NON-COVALENT INTERACTIONS IN THE SYNTHESIS AND DESIGN OF NEW COMPOUNDS

NON-COVALENT INTERACTIONS IN THE SYNTHESIS AND DESIGN OF NEW COMPOUNDS

Edited by

ABEL M. MAHARRAMOV

KAMRAN T. MAHMUDOV

MAXIMILIAN N. KOPYLOVICH

ARMANDO J. L. POMBEIRO

WILEY

Published by John Wiley & Sons, Inc., Hoboken, New Jersey
Published simultaneously in Canada

Library of Congress Cataloging-in-Publication Data:

Names: Maharramov, A. M., editor. | Mahmudov, Kamran T., 1980– editor. | Kopylovich, Maximilian N., 1970– editor. | Pombeiro, A. J. L. (Armando J. L.), editor.
Title: Non-covalent interactions in the synthesis and design of new compounds / edited by Abel M. Maharramov, Kamran T. Mahmudov, Maximilian N. Kopylovich, Armando J. L. Pombeiro.
Description: Hoboken, New Jersey : John Wiley & Sons, 2016. | Includes bibliographical references and index.
Identifiers: LCCN 2015040502 (print) | LCCN 2015047682 (ebook) | ISBN 9781119109891 (cloth) | ISBN 9781119113850 (Adobe PDF) | ISBN 9781119113867 (ePub)
Subjects: LCSH: Supramolecular chemistry. | Valence (Theoretical chemistry)
Classification: LCC QD878 .N66 2016 (print) | LCC QD878 (ebook) | DDC 541/.226–dc23
LC record available at http://lccn.loc.gov/2015040502

10 9 8 7 6 5 4 3 2 1

CONTENTS

Notes on Editors ix

List of Contributors xi

Preface xv

PART I ORGANIC SYNTHESIS 1

1 Activation of Covalent Bonds Through Non-Covalent Interactions 3
Abel M. Maharramov, Kamran T. Mahmudov, Maximilian N. Kopylovich, M. Fátima C. Guedes da Silva, and Armando J. L. Pombeiro

2 Boron–Nitrogen Bond: A Useful Molecular Construction Tool 23
João P. M. Antonio, Gonçalo D. V. Farias, Fábio M. F. Santos, Rudi Oliveira, Pedro M. S. D. Cal, and Pedro M. P. Gois

3 Influence of Steric, Electronic, and Molecular Preorganization Effects in the Reactivity of β-Phenylethylamines with Nonenolizable Aldehydes 49
Rodolfo Quevedo

4 Non-Covalent Interactions in the Synthesis of Macrocycles 63
Evgeny A. Kataev

PART II INORGANIC, COORDINATION AND ORGANOMETALLIC SYNTHESES 83

5 Non-Covalent Interactions of Water with Metal Complexes in Solution 85
Antonio Romerosa and Franco Scalambra

6 π–π Interaction Directed Applications of Metal Complexes 101
Manas Sutradhar and Armando J. L. Pombeiro

7 **Non-Covalent Stabilization in Transition Metal Coordination and Organometallic Complexes** 115
Predrag Petrović, Jean-Pierre Djukic, Andreas Hansen, Christoph Bannwarth, and Stefan Grimme

8 **Halogen Bonding in the Synthesis and Design of Coordination and Organometallic Compounds** 145
Abel M. Maharramov, Namiq Q. Shixaliyev, Atash V. Gurbanov, Kamran T. Mahmudov, Valentine G. Nenajdenko, Armando J. L. Pombeiro, and Maximilian N. Kopylovich

9 **The Influence of Non-Covalent Interactions in the Structure and Dimensionality of Hybrid Compounds and Coordination Polymers** 163
Ferdinando Costantino, Andrea Ienco, and Marco Taddei

PART III CRYSTAL DESIGN AND HOST–GUEST COMPOUNDS 185

10 **Dihalogens as Halogen Bond Donors** 187
Matti Haukka, Pipsa Hirva, and Kari Rissanen

11 **Construction of Supramolecular Assemblies Based on Anion–π Interactions** 199
Antonio Bauzá and Antonio Frontera

12 **Asymmetric Azamacrocycles as Chiral Solvating Agents** 213
Koichi Tanaka

13 **New Strategies for the Design of Inclusion Compounds with Cucurbituril Hosts** 229
Na'il Saleh

14 **Parallel Interactions of Aromatic Molecules at Large Horizontal Displacements** 241
Dušan P. Malenov and Snežana D. Zarić

15 **Selective Molecular Binding and Nanosupramolecular Assembly of *p*-Sulfonatocalix[*n*]arenes** 261
Yu Liu and Yi-Xuan Wang

16 **Synthesis, Design, Characterization, and Application of Metallo-Supramolecular Polymers** 283
Lipeng He and Weifeng Bu

PART IV CATALYSIS 313

17 **Catalytic Applications of Metal Complexes Immobilized by Non-Covalent Interactions onto Chemically Derived Graphenes and Related Materials** 315
Sara Sabater and Jose A. Mata

18 Cooperation of Non-Covalent Interactions and Coordination in Catalysis 327
Abel M. Maharramov, Kamran T. Mahmudov, Maximilian N. Kopylovich, Rafiga A. Aliyeva, and Armando J. L. Pombeiro

19 Hybrid Ordered Mesoporous Materials as Supports for Permanent Enzyme Immobilization Through Non-Covalent Interactions 345
Victoria Gascón, Carlos Márquez-Alvarez, Isabel Díaz, and Rosa M. Blanco

PART V BIORELEVANT SYNTHESES 361

20 Modulation of Biorelevant Radical Reactions by Non-Covalent Interactions 363
Riccardo Amorati and Luca Valgimigli

21 Current Understanding of π–π Interactions and the Applications in Protein Design 381
Jianmin Gao and Azade S. Hosseini

22 Non-Covalent Interactions for the Preparation of Pseudopeptidic Synthetic Compounds and Materials 391
Jordi Solà and Ignacio Alfonso

PART VI MATERIAL CHEMISTRY 413

23 Non-Covalent Exfoliation of Graphite to Produce Graphene 415
Yingkui Yang, Dean Shi, and Tao Jiang

24 Electrostatic Interactions in the Design of Polymeric Products 431
Lorenzo M. Polgar and Francesco Picchioni

25 Supramolecular Steric Hindrance at Bulky Organic/Polymer Semiconductors and Devices 443
Ling-Hai Xie and Wei Huang

Index 457

NOTES ON EDITORS

Abel M. Maharramov graduated in chemistry from Baku State University, Azerbaijan, in 1971. He obtained his Ph.D. degree from Moscow State University in 1976. Since 1991, he has been a Full Professor at Baku State University and served as the Head of the Department of Chemistry from 1993 to 1999; now he is the Rector of this university (from 1999). He is a Full Member of the Azerbaijan National Academy of Sciences (from 2007) and Georgian National Academy of Sciences (2013). He authored 71 books, (co-)authored 650 research publications (in International Scientific Indexing journals), held 55 patents, and presented 45 lectures at international conferences. He has supervised 52 Ph.D. students and Doctors of Science. His research group focuses on organic, coordination, and supramolecular chemistry and, in particular, on the development of non-covalent interactions in the synthesis, catalysis, and crystal engineering. Among his many honors are Order of Glory (2009, Azerbaijan), Gold Medal of the Economic Cooperation Organization (2000), Honor "Golden Star" COMANDOR level Order of Romania (2004), and the ISESCO Prize on Science and Technology (2012). He is the Honorary Doctor of the leading universities of Turkey, Romania, Kazakhstan, Ukraine, China, and Albania, and also the Honorary Professor of Lomonosov *Moscow State University.* He is the Vice-President of the Eurasian Association of Universities. He is also the Honored Scientist of Azerbaijan. He was elected as Member of Parliament of the Republic of Azerbaijan in 2005 and 2010.

Kamran T. Mahmudov was born in Tovuz, Azerbaijan, and received his B.Sc. (2001), M.Sc. (2003), Ph.D. (2007), and Habilitation (2012) (scientific advisers A.M. Maharramov and Prof. R.A. Aliyeva) degrees at the Baku State University, where he worked as an Assistant Professor (2006–2008). Since 2009, he has been conducting a postdoctoral research under the supervision of A.J.L. Pombeiro at the University of Lisbon. His publication record contains more than 75 peer-reviewed publications in International Scientific Indexing journals, two book chapters, and two reviews. His main research interests are related to the synthesis of organic and coordination compounds, *E*/*Z* isomerization, homogeneous catalysis, and activation of covalent bonds by non-covalent interactions (in particular, resonance-assisted hydrogen bonding) or by their cooperation with coordination. He is married to Farida Abdiyeva who is a Journalist.

Maximilian N. Kopylovich graduated in Chemical Engineering (1993) from Byelorussian State Technological University, and he completed his Ph.D. degree in Chemistry (1998) at the same institution. Then he worked as a post-doctoral fellow (2000–2007), auxiliary (2008–2013), and senior (2014–present) researcher at the Centro de Química Estrutural, Instituto Superior Técnico (IST), Universidade de Lisboa. His scientific interests lay mainly within synthetic coordination chemistry and catalysis. In particular, he participated in the development of new methods of synthesis of triazapentadienato, aminopolyalcoholato, arylhydrazo, iminoisoindolinone, and tetrazolato complexes of metals from virtually all groups of the periodic table. His current research interests are related to the metal-mediated (template) and hydrogen(halogen)-assisted syntheses of new compounds and supramolecular assemblies.

Armando J. L. Pombeiro is Full Professor at the Instituto Superior Técnico, Universidade de Lisboa, President of the Centro de Química Estrutural and Coordinator of its Synthesis and Catalysis thematic line, Director of the Catalysis and Sustainability (CATSUS) PhD program, Full Member of the Academy of Sciences of Lisbon (in charge of various directive and representative positions) and former President of the Portuguese Electrochemical Society. His research group investigates the activation of small molecules with industrial, environmental or biological significance, including metal-mediated synthesis and catalysis (e.g., functionalization of alkanes under mild conditions), crystal engineering of coordination compounds, design and self-assembly of polynuclear and supramolecular structures, molecular electrochemistry and theoretical studies. He was Chairman of the 25th ICOMC and member of organizing/scientific committees of 40 international conferences or schools. He authored one book (plus four as editor), (co-)authored over 600 research publications, 33 patents, and presented 100 invited lectures at international conferences. His work has received ca. 14,000 citations, h-index = 56 (Web of Science). Among his honors, he was awarded the Madinabeitia-Lourenço Prize from the Spanish Royal Chemical Society and the Prizes of the Portuguese Chemical and Electrochemical Societies.

LIST OF CONTRIBUTORS

Ignacio Alfonso, Department of Biological Chemistry and Molecular Modeling, Institute of Advanced Chemistry of Catalonia (IQAC-CSIC), Barcelona, Spain

Rafiga A. Aliyeva, Department of Chemistry, Baku State University, Baku, Azerbaijan

Riccardo Amorati, Department of Chemistry "G. Ciamician", University of Bologna, Bologna, Italy

João P. M. Antonio, Instituto de Investigação do Medicamento (iMed.ULisboa), Faculdade de Farmácia, Universidade de Lisboa, Lisbon, Portugal

Christoph Bannwarth, Mulliken Center for Theoretical Chemistry, Institut für Physikalische und Theoretische Chemie, Universität Bonn, Bonn, Germany

Antonio Bauzá, Departament de Química, Universitat de les Illes Balears, Palma de Mallorca (Baleares), Spain

Rosa M. Blanco, Instituto de Catálisis y Petroleoquímica, CSIC, Madrid, Spain

Weifeng Bu, Key Laboratory of Nonferrous Metals Chemistry and Resources Utilization of Gansu Province, State Key Laboratory of Applied Organic Chemistry, College of Chemistry and Chemical Engineering, Lanzhou University, Lanzhou, China

Pedro M. S. D. Cal, Instituto de Medicina Molecular (IMM), Faculdade de Medicina da Universidade de Lisboa, Lisbon, Portugal

Ferdinando Costantino, Dipartimento di Chimica, Biologia e Biotecnologie, University of Perugia, Perugia, Italy; Department at Consiglio Nazionale delle Ricerche, Institute of Chemistry of Organometallic Compounds (CNR–ICCOM), Florence, Italy

Isabel Díaz, Instituto de Catálisis y Petroleoquímica, CSIC, Madrid, Spain

Jean-Pierre Djukic, Institut de Chimie de Strasbourg, CNRS UMR 7177, Université de Strasbourg, Strasbourg, France

Gonçalo D. V. Farias, Instituto de Investigação do Medicamento (iMed.ULisboa), Faculdade de Farmácia, Universidade de Lisboa, Lisbon, Portugal

Antonio Frontera, Departament de Química, Universitat de les Illes Balears, Palma de Mallorca (Baleares), Spain

Jianmin Gao, Department of Chemistry, Merkert Chemistry Center, Boston College, Chestnut Hill, MA, USA

Victoria Gascón, Instituto de Catálisis y Petroleoquímica, CSIC, Madrid, Spain

Pedro M. P. Gois, Instituto de Investigação do Medicamento (iMed.ULisboa), Faculdade de Farmácia, Universidade de Lisboa, Lisbon, Portugal

Stefan Grimme, Mulliken Center for Theoretical Chemistry, Institut für Physikalische und Theoretische Chemie, Universität Bonn, Bonn, Germany

M. Fátima C. Guedes da Silva, Centro de Química Estrutural, Instituto Superior Técnico, Universidade de Lisboa, Lisbon, Portugal

Atash V. Gurbanov, Department of Chemistry, Baku State University, Baku, Azerbaijan

Andreas Hansen, Mulliken Center for Theoretical Chemistry, Institut für Physikalische und Theoretische Chemie, Universität Bonn, Bonn, Germany

Matti Haukka, Department of Chemistry, University of Jyvaskyla, Jyvaskyla, Finland

Lipeng He, Key Laboratory of Nonferrous Metals Chemistry and Resources Utilization of Gansu Province, State Key Laboratory of Applied Organic Chemistry, College of Chemistry and Chemical Engineering, Lanzhou University, Lanzhou, China

Pipsa Hirva, Department of Chemistry, University of Eastern Finland, Joensuu, Finland

Azade S. Hosseini, Department of Chemistry, Merkert Chemistry Center, Boston College, Chestnut Hill, MA, USA

Wei Huang, Center for Molecular System & Organic Devices (CMSOD), Key Laboratory for Organic Electronics and Information Displays & Institute of Advanced Materials (IAM), Jiangsu National Synergetic Innovation Center for Advanced Materials (SICAM), Nanjing University of Posts & Telecommunications, Nanjing, China; Key Laboratory of Flexible Electronics (KLOFE) & Institute of Advanced Materials (IAM), Jiangsu National Synergetic Innovation Center for Advanced Materials (SICAM), Nanjing Tech University (NanjingTech), Nanjing, China

Andrea Ienco, Department at Consiglio Nazionale delle Ricerche, Institute of Chemistry of Organometallic Compounds (CNR–ICCOM), Florence, Italy

Tao Jiang, MOE Key Laboratory for Green Preparation and Application of Functional Materials, Hubei Collaborative Innovation Center for Advanced Organic Chemical Materials, and Faculty of Materials Science and Engineering, Hubei University, Wuhan, China

Evgeny A. Kataev, Institute of Chemistry, Faculty of Natural Sciences, Technische Universität Chemnitz, Chemnitz, Germany

Maximilian N. Kopylovich, Centro de Química Estrutural, Instituto Superior Técnico, Universidade de Lisboa, Lisbon, Portugal

Yu Liu, Department of Chemistry, State Key Laboratory of Elemento-Organic Chemistry, Collaborative Innovation Center of Chemical Science and Engineering (Tianjin), Nankai University, Tianjin, China

Abel M. Maharramov, Department of Chemistry, Baku State University, Baku, Azerbaijan

Kamran T. Mahmudov, Department of Chemistry, Baku State University, Baku, Azerbaijan; Centro de Química Estrutural, Instituto Superior Técnico, Universidade de Lisboa, Lisbon, Portugal

Dušan P. Malenov, Innovation Center of the Faculty of Chemistry, University of Belgrade, Belgrade, Serbia

Carlos Márquez-Alvarez, Instituto de Catálisis y Petroleoquímica, CSIC, Madrid, Spain

Jose A. Mata, Departamento de Química Inorgánica y Orgánica, Institute of Advanced Materials (INAM), Universitat Jaume I, Castellón de la Plana, Spain

Samira Nahim-Granados, Área de Química Inorgánica-CIESOL, University of Almeria, Almeria, Spain

Valentine G. Nenajdenko, Department of Chemistry, Moscow State University, Moscow, Russia

Rudi Oliveira, Instituto de Investigação do Medicamento (iMed.ULisboa), Faculdade de Farmácia, Universidade de Lisboa, Lisbon, Portugal

Predrag Petrović, Department of Chemistry, Innovation Center, University of Belgrade, Belgrade, Serbia

Francesco Picchioni, Department of Chemical Engineering, University of Groningen, Groningen, the Netherlands

Lorenzo M. Polgar, Department of Chemical Engineering, University of Groningen, Groningen, the Netherlands

Armando J. L. Pombeiro, Centro de Química Estrutural, Instituto Superior Técnico, Universidade de Lisboa, Lisbon, Portugal

Rodolfo Quevedo, Departamento de Química, Facultad de Ciencias, Universidad Nacional de Colombia, Bogotá, Colombia

Kari Rissanen, Department of Chemistry, University of Jyvaskyla, Jyvaskyla, Finland; Nanoscience Center, University of Jyvaskyla, Jyvaskyla, Finland

Antonio Romerosa, Área de Química Inorgánica-CIESOL, University of Almeria, Almeria, Spain

Sara Sabater, Departamento de Química Inorgánica y Orgánica, Institute of Advanced Materials (INAM), Universitat Jaume I, Castellón de la Plana, Spain

Na'il Saleh, Department of Chemistry, College of Science, United Arab Emirates University, Al-Ain, United Arab Emirates

Fábio M. F. Santos, Instituto de Investigação do Medicamento (iMed.ULisboa), Faculdade de Farmácia, Universidade de Lisboa, Lisbon, Portugal

Franco Scalambra, Área de Química Inorgánica-CIESOL, University of Almeria, Almeria, Spain

Dean Shi, MOE Key Laboratory for Green Preparation and Application of Functional Materials, Hubei Collaborative Innovation Center for Advanced Organic Chemical Materials, and Faculty of Materials Science and Engineering, Hubei University, Wuhan, China

Namiq Q. Shixaliyev, Department of Chemistry, Baku State University, Baku, Azerbaijan

Jordi Solà, Department of Biological Chemistry and Molecular Modeling, Institute of Advanced Chemistry of Catalonia (IQAC-CSIC), Barcelona, Spain

Manas Sutradhar, Centro de Química Estrutural, Instituto Superior Técnico, Universidade de Lisboa, Lisbon, Portugal

Marco Taddei, Laboratory for Catalysis and Sustainable Chemistry, Paul Scherrer Institut (PSI), Villigen, Switzerland

Koichi Tanaka, Department of Chemistry and Materials Engineering, Faculty of Chemistry, Materials and Bioengineering, Kansai University, Osaka, Japan

Luca Valgimigli, Department of Chemistry "G. Ciamician", University of Bologna, Bologna, Italy

Yi-Xuan Wang, Department of Chemistry, State Key Laboratory of Elemento-Organic Chemistry, Collaborative Innovation Center of Chemical Science and Engineering (Tianjin), Nankai University, Tianjin, China

Ling-Hai Xie, Center for Molecular System & Organic Devices (CMSOD), Key Laboratory for Organic Electronics and Information Displays & Institute of Advanced Materials (IAM), Jiangsu National Synergetic Innovation Center for Advanced Materials (SICAM), Nanjing University of Posts & Telecommunications, Nanjing, China

Yingkui Yang, MOE Key Laboratory for Green Preparation and Application of Functional Materials, Hubei Collaborative Innovation Center for Advanced Organic Chemical Materials, and Faculty of Materials Science and Engineering, Hubei University, Wuhan, China

Snežana D. Zarić, Department of Chemistry, University of Belgrade, Belgrade, Serbia; Department of Chemistry, Texas A&M University at Qatar, Doha, Qatar

PREFACE

The importance of non-covalent interactions has been recognized since a long time ago [1], and the field continues to attract a wide attention in different areas of chemistry and technology. Generally, non-covalent interactions concern atomic or molecular contacts which do not involve electron sharing or loss. As any simplification, this definition is rather debatable, but it still allows to separate covalent, coordination, metallic, or ionic bonds from the subtle (non-covalent) interactions. The latter, in particular, include the so-called hydrogen, halogen, chalcogen and pnictogen bonds, van der Waals forces, π-interactions, and so on.

It should be mentioned that in different areas of chemistry, distinct types of bonding can be considered as the "non-covalent" ones. For instance, in organic chemistry the coordination bond can be regarded as a non-covalent interaction, while the reversible dative B—N bond of a similar nature and low energy can be considered as a covalent one. Nevertheless, some common ground can be found. Thus, if one observes an interaction, which manifests itself in a significant reduction of the interatomic distance if compared with the sum of van der Waals radii; possesses a certain directionality; and involves particular types of atoms and at the same time is of much lower energy than a typical "covalent" bond (i.e., <100 kJ/mol), then one can consider such bond as a non-covalent one.

The subtle nature and diversity of non-covalent interactions ensure their prominent role in various applications, for example, in the syntheses of organic, inorganic, coordination and organometallic compounds, and supramolecular assemblies. In those cases, they can organize reagents in the space in a favorable fashion, provide a suitable environment, and drive the energetic pattern of the process, facilitating its course. For instance, the formation of multiple halogen bonds between the reaction products can be one of main driving forces, giving rise to the so-called *spider effect* in the synthesis of halogen-rich organic and coordination compounds. Although sometimes we can analyze and predict the result of the influence of non-covalent interactions on a reaction, usually the synthesis remains a matter of intuition, good luck, and serendipity.

This book is intended to rationalize, at least in part, reaction patterns and driving forces of some selected reactions, driven by non-covalent interactions. The role of these interactions in related fields of material design and biorelevant chemistry is also highlighted. It should be noted that, in contrast to the commonly separated courses on the main fields of chemistry, the book is thought of converging different chemical fields and approaches. In this view, it gathers scientists of different generations and backgrounds, from all over the world, whose research interests and work, however, are directly influenced by those innumerable tiny forces around us.

Although prominent books have been published recently on different types of non-covalent interactions [2–9], new facts and concepts continue to appear. As a result, several chapters of the books are devoted to recent advances in the theory of non-covalent interactions. On the other hand, the current progress in structural and computer analyses gives rise to a statistical approach to non-covalent interactions in compounds [3–5, 10]. This allows to distinguish the most relevant and widely spread tendencies and, at the same time, drives attention to some earlier unnoticed interactions, which, however, could be of great importance for many applications.

It is already well established that non-covalent contacts play a relevant role in the synthesis of new compounds and advanced materials. The catalytic chemical transformations driven by non-covalent interactions or by their synergetic action with coordination (e.g., in cooperative catalysis) constitute one of the most promising strategies in organic synthesis [2]. Furthermore, non-covalent interactions can be essential for understanding the action of drugs and functioning of catalytic sites within biological systems [11]. In this respect, many biomedical and other related areas would significantly benefit from a deeper understanding and conscious application of these interactions. In particular, non-covalent interactions can provide vehicles for encapsulation and targeted delivery of therapeutic agents or bioactive materials.

In view of the aforementioned considerations, this book is formally divided into several sections, related to organic, inorganic and organometallic syntheses, catalysis, crystal design, and bio-relevant and material chemistries. Although this division is rather artificial, it points out some areas of current interest and assists the general orientation within the book.

Its first part is dedicated to the role of non-covalent interactions in organic synthesis. Hydrogen, halogen, chalcogen, and pnictogen bond-assisted activation of saturated and unsaturated covalent bonds, synthesis of macrocycles, and molecular preorganization are among the discussed topics within this section. The following part of the book discusses how the non-covalent interactions influence the preparation, design, and applications of inorganic, coordination, and organometallic compounds; while the third section is mainly related to crystal design and organization of host–guest assemblies. Dihalogens as halogen bond donors, parallel interactions of aromatic molecules, anion–π interactions as a driving force in the construction of supramolecular assemblies, and design of inclusion compounds (e.g., of cucurbituril or calixarene hosts), in chiral azamacrocycles and in molecular sensing, are particularly discussed in the third part.

In the fourth section of the book, some possible catalytic applications are overviewed, and this somehow relates to the fifth section, where the role of non-covalent forces in bio-relevant radical reactions and in the preparation of peptide-like compounds is reported. The final sixth section is related to material chemistry, in particular, to the fabrication and design of new graphene-based, polymeric and hybrid materials and devices.

Therefore, various fields of chemistry, theoretical and practical approaches, are covered within the general concept of non-covalent interactions, hopefully providing an integrated overview of their known and envisioned applications. We hope that this book will be helpful for the readers interested in the design and synthesis of new useful materials and will inspire new studies in the field.

We acknowledge the authors of the chapters for their valuable contributions. Particular thanks are due to Anita Lekhwani (Senior Acquisitions Editor at John Wiley & Sons, Inc.) and the team at Wiley at Hoboken for their editorial assistance. The support of the Baku State University, Azerbaijan, as well as Foundation for Science and Technology (FCT), Portugal, should also be mentioned.

Abel M. Maharramov
Kamran T. Mahmudov
Maximilian N. Kopylovich
Armando J. L. Pombeiro

REFERENCES

1. J. D. van der Waals, Over de Continuiteit van den Gas- en Vloeistoftoestand [on the continuity of the gas and liquid state]. Doctoral Dissertation, Universiteit Leiden, Leiden, 1873.
2. P. M. Pihko, Hydrogen Bonding in Organic Synthesis, Wiley-VCH: Weinheim, 2009.
3. P. Gilli, G. Gilli, The Nature of the Hydrogen Bond, Oxford University Press: Oxford, 2009.
4. P. Hobza, K. Müller-Dethlefs, Non-covalent Interactions: Theory and Experiment, Royal Society of Chemistry: Cambridge, 2010.
5. S. Scheiner, Non-covalent Forces (Challenges and Advances in Computational Chemistry and Physics volume *19*), Springer, Heidelberg, 2015.
6. P. A. Gale, J. W. Steed, Supramolecular Chemistry: From Molecules to Nanomaterials, John Wiley & Sons, Ltd: Chichester, 2012.
7. A. K. Baev, Specific Intermolecular Interactions of Organic Compounds, Springer: Heidelberg, 2012.
8. A. K. Baev, Specific Intermolecular Interactions of Nitrogenated and Bioorganic Compounds, Springer: Heidelberg, 2014.
9. P. Metrangolo, G. Resnati, Halogen Bonding: Fundamentals and Applications (Stucture and Bonding); Springer: Heidelberg, 2010.
10. S. Scheiner, *Acc. Chem. Res.* 2013, *46*, 280.
11. A. Karshikoff, Non-covalent Interactions in Proteins, Imperial College Press: Singapore, 2006.

PART I

ORGANIC SYNTHESIS

1

ACTIVATION OF COVALENT BONDS THROUGH NON-COVALENT INTERACTIONS

ABEL M. MAHARRAMOV[1], KAMRAN T. MAHMUDOV[1,2,*], MAXIMILIAN N. KOPYLOVICH[2], M. FÁTIMA C. GUEDES DA SILVA[2], AND ARMANDO J. L. POMBEIRO[2,†]

[1] *Department of Chemistry, Baku State University, Baku, Azerbaijan*
[2] *Centro de Química Estrutural, Instituto Superior Técnico, Universidade de Lisboa, Lisbon, Portugal*

1.1 INTRODUCTION

A large number of fundamental chemical processes involve cleavage and/or transformation of strong covalent bonds, which can require a high energy consumption to fulfill the thermodynamic and kinetic profile of the reaction. The finding of pathways where intermediates with a lower energy consumption are formed is a common tool to facilitate a reaction, and non-covalent interactions can play an important role in the formation and stabilization of intermediates. Usually the energy of non-covalent interactions (hydrogen and halogen bonding, van der Waals forces, π-effects, etc.) is much lower than that of typical covalent bonds [1–6], but, on account of their multiplicity and facile transformation, their overall influence on the course of a reaction can be decisive. Thus, the "soft force" of multiple low-energy and low-level interactions can play a significant role in the activation of covalent bonds, directing the "brute action" of pronounced powers such as heat, current, light, etc. As a result, the cooperation of these "strong" external forces and "weak" non-covalent interactions can be relevant for an effective activation of covalent bonds.

For instance, if hydrogen bonds are under consideration, one can surmise that they influence the reactivity of substrates in several ways (Scheme 1.1). When the oxygen atom of a carbonyl group or the sulfur atom of a thioketone is involved in a strong bifurcated hydrogen bonding, the electrophilic character of the corresponding carbon atom increases toward nucleophilic attack when compared with a simple H-bonding (compare **II** with **I** and **IV** with **III**). In addition, a nucleophilic attack to the carbon atom in a carbonyl is usually more favorable than in a thioketone (compare **I** with **III** and **II** with **IV**). Hence, one can use these considerations to perform a regiospecific activation of a particular moiety.

Similar approaches can be used for the activation of a carbon atom in imines (**V** and **VI**), alkenes (**VII** and **VIII**), and alkynes (**IX** and **X**). Another approach can involve the creation of flexible or cyclic hydrogen bonding systems in a near proximity to the target covalent bond to promote the desired nucleophilic attack (**XI–XVI**). This allows to perform a proper orientation in space of the electron-poor carbon atoms; and an easy attack by an electron-rich nucleophilic group occurs smoothly. In many cases the creation of such supporting non-covalent bondings can be rationalized, but often it is not straightforward to predict the activation effect. Hence, more efforts should be devoted to the study and systematization of this field.

Therefore, the use of non-covalent interactions in synthesis is becoming an active area of research with potential applications in separation processes, medicine, catalysis, and biomolecular systems. Most of the known applications in synthesis are related to the highly widespread hydrogen bonds [1, 5, 6]. However, other types of interactions, such as O···B, O···Si, O···S, N···B, N···N, N···P, N···S, or halogen

* Corresponnding author Kamran T. Mahmudov (kamran_chem@mail.ru)
† Armando J.L. Pombeiro (pombeiro@tecnico.ulisboa.pt)

Non-Covalent Interactions in the Synthesis and Design of New Compounds, First Edition.
Edited by Abel M. Maharramov, Kamran T. Mahmudov, Maximilian N. Kopylovich, and Armando J. L. Pombeiro.

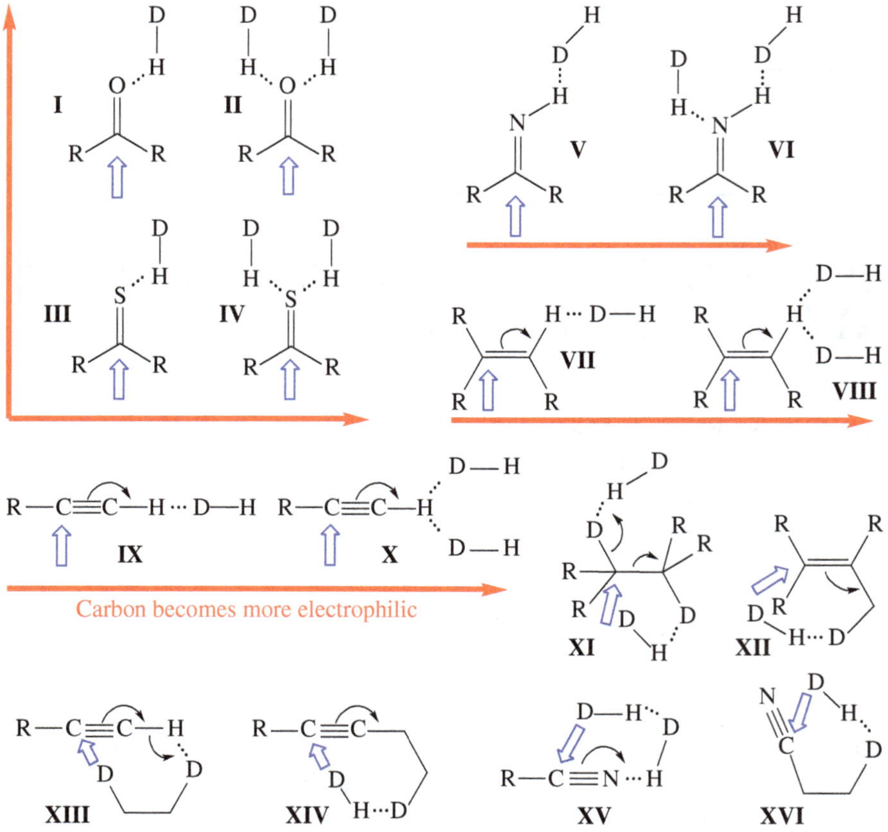

SCHEME 1.1 Some modes of covalent bond activation by hydrogen bonds.

bonding, have also grown from a scientific curiosity to a real tool to promote a number of organic and inorganic transformations and to construct supramolecular assemblies [7, 8]. This chapter provides some examples of activation of different types of covalent bonds (C—C, C=C, C≡C, C—N, C=N, C≡N, C—O, C=O, C—S, and C=S) assisted by non-covalent interactions. We hope it will help the reader to understand some key principles, which can be used for elaboration of new synthetic pathways to target compounds.

1.2 EXAMPLES OF HYDROGEN BOND-ASSISTED ACTIVATIONS

1.2.1 Activation of C—C Bond

The fission of C—C bonds by incorporation of oxygen (oxidative cleavage) or water (hydrolytic cleavage) is a crucial step in many biochemical pathways through which microorganisms assimilate xenobiotics and environmentally toxic compounds [9]. These reactions are of considerable interest due to their broad array of applications, including fuel production from biomass, removal of organic contaminants, or treatment of diseases. In many cases, the activation of a C—C bond adjacent to a carbonyl moiety is an essential step to reach the goal. For instance, hydrolytic cleavage of C—C bonds catalyzed by hydrolase enzymes is important for the bacterial degradation of aromatic compounds. The proposed mechanism for C—C cleavage in 2-hydroxy-6-keto-6-phenyl-hexa-2,4-dienoic acid, catalyzed by a Ser–His–Arg system (Scheme 1.2), commences by the keto–enol tautomerization of the substrate to give a keto intermediate (**I**), proved by stopped flow kinetic analysis and deuterium isotope exchange studies [10, 11]. Hydrogen bond-assisted nucleophilic attack of H_2O on the C6 carbonyl is then followed by a stereospecific C—C fragmentation (intermediate **II**) to give (*E*/*Z*)-2-hydroxypenta-2,4-dienoic and benzoic acids [10, 11].

Artificial systems for hydrogen bond-assisted C—C bond activation were elaborated for β-diketones, important precursors for several widely spread heterocyclic compounds (pyrazoles, triazoles, etc.). The cleavage of C—C bonds in β-diketones, also known as retro-Claisen reaction, is of interest due to its relevance to metabolism of aromatics and terpenes, bioremediation, and preparative biocatalysis. Acetylacetone dioxygenase (Dke1) performs oxidative fission of the C—C bond by consecutive coordination of β-diketone to an iron(II) center, dioxygen incorporation with C—O bond formation, and nucleophilic attack of the peroxidate moiety on a carbonyl group (Scheme 1.3a) [9]. A number of retro-Claisen catalytic transformations are known, involving, as catalysts, indium, iron, palladium, copper salts, and complexes [9]. However, a new metal-free C—C cleavage in β-diketones with assistance of, for example, hydrogen bonding can significantly improve the ecological potential of the reaction.

To create an intramolecular hydrogen bond system, which can assist specific nucleophilic attacks to the carbonyl moieties, a triazine moiety was introduced into β-diketones (Scheme 1.3b) [12]. Protonation of the triazine derivatives of

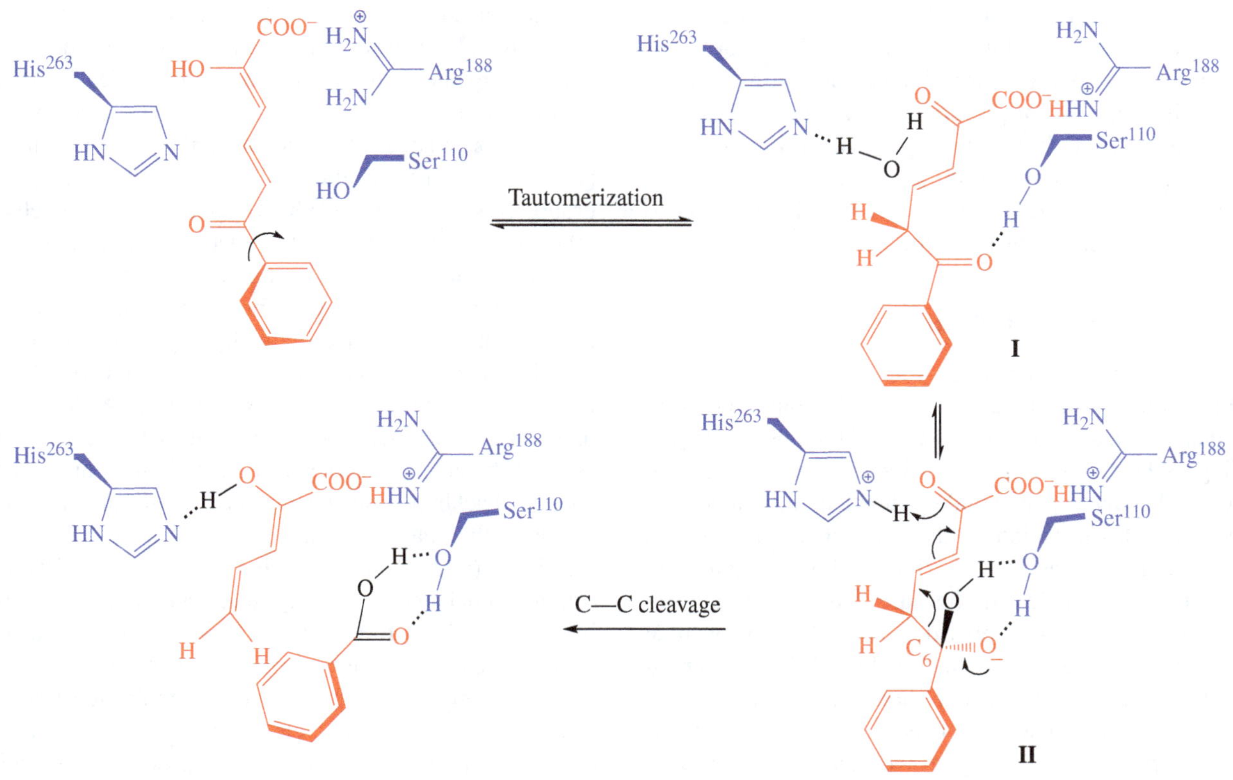

SCHEME 1.2 C—C cleavage in 2-hydroxy-6-keto-6-phenyl-hexa-2,4-dienoic acid [10, 11].

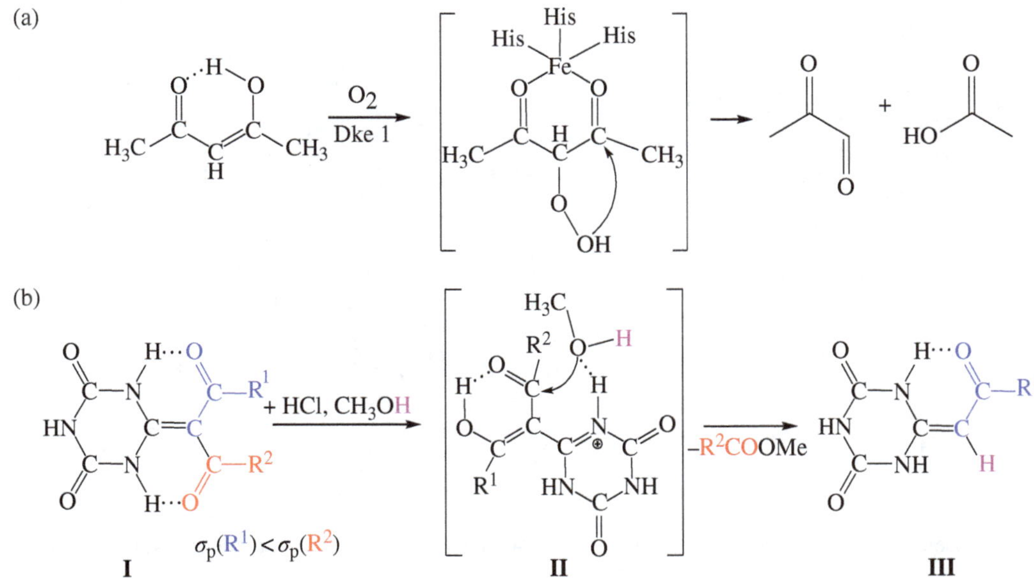

SCHEME 1.3 Acetylacetone dioxygenase (Dke1)-catalyzed oxidative C—C cleavage in acetylacetone (*retro*-Claisen reaction) (a) [9]; regioselective H-bond-assisted C—C cleavage in β-diketones (b) [12].

β-diketone (**I**) leads to the enolic form (**II**). Hydrogen bond-assisted nucleophilic attack of MeOH on this enolic form with liberation of the R^2COOMe ester leads to the formation of the unsymmetrical ketone product **III**. Although the electron donor/acceptor character of substituents within the β-diketone moiety can be rather close [the values of Hammett's σ_p constant are −0.13 for C_3H_7, −0.17 for CH_3, −0.24 for OC_2H_5, −0.27 for OCH_3, and −0.83 for $N(CH_3)_2$], in all the studied cases, the C—C bond breaks selectively at the side of the weaker electron-releasing substituent; the regioselectivity is possibly amplified by the formation of a hydrogen bonding system.

A similar approach was used for the regioselective activation of C—C bond in 4,4,4-trifluoro-1-(thiophen-2-yl)butane-1,3-dione, where an arylhydrazone moiety was introduced to create an intramolecular hydrogen bond at a desired position [13]. In this case, due to the stability of intramolecular hydrogen bonding, the C—C cleavage occurs at the more electron-accepting part of the substituted 1,3-dione (Scheme 1.4). Such hydrogen bond-assisted C—C bond cleavages with elimination of an acetyl group do not require any catalyst and proceed smoothly under ambient conditions, which contrasts to other known reactions of this type.

1.2.2 Activation of C=C Bond

Alkenes are known substrates for many reactions, and economic ways for activation of the C=C double bond, in particular with assistance of the "cost-of-nothing" hydrogen bonding, is becoming a matter of increasing attention. For instance, the hydrogen bonding-directed intermolecular Cope-type hydroamination of alkenes (Scheme 1.5) allows an atom-economical access to various highly functionalized vicinal diamines [14, 15]. In this case, the formed intermolecular hydrogen bonding provides the buildup of positive charges at both N atoms of intermediate **I**. The observed exclusive formation of vicinal diamines **II** most likely originates from the synergy between the H-bonding effect and the propensity of Cope-type hydroaminations to favor Markovnikov addition products.

Michael addition of a nucleophile to electron-deficient olefins is a classical way to construct C—C or C—heteroatom bonds in organic synthesis. A highly enantioselective Michael addition of diethyl malonate with chalcones is catalyzed by cinchona-originated bifunctional tertiary amine–thioureas [16]. The functionalization of the catalyst with hydroxyl group gives rise to multiple hydrogen bonds and thereby significantly enhances the catalytic activity and rigidity of the envisaged transition state, leading to high enantioselectivity and activity (Scheme 1.6). Thus, the carbonyl group in intermediate **I** is activated by the hydrogen bonding interactions between the oxygen atom of the carbonyl with the thiourea moiety and the extra hydroxyl group of **A**. At the same time, diethyl malonate is deprotonated by the basic nitrogen atom of the cinchona moiety. The chiral environment of the alkaloid backbone and the matched amino alcohol moiety enable diethyl malonate to attack the activated chalcone from the *Si*-face (intermediate **I**), giving the *R*-product. If the interaction of chalcone and diethyl malonate is catalyzed by **B**, intermediate **II** is formed and a *Re*-face attack is favored, leading to the *S*-configured product (Scheme 1.6).

Cyclopropanes are particularly suitable substrates for synthetic applications, and electronic effects of substituents support their specific activation and allow to prepare a high versatility of products with multiple stereocenters [17, 18].

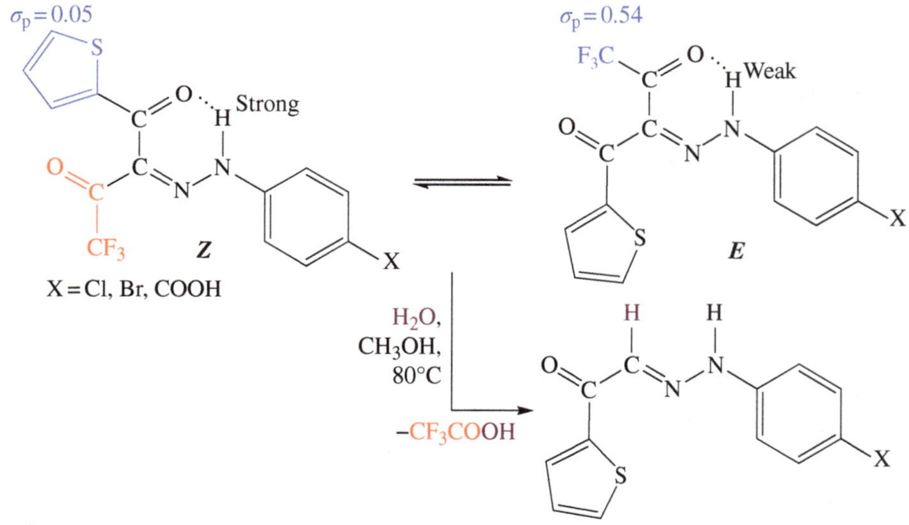

SCHEME 1.4 Regioselective C—C cleavage in (*E,Z*)-2-(2-(4-substitutedphenyl) hydrazono)-4,4,4-trifluoro-1-(thiophen-2-yl) butane-1,3-dione [13].

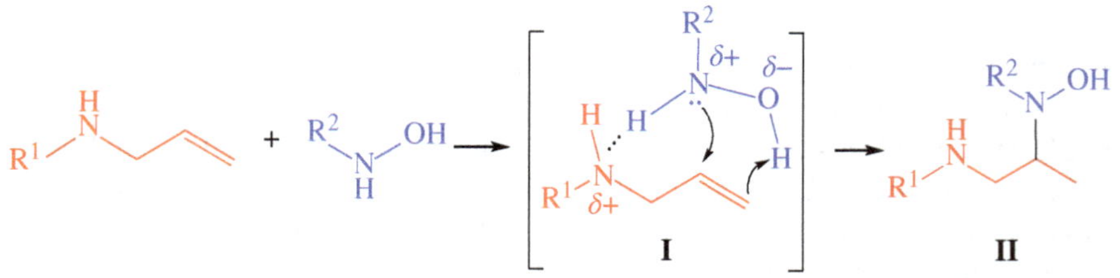

SCHEME 1.5 Cope-type hydroamination of alkenes [14, 15].

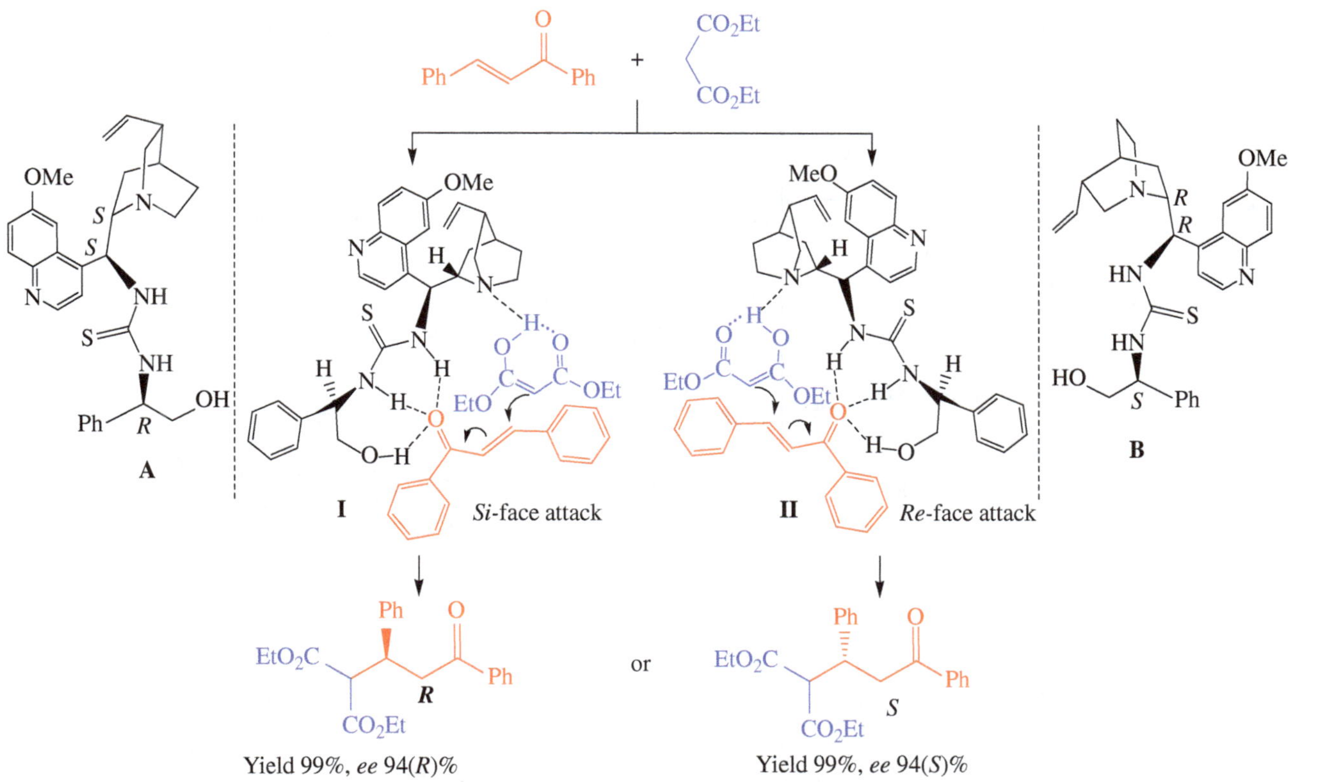

SCHEME 1.6 Asymmetric Michael reaction between chalcone and diethyl malonate [16].

Manifold synthetic methods have been developed for the synthesis of asymmetric vinylcyclopropanes [17], including a hydrogen bond-assisted activation of C=C in methyl cinnamate (Scheme 1.7) [18]. The starting ylide has two conformations, with the vinyl group either *syn* (e.g., in **I** and **IV**; in **V** and **VIII**) or *anti* (e.g., in **II** and **III**; in **VI** and **VII**) to the sulfur lone pair. Hence, theoretically eight hydrogen bond-assisted intermediates can be formed for each ylide (Scheme 1.7). The proposed model with the formation of 10-membered hydrogen bonding ring provides an explanation for the opposite enantioselectivities of *exo* and *endo* sulfur ylide cyclopropanation reactions. Thus, due to the ring constraint and steric CH–O and CH–S hindrances with the norbornyl group, the intermediates **II**, **IV**, **V**, and **VII** are forced to distort. Due to the directionality of hydrogen bonding, the hydroxyl group plays a critical role in the control of enantioselectivity (98 or 99%) and diastereoselectivity (99/1). The solvent also plays an important role in enhancing both diastereo- and enantioselectivity, possibly due of its participation in non-covalent interactions with intermediates [18].

1.2.3 Activation of C≡C Bond

Alkynes can be used as nucleophiles or electrophiles, which provide a wide diversity of reactions in which they can be involved. For instance, intermolecular cycloadditions to alkynes can lead directly to a variety of carbocyclic compounds, but metal-containing catalysts are usually necessary [19]. On the other hand, many substrates in reactions with alkynes can be activated by amines or ammonium salts in combination with inorganic bases; the non-covalent interactions can play an important role in those cases [20]. Thus, hydrogen bond-assisted activation of the C≡C triple bond in acetylenic β-diketones can lead to enantioselective carbocyclization (Conia-ene reaction; Scheme 1.8) [21]. This reaction represents one of the most direct ways to carbocycles and is particularly attractive for the preparation of cyclopentane derivatives. The H-bond acceptor catalyst 3,5-bis(pentafluorosulfanyl)phenylboronic acid is more active in a nonpolar solvent such as toluene and 1,1,1,3,3-pentafluorobutane in comparison to 3-nitro phenylboronic acid, which indicates that the SF_5 function can be a lipophilic and sterically demanding alternative to the NO_2 group in catalyst design. The possible mechanism of this reaction involves H-bond-assisted enolization of starting 2-alkynic β-diketones; a subsequent concerted ene reaction of the enol form affords the product (Scheme 1.8).

H-bond-assisted enantioselective isomerization of 3-alkynoates to chiral allenoates (Scheme 1.9) [22, 23] with benzothiadiazine-1,1-dioxide-type catalysts is of particular importance for the synthesis of many biologically active compounds. The Brønsted base-catalyzed 1,3-proton shift allows to promote the isomerization of alkynoates to allenoates with a high enantioselectivity [22]. The enantioselectivity of the reaction can be explained by the H-bond-assisted deprotonation of both alkynoate enantiomers in intermediates **A** and **B**; this enables their interconversion.

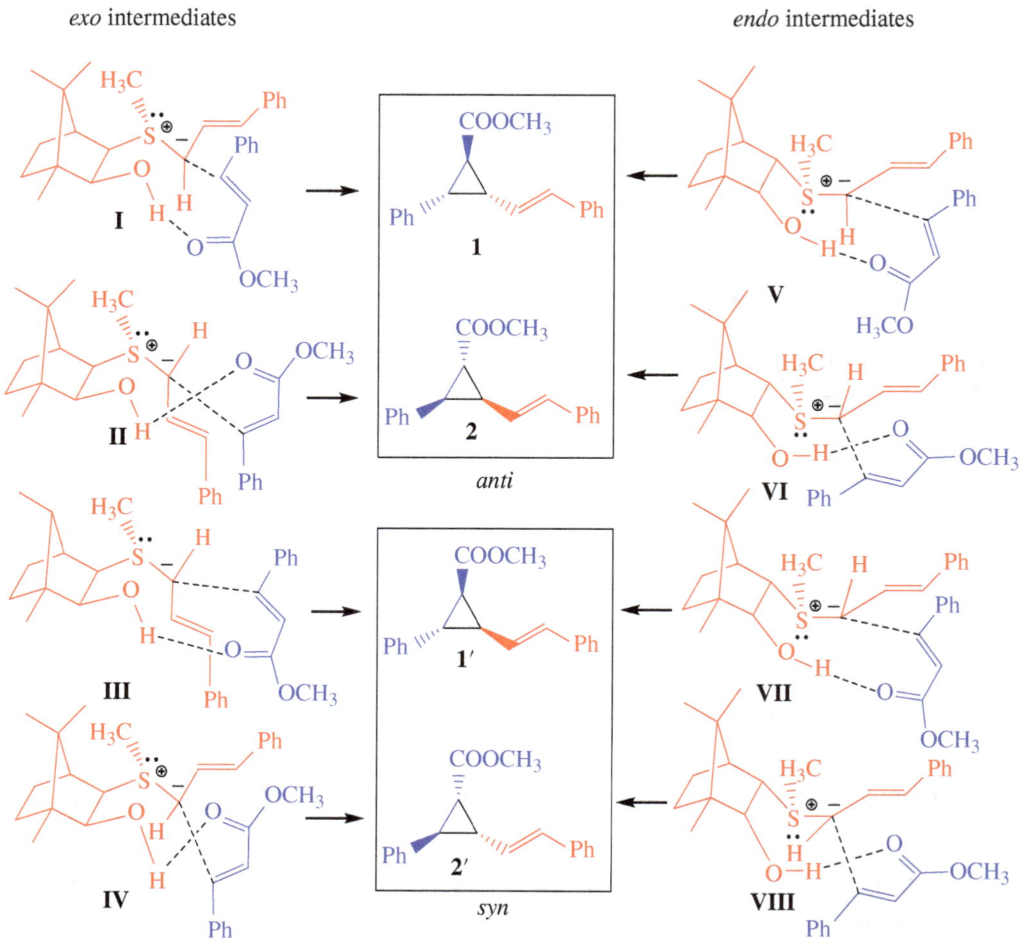

Yield 77%, **1/1′** (>99/1); 99% ***ee***

Yield 86%, **2/2′** (>99/1); 98% ***ee***

SCHEME 1.7 Hydrogen bond-assisted enantio- and diastereoselective synthesis of vinylcyclopropanes [18].

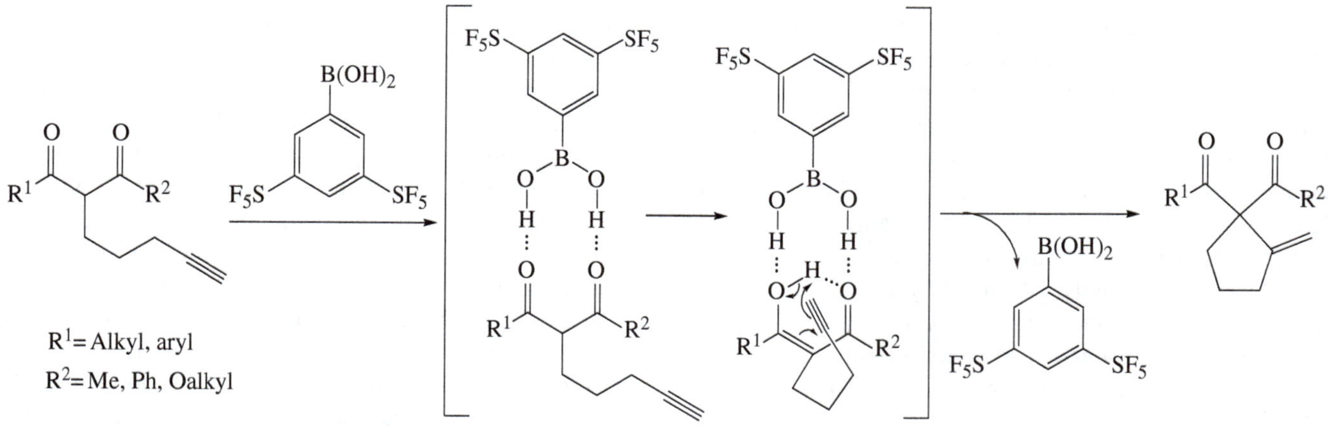

SCHEME 1.8 Enantioselective cyclization of acetylenic β-dicarbonyl compounds [21].

However, the intermediate **C** (*E*-enol), derived from **A**, does not isomerize into (*R*)-allenoate **D**, because the ammonium proton cannot reach the γ-sp-carbon of the alkynoate, while the isomerization of **B** into (*S*)-allenoate **F** occurs smoothly via the H-bond-assisted intermediate *Z*-enol **E** to give the intermediate **F** and finally the targeted allene (Scheme 1.9).

1.2.4 Activation of C—N Bond

To activate the C—N bond in amines and related compounds, different types of catalysts have been used, and in many cases the activation processes occur via an intermediate with hydrogen bonding. For instance, a H-bond-assisted mechanism is

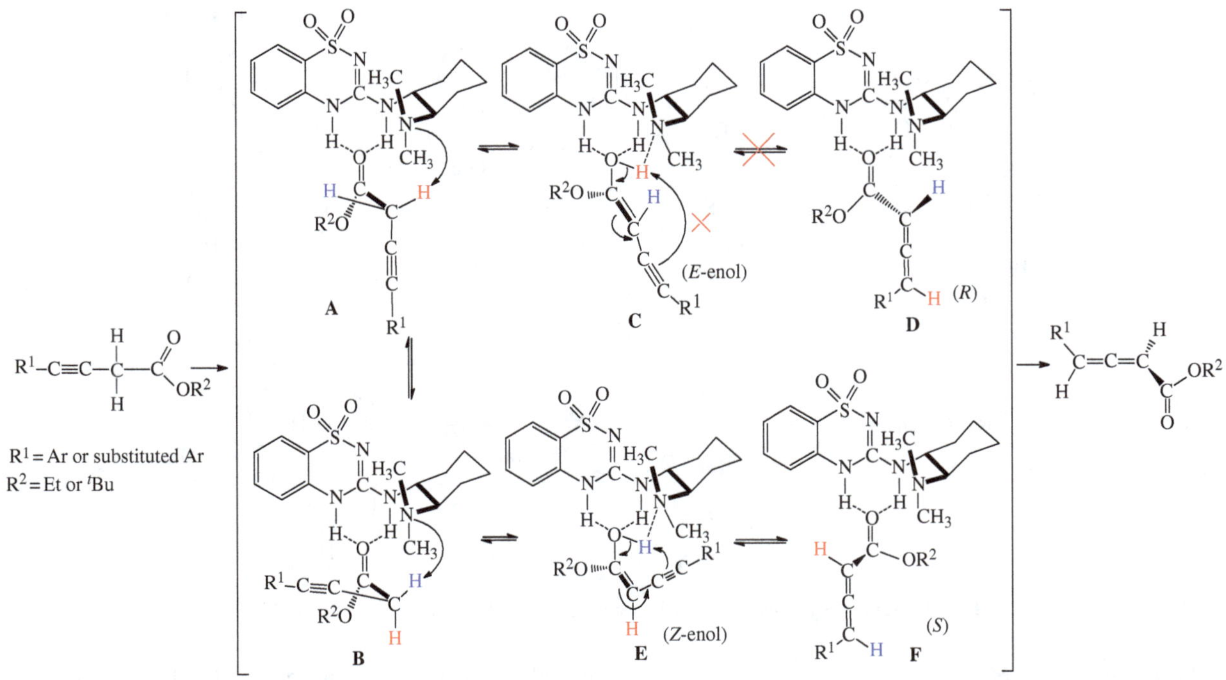

SCHEME 1.9 Activation/isomerization of alkynoates [22].

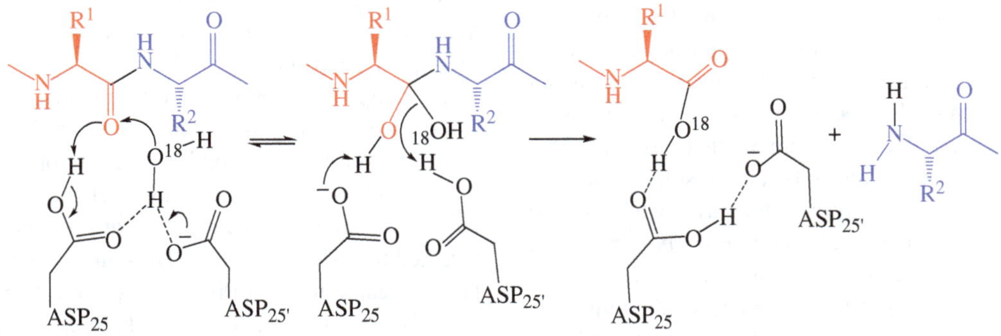

SCHEME 1.10 Polypeptide hydrolysis within aspartic proteases [24].

considered most likely for polypeptide hydrolysis catalyzed by aspartic proteases (Scheme 1.10). According to a proposed mechanism [24], one aspartate residue of the enzyme undergoes protonation, while the second one is ionized. Indeed, there is experimental evidence that the two aspartates share one proton at a physiological pH [24]. Thus, one aspartate activates a water molecule by abstracting a proton; this enables water to attack the carbonyl carbon of the substrate scissile bond, generating a tetrahedral zwitterionic intermediate; its reorganization leads to protonation of the scissile amide. An additional water molecule interconnects the substrate and the main chain amide groups of the enzyme via H-bond and is thought to twist the scissile peptide bond out of planarity, thereby facilitating its cleavage.

Another example of hydrogen bond-assisted activation of a C—N bond is provided by the hydrolysis of carboxyamides to carboxylic acids. Usually, one-step hydrolysis of a nitrile to a carboxylic acid occurs through the formation of an amide as an intermediate and requires harsh acidic or basic reaction conditions; however, a Glu–Lys–Cys catalytic triad can be used as an organocatalyst for the C≡N bond hydrolysis and for the subsequent C—N bond cleavage in nitriles (Scheme 1.11) [25]. The proposed mechanism suggests a nucleophilic attack on the nitrile carbon atom by a conserved cysteine residue of the catalyst; the formed thioimidate affords subsequently a tetrahedral intermediate with the H-bond-assisted addition of water (intermediates **I**→**II**→**III**). In **III**, glutamate acts as a base, while the lysine residue is involved in stabilization of a tetrahedral transition state. Addition of the water molecule leads to the H-bonding intermediate **IV**. Elimination of NH_3 from this intermediate gives a thioester **V**, which reacts with a second water molecule to afford carboxylic acid.

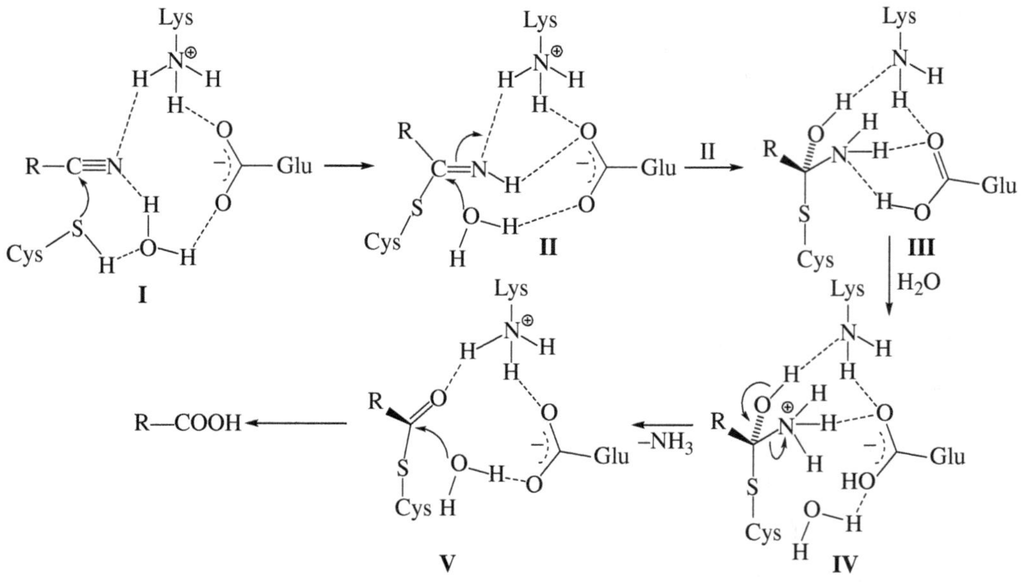

SCHEME 1.11 Hydrolysis of nitriles with the Glu–Lys–Cys catalytic triad [25].

1.2.5 Activation of C=N Bond

The activation of C=N double bonds of imines provides convenient and versatile routes to many organic compounds such as cucurbituril, oxaziridines, optically active amines bearing a stereogenic center, etc. [26]. Recently much attention has been devoted to the development of new methods for stereoselective generation of nitrogen-bonded chiral centers, among the most thoroughly investigated being addition reactions to C=N functionalities [27]. For instance, enantio- and diastereoselective Mannich reactions between a chiral Ni(II) complex of glycine and α-amino sulfones such as *tert*-butyl(phenyl(phenylsulfonyl)methyl)carbamate involve the H-bond promoted creation of a carbon–carbon bond and two stereogenic centers in a single run (Scheme 1.12) [28]. This method represents an attractive route to the synthesis α,β-diamino acids, versatile chiral auxiliaries and ligands for asymmetric synthesis, and medicinal and peptide/peptidomimetic chemistry [28, 29]. The proposed mechanism (Scheme 1.12) involves three steps: (i) enolization of a (*S*)-Ni(II) complex, promoted by 1,8-diazabicyclo[5.4.0]-undec-7-ene, which creates a hydrogen bond with substrate to form the intermediate **A** (benefiting 9.46 kcal/mol); (ii) reaction of Ni(II)-enolate with *tert*-butyl(phenyl(phenylsulfonyl) methyl)carbamate to give *syn* and *anti* diastereoisomers of compound **B**; and (iii) release of an α,β-diamino acid as a HCl adduct and $NiCl_2$ with the intramolecular hydrogen bond-assisted recovery of the ligand precursor.

The enantioselective hydrogenation of imines is used as one of the most efficient and convenient methods for the preparation of chiral amines and their derivatives. Thus, asymmetric synthesis of nonracemic 1,3-diamines by hydrogen bond-directed diastereoselective reduction of enantiopure *N-tert*-butanesulfinylketimines (Scheme 1.13) involves *E*/*Z* isomerization of *N-tert*-butanesulfinylketimines, which was found to depend on the solvent [30]. The correlation between facial selectivity of the reduction and *E* or *Z* geometry of the starting ketimines suggests the involvement of a cyclic transition state stabilized with hydrogen bond. The *ortho*-substituent participates in hydrogen bonding thus controlling the geometry of *N-tert*-butanesulfinylketimines: *E*-imines are reduced to give *R* configuration at the newly created chiral center, while the *Z*-isomer affords the *S* configuration.

Nucleophilic addition of nitroalkanes to imines to give β-nitroamine derivatives, known as aza-Henry reaction, is another useful C=N bond activation process [31]. Many organocatalysts are applied to reach higher diastereo- and enantioselectivities, which, in many cases, are essentially directed by H-bonding [32]. Thus, upon the asymmetric aza-Henry reaction of ketimines derived from isatins with nitroalkanes using cinchona alkaloids as organocatalysts (Scheme 1.14), the tertiary amine group of the catalyst deprotonates the α-proton of nitroethane, activating it for nucleophilic attack on the *Re*-face of ketimines, which, in their turn, are activated through hydrogen bonding with the —OH group of the catalyst, providing the (*R*)-enantiomer of the product [33]. Hence, high diastereo- and enantioselectivities were obtained in this reaction through the synergistic activation of ketimines and nitroalkanes by a bifunctional alkaloid organocatalyst.

1.2.6 Activation of C≡N Bond

Reactions of electrophiles or nucleophiles with organonitriles are among the most significant ones in organic synthesis [34, 35]. Many methods have been developed to activate the C≡N bond, namely, involving the use of an electron-withdrawing R group in the R—C≡N nitrile molecule, protonation or alkylation of the nitrile N atom, or coordination to a metal center [34]. These methods suffer from various drawbacks,

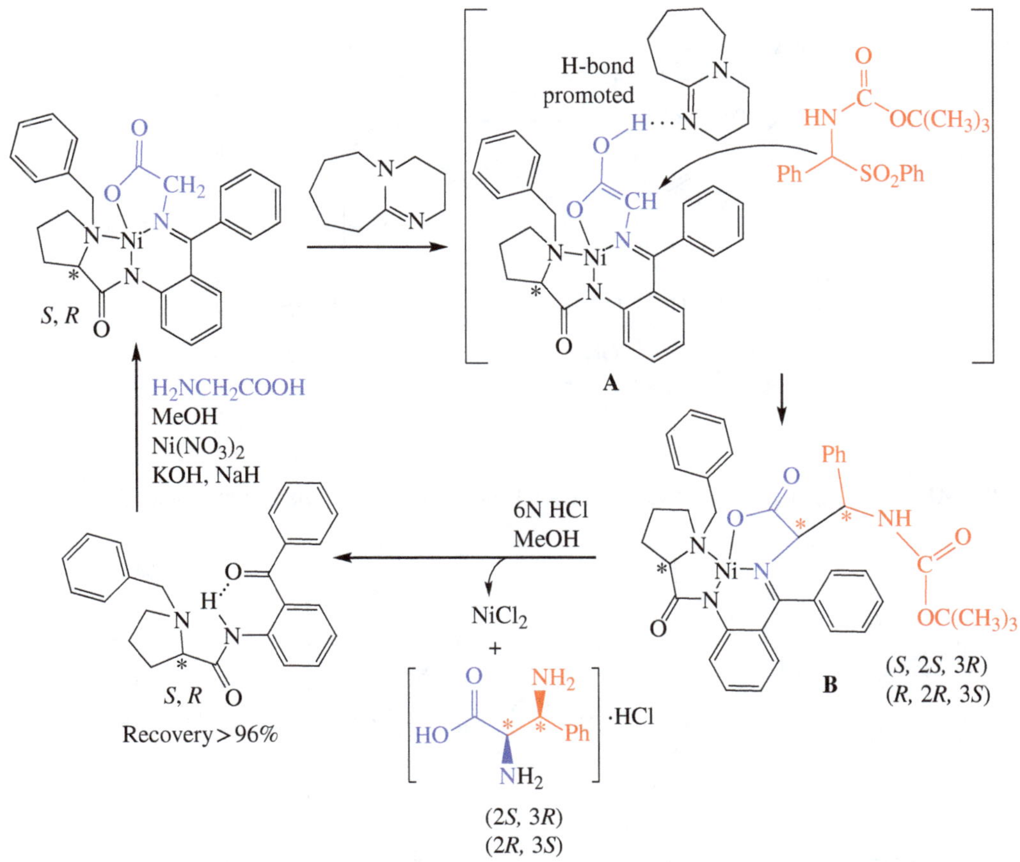

SCHEME 1.12 Stereoselective synthesis of enantioenriched α,β-diamino acids via a chiral imine complex of Ni(II) with sulfones [28].

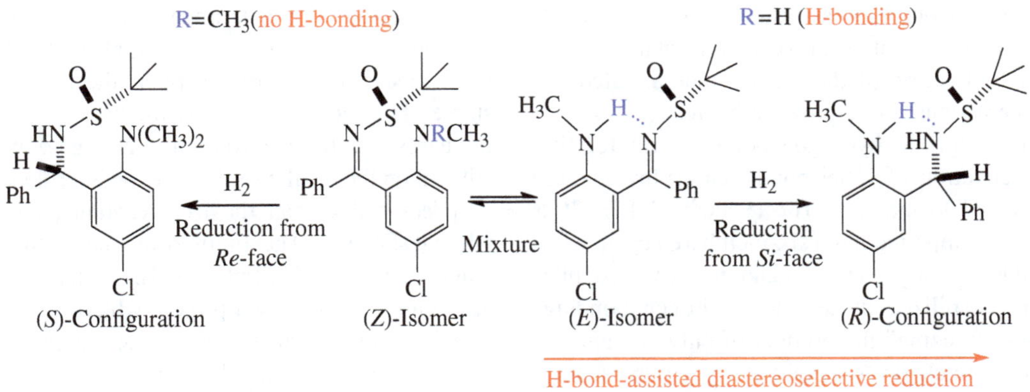

SCHEME 1.13 Diastereoselective reduction of diaryl *N-tert*-butanesulfinylketimines to nonracemic 1,3-diamines [30].

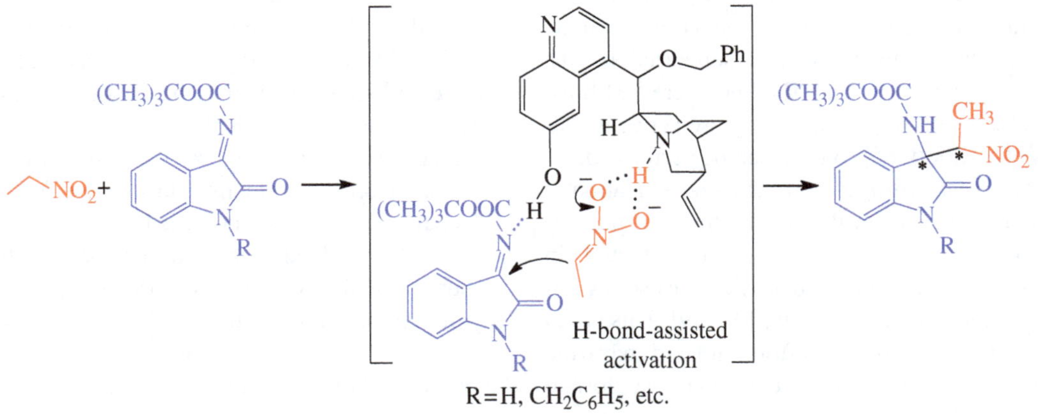

SCHEME 1.14 Proposed dual H-bond-assisted activation for the asymmetric aza-Henry reaction of nitroethane with ketimines [33].

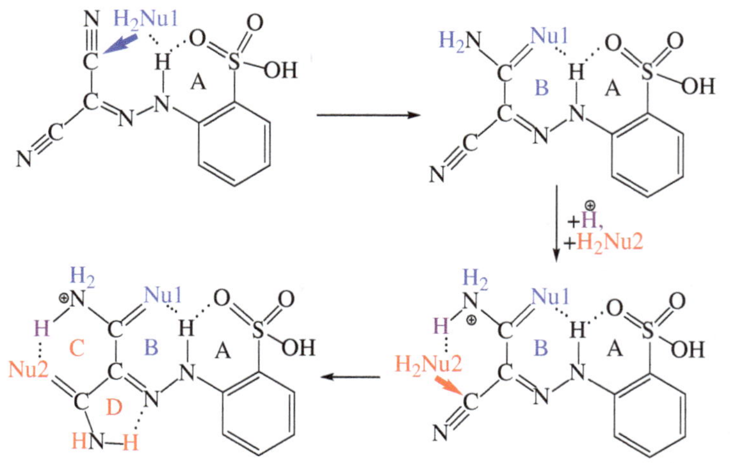

SCHEME 1.15 Regioselective activation of C≡N bonds in an arylhydrazone of malononitrile [36].

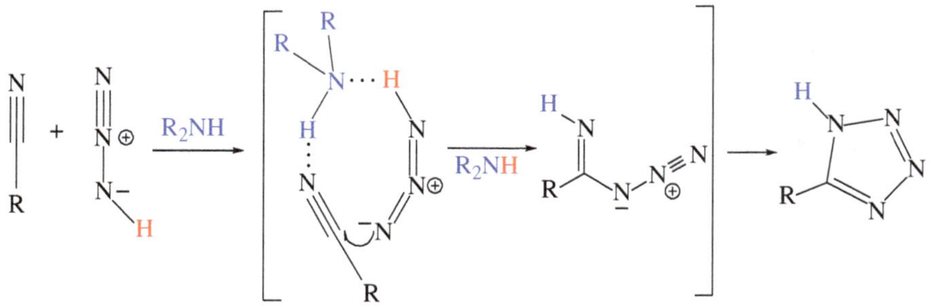

SCHEME 1.16 Proposed [37] mechanism for the direct synthesis of tetrazoles.

for example, high cost of the catalysts, difficulties in the separation of the product and catalyst, the use of an inert atmosphere for handling air-sensitive metal catalysts, etc. As a result, the activation of the C≡N bonds assisted by non-covalent interactions is of a growing interest.

To create a supporting hydrogen bond system for the regioselective activation of C≡N bonds, an arylhydrazone unit was introduced to malononitrile (Scheme 1.15) [36]. The thus obtained compound has a six-membered cycle with hydrogen bonding (A); the product of addition of a nucleophile (Nu1) is further stabilized by another hydrogen bonding system (B), thus "freezing" the product of only one attack. The nucleophilic attack on a second C≡N group is only possible after acidification of this product: addition of an acid leads to protonation of the nitrogen atom of the amino group, promoting the formation of another hydrogen bond, which assists nucleophilic attack by the second nucleophile (Nu2). Stabilization of this product is supported by the formation of two more six- and five-membered H-bond cycles (C and D). Hence, a convenient regioselective metal-free syntheses of a variety of products can be performed.

Another example of hydrogen bond-assisted activation of C≡N bonds is in field of the synthesis of tetrazoles, popular compounds as catalysts, propellants, explosives, ligands in coordination chemistry, and nonclassical isosteres of carboxylic acids in medicinal chemistry [37–39]. This broad utility has prompted significant efforts toward tetrazole synthesis; the most direct method is via the formal cycloaddition of azides and nitriles. Thus, interaction of azide salts with a nitrile under certain conditions produces the corresponding tetrazoles in high yields (Scheme 1.16) [38]. Mechanistic studies support the formation of a quite stable H-bond-assisted intermediate [37]. Its stability increases with the electron-withdrawing potential of the substituent on the nitrile: for R = Me the intermediate is 3.3 kcal/mol less stable than the free reactants, HN_3 and MeCN, while it shows 11.4 kcal/mol lower energy than the free reactants, when R = $CH_3S(=O)_2$—. The transition state involves activation of nitrile by a proton, which facilitates the attack of azide on the carbon of nitrile, and then a 1,5-cyclization occurs to give the tetrazole [37].

Hydrogen bond-assisted interaction of nitrile oxide with aliphatic or electron-rich aromatic nitriles leads to 3-functionalized 1,2,4-oxadiazoles (Scheme 1.17); the reaction proceeds under mild conditions even in the absence of a Lewis acid [40]. Among functionalized 1,2,4-oxadiazoles, carbamoyl-substituted ones have attracted much attention because of their biological activity and utility for diabetes treatment [40]. Furthermore, the carbamoyl group can be used as a precursor to introduce other functional groups. Mechanistic studies support the inverse electron-demanding 1,3-dipolar cycloaddition and suggest that the carbamoyl group of nitrile oxide activates the dipolarophilic nitrile by hydrogen bonding (Scheme 1.17). Thus, the carbamoyl group of nitrile oxide plays an important role not only as an electron-withdrawing group but also as a H-donor to form a

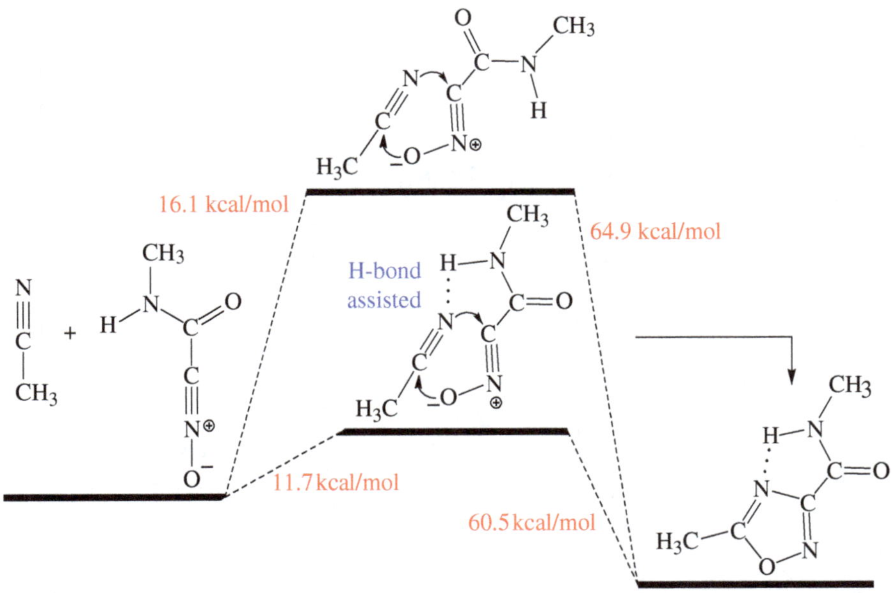

SCHEME 1.17 Possible ways for the 1,3-dipolar cycloaddition of nitrile oxides to acetonitrile [40].

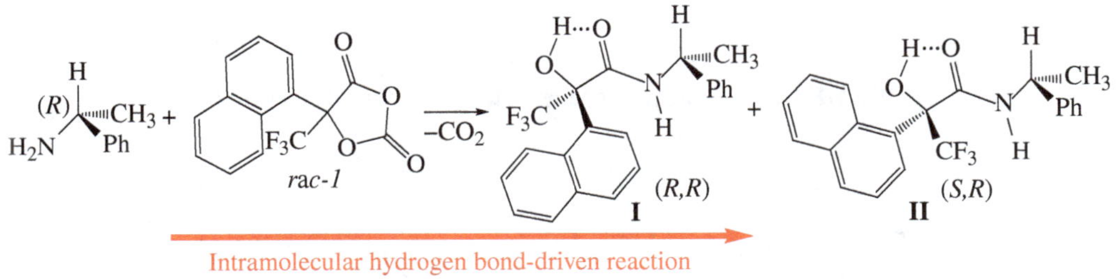

SCHEME 1.18 Reaction of *rac*-(1-naphthyl)(trifluoromethyl) *O*-carboxy-anhydride with (*R*)-α-methylbenzylamine [41].

hydrogen bond with a dipolarophilic nitrile. In order to probe the effect of the hydrogen bond, both intermediates, in which only the orientation of the carbamoyl group is different, were considered (Scheme 1.17). The H-bond-assisted intermediate was found to be more stable than non-H-bonded one in 4.4 kcal/mol; hence, the formation of a hydrogen bond with the carbamoyl group of nitrile oxide facilitates the inverse electron-demanding cycloaddition.

1.2.7 Activation of C—O Bond

The hydrogen bond-driven reaction between *rac*-(1-naphthyl) (trifluoromethyl) *O*-carboxy-anhydride and (*R*)-α-methylbenzylamine afforded an equimolar mixture of α-hydroxyamides **I** and **II** (Scheme 1.18) [41]. Due to strong intramolecular hydrogen bonding, high $\Delta\delta^{RS}$ values were observed in the ^{1}H NMR spectra of the obtained compounds and have been correlated with a marked preference of the corresponding α-hydroxyamides for the *eclipsed conformation*. This fact can be explained by maximization of the anisotropic effect of the naphthyl group. For comparison, the related O-methylated amides are shown to adopt staggered conformations, which substantiate the critical role of intramolecular hydrogen bonding in maximizing the anisotropic effect. Accordingly, this *O*-carboxy-anhydride can be considered a promising chiral derivatizing agent; hence, further improvements can be expected in both reactivity and resolution efficiency by varying the substitution pattern.

Another example of the hydrogen bond-assisted C—O bond activation is the hydrolysis of esters to acids and alcohols, catalyzed by the ionic liquid cholinium hexanoate, $[(CH_3)_3NCH_2CH_2OH]^+[C_5H_{11}COO]^-$ (Scheme 1.19) [42].

A mechanism of the reaction involves two steps. (i) The hydroxyl group of the cholinium cation establishes a hydrogen bond with the oxygen atom of the carbonyl group of the ester, favoring the nucleophilic attack of water at the carbon (intermediate **I**). As a result, a proton of water molecule shifts to the $C_5H_{11}COO^-$ anion (intermediate **II**). (ii) Intramolecular proton transfer leads to six-membered H-bond system, weakening the C—O bond in intermediate **III**. Finally, the products, the acid and the alcohol, are formed, and the ionic liquid is regenerated, closing the H-bond-assisted catalytic cycle.

1.2.8 Activation of C=O Bond

Activation of the C=O double bond in carbonyl compounds is a strategically important method for the synthesis of a large number of building blocks, such as β-hydroxy carbonyl

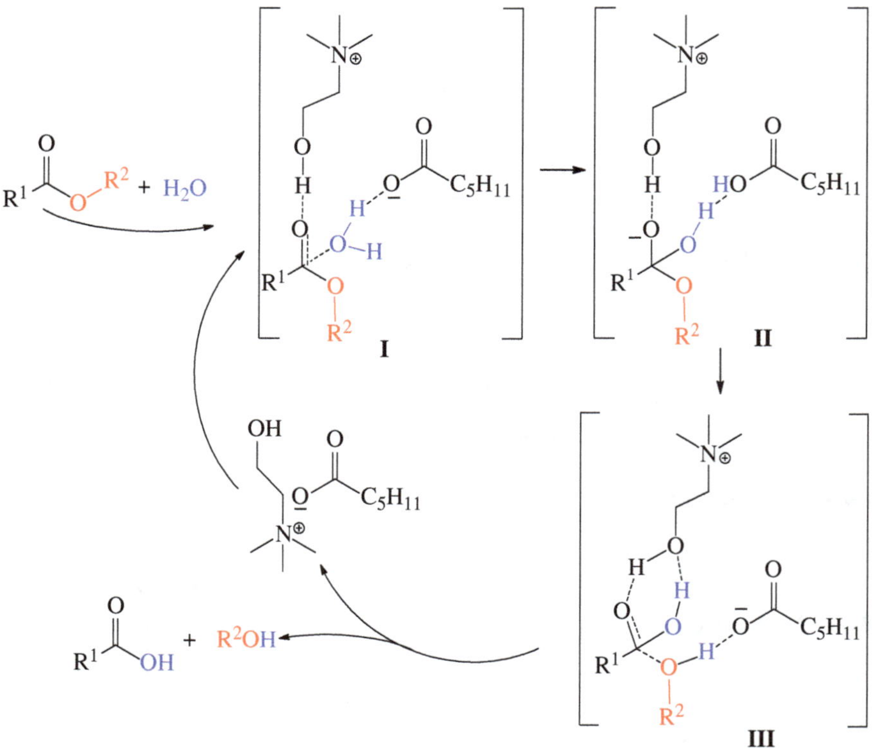

SCHEME 1.19 Hydrogen bonding in the hydrolysis of ethers [42].

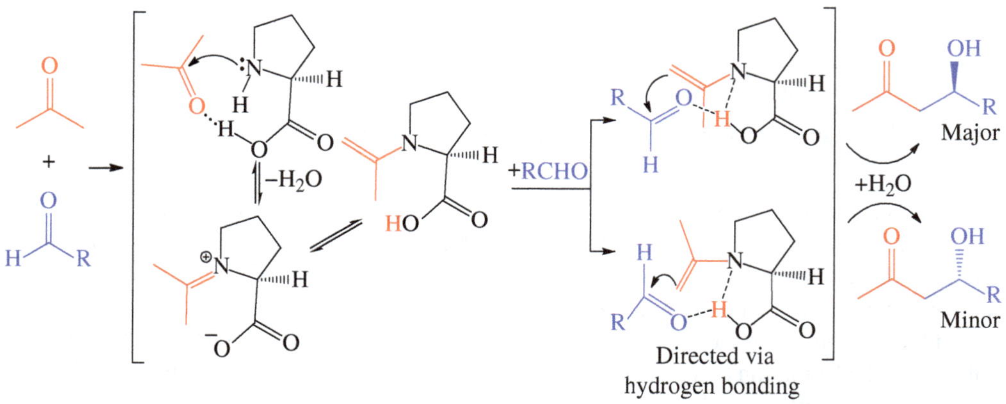

SCHEME 1.20 Proline-catalyzed asymmetric aldol reaction [45].

compounds (Aldol reaction), β-nitro alcohols (Henry or nitro-aldol reaction), thioacetals, hemiacetals, hemiketals, cyanohydrins, etc. [43]. Numerous methodologies, including metal-mediated approaches, for the C=O bond activation have been developed [43, 44]; the viability of non-covalent interactions in the aldol and related reactions has also been proved [45–48]. In those cases, non-covalent interactions weaken the C=O bond of carbonyl compound via hydrogen bonding with the lone pair of the oxygen atom and make it more susceptible for nucleophilic attack. It is not surprising that a plethora of carbonyl chemistry can be performed simply by adding a catalytic quantity of a strong protic acid (proton donor) to a mixture of carbonyl compound and a nucleophile. The formation of β-nitro alcohols from carbonyl compounds and nitroalkanes serves as an example.

Asymmetric aldol reaction involves the C=O bond activation and allows to build a variety of pharmaceutically relevant substances, in particular those with polyoxygenated subunits. Thus, the proline-catalyzed asymmetric aldol reaction is supposed to occur via an enamine intermediate (Scheme 1.20) [45]. In this case, the intermolecular hydrogen bond assists the nucleophilic attack of the amino group of pyroline to the C=O carbon resulting in the pyrrolidinum carboxylate formation and elimination of a water molecule. Hence, proline plays a dual role, forming, as secondary amine, a key nucleophilic enamine intermediate (activation of acetone), while the carboxylic acid group activates the aldehyde toward nucleophilic attack of the enamine. The enantioselectivity of the reaction can be related to the stability of the H-bond-assisted intermediates, which lead to the major product [45].

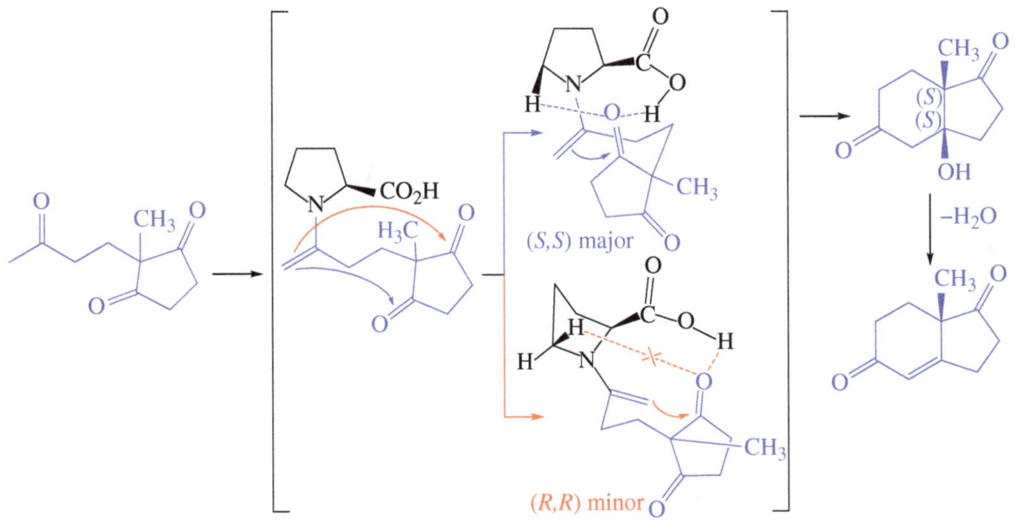

SCHEME 1.21 C—H···O interaction-directed asymmetric Hajos–Parrish reaction [47].

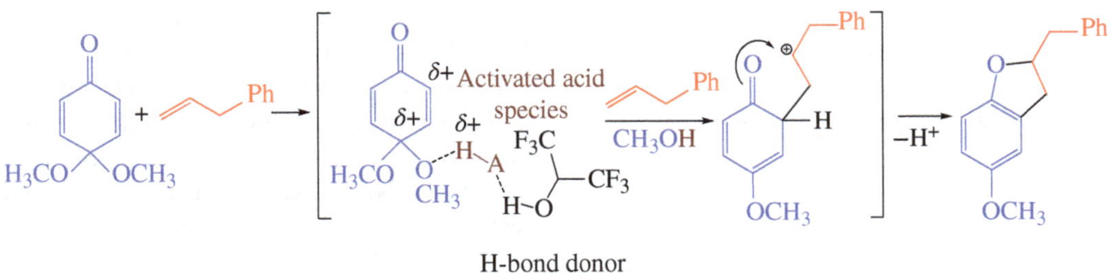

SCHEME 1.22 Synthesis of benzofurans based on non-covalent interactions of Brønsted acids with hexafluoroisopropanol [49].

Although the C—H···O/N contacts are weaker than typical hydrogen bonds, they still provide enough stabilization to render control of selectivity [46]. The C—H···O interactions were also found to be important in controlling the stereoselectivity in the Hajos–Parrish-type aldol and Mannich reactions. Thus, proline-catalyzed intramolecular Hajos–Parrish transformation of 2-methyl-2-(3-oxobutyl) cyclopentane-1,3-dione (Scheme 1.21) was shown to be directed by C—H···O interactions [47].

The obtained high stereoselectivity is explained by the following facts: (i) the greater iminium planarity distortion in one H-bonded (*R*,*R*) *syn*-enamine compared to two H-bonded *anti* (*S*,*S*) ones and (ii) the ability of a prolinyl C(sp^3)—H to stabilize the developing negative charge on the carbonyl oxygen, the major (*S*,*S*) product exhibiting a shorter C—H···O interaction and a lower energy (by 3.4 kcal/mol) than the minor (*R*,*R*) product where this interaction is more distant and presumably weaker.

Quinone monoacetals possess both α,β-unsaturated carbonyl and allyl acetal functionalities in one skeleton and are interesting due to their broad utilities in organic transformations, as intermediates and important building blocks for the synthesis of natural products [48]. The reactions of nucleophilic attack on quinone monoacetal carbons include, for instance, addition to the carbonyl carbon (i.e., 1,2-additions), conjugated addition to the enone moiety (e.g., 1,4-additions), and cyclizations involving those processes. Regarding the chemo- and regioselective reactions, many applied organic and inorganic catalysts relay on cooperative action of non-covalent interactions, coordination, etc. [48]. The strategy for the [3+2] coupling of quinone acetal with a series of alkene nucleophiles (Scheme 1.22) utilizes a Brønsted acid activated upon hydrogen bond with the donor solvent, hexafluoroisopropanol [49]. This polar solvent has a high hydrogen bond donor ability and does not behave as a hydrogen bond acceptor. The H-bonded intermediate allows cyclization of the carbonyl moiety in the keto-type tautomer, which is accompanied by aromatization as an additional driving force to afford the formal [3+2] coupling product.

1.2.9 Activation of C—S Bond

The transformations of the thiol (R—SH) or thioether (R—S—R′) groups and their derivatives are key steps of many enzymatic reactions and metabolic processes. For instance, coenzyme A (HSCoA) is a cocatalyst for activations of a number of substrates in organisms; as an example, the conversion of (*S*)-citryl-SCoA to citrate in the Krebs cycle (Scheme 1.23) can be mentioned. The reaction occurs by H-bond-assisted nucleophilic attack of enolate or thioester on the α-carbon of citrylate [50]. The hydrolysis

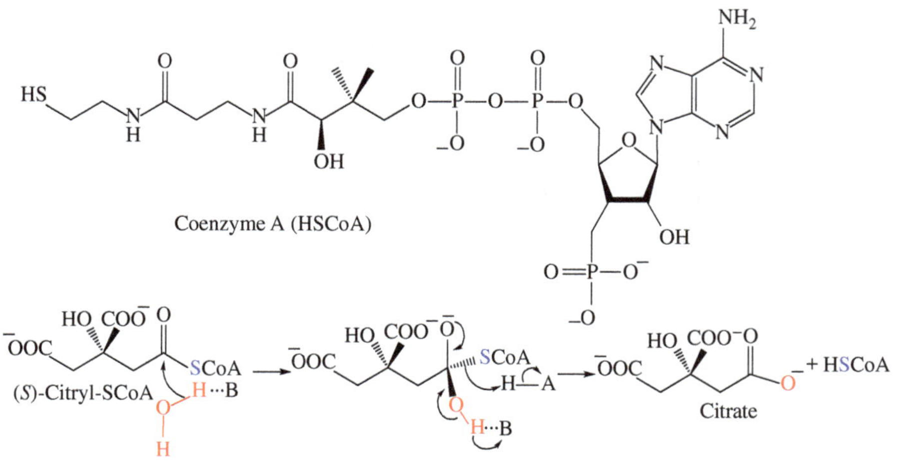

SCHEME 1.23 Hydrolysis of thioether to citrate in the Krebs cycle.

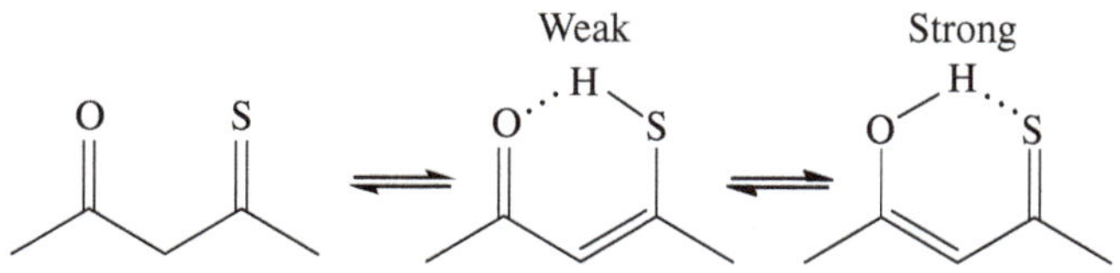

SCHEME 1.24 Keto–enol tautomerism in 4-thioxopentan-2-one.

of the thioester (citryl-SCoA) formed upon aldol condensation provides a significant free energy reduction ($\Delta G^{\circ\prime}=-31.5$ kJ/mol or -7.5 kcal/mol), which drives forward the net reaction.

1.2.10 Activation of C=S Bond

Stereospecific nucleophilic attack on the carbon atom of C=S is a simple and versatile way to construct stereocenters next to heteroatoms with an overall inversion of stereochemistry [51]. In many cases, intermolecular and intramolecular hydrogen bonds not only control the stereochemistry but also influence the reactivity and other properties of thioketones or thioaldehydes. The C=S bond in thioketones is longer than the corresponding C=O bond in ketones, and its length also depends on the substituents. The sulfur atom is a weaker hydrogen bond acceptor to oxygen, and accordingly, the intramolecular O—H···S hydrogen bonding is preferable to the S—H···O one (Scheme 1.24) [52]. This effect is manifested in the reactivity of 4-thioxopentan-2-one: only the regioselective thioketo transformation is found in its condensation with phenylhydrazine (Scheme 1.25) [53].

1.3 HALOGEN BOND-ASSISTED ACTIVATIONS

Halogen bonding concerns non-covalent interactions between terminal halogen atoms in compounds of the type R—X (X=Cl, Br, I) and Lewis bases [7, 54]. Stronger halogen bonds are formed when R is highly electronegative, for example, in the case of polyfluorinated alkyl or phenyl substituents. Several primary differences between halogen and hydrogen bondings can be mentioned [54]: (i) halogen bonds tend to be more directional than (single) hydrogen bonds; (ii) the strength of a halogen bonding can be easily tuned, while in the case of a hydrogen bond, considerable changes are required into the residue the donor site is bound to; (iii) halogen bonds are hydrophobic, whereas hydrogen bonds are hydrophilic; (iv) the size of the bond donor (halogen) atom is significantly larger than hydrogen. The halogen bonding has found some useful applications in crystal design and engineering, molecular recognition, synthesis, catalysis, molecular conductors, liquid crystals, and bioorganic chemistry [7, 54]. The role of halogen bonding in organic synthesis is also becoming more prominent.

Thus, the fluoronium cation F^+ derived from an *N*-fluoroheterocyclic salt plays the role of a convenient and effective catalyst to mediate the interaction of *N*-substituted imines and ethyl diazoacetate affording *N*-substituted aziridines (Scheme 1.26) [55]. The highly electrophilic nature of F^+ allows easy imine activation and subsequent nucleophilic attack by ethyl diazoacetate.

The N···X halogen bonding (X=Br, I) can be also applied for the activation of C=N bonds by haloalkanes toward electrophilic hydride additions. In particular, 2-phenylquinoline was reduced by using a Hantzsch ester (diethyl 2,6-dimethyl-1,4-dihydropyridine-3,5-dicarboxylate) as a reductant (Scheme 1.27) [56]. While there was no reaction in the absence of a haloalkane, yields up to 98% were achieved in the presence of 10 mol% perfluoroiodooctane, $CF_3(CF_2)_7I$. No signal shifts were found in the ^{1}H NMR spectra; this contrasted to the ^{13}C NMR behavior, where the quinoline signals were slightly shifted to lower field (by 0.01–0.06 ppm). This is indicative of a weak interaction between the quinoline nitrogen and iodide of the perfluoro compound, which is supported by the ^{19}F NMR spectra, in which the signal for the CF_2I group resonated at a lower magnetic field ($\Delta\delta=0.1$ ppm), and all

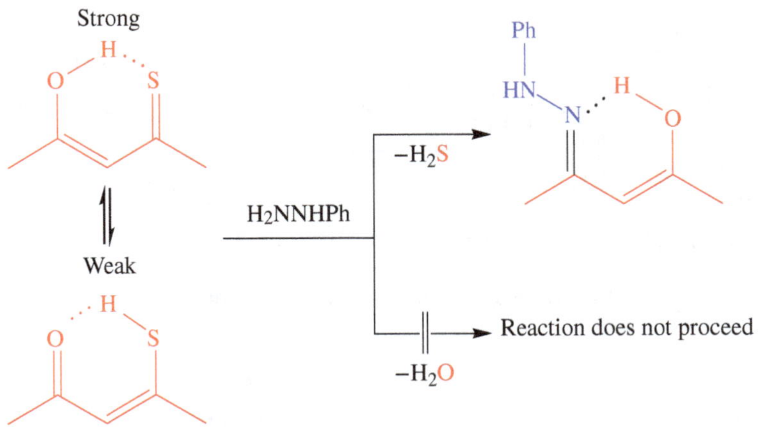

SCHEME 1.25 Regioselective synthesis of (2*Z*,4*E*)-4-(2-phenylhydrazono)pent-2-en-2-ol [53].

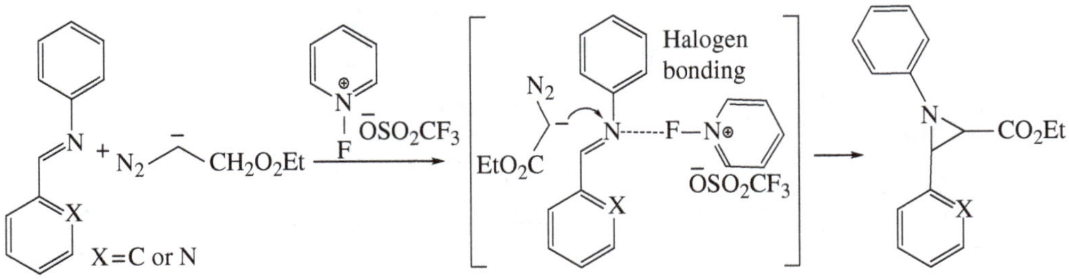

SCHEME 1.26 Synthesis of *N*-arylaziridine [55].

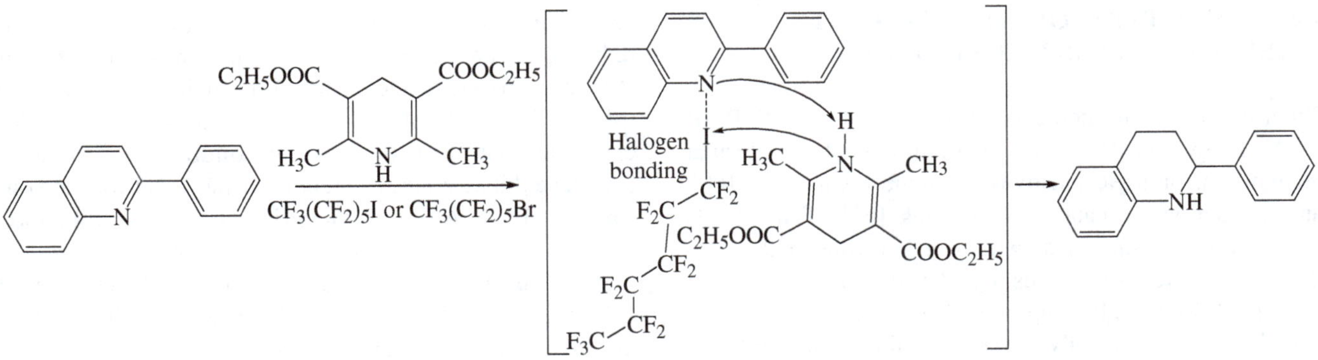

SCHEME 1.27 Reduction of 2-phenylquinoline [56].

chemical shifts are strongly dependant on concentration. The obtained results can be related to the fact that the fluorinated iodoalkanes, such as $CF_3(CF_2)_7I$ or $CF_3(CF_2)_7Br$, form strong halogen bonds with the sp^2-type nitrogen atoms [56].

Generally, the electrophilic character of halogen atoms in halogen bonding is significant when they are bonded to a strong electron-withdrawing group, such as a fluorocarbon chain, as in $CF_3(CF_2)_7I$. This high halogen-bonding ability of perfluoroalkyl halides can be used as an alternative to covalent perfluorinated ponytails to drag hydrocarbon reactants from an organic phase [57]. Thus, the halogen-bonded adduct between 1,4-diazabicyclo[2.2.2]octane (DABCO) and two molecules of perfluorooctyl iodide (DABCO·$(C_8F_{17}I)_2$) was applied as a recyclable supramolecular fluorous organocatalyst for the addition of aromatic aldehydes to methyl acrylate (Morita–Baylis–Hillman reaction; Scheme 1.28) [58]. The adduct DABCO·$(C_8F_{17}I)_2$ was formed through halogen bonding; after the catalytic reaction, DABCO·$(C_8F_{17}I)_2$ is easily recoverable by filtration.

A halogen-containing catalyst also activates the C—Br bond of benzhydryl bromide in its reaction with deuterated acetonitrile (Scheme 1.29) [59]. Hydrolysis of the resulting nitrilium intermediate by traces of water yielded *N*-benzhydryl acetamide in good yield. Without the catalyst or in the presence of its H-analogue, the yield of the reaction drops to less than 5%. Hence, the halogen-bonding promoter does successfully accelerate this reaction, the best activation being observed in the presence of the BF_4^- anion (yield 97%), while the reaction with CF_3SO_3H afforded 25% yield of the product.

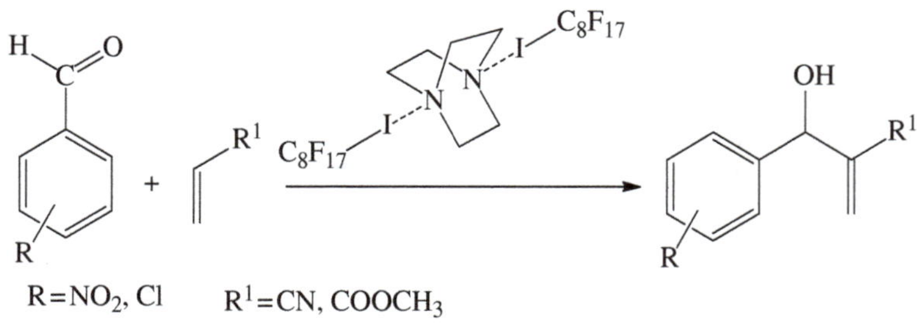

SCHEME 1.28 Baylis–Hillman reaction [58].

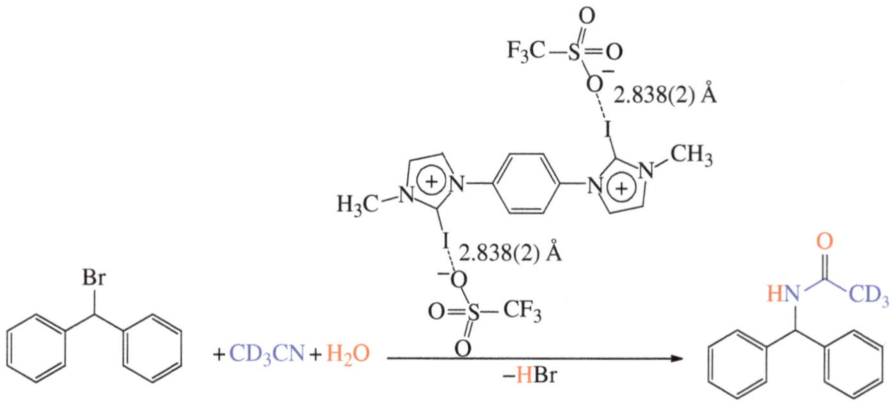

SCHEME 1.29 Synthesis of *N*-benzhydryl acetamide [59].

1.4 OTHER TYPES OF NON-COVALENT INTERACTIONS AND PERSPECTIVES

Other non-covalent interactions such as S···O, S···N, P···P, P···N, B···N, and B···O are also important for molecular recognition, biological activity of some compounds [60], and regulation of catalytic functions [61]. Thus, S···O interactions are supposed to play a key role in the synthesis of chiral sulfoxides by the Pummerer reaction (Scheme 1.30) [62]. It is supposed that coordination of acetic anhydride to trimethylsilyl triflate in amide (or lactam) generates an intermediate **I**. In parallel, an interaction of chiral sulfoxide with amide (or lactam) leads to intermediates **II** and **III** bearing the inter- and intramolecular S···O interactions. It appears that structure **II** is more stable than **III** due to the steric repulsion between the aryl group and the carbonyl group in **III**. Subsequently, acetylation of the sulfonyl oxygen atom of **II** with acetic anhydride in the presence of active species **I** affords a chiral sulfurane-type intermediate **IV**. The subsequent abstraction of the methylene α-hydrogen atom by $CF_3S(=O)_2O^-$ from **IV** generates a rigid ylide intermediate **V** supported by both inter- and intramolecular S···O interactions. The stereoselective 1,2-acetoxy transfer in **V** proceeds via three kinds of plausible modes, intimate ion pair, sliding, and cyclic, to furnish the desired chiral Pummerer product.

Intramolecular nonconventional S···X (X = O or S) interactions were found in a large number of organosulfur compounds, such as acetazolamides, thiadiazolines, natural antitumor antibiotic leinamycin, cyclic sulfilimine, or 1,3-dithiole, and have shown to be important in controlling their structural properties and chemical reactivity. For instance, a role of intramolecular S···O interactions in the regiochemical control of the 4-oxothiazolidine → 1,2-dithiole-3-ylidene thione rearrangement (Scheme 1.31) was demonstrated [63, 64]. A ring-opening/ring-closing transformation of selected 4-oxothiazolidine enaminoketones in the presence of Lawesson's reagent produces functionalized 1,2-dithiole-3-ylidene thiones. This reaction is initiated by the directional non-covalent 1,5-type S···O interaction in the 4-oxothiazolidines. The S···S distances, being 2.3374(5) and 2.3408(5) Å in 1,2-dithiole-3-ylidene thiones, are shorter than the sum of the van der Waals radii but relatively long in comparison to a S—S covalent bond (2.08 Å) and not exactly of equal length, as the two fused rings differ (see isomerization of 1,2-dithiole-3-ylidene thiones in Scheme 1.31).

Another sort of non-covalent interaction, the pnicogen bond, concerns group 15 elements (N, P, or As, i.e., a pnictogen element) and a Lewis base and has also drawn attention [8]. Different from the hydrogen bond but similar to halogen and chalcogen bonds, a pnicogen bond (X)A(H)···D is formed with the electrons transferred from the Lewis base D to the pnicogen atom A, which acts as a Lewis acid and shows a high degree of anisotropy. X represents substituent group(s) and H is hydrogen atom(s) bonded to A but not directly interacting with D. For instance, B···N, P···N, As···N, S···N, and P···P interactions are recognized [8] but still without respective applications in synthesis.

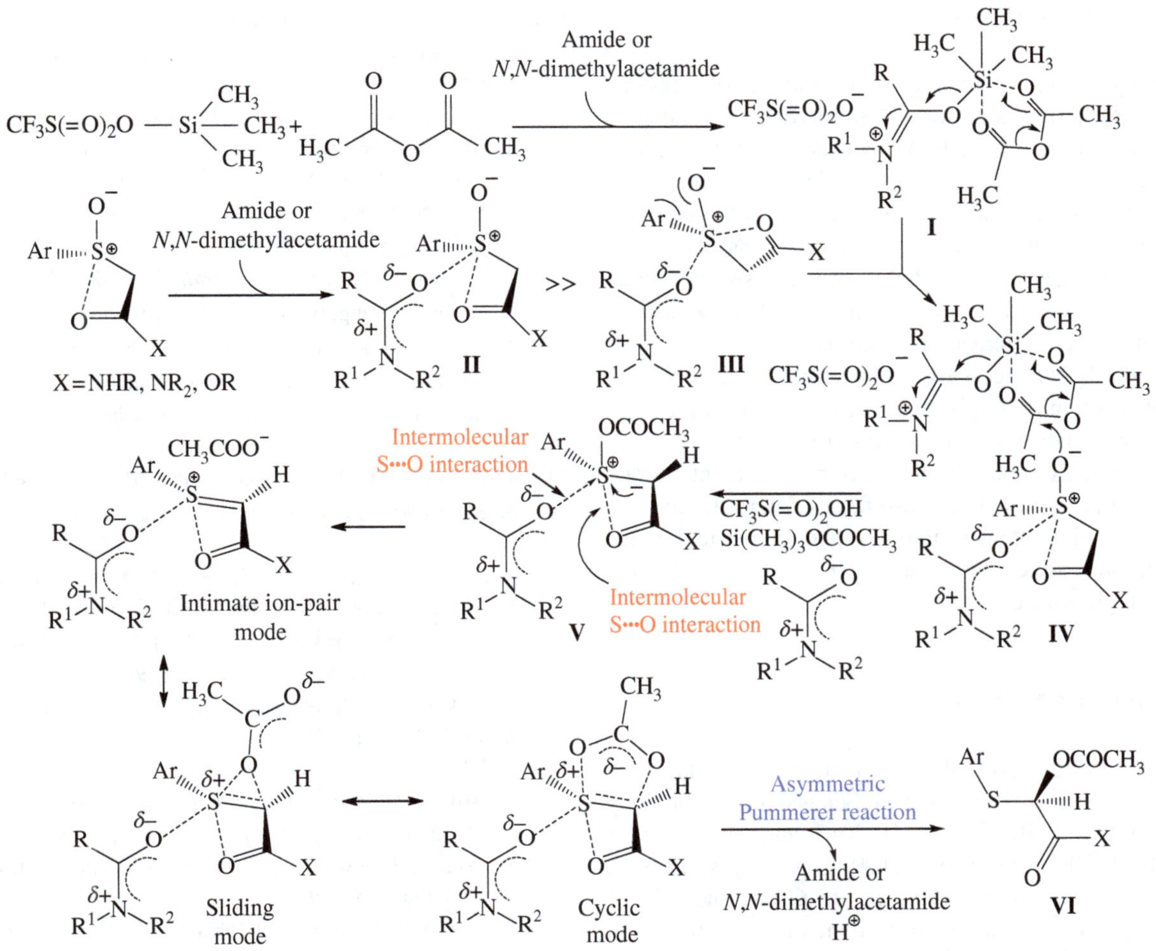

SCHEME 1.30 Asymmetric Pummerer reaction [62].

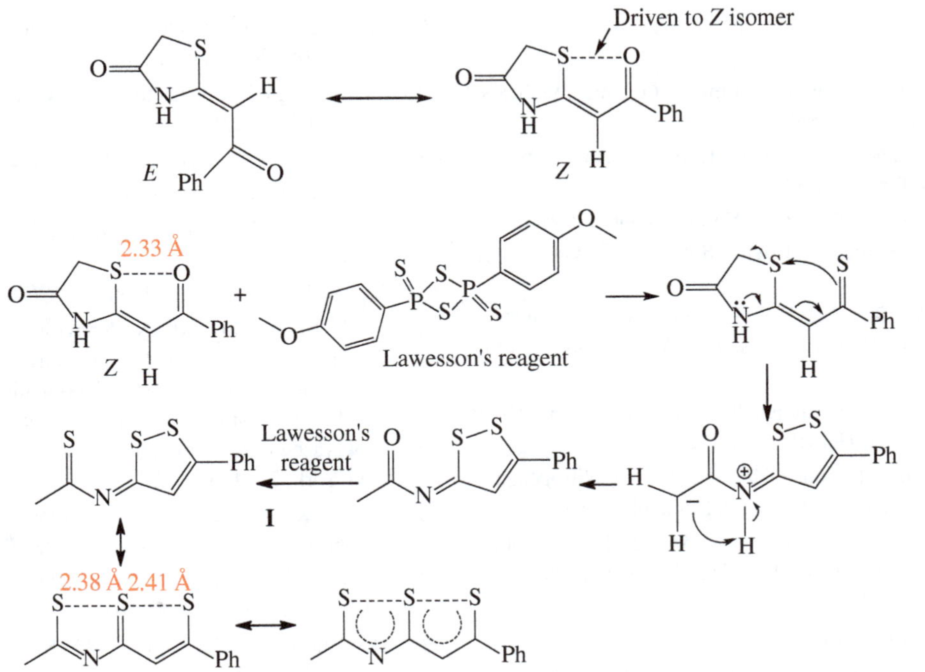

SCHEME 1.31 Selective synthesis of (*E,Z*)-*N*-(5-phenyl-3H-1,2-dithiol-3-ylidene)ethan-ethioamide (I) and its isomerization (II) [63, 64].

1.5 FINAL COMMENTS

In summary, non-covalent interactions have a rich history, being largely responsible for the properties of condensed phases, solutions, and crystals. Many examples of their use as a tool in activation of covalent bonds and thus in synthesis are reported. Increasing attention has been paid to hydrogen bonds and, to a lower extent, to halogen bond interactions, due to their extensive potential in many fields of catalysis, molecular recognition, and supramolecular, material, and bio-chemistries. Other non-covalent interactions, such as S···O, S···N, P···P, P···N, B···N, B···O, etc., have also been recognized, but still their application in synthesis remains to be developed (mainly the pnictogen bonds, involving a group 15 elements).

The selected examples on the use of non-covalent interactions in synthesis show that these forces determine the fate of many chemical transformations, and in particular, the regio- and stereoselective control of reactions is a current topic of a high prospect.

ACKNOWLEDGMENTS

This work has been partially supported by the Baku State University as well as Foundation for Science and Technology (FCT), Portugal [UID/QUI/00100/2013 and "Investigador 2013" (IF/01270/2013/CP1163/CT0007) programs]. M.N.K. thanks FCT for the working contract. M.F.C.G.S. and A.J.L.P. acknowledge the Russian Science Foundation for support in the structural analysis and help in organizing the International Laboratory (grant 14-43-00017)

REFERENCES

1. P. M. Pihko (Ed.), Hydrogen Bonding in Organic Synthesis, Wiley-VCH, Weinheim, 2009.
2. P. Gilli, G. Gilli, The Nature of the Hydrogen Bond, Oxford University Press, Oxford, 2009.
3. P. Hobza, K. Müller-Dethlefs, Non-covalent Interactions: Theory and Experiment, Royal Society of Chemistry, Cambridge, UK, 2010.
4. P. A. Gale, J. W. Steed (Eds.), Supramolecular Chemistry: From Molecules to Nanomaterials, John Wiley & Sons, Ltd, Chichester, 2012.
5. A. K. Baev, Specific Intermolecular Interactions of Organic Compounds, Springer, Heidelberg, 2012.
6. A. K. Baev, Specific Intermolecular Interactions of Nitrogenated and Bioorganic Compounds, Springer, Heidelberg, 2014.
7. P. Metrangolo, G. Resnati, Halogen Bonding: Fundamentals and Applications (Structure and Bonding); Springer, Heidelberg, 2010.
8. S. Scheiner, *Acc. Chem. Res.* 2013, *46*, 280.
9. C. J. Allpress, L. M. Berreau, *Coord. Chem. Rev.* 2013, *257*, 3005.
10. J.-J. Li, T. D. H. Bugg, *Chem. Commun.* 2005, *5*, 130.
11. J.-J. Li, T. D. H. Bugg, *Org. Biomol. Chem.* 2007, *5*, 507.
12. M. N. Kopylovich, K. T. Mahmudov, M. Haukka, A. J. L. Pombeiro, *New J. Chem.* 2014, *38*, 495.
13. R. Solhnejad, F. S. Aliyeva, A. M. Maharramov, R. A. Aliyeva, F. M. Chyragov, A. V. Gurbanov, K. T. Mahmudov, M. N. Kopylovich, *J. Mol. Struct.* 2013, *1050*, 180.
14. S.-B. Zhao, E. Bilodeau, V. Lemieux, A. M. Beauchemin, *Org. Lett.* 2012, *14*, 5082.
15. N. J. Cooper, D. W. Knight, *Tetrahedron* 2004, *60*, 243.
16. X. Wang, X. Wang, W. He, *Org. Biomol. Chem.* 2014, *12*, 3163.
17. H.-U. Reissig, R. Zimmer, *Chem. Rev.* 2003, *103*, 1151.
18. X.-M. Deng, P. Cai, S. Ye, X.-L. Sun, W.-W. Liao, K. Li, Y. Tang, Y.-D. Wu, L.-X. Dai, *J. Am. Chem. Soc.* 2006, *128*, 9730.
19. R. Chinchilla, C. Najera, *Chem. Rev.* 2014, *114*, 1783.
20. R. Salvio, M. Moliterno, M. Bella, *Asian J. Org. Chem.* 2014, *3*, 340.
21. S. Suzuki, E. Tokunaga, D. S. Reddy, T. Matsumoto, M. Shiro, N. Shibata, *Angew. Chem. Int. Ed.* 2012, *51*, 4131.
22. T. Inokuma, M. Furukawa, Y. Suzuki, T. Kimachi, Y. Kobayashi, Y. Takemoto, *ChemCatChem* 2012, *4*, 983.
23. T. Inokuma, M. Furukawa, T. Uno, Y. Suzuki, K. Yoshida, Y. Yano, K. Matsuzaki, Y. Takemoto, *Chem. Eur. J.* 2011, *17*, 10470.
24. A. Brik, C.-H. Wong, *Org. Biomol. Chem.* 2003, *1*, 5.
25. B. C. M. Fernandes, C. Mateo, C. Kiziak, A. Chmura, W. Wacker, F. Rantwijk, A. Stolz, R. A. Sheldon, *Adv. Synth. Catal.* 2006, *348*, 2597.
26. A. Erkkilä, I. Majander, P. M. Pihko, *Chem. Rev.* 2007, *107*, 5416.
27. S. Kobayashi, Y. Mori, J. S. Fossey, M. M. Salter, *Chem. Rev.* 2011, *111*, 2626.
28. J. Wang, T. Shi, G. Deng, H. Jiang, H. Liu, *J. Org. Chem.* 2008, *73*, 8563.
29. A. Viso, R. Fernández de la Pradillla, A. García, A. Flores, *Chem. Rev.* 2005, *105*, 3167.
30. M. Martjuga, D. Shabashov, S. Belyakov, E. Liepinsh, E. Suna, *J. Org. Chem.* 2010, *75*, 2357.
31. A. Noble, J. C. Anderson, *Chem. Rev.* 2013, *113*, 2887.
32. T. Cheng, S. Meng, Y. Huang, *Org. Lett.* 2013, *15*, 1958.
33. A. Kumar, J. Kaur, S. S. Chimni, A. K. Jassal, *RSC Adv.* 2014, *4*, 24816.
34. V. Y. Kukushkin, A. J. L. Pombeiro, *Chem. Rev.* 2002, *102*, 1771.
35. K. T. Mahmudov, M. N. Kopylovich, A. J. L. Pombeiro, Arylhydrazones of methylene active nitriles as promising ligands and starting materials for organic synthesis, in Ligands: Synthesis, Characterization and Role in Biotechnology, P. Gawryszewska, P. Smoleński (eds.), Nova Science Publishers, New York, 2014, pp. 177–198.
36. M. N. Kopylovich, K. T. Mahmudov, A. Mizar, A. J. L. Pombeiro, *Chem. Commun.* 2011, *47*, 7248.
37. F. Himo, Z. P. Demko, L. Noodleman, K. B. Sharpless, *J. Am. Chem. Soc.* 2002, *124*, 12210.
38. D. Cantillo, B. Gutmann, C. O. Kappe, *J. Am. Chem. Soc.* 2011, *133*, 4465.
39. P. B. Palde, T. F. Jamison, *Angew. Chem. Int. Ed.* 2011, *50*, 3525.

40. N. Nishiwaki, K. Kobiro, S. Hirao, J. Sawayama, K. Saigo, Y. Ise, Y. Okajima, M. Ariga, *Org. Biomol. Chem.* 2011, *9*, 6750.

41. O. T. du Boullay, A. Alba, F. Oukhatar, B. Martin-Vaca, D. Bourissou, *Org. Lett.* 2008, *10*, 4669.

42. R. Ferreira, H. Garcia, A. F. Sousa, M. Guerreiro, F. J. S. Duarte, C. S. R. Freire, M. J. Calhorda, A. J. D. Silvestre, W. Kunz, L. P. N. Rebelo, C. S. Pereira, *RSC Adv.* 2014, *4*, 2993.

43. A. M. R. Smith, K. K. (Mimi) Hii, *Chem. Rev.* 2011, *111*, 1637.

44. M. Yus, J. C. Gonzáez-Gómez, F. Foubelo, *Chem. Rev.* 2013, *113*, 5595.

45. S. Bahmanyar, K. N. Houk, H. J. Martin, B. List, *J. Am. Chem. Soc.* 2003, *125*, 2475.

46. G. G. R. Desiraju, T. Steiner, The Weak Hydrogen Bond in Structural Chemistry and Biology, Oxford University Press, Oxford, 1999.

47. R. C. Johnston, P. H.-Y. Cheong, *Org. Biomol. Chem.* 2013, *11*, 5057.

48. D. Magdziak, S. J. Meek, T. R. R. Pettus, *Chem. Rev.* 2004, *104*, 1383.

49. T. Dohi, Y. Hu, T. Kamitanaka, N. Washimi, Y. Kita, *Org. Lett.* 2011, *13*, 4814.

50. J. M. Berg, J. L. Tymoczko, L. Stryer, Biochemistry, W. H. Freeman, New York, 2002.

51. M. Esseffar, M. El Messaoudi, S. Azzouzi, R. Jalal, J. A. Sáez, L. R. Domingo, J. Latorre, M. Liu-González, *J. Phys. Org. Chem.* 2009, *22*, 31.

52. G. Buemi, *J. Chem. Soc. Faraday Trans.* 1990, *86*, 2813.

53. S. V. Kumar, S. K. Yadav, B. Raghava, B. Saraiah, H. Ila, K. S. Rangappa, A. Hazra, *J. Org. Chem.* 2013, *78*, 4960.

54. A. Priimagi, G. Cavallo, P. Metrangolo, G. Resnati, *Acc. Chem. Res.* 2013, *46*, 2686.

55. S. P. Bew, S. A. Fairhurst, D. L. Hughes, L. Legentil, J. Liddle, P. Pesce, S. Nigudkar, M. A. Wilson, *Org. Lett.* 2009, *11*, 4552.

56. A. Bruckmann, M. A. Pena, C. Bolm, *Synlett* 2008, *6*, 900.

57. S. Dordonne, B. Crousse, D. Bonnet-Delpon, J. Legros, *Chem. Commun.* 2011, *47*, 5855.

58. S. M. Walter, F. Kniep, E. Herdtweck, S. M. Huber, *Angew. Chem. Int. Ed.* 2011, *50*, 7187.

59. S. Zhu, C. Xing, W. Xu, Z. Li, *Tetrahedron Lett.* 2004, *45*, 777.

60. M. Iwaoka, S. Takemoto, M. Okada, S. Tomoda, *Chem. Lett.* 2001, *132*.

61. C. Dugave, L. Demange, *Chem. Rev.* 2003, *103*, 2475.

62. Y. Nagao, S. Miyamoto, M. Miyamoto, H. Takeshige, K. Hayashi, S. Sano, M. Shiro, K. Yamaguchi, Y. Sei, *J. Am. Chem. Soc.* 2006, *128*, 9722.

63. A. Rašović, P. J. Steel, E. Kleinpeter, R. Marković, *Tetrahedron* 2007, *63*, 1937.

64. A. Rašović, A. Koch, E. Kleinpeter, R. Marković, *Tetrahedron* 2013, *69*, 10849.

2

BORON–NITROGEN BOND: A USEFUL MOLECULAR CONSTRUCTION TOOL

João P. M. Antonio[1], Gonçalo D. V. Farias[1], Fábio M. F. Santos[1], Rudi Oliveira[1], Pedro M. S. D. Cal[2], and Pedro M. P. Gois[1]

[1] *Instituto de Investigação do Medicamento (iMed.ULisboa), Faculdade de Farmácia, Universidade de Lisboa, Lisbon, Portugal*

[2] *Instituto de Medicina Molecular (IMM), Faculdade de Medicina, Universidade de Lisboa, Lisbon, Portugal*

2.1 INTRODUCTION

Boronic acids are organic compounds that have a sp^2-hybridized boron atom attached to one alkyl, vinyl, or aryl substituent, two hydroxyl groups, and a vacant p-orbital as a consequence of having only six valence electrons (**1**; Scheme 2.1). Generally, boronic acids present trigonal planar geometry, and their hydroxyl groups are in one plane to the aryl substituent, while the empty orbital is perpendicular. The exception for this rule is the case of *ortho*-substituted aromatic moieties, where the aromatic ring is perpendicular to minimize the steric strain.

However, in order to complete boron's octet, boronic acids can also coordinate basic molecules and exist as stable tetracoordinated adducts with tetrahedral geometry. This coordination caused by boron's deficient valence results in a negative charge that is formally drawn on this atom (**2**; Scheme 2.1). This complexation explains boronic acids' unique properties as mild organic Lewis acids in opposition to a usual Brønsted base formed in the equilibrium of hydroxyl groups in water. In other words, the boronic acids ionize water and form hydronium ions by "indirect" proton transfer instead of direct proton removal from their hydroxyl groups [1]. This protonation behavior is crucial for the binding to vicinal Lewis bases such as catechols, carbohydrates, hydroxamic acids, or even some proteins.

Scheme 2.1 exemplifies the addition of a vicinal diol to a boronic acid, which results in the formation of the cyclic boronate ester **3**, with concomitant ionization of boron to maintain the acid–base equilibrium. This feature decreases the apparent pK_a of the boronic acid **1** and promotes its consumption in the reaction.

A novel structure developed to improve this binding ability of boronic acids at neutral pH was reported by Wulff, which was achieved by lowering the boronic acid's pK_a value through the introduction of a neighboring amino group capable of promoting an intramolecular BN interaction [2]. This dative bond induces also a contraction of the O–B–O bond angles, contributing to a stabilizing effect of boronate esters at neutral pH, expanding their utility to many biological applications. Furthermore, the strength of the BN interaction in tetracoordinated boron can also influence the fluorescence of neighboring fluorophores [3]. In this aspect, a particular type of these compounds featuring an *ortho* dialkylaminomethyl to the boronic acid (e.g., compound **4**) assumed an essential role in the development of various chemosensors and became known as Wulff-type boronic acids [4, 5]. These compounds will be discussed in more detail later in this chapter.

In relation to carbon, the boron atom is electron deficient with one less electron, and nitrogen is electron rich with one more electron. Therefore, it is conceivable that a covalent BN bond is isoelectric to a carbon–carbon (CC) bond. The major difference lies in the heterogeneous/ionic character of the bond, which provides different electronic and optical properties in comparison to the

Non-Covalent Interactions in the Synthesis and Design of New Compounds, First Edition.
Edited by Abel M. Maharramov, Kamran T. Mahmudov, Maximilian N. Kopylovich, and Armando J. L. Pombeiro.

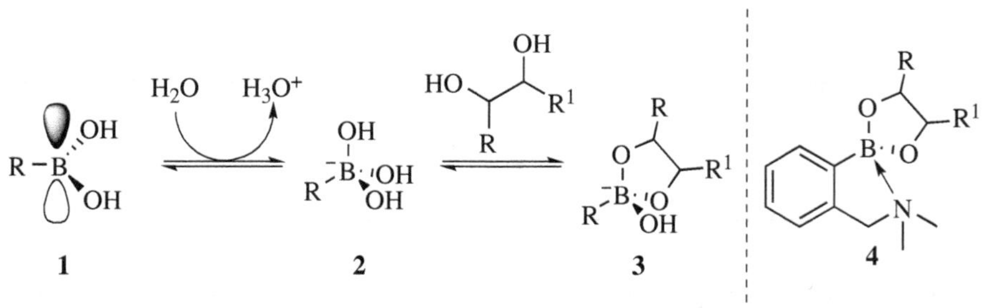

SCHEME 2.1 Ionization equilibrium of boronic acid in water.

corresponding CC compounds [6]. As a result, covalent BN bond has been used as a CC isostere for many decades in the search for compounds with improved properties.

This chapter highlights the relevance of dative and covalent BN bonds in various applications, with emphasis in the structural variability of BN-containing compounds, starting from CC isosterism in simple heterocycles to the integration in supramolecular structures and chemosensors.

2.2 BN AROMATIC HETEROCYCLES

Ever since its discovery by Faraday in 1825 [7], benzene, **5**, became one of the most studied molecules in chemistry. Its particular electron conjugation characteristics result in high stability and reactivity, which was very surprising at the time. Aromatic compounds assumed a key position in science with applications ranging from biomedical industry and drug design [8, 9], material and polymer industry [10], organic electronics [11], agriculture [12], and many others. And so, due to the aforementioned CC/BN isosterism, BN-containing aromatic molecules are envisioned as promising compounds. In this section, focus will be given to borazine, **6**, and azaborine, **7**, and their polycyclic derivatives, structural characteristics, reactivity, and applications will be discussed (Fig. 2.1).

2.2.1 Borazine

The first report of a benzene BN-containing analogue was reported by Stock and Pohland [13] who synthesized borazine **6** in 1926. Borazine is commonly called "inorganic benzene" due to its isoelectronic and isostructural similarities to benzene (Scheme 2.2). However, despite an apparent similarity to benzene, borazine's aromaticity is still a matter of discussion. On the early days, its aromaticity seemed obvious as borazine has the classical standard aromatic properties: $4n+2$ π-electrons, planar structure, and uniform bond length. The bond length (1.432 Å) [14] is in between the lengths of a single and double BN bond (1.58 and 1.40 Å, respectively) [15], just as in its organic counterpart, pointing even more toward aromaticity.

Nevertheless, due to the electronegativity differences between nitrogen and boron, the aromaticity is expected to be lower in borazine. This is supported by the preferential addition reactions suffered by borazine instead of the typical aromatic substitutions undertaken by benzene. In fact, the first electrophilic aromatic substitution was only reported by the Fornarini group in 1999 [16]. More recently, new procedures to evaluate aromaticity standards have been developed and, when applied to borazine, present more conclusive evidence [17]. These results have been compiled by Merino *et al.* and point to interesting conclusions [18]. Aromatic stabilization energies [19, 20], valence bond calculations [21], energy decomposition analysis [22], nucleus-independent chemical shift values [23], and other magnetic experiments [24] all point toward low aromaticity or no aromaticity at all for borazine. Due to the more electronegative nitrogen atoms, the electron dispersion is not uniform, suggesting a more localized π-electron feature around those atoms. Even the more recent experiments are unable to fully clarify the aromatic character of borazine [25–31]. What is obvious is that its characteristics are clearly different from the benzene ones.

Regarding its applications, borazine has been proposed as a potential participant in hydrogen storage materials [32–34]. Borazine is also an important intermediate for the production of boron nitride ceramics and nanotubes. In fact, borazine presents the ideal BN ratio for this application. Boron nitride is a binary compound possessing one atom of boron and one of nitrogen that, due to its similarity to CC compounds, can create interesting polymers resembling graphite and diamond (hexagonal boron nitride and cubic boron nitride, respectively). The cubic structure's hardness is only surpassed by diamond itself, although the boron nitride structure presents higher thermal and chemical stability. The hexagonal structure is the most stable crystalline form and is organized with a series of graphene-like hexagonal layers containing covalently bound nitrogen and boron atoms (Fig. 2.2) [35]. The layers are held together through van der Waals attractive forces, just as in graphite. Its interest emanates from its particular characteristics such as high thermal conductivity and stability, high resistance to corrosion and oxidation as well as a strong UV emission, superhydrophobicity, no toxicity, good environmental compatibility, high dielectric breakdown strength, high volume resistivity, and many other valuable assets [35]. In addition to all this, boron nitride complexes offer the possibility to modulate those properties simply by altering their synthetic route. The different procedures, polymer structures, characteristics, and

Organic Inorganic Hybrid

5 Benzene; 6 Borazine; 7 1,2-Azaborine; 8 1,3-Azaborine; 9 1,4-Azaborine

FIGURE 2.1 Representation of benzene's inorganic analogue borazine, **6**, and the hybrid azaborines **7**, **8**, and **9**.

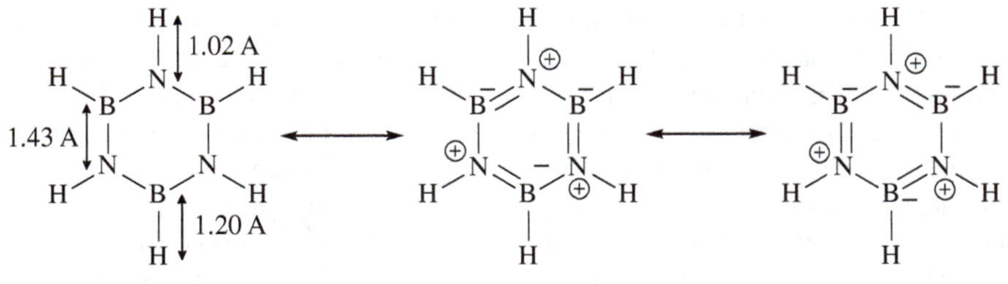

6

SCHEME 2.2 Borazine representation: bond lengths and resonance structures showing the delocalization of the nitrogen electrons.

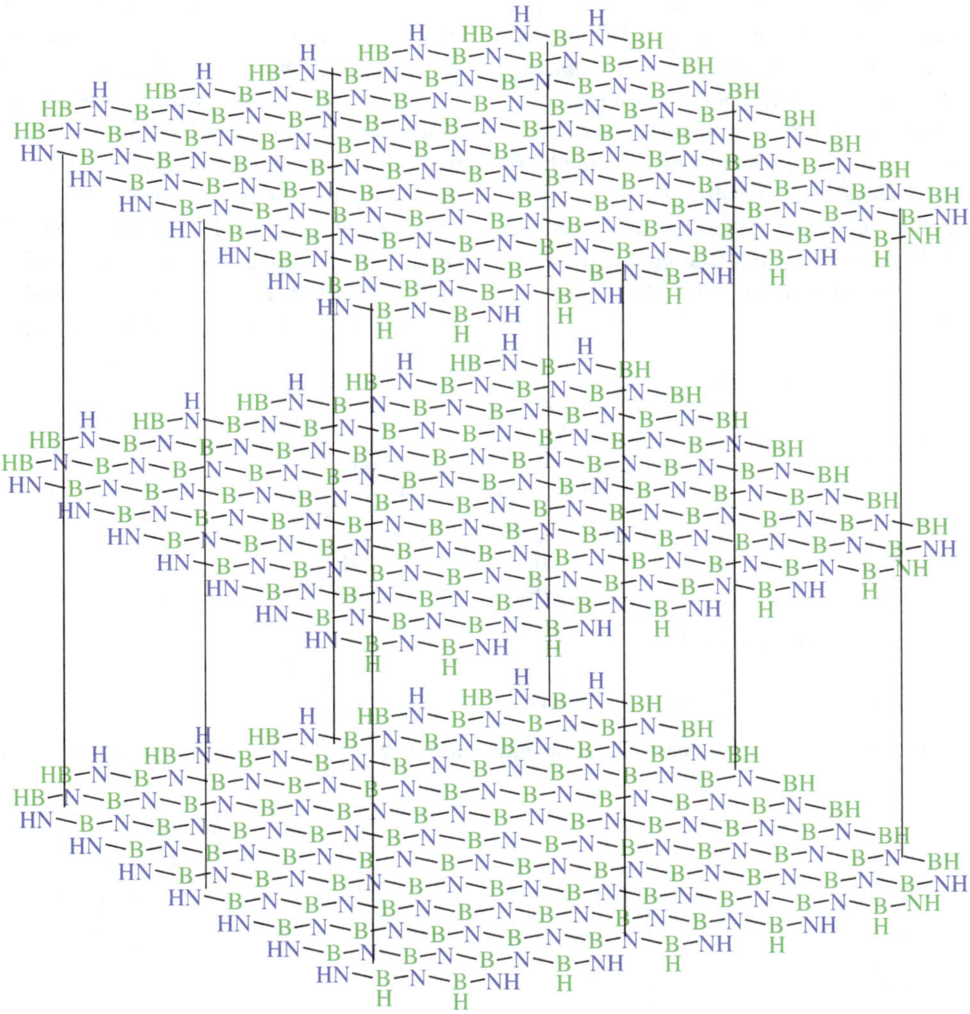

FIGURE 2.2 Hexagonal multilayer structure of h-boron nitride; resembles graphite structure but with the added properties of the polarization of the BN bond.

detailed applications can be accessed on the review made by Miele and Bernard [35].

It is important to mention that most of the starting materials for the production of boron nitride come from borazine and its derivatives, so it is essential to understand their reactivity. B-azide, B-nitro, and B-nitramine derivatives have also been produced [36] and have interesting application as energetic and explosive materials. Particularly, borazines **11** and **12** seem to present equal and better detonation performances than the standard explosive 4,6-diazido-*N*-nitro-1,3,5-triazine-2-amine, **10** (Fig. 2.3) [37]. The Lewis basicity of the nitrogen atoms allow the formation of adducts with aluminum, gallium, and tin [38]. This bonding, just as with other bulky B-substituents, might promote the loss of planarity of the borazine ring [39]. Some substituents on the other hand, such as fluorine, might increase the ring stability by increasing the electron delocalization. Therefore, it is important to take into account the substituent's effect on borazine physical and chemical properties.

2.2.2 Azaborine and Polycyclic Derivatives

A hybrid organic–inorganic molecule mixing benzene and borazine was thought to present an interesting and promising mix of characteristics. This can be obtained by replacing only one CC bond by the isosteric BN bond. The first report of a single CC/BN replacement in an aromatic compound is credited to Dewar *et al.* who, in 1958, synthesized the phenanthrene analogue **14** (Scheme 2.3) [40]. Dewar worked further on aromatic analogues and synthesized an interesting library of mono- [41] and polycyclic [42] compounds. This work promoted extensive research in this class of compounds [15, 43, 44].

2.2.2.1 Monocyclic Compounds The representative molecule of this class is the 1,2-dihydro-1,2-azaborine, **7**. Other azaborines with boron and nitrogen in different positions (**8** and **9**) have been obtained and studied, but since there is no formal bonding between boron and nitrogen, these are not part of the scope of this chapter. Thus, from now on, 1,2-dihydro-1,2-azaborine will be commonly referred as azaborine. Despite many advances in azaborine chemistry, with many reported B- and N-substitutions [43], the parent compound, 1,2-azaborine, had not been synthesized yet. That problem was only solved in 2009 by Liu *et al.* who reported its isolation, although in very low yields. Four years later, in 2013, Liu once again described a protection group-free gram-scale synthesis of compound **7**, which finally allowed an extensive characterization [45].

Preliminary reports indicated classical aromaticity properties from azaborine such as its availability to undergo electrophilic substitutions and delocalized structures [46, 47]. X-ray data and microwave spectroscopy analysis demonstrate ring planarity with all bond lengths (CC, CN, CB, and BN) between single and double bonds [47, 48]. Magnetic properties [49], resonance stabilization energies (theoretical and experimental) [50, 51], nucleus-independent chemical shielding, harmonic oscillator model of aromaticity, *para* delocalization index, and aromatic fluctuation index [52] all point toward a lower aromaticity than benzene, but significantly higher than borazine. A curious finding shows that 1,3-azaborine is more aromatic, yet less stable, than 1,2-azaborine. It should be reminded that stability is a global property of a molecule, whereas aromaticity is a stabilizing effect that arises from cyclic electron delocalization [53]. Azaborine

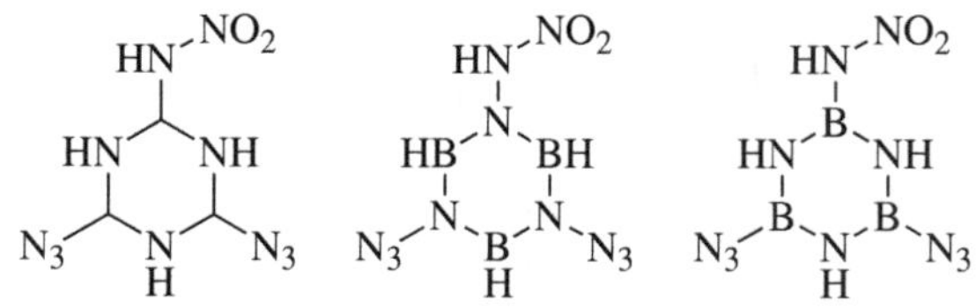

	10	11	12
Detonation pressure (kbar)	330	335	364
Detonation velocity (m/s)	8472	9620	9734

FIGURE 2.3 Structure and detonation performance of diazidonitroamine borazines **11** and **12** in comparison to a standard explosive **10**.

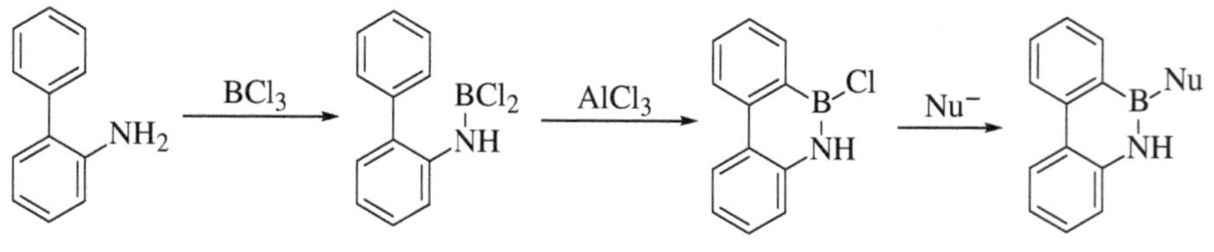

13 **14**

SCHEME 2.3 First reported synthesis of a BN isostere of an aromatic compound.

seems to be stable in water and resistant to oxidation, with the degree of resistance changing upon the substituents in the ring, mostly in boron and C_3 [54].

Azaborine, being an isostere of benzene, can virtually have all the same application as its organic counterpart while adding boron's useful elemental and chemical features, which include nuclear spin, large cross section for neutron capture, and Lewis acidity [55]. It has been shown that azaborine **15** (Fig. 2.4) binds to a hydrophobic pocket of T4 lysozyme, which is an interesting concept in azaborine's biological applications. On the other hand, diazaborine **16** is a well-known antibacterial known to inhibit fatty acid synthesis [56]. However, new reports also demonstrate antifungal activity for compound **17** by inhibition of rRNA's maturation for the large ribosomal subunit (Fig. 2.4) [57].

The development of benzodiazaborines (e.g., compounds **16** and **17**) as structural analogues of the antituberculosis drug isoniazid (INH) is a classic example of this isosterism [56, 58, 59]. The target for these drugs and INH is a NAD(P)H-dependent enoyl-acyl carrier protein reductase (ENR), which catalyzes the final step of fatty acid synthase (FAS) [60]. In benzodiazaborines, the Lewis acid character of boron allows the attack of the OH group from the ribose unit of an NAD cofactor of ENR [58, 59]. Other microorganisms like Gram-negative bacteria (e.g., *Escherichia coli*) [61] and *Plasmodium* parasites [62] are also sensitive to this class of compounds, since these have similar enzymes to ENR in their fatty acid biosynthesis.

*2.2.2.2 **Polycyclic Compounds*** The first example of a polycyclic compound was already mentioned before, and it was the synthesis of phenanthrene **14** by Dewar *et al.* [40]. N-derivatives can also be synthesized by nucleophilic substitution [63]. Besides phenanthrenes and azaborines, Dewar's legacy also includes triphenylene **18** [41], pyrenes **19–21** [64], chrysene **22**, benzo[*a*]anthracenes **23** and **24**, and naphthalene **25** (Fig. 2.5) [42].

The purpose of this library of polycyclic compounds returned with the advent of electronics, due to their interesting optical properties. With that, there were significant advances in synthetic methodologies that were summarized by Pei *et al.* [44]. The applications of these compounds are related to their optical and electronic properties. Compound **26** presents high thermal stability (5% weight loss in the range of 340–400°C) and is stable to oxygen and moisture, which are the ideal characteristics for electronic devices. Field-effect transistors with this compound have been synthesized with success in a bottom-gate/top-contact device configuration [65]. On the other hand, due to the immense electron conjugation of these systems, some compounds are thought to own interesting optical properties. Compound **27** shows deep-blue fluorescence [66], which inspired the creation of organic light-emitting diodes using compounds **28** and **29** (Fig. 2.6) [67, 68].

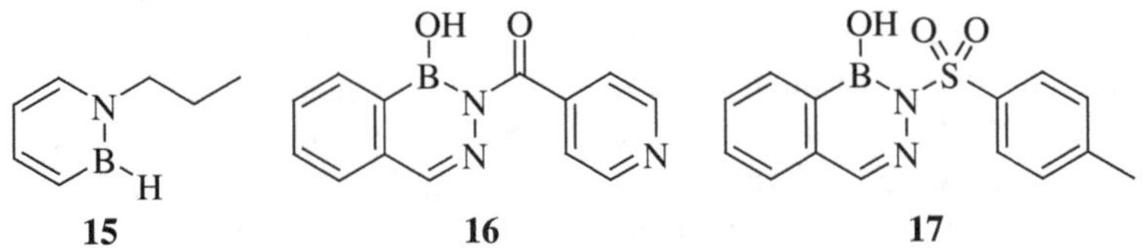

FIGURE 2.4 Examples of biologically active boron compounds.

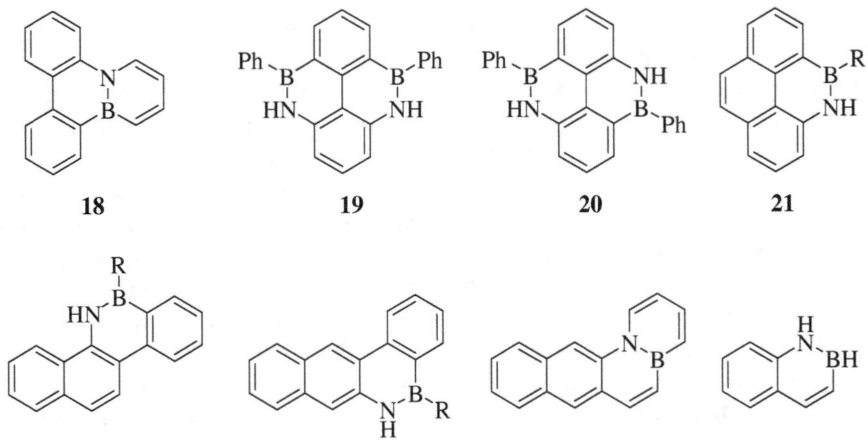

FIGURE 2.5 Library of aromatic analogues synthesized by Dewar *et al.* [42].

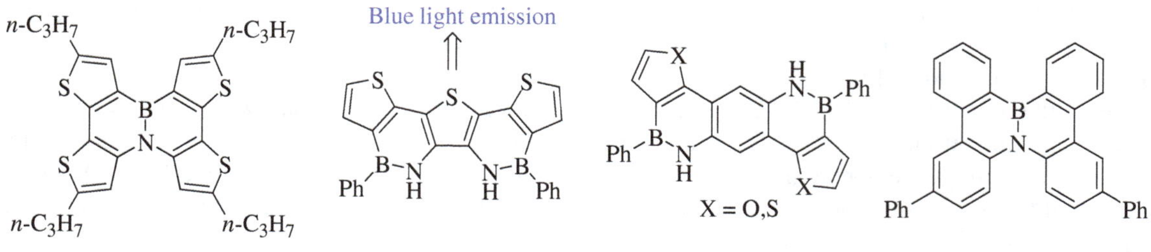

FIGURE 2.6 Electronic and optically active compounds comprising an azaborine moiety in their structure.

2.3 BN NONAROMATIC HETEROCYCLES

2.3.1 Boron Protecting Groups and Ligands

Presently, boronic acids are very useful compounds in organic and inorganic chemistry, with special emphasis in the area of CC bond formation (e.g., Suzuki–Miyaura reactions) [1, 69]. Having a key function in these reactions, the degradation of boronic acids in reaction conditions (protodeboronation, oxidation, and palladium-catalyzed homocoupling) is a considerable limitation [70]. In addition, the inherent instability of some boronic acids complicates their use in the laboratory, namely, in usual handling conditions (air, moisture, oxygen, and temperature) and purification techniques (silica gel) [71]. To overcome these issues, a variety of protecting groups were developed to cover a range of physical, chemical, and reactivity characteristics that can be used in different situations [1, 69].

Two types of BN heterocycles have been described as protecting groups for boronic acids: N-coordinates boronates and boronamides (Fig. 2.7). In the former, boron is covalently bonded to two oxygen substituents (hydroxyl or carboxylate) and forms a dative bond between the Lewis basic lone pair of the nitrogen and the empty p-orbital of the Lewis acidic boron, which increases the stability of the complex when compared with simple boronate esters (Fig. 2.7). The dative bond causes boron to hybridize from sp^2 to sp^3, creating a zwitterionic neutral complex with a tetrahedral geometry. In diethanolamine (DEA) complexes, the dative bond is labile at room temperature [72], so the deprotection is achieved in acidic conditions, just like boronate esters. With the *N*-methyliminodiacetic acid (MIDA) group, the dative bond is stronger (up to 150°C), so the stability is dependent on the hydrolysis of the carboxylate groups in the presence of hard nucleophiles or basic aqueous conditions.

Although these ligands were first intended to protect boronic acids, in certain reactions the complexes can be used directly as reactant to yield better results. For example, DEA [73] and MIDA [71] boronates showed better results than the unstable boronic acids in Suzuki–Miyaura reactions. In both DEA and MIDA, the nitrogen can accommodate different substituents, including asymmetric moieties to allow the enantioselective control of reactions in distal functional groups [74] or improve the optical purity of hydroboration intermediates through selective crystallization [75].

Boronamides are characterized with a neutral sp^2-hybridized boron bonded to two nitrogen substituents. Lone-pair donation of the Lewis basic nitrogen atoms reduces the Lewis acidity of boron, making the compounds less reactive and thus more stable. The ligand 1,8-diaminonaphthalene (DAN), also known as the core of the "proton sponge," has a higher Lewis base character than the ligands anthranilamide (AAM) and 2-(pyrazol-5-yl)aniline (PZA), thus forming more stable boronamides. Common to the three ligands is their stability in basic aqueous Suzuki–Miyaura conditions [76–78]. As these conditions tend to increase the rate of the Suzuki–Miyaura reaction, boronamides offer an advantage over MIDA boronates in iterative cross-coupling reactions [79, 80]. Removal of this type of ligands is achieved in acidic conditions *via* protonation of the amines.

Benzoxaboroles, chemical analogues of boronic acids, can also be protected with a derivative of the "proton sponge" (Fig. 2.7), just like B(DAN) [81]. The protecting group is stable in pH over 7 and reducing conditions ($LiAlH_4$) but labile at lower pH and oxidative conditions.

Aside from protecting groups, several other ligands have been developed to stabilize boron species. The resulting complexes have been used to improve the stereoselectivity of reactions [82–85], promote polymerization [86–88], self-assembly reactions (detailed in supramolecular BN architectures) [88–90], and study metal stereogenic centers [91, 92].

Contreras *et al.* and later Farfán *et al.* compiled a number of reagents/ligands that stabilized boron by promoting the stable tetrahedral geometry [93, 94]. The stabilization of the complex can be quantified using the distance between boron and nitrogen atoms [95], which for this type of ligands is usually between 1.6 and 1.7Å. Salicylaldehyde is the basis of a group of ligands for boronic acids named salen. Depending on the type, the ligands can be tri- or tetradentate (Fig. 2.8). In addition, the substituents and reaction conditions can influence the structure of the ligand–boronic acid complexes, namely, the complexation of more than one boronic acid *per* ligand, or the formation of dimers and polymers. On the other hand, a recurring characteristic of these complexes is the *cis* addition of the hydroxyl arms of the ligand to boron.

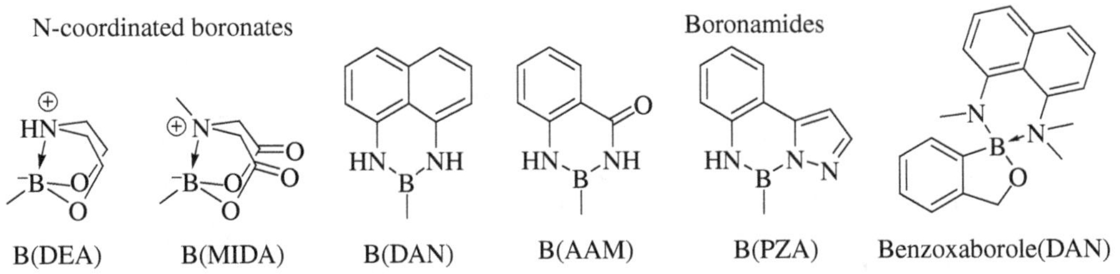

FIGURE 2.7 Protecting groups for boronic acids and benzoxaborole.

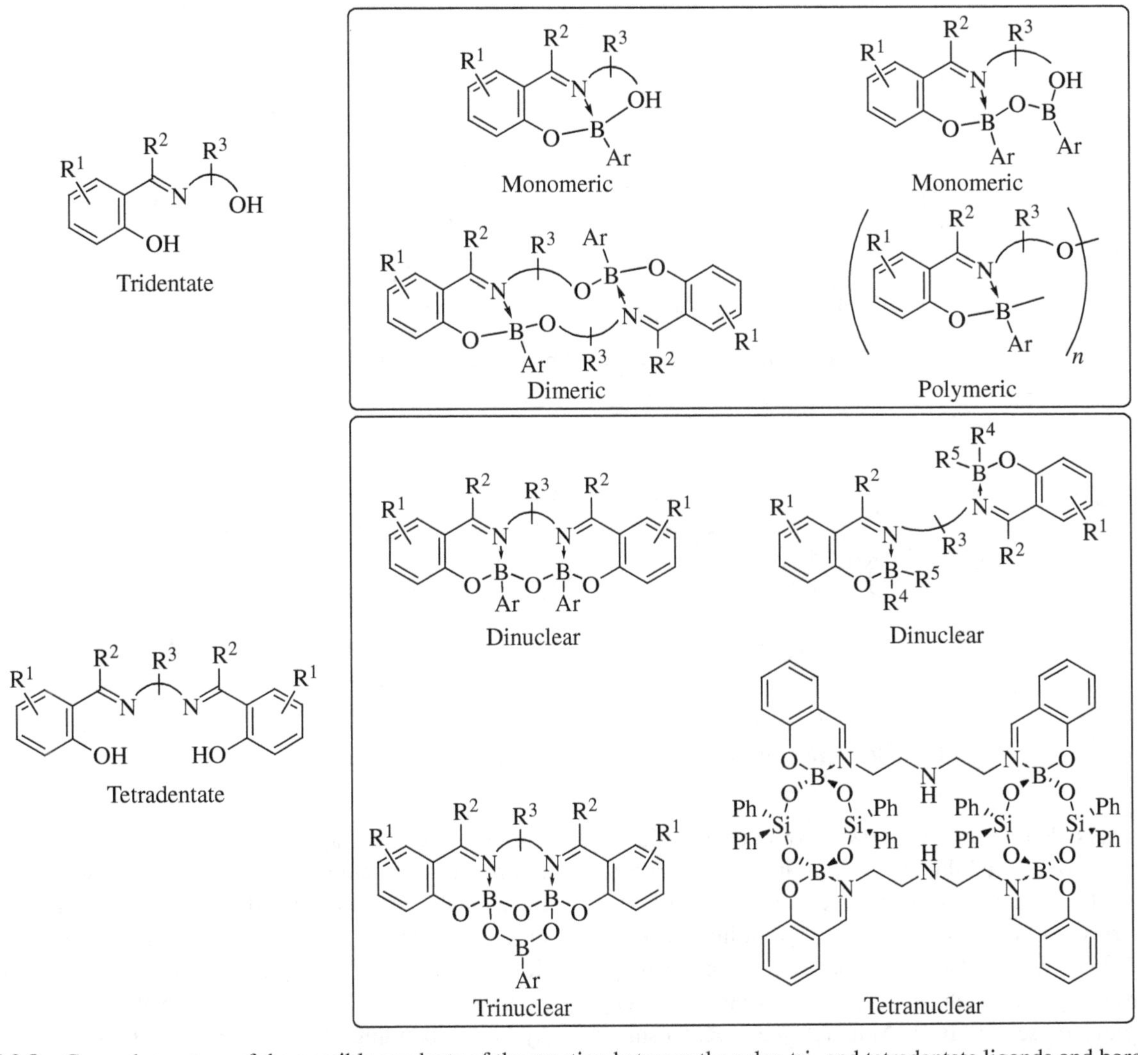

FIGURE 2.8 General structure of the possible products of the reaction between the salen tri- and tetradentate ligands and boronic acid.

2.3.2 Bioactive Heterocycles

A wide range of BN heterocycles has been associated with various biological activities. Many of these cases result from the direct substitution of CC bonds by BN in compounds known to induce a biological effect. For instance, Gois *et al.* prepared isosteres of bioactive natural products (e.g., compound **30**), exploring the known *cis* addition of tridentate ligands to boronic acids (Fig. 2.9) [96]. Later, the same strategy was used with a salen-type ligand and an amino acid precursor to yield bioactive boron complexes that acted as inhibitors of human neutrophil elastase (e.g., compound **31**) [97] and others as phenylalanine hydroxylase modulators (e.g., compound **32**) [98].

Srebnik *et al.* used amino alcohols to prepare oxazaborolidines (e.g., compound **33**) with activity against *Streptococcus mutans* [99]. These compounds were selected due to their structural resemblance to the autoinducer 2 (AI-2), a universal signaling molecule in bacteria [100]. The BN bond was thought to contribute to the hydrogen bonding with the target proteins known as AI-2–Lux proteins [100]. Blackmore

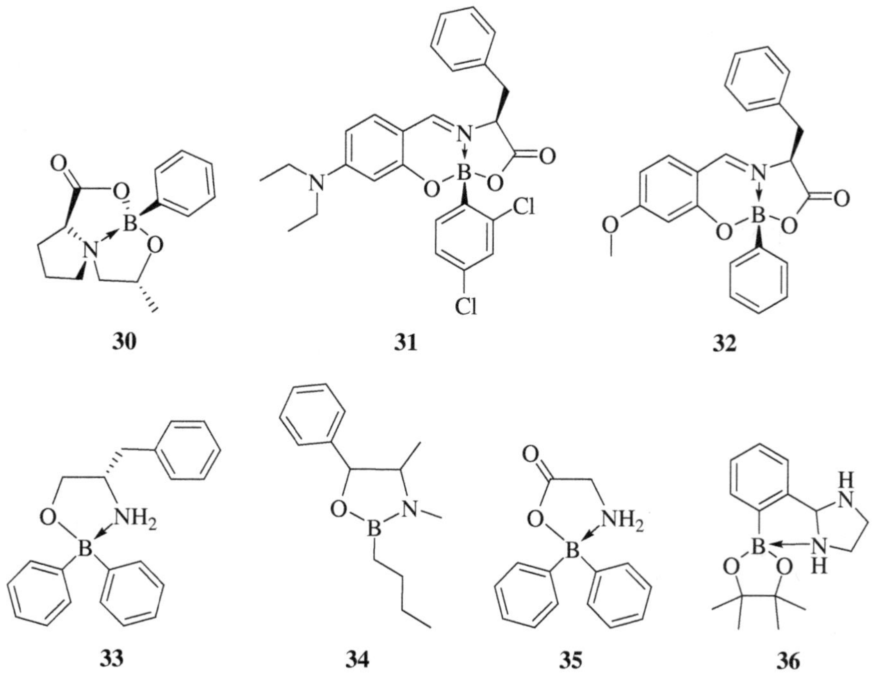

FIGURE 2.9 Examples of bioactive BN heterocycles.

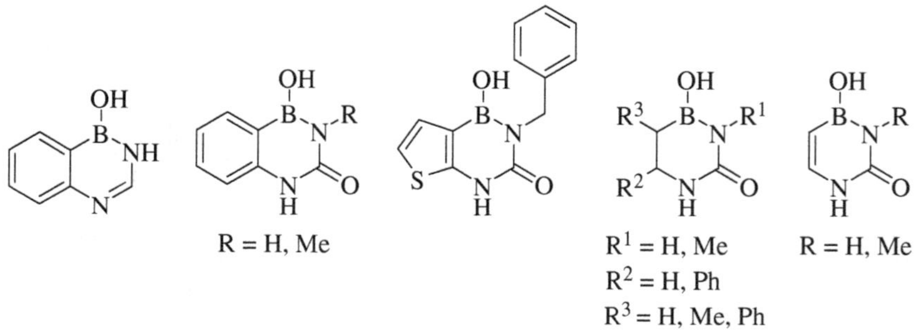

FIGURE 2.10 BN analogues of nucleic acid bases for application in BNCT.

and coworkers demonstrated that similar oxazaborolidines (e.g., compound **34**) inhibited calcium influx in platelets (Fig. 2.9) [101]. Later, Sánchez-Torres *et al.* published a similar series of compounds—oxazaborolidinones (e.g., compound **35**)—with a dative BN bond. This set of drugs derived from glycine displayed improved cytotoxic activity over the previous series [102]. Although the mechanism of action is yet to be confirmed, some observations were already reported, such as the reduction of transmembrane potential, caspase-3 activation, and DNA fragmentation, all signs of apoptosis [102].

Simple pinacol boronates (e.g., compound **36**) with a dative BN bond showed moderate antifungal activity, yet no mechanism of action has been identified [103].

BN heterocycles are also important compounds for boron neutron capture therapy (BNCT) [104, 105]. This therapeutic approach consists in the irradiation of nonradioactive boron-10 (^{10}B) with low-energy thermal neutrons, causing the release of high-energy particles that destroy cells within a very tight range (5–9 μm). Special importance is given to the selective targeting of cancer cells versus healthy cells for improved therapeutic effect. Several reviews have compiled the available techniques to deliver ^{10}B compounds more specifically to cancer cells [105–107]. A popular strategy is to develop stable BN analogues of nucleic acids that mimic the natural structure and preserve the pattern of Watson–Crick hydrogen bonding. The structures of these compounds are displayed in Figure 2.10 [108].

2.3.3 Catalysis

Molecules holding a Lewis acid in equilibrium with a Lewis base are known to activate a variety of strong bonds, such as alkynes, CO, CO_2, NO, and N_2O, among others [109, 110]. Then, the bulky moieties surrounding the Lewis acid and base favor the disruption of the mentioned bonds. These so-called frustrated Lewis pairs are therefore useful in many organic transformations. Compounds **37** [110] and **38** [111] are examples of BN frustrated Lewis pairs, although many others are known using specially the pair boron–phosphorus (Fig. 2.11) [109].

Heterocyclic borenium ions are used as catalyst in a variety of reactions. The facile route of synthesis and the high range of derivatization were important advantages of these compounds when optimizing the catalysts.

Most borenium heterocycles have a tridimensional structure centered on boron, making it useful in the enantioselective control of several reactions. Moreover, Schrimpf *et al.* reported that boron can return to the same asymmetric conformation after losing temporarily the tetrahedral geometry [112, 113]. Figure 2.12 shows some of these borenium catalysts that where optimized for the following enantioselective reactions: ketone reduction (e.g., catalysts **39** and **40**) [114, 115], Diels–Alder (e.g., catalysts **41** and **42**) [116, 117], asynchronous [2+2] cycloaddition (e.g., catalyst **43**) [118], cyanosilylation of ketones (e.g., catalyst **44**) [119, 120], and ethynylation of aldehydes (e.g., catalyst **45**) [121].

2.3.4 Boron Radical

In 2012, Nozaki and coworkers reported the isolation and characterization of the first neutral boryl radical **46** [122]. The key to modulate the stability and reactivity of boryl radicals is to tune the structure of the Lewis base ligand and corresponding substituents surrounding the boron atom [123]. This is intended to decrease the spin density on boron and improve stability. Nitrogen substituents are often used as stabilizing groups due to their electron-donating effect to boron.

In parallel with other radicals formed in nonheterocyclic BN compounds [124], boryl radicals **46** and **47** [125] (Fig. 2.13) have the potential to promote radical photopolymerization and dehalogenation reactions [125, 126] and can also be used as polarity reversal catalysts [127].

2.3.5 Luminescence

The interest in luminescent materials is common to many scientific disciplines. The family of 4,4′-difluoro-4-bora-3*a*,4*a*-diaza-*s*-indacene (BODIPY) compounds evolved to be an important class of luminescent dyes due to their outstanding chemical and photophysical properties, together with the wide range of structural variability. Ziessel *et al.* proposed the division of BODIPYs and derivatives according to the nature of the boron ligand (N,N and N,O) and number and size of the aromatic systems associated to boron (Fig. 2.14) [128]. The complete array of luminescent BN heterocycles is distributed in recent reviews [128–132].

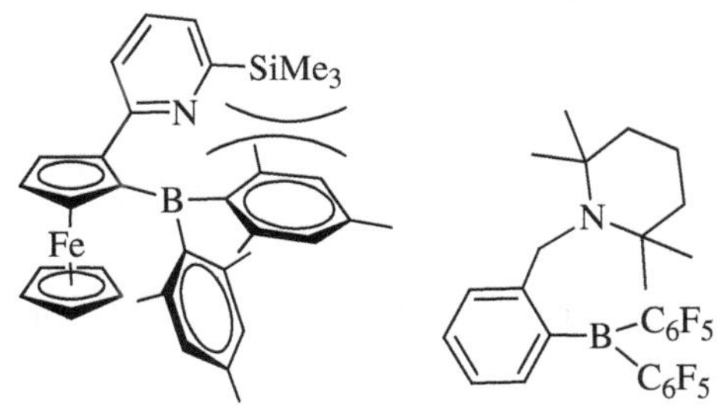

37 **38**

FIGURE 2.11 Examples of BN frustrated Lewis pairs.

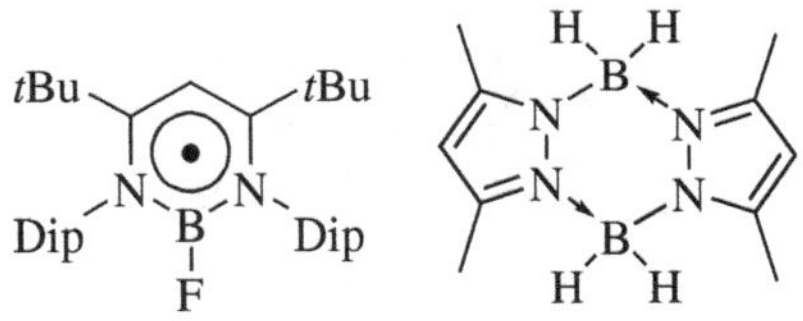

46 **47**

FIGURE 2.13 Boryl radicals **46** and **47** with benzoquinone and benzoyl peroxide.

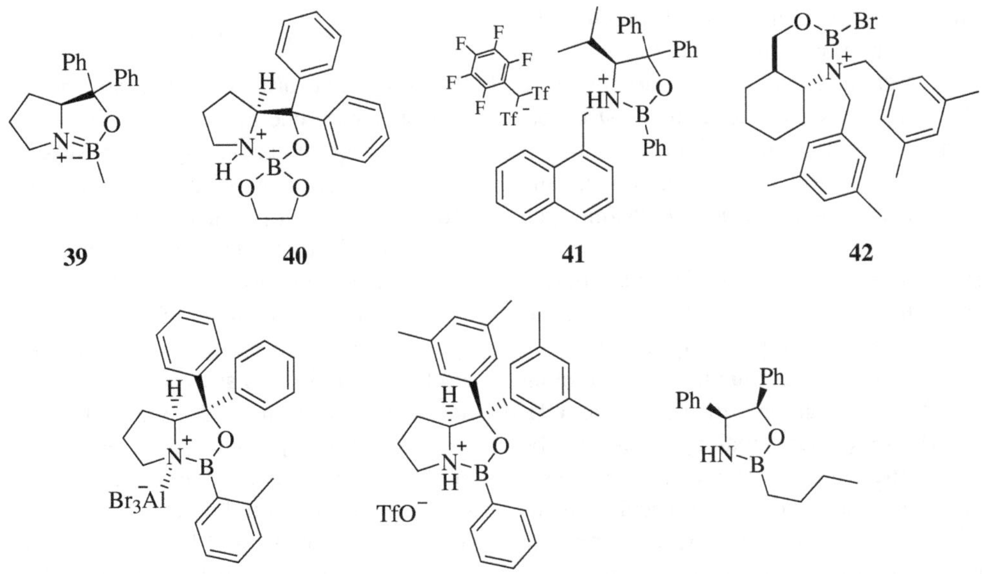

FIGURE 2.12 Heterocyclic borenium ion catalysts.

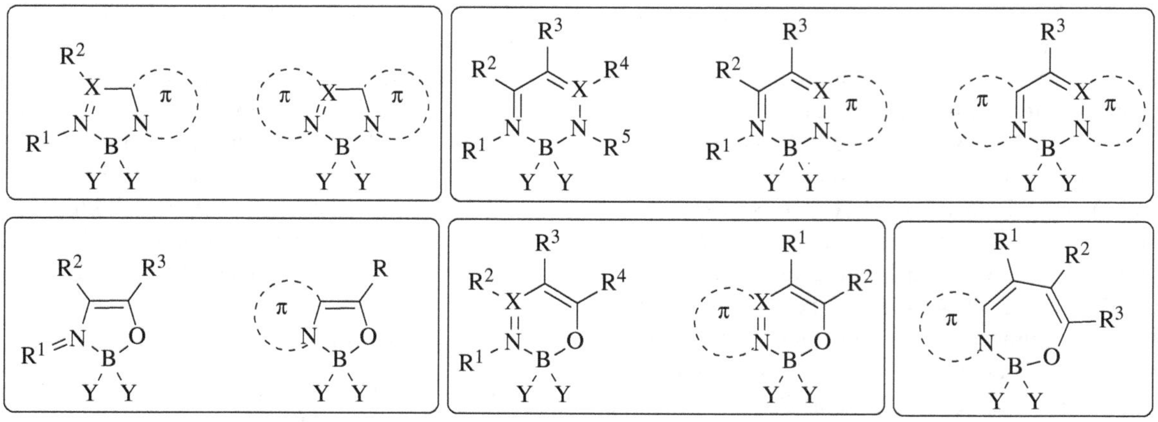

X = C or N
Y = F, Ar, OR

FIGURE 2.14 General structures of BODIPYs and analogues [128].

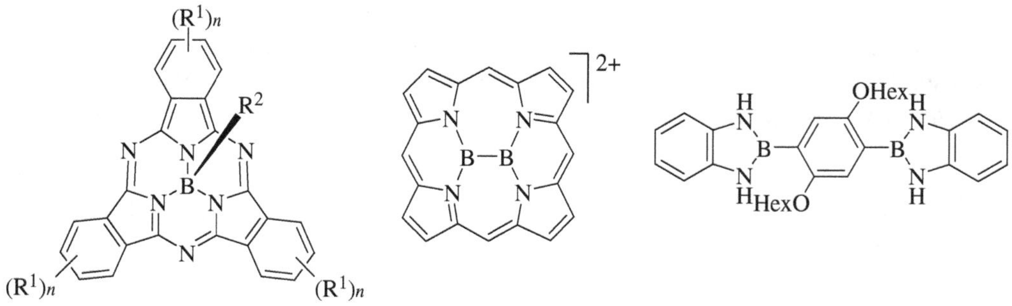

48 **49** **50**

FIGURE 2.15 General structure of boron subphthalocyanines **48** and examples of other luminescent BN heterocycles: boron porphyrin complex, **49**, and 1,3,2-diazaboroline, **50**.

Directed modification of the structure surrounding the boron atom harvested the luminescent properties for various applications. For instance, the increase of conjugation in the molecule is often used to obtain compounds that emit in the near-infrared (NIR) region, a desirable characteristic for *in vivo* imaging [133]. Attaching specific recognition moieties to a trigger BODIPY core yielded various indicators/sensors for ions (e.g., fluoride, cyanide, and calcium), various pH values, biomolecules (e.g., toxins and proteins), and oxidative stress (reactive oxygen species (ROS)/nitrogen species) [133, 134].

Subphthalocyanines (general structure **48**) are a different set of luminescent BN heterocycles (Fig. 2.15). These compounds have a peculiar bowl-shaped geometry formed by an aromatic core of 14 π-electrons surrounding a boron atom that is completed with an axial anionic ligand (halide, OH, alkoxy, or aryl) [131, 135]. Subphthalocyanines can be applied in nonlinear optics, organic semiconductor, light-emitting diodes, anion sensing, and supramolecular chemistry [136]. The luminescent properties result from the conjugation of the vacant p-orbital on the boron center with the π^* orbital of the attached organic π-system. This effect is also present in other luminescent BN heterocycles, namely, porphyrin derivatives (e.g., compound **49**) [131] and 1,3,2-diazaborolines (e.g., compound **50**) [137].

2.3.6 Hydrogen Storage

BN compounds are among the most promising materials to deliver clean energy *via* hydrogen storage. The great advantage of molecular hydrogen versus petroleum is its higher energy density that can be extracted with great efficiency without producing greenhouse gases. The importance of this fuel alternative is evidenced in the several reviews published over the years [138–143].

1,2-BN-cyclohexane, **51**, and 3-methyl-1,2-BN-cyclopentane, **53** (Scheme 2.4), are the main BN heterocycles studied as hydrogen storage molecules. Here, the Lewis acidic boron atom holds two hydrides, while the adjacent nitrogen captures two more hydrogen atoms [142]. With suitable activation, both electron-rich and electron-poor hydrogen nuclei combine to form molecular hydrogen. Activation of 1,2-BN-cyclohexane is achieved using mild thermal activation (150°C); however, 3-methyl-1,2-BN-cyclopentane can release hydrogen at much lower temperatures (≤80°C) using low-cost metal halide catalysts such as $FeCl_2$ and $NiCl_2$. Both systems can be regenerated using high-energy hydrides [144, 145], although this pathway becomes expensive in large-scale application. Currently, linear BN compounds (e.g., NH_3—BH_2Me and $MeNH_2$— BH_2Me) are gaining ground as alternatives to the heterocyclic counterparts for hydrogen storage [143].

2.4 BN INTERACTION: SUPRAMOLECULAR ARCHITECTURES

Supramolecular architectures containing BN bonds have been increasing in recent years. BN interactions reveal to be of great importance in the self-assembly of small molecules in order to construct supramolecular entities, such as macrocycles, cages, polymeric structures, and organogels [1, 90, 146–149]. Furthermore, due to the reversible nature of BN bond, it can be applied in the construction of bioconjugated compounds composed by proteins, vitamins, peptides, and RNA [150–152].

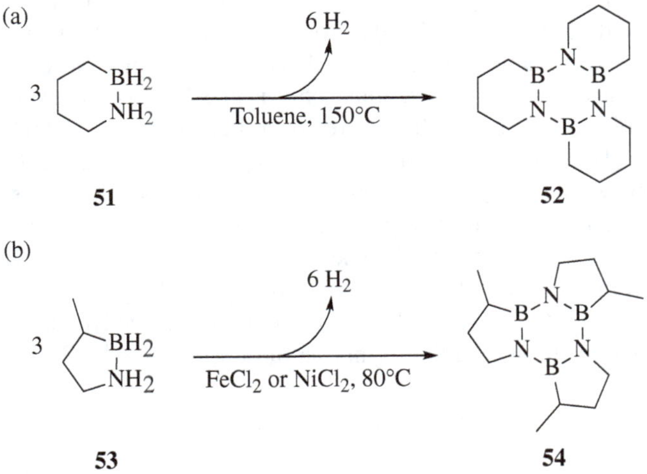

SCHEME 2.4 Hydrogen release conditions for 1,2-BN-cyclohexane (a) and 3-methyl-1,2-BN-cyclopentane (b).

2.4.1 Macrocycles/Cages

Barba, Farfan and Höpfl were pioneers in the development of boronic acid macrocycles. Most of these structures were designed with the main goal of incorporating small organic molecules in their cavities.

The preparation of BN macrocycles and cages is usually made through boronic acid–diol [2+2] condensation, assisted by BN interactions [153–159]. However, [3+3] trimeric and [4+4] tetrameric macrocyclic compounds have also been reported. For instance, the trimeric macrocycle **55** (Fig. 2.16) is arranged in a calix-like shape, which was found to accommodate one molecule of dichloromethane in its cavity [153, 155, 157, 160–162]. Barba and Betanzos have also prepared a hemicarcerand architecture **56** (Fig. 2.16), which was able to incorporate two benzene molecules [163]. Severin *et al.* reported trigonal prismatic cages **57** (Fig. 2.16) that resulted from the assembly of diboronic acids, catechols, and tripyridyl linkers [164–169]. The size of the cage could be adjusted by changing the length of diboronic acid, and depending on such aspect, supramolecular structures can encapsulate different polyaromatic compounds, for example, coronene or triphenylene [168, 169].

BN cages can also be applied in the construction of complex 3D molecular networks. Höpfl *et al.* reported the synthesis of a pentameric boroxine cage **58**, which was further used in the preparation of a supramolecular 3D structure. The boroxine core was constructed from the self-assembly of 15 molecules of 3-pyridineboronic, and then the BN interactions promoted the formation of the BN cage **58** (Fig. 2.17) [170].

James *et al.* developed a different self-assembly system composed by boronic acids, amines, and diols. These structures were used to determine the enantiomeric excess of either chiral amines or chiral diols using NMR spectroscopy and electrochemical methods [171–178]. This methodology was later applied for the construction of a library of Schiff base BN boracycles **59** and **60** from the condensation of boronic acids with different 1,2-amino alcohols (Scheme 2.5) [179].

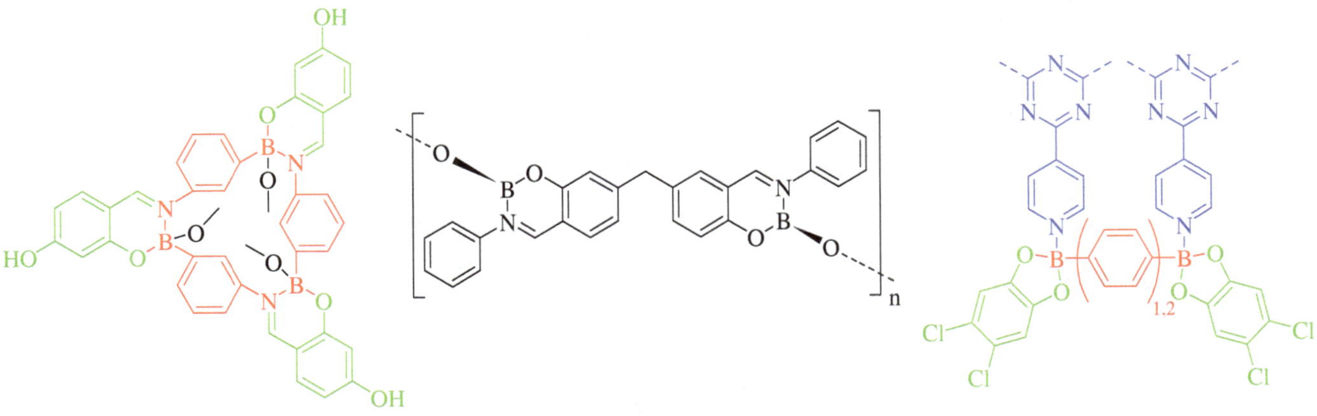

55 56 57

FIGURE 2.16 BN macrocycles and cages **55–57**.

2.4.2 Polymeric Structures

As far as our knowledge goes, the first polymers containing BN bonds were constructed in the 1960s and were made of different cores, such as borono-hydrazones [180], diazaboroles [181], and borazines [182]. In the 1990s, several research groups reported different BN polymers **61–64** with diverse properties as described in Figure 2.18 [183–187].

More recently, Severin *et al.* have done an extensive work in the preparation of BN polymers [166, 167, 169, 188–190]. Using a three-component reaction of aryl boronic acids with 1,2,4,5-tetrahydroxybenzene and 1,2-bis(4-pyridyl)ethylene or 4,4′-bipyridine, BN polymers **65** were constructed (Fig. 2.19), which were deeply colored due to efficient charge transfer transitions [188]. Four years later, two-dimensional (2D) BN polymers were developed using the aforementioned three-component methodology, but in this case the building blocks used were triboronic acid, 4-*tert*-butylcatechol, and 4,4′-bipyridine or 1,2-di(4-pyridyl) ethylene [189].

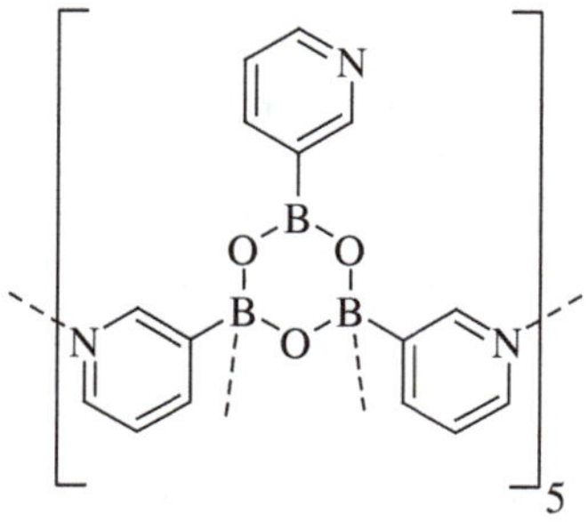

58

FIGURE 2.17 Pentameric boroxine cage **58**.

Höpfl *et al.* reported the synthesis of BN polymers from N-containing boronic acids 5-isoquinolineboronic acid or 4-pyridineboronic acid and pentaerythritol. X-ray crystallography revealed that polymer **66** has a 2D layered structure, while polymer **67** adopts a 3D structure (Fig. 2.19) [191]. This demonstrated that a simple change in the aromatic core the boronic acid (isoquinoline to pyridine) was sufficient to alter the construction of the polymer, from a 2D to a 3D structure.

2.4.3 Organogels

Severin *et al.* were also involved in the synthesis of BN supramolecular organogels. In 2011, an orange organogel **68** was prepared from the condensation between an extended triboronic acid, 4-*tert*-butylcatechol, and 4,4′-bipyridine. It was concluded that the use of this triboronic acid and the formation of BN interactions were fundamental for the gelation process. In fact, heating caused this organogel to become liquid, due to the disruption of the dative BN bonds [169, 189].

With the goal to prepare organogels, the same group performed recently a screening using different arylboronic acids containing a formyl group, arylboronic acids with a primary amine group, catechols, and ditopic N-donor ligands. The efforts resulted only in the BN organogel **69** (Fig. 2.20), as the others were not sufficiently stable [192].

Iwasawa *et al.* reported another organogel **70** (Fig. 2.20), in which the gelation was also promoted by BN bonds. The preparation of organogel **70** was done in two stages. First, it constructed a macrocycle from a bis(1,2-diol) compound and 2,5-difluoro-1,4-phenylenediboronic acid. Then, after addition of 1,3-diaminopropane, the BN interactions promoted the formation of organogel **70** [193].

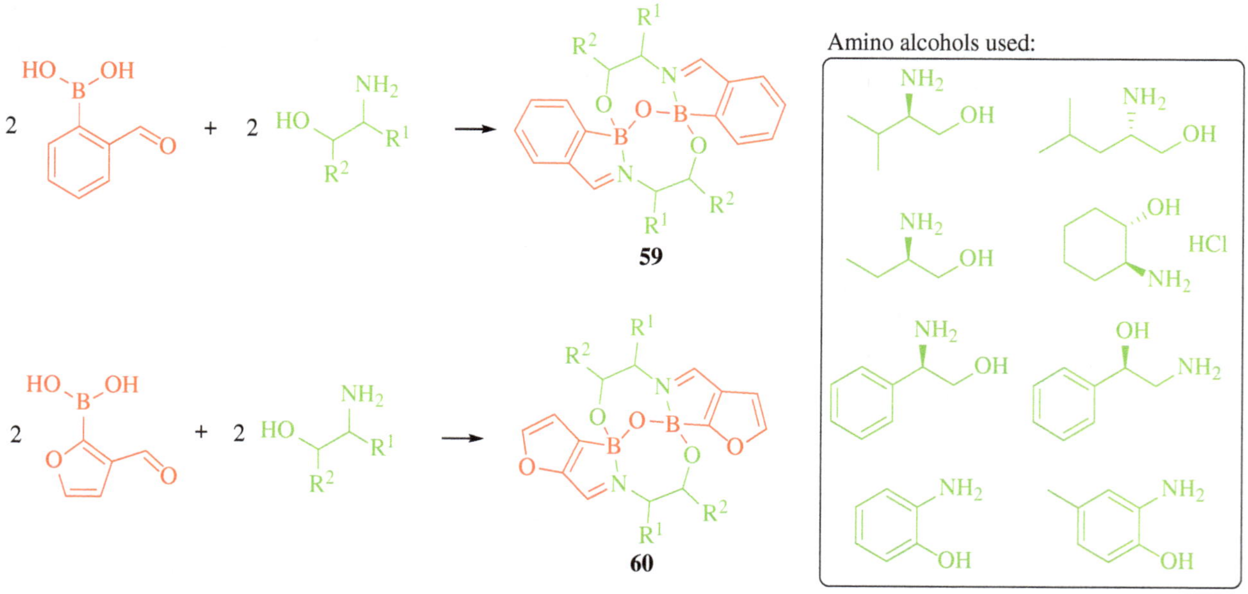

SCHEME 2.5 Synthesis of macrocyclic boracycles **59** and **60**.

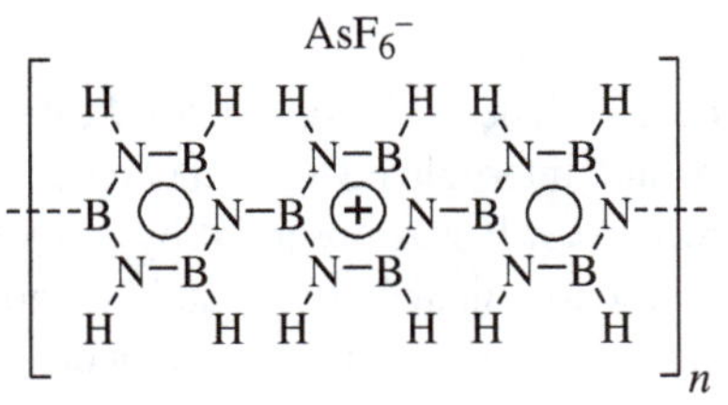

61

Borazine polymer—conducting properties

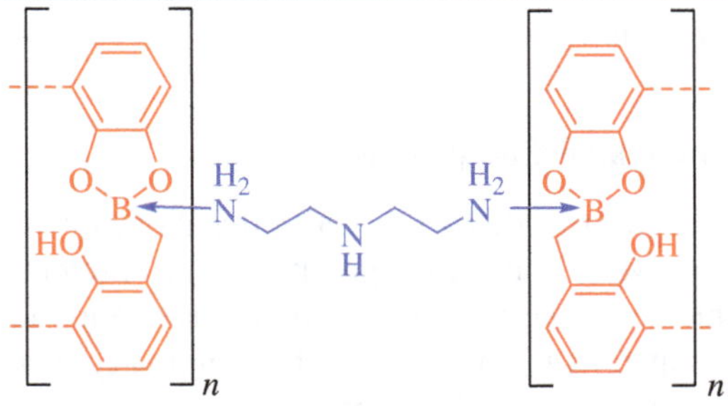

62

Urushiol/diethylene triamine polymer—anticorrosive

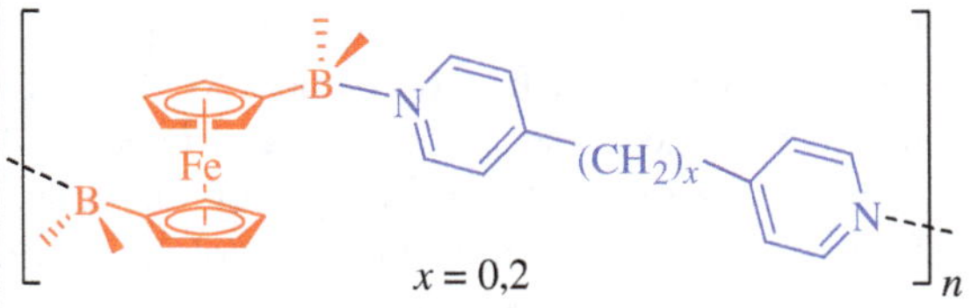

63

Ferrocenylborane/pyridine polymer—electron donor in organometallic redox chemistry

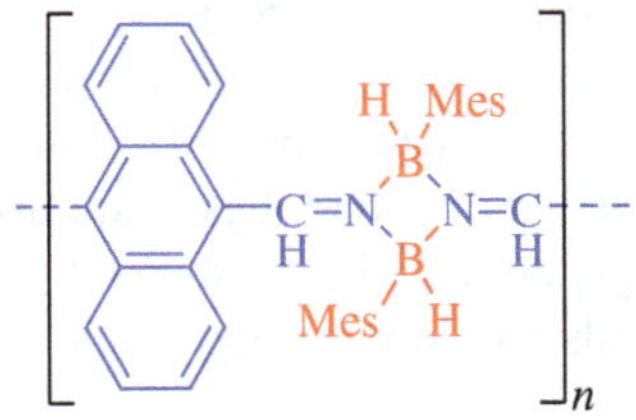

64

Cyclodiborazane polymer—broad emission in the green region (494 nm)

FIGURE 2.18 BN polymers **61–64** and their applications.

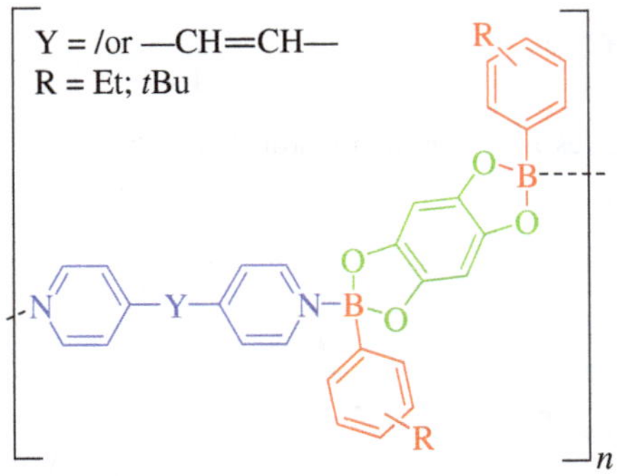

65

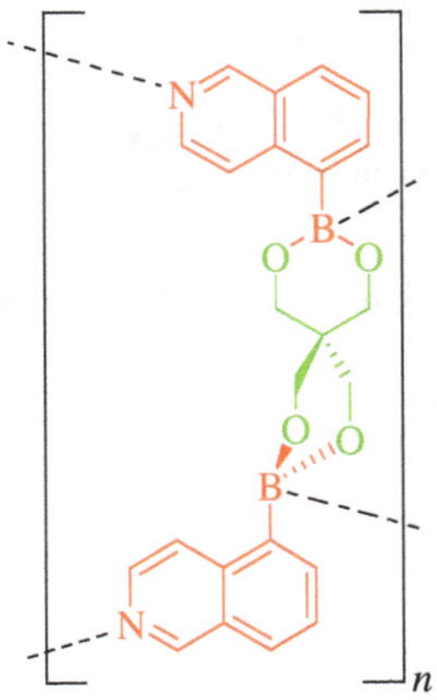

66

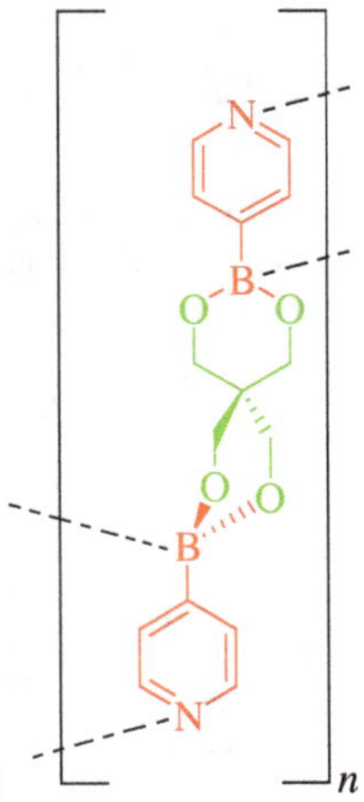

67

FIGURE 2.19 BN polymers **65–67**.

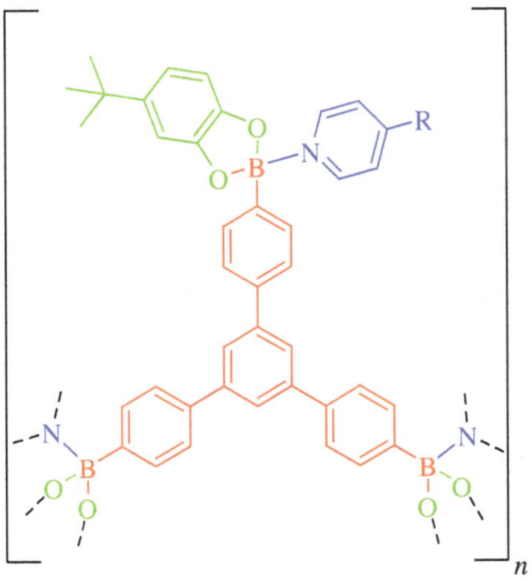

68

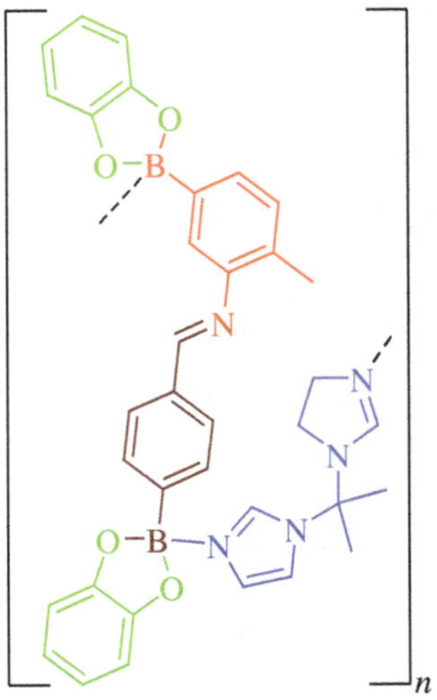

69

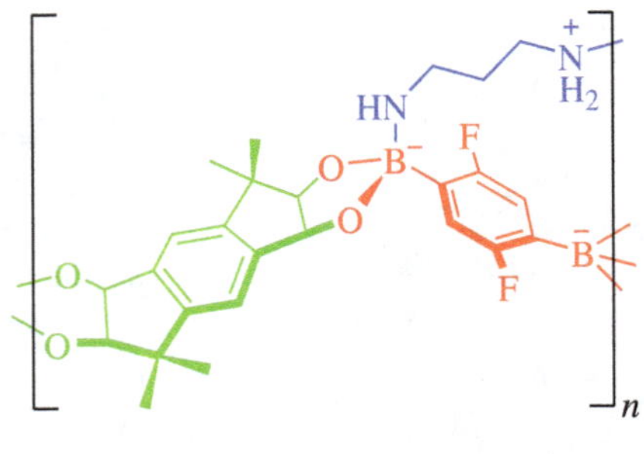

70

FIGURE 2.20 BN organogels **68–70**.

2.4.4 Bioconjugated Compounds

In 2012, Gois *et al.* reported a new bioconjugation strategy to modify the lysine's ε-amino group and the protein's N-terminal moiety based on the formation of stable iminoboronates in aqueous media. Various bioconjugates were constructed using this novel bioconjugation approach, in which different boronic acids were conjugated with several biomolecules such as somatostatin, lysozyme, cytochrome C, ribonuclease A, and myoglobin (Scheme 2.6). The most significant feature of these bioconjugated structures relied on their stability toward hydrolysis, which was a direct consequence of the dative BN bond. Upon addition of dopamine, fructose, and glutathione, it was shown that these bioconjugated compounds would decompose due to BN bond dissociation [150].

In a subsequent study, Gois *et al.* reported the synthesis of fluorescent bioconjugated compounds **72**, in which fluorescent 2-acetylbenzeneboronic acids underwent BN promoted conjugation with *N*-(2-aminoethyl) folic acid (Scheme 2.7). These bioconjugates **72** were selectively recognized and internalized by cancer cells that overexpress folate receptors. It was observed by a fluorescent assay that the same bioconjugated compounds without the iminoboronate function were too unstable and were not detected inside the cancer cells [151].

Inspired by the reversible nature of the boronic acid–salicylhydroxamic acid (BA–SHA) system, Weil *et al.* developed a new supramolecular protecting system for enzymes. The result was a hybrid dendrimer–enzyme complex **73**, made through the insertion of a pH-responsive dendrimer onto the enzyme surface (Scheme 2.8). The enzymes were inactive in the presence of the dendrimer shell, but their activity was restored at acidic pH by dissociation of BA–SHA system and subsequent removal of the dendrimer structure. The inactive complex **73** was efficiently taken up by A549 cells releasing the active enzyme after exposure to the acidic compartments [194].

The BA–SHA system was also explored by Wiley *et al.* in the construction of bioconjugates **74** through protein

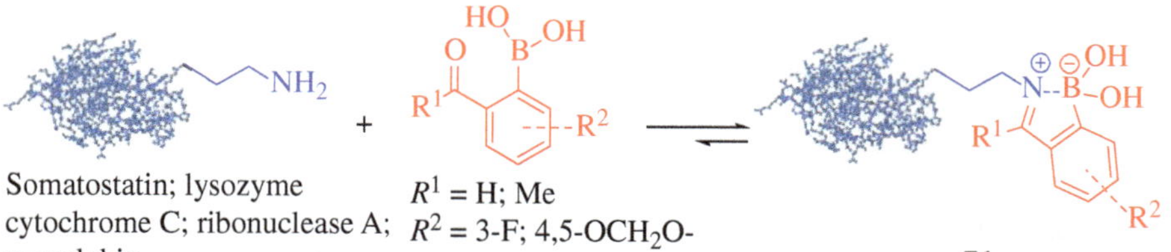

71

SCHEME 2.6 Construction of bioconjugates **71** *via* an iminoboronate function.

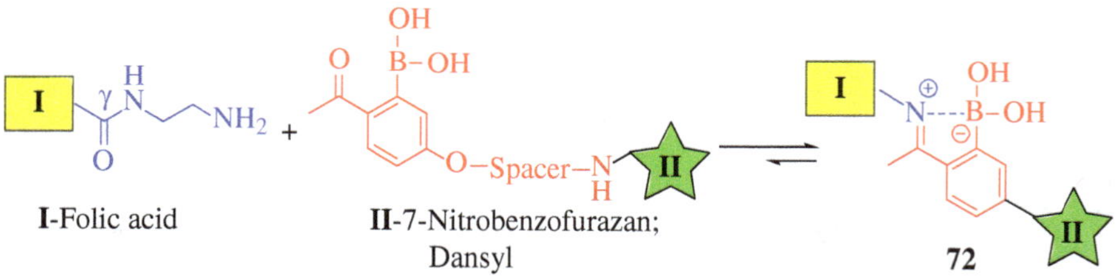

72

SCHEME 2.7 Folic acid modification with fluorescent boronic acids. Construction of BN-based cancer cell targeting fluorescent conjugates **72**.

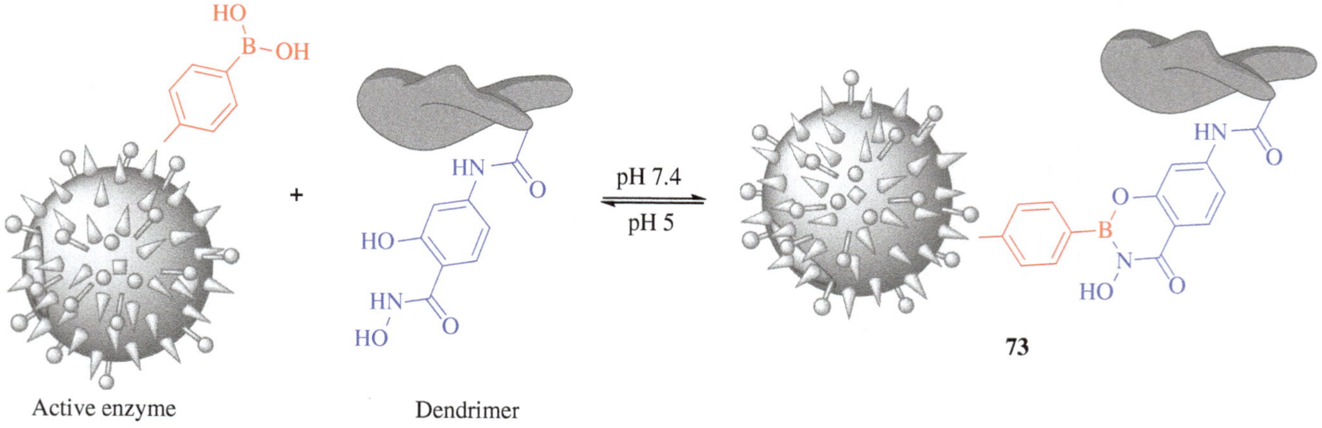

73

SCHEME 2.8 Supramolecular protecting system for enzymes based on the reversible nature of the BA–SHA system.

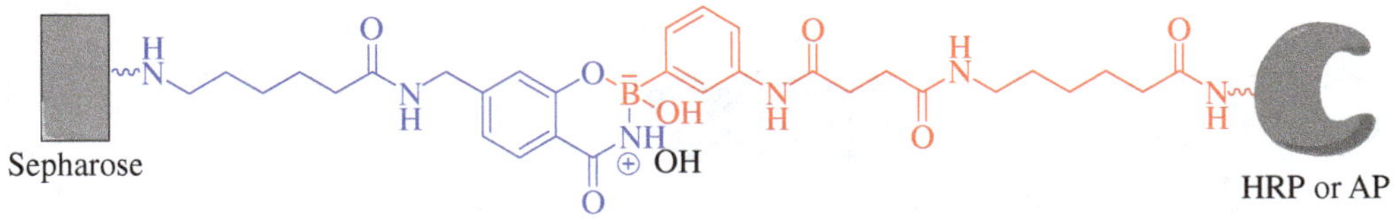

74

FIGURE 2.21 Construction of bioconjugates **74** from immobilization of HRP or AP into a chromatographic support (sepharose).

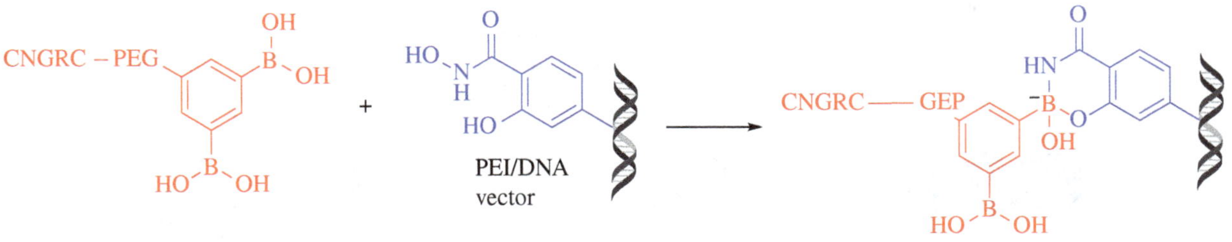

75

SCHEME 2.9 Complexation of the phenyldiboronic acid–PEG-linked peptide (CNGRC) and the salicylhydroxamate polyethylenimine/DNA-bgal vector to construct vector **75**.

immobilization on a chromatographic support (Fig. 2.21). Alkaline phosphatase (AP) and horseradish peroxidase (HRP) were modified with a boronic acid moiety, and the salicylhydroxamic acid group was inserted in the chromatographic support (sepharose). This immobilization process was used to purify boronic acid-derivatized proteins, since the BA–SHA complexation is reversible at acidic pH [195, 196].

Cristiano *et al.* developed a new polycation-based vector (polyethyleneimine (PEI)/DNA), which was coupled with a specific peptide (CNGRC) *via* BA–SHA complexation, ensuring the structural integrity of the final vector **75** (Scheme 2.9). The constructed vector **75** was capable of efficient tumor-specific delivery, and the future goal of their work is the delivery of therapeutic genes into solid tumors [197].

2.5 BN INTERACTION: CHEMOSENSING

Due to the dynamic properties of BN interactions, numerous studies have been performed to attempt a triggered response based on the boronic acid's selective diol recognition [198]. Therefore, different sensors were developed to display measurable parameters upon establishment or disruption of a BN interaction in the presence of specific species. Several examples of chemosensing applications will be discussed in further detail.

2.5.1 Fluorescent Receptors

Fluorescence is one of the most effective methods to evaluate chemosensing. Taking into account that BN interaction can be modulated for an "off–on" response upon recognition of a saccharide molecule (for instance) [199], it is not surprising that the majority of applications of chemosensing rely on this spectrophotometric technique. To take advantage of this system, several fluorescent aminoboronated sensors were developed throughout the years, which can be divided in photoinduced electron transfer (PET) sensors or non-PET sensors.

PET sensors comprise a photoexcitable molecule that transfers an excited electron from a donor to an acceptor. A change in the redox potential of the donor can modify the PET process and thus vary the fluorescence [200].

The first strategy to modulate fluorescence variation (PET quenching) through a mechanism dependable on BN interaction was suggested by Shinkai *et al.* in 1994 [201]. Such approach is based on the known fact that anthracene fluorescent quenching is related to the lone-pair electrons on the benzylic amine moiety. In this case, the fluorescence intensity is controlled according to the presence of saccharide molecules that generate boronate esters and modify the amine's availability.

There are two different mechanisms to explain the variations in fluorescence intensity of these Wulff-type boronic acids (Scheme 2.10). One was suggested by Shinkai *et al.* while reporting the amino anthracene compounds, which assumed that the fluorescence intensity increase upon saccharide binding was due to the formation or strengthening of the BN interaction. Such claim was based on the increased acidity of the boronate ester that enhances BN binding, thus preventing fluorescence quench through PET [201, 202].

On the other hand, Wang *et al.* reported an alternative mechanism denominated by hydrolysis or solvent insertion mechanism, which is based on the formation of a zwitterionic species upon saccharide binding, since the increased acidity of boronate ester promotes its reaction with a water

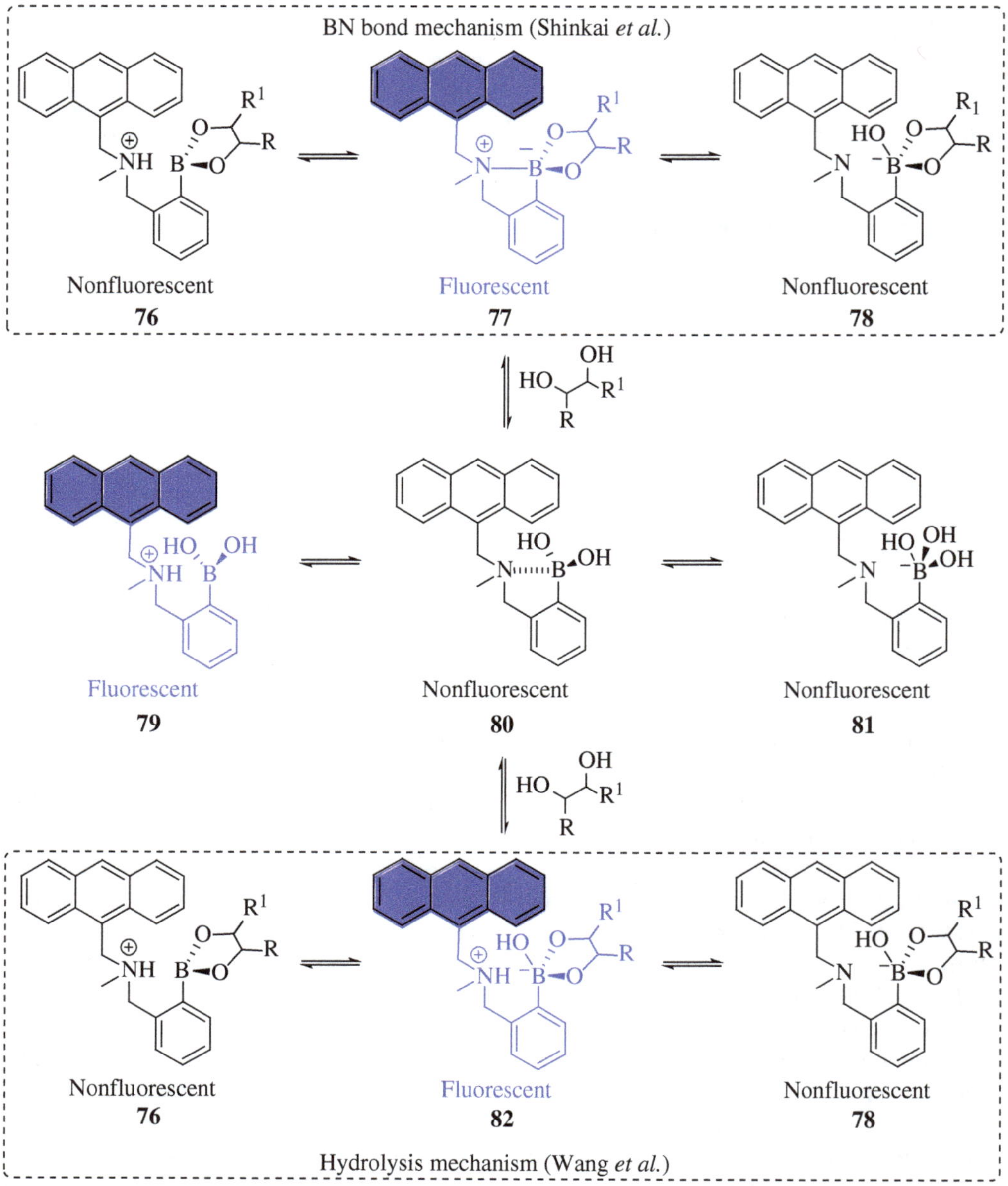

SCHEME 2.10 Mechanisms for the fluorescence intensity changes for the amino anthracene fluorophore.

molecule [203]. Therefore, upon saccharide binding, the conversion of the weak BN bond (compound **80**) into a protonated amine (compound **82**) is favored, leading to an increased fluorescence intensity through PET quenching.

Taking into account the experimental data collected by Wang *et al.*, the hydrolysis mechanism has been suggested as the most probable mechanism to explain the fluorescence intensity alteration of these Wulff-type anthracene boronic acids.

Once this anthracene PET sensor was described, several modifications of its structure have been performed in order to modulate the selectivity toward specific saccharide molecules. One of those attempts was to develop Wulff-type aza-crown ether boronic acids (**83**; Fig. 2.22), which showed D-glucosamine-responsive selective fluorescence [204, 205]. The authors suggest that the primary amine of D-glucosamine is involved in a hydrogen bond network with the benzylic nitrogen from the aza-crown ether in order to suppress PET. This hypothesis was confirmed by studying the same interactions between the saccharide molecules and two synthesized compounds, namely, compounds **83** bearing no Wulff-type boronic acid and also a derivative with no *para*-substituted aza-crown (compound **80** mentioned in Scheme 2.10).

Apart from such attempts of modulating selectivity with chemical substitutions on compound **80**, another common strategy was to synthesize different diboronic acids bearing fluorophores that are able to chelate to specific saccharide molecules. For instance, Shinkai *et al.* tried to alter recognition selectivity by synthesizing a diboronic acid PET sensor based on the previously reported anthracene monomer. This PET sensor retained the same "on–off" fluorescent properties and promoted an effective binding pocket for glucose as depicted in complex **84** of Figure 2.23 [206, 207]. Another example of this higher control over saccharide binding was obtained by James and Shinkai through the development of an allosteric diboronic acid with two aza-crown structures (**85**; Fig. 2.23). In this case, the "on–off" PET sensor had the advantage of having a cause–effect relationship of saccharide release upon the formation of a metal crown sandwich [208].

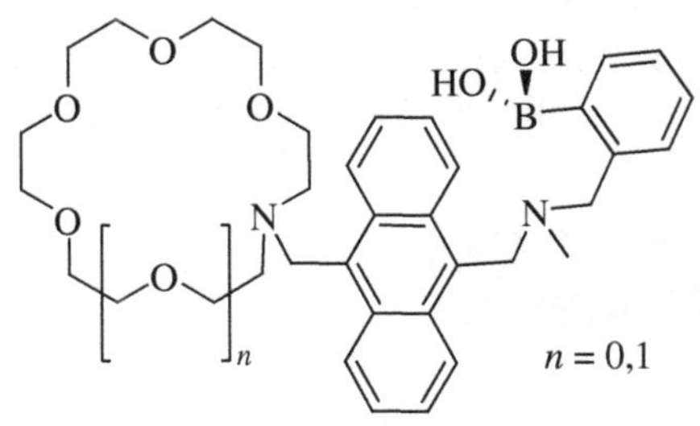

83

FIGURE 2.22 Wulff-type aza-crown ether boronic acids with specific fluorescence increase upon binding with D-glucosamine.

Pursuing enhanced saccharide recognition selectivity, several other examples of Wulff-type boronic acids have been developed with different fluorophores for varied applications. A D-glucuronic acid-selective fluorescent system was developed bearing a metal-chelating fluorophore (phenanthroline, **86**; Fig. 2.24) [209, 210]; a diboronic acid bearing the same metal ligand was also reported, but it presented issues regarding selectivity and sensitivity due to the large spacer between the Wulff-type boronic acids (**87**; Fig. 2.24) [211]. Another diboronic acid bearing two pyrene groups that composed a 2D PET sensor was also described (**88**; Fig. 2.25) [212], and, lastly, a chiral binaphthyl building block was synthesized to design a chiral recognition of saccharides using steric and electronic factors (**89**; Fig. 2.25) [213].

Separately, the Wulff-type anthracene boronic acid (compound **80**) was also further developed for diverse applications, namely, as a dual chemosensor for the detection of both saccharide molecules and ROS, which are relevant

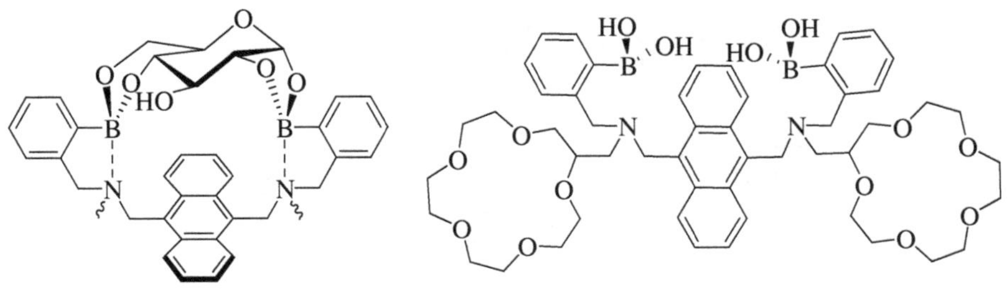

84 **85**

FIGURE 2.23 Wulff-type diboronic acids: compound **84**, diboronic acid PET sensor chelating a glucose molecule; compound **85**, metal-chelating diboronic PET sensor.

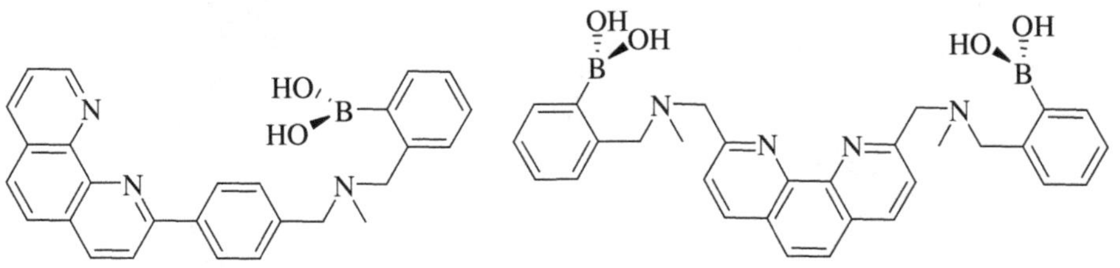

86 **87**

FIGURE 2.24 Compound **86**, D-glucuronic acid-selective fluorescent system and metal chelate; compound **87**, diboronic acid with a phenanthroline spacer.

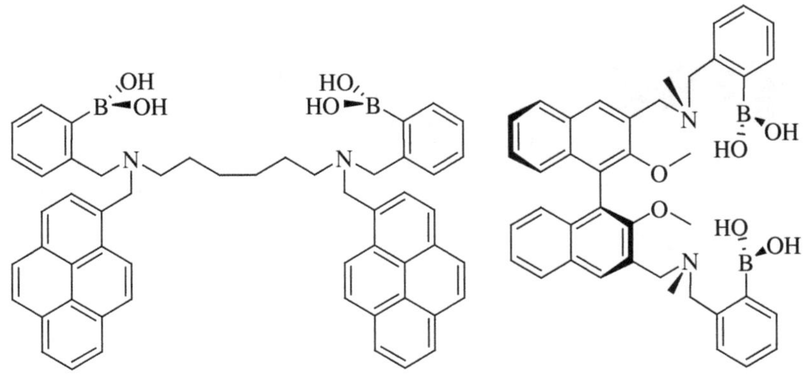

88 **89**

FIGURE 2.25 Compound **88**, two-dimensional PET sensor; compound **89**, chiral recognition by a di-Wulff-type boronic acid.

biomarkers for a large number of pathologies associated with oxidative cellular damage (Scheme 2.11) [214], and as a chemosensor for quantification of trace amount of water in organic solvents (Scheme 2.12) [215] and also for dopamine sensing (Scheme 2.13) [216].

Another PET sensor based on BN interaction was developed by Mohr *et al.*, which used naphthalimide fluorophore as basis for a similar "on–off" probe, as depicted in Scheme 2.14 [217]. Based on Shinkai's work, it is once again suggested that, upon boronic acids' binding to saccharide molecules, the increased acidity of boron strengthened BN interaction leading to PET fluorescence quenching and thus to increase in fluorescence intensity. In this case *ortho* boronic acid position is compared with *meta* and *para* derivatives to exhibit the unique performance in displaying an increased fluorescence intensity.

Following this work, Wang *et al.* described the synthesis of naphthalimide PET sensors, but without any

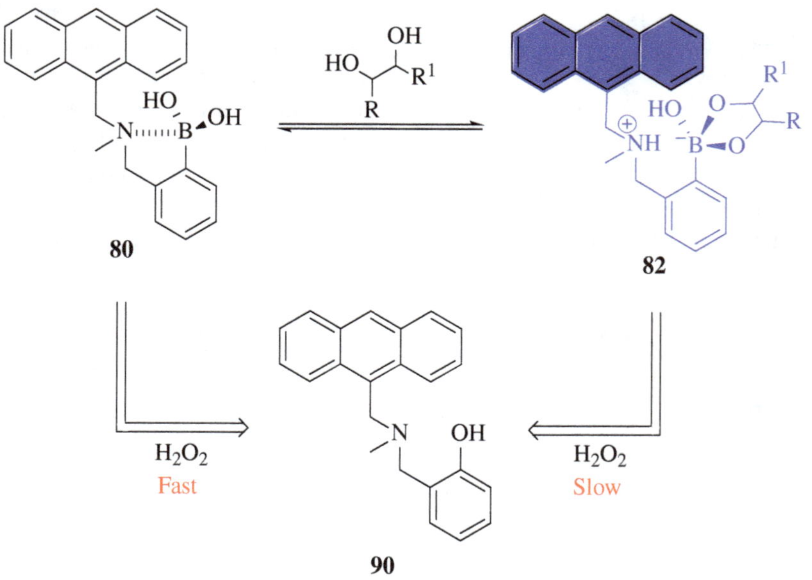

SCHEME 2.11 ROS detection based on the PET sensor.

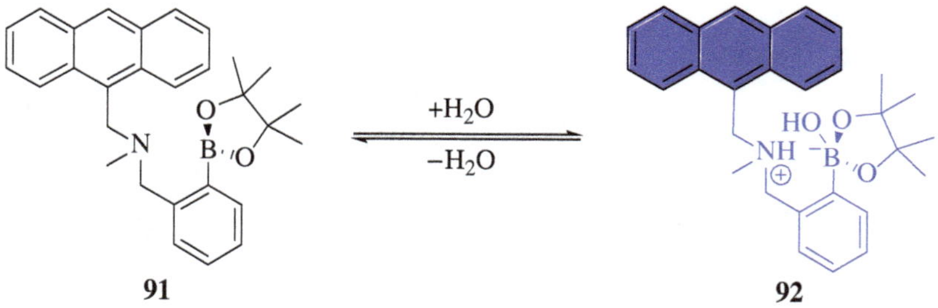

SCHEME 2.12 Water detection in organic solvents using a PET sensor.

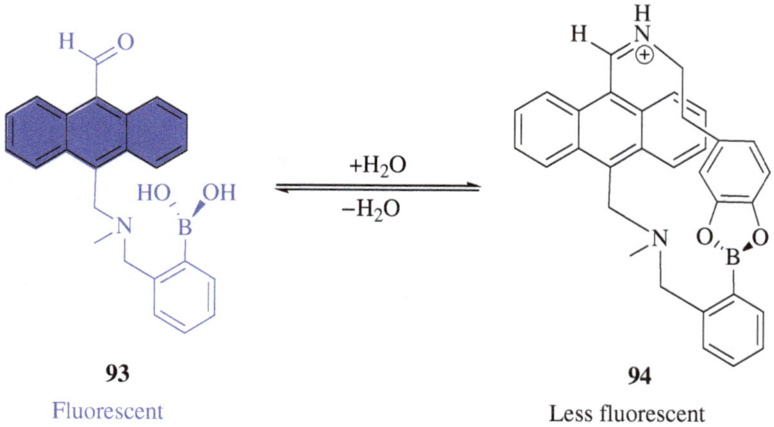

SCHEME 2.13 Dopamine detection using a PET sensor.

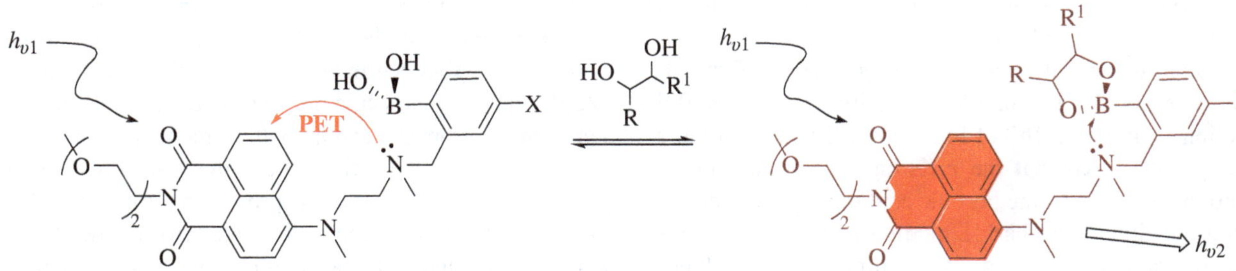

SCHEME 2.14 Fluorescence intensity changes for the naphthalimide fluorophore.

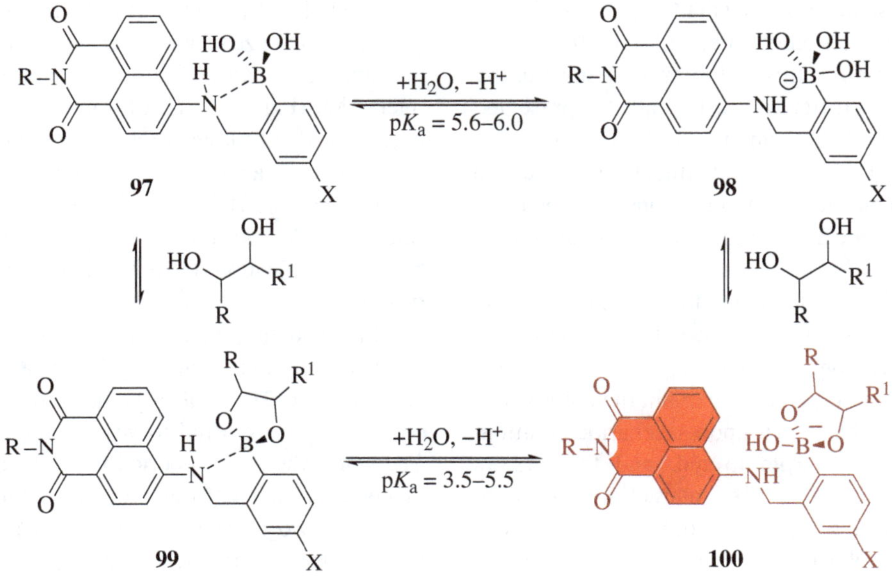

SCHEME 2.15 Wang *et al.*'s study of the naphthalimide fluorophore.

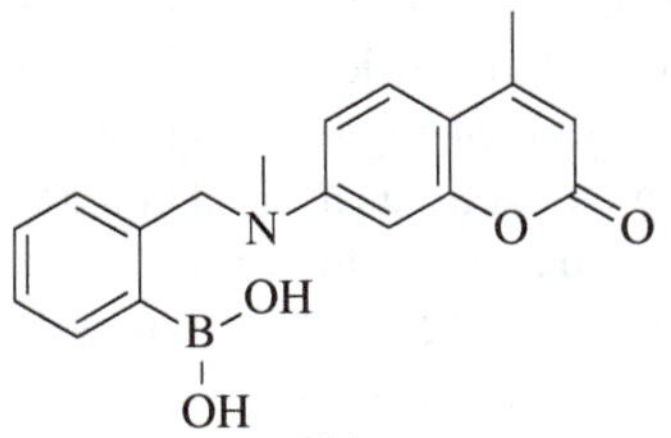

101

FIGURE 2.26 ICT sensor–aminocoumarin chromophore.

spacer between the fluorophore and the benzylboronic acid moiety (Scheme 2.15). First, the authors evaluate the effect of *N*-substitutions of the naphthalimide core on fluorescence variation (R position is either unsubstituted, methylated, or benzylated), concluding that upon saccharide binding the increase of steric hindrance lowers the fluorescent response of the PET sensor [218]. On another work, the authors evaluate the effect of using electron-donating and electron-withdrawing substituents at the *para* position of the phenylboronic acid moiety (X position depicted in Scheme 2.15 is either an ether, an ester, or a fluorine atom) and the pH effect on fluorescence intensity. First, it was found that such *para* substitutions had negligible effect on both the binding affinity and spectroscopic properties, while basic pH was described as crucial to disrupt the BN interaction by generating an anionic boronate saccharide ester that decreases the rotational freedom around CN bond, which, in turn, justifies the increased fluorescence intensity [219].

In 1995, Shinkai *et al.* also described the synthesis of a non-PET sensor based on internal charge transfer (ICT) depicted in Figure 2.26 [220]. In the case of this aminocoumarin (compound **101**), fluorescence intensity and wavelength are affected by the relationship between BN interaction and saccharide sensing, due to the fact that the nitrogen is directly linked to the chromophore. However, these properties showed only slight deviations between the free boronic acids and the saccharide boronate esters, which were unfortunate prepositions for this type of methodology [200].

2.5.2 Colorimetric Receptors

Another common method to evaluate chemosensing is based on color change detection, which can be achieved by having a recognition element covalently attached to a chromophore. Therefore, modulation of BN interaction upon recognition of a saccharide molecule can act as the trigger to perturb the chromophore's electronic transitions. In order to develop this methodology, several colorimetric aminoboronated sensors were developed with different chromophores.

In 1994, Sandanayake and Shinkai reported the first colorimetric sensor for saccharide recognition based on BN interaction (compound **102** of Fig. 2.27) [221]. The authors rationally considered that the electron-rich tertiary amine interaction would make the boronic acid moiety more acidic and, thus, facilitate saccharide binding at neutral aqueous media. In fact, as expected, the anilinic electron donor moiety of this ICT chromophore changed upon saccharide binding, which was detected spectrophotometrically.

However, this system showed small variations in the absorption bands of the chromophore, and, therefore, years later, James *et al.* optimized this ICT sensor by substituting the sulfonyl moiety with different substituents that presented diverse electronic properties (compound **103a–e** of Fig. 2.27) [222, 223]. The most effective substituent to produce a change in color from purple to red in aqueous solution upon the presence of saccharide molecules was the derivative with a nitro substituent.

According to James *et al.* the main difference responsible for an enhanced response of the reported dye (compound **103a**) is the anilinic hydrogen, which led the authors to propose different equilibrium species to explain the color variation (Scheme 2.16). At basic pH, upon saccharide addition, the solution turns from purple to red, and the authors associate this change to a stronger BN interaction and a more acidic NH proton, which, after deprotonation, generates compound **108**. This color change is justified according to the differences of electronic environment of the anilinic nitrogen that is linked to the azo chromophore, causing changes in the absorption energy and wavelength.

Later, the same group developed another Wulff-type boronic acid (**109**; Fig. 2.28), which applied the same concept in a neutral aqueous solution. In this case, the ICT sensor, upon saccharide recognition, changed its color from purple to pink [224].

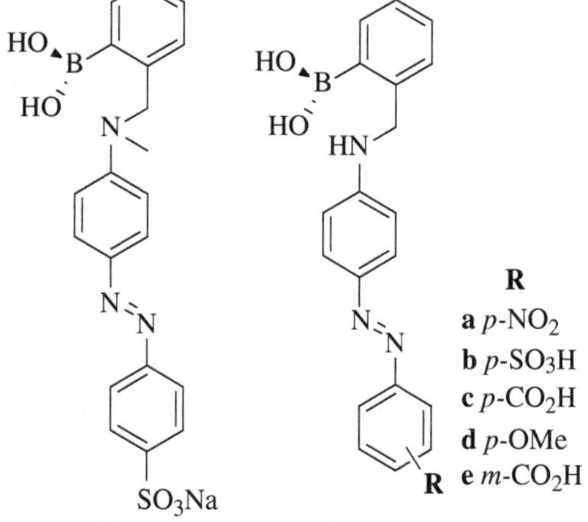

FIGURE 2.27 Colorimetric sensor based on ICT chromophore.

In 1995, Shinkai *et al.* reported another colorimetric sensor for saccharide molecules based on BN interaction, which was based on a spirobenzopyran boronic acid core (Scheme 2.17) [225]. This core molecule presents a tautomeric equilibrium between a spiropyran and a zwitterionic merocyanine form. In the absence of the carbohydrate, this last form is predominant due to a stabilizing effect promoted through a hydrogen bonding network between the boronic acid and the phenolate anion (compound **111**), presenting a purple color. Upon the presence of saccharide molecules, the authors suggest that an enhanced BN interaction stabilizes the spiropyran form (compound **112**), and this results in a color change.

Later, azobenzene boronic acid compounds were developed as colorimetric sensor for saccharide molecules based on BN interaction (Fig. 2.29) [226]. Egawa *et al.* synthesized *ortho* boronic acid-substituted azobenzenes based on the assumption that a boronic acid group adjacent to a diazo chromophore would affect the BN interaction upon binding with a saccharide molecule and, consequently, the sensor's color. In fact, both pH basification (from 7 to 13) and addition of D-fructose (up to 100 mM) caused a shift in the UV-visible absorption spectra, validating the premises of this work. Furthermore, in order to improve water solubility, a disulfonyl derivative (compound **115**) was also synthesized, and it showed a severe color change (from red to yellow) upon saccharide binding. The work was also supported with ^{15}N and ^{11}B NMR experiments showing interesting properties of the BN interaction.

Lastly, two literature reports using intermolecular interactions between boronic acids and dyes that involve a BN interaction will be described. The first uses an azo dye that switches its color from yellow to orange when the BN interaction is established, and, then, upon saccharide binging, it shifts to red due to a stronger BN interaction (**116**; Fig. 2.30) [227]. In the second work, Anslyn *et al.* reported a specific colorimetric chemosensing ensemble for the analysis of tartrate/malate in beverages, which is a common natural product present in several grape-derived beverages [228]. This sensor is composed by a mixture of a compound with a BN interaction and alizarin complexone dye (**117**; Fig. 2.30), which causes the color changes between the natural color of alizarin (burgundy) and the yellow/orange color that the complex generates. In this case, due to a higher selectivity toward tartrate, the saccharide molecule induces a displacement of the dye, changing the color back to burgundy upon the presence of the recognizable moiety.

2.5.3 Other Receptors

Apart from the fluorescent and colorimetric sensors that can be evaluated with spectrophotometric parameters, another useful saccharide sensor assesses hydrosolubility as a way to evaluate boronate formation and the BN interaction. In 2009, van Hest *et al.* reported Wulff-type boronic acids in polymers that bind with saccharides due to the presence of a BN interaction that lowers the pK_a of the boronic acid (Scheme 2.18) [4]. Due to this interaction,

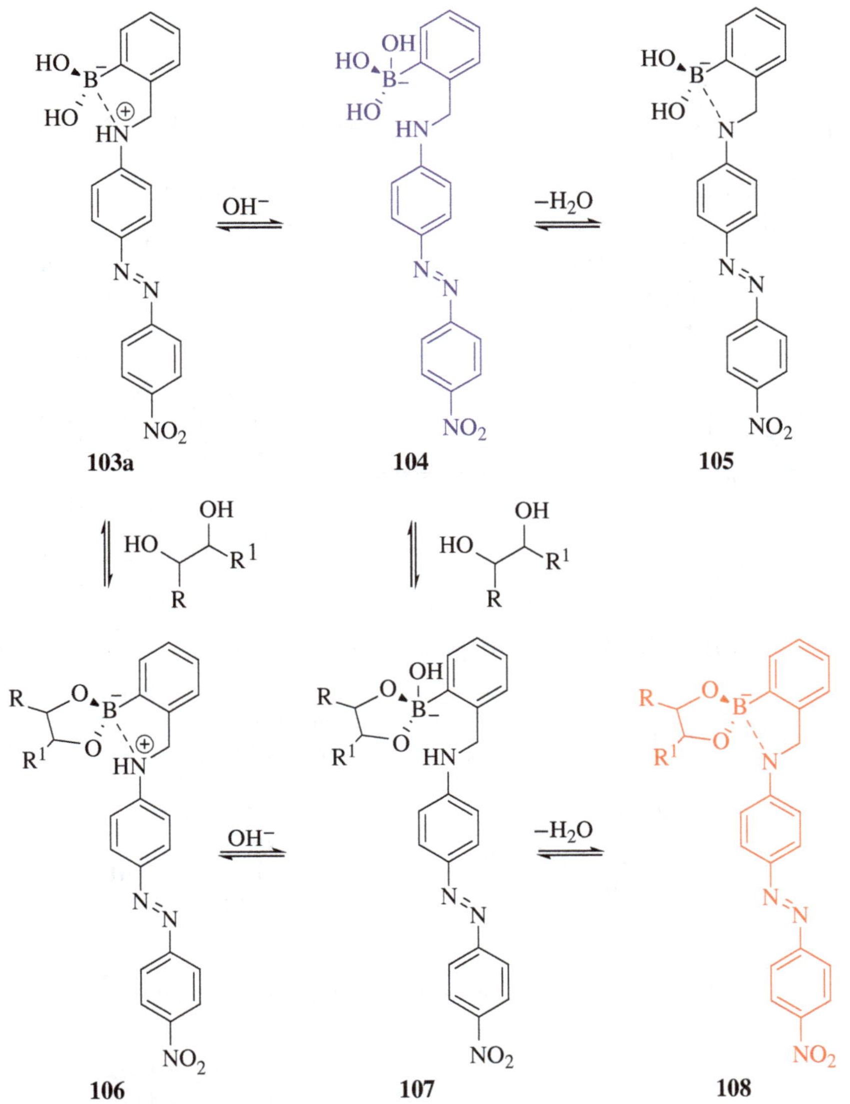

SCHEME 2.16 Proposed equilibria for the ICT sensor.

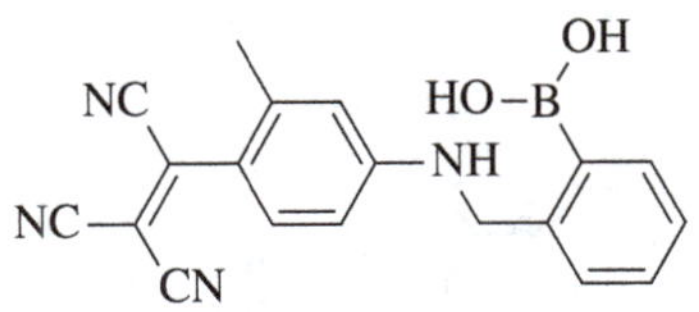

109

FIGURE 2.28 Saccharide recognition–response ICT sensor in aqueous methanol at pH 8.

these styrenic polymers present an increase of hydrosolubility at neutral pH conditions [229].

2.6 FINAL COMMENTS

The broad array of applications hereby presented clearly state the impact of covalent and dative BN bonds as chemical tools and one of the most promising areas of chemistry.

The endless argument on BN analogues' aromaticity has indirectly been a valuable asset in understanding the true foundations of aromaticity itself. Replacing covalent CC bonds by BN bonds has provided in many cases better properties to compounds already used for different biologic and electronic applications. The heterogeneity of the BN bond was also essential to create new compounds for current demands such as hydrogen storage, polymers, and catalysts. The incorporation of BN bonds for these applications has benefited from the immense development of boron chemistry in the last decades.

Comprehensive knowledge of the strong BN dative interactions paved the way for self-assembly strategies and formation of macrostructures. On the other hand, the reversibility of these interactions plays a critical role in the preparation of various bioconjugates and chemosensors.

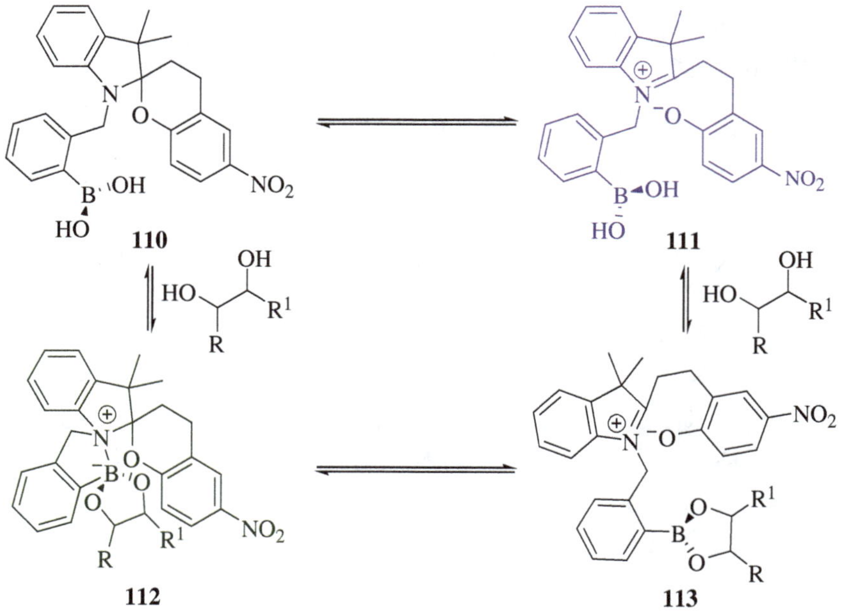

SCHEME 2.17 Spirobenzopyran boronic acid colorimetric sensor.

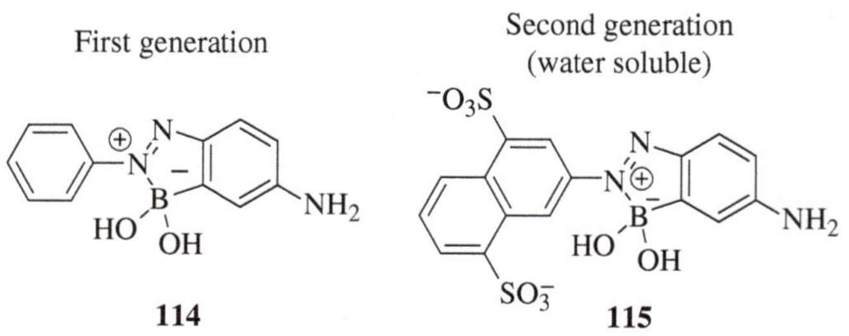

FIGURE 2.29 *Ortho* boronic acid-substituted azobenzene.

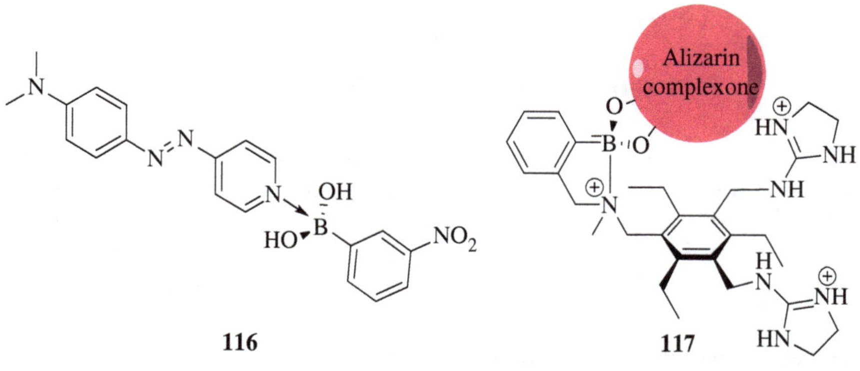

FIGURE 2.30 Compound **116**, intermolecular BN interaction with a color change upon saccharide binding; compound **117**, colorimetric sensor complex bearing a compound with a BN interaction and alizarin complexone.

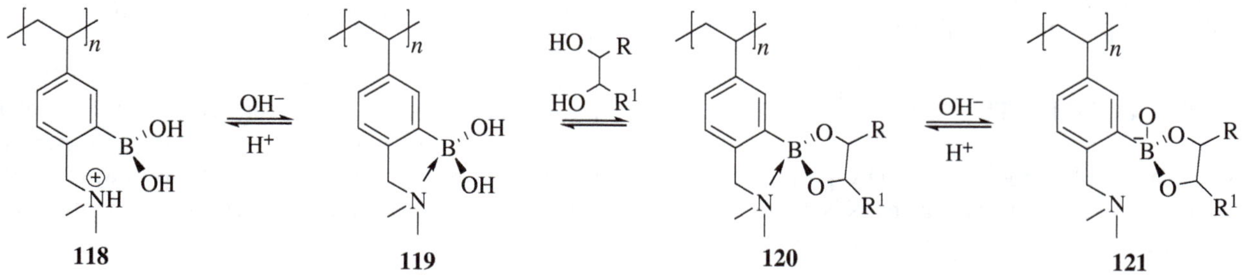

SCHEME 2.18 Interaction between the reported polymer **118** and saccharides in different pH environments.

ACKNOWLEDGMENTS

Fundação para a Ciência e Tecnologia (SFRH/BD/94779/2013; SFRH/BPD/103172/2014; PTDC/QUI-QUI/118315/2010; PTDC/QUI-QUI/119823/2010; Pest-OE/SAU/UI4013/2011). P.M.P. Gois, a FCT Investigator, thanks for financial support funded, in part, by iMed.ULisboa (UID/DTP/04138/2013).

REFERENCES

1. D. G. Hall, Boronic Acids: Preparation, Applications in Organic Synthesis and Medicine, Wiley-VCH, Weinheim, 2005.
2. G. Wulff, *Pure Appl. Chem.* 1982, *54*, 2093.
3. F. Schulz, C. Gelhaus, B. Degel, R. Vicik, S. Heppner, E. Breuning, M. Leippe, J. Gut, P. J. Rosenthal, T. Schirmeister, *ChemMedChem* 2007, *2*, 1214.
4. T. K. Kyoung, J. J. L. M. Cornelissen, R. J. M. Nolte, J. C. M. Van Hest, *J. Am. Chem. Soc.* 2009, *131*, 13908.
5. X. Wang, N. Xia, L. Liu, *Int. J. Mol. Sci.* 2013, *14*, 20890.
6. Z. Liu, T. B. Marder, *Angew. Chem. Int. Ed.* 2008, *47*, 242.
7. M. Faraday, *Philos. Trans. R. Soc. Lond.* 1825, *115*, 440.
8. R. P. Gangwal, M. V. Damre, N. R. Das, G. V. Dhoke, A. Bhadauriya, R. A. Varikoti, S. S. Sharma, A. T. Sangamwar, *J. Mol. Graph. Model.* 2015, *57*, 89.
9. A. Barnard, K. Long, H. L. Martin, J. A. Miles, T. A. Edwards, D. C. Tomlinson, A. Macdonald, A. J. Wilson, *Angew. Chem. Int. Ed.* 2015, *54*, 2960.
10. N. Awang, A. F. Ismail, J. Jaafar, T. Matsuura, H. Junoh, M. H. D. Othman, M. A. Rahman, *React. Funct. Polym.* 2015, *86*, 248.
11. K. Takimiya, I. Osaka, M. Nakano, *Chem. Mater.* 2014, *26*, 587.
12. C. Postigo, D. Barceló, *Sci. Total Environ.* 2015, *503–504*, 32.
13. A. Stock, P. Borwasserstoffe, *Ber. Dtsch. Chem. Ges.* 1926, *59*, 2210.
14. W. Shen, M. Li, Y. Li, S. Wang, *Inorg. Chim. Acta* 2007, *360*, 619.
15. M. J. D. Bosdet, W. E. Piers, *Can. J. Chem.* 2009, *87*, 8.
16. B. Chiavarino, M. E. Crestoni, S. Fornarini, *J. Am. Chem. Soc.* 1999, *121*, 2619.
17. M. K. Cyrański, *Chem. Rev.* 2005, *105*, 3773.
18. R. Islas, E. Chamorro, J. Robles, T. Heine, J. C. Santos, G. Merino, *Struct. Chem.* 2007, *18*, 833.
19. W. H. Fink, J. C. Richards, *J. Am. Chem. Soc.* 1991, *113*, 3393.
20. P. V. R. Schleyer, H. Jiao, N. J. R. V. E. Hommes, V. G. Malkin, O. L. Malkina, *J. Am. Chem. Soc.* 1997, *119*, 12669.
21. D. Benker, T. M. Klapötke, G. Kuhn, J. Li, C. Miller, *Heteroat. Chem.* 2005, *16*, 311.
22. I. Fernández, G. Frenking, *Faraday Discuss.* 2007, *135*, 403.
23. P. W. Fowler, E. Steiner, *J. Phys. Chem. A* 1997, *101*, 1409.
24. E. D. Jemmis, B. Kiran, *Inorg. Chem.* 1998, *37*, 2110.
25. H. Jacobsen, *Chem. Phys. Lett.* 2013, *582*, 144.
26. M. Baranac-Stojanović, M. Stojanović, *RSC Adv.* 2013, *3*, 24108.
27. A. K. Phukan, A. K. Guha, B. Silvi, *Dalton Trans.* 2010, *39*, 4126.
28. S. N. Steinmann, Y. Mo, C. Corminboeuf, *Phys. Chem. Chem. Phys.* 2011, *13*, 20584.
29. M. Baranac-Stojanović, *J. Phys. Chem. A* 2013, *117*, 11540.
30. M. Baranac-Stojanović, A. Koch, E. Kleinpeter, *ChemPhysChem* 2012, *13*, 3803.
31. A. S. Lisovenko, A. Y. Timoshkin, *Russ. J. Gen. Chem.* 2011, *81*, 831.
32. H. A. Dabbagh, M. Shahraki, H. Farrokhpour, *Phys. Chem. Chem. Phys.* 2014, *16*, 10519.
33. T. B. Marder, *Angew. Chem. Int. Ed.* 2007, *46*, 8116.
34. F. H. Stephens, V. Pons, R. Tom Baker, *Dalton Trans.* 2007, *2*, 2613.
35. S. Bernard, P. Miele, *Materials (Basel)* 2014, *7*, 7436.
36. R. T. Paine, W. Koestle, T. T. Borek, G. L. Wood, E. A. Pruss, E. N. Duesler, M. A. Hiskey, *Inorg. Chem.* 1999, *3*, 3738.
37. M. Zamani, M. Hossein, *Comput. Mater. Sci.* 2015, *97*, 295.
38. B. Gemünda, B. Günthera, H. Nöthb, *Arkivoc* 2008, *2008*, 136.
39. B. Anand, H. Nöth, H. Schwenk-Kircher, A. Troll, *Eur. J. Inorg. Chem.* 2008, 3186.
40. M. J. S. Dewar, V. P. Kubba, R. Pettit, *J. Chem. Soc.* 1958, 3073.
41. G. C. Culling, M. J. S. Dewar, P. A. Marr, *J. Am. Chem. Soc.* 1964, *86*, 1125.
42. M. J. S. Dewar, R. Dietz, *J. Chem. Soc.* 1959, 65.
43. P. G. Campbell, A. J. V Marwitz, S. Y. Liu, *Angew. Chem. Int. Ed.* 2012, *51*, 6074.
44. X.-Y. Wang, J.-Y. Wang, J. Pei, *Chem. Eur. J.* 2015, *21*, 3528.
45. E. R. Abbey, A. N. Lamm, A. W. Baggett, L. N. Zakharov, S. Liu, *J. Am. Chem. Soc.* 2013, *135*, 12908.
46. J. Pan, J. W. Kampf, A. J. Ashe III, *Org. Lett.* 2007, *9*, 679.
47. E. R. Abbey, L. N. Zakharov, S. Liu, *J. Am. Chem. Soc.* 2008, *130*, 7250.
48. C. Tanjaroon, A. Daly, A. J. V. Marwitz, S.-Y. Liu, S. Kukolich, *J. Chem. Phys.* 2009, *131*, 224312.
49. R. Carion, V. Liégeois, B. Champagne, D. Bonifazi, S. Pelloni, P. Lazzeretti, *J. Phys. Chem. Lett.* 2010, *1*, 1563.
50. J. Del Bene, M. Yáñez, *J. Chem.* 2009, *5*, 2239.
51. P. G. Campbell, E. R. Abbey, D. Neiner, D. J. Grant, D. A. Dixon, S. Y. Liu, *J. Am. Chem. Soc.* 2010, *132*, 18048.
52. D. Ghosh, G. Periyasamy, S. K. Pati, *Phys. Chem. Chem. Phys.* 2011, *13*, 20627.
53. M. Baranac-Stojanović, *Chem. Eur. J.* 2014, *20*, 16558.
54. A. N. Lamm, S.-Y. Liu, *Mol. Biosyst.* 2009, *5*, 1303.
55. P. Bendel, *NMR Biomed.* 2005, *18*, 74.
56. D. Kanichar, L. Roppiyakuda, E. Kosmowska, M. A. Faust, K. P. Tran, F. Chow, E. Buglo, M. P. Groziak, E. A. Sarina, M. M. Olmstead, *et al.*, *Chem. Biodivers.* 2014, *11*, 1381.
57. B. Pertschy, G. Zisser, H. Schein, R. Köffel, G. Rauch, K. Grillitsch, C. Morgenstern, M. Durchschlag, G. Högenauer, H. Bergler, *Mol. Cell. Biol.* 2004, *24*, 6476.
58. C. Baldock, J. B. Rafferty, S. E. Sedelnikova, P. J. Baker, A. R. Stuitje, A. R. Slabas, T. R. Hawkes, D. W. Rice, *Science* 1996, *274*, 2107.
59. C. W. Levy, C. Baldock, A. J. Wallace, S. Sedelnikova, R. C. Viner, J. M. Clough, A. R. Stuitje, A. R. Slabas, D. W. Rice, J. B. Rafferty, *J. Mol. Biol.* 2001, *309*, 171.
60. C. Baldock, G. J. De Boer, J. B. Rafferty, A. R. Stuitje, D. W. Rice, *Biochem. Pharmacol.* 1998, *55*, 1541.

61. G. Hogenauer, M. Woisetschlager, *Nature* 1981, *293*, 662.
62. N. Surolia, S. P. Ramachandrarao, A. Surolia, *Bioessays* 2002, *24*, 192.
63. M. J. S. Dewar, P. M. Maitlis, *Tetrahedron* 1959, *7*, 213.
64. S. S. Chissick, M. J. S. Dewar, P. M. Maitlis, *Tetrahedron Lett.* 1960, *23*, 8.
65. X. Y. Wang, H. R. Lin, T. Lei, D. C. Yang, F. D. Zhuang, J. Y. Wang, S. C. Yuan, J. Pei, *Angew. Chem. Int. Ed.* 2013, *52*, 3117.
66. M. Lepeltier, O. Lukoyanova, A. Jacobson, S. Jeeva, D. F. Perepichka, *Chem. Commun.* 2010, *46*, 7007.
67. S. Hashimoto, T. Ikuta, K. Shiren, S. Nakatsuka, J. Ni, M. Nakamura, T. Hatakeyama, *Chem. Mater.* 2014, *26*, 6265.
68. X. Wang, F. Zhang, J. Liu, R. Tang, Y. Fu, D. Wu, Q. Xu, X. Zhuang, G. He, X. Feng, *Org. Lett.* 2013, *15*, 5714.
69. A. J. J. Lennox, G. C. Lloyd-Jones, *Chem. Soc. Rev.* 2014, *43*, 412.
70. A. J. J. Lennox, G. C. Lloyd-Jones, *Isr. J. Chem.* 2010, *50*, 664.
71. D. M. Knapp, E. P. Gillis, M. D. Burke, *J. Am. Chem. Soc.* 2009, *131*, 6961.
72. E. P. Gillis, M. D. Burke, *J. Am. Chem. Soc.* 2008, *130*, 14084.
73. H. Bonin, D. Delbrayelle, P. Demonchaux, E. Gras, *Chem. Commun.* 2010, *46*, 2677.
74. J. Li, M. D. Burke, *J. Am. Chem. Soc.* 2011, *133*, 13774.
75. H. C. Brown, J. V. N. V. Prasad, *J. Org. Chem.* 1986, *51*, 4526.
76. H. Noguchi, K. Hojo, M. Suginome, *J. Am. Chem. Soc.* 2007, *129*, 758.
77. N. Xia, M. Taillefer, *Angew. Chem. Int. Ed. Engl.* 2009, *48*, 337.
78. H. Ihara, M. Koyanagi, M. Suginome, *Org. Lett.* 2011, *13*, 2662.
79. N. Iwadate, M. Suginome, *Org. Lett.* 2009, *11*, 1899.
80. E. P. Gillis, M. D. Burke, *Aldrichimica Acta* 2009, *42*, 17.
81. B. Vanveller, M. R. Aronoff, R. T. Raines, *RSC Adv.* 2013, *44*, 21331.
82. V. Barba, D. Cuahutle, R. Santillan, N. Farfán, *Can. J. Chem.* 2001, *79*, 1229.
83. T. Mancilla, M. de los Ángeles Calixto Romo, L. A. Delgado, *Polyhedron* 2007, *26*, 1023.
84. C. C. Musonda, G. A. Whitlock, M. J. Witty, R. Brun, M. Kaiser, *Bioorg. Med. Chem. Lett.* 2009, *19*, 481.
85. J. M. Rivera, S. Rincón, N. Farfán, R. Santillan, *J. Organomet. Chem.* 2011, *696*, 2420.
86. P. Wei, D. A. Atwood, *Inorg. Chem.* 1998, *37*, 4934.
87. M. Sánchez, O. Sánchez, H. Höpfl, M. E. Ochoa, D. Castillo, N. Farfán, S. Rojas-Lima, *J. Organomet. Chem.* 2004, *689*, 811.
88. G. Vargas-Díaz, H. Höpfl, *J. Organomet. Chem.* 2009, *694*, 3660.
89. P. V. V. N. Kishore, A. K. Jami, V. Baskar, *Inorg. Chim. Acta* 2011, *372*, 321.
90. R. Nishiyabu, Y. Kubo, T. D. James, J. S. Fossey, *Chem. Commun.* 2011, *47*, 1124.
91. R. Contreras, A. Flores-Parra, H. C. López-Sandoval, N. Barba-Behrens, *Coord. Chem. Rev.* 2007, *251*, 1852.
92. C. J. Whiteoak, G. Salassa, A. W. Kleij, *Chem. Soc. Rev.* 2012, *41*, 622.
93. A. Flores-Parra, R. Contreras, *Coord. Chem. Rev.* 2000, *196*, 85.
94. A. Abreu, S. J. Alas, H. I. Beltrán, R. Santillan, N. Farfán, *J. Organomet. Chem.* 2006, *691*, 337.
95. H. Höpfl, *J. Organomet. Chem.* 1999, *581*, 129.
96. F. Montalbano, N. R. Candeias, M. T. Duarte, M. R. Bronze, R. Moreira, P. M. P. Gois, *Org. Lett.* 2012, *14*, 988.
97. F. Montalbano, P. M. S. D. Cal, M. A. B. R. Carvalho, L. M. Gonçalves, S. D. Lucas, R. C. Guedes, L. F. Veiros, R. Moreira, P. M. P. Gois, *Org. Biomol. Chem.* 2013, *11*, 4465.
98. F. Montalbano, J. Leandro, G. D. V. F. Farias, P. R. Lino, R. C. Guedes, J. B. Vicente, P. Leandro, P. M. P. Gois, *RSC Adv.* 2014, *4*, 61022.
99. A. Jabbour, D. Steinberg, V. M. Dembitsky, A. Moussaieff, B. Zaks, M. Srebnik, *J. Med. Chem.* 2004, *47*, 2409.
100. X. Chen, S. Schauder, N. Potier, A. Van Dorsselaer, I. Pelczer, B. L. Bassler, F. M. Hughson, *Nature* 2002, *415*, 545.
101. Y. Dobrydneva, C. J. Abelt, B. Dovel, C. M. Thadigiri, R. L. Williams, P. F. Blackmore, *Mol. Pharmacol.* 2006, *69*, 247.
102. B. Velasco, J. G. Trujillo-Ferrara, L. H. F. Castillo, R. Miranda, L. E. Sánchez-Torres, *Life Sci.* 2007, *80*, 1007.
103. A. M. Irving, C. M. Vogels, L. G. Nikolcheva, J. P. Edwards, X.-F. He, M. G. Hamilton, M. O. Baerlocher, F. J. Baerlocher, A. Decken, S. A. Westcott, *New J. Chem.* 2003, *27*, 1419.
104. A. H. Soloway, W. Tjarks, B. A. Barnum, F.-G. Rong, R. F. Barth, I. M. Codogni, J. G. Wilson, *Chem. Rev.* 1998, *98*, 1515.
105. R. F. Barth, J. A. Coderre, M. G. H. Vicente, T. E. Blue, *Clin. Cancer Res.* 2005, *11*, 3987.
106. W. Chen, S. C. Mehta, D. R. Lu, *Adv. Drug Deliv. Rev.* 1997, *26*, 231.
107. H. Yanagië, A. Ogata, H. Sugiyama, M. Eriguchi, S. Takamoto, H. Takahashi, *Expert Opin. Drug Deliv.* 2008, *5*, 427.
108. J. Nizio, W. Rode, T. Ruman, *Mini-Rev. Org. Chem.* 2012, *9*, 418.
109. D. W. Stephan, G. Erker, *Angew. Chem. Int. Ed.* 2010, *49*, 46.
110. J. Chen, R. A. Lalancette, F. Jäkle, *Chem. Eur. J.* 2014, *20*, 9120.
111. V. Sumerin, F. Schulz, M. Atsumi, C. Wang, M. Nieger, M. Leskelä, T. Repo, P. Pyykkö, B. Rieger, *J. Am. Chem. Soc.* 2008, *130*, 14117.
112. E. Vedejs, S. C. Fields, R. Hayashi, S. R. Hitchcock, D. R. Powell, M. R. Schrimpf, *J. Am. Chem. Soc.* 1999, *121*, 2460.
113. T. S. De Vries, A. Prokofjevs, E. Vedejs, *Chem. Rev.* 2012, *112*, 4246.
114. S. Itsuno, Y. Sakurai, K. Ito, A. Hirao, S. Nakahama, *Bull. Chem. Soc. Jpn* 1987, *60*, 395.
115. E. J. Corey, R. K. Bakshi, S. Shibata, C. P. Chen, V. K. Singh, *J. Am. Chem. Soc.* 1987, *109*, 7925.
116. Y. Hayashi, J. J. Rohde, E. J. Corey, *J. Am. Chem. Soc.* 1996, *118*, 5502.
117. J. N. Payette, H. Yamamoto, *J. Am. Chem. Soc.* 2007, *129*, 9536.
118. E. Canales, E. J. Corey, *J. Am. Chem. Soc.* 2007, *129*, 12686.
119. D. H. Ryu, E. J. Corey, *J. Am. Chem. Soc.* 2004, *126*, 8106.
120. D. H. Ryu, E. J. Corey, *J. Am. Chem. Soc.* 2005, *127*, 5384.
121. E. J. Corey, K. A. Cimprich, *J. Am. Chem. Soc.* 1994, *116*, 3151.
122. Y. Aramaki, H. Omiya, M. Yamashita, K. Nakabayashi, S. I. Ohkoshi, K. Nozaki, *J. Am. Chem. Soc.* 2012, *134*, 19989.
123. D. Lu, C. Wu, P. Li, *Chem. Eur. J.* 2014, *20*, 1630.
124. AS. H. Ueng, A. Solovyev, X. Yuan, S. J. Geib, L. Fensterbank, E. Lacôte, M. Malacria, M. Newcomb, J. C. Walton, D. P. Curran, *J. Am. Chem. Soc.* 2009, *131*, 11256.

125. M. A. Tehfe, S. Schweizer, A. C. Chany, C. Ysacco, J. L. Clément, D. Gigmes, F. Morlet-Savary, J. P. Fouassier, M. Neuburger, T. Tschamber, *et al.*, *Chem. Eur. J.* 2014, *20*, 5054.

126. J. Lalevée, S. Telitel, M. A. Tehfe, J. P. Fouassier, D. P. Curran, E. Lacôte, *Angew. Chem. Int. Ed.* 2012, *51*, 5958.

127. X. Pan, E. Lacôte, J. Lalevée, D. P. Curran, *J. Am. Chem. Soc.* 2012, *134*, 5669.

128. D. Frath, J. Massue, G. Ulrich, R. Ziessel, *Angew. Chem. Int. Ed.* 2014, *53*, 2290.

129. A. Loudet, K. Burgess, *Chem. Rev.* 2007, *107*, 4891.

130. G. Ulrich, R. Ziessel, A. Harriman, *Angew. Chem. Int. Ed.* 2008, *47*, 1184.

131. P. J. Brothers, *Inorg. Chem.* 2011, *50*, 12374.

132. N. Boens, V. Leen, W. Dehaen, *Chem. Soc. Rev.* 2012, *41*, 1130.

133. H. Lu, J. Mack, Y. Yang, Z. Shen, *Chem. Soc. Rev.* 2014, *43*, 4778–4823.

134. T. W. Hudnall, F. P. Gabbaï, *Chem. Commun.* 2008, 4596.

135. G. E. Morse, T. P. Bender, *ACS Appl. Mater. Interfaces* 2012, *4*, 5055.

136. D. González-Rodríguez, C. G. Claessens, T. Torres, *J. Porphyrins Phthalocyanines* 2009, *13*, 203.

137. L. Weber, *Coord. Chem. Rev.* 2008, *252*, 1.

138. C. W. Hamilton, R. T. Baker, A. Staubitz, I. Manners, *Chem. Soc. Rev.* 2009, *38*, 279.

139. A. Staubitz, A. P. M. Robertson, I. Manners, *Chem. Rev.* 2010, *110*, 4079.

140. N. C. Smythe, J. C. Gordon, *Eur. J. Inorg. Chem.* 2010, 509.

141. P. Wang, *Dalton Trans.* 2012, *41*, 4296.

142. Z. Huang, T. Autrey, *Energy Environ. Sci.* 2012, *5*, 9257.

143. Q.-L. Zhu, Q. Xu, *Energy Environ. Sci.* 2015, *8*, 478.

144. P. G. Campbell, L. N. Zakharov, D. J. Grant, D. A. Dixon, S. Y. Liu, *J. Am. Chem. Soc.* 2010, *132*, 3289.

145. W. Luo, D. Neiner, A. Karkamkar, K. Parab, E. B. Garner III, D. A. Dixon, D. Matson, T. Autrey, S.-Y. Liu, *Dalton Trans.* 2013, *42*, 611.

146. P. M. Mitrasinovic, *Curr. Org. Synth.* 2012, *9*, 233.

147. N. Fujita, S. Shinkai, T. D. James, *Chem. Asian J.* 2008, *3*, 1076.

148. Y. Kubo, R. Nishiyabu, T. D. James, *Chem. Commun.* 2015, *51*, 2005.

149. M. L. Saha, S. De, S. Pramanik, M. Schmittel, *Chem. Soc. Rev.* 2013, *42*, 6860.

150. P. M. S. D. Cal, J. B. Vicente, E. Pires, A. V. Coelho, L. F. Veiros, C. Cordeiro, P. M. P. Gois, *J. Am. Chem. Soc.* 2012, *134*, 10299.

151. P. M. S. D. Cal, R. F. M. Frade, V. Chudasama, C. Cordeiro, S. Caddick, P. M. P. Gois, *Chem. Commun.* 2014, *50*, 5261.

152. M. Arzt, C. Seidler, D. Y. W. Ng, T. Weil, *Chem. Asian J.* 2014, *9*, 1994.

153. H. Höpfl, N. Farfán, *J. Organomet. Chem.* 1997, *547*, 71.

154. H. Höpfl, M. Sánchez, N. Farfán, V. Barba, *Inorg. Chem.* 1998, *37*, 1679.

155. N. Farfán, H. Höpfl, V. Barba, M. Eugenia Ochoa, R. Santillan, E. Gómez, A. Gutiérrez, *J. Organomet. Chem.* 1999, *581*, 70.

156. V. Barba, R. Luna, D. Castillo, *J. Organomet. Chem.* 2000, *604*, 273.

157. V. Barba, E. Gallegos, R. Santillan, N. Farfán, *J. Organomet. Chem.* 2001, *622*, 259.

158. H. Hopfl, Structure and Bonding in Boron-Containing Macrocycles and Cages – Comparison to Related Structures with Other Elements Including Organic Molecules, in Structure and Bonding, Eds. H. W. Roesky & D. A. Atwood, Berlin: Springer Verlag, vol. *103*, 2002, pp. 1–56.

159. M. Sánchez, R. Santillan, S. Rojas-lima, *Chem. Eur. J.* 2002, *8*, 612.

160. V. Barba, R. Villamil, R. Luna, C. Godoy-Alcántar, H. Höpfl, H. I. Beltran, L. S. Zamudio-Rivera, R. Santillan, N. Farfán, *Inorg. Chem.* 2006, *45*, 2553.

161. V. Barba, H. Höpfl, N. Farfán, R. Santillan, H. I. Beltran, L. S. Zamudio-Rivera, *Chem. Commun.* 2004, 2834.

162. V. Barba, R. Hernández, H. Höpfl, R. Santillan, N. Farfán, *J. Organomet. Chem.* 2009, *694*, 2127.

163. V. Barba, I. Betanzos, *J. Organomet. Chem.* 2007, *692*, 4903.

164. N. Christinat, R. Scopelliti, K. Severin, *Chem. Commun.* 2004, 1158.

165. N. Christinat, R. Scopelliti, K. Severin, *J. Org. Chem.* 2007, *72*, 2192.

166. B. Icli, E. Solari, B. Kilbas, R. Scopelliti, K. Severin, *Chem. Eur. J.* 2012, *18*, 14867.

167. E. Sheepwash, K. Zhou, R. Scopelliti, K. Severin, *Eur. J. Inorg. Chem.* 2013, 2558.

168. B. Icli, E. Sheepwash, T. Riis-Johannessen, K. Schenk, Y. Filinchuk, R. Scopelliti, K. Severin, *Chem. Sci.* 2011, *2*, 1719.

169. E. Sheepwash, B. Icli, K. Severin, *Chimia (Aarau).* 2012, *66*, 212.

170. D. Salazar-Mendoza, J. Guerrero-Alvarez, H. Höpfl, *Chem. Commun.* 2008, 6543.

171. Y. Pérez-Fuertes, A. M. Kelly, A. L. Johnson, S. Arimori, S. D. Bull, T. D. James, *Org. Lett.* 2006, *8*, 609.

172. A. M. Kelly, Y. Pérez-Fuertes, J. S. Fossey, S. L. Yeste, S. D. Bull, T. D. James, *Org. Lett.* 2006, *8*, 1971.

173. Y. Pérez-Fuertes, A. M. Kelly, A. L. Johnson, S. Arimori, S. D. Bull, T. D. James, *Nat. Protoc.* 2008, *3*, 210.

174. A. M. Kelly, Y. Pérez-Fuertes, J. S. Fossey, S. L. Yeste, S. D. Bull, T. D. James, *Nat. Protoc.* 2008, *3*, 215.

175. A. M. Kelly, S. D. Bull, T. D. James, *Tetrahedron Asymmetry* 2008, *19*, 489.

176. M. E. Powell, A. M. Kelly, S. D. Bull, T. D. James, *Tetrahedron Lett.* 2009, *50*, 876.

177. S. L. Yeste, M. E. Powell, S. D. Bull, T. D. James, *J. Org. Chem.* 2009, *74*, 427.

178. G. Mirri, S. D. Bull, P. N. Horton, T. D. James, L. Male, J. H. R. Tucker, *J. Am. Chem. Soc.* 2010, *132*, 8903.

179. E. Galbraith, A. M. Kelly, J. S. Fossey, G. Kociok-Kohn, M. G. Davidson, S. D. Bull, T. D. James, *New J. Chem.* 2009, *33*, 181.

180. C. Ungurenaşu, S. Cihodaru, I. Popescu, *Tetrahedron Lett.* 1969, *10*, 1435.

181. J. E. Mulvaney, J. J. Bloomfield, C. S. Marvel, *J. Polym. Sci.* 1962, *62*, 59.

182. R. I. Wagner, J. L. Bradford, *Inorg. Chem.* 1962, *1*, 99.

183. E. W. Chang, R. B. Kaner, *Synth. Met.* 1991, *41*, 3075.

184. L. Jinhuo, C. Wending, *Chin. J. Polym. Sci.* 1997, *15*, 305.

185. M. Fontani, F. Peters, W. Scherer, W. Wachter, M. Wagner, P. Zanello, *Eur. J. Inorg. Chem.* 1998, *10*, 1453.
186. N. Matsumi, K. Naka, Y. Chujo, *Macromolecules* 1998, *31*, 8047.
187. C. D. Entwistle, T. B. Marder, *Angew. Chem. Int. Ed.* 2002, *41*, 2927.
188. N. Christinat, E. Croisier, R. Scopelliti, M. Cascella, U. Röthlisberger, K. Severin, *Eur. J. Inorg. Chem.* 2007, *2007*, 5177.
189. E. Sheepwash, V. Krampl, R. Scopelliti, O. Sereda, A. Neels, K. Severin, Angew. *Chem. Int. Ed.* 2011, *50*, 3034.
190. E. Sheepwash, N. Luisier, M. R. Krause, S. Noe, K. Severin, *Chem. Commun.* 2012, *48*, 7808.
191. J. Cruz-Huerta, D. Salazar-Mendoza, J. Hernández-Paredes, I. F. Hernández Ahuactzi, H. Höpfl, *Chem. Commun.* 2012, *48*, 4241.
192. N. Luisier, K. Schenk, K. Severin, *Chem. Commun.* 2014, *50*, 10233.
193. S. Ito, H. Takata, K. Ono, N. Iwasawa, *Angew. Chem. Int. Ed.* 2013, *52*, 11045.
194. D. Y. W. Ng, M. Arzt, Y. Wu, S. L. Kuan, M. Lamla, T. Weil, *Angew. Chem. Int. Ed.* 2014, *53*, 324.
195. M. L. Stolowitz, C. Ahlem, K. A. Hughes, R. J. Kaiser, A. Kesicki, G. Li, K. P. Lund, S. M. Torkelson, J. P. Wiley, E. A. Kesicki, *Bioconjug. Chem.* 2001, *12*, 229.
196. J. P. Wiley, K. A. Hughes, R. J. Kaiser, E. A. Kesicki, K. P. Lund, M. L. Stolowitz, *Bioconjug. Chem.* 2001, *12*, 240.
197. S. Moffatt, S. Wiehle, R. J. Cristiano, *Hum. Gene Ther.* 2005, *16*, 57.
198. A. Wilson, G. Gasparini, S. Matile, *Chem. Soc. Rev.* 2014, *43*, 1948.
199. S. D. Bull, M. G. Davidson, J. M. H. Van Den Elsen, J. S. Fossey, A. T. A. Jenkins, Y. B. Jiang, Y. Kubo, F. Marken, K. Sakurai, J. Zhao, *Acc. Chem. Res.* 2013, *46*, 312.
200. T. D. James, S. Shinkai, *Top. Curr. Chem.* 2002, *218*, 159.
201. T. D. James, K. R. A. S. Sandanayake, S. Shinkai, *J. Chem. Soc. Chem. Commun.* 1994, 477.
202. J. Yan, H. Fang, B. Wang, *Med. Res. Rev.* 2005, *25*, 490.
203. W. Ni, G. Kaur, G. Springsteen, B. Wang, S. Franzen, *Bioorg. Chem.* 2004, *32*, 571.
204. C. R. Cooper, T. D. James, *Chem. Commun.* 1997, 1419.
205. C. R. Cooper, T. D. James, *J. Chem. Soc., Perkin Trans. 1* 2000, 963.
206. T. D. James, K. R. A. S. Sandanayake, S. Shinkai, *Angew. Chem. Int. Ed.* 1994, *33*, 2207.
207. T. D. James, K. R. A. S. Sandanayake, R. Iguchi, S. Shinkai, *J. Am. Chem. Soc.* 1995, *117*, 8982.
208. T. D. James, S. Shinkai, *J. Chem. Soc. Chem. Commun.* 1995, *3*, 1483.
209. M. Yamamoto, M. Takeuchi, S. Shinkai, *Tetrahedron* 1998, *54*, 3125.
210. M. Takeuchi, M. Yamamoto, S. Shinkai, *Chem. Commun.* 1997, 1731.
211. P. Linnane, T. D. James, S. Imazu, S. Shinkai, *Tetrahedron Lett.* 1995, *36*, 8833.
212. K. R. A. S. Sandanayake, T. D. James, S. Shinkai, *Chem. Lett.* 1995, 503.
213. T. D. James, K. R. A. S. Sandanayake, S. Shinkai, *Nature* 1995, *374*, 345.
214. X. Sun, S.-Y. Xu, S. E. Flower, J. S. Fossey, X. Qian, T. D. James, *Chem. Commun.* 2013, *49*, 8311.
215. Y. Ooyama, K. Uenaka, A. Matsugasako, Y. Harima, J. Ohshita, *RSC Adv.* 2013, *3*, 23255.
216. J. J. Yun, H. J. Ji, K. M. K. Swamy, K. Nakamura, S. K. Hwa, J. Y. Yeo, J. Yoon, *Bull. Kor. Chem. Soc.* 2005, *26*, 2041.
217. S. Trupp, A. Schweitzer, G. J. Mohr, *Org. Biomol. Chem.* 2006, *4*, 2965.
218. J. Wang, S. Jin, S. Akay, B. Wang, *Eur. J. Org. Chem.* 2007, 2091.
219. S. Jin, J. Wang, M. Li, B. Wang, *Chem. Eur. J.* 2008, *14*, 2795.
220. K. R. A. S. Sandanayake, S. Imazu, T. D. James, M. Mikami, S. Shinkai, *Chem. Lett.* 1995, 139.
221. K. R. A. S. Sandanayake, S. Shinkai, *J. Chem. Soc. Chem. Commun.* 1994, 1083.
222. C. J. Ward, P. Patel, P. R. Ashton, T. D. James, *Chem. Commun.* 2000, 229.
223. C. J. Ward, P. Patel, T. D. James, *J. Chem. Soc., Perkin Trans. 1* 2002, 462.
224. C. J. Ward, P. Patel, T. D. James, *Org. Lett.* 2002, *4*, 477.
225. H. Shinmori, M. Takeuchi, S. Shinkai, *J. Chem. Soc., Perkin Trans. 2* 1996, 1.
226. Y. Egawa, R. Miki, T. Seki, *Materials (Basel)* 2014, *7*, 1201.
227. K. Koumoto, M. Takeuchi, S. Shinkai, *Supramol. Chem.* 1998, *9*, 203.
228. S. L. Wiskur, H. Ait-Haddou, J. J. Lavigne, E. V. Anslyn, *Angew. Chem. Int. Ed.* 1999, *38*, 3666.
229. J. N. Cambre, B. S. Sumerlin, *Polymer (Guildf)* 2011, *52*, 4631.

3

INFLUENCE OF STERIC, ELECTRONIC, AND MOLECULAR PREORGANIZATION EFFECTS IN THE REACTIVITY OF β-PHENYLETHYLAMINES WITH NONENOLIZABLE ALDEHYDES

RODOLFO QUEVEDO

Departamento de Química, Facultad de Ciencias, Universidad Nacional de Colombia, Bogotá, Colombia

3.1 INTRODUCTION

The β-phenylethylamines are compounds whose structure is formed by an amino group bound to an aromatic ring by means of an ethylene; they are known for their effects on the central nervous system where they act as neurotransmitters (i.e., dopamine, noradrenaline, and epinephrine), for their psychotropic effects (i.e., amphetamine, methamphetamine, and 3,4-methylenedioxyamphetamine (MDMA)), and for being precursors in the synthesis and biosynthesis of nitrogenous metabolites having a broad spectrum of biological activity (i.e., aporphines and isoquinolines) (Scheme 3.1) [1, 2].

The alkaloids having one or more β-phenylethylamine units in their structure are related to protecting plants against the predatory acts of insects and herbivores and may be toxic for mammals in some cases [3–6]. It has been suggested that some of them participate in vegetable growth, due to their ability to form chelates or intervene in oxide reduction phenomena. These alkaloids' nitrogen atom comes originally from tyrosine, and this amino acid's carbon structure is kept intact in the alkaloid's structure, while the carbon from carboxylic acid undergoes decarboxylation.

However, when a molecule has two or more reactive sites and a reaction occurs preferentially at one of these sites, then one is talking about selective processes; such selective processes' directionality is due to both electronic effects and steric factors or a combination of both. This chapter analyzes the importance of electronic effects and steric factors on the course of phenylethylamines' reaction with nonenolizable aldehydes and presents 4-hydroxyphenylethylamine molecules' molecular preorganization as a third determinant factor regarding the course of this reaction and its use in the synthesis of a novel group of tyrosine-derived azacyclophane macrocyclic compounds.

3.2 DOPAMINE REACTION WITH NONENOLIZABLE ALDEHYDES: THE PICTET–SPENGLER REACTION

The Pictet–Spengler reaction is the most powerful methodology for the synthesis of biologically important tetrahydroisoquinolinic alkaloids [7–9]. This reaction involves adding a β-phenylethylamine to a carbonyl compound to form the respective Schiff base. The Schiff base activated by adding an acid catalyst promotes an electrophilic aromatic substitution, thereby forming the respective 1,2,3,4-tetrahydroisoquinoline (Scheme 3.2).

The Pictet–Spengler reaction only happens on strongly activated aromatic rings; it traditionally occurs in an aprotic solvent in the presence of an acid catalyst, usually acetic or trifluoroacetic acid. A mixture of the two possible regioisomers is obtained when the starting phenylethylamine has more than one activated position, the cyclization product prevailing in the less hindered *ortho* position in most cases; only a regioisomer is obtained in high yield when the reaction involves dopamine hydrobromide using methanol as solvent and acetic acid as catalyst (Scheme 3.3) [10, 11].

Non-Covalent Interactions in the Synthesis and Design of New Compounds, First Edition.
Edited by Abel M. Maharramov, Kamran T. Mahmudov, Maximilian N. Kopylovich, and Armando J. L. Pombeiro.

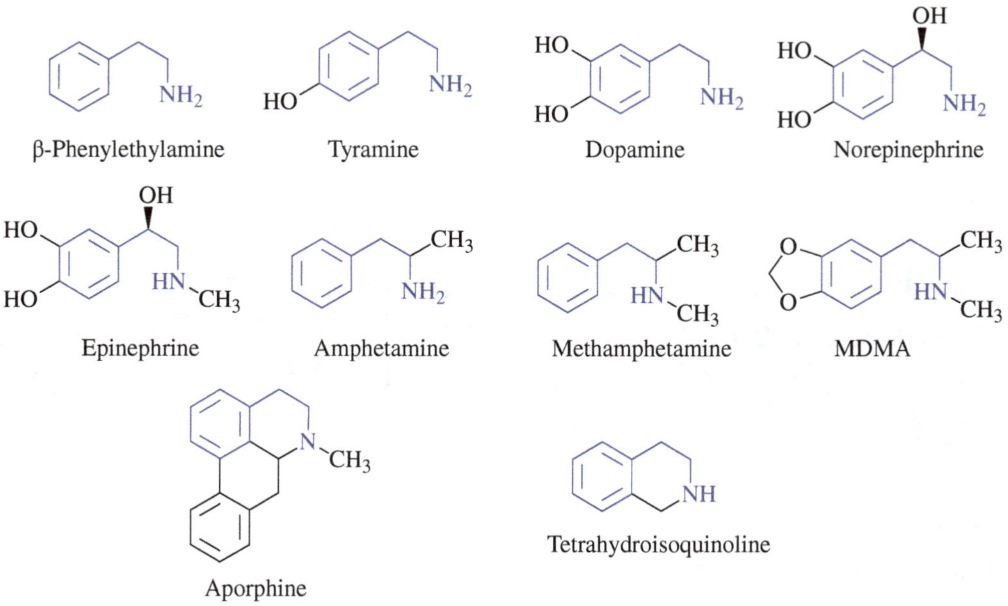

SCHEME 3.1 β-phenylethylamines.

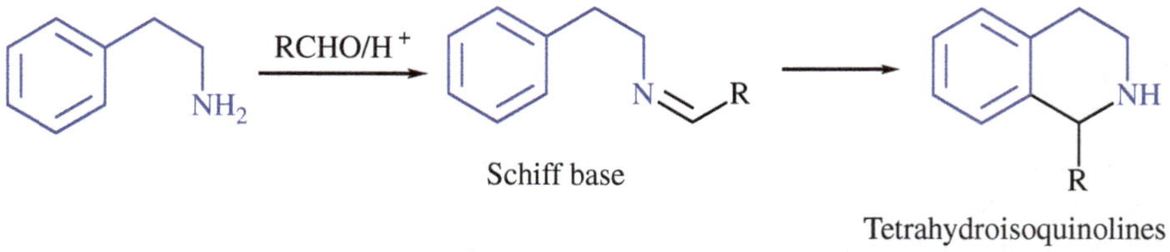

SCHEME 3.2 The Pictet–Spengler reaction.

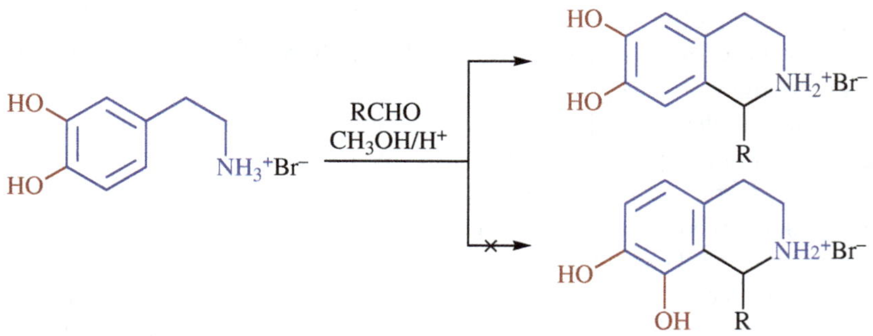

SCHEME 3.3 Tetrahydroisoquinoline synthesis from dopamine.

Pictet–Spengler reaction regioselectivity between *m*-tyramine or dopamine and aromatic aldehydes can be explained by electronic factors; the hydroxyl located on aromatic ring carbon 3 activates positions 2 and 6 due to the effect of resonance and directs electrophilic aromatic substitution toward these carbons (Scheme 3.3). The influence of electronic effects on Pictet–Spengler reaction regioselectivity has been demonstrated by calculating Mulliken atomic charge distribution for intermediary Schiff base for the reaction between dopamine and benzaldehyde. Carbon 6 has greater electronic density (−0.0559) than carbon 2 (−0.0262); such distribution of charge shows that both carbons are strongly activated but also that the Pictet–Spengler reaction is electronically favored toward carbon 6 on the aromatic ring of dopamine [11].

From the electronic point of view, it is clear why the Pictet–Spengler reaction is regioselective and explains why the mixture of both regioisomers is obtained in different proportions; however, total regioselectivity is observed when the reaction is done in methanol as only one of the regioisomers is formed. Total regioselectivity results from using protic solvents that increase steric effects due to the solvation of the polar groups on the phenylethylamine structure;

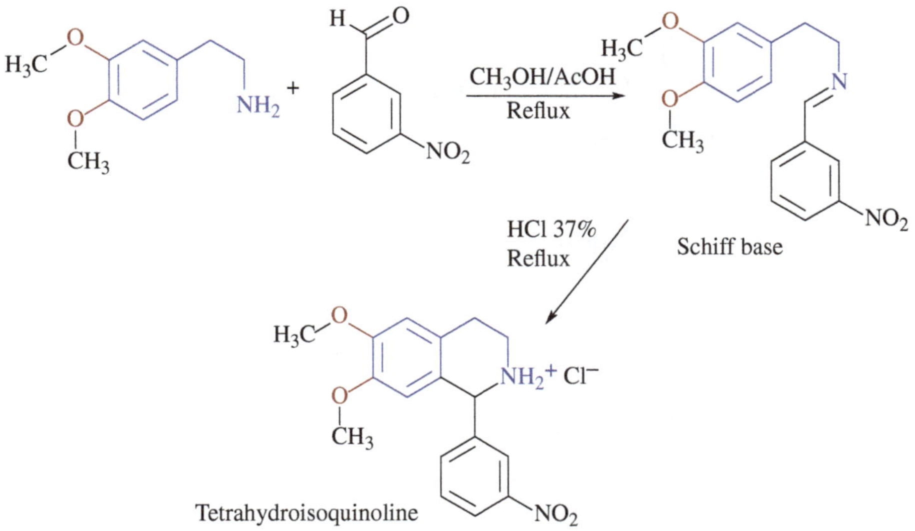

SCHEME 3.4 Reaction of 6,7-dimethoxiphenylethylamine with 3-nitrobenzaldehyde.

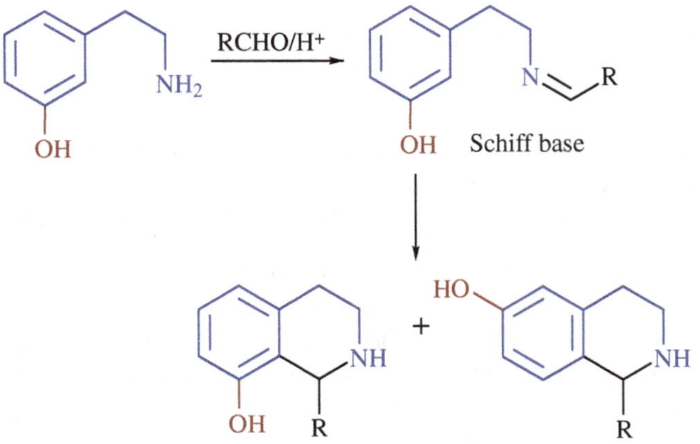

SCHEME 3.5 Reaction of *m*-tyramine with aldehydes.

the steric hindrance produced by such solvation directed cyclization toward the less hindered carbon (carbon 6). The reaction between 6,7-dimetoxyphenylethylamine (dimethyl dopamine) and 3-nitrobenzaldehyde (again using methanol as solvent and acetic acid as catalyst) stops at the intermediary Schiff base, showing that decreased activation of the ring by the methoxyls does not favor electrophilic aromatic substitution and that cyclization only occurs when drastic experimental conditions are used (37% HCl) [10, 11]. The differences regarding electronic effects between dopamine and dimethyl dopamine influence the activation of the ring, but (like reactions involving dopamine) the steric effects due to solvation induce total regioselectivity to the isoquinoline during cyclization (Scheme 3.4).

Electronic effects due to the presence of strongly activating groups on the aromatic ring and steric factors induced by solvation with protic solvents produce total regioselectivity in the Pictet–Spengler reaction. Using solid catalysts, such as bentonites, vermiculites, and mixed oxides in methanol, in the Pictet–Spengler reaction between dopamine and aromatic aldehydes leads to greater yields but less regioselectivity than that obtained with acetic acid as catalyst; deficient regioselectivity with some solid catalysts is related to solvation difficulties due to the catalyst's textural properties [12].

3.3 TYROSINE DERIVATIVES REACTION WITH NONENOLIZABLE ALDEHYDES

3.3.1 The L-Tyrosine Ethyl Ester Reaction with Aldehydes

The electronic effects caused by hydroxyls or methoxyls and the steric factors produced by solvation, when protic solvents are used, have a direct effect on the course of the Pictet–Spengler reaction between dopamine and nonenolizable aldehydes. How does the number of phenolic hydroxyls influence the course of the reaction? The reaction between *m*-tyramine and aromatic aldehydes has been studied recently using dichloromethane as solvent (Scheme 3.4). The hydroxyl in position 3 of the aromatic ring activates the *ortho* position and favors cyclization via the Pictet–Spengler reaction. This reaction occurs with low yield due to the low activation of the ring. Using aprotic solvents does not produce steric effects due to solvation but does lead to the mixture of regioisomers (Scheme 3.5) [10].

No product formation can be observed when the reaction between tyrosine esters and aromatic aldehydes takes place under conditions usually used for the Pictet–Spengler reaction. The reaction with formaldehyde follows a different pattern and leads to the formation of macrocyclic compounds. These macrocycles are the product of aromatic Mannich condensation between two units of L-tyrosine alkyl ester and four formaldehydes. Macrocyclic formation by means of the L-tyrosine alkyl ester reaction with

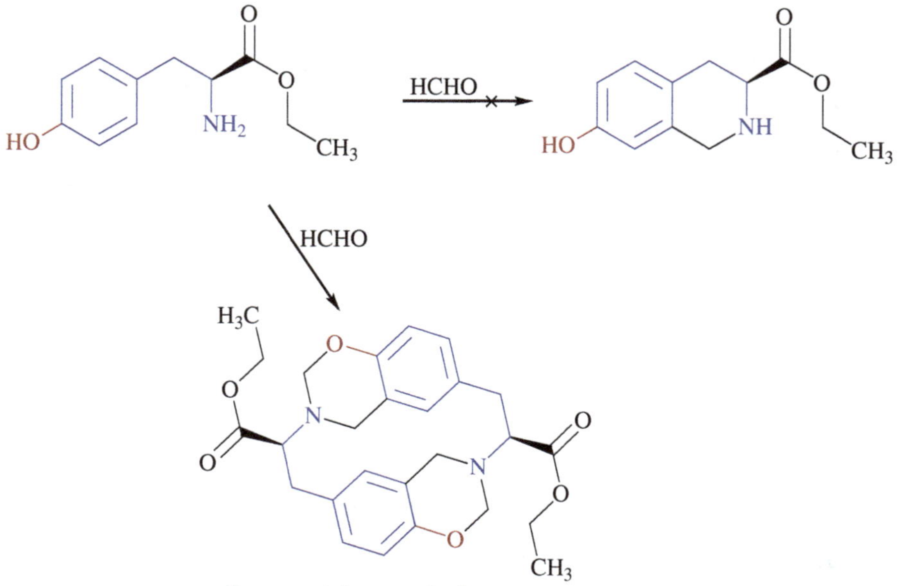

Benzoxazinic azacyclophane
benzoxazinephane

SCHEME 3.6 Reaction of L-tyrosine ethyl ester with formaldehyde.

formaldehyde competes with reactions involving oligomer formation and ester hydrolysis. Macrocyclization reaction yield becomes improved when such reaction involves using an excess of formaldehyde and absence of acid catalysts (Scheme 3.6) [13–15].

The formation of macrocyclic compounds in the reaction between formaldehyde and tyrosine derivatives shows the importance of electronic factors regarding the course and directionality of the reaction. Compounds having rings with strongly activated positions 2 and 6, such as 3-hydroxyphenylethylamine or dopamine, react with aldehydes for producing the respective tetrahydroisoquinolines via the Pictet–Spengler reaction. The Pictet–Spengler reaction does not occur when the reaction is run with 4-hydroxyphenylethylamines, such as tyrosine alkyl esters or tyramine compounds in which carbons 3 and 5 are the positions having the greatest electron density, and macrocyclic products are obtained having a pentacyclic skeleton formed by two units of 3,4-dihydro-2*H*-1,3-benzoxazine joined by two ethylene groups by means of aromatic Mannich condensation (Scheme 3.6). These macrocycle formation is also conformationally favored; the lowest energy conformation has two benzoxazinic nuclei located face to face with the hydrogen on the chiral carbon in *pseudo*-axial position and the ethyloxycarbonyl group in *pseudo*-equatorial position. The optimized structure also shows that the oxazinic ring adopts a half-chair conformation with the nitrogens located outside the plane of the aromatic ring toward the azacyclophane cavity and the carbons between the oxygen and the nitrogen on the oxazinic ring are located outside the plane of the aromatic ring pointing outside the cavity (Scheme 3.6).

3.3.2 Molecular Preorganization by Means of Non-covalent Interactions and Their Participation in the Synthesis of Azacyclophanes from Tyrosine Derivatives

The cyclophanes are molecules formed by aromatic rings linked by short spacers. They have structural characteristics allowing them to act as hosts due to their ability to trap guests in their cavity by means of non-covalent interactions; this ability can be seen with both organic molecules and metal cations and anions [16–19].

The cyclophanes are usually synthesized by macrocyclization processes involving several stages, using protector groups or diluted solutions or through the formation of templates at metal ions and presynthesized subunits [20–24]. The non-covalent interaction-driven methodology for the synthesis of azacyclophanes from tyrosine derivatives does not involve any of these aspects; it occurs in just one step, uses high concentration aqueous solutions, and does not require the transition metal templates. This reaction's directionality cannot be satisfactorily explained by analyzing electronic effects and steric factors; a Mannich reaction between tyrosine derivatives and formaldehyde could lead to the formation of linear polymers, as reported in the last century [25, 26] (Scheme 3.7).

Formally, electronic effects and steric factors regarding L-tyrosine derivative ester and tyramine reaction with formaldehyde should lead to the formation of linear oligomers, as represented in Scheme 3.7. But what favors macrocyclization for the formation of azacyclophanes from tyrosine derivatives? As a possible response to the patterns observed, a preorganization of the molecules in solution via intermolecular hydrogen bonds between the amino groups and the phenolic hydroxyls of the two molecules of the respective L-tyrosine derivative has been proposed, due to the amphoteric nature of the L-tyrosine derivative esters (Scheme 3.8).

3.3.3 Dimerization Constant for Tyrosine Isopropyl Ester in Solution

When a chemical species is associated via intermolecular hydrogen bonds, the protons directly involved in such interaction experience different environments resulting in two different chemical displacements in ^{1}H NMR spectra: δ_{free} and δ_{bound}; however, for a system experiencing rapid hydrogen exchange (i.e., having atomic lifetimes in milliseconds or less), the magnetic setting experienced by the proton is averaged between free and bound state settings, and a broad singlet can be observed at a frequency that is the weighted average of δ_{free} and δ_{bound}. The ^{1}H NMR spectrum for tyrosine isopropyl ester in $CDCl_3$ shows exchangeable protons at 2.72 ppm at 18 mM concentration (broad singlet); this signal corresponds to the protons involved in intermolecular hydrogen bonds (OH and NH) and is sensitive to any change in concentration. This signal becomes displaced to the low field by an increase in concentration due to the protons becoming deshielded by the formation of hydrogen bonds; greater concentration leads to greater dimer template formation [27–30].

Analyzing chemical displacement data regarding tyrosine isopropyl ester concentration has established that the association between ester molecules fits a dimerization model with constant $K_d = 38\,M^{-1}$, such value indicating that this is favored in the equilibrium, around 98% of the ester being in dimer form [15, 31].

Dimer formation in solution is also evident in mass spectra when using the electrospray technique $[M–H]^-$ at m/z 222.1088 and $[2M–H]^-$ at m/z 445.2256, ions corresponding to the L-tyrosine isopropyl ester monomer and dimer. The absence of larger m/z values rules out the formation of complexes having more than two ester units [31].

Analyzing the ^{1}H NMR and ESI-MS spectra reveals the association by hydrogen bonds for dimer formation; however, such dimer could be either the proposed template (Scheme 3.9a) or a linear arrangement (Scheme 3.9b). Computational calculations using semiempirical methods regarding both possible dimer arrangements show that it is more probable that their structure is stabilized by two hydrogen bonds between the phenolic hydroxyl group of one unit and the amino group of the other unit (O—H···N). This arrangement is also lower in energy compared to that for the two separate molecules. The dimeric arrangement has a suitable formation for geometrically assisting the macrocyclization reaction between L-tyrosine ester derivatives and formaldehyde [32–36].

The formation of these templates is entropically unfavorable but enthalpically favorable, and the overall free energy drives the dimer formation. In spite of the entropic cost regarding tyrosine ester molecule orientation toward template formation, the energy of the hydrogen bonds so formed

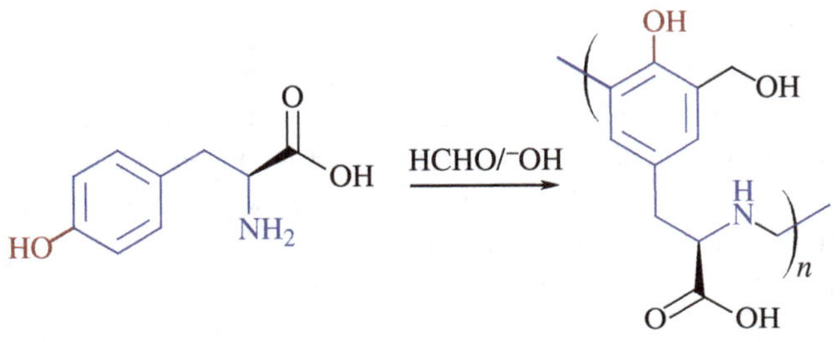

SCHEME 3.7 Mannich-type reaction of L-tyrosine with formaldehyde in basic medium.

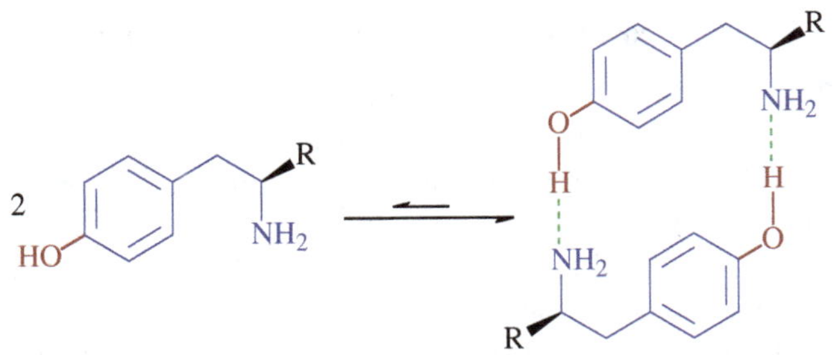

SCHEME 3.8 Molecular preorganization by self-assembly of L-tyrosine derivatives via intermolecular hydrogen bonds.

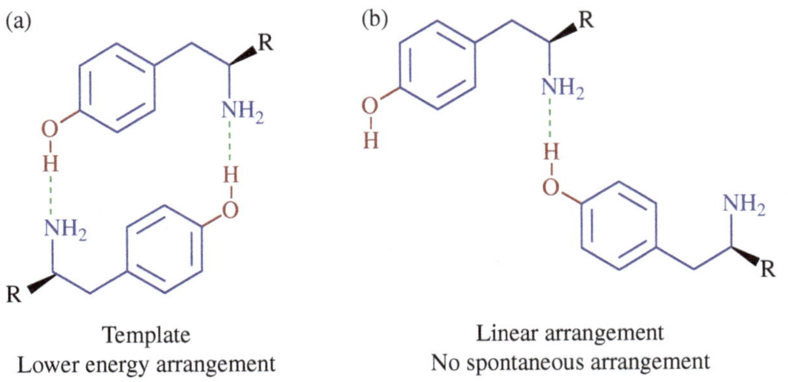

SCHEME 3.9 Hydrogen bonds for dimer formation.

is responsible for the spontaneity of the process. The formation of the linear arrangement is not spontaneous because the entropic cost of binding two ester molecules in a linear disposition is greater than the energy provided by a single hydrogen bond.

Both spectroscopic patterns and computational calculations show that the directionality of the reaction between tyrosine derivatives and formaldehyde is induced by molecular preorganization via hydrogen bonds between the amino groups and the phenolic hydroxyls from two L-tyrosine derivative units. The molecular preorganization between the two molecules keeps the amino group from one molecule at a short distance from the phenolic hydroxyl group of the other, and when formaldehyde is added, its highly electrophilic carbon reacts with the two closest nucleophilic sites (i.e., the nitrogen from one of the L-tyrosine derivative units and the aromatic ring from the other). The template formed by hydrogen bonds impedes condensation to linear products (oligomers and polymers); the macrocycle so formed is thus a kinetic product and the molecules' preorganization provides a kinetic template effect in course of the reaction.

To summarize, the L-tyrosine derivative macrocyclization is induced by the spontaneous formation of an energetically favored dimer that acts as template, followed by covalent modification. As the macrocyclization reaction involves intermolecular condensation, high-dilution conditions are not required in the synthesis. Preorganization by hydrogen bonds in this type of molecule determines the course of the reaction with formaldehyde. The molecules interact by means of complementary binding sites (recognition) in this template and become organized in spatial terms, thereby modifying their reactivity (organization), and information is transferred from the spatial disposition to the product (information transfer).

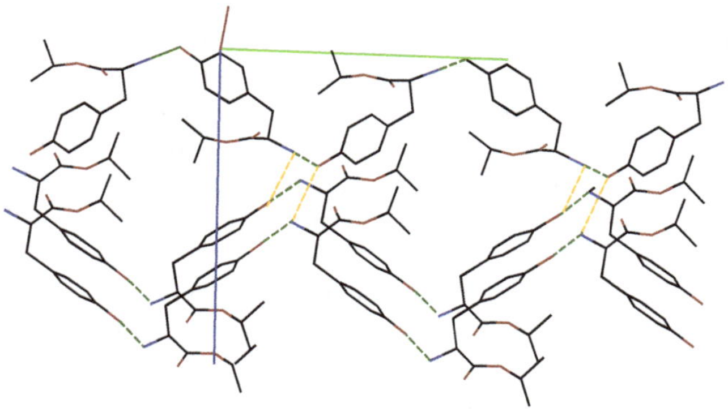

SCHEME 3.10 Crystal packing of L-tyrosine isopropyl ester. The O—H···N hydrogen bonds are shown as dashed green lines, and N—H···O hydrogen bonds are shown as dashed lines.

3.3.4 Solid-State Analysis

Analyzing the crystal structure of tyrosine alkyl esters by single-crystal X-ray diffraction has shown that this type of compound adopts a folded conformation called U-shaped or scorpion conformation (Scheme 3.10). Molecules bound by strong intermolecular hydrogen bonds can be observed in the crystal structure, with phenolic hydroxyl as proton donor and the amino group as proton acceptor (O—H···N), forming one-dimensional helical infinite zigzag chains. Weak hydrogen bonds can be observed between the chains, with the amino group acting as proton donor and phenolic hydroxyl as proton acceptor (N—H···O). Intermolecular interactions (C—H···O) may also be seen between the hydrogen bonded to chiral carbon and carbonyl oxygen; no π-stacking interactions have been observed in these structures [37–40]. As can be seen, the proposed dimeric template is not observed in solid state; however, the ester molecules in solution may reorganized by strong hydrogen bond interactions to form the template.

3.3.5 Non-covalent Interactions Involving Tyramine

Tyramine is an L-tyrosine derivative that leads to the formation of the respective azacyclophane in high yields and provides the simplest model for studying preorganization by means of hydrogen bonds (Scheme 3.11).

Semiempirical calculations have shown that the formation of templates in vacuum via hydrogen bonds between the phenolic hydroxyl from one unit of tyramine and the amino group from the other unit (O—H···N) is energetically favored, confirming this compound's tendency to preorganization by means of hydrogen bonds. Calculations using water and ethanol as solvent have shown the same tendency

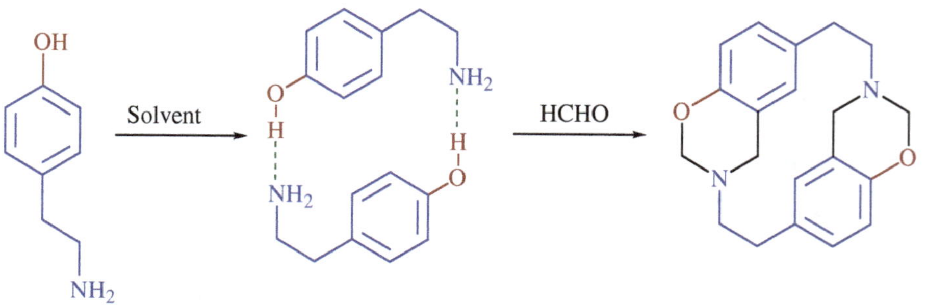

SCHEME 3.11 Reaction of tyramine with formaldehyde.

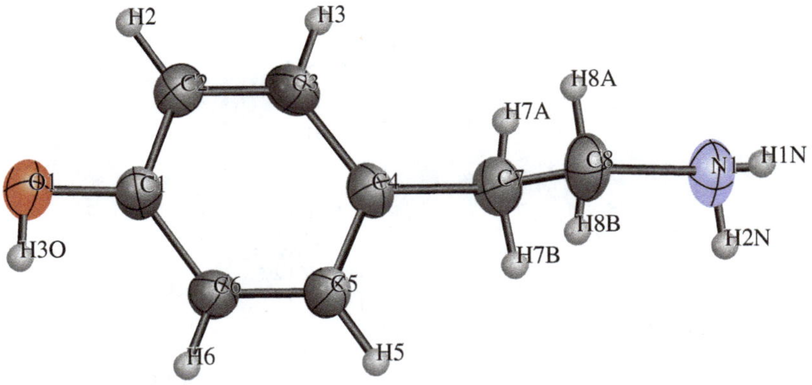

SCHEME 3.12 ORTEP diagram for tyramine.

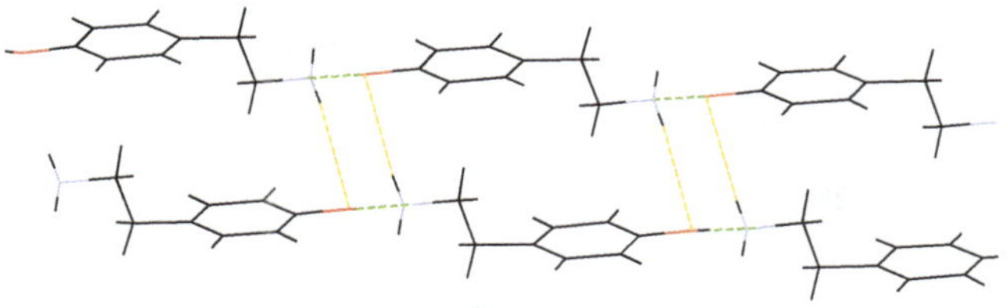

SCHEME 3.13 The extended hydrogen bond network in tyramine crystal structure. The O—H···N hydrogen bonds are shown as dashed lines, and N—H···O hydrogen bonds are shown as dashed yellow lines.

in vacuum. Such results suggest that the solvent does not participate in the molecular preorganization of tyramine. The ESI-HRMS spectrum for tyramine in positive mode has revealed the following ions: $[M+H]^+$ (138.0916 *m/z*) and $[2M+H]^+$ (275.1744 *m/z*). The $[2M+H]^+$ ion corresponds to the tyramine dimer [41].

Single-crystal X-ray diffraction analysis of tyramine as free base (Scheme 3.12) has revealed dihedral angles C_8—C_7—C_4—C_3 (102.14°), N_1—C_8—C_7—C_4 (176.39°), and H_{1N}—N_1—C_8—C_7 (−177.45°), thereby indicating that the ethylamine side chain has an anti conformation regarding rotation around the C_8—C_7 bond. The anti conformation is asymmetric according to rotation around the N_1—C_8 bond; the asymmetric anti conformation is *cis* due to the relative orientation of the NH_2 and OH groups.

Molecular packing involves a linear arrangement of tyramine molecules connected head to tail by strong intermolecular hydrogen bonds all along the *b* axis, with phenolic hydroxyl as proton donor and the amino group as proton acceptor (Scheme 3.13). The linear arrangements are interconnected, thereby forming double chains that are stabilized by weak head-to-tail hydrogen bonds, with the amino group acting as proton donor and phenolic hydroxyl as proton acceptor (Scheme 3.13). The hydrogen bond interactions order the molecules to form double chains stabilized by cyclic arrangements of hydrogen bonds between neighboring tyramine units (Scheme 3.13). Such organization between two chains in solid state geometrically may favor the formation of templates in solution [41].

3.4 NON-COVALENT INTERACTIONS REGARDING L-TYROSINE AND ITS TETRABUTYLAMMONIUM SALT

3.4.1 L-Tyrosine

The reaction between L-tyrosine and formaldehyde was first studied during the last century, thereby establishing that L-tyrosine reacts with two equivalents of formaldehyde in alkaline medium to form high molecular weight compounds (Scheme 3.7) [25, 26]. Recent studies concerning this reaction have evaluated the influence of L-tyrosine concentration, the number of base equivalents, the amount of formaldehyde, and the reaction time. Such reactions usually led to mixtures of oligomer products of variable composition. The only fully characterized product was linear oligomer formed by three tyrosine units bound by two methylene groups (Scheme 3.14) [42].

The macrocyclization of L-tyrosine derivatives does not occur with L-tyrosine due to the impossibility of L-tyrosine molecules to preorganize themselves via hydrogen bonds in basic aqueous medium. One or two equivalents of NaOH must be added to the reaction medium due to tyrosine's low solubility; the tyrosine keeps its net negative charge, and this probably causes the anions formed in aqueous medium to prefer being solvated by water molecules and separated. This would mean that the template proposed for explaining the macrocyclization reaction observed between L-tyrosine ester derivatives and formaldehyde in basic aqueous medium does not favor L-tyrosine (Scheme 3.15) [42].

3.4.2 L-Tyrosine Tetrabutylammonium

Amino acid quaternary ammonium salts are soluble in organic solvents (especially tetra-*n*-butylammonium salts), and the amino group is free (Scheme 3.16) [43].

L-tyrosine tetrabutylammonium salt reaction with formaldehyde produces a mixture of oligomers having two and three units of L-tyrosine bound by methylene groups. This reaction favors the formation of an L-tyrosine linear dimer; such pattern is due to tetrabutylammonium salt having an ion pair

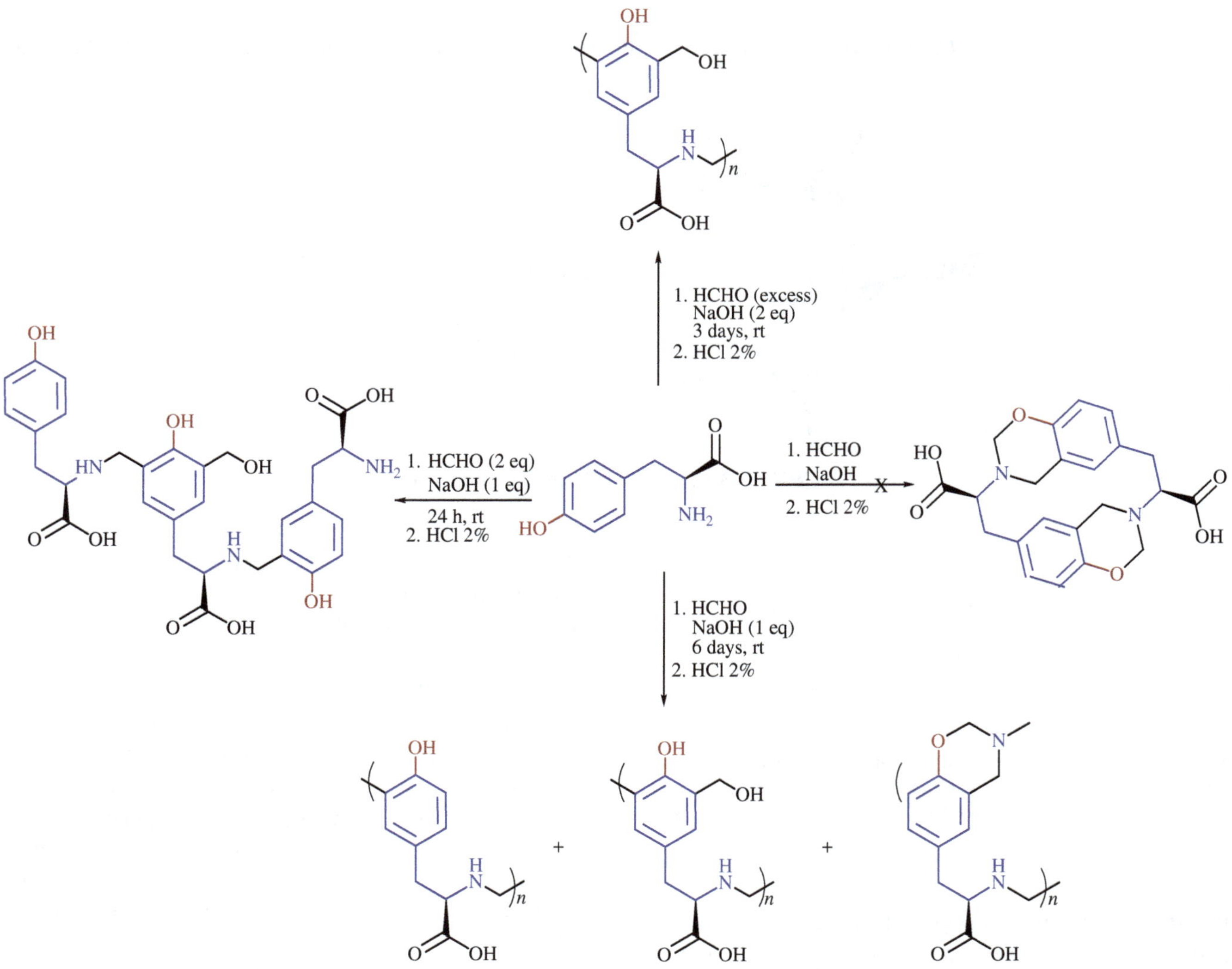

SCHEME 3.14 L-Tyrosine reaction with formaldehyde in basic medium.

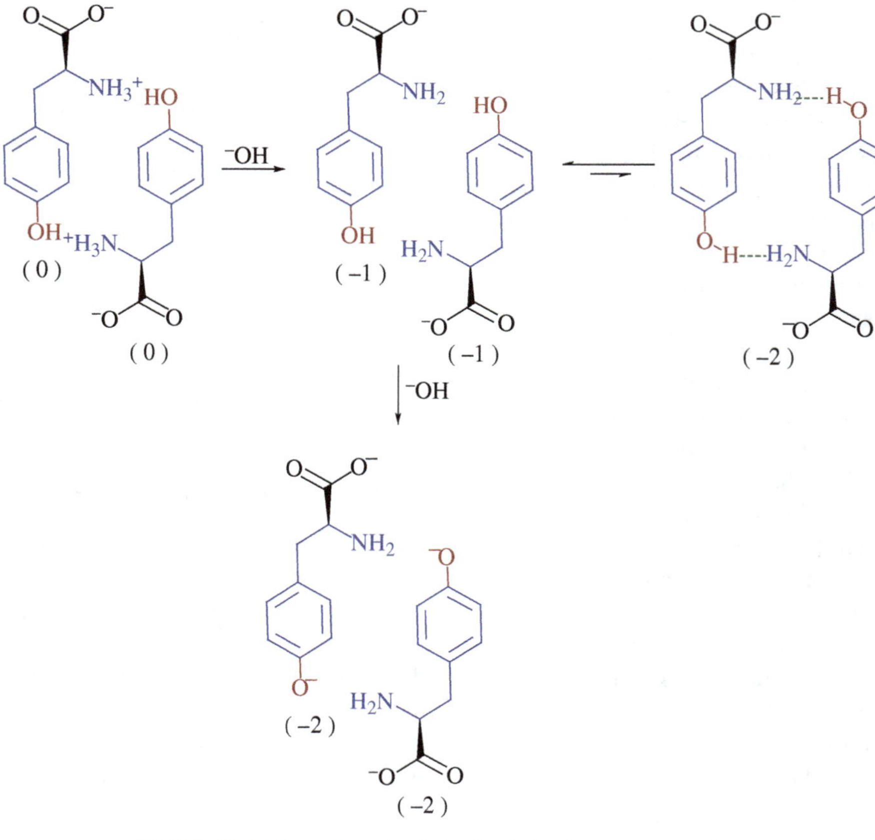

SCHEME 3.15 L-Tyrosine behavior in basic aqueous medium.

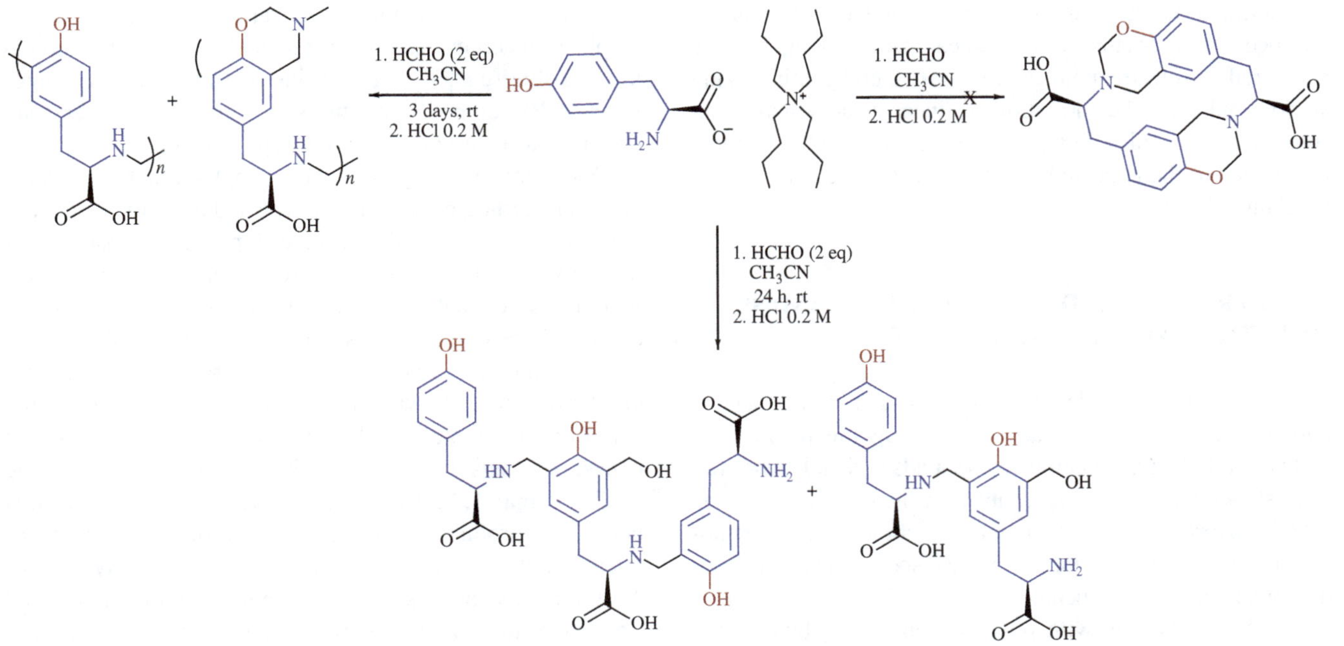

SCHEME 3.16 L-Tyrosine tetrabutylammonium reaction with formaldehyde.

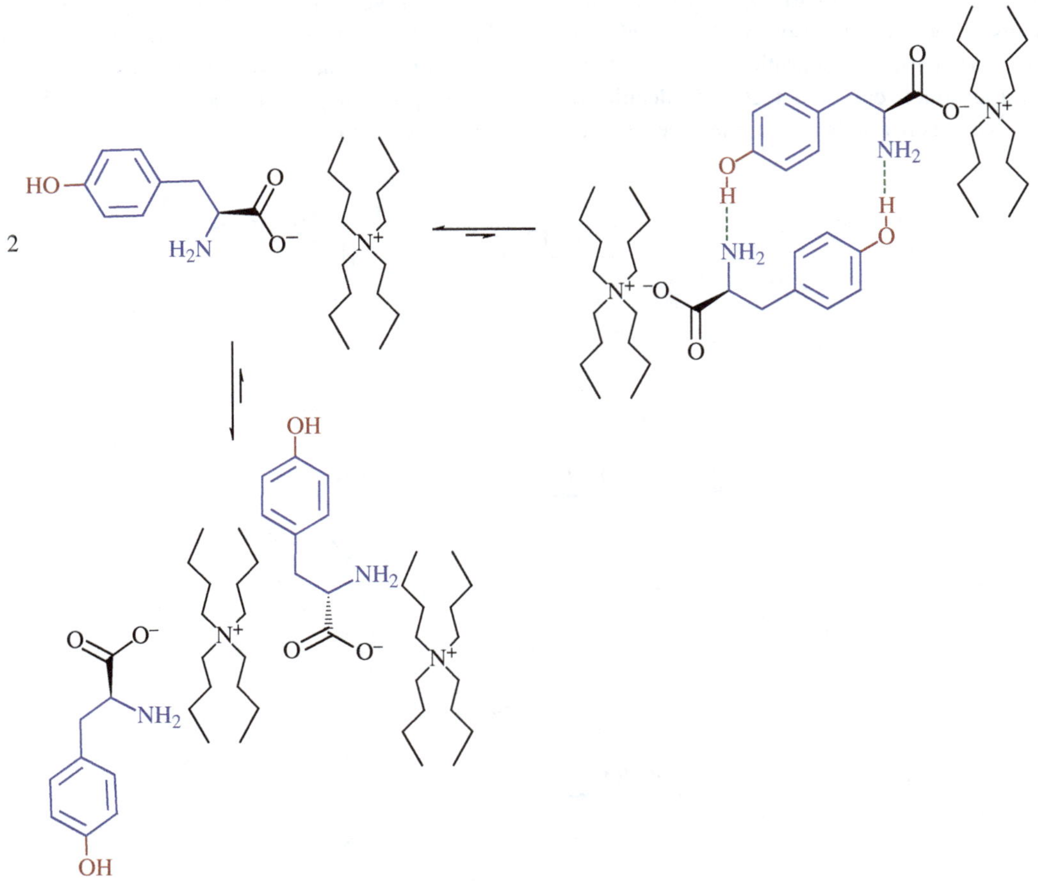

SCHEME 3.17 L-Tyrosine tetrabutylammonium behavior in acetonitrile.

in solution and the bulky cation producing steric hindrance, thereby slowing down L-tyrosine units' reaction with formaldehyde. Steric factors acting on tetrabutylammonium cation directed the reaction toward the formation of linear products and impede molecular preorganization (Scheme 3.16) [42].

Just as the reaction involving L-tyrosine, the formation of linear oligomers having the overall structure shown in Scheme 3.16 is favored, and macrocyclic formation is never observed. This pattern can be explained in terms of the preorganization of the molecules under dissolution (Scheme 3.17).

The tetrabutylammonium salt forms ion pairs in solution that become associated to form linear arrangements that are stabilized by electrostatic interactions, thereby minimizing steric repulsion. The formation of linear arrangements between the molecules favors condensation to afford linear products, but the expected macrocyclization product cannot be obtained.

3.5 TYRAMINE AND β-PHENYLETHYLAMINE REACTION WITH FORMALDEHYDE

The importance of electronic effects and steric factors and molecular preorganization regarding the reaction of β-(hydroxyl)phenylethylamine compounds with aldehydes has been shown throughout this chapter. In the case of dopamine β-(3,4-dihydroxyphenylethylamine), electronic factors regioselectively lead to the formation of the respective tetrahydroisoquinoline via a Pictet–Spengler reaction, whereas steric effects induced by solvation with polar solvents contribute to the reaction's directionality and allow the respective tetrahydroisoquinolines to be obtained with total regioselectivity.

When working with 4-hydroxyphenylethylamines, electronic effects direct the reaction with formaldehyde to Mannich-type aromatic condensation, and no Pictet–Spengler reaction products can be detected. Molecular preorganization by hydrogen bonds determines the course of this reaction, favoring macrocyclization that leads to azacyclophane macrocycles. When molecular preorganization is not possible (e.g., with L-tyrosine and its tetrabutylammonium salt), complex mixtures of linear oligomers are obtained, as result of a Mannich-type reaction.

Dopamine (3,4-dihydroxyphenylethylamine) has the same pattern regarding both formaldehyde and aromatic aldehydes that are less electrophilic. Unlike dopamine, tyrosine ester derivatives, having a less activated aromatic ring, do not react with aromatic aldehydes, and the starting ester and tyrosine are usually recovered as products of ester hydrolysis.

Computational studies involving tyramine and phenylethylamine have shown that the amino group represents the site with the greatest nucleophilicity in phenylethylamine; hence, it reacts with formaldehyde, and it produces the cyclic aminal 1,3,5-triphenylhexahydro-1,3,5-triazine, this being the product from the reaction of three moles of phenylethylamine and three moles of formaldehyde [44]. Tyramine has carbons *ortho* to phenolic hydroxyl (C3 and C5) as the sites having the greatest nucleophilicity; the reaction with formaldehyde only occurs on one of these carbons by means of a Mannich-type aromatic condensation to produce azacyclophane-type macrocyclic systems. Dopamine has carbon 6 as the position having greatest activation and, regarding formaldehyde, produces the respective tetrahydroisoquinoline by means of a Pictet–Spengler reaction (Scheme 3.18).

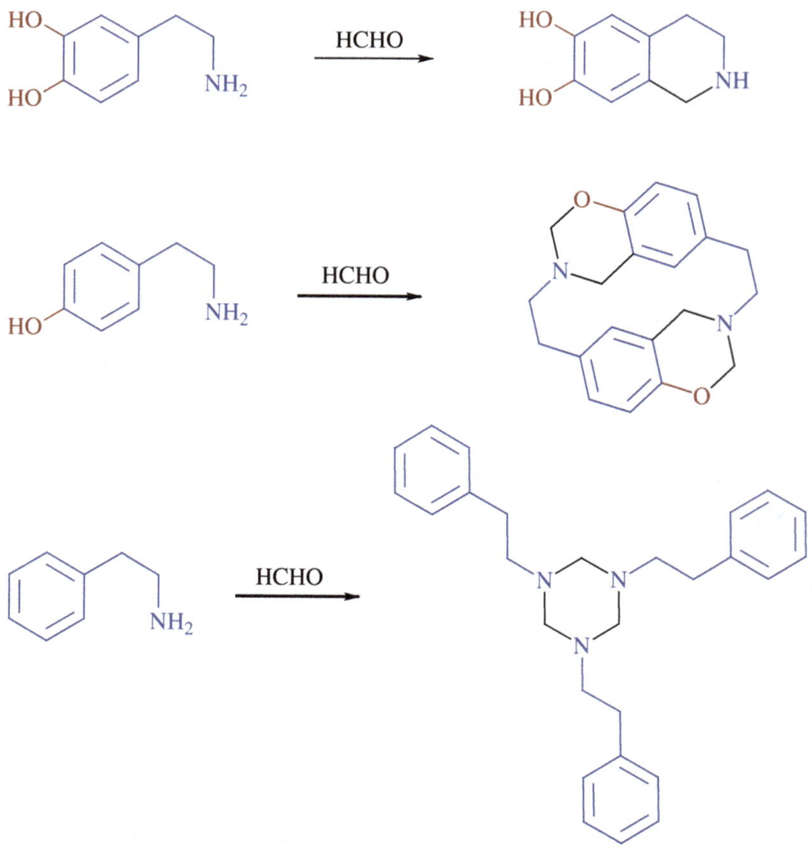

SCHEME 3.18 β-Phenylethylamine reaction with formaldehyde.

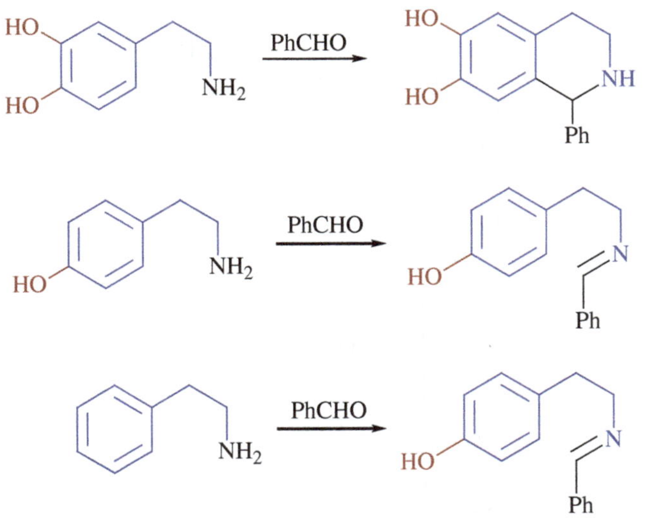

SCHEME 3.19 β-Phenylethylamine reaction with aromatic aldehydes.

As mentioned above, dopamine reacts with aromatic aldehydes to produce tetrahydroisoquinolines by means of a Pictet–Spengler reaction, as occurs with formaldehyde. The same does not happen when the reaction involves *trans* cinnamaldehyde; the corresponding intermediary Schiff base is obtained in this case. Tetrahydroisoquinoline formation is not favored in this case due to the conjugation that reduces carbon electrophilicity and hinders nucleophilic addition.

Tyramine and phenylethylamine reaction with aromatic aldehydes leads to the corresponding Schiff bases. Yields are not significantly affected by electron acceptor or electron donor groups in the aldehyde; however, phenylethylamine gives better yields due to greater nitrogen nucleophilicity. Schiff bases are obtained because the aromatic rings' nucleophilicity in these two phenylethylamines is low and does not favor the aromatic electrophilic substitution needed for tetrahydroisoquinoline formation. Likewise, Schiff base electrophilicity is low for producing aminals or azacyclophanes by means of reactions involving nucleophilic addition, as happens in reactions involving formaldehyde (Scheme 3.19).

The β-phenylethylamines (phenylethylamine, tyramine, and dopamine) react differently with formaldehyde and produce a cyclic aminal, an azacyclophane, and a tetrahydroisoquinoline (respectively); such differences regarding reactivity are due to differing degrees of aromatic ring activation and different nitrogen nucleophilicity, together with the effects of molecular preorganization in tyramine. The reaction with other less reactive nonenolizable aldehydes only occurs with the amine from the ring having the greatest activation (dopamine), thereby producing the respective tetrahydroisoquinolines; the β-phenylethylamine reaction with the least activated rings only gives the respective Schiff bases.

3.6 BENZYLAZACYCLOPHANE SYNTHESIS

The amount and position of hydroxyls in the aromatic ring play an important role in the course of phenylethylamine reaction with formaldehyde because they directly affect the degree and position of ring activation, modify nitrogen nucleophilicity, and govern the course of the reaction with formaldehyde by means of non-covalent interactions. Do substituents on nitrogen affect template formation by hydrogen bonds?

Semiempirical computational calculations regarding *N*-benzyltyramine intermolecular associations by O—H···N hydrogen bonds have shown that cyclic dimer formation is energetically favored. The presence of protic solvents (water and ethanol) favors template formation, thereby increasing the hydrogen bonds' ionic nature (greater O—H bond length and shorter H···N distance).

The reaction between *N*-benzyltyramine and formaldehyde produces the corresponding *N*-benzylazacyclophane macrocycle formed by two units of *N*-benzyltyramine bound by two methylene groups (Scheme 3.20) [45].

Studying another two *N*-benzyltyramines (the first having an electron donor substituent (*N*-(4-methoxybenzyl) tyramine) and the second an electron acceptor group (*N*-(3-nitrobenzyl)tyramine)) has revealed significant differences concerning their behavior regarding formaldehyde. The former leads to the formation of the respective *N*-benzylazacyclophane with good yield (Scheme 3.20), while the latter produces a complex mixture of oligomers, mostly two N-methylated linear compounds (*N*-methylated benzylamine) and a linear dimer being the product of the reaction of two units of benzylamine with one unit of formaldehyde (Scheme 3.21). Methylation also occurs on one of the dimer's nitrogens; N-methylated product formation follows an Eschweiler–Clarke reaction involving the participation of formic acid as reducing agent formed *in situ* by formaldehyde oxidation [46]. The presence of electron donor substituents in the starting *N*-benzyltyramine increases macrocyclization yield with formaldehyde, while the electron attractor substituent favors linear oligomerization and N-methylation.

Thus, the reaction of *N*-benzyltyramines shows that the presence of electron donor groups in the *N*-benzyl rings favors macrocyclization reaction because they increase nitrogen nucleophilicity and basicity in the starting *N*-benzyltyramine: increased nucleophilicity favors nucleophilic addition to formaldehyde, and increased basicity favors hydrogen bond interaction with the phenolic hydroxyl needed for macrocyclization (Scheme 3.20). The presence of electron attractor substituents reduces nucleophilicity and basicity: reduced nucleophilicity affects reactivity regarding formaldehyde, and reduced basicity hinders the preorganization needed for benzylazacyclophane synthesis. The electronic effects caused by the presence of the electron attractor substituent favor side-chain reactions such as Eschweiler–Clarke methylation (Scheme 3.21) [46].

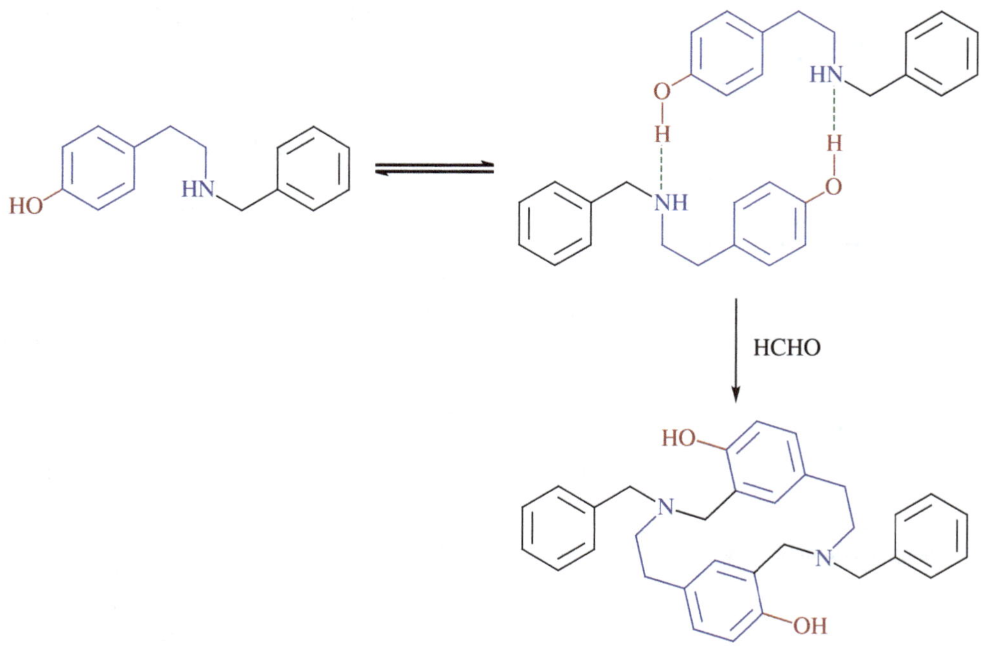

SCHEME 3.20 *N*-Benzylazacyclophane synthesis.

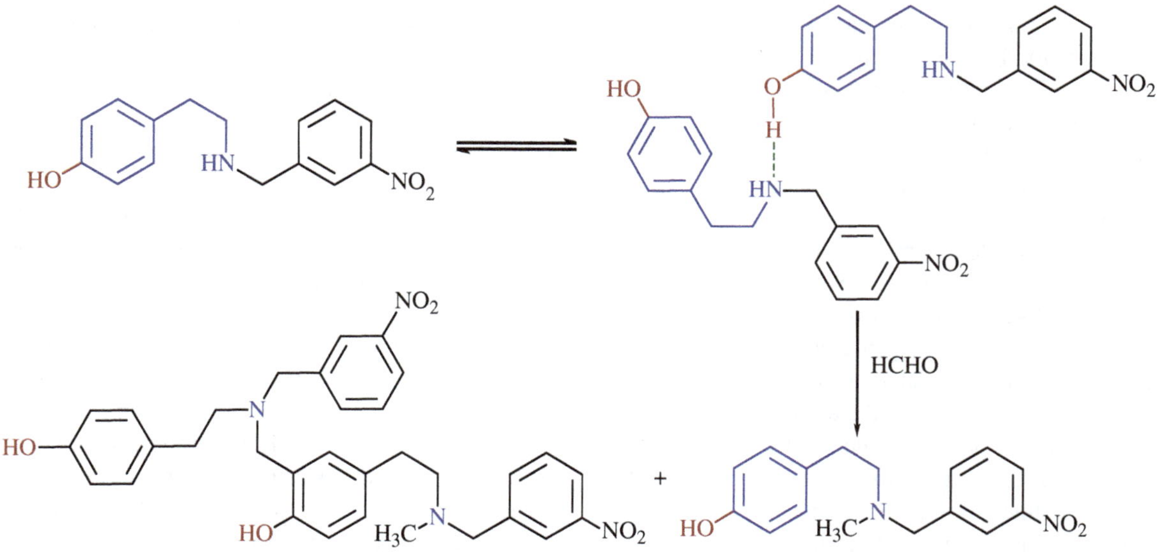

SCHEME 3.21 *N*-(3-Nitrobenzyl)tyramine reaction with formaldehyde.

3.7 FINAL COMMENTS

The reaction between dopamine and nonenolizable aldehydes leads to the formation of tetrahydroisoquinolines by means of a Pictet–Spengler reaction; such reaction is regioselective due to electronic effects, and total regioselectivity is achieved by steric factors due to the solvation of the polar groups on the aromatic ring by solvent.

The reaction of tyramine or L-tyrosine ester derivatives with formaldehyde leads to azacyclophane formation involving two benzoxazinic nuclei connected by two ethylene groups. The course of the azacyclophane formation reaction (aromatic Mannich reaction) is directed by electronic effects, and macrocyclization is driven by non-covalent interactions by hydrogen bonds between the amino groups and phenolic hydroxyls from two tyrosine units. The macrocyclization is induced by the spontaneous formation of an energetically favored dimer (thermodynamic minimum), acting as template, followed by covalent modification to fix the macrocycle (kinetic product). Such synthetic tool provides a new type of macrocyclization.

As macrocyclization reaction is a process of intermolecular condensation and molecular preorganization directs the reaction toward the macrocyclic product, high dilution is not required for the synthesis. Hence, non-covalent interactions in the course of macrocyclization provide operational simplicity and chemical and conformational stability, simplify purification procedures, favor high yields, and do not affect the environment.

The reaction of L-tyrosine and its tetrabutylammonium salt with formaldehyde does not lead to macrocyclic azacyclophane compounds. Linear oligomers having two or more L-tyrosine units bound by methylene groups are obtained in all cases. Such 4-hydroxyphenylethylamine structural characteristics do not allow template formation by hydrogen bonds but do favor linear compound formation.

The differences in the reactivity of β-phenylethylamine (phenylethylamine, tyramine, and dopamine) with formaldehyde lead to obtaining a cyclic aminal, an azacyclophane, and a tetrahydroisoquinoline, respectively; such differences regarding reactivity result from different degrees of aromatic ring activation and differing nitrogen nucleophilicity, together with the ability (or inability) to form templates in solution. The reaction with other less reactive nonenolizable aldehydes only occurs with the amine presented by the ring having the greatest activation (dopamine), thereby affording the respective tetrahydroisoquinolines; the β-phenylethylamine reaction with less activated rings only leads to the respective Schiff bases. Computational calculations have highlighted a greater degree of hydroxylation on the ring and less nitrogen nucleophilicity.

N-Benzyltyramine reaction with formaldehyde shows that the presence of electron donor groups on *N*-benzyl rings favors the macrocyclization reaction because they increase nitrogen nucleophilicity and basicity in the starting *N*-benzyltyramine: increased nucleophilicity favors nucleophilic addition to formaldehyde, and increased basicity favors the interaction involving hydrogen bonds with the phenolic hydroxyl needed for macrocyclization. The presence of electron attractor substituents reduces nucleophilicity and basicity: reduced nucleophilicity affects reactivity regarding formaldehyde, and reduced basicity hinders the preorganization necessary for benzylazacyclophane synthesis.

ACKNOWLEDGMENTS

I would like to thank the Universidad Nacional de Colombia for funding this work (DIB research project no. 28217) and acknowledge the following students' contribution to this work: Edwin Baquero, Marlon González, Nelson Nuñez, Christian Díaz, and Yovanny Quevedo.

REFERENCES

1. S. Freeman, J. F. Alder, *Eur. J. Med. Chem.* 2002, *37*, 527.
2. K. W. Bentley, *Nat. Prod. Rep.* 2006, *23*, 444.
3. J. L. Martinez, R. Torres, M. A. Morales, *Phytother. Res.* 1997, *11*, 246.
4. P. Schiff Jr., *J. Nat. Prod.* 1997, *60*, 934.
5. R. Quevedo, K. Valderrama, B. Moreno, M. Laverde, V. Fajardo, *Biochem. Syst. Ecol.*, 2008, *36*, 812.
6. R. Quevedo, L. Nuñez, B. Moreno, *Nat. Prod. Res.* 2011, *25*, 934.
7. A. Pictet, T. Spengler, *Berichte*, 1911, *44*, 2030.
8. K. Manabe, D. Nobutou, S. Kobayashi, *Bioorg. Med. Chem. Lett.* 2005, *13*, 5154.
9. P. Mayer, J. Zhang, *Tetrahedron Lett.* 1996, *37*, 5633.
10. M. Zhong, M. Villani, R. Marzouq, *Org. Process Res. Dev.* 2007, *11*, 463.
11. R. Quevedo, E. Baquero, M. Rodriguez, *Tetrahedron Lett.* 2010, *51*, 1774.
12. R. Quevedo, C. Perdomo, S. Moreno, *J. Chem.* 2013, ID 125302, 5.
13. R. Quevedo, B. Moreno-Murillo, *Tetrahedron Lett.* 2009, *50*, 936.
14. R. Quevedo, M. González, C. Díaz-Oviedo, *Tetrahedron Lett.* 2012, *53*, 1595.
15. N. Nuñez-Dallos, A. Reyes, R. Quevedo, *Tetrahedron Lett.* 2012, *53*, 530.
16. H. Hopf, R. Gleiter, Modern Cyclophane Chemistry; Wiley-VCH, Chichester, 2004.
17. K. Gloe, Macrocyclic Chemistry; Springer, Dordrecht, 2005.
18. A. Bordunov, N. Lukyanenko, V. Pastushok, K. Krakowiak, J. Bradshaw, K. Dalley, X. Kout, *J. Org. Chem.* 1995, *60*, 4912.
19. P. Rajakumar, M. Rasheed, *Tetrahedron* 2005, *61*, 5351.
20. J. W. Steed, J. L. Atwood, Molecular Guests in Solution. In Supramolecular Chemistry, John Wiley & Sons, Ltd: Chichester, 2009; pp. 307–384.
21. T. Gulder, P. S. Baran, *Nat. Prod. Rep.* 2012, *29*, 899.
22. C. Thilgen, V. A. Azov, Cyclophanes: Definition and Scope. In Encyclopedia of Supramolecular Chemistry, J. L. Atwood, and J. W. Steed (eds), Marcel Dekker: New York, 2004; vol. *1*, pp. 414–423.
23. S. I. Hauck, K. V. Lakshmi, J. F. Hartwig, *Org. Lett.* 1999, *1*, 2057.
24. J. W. Steed, J. L. Atwood, Cation-Binding Hosts. In Supramolecular Chemistry, John Wiley & Sons, Ltd: Chichester, 2009; pp. 105–222.
25. A. E. Brown, *J. Am. Chem. Soc.* 1946, *68*, 1011.
26. H. J. Olcott, *J. Am. Chem. Soc.* 1955, *77*, 5434.
27. E. J. Billo, Calculation of Binding Constants. In Excel® for Chemists, John Wiley & Sons, Ltd: New York, 2001, pp. 223–372.
28. K. Hirose, Determination of Binding Constants. In Analytical Methods in Supramolecular Chemistry, Editor-in-Chief: Schalley, C., Wiley-VCH Verlag GmbH & Co. KGaA: Weinheim, 2007, pp. 17–54.
29. W. J. Chu, Y. Yang, C. F. Chen, *Org. Lett.* 2010, *12*, 3156.
30. P. Zhang, H. Chu, X. Li, W. Feng, P. Deng, L. Yuan, B. Gong, *Org. Lett.* 2011, *13*, 54.
31. N. Nuñez, Estudio de la reacción entre derivados del ácido (2S)-2-amino-3-(4-hidroxifenil)propanoico y formaldehído, Master's Thesis, Universidad Nacional de Colombia, Colombia, 2012.

32. MOPAC2009, J. J. P. Stewart, Stewart Computational Chemistry, Version 11.366W web: http://OpenMOPAC.net (Accessed on October 20, 2015).
33. M. Korth, *J. Chem. Theory Comput.* 2010, *6*, 3808.
34. J. Řezáč, P. Hobza, *J. Chem. Theory Comput.* 2012, *8*, 141.
35. F. Avila-Salas, C. Sandoval, J. Caballero, S. Guiñez-Molinos, L. S. Santos, R. E. Cachau, F. D. González-Nilo, *J. Phys. Chem. B* 2012, *116*, 2031.
36. L. Leclercq, I. Suisse, P. Roussel, F. Agbossou-Niedercorn, *J. Mol. Struct.* 2012, *1010*, 152.
37. N. Nuñez-Dallos, K. Wurst, R. Quevedo, *Acta Crystallogr. E* 2012, *68*, o3173.
38. S. S. Qian, H. L. Zhu, E. R. T. Tiekink, *Acta Crystallogr. E* 2006, *62*, o882.
39. B. Nicolaï, N. Mahé, R. Céolin, I. Rietveld, M. Barrio, J. L. Tamarit, *Struct. Chem.* 2011, *22*, 649.
40. I. B. Rietveld, M. Barrio, J. L. Tamarit, B. Nicolaï, J. Van de Streek, N. Mahé, R. Ceolin, B. Do, *J. Pharm. Sci.* 2011, *100*, 4774.
41. R. Quevedo, N. Nuñez-Dallos, K. Wurst, A. Duarte-Ruiz, *J. Mol. Struct.* 2012, *1029*, 175.
42. N. Nuñez-Dallos, C. Díaz-Oviedo. R. Quevedo, *Tetrahedron Lett.* 2014, *55*, 4216.
43. A. Sudo, R. Kudoh, T. Endo, *J. Polym. Sci. A Polym. Chem.* 2011, *49*, 1724.
44. R. Quevedo, C. Díaz-Oviedo, Y. Quevedo-Acosta, *Res. Chem. Intermed.* 2015, *41*, 9835.
45. C. Díaz-Oviedo, R. Quevedo, *Tetrahedron Lett.* 2014, *55*, 6571.
46. J. Harding, J. Jones, S. Lu, R. Wood, *Tetrahedron Lett.* 2002, *43*, 9487.

4

NON-COVALENT INTERACTIONS IN THE SYNTHESIS OF MACROCYCLES

Evgeny A. Kataev

Institute of Chemistry, Faculty of Natural Sciences, Technische Universität Chemnitz, Chemnitz, Germany

4.1 INTRODUCTION

Synthesis of macrocycles by using non-covalent interactions between a precursor and a template or by intramolecular interactions has attracted considerable attention in recent years, since it has opened an access to large molecules through multicomponent reactions [1]. One of the most often used approaches, which gives excellent yields of macrocycles, is the template synthesis. A template interacts with starting materials (precursor or building blocks) by means of hydrogen bonds, π–π interactions, and ion–dipole, ion–ion, or dipole–dipole interactions and drives the assembly of starting materials to a macrocycle. For nature the use of templates is a major tool to transfer the information, for example, in DNA replication, enzyme synthesis, and signal transduction, where recognition often occurs through weak interactions. With this approach, a number of selective macrocyclic receptors have been discovered using a dynamic covalent chemistry. In this case, mainly reversible covalent bonds have been employed to connect building blocks in the presence of anionic, neutral, and cationic species under thermodynamic control [2–7]. The use of irreversible bonds in the synthesis of macrocycles is still scarce. However, in most cases these reactions give much better yields because they proceed under kinetic control. Historically important examples go back to the work of Pederson in 1967, who has accidentally discovered a template effect of cations in the formation of crown ethers [8]. Later, in 1976 Dale observed the formation of crown ether during the polymerization of ethylene oxide [9]. The field of cation-templated reactions was further developed by the research work of Busch [10]. Anion-templated macrocyclization through irreversible covalent bonds was initially studied by Shur [11] and Hawthorne [12].

The synthesis of macrocycles through irreversible bonds is of high interest because the reaction products are usually robust and offer a number of interesting properties such as recognition of small molecules, similar to the natural cyclic compound bioactivity and an ability for self-assembly in solution or in the solid state. One of the most used covalent bonds to produce macrocycles is the amide bond. The progress in the synthesis of macrocycles without template effects has been already summarized in the literature and includes the synthesis of peptides in nature [13], peptide cyclization methods [14], and synthetic cyclic peptides for supramolecular chemistry [15, 16]. The synthesis of macrocycles from rigid or preorganized starting compounds is not included in this review because the geometry of the reactants and the position/orientation of reacting groups determine the structure of macrocyclic products [17]. We also omit methods of macrocycle synthesis, in which transition metal coordination bonds are used to drive the reaction toward cyclization products [18, 19]. The synthesis of catenanes by using hydrogen bonds has been reviewed recently [20, 21]; thus hydrogen bond-directed synthesis is narrowed to cyclo-oligomerization reactions.

In this chapter, we will discuss the syntheses of macrocycles, which proceed with the help of inter- or intramolecular non-covalent interactions. In particular, we consider three major groups of interactions: precursor–anionic template, precursor–cationic template, and intramolecular hydrogen bonds formed during the reaction. These interactions have been suggested or found to have a decisive role in determining the product distribution in macrocyclization

Non-Covalent Interactions in the Synthesis and Design of New Compounds, First Edition.
Edited by Abel M. Maharramov, Kamran T. Mahmudov, Maximilian N. Kopylovich, and Armando J. L. Pombeiro.

reactions. We focus in particular on the reactions, which in principle can give a broad range of cyclic oligomers. The presence of non-covalent interactions either intramolecular or with a template usually narrows the range of possible products. Under the term "irreversible bonds," we consider those covalent bonds that are formed under kinetic control and that are not reversible under conditions studied. However, as it can be seen throughout this chapter, some of the reactions, which are usually considered as irreversible, under certain conditions become reversible, and the formation of the product was found to occur under thermodynamic control.

4.2 ANION-TEMPLATED SYNTHESIS

Anions are known to form strong hydrogen bonds and they have often been used as templates in anion-templated self-assembly [5, 22, 23], macrocyclization reactions [6, 24, 25], and synthesis of molecular imprinted polymers [26]. In 1991 Hawthorne and coworkers [12] reported on Hg(II) complexes and postulated that one product was formed due to the strong "template effect with halide ion serving in that role" (Scheme 4.1). The synthesis led to the formation of a macrocycle, which consist of four repeating units, and in 1993 [27] the authors obtained a crystal structure of the iodide complex (**1**). Such compounds were named as "anticrowns" because they could coordinate anions in a similar manner as crown ether coordinates cations [28]. Almost at the same time Sessler and coworkers observed an anion template effect in the synthesis of expanded porphyrins [29]. The porphyrin could be prepared by using an UO_2^{2+} cation with a high yield (75%). However, the use of nitric acid in the condensation reaction yielded the macrocycle quantitatively. Interestingly, hydrochloric acid did not give good yields. These results "led to an aside to suggest that these types of Schiff base condensation reaction could be a subject to a general anion template effect." The subsequent work on the bipyrrole [30] and terpyrrole [31] dialdehyde has shown similar behavior of the pyrrole building blocks in the macrocyclization with *o*-phenylenediamine, namely, high yields of products were observed in the presence of nitric acid as compared to other acids.

These reactions can be considered among those that initiated the synthesis of macrocycles with the help of anionic templates and the resulting macrocycles served as selective host for these templates.

Highly efficient synthesis of cyclooctapyrrole through a C—C bond forming reaction was reported by Sessler and coworkers [32, 33]. They found that 2,2′-bipyrrole could be cyclized into compound **2** by means of an oxidative coupling, and the type of acid used was decisive. Only in the

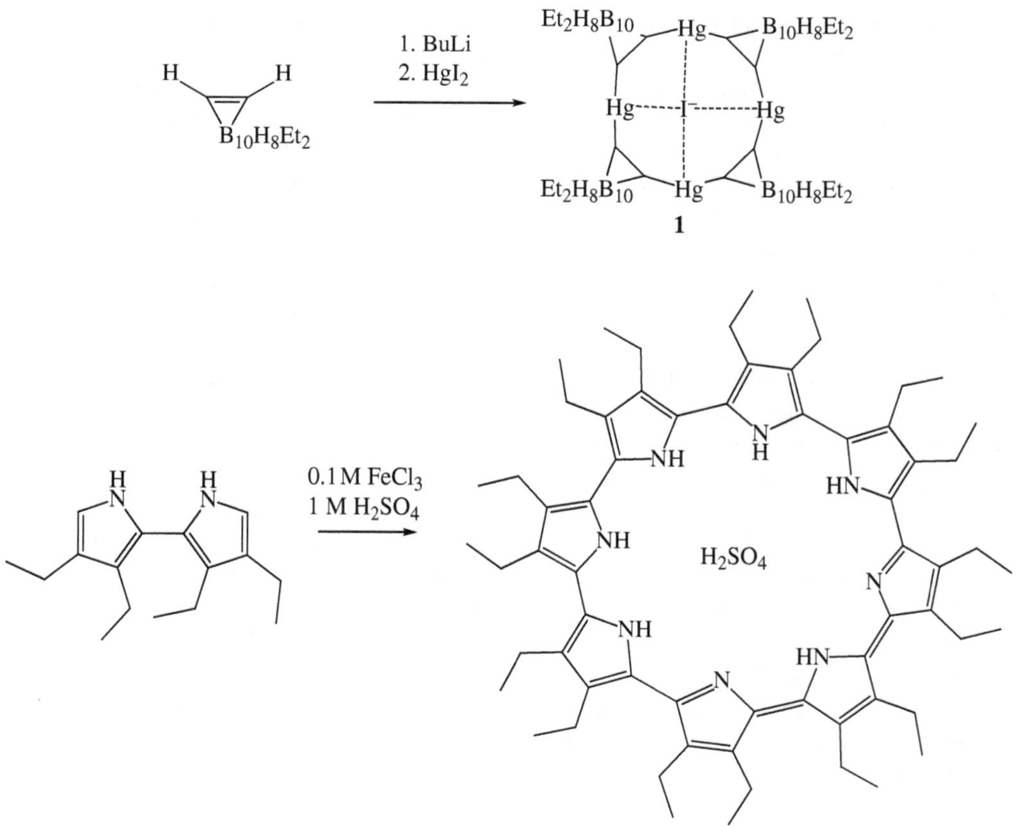

1

2

SCHEME 4.1 Iodide-templated synthesis of Hg(II) complex **1** and the synthesis of cyclooctapyrrole **2** in the presence of sulfate anions.

presence of sulfuric acid the product could be obtained in more than 70% yield. It was proposed that sulfate anions serve as efficient templates for the formation of macrocycle **2** (Scheme 4.1). The synthesis of calix[*n*]pyrroles can be also assigned to the anion-templated synthesis [34]. For instance, trihalogenated acids can act as effective and selective templates in the synthesis of *meso*-hexaphenyl-calix[6] pyrrole [35].

Cyclic ureas are often synthesized by using non-covalent interactions of starting compounds with an anion. Böhmer and coworkers discovered a template effect of a chloride anion, which gives rise to the formation of large hexaureas (Scheme 4.2) [36]. A number of oligourea macrocyclic compounds were investigated in terms of anion-assisted synthesis and anion binding properties [37]. The macrocyclic compounds were synthesized from starting diamines (**3** and **4**) and diisocyanates (**5** and **6**). The major products of reactions between diamines and diisocyanates were cyclic oligoureas with either [1+1] or [2+2] composition (Scheme 4.2). Acetate and chloride were tested as templates in the macrocyclization reaction, and their influence on the product distribution and yield of the reactions is gathered in Table 4.1. The most striking reaction appeared to be the reaction between compounds **3** and **6**. The presence of chloride totally reverts the ratio of [1+1] and [2+2] products, indicating that there is a preference for two chloride anions to induce the formation of a larger macrocycle. This suggestion was obtained from anion binding studies of the [2+2] macrocycle, which showed that this large ring can bind two chloride anions stepwise ($\log K_{11}=6.16$, $\log K_{12}=5.25$). According to the crystal structure analysis of complex **10**•($2Cl^-$), the hexaurea binds two chloride anions and shows high order of complementarity between the host and two chlorides. The macrocycle wraps around two chloride anions forming an 8-shaped structure. A separate experiment with the bromide anion demonstrated that such an amplification

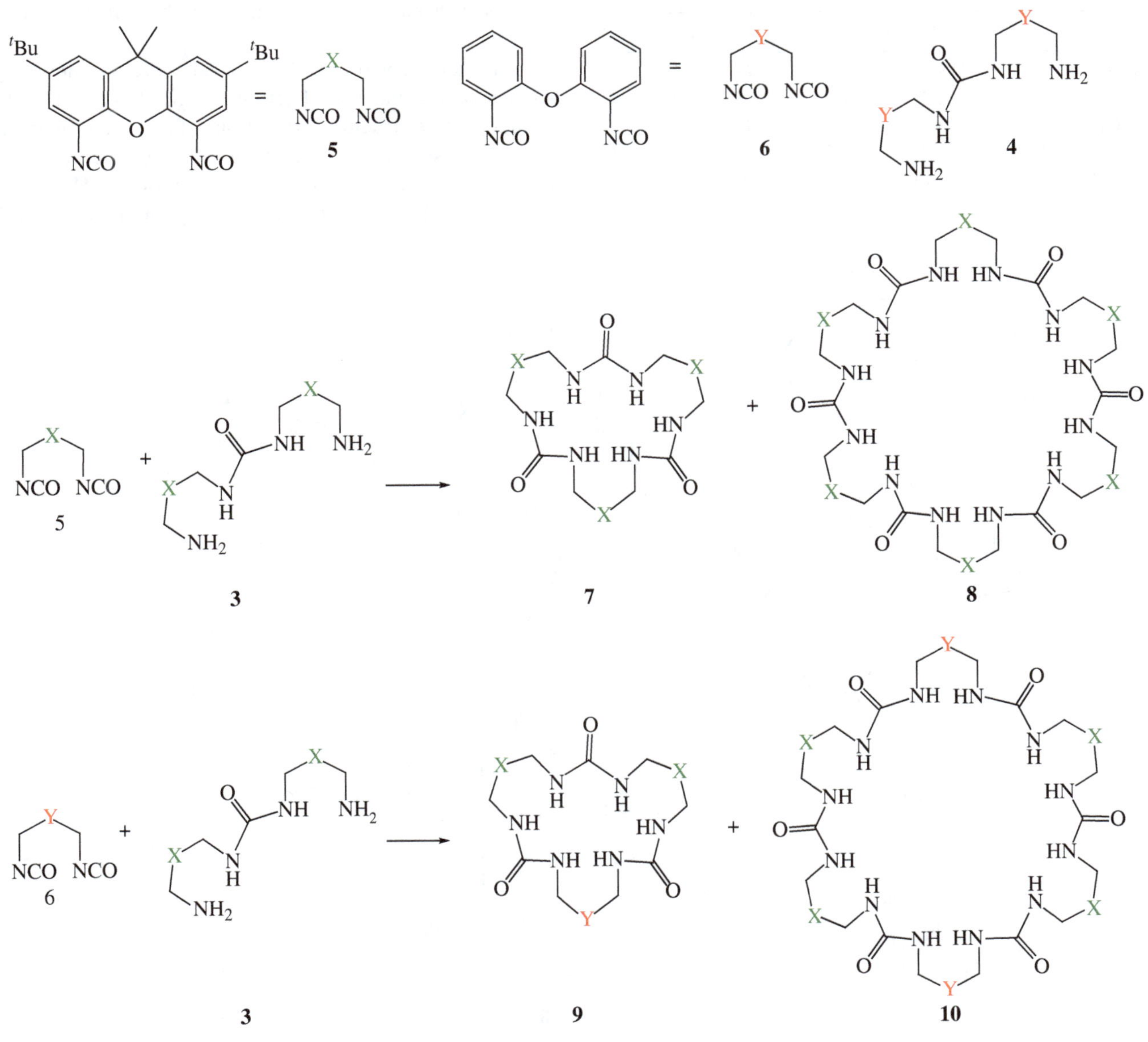

SCHEME 4.2 Synthesis of polyuria macrocycles **8–10**.

of a larger product is less efficient, likely because of the smaller affinity of the product or starting material for bromide [36]. Two interesting facts were gathered from investigations of these macrocycles. As a rule, in polar solvents, like tetrahydrofuran (THF), the larger structure was formed in low yields, while in apolar solvents, like dichloromethane, the yield was considerably higher. In the anion-templated synthesis, the anion had greater influence on the product distribution and the yield of the reaction if the starting building blocks were rigid.

Effect of chloride on the cyclization process was observed by Alcade and coworkers in the synthesis of imidazoliophanes by a [3+1] reaction (Scheme 4.3) [38, 39]. Macrocycles **11**, **12**, and **13** could be prepared by the reaction between corresponding starting materials **14–17** with

TABLE 4.1 Effect of an Anion on the Product Distribution of the Reaction between Diamines 3 and 4 and Diisocyanates 5 and 6

Salt	Type of Product	3+5	3+6	4+5	4+6
No anion	[1+1]	—	75	46	60
	[2+2]	69	15	29	20
TBA^+Cl^-	[1+1]	—	8	86	63
	[2+2]	55	82	—	37
$TBA^+CH_3COO^-$	[1+1]	ca. 100	40	71	51
	[2+2]	—	—	—	4

high yields in the presence of tetrabutylammonium chloride as a template. For instance, the yield of product **11** increases from 42 to 70% when the template was used in the reaction. It was also found that the generated chloride during the reaction works as a template. Thus, the reaction can be considered as a self-templating reaction. A series of different inorganic anions were studied: chloride, bromide, iodide, phosphate, nitrate, sulfate, fluoride, tetrafluoroborate, and hexafluorophosphate. However, only chloride and bromide demonstrated best templating efficiency. Detailed kinetic studies of the chloride-templated reaction starting from acyclic precursor **18** were conducted [40]. Instead of chloride, compound **18** contains picrate as a leaving group and as a counteranion to avoid the self-templating reactions. The kinetic studies were carried out with UV-vis spectroscopy at 0.0007 M concentration of compound **18** and 0.04 M concentration of tetrabutylammonium chloride. The suggested kinetic model delivers the association constant between **18** and chloride, which was calculated as $72\,M^{-1}$. Thus, it was suggested that chloride stabilizes the transition state in the cyclization reaction.

Luis and coworkers conducted several interesting studies of the macrocyclization reactions leading to the pseudopeptide structures. Reactions of diamines **19** and **20** with 1,3-bis(bromomethyl)benzene led to the formation of the corresponding macrocyclic products **21** and **22** (Scheme 4.4) [41]. Different tetrabutylammonium (TBA)

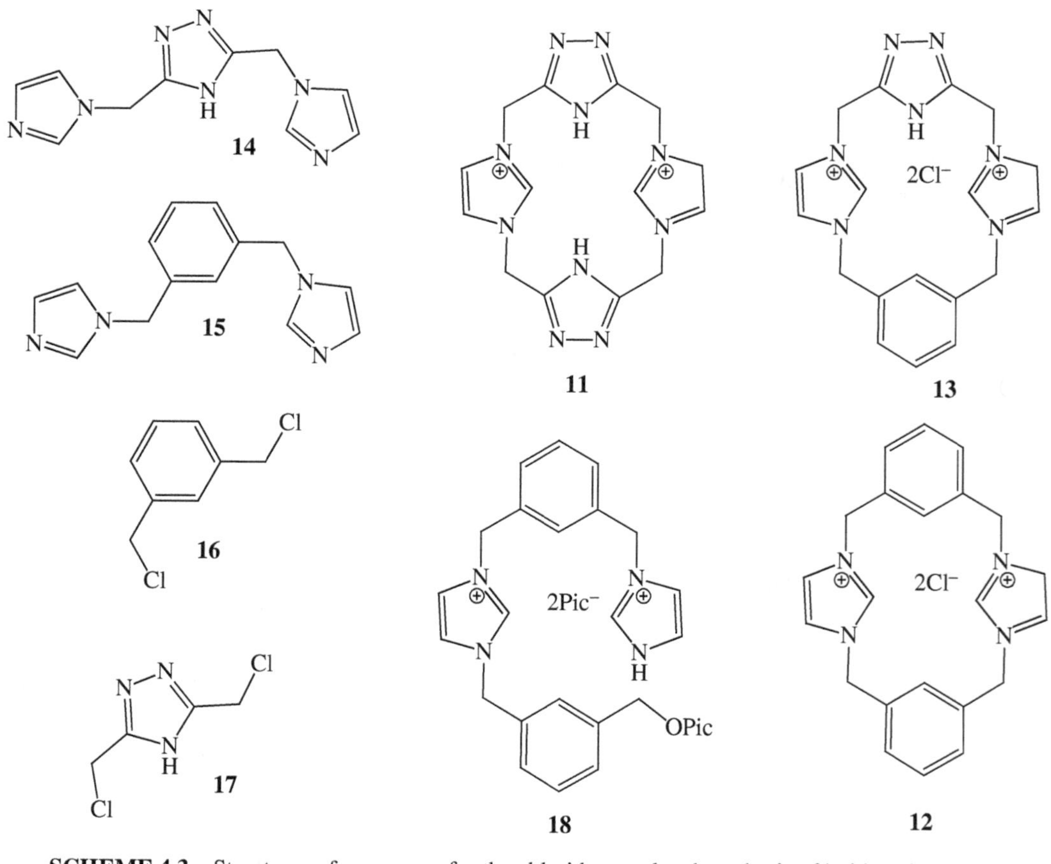

SCHEME 4.3 Structures of precursors for the chloride-templated synthesis of imidazoliophanes.

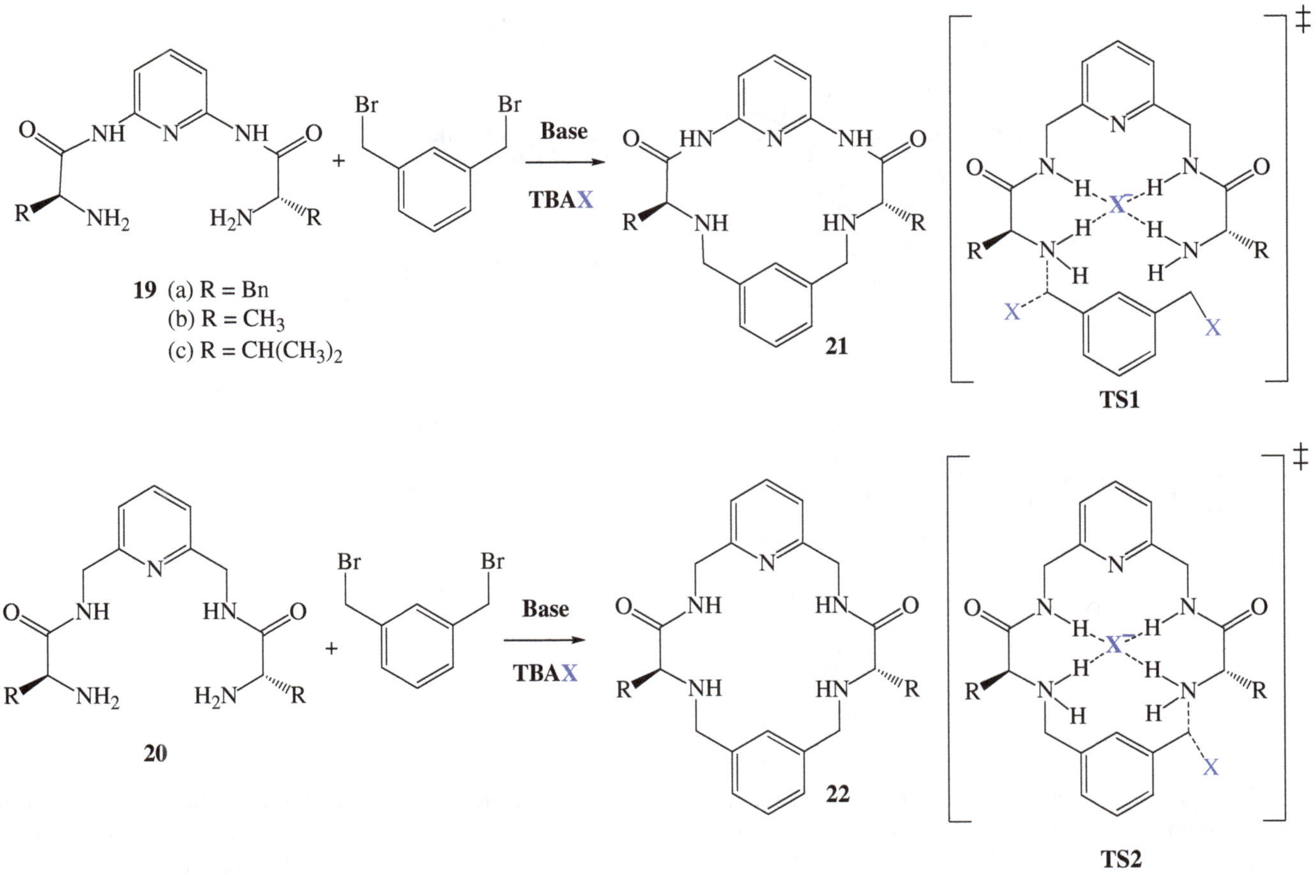

SCHEME 4.4 Synthesis of macrocycles **21** and **22** and the proposed transition states.

salts were used as phase transfer catalysts in the presence of potassium carbonate or diisopropylethylamine (DIPEA) as a base. Interestingly, the formation of **22** was completed with highest yields (93–96%; Table 4.2) in the presence of tetrabutylammonium bromide (TBABr) when compared with TBA salts of other inorganic acids. Apparently, bromide was the best template for the cyclization reaction. The kinetic studies showed that the presence of bromide increases both the yield and the rate of the reaction. Based on quantum chemical calculations, the authors proposed two transition states **TS1** and **TS2** (Scheme 4.4), which are stabilized in the presence of a template. The calculations predicted that the best anionic template is that that stabilizes **TS2** greater than **TS1**. In this case, the macrocyclization step proceeds much faster than the oligomerization reaction. By using the suggested model, the authors predicted that chloride is the best template for the macrocyclization of **20** to **22**. This fact was successfully confirmed by an experiment, and compound **22** was formed in 30% yield. Several kinetic models were developed to describe the reaction [42]. They allowed one to optimize the reaction conditions in order to form the desired product with a high yield.

The effect of substituents on the reaction kinetics was also investigated [43]. It appears that the yield of the reaction not only depended on the presence of a template but also on

TABLE 4.2 Reaction Yields in the Synthesis of 22 Templated by Different Anions

Entry	Solvent	Base	TBAX	Yield
1	CH_3CN	K_2CO_3	Br	93
2	CH_3CN	DIPEA	—	82
3	CH_3CN	DIPEA	$OCOCH_3$	69
4	CH_3CN	DIPEA	Br	96
5	CH_3CN	DIPEA	Cl	81
6	CH_3CN	DIPEA	F	30
7	CH_3CN	—	—	58

preorganization of an intermediate complex for the cyclization reaction. Thus, a series of diamines was investigated in the reaction with 1,3-bis(bromomethyl)benzene yielding macrocycles **23a–23i** (Scheme 4.5). The cyclohexylalanine-containing precursor showed the best yield in cyclization reactions. Computational analysis confirmed that the most stable conformer of the intermediate is stabilized by a template (bromide) and has a proper conformation for cyclization.

The anion-templated synthesis was further used by the authors in construction of new pseudopeptidic tripodal small cages **26** and **27** for binding of chloride anion. TBABr was used in catalytic quantities to template the macrocyclization step as depicted in Scheme 4.5 [44].

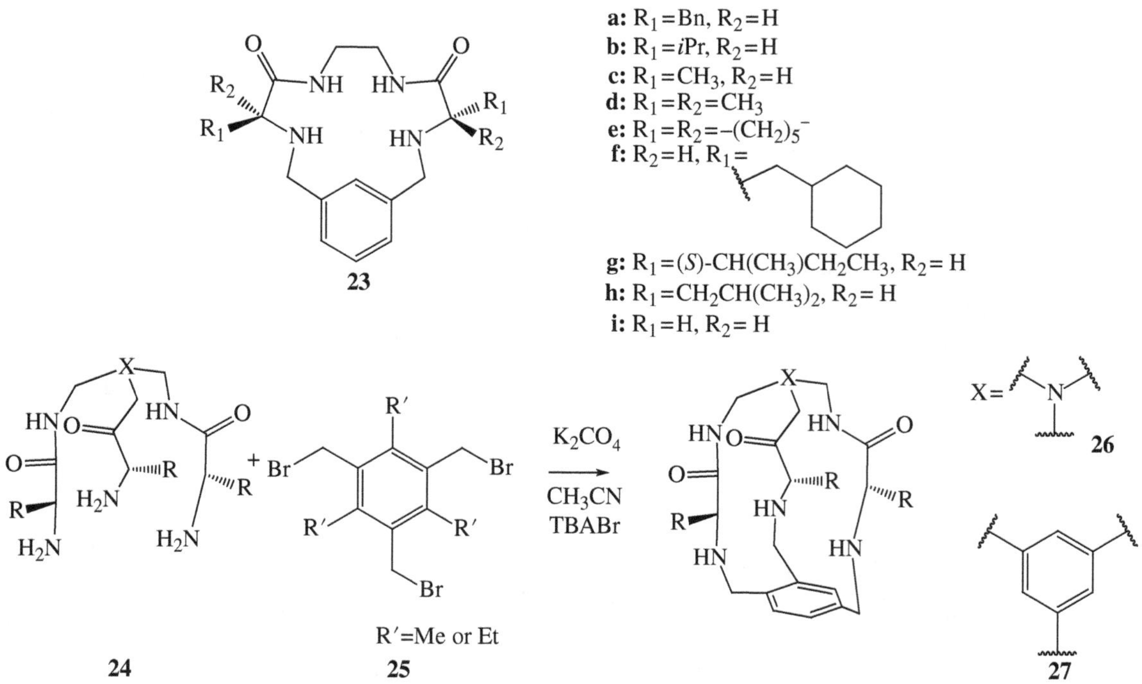

SCHEME 4.5 Synthesis of macrocycles **23**, **26**, and **27**.

Self-templating processes in macrocyclization reactions were observed by the author of this chapter and coworkers when diacid dichlorides (e.g., **29**) were reacted with diamines (e.g., **28**) [45, 46] (Scheme 4.6). The presence of other anions had no influence on neither the structure of the product nor the yield, which was always very good (>60%). The leaving chloride anion was found in all cases in products as a guest (**30a,b**). This behavior can be explained in terms of high affinity of generated intermediates and the product for chloride, as well as rigidity of the building blocks. However, a disadvantage of using acid chlorides as reactive species for coupling with amines is that the reaction rate is very high and lowering the temperature led to better yields because of stronger host–guest interactions.

In the next work, a number of different activation groups were screened, which (i) allow the amide bond formation under mild conditions and (ii) do not interact with building blocks in organic solvents. The best results were obtained for 2-mercaptothiazoline derivative **32** shown in Scheme 4.7. It was possible to observe different distributions of macrocycles from [1+1] to [4+4] condensation products in the presence of various anions. Interestingly, intramolecular hydrogen bonding prevails in the absence of a template, and the reaction led to a [2+2] macrocycle **33** with a high yield. The use of an anionic template shifted the reaction to a smaller [1+1] macrocycle **34**. Although 2,2′-bipyrrole-based diamine **31** is difficult to synthesize, it was used because of good fluorescent properties. Fluorescence spectroscopy appeared to be an excellent method to follow the macrocyclization reaction and assess the reaction rates and kinetics. By using fluorescence, it was found that the formation of a [2+2] macrocycle **33** in the absence of a template proceeds through an intermediate **36**. If activated isophthalic acid was used in the reaction, no intermediates were observed. It was suggested that the reaction proceeds through intermediate **35** when tetrabutylammonium chloride was used as a template. The template brakes intramolecular hydrogen bonds; this leads to formation of a kinetically favorable [1+1] macrocycle. This difference is clearly seen in Scheme 4.7. Fluorescence measurements allowed the authors to investigate the influence of the template concentration on the reaction rate [47].

Flood and coworkers developed a high-yielding and multigram-scale synthesis of C5-symmetric cyanostar macrocycle **37** (Scheme 4.8). The reaction employs Knoevenagel condensations of monomer **38** to a macrocycle in the presence of a catalytic quantity (5%) of cesium carbonate. The authors suggested that carbonate may act as a template or the deprotonated benzylic anion may help the cyclization. This conclusion was supported by the experiments with other bases such as pyridine, piperidine, triethylamine, and 1,8-diazobicycloundec-7-ene, which led to very low yields of the macrocycle even at elevated temperature and longer reaction times. Interestingly, the cyanostar macrocycle demonstrated relatively high affinities toward hydrophobic anions in polar solvent mixture 40% methanol/dichloromethane. The synthetic receptor has a property to bind anions by using CH hydrogen bonding units [48].

The synthesis of cucurbiturils is known to be directed by anions or other compounds present in the reaction mixture. Since the mechanistic investigation on the synthesis of cucurbiturils was recently reviewed by Isaacs [40, 49], we focus here only on the synthesis of hemicucurbiturils and bambusurils [50], which have attracted recently great attention as new anion receptors.

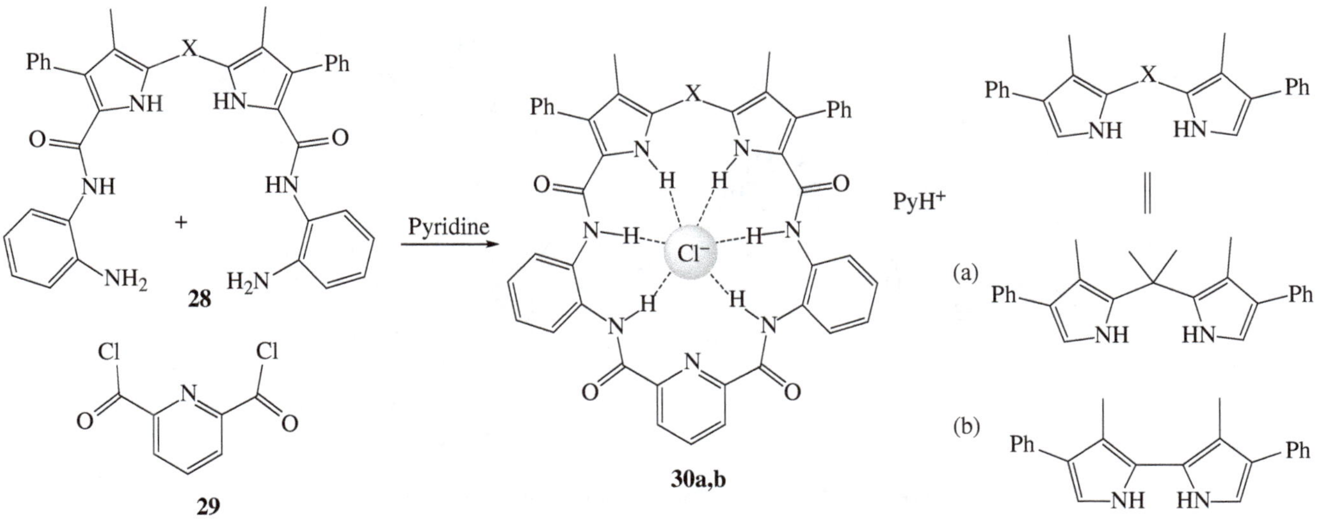

SCHEME 4.6 Self-templating macrocyclization producing chloride complexes **30a** and **b**.

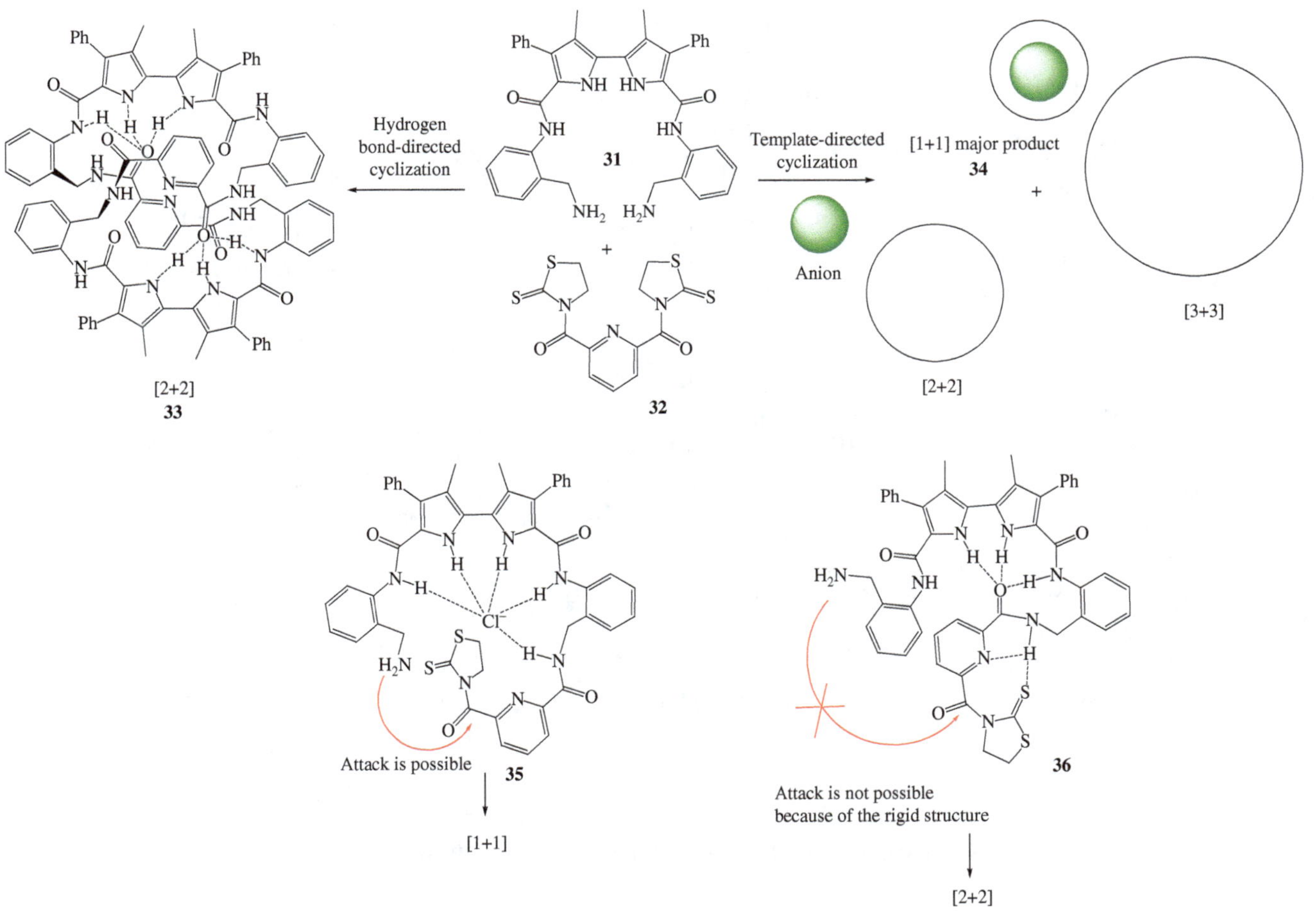

SCHEME 4.7 Formation of macrocycles **33** and **34** without and with the chloride anion, respectively. Structures of the proposed intermediates **35** and **36** are also shown.

Hemicucurbiturils were introduced and synthesized by Miyahara with the help of hydrochloric acid-catalyzed Mannich reaction of ethyleneurea **39** with 37% aqueous solution of formalin [51]. Hemicucurbiturils[6] (**40**) and hemicucurbiturils[12] were obtained with high yields (Scheme 4.9). Hemicucurbit[6]uril was found to form in 1N HCl at room temperature (r.t.) with 94% yield, while hemicucurbit[12]urils were prepared in 4N HCl at 55°C with 93% yield. The X-ray crystal structure of the smaller oligomer revealed the presence of the chloride anion bound in

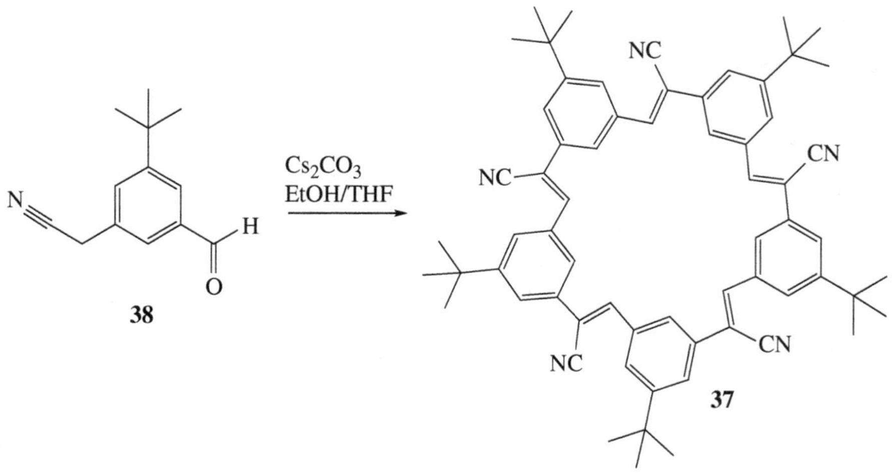

SCHEME 4.8 Synthesis of C5-symmetric cyanostar macrocycle **37**.

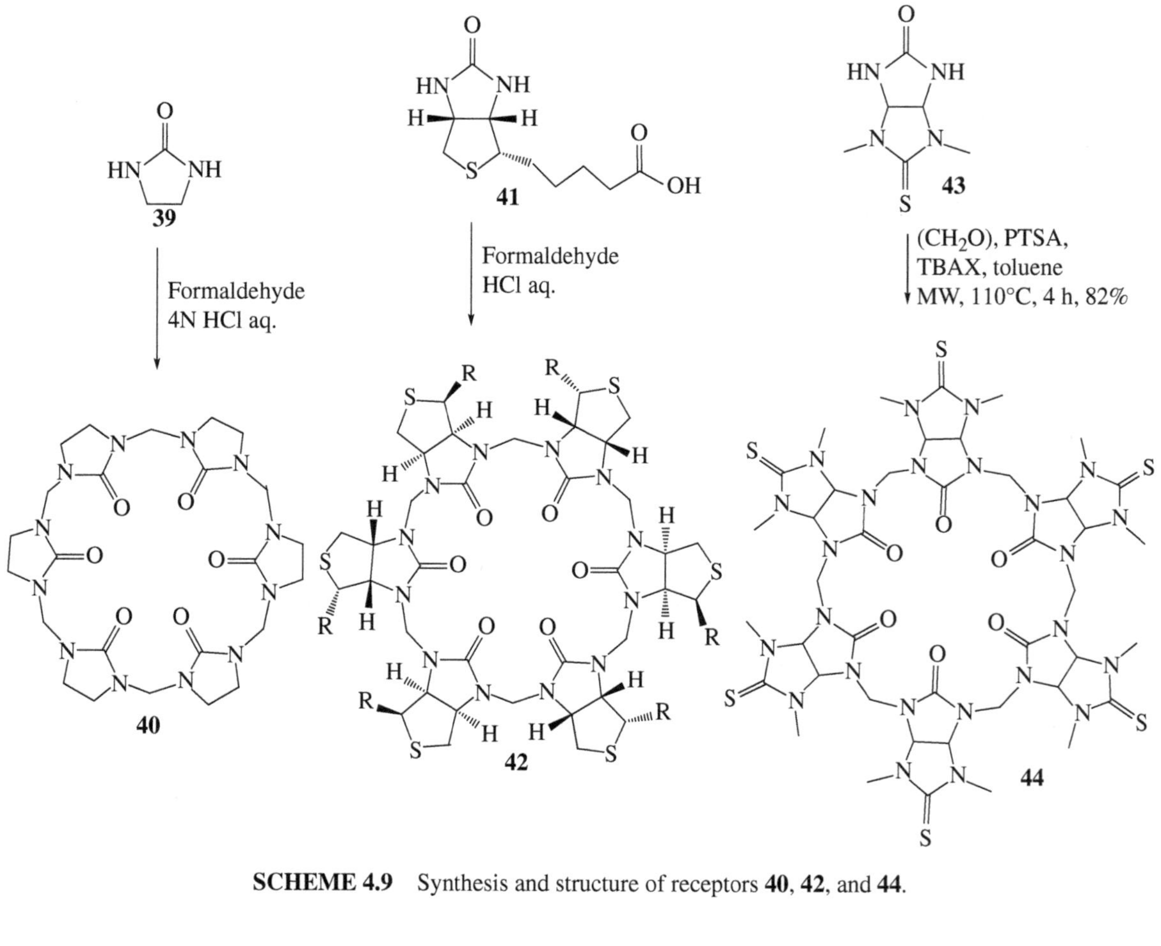

SCHEME 4.9 Synthesis and structure of receptors **40**, **42**, and **44**.

the center of the cavity and forming hydrogen bonds with water molecules situated outside the cavity. Thus, chloride was suggested to work as a template in cyclization reactions. Interestingly, the macrocycle was found in the "alternate" conformation. This fact could be the reason why hemicucurbiturils do not bind cations. The solubility of hemicucurbit[6]uril was dramatically increased upon addition of ammonium thiocyanate. Buschmann and coworkers explained this behavior by determining binding affinities of hemicucurbit[6]uril toward different salts in aqueous solution [52]. Among all tested salts thiocyanate demonstrated the best stability constants (e.g., log$K$ for NH_4SCN is 2.15). Only water molecules were found in the cavity of the larger macrocycle—hemicucurbit[12]urils. There was no exact mechanism suggested for the formation of this big structure in such a high yield.

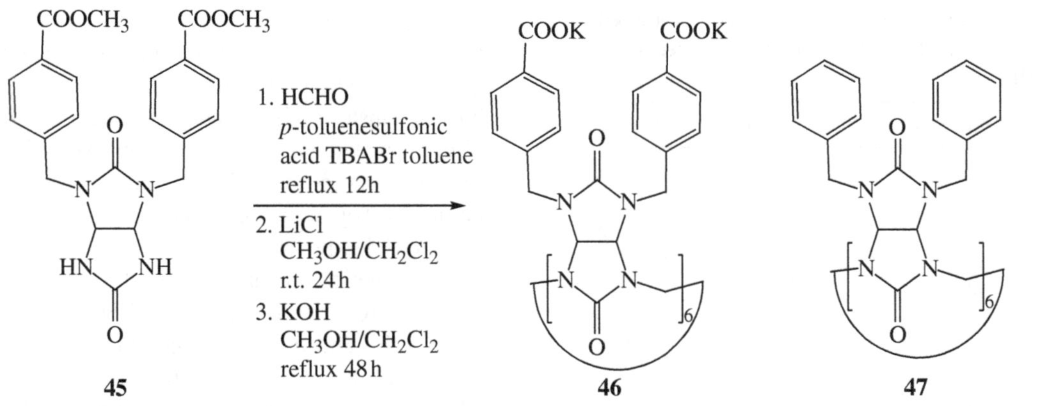

SCHEME 4.10 Synthesis and structures of bambusurils, which bind anions with high affinity.

Similar halide-templated synthesis for the construction of hemicucurbiturils was used by Pittelkow and coworkers to construct a [6+6] macrocycle from biotin **41** and formaldehyde (Scheme 4.9) [53]. Large macrocycle **42** was obtained as HCl adduct under heating the starting materials in 7N HCl at 60°C. If the reaction was carried out in other acids, for example, sulfuric acid, then the reaction resulted in a mixture of noncyclic oligomers. When chloride, bromide, or iodide salts were added together with sulfuric acid, the product was again observed. The best yield of 63% was achieved by using a mixture of 3.5 M H_2SO_4 and 5 M NaBr. According to the NMR and ITC titrations, the macrocycle showed high selectivity for iodide (K_a from NMR titrations: Cl^-, 59 M^{-1}; Br^-, 540 M^{-1}; I^-, 2200 M^{-1}).

Reany and coworkers have reported an anion-templated synthesis of semithiobambusuril BU[6] **44** with 82% yield starting from **43** [54]. Optimization of reaction conditions revealed that not only halide anions can act as a template for the formation of the hexamer but also perchlorate. It appears that the influence of a countercation in this template reaction is negligible, for example, Et_4NClO_4, Ph_4PBr, CsI, and Me_4NI gave similar yields of macrocycles. The prepared receptors demonstrated good solubility in organic solvent in the presence of anions. Interestingly, the anion binding properties of semithiobambus[6]uril is slightly different from bambus[6]uril. The Semithiomacrocycle has almost similar affinities for chloride, bromide, and iodide, while for bambus[6]uril the affinity constants for these anions range over three orders of magnitude. This difference was attributed by the authors to the significantly distorted cavity of the semithiomacrocycle, which was observed in the X-ray crystal structure.

Similar approach for the synthesis of hemicucurbit[6]uril was used by Aav and coworkers. They obtained the product in 85% yield by using HCl and in 64% yield by using HBr as a catalyst and a template at the same time [55].

Sindelar and coworkers used a template method to synthesize a range of bambus[6]urils from monomer **45** by an acid-catalyzed reaction in the presence of TBABr (Scheme 4.10) [56]. The carboxylate-functionalized bambusuril **46** can coordinate hydrophobic anion with unprecedented affinity reaching $10^7 M^{-1}$ for perchlorate and iodide in pure water (D_2O, 20 mM K_2DPO_4, pD 7.1). Interestingly, in a mixture of 5% water in DMSO, macrocycle **47** binds anions with low exchange rate. This property allowed the authors to differentiate several different anions in one mixture [57].

4.3 CATION-TEMPLATED SYNTHESIS

One of the first interesting cyclo-oligomerization reactions was observed by Dale and coworkers. The authors discovered the effect of a cation on polymerization of ethylene oxide and investigated the cyclo-oligomerization reaction of ethylene oxide in dioxane in the presence of BF_3 as a catalyst and metal templates [9]. This is an important contribution in the field of crown ether synthesis after the Pederson work. Fluoroborates, fluorophosphates, and fluoroantimonates of alkali metals, alkaline earth, and transitions metals were used as templates in this study. All templates functioned well yielding macrocycles $(\text{—CH}_2\text{—CH}_2\text{—O—})_n$ with $n=4$–6 as major products. Common salts, like sulfates and halides, could not be used in this reaction because the anions, which are basic in nature, neutralized Lewis acid catalysts.

It was found that the reaction is actually reversible under the conditions studied. When deuterated dioxane was used as a solvent, the deuterium was found in the growing chain. The distribution of major products is shown in Table 4.3. Interestingly, the hexamer, which is 18-crown-6, was formed quantitatively in the presence of Cs^+ and Rb^+ cations. The pentamer was formed with best yields in the presence of Li^+ and transition metals, such as Ni^{2+}, Cu^{2+}, and Zn^{2+}. Later Mandolini did a thorough investigation of the effect of cations (Ba^{2+}, Sr^{2+}, K^+, Na^+, and Li^+) on the cyclization rate of benzo-18-crown-6 in water solution and found that the efficiency of a template repeats the order of affinities of crown ether for alkali metals [58].

A simple proton can serve as a cationic template in a cyclo-oligomerization reaction. This idea was postulated by Kanbara and coworkers, who prepared monoprotonated azacalix[3]pyridines **48** with the help of Ullmann coupling reaction [59]. Starting compounds **49a–c** were converted to macrocycles **48a–c** (HBr salts) with 80% yield in one step. High yields were attributed to the template effect of proton. In the coupling

TABLE 4.3 Distribution of Cyclic Oligomers Obtained by Oligomerization of Ethylene Oxide in the Presence of Anhydrous Salts

Salt	Tetramer (%)	Pentamer (%)	Hexamer (%)
$LiBF_4$	30	70	—
$NaBF_4$	25	50	25
KBF_4	—	50	50
KPF_6	20	40	40
$KSbF_4$	40	20	40
$RbBF_4$	—	—	100
$CsBF_4$	—	—	100
$Ca(BF_4)_2$	50	50	—
$Sr(BF_4)_2$	10	45	45
$Ba(BF_4)_2$	10	30	60
$Ag(BF_4)$	35	30	35
$Hg(BF_4)_2$	20	70	10
$Ni(BF_4)_2$	20	80	—
$Cu(BF_4)_2$	5	90	5
$Zn(BF_4)_2$	5	90	5

reaction, hydrogen bromide is produced as a by-product and induces the formation of the macrocycle. Interestingly, if a strong base was used in the reaction, for example, sodium tert-butoxide instead of potassium carbonate, linear aminopyridine oligomers and larger macrocycles were formed. This fact indicates that stronger bases react immediately with eliminating hydrogen bromide and inhibit the template effect of proton (Scheme 4.11).

Investigations on ammonium-directed cyclo-oligomerization were reported by Jolliffe and Pattenden in the synthesis of oxazole- and thiazole-based cyclopeptides. They showed that pentafluorophenyl diphenylphosphinate (FDPP) reagent allows one to increase the yield of cyclo-oligomerization of amino acid **50** up to 60% in DMF [60–62]. The template effect of cations on the cyclo-oligomerization reaction using FDPP was investigated [63]. The best amplification of trimer **51** was observed in the presence of the ammonium cation. This template effect was explained in terms of hydrogen bonds that stabilized host–guest complex **51**•NH_4^+ between the macrocycle and the ammonium cation (Scheme 4.12).

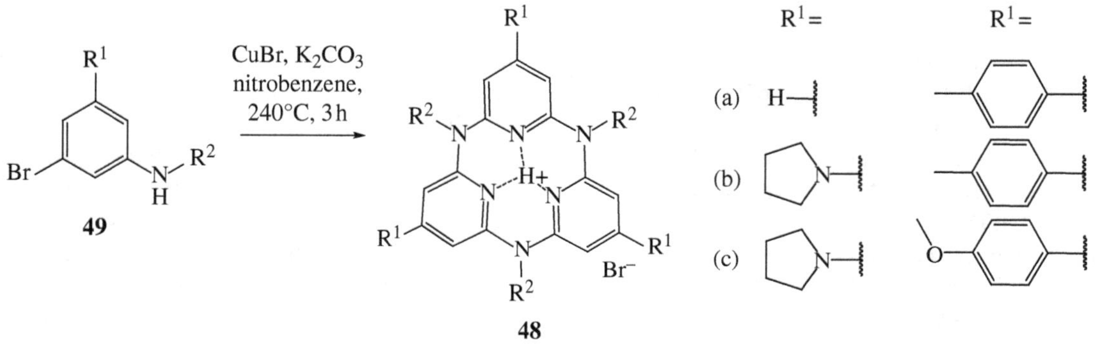

SCHEME 4.11 Proton-templated synthesis of macrocycles **48a–c**.

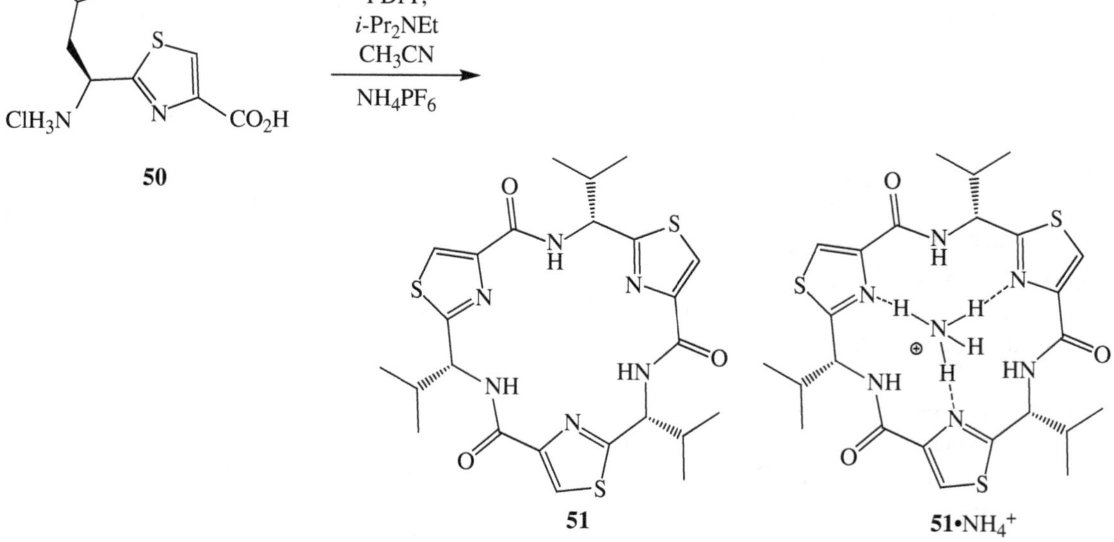

SCHEME 4.12 Synthesis of cyclopeptide **51**.

Several reports were published, which deal with the synthesis of sugar-containing macrocycles. Sugars were connected through ether or ester bonds and cyclized to produce large macrocycles, which can bind cations. Thus, Marsura and coworkers investigated the template effect of Cs^+ and Na^+ as their carbonate salts in the synthesis of macrocycles from diamine **52** and diazides **53** and **54** in the presence of PPh_3 (Scheme 4.13) [64]. The formation of macrocycles [1 + 1] and [2 + 2] was observed when the reaction was carried out without a template. The presence of sodium carbonate induces the formation of a [1 + 1] macrocycle (**55**) as a major product, while cesium carbonate led to the formation of a [2+2] macrocycle (**56**). Similar results were observed for diazide **54**.

The *cis*-2-alkyl-3-oxy-tetrahydropyran was used in the work of Martin and coworkers as a novel building block for the design of cation receptors [65]. The building blocks were combined through ester bonds into macrocycles of different sizes. The template effect of Li^+, K^+, and Na^+ cations was investigated in the synthesis of receptors. The macrocycles could be prepared either under thermodynamic or kinetic control depending on the reagents used to form ester bonds. However, under thermodynamic control the reaction of compound **57** with potassium or sodium methoxide did not give desirable macrocycles **58** and **59** (Scheme 4.14). Under the Yamaguchi conditions compound **60** reacted to form a mixture of **58** and **59** with different ratios depending on the template used. If the cyclization of **60** was conducted at r.t. without the template, then a 40:60 (**58**:**59**) ratio was observed. After addition of 20 equivalents of potassium carbonate, the reaction mixture changed the product ratio to 87:13. For the cyclization of **61**, similar product ratio was observed in the presence of potassium carbonate, while the use of DMAP instead of the template led to the formation of only larger macrocycle **59**. Interestingly, the sodium cation, which was found to be the best host for receptor **58**, was not the best template. The best template in all these reactions was the potassium cation. This may be explained in terms of different cavity size of the intermediate of the transition state, which lead to compound **58**.

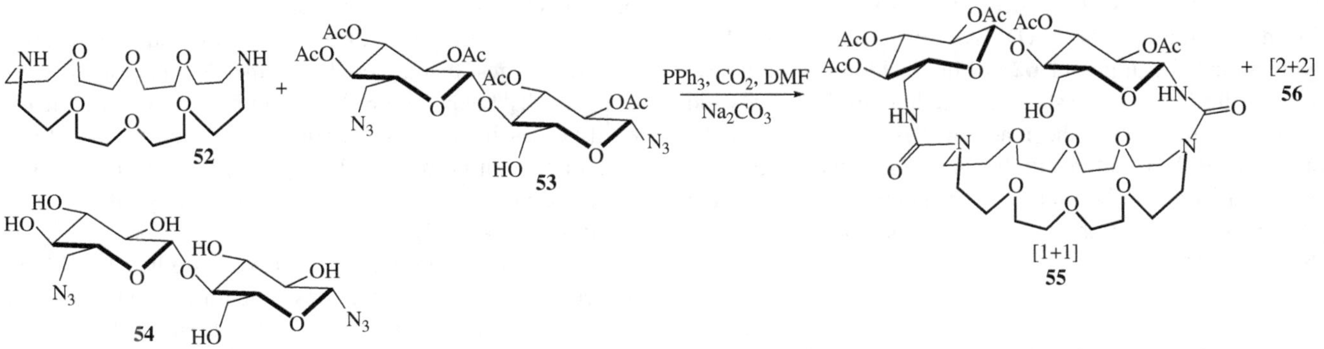

SCHEME 4.13 Synthesis of sugar-containing macrocycles **55** and **56**.

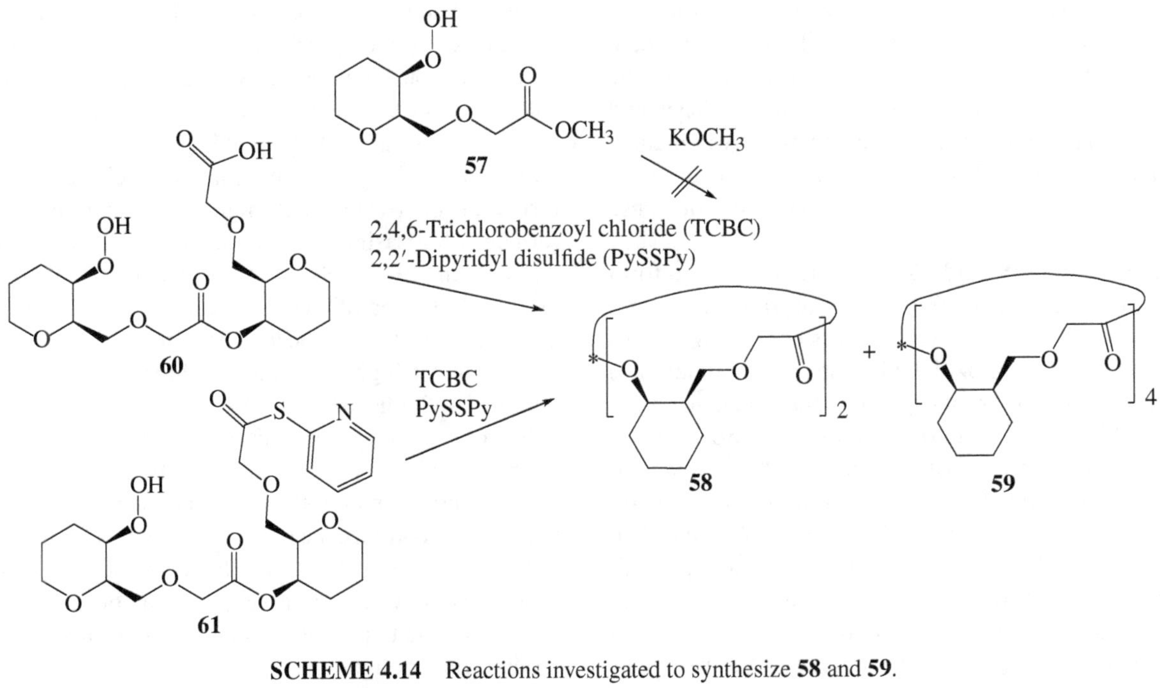

SCHEME 4.14 Reactions investigated to synthesize **58** and **59**.

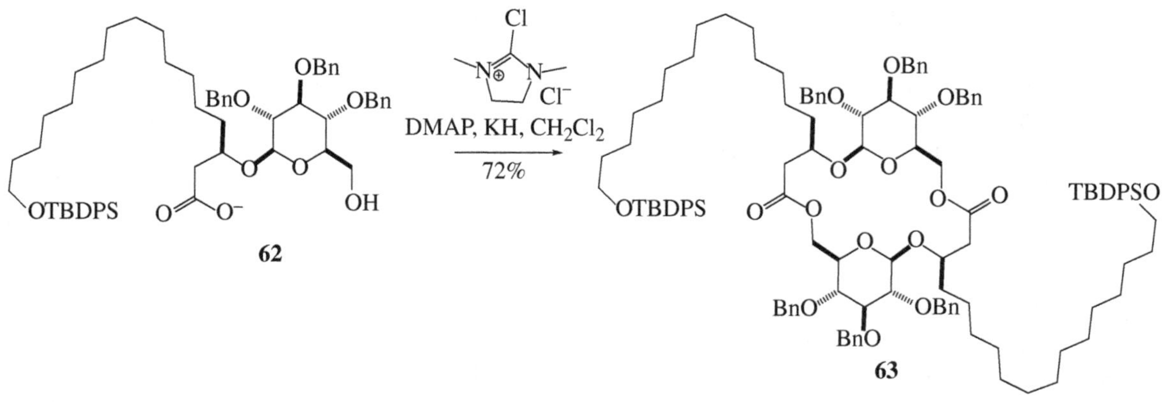

SCHEME 4.15 Synthesis of sugar-containing macrocycle **63**.

Another interesting example of a template-induced synthesis was reported by Fürstner and coworkers, who prepared natural compound cycloviracin B_1 [66]. The C2-symmetrical core structure **63** of this natural compound was synthesized by a template-directed macrodilactonization. Dicyclohexylcarbodiimide (DCC) together with 4-dimethylaminopyridine (DMAP) was unsuccessful in cyclization of compound **62**, while 2-chloro-1,3-dimethylimidazolinium chloride showed excellent results (72% yield of the product) when the reaction was carried out in the presence of KH. Interestingly, K^+ was a better template in comparison with Na^+ and Cs^+. The potassium cation could preorganize precursor **62** for the macrodilactonization (Scheme 4.15).

4.4 HYDROGEN BOND-DIRECTED SYNTHESIS

This kind of macrocyclization reactions is also known as a self-templated cyclization because the formed intramolecular hydrogen bonds in intermediates drive further chain growth and finally lead to a macrocyclization step. The review of Zeng and coworkers describes the synthesis and properties of particularly pentameric macrocycles obtained in his laboratory [67]. Thus, in this part we focus more on the mechanistic aspects of the synthesis of these macrocycles.

Amino acids **64** [58] and **65** [17] were reported to form macrocyclic compounds after an appropriate activation (Scheme 4.16). The amino acid **64** was converted to pentamer **66** with (benzotriazol-1-yloxy)tris(dimethylamino) phosphonium hexafluorophosphate (BOP reagent) with the best yield of 25%. Amino acids **65a** and **65b** could be converted to macrocycles with the help of $POCl_3$ in acetonitrile solution with much better efficiency. The best reaction conditions for the formation of pentamer **67a** and **67b** were 40°C and 1 M concentration of the starting material. The pentamer was formed in yields reaching 45% under the optimized conditions. It was found that at elevated temperatures the yields of the hexamer and the heptamer were also much higher as compared to those of the reactions under r.t. The cyclization of starting material **65b** was anticipated to give more hexamer and heptamer macrocycles because of the bulky nature of the ethoxy group. However, pentamer **67b** was again the major product of the cyclo-oligomerization reaction. Additional theoretical modeling of structures of macrocycles and the reaction kinetics suggest that although hexamer **68b** is more thermodynamically favorable product, pentamer **67b** is kinetically easily accessible. The same authors also demonstrated that pentamers can be prepared by [5+1] or [3+2] approach, which gave rise to macrocycles consisting of different building blocks [68].

Unusual isomeric cyclic ureas were studied by Böhme and coworkers. The macrocycles were formed in the reaction between 2,6-diaminopyridine and *N,N′*-carbonyldiimidazole (CDI) [69]. Heating the starting materials at 80°C in DMSO solution led to the formation of two isomeric trimers and one tetramer, in which structures are depicted in Scheme 4.17. Trimer **70** appeared to be not soluble and precipitated from the reaction mixture, while trimer **69** had a good solubility. If the reaction temperature was raised to 180°C, then the soluble trimer **69** was formed in almost quantitative yields, as a thermodynamically favorable product. Likely, the intramolecular hydrogen bonds play a decisive role in the formation of a trimer and higher oligomers could not be formed. Surprisingly, the same products and their distributions were obtained by using 1,3-bis-(6-aminopyridin-2-yl)-urea (dimer) with carbonyldiimidazole (CDI). This fact indicates that the urea bond formation is a reversible process under high temperatures. Indeed, a mixture of **69**, **70**, and **71** after heating at 180°C for 8 h gave trimer **69** as the only product. The authors proposed that the exchange reaction might take place between urea and the amino group of the dimer. Understanding the reaction mechanism was very important to optimize the reaction conditions for the synthesis of linear tetramer, which has a helical structure. Under mild conditions (in acetone or in THF under ambient temperature), it was possible to achieve kinetic control of the reaction and produce linear oligomers.

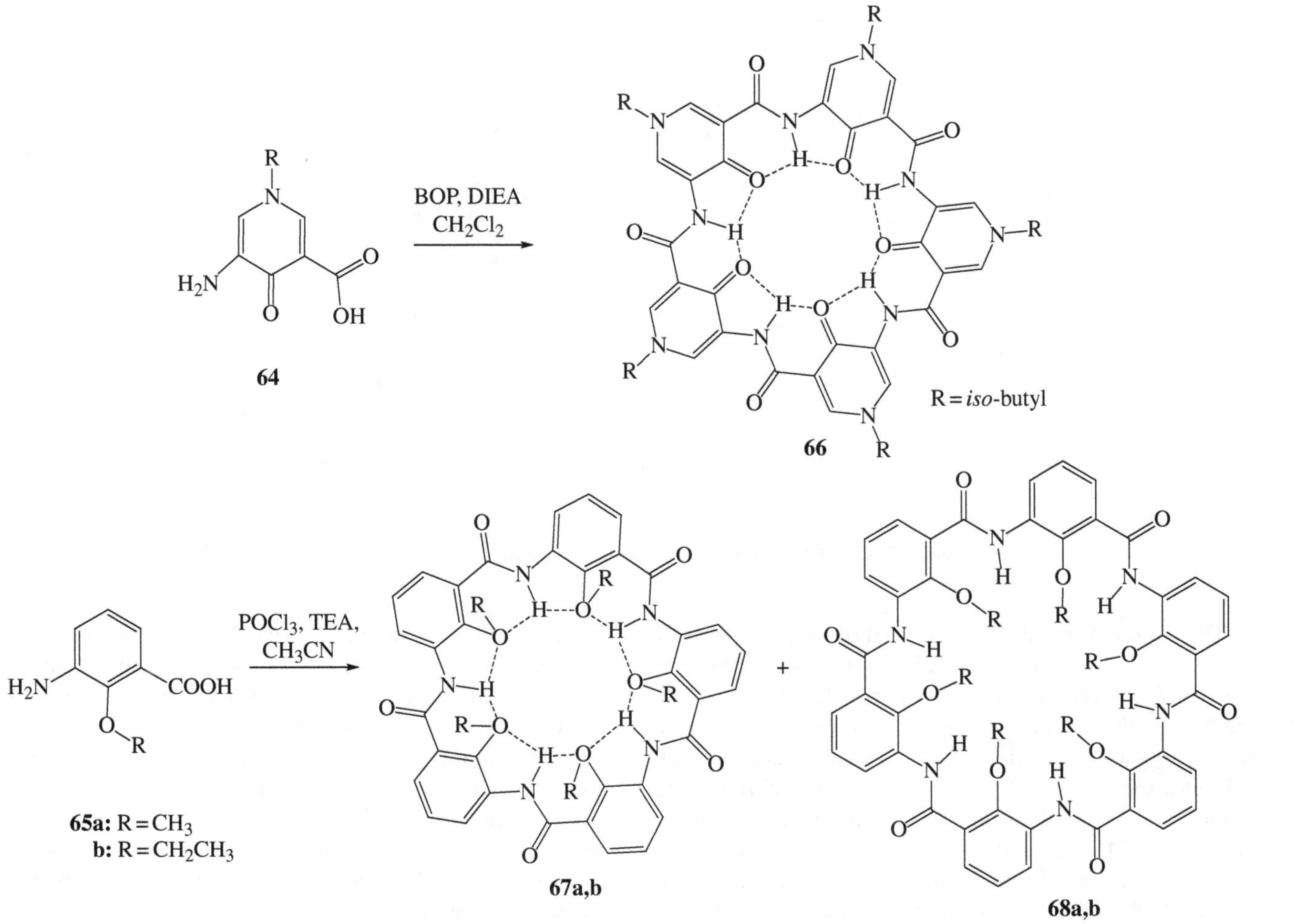

SCHEME 4.16 Cyclo-oligomerization of amino acids **64** and **65**.

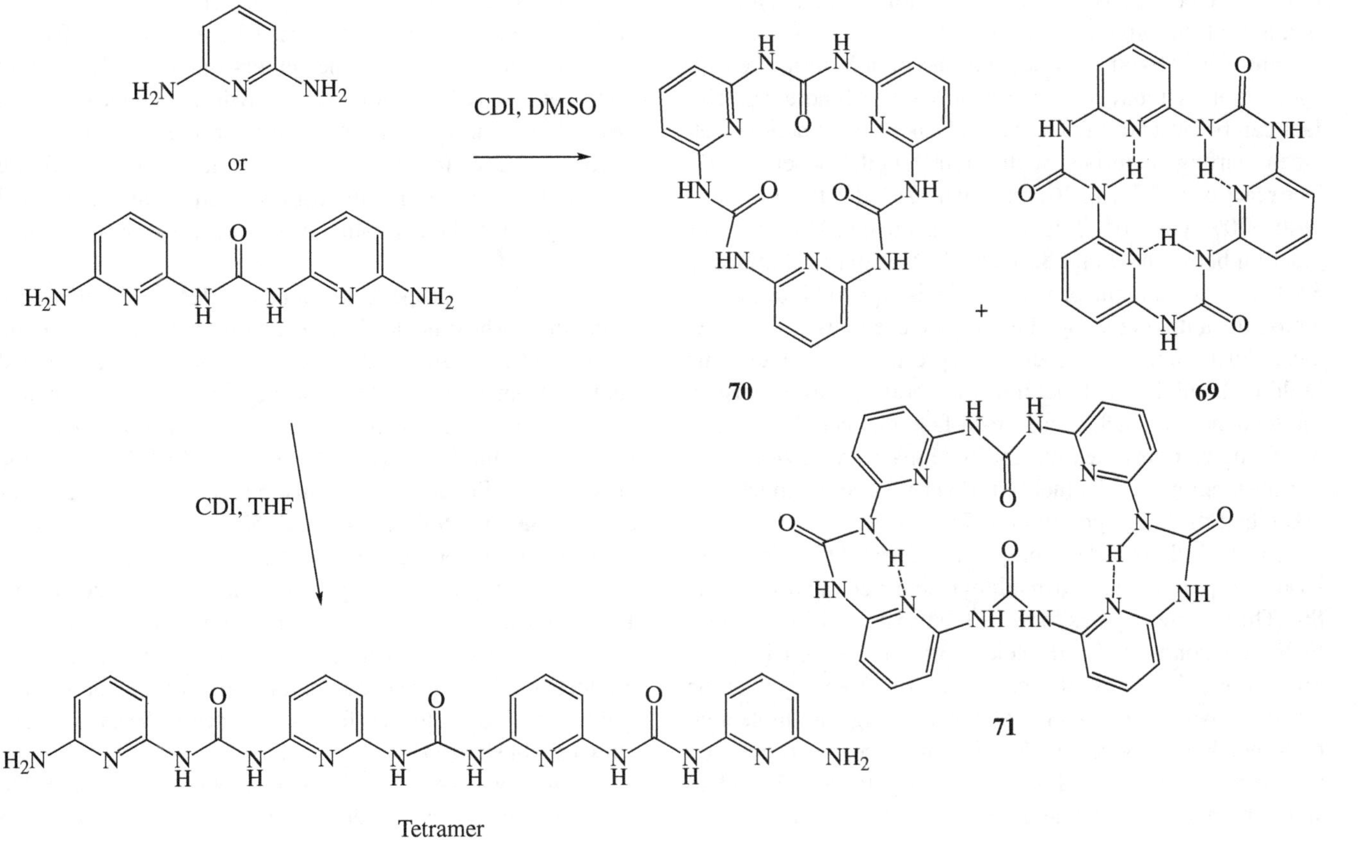

SCHEME 4.17 Synthesis of cyclic ureas **69–71**.

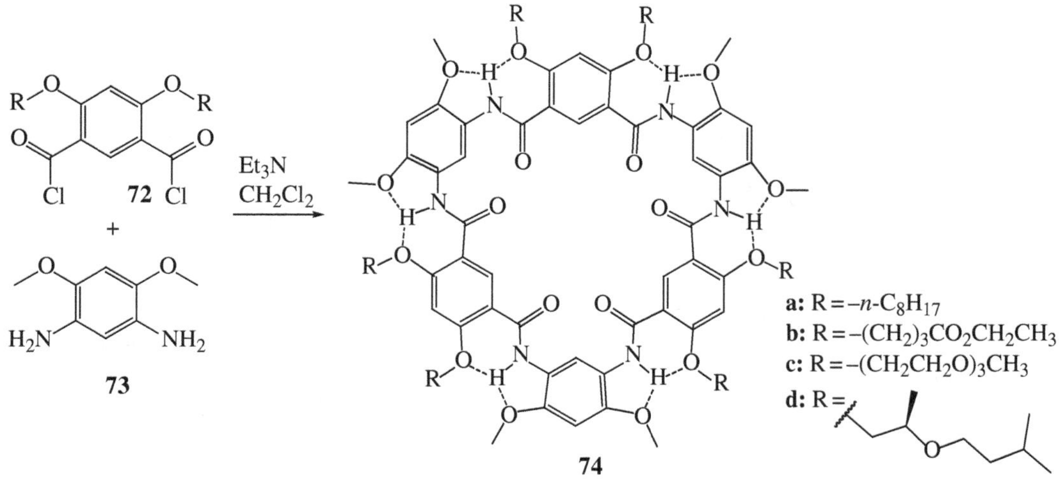

SCHEME 4.18 Hydrogen bond-directed synthesis of macrocycle **74**.

The importance of the structure and rigidity of intermediates formed in kinetic cyclization was demonstrated by Gong and coworkers. Intramolecular hydrogen bonds that are formed during the cyclization reaction were the reason for high yields in multicomponent reactions between acid chlorides and amines [70]. For instance, the cyclization between acid chloride **72** and amine **73** leads to the formation of macrocycle **74** with greater than 80% yield (Scheme 4.18). However, theoretical simulations of the reaction kinetics predicted only 36% of a maximum allowed yield. Detailed mechanistic investigations showed that either a chain growth mechanism or a reaction of smaller oligomers with each other could be the case. The present intramolecular hydrogen bonds fold the oligomers in the helix-like structures. If the length of an oligomer is beyond one helical turn, then the cyclization is retarded because of a steric hindrance. This fact can be observed from the analysis of the reactions between starting materials of different length (Scheme 4.19) The reaction of **75** with **76** and **77** with **78** led to the formation with >70% yield of **79** and **80**, respectively. However, the reaction between **81** and **82** yielded only small quantities of **83**. Understanding the principles of this type of kinetic cyclization the authors managed to prepare even larger macrocycles. Such large macrocycle represents an interesting building block for artificial transmembrane pores because of the tendency to stack in a face-to-face manner. Thus, Shao and Gong demonstrated the ability of macrocycle **74** to form a transmembrane channel with very large conductance, about half of that of gramicidin [71].

Cuccia and coworkers prepared a series of macrocycles based on pyrizadine and naphthyridine derivatives **84** and **85**. These heterocycles have an intrinsic ability to form hydrogen bonds with NH acidic protons located in close proximity [72]. Thus, the efficiency of [3+3] cyclization reactions was high because of the forming intramolecular hydrogen bonds (Scheme 4.20). For instance, macrocycle **86** was formed in 64% yield, while macrocycles **87**, **88**, and **89** were obtained in 75, 64, and 46% yield, respectively.

Rigid aromatic amino acids can be converted into cyclic products with the help of coupling reagents. Huc and coworkers discovered that starting building block **90** can be converted to trimer **91** and tetramer **92**, formed in 20% yield each, by heating the starting compound with triphenyl phosphite (Scheme 4.21) [21]. Analysis of structures of the corresponding linear trimer and tetramer, which could serve as a precursor for cyclization, revealed that intramolecular hydrogen bonds fold the structures into a helix structure. From the X-ray structure analysis, it was clear that one turn consists of three units. This could be the reason for the easy formation of the trimer macrocycle. The cyclic tetramer was not expected because the noncyclic tetramer forms helix, which extends to over 1.5 turns. To explain the fact of formation of macrocycles, the authors suggested that at high temperatures two intramolecular hydrogen bonds can be easily broken in the noncyclic tetramer and the macrocycle is formed. Carrying out the reaction at lower temperature proved this suggestion—the trimer macrocycle was formed in 50% yield, and the tetramer macrocycle was only obtained in trace amounts.

Synthesis and properties of macrocycles based on aminoquinoline carboxylic acid were investigated also by Jiang and coworkers [73]. The macrocyclic peptides were obtained starting from **93a** and **93b** bearing methoxy group, which can participate in hydrogen bonding interactions with the neighboring amide group. Under reflux in dry THF and in the presence of PPh_3Cl_2, starting materials react to form macrocycles **94a** and **94b** in 46 and 53% yield, respectively. Interestingly, other reagents used to activate the acid, such as thionyl chloride, oxalyl chloride, and benzotriazol-1-yl-oxytripyrrolidinophosphonium hexafluorophosphate (PyBOP), failed to generate the desired macrocycles. The products tend to aggregate in chloroform solution and in the solid phase. The molecules form stable face-to-face π–π stacks with a slight slippage.

Li and coworkers found that macrocycles, in which the conformation is rigidified by intramolecular hydrogen

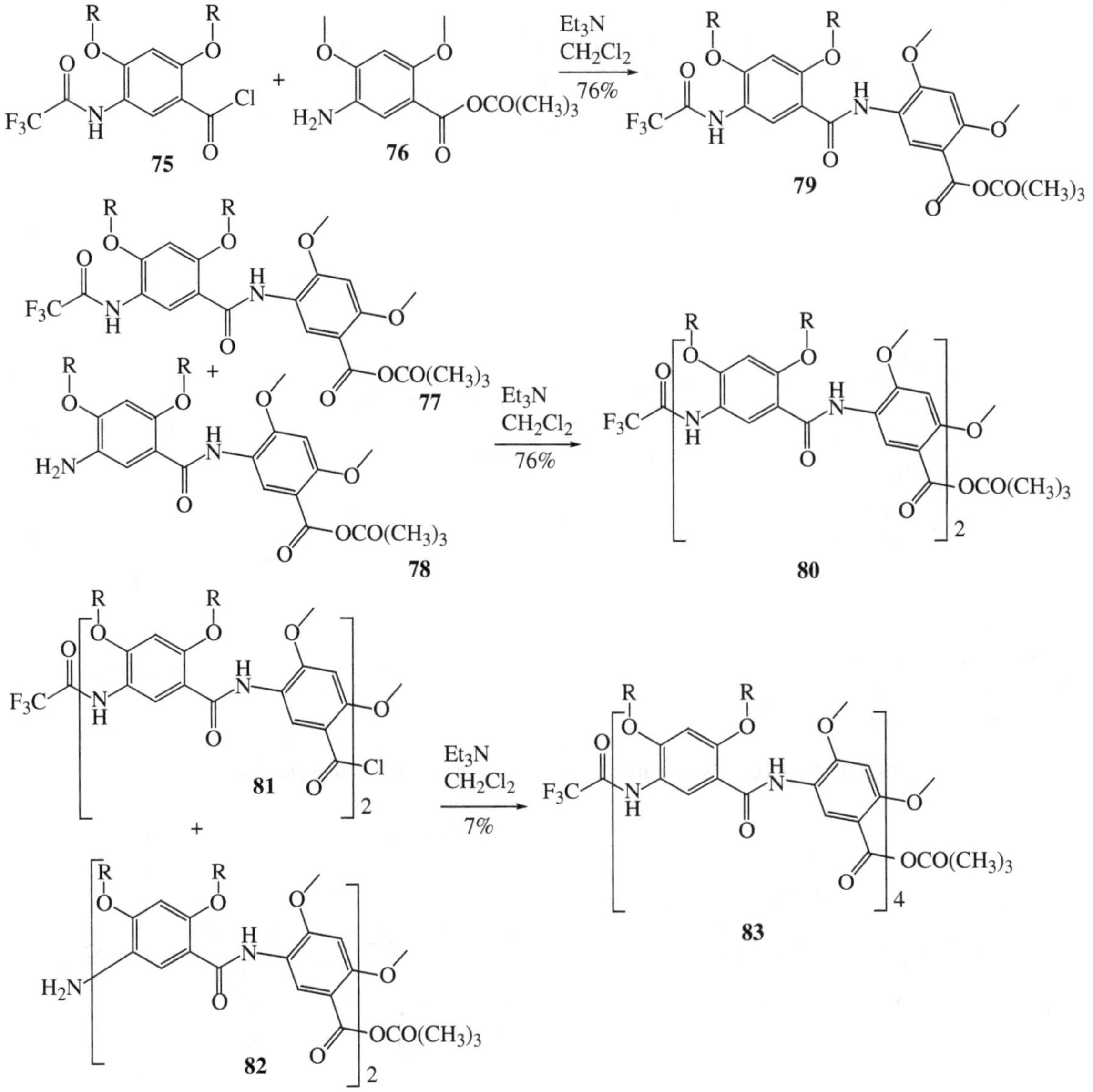

SCHEME 4.19 Reaction of monomers and dimers to form oligomers.

bonds, can interact with fullerene and coronene in chloroform solution [65]. The macrocycles were prepared by the reaction of diamine **95** with diacid **96** by simple mixing the starting materials in THF in the presence of a base. Compound **96** reacted with **95** producing a [2+2] (**97**) and a [3+3] (**98**) products of condensation. In the case of the reaction between **99** and **100**, only a [3+3] product (**101**) was isolated (Scheme 4.22).

4.5 OTHER MACROCYCLIZATIONS

Macrocycles can be also synthesized by using other non-covalent interactions, for example, van der Waals and other forces. For instance, Shanzer and coworkers found that cyclic amides can be prepared by template synthesis through the preparation of diazasilolidines with sequential acylation by diacid dichlorides [74]. The reactions give good yields due to the interaction between silicon and oxygen of the carbonyl group [75]. It was suggested that the reaction between **102** and **103** proceeds through intermediate **104**, in which non-covalent interactions between silicon and oxygen stabilize an appropriate conformation for the formation of macrocycle **105** (Scheme 4.23). By using similar approach, it was possible to prepare lactones through tin ethers of thioalcohols upon acylation by diacid dichlorides [76].

In recent years pillar[n]arenes found broad applications in supramolecular chemistry. It was found that the major product of cyclization between 1,4-disubstituted hydroquinone and formaldehyde under Friedel–Crafts conditions—pillar[5]arene—is a kinetic product. Nierengarten reported that pillar[5]arene is actually a thermodynamic product because the condensation reaction occurs under reversible bond formation [77]. This evidence was obtained after an attempt to prepare pillar[6]arene from 2,5-bis(bromomethyl)-1,4-dialkoxybenzenes and

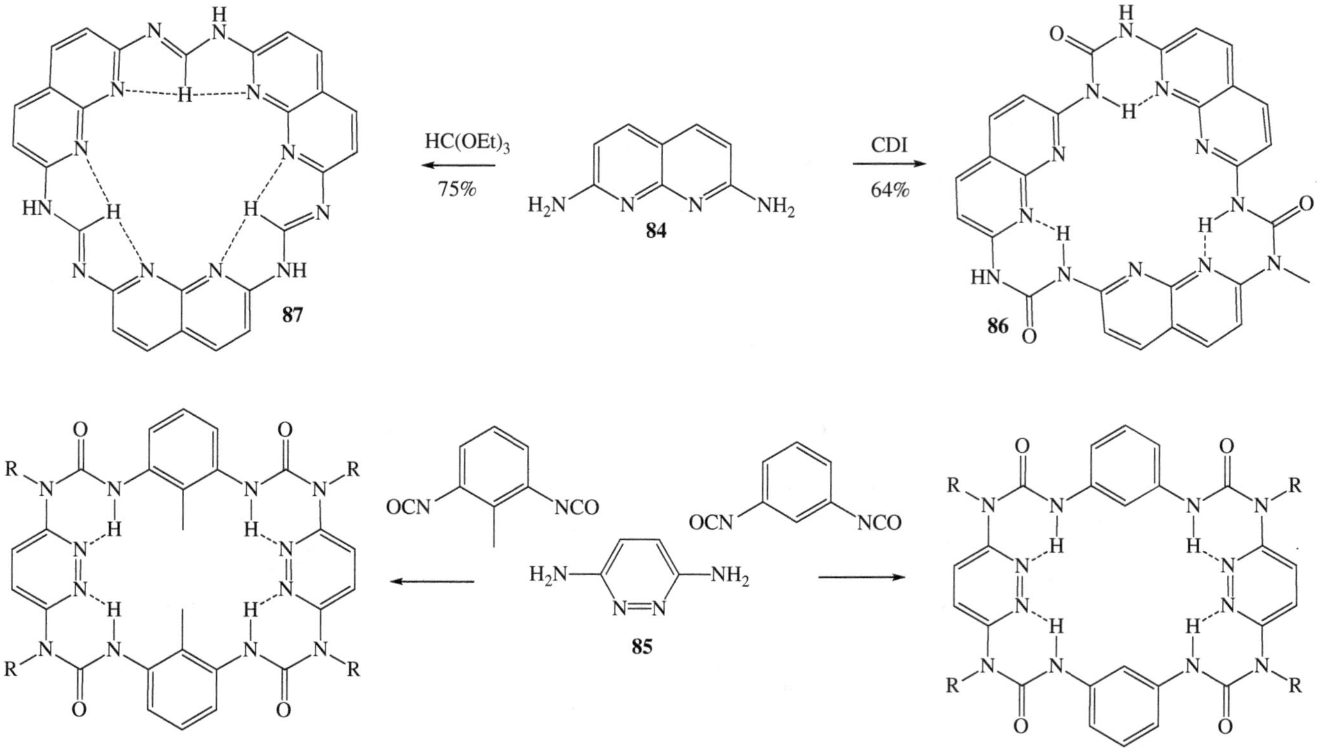

88a: R = Ph-$C_{10}H_{21}$, 64%
88b: R = *n*-$C_{12}H_{25}$, 67%
88c: R = *i*-C_4H_9, 64%

89, R = *i*-C_4H_9, 46%

SCHEME 4.20 Hydrogen bond-directed synthesis of macrocycles **86–89**.

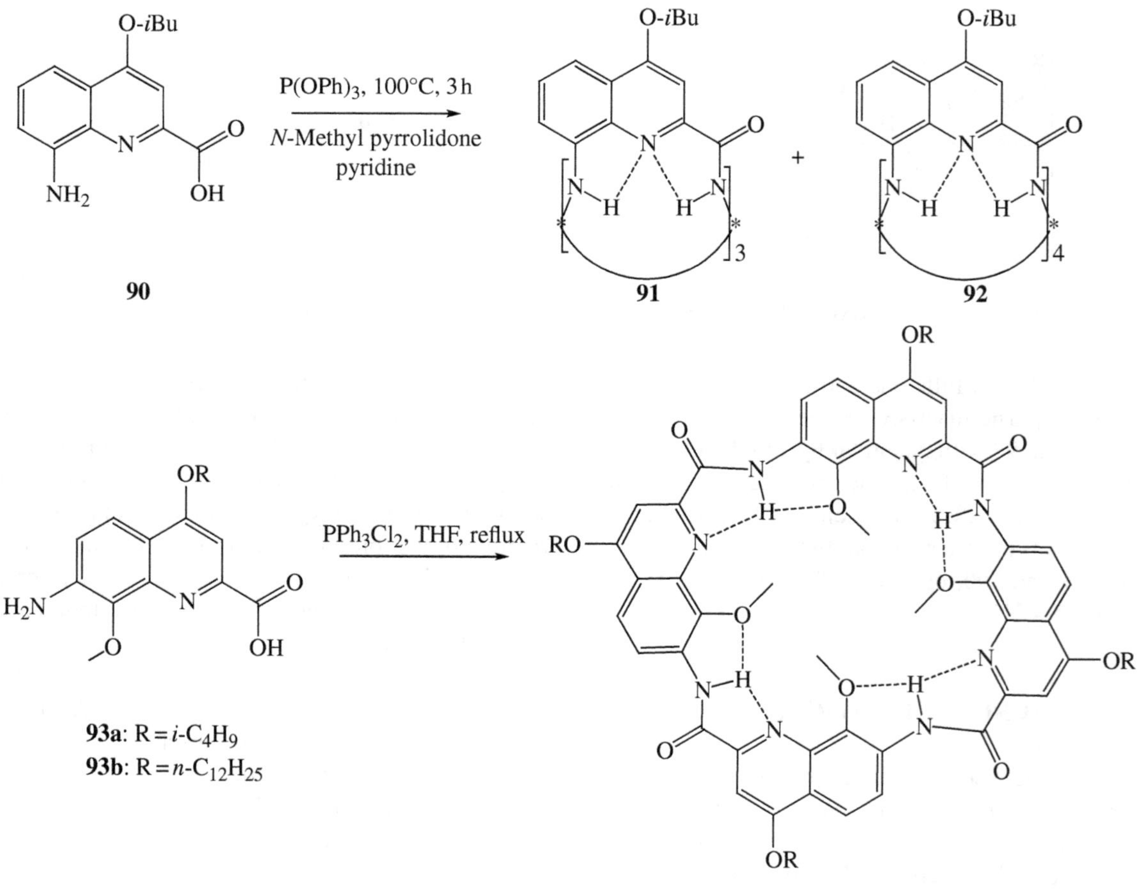

93a: R = *i*-C_4H_9
93b: R = *n*-$C_{12}H_{25}$

94a: R = *i*-C_4H_9
94b: R = *n*-$C_{12}H_{25}$

SCHEME 4.21 Synthesis of rigid cyclic peptides.

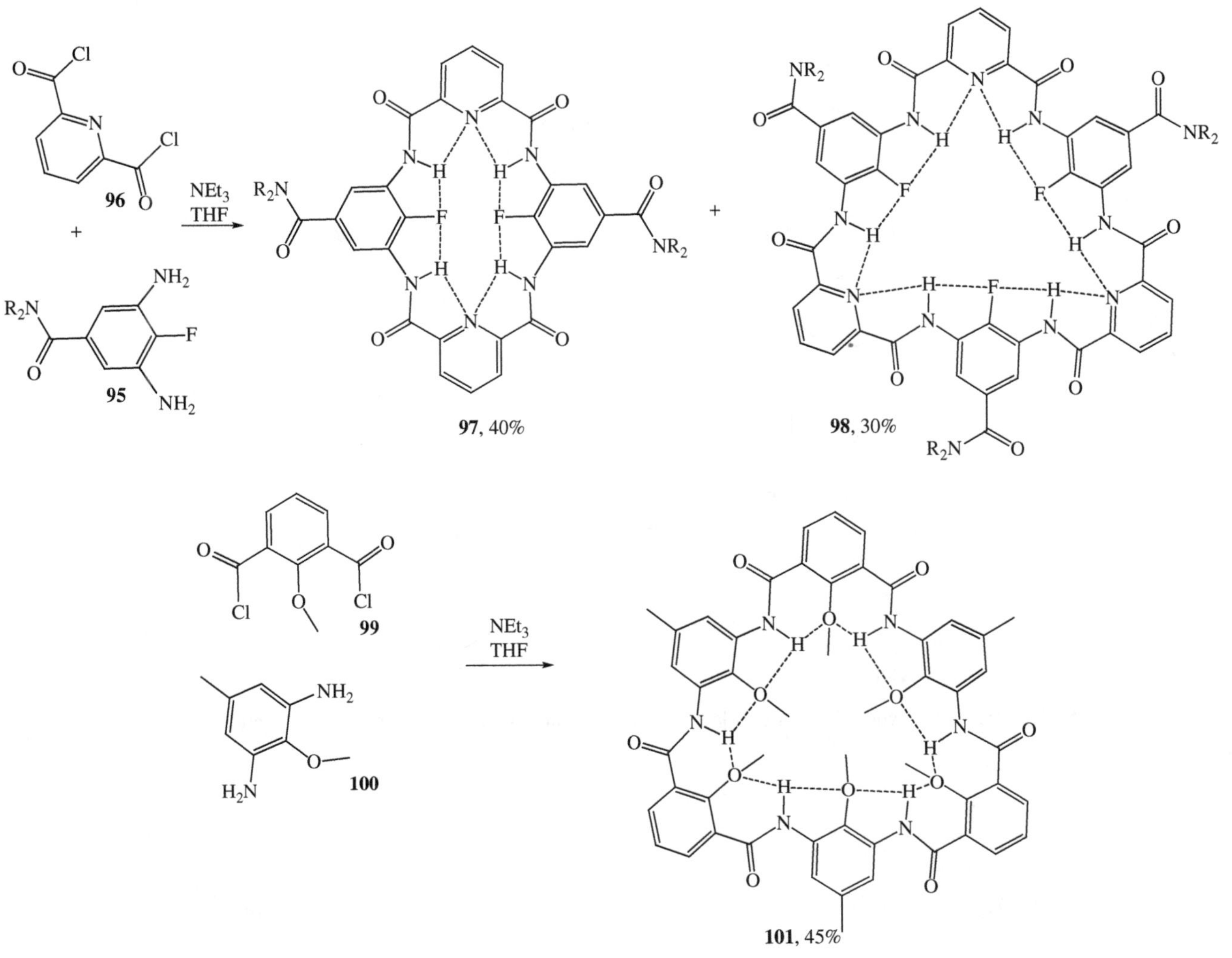

SCHEME 4.22 Synthesis of macrocycles **97**, **98**, and **101** from diacids and diamines.

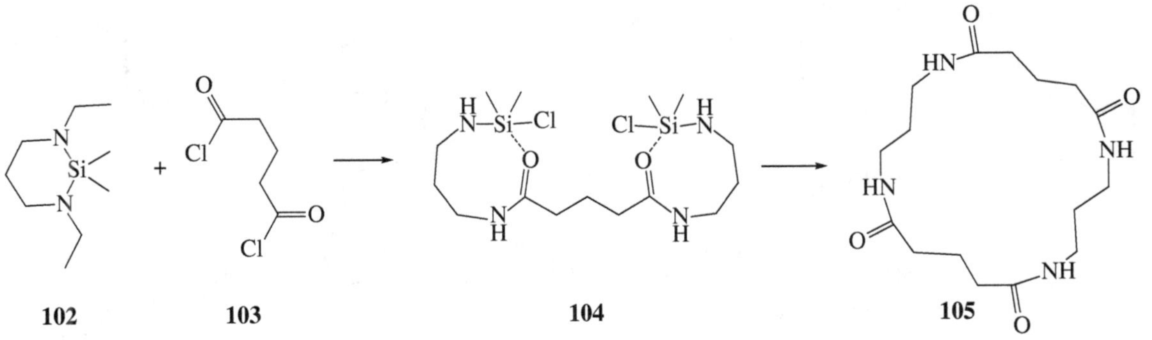

SCHEME 4.23 Proposed pathway for the formation of macrocycle **105**.

1,4-dialkoxybenzenes. Instead of the expected product, pillar[5] arene was the major product. Interestingly, Ogoshi and coworkers suggested similar mechanism in the synthesis of pillar[5]arenes [78]. The reaction of **106** with formaldehyde and a Lewis acid was proposed to be reversible and proceeds under thermodynamic control in dichloroethane solution yielding 71% of product pillar[5]arene (Scheme 4.24). It appears that the solvent—dichloroethane—works as a template in this reaction. In contrast, the reaction of 1,4-diethoxybenzene with paraformaldehyde in chloroform as a solvent gives a mixture of pillar[5–10] arenes **107** with P[5] and P[6] being isolated only in 20 and 15% yield, respectively. Understanding the importance of a solvent in the cyclization reaction, the authors found that chlorocyclohexane is an excellent template and a solvent for the synthesis of pillar[6]arene. The latter compound could be obtained in 87% yield. To prove the fact that macrocycles are formed under reversible condition, pillar[6]arene **108** was heated with a Lewis acid in dichloroethane, while pillar[5]arene **109** was heated with the a Lewis acid in chlorocyclohexane (Scheme 4.24). The conversion of one macrocycle into the other was detected in

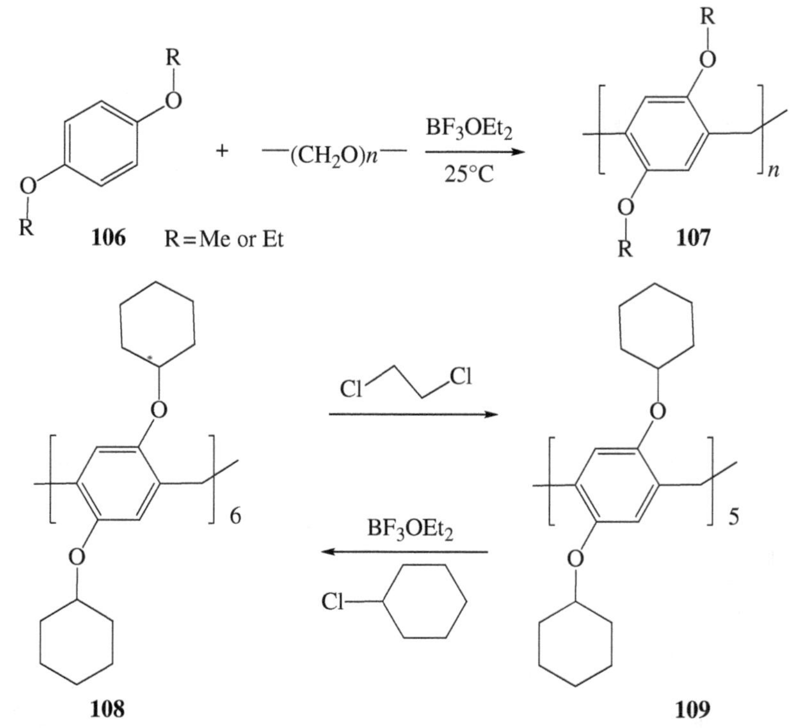

SCHEME 4.24 Synthesis of pillar[n]arenes and their interconversion under thermodynamic control.

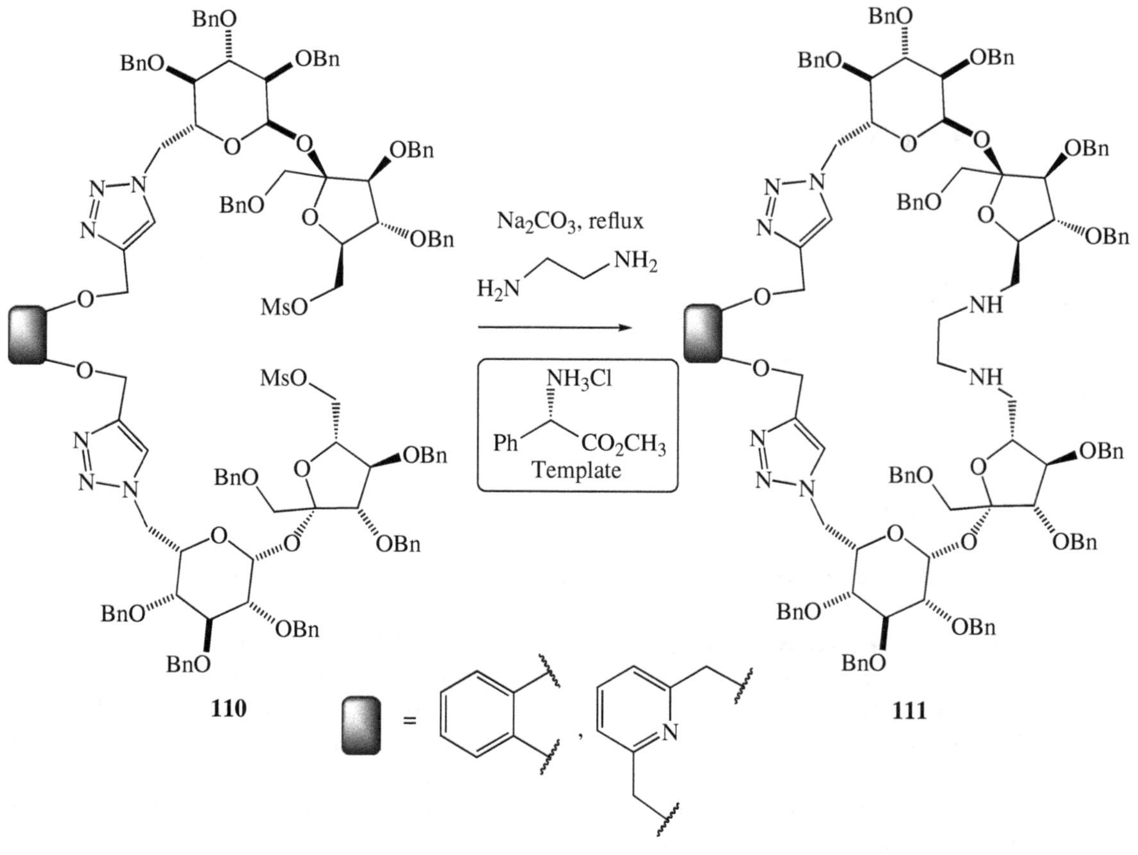

SCHEME 4.25 Amino acid-templated synthesis of macrocycle **111**.

both reactions, indicating that under these conditions the reaction is indeed reversible and that the macrocycles are thermodynamically favorable products.

Surprisingly high yields for the formation of [2+2] macrocycles were reported by Wen and coworkers in the synthesis of oxocalixarenes [79]. The connection of building blocks was performed through ether bonds from dihalide- and dialcohol-containing compounds. It was suggested that under the studied conditions (Cs_2CO_3, DMSO, r.t.), the formation of ether bonds was under thermodynamic control. However, detailed studies of the mechanism of the reaction were not carried out.

Multiple hydrogen bonds between a template and a precursor for the formation of a macrocycle were suggested by Jarocz and coworkers in the synthesis of sucrose-derived macrocycles. The reactions were carried out with the help of an amino acid as a template in the last macrocyclization step [80]. The compound **110** was converted in **111** without a template only with a 5% overall yield. Neither metal cations nor simple aromatic compounds were effective as templates. Only application of L-enantiomer of phenylglycine methyl ester hydrochloride gave rise to product **111** with 25% yield. It was suggested that this amino acid preorganizes the conformation of the precursor for the formation of the macrocycle (Scheme 4.25).

4.6 FINAL COMMENTS

In summary, non-covalent interactions play an important role in the synthesis of macrocyclic compounds. As it has been demonstrated in this chapter, non-covalent interactions between precursors and a template are not only decisive for macrocyclizations under thermodynamic control but also under kinetic conditions, that is, the reactions that proceed through irreversible bonds, like amides, ethers, esters, and C—C bonds. Another driving force for efficient macrocyclization is intramolecular hydrogen bonding. Appropriately designed precursors can be cyclized in a macrocycle with a high yield thanks to intramolecular hydrogen bonds that are formed during oligomerization reactions. These non-covalent interactions rigidify the resulting structure and drive the reaction to a kinetically favorable macrocycle. Several examples have been given that show the importance of understanding of mechanisms of macrocyclization reactions. This understanding has allowed one to optimize the reaction condition and obtain macrocycles in almost quantitative yields. Realizing a great potential of using non-covalent interactions, we mimic the tricks of self-assembly, which nature uses to construct beautiful structures. Thus, programming the molecules and templates in view of their non-covalent interactions in course of reaction might be a promising approach to design new structure with desired properties.

ACKNOWLEDGMENTS

This work was supported by Fonds der Chemischen Industrie, Deutsche Bundesstiftung Umwelt, Deutsche Forschungsgemeinschaft (KA 3444/7-1), and Technische Universität Chemnitz.

REFERENCES

1. F. Diederich, P. J. Stang, R. R. Tykwinski, Modern Supramolecular Chemistry: Strategies for Macrocycle synthesis, Wiley-VCH, Weinheim/Chichester, 2008.
2. S. Otto, *Acc. Chem. Res.* 2012, *45*, 2200.
3. S. J. Rowan, S. J. Cantrill, G. R. L. Cousins, J. K. M. Sanders, J. F. Stoddart, *Angew. Chem. Int. Ed.* 2002, *41*, 898.
4. J.-M. Lehn, *Chem. Eur. J.* 1999, *5*, 2455.
5. M. S. Vickers, P. D. Beer, *Chem. Soc. Rev.* 2007, *36*, 211.
6. R. Vilar, *Angew. Chem. Int. Ed.* 2003, *42*, 1460.
7. R. Vilar, *Eur. J. Inorg. Chem.* 2008, 357.
8. C. J. Pedersen, *J. Am. Chem. Soc.* 1967, *89*, 7017.
9. J. Dale, K. Daasvatn, *J. Chem. Soc. Chem. Commun.* 1976, 295.
10. D. H. Busch, *Top. Curr. Chem.* 2005, *249*, 1.
11. V. B. Shur, I. A. Tikhonova, *Russ. Chem. Bull.* 2003, *52*, 2539.
12. X. Yang, C. B. Knobler, M. F. Hawthorne, *Angew. Chem. Int. Ed.* 1991, *30*, 1507.
13. J. Grünewald, M. A. Marahiel, *Microbiol. Mol. Biol. Rev.* 2006, *70*, 121.
14. C. J. White, A. K. Yudin, *Nat. Chem.* 2011, *3*, 509.
15. K. A. Jolliffe, *Supramol. Chem.* 2005, *17*, 81.
16. S. Kubik, Synthetic Peptide-Based Receptors, in Supramolecular Chemistry: From Molecules to Nanomaterials (Eds.: J. W. Steed, P. A. Gale), John Wiley & Sons, Inc., Hoboken, 2012.
17. J. Blankenstein, J. Zhu, *Eur. J. Org. Chem.* 2005, 1949.
18. M. C. O'Sullivan, J. K. Sprafke, D. V. Kondratuk, C. Rinfray, T. D. W. Claridge, A. Saywell, M. O. Blunt, J. N. O'Shea, P. H. Beton, M. Malfois, H. L. Anderson, *Nature* 2011, *469*, 72.
19. D. V. Kondratuk, J. K. Sprafke, M. C. O'Sullivan, L. M. A. Perdigao, A. Saywell, M. Malfois, J. N. O'Shea, P. H. Beton, A. L. Thompson, H. L. Anderson, *Chem. Eur. J.* 2014, *20*, 12826.
20. C. A. Schalley, W. Reckien, S. Peyerimhoff, B. Baytekin, F. Vogtle, *Chem. Eur. J.* 2004, *10*, 4777.
21. H. Jiang, J.-M. Léger, P. Guionneau, I. Huc, *Org. Lett.* 2004, *6*, 2985.
22. H. Juwarker, K. S. Jeong, *Chem. Soc. Rev.* 2010, *39*, 3664.
23. M. D. Lankshear, P. D. Beer, *Coord. Chem. Rev.* 2006, *250*, 3142.
24. R. Vilar, *Struct. Bond.* 2008, *129*, 175.
25. M. Wenzel, J. R. Hiscock, P. A. Gale, *Chem. Soc. Rev.* 2012, *41*, 480.
26. S. L. Ewen, J. H. G. Steinke, *Recognit. Anions* 2008, *129*, 207.
27. Z. Zheng, X. Yang, C. B. Knobler, M. F. Hawthorne, *J. Am. Chem. Soc.* 1993, *115*, 5320.
28. X. Yang, C. B. Knobler, Z. Zheng, M. F. Hawthorne, *J. Am. Chem. Soc.* 1994, *116*, 7142.
29. J. L. Sessler, T. D. Mody, V. Lynch, *Inorg. Chem.* 1992, *31*, 529.
30. J. L. Sessler, T. D. Mody, V. M. Lynch, *J. Am. Chem. Soc.* 1993, *115*, 3346.
31. S. Meyer, B. Andrioletti, J. L. Sessler, V. Lynch, *J. Org. Chem.* 1998, *63*, 6752.
32. J. L. Sessler, D. Seidel, *Angew. Chem. Int. Ed.* 2003, *42*, 5134.
33. D. Seidel, V. Lynch, J. L. Sessler, *Angew. Chem. Int. Ed.* 2002, *41*, 1422.

34. J. L. Sessler, J. A. Shriver, K. Jursíková, V. M. Lynch, M. Marquez, *J. Am. Chem. Soc.* 2000, *122*, 12061.
35. B. Turner, A. Shterenberg, M. Kapon, K. Suwinska, Y. Eichen, *Chem. Commun.* 2002, 404.
36. D. Meshcheryakov, V. Böhmer, M. Bolte, V. Hubscher-Bruder, F. Arnaud-Neu, H. Herschbach, A. Dorsselaer, I. Thondorf, W. Mögelin, *Angew. Chem. Int. Ed.* 2006, *45*, 1648.
37. D. Meshcheryakov, V. Böhmer, M. Bolte, V. Hubscher-Bruder, F. Arnaud-Neu, *Chem. Eur. J.* 2009, *15*, 4811.
38. E. Alcalde, S. Ramos, L. Pérez-García, *Org. Lett.* 1999, *1*, 1035.
39. S. Ramos, E. Alcalde, G. Doddi, P. Mencarelli, L. Pérez-García, *J. Org. Chem.* 2002, *67*, 8463.
40. L. Isaacs, *Isr. J. Chem.* 2011, *51*, 578.
41. V. Martí-Centelles, M. I. Burguete, S. V. Luis, *Chem. Eur. J.* 2012, *18*, 2409.
42. V. Marti-Centelles, M. I. Burguete, S. V. Luis, *Sci. World J.*, 2012, 10.1100/2012/748251.
43. V. Martí-Centelles, M. I. Burguete, C. Cativiela, S. V. Luis, *J. Org. Chem.* 2014, *79*, 559.
44. I. Martí, J. Rubio, M. Bolte, M. I. Burguete, C. Vicent, R. Quesada, I. Alfonso, S. V. Luis, *Chem. Eur. J.* 2012, *18*, 16728.
45. E. A. Katayev, N. V. Boev, E. Myshkovskaya, V. N. Khrustalev, Y. A. Ustynyuk, *Chem. Eur. J.* 2008, *14*, 9065.
46. E. A. Katayev, E. N. Myshkovskaya, N. V. Boev, V. N. Khrustalev, *Supramol. Chem.* 2008, *20*, 619.
47. E. A. Kataev, G. V. Kolesnikov, R. Arnold, H. V. Lavrov, V. N. Khrustalev, *Chem. Eur. J.* 2013, *19*, 3710.
48. S. Lee, C.-H. Chen, A. H. Flood, *Nat. Chem.* 2013, *5*, 704.
49. J. Lagona, P. Mukhopadhyay, S. Chakrabarti, L. Isaacs, *Angew. Chem. Int. Ed.* 2005, *44*, 4844.
50. J. Rivollier, P. Thuéry, M.-P. Heck, *Org. Lett.* 2013, *15*, 480.
51. Y. Miyahara, K. Goto, M. Oka, T. Inazu, *Angew. Chem. Int. Ed.* 2004, *43*, 5019.
52. H. J. Buschmann, E. Cleve, E. Schollmeyer, *Inorg. Chem. Commun.* 2005, *8*, 125.
53. M. Lisbjerg, B. M. Jessen, B. Rasmussen, B. E. Nielsen, A. O. Madsen, M. Pittelkow, *Chem. Sci.* 2014, *5*, 2647.
54. M. Singh, E. Solel, E. Keinan, O. Reany, *Chem. Eur. J.* 2015, *21*, 536.
55. R. Aav, E. Shmatova, I. Reile, M. Borissova, F. Topić, K. Rissanen, *Org. Lett.* 2013, *15*, 3786.
56. M. A. Yawer, V. Havel, V. Sindelar, *Angew. Chem. Int. Ed.* 2015, *54*, 276.
57. V. Havel, M. A. Yawer, V. Sindelar, *Chem. Commun.* 2015, *51*, 4666.
58. L. Mandolini, B. Masci, *J. Am. Chem. Soc.* 1977, *99*, 7709.
59. N. Uchida, R. Zhi, J. Kuwabara, T. Kanbara, *Tetrahedron Lett.* 2014, *55*, 3070.
60. K. A. Fairweather, N. Sayyadi, I. J. Luck, J. K. Clegg, K. A. Jolliffe, *Org. Lett.* 2010, *12*, 3136.
61. A. Bertram, A. J. Blake, F. G.-L. de Turiso, J. S. Hannam, K. A. Jolliffe, G. Pattenden, M. Skae, *Tetrahedron* 2003, *59*, 6979.
62. L. Dudin, G. Pattenden, M. S. Viljoen, C. Wilson, *Tetrahedron* 2005, *61*, 1257.
63. A. J. Blake, J. S. Hannam, K. A. Jolliffe, G. Pattenden, *Synlett* 2000, 1515.
64. S. Porwanski, A. Marsura, *Eur. J. Org. Chem.* 2009, 2047.
65. R. Carrillo, V. S. Martín, M. López, T. Martín, *Tetrahedron* 2005, *61*, 8177.
66. A. Furstner, M. Albert, J. Mlynarski, M. Matheu, E. DeClercq, *J. Am. Chem. Soc.* 2003, *125*, 13132.
67. H. Fu, Y. Liu, H. Zeng, *Chem. Commun.* 2013, *49*, 4127.
68. B. Qin, C. Sun, Y. Liu, J. Shen, R. J. Ye, J. Zhu, X. F. Duan, H. Q. Zeng, *Org. Lett.* 2011, *13*, 2270.
69. A. Gube, H. Komber, K. Sahre, P. Friedel, B. Voit, F. Böhme, *J. Org. Chem.* 2012, *77*, 9620.
70. W. Feng, K. Yamato, L. Yang, J. S. Ferguson, L. Zhong, S. Zou, L. Yuan, X. C. Zeng, B. Gong, *J. Am. Chem. Soc.* 2009, *131*, 2629.
71. A. J. Helsel, A. L. Brown, K. Yamato, W. Feng, L. Yuan, A. J. Clements, S. V. Harding, G. Szabo, Z. Shao, B. Gong, *J. Am. Chem. Soc.* 2008, *130*, 15784.
72. L. Xing, U. Ziener, T. C. Sutherland, L. A. Cuccia, *Chem. Commun.* 2005, *46*, 5751.
73. F. Li, Q. Gan, L. Xue, Z.-M. Wang, H. Jiang, *Tetrahedron Lett.* 2009, *50*, 2367.
74. E. Schwartz, H. E. Gottlieb, F. Frolow, A. Shanzer, *J. Org. Chem.* 1985, *50*, 5469.
75. Y. Tor, J. Libman, F. Frolow, H. E. Gottlieb, R. Lazar, A. Shanzer, *J. Org. Chem.* 1985, *50*, 5476.
76. A. Shanzer, J. Libman, F. Frolow, *Acc. Chem. Res.* 1983, *16*, 60.
77. M. Holler, N. Allenbach, J. Sonet, J.-F. Nierengarten, *Chem. Commun.* 2012, *48*, 2576.
78. T. Ogoshi, N. Ueshima, T. Akutsu, D. Yamafuji, T. Furuta, F. Sakakibara, T. A. Yamagishi, *Chem. Commun.* 2014, *50*, 5774.
79. W.-J. Hu, X.-L. Zhao, M.-L. Ma, F. Guo, X.-Q. Mi, B. Jiang, K. Wen, *Eur. J. Org. Chem.* 2012, *2012*, 1448.
80. B. Lewandowski, S. Jarosz, *Org. Lett.* 2010, *12*, 2532.

PART II

INORGANIC, COORDINATION AND ORGANOMETALLIC SYNTHESES

5

NON-COVALENT INTERACTIONS OF WATER WITH METAL COMPLEXES IN SOLUTION

ANTONIO ROMEROSA AND FRANCO SCALAMBRA
Área de Química Inorgánica-CIESOL, University of Almeria, Almeria, Spain

5.1 INTRODUCTION

It is said that atoms are connected by a non-covalent interaction when they stick together through an assortment of weak bonding interactions that does not involve sharing electrons. Non-covalent interactions are generally classified into four categories: electrostatic interactions, ionic bonding, π-effects, and hydrophobic effects [1]. The energy of non-covalent interactions is much lower than that of typical covalent bonds, but due to their multiplicity and ubiquity their overall influence on the course of a reaction can be decisive. Water is able of interacting with molecules by either of the precedent non-covalent interactions and frequently more than one of them at the time. The hydrophobic effect is defined as the propensity of nonpolar substances to aggregate in aqueous solution with exclusion of water molecules; it is not well understood yet, being mainly produced by the hydrogen bonds among water molecules, and therefore it will not be discussed in this chapter. The π-effect, although being very important to understand the interaction among water and aromatic compounds, will not be also discussed. Intermolecular forces are a subset of non-covalent interactions, simply because non-covalent interactions can be both intermolecular, that is, interaction between water molecules, and intramolecular, that is, protein folding. Intermolecular forces, on the other hand, are defined as interactions between molecules and not within molecules.

The most important and known non-covalent interactions of water with itself and other molecules are two subsets of electrostatic interactions: hydrogen bonding and van der Waals forces. This chapter is focused mainly on describing the interaction by these two forces among metal complexes and water molecules in solution.

Just after the end of the World War II, in 1945, Frank and Evans [2] subjected that the hydration in water of metal ions is produced by the strong electrostatic interactions due to the charge borne that induces a series of concentric shells. Normally only two hydration shells are well defined, and in those beyond them the bulk water properties are recovered. Some highly charged metal cations could induce the formation of a third shell. A further step of this model recognized that the first-shell water molecules could be envisaged as ligands intimately joined or coordinated to highly charged metal cations, since these water molecules are different from the bulk water molecules [3, 4]. The oxygen–metal bond of the first shell is covalent, but the interaction among the water molecules of the second shell with the first-shell ones is mainly non-covalent. A correct description of the solvation shell is of main importance for understanding the molecular dynamics (MD) of metal complexes in water.

The study of the evolution of molecular geometry and electronic structure during ionic solvation by water is fundamental to understanding the behavior of ions in chemical, biological, and industrial systems and processes [5–7]. For this reason, interactions of water molecules with ions, molecules, and interfaces have been intensively studied [8–12]. It is generally thought that the presence of ions can either enhance or weaken the local hydrogen bond network, depending on the ionic charge and size, but quantification of these effects and their principle causes is difficult. It was

Non-Covalent Interactions in the Synthesis and Design of New Compounds, First Edition.
Edited by Abel M. Maharramov, Kamran T. Mahmudov, Maximilian N. Kopylovich, and Armando J. L. Pombeiro.

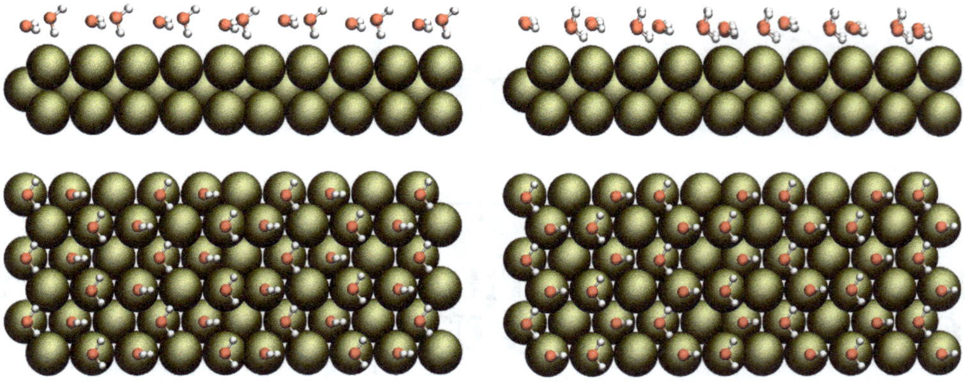

FIGURE 5.1 Ordered ice-like structures for 2/3 ML of water on Pt(111). The left figure is the traditional bilayer structure (denoted as H-up bilayer) with half of the water parallel to the surface and bonded through the O lone pair directly to a Pt atom. The right figure shows the H-down bilayer structure in which the perpendicular waters (HOH) are in H-down configuration forming an agostic Pt—HOH bond.

also observed that both the X^-–O and Y^+–O distances increase with increasing ionic radius in the alkali metal halide series ((X^-) halide anion, (Y^+) cation) and that water molecules preferentially orient with a single hydrogen directed toward an anion but with the hydrogens directed away from a cation [13]. Since water is present on nearly every real surface and plays an important role in electrochemical reactions on metals, the structures of the water layers on metals have been studied thoroughly. The most widely accepted model for the saturated first water layer on hexagonal metal surfaces has been the H-up bilayer structure (Fig. 5.1a) by Doering and Madey [14]. This has half of the molecules (denoted as H_2O) binded directly to the surface using one of the lone pair orbital of the oxygen. The other half of the water molecules is shifted away from the surface and stabilized by hydrogen bonds to the surface-adsorbed water molecules.

Therefore, liquid water has a very complex hydrogen bond network, and its nature is still in the process of being clarified using spectroscopic techniques and computational methods [15]. Gas-phase studies of hydrated ions can reveal detailed information about both direct ion–solvent interactions and solvent shell structures [16]. Important data about hydrogen bonds have been obtained from crystal structures [17–20]. Several review papers about the interactions of water with different systems have been published [21–24]. In this chapter will be presented the last findings in how water interacts with metal complexes in solution, the actual techniques and methods used to understand how the interaction occurs, and its consequences. Additionally, the concepts of hydrogen and van der Waals bonds have been changed as a deep knowledge of them have been obtained along the years, in some aspects far from the classical definitions. To have a more precise understanding of how water molecules interact with metal complexes, an overview of the new definitions and properties of hydrogen bonding and van der Waals interactions is mandatory.

5.2 HYDROGEN BONDING AND VAN DER WAALS INTERACTIONS

5.2.1 Hydrogen Bonding

Although hydrogen bonding and halogen bonding are both forms of dipole–dipole interactions, these are typically not classified as van der Waals forces by convention. The concept of hydrogen bond was introduced almost 100 years ago [25], but it is still and even more a topic of vital scientific research. The earlier classical view on hydrogen bonding considers the directional interaction between water molecules as the prototype of all hydrogen bonds. Despite significant charge transfer in the hydrogen bond, the total interaction is dominantly electrostatic, which leads to pronounced flexibility in the bond length and angle. The dissociation energy is around 3–5 kcal mol^{-1}. The large difference in electronegativity between the H and O atoms makes the O—H bonds of a water molecule inherently polar, with partial atomic charges of around +0.4 on each H atom and −0.8 on the O atom. Neighboring water molecules orient in such a way that local dipoles O—H point at negative partial charges on the lone pairs of the filled p orbitals of the O atoms. For the hydrogen-bonded pairs, one finds a maximum in the distribution of the distances at about 2.9 Å and intermolecular contacts disregarding hydrogen-containing atoms give a van der Waals peak at around 3.3 Å.

After more than a century, it is known that the so-called the *classical hydrogen bond*, though abundant and important one, is just one among many. There are hydrogen bonds that are so strong that they resemble covalent bonds in most of their properties and others so weak that they can hardly be distinguished from van der Waals interactions. The electrostatic dominance of the hydrogen bond is true only for some of the occurring configurations. The hydrogen bond distance is not ever shorter than the sum of the van der Waals radii, and for constituting a hydrogen bond, one of the atom does not need to be very electronegative; it is only necessary that the bond to be at least slightly polar. The energy range for dissociation of hydrogen bonds covers from about 0.2 to 40 kcal mol^{-1}, and the possible functions of a particular type of hydrogen bond depend on its location on this scale.

A general and flexible definition for hydrogen bonding was proposed by Pimentel and McClellan; they suggest that "...a hydrogen bond exists if there is evidence of a bond and there is evidence that this bond sterically involves a hydrogen atom already bonded to another atom" [26]. This definition in the strict sense includes pure van der Waals contacts with H atoms, and it also includes three-center two-electron interactions (agostic interactions) where electrons of an X—H bond are donated sideways to an electron-deficient center. It was proposed [19] to modify the second condition by: "An X—H—A interaction is called a hydrogen bond, if it constitutes a local bond and X—H acts as proton donor."

Chemical variation of donor and/or acceptor atoms produced by molecular changes or by the environment can progressively modify a hydrogen bond to another interaction type. The hydrogen bond has broad transition regions—gray areas—with the covalent bond, van der Waals, ionic, and cation–π interactions. The transition to pure van der Waals interaction is very common. The reduction of the polarity of the bonds in the hydrogen bonding decreases the electrostatic interaction of components whereas the van der Waals part is much less affected, as a result the angular characteristics gradually change from directional to isotropic [27]. The main structural feature distinguishing the hydrogen bond from the van der Waals interaction is its preference for linearity. In solid state, hydrogen bond distances and angles vary over wide ranges even within the same donor–acceptor combination.

Furthermore, there is also a continuous transition to covalent bonding [28]. In the so-called symmetric hydrogen bonds where an H atom is equally shared between two chemically identical atoms X, no distinction can be made between a donor and an acceptor and therefore between a covalent and non-covalent bond. It is important to stress that hydrogen bonding could have an important effect on the covalent geometry of the molecules as described as early as in the 1950s [29]. Moreover, a gradual transition from hydrogen bonding to ionic interactions exists. If the charges on the atoms are large, the ionic behavior may become the most important, and for fully charged hydrogen bond partners, the bond is mainly formed by a Coulombic interaction between the charge centers but the total interaction still remains directional. Finally, there is a transition region between the hydrogen bond and the cation–π interaction.

Therefore, hydrogen bonds are able to have a continuum of strengths. For practical reasons a classification for hydrogen bonding is used: weak, strong, and in between. Jeffrey [30] defined the hydrogen bonds as moderate or normal when they resemble those between water molecules or carbohydrates (associated energies in the range of 4–15 kcal·mol^{-1}). Hydrogen bonds with energies above and below this range are called, respectively, strong and weak, with not clear cutoff limits between them. Some examples for these three hydrogen bonds categories are displayed in Table 5.1.

The called Ubbelohde effect, which happens when D exchanges the H, leads to a slightly increasing of the hydrogen bond lengths upon deuteration [31]. It is thought that the O—D bond is more stable than the O—H bond due to its lower zero-point vibrational energy. Therefore D is more difficult to be abstracted from O than H, and the hydrogen bonds are weaker. It is much smaller for long hydrogen bonds and about zero for very short ones (possibly, even negative values are allowed). The details of the effect remain yet unexplained. Isotope exchange occasionally leads to more severe structural changes.

If a polar molecule forms a hydrogen bond, it becomes more polar. The same occurs whether it accepts a hydrogen bond, and thus, in a chain with two hydrogen bonds, both of them become stronger (around 20% relative to isolated interactions) [32]. The effect is often called "bond cooperativity," but the terms "polarization-enhanced hydrogen bonding" [33] and "polarization-assisted hydrogen bonding" (as opposed to "resonance assisted hydrogen bonding") [34] have also been proposed. If double donors (such as H_2O) and/or

TABLE 5.1 Hydrogen Bonds According to Jeffrey

Interaction Type	Strong Strongly Covalent	Moderate Mostly Electrostatic	Weak Electrostatic/Dispersive
Bond lengths H···A (Å)	1.2–1.5	1.5–2.2	>2.2
Lengthening of H—X (Å)	0.08–0.25	0.02–0.08	<0.02
H—X/H···A comparisons	H—X ≈ H···A	H—X < H···A	H—X ≪ H···A
Distances H—A (Å)	2.2–2.5	2.5–3.2	>3.2
Directionality	Strong	Moderate	Weak
Bond angles (°)	170–180	>130	>90
Bond energies (kcal·mol^{-1})	15–40	4–15	<4
Relative IR shift $\Delta\bar{\nu}_{HX}$ (cm^{-1})	25%	10–25%	<10%
^1H downfield shift	14–22	<14	

The numerical data are guiding values only.

double acceptors are involved, they can interconnect chains and rings to form complex networks. A polar molecule or group may also be polarized by charge flow through π bonds, and the effect is called π-bond cooperativity. Nevertheless, hydrogen bonds may also reduce the strengths of each other. This occurs where two roughly parallel donor dipoles repel each other. This effect is probably responsible for the preferences of homodromic over antidromic cycles of hydrogen bonds [35].

Actually, predicting a hydrogen bond between a potential donor X—H and a potential acceptor A is sometimes very difficult and defining a probability of formation is only possible. The probability of formation of a hydrogen bond is defined as the fraction of X—H···A hydrogen bonds among the number of such hydrogen bonds that could be formed in principle. Therefore, this is averaged over all chemical and structural situations.

5.2.2 van der Waals Interactions

As indicated in the introduction, the van der Waals forces are a subset of electrostatic interactions involving permanent or induced dipoles (or multipoles), being classified in permanent dipole–dipole interactions (Keesom force), dipole–induced dipole interactions (Debye force), and induced dipole–induced dipole interactions (London dispersion forces). It is commonly indicated that there is a van der Waals bond between molecules or atoms when the distance among them is in the range of the van der Waals radii. This radius is also extensively used for studying properties such as crystal packing and supramolecular interaction analysis [36, 37].

Different size measures should be adopted for ions, atoms within molecules, and atoms connected through van der Waals forces. We accept that an atom preserves its size even whether its number of valence electrons may be quite different as witnessed by the widely variable oxidation states in transition metals. These sizes, ionic, covalent, and van der Waals radii [38], have been traditionally deduced mainly from experimental interatomic distances. In spite of the extensive use made of such empirical parameters, only in recent years consistent covalent radii were calculated [39]. Rowland and Taylor deduced van der Waals radii for a few main group elements using a statistical analysis of intermolecular contacts in X-ray crystal structures, which were found to be consistent with those proposed earlier by Bondi [40]. In a remarkable study, S. Álvarez [41] applied the methodology proposed earlier by Rowland and Taylor to establish representative van der Waals radii of as many elements as possible by using the structural data on intermolecular interactions present in the Cambridge Structural Database (CSD) [42]. The atoms were considered as spherical, and in general such fine details as the oxidation state and coordination number of the element under consideration was not considered [43–45].

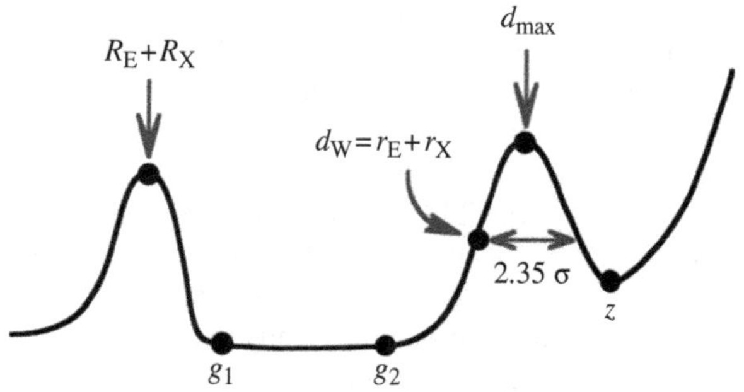

FIGURE 5.2 Distance distribution from the central atom.

The adopted approach consisted in calculating the van der Waals radius for an element $E(r_E)$ from the experimental distances of its intermolecular contacts with a reference element X, whose precise van der Waals radius, r_X, is known. In a E–X distance distribution map, several distinct regions are observed. The short distances correspond to covalent bonds (RE + RX), while a range of distances at which practically no structures are found (Fig. 5.2) form a van der Waals gap (g1 – g2) between the intra- and intermolecular distances. Further, a peak that corresponds to a maximum in the distribution function, which can be assigned to the van der Waals interactions, is referred to the van der Waals peak (d_{max}). Finally, a region of randomly distributed noninteracting of atoms follows. The noninteracting and van der Waals regions often overlap. Special caution must be taken for excessive proximity between atoms, particularly the presence of hydrogen bonding. In general for all atoms studied, each van der Waals radius is on average 0.9Å longer than the corresponding covalent one but with a differences ranging from 0.6 to 1.2Å.

The obtained van der Waals radii for metals interacting with O provide interesting conclusions. Alkaline elements show distributions with two peaks for the M···O contact distances, the shortest distance peaks being attributed to chemical bond. The centers of those peaks shift from Li to Rb (from 2.5 to 5.4Å). In contrast with the alkaline elements, the lighter alkaline earths present sharp van der Waals gaps that turn into a pseudogap for Ba. The transition metal atom van der Waals radii reveals a periodic dependence roughly parallel to that shown by the covalent radii. In contrast with the known smooth variation in the same period of the covalent radii from transition metals to p-block elements, a sharp drop of the van der Waals radii is observed. Additionally, the van der Waals radii of the 3d and 4d transition elements experience lesser contraction along the period than the covalent radii. It is important to point out that there are no significant differences between the van der Waals regions for high- and low-spin complexes. Additionally it is interesting that the van der Waals sizes vary very little down at group. This fact, together with the small variation of the radii of the 4f elements, indicates the absence of a lanthanide contraction in the van der Waals radii.

It is clear that the best way to address the analysis of the E···X interaction in a particular compound would be to obtain the specific E–X van der Waals map. Nevertheless, such a map confection requires significant effort. For this reason, generally the sum of the van der Waals radii is used to try to decide which of these bonding situations are compatible with the E–X distance. Although this strategy is clearly a severe simplification of the problem, the qualitative information obtained is often so useful that it is worthwhile considered suitable to describe approximately the van der Waals radii. In general, all obtained E–X distances comprised between 0.7Å less and 0.7Å more than the van der Waals radii sum (*dW*) could be considered, while longer distances indicate noninteracting atoms. Distances shorter than the van der Waals radii sum by more than 1.3Å correspond most likely to a chemical bond, and those 0.7–1.3Å shorter fall within the van der Waals gap, thus suggesting a special bonding situation that asks for a deeper analysis.

It is important to stress that a revelation of the last studies is that the van der Waals radii sum should not be used as a cutoff distance as the radius does not correspond to the shortest approach interaction. The radii proposed point to the leading edge of the van der Waals peak a few tenths of Angstrom above the cutoff for intermolecular contacts. Also, the crystal packing and solution distance analysis should not be limited to distances smaller than the van der Waals radii sum and that the majority of van der Waals contacts are outside that distance. It is evident also as an interesting conclusion that there are van der Waals bonds of varying strengths depending on the atom pair considered. Therefore, distances 0.7Å shorter or longer than the radii sum are likely to correspond to van der Waals interactions, but only distances approximately 1.3Å shorter than the radii sum can unequivocally be assigned to chemical bonds based on a distance criterion. Finally, when there are not enough available structural data to obtain the van der Waals map of an element, its van der Waals radius could be reasonably estimated as the product of the addition of 0.9Å to its covalent radius.

5.3 NEUTRON AND X-RAY METHODS TO INVESTIGATE STRUCTURAL ASPECTS OF METAL ION AND METAL COMPLEX SOLVATION AND THEIR SOLUTION STRUCTURE

Despite of the importance of obtaining as much information as possible for the interaction of water with itself and other molecules, in particular in solution, it is not easily affordable. Techniques such as IR and NMR spectroscopy provide partial valuable information on the interaction of water with dissolved molecules that were very useful for obtaining most of the actual knowledge on the solvation of metal ions and complexes. The last findings in X-ray and neutron diffraction have significantly increased the options for obtaining a picture of how the water molecules interact with solvates. In parallel, simultaneous developments in advanced computational methods have facilitated to extract the maximum information from the resulting experimental data. Among the new techniques, X-ray diffraction (XRD) and neutron diffraction (ND), with highly complementary characteristics, are arising as the most powerful as they probe the entire structure of a solution. There are interesting papers, reviews, and books that describe deeply and extensively both techniques [46–48]. In this chapter will be offered a cursory description of how it is possible to obtain information on the molecules solvation from XRD and ND.

The diffuse diffraction pattern of a liquid only contains information about the pair correlations between the system's constituent atoms. To model the structure of liquids from diffraction data, it is mandatory to calculate the atomic pair distribution function [49]. This function could be obtained by using the formalism for X-ray absorption fine structure spectroscopy (EXAFS) that has the advantage that the atomic pair distribution function includes the primary structural observable:

$$\gamma_{\alpha\beta}^{(2)} \cong A_{\alpha\beta}(k,r)\sin\left(2kr+\Phi_{\alpha\beta}(k,r)\right)$$

This function is the photoelectron scattering signals associated with a single atom, β, situated at a distance r from the photoabsorber, α. $A_{\alpha\beta}(k,r)$ is the amplitude term which incorporates factors such as the backscattering amplitude, and $\Phi_{\alpha\beta}(k,r)$ that is the total phase factor. This equation could be used to a certain extent for other experimental techniques such as XRD and ND. One procedure includes first-order difference neutron diffraction with isotopic substitution [50]. This method only is possible to be applied if isotopes of an element with neutron scattering length differences, Δb, greater than ≈1 fm can be obtained. Two neutron diffraction experiments on identical solutions are performed where only the isotopic composition of the system has been changed. Thus, by the composite partial pair correlation function it is possible to directly access to the specie of interest and to the other atoms present in solution. In most aqueous solutions the partial pair correlation function will be determined mainly in the near neighbor distance range by the ion–water contributions. The anomalous difference X-ray scattering technique is the analog method in X-ray diffraction [51]. This X-ray technique makes use of the change in the X-ray scattering form factor for an element when the incident photon energy is tuned to slightly different values close to its absorption edges. Neutron and X-ray methods are technically very challenging to execute, since the differences in the scattering properties of an element are, barring a few very favorable cases, extremely small. In both methods corrections for contributions such as absorption of the radiation in the sample, multiple scattering, inelastic scattering, and cell and instrument background scattering have to be done and, therefore, they are only really suitable for quite concentrated systems (at least 0.1M) [52, 53].

Therefore, it is possible to obtain sufficient information to deduce the structures of most simple ionic solutions and enough soluble metal complexes through the standard X-ray diffraction and H/D isotopic substitution neutron diffraction (and sometimes EXAFS spectroscopy) experiment combination. Nevertheless, the experimental results obtained by combination of these techniques are not enough for solving directly the pairwise structure of the aqueous solution. The experimental data must be analyzed by computational approaches for incorporating additional constraints and producing model solutions that can be tested for experimental compatibility.

The empirical potential structure refinement (EPSR) technique is one of the most successful computational methods due to its flexibility, large possibilities, and relatively ease [54–57]. The first step in this method involves a classical Monte Carlo simulation of a fluid that generates a model based on a set of interatomic reference potentials. These classical potentials allow the model to incorporate reasonable but fundamental physicochemical constraints, such as the sizes of atoms, ionic interactions, the atomic density of the fluid, and basic molecular structures. The final model is used to calculate the structure factors that have been measured in neutron and X-ray diffraction experiments. A perturbation potential is generated from the difference between the experimental and model diffraction patterns, which is then added to the reference potential. The simulation procedure continues until the positions of the atoms and molecules in the model fluid are driven into consistency with the pair correlation functions measured in the diffraction experiments. Once the model is in agreement with the experimental data, average structural information can be generated. Therefore, at the end of the process, the obtained configurations of atoms and molecules are consistent with the information in the experimentally determined structure factors. Actual computational limits constringe the standard maximum length scale over which the potentials operate at 12Å [58]. A large length scale is possible but with a noticeable reduction of the refinement process speed. The final results are formally the experimental data with imposed reasonable constrains and therefore other solutions for the model are possible. For this reason, to support the resulting model with experimental techniques as much as possible is mandatory to increase the confidence in it.

5.4 WATER, METAL IONS, AND METAL COMPLEXES IN SOLUTION

Water is a very exotic molecule that is able to interact with metal and its ions in solid state and in solution. As commented in the introduction, there are interesting reviews on the non-covalent interactions of water with metal complexes in solid state but few examples of experimental studies about how water interacts—directly or through the ligands—with the metal of a complex. Before discussing the last findings on non-covalent interaction of water with metal complexes, a previous findings on how water behaves in the presence of metal ions is needed to be mentioned.

Modern neutron diffraction combined with X-ray techniques leads to three-dimensional atomistic models of water solutions that provide the orientational correlations between molecular units [59]. A set of coordinate reference axes for each molecule type in the system is needed for obtaining a three-dimensional solution model. Through the analysis of the resulting three-dimensional models, it is possible to obtain valuable information on the nature of the molecular interactions in solution such as distances and angles between molecules. The resulting model can be used as starting point for ab initio calculation that is able to provide electronic information on the solved molecules and solvent.

5.4.1 Interaction of Water Molecules with Metal Ions

Positively charged ions in solution play key roles in earth, biological, and industrial processes, such as mineral formation and extraction [60, 61], stability of surfaces and nanoparticles [62, 63], and in partially solvated ions (cofactors) activating biochemical reactions [64]. The chemistry of cations in solution has mainly been rationalized by using such concepts as Pearson hardness [65], ion charge to size ratios [66], electrophilicity, nucleophilicity, and nucleofugality [67]. However, these concepts are fundamentally incomplete since their background mostly comes from properties predicted on isolated systems, and they do not take into account the important chemical changes associated with the strong interaction between the ions and the water molecules. Due to their complexity, most of the aspects related to how water molecules interact with metal cations in solution can only correctly be understood through comprehensive three-dimensional models of the solution structures. The recent results obtained by neutron and X-ray diffraction data [68] illustrate what is the actual understanding on how is the interaction of water molecules with metal ions in solution.

The geometry adopted by the water molecules in the first hydration shell of metal cations was one of the major questions in aquation chemistry literature over the past 50 years [69]. This type of structural information is very weakly encoded in the pair distribution functions that are accessible by the primary structure characterization techniques applicable to liquids. Structures are generally inferred by comparison of experimentally derived coordination numbers and bond lengths with the environments found in crystalline analogues of the system and then by testing the resulting assertion by advanced, very challenging, electronic structure modeling methods to discriminate between hypotheses [70]. Better results could be obtained by the EPSR of atomistic models against multiple diffraction and EXAFS data. Thus Figure 5.3 illustrates how using the bond angle distributions is possible to distinguish between different hydration shell geometries around the ion. Loosely coordinated ions with no favored hydration shell geometries have relatively unstructured bond angle distributions, while tetrahedral configurations favor OW–cation–OW angles around 109°, octahedral configurations require 90° and 180° angles, square antiprisms favor 72° and 141°, and trigonal tricapped prismatic forms have significant features at 68.5° and 136.5°. The

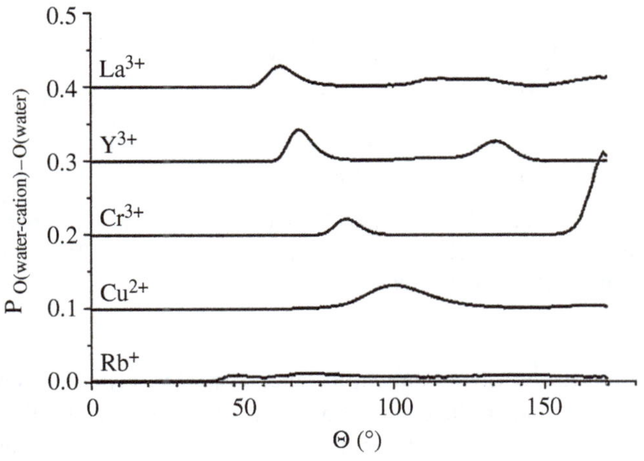

FIGURE 5.3 Water oxygen–cation–water oxygen bond angle distributions calculated for aqueous solutions of Rb^+ (0.1 M RbBr), Cu^{2+} (0.5 M $Cu(ClO_4)_2$), Cr^{3+} (1.0 M $Cr(NO_3)_3$), Y^{3+} (1.0 M YCl_3), and La^{3+} (1.0 M $LaCl_3$). These distributions correspond to hydration shells of no specific geometry for Rb^+, tetrahedral (plus components of trigonal bipyramidal and distorted octahedral) geometry for Cu^{2+}, octahedral geometry for Cr^{3+}, square antiprism geometry for Y^{3+}, and tricapped trigonal prism geometry for La^{3+}.

structural disorder of the hydration shell is reflected in the angular widths of the favored correlations.

The evolution of molecular geometry and electronic structure during ionic solvation by liquid water is fundamental for understanding the behavior of ions in chemical and biological systems and processes. It is generally thought that the presence of ions can either enhance or weaken the local hydrogen bond network, depending on the ionic charge and size, but quantification of these effects and their fundamental causes is difficult. The structure of the solvent medium itself and how this is perturbed by the presence of ions could be obtained from combining X-ray and neutron diffraction data since the chemically specific X-ray absorption information is focused entirely on the solute. X-ray diffraction and neutron scattering studies provide information on the local atomic arrangement of water molecules around an ion, demonstrating, for example, that in the alkali metal halide series both the X^-–O and Y^+–O distances increase with increasing ionic radius and the water molecules in the first (or coordination) shell differently orient around the anion and the cation: one single hydrogen is directed toward the anion while both hydrogens are directed away from the cation [13].

Despite the solvent support of the chemical processes that take place within it, the study of its structure around solutes is often neglected. Bernal and Fowler [71] propose terms kosmotropes (structure makers) or chaotropes (structure breakers) [72] for referring the effects of ions upon water around structure. The effect of ions in solution on the structure of water was rationalized through the concept of an induced structural temperature that was either higher or lower depending on whether an ion was perceived to have disordered or ordered the structure of the solvent. In 1995

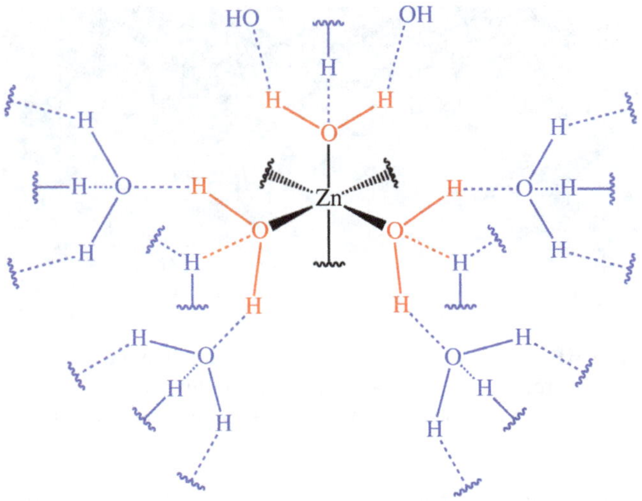

FIGURE 5.4 Trigonal first-shell structure and mostly tetrahedral structure in the second shell and beyond. A limited number of waters drawn for clarity.

Leberman and Soper suggested an analogy between the restructuring of bulk water by the presence of ions in solution and the effect of applied pressure to the pure liquid [73]. More recently the more advanced analysis using structure refinement methods has additionally established that the restructuring phenomenon extends beyond the immediate ion environment into the bulk solvent and is dominated by reorganization of the water oxygen network [74]. X-ray absorption spectrum (XAS) of aqueous alkali metal halide solutions provides information on the local electronic environment of atomic sites within molecules by probing the nature of the unoccupied orbitals. It was proved that the electronic structure of water is very sensitive to the specific hydrogen bonding configurations of the water molecules. Additionally, some new features were observed that were attributed to mixing between the molecular orbitals of water molecules and the open d-shell of the transition metal cation.

Most recently an interesting study showed how Zn^{2+} ion changes the distribution of the water molecules [75]. Ab initio molecular dynamics (AIMD) simulations of the dynamics of water molecules in the hydration shells surrounding the Zn^{2+} ion were compared to those simulations combining quantum and classical molecular dynamics (MD), molecular mechanical, approach. The hydration structures predicted from the AIMD and AIMD/MM simulations were found to agree in detail with each other and with the structural results from X-ray data despite the very limited QM region in the AIMD/MM simulation, and they also agreed in remarkable detail with EXAFS data.

The simulations of the experimental data agreed with a stable octahedral structure containing six water molecules for the first hydration shell (Fig. 5.4). The highly structured second hydration shell is found to contain, on the average, 14 water molecules trigonally coordinated via hydrogen bonds to the six first-shell water molecules. Beyond the second shell, the bonding pattern is found to substantially return

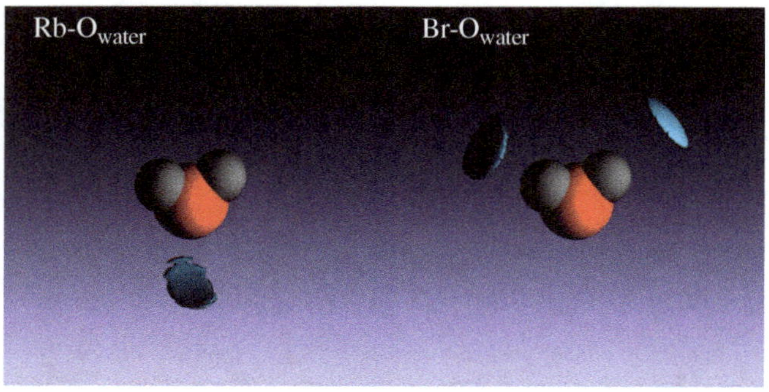

FIGURE 5.5 Spatial density functions highlighting the most probable regions around a water molecule to find an Rb^+ cation or a Br^- anion in a 0.002 mole fraction aqueous solution of RbBr.

to the tetrahedral structure of bulk water. Molecules of water transfer into and out of the second solvation shell are also observed. These exchanges are found to occur on a picosecond timescale via dissociative and associative mechanisms. In these calculations, the full d^{10} electronic structure of the valence shell of the Zn^{2+} ion is retained. The simulations showed substantial and complex charge relocation on both the Zn^{2+} ion and the first hydration shell. Little polarization is found for the water in the second hydration shell, and no exchanges were seen between the first and the second hydrations shells; however, many water transfers between the second hydration shell and the bulk were observed [75].

Finally, a convenient way to have a view of the distribution of ions about solvent molecules is using the spatial density function (SDF) [76]. This approach maps out spatial volumes in the coordinate reference frame of the molecule placed at the origin, which correspond to occupation probabilities of neighboring species. As example, Figure 5.5 shows the SDFs for Rb^+ cations and Br^- anions in the first neighbor region of water molecules. The highlighted regions clearly show that the cation prefers to be solvated by the negatively polarized oxygen atoms of the water molecules whereas the anions are solvated by the positively polarized water hydrogen atoms.

5.4.2 Interaction of Water Molecules with the Complex Metal Center

The discovery of the anticancer properties of *cis*platin ($[PtCl_2(NH_3)_2]$) in the sixties by Rosenberg [77] proved that transition metal complexes could be toxic for tumor cells. Transition metal complexes designed to bind to DNA have been extensively studied in the last few decades [78] and a huge number of metal compounds of wide structural diversity have been tested, some of them being active therapeutic agents for cancer treatment. Still now *cis*platin and its analogues are among the most widely used chemotherapeutic agents [79]. A correct description of the solvation shell is of fundamental importance for enabling reliable MD simulations of platinum complexes. Classical MD and AIMD

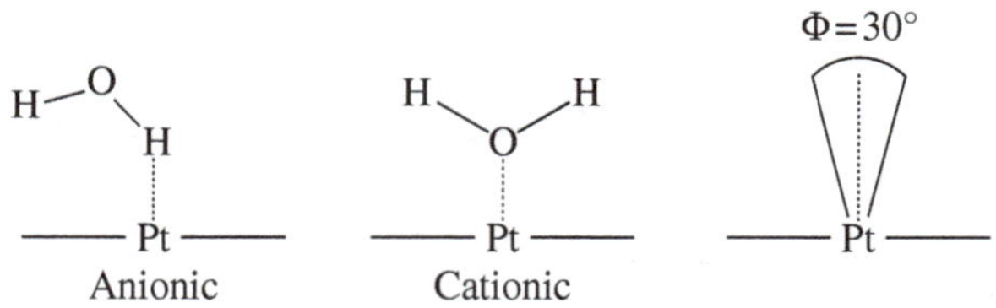

SCHEME 5.1 Possible anionic and cationic interaction modes between a water molecule and a square-planar Pt^{2+} center. Definition of the axial region used in the analysis.

simulations of solvated platinum complexes have been instrumental in the evaluation of the structure of platinum–DNA cross-links that are formed by platinum anticancer drugs [80]. AIMD simulations have been also used for the calculation of ^{195}Pt NMR chemical shifts [81]. Despite of the large interest in having as much information as possible on how *cis*platin is dissolved in the physiological fluid and therefore on its interaction with water molecules, a direct experimental information was only recently provided.

The neutron diffraction structure of the relayed to *cis*platin complex *trans*-$[PtCl_2(NH_3)(N\text{-glycine})]\cdot H_2O$ constituted the first experimental evidence for a hydrogen bonding-like interaction between a water molecule and a d^8 metal ion and revealed two nonclassical intermolecular hydrogen bonds involving Pt: one with the NH_3 ligand of a neighboring complex and another with a water molecule [82, 83]. These results suggested that similar interactions should exist in solution too. Previous AIMD simulations on Pt^{2+} compounds showed that both anionic (Pt···HOH) and cationic configurations (Pt···OH_2; Scheme 5.1) are possible [84]. Such "inverse hydration" [85] of d^8 complexes could represent an impediment for ligand exchange with water (aquation) where the water molecule attacks the metal center along the *z*-axis with its oxygen atom. Additionally, it was evidenced that the nature of the ligands strongly influences the hydration shell structure and the number of water molecules in the axial region [84], and therefore the axial hydration pattern cannot be generalized and should be investigated for each complex.

The simulations of *trans*-$[PtCl_2(NH_3)(N\text{-glycine})]$ in solution confirmed the formation of H-bond interactions between the Pt^{2+} center and water molecules [86]. However, the axial hydration pattern turned to be asymmetric with respect to the plane of the complex, different from what may be inferred from calculations on models of reduced size. Nevertheless, the calculations did not reveal the role that anionic interactions play in the reactivity of the Pt^{2+} complex, and the obtained solvation shell model was not very realistic. However, the important conclusion was that d^8 ions such as Pt^{2+} can act as hydrogen bonding acceptor. This finding implies that electrically neutral Pt^{2+} complexes could be hydrated in the "inverse" (or "anionic") fashion and suggests that the general view of Pt^{2+} hydration should be revised. This fact has an important consequence, as in the cellular

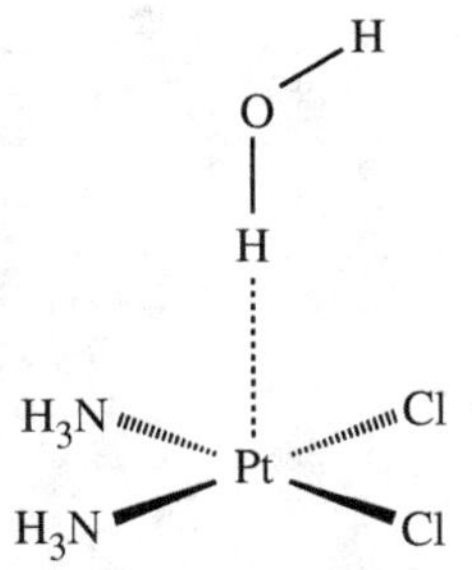

FIGURE 5.6 *cis*-$[PtCl_2(NH_3)_2]\cdot H_2O$ complex in the "H-ahead" orientation for d(Pt—O) = 3.4 Å.

processing of the antitumor drug *cis*platin, the "inverse hydration" could be the reason that justifies the observed unexpectedly slow first rate-determining aquation step. Platinum(II) complexes display a square planar coordination geometry, and, in aqueous solutions, water molecules could either approach the platinum atom along the *z*-axis (perpendicular to the coordination plane) or they can contact the ligands in an in-plane fashion.

It is important to stress that the measurement of the aquation rate constants for *cis*platin supports the existence of the O—H···Pt hydrogen bonding. The *cis*platin is hydrolyzed in water in two steps, the two aquation rate constants being nearly equal [87]. This is a strange result, the second aquation rate should be considerably lower than the first one due to the impediment produced by attraction on the outgoing Cl^- of the monoaquated complex positive charge. This fact can be interpreted as evidence for the "inverse hydration" of the neutral dichloride form of *cis*platin, which orients the axial molecule unfavorably for nucleophilic substitution (Fig. 5.6).

In a recent paper Bergès et al. carried out an extended analysis of the nonconventional H-bond that forms when an uncharged Pt(II) complex is axially approached by a water molecule [88]. The obtained results allowed establishing four descriptors of the nonconventional OH···Pt bond, one energetic and three of topological nature. The important energetic feature appears to be the dispersion-driven character of the "H-ahead" interaction, which is modulated by the nature and the configuration of the polar ligands in vacuo. In the "H-ahead" approach, the charge transfer between all the neutral platinum complexes and the water molecule (0.04 e at equilibrium) displays the same order of magnitude as in other classical H-bonded compounds. This charge transfer reflects a small orbital interaction that increases at shorter distances. In contrast, no charge transfer was detected for the "lone pair-ahead" approach down to 2.4 Å. The charge transfer values were not modified in the solvent. The variations, as a function of the Pt—O distance, of the electron density and its Laplacian at the bond critical point consolidate the classification of this interaction as an H-bond but also point out the influence of the nature of the ligands in vacuum as well as in the solvent. The "H-ahead" approach is accompanied by an increase of the dipolar polarization of the interacting O—H bond. Conversely, the polarization of the noninteracting O—H bond decreases slightly, which differentiates the interacting O—H bond from the noninteracting one. It is noteworthy that the variations of the dipolar polarization and Laplacian are similar, the position of their maxima being close to each other. The dipolar polarization of the O—H bonds increases in the aqueous medium, but the ligand selectivity is retained.

The first wide angle X-ray scattering (WAXS) [89, 90] and extended EXAFS [91] studies on the hydration structure of the Pt^{2+} did not find evidences of axial water molecules. Martinez et al. [92] using classical MD simulations of a highly diluted Pd(II) aqueous solution detected a new hydration shell in the axial regions in which water molecules show structural and dynamical properties noticeably different from those of the first and second shell. The properties of this new shell, called meso-shell, are found to be a compromise among first and second shell and even the bulk water.

A later classical MD study of the Pt^{2+} case [93] gave evidence for a less strongly bound meso- shell at ≈3 Å, which is not possible to be detected by EXAFS because of the structural and dynamic disorder of the water molecules. Further experimental and theoretical studies of Pt^{2+} aqueous solutions have supplied different qualitative and quantitative descriptions on the number and relative orientations of the axial water molecules [94]. Different results were found depending of the used technic. Despite the Pt^{2+} and Pd^{2+} ions display somewhat different behavior in water, they are considered almost similar metals, and the results found for one of them could be probably used for foreseeing the behavior of the other one. The Pd^{2+} ion is stable and soluble in water and therefore is a useful metal to be studied with neutron and X-ray diffraction combined with EPSR simulations [95].

The obtained results showed an unambiguous experimental evidence at the existence of the meso-shell around Pd^{2+} in water solution. The axial hydration was found centered at 2.85 Å containing on average two water molecules, each of them located at one side of the molecular plane. There is then an intermediate shell between the strongly bound water molecules of the first shell (2.04 Å) (coordinated water molecules) and the equatorial water molecules forming the second hydration shell (4.2 Å). The experiments were reformed in a nondiluted aqueous solutions at acid pH (H_2ClO_4) to prevent the hydrolysis of the $[Pd(H_2O)_4]^{2+}$ aqua ion. The analysis of the resulting data showed that a perchlorate anion distribution is significant in the close environment of the aqua ion and appears to display a similar axial coordination pattern as adopted by the meso-shell water molecules (Fig. 5.7).

This evidence highlights the fact that the meso-shell is not only a hydration structure but rather a coordination region built by the cooperative effects of the square-planar cation and the water structure around it. Additionally, ligands with coordinative and topological properties may be involved in the meso-shell together with water molecules. Additionally,

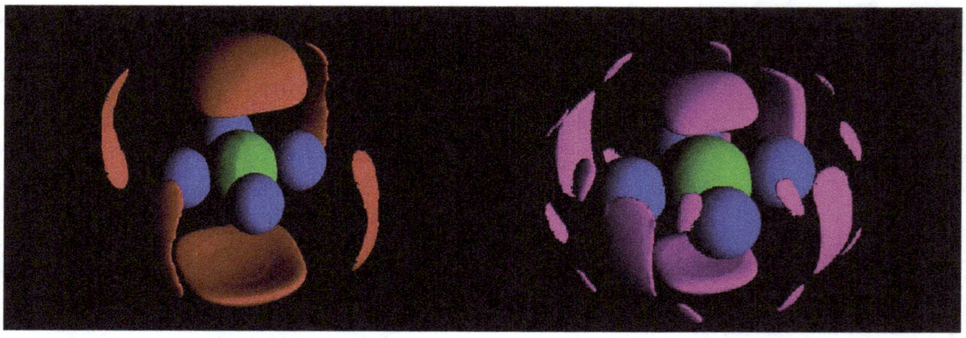

FIGURE 5.7 Spatial distribution functions around the $[Pd(OH_2)_4]^{2+}$. Pd (green sphere), coordinated oxygen atoms (blue spheres). Right, water oxygen atoms (red surfaces); left, perchlorate chlorine atoms (purple surfaces).

the obtained results show that initial steps for coordination of a ligand through the axial region may also be affected by the presence of counterions and charged species, which may impose a different behavior on the entering reactant. In conclusion, depending of the counterion chosen for the metallo-precursor salt, the subsequent reaction mechanisms could be affected. This specificity is going beyond the classical Brönsted–Bjerrum behavior [95, 96] where the ion effects on chemical reactions are quantified by the ionic strength. This is a particular effect that is not depending on the intrinsic chemical interaction of anion and reactants but on the combination of the hydration structure and the involved reactants. Therefore, this chemical effect is specific and based on a collective medium behavior. Extending these results to other squared-planar metals like Pt^{2+} complexes means that a particular attention is needed to be paid to the specific effects that counterions may produce in the media where reactions occur.

5.4.3 Interaction of Water Molecules with Ligands within Metal Complexes

The non-covalent interaction of water molecules with metal complex ligands was observed when the first crystal structures of metal complexes were obtained. There is a long tradition of analyzing and comparing the non-covalent interaction of water and other molecules with ligands in solid state, most of them being summarized in numerous reviews [97–99]. Until not long ago the possible non-covalent interaction of water with ligands in water solution was mainly studied by theoretical calculations. The partial correspondence to experimental data from techniques such as IR and NMR spectroscopy is normally considered to be enough for supporting the obtained theoretical results [100–103].

In a previous paper, B. Frost et al. [104] showed the pH-dependent speciation of aqueous solutions of $[CpRu(PTA)_2H]$ (PTA = 1,3,5-triaza-7-phosphaadamantane) under a pressure of hydrogen. In this conditions one or both PTA ligands can be protonated at one nitrogen atom, giving rise the complex $[CpRu(PTA)\{PTA(H)\}H]^+$ and $[CpRu\{PTA\text{-}(H)\}_2H]^{2+}$ as observed by NMR (Scheme 5.2) [105–107]. The study made by DFT calculations on the relative energies of the various protonated species of $[CpRu(PTA)_2H]$ in the gas phase did not provide, obviously, any information on the interaction with a noninnocent molecule as water, in which solvent the reaction took place. The DFT computational study of the H—H splitting in water mediated by $[CpRu(PTA)_2Cl]$ showed that two pathways are possible (Scheme 5.2): the homolytic (oxidative addition) and the heterolytic ones.

The Gibbs energy profile in water for $[CpRu(PTA)_2Cl]$, $[CpRu(PTA)_2(H_2O)]$, and the dihydrogen complex $\{[CpRu(PTA)_2(\eta^2\text{-}H_2)]^+\cdot(H_2O)_3\}\{5\cdot(H_2O)_3\}$ is showed in Figure 5.8. The initial step consists of the heterolytic Ru—Cl bond cleavage in $[CpRu(PTA)_2Cl]$, yielding the unsaturated $[CpRu(PTA)_2]^+$ fragment and releasing the Cl^- anion, a process favored in water. The high solvation Gibbs energy of the chloride anion in water $\Delta G_{solv}(Cl^-) = -78.5$ kcal mol^{-1}) [107] leads to a rather low energy Ru—Cl hydrolysis in water (29.1 kcal mol^{-1}). The lowest minimum for the three complexes were found for an arrangement where three water molecules build a hydrogen-bonded chain that connects two basic centers of the molecule (the chloride, H_2O and H_2, and a nitrogen atom from one PTA ligand) (Fig. 5.8).

The presence of water strongly favors chloride dissociation and subsequent formation of the aquo complex $\{[CpRu(PTA)_2(H_2O)]^+\cdot(H_2O)_2\}$. The stability of the monoaquo complex $[CpRu(PTA)_2(H_2O)]^+$ agrees with its experimental NMR detection by dissolving $[CpRu(PTA)_2Cl]$ in water [105]. Displacement of the coordinated water by H_2, resulting in the complex $\{[CpRu(PTA)_2(\eta^2\text{-}H_2)]^+\cdot(H_2O)_3\}\{5\cdot(H_2O)_3\}$, is endergonic. The dihydrogen complex has not ever been observed experimentally. However, the formation of this key initial intermediate from which the H—H activation starts can be driven by the high H_2 pressures at which the reaction is being performed even if it is reasonable that, under these conditions, it rapidly evolves to more stable species.

As indicated, the H—H bond in $\{\{[CpRu(PTA)_2(\eta^2\text{-}H_2)]^+\cdot(H_2O)_3\}\{5\cdot(H_2O)_3\}$ can be activated by either a homolytic or a heterolytic pathway. The homolytic splitting does not require any direct solvent participation to occur and it can be seen as a formal oxidative addition of the H—H bond to Ru^{2+} affording a Ru^{4+} dihydride. It was not possible to locate a minimum for a *cis*-dihydride species and therefore

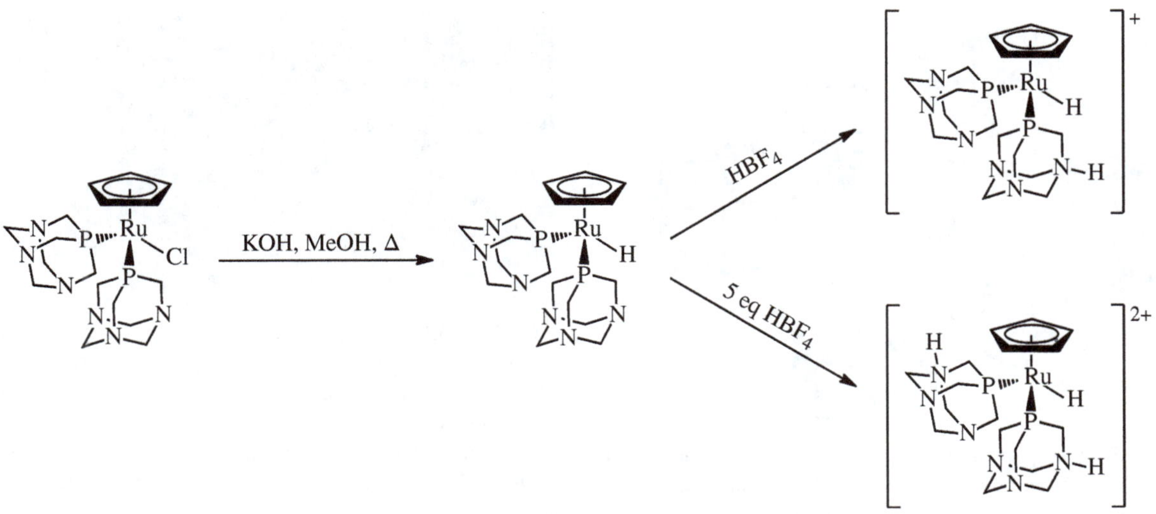

SCHEME 5.2 Possible pathways for the H—H splitting mediated by $[CpRu(PTA)_2Cl]$.

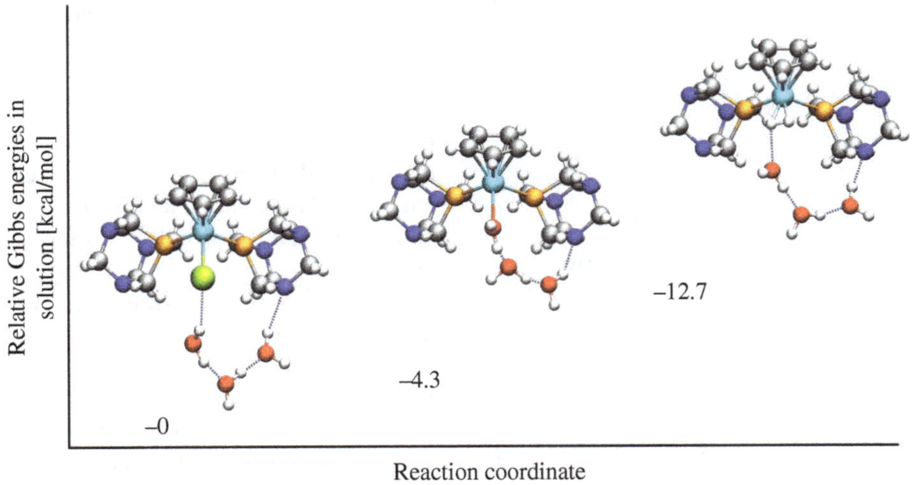

FIGURE 5.8 Gibbs energy profile in water for the chloride by H_2 ligand exchange, starting from $[CpRu(PTA)_2Cl]$ and leading to the formation of the dihydrogen complex $\{[CpRu(PTA)_2(\eta_2\text{-}H_2)](H_2O)_3\}$.

the homolytic cleavage of the H—H bond should be accompanied by a *cis–trans* rearrangement, involving the migration of one H atom from one side of the molecule to the opposite side. In contrast, the heterolytic H—H splitting involving a direct water effect was possible to be modeled containing the three water molecules effectively bridged between the coordinated dihydrogen/hydride ligand and a nitrogen atom of one of the PTA ligands. The three solvent molecules behave as an efficient "proton carrier" to assist the intramolecular process eventually resulting in the heterolytic splitting of the molecular hydrogen ligand.

The present DFT study on the role of water in the activation of dihydrogen by $[CpRu(PTA)_2Cl]$ evidenced that water molecules take an active part in assisting the heterolytic splitting of the H—H bond. Compared to other mechanisms where no direct involvement of the solvent is proposed, the ability of water to form extensive hydrogen bonding is essential to lower the energy barriers in this reaction. It is important to stress that theoretical study shows that ruthenium-coordinated PTA ligand efficiently assists the heterolytic splitting by taking up a proton from H_2 via N-quaternization and confirm that the formation of a monohydrido species is favored over the *trans*-dihydride, as suggested by experimental results.

The first experimental determination of the interaction of water molecules with an aqua-soluble organometallic complex in solution was published at the end of 2014 [106]. The large solubility ($S_{25°C} = 320\,mg\,ml^{-1}$) and stability of solution of the complex $[RuCp(PTA)_2\text{-}\mu\text{-}CN\text{-}1\kappa C{:}2\kappa^2 N\text{-}RuCp(PTA)_2](CF_3SO_3)$ (Fig. 5.9), still in contact with metals like Ti and Al, made it ideal for evaluating whether X-ray and neutron diffraction methods could be used for determining experimentally how water molecules interact with metal complexes in water solution. Neutron diffraction enhanced by hydrogen/deuterium isotopic substitution (NDIS) and white-beam (Ag source) X-ray diffraction were used to study a 0.1 M solution

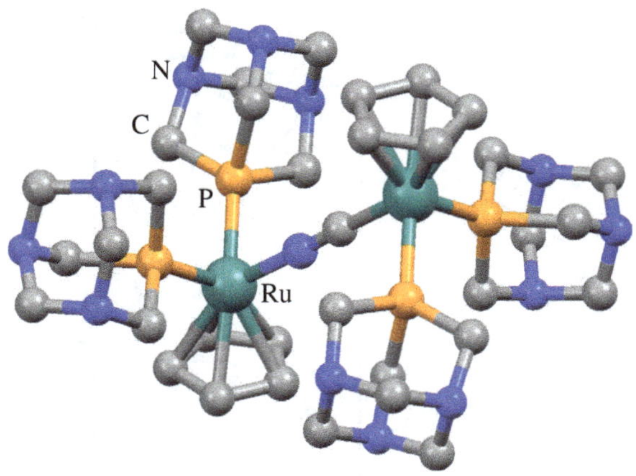

FIGURE 5.9 Complex unit single crystal X-ray structure of $[RuCp(PTA)_2\text{-}\mu\text{-}CN\text{-}1\kappa C{:}2\kappa^2 N\text{-}RuCp(PTA)_2]^+$, including the atomic labelling scheme. For clarity hydrogen atoms are not included.

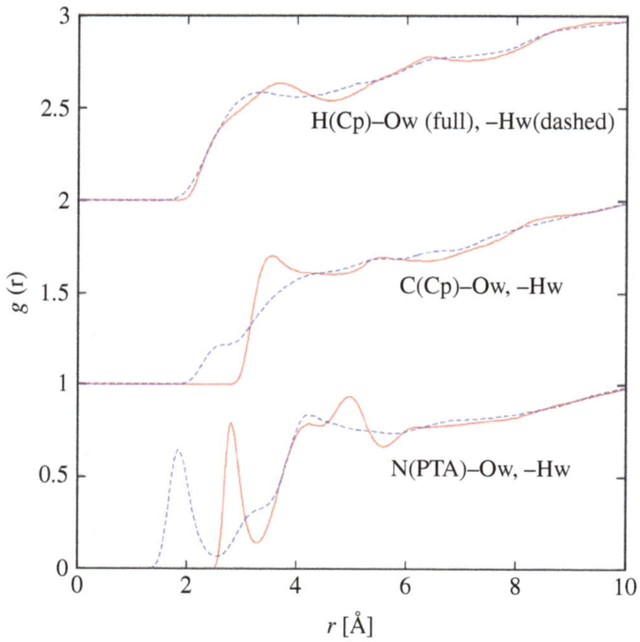

FIGURE 5.10 Pair correlation functions extracted through EPSR simulation for water around $[RuCp(PTA)_2\text{-}\mu\text{-}CN\text{-}1\kappa C{:}2\kappa^2 N\text{-}RuCp(PTA)_2]^+$. These functions express the probability of finding an atom at a given distance from another atom. Probability to find water oxygen (Ow, full line) or water hydrogen (Hw, dashed line) around three atoms belonging to $[RuCp(PTA)_2\text{-}\mu\text{-}CN\text{-}1\kappa C{:}2\kappa^2 N\text{-}RuCp(PTA)_2]^+$, namely, H and C on the Cp ligand and N on the PTA ligand. Lines have been shifted vertically for clarity.

of this complex in water. The experimental spectra are used to constrain a computer EPSR simulation. Neutron data provided a strong constraint on the solvent structure while the X-ray data complemented the information on the solute–solvent interaction.

A selection of the radial atom–atom correlation functions $g(r)$ extracted through EPSR simulation for water around dimeric Ru complex (Fig. 5.10) showed the shorter and

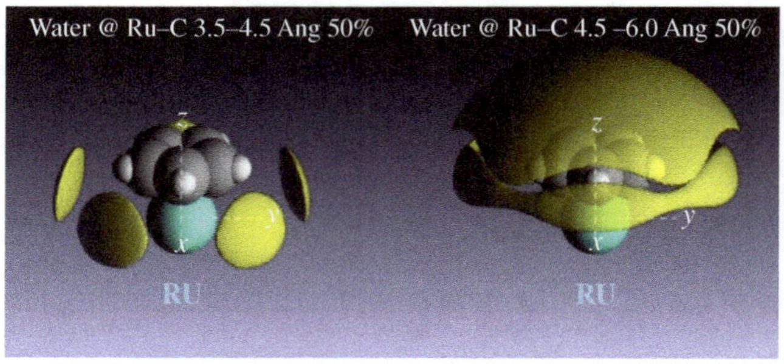

FIGURE 5.11 Spatial distribution functions for water molecules around cyclopentadiene ligand [107, 108]. The yellow cloud represents the region where the probability of finding a water molecule exceeds 50% (regardless of its orientation). The left panel is restricted to distance 3.5–4.5 Å from the central Ru atom, while the right-hand panel explores the region between 4.5 and 6.0 Å.

better-defined correlation is around the N_{PTA} acceptors for the water protons.

Surprisingly, interaction of water molecules with the hydrophobic cyclopentadiene (Cp) group was also observed. These interactions do not produce a well-defined peak but a nonnegligible intensity at a distance as short as 2.5 Å (C–Hw), and 3.5 Å (C–Ow) is a strong indication for water molecules and the Cp ligand interacting (Fig. 5.11).

The electronic structure and UV/optical response of dimeric ruthenium complex both in the gas phase and in the presence of a surrounding shell of water molecules, using density functional theory (DFT) and time-dependent density functional theory (TD-DFT), was calculated. For the solvated complex, the atomic configurations obtained from the EPSR simulations was considered for one complex molecule and up to 50 water molecules to represent the first solvation shell. It is important to stress that combining the output of a full molecular Monte Carlo simulation with a smaller-scale ab initio simulation is an original approach, requiring a careful selection of the parameter coupling the two simulation techniques. The properties of the solvated complex were obtained by averaging over different configurations to simulate the effects of the dynamical disorder in the liquid phase. The simulated TD-DFT and experimental UV absorption spectra are shown in Figure 5.12.

In the gas phase, the UV response is dominated by two main groups of peaks, at 300–400 nm (localized Ru-4d—4d transitions) and at 220–300 nm (partial charge-transfer character from Ru-4d to P atoms of PTA). In water solution, the Ru-4d—4d transitions are shifted to higher wavelengths (bathochromic shift). An analysis of the atomic charge distribution indicates that this red shift in absorption energies is correlated to electronic charge redistribution on the Cp groups caused by the interaction of the ligand with

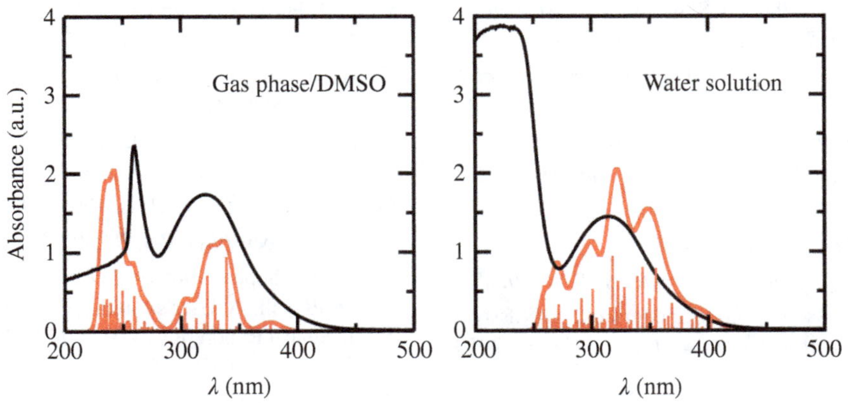

FIGURE 5.12 Experimental and TD-DFT UV-vis absorption spectra (5×10^{-4} M). The thin vertical lines represent the energies of the individual TD-DFT excitations.

surrounding water molecules. The bathochromic shift in the UV spectrum is therefore largely an indirect effect which can be interpreted as a solvation-induced change in the strength of the ligand field induced on the Ru atoms by the presence of the coordinating Cp groups.

5.5 FINAL COMMENTS

Non-covalent interaction of water molecules with metal complexes is largely known and discussed in solid state but shortly in solution despite of these interactions produce radical changes in the electronic distribution of the complexes and therefore in their properties. This panorama was the result of the lack of useful techniques to investigate how water molecules interact with metal complexes. In the last years, the X-ray and neutron diffraction techniques arrived to provide powerful methods for understanding the chemical interactions of water molecules with water-soluble complexes at an unprecedented level of detail. It is quite obvious that in this moment the neutron techniques alone cannot realistically be applied as a routine characterization method. Nevertheless, continuous efforts are targeted to address the optimization of X-ray and neutron diffraction techniques in connection with theoretical tools (EPSR, DFT, and TD-DFT) to extend the range of applicability of these techniques. As showed by the results obtained from studied systems containing metal complexes in water, it is imaginable that if the methodology is used on carefully selected model compounds, the depth of understanding of interaction at the molecular level should shed light on a number of similar compounds with interesting catalytic and biological properties. This progress with the redefinition of the meaning of non-covalent interaction and the better understanding of the structure of bulk water is providing a new view of how water interacts with metal complexes, showing that with this kind of compounds everything is possible.

ACKNOWLEDGEMENTS

This work has been partially supported by the projects CTQ2010-20952 and P09-FQM-5402. Thanks are also given to Junta de Andalucía-PAI (FQM-317) and COST Action CM1302 (WG1, WG2). F. S. to UAL for a predoctoral grant. Cofinanced by the EU FEDER. Thanks are given to Samira Nahim Granados.

REFERENCES

1. C. E. Housecroft, A. G. Sharpe, Inorganic Chemistry, Fourth ed. Pearson/Prentice Hall, New York, 2012.
2. H. S. Frank, M. W. Evans, *J. Chem. Phys.* 1945, *13*, 507.
3. H. J. Taube, *Phys. Chem. Chem. Phys.* 1954, *58*, 523.
4. G. K. Schweitzer, L. L. Pesterfield. The Aqueous Chemistry of the Elements. Oxford University Press. 2010.
5. M. A. Elizondo-Riojas, J. Kozelka, *J. Mol. Biol.* 2001, *314*, 1227, and references therein.
6. D. Bhattacharyya, S. Ramachandran, S. S. Sharma, W. Pathmasiri, C. L. King, I. Baskerville-Abraham, G. Boysen, J. A. Swenberg, S. L. Campbell, N. V. Dokholyan, *PLoS One* 2011, *6*, 23582.
7. P. Carloni, M. Sprik, W. Andreoni, *J. Phys. Chem.* 2000, *104*, 823.
8. G. R. Desiraju, *Angew. Chem. Int. Ed.* 2010, *49*, 2.
9. S. Yamada, Y. Nojiri, *Chem. Commun.* 2011, *47*, 9143.
10. J. S. Prell, J. T. O'Brien, E. R. Williams, *J. Am. Chem. Soc.* 2011, *133*, 4810.
11. N. Guillou, F. Millangea, R. I. Walton, *Chem. Commun.* 2011, *47*, 713.
12. B. D. Ostojić, G. V. Janjić, S. D. Zarić, *Chem. Commun.* 2008, *48*, 6546.
13. G. W. Neilson, P. E. Mason, S. Ramos, D. Sullivan, *Philos. Trans. R. Soc. Lond. A* 2001, *359*, 1575.
14. D. L. Doering, T. E. Madey, *Surf. Sci.* 1982, *305*, 123.
15. N. Agmon, *Acc. Chem. Res.* 2012, *45*, 63.
16. M. Bustamante, I. Valencia, M. Castro, *J. Phys. Chem. A* 2011, *115*, 4115.

17. U. S. Raghavender, B. Kantharaju, S. Aravinda, N. Shamala, P. Balaram, *J. Am. Chem. Soc.* 2010, *132*, 1075.
18. D. Ž. Veljković, G. V. Janjić, S. D. Zarić, *CrystEngComm* 2011, *13*, 5005.
19. T. Steiner, *Angew. Chem. Int. Ed.* 2002, *41*, 48.
20. D. Natale, J. C. Mareque-Rivas, *Chem. Commun.* 2008, 425.
21. L. R. Falvello, *Angew. Chem. Int. Ed.* 2010, *49*, 10045.
22. M. D. Fayer, *Acc. Chem. Res.* 2012, *45*, 3.
23. D. Paschek, R. Ludwig, *Angew. Chem. Int. Ed.* 2011, *50*, 352.
24. T. Schiros, K. J. Andersson, L. G. M. Pettersson, A. Nilssona, H. Ogasawaraa, *J. Electron Spectrosc. Relat. Phenom.* 2010, *177*, 85.
25. G. R. Desiraju, T. Steiner, The Weak Hydrogen Bond in Structural Chemistry and Biology, Oxford University Press, Oxford, 1999.
26. G. C. Pimentel, A. L. McClellan, The Hydrogen Bond, Freeman, San Francisco, 1960.
27. T. Steiner, G. R. Desiraju, *Chem. Commun.* 1998, 891.
28. P. Gilli, V. Ferretti, V. Bertolasi, G. Gilli, *J. Am. Chem. Soc.* 1994, *116*, 909.
29. K. Nakamoto, M. Margoshes, R. E. Rundle, *J. Am. Chem. Soc.* 1955, *77*, 6480.
30. G. A. Jeffrey, An Introduction to Hydrogen Bonding, Oxford University Press, Oxford, 1997.
31. A. R. Ubbelohde, K. J. Gallagher, *Acta Crystallogr.* 1955, *8*, 71.
32. S. Scheiner, Hydrogen Bonding: A Theoretical Perspective, Oxford University Press, Oxford, 1997.
33. G. A. Jeffrey, *Crystallogr. Rev.* 1995, *3*, 213.
34. G. Gilli, P. Gilli, *J. Mol. Struct.* 2000, *552*, 1.
35. W. Saenger, *Nature* 1979, *279*, 343.
36. D. M. P. Mingos, A. L. Rohl, *J. Chem. Soc. Dalton Trans.* 1991, 3419
37. A. Gavezzotti, *J. Am. Chem. Soc.* 1983, *105*, 5220.
38. L. Pauling, The Nature of the Chemical Bond, Cornell University Press, Ithaca, 1960.
39. B. Cordero, V. Gómez, A. E. Platero-Prats, M. Revés, J. Echeverría, E. Cremades, F. Barragán, S. Alvarez, *Dalton Trans.* 2008, 2832.
40. R. S. Rowland, R. Taylor, *J. Phys. Chem.* 1996, *100*, 7384.
41. S. Alvarez, *Dalton Trans.* 2013, *42*, 8617.
42. F. H. Allen, *Acta Crystallogr., Sect. B: Struct. Sci.* 2002, *58*, 380.
43. J. P. M. Lommerse, A. J. Stone, R. Taylor, F. H. Allen, *J. Am. Chem. Soc.* 1996, *118*, 3108.
44. S. C. Nyburg, C. H. Faerman, *Acta Crystallogr., Sect. B: Struct. Sci.* 1985, *41*, 274
45. T. N. G. Row, R. Parthasarathy, *J. Am. Chem. Soc.* 1981, *103*, 477.
46. ISIS neutron facilities (http://www.isis.stfc.ac.uk/instruments/instruments2105.html) (Accessed October 26, 2015).
47. D. T. Bowron, S. Diaz-Moreno, *Coord. Chem. Rev.* 2014, *277–278*, 2.
48. D. A. Keen, *J. Appl. Crystallogr.* 2000, *34*, 172.
49. S. Diaz-Moreno, S. Hayama, M. Amboage, A. Freeman, G. Duller, J. Sutter, *J. Phys. Conf. Ser.* 2009, *190*, 012038.
50. J. E. Enderby, *Chem. Soc. Rev.* 1995, 159.
51. P. H. Fuoss, P. Eisenberger, W. K. Warburton, A. Bienenstock, *Phys. Rev. Lett.* 1981, *46*, 1537.
52. A. K. Soper, G. W. Neilson, J. E. Enderby, R. A. Howe, *J. Phys. C Solid State Phys.* 1977, *10*, 1793.
53. S. Ramos, A. C. Barnes, G. W. Neilson, D. Thiaudiere, S. Lequien, *J. Phys. Condens. Matter* 1999, *11*, A203.
54. A. K. Soper, *Chem. Phys.* 1996, *202*, 295.
55. A. K. Soper, *Phys. Rev. B* 2005, *72*, 104204.
56. A. K. Soper, Empirical Potential Structure Refinement—EPSRshell: A User's Guide. Rutherford Appleton Laboratory Technical Report, RAL-TR-2011-012, Science and Technology Facilities Council, UK, 2011.
57. D. T. Bowron, S. Diaz-Moreno, *J. Phys. Chem. B* 2007, *111*, 11393.
58. M. P. Allen, D. J. Tildesley, Computer Simulation of Liquids, Clarendon Press, Oxford, 1987.
59. A. K. Soper, *J. Chem. Phys.* 1994, *101*, 6888.
60. M. F. Hochella, S. K. Lower, P. A. Maurice, R. L. Penn, N. Sahai, D. L. Sparks, B. S. Twining, *Science* 2008, *319*, 1631.
61. J. H. Weare, *Rev. Mineral.* 1987, *17*, 143.
62. S. Kerisit, D. J. Cooke, D. Spagnoli, S. C. Parker, *J. Mater. Chem.* 2005, *15*, 1454.
63. D. Spagnoli, D. J. Cooke, S. Kerisit, S. C. Parker, *J. Mater. Chem.* 2006, *16*, 1997.
64. M. Valiev, J. Yang, J. A. Adams, S. S. Taylor, J. H. Weare, *J. Phys. Chem. B* 2007, *111*, 13455.
65. R. G. Parr, R. G. Pearson, *J. Am. Chem. Soc.* 1983, *105*, 7512.
66. D. T. Richens, The Chemistry of Aqua Ions: Synthesis, Structure, and Reactivity: A Tour Through the Periodic Table of the Elements, John Wiley & Sons, Ltd, Chichester, 1997.
67. A. D. McNaught, A. Wilkinson, IUPAC. Compendium of Chemical Terminology (the "Gold Book"), Blackwell Science, Oxford, 1997.
68. D. T. Bowron, S. Diaz-Moreno, *J. Phys. Chem. B* 2009, *113*, 11858.
69. I. Persson, *Pure Appl. Chem.* 2010, *82*, 1901.
70. T. Ikeda, M. Hirata, T. Kimura, *J. Chem. Phys.* 2005, *122*, 244507.
71. J. D. Bernal, R. H. Fowler, *J. Chem. Phys.* 1933, *1*, 515.
72. B. Hribar, N. T. Southall, V. Vlachy, K. A. Dill, *J. Am. Chem. Soc.* 2002, *124*, 12302.
73. R. Leberman, A. K. Soper, *Nature* 1995, *378*, 364.
74. R. Mancinelli, A. Botti, F. Bruni, M. A. Ricci, A. K. Soper, *Phys. Chem. Chem. Phys.* 2007, *9*, 2959.
75. E. Cauët, S. Bogatko, J. H. Weare, J. L. Fulton, G. K. Schenter, E. J. Bylaska, *J. Chem. Phys.* 2010, *132*, 194502.
76. I. M. Svishchev, P. G. Kusalik, *J. Chem. Phys.* 1993, *99*, 3049.
77. B. Rosenberg, L. van Camp, T. Krigas, *Nature* 1965, *205*, 698.
78. S. P. Foxon, T. Phillips, M. R. Gill, M. Towrie, A. W. Parker, M. Webb, J. A. Thomas, *Angew. Chem. Int. Ed.* 2007, *46*, 3686.
79. E. R. Jamieson, S. J. Lippard, *Chem. Rev.* 1999, *99*, 2467.
80. S. G. Chaney, S. L. Campbell, E. Bassett, Y. Wu, *Crit. Rev. Oncol. Hematol.* 2005, *53*, 3.
81. L. A. Truflandier, J. Autschbach, *J. Am. Chem. Soc.* 2010, *132*, 3472.
82. S. Rizzato, J. Berges, S. A. Mason, A. Albinati, J. Kozelka, *Angew. Chem.* 2010, *122*, 7602.

83. S. Rizzato, J. Bergès, S. A. Mason, A. Albinati, J. Kozelka, *Angew. Chem. Int. Ed.* 2010, *49*, 7440.

84. L. A. Truflandier, K. Sutter, J. Autschbach, *Inorg. Chem.* 2011, *50*, 1723.

85. J. Bergès, J. Caillet, J. Langlet, J. Kozelka, *Chem. Phys. Lett.* 2001, *344*, 573.

86. P. Vidossich, M. Á. Ortuño, G. Ujaque, A. Lledós, *Chem. Phys. Chem.* 2011, *12*, 1666.

87. J. Vinje, E. Sletten, J. Kozelka, *Chem. Eur. J.* 2005, *11*, 3863.

88. J. Bergés, I. Fourré, J. Pilmé, J. Kozelka, *Inorg. Chem.* 2013, *52*, 1217.

89. B. Hellquist, L. Bengtsson, B. Holmberg, B. Hedman, I. Persson, L. I. Elding, *Acta Chem. Scand.* 1991, *45*, 449.

90. R. J. Deeth, L. Elding, *Inorg. Chem.* 1996, *35*, 5019.

91. R. Ayala, E. Sanchez Marcos, S. Daz-Moreno, V. A. Sole, A. J. Muñoz-Paez, *J. Phys. Chem. B* 2001, *105*, 7588.

92. J. M. Martínez, F. Torrico, R. R. Pappalardo, E. J. S. Marcos, *Phys. Chem. B* 2004, *108*, 15851.

93. F. Torrico, R. R. Pappalardo, E. J. Sanchez Marcos, J. M. Martínez, *Theor. Chem. Accounts* 2006, *115*, 196.

94. K. Kamienska-Trela, J. Wójcik, *Nucl. Magn. Reson.* 2013, *42*, 97.

95. D. T. Bowron, E. C. Beret, E. Martin-Zamora, A. K. Soper, E. Sánchez Marcos, *J. Am. Chem. Soc.* 2012, *134*, 962.

96. N. Z. Bjerrum, *Phys. Chem. Chem. Phys.* 1925, *118*, 251.

97. J. N. Z. Bronsted, *Phys. Chem. Chem. Phys.* 1925, *115*, 337.

98. S. Zuluaga, P. Canepa, K. Tan, Y. J. Chabal, T. Thonhauser, *J. Phys. Condens. Matter* 2014, *26*, 133002.

99. C. W. Bock, A. K. Katz, J. P. Glusker, *J. Am. Chem. Soc.* 1995, *13*, 3754.

100. D. S. Auld, *Biometals* 2001, *14*, 271.

101. A. Rossin, L. Gonsalvi, A. D. Phillips, O. Maresca, A. Lledós, M. Peruzzini, *Organometallics* 2007, *26*, 3289.

102. B. J. Frost, C. A. Mebi, *Organometallics* 2004, *23*, 5317.

103. C. A. Mebi, B. J. Frost, *Organometallics* 2005, *24*, 2339.

104. C. A. Mebi, R. P. Nair, B. J. Frost, *Organometallics* 2007, *26*, 429.

105. This value is in very good agreement with accurate evaluations of the chloride solvation Gibbs energy recently reported: C. P. Kelly, C. J. Cramer, D. G. Truhlar, *J. Phys. Chem. B* 2006, *110*, 16066.

106. M. Serrano-Ruiz, S. Imberti, L. Bernasconi, N. Jadagayeva, F. Scalambra, A. Romerosa, *Chem. Commun.* 2014, *50*, 11587.

107. A. K. Soper, E. R. Barney, *J. Appl. Crystallogr.* 2011, *44*, 714.

108. A. K. Soper, Rutherford Appleton Laboratory Technical Report, RAL–TR–2011–013, 2011.

6

π–π INTERACTION DIRECTED APPLICATIONS OF METAL COMPLEXES

MANAS SUTRADHAR* AND ARMANDO J. L. POMBEIRO†

Centro de Química Estrutural, Instituto Superior Técnico, Universidade de Lisboa, Lisbon, Portugal

6.1 INTRODUCTION

Non-covalent interactions involving π-systems concern a well-studied fundamental topic in chemistry and biochemistry. Among the various types of π-interactions, the most common ones involve H–π, π–π, cation–π, anion–π, lone pair–π, π^+–π, and anion–π–π interactions [1–3]. The non-covalent interactions between aromatic rings (π–π stacking) involving π-systems, that is, π-stacking interactions between the π-electron clouds of aromatic systems, are widely explored aromatic π-interactions in many fields of chemistry [4–10] and biochemistry [11–14]. They are important in the structures of biological macromolecules [15, 16] and are useful to study the intercalation of drugs into DNA [17]. They are also involved in template-directed synthesis, molecular recognition, molecular aggregates, and supramolecular modification [18–27]. Depending on the donor/acceptor properties of the aromatic system, the stability order of π–π stacking follows the trend: π-electron deficient–π-electron deficient>π-electron deficient–π-electron rich>π-electron rich–π-electron rich [28]. Transition metal complexes having planar chelate rings with delocalized π-bonds also exhibit an aromatic character and can involve stacking interactions similar to typical π–π interactions found in aromatic organic molecules [29–31].

In this context, we illustrate herein some important applications of π–π interactions in metal complexes, such as anchoring catalysts via non-covalent π–π stacking, magnetic coupling via π–π interactions, and photoluminescent properties directed by π–π stacking.

6.2 MODEL AND SOME RULES OF π–π STACKING

In 1990 Hunter and Sanders proposed a simple model to explain experimental observations of aromatic interactions based on the charge distribution in a π-system [32]. This model considers the σ-framework and the π-system separately and the overall π–π interactions are the result of π–σ attractions that overcome π–π repulsions. The electrostatics of the model can be visualized by describing the aromatic ring as two regions of negatively charged π-electron density surrounded by a positively charged σ-framework (Fig. 6.1).

Generally, three stacking patterns for phenyl–phenyl π–π stacking interactions can be considered: face-to-face (sandwich), edge-to-face (T-shaped), and offset face-to-face (parallel-displaced) (Fig. 6.2). On account of the electrostatic effects, the latter two stacking patterns are stable in energy, while the face-to-face stacking is unstable and rarely observed because of the repulsion between the two stacking phenyl rings.

From above findings, Hunter and Sanders described some rules for nonpolarized π-systems [32]:

Rule 1: π–π repulsion dominates in a face-to-face π-stacked geometry.

Rule 2: π–σ attraction dominates in an edge-on geometry.

Rule 3: π–σ attraction dominates in an offset π-stacked geometry.

A substituent or a heteroatom in the aromatic ring can polarize the aromatic system and thus the stacking interaction

* Corresponding author Manas Sutradhar (manaschem@yahoo.co.in)

† Corresponding author Armando J. L. Pomberio (pombeiro@tecnico.ulisboa.pt)

Non-Covalent Interactions in the Synthesis and Design of New Compounds, First Edition.
Edited by Abel M. Maharramov, Kamran T. Mahmudov, Maximilian N. Kopylovich, and Armando J. L. Pombeiro.

can be affected. An electron-withdrawing substituent reduces the electron density in the ring and hence decreases the π–π repulsion between the two rings. Therefore, the stability of π–π interactions increases. Electron-donating substituents increase the electron density in the aromatic ring and hence have the reverse effect. Therefore, for polarized π-systems they proposed the following rules [32]:

Rule 4: For interactions between highly charged atoms, charge–charge interactions dominate.

Rule 5: A favorable interaction with a neutral or weakly polarized site requires the following π-polarization:

a. A π-deficient atom in a face-to-face geometry
b. A π-deficient atom in the vertical T-group in the edge-on geometry
c. A π-rich atom in the horizontal T-group in the edge-on geometry

Rule 6: A favorable interaction with a neutral or weakly polarized site requires the following σ-polarization:

a. A positively charged atom in a face-to-face geometry
b. A positively charged atom in the vertical T-group in the edge-on geometry
c. A negatively charged atom in the edge-on geometry

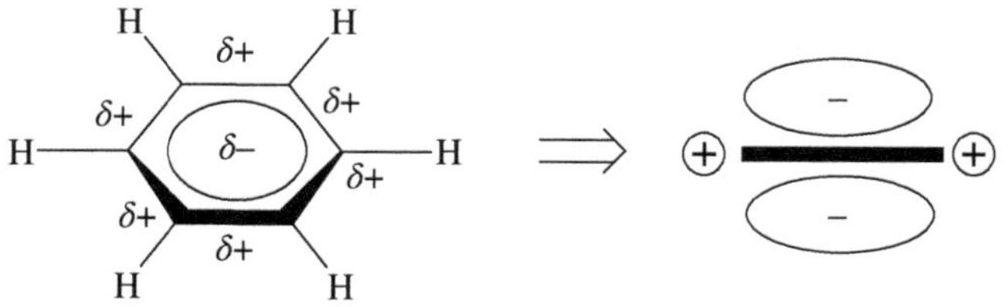

FIGURE 6.1 Electrostatic model of the quadrupole moments of benzene.

6.3 CATALYSIS

Nowadays, improvements in the catalyst development have focused *inter alia* on the challenges of cost-effective, more efficient, and environmentally benign greener systems. Although homogeneous catalysts can exhibit a high activity and selectivity in organic reactions, they commonly have

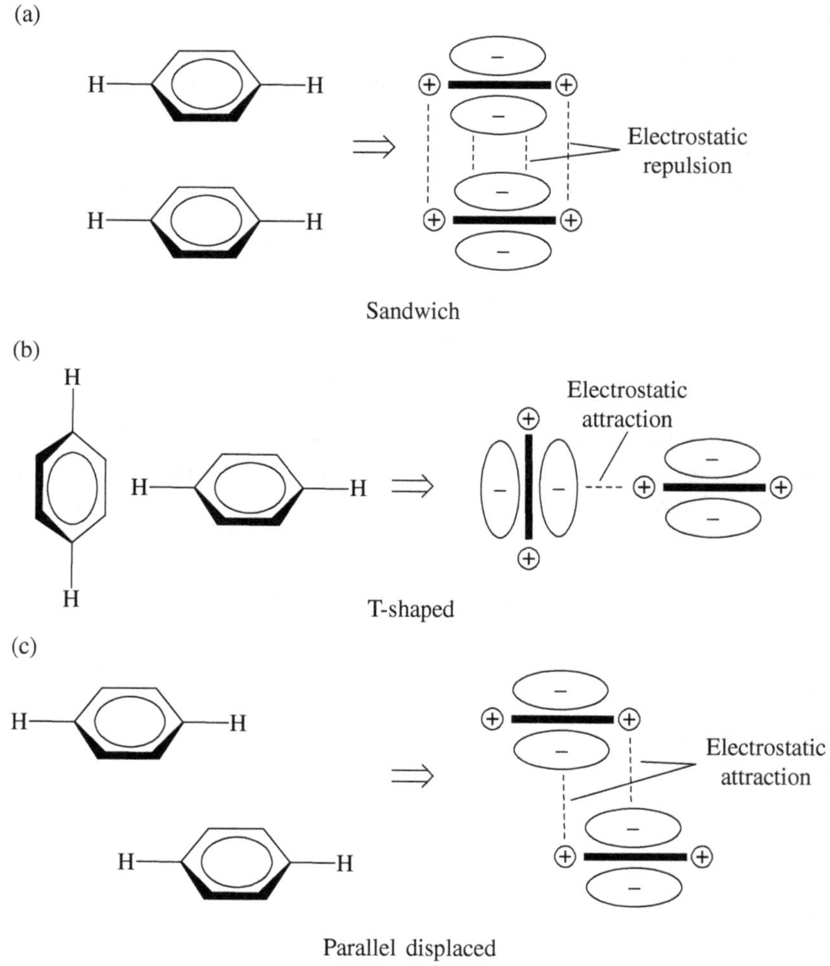

FIGURE 6.2 Stacking patterns in benzene rings: (a) face-to-face (sandwich), (b) edge-to-face (T-shaped), and (c) offset face-to-face (parallel-displaced).

limitations due to catalyst instability and difficulty in catalyst/product separation. Immobilization of homogeneous catalysts on a solid support has distinct advantages in terms of reusability and sometimes also of stability and activity, which improve the catalytic performance and provide green chemistry strategies for chemical synthesis.

Several methods for immobilizing catalysts using a non-covalent method via electrostatic interactions have been reported [33–35]. They are simpler and more convenient for recycling and reuse of the supported materials compared to traditional covalent bonding immobilization methods [36–40] and have been successfully used for the immobilization of transition metal catalysts. Recently, single-walled carbon nanotubes (SWCNTs) have been used as an ideal solid support material for anchoring catalysts via non-covalent π–π stacking [41].

6.3.1 Pyrene-Tagged (and Other π–π) Ruthenium(II) and Rhodium(I) Complexes

A pyrene-tagged ruthenium carbene complex **1** (Scheme 6.1) was prepared and anchored on the surface of a SWCNT (**2**) via π–π interactions (Scheme 6.2) between the pyrene and the SWCNT [42]. This anchored complex was used for ring-closing metathesis of selected substrates (Scheme 6.3). The π–π stacking interaction between the pyrene and the SWCNT is reversible and is affected by the polarity of the solvent and by temperature. Both the ruthenium catalysts and the supporting SWCNTs can be recycled [42]. This non-covalent immobilization was conducted in CH_2Cl_2 and the SWCNTs can be easily recycled by washing with tetrahydrofuran. Desorption of **1** can be accelerated by sonication (~98% in 4 h) [42].

In situ reaction of $[Rh(COD)_2]BF4$ (COD = 1,5-cyclooctadiene) with the pyrene-modified pyrphos ligand **3** (Scheme 6.4), followed by treatment wi th multiwalled carbon nanotubes (MWCNTs), generates the chiral Rh(I) catalyst via π–π stacking (similar to Scheme 6.2) [43]. The chiral catalyst was used for asymmetric hydrogenation of α-dehydroamino esters (Scheme 6.5) and was recovered via general solid/liquid separation simply by changing the solvent (here EtOAC) [43].

A monomeric ruthenium catalyst **4** (Scheme 6.6) with a pyridine-bearing pyrene ligand at the axial position was bound to MWCNTs by non-covalent π–π stacking

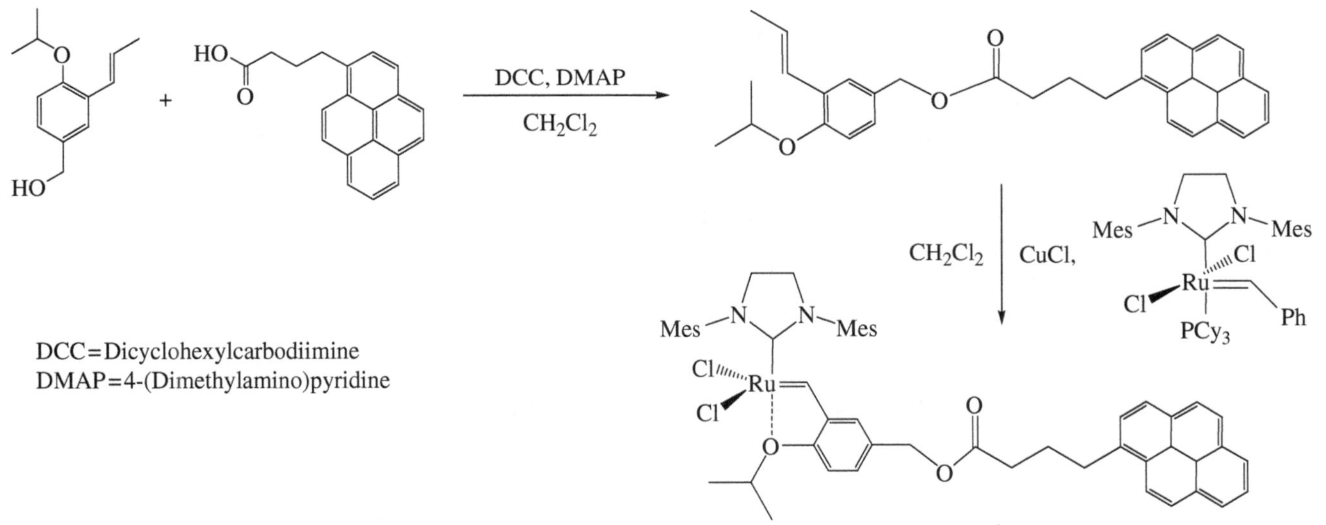

1

SCHEME 6.1 Synthesis of a pyrene-tagged ruthenium carbene complex **1** [42].

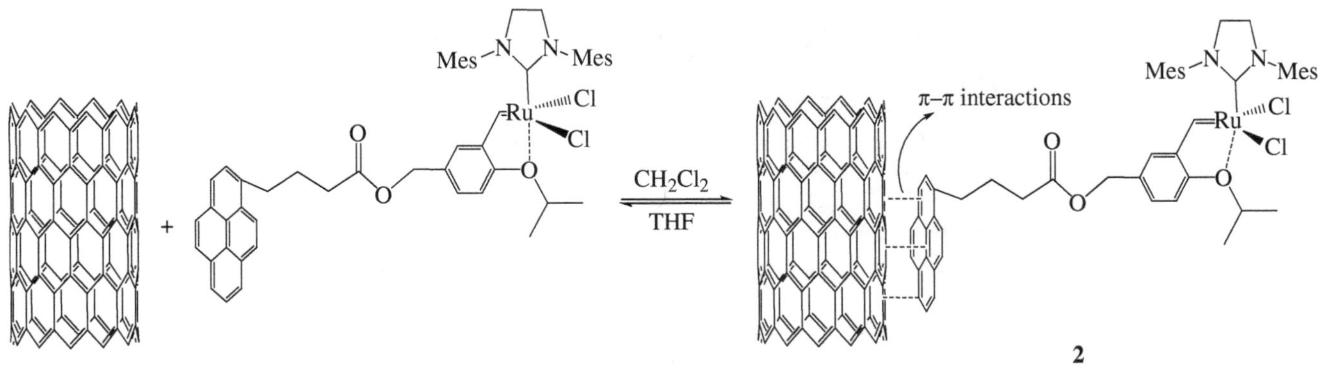

2

SCHEME 6.2 Pyrene-tagged ruthenium carbene complex **1** anchored on the surface of SWCNTs via π–π interactions [42].

interactions and was used for the electrocatalytic oxidation of water (Scheme 6.6) [44]. The electrolysis was carried out in neutral aqueous solution containing 0.1 m sodium sulfate in a three-electrode cell equipped with a MWCNT/ITO working electrode (ITO = indium tin oxide or tin-doped indium oxide) with immobilized complex **4**, a Pt mesh as counter electrode, and a Ag/AgCl reference electrode [44]. The water oxidation process occurs efficiently at a relatively low applied potential in nonbuffered neutral aqueous solution, and the molecular nature of the catalyst retains at the heterogeneous surface.

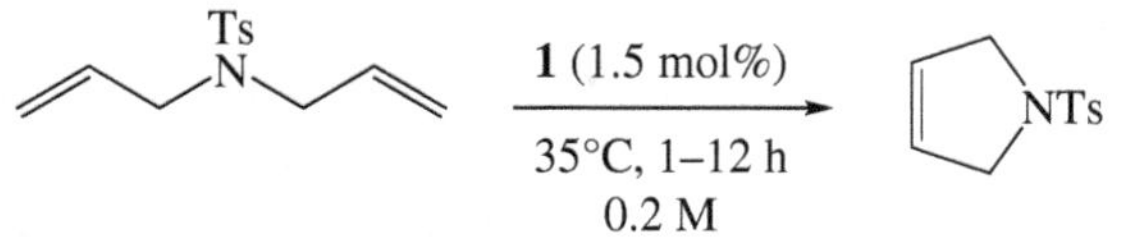

SCHEME 6.3 Ring-closing metathesis reaction catalyzed by **1**@SWCNT [42].

An intramolecular edge-to-face aromatic π–π interaction was observed by X-ray crystallography in optically active diruthenium–allenylidene compounds (complex **5**, Scheme 6.7) bearing optically active thiolate moieties [45]. This interaction between the Ru catalyst (**Ru–Cat**) and the propargylic alcohol-derived ligand plays a crucial role in achieving the high enantioselectivity in the ruthenium-catalyzed enantioselective propargylic substitution reactions of propargylic alcohols with nucleophiles (Scheme 6.8) [45].

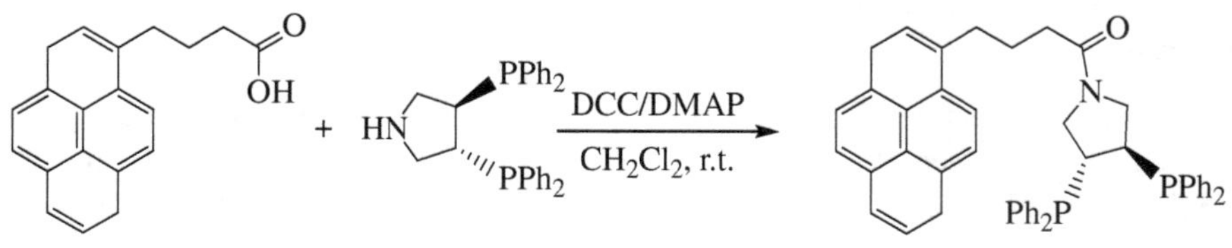

3

SCHEME 6.4 Synthesis of pyrene-modified pyrphos ligand **3** [43].

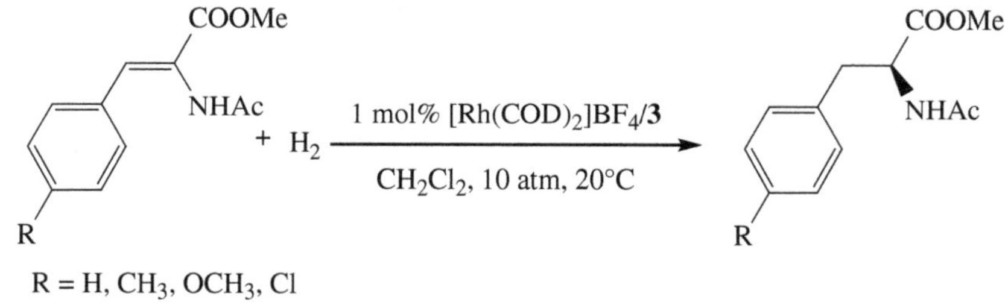

SCHEME 6.5 Asymmetric hydrogenation of α-dehydroamino esters catalyzed by $[Rh(COD)_2]BF_4$@**3** [43].

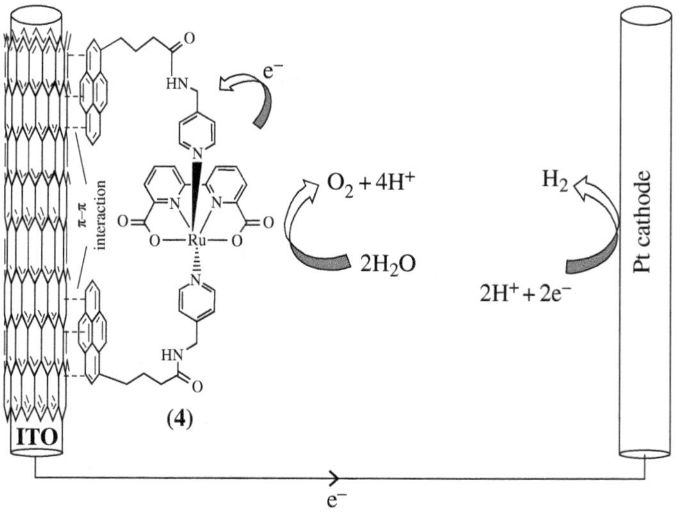

SCHEME 6.6 Electrocatalytic oxidation of water by a monomeric ruthenium catalyst **4**@MWCNT [44].

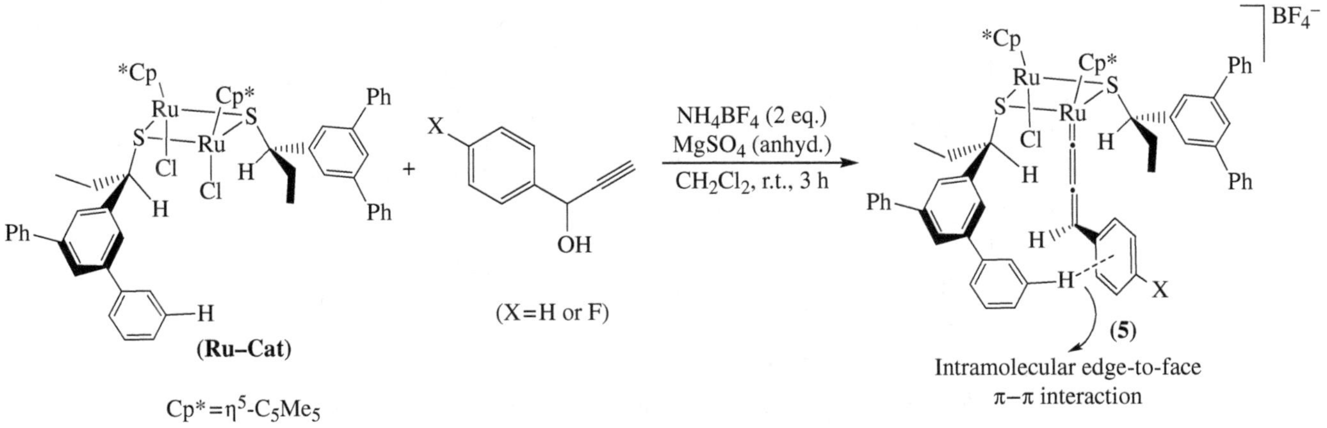

SCHEME 6.7 Edge-to-face aromatic π–π interaction in an optically active diruthenium–allenylidene complex **5** [45].

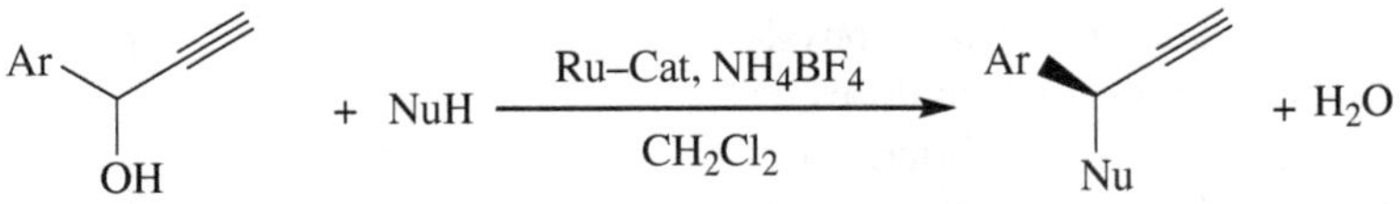

SCHEME 6.8 Enantioselective propargylic substitution reactions of propargylic alcohols with nucleophiles catalyzed by **Ru–Cat** [45].

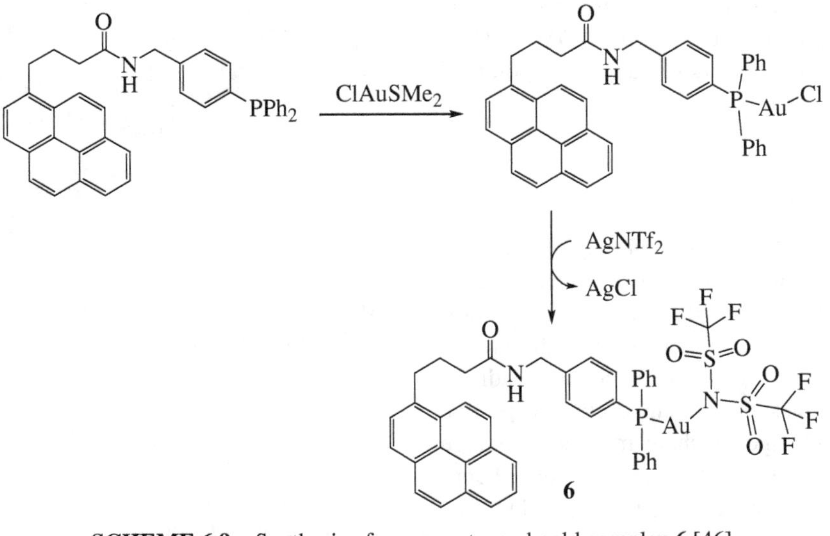

SCHEME 6.9 Synthesis of a pyrene-tagged gold complex **6** [46].

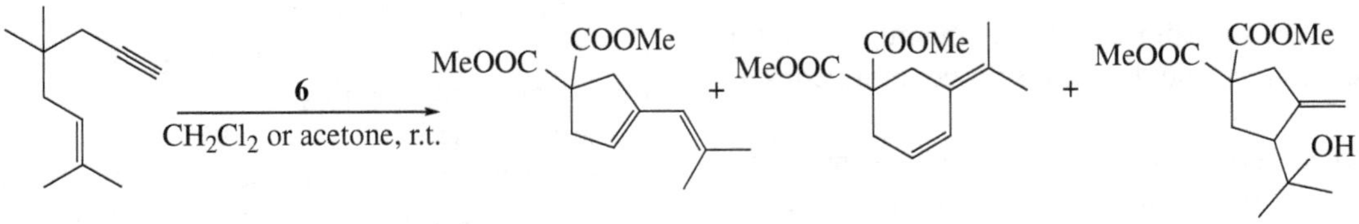

SCHEME 6.10 Cycloisomerization of enyne catalyzed by **6**@MWCNT [46].

6.3.2 Pyrene-Tagged Gold(I) Complex

A pyrene-tagged gold complex **6** (Scheme 6.9) has been synthesized and immobilized onto MWCNTs through π–π interactions [46]. Complex **6** shows a reversible immobilization similar to that of **1**, dependent on the polarity of the solvents and temperature, and behaves as a supported homogeneous "boomerang" catalyst. The immobilized catalyst shows a high activity and selectivity in cycloisomerization of enyne (Scheme 6.10), as its homogeneous counterpart [46]. Recycling was carried out at low temperatures.

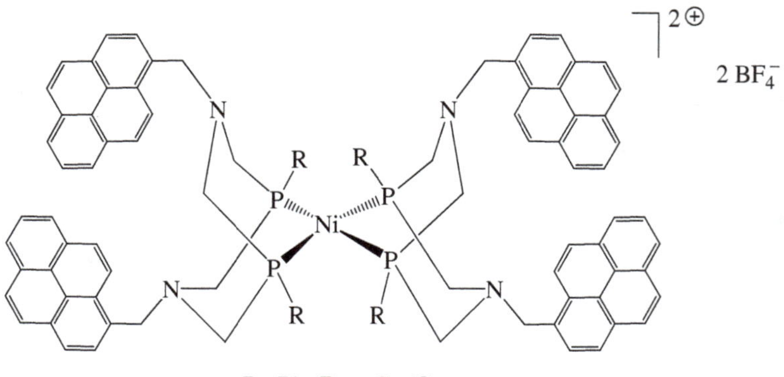

R = Ph (**7**) or Cy (**8**)

SCHEME 6.11 Ni(II) complexes **7** and **8** with pyrene-based ligand [47].

6.3.3 Pyrene-Tagged Nickel(II) Complexes

Two Ni(II) complexes (**7** and **8**, Scheme 6.11) were physisorbed on MWCNTs deposited onto commercial gas diffusion layers (GDL), a supporting material in a membrane electrode assembly, through π–π stacking interactions between the pyrene moieties and graphene motifs [47]. The Ni-functionalized MWCNTs/GDL electrodes display electrocatalytic activities for H_2 evolution as well as for H_2 oxidation from protonated *N,N*-dimethylformamide ([DMFH]OTf) in CH_3CN with similar catalytic rates and the overpotential of 0.1 V [47].

6.4 MAGNETISM

In molecule-based magnets there exist a variety of magnetic interactions between the spin carriers, such as metallic ions and radicals, which lead to interesting magnetic properties. The coupling spin carriers can interact through various exchange mechanisms, such as direct exchange, indirect exchange, double exchange, super exchange, etc. Irrespective of the nature of the exchange mechanisms, the exchange interactions are usually quantified through the Heisenberg–Dirac–van Vleck (HDVV) spin Hamiltonian (Eq. 6.1) [48]:

$$\hat{H} = -\sum_{i<j} J_{ij}\hat{S}_i \cdot \hat{S}_j \tag{6.1}$$

where $\hat{S}_i$ and $\hat{S}_j$ are the spin angular momentum operators on spin carriers *i* and *j* and J_{ij} is the spin coupling constant between them.

In the magnetic coupling field, π–π stacking interaction also plays an important role and some examples are discussed in the following sections.

6.4.1 Cu(II) Complexes

A binuclear Cu(II) complex $[Cu_2(\mu_2\text{-}OOCCH_3)_2(\text{bpydiol-H})_2(H_2O)_2]$ (**9**, Fig. 6.3) (bpydiol-H = monodeprotonated 2,2′-bypyridine-3,3′-diol), bridged by two acetate ligands and with a π–π stacking between the adjacent pyridine rings, shows an anomalous magnetic interaction [49]. A weak magnetic coupling with experimental fitting $2J = -59.61\ cm^{-1}$ is observed [49]. Based on theoretical calculations, the acetate bridge ligand in **9** leads to an antiferromagnetic coupling with $2J = -166.72\ cm^{-1}$, whereas the π–π interaction results into a ferromagnetic interaction with $2J = 21.0\ cm^{-1}$. Therefore, the deviation observed in the experimental fitting can be considered due to the offset from the antiferromagnetic coupling through the acetate bridge ligand and the ferromagnetic interaction by π–π stacking [49].

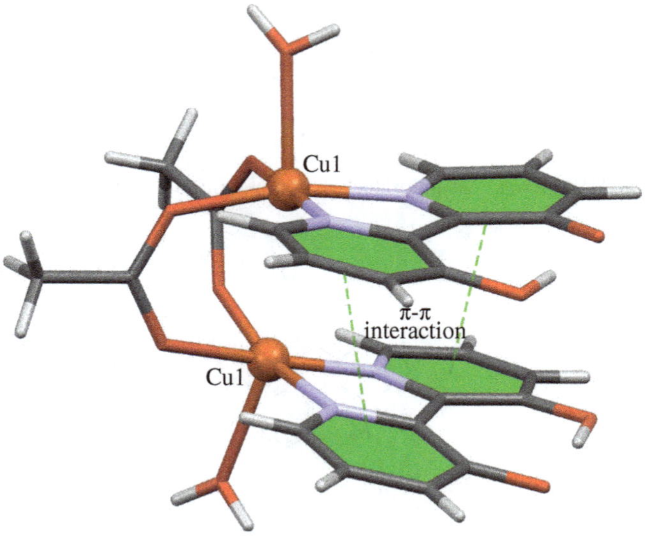

FIGURE 6.3 π–π stacking interaction between the adjacent pyridine rings in **9** [49].

An effect of π–π stacking interaction was also observed in the mononuclear Cu(II) complex $[CuBr(PMP)]ClO_4$ (**10**, Fig. 6.4) (PMP = 2,9-bis(pyridin-2-methoxyl)-1,10-phenanthroline) [50]. In its crystal structure there exist three types of π–π stackings involving symmetrically related 1,10-phenanthroline rings slipped in relevant distances with symmetry codes: (1−*x*, −*y*, 1−*z*), (−*x*, 1−*y*, 1−*z*), and (1−*x*, 1−*y*,1−*z*). A weak antiferromagnetic coupling between adjacent Cu(II) ions with the Weiss constant $\theta = -2.99$, $K = -2.08\ cm^{-1}$ is observed in the variable temperature magnetic susceptibility measurement [50].

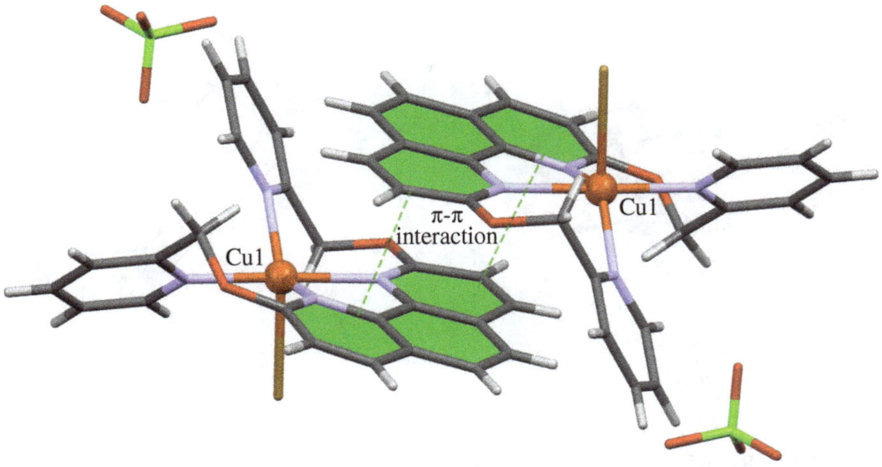

FIGURE 6.4 π–π stacking interaction in **10** [50].

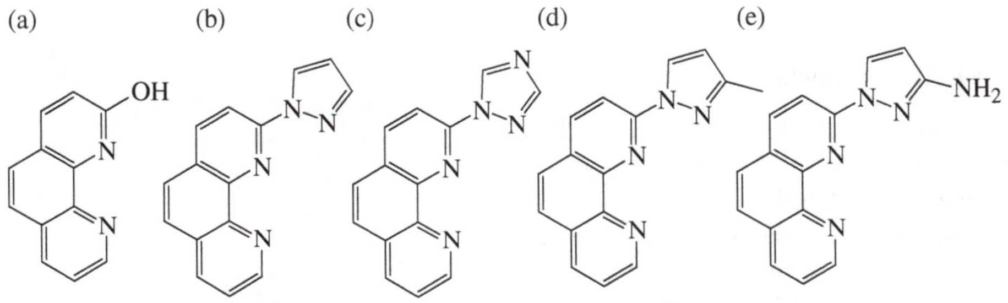

SCHEME 6.12 Ligands used in the syntheses of Cu(II) complexes **11–17**; (a) 1,10-phenanthrolin-2-ol (PhenOH) (b) 2-(1*H*-pyrazol-1-yl)-1, 10-phenanthroline (PhenP) (c) 2-(1*H*-1,2,4-triazol-1-yl)-1,10-phenanthroline (PhenTA) (d) 2-(3-methyl-pyrazol-1*H*-yl)-1,10-phenanthroline (PhenMP) (e) 2-(3-amino-pyrazol-1*H*-yl)-1,10-phenanthroline (PhenAP) [55].

A theoretical study of the magnetic interaction by density function calculations was undertaken according to Equation 6.2, assuming spin Hamiltonian (Eq. 6.1), in the three types of π–π stacking:

$$J = \frac{E_{BS} - E_{HS}}{4S_1S_2} \quad (6.2)$$

where E_{HS} = high-spin state and E_{BS} = broken-symmetry state [51–53]. The first two types resulted in antiferromagnetic couplings with $2J = -12.40\,cm^{-1}$ and $2J = -9.74\,cm^{-1}$, and the third type led to a weak ferromagnetic interaction with $2J = -4.28\,cm^{-1}$ [50]. Therefore, the magnitude of the overall antiferromagnetic coupling is larger than that of the ferromagnetic coupling. In this case, only the sign of the ferromagnetic coupling accords with McConnell I spin polarization [54], but the antiferromagnetic coupling signs do not. McConnell I spin-polarization mechanism considers that a global ferromagnetic coupling develops from the interaction between spin densities of opposite signs, whereas an antiferromagnetism dominates when the coupling arises from the interaction between spin densities of the same sign [54].

Five N-donor ligands (Scheme 6.12) were used to synthesize a series of seven mononuclear Cu(II) complexes, [CuCl(OH)(PhenOH)] (**11**), $[Cu(ONO_2)_2(PhenP)]$ (**12**), $[Cu(PhenTA)_2](ClO_4)_2{\cdot}CH_3CH_2OH$ (**13**), $[CuCl_2$ (PhenMP)] (**14**), $[CuBr_2$(PhenMP)] (**15**), $[Cu(ONO_2)(PhenAP)(HOCH_3)]NO_3$(**16**), and $[CuCl_2$(PhenAP)] (**17**), exhibiting π–π stacking interactions as the major intermolecular forces [55]. The spin densities of these seven complexes were calculated and their intermolecular magnetic coupling strengths were studied.

It was observed that the spin density is one of the major factors that contribute to the magnetic coupling strength in the π–π stacking systems, as proposed by McConnell [54]. In **11–17**, the smaller spin densities resulted in weak π–π stacking magnetic coupling strengths. Moreover, in addition to spin density, the distance of the short contact between the π–π stacking planes is another major factor that contributes to the magnetic coupling strength of the π–π stacking system [55]. A decrease in the separation of the π–π stacking planes results in an increase in spin densities on the relevant short contact atoms. Therefore, in order to prepare compounds with strong π–π stacking magnetic coupling strength, it may be necessary to

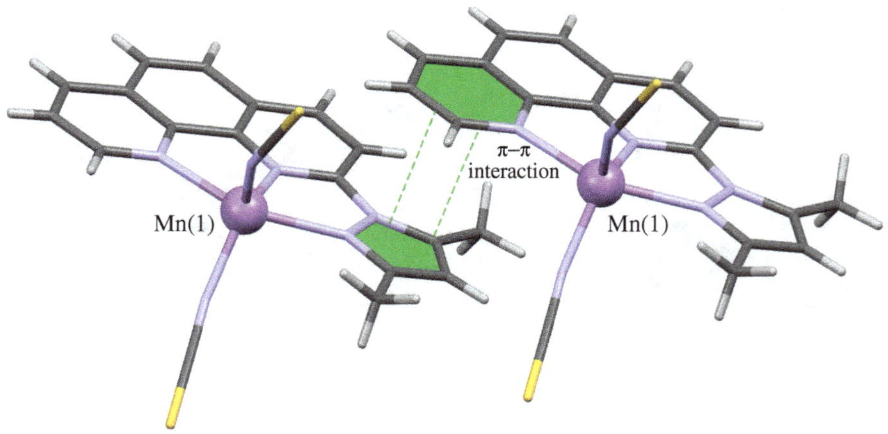

FIGURE 6.5 π–π stacking in mononuclear manganese(II) complex **18** [56].

increase the spin density on the short contact atoms and decrease the short contact distance.

6.4.2 Mn(II) and Mn(III) Complexes

The mononuclear manganese(II) complex [Mn(NCS)$_2$(DPP)] (**18**, Fig. 6.5) (DPP = 2-(3,5-dimethyl-1*H*-pyrazol-1-yl)-1, 10-phenanthroline) exhibits π–π stacking between adjacent molecules involving a pyridyl ring and the symmetry-related pyrazolyl ring, which leads to the formation of a supramolecular one-dimensional (1D) chain [56]. Variable temperature magnetic susceptibility measurements with the experimental fitting $J = -0.11\,\text{cm}^{-1}$ indicate that there is a weak antiferromagnetic coupling between the adjacent mononuclear Mn(II) molecules via π–π stacking pathway [56].

Theoretical calculations based on density functional theory (DFT) also suggested a weak antiferromagnetic coupling between the adjacent Mn(II) ions with $J = -0.02\,\text{cm}^{-1}$, which can be explained by the spin density population and McConnell I spin-polarization mechanism [54]. It was suggested that the distance between the atoms involving π–π stacking plays an important role that dominates the magnitude of the magnetic coupling.

An effect of π–π interactions in the magnetic properties is also observed in the mononuclear manganese(III) complex [Mn(O$_2$CMe)(3-OEtsalen)(H$_2$O)] **19**·CH$_3$CN (Fig. 6.6) and in the binuclear one [{Mn(3-OEtsalen)(H$_2$O)}$_2$(μ-O$_2$CMe)](ClO$_4$) **20**·4CH$_3$CN (Fig. 6.7), where 3-OEtsalenH$_2$ represents *N,N′*-ethylenebis(3-ethoxysalicylideneaminate) [57]. Intermolecular hydrogen bonding in both **19** and **20** are also observed. Such dual non-covalent interactions lead to dimerization of **19** and formation of a *syn–anti* 1D chain for **20**. The observed small negative *J* values (*ca.* $-0.6\,\text{cm}^{-1}$) for both **19** and **20** suggest very weak antiferromagnetic intrachain exchange coupling between the Mn(III) spin centers [57]. As there is no covalent linking between the mononuclear species in **19**, the exchange interactions are expected to occur through π–π stacking interactions and/or hydrogen bonding. The magnetic properties are similar in

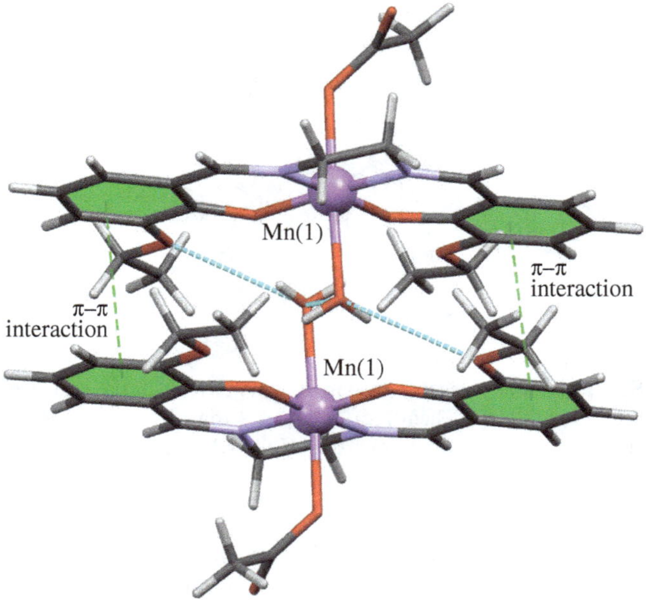

FIGURE 6.6 π–π interaction and hydrogen bonding in a mononuclear manganese(III) complex **19** (solvent molecules are omitted) [57].

both Mn(III) complexes and thus the bridging acetate group in **20** doesn't play a significant role for the exchange coupling pathway. The overall exchange coupling interaction in **20** is due to an effective π–π stacking interaction as well as to H-bonding [57].

6.4.3 Ni(II) Complex

Intermolecular π–π interactions are exhibited by the mononuclear Ni(II) complex [NiL$_2$]·0.5Me$_2$CO (**21**, Fig. 6.8) (HL = *N*-(8-quinolyl)salicylaldimine) [58]. They lead to a π-dimer, which further connects with others via H-bonding interactions forming a 1D chain. Magnetic susceptibility measurements indicate that there is a very weak magnetic coupling interaction within the π-dimer and the zero field splitting dominates the magnetic behavior. The expected stronger magnetic coupling interactions via π-orbital

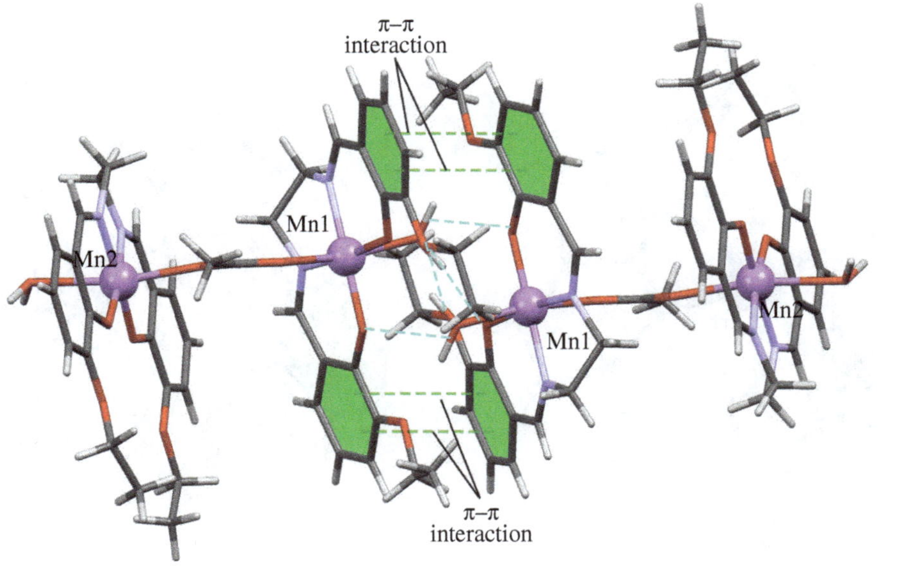

FIGURE 6.7 π–π interaction and hydrogen bonding in the binuclear manganese(III) complex **20** (perchlorate ions and solvent molecules are omitted) [57].

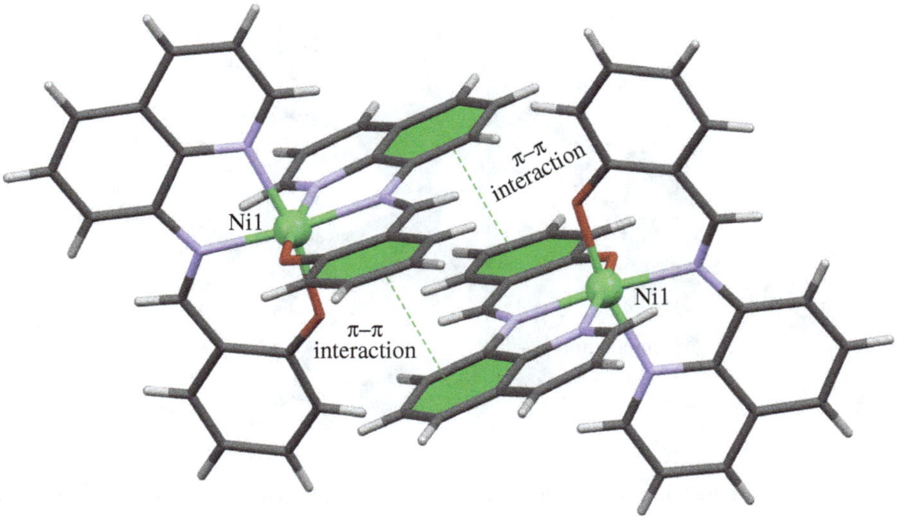

FIGURE 6.8 π–π interactions exhibited by $[NiL_2]\cdot 0.5Me_2CO$ (**21**, solvent molecules are omitted) [58].

overlapping of the ligands and spin-polarization effect were not observed in **21** since the unpaired electrons are mostly localized on the Ni(II) ion and do not delocalize to the ligands, which is in consistence with the DFT calculation [58].

6.5 PHOTOPHYSICAL PROPERTIES

The exploration of novel electroluminescent (EL) materials with specific optoelectronic properties is a matter of current interest. π–π interactions can control various physical and chemical properties, which involve the functions of organic EL devices, including light-emitting electrochemical cells [59–62] and organic light-emitting diodes [63–65]. Examples of some photoluminescent complexes where π–π interactions play an important role on their photophysical properties are given in the following sections.

6.5.1 Iridium Complexes

Intramolecular π–π stacking interactions are observed in the blue-green-emitting cationic iridium(III) complex $[Ir(ppy)_2(phpzpy)]PF_6$ (**22**, Fig. 6.9a), with two deprotonated 2-phenylpyridine (Hppy) cyclometalated ligands and 2-(1-phenyl-1*H*-pyrazol-3-yl) pyridine (phpzpy) as the ancillary ligand [67]. The similar complexes $[Ir(ppy)_2(F_2phpzpy)]PF_6$ (**23**, Fig. 6.9b) and $[Ir(ppy)_2(F_5phpzpy)]PF_6$ (**24**, Fig. 6.9c) (where F_2phpzpy is 2-(1-(3,5-difluorophenyl)-1Hpyrazol-3-yl)pyridine and F_5phpzpy is 2-(1-pentafluorophenyl-1*H*-pyrazol-3-yl)-pyridine), having two and five substituent fluorine atoms at the pendant phenyl rings on the ancillary

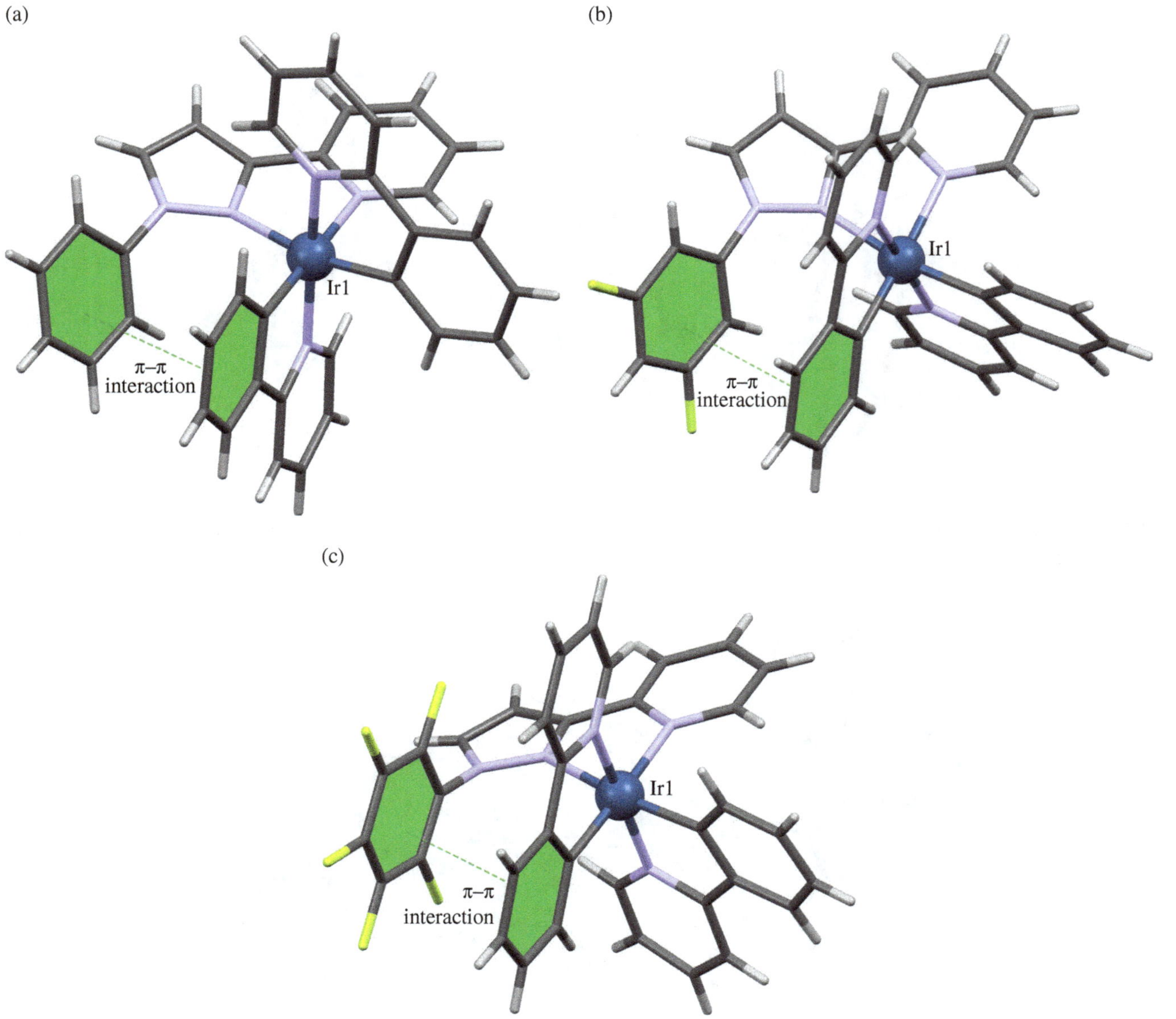

FIGURE 6.9 Intramolecular π–π stacking interaction in (a) complex **22**, (b) complex **23**, and (c) complex **24** (hexafluorophosphate anions are omitted) [66].

ligands, have also been synthesized to compare the photoluminescent efficiency in solution with that of **22** [67]. The fluorine-substituted pendant phenyl rings on the ancillary ligands (in **23** and **24**) stack more parallel and closer to the phenyl rings of the ppy ligands resulting in a significantly reinforced intramolecular face-to-face π–π stacking interaction and a high luminescent efficiency in solution. By theoretical calculations, it is found that the intramolecular π–π stacking interaction remains on both the ground and the excited triplet states in **24**. Among the three complexes, **24** exhibits the smallest structural deviations between the ground and the lowest triplet states and shows a significantly enhanced photoluminescent efficiency in solution [66].

The robust cationic iridium(III) complexes $[Ir(dfppz)_2(phsp)]PF_6$ (**25**) and $[Ir(dfppz)_2(diphsp)]PF_6$ (**26**, Fig. 6.10) (where dfppz = cyclometalated anionic dfppz 1-(2,4-difluorophenyl) pyrazole, phsp = 3-phenyl-4,5-diaza-9,9′-spirobifluorene, and diphsp = 3,6-diphenyl-4,5-diaza-9,9′-spirobifluorene) show a strong intramolecular face-to-face π-stacking between the pendant phenyl ring of phsp or diphsp and the difluorophenyl ring of the cyclometalated anionic dfppz ligand [68]. The presence of such interaction minimizes the expansion of the metal–ligand bonds in the excited state, resulting in a longer device lifetime in the light-emitting electrochemical cell (LEC) devices. Much suppressed current densities were observed by the LEC devices based on **25** and **26** because of densely extended π-stacking structures. The lifetime of the devices based on these complexes is 585 and 102 min, respectively [68]. The enhancement on the device lifetime of **25** as compared to that of **26** has benefited by the less current flux and enhanced molecular stability.

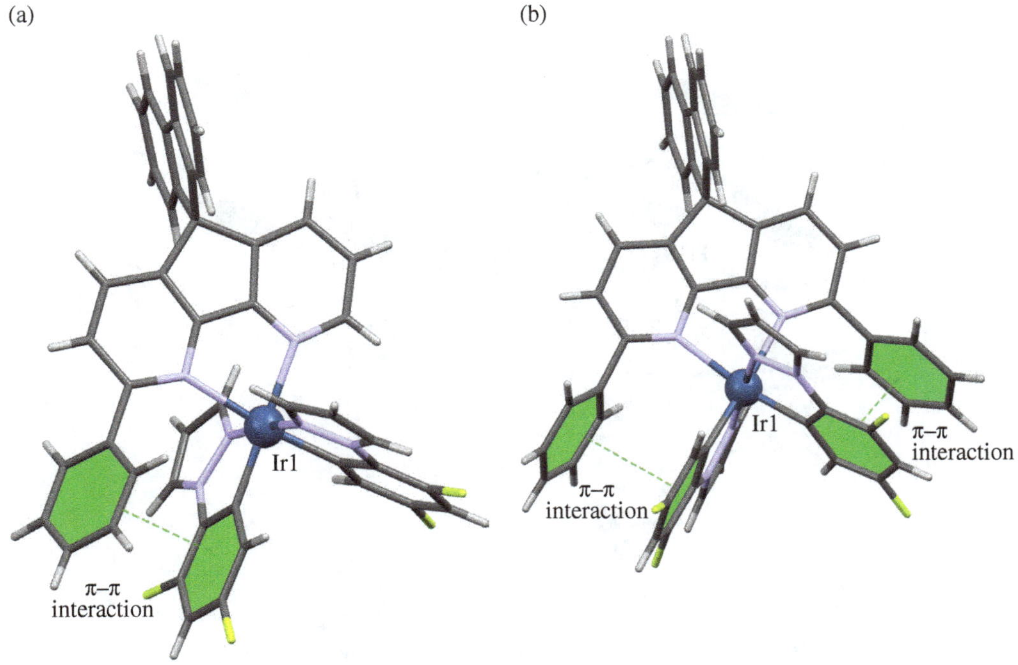

FIGURE 6.10 Intramolecular face-to-face π-stacking in (a) complex **25** and (b) complex **26** (hexafluorophosphate anions are omitted) [68].

6.5.2 Platinum Complexes

Like iridium(III) complexes, platinum(II) complexes also show promising luminescence properties. Slaughter et al. [69] examined the luminescent properties of a series of isocyanide platinum(II) complexes of general formula *cis*-[$PtCl_2(CNAr)_2$] (Ar = substituted phenyl). Extended Pt···Pt chain structures and π–π interactions are observed in [$PtCl_2(CNC_6H_4$-p-F$)_2$] (**27**), [$PtCl_2(CNC_6H_5)_2$] (**28**), and [$PtCl_2(CNC_6H_4$-*p*-$CH_3)_2$] (**29**) (Fig. 6.11). All the three complexes display Pt···Pt chain-centered luminescence assignable to ($d\sigma^* \rightarrow p\sigma$) excited states, demonstrating the existence of substantial orbital communication along the metal–metal chains. Variable temperature structural studies showed that both M···M and π–π interactions contract upon cooling and the synergy with π–π and other interactions (CF_3···π and CH_3···π) is necessary to stabilize the extended M···M chain structures [69].

The neutral Pt(II) complex **30** (Fig. 6.12), containing the N^N^N azole-based tridentate chromophoric ligand 2,6-bis(3-(trifluoromethyl)-1*H*-1,2,4-triazol-5-yl)pyridine (pyC5-CF3-tzH_2) and 4-amylpyridine as the ancillary ligand, spontaneously assembles into highly ordered structures (fibers) through metal···metal and π–π interactions. Selected area electron diffraction (SAED) and small-angle X-ray scattering/wide-angle X-ray scattering (SAXS/WAXS) experiments showed an intermolecular distance of 3.44 and 3.28 Å, respectively, attributable to the $d_z{}^2\pi \cdots d_z{}^2\pi$ interactions along the growth axis of the fibers in **30**. These discrete crystalline fibers efficiently emit polarized yellow light with a very high photoluminescence quantum yield (PLQY) value of 74% [70].

The related water-soluble neutral Pt(II) complex **31** consisting of the tridentate chromophoric ligand 2,6-bis (tetrazole-5-yl)pyridine and an ancillary pyridine ligand with two tetraethylene glycol (TEG) chains was also reported [71]. The TEG chains on the ancillary ligand promote its solubilization in water. **31** aggregates more efficiently in aqueous solutions than in an organic solvent (CH_2Cl_2) and shows luminescent properties. The aromatic (hydrophobic) part is mainly responsible for the aggregation process induced by π–π interactions, which leads to a clear blueshift of the luminescence when going from CH_2Cl_2 to water. A fourfold enhancement of the luminescence quantum yield is observed in water, as compared to CH_2Cl_2 [71]. **31** can form phosphorescent hydrogels upon interaction (host–guest) of the TEG chains with cyclodextrins (CDs), and the luminescence of the resulting hydrogels can be tuned by changing the size of the CD host (α- or β-CD) (Fig. 6.13) [71].

6.6 CONCLUSION

The role of π–π interactions in metal complexes to rationalize structural features and address catalytic, magnetic, and photoluminescent properties is illustrated with suitable examples. In the area of catalysis, such interactions were applied for anchoring catalysts by using pyrene-tagged systems. The anchored pyrene-tagged catalyst on the surface of carbon nanotubes can be easily recycled, and the catalyst immobilization is affected by the polarity of the solvents and by temperature. Concerning magnetism, the magnetic coupling via π–π stacking affects the magnetic properties of metal

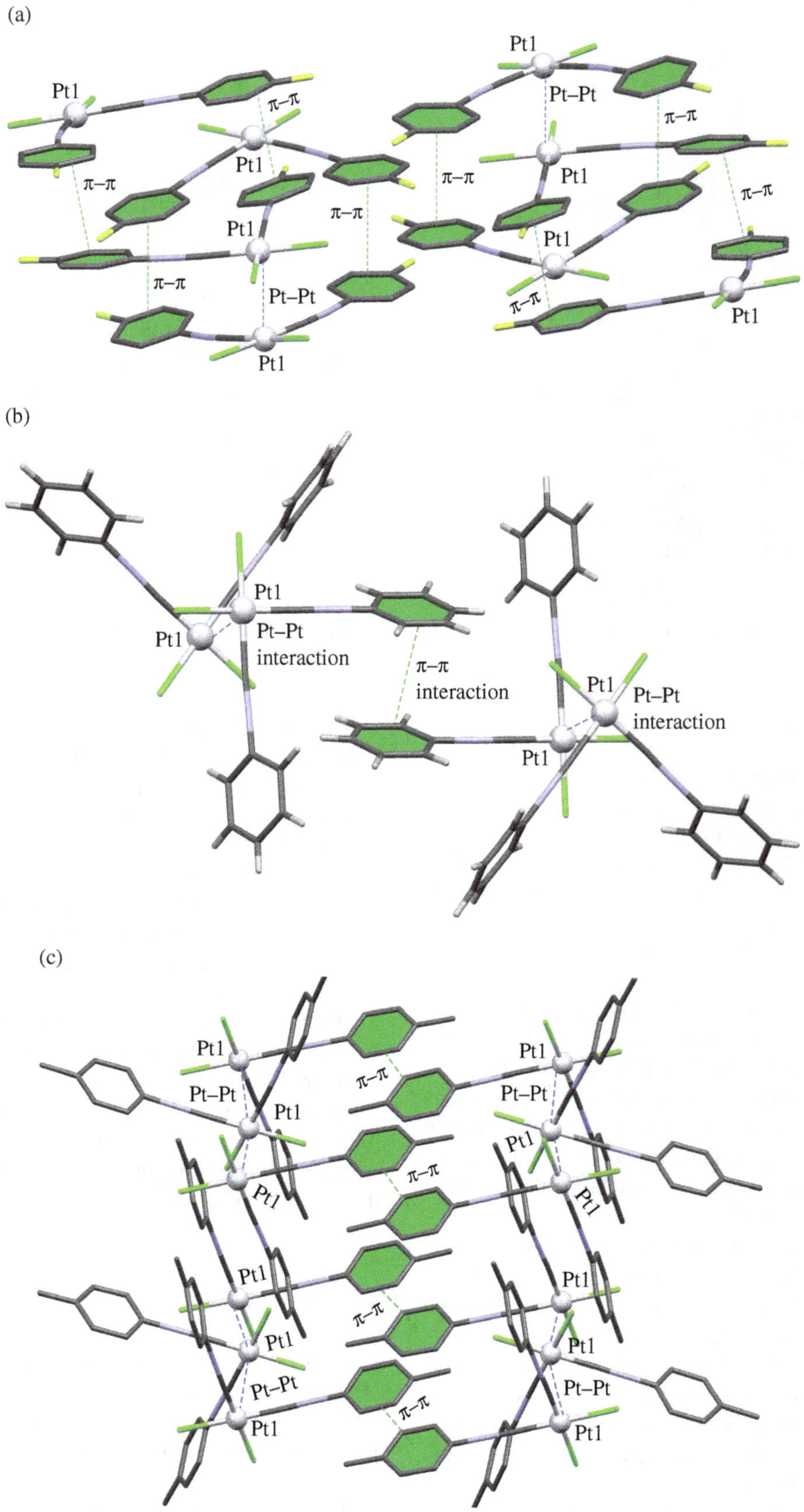

FIGURE 6.11 Pt···Pt chain structures and π–π interactions in (a) complex **27** (H-atoms are omitted), (b) complex **28**, and (c) complex **29** (H-atoms are omitted) [69].

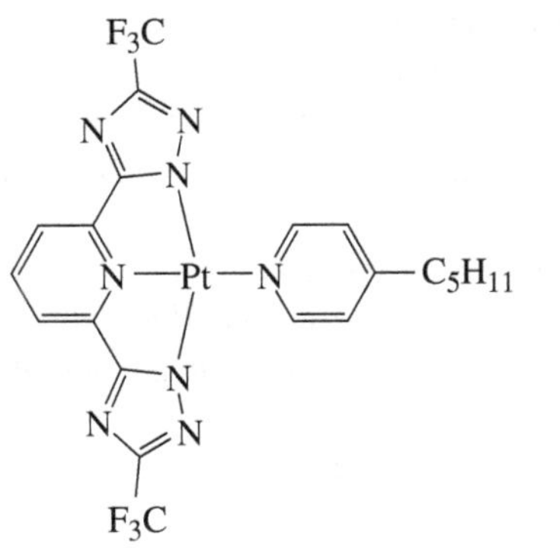

FIGURE 6.12 Molecular structure of the platinum(II) complex **30** [70].

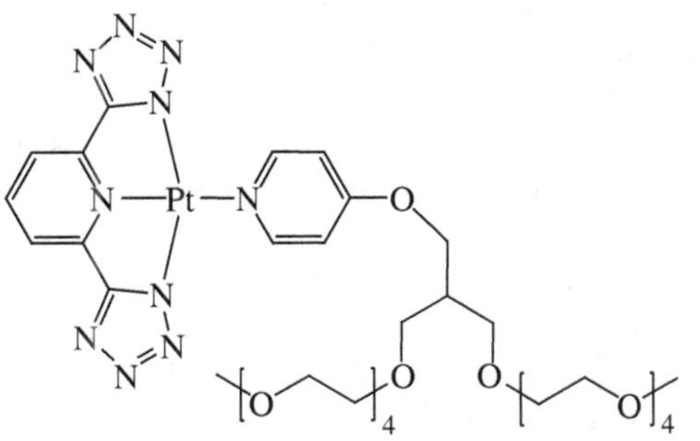

FIGURE 6.13 Molecular structure of the platinum(II) complex **31** [70].

complexes as we discussed for Cu(II), Mn(II), Mn(III), and Ni(II) complexes. DFT calculations also support the existence of such effects. The presence of intramolecular π–π stacking in Ir(III) complexes and intermolecular π–π stacking in Pt(II) complexes enhances the photoluminescence properties.

Such responsive effects of π–π interactions are useful to tune the physical and chemical properties of organic and inorganic compounds, and we can expect further important applications in the future.

ACKNOWLEDGMENTS

The authors are grateful to the Foundation for Science and Technology (FCT) (project UID/QUI/00100/2013), Portugal, for financial support. M.S. acknowledges the FCT, Portugal, for a postdoctoral fellowship (SFRH/BPD/86067/2012).

REFERENCES

1. L. M. Salonen, M. Ellermann, F. Diederich, *Angew. Chem. Int. Ed.* 2011, *50*, 4808.
2. K. Ishihara, M. Fushimi, M. Akakura, *Acc. Chem. Res.* 2007, *40*, 1049.
3. R. Gyepes, S. Pacigová, J. Tatiersky, M. Sivák, *J. Mol. Struct.* 2013, *1041*, 113.
4. E. Despagnet-Ayoub, S. Schigand, L. Vendier, M. Etienne, *Organometallics* 2009, *28*, 2188.
5. N. Schultheiss, D. R. Powell, E. Bosch, *Inorg. Chem.* 2003, *42*, 5304.
6. T. M. Swager, *ACS Macro Lett.* 2011, *1*, 3.
7. S. Tsuzuki, K. Honda, R. Azumi, *J. Am. Chem. Soc.* 2002, *124*, 12200.
8. G. Zeni, R. C. Larock, *Chem. Rev.* 2004, *104*, 2285.
9. T.-H. Huang, M.-H. Zhang, *Inorg. Chim. Acta* 2014, *410*, 150.
10. T.-H. Huang, M.-H. Zhang, C.-Y. Gao, L.-T. Wang, *Inorg. Chim. Acta* 2013, *408*, 91.
11. H. Baruah, M. W. Wright, U. Bierbach, *Biochemistry* 2005, *44*, 6059.
12. A. Migliore, *J. Chem. Theory Comput.* 2011, *7*, 1712.
13. W. Yan, L. Zhang, D. Xie, J. Zeng, *J. Phys. Chem. B* 2007, *111*, 14055.
14. X. Zhang, I. Lee, A. J. Berdis, *Biochemistry* 2005, *44*, 13101.
15. W. T. Góźdź, *Langmuir* 2008, *24*, 12458.
16. A. S. Mahadevi, G. N. Sastry, *Chem. Rev.* 2012, *113*, 2100.
17. Y. Shi, M. J. Steenbergen, E. A. Teunissen, L. S. Novo, S. Gradmann, M. Baldus, C. F. Nostrum, W. E. Hennink, *Biomacromolecules* 2013, *14*, 1826.
18. B. Nohra, Y. Yao, C. Lescop, R. Réau, *Angew. Chem. Int. Ed.* 2007, *46*, 8242.
19. X.-J. Wang, L.-C. Gui, Q.-L. Ni, Y.-F. Liao, X.-F. Jiang, L.-H. Tang, Z. Zhang, Q. Wu, *CrystEngComm.* 2008, *10*, 1003.
20. J.-P. Sauvage, J.-P. Collin, J.-C. Chambron, S. Guillerez, C. Coudret, *Chem. Rev.* 1994, *94*, 993.
21. A. R. Paital, A.-Q. Wu, G.-C. Guo, G. Aromí, J. Ribas-Ariño, D. Ray, *Inorg. Chem.* 2007, *46*, 2947.
22. J. G. Planas, C. Masalles, R. Sillanpää, R. Kivekäs, F. Teixidor, C. Viñas, *CrystEngComm* 2006, *8*, 75.
23. D. Gut, A. Rudi, J. Kopilov, I. Goldberg, M. Kol, *J. Am. Chem. Soc.* 2002, *124*, 5449.
24. M. Yamakawa, I. Yamada, R. Noyori, *Angew. Chem. Int. Ed.* 2001, *40*, 2818.
25. C. E. Smith, P. S. Smith, R. L. Thomas, E. G. Robins, J. C. Collings, C. Dai, A. J. Scott, S. Borwick, A. S. Batsanov, S. W. Watt, S. J. Clark, C. Viney, J. A. K. Howard, W. Clegg, T. B. Marder, *J. Mater. Chem.* 2004, *14*, 413.
26. X. Zhang, P. Xie, Z. Shen, J. Jiang, C. Zhu, H. Li, T. Zhang, C. C. Han, L. Wan, S. Yan, R. Zhang, *Angew. Chem. Int. Ed.* 2006, *45*, 3112.
27. C. Janiak, *J. Chem. Soc. Dalton Trans.* 2000, 3885.
28. Q.-L. Ni, X.-F. Jiang, L.-C. Gui, X.-J. Wang, K.-G. Yang, X.-S. Bi, *New J. Chem.* 2011, *35*, 2471.
29. D. Sredojević, G. A. Bogdanović, Z. D. Tomić, S. D. Zarić, *CrystEngComm* 2007, *9*, 793.
30. X.-J. Wang, H.-X. Jian, Z.-P. Liu, Q.-L. Ni, L.-C. Gui, L.-H. Tang, *Polyhedron* 2008, *27*, 2634.
31. H. H. Monfared, Z. Kalantari, M.-A. Kamyabi, C. Janiak, Z. *Anorg. Allg. Chem.* 2007, *633*, 1945.
32. C. A. Hunter, J. K. M. Sanders, *J. Am. Chem. Soc.* 1990, *112*, 5525.
33. A. Michrowska, K. Mennecke, U. Kunz, A. Kirschning, K. Grela, *J. Am. Chem. Soc.* 2006, *128*, 13261.

34. F. M. de Rege, D. K. Morita, K. C. Ott, W. Tumas, R. D. Broene, *Chem. Commun.* 2000, *18*, 1797.
35. P. Scovazzo, A. E. Visser, J. H. Davis Jr., R. D. Rogers, C. A. Koval, D. L. DuBois, R. D. Noble, Supported Ionic Liquid Membranes and Facilitated Ionic Liquid Membranes. In Ionic Liquids: Industrial Applications to Green Chemistry; R. D. Rogers, K. R. Seddon (Eds), ACS Symposium Series *818*; American Chemical Society: Washington, DC, 2002; p 68.
36. J. Horn, F. Michalek, C. C. Tzschucke, W. Bannwarth, *Top. Curr. Chem.* 2004, *242*, 43.
37. A. Kirschning, G. Jas, *Top. Curr. Chem.* 2004, *242*, 209.
38. G. Jas, A. Kirschning, *Chem. Eur. J.* 2003, *9*, 5708.
39. P. D. I. Fletcher, S. J. Haswell, E. Pombo-Villar, B. H. Warrington, P. Watts, S. Y. F. Wong, X. Zhang, *Tetrahedron* 2002, *58*, 4735.
40. A. Kirschning, W. Solodenko, K. Mennecke, *Chem. Eur. J.* 2006, *12*, 5972.
41. E. W. McQueen, J. I. Goldsmith, *J. Am. Chem. Soc.* 2009, *131*, 17554.
42. G. Liu, B. Wu, J. Zhang, X. Wang, M. Shao, J. Wang, *Inorg. Chem.* 2009, *48*, 2383.
43. L. Xing, J.-H. Xie, Y.-S. Chen, L.-X. Wang, Q.-L. Zhou, *Adv. Synth. Catal.* 2008, *350*, 1013.
44. F. Li, B. Zhang, X. Li, Y. Jiang, L. Chen, Y. Li, L. Sun, *Angew. Chem. Int. Ed.* 2011, *50*, 12276.
45. K. Kanao, Y. Tanabe, Y. Miyake, Y. Nishibayashi, *Organometallics* 2010, *29*, 2381.
46. C. Vriamont, M. Devillers, O. Riant, S. Hermans, *Chem. Eur. J.* 2013, *19*, 12009.
47. P. D. Tran, A. L. Goff, J. Heidkamp, B. Jousselme, N. Guillet, S. Palacin, H. Dau, M. Fontecave, V. Artero, *Angew. Chem. Int. Ed.* 2011, *123*, 1407.
48. J. H. Van Vleck, The Theory of Electric and Magnetic Susceptibilities; Oxford University Press: Oxford, 1932.
49. C. Hou, J.-M. Shi, Y.-M. Sun, W. Shi, P. Cheng, L.-D. Liu, *Dalton Trans.* 2008, 5970.
50. H. Li, S.-G. Zhang, L.-M. Xie, L. Yu, J.-M. Shi, *J. Coord. Chem.* 2011, *64*, 3595.
51. L. Noodleman, *J. Chem. Phys.* 1981, *74*, 5737.
52. L. Noodleman, E. J. Baerends, *J. Am. Chem. Soc.* 1984, *106*, 2316.
53. L. Noodleman, D. A. Case, *Adv. Inorg. Chem.* 1992, *38*, 423.
54. H. M. McConnell, *J. Chem. Phys.* 1963, *39*, 1910.
55. Y.-H. Chi, J.-M. Shi, H.-N. Li, W. Wei, E. Cottrill, N. Pan, H. Chen, Y. Liang, L. Yu, Y.-Q. Zhang, C. Hou, *Dalton Trans.* 2013, *42*, 15559.
56. L. Yu, J.-M. Shi, Y.-Q. Zhang, Y.-Q. Wang, Y.-N. Fan, G.-Q. Zhang, W. Shi, P. Cheng, *J. Mol. Struct.* 2011, *987*, 138.
57. M. Sutradhar, L. M. Carrella, E. Rentschler, *Polyhedron* 2012, *38*, 297.
58. H. Xu, Z. P. Ni, X. M. Ren, Q. J. Meng *J. Mol. Struct.* 2005, *752*, 153.
59. Q. B. Pei, G. Yu, C. Zhang, Y. Yang, A. J. Heeger, *Science* 1995, *269*, 1086.
60. P. Matyba, K. Maturova, M. Kemerink, N. D. Robinson, L. Edman, *Nat. Mater.* 2009, *8*, 672.
61. J. D. Slinker, A. A. Gorodetsky, M. S. Lowry, J. J. Wang, S. Parker, R. Rohl, S. Bernhard, G. G. Malliaras, *J. Am. Chem. Soc.* 2004, *126*, 2763.
62. J. D. Slinker, J. Rivnay, J. S. Moskowitz, J. B. Parker, S. Bernhard, H. D. Abruna, G. G. Malliaras, *J. Mater. Chem.* 2007, *17*, 2976.
63. E. A. Plummer, A. van Dijken, H. W. Hofstraat, L. De Cola, K. Brunner, *Adv. Funct. Mater.* 2005, *15*, 281.
64. L. He, L. Duan, J. Qiao, D. Q. Zhang, G. F. Dong, L. D. Wang, Y. Qiu, *Org. Electron.* 2009, *10*, 152.
65. L. He, L. A. Duan, J. A. Qiao, D. Q. Zhang, L. D. Wang, Y. Qiu, *Org. Electron.* 2010, *11*, 1185.
66. L. He, D. Ma, L. Duan, Y. Wei, J. Qiao, D. Zhang, G. Dong, L. Wang, Y. Qiu, *Inorg. Chem.* 2012, *51*, 4502.
67. L. He, L. Duan, J. Qiao, D. Zhang, L. Wang, Y. Qiu, *Chem. Commun.* 2011, *47*, 6467.
68. H.-F. Chen, W.-Y. Hung, S.-W. Chen, T.-C. Wang, S.-W. Lin, S.-H. Chou, C.-T. Liao, H.-C. Su, H.-A. Pan, P.-T. Chou, Y.-H. Liu, K.-T. Wong, *Inorg. Chem.* 2012, *51*, 12114.
69. I. M. Sluch, A. J. Miranda, O. Elbjeirami, M. A. Omary, L. M. Slaughter, *Inorg. Chem.* 2012, *51*, 10728.
70. M. Mauro, A. Aliprandi, C. Cebrián, D. Wang, C. Kübelc, L. De Cola, *Chem. Commun.* 2014, *50*, 7269.
71. N. K. Allampally, M. Bredol, C. A. Strassert, L. De Cola, *Chem. Eur. J.* 2014, *20*, 16863.

7

NON-COVALENT STABILIZATION IN TRANSITION METAL COORDINATION AND ORGANOMETALLIC COMPLEXES

PREDRAG PETROVIĆ[1], JEAN-PIERRE DJUKIC[2], ANDREAS HANSEN[3], CHRISTOPH BANNWARTH[3], AND STEFAN GRIMME[3]

[1] *Department of Chemistry, Innovation Center, University of Belgrade, Belgrade, Serbia*
[2] *Institut de Chimie de Strasbourg, CNRS UMR 7177, Université de Strasbourg, Strasbourg, France*
[3] *Mulliken Center for Theoretical Chemistry, Institut für Physikalische und Theoretische Chemie, Universität Bonn, Bonn, Germany*

7.1 INTRODUCTION

What would nature look like if, by some means, one could switch non-covalent interactions (NCIs) off? Nature would surely lose its greatest wealth, that is, flexibility, adaptability, and diversity, which makes, above all, life possible. Chemistry would be reduced to the art of establishing still a wide range of covalent bonds.

The London force, also known as dispersion [1], is generally considered today as one of the most essential components of the van der Waals NCIs [2], which until the late 1920s were essentially considered to be a combination of the Debye and Keesom interactions [3]. Introduction of the London force as a new concept within the general theory of chemical bonding and molecular forces was a major fundamental breakthrough, which unlocked the resolution of the ancient enigma [4] of condensed phase formation [5]. The "cohesion of matter" was a major source of sustained and often fierce debates during the 1920s [6].

The dispersion force formalized by F. London and R. Eisenschitz in 1930 [7] using the theoretical framework of quantum physics [8] can be defined as a permanently attractive interaction that arises upon *mutual action at a distance* between instantaneous induced electronic dipoles arising at polar or apolar molecules (or fragments of them) or at monoatoms [1, 5, 9]. Dispersion is an attractive interaction caused by long-range electron correlation. Its influence on individual covalent bonds is rather weak compared to the more local correlation, exchange interactions, and Pauli repulsion. The strength of dispersion lies in its collective action at distance, which becomes more energetically favorable as the polarizability and the size of the considered fragments increase. Dispersion was swiftly assimilated by physical chemists and incorporated into the then nascent field of reaction thermodynamics. Already in 1932 [10], H. Eyring includes dispersion as an essential energy component of his kinetic theory of collision between atoms for the calculation of collision energy potentials. In 1955 [11], Pitzer attempts a first evaluation of dispersion contribution to the bond energies of diatomic molecules. Noteworthy are the 1957s conjectures of Bunnett [12] on the possible correlation of nonlinearities in the kinetics of nucleophilic substitution reactions due to the dispersion stabilization of transition states arising from interactions of high polarizability reactants. In 1959 [13], Spinner echoes Bunnett's conjecture by stressing the need for including dispersion in the calculation of the "cohesion energies" of transition states of bimolecular nucleophilic displacement reactions of halide ion with alkyl halide.

If, by its novelty, dispersion attracted a great deal of interest in the first half of the twentieth century, its assimilation as an essential contribution to chemical bonding remained purely that of an abstract albeit physically grounded concept, difficult to compute on polyatomic molecules, hardly observable experimentally [14], and for this

Non-Covalent Interactions in the Synthesis and Design of New Compounds, First Edition.
Edited by Abel M. Maharramov, Kamran T. Mahmudov, Maximilian N. Kopylovich, and Armando J. L. Pombeiro.

reason subject to persistent skepticism among experimentalists about its actual energetic relevance in "real-world" chemistry. This skepticism did not preclude the rise of a new field, that is, supramolecular chemistry [15], and of new concepts that are based on the reversible action of NCIs, such as dynamic combinatorial chemistry [16].

In the field of coordination and organometallic chemistries of d-block transition metals, the role of dispersion and more generally NCIs has long been overlooked though.

Like proposed herein, a new look at this essential physical effect may indeed provide some answers to old issues around the stabilization of complexes that apparently violate the Langmuir–Sidgwick [17] 18 valence electron rule (abbr. as LS rule) [18], which is the extension to transition metals of the empirical octet rule proposed by Lewis for main group elements. The LS rule, as a rather consistent empirical system for predicting the viability of metal complexes, has had and still has a major role in the development of d-block metals' chemistry, notwithstanding permanent controversies on its universal pertinence that appear cyclically in the literature, revealing schismatic trends among experimentalists and theoreticians [19]. Such a look over NCIs may also unveil new paradigms for the engineering of advanced functional molecules.

Since dispersion cannot be dissociated from other NCIs, complete description of its role in bonding has been conditioned by the wake of new theoretical methods capable of embracing the complexity of real transition metal-containing molecular systems. Such an endeavor was initiated by experimentalists and theoreticians investigating the so-called aurophilic effect [20], other "metallophilic" interactions [21], and other so-called d–d interactions [22], for which direct non-covalent metal–metal interactions and relativistic effect are thought to make most of the complexes' cohesion. Recent advances in the development of dispersion-corrected theoretical methods pave the way for a comprehensive treatment of the role of NCIs in transition metal chemistry.

This chapter provides an insight on past and present cases where NCIs allow the stabilization of d-block transition metal complex assemblies or complexes that are obviously challenging the reductionist LS rule.

The scope is deliberately restrained to solution-persistent molecular situations of significant fundamental importance, which still represent a limited number of cases among the large corpus of articles published in the fields of coordination and organometallic chemistries in the past decades.

Solvation of any molecular system [23] and particularly transition metal complexes may challenge the attractive and stabilizing effect of NCIs acting between molecules or within molecules. Therefore, this chapter will not cover the wealth of cases reporting on d–d interactions with oligomers of d^{8-10} metal complexes that were mostly observed in the crystal state and only in rare cases evidenced *in solutio*, even though a steadily growing interest is focused on investigating the use of this intermetal interaction for the elaboration of new functional molecules [24]. So-called agostic C—H···M interactions, which are borderline cases between covalent and non-covalent bonds, will not be treated therein neither [25].

In some cases treated herein, Yang [26] and Piquemal's [27] density reduced gradient method of visualization of local NCIs is used to materialize in a reader-friendly and intuitive way the local NCIs in molecules that for decades defied the understanding of chemists. This method of detection of NCIs in the real space is based on electron density ρ and its derivatives. Very much complementary to analyses based on the Quantum Theory of Atoms in Molecule (QTAIM) [28], NCI region analysis provides an intuitive 3D isosurface representation of local steric (Pauli) repulsion, van der Waals interactions, and other local attractive NCIs. By the choice of an appropriate cutoff value for s, the strong dependence of the so-called reduced density gradient s on the signed value of the density $\lambda_2\rho$ allows the separation of non-covalent from covalent interactions, the former being further sorted into bonded and nonbonded classes depending on the sign of the second electron density Hessian eigenvalue λ_2. The sign of the latter is purportedly positive for nonbonded and repulsive interactions and negative for a range of attractive NCIs. Indeed, NCI analysis provides a striking insight into cases of the past and over bonding relationships. However it does not provide any clues about the origin of the observed stereospecificity of reactions that exclusively lead to a single product.

7.2 THEORETICAL CHALLENGES IN ADDRESSING NCIs AND DISPERSION IN TRANSITION METAL COMPLEXES

Modern *ab initio*-type methods of quantum chemistry together with the latest hard- and software developments are able to calculate many properties of molecules and their reactions with high accuracy. For a recent overview of (non-relativistic) contemporary wave function theory (WFT) and density functional theory (DFT), see, for example, Ref. [29]. This also holds more and more for solids and surfaces although here the arsenal of quantum chemical (QC) methods compared to the isolated molecule situation is somewhat limited. Chemical accuracy for energetic properties (errors < 1 kcal/mol, < 1–2% relative error) can be obtained routinely by canonical coupled-cluster (CC)-based methods for small- and medium-sized molecules (<20–30 atoms). With recent local CC implementations employing pair-natural orbitals and domain-based techniques (DLPNO), CCSD(T) reference calculations for about 100–150 atoms with reasonable triple-zeta AO basis sets are possible [30]. This technique is well suited to systematically study the thermochemistry of transition metal complexes.

For medium-sized systems (<50 atoms), the best contemporary dispersion-corrected DFT methods on average yield energies with errors of only 1–2 kcal/mol (<5%) [31]. In addition to the calculation of equilibrium properties, the theoretical description of transition states (barrier heights) has become satisfactory. However, the accuracy of DFT for larger systems (200–300 atoms) or solids slightly diminishes [32]. Structural DFT predictions for isolated molecules [33] and even for molecular crystals [34] are accurate and in fact often used in a complementary manner for interpreting experimental data. In this context, it seems desirable to provide accurate thermochemical reference data for larger complexes of chemical interest in order to assess the performance of different density functionals. Regarding the general chemical nature of the systems, which can be investigated, and their binding modes, there are practically no restrictions, and, for example, Lewis acid–base pairs, hydrogen bonds, or metal–ligand interactions, can be treated without specific adjustments of the theoretical methods.

The description of NCIs in DFT has advanced extraordinarily in recent years by the accurate account of the long-range London dispersion forces. Within the atom-pairwise D3 method [35] proposed in 2010, London dispersion forces are computed in a semiclassical fashion between atom pairs. Compared to different precursors (e.g., D2; see Ref. [32]), it includes geometrical information about the molecular environment of the atoms making it more accurate and less empirical. Furthermore, it is consistently extended to the whole periodic table of elements. Due to its high accuracy, simplicity, and availability of cheap gradients, it is the de facto standard in dispersion-corrected DFT calculations on molecules and is used routinely in periodic solid state calculations as well [36]. Van der Waals complexes, adsorption phenomena, or in general the thermochemistry of many nonmetallic condensed matter systems do not pose a fundamental challenge to electronic structure theory nowadays. However, describing the electronically more difficult transition metal complex chemistry, specifically in solution, remains challenging, and we will discuss a few examples in this section. As an alternative to account for London dispersion effects, the density dependent nonlocal (NL) VV10 correlation functional of Vydrow and van Voorhis [37] has been implemented into the ORCA [38] quantum chemistry program package, and successful applications (including direct comparisons to D3) have been published [39]. This DFT-NL approach is suggested as a viable alternative for more complicated electronic structures (open shell, metallic), and differences between D3 and NL approaches (for the same short-range exchange–correlation functional) can be used to estimate the error of the D3 dispersion energy. This sanity check is important for transition metal complexes because the oxidation state dependence of the dispersion coefficients [40] is not accounted for in D3 but in the NL scheme.

The importance of London dispersion forces in transition metal complexes was previously demonstrated for $[(PhNC)_4Rh]^+$ [24e] (*vide infra*). Despite of the fact that this complex is positively charged, the formation of dimers was experimentally observed. Computations mostly at the dispersion-corrected DFT level clearly showed that the repulsive Coulomb force between the two complexes is overwhelmed by attractive London dispersion forces mostly between the ligands. This counterintuitive dimerization of two cationic species could be interpreted by the applied QC methods since they allow a correct description of the repulsive Coulomb (and Pauli exchange) forces on the one hand and the attractive London dispersion forces on the other.

However, two pitfalls still remain if reliable predictions of thermochemical properties of transition metal complexes are in the focus. The first one is the lack of comprehensive accurate thermodynamic data for ligand–metal binding, which could serve as reference for the evaluation of approximate QC methods. The second, intimately bound to the improvement of the former, is rooted in the known limitations of continuum-based solvation models, which is corroborated by three examples discussed in the following.

The first example for the performance of the DFT-D3 approach on transition metal complexes "in solution" addresses the trapping of reactive organometallic intermediates in water oxidation catalyzed by iridacycles in the presence of sacrificial oxidant (Ce(IV)) [41]. Although it was known that in the presence of a sacrificial oxidant $Cp^*Ir(Cl)[C,N]$ metallacycles would unavoidably decompose into IrO_2 nanoparticles giving eventually unreactive colloids, we were interested in uncovering the nature of the first organometallic products of catalyst's hydrolytic oxidation such as the radical species $Cp^*Ir^{IV}(OH_2)_2(O)$. Kinetic monitoring of the production of O_2 with cucurbit[7]uril used as a non-covalent scavenger of any Cp^*Ir(aquo) species showed obvious inhibition of gas production. Isothermal titration calorimetry (ITC) [42] measurements of the association of $Cp^*Ir(H_2O)_3^{2+}$ with cucurbit[7]uril gave an enthalpy of association of about −3 kcal/mol in water (25°C). DFT-D3 assisted with a COnductor-like Screening MOdel (COSMO) [43] treatment including explicit counteranions produced values of the enthalpy of association in reasonable agreement with experiment, not only underlining the essential role of the dispersion correction but also stressing that the main discrepancy in theory may arise from insufficient accuracy in modeling solvation.

However, using the COSMO-RS [44] (COSMO for Real Solvents) model, which is based on a statistic evaluation of the distributions of the charge density segments of the solute over charge densities multiplied by the surface area of the molecular volume, more accurate thermochemical data can be reproduced or even predicted by theory [41, 45]. This latter model represents an acceptable compromise between classical molecular dynamics solvation energy averaging and the reductionist models (COSMO or the polarizable continuum model PCM [46]). In critical cases, explicit (strongly interacting) solvent molecules or counterions should be included in addition to the implicit COSMO solvation model. A continuum model-based alternative is, for example, Truhlar's "density-based" solvation model (SMD) [47] (as implemented in GAMESS [48]), but more experience needs to be gained on its performance for "real-life" systems. This is even more the case for the embedded cluster reference interaction site model (EC-RISM) [49] approach, which might be an alternative in the future and is simply mentioned at this point for completeness.

Together with the above mentioned accurate QC methods, a robust standard thermochemistry protocol was developed [50] and successfully applied to a variety of main group chemistry problems [51]. In this protocol, reaction free energies are computed in an additive scheme by

$$\Delta G = \Delta E + \Delta G_{\text{RRHO}} + \Delta\delta G_{\text{solv}}.$$

Here, ΔE is the reaction energy (difference in electronic energies as well as in nuclear repulsion and London dispersion for products and reactants) at 0 K computed by a high-level QC method, that is, dispersion-corrected DFT or correlated WFT. ΔG_{RRHO} is the difference in ro-vibrational contributions (including zero-point vibrational energy) to the free energy at a given temperature. The last term $\Delta\delta G_{\text{solv}}$ accounts for changes in the free energy of solvation as computed by either one of the above mentioned approaches. In our case, this contribution is computed within the COSMO-RS approach.

It will be demonstrated in the course of the second example for dispersion-driven transition metal reactions [52] discussed in the following and depicted in Scheme 7.1. Here, the Pd–Cl_2–Pd bridge is quenched by a phosphane ligand PR_3 (R = phenyl (Ph) or cyclohexyl (Cy)) yielding the corresponding monopalladium complex. The ITC measurements in solution (room temperature; solvent: chlorobenzene (PhCl)) revealed that the reaction with R = Cy (ΔH = −25.3 kcal/mol) is slightly more exothermic than that with R = Ph (ΔH = −23.6 kcal/mol). This is in agreement with chemical intuition since Cy is a stronger donor than Ph, which increases the Lewis base character of the phosphorus atom. Moreover, there is no strong steric cluttering around the Pd center thus it is easily accessible for both ligands. The measured reaction enthalpies are, however, not directly comparable to results from a QC calculation since the single point energies (SPEs) of the molecules in question represent pure electronic energies in the gas phase at 0 K. Hence, solvent and vibrational effects as well as their temperature dependence have to be properly taken into account.

At first sight, calculating reaction enthalpies from reaction energies in an analogous fashion as in our thermochemical protocol for free energies (see above) seems to be the most straightforward way to compare the experimental and calculated values. Unfortunately, COSMO-RS is parameterized for an accurate description of Gibbs free energies and not for enthalpies. Thus, the measured reaction enthalpies were converted to Gibbs free energies by applying a fitting procedure. Since we like to focus on the performance of the applied QC methods, the original thermochemical protocol was modified to obtain an "experimental" reaction energy by "back-correcting" the experimentally determined reaction free energy with solvation (COSMO-RS) and ro-vibrational corrections to the Gibbs free energy. The latter are calculated within a modified rigid rotor–harmonic oscillator (RRHO) approximation [50] at room temperature using analytic PBE [53] –D3/def2-TZVP [54] frequencies. SD(28,MWB) effective core potentials, and matching basis sets were employed for the Pd atoms [55]. By applying this scheme, an "experimental" reaction energy is obtained that can serve as a reference for approximative QC methods.

The corrections (as the SPEs itself) do, of course, depend on the quality of the optimized structures and on how well these resemble the actual geometries of the molecules in solution (Fig. 7.1). The effect of different geometries on the (back-corrected) experimental reaction energies could be estimated to about 1 kcal/mol.

However, the largest source of error stems from the free solvation energy. Checking the COSMO-RS results against SMD revealed some differences from which we estimate an error of ±2 kcal/mol in the solvation free energy. The final best estimate experimental reference values for the quenching reactions with both ligands are −32 ± 3 kcal/mol and −36 ± 3 kcal/mol for R = Ph and R = Cy, respectively. The relatively large uncertainty also reflects the size and complexity of the involved chemical species and hence the term "chemical accuracy," which is defined as 1 kcal/mol error for small molecules, needs to be adjusted for large systems to a more appropriate target of about 5% relative error.

Several DFT and WFT methods were tested concerning their ability to reproduce the experimental ΔE reference values for the investigated reaction. In the following, only a selection of all results will be discussed (for details see

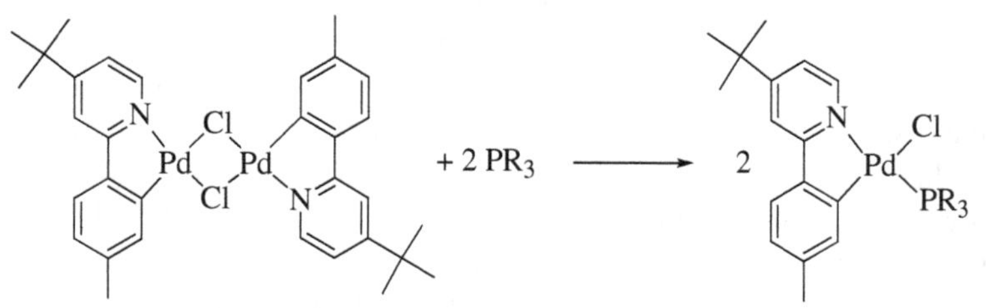

(R = Ph, Cy)

SCHEME 7.1 Reaction of phosphanes with a μ-chloro-bridged palladacycle derived from 2-phenylpyridine [50].

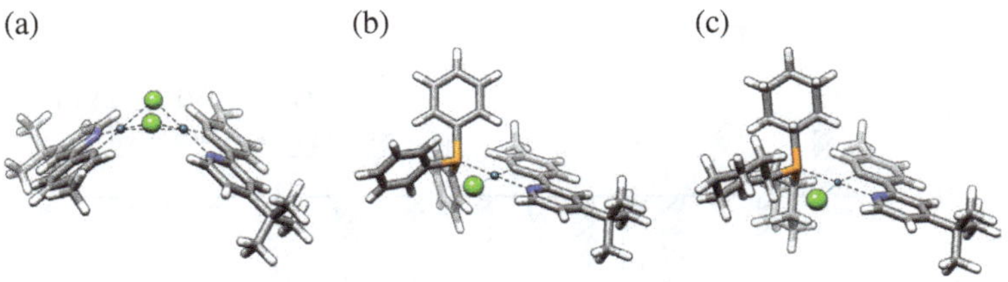

FIGURE 7.1 Singlet ground state geometries (PBE-D3/def2-TZVP) of the palladacyclic dimer (a) and its associated phosphane adducts (b: R=Ph, c: R=Cy). Note that the dimer in the solvent cavity as well as in the gas phase is not planar at its energy minimum but slightly folded. (a) The starting dimer; (b) the PPh_3 adduct; (c) the PCy_3 adduct. Source: Reproduced from Ref. [52] with permission from Wiley.

Ref. [52]). The DFT calculations (TPSS [56], B3LYP [57], M06-2X [58], M06 [59], PW6B95 [60], B2PLYP [60]) were performed with the ORCA program (Ver. 3.0.0) [38] using either the def2-TZVP [54] or the def2-QZVP [54] basis sets and, if applicable, corresponding auxiliary basis sets (def2-TZVP/C [54], def2-QZVP/C [54]) as well as SD(28,MWB) ECPs [55] for Pd. The D3 dispersion correction [35] (or D3M+, vide infra) with Becke–Johnson damping [35] was applied for all functionals except M06 and M06-2X, where the D3 correction with "zero"-damping (D3(0)) was used since medium-range correlation effects are already captured by these functionals. The NL van der Waals functional approach (VV10 [37], DFT-NL [39]) was employed non-self-consistently for B3LYP, that is, calculating the NL energy only once after the SCF iterations have converged. Hartree–Fock (HF), RI-MP2, and RI-SCS-MP2 [61] calculations were done with Turbomole 6.4 [62]. Extrapolation to the complete basis set limit (CBS(TQ)) was carried out according to Helgaker's two-point extrapolation scheme separately for DFT/HF energies [63] and MP2 correlation energies [63]. Furthermore, the recently developed DLPNO–CCSD(T) [30] method was applied using two different values for the electron pair cutoff, the default value of $T_{CutPairs} = 10^{-4}\,E_h$, and a tighter value of $T_{CutPairs} = 10^{-5}\,E_h$. The DLPNO–CCSD(T)/def-TZVP [64] correlation energies were extrapolated to the basis set limit using an additive correction at the MP2 level (ΔCBS) [65]. All correlation calculations were done with the frozen-core approximation (chemical core, i.e., only the valence electrons were correlated). The results are summarized in Table 7.1.

The HF method predicts both reaction energies qualitatively wrong (endothermic), which clearly shows that the driving force for the investigated reactions is electron correlation and, in particular, London dispersion, as can be seen from Figure 7.2. However, it is not only the relatively large medium-range correlation but the significant long-range contribution to the dispersion energy that renders this reaction particularly difficult. Hence, not surprisingly, DFT methods without dispersion correction are not able to reproduce the experimental data. The hybrid functional B3LYP performs even worse than the meta-GGA TPSS. D3 corrects for a large part this error, but the reaction energies are too low, which indicates a slight overestimation of the dispersion energy. This can be understood since in typical organometallic compounds, the metal atoms usually carry a positive partial charge, which is not taken into account within the D3 framework.

TABLE 7.1 Reaction Energies (see Scheme 7.1) Computed at Various Levels of Theory

	E_{reac} (kcal/mol)	
Method/Basis Set	R=Ph	R=Cy
Exptl. Ref. ($E \rightarrow G$)	−32±3	−36±3
HF/CBS(TQ)	4.2	4.4
TPSS/CBS(TQ)	−17.7	−18.8
TPSS-D3/CBS(TQ)	−42.3	−44.2
TPSS-D3M+/CBS(TQ)	−39.9	−41.7
B3LYP/CBS(TQ)	−11.8	−12.7
B3LYP-D3/CBS(TQ)	−41.4	−44.3
B3LYP-D3M+/CBS(TQ)	−38.5	−40.1
B3LYP-NL/CBS(TQ)	−34.7	−38.3
M06-2X/CBS	−20.2	−23.6
M06/CBS	−28.5	−31.0
M06-D3M+(0)/CBS(TQ)	−34.6	−38.1
PW6B95-D3M+/CBS(TQ)	−33.7	−35.0
B2PLYP-D3M+/CBS(TQ)	−42.4	−43.2
MP2/CBS(TQ)	−54.4	−54.4
SCS-MP2/CBS(TQ)	−41.1	−41.1
DLPNO-CCSD(T)/ΔCBS(TQ) ($T_{CutPairs} = 10^{-4}\,E_h$)	−36.3	−36.7
DLPNO-CCSD(T)/ΔCBS(TQ) ($T_{CutPairs} = 10^{-5}\,E_h$)	−32.1	−33.8

The dispersion forces of neutral and (partially) charged metal are expected to be somewhat different. In the D3 treatment, the C_6 coefficients are determined for different coordination numbers of an element, that is, using the (neutral) free atoms and different hydrides as model systems. In the case of metals, the local electron density of the metal atom in these model systems will be larger than for typical organometallic compounds. This leads to too large C_6 coefficients and will most likely overestimate the corresponding London dispersion forces. Due to the presence of Pd atoms and the importance of dispersion forces in the systems discussed here, we calculated new C_6 coefficients based on model systems with a net charge of +1. This modified version of the D3 dispersion correction is denoted D3M+. The TPSS-D3M+ and B3LYP-D3M+ reaction

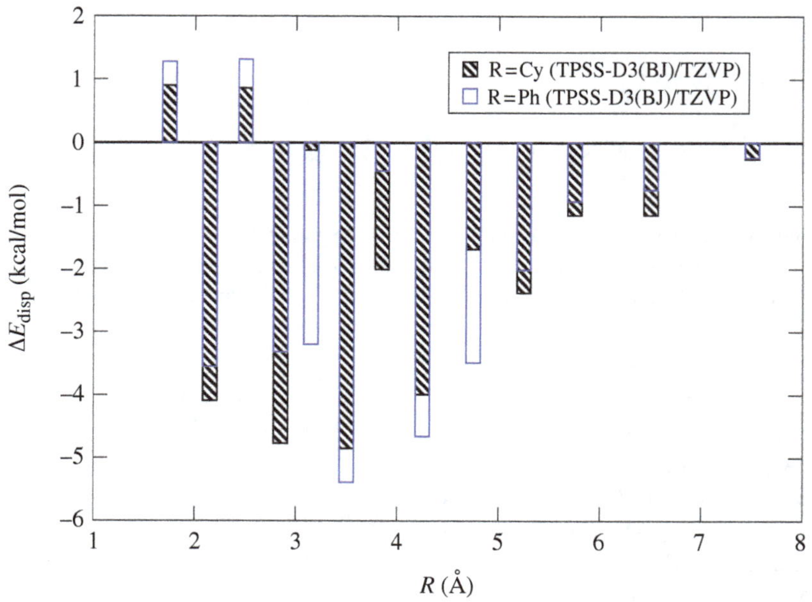

FIGURE 7.2 London dispersion contribution to the reaction energy (from TPSS-D3/TZVP [64]) with respect to the interatomic pair distance. Reproduced from Ref. [52] in slightly modified form with permission from Wiley.

energies are in better agreement with the experimental reference values. Moreover, the density-dependent NL correction employed with the B3LYP functional leads to reaction energies in reasonable agreement with the corresponding D3M+ results, thus indicating that the new dispersion coefficients and their nonempirical generation is physically sound. The M06 and M06-2X functionals are constructed in a way that medium-range correlation is already included. However, the long-range part, which is responsible for the correct asymptotic decay of the London dispersion energy, is missing, and only after correcting for this by means of D3M+(0), the reaction energies are reproduced correctly with M06. The best performance of all tested DFT methods shows the (meta-)hybrid PW6B95 with D3M+ correction, which is able to reproduce both reaction energies quantitatively correctly. The double hybrid B2PLYP suffers from the overestimation of the reaction energy on the MP2 level, and hence it is not recommended in general for such systems. MP2 and to a much lesser extent SCS-MP2 overshoot the reaction energy due to the strong overestimation of π–π interactions. Since the DLPNO methods make use of an MP2 error estimate for the correlation energy of the neglected pairs and truncated PNO expansion, the DLPNO–CCSD(T) reaction energies are also slightly too large if the default value for the pair cutoff is used. Tightening of this threshold by one order of magnitude, however, yields reaction energies, which are in excellent agreement with the experimental reference values and hence qualify this method as reference for thermochemical properties of larger molecules including closed-shell transition metals.

As a last example, we will briefly discuss a recent study of the accurate prediction of Co—C bond dissociation energies (BDEs) for the biologically important and sizeable cobalamin derivatives in solution and in gas phase [66]. The apparent discrepancy between experimental and theoretical gas phase BDEs for adenosylcobinamide could be rebutted with DFT-D3 and open-shell DLPNO–CCSD calculations, which both yielded fully consistent and physically sound results. The Co—C BDE in the complex with the bulky adenosyl ligand is enhanced by about 10 kcal/mol due to several ligand–host NCIs compared to the BDE for the methyl ligand in the gas phase. The additional hydrogen bonds and intramolecular dispersion interactions are not fully quenched in aqueous solution. The application of our thermochemistry protocol with COSMO-RS yielded results in full accordance with experimental data in solution and further combined experimental/theoretical studies of this important system [67]. Hence, it can be considered as another successful application of dispersion-corrected DFT methods in theoretical organometallic chemistry involving sizeable and electronically difficult molecules.

It is encouraging to see that describing the physics correctly, that is, proper treatment of medium- and long-range correlation effects, both DFT and WFT methods can produce the experimental reaction energies sufficiently accurate, thus giving the right answer for the right reason. This finding is further supported by a previous study about ligand substitution energies for transition metal complexes [52]. Thus, we are confident that the available computational methods are well suited for studying cohesion effects in transition metal chemistry, but further exploration of the applicability and accuracy of this approach for a greater variety of transition metal complexes in solution is still needed. The question whether three-body dispersion ($E^{(3)}_{disp}$) [68] or higher-order many-body effects, which are important in solid and supramolecular systems [36, 51], should also be considered for transition metal cohesion will be addressed in future studies. The major part of the remaining error, however, stems from the free solvation enthalpies, and hence it is of great importance to find more accurate descriptions of the latter. Nevertheless, particularly due to the possibility of physically correctly accounting for London dispersion forces efficiently by efficient DFT-D3 approach, the future for realistic predictions of thermochemical properties for transition metal complexes is bright.

7.3 INTRAMOLECULAR NCIs AS A STABILIZING FACTOR

The coordination of aromatic ligands has attracted a great interest starting from the late 1960s. It contributed significantly to establishing the bases of organometallic chemistry. One of the most striking aspects of these early studies is the intense recourse to structural X-ray diffraction (XRD) analyses that were often used to characterize the bonding mode of metals to their ligands. In many studies of this period, authors have concluded briefly on the existence of a bond by applying a deductionist principle, that is, the comparison to similar situations where the quality of a given bond or interaction was apparently firmly established. Only with the popularization of DFT methods in the 1980–1990s, a more complete description of bonding relationships was at reach for experimentalists who could hence proceed to a substantiated analysis of bonding. This section treats cases of stereoselective reactions involving d-block transition metal complexes. Some of these cases are true *left-outs of fundamental research in chemistry*, for no rationalization of the stereospecificity of their formation was attempted. This chapter outlines some *terrae incognitae* of transition metal chemistry that remain to be explored. The most interesting aspect of this chemistry lies in the consequences of the diffused character of valence density of metals, which generally leads to metal–metal interactions that can be rarely depicted as shared (covalent) interactions owing to the large delocalization of the electron density, according to Macchi and Sironi [69]. If this trend is relatively well known, though still being subject of intense research with the contribution of XRD multipolar refinement techniques and theoretical analyses of the electron density topology, it has been only rarely applied intentionally to the engineering of new classes of molecules.

7.4 FACE-SELECTIVE METAL COORDINATION TO AROMATIC LIGANDS

7.4.1 The Face-Selective Thermolytic Binding of Metalcarbonyls to Polyaromatics

In the late 1950s, Wilkinson *et al.* reported the synthesis of bimetallic complexes of azulene by treating the latter ligand thermolytically with metalcarbonyls [70]. Typical azulene can be seen as an heteroditopic ligand where the fused 5- and 7-membered rings can express various types of hapticities depending on the bound metals (Scheme 7.2), spanning often from η^5 to η^1 on both rings. This flexibility in binding capabilities was thoroughly investigated from the early 1950s to the late 1980s with a range of other fused aromatic and cyclic polyenes [71]. The first two cases reported by Wilkinson *et al.* were derivatives of molybdenum and iron carbonyls both bearing the metal carbonyl moieties on the same side of the arene; in the former case the $Mo(CO)_3$ fragment originated from $Mo(CO)_6$ [70a], whereas in the latter the source of ironcarbonyl moieties, that is, $Fe(CO)_3$ and $Fe(CO)_2$, was $Fe_2(CO)_9$ [70b].

Later in the 1960s Churchill *et al.* investigated further the coordination chemistry of azulenes and noted that depending on the nature of metal carbonyl reactants, the stereochemical course of the reaction with azulenes could differ, leading either to syn-facial or to anti-facial bimetallic complexes [72]. Quite common in this period, the existence of bonds between metals in syn-facial complexes was directly inferred from structural considerations based on experimental crystallographic data. As a direct consequence, the formulation of the Lewis structure included the assumed metal–metal bond and a consistent bonding relationship to the π-bonded azulene.

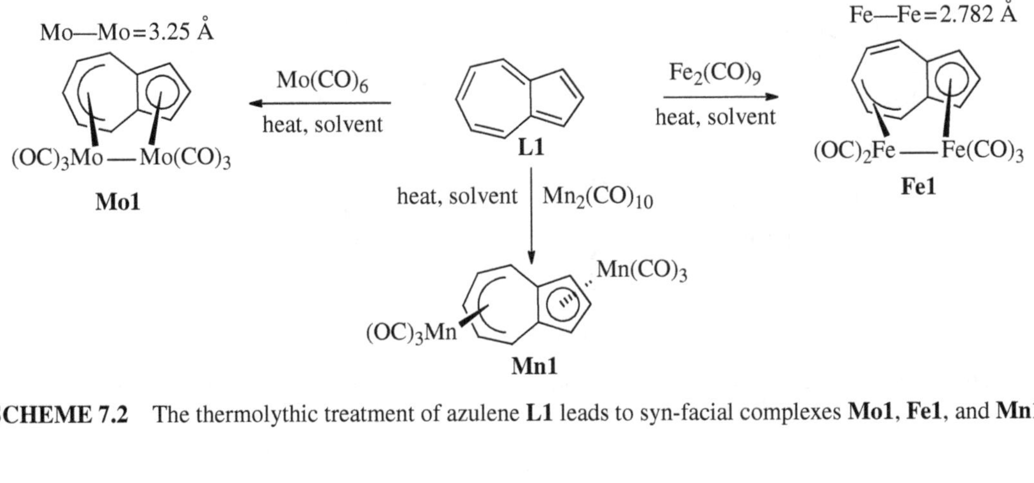

SCHEME 7.2 The thermolythic treatment of azulene **L1** leads to syn-facial complexes **Mo1**, **Fe1**, and **Mn1**.

Quite surprisingly, the preference for either syn- or anti-facial forms was never explicitly addressed by Churchill. The works of Behrens *et al.* [73] further documented the propensity of various metal fragments to coordinate monometallic complexes of azulene. Behrens showed, for instance, that syn-facial heterobimetallic complexes of azulenes (Scheme 7.3) could also be synthesized by treating at relatively high temperatures in noncoordinating solvents (*ca.* 100°C) azulene Mo and Cr complexes with labile Cr-, Rh-, and Fe-containing substrates such as $(MeCN)_3Cr(CO)_3$, $[RhCl(CO)_2]_2$, and $Fe_2(CO)_9$, which would respectively transfer a $Cr(CO)_3$, a $Rh(CO)_2^+$, and a $Fe(CO)_3$ fragment to the 7-membered ring of the azulene ligand. Such results led the authors to express deep doubts about the existence of effective metal—metal bonds in many instances due to the overly long intermetal distances observed in XRD structures (d>2.8Å, i.e., longer than the sum of empirical van der Waals radii for d-block transition metals). Quite interesting is the case of the **Cr3** complex (Scheme 7.4) that was first formulated by Behrens as possessing a Cr—Cr bond, following in that Wilkinson's initial formulation for his Mo analogue. Behrens *et al.*, however, pointed out the abnormal similarity of the Cr—Cr interatomic distance with that reported for Wilkinson's Mo–Mo isomer, which in other terms was questioning the reality of a metal–metal bond. Worthy to note, the fluxionality of **Rh1** (Scheme 7.3) was established from ^1H and ^{13}C NMR spectra and considered as symptomatic of a rather weak metal–metal interaction [73b]: this dynamic behavior was assigned to the excursion of the $Rh(CO)_2$ moiety along the 7-membered ring C–C double bonds with a Gibbs enthalpy of activation of *ca.* 10 kcal/mol. All other complexes bearing the (η^6-benzene)molybdenum moiety displayed akin dynamic behavior in solution ^1H and ^{13}C NMR spectroscopy [73a].

Another case study leading to similar observations is that of O'Hare *et al.*, on the reaction of 1,2,3,4,5-pentamethyl-2-methylene-1,2-dihydropentalene with $Fe_2(CO)_9$ and $Co_2(CO)_8$ under thermolytic conditions [74]. In both cases the reaction led, via a C—H bond activation, to the formation of syn-facial permethylpentalene complexes of biscobalt and bisiron carbonyl exclusively (Scheme 7.5). The two complexes displayed a dynamic behavior in solution. For

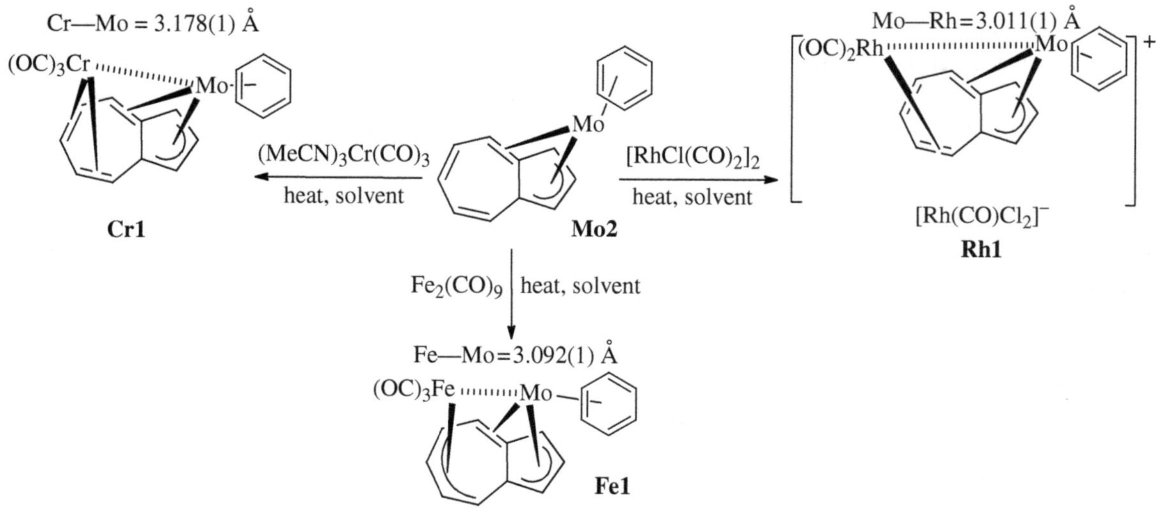

SCHEME 7.3 Thermolytic treatment of **Mo2** with $(Me_3CN)_3Cr(CO)_3$, $[RhCl(CO)_2]_2$, and $Fe_2(CO)_9$ leads to syn-facial heterobimetallics **Cr1**, **Rh1**, and **Fe1**.

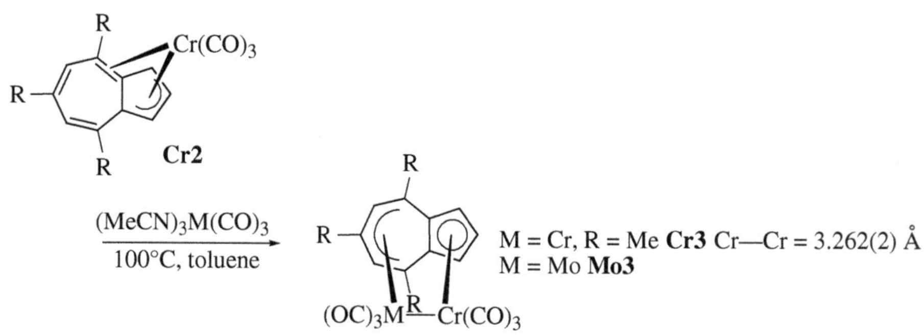

SCHEME 7.4 The thermolytic treatment of **Cr2** with various labile $(MeCN)_3M(CO)_3$ complexes produces the syn-facial complexes **Cr3** and **Mo3**.

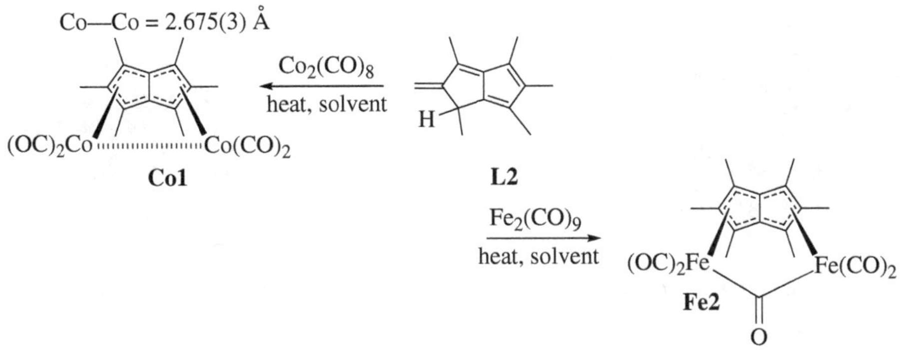

SCHEME 7.5 The thermolytic treatment of **L2** produces the syn-facial homometallic complexes **Co1** and **Fe2**.

Fe2, the rate of exchange of the carbonyl ligands related to the rotation of the iron carbonyl fragments, and the exchange of the bridging carbonyl ligand was estimated to be in the order of 1 MHz at room temperature. In the case of the bisiron complex **Fe2**, DFT calculations and QTAIM investigations carried out at the ZORA-BP86/TZ2P (frozen cores at O, C, Fe) concluded to the absence of a Fe—Fe bond and to existence of a rather repulsive interaction between the two metal centers.

In the case of the biscobalt complex **Co1**, the very low partial bond order and the absence of a bond path according to QTAIM for the Co—Co segment were in favor of the absence of any significant covalent Co—Co bonding. Noteworthy, the authors stated that the existence of such a bond was further ruled out by the quasi isoenergetic character of the fictitious anti-facial isomer, which was found to be by only *ca.* 2 kcal/mol less stable than syn-facial **Co1**. The dissociation enthalpy of $Co_2(CO)_8$ being estimated to be between 15 and 35 kcal/mol, the computed tiny difference of energy between the syn and anti isomers, was sufficient for the authors to rule out the existence of any significant intermetal covalent bond in **Co1**. This conclusion echoes earlier considerations of Reinhold *et al.* [19a] on the electronic structure of $Co_2(CO)_8$, which confirmed the lack of Co—Co covalent bond, that is, the lack of a bond path and bond critical point according to the QTAIM in the Co—Co segment.

The use of solution dynamic NMR spectroscopy since the 1970s has provided an insight into the dynamic behavior of complexes that complements the static and somewhat partial views based on structural XRD analyses. Any dynamic behavior of a π-bonded metal in the compounds treated here raises caution about the Lewis formulation of the entire species and questions the pertinence of the LS rule applied to the static model suggested by XRD.

Nonetheless, close inspection of signed reduced gradient NCI isosurfaces, a computation that can be readily done nowadays using standard DFT codes, provides immediate clues. The models used to compute the NCI isosurfaces displayed in Figure 7.3 are gas phase singlet state geometries of structures either deposited with the CSDB or rebuilt from literature, which were optimized with ADF [75] at the dispersion-corrected scalar ZORA [76] -TPSS [56] -D3(BJ) [35]/all electron TZ2P [77] level. In most of the cases treated here, NCI plots tend to rule out the existence of significant metal–metal covalent bonds (Fig. 7.3). It is readily visible that, in most cases, strong areas of bonding (attractive) NCIs (red colored) are established in the intermetal space that are surrounded with large banks of either Pauli repulsion or non-bonding van der Waals contacts (blue colored).

Figure 7.4 displays the gas phase optimized structures of **Co1** and $Co_2(CO)_8$ as computed at the above mentioned level, using the XRD data as starting geometries [74, 78]. The NCI isosurfaces are neatly defined and essentially composed of attractive domains (red-colored isosurfaces) that are positioned ideally where common intuition would place a bent covalent Co—Co bond.

The computations showed herein raise several questions relative to the nature of those NCIs establishing between the metal centers. This counterintuitive set of observations serves the hypothesis of a major thermodynamic drive toward syn-facial bimetallics operated by attractive NCIs between metallic fragments.

7.4.2 The Face-Selective Binding of Rh and Ir(I) Moieties to Tricarbonyl(η^6-Indenyl)Chromium Anions

Arenetricarbonylchromium complexes have been long studied for their chemical reactivity toward nucleophiles, electrophiles, and bases [79]. The $Cr(CO)_3$ moiety induces a significant electron-withdrawing effect upon the bound arene, which creates a major electrostatic unbalance with the buildup of charge density at the chromium and the terminal carbonyl oxygen atoms [80]. This peculiar property is partly responsible for the enhanced electrophilic character of the arene ligand and for its ability to stabilize benzylic and pseudo-benzylic anions. It is generally observed that the $Cr(CO)_3$ moiety operates a stereoelectronic control mostly of steric nature over the chemical transformations that aim the arene ligand. However, like pointed out by Lewis [81] and Magomedov [82], the $Cr(CO)_3$ moiety can also express some Lewis-type donor ability by capturing Lewis acidic metals to produce persistent molecular complexes [83]. These earlier findings on the coordination chemistry of arenetricarbonylchromium

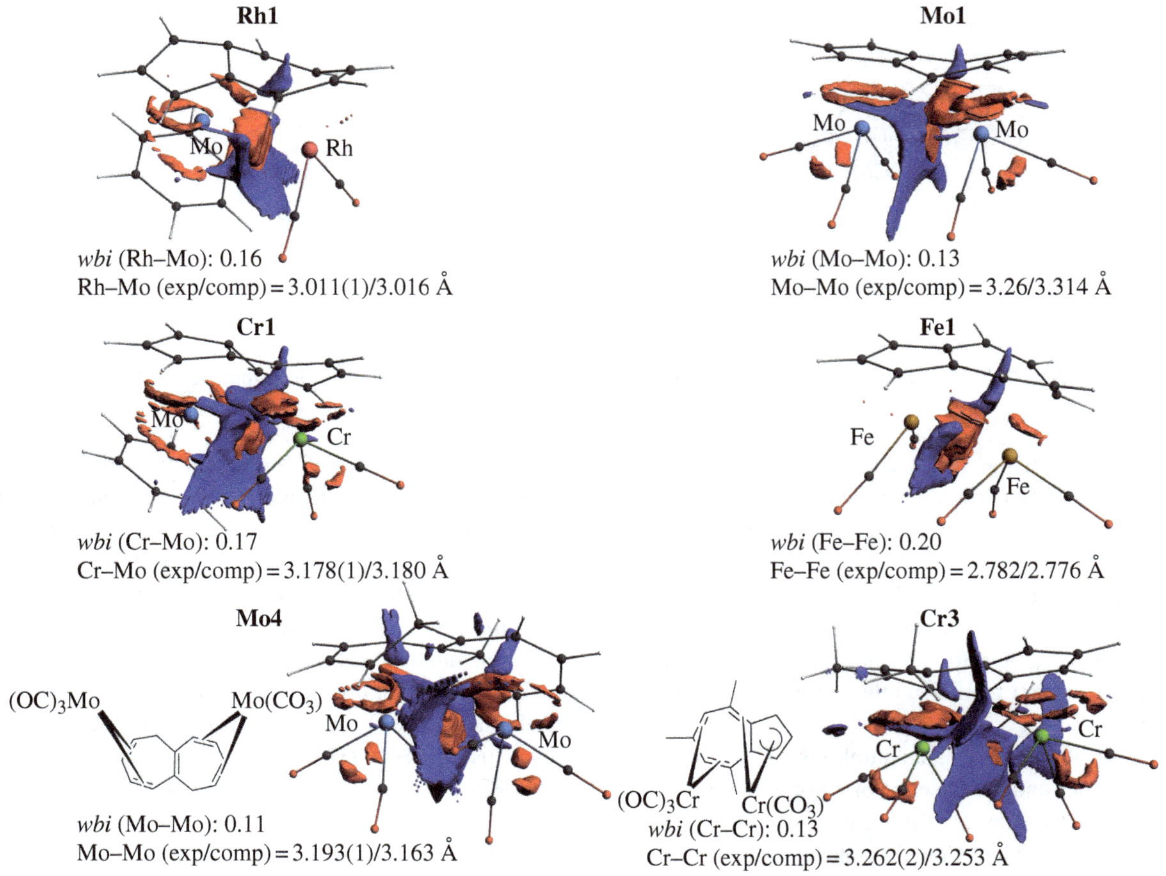

FIGURE 7.3 ADFview2013 plots of non-covalent interaction (NCI) regions materialized by reduced density gradient isosurfaces (cutoff value $s = 0.02$ a.u., $\rho = 0.05$ a.u.) colored according to the sign of the signed density $\lambda_2\rho$ (red and blue colors are associated to negatively and positively signed terms) for gas phase relaxed singlet ground state models of **Rh1**, **Mo1**, **Cr1**, **Fe1**, **Mo4**, **Cr3**. Calculations were performed with the gas phase singlet ground state geometry optimized at the ZORA-TPSS-D3(BJ)/all electron TZ2P level within the ADF2013 package [75a]. Experimental and computed intermetal distances are provided along with the computed Wiberg bond indices.

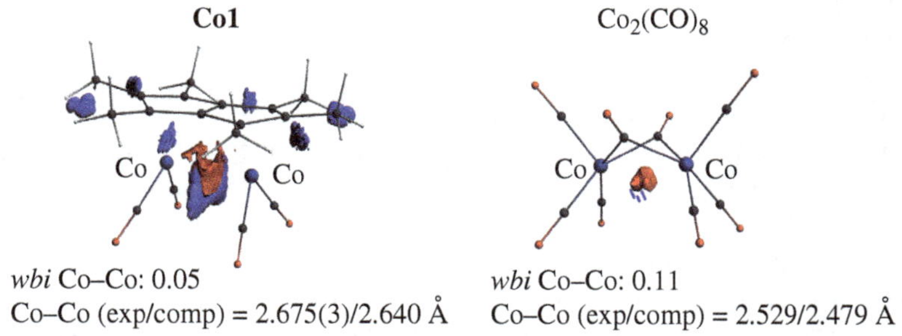

FIGURE 7.4 ADFview2013 plots of non-covalent interaction (NCI) regions materialized by reduced density gradient isosurfaces (cutoff value $s = 0.02$ a.u., $\rho = 0.05$ a.u.) for gas phase relaxed singlet ground state models of **Co1** and $Co_2(CO)_8$. Calculations were performed with the gas phase singlet ground state geometry optimized at the ZORA-TPSS-D3(BJ)/all electron TZ2P level within the ADF2013 package [75a]. Experimental and computed intermetal distances are provided along with the computed Wiberg bond indices.

complexes did not actually attract much interest as most of the research on this class of organometallics was devoted to organic synthesis-oriented applications.

The works of Ceccon *et al.* exemplify the duality of the $Cr(CO)_3$ moiety as a sterically bulky rotor and as a noninnocent electron-withdrawing group. The main motivation of Ceccon's studies was to probe the catalytic activity of Rh(I) η^5-indene complexes and to establish [84] the role of the $Cr(CO)_3$ moiety in alkyne cyclotrimerization [85]. In this endeavor that lasted about a decade, a host of information was gathered on a challenging molecular system. The reaction of anion **Cr4** with $[RhCl(COD)]_2$ (COD: cyclooctadiene) [86] and with $[RhCl(NBD)]_2$ (NBD: norbornadiene) [87] at subambient temperature provided single products with opposite facial distribution of the metals at the indenyl ligand (Scheme 7.6). In compound **Rh2** [86], the anti-facial heterobimetallic complex (Scheme 7.6), the Rh(COD) unit was seemingly η^5-bonded to the 5-membered ring. In **Rh3** [87], however, the inferred picture was that of an η^3-bonding of the Rh atom to the 5-membered ring because of a significant puckering of the indenyl ligand at the benzylic positions. **Rh3** served as a starting material for the synthesis of other similar syn-facial complexes by selective ligand displacement at the Rh [88]. For instance, addition of CO would yield quantitatively syn-facial **Rh4** [87], which then could be treated with excess COD to afford **Rh5** [89], which was not obtained in the direct reaction of **Cr4** with $[RhCl(COD)]_2$ (Scheme 7.6).

Critical inspection of the published structures of syn-facial complexes **Rh3–5** indicates that the Rh—Cr distance reflects more the steric demand of the Rh-bound ligand than the strength of any Rh—Cr bond. The distances lie around a mean value of 3.05 Å.

Key information on the thermodynamic bias favoring syn-facial arrangements rather than anti-facial were produced when complex **Rh7**, produced from **Rh6** by a ligand exchange reaction, was found to convert *in solutio* into **Rh3** in the presence of catalytic amounts (*ca.* 5% mol) of $[Rh(NBD)]BF_4$ [90] (Scheme 7.7).

The authors reported that the anti/syn isomerization was likewise promoted by other cations such as $[Rh(COD)]^+$, $[Rh(COD)_2]^+$, and $[Ir(COD)]^+$ without scrambling of either ligands or metals. The enthalpy of activation of this reaction was evaluated to *ca.* 12 kcal/mol, which represents a rather low value, of the order of magnitude of the barrier to rotation of a Rh(COD) moiety measured in **Rh5** [89]. The authors concluded that the mechanism is a nondissociative process implying at least two 1,3 hydrogen atom shifts to unlock the interfacial migration of the Rh(NBD) moiety (Scheme 7.8).

Worthy to note, the reproduction of similar chemistry with Ir(I)-based moieties proved to be a hard issue due to the exclusive formation of anti-facial η^1-bonded Ir(I) complexes. The synthesis of **Ir2** implied the treatment of dicarbonyl (η^5-indenyl)iridium **Ir1** with $(MeCN)_3Cr(CO)_3$ [91] (Scheme 7.9).

The stability of syn-facial complexes motivated a first theoretical investigation [92] using the Extended Hückel molecular orbital method (EHMO), which concluded to the "heterodox" nature [93] of the Cr—Rh bond for the criterion for a covalent bond was not met owing to the very low orbital overlap population between the two metal centers and to the repulsive character of the 4-electron 2-orbital Cr—Rh interaction. It was however inferred that in syn-facial complexes of this type, the Rh(I) atom displays enhanced Lewis acidic character, which should find the basis for a weak Cr—Rh donor–acceptor (DA) interaction.

More recently, a complete study was undertaken pointing out explicitly the role of electron correlation in the stabilization of the syn-facial isomers [94]. In this report by Schwabe *et al.*, the energy differences of syn- and anti-facial isomers were compared at different levels of the wave function

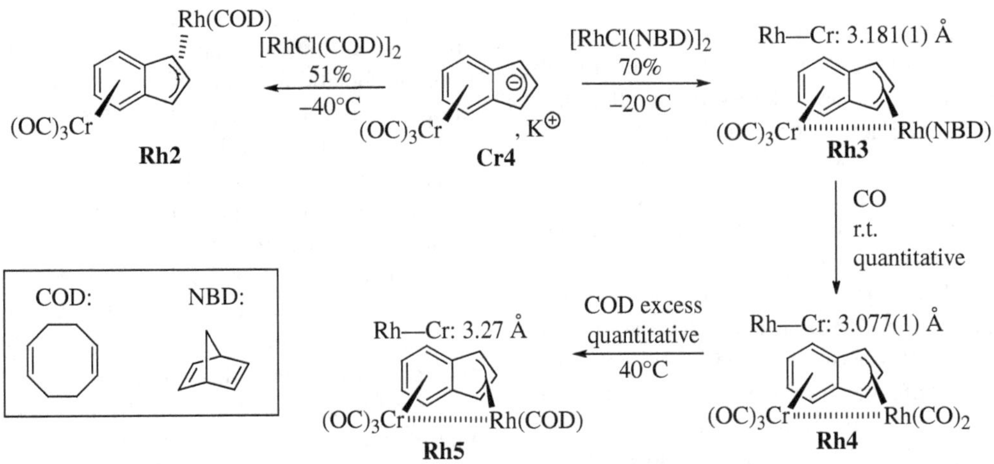

SCHEME 7.6 Synthesis of Ceccon's heterobimetallic complexes **Rh2–5** from **Cr4**.

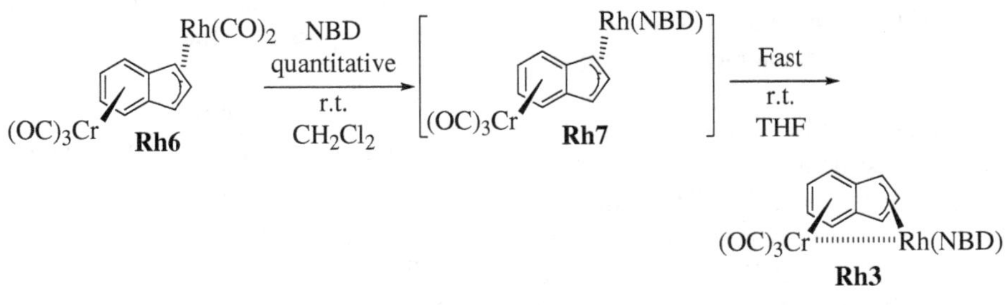

SCHEME 7.7 Ligand exchange at **Rh6** that leads to **Rh3** upon isomerization.

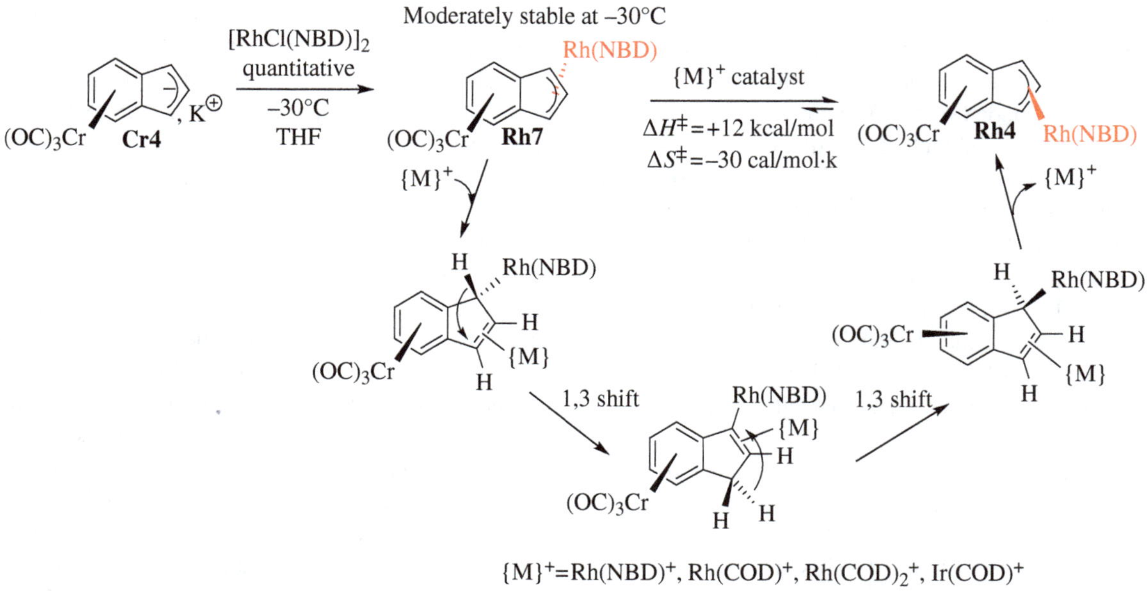

SCHEME 7.8 Proposed mechanism of isomerization of complex **Rh7** into **Rh4** catalyzed by Lewis acidic rhodium(I) and iridium(I) salts.

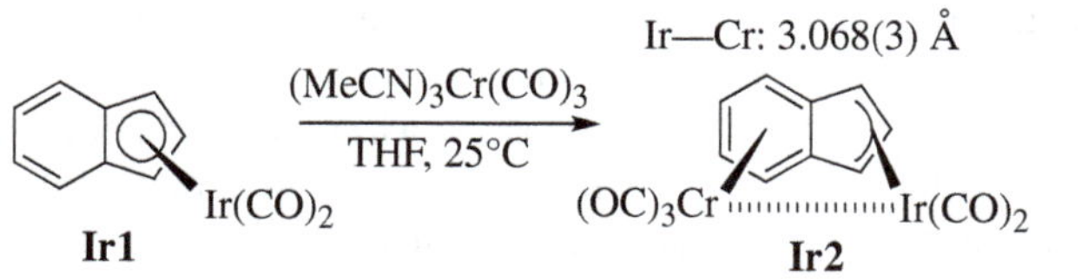

SCHEME 7.9 The treatment of **Ir1** at room temperature with $(MeCN)_3Cr(CO)_3$ affords syn-facial complex **Ir2**.

(HF and SCS-MP2) and density functional theories. For the latter, the performance of DFT versus DFT-D2 functionals, such as the meta-GGA TPSS, was addressed. Calculations at the correlation-exempt HF level were found to favor the anti-facial isomers, whereas SCS-MP2 reproduced correctly the experimental trends favoring the syn-facial arrangements, particularly for **Rh5** wherein the attractive effect of dispersion barely compensates Pauli repulsion. In that aspect the TPSS-D2 functional matched the results of the SCS-MP2 calculations. Using the Pipek–Mezey localization of molecular orbitals (LMO) between the syn- and anti-facial complexes **Rh4** and **Rh6**, interfragment correlation energies were estimated to reach maximum values for the Cr—Rh interaction (−6 kcal/mol) and for the Rh-bound CO–to–Cr interactions. Ligand–ligand correlation energies were also found essential in determining the thermodynamic bias in favor of the syn-facial isomers. Similar calculations carried out with the Ir analogues of **Rh4/Rh6**, that is, **Ir1** and fictitious **Ir2**, gave a Cr—Ir correlation interaction energy of −11 kcal/mol.

Yang's NCI plots for a methyl-substituted isomer of **Rh4**, that is, **Rh8**, synthesized and structurally characterized by Werlé *et al.* [95] are shown in Figure 7.5. The attractive domain of NCI is exactly seated in the Cr—Rh segment and largely surrounded by a bank of van der Waals contacts and repulsive Pauli interactions. The main teasing conclusion of this theoretical study was that the drive to syn-facial isomers was governed by electron correlation and dispersion.

7.4.3 Hemichelation

Chelation is generally conceived as the entropy-promoted coordination of a ditopic ligand to a metal center, where the coordination bonds are both of covalent character. The term hemichelation [96] has emerged to rationalize peculiar situations where the coordination only fulfills partly the conditions of chelation on one of the two bonded centers of the ditopic ligand, providing nonetheless a stabilization of the metal, thanks to NCIs (Fig. 7.6).

Hemichelation [96] is therefore a peculiar bonding relationship between a sequestrated metal center and a ligand; it entails the contribution of a new type of non-covalent donor ligand that occupies explicitly a free coordination site at the hemichelated metal. In some aspects, it can be paralleled with the concept of "auto-solvation" proposed by P. Gyula [97] wherein an unsaturated metal center in violation of the 18 electron LS rule is stabilized by dispersion-based intramolecular interactions with a ligand of its coordination retinue.

In the examples listed in this section, the central role of the $Cr(CO)_3$ moiety emerges under a new light. Of course the chemistry shown in the following sections could in principle be extended to ligands containing a different metalcarbonyl moiety that possesses the same electronic features, namely:

- Electron-withdrawing properties
- Capability to concentrate charge density on terminal oxygen atoms and at the metal
- Ability to behave as a fast conformational rotor under normal conditions
- Relative resistance to oxidation by the hemichelated metal

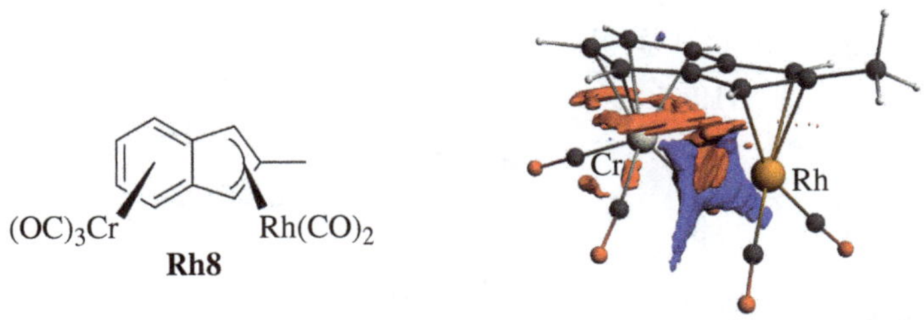

FIGURE 7.5 ADFview2013 plot of non-covalent interaction (NCI) regions materialized by reduced density gradient isosurfaces (cutoff value $s=0.02$ a.u., $\rho=0.05$ a.u.) for gas phase relaxed singlet ground state models of **Rh8**. Calculations were performed with the gas phase singlet ground state geometry optimized at the ZORA-TPSS-D3(BJ)/all electron TZP level within the ADF2013 package [75a]. Experimental and computed intermetal distances are provided along with the computed Wiberg bond indices.

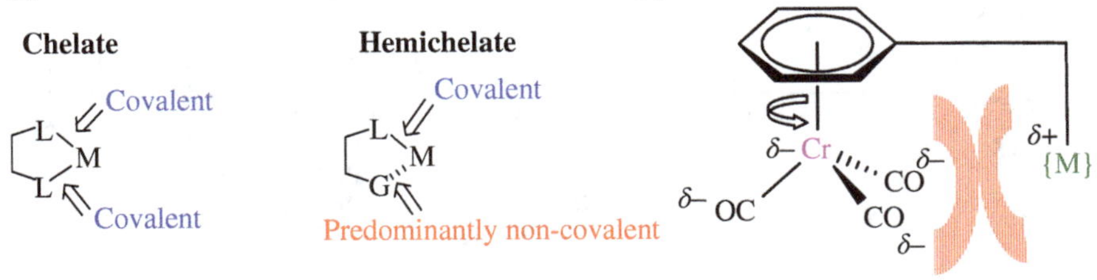

FIGURE 7.6 (a) Schematic definition of the concept of hemichelation; (b) the stabilizing non-covalent interactions in hemichelates based on the tricarbonyl(η^6-benzyl)chromium anion operate in solution as well as in the solid state mostly through the support of attractive coulombic interactions.

For instance, the benzylic anion derived from tricarbonyl (η^6-toluene)chromium can be considered as an anionic ligand capable of expressing hemichelating behavior; the benzylic carbon officiates the covalent anchoring of the metal, while the $Cr(CO)_3$ moiety, which concentrates quite a significant charge density, is the non-covalent anchoring point (Fig. 7.6b). Literature shows that such benzylic anion moiety represents one of the leading motives of hemichelating ligands (*vide infra*).

Hemichelation, because of the frail albeit stabilizing character of NCIs, is generally associated with dynamic behavior in solution. It can be defined as a mean to stabilize electron-unsaturated metal centers and isolate them under a persistent form. It provides a convenient framework recognizing the significant stabilizing effect of NCIs.

First cases of hemichelates of Pd(II) appeared in Kalinin's report of the early 1990s. The authors demonstrated that benzylic anions stabilized under the form of organozinc reagents **Zn1** [98] and **Zn2** [99] could react with the η^3-allylic Pd(II) complex $[(\eta^3\text{-}C_3H_5)PdCl]_2$ to produce, upon transmetallation, heterobimetallic complexes **Pd1** and **Pd2**, which both displayed a quite unusual syn-facial geometry (Scheme 7.10). ^{13}C NMR investigation pointed out the hindered rotation of the $Cr(CO)_3$ rotor, which would produce three distinct singlets at −55°C. **Pd2** was shown to exist as a mixture of two diastereomers, the ratio of which would vary from 1:1 in solution to *ca.* 4:1 in the crystal. The relatively short Pd–Cr distance of *ca.* 2.7 Å led the authors to propose the existence of a covalent bond: such a bond would have entailed the formal oxidation of the Cr(0) center and the reduction of the Pd(II), an assumption that was not matching to the chemical propensity of **Pd1** to produce the Cr(0) complex **Cr4**, for example.

It is about 25 years later that it was proposed that such Pd complexes are most probably hemichelates of a Pd(II) center bearing a formal 14 electron configuration [96]. Anion **Cr5** (Scheme 7.11) reacted readily with a series of η^3-allylic Pd(II) μ-chlorido-bridged complexes to afford highly fluxional complexes **Pd3–5** [100] akin to those reported by Kalinin. The fluxionality was associated with the fast migration of the (η^3-allyl)Pd moiety from one benzylic position of the indenyl to the other with exchange rates estimated to be of the magnitude of a few kilohertz. Complexes **Pd3–5** proved to be rather difficult to handle in the solid state because of their inherent thermal instability. In solution, in the presence of O_2 **Pd3–5** would readily decompose by loss of Pd(0) black to afford C–C coupling products [100].

Improved stability was observed in similar hemichelates of Pd(II) prepared from dimers of μ-chlorido-bridged palladacycles [96]. Complexes **Pd6–10** (Scheme 7.12) all displayed the same syn-facial arrangement of the metals and a *cis* relationship between the benzylic carbon donor atom of the palladacycle. Only in one case, that is, the reaction leading to major product **Pd7**, where the authors isolated minute amounts of the material that precipitated out of the

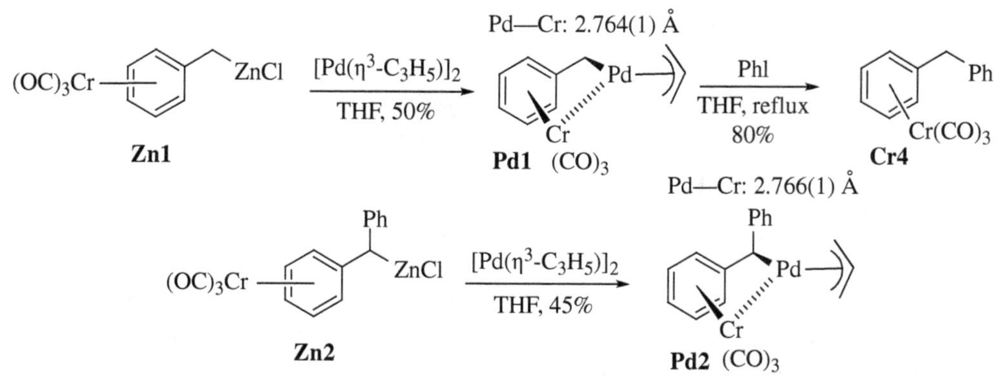

SCHEME 7.10 Kalinin's synthesis of syn-facial complexes **Pd1** and **Pd2** from **Zn1** and **Zn2**, respectively [98].

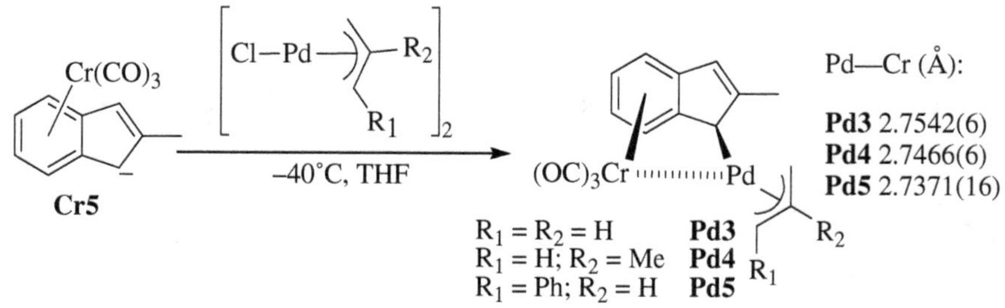

SCHEME 7.11 Synthesis of hemichelates **Pd3–5** from the reaction of anion **Cr5** with μ-chlorido-bridged palladium(II) allyl complexes [100].

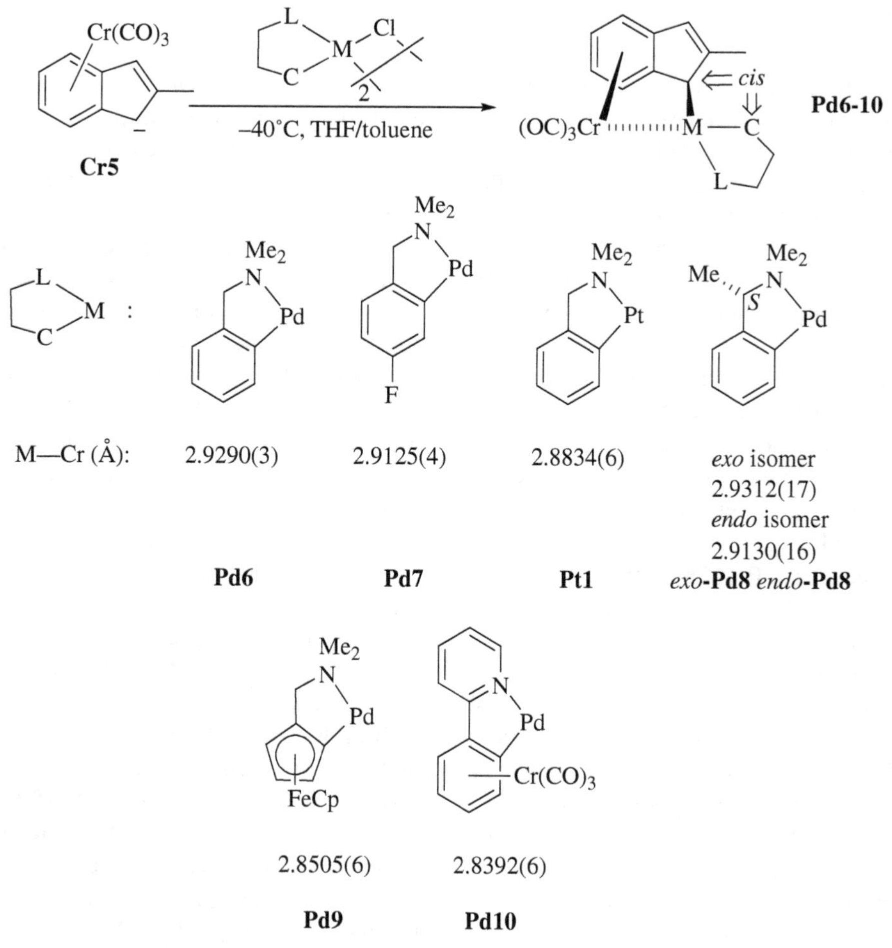

SCHEME 7.12 Synthesis of hemichelates **Pd6–10** and **Pt1** from anion **Cr5** and a range of μ-chlorido-bridged pallada- and platinacycles [96].

reaction mixture as the anti-facial isomer wherein the Pd(II) center was η[3]-bonded to indenyl's 5-membered ring.

Although fluxionality of **Pd6–10** was suspected to follow a mechanism similar to that proposed for **Pd3**, it was not firmly established spectroscopically but rather by chemical means applying the Curtin–Hammett principle in the case of **Pd8**, which bears a chiral palladacycle capable of inducing two planar chiral configurations at the indenyl ligand depending on the position occupied by the palladacycle [96].

Under acidic treatment of **Pd8** planar chiral complex, **Cr6** would be released with temperature-dependent enantiomeric excesses, thus proving that the chiral palladacycle in **Pd8** is subject to rapid metal migration expressed by a swift mutual interconversion of *exo*- and *endo*-**Pd8** in solution (Scheme 7.13).

Theoretical investigations based on the DFT showed in all cases, that is, **Pd3–10** [96, 100], that the *syn*-facial arrangement of the metals, which entails an η[1] bonding of the hemichelated metal to the indenyl ligand, was by *ca.* 10–15 kcal/mol privileged over any anti-facial arrangement. Furthermore, calculations showed that syn-facial situations, where the $Cr(CO)_3$ formally occupies the fourth coordination site of Pd (or Pt like in **Pt1**), were more stable by *ca.* 15 kcal/mol than the alternative syn-facial η[3] bonding of the Pd to the indenyl ligand. Weak covalent interaction was assigned to a short contact between a Cr-bound CO ligand and the Pd atom in **Pd6**: it was defined as an incipient bridging interaction of insufficient strength to halt the rotation of the $Cr(CO)_3$ moiety in solution. The nature of the thermodynamic drive toward syn-facial geometries was assigned to a major contribution of attractive coulombic interaction combined to minor contributions of covalence and dispersion. Of course, these considerations have to be tempered by the fact that **Pd3–10** have a dynamic behavior in solution, XRD structures being therefore more likely considered as energy minimum structures within a shallow potential energy map.

Two rhodium(I) hemichelates **Rh9** and **Rh10**, very similar to the above mentioned Pd and Pt complexes, were synthesized by opposing anion **Cr7** to $[RhCl(NBD)]_2$ and $[RhCl(CO)_2]_2$ (Scheme 7.14). The fluorenyl anion displayed a different reactivity than **Cr5** in that the RhL_2 moiety was less prone to bind the anionic aromatic ligand in an η[3] fashion due to the lower charge density available from the benzo group in **Cr7** compared to **Cr5**, as a consequence of the conjugate electron-withdrawing action of the two $Cr(CO)_3$ moieties. In both complexes, the Rh atom is η[1] bonded to the benzylic position and distant by about 2.8 Å from the Cr center. The slightly hindered $Cr(CO)_3$ rotor is positioned ideally in the vacant fourth coordination site of the Rh center. Theoretical analyses based on NCI and QTAIM calculations

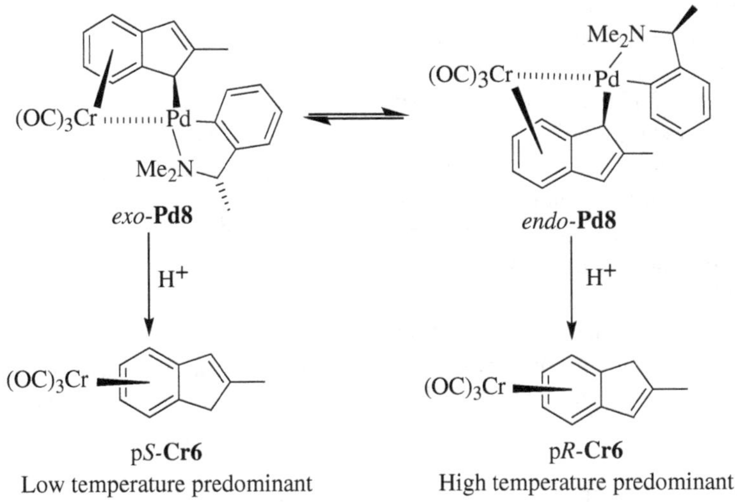

SCHEME 7.13 The acidic demetallation of fast exchanging **Pd8** produces enantioenriched **Cr6**, of which the enantiomeric excess is temperature dependent [96].

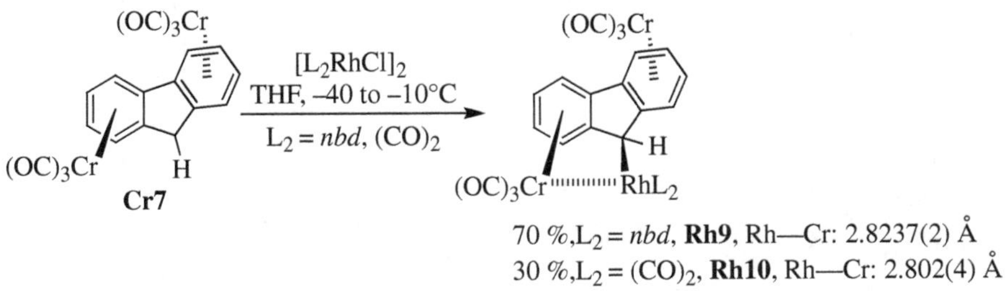

SCHEME 7.14 Synthesis of Rh(I) hemichelates **Rh9** and **Rh10** from the reaction of anion **Cr7** with $[(NBD)RhCl]_2$ and $[(CO)_2RhCl]_2$, respectively [95].

suggested that no covalent bond exists between the two metallic atoms and that most of the stability of the two species stems from attractive coulombic interactions between $Cr(CO)_3$ and the RhL_2 moieties.

The concept of hemichelation [96] may also apply to electron-unsaturated metals placed within a pseudo octahedral coordination geometry such as the Mn(I) center in **Mn6–7** (Scheme 7.15). The main examples of such hemichelates were first reported in 1998 [101]. The complexes in question can be prepared by the insertion of a Fischer-type carbene ligand into the aromatic carbon–metal bond of manganese(I) and rhenium(I) metallacycles. The carbene ligand can be formed by two different methods.

The first one consists of the conventional electrophilic trapping of an acyl-metallate [102] such as **Mn3** prepared *in situ* by the selective nucleophilic addition of an aryl or alkyllithium reagent to a tetracarbonylmetal (metal = Mn(I), Re(I)); the transient Fischer carbene (*cf.* e.g., **Mn4**) rearranges swiftly to yield the syn-facial Cr, Mn product [103]. An alternative method consists of the thermolytic treatment above 60°C of a tetracarbonylmetallacycle with a diazoalkane, which leads to the departure of one CO ligand, the fast coordination of the carbene ligand upon release of dinitrogen [104], and a *cis*-migration of the vicinal aromatic carbon with release of one coordination site at the metal to produce a new electron-unsaturated 6-membered metallacycle (*cf.* **Mn7**).

When no $Cr(CO)_3$ moiety binds the vicinal metallacycle, the metal center fulfills its valence electron shell by binding the closest π-system, thus producing an η^3-benzylic metal complex [105] (*cf.* **Mn5**, Scheme 7.15). The newly formed 6-membered metallacycles are air stable and solution persistent. Structurally, the $Cr(CO)_3$ moiety apparently occupies the sixth vacant coordination site at the metal without any major bonding to the Mn(I) or Re(I). Furthermore, the $Cr(CO)_3$ moiety has a hindered rotation with a Gibbs barrier to rotation of *ca.* 15 kcal/mol [106].

Most of such Mn and Re hemichelates derived from 2-phenylpyridine display similar structural features, with average Cr–Mn(Re) distances of *ca.* 3.04 Å. The 6-membered metallacycle is generally folded and rather rigid when it is derived from 2-phenylpyridines or similar azaheterocycles [107].

With a derivative of 2-phenyloxazoline, the resulting hemichelate **Mn8** (Scheme 7.16) displays dynamic structural changes in solution, that is, a combined chemical and conformational exchange wherein the hemichelate converts into a planar 6-membered manganacycle stabilized by an η^3-benzylic interaction with a proximal phenyl group, for example, **Mn9** (Scheme 7.16). This exchange is purportedly enabled by the ability of the oxazolyl to absorb the ring strains induced by the flattening of the 6-membered manganacycle. The **Mn8–9** dyad represents a rare example of mutual interconversion between isoenergetic formal 36 and 34 electron complexes.

In a DFT-D1 theoretical study of Hyla-Kryspin *et al.* [107], the nature of the interaction between the Cr(0) center and the Mn(I) was identified to be a mutually compensating combination of a Cr → Mn DA 3-orbital 4-electron interactions and its Mn → Cr back-bonding counterpart sustained by dispersion and electron correlation. It was found that hypothetical anti-facial isomer would be less stable by *ca.* 15 kcal/mol than the syn-facial arrangement. Figure 7.7 displays the NCI regions for complex **Mn10** [104c], which shows a large attractive domain in the Cr—Mn interatomic space.

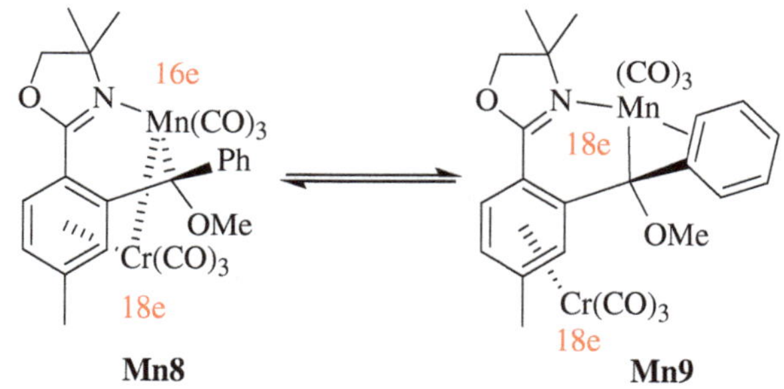

SCHEME 7.16 The electron-unsaturated complex **Mn8** equilibrates with complex **Mn9** in solution.

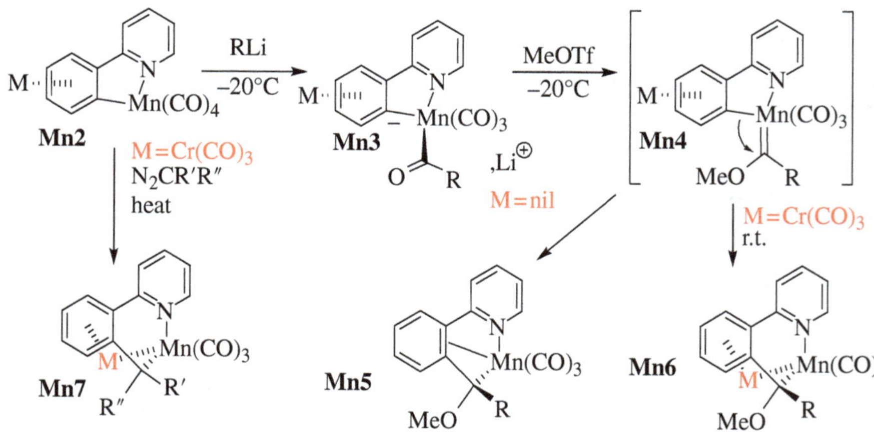

SCHEME 7.15 Synthesis of Mn(I) hemichelates **Mn6–7** by the derivation of tetracarbonyl manganacycle derived from 2-phenylpyridine **Mn2** by insertion of a Fischer-type carbene into the C_{Ar}—Mn bond.

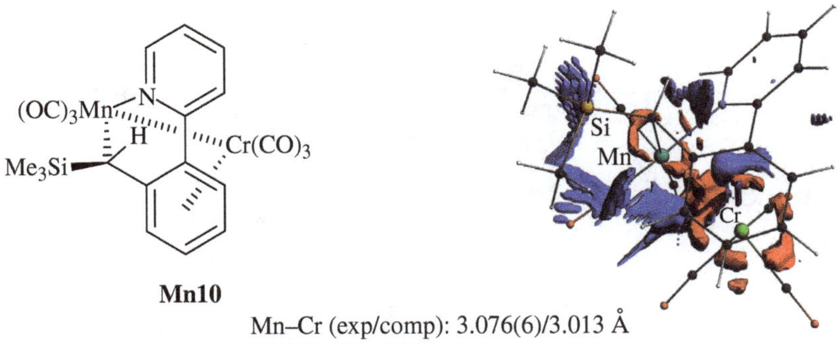

Mn10

Mn–Cr (exp/comp): 3.076(6)/3.013 Å
wbi Mn–Cr: 0.11

FIGURE 7.7 ADFview2013 plot of non-covalent interaction (NCI) regions materialized by reduced density gradient isosurfaces (cutoff value $s = 0.02$ a.u., $\rho = 0.05$ a.u.) for gas phase relaxed singlet ground state models of **Mn10**. Calculations were performed with the gas phase singlet ground state geometry optimized at the ZORA-TPSS-D3(BJ)/all electron TZP level within the ADF2013 package [75a]. Experimental and computed intermetal distances are provided along with the computed Wiberg bond indices.

7.5 INTERMOLECULAR NCIs IN METAL–METAL DA COMPLEXES AND OTHER AGGREGATES

Metallophilicity is probably the most widely known concept that can be associated with the affinity of two metal centers for each other. The most widespread way this concept is used is through the term "aurophilicity" that was introduced in the late 1980s and then defined in the 1990s by Schmidbaur *et al.* [108]. This term was coined to express "the specificity of the unexpected attraction between two or more gold centers in compounds where the metals are already chemically saturated according to conventional valence concepts: gold atoms always interact with the gold atoms in neighboring functions or molecules, ignoring all other potential donor or acceptor sites" [109]. In other words, two metal centers are separated by a distance shorter than the sum of their van der Waals radii.

In the second half of the twentieth century, numerous articles [108a, 110] have addressed closed-shell attractive interactions involving 4d, 5d, and late-transition metal centers in solid state. Interaction between metal centers was studied by analyzing steric and electronic effects of both ligands and metals [111]. The attractive nature of these interactions was attributed to the combination of various effects, where dispersion between metal centers and coordinated ligands could play a role [20a, 110j, 112].

7.5.1 DA Complexes Sustained by Dispersion

If the influence of dispersion is taken into consideration, dative metal–metal bonds should be regarded as borderline cases, since the DA interaction between electron-deficient and saturated metal centers retains a significant amount of covalent character [113]. Pomeroy's group has reported syntheses of unbridged complexes of Ru and Os with group 15 ligands where dative bond was formed between metal and Sb, As, and P atoms in relatively small yields, except for $Os(CO)_4(PPh_3)$ (Scheme 7.17) [114].

These compounds inspired the synthesis of complexes with a nonbridged DA bond between metals and 18-electron complexes of Os, where the latter would act as a donor ligand for the second Os atom [115] (Scheme 7.18). Coordination around each Os atom is approximately octahedral. This compound is very stable, even after heating in solution. Possible mechanism for formation of this complex is shown in Scheme 7.18.

Pomeroy *et al.* discovered that the best suited metal donor for the synthesis of DA complexes were 18-electron, quasi-square-pyramidal pentacoordinated Os(0) complexes that formally act as two-electron Lewis donors. In turn, labile solvato complexes of W, Cr, or Mn produced photolytically under UV light in tetrahydrofurane were used to play the role of Lewis acceptors (Scheme 7.19) [116].

The concept of dative bonds between different closed-shell metals was further validated experimentally in numerous cases by Pomeroy *et al.* [117]. Among these, one particularly stands out [117a], where a formal dative bond between Ir and W metals was formed, that is, **Ir3** (Scheme 7.20). Complex $(\eta^5\text{-}C_5Me_5)Ir(CO)_2$ was selected to act as a donor because it contains a five-coordinate metal atom that is expected to readily expand its coordination number to six and also because third-row metals form stronger dative bonds with neutral transition metal donors. It was noticed that **Ir3** complex would decompose into starting compounds in a period of several hours in solution.

First theoretical analysis of dative metal–metal bond in nonbridged complexes was done by Nakatsuji *et al.* [118]. Using restricted *ab initio* calculations at the HF level on simplified Pomeroy's complexes (as shown on Scheme 7.19), they have provided evidence for the dative character of metal–metal bonds where electron transfer is not restricted only to the metal–metal region but is spread all over interacting fragments.

More recently, a detailed study on the DA interactions in Pomeroy's compound was conducted [119]. Nakatsuji *et al.*'s assumptions on the electronic structure of considered

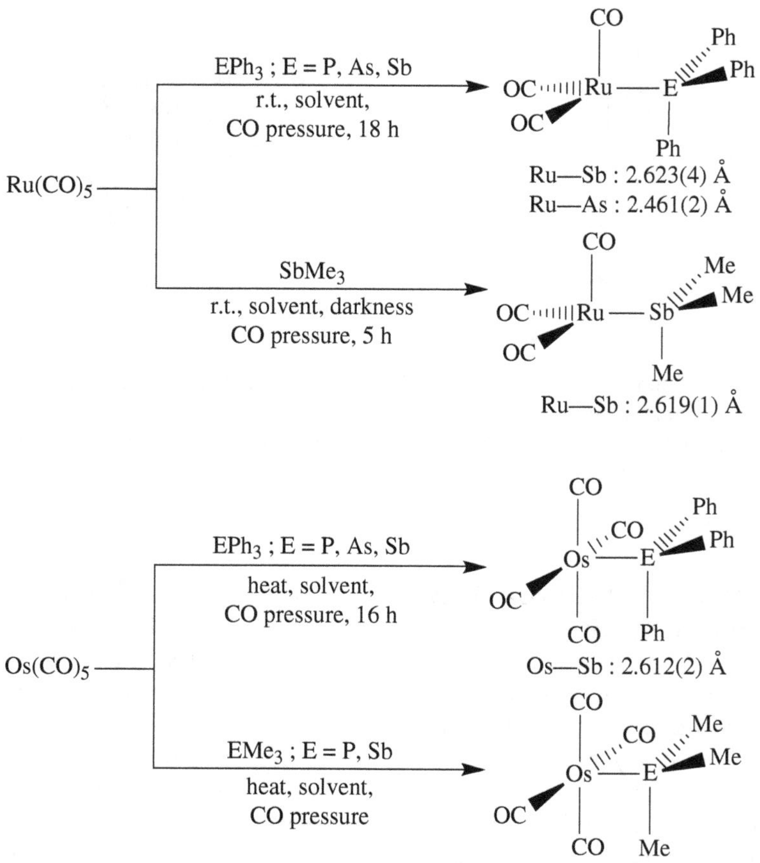

SCHEME 7.17 Synthesis of Ru and Os complexes with M—E dative bond; M = Ru, Os; E = P, Sb, As.

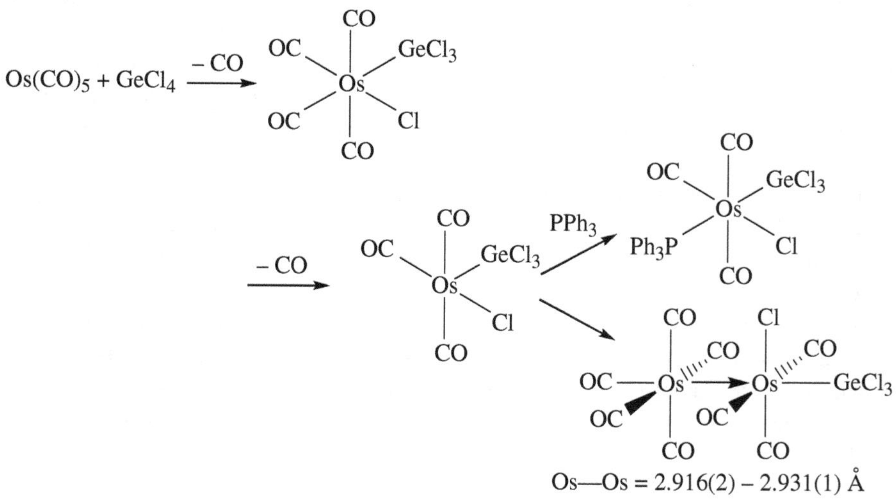

SCHEME 7.18 Possible mechanism of $(PPh_3)(CO)_4Os–Os(Cl)(CO)_3GeCl_3$ complex formation.

compounds were confirmed using semilocal DFT approximations like Becke–Perdew or TPSS, the double-hybrid B2PLYP functional as well as the corresponding dispersion-corrected methods, that is, TPSS-D2 and B2PLYP-D2 functionals and the wave function-based spin-component-scaled second-order perturbative Møller–Plesset theory (SCS-MP2). The so-called dative bond results from a DA interaction between the highest occupied molecular orbital (HOMO) of the $L(CO)_4Os$ fragment and the lowest unoccupied molecular orbital (LUMO) of the $M(CO)_5$ fragment. Analysis of Wiberg bond indices values revealed poor contribution of covalence in the metal–metal interaction (values <0.1). The plot of the coulombic (electrostatic) potential over the isosurface of the SCF density conspicuously shows (Fig. 7.8a) very low accumulation of charge densities around the two metal centers and slight differences in potential between the two vicinal metal centers.

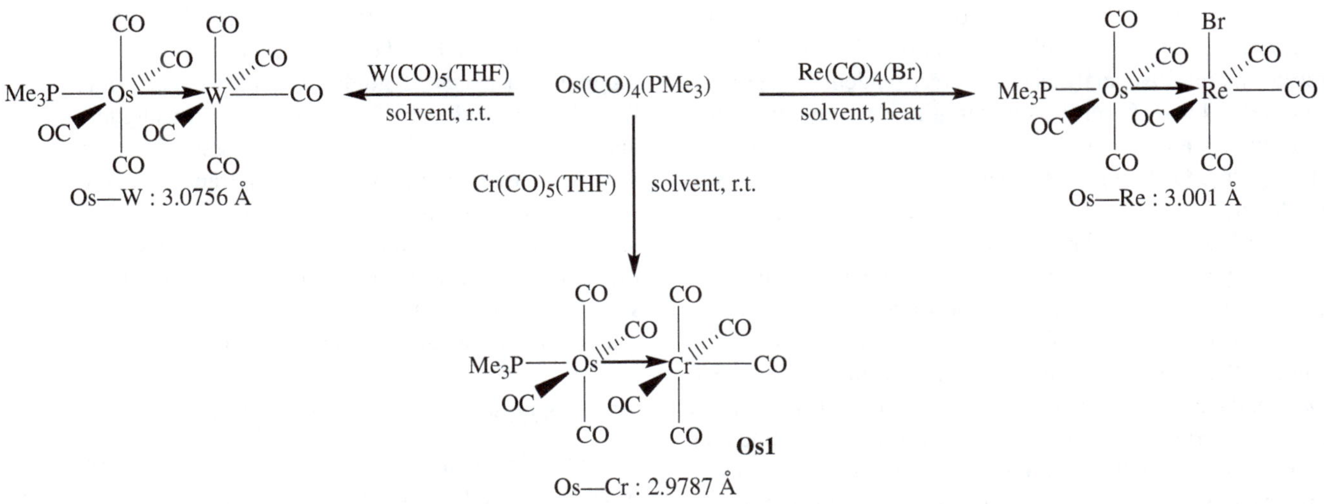

SCHEME 7.19 Synthesis of Pomeroy's compounds with formation of dative Os—M bond; M = Re, Cr, W.

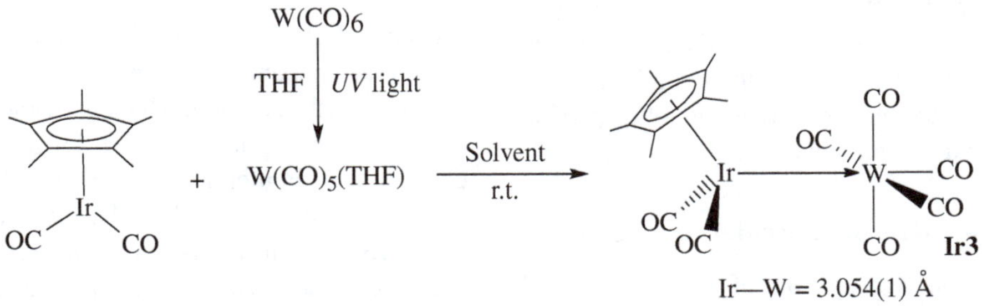

SCHEME 7.20 Synthesis of (η^5-C_5Me_5)$(CO)_2$Ir-W$(CO)_5$ complex **Ir3**.

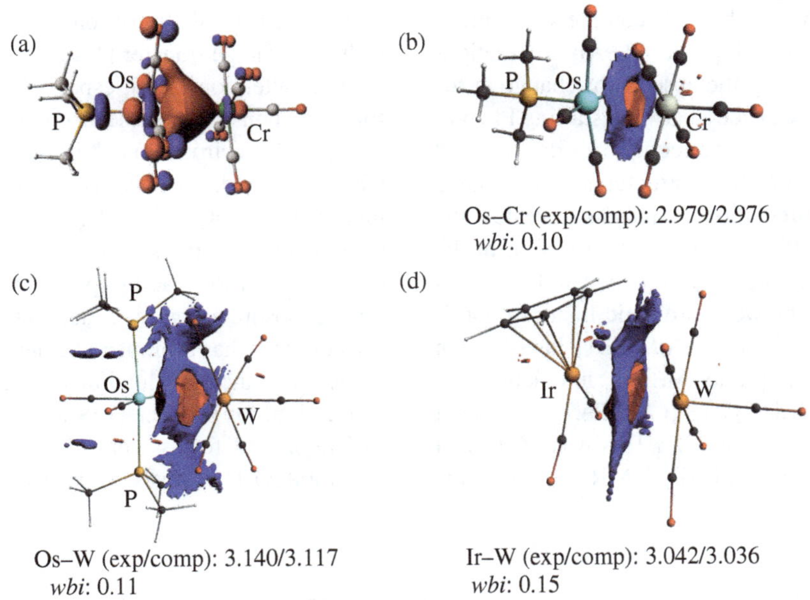

FIGURE 7.8 (a) ADFview2008 drawings of interacting resulting bonding HOMO Kohn–Sham orbitals of Os–Cr complex **Os1**. Orbitals related to the complex are depicted here with an isosurface contour value of 0.04 e bohr^{-3}(ZORA-BP-D2/all electron TZP level) [119]. (b–d) ADFview2013 [75a] plots of non-covalent interaction (NCI) regions materialized by reduced density gradient isosurfaces (cutoff value $s = 0.02$ a.u., $\rho = 0.05$ a.u.) colored according to the sign of the signed density $\lambda_2\rho$ (red and blue colors are associated to negatively and positively signed terms) for gas phase relaxed singlet ground state ZORA-TPSS-D3(BJ)/all electron TZ2P level models of **Os1** (b), **Os2** (c), and **Ir3** (d). Experimental and computed intermetal distances are provided along with the computed Wiberg bond indices.

As mentioned above, Yang and Piquemal's [27] density reduced gradient method of visualization of local NCIs represents an intuitive analysis of the electron density topology. The red-colored isosurfaces materialize the attractive intermolecular interactions between metal centers in **Os1** (Fig. 7.8b), **Os2** (Fig. 7.8c), and **Ir3** complex (Fig. 7.8d), while the blue isosurfaces are assigned to van der Waals interactions or to Pauli repulsion.

Energy decomposition analysis in combination with SCS-MP2 calculations gave a clear demonstration of the significant role of (London) dispersion forces in stabilization of the studied Pomeroy's complexes [116], wherein the Os–metal DA bond bears a rather low covalent character. It was shown that for shorter M—M distances, the dispersion interactions between the spatially close-lying metal shells become more important, while the M—L correlations are stronger at longer distances. Another important conclusion that arose from this study is that nondispersion corrected semilocal DFT methods (e.g., Becke–Perdew or TPSS) underestimate metal–metal dissociation energy by about 50% in considered examples (*ca.* 15–20 kcal/mol) and that the contribution of dispersion should rise with the size of the substituents in the complexes, thus emphasizing the importance of proper treatment of dispersion in all studies of NCI in transition metal complexes.

7.5.2 Intermolecular NCIs in Molecular Aggregates

Most of the articles on aggregation of transition metal complexes deal with the formation of dimers or higher oligomers in solid state. However, there are several reports that either hint on or show existence of aggregates even in solution. Doerrer *et al.* [120] have reported the synthesis and characterization of Pt—Au double salt complexes where it was shown that direct metallophilic Pt—Au interactions vanish in solution, justified by the absence of change in the electronic and fluorescence spectra between starting Pt complex and Pt—Au double salt. In these complexes direct Pt—Pt interactions are inferred from XRD structures, in oligomers and polymers. DFT computational analysis revealed the influence of two forces effective on the stabilization of the aggregates: metallophilic interactions between Pt centers and stacking interactions between aromatic ligands coordinated to Pt. López-de-Luzuriaga *et al.* [121] have shown strong evidence of metallophilic Au—Ag interactions in polymeric $[AuAg_3(C_6F_5)(CF_3CO_2)_3{-}(CH_2PPh_3)]_n$ persistent even in concentrated solutions of THF. This was confirmed by pulsed-gradient spin-echo (PGSE) NMR spectroscopy, extended X-ray absorption fine structure (EXAFS) analysis, time-dependent DFT (TD-DFT), and *ab initio* calculations. Aggregation states of gold(I) complexes with phosphanyl ligands were studied by Diversi *et al.* [122] using mass (ESI) and diffusion-ordered NMR (DOSY) spectroscopies. In the solid state, these complexes exist as tetranuclear compounds stabilized by aurophilic interactions between four gold centers. Some of these compounds maintain their structure even in solution. Several other articles report on gold complexes supported by bidentate phosphanes where short Au(I)—Au(I) interactions were partially maintained in solution [123].

7.5.3 Aggregation of Rh(I) Complexes

Studies of the chemical and physical properties of tetrakis(alkyl or arylisonitrile)Rh(I) complexes have been highlighted by a large number of reports in last three decades [124]. These studies illustrated the physical properties of Rh(I) complexes as well as their unusual propensity to spontaneously form adducts or oligomers in solution and in the solid state (Scheme 7.21) [125]. The formation of these oligomers is always accompanied with major changes in the electronic spectrum of the solution, which is characterized by the growth of a new absorption band at longer wavelengths as the concentration in monomer increases [126]. This typical spectroscopic signature [127] of the dimeric and trimeric forms expresses the sensible perturbations caused to the electronic structure of the system by the self-association of two identical square planar d^8 fragments in a face-to-face arrangement [128].

This property [129] has originally sparked conjectures on a specific "d^8–d^8" interaction that would act somewhat as a driving force in the process of so-called self-assembly [130]. The new electronic transitions, which originate from metal-to-ligand charge transfer [130b, 131], have been the subject of great attention [132], especially because they were also found responsible for the peculiar photochemical/physical [133] and electrochemical properties of those oligomers, which clearly differ from the behavior of the corresponding monomers [134]. The oligomers' propensity to readily undergo electron transfer reactions [134a, c, 135] and oxidations [134a, 136] leading to M—M-bound complexes [137] upon exposure to light was demonstrated in many instances. This property has prompted the development of bridged analogues and their evaluation as catalysts for solar energy conversion, for instance [138].

Oligomeric forms were characterized in the crystal by structural XRD analysis [130b, 139]. Numerous examples of

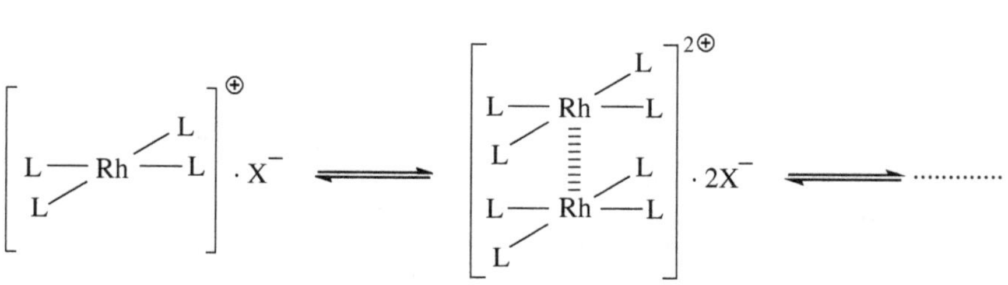

SCHEME 7.21 Formation of the oligomers in the Rh(I) square-planar complexes.

dimers [140] and trimers [141] can be found in structural databases, which clearly demonstrate the reality of the propensity of cationic Rh(I) tetrakis(isonitrile) complexes to self-aggregate in the solid state. Self-aggregation with metal–metal contacts is precluded by steric effects created by bulky isonitrile ligands [142]. In the most common cases, Rh(I)—Rh(I) dimers display an average metal–metal distance of *ca.* 3.1 Å [143]. The metal–metal interaction was studied in so-called bridged systems whereby the two metal fragments are somewhat forced to remain in close vicinity [139a, 144]. It was demonstrated that the Rh—Rh distance would be significantly shortened by *ca.* 0.85 Å in a bridged complex brought to its triplet excited state upon exposure to the radiation of a pulsed laser beam [143a], thus confirming earlier results inferred from the spectroscopic study of these bridged species [145]. If great efforts have been invested in investigating the photophysical properties of these binuclear Rh(I)—Rh(I) assemblies [143a, 146], not much knowledge about the real origin of the interactions that govern the process of self-aggregation of the cationic monomers was sought [147]. Structure, synthesis, and chemical properties of $[(PhCN)_4Rh]^+$ (**Rh11**) complex reported by Gray *et al.* [125c] constitute a prototypical case that was the focus of a recent report [24e] addressing experimentally and theoretically the propensity of **Rh11** to self-aggregate in the gas phase and solution. Experimental evidence for the existence of dimer $[\mathbf{Rh11}]_2$ in solution was provided by ^{1}H DOSY NMR spectroscopy, whereas ESI mass spectroscopy suggested the absence of oligomers in the gas phase. Theoretical calculations were performed using DFT-D3 and wave function-based SCS–MP2 methods. Due to shallow energy minima, conformational search for low-energy gas phase structures of **Rh11** complex yielded three minima (Fig. 7.9).

The influence of the dispersion correction on the computed structures of the dimers was found much greater than the choice of the density functional. For the Rh—Rh distance in the dimers, the dispersion correction was found to have a shortening effect of 0.15–0.34 Å. Furthermore, the electrostatic potential (ESP) map showed the absence of buildup of electronic charge density at the two metal centers, while the positive charge was spread over whole of the monomer fragments. Influence of dispersion can be seen in the two cases (Fig. 7.10) showing the resulting geometries optimized with and without dispersion correction.

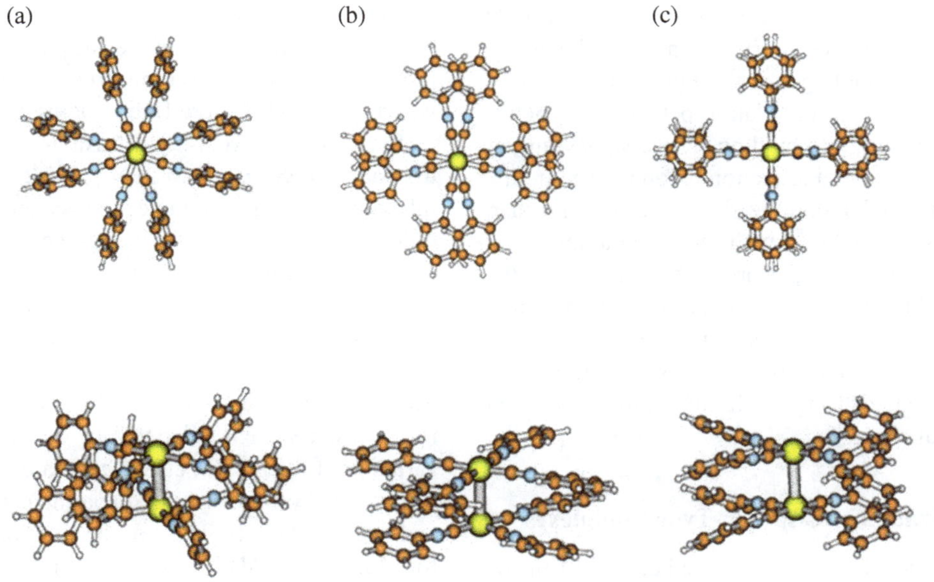

FIGURE 7.9 Optimized singlet ground state structures for the three conformers (a–c) of $[\mathbf{Rh11}]_2^{2+}$ [24e]. Reprinted (adapted) with permission from Ref. [24e]. Copyright 2011 American Chemical Society.

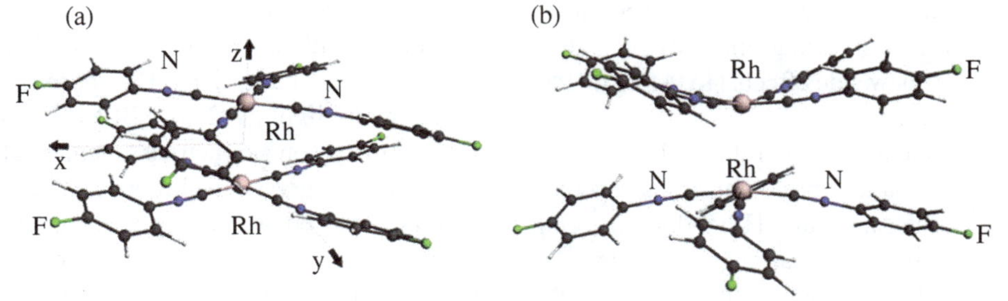

FIGURE 7.10 Singlet ground state DFT geometries of $[\mathbf{Rh11}]_2^{2+}$ computed with (a) and without (b) dispersion correction [24e]. Reprinted (adapted) with permission from Ref. [24e]. Copyright 2011 American Chemical Society.

Distortion of the geometry in Figure 7.10b has two origins. The first is the strong Pauli and electrostatic repulsion between the ligands that carry most of the Coulombic repulsive energy term, with the Rh atoms being almost neutral. The second is the attractive short-range correlation-based d^8–d^8 interactions that are accounted for by the dispersion correction devoid native DFT functional. Analysis of bonding energies shows bonding situation only when TPSS-D3/COSMO level of theory was used, with a large contribution of the dispersion correction to binding energy (*ca.* −43 kcal/mol for the most stable geometry—A of $[\mathbf{Rh11}]_2^{2+}$). This shows that even if the electrostatic repulsion is screened by solvation, the molecular state becomes unbound when dispersion is neglected. Calculated Gibbs free energy for dimer formation (e.g., −7.3 kcal/mol for geometry A of $[\mathbf{Rh11}]_2^{2+}$) compares reasonably well with the experimental value of −2.1 kcal/mol inferred from the original report published by Gray *et al.* Analysis of the dispersion contribution between different interacting fragments using DFT-D3 methods revealed that long-range dispersion effects between the metals are relatively unimportant (*ca.* 4 kcal/mol compared to the total dispersion energy of more than 40 kcal/mol near the equilibrium distance). This was further confirmed by SCS-LMP2 calculations. According to the SCS-LMP2 calculations, d^8–d^8 valence correlation contributes by about −3.5 kcal/mol to binding near the equilibrium distance, while core and core-valence correlation between the metals adds another −1.3 kcal/mol to binding. These values further explain why some intermetal attraction is found even in dispersion uncorrected DFT-optimized geometries. Further, when phenyl rings are replaced with hydrogen atoms, total interfragment dispersion energies are much smaller. All the analyses converged to the conclusion that when many bulky groups come spatially close and can arrange in an optimal fashion, their combined attractive dispersion interactions can overcompensate the strong Coulomb repulsion between two positive formal charges.

7.5.4 Oligomerization of Cisplatin-Type Complexes

The discovery of the cytotoxicity of cisplatin [148] in the 1960s has introduced the use of coordination complexes for the therapeutic treatment of diseases such as cancers. Due to the toxicity of cisplatin, new anticancer agents such as oxaliplatin (Eloxatin™ drug) and carboplatin (Fig. 7.11) have been developed. This class of clinically approved drugs shows a range of activities on various lineages of malign cells [150] as well as some new side effects [151] that dictate cautious formulations [152].

Available crystallographic information shows that the solid state structuring of these cisplatin-type complexes is governed by the directing effect of intermolecular H-bonding, favoring various modes of intermolecular arrangements (stacks, strands) [153]. However, the question of the nature of the forces driving these complexes to form oligomers of sufficient persistence in solution or in the gas phase remains open.

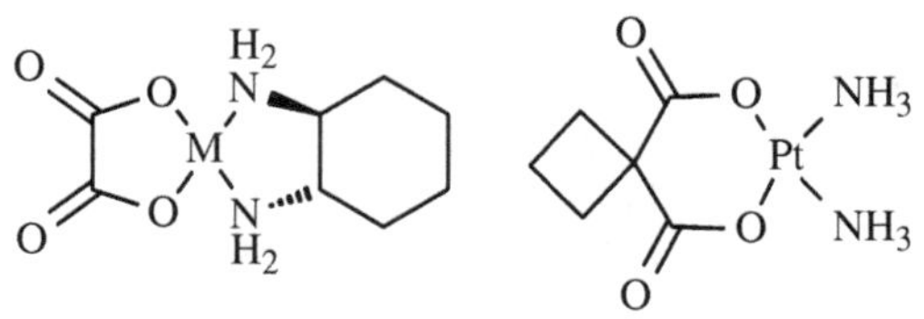

FIGURE 7.11 Formulas of complexes **Pt2** (oxaliplatin), **Pd11**, and **Pt3** (carboplatin). Reproduced from Ref. [149] with permission from the PCCP Owner Societies.

Unusual characteristic of these complexes, that is, long shelf stability in the form of concentrated infusions, was investigated by Dabrowiak *et al.* [154]. They proposed, on the basis of NMR spectroscopic analyses, that these complexes exist as dimers in relatively concentrated solutions. Aggregation of the complexes would also explain why their hydrolysis to bis–aquo complexes [155], a process likely happening in dilute neutral and basic aqueous preparations [156], is apparently precluded in concentrated infusions. This proposal was supported by striking spectroscopic evidence of oligomer formation in the gas phase, thanks to electrospray ionization (ESI) mass spectroscopy [154]. Such behavior in solution and in the gas phase was already known [157] for other cisplatin-type structures, but the reason for it was never sought before Dabrowiak's paper.

In a recent report [149], diffusion-ordered ^{1}H NMR spectroscopy (DOSY), which may provide important structural information on small molecular systems [24e, 158], was used for the investigation of aggregation of cisplatin-type drugs in solution. However, in the case of solutions of oxaliplatin (**Pt2**), oxalipalladium (**Pd11**), and carboplatin (**Pt3**), the formation of oligomers was not detected. Measuring the variations of absorbance A of the UV–Vis absorption spectrum of **Pt2** as a function of analyte concentration c was also inconclusive since it failed to provide the expected nonlinear variation of the $A = f(c)$ curve, generally symptomatic of the formation of oligomers. ITC experiments gave no noticeable endothermic heat flow above the dilution background trace for **Pt2**. For **Pd11**, an endothermic trace typical for aggregation was recorded giving enthalpy of dimer formation ΔH_f of −0.5 kcal/mol. Titrations of compounds **Pt2** and **Pt11** gave responses that could suggest possible mutual interaction with an enthalpy of the association of −0.6 kcal/mol.

The aggregation of **Pt2**, **Pd11**, and **Pt3** in either the gas phase or solution was modeled by taking advantage of the first-principles-based DFT-D3 approach [35a], using computationally light (ZORA) BLYP-D3(0)/all electron-TZP level of theory on geometries schematically represented with their relative energies on Figure 7.12. Calculations indicated that the stacked arrangement β, observed in one experimental structure of a similar complex by Galanski *et al.* [153b], leads to optimal stabilization (Table 7.2), which further points to the existence of the dimers in the gas phase, already confirmed by ESI mass spectroscopy.

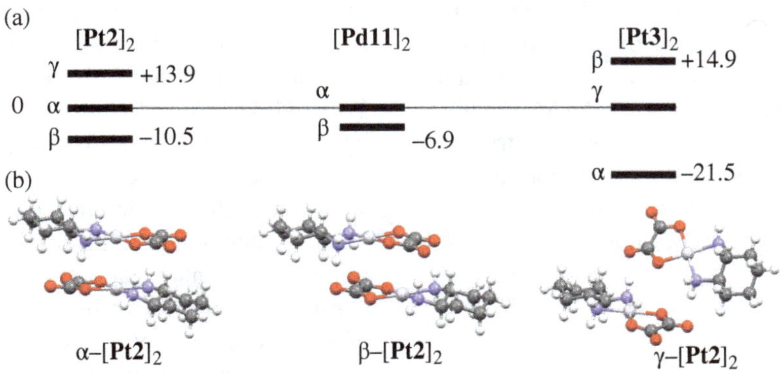

FIGURE 7.12 (a) Relative total bonding energies for α, β, and γ geometries of dimers of **Pt2**, **Pd11**, and **Pt3**. (b) Explicit geometries of α, β, and γ arrangements for $[\mathbf{Pt2}]_2$. Reproduced from Ref. [149] with permission from the PCCP Owner Societies.

TABLE 7.2 Calculated Thermochemical Parameters for Various Combinations of Monomers, in the Gas Phase as well as with a COSMO Solvation Treatment (Basis Set Superposition Error Was not Accounted For)

	ΔH_f^{298} (kcal/mol)	ΔS_f^{298} (kcal/K·mol)	ΔG_f^{298} (kcal/mol)
exptl $[\mathbf{Pt2}]_2$ (water)[a]	NA	NA	NA
α-$[\mathbf{Pt2}]_2$	−53.4	−0.054	−37.4
β-$[\mathbf{Pt2}]_2$	−63.3	−0.057	−46.5
β-$[\mathbf{Pt2}]_2$-(DMSO)	−20.0	−0.042	−7.4
α-$[\mathbf{Pt2}]_2$-(water)	−15.6	−0.060	2.2
β-$[\mathbf{Pt2}]_2$-(water)	−18.9	−0.037	−7.7
exptl $[\mathbf{Pd11}]_2$ (water)[a]	−0.5	−0.017	+4.6
α-$[\mathbf{Pd11}]_2$	−50.6	−0.050	−35.6
β-$[\mathbf{Pd11}]_2$	−56.8	−0.042	−44.0
β-$[\mathbf{Pd11}]_2$-(DMSO)	−17.0	−0.053	−1.3
β-$[\mathbf{Pd11}]_2$-(water)	−16.4	−0.051	−1.1
exptl [**Pt2•Pd11**] (water)[a]	−0.6	−0.017	+4.5
β-[**Pt2•Pd11**]	−60.2	−0.051	−45.1
β-[**Pt2•Pd11**] (water)	−18.5	−0.043	−5.6

NA, not available.
[a] From ITC experiments.

Calculations showed that theory overestimates the propensity of the system to self-assemble. Difference in theoretical and experimental results was deemed to probably result from lack of consideration for the specific interactions of water with the solute by the COSMO model and to the neglected basis set superposition error. Energy decomposition analysis shows that significant portion of total interaction energy goes to dispersion, for example, up to 50% of $\Delta E_{int} = -60.1$ kcal/mol in α-$[\mathbf{Pt2}]_2$. H-bonding constitutes a major effect responsible for the cohesion of those dimers in the gas phase. ETS–NOCV [159] analysis provided a clear picture of the contribution of H-bonding in the cohesion of the gas phase dimer and confirmed the absence of any significant metal–metal orbital interaction (Fig. 7.13a), while the anchoring effect of H-bonding can further be visualized by NCI regions (Fig. 7.13b).

Another instance, where the proper treatment of dispersion and reliable experimental thermochemical data play a great role in describing NCI, is in the inclusion of transition metal complexes into lipophilic cavity of cucurbit[*n*]urils (**CB**[*n*]), a process well documented in the literature. Room-temperature ITC experiments on the inclusion of **Pt2** into **CB**[7] afforded an enthalpy of inclusion ΔH_{mod} of −8.7(1) kcal/mol and ΔH_{mod} of −6.6(5) kcal/mol for the inclusion of **Pd11** into **CB**[7]. Estimates of thermochemical parameters obtained by DFT-D3 calculations with COSMO treatment of solvation are summarized in the Table 7.3. The computed correct sign and right order of magnitude for ΔG compared to experimental values is a promising result, knowing the limitations of the implicit solvent model.

7.6 FINAL COMMENTS

NCIs play a significant role in the chemistry of transition metals. The present state of advancement of theoretical chemistry finds the bases for a new insight based on an analytical approach of bonding as well as in the reproduction of observables such as the enthalpies of reactions in solution. Most molecules treated herein display dynamic behavior that challenges the conventional methodology for assigning the quality of a bond from static structures obtained by XRD analysis. Several cases reported in the period spanning 1960–1990 would deserve a new interest from the point of view of the forces that drive with high selectivity the resulting bimetallic product to be as compact as possible by adopting a syn-facial arrangement. We have provided several hints (NCI region plots and values of Wiberg bond indices) that strongly support a major role of NCIs in molecular cohesion. It appears that NCIs generally support weak DA interactions that otherwise would not be strong enough to maintain the cohesion of a metal–metal bonding interaction. This topic of research is of great importance; it may lead to new synthetic strategies for the construction of electron-deficient complexes that could be of use in various applications.

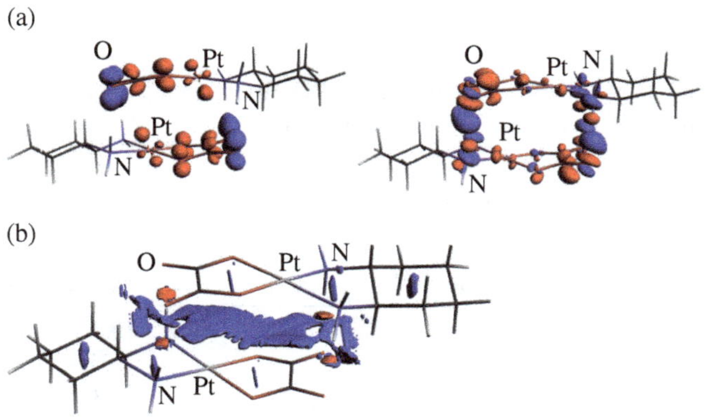

FIGURE 7.13 (a) NOCV deformation density $\Delta\rho_1$ (left—ΔE_{orb} = −5.84 kcal/mol and ω = 15%), $\Delta\rho_2$ (right—ΔE_{orb} = −4.07 kcal/mol; ω = 11%); ΔE_{orb}—orbital stabilization energy for corresponding deformation density ΔE_{torb}—energy of total orbital interactions in the corresponding dimer ω—percentage of ΔE_{orb} in the ΔE_{torb}. (b) ADFview2013 plots of non-covalent interaction (NCI) regions indicated by *reduced density gradient* isosurfaces (cutoff value s = 0.02 a.u., ρ = 0.05 a.u.) colored according to the sign of the signed density $\lambda_2\rho$ (red and blue colors are associated to negatively and positively signed terms) for the gas phase relaxed singlet ground state model of Pt-containing complexes β-[**Pt2**]$_2$. All calculations were performed with gas phase singlet ground state optimized geometries at the ZORA-BLYP-D3/all electron TZP level. Non-covalent N—H···O bonds are materialized by attractive non-covalent red-colored isosurfaces. Blue isosurfaces are assigned to van der Waals interactions or to Pauli repulsion. Reproduced from Ref. [149] with permission from the PCCP Owner Societies.

The concept of hemichelation has emerged as a new formulation of the relationship that exists between the intrinsic unbalance of charge density within polymetallic species and the stabilization of valence-unsaturated metal centers. Hemichelation is the name given to a type of coordination in complexes; it summarizes within a single term the unexpected stabilizing conjunction of coordination and non-covalent bonding, which can be traced in several reports of the past for molecules that escaped understanding. Recent reports suggest that hemichelation can be privileged over other conventional modes of coordination by a reasoned engineering of ligands. This endeavor requires a proper knowledge of the electronic structure of the target ligand and the anticipation of its binding properties by DFT methods, which creates a new alliance between theoretical chemistry and synthesis.

ACKNOWLEDGMENTS

This work was supported by the LABEX Complex Chemical Systems, the Centre National de la Recherche Scientifique, the University of Strasbourg, the joint funds of the Agence Nationale de la Recherche (ANR), and the Deutsche Forschungsgemeinschaft (DFG) through project COCOORDCHEM. JPD thanks Prof. Dr Hélène Michel (SAGE, UMR 7363 CNRS) for fruitful and inspiring discussions.

TABLE 7.3 Estimate of the Thermochemical Parameters for the Inclusion of Pt2 and Pd11 into CB[7] in the Gas Phase (ZORA-BLYP-D3/All Electron TZP) and in Water (COSMO Geometries, Treated with COSMO-RS) (Energies in kcal/mol; in case of ΔS_f Energy is in cal/(K·mol))

Gas Phase	ΔE_f	ΔH_f	ΔS_f	ΔG_f^{298}
Pt2@CB7	−49.6	−48.9	−0.065	−29.4
Pd11@CB7	−48.9	−47.6	−0.058	−30.2

COSMO-RS (Water)	ΔE_g	ΔG_{RRHO}	$\Delta\delta G_{solv}$	$\Delta E_{disp}^{(3)}$	BSSE	ΔG_f^{298}
Pt2@CB7	−47.4	20.2	8.5	4.6	1.6	−12.3
Pd11@CB7	−47.4	18.7	8.9	4.7	1.5	−13.6

ΔE_g, is the host–guest association energy for COSMO-relaxed geometries taken in the gas phase; ΔG_{RRHO}, is the rigid rotor harmonic oscillator energy correction from gas phase energy E to free Gibbs energy G, taken for gas phase calculations; $\Delta E_{disp}^{(3)}$ is the three-body dispersion energy term; ΔE_{BSSE} is the energy correction for the basis set superposition error.

REFERENCES

1. F. London, *Trans. Faraday* 1937, *33*, 8.
2. M. Kolar, T. Kubar, P. Hobza, *J. Phys. Chem. B* 2011, *115*, 8038.
3. F. London, *Z. Phys.* 1930, *63*, 245.
4. J. Rowlinson, Cohesion, a Scientific History of Intermolecular Forces, Cambridge University Press, Cambridge, 2004.
5. F. London, *Z. Phys. Chem.* 1930, *B11*, 222.
6. (a) H. Chatley, *Nature* 1921, *107*, 786; (b) W. Taylor, *Nature* 1922, *109*, 10; (c) anonymous, *Nature* 1923, *112*, 773; (d) anonymous, *Nature* 1931, *128*, 462.
7. R. Eisenschitz, F. London, *Z. Phys.* 1930, *60*, 491.
8. W. Heitler, F. London, *Z. Phys.* 1927, *44*, 455.
9. C. A. Coulson, *Nature* 1962, *195*, 744.
10. H. Eyring, *J. Am. Chem. Soc.* 1932, *54*, 3191.
11. K. S. Pitzer, *J. Chem. Phys.* 1955, *23*, 1735.
12. (a) J. F. Bunnett, *J. Am. Chem. Soc.* 1957, *79*, 5969; (b) J. F. Bunnet, J. D. Reinheimer, *J. Am. Chem. Soc.* 1962, *84*, 3284; (c) D. L. Dalrymple, J. D. Reinheimer, D. Barnes, R. Baker, *J. Org. Chem.* 1964, *29*, 2647; (d) A. J. Sisti, J. P. Sawinski, *J. Org. Chem.* 1976, *41*, 2746; (e) G. Bartoli, P. E. Todesco, M. Fiorentino, *J. Am. Chem. Soc.* 1977, *99*, 6874.
13. E. Spinner, *Aust. J. Chem.* 1960, *13*, 218.
14. P. Hobza, K. Müller-Dethlefs, Non-Covalent Interactions: Theory and Experiments, Royal Society of Chemistry, Cambridge, 2010.
15. H. J. Schneider, A. K. Yatsimirsky, Principles and Methods in supramolecular Chemistry, John Wiley & sons, Inc., New York, 2000.
16. J. N. H. Joost, S. Otto, Dynamic Combinatorial Chemistry, Wiley-VCH Verlag GmbH, Weinheim, 2010.
17. W. B. Jensen, *J. Chem. Educ.* 2005, *82*, 28.
18. (a) P. R. Mitchell, R. V. Parish, *J. Chem. Educ.* 1969, *46*, 811; (b) C. A. Tolman, *Chem. Soc. Rev.* 1972, *1*, 337; (c) D. M. P. Mingos, *J. Chem. Soc. Dalton Trans.* 1977, 31.
19. (a) O. Kluge, M. Finger, J. Reinhold, *Inorg. Chem.* 2005, *44*, 6494; (b) M. Ahlquist, P. Fristrup, D. Tanner, P.-O. Norrby, *Organometallics* 2006, *25*, 2066; (c) R. Ponec, G. Lendvay, J. Chaves, *J. Comput. Chem.* 2008, *29*, 1387; (d) R. Ponec, G. Lendvay, M. R. Sundberg, *J. Phys. Chem. A* 2008, *112*, 9936; (e) X. Zhang, Q.-S. Li, Y. Xie, R. B. King, H. F. Schaefer, III, *J. Mol. Struct.* 2008, *890*, 184; (f) A. D. Brathwaite, J. A. Maner, M. A. Duncan, *Inorg. Chem.* 2014, *53*, 1166; (g) G. W. Waldhart, A. J. Webster, B. M. Schreiber, R. B. Siedschlag, J. S. D'Acchioli, *J. Inorg. Organomet. Polym. Mater.* 2014, *24*, 87.
20. (a) S.-G. Wang, W. H. E. Schwarz, *J. Am. Chem. Soc.* 2004, *126*, 1266; (b) H. Fang, S.-G. Wang, X.-G. Zhang, *Int. J. Quantum Chem.* 2008, *109*, 526; (c) H. Fang, X.-G. Zhang, S.-G. Wang, *Phys. Chem. Chem. Phys.* 2009, *11*, 5796; (d) F. Mendizabal, S. Miranda-Rojas, L. Barrientos-Poblete, *Comput. Theor. Chem.* 2015, *1057*, 74; (e) H. Schmidbaur, A. Schier, *Chem. Soc. Rev.* 2012, *41*, 370; (f) H. Schmidbaur, A. Schier, *Chem. Soc. Rev.* 2008, *37*, 1931; (g) P. Pyykkö, *Angew. Chem. Int. Ed.* 2004, *43*, 4412.
21. (a) L. H. Doerrer, *Comments Inorg. Chem.* 2008, *29*, 93; (b) A. Otero-de-la-Roza, J. D. Mallory, E. R. Johnson, *J. Chem. Phys.* 2014, 140, 18A504/1.
22. (a) C.-M. Che, S.-W. Lai, *Coord. Chem. Rev.* 2005, *249*, 1296; (b) S. Sculfort, P. Braunstein, *Chem. Soc. Rev.* 2011, *40*, 2741.
23. (a) D.-W. Fong, E. Grunwald, *J. Phys. Chem.* 1969, *73*, 3909; (b) E. Grunwald, R. L. Lipnick, E. K. Ralph, III, *J. Am. Chem. Soc.* 1969, *91*, 4333; (c) G. R. Luckhurst, M. Setaka, *Mol. Cryst. Liq. Cryst.* 1973, *19*, 279; (d) J. Mahanty, C. N. R. Rao, *Z. Naturforsch., A* 1976, *31A*, 1094; (e) V. L. Vilker, G. P. Uyeno, W. G. McMillan, *J. Phys. Chem.* 1981, *85*, 2013; (f) T. A. Hagen, G. L. Flynn, *J. Pharm. Sci.* 1983, *72*, 409; (g) K. Yoshida, T. Shigeoka, F. Yamauchi, *Ecotoxicol. Environ. Saf.*, 1983, *7*, 558; (h) U. Berg, W. P. Jencks, *J. Am. Chem. Soc.* 1991, *113*, 6997; (i) K. L. Dillman, K. R. Shelly, W. F. Beck, *J. Phys. Chem. B* 2009, *113*, 6127.
24. (a) D. C. Boyd, R. Szalapski, K. R. Mann, *Organometallics* 1989, *8*, 790; (b) T. J. Wadas, Q.-M. Wang, Y.-J. Kim, C. Flaschenreim, T. N. Blanton, R. Eisenberg, *J. Am. Chem. Soc.* 2004, *126*, 16841; (c) C.-K. Koo, B. Lam, S.-K. Leung, M. H.-W. Lam, W.-Y. Wong, *J. Am. Chem. Soc.* 2006, *128*, 16434; (d) J. E. Bercaw, A. C. Durrell, H. B. Gray, J. C. Green, N. Hazari, J. A. Labinger, J. R. Winkler, *Inorg. Chem.* 2010, *49*, 1801; (e) S. Grimme, J.-P. Djukic, *Inorg. Chem.* 2011, *50*, 2619; (f) J. Luo, J. R. Khusnutdinova, N. P. Rath, L. M. Mirica, *Chem. Commun.* 2012, *48*, 1532; (g) M. Weber, J. E. M. N. Klein, B. Miehlich, W. Frey, R. Peters, *Organometallics* 2013, *32*, 5810.
25. (a) D. Braga, F. Grepioni, E. Tedesco, K. Biradha, G. R. Desiraju, *Organometallics* 1997, *16*, 1846; (b) T. W. Hambley, *Inorg. Chem.* 1998, *37*, 3767; (c) P. L. A. Popelier, G. Logothetis, *J. Organomet. Chem.* 1998, *555*, 101; (d) S. Pillet, G. Wu, V. Kulsomphob, B. G. Harvey, R. D. Ernst, P. Coppens, *J. Am. Chem. Soc.* 2003, *125*, 1937; (e) S. H. Crosby, G. J. Clarkson, R. J. Deeth, J. P. Rourke, *Organometallics* 2010, *29*, 1966; (f) V. Tognetti, L. Joubert, C. Adamo, *J. Chem. Phys.* 2010, *132*, 211101; (g) J. Sassmannshausen, *Dalton Trans.* 2012, *41*, 1919.
26. E. Johnson, S. Keinan, P. Mauri-Sanchez, J. Contreras-Garcia, A. J. Cohen, W. Yang, *J. Am. Chem. Soc.* 2010, *132*, 6498.
27. J. Contreras-Garcia, E. R. Johnson, S. Keinan, R. Chaudret, J. P. Piquemal, D. N. Beratan, W. Yang, *J. Chem. Theory Comput.* 2011, *7*, 625.
28. (a) R. F. W. Bader, In Atoms in Molecules: A Quantum Theory, Clarendon, Oxford, 1990; (b) P. Popelier, Atoms in Molecules, An Introduction, Prentice Hall, Harlow, 2000.
29. (a) C. Hättig, W. Klopper, A. Köhn, D. P. Tew, *Chem. Rev.* 2012, *112*, 4; (b) P. G. Szalay, T. Müller, G. Gidofalvi, H. Lischka, R. Shepard, *Chem. Rev.* 2012, *112*, 108; (c) D. I. Lyakh, M. Musial, V. F. Lotrich, R. J. Bartlett, *Chem. Rev.* 2012, *112*, 182; (d) A. J. Cohen, P. Mori-Sánchez, W. Yang, *Chem. Rev.* 2012, *112*, 289.
30. C. Riplinger, B. Sandhoefer, A. Hansen, F. Neese, *J. Chem. Phys.* 2013, *139*, 134101.
31. (a) L. Goerigk, S. Grimme, *Phys. Chem. Chem. Phys.* 2011, *13*, 6670; (b) L. Goerigk, S. Grimme, *J. Chem. Theory Comput.* 2011, *7*, 291; (c) M. Steinmetz, A. Hansen, S. Ehrlich, T. Risthaus, S. Grimme, *Top. Curr. Chem.* 2015, *365*, 1.
32. R. Huenerbein, B. Schirmer, J. Moellmann, S. Grimme, *Phys. Chem. Chem. Phys.* 2010, *12*, 6940.
33. (a) S. Grimme, M. Steinmetz, *Phys. Chem. Chem. Phys.* 2013, *15*, 16031; (b) T. Risthaus, M. Steinmetz, S. Grimme, *J. Comp. Chem.* 2014, *35*, 1509.

34. (a) T. Bucko, J. Hafner, S. Lebegue, J. G. Angyan, *J. Phys. Chem. A* 2010, *114*, 11814; (b) J. Moellmann, S. Grimme, *Phys. Chem. Chem. Phys.* 2010, *12*, 8500; (c) J. Moellmann, S. Grimme, *Organometallics* 2013, *32*, 3784.
35. (a) S. Grimme, J. Antony, S. Ehrlich, H. Krieg, *J. Chem. Phys.* 2010, *132*, 154104; (b) S. Grimme, S. Ehrlich, L. Goerigk, *J. Comput. Chem.* 2011, *32*, 1456; (c) S. Grimme, *WIREs Comput. Mol. Sci.* 2011, *1*, 211.
36. (a) J. G. Brandenburg, S. Grimme, *Theor. Chem. Acc.* 2013, *132*, 1399; (b) J. G. Brandenburg, S. Grimme, *Top. Curr. Chem.* 2014, *345*, 1.
37. O. A. Vydrov, T. V. Voorhis, *J. Chem. Phys.* 2010, *132*, 164113.
38. (a) F. Neese, *Wiley Interdiscip. Rev.: Comput. Mol. Sci.* 2012, *2*, 73; (b) F. Neese, ORCA - An Ab Initio, Density Functional and Semiempirical Program Package 2014, version 3.0.0; Bonn University: Bonn, 2014.
39. (a) W. Hujo, S. Grimme, *Phys. Chem. Chem. Phys.* 2011, *13*, 13942; (b) S. Grimme, W. Hujo, B. Kirchner, *Phys. Chem. Chem. Phys.* 2012, *14*, 4875; (c) W. Hujo, S. Grimme, *J. Chem. Theory Comput.* 2013, *9*, 308; (d) W. Hujo, S. Grimme, *J. Chem. Theory Comput.* 2011, *7*, 3866.
40. S. Ehrlich, J. Moellmann, W. Reckien, T. Bredow, S. Grimme, *ChemPhysChem* 2011, *12*, 3414.
41. W. Iali, P. Petrovic, M. Pfeffer, S. Grimme, J.-P. Djukic, *Dalton Trans.* 2012, *41*, 12233.
42. M. E. M. de Piedade, J. A. M. Simoes, Molecular Energetics: Condensed-Phase Thermochemical Techniques, Oxford University Press, New York, 2006.
43. A. Klamt, G. Schüürmann, *J. Chem. Soc., Perkin Trans.* 1993, *2*, 799.
44. (a) A. Klamt, *WIREs Comput. Mol. Sci.* 2011, *1*, 699; (b) F. Eckert, A. Klamt, COSMOtherm, Version C2.1, Release 01.11 ed.; COSMOlogic GmbH & Co. KG, Leverkusen, 2010.
45. S. Grimme, *ChemPhysChem* 2012, *13*, 1407.
46. J. Tomasi, B. Mennucci, R. Cammi, *Chem. Rev.* 2005, *105*, 2999.
47. A. V. Marenich, C. J. Cramer, D. G. Truhlar, *J. Phys. Chem. B* 2009, *113*, 6378.
48. M. W. Schmidt, K. K. Baldridge, J. A. Boatz, S. T. Elbert, M. S. Gordon, J. H. Jensen, S. Koseki, N. Matsunaga, K. A. Nguyen, S. Su, T. L. Windus, M. Dupuis, J. A. Montgomery, *J. Comput. Chem.* 1993, *14*, 1347.
49. T. Kloss, J. Heil, S. M. Kast, *J. Phys. Chem. B* 2008, *112*, 4337.
50. S. Grimme, *Chem. Eur. J.* 2012, *18*, 9955.
51. (a) T. Risthaus, S. Grimme, *J. Chem. Theory Comput.* 2013, *9*, 1580; (b) R. Sure, J. Antony, S. Grimme, *J. Phys. Chem. B* 2014, *118*, 3431; (c) J. Antony, R. Sure, S. Grimme, *Chem. Commun.* 2015, *51*, 1764; (d) C. Bannwarth, A. Hansen, S. Grimme, *Isr. J. Chem.* 2015, *55*, 235.
52. A. Hansen, C. Bannwarth, S. Grimme, P. Petrović, C. Werle, J.-P. Djukic, *ChemistryOpen* 2014, *3*, 177.
53. (a) J. P. Perdew, K. Burke, M. Ernzerhof, *Phys. Rev. Lett.* 1996, *77*, 3865; (b) J. P. Perdew, K. Burke, M. Ernzerhof, *Phys. Rev. Lett.* 1997, *78*, 1396.
54. (a) F. Weigend, R. Ahlrichs, *Phys. Chem. Chem. Phys.* 2005, *7*, 3297; (b) F. Weigend, M. Häser, H. Patzelt, R. Ahlrichs, *Chem. Phys. Lett.* 1998, *294*, 143; (c) A. Hellweg, C. Hättig, S. Höfener, W. Klopper, *Theor. Chem. Acc.* 2007, *117*, 587.
55. D. Andrae, U. Haeussermann, M. Dolg, H. Stoll, H. Preuss, *Theor. Chim. Acta* 1990, *77*, 123.
56. J. Tao, J. P. Perdew, V. N. Staroverov, G. E. Scuseria, *Phys. Rev. Lett.* 2003, *91*, 146401.
57. (a) A. D. Becke, *J. Chem. Phys.* 1993, *98*, 5648; (b) C. Lee, W. Yang, R. G. Parr, *Phys. Rev. B* 1988, *37*, 785.
58. Y. Zhao, D. G. Truhlar, *J. Phys. Chem. A* 2005, *109*, 5656.
59. Y. Zhao, D. G. Truhlar, *Theor. Chem. Acc.* 2008, *120*, 215.
60. S. Grimme, *J. Chem. Phys.* 2006, *124*, 034108.
61. S. Grimme, L. Goerigk, R. F. Fink, *WIREs Comput. Mol. Sci.* 2012, *2*, 886.
62. TURBOMOLE, TURBOMOLE, 6.4 ed., TURBOMOLE GmbH, University of Karlsruhe and Forschungszentrum Karlsruhe GmbH, 2012. Available from http://www.turbomole.com. Accessed 24 June, 2015.
63. (a) A. Halkier, T. Helgaker, P. Jørgensen, W. Klopper, H. Koch, J. Olsen, A. K. Wilson, *Chem. Phys. Lett.* 1998, *286*, 243; (b) A. Halkier, T. Helgaker, P. Jørgensen, W. Klopper, J. Olsen, *Chem. Phys. Lett.* 1999, *302*, 437.
64. A. Schäfer, C. Huber, R. Ahlrichs, *J. Chem. Phys.* 1994, *100*, 5829.
65. A. G. Csaszar, W. D. Allen, H. F. Schaefer, *J. Chem. Phys.* 1998, *108*, 9751.
66. Z.-W. Qu, A. Hansen, S. Grimme, *J. Chem. Theory Comput.* 2015, *11*, 1037.
67. (a) P. M. Kozlowski, M. Kumar, P. Piecuch, W. Li, N. P. Bauman, J. A. Hansen, P. Lodowski, M. Jaworska, *J. Chem. Theory Comput.* 2012, *8*, 1870; (b) K. P. Kepp, *J. Phys. Chem. A* 2014, *118*, 7104.
68. B. M. Axilrod, E. Teller, *J. Chem. Phys.* 1943, *11*, 299.
69. P. Macchi, A. Sironi, *Coord. Chem. Rev.* 2003, *238–239*, 383.
70. (a) R. Burton, G. Wilkinson, *Chem. Ind.* 1958, 1205; (b) R. Burton, M. L. H. Green, E. W. Abel, G. Wilkinson, *Chem. Ind.* 1958, 1592.
71. P. F. Lindsey, O. S. Mills, *J. Chem. Soc. A* 1969, 1286.
72. (a) M. R. Churchill, P. H. Bird, *Inorg. Chem.* 1968, *7*, 1545; (b) M. R. Churchill, *Inorg. Chem.* 1967, *6*, 190; (c) M. R. Churchill, P. H. Bird, *Chem. Commun.* 1967, 746.
73. (a) S. Töfke, U. Behrens, *Angew. Chem. Int. Ed. Engl.* 1987, 26; (b) S. Töfke, U. Behrens, *J. Organomet. Chem.* 1988, *338*, 29; (c) F. Edelmann, S. Töfke, U. Behrens, *J. Organomet. Chem.* 1986, *308*, 27.
74. A. E. Ashley, G. Balázs, A. R. Cowley, J. C. Green, D. O'Hare, *Organometallics* 2007, *26*, 5517.
75. (a) Amsterdam Density Functional ADF2013, SCM, Theoretical Chemistry, Vrije Universiteit, Amsterdam, The Netherlands, 2013. Available from http://www.scm.com. Accessed October 31, 2015; (b) G. te Velde, F. M. Bickelhaupt, E. J. Baerends, C. Fonseca Guerra, S. J. A. van Gisbergen, J. G. Snijders, T. Ziegler, *J. Comp. Chem.* 2001, *22*, 931.
76. E. van Lenthe, A. E. Ehlers, E. J. Baerends, *J. Chem. Phys.* 1999, *110*, 8943.
77. E. van Lenthe, E. J. Baerends, *J. Comp. Chem.* 2003, *24*, 1142.
78. P. C. Leung, P. Coppens, *Acta Crystallogr.* 1983, *B39*, 535.
79. V. N. Kalinin, *Russ. Chem. Soc.* 1987, *56*, 682.
80. A. Solladié-Cavallo, *Polyhedron* 1985, *4*, 901.
81. K. Edgar, B. F. G. Johnson, J. Lewis, S. B. Wild, *J. Chem. Soc. A* 1968, 2851.

82. G. I. Magomedov, V. G. Syrkin, A. S. Frenkel, O. A. Zakharchenko, *Zh. Obshch. Khim.* 1975, *45*, 2530.

83. B. V. Lokshin, E. B. Rusach, N. E. Kolobova, Y. V. Makarov, N. A. Ustynyuk, V. I. Zdanovich, A. Z. Zhakaeva, V. N. Setkina, *J. Organomet. Chem.* 1976, *108*, 353.

84. C. Bonifaci, A. Ceccon, A. Gambaro, F. Manoli, L. Mantovani, P. Ganis, S. Santi, A. Venzo, *J. Organomet. Chem.* 1998, *577*, 97.

85. (a) A. Ceccon, A. Gambaro, S. Santi, A. Venzo, *J. Mol. Cat.* 1991, *69*, L1; (b) L. Mantovani, A. Ceccon, A. Gambaro, S. Santi, P. Ganis, A. Venzo, *Organometallics* 1997, *16*, 2682.

86. A. Ceccon, A. Gambarro, S. Santi, G. Valle, A. Venzo, *J. Chem. Soc., Chem. Commun.* 1989, 51.

87. C. Bonifaci, A. Ceccon, A. Gambaro, P. Ganis, S. Santi, G. Valle, A. Venzo, *Organometallics* 1993, *12*, 4211.

88. C. Bonifaci, G. Carta, A. Ceccon, A. Gambaro, S. Santi, A. Venzo, *Organometallics* 1996, *15*, 1630.

89. C. Bonifaci, A. Ceccon, A. Gambaro, P. Ganis, S. Santi, G. Valle, A. Venzo, *J. Organomet. Chem.* 1995, *1995*, 35.

90. C. Bonifaci, A. Ceccon, A. Gambaro, P. Ganis, S. Santi, A. Venzo, *Organometallics* 1995, *14*, 2430.

91. P. Cecchetto, A. Ceccon, A. Gambaro, S. Santi, P. Ganis, R. Gobetto, G. Valle, A. Venzo, *Organometallics* 1998, *17*, 752.

92. C. Bonifaci, A. Ceccon, S. Santi, C. Mealli, R. W. Zoellner, *Inorg. Chim. Acta* 1995, *240*, 541.

93. S. Alvarez, P. Alemany, G. Aullon, A. A. Palacios, J. J. Novoa, Heterodox bonding effects between transition metal atoms, in The Synergy between Dynamics and Reactivity at Clusters and Surfaces, Vol. *C465* (Ed.: L. J. Farugia), Kluwer Academic Publishers, Dordrecht, 1995, pp. 241.

94. T. Schwabe, S. Grimme, J. P. Djukic, *J. Am. Chem. Soc.* 2009, *131*, 14156.

95. C. Werlé, C. Bailly, L. Karmazin-Brelot, X. F. L. Goff, M. Pfeffer, J. P. Djukic, *Angew. Chem. Int. Ed.* 2014, *53*, 9827.

96. C. Werlé, C. Bailly, L. Karmazin-Brelot, X. F. L. Goff, L. Ricard, J. P. Djukic, *J. Am. Chem. Soc.* 2013, *135*, 17839.

97. G. Palyi, *Transit. Met. Chem.* 1977, *2*, 273.

98. V. N. Kalinin, I. A. Cherepanov, S. K. Moiseev, A. S. Batsanov, Y. T. Struchkov, *Mendeleev Commun.* 1991, 77.

99. (a) S. K. Moiseev, I. A. Cherepanov, P. V. Petrovskii, M. G. Ezernitskaya, H. Butenschoen, M. Strotmann, V. N. Kalinin, *Inorg. Chim. Acta* 1998, *280*, 71; (b) V. N. Kalinin, I. A. Cherepanov, S. K. Moiseev, F. M. Dolgushin, A. I. Yanovsky, Y. T. Struchkov, *Acta Crystallogr. Sect. C: Cryst. Struct. Commun.* 1993, *C49*, 805.

100. C. Werlé, M. Hamdaoui, C. Bailly, X.-F. Le Goff, L. Brelot, J.-P. Djukic, *J. Am. Chem. Soc.* 2013, *135*, 1715.

101. J.-P. Djukic, A. Maisse, M. Pfeffer, K.-H. Doetz, M. Nieger, *Eur. J. Inorg. Chem. 1998*, 1781.

102. J.-P. Djukic, A. Maisse, M. Pfeffer, K. H. Doetz, M. Nieger, *Organometallics* 1999, *18*, 2786.

103. J.-P. Djukic, A. Maisse-Francois, M. Pfeffer, K. H. Doetz, A. De Cian, J. Fischer, *Organometallics* 2000, *19*, 5484.

104. (a) J.-P. Djukic, C. Michon, A. Maisse-Francois, R. Allagapen, M. Pfeffer, K. H. Dotz, A. De Cian, J. Fischer, *Chem. Eur. J.* 2000, *6*, 1064; (b) C. Werle, X.-F. Le Goff, J.-P. Djukic, *J. Organomet. Chem.* 2014, *751*, 754; (c) J. P. Djukic, C. Michon, A. Berger, M. Pfeffer, A. DeCian, A. Kyritsakas-Gruber, *J. Organomet. Chem.* 2006, *691*, 846.

105. J. P. Djukic, K. H. Dötz, M. Pfeffer, A. De Cian, J. Fischer, *Organometallics* 1997, *16*, 5171.

106. J.-P. Djukic, M. Pfeffer, K. H. Dotz, *C. R. Acad. Sci., Ser. IIc: Chim.* 1999, *2*, 403.

107. I. Hyla-Kryspin, S. Grimme, J. P. Djukic, *Organometallics* 2009, *28*, 1001.

108. (a) H. Schmidbaur, *Gold Bull. Gold Put. Dig.* 1990, *23*, 11; (b) H. Schmidbaur, *Chem. Soc. Rev.* 1995, *24*, 391; (c) F. Scherbaum, A. Grohmann, G. Mueller, H. Schmidbaur, *Angew. Chem.* 1989, *101*, 464; (d) F. Scherbaum, A. Grohmann, B. Huber, C. Krueger, H. Schmidbaur, *Angew. Chem.* 1988, *100*, 1602; (e) F. Scherbaum, B. Huber, G. Mueller, H. Schmidbaur, *Angew. Chem.* 1988, *100*, 1600.

109. H. Schmidbaur, *Gold Bull.* 2000, *33*, 3.

110. (a) N. W. Alcock, *Adv. Inorg. Chem. Radiochem.* 1972, *15*, 1; (b) H. A. Bent, *Chem. Rev.* 1968, *68*, 587; (c) G. R. Desiraju, Crystal Engineering: The Design of Organic Solids, Elsevier, Amsterdam, 1989; (d) J. P. Fackler, Jr., *Polyhedron* 1997, *16*, 1271g; (e) C. Glidewell, *Inorg. Chim. Acta* 1975, *12*, 219; (f) K. I. Grandberg, V. P. Dyadchenko, *J. Organomet. Chem.* 1994, *474*, 1; (g) C. E. Housecroft, *Coord. Chem. Rev.* 1992, *115*, 117; (h) C. E. Housecroft, *Coord. Chem. Rev.* 1993, *127*, 187; (i) C. E. Housecroft, *Coord. Chem. Rev.* 1995, *142*, 101; (j) M. Jansen, *Angew. Chem.* 1987, *99*, 1136; (k) P. G. Jones, *Gold Bull.* 1981, *14*, 159; (l) P. G. Jones, *Gold Bull.* 1981, *14*, 102; (m) M. Melnik, R. V. Parish, *Coord. Chem. Rev.* 1986, *70*, 157; (n) D. M. P. Mingos, *J. Chem. Soc. Dalton Trans.* 1996, 561; (o) D. M. P. Mingos, M. J. Watson, *Adv. Inorg. Chem.* 1992, *39*, 327; (p) P. J. Sadler, *Struct. Bond.* 1976, *29*, 171; (q) G. van Koten, S. L. James, J. T. B. H. Jastrzebski, Copper and silver, in Comprehensive Organometallic Chemistry II (Eds.: E. W. Abel, F. G. A. Stone, G. Wilkinson), Elsevier, Oxford, 1995, pp. 57; (r) C.-M. Che, S.-W. Lai, *Coord. Chem. Rev.* 2005, *249*, 1296.

111. (a) L. H. Doerrer, *Dalton Trans.* 2010, *39*, 3543; (b) P. Pyykkö, *Chem. Rev.* 1997, *97*, 597; (c) K. Uehara, S. Hikichi, A. Inagaki, M. Akita, *Chem. Eur. J.* 2005, *11*, 2788; (d) J.-M. Poblet, M. Bénard, *Chem. Commun.* 1998, 1179; (e) H. L. Hermann, G. Boche, P. Schwerdtfeger, *Chem. Eur. J.* 2001, *7*, 5333; (f) B.-H. Xia, H.-X. Zhang, Y.-Q. Jiao, Q.-J. Pan, Z.-S. Li, C.-C. Sun, *J. Chem. Phys.* 2004, *120*, 11487.

112. (a) P. Coppens, I. V. Novozhilova, *Int. J. Quantum Chem.* 2005, *101*, 611; (b) Z.-N. Chen, N. Zhao, Y. Fan, J. Ni, *Coord. Chem. Rev.* 2009, *253*, 1.

113. L. W. Arndt, M. Y. Darensbourg, T. Delord, B. T. Bancroft, *J. Am. Chem. Soc.* 1986, *108*, 2617.

114. (a) L. R. Martin, F. W. B. Einstein, R. K. Pomeroy, *Inorg. Chem.* 1983, *22*, 1959; (b) L. R. Martin, F. W. B. Einstein, R. K. Pomeroy, *Inorg. Chem.* 1985, *24*, 2777.

115. F. W. B. Einstein, R. K. Pomeroy, P. Rushman, A. C. Willis, *Chem. Commun.* 1983, 854.

116. F. W. B. Einstein, T. Jones, R. K. Pomeroy, P. Rushman, *J. Am. Chem. Soc.* 1984, *106*, 2707.

117. (a) F. W. B. Einstein, R. K. Pomeroy, P. Rushman, A. C. Willis, *Organometallics* 1985, *4*, 250; (b) R. J. Batchelor, H. B. Davis, F. W. B. Einstein, R. K. Pomeroy, *J. Am. Chem. Soc.* 1990, *112*, 2036; (c) H. B. Davis, F. W. B. Einstein,

P. G. Glavina, T. Jones, R. K. Pomeroy, P. Rushman, *Organometallics* 1989, *8*, 1030; (d) J. A. Shipley, R. J. Batchelor, F. W. B. Einstein, R. K. Pomeroy, *Organometallics* 1991, *10*, 3620; (e) F. Jiang, H. A. Jenkins, K. Biradha, H. B. Davis, R. K. Pomeroy, M. J. Zaworotko, *Organometallics* 2000, *19*, 5049; (f) F. Jiang, J. L. Male, K. Biradha, W. K. Leong, R. K. Pomeroy, M. J. Zaworotko, *Organometallics* 1998, *17*, 5810.

118. H. Nakatsuji, M. Hada, A. Kawashima, *Inorg. Chem.* 1992, *31*, 1740.

119. S. Grimme, J.-P. Djukic, *Inorg. Chem.* 2010, *49*, 2911.

120. V. Phillips, K. J. Willard, J. A. Golen, C. J. Moore, A. L. Rheingold, L. H. Doerrer, *Inorg. Chem.* 2010, *49*, 9265.

121. E. J. Fernández, C. Hardacre, A. Laguna, M. C. Lagunas, J. M. López-de-Luzuriaga, M. Monge, M. Montiel, M. E. Olmos, R. C. Puelles, E. Sánchez-Forcada, *Chem. Eur. J.* 2009, *15*, 6222.

122. F. Balzano, A. Cuzzola, P. Diversi, F. Ghiotto, G. Uccello-Barretta, *Eur. J. Inorg. Chem.* 2007, *2007*, 5556.

123. (a) R. Narayanaswamy, M. A. Young, E. Parkhurst, M. Ouellette, M. E. Kerr, D. M. Ho, R. C. Elder, A. E. Bruce, M. R. M. Bruce, *Inorg. Chem.* 1993, *32*, 2506; (b) A. Pintado-Alba, H. de la Riva, M. Nieuwhuyzen, D. Bautista, P. R. Raithby, H. A. Sparkes, S. J. Teat, J. M. Lopez-de-Luzuriaga, M. C. Lagunas, *Dalton Trans.* 2004, 3459; (c) H. de la Riva, A. Pintado-Alba, M. Nieuwenhuyzen, C. Hardacre, M. C. Lagunas, *Chem. Commun.* 2005, 4970; (d) A. Deák, T. Megyes, G. Tárkányi, P. Király, L. Biczók, G. Pálinkás, P. J. Stang, *J. Am. Chem. Soc.* 2006, *128*, 12668.

124. (a) A. L. Balch, J. Miller, *J. Organomet. Chem.* 1971, *32*, 263; (b) P. R. Branson, M. Green, *Dalton Trans.* 1972, 1303; (c) J. W. Dart, M. K. Lloyd, J. A. McCleverty, R. Mason, *Chem. Commun.* 1971, 1197.

125. (a) A. L. Balch, M. M. Olmstead, *J. Am. Chem. Soc.* 1976, *98*, 2354; (b) N. S. Lewis, K. R. Mann, J. G. Gordon, II, H. B. Gray, *J. Am. Chem. Soc.* 1976, *98*, 7461; (c) K. R. Mann, J. G. Gordon, II, H. B. Gray, *J. Am. Chem. Soc.* 1975, *97*, 3553; (d) J. A. McCleverty, J. Williams, *Transit. Met. Chem.* 1978, *3*, 205.

126. (a) V. M. Miskowski, G. L. Nobinger, D. S. Kliger, G. S. Hammond, N. S. Lewis, K. R. Mann, H. B. Gray, *J. Am. Chem. Soc.* 1978, *100*, 485; (b) S. Miya, Y. Yamamoto, H. Yamazaki, *Inorg. Chem.* 1982, *21*, 1486.

127. I. S. Sigal, H. B. Gray, *J. Am. Chem. Soc.* 1981, *103*, 2220.

128. A. L. Balch, V. J. Catalano, *Inorg. Chem.* 1992, *31*, 3934.

129. (a) V. M. Miskowski, T. P. Smith, T. M. Loehr, H. B. Gray, *J. Am. Chem. Soc.* 1985, *107*, 7925; (b) S. F. Rice, H. B. Gray, *J. Am. Chem. Soc.* 1981, *103*, 1593.

130. (a) C.-K. Koo, K.-L. Wong, K.-C. Lau, W.-Y. Wong, M. H.-W. Lam, *Chem. Eur. J.* 2009, *15*, 7689; (b) K. R. Mann, J. A. Thich, R. A. Bell, C. L. Coyle, H. B. Gray, *Inorg. Chem.* 1980, *19*, 2462; (c) G. S. Rodman, K. R. Mann, *Inorg. Chem.* 1988, *27*, 3338; (d) A. E. Stiegman, S. F. Rice, H. B. Gray, V. M. Miskowski, *Inorg. Chem.* 1987, *26*, 1112; (e) J. R. Winkler, J. L. Marshall, T. L. Netzel, H. B. Gray, *J. Am. Chem. Soc.* 1986, *108*, 2263.

131. G. L. Geoffroy, M. S. Wrighton, G. S. Hammond, H. B. Gray, *J. Am. Chem. Soc.* 1974, *96*, 3105.

132. S. F. Rice, S. J. Milder, H. B. Gray, R. A. Goldbeck, D. S. Kliger, *Coord. Chem. Rev.* 1982, *43*, 349.

133. (a) S. K. Doorn, K. C. Gordon, R. B. Dyer, W. H. Woodruff, *Inorg. Chem.* 1992, *31*, 2284; (b) S. J. Milder, *Inorg. Chem.* 1985, *24*, 3376; (c) S. J. Milder, D. S. Kliger, *J. Phys. Chem. B* 1985, *89*, 4170; (d) S. Oberneder, G. Gliemann, *J. Phys. Chem.* 1989, *93*, 4487; (e) P. V. Yaneff, J. Powell, *J. Organomet. Chem.* 1979, *179*, 101.

134. (a) S. Fukuzumi, N. Nishizawa, T. Tanaka, *Bull. Chem. Soc. Jpn.* 1982, *55*, 2892; (b) S. Fukuzumi, N. Nishizawa, T. Tanaka, *Bull. Chem. Soc. Jpn.* 1982, *55*, 3482; (c) M. G. Hill, K. R. Mann, *Catal. Lett.* 1991, *11*, 341.

135. M. G. Hill, J. P. Bullock, T. Wilson, P. Bacon, C. A. Blaine, K. R. Mann, *Inorg. Chim. Acta* 1994, *226*, 61.

136. M. R. Rhodes, K. R. Mann, *Inorg. Chem.* 1984, *23*, 2053.

137. A. L. Balch, M. M. Olmstead, *J. Am. Chem. Soc.* 1979, *101*, 3128.

138. (a) H. B. Gray, V. M. Miskowski, S. J. Milder, T. P. Smith, A. W. Maverick, J. D. Buhr, W. L. Gladfelter, I. S. Sigal, K. R. Mann, Mechanistic aspects of solar energy storage reactions involving polynuclear rhodium isocyanide complexes. Preparation of new binuclear isocyanide complexes of iridium, cobalt, nickel, and ruthenium, in Fundamental Research in Homogeneous Catalysis, (Ed.: M. Tsutsui), Springer, New York, 1979, pp. 819; (b) K. R. Mann, R. A. Bell, H. B. Gray, *Inorg. Chem.* 1979, *18*, 2671; (c) K. R. Mann, M. J. DiPierro, T. P. Gill, *J. Am. Chem. Soc.* 1980, *102*, 3965; (d) I. S. Sigal, K. R. Mann, H. B. Gray, *J. Am. Chem. Soc.* 1980, *102*, 7252.

139. (a) A. W. Maverick, T. P. Smith, E. F. Maverick, H. B. Gray, *Inorg. Chem.* 1987, *26*, 4336; (b) X. Shen, Y. Yamaguchi, K. Sakata, M. Hashimoto, *Anal. Sci.: X-Ray Struct. Anal. Online* 2005, *21*, x75.

140. (a) H. Endres, N. Gottstein, H. J. Keller, R. Martin, W. Rodemer, W. Steiger, *Z. Naturforsch. B: Chem. Sci.* 1979, *34*, 827; (b) M. M. Olmstead, A. L. Balch, *J. Organomet. Chem.* 1978, *148*, C15.

141. (a) N. T. Tran, J. R. Stork, D. Pham, C. J. Chancellor, M. M. Olmstead, J. C. Fettinger, A. L. Balch, *Inorg. Chem.* 2007, *46*, 7998; (b) N. T. Tran, J. R. Stork, D. Pham, M. M. Olmstead, J. C. Fettinger, A. L. Balch, *Chem. Commun.* 2006, 1130.

142. (a) D. T. Plummer, B. A. Karcher, R. A. Jacobson, R. J. Angelici, *J. Organomet. Chem.* 1984, *260*, 347; (b) Y. Yamamoto, K. Aoki, H. Yamazaki, *Inorg. Chem.* 1979, *18*, 1681.

143. (a) P. Coppens, O. Gerlits, I. I. Vorontsov, A. Y. Kovalevsky, Y.-S. Chen, T. Graber, M. Gembicky, I. V. Novozhilova, *Chem. Commun.* 2004, 2144; (b) I. V. Novozhilova, A. V. Volkov, P. Coppens, *Inorg. Chem.* 2004, *43*, 2299.

144. (a) C. L. Exstrom, D. Britton, K. R. Mann, M. G. Hill, V. M. Miskowski, W. P. Schaefer, H. B. Gray, W. M. Lamanna, *Inorg. Chem.* 1996, *35*, 549; (b) M. Fetterolf, A. E. Friedman, Y. Y. Yang, H. Offen, P. C. Ford, *J. Phys. Chem.* 1988, *92*, 3760.

145. (a) R. F. Dallinger, M. J. Carlson, V. M. Miskowski, H. B. Gray, *Inorg. Chem.* 1998, *37*, 5011; (b) O. Gerlits, A. Y. Kovalevsky, P. Coppens, *Dalton Trans.* 2004, 3955; (c) V. M. Miskowski, S. F. Rice, H. B. Gray, R. F. Dallinger, S. J. Milder, M. G. Hill, C. L. Exstrom, K. R. Mann, *Inorg. Chem.* 1994, *33*, 2799.

146. (a) M. G. Hill, K. R. Mann, *Inorg. Chim. Acta* 1996, *243*, 219; (b) D. C. Smith, V. M. Miskowski, W. R. Mason, H. B. Gray, *J. Am. Chem. Soc.* 1990, *112*, 3759.

147. J. J. Novoa, G. Aullon, P. Alemany, S. Alvarez, *J. Am. Chem. Soc.* 1995, *117*, 7169.

148. B. Lippert, Cisplatin: Chemistry and Biochemistry of a Leading Anticancer Drug, Verlag Helvetica Chimica Acta, Zurich, 1999.

149. P. V. Petrovic, S. Grimme, S. D. Zaric, M. Pfeffer, J.-P. Djukic, *Phys. Chem. Chem. Phys.* 2014, *16*, 14688.

150. (a) S. G. Chaney, S. L. Campbell, E. Bassett, Y. Wu, *Crit. Rev. Oncol. Hematol.* 2005, *53*, 3; (b) I. Ott, R. Gust, *Anticancer Agents Med Chem.* 2007, *7*, 95.

151. J. Peng, R. Mandal, M. Sawyer, X.-F. Li, *Clin. Chem.* 2005, *51*, 2274.

152. D. M. Kweekel, H. Gelderblom, H. J. Guchelaar, *Cancer Treat. Rev.* 2005, *31*, 90.

153. (a) M. A. Bruck, R. Bau, M. Noji, K. Inagaki, Y. Kidani, *Inorg. Chim. Acta* 1984, *92*, 279; (b) M. Galanski, A. Yasemi, S. Slaby, M. A. Jakupec, V. B. Arion, M. Rausch, A. A. Nazarov, B. K. Keppler, *Eur. J. Med. Chem.* 2004, *39*, 707.

154. A. J. Di Pasqua, D. J. Kerwood, Y. Shi, J. Goodisman, J. C. Dabrowiak, *Dalton Trans.* 2011, *40*, 4821.

155. (a) M. Lederer, E. Leipzig-Pagani, *Int. J. Pharm.* 1998, *167*, 223; (b) M. F. A. Lucas, M. Pavelka, M. E. Alberto, N. Russo, *J. Phys. Chem. B* 2009, *113*, 831; (c) M. Pavelka, M. F. A. Lucas, N. Russo, *Chem. Eur. J.* 2007, *13*, 10108.

156. P. Videhult, J. Yachnin, E. Jerremalm, R. Lewensohn, H. Ehrsson, *Cancer Lett.* 2002, *180*, 191.

157. R. B. Burns, R. W. Burton, S. P. Albon, L. Embree, *J. Pharm. Biomed. Anal.* 1996, *14*, 367.

158. S. E. Harding, *Biophys. Chem.* 1995, *55*, 69.

159. M. Mitoraj, A. Michalak, T. Ziegler, *J. Chem. Theory Comput.* 2009, *5*, 962.

8

HALOGEN BONDING IN THE SYNTHESIS AND DESIGN OF COORDINATION AND ORGANOMETALLIC COMPOUNDS

Abel M. Maharramov[1], Namiq Q. Shixaliyev[1], Atash V. Gurbanov[1], Kamran T. Mahmudov[1,2,*], Valentine G. Nenajdenko[3], Armando J. L. Pombeiro[2], and Maximilian N. Kopylovich[2,†]

[1] *Department of Chemistry, Baku State University, Baku, Azerbaijan*
[2] *Centro de Química Estrutural, Instituto Superior Técnico, Universidade de Lisboa, Lisbon, Portugal*
[3] *Department of Chemistry, Moscow State University, Moscow, Russia*

8.1 INTRODUCTION

Halogen bonding is the collective term for electron density donation-based interactions of halogen atoms and neutral or anionic Lewis bases, such as N, P, O, S, halogen functionalities, and π electron donors, and has been recently recognized to play a significant role in crystal engineering, anion recognition, coordination, control of properties of halogenated materials, etc. [1–7]. The halogen–halogen interactions are believed to occur due to anisotropy of electrostatic potentials on the halogen atoms and, as consequence, electrostatic attraction between electron-rich and positive areas (σ-holes) on the surfaces of halogen atoms [1]. The direction and strength of halogen bonding generally dependent on the nature of the halogen donor and acceptor and can be tuned by introduction of the substituents and cooperative effects [4, 5]. It should be noted that fluorine seldom participates in halogen bonding [8] because in most molecules the electrostatic potential of fluorine is totally negative. However, when fluorine is bonded to a highly electron-attracting framework, it can form weak halogen bonds [8].

In conventional textbooks, for example, [9], a partial negative charge is assigned to the covalently bound halogens (Scheme 8.1a), and the description to a certain degree explains the electrophilic behavior of halogenated carbon atoms or the hydrogen bonding ability of halogen moieties. However, this model is not appropriate for a correct interpretation of experimental data on some other secondary interactions. Thus, a more than 5% reduction of the sum of van der Waals radii of interacting polarized halogen and acceptor atoms is formally an indication of a non-covalent bonding [1], and hence, these contacts should be also considered for the sake of correctness. As a consequence, a corrected model, based on anisotropic distribution of the electrons in halogens (Scheme 8.1b), was proposed [5, 6].

Within this concept, halogens can act as Lewis acids or bases depending on the interaction partners and geometry of the non-covalent bonding. Thus, upon interaction with Lewis acids, the halogen moieties act as respective bases, donating electrons from their nonbonding orbitals with typical Y—X···LA$^+$ angles of 90–120°, where Y stands for any type of atom, X for halogen, and LA$^+$ for a Lewis acid [6]. A typical example is the hydrogen bonding, with interaction energies up to 140 kJ mol^{-1} and the F>Cl>Br>I trend, if bond strength is under consideration. When acting as Lewis acids, the Y—X···LB$^-$ interactions (LB$^-$ stands for a Lewis base) are generally linear with strengths commonly of 5–30 kJ mol^{-1} (in exceptional cases up to 180 kJ mol^{-1}) and

* Kamran T. Mahmudov (kamran_chem@mail.ru)
† Maximilian N. Kopylovich (maximilian.kopylovich@tecnico.ulisboa.pt)

Non-Covalent Interactions in the Synthesis and Design of New Compounds, First Edition.
Edited by Abel M. Maharramov, Kamran T. Mahmudov, Maximilian N. Kopylovich, and Armando J. L. Pombeiro.

opposite to the former trend, namely, I > Br > Cl > F [6]. This latter type of interaction has received the name "halogen bonding" and during the past decades has been increasingly investigated using X-ray and computational methods.

The majority of the studies on halogen bonding are focused on the non-covalent interactions between/within organic molecules, possibly due to available extensive experimental data and relatively easy theoretical calculations. Thus, two different intermolecular non-covalent interactions, that is, N···I halogen bond and $^+$N—H···F hydrogen bond, were identified upon self-assembly of new supramolecular systems from monoiodoperfluorocarbon $C_6F_5CF_2I$ and *N,N,N′,N′*-tetramethylethylenediamine (Scheme 8.2) [10]. Single-crystal X-ray analysis of **1** revealed a strong N···I halogen bond with the N···I—C bond distance of 2.799 Å, which is substantially shorter than the sum of the van der Waals radii of nitrogen (1.55 Å) and iodine (1.98 Å). An interaction of **1** with the water of air leads to the hydrogen bond-assisted associate **2** (Scheme 8.2) with the distance between H and F of 1.795 Å, while the distance between N and F is 2.660 Å [10]. The formed halogen and hydrogen bonding most probably play a significant role in the course of hydrolysis.

In another example, the N···I halogen bonding was applied to drive intermolecular recognition of (*E*)-1,2-diiodo-1,2-difluoroethene (bidentate electron acceptor) by *N*^{1},*N*^{1},*N*^{2},*N*^{2}-tetramethylethane-1,2-diamine or 4,4′-bipyridine (bidentate

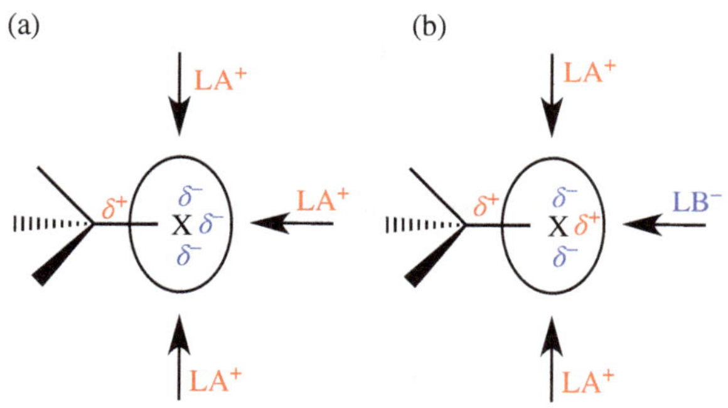

SCHEME 8.1 Schematic description of the electron density distribution of covalently bound halogens (a) and the expected halogen bonding intermolecular interactions (b) [6].

electron donors), with the formation of 1D infinite chains in both cocrystals (**3** and **4**, Scheme 8.3) [11]. The N···I distances in the halogen-bonded associates of 2.805(2) and 2.757(1) Å, respectively, are substantially shorter than the sum of van der Waals radii values and hence prove the presence of strong attractive interactions between the atoms involved.

Multiple halogen bond-driven interactions of (1*E*,1′*E*)-((perfluoro-1,4-phenylene)bis(methanylylidene))bis(hydrazine) with perbromomethane, bromotrichloromethane, and tribromofluoromethane in aqueous ammonia and DMSO in the presence of CuCl lead to 1,4-bis(2,2-dibromovinyl)-2,3,5,6-tetrafluorobenzene (**5a**), 1,4-bis(2,2-dichlorovinyl)-2,3,5,6-tetrafluorobenzene (**5b**), and 1,4-bis((*E*)-2-bromo-2-fluorovinyl)-2,3,5,6-tetrafluorobenzene (**5c**) giving rise to a spider-on-net-looking superstructure (Scheme 8.4). In structures of **5a–c**, each unit interacts with its neighbors via extensive halogen bonds (Scheme 8.4), which evidently contribute to the overall reaction energetics and facilitate the formation of the products **5a–c**. **5a–c** crystallize in space groups *P* 21/*n*, *C*2/*c*, and *P*21/*n*, respectively, [CCDC 1407946 (**5a**), 1401867 (**5b**), and 1062063 (**5c**)], while the Hal–Hal distances in **5a–c** are in the range of 2.771–3.633 Å. The halogen/halogen interactions link molecules into two-dimensional networks in parallel to *bc*-plane.

Apart from the major application of halogen bonding for supramolecular arrangement of organic materials, some recent publications point out a significant potential of this type of non-covalent interactions for the preparation of coordination compounds [12, 13]. For instance, halogen bonding plays a significant role in the solid-state architecture of many coordination and organometallic compounds, where non-covalent interactions between ionic counterparts occur. The cocrystallization outcome is related to the structural fitting of the components and provides numerous opportunities for the specific arrangement of, for example, metal- and halogen-containing pharmaceutical substances with a variety of inorganic counterparts [3].

As can be foreseen, the application of magnetic, optical, redox, and catalytic properties of metal complexes with adjustable halogenated organic frameworks has a great

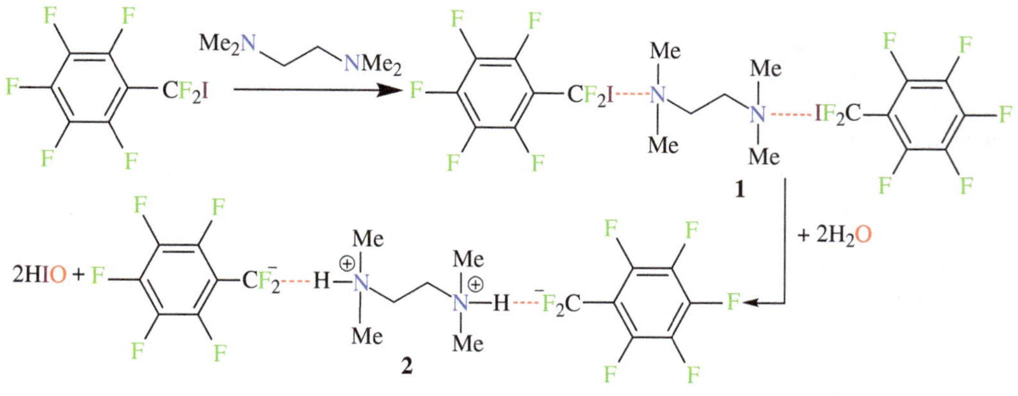

SCHEME 8.2 Synthesis of halogen bond-assisted adduct and its hydrolysis [10].

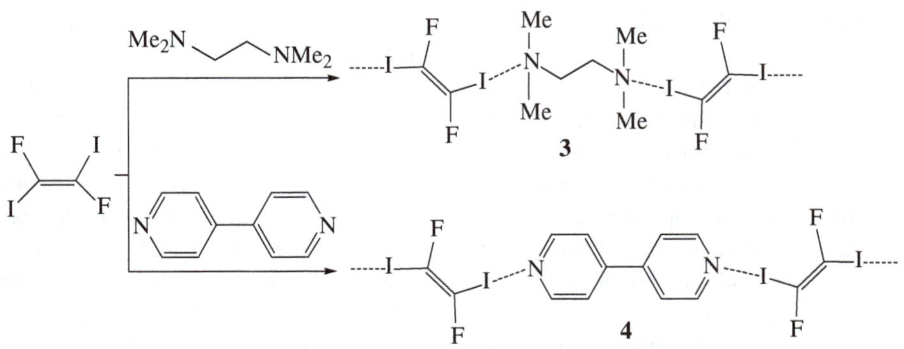

SCHEME 8.3 Formation of 1D infinite networks through halogen bonding-driven self-assembly [11].

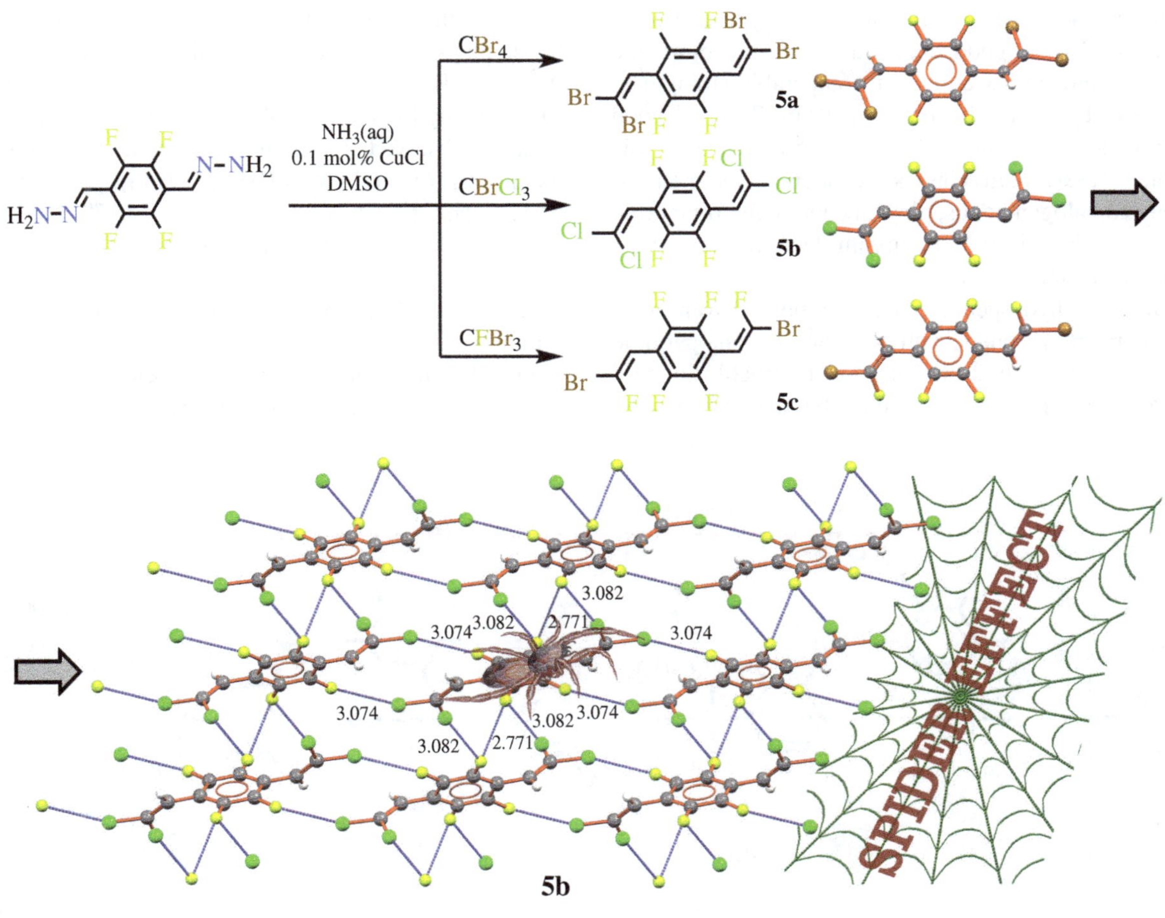

SCHEME 8.4 Halogen bond-assisted synthesis of **5a**; multiple halogen bond interactions in **5b**.

potential for material chemistry and catalysis. And *vice versa*, an introduction of metal-containing moieties into the halogen bonding-driven supramolecular architectures can significantly influence their geometry and stability. This influence can be performed in different ways, for example, through polarization of halogen atoms, which, as known, usually leads to stronger halogen bonds. One of the effective ways to polarize halogen atoms is the introduction of conjugated systems in the nearest proximity, and metallacycles provide a simple and convenient example of such charged conjugated electronic structures [12]. Hence, the formation of coordination compounds with metallacycles can significantly influence the strength and directionality of a halogen bond network. From another perspective, the number and energy of halogen bonds can grow significantly in the course of complex formation, and this effect can significantly influence the reaction outcome and its yield [12].

8.2 HALOGEN–HALOGEN BONDING

Generally, in the halogen-assisted syntheses, metal complexes can be employed in two different ways: (i) as acceptors and (ii) as donors. In the former case, they act as acceptor nodes through their ligands, thereby dictating the directionality and dimensionality of supramolecular networks. In the latter case, metal coordination has been shown to increase the electrophilicity of halogen atoms in ligands [13–15]. It was stated that halogen bonds in the produced metal complexes are weaker than those reported for purely organic binding partners [13].

Interaction enthalpies and entropies of C—X···X′—M halogen bonds were determined by titration of a nickel fluoride complex **6** with contradiction C_6F_5I (Scheme 8.5) [16]. The studies reveal that the C_6F_5I unit, a known hydrogen bond acceptor, is also a good halogen bond acceptor: the C—I···F—Ni halogen bonds are of comparable strength to N—H···F—Ni hydrogen bonds formed in **7** by a good hydrogen bond donor, indole (Scheme 8.5). Hence, in this and related cases, both classes of interactions, namely, hydrogen and halogen bonds, play an equally important role in crystal engineering and supramolecular control of transition metal catalysts [17].

Coordination, hydrogen, and halogen bond-driven interactions of copper(II) bromide with 2-, 3-, and 4-bromopyridine in the presence of hydrobromic or hydrochloric acid lead to a series of bromopyridinium tetrahalocuprates(II) (Scheme 8.6) [18]. In fact, the structures of **8–10** are dictated by N—H···X—Cu and C—Br···X—Cu (X = Cl, Br) halogen bonds with Br···Cl distances of 3.27–3.36 Å and Br···Br separations being from 3.39 to 3.45 Å (what accounts for the 7–9% diameter reduction) [19]. In similar 3-halopyridinium tetrachlorocobaltates(II) with 3-chloropyridinium, 3-bromopyridinium, and 3-iodopyridinium cations, N—H···Cl—Co and C—X···Cl—Co (X = Cl, Br, I) contacts were identified based on the X···Cl distances reduction, which is less pronounced in the series I > Br > Cl [20].

Cooperative action of coordination, hydrogen, and halogen bonding in one-pot template condensation of $CCl_3C{\equiv}N$ with ammonia on a metal source led to tris{2,4-bis (trichloromethyl)-1,3,5-triazapentadienato}-M(III) complexes (Scheme 8.7) [12]. The used DMSO solvent actively participates in the formation of the non-covalent network and hence plays an important role in the course of the reaction. Additionally, due to stabilization of the formed products **11–13** by hydrogen and halogen bonds, a high exothermic effect is observed. It was indicated that ionic radii of the metal ions [68.5 (Co) < 69 (Fe) < 72 (Mn), pm] relate to the corresponding Cl···Cl distances [3.178 (**13**) > 3.155 (**12**) > 3.133 (**11**) Å] and hence the strength of the halogen bonding.

The crystal packings of **11–13** are similar and reveal infinite strands of the complexes assembled by the Cl···Cl intermolecular interactions between chlorine atoms (Scheme 8.8). Intermolecular N···Cl interactions are also

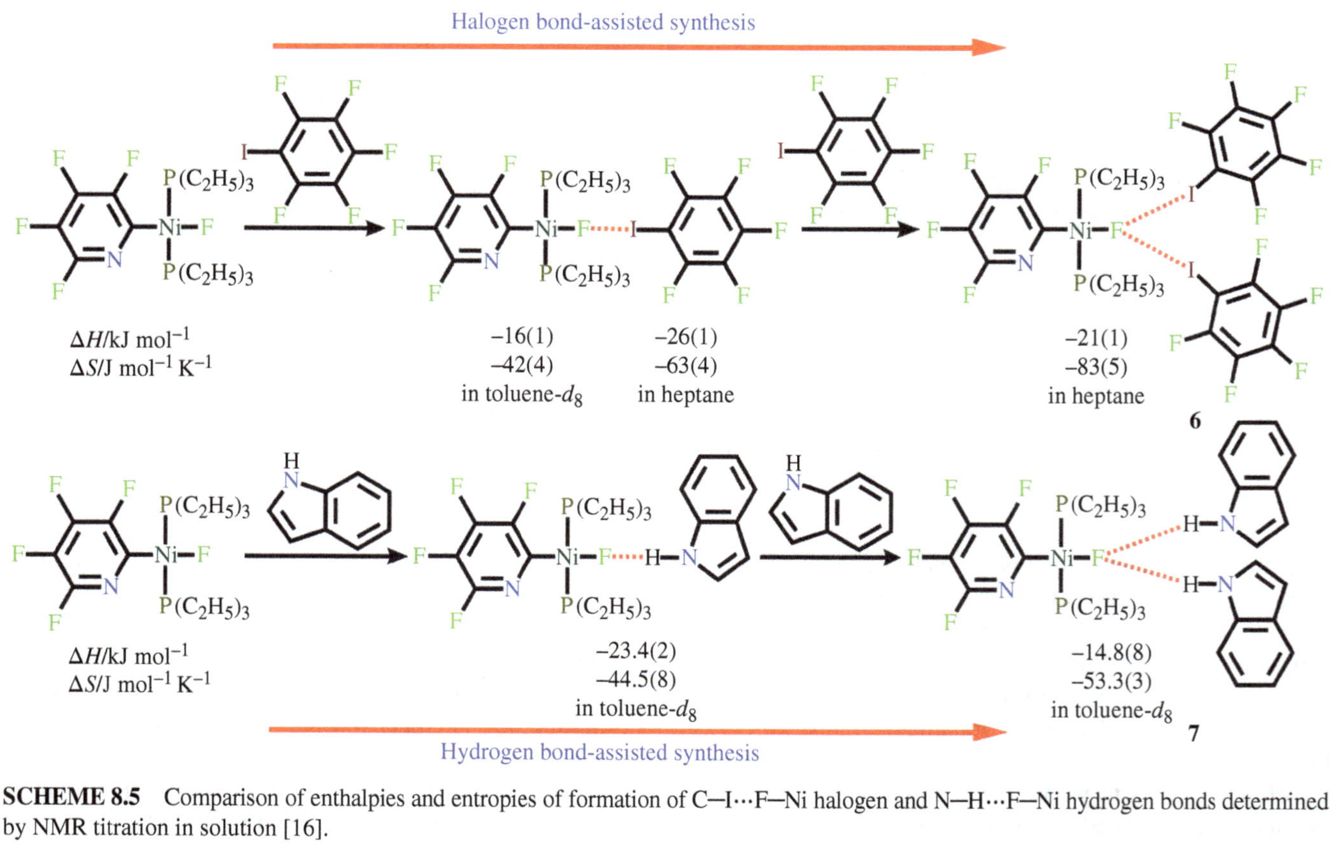

SCHEME 8.5 Comparison of enthalpies and entropies of formation of C—I···F—Ni halogen and N—H···F—Ni hydrogen bonds determined by NMR titration in solution [16].

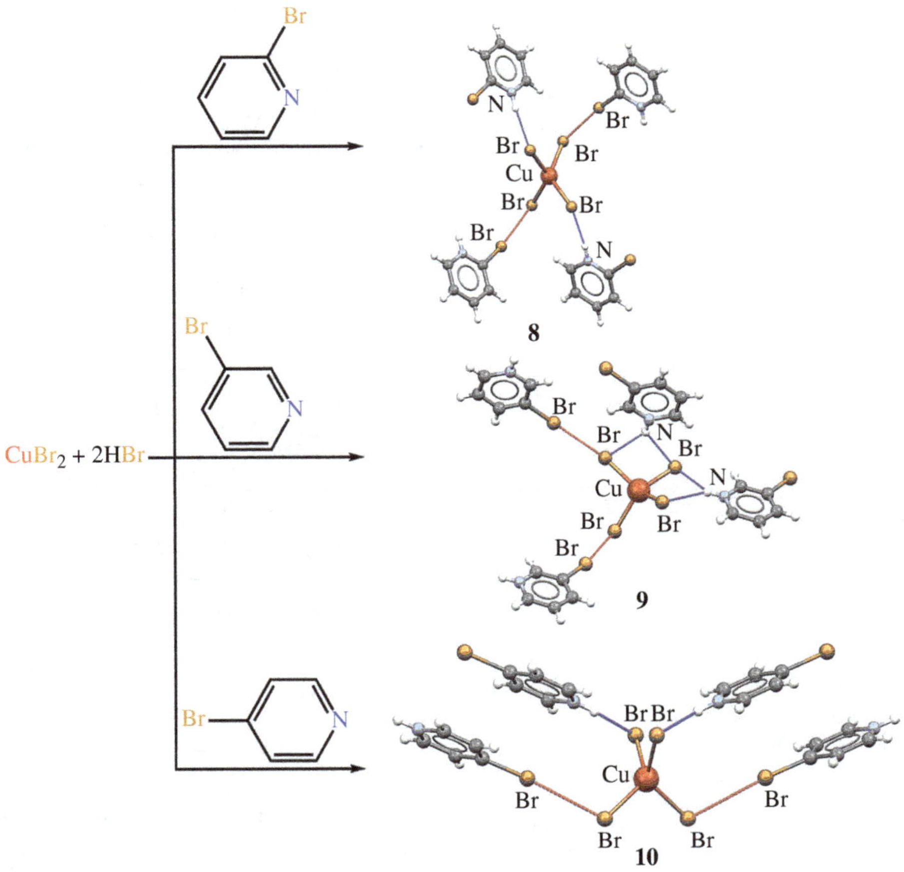

SCHEME 8.6 Synthesis and synthon interactions in **6–10** [18].

observed between the central nitrogen atoms of the six-membered metallacycles and the Cl atoms of $-CCl_3$ groups with distances of 3.221(6)–3.225(6) Å. In overall, the cooperation of six-membered hydrogen bonding systems and Cl···Cl interactions leads to 3D supramolecular architectures.

Similarly to the above examples, the cooperative coordination, hydrogen, and halogen bondings in the one-pot template condensation of $CCl_3C{\equiv}N$ with ammonia on a metal source in DMSO resulted in supramolecularly stabilized bis{2,4-bis(trichloromethyl)-1,3,5-triazapentadienato}-M(II) complexes with a high exothermic effect (Scheme 8.9a) [21]. The formation of Cl···Cl interactions between complex molecules is one of major driving forces of this reaction, which can be called as a *spider effect* (Scheme 8.9b). The (N)H atoms of the ligands interact with oxygen atoms of $(CH_3)_2S{=}O$ with formation of six-membered intermolecular hydrogen bonding rings, while Cl atoms of the CCl_3 groups participate in Cl···Cl interactions leading to an overall supramolecular assembly of a distinctive geometry. The structures of **14–16** are similar and reveal infinite strands of the complexes assembled by the Cl···Cl intermolecular interactions between four chlorine atoms (Scheme 8.9). The Cl···Cl distances within the strands are related to the nature of metal ions and follow the order: 3.342 (Cu) < 3.346 (Ni) < 3.387 Å (Pd).

The series of halogen bond-assembled complexes $[CuLCl_2]$ (**17**), $[Cu_2(\mu_2\text{-}Cl)_2L_2(H_2O)_2]Cl_2{\cdot}2(CH_3)_2CO$ (**18**), and $[Cu_3(\mu_2\text{-}Cl)_6L_2]{\cdot}(CH_3)_2CO$ (**19**) (Scheme 8.9) is obtained from reactions of $CuCl_2{\cdot}2H_2O$ with bis(3,5-dimethyl-4-iodo-pyrazol-1-yl)methane (L) in the presence of HCl, acetone, and ethanol [22]. The supramolecular assemblies are interconvertible, kinetic, and thermodynamic factors taking the lead in sequence: upon slow evaporation of the reaction mixture, firstly the red complex **17** precipitates along with green crystals of **18**, which disappear with time, slowly transforming into the brown crystals of **19** (Scheme 8.10).

It seems that in **17** and **18**, the non-covalent bonds are formed faster than in **19**, at the same time giving lesser thermodynamic stability. In the kinetic products **17** and **18**, the C—I···Cl^- interactions with coordinated and noncoordinated Cl^- play a major role and drive the superstructure stabilization, while the structure of **19** is based on the stable cluster of three copper atoms, combined by three μ-chlorine bridges, with the chelating L ligands coordinated to the terminal copper atoms. Apparently, the formation of these clusters is slower but affords a high stability of the overall system. Concerning non-covalent interactions, two similar symmetrically independent contacts occur in **19** (I1···Cl3 and I2···Cl1, with lengths of 3.299(2) and 3.439(2) Å, respectively), which are substantially shorter than the sum of van

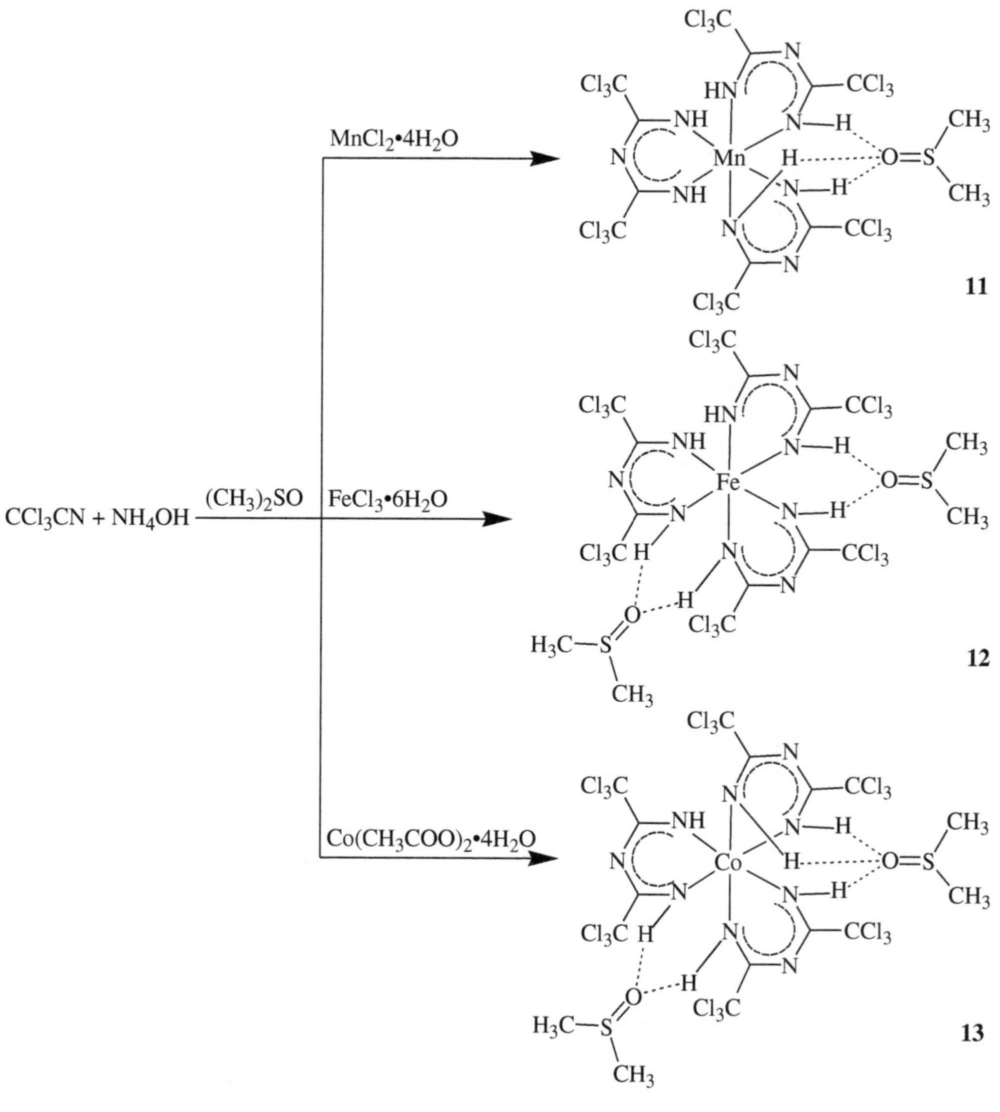

SCHEME 8.7 Cooperation of coordination, hydrogen, and halogen bonding, in the synthesis of **10–12** [12].

der Waals radii of iodine and chlorine atoms. With regard to the shorter I1···Cl3 contact, the structure resembles a polymeric ribbon. The ribbons are further cross-linked through the longer I2···Cl1 contacts into a 2D network. Additionally, the noncoordinated chloride anions are also engaged in hydrogen bonding with coordinated water molecules. The formation of all the mentioned bonds and their space arrangement takes a considerable time but finally provides a multilevel stabilization of the structure of **19**.

8.3 HALOGEN–NITROGEN INTERACTIONS

Several approaches can be used to increase the positive electrostatic potential along the C—X axis in supramolecular associates with the halogen–nitrogen non-covalent bonding, for example, protonation and coordination to metal ions [18, 20]. Another approach involves alkylation of the nitrogen atom in the pyridine moiety [23]. The latter method enables preparation of halogen-bonded supramolecular structures through ion exchange between *N*-alkylhalopyridinium salts and salts with anions of interest, for example, *N*-methylhalopyridinium hexafluorophosphate and $[(Ph)_3P{=}N{=}P(Ph)_3]_2[Ru(bipy)(CN)_4]$ giving compound **20** (Scheme 8.11) [24].

Although the study of the complex formation in solution questions the role of halogen bonding [24], the crystal structures of the solid adducts are clearly strongly influenced by the halogen–nitrogen non-covalent interactions (Scheme 8.11). The individual halogen bonding contacts differ significantly in geometry: there are three different C—I···N—C contacts with I···N distances being within the 2.74–2.97 Å range. Although the angles of the C—I···N bonds lie within a relatively small range (166.4–174.0°), the I···N—C angles differ significantly (from 152.7 and 144.0° to 104.9°). This can be related to the increased contribution from the π(CN) electron donors if compared to the nitrogen lone pair, also giving an elongation of the I···N interaction [24].

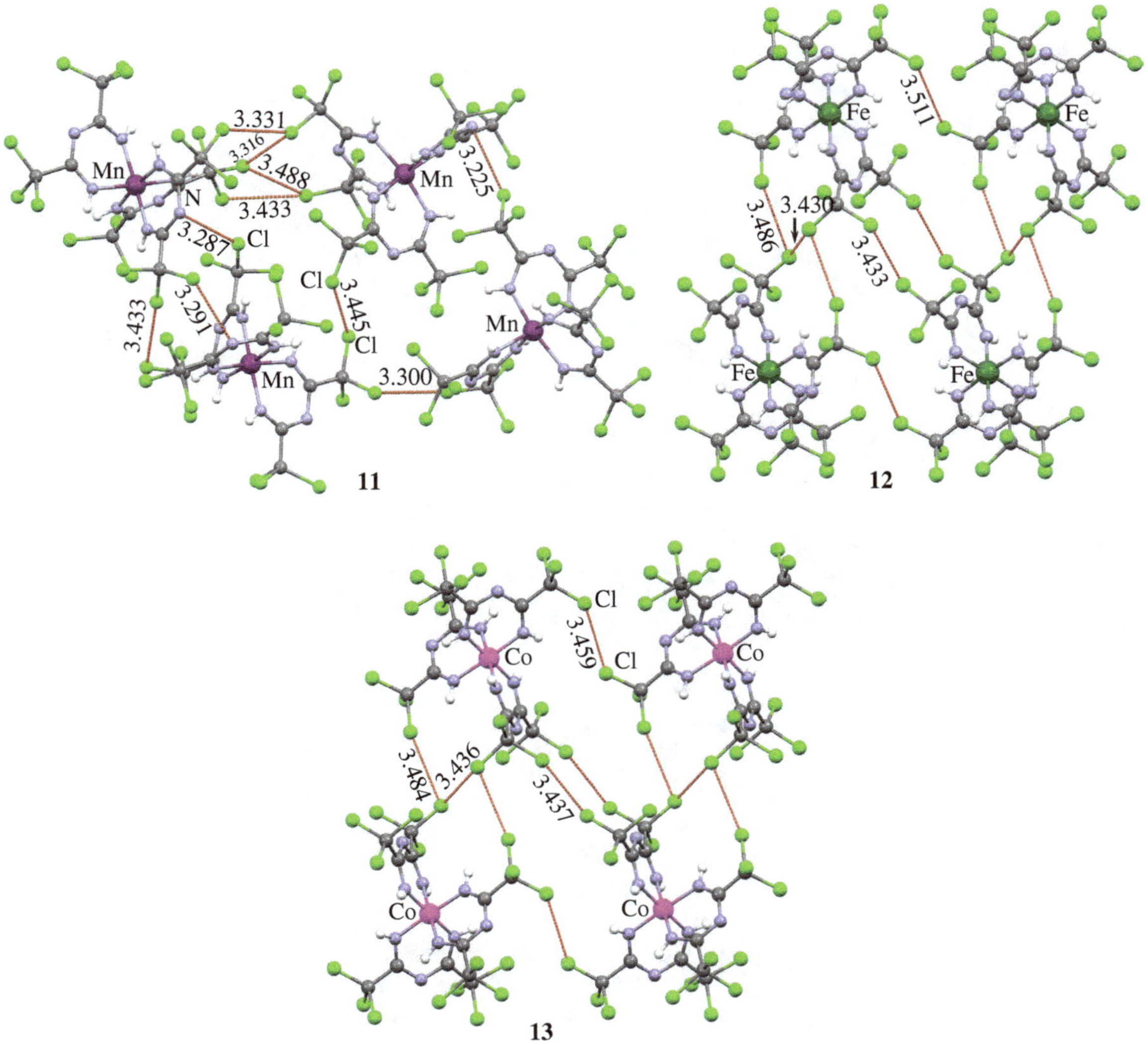

SCHEME 8.8 Cl···Cl interactions in **11–13**; DMSO molecules are omitted for clarity [12].

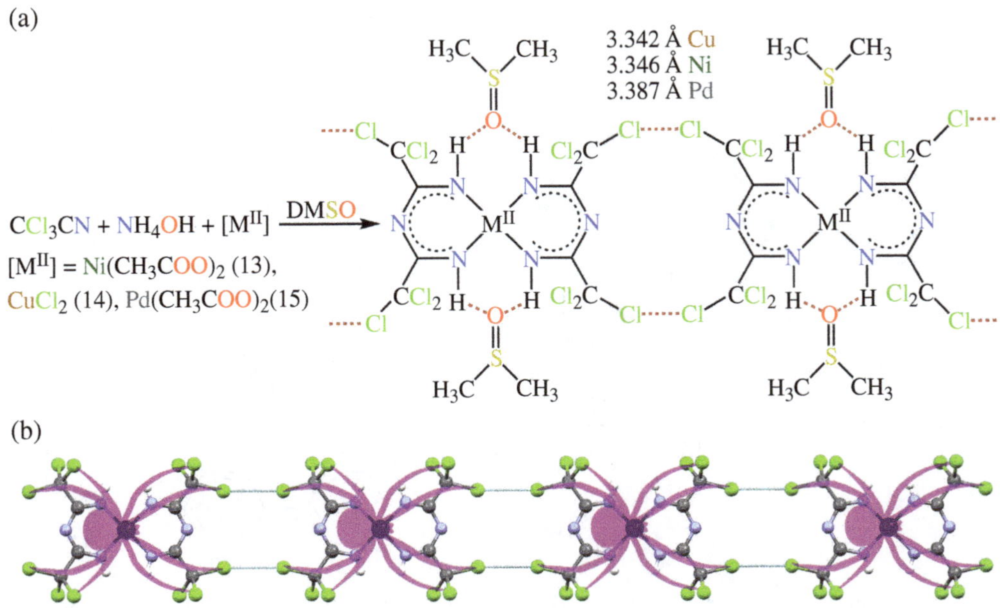

SCHEME 8.9 Coordination, hydrogen, and halogen bond-assisted synthesis of **14–16** [21].

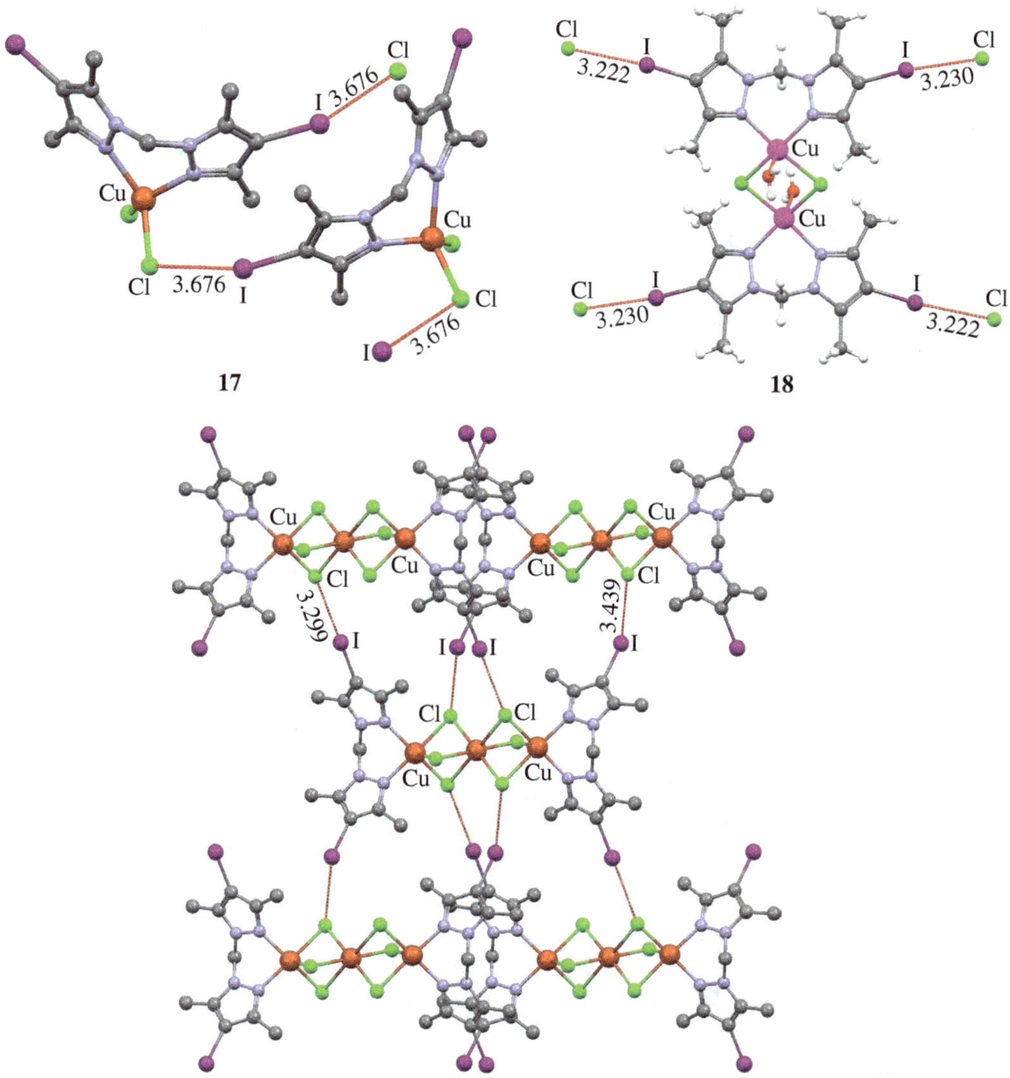

19

SCHEME 8.10 Halogen bonding in **17–19** [22].

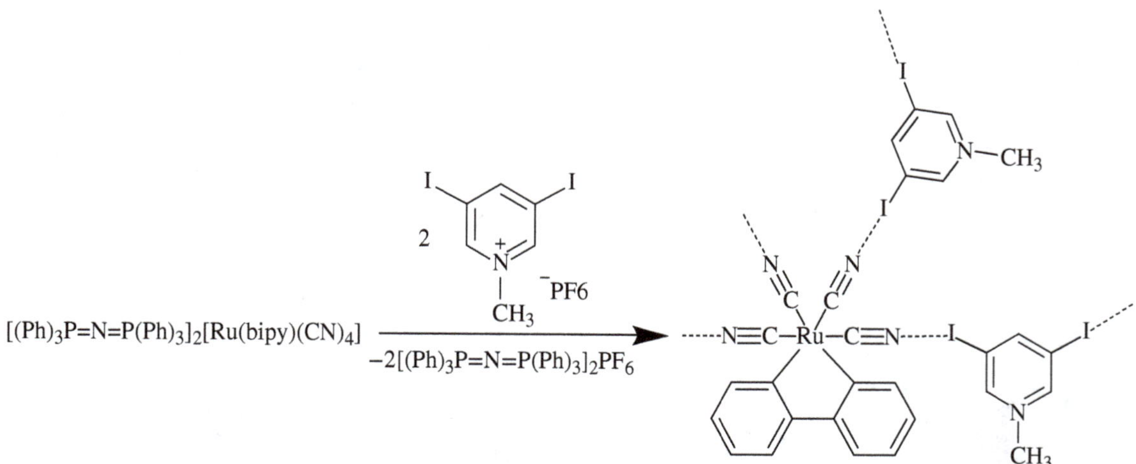

20

SCHEME 8.11 Supramolecular chains in **20** consisting of *N*-methyl-3,5-diiodopyridinium cations and $[Ru(bipy)(CN)_4]^{2-}$ anions connected by C–I···N(cyano) halogen bonds [24].

Two families of halopyridinium hexacyanometallate salts, $(3\text{-XpyMe})_3[M(CN)_6]$ and $(3,5\text{-}X_2\text{pyMe})_3[M(CN)_6]$ (X = I, Br; M = Cr, Fe, Co), are derived from halogen bond-assisted interactions of $[(Ph)_3P{=}N{=}P(Ph)_3]_3[M(CN)_6]$ (M = Fe, Co, Cr) with *N*-methyl-3-halo- and *N*-methyl-3,5-dihalopyridinium ($3\text{-}XC_5H_4NCH_3$ and $3,5\text{-}X_2C_5H_3NCH_3$ with X = Br, I) salts [25]. The close structural relationships between the studied compounds provide an opportunity to examine halogen bonding in interactions involving cyanide ligands as halogen bond acceptors, that is, C—X···NC(M). Halogen bonds become shorter across each isostructural series for Cr > Fe > Co.

All halogens participate in halogen bonding, forming $(3,5\text{-}Br_2C_5H_3NCH_3)_3[Cr(CN)_6]\cdot 4H_2O$ (compound **21**, Scheme 8.12) with C—X···NC(M) and in one case C—Br···OH_2 (2.86 Å) bonds [25]. Longer halogen bonds (Br···N distance of 3.041(6)–3.380(6)) are formed in **21** due to competition with the O—H···N hydrogen bonding. All the halogen bonds have close to linear geometries (most have ∠C—X···N > 165°, with the smallest one being 154.1(3)°). The geometry of interaction of the halogen bond donor (C—X) with the cyanide ligand suggests predominant interaction with the exo lone pair of the nitrogen atom (C≡N···X > 145°) or involvement of the C≡N π-bond (C≡N···X < 105°) as an acceptor.

Pseudohalides can also assist the formation of metal–organic frameworks. Thus, the coordination polymer $[Cu(N_3)_2(4\text{-bromopyridine})]_n$ (**22**) (Scheme 8.13a) was prepared from NaN_3, $Cu(NO_3)_2\cdot 3H_2O$, and 4-bromopyridine [26]. **22** consists of two-dimensional sheets formed by Cu—N(azide) sewed by 4-bromopyridine molecules. These sheets are then assembled through C—Br···NNN halogen

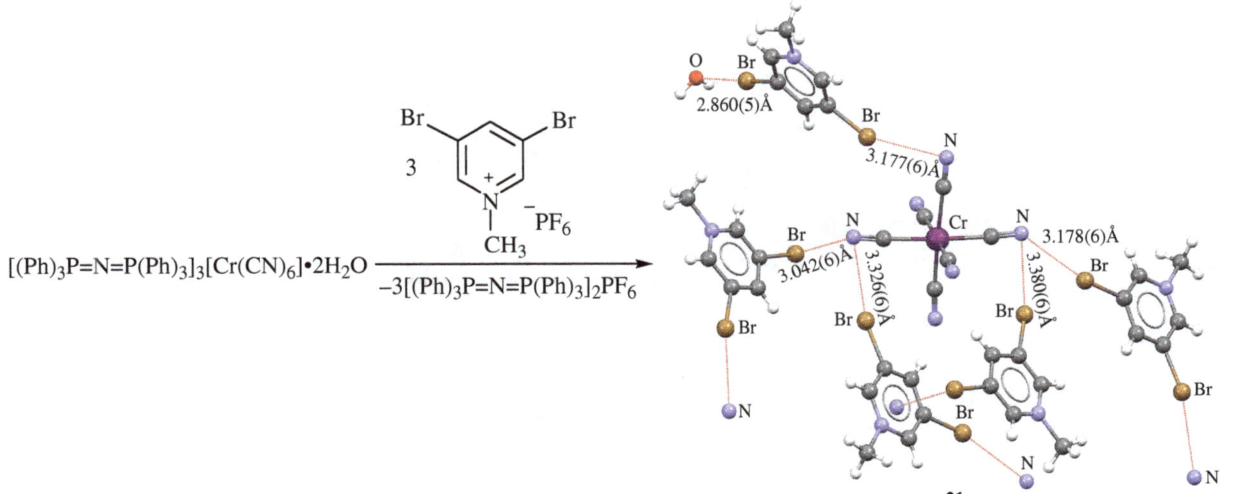

21

SCHEME 8.12 Synthesis and structure of **21** showing C—Br···N and C—Br···O halogen bonds [25].

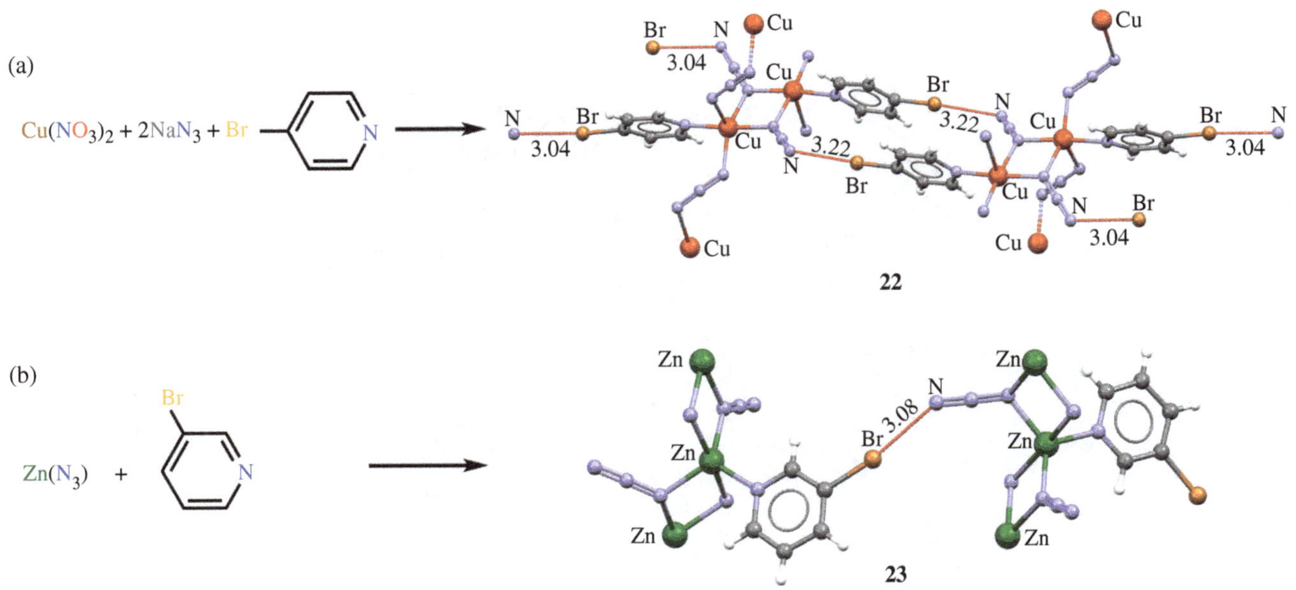

SCHEME 8.13 Synthesis of coordination polymers **22** and **23** [26, 27].

bonds (Br···N distances of 3.04 and 3.22 Å) between the 4-bromopyridine ligands and the terminal nitrogen atoms of the bridging azide ligands. Similarly, supramolecular chains assembled by C—Br···NNN halogen bonds (3.08 Å) are found in $[Zn(N_3)_2(3\text{-bromopyridine})]_n$ **23** (Scheme 8.13b) [27].

Halogen bonding was also explored to prepare noncentrosymmetric solid phases based on porphyrins functionalized with iodophenyl and pyridyl groups [28]. For that, $M(OAc)_2$ hydrate (where M = Cu(II) and Zn(II)) was reacted with porphyrins, giving the respective metal complexes, which then were assembled with pyridine in DMF to form [Zn(py)-PyTIPP] (**24**) and a pyridine solvate of [Cu-PyTIPP] (**25**), where PyTIPP = 5-(4′-pyridyl)-10,15,20-tris(4′-iodophenyl) porphyrin (Scheme 8.14). Both compounds are composed of flat supramolecular layers of porphyrin units (irrespective of their metallation state, molecular conformation, and unit cell content) interconnected to one another by head-to-tail N···I halogen bonds with 2.87–3.06 Å distances between adjacent molecules and aligned along straight lines [28]. The *trans* pyridyl and iodophenyl groups of the porphyrin moiety are involved in these interactions at distances that are considerably shorter than the sum of the corresponding van der Waals radii of N and I atoms (*ca.* 3.5–3.8 Å). The neighboring chains extend further via secondary I···π interactions (with distances within 3.4–3.6 Å and with the sideways-directed C—I bonds of one chain being oriented nearly perpendicular to the iodophenyl and pyridyl aryl groups of adjacent N···I halogen-bonded chains). In **24**, attractive interchain I···I contacts (3.68–3.73 Å) are also apparent.

The ditopic character of 3-(4-pyridyl)-2,4-pentanedionate (acacPy) together with its nucleophilic and basic sites provides a high coordination ability and drives the halogen bond-assisted links between complex molecules upon interaction of $[Fe(acacPy)_3]$ or $[Al(acacPy)_3]\cdot 3H_2O$ with

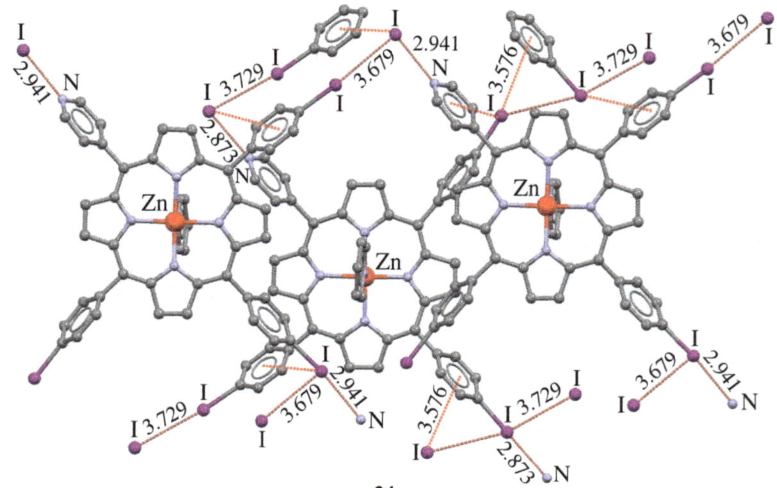

24

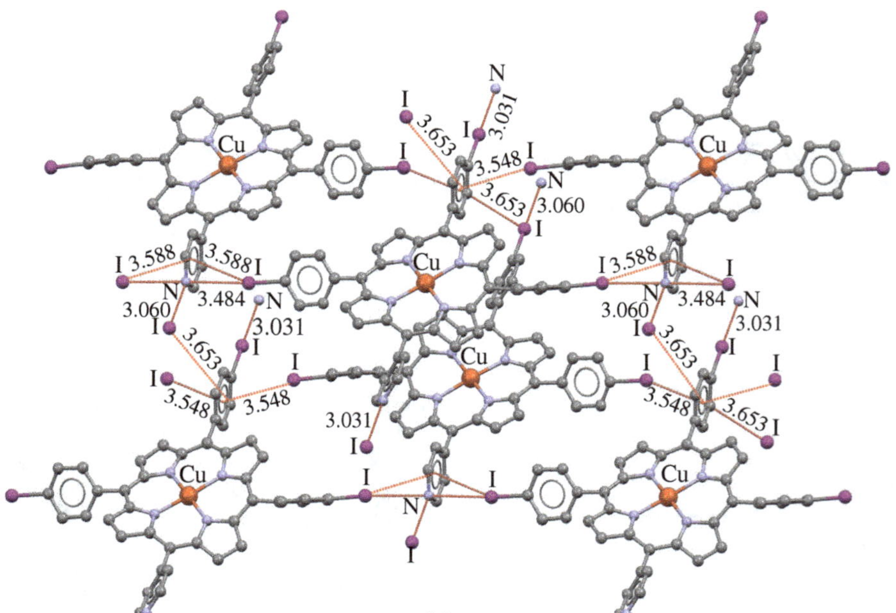

25

SCHEME 8.14 Halogen bonding in **24** and **25** [28].

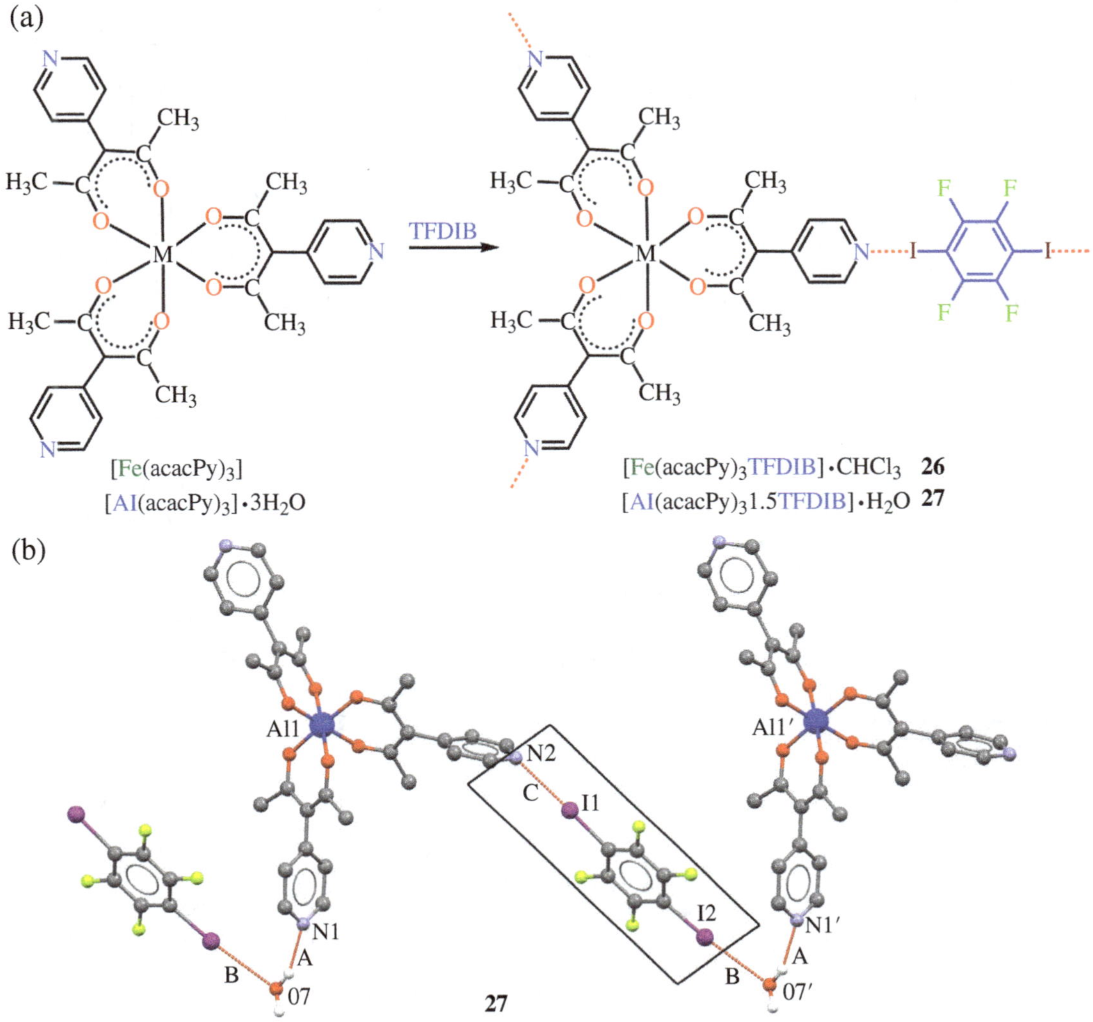

SCHEME 8.15 (a) Schematic representation of the self-assembly of TFDIB with [Fe(acacPy)$_3$] and [Al(acacPy)$_3$]·3H$_2$O; (b) Shortest intermolecular contacts in **27** are due to hydrogen bonds (**A**) and O···I (**B**) and N···I (**C**) interactions [29].

tetrafluorodiiodobenzene (TFDIB) (Scheme 8.15a), thus leading to [Fe(acacPy)$_3$TFDIB]·CHCl$_3$ (**26**) and [Al(acacPy)$_3$·1.5TFDIB]·H$_2$O (**26**) complexes, respectively [29]. In **26**, a deprotonated acacPy is involved in metal coordination via its chelating acetylacetonate function and in halogen bonding. Two of the three pyridine N atoms of the neutral [Fe(acacPy)$_3$] complex act as nucleophiles and approach iodine substituents of TFDIB, with N···I distances of slightly less than 2.8 Å. In **26**, the extended network is based on coordination and halogen bond-assisted building blocks, while an additional interaction becomes relevant in **27**: an acacPy nitrogen atom in the [Al(acacPy)$_3$] unit is engaged in a conventional hydrogen bond to a water molecule (interaction **A** in Scheme 8.15b). The O atom of this water molecule is the nucleophilic partner in a halogen bond to TFDIB (interaction **B**), which in turn acts as the electrophile toward N2 of a translationally equivalent [Al(acacPy)$_3$] complex (**C** in Scheme 8.15b). In the latter case, the coexisting covalent, coordinative, hydrogen, and halogen bonds contribute to the formation of the extended crystal structure. Thus, a combination of halogen bonds and hydrogen bonds contribute to the heteromeric chains assembly.

8.4 HALOGEN–CHALCOGEN BONDING

Halogen bonding and coordination-driven reaction of FeCl$_2$·4H$_2$O with 3-bromo-1,10-phenanthroline (3-Br-phen) in the presence of NH$_4$SCN leads to [Fe(3-Br-phen)$_2$(NCS)$_2$]·2CH$_2$Cl$_2$, **28** (Scheme 8.16) [30]. Auxiliary thiocyanate ligands also take part in halogen bonding; neutral complex moieties form chains through pairs of C—Br···SCN halogen bonds (Br···S distance of 3.32 Å and Br···S═C angle of 86°, Scheme 8.16), which then assemble through C—H···π and π···π interactions. The geometrical parameters of the C—Br···SCN halogen bonds indicate a contribution of π(S═C) electrons to the halogen bonding and reflect difference in the electron density distribution around the sulfur atom of the SCN ligand compared with the nitrogen of the CN moiety.

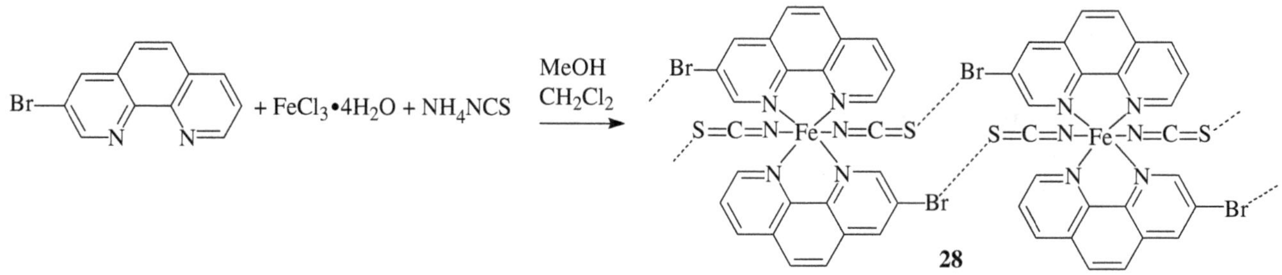

SCHEME 8.16 Synthesis and supramolecular chains in the structure of **28** formed through C—Br···SCN halogen bonds [30].

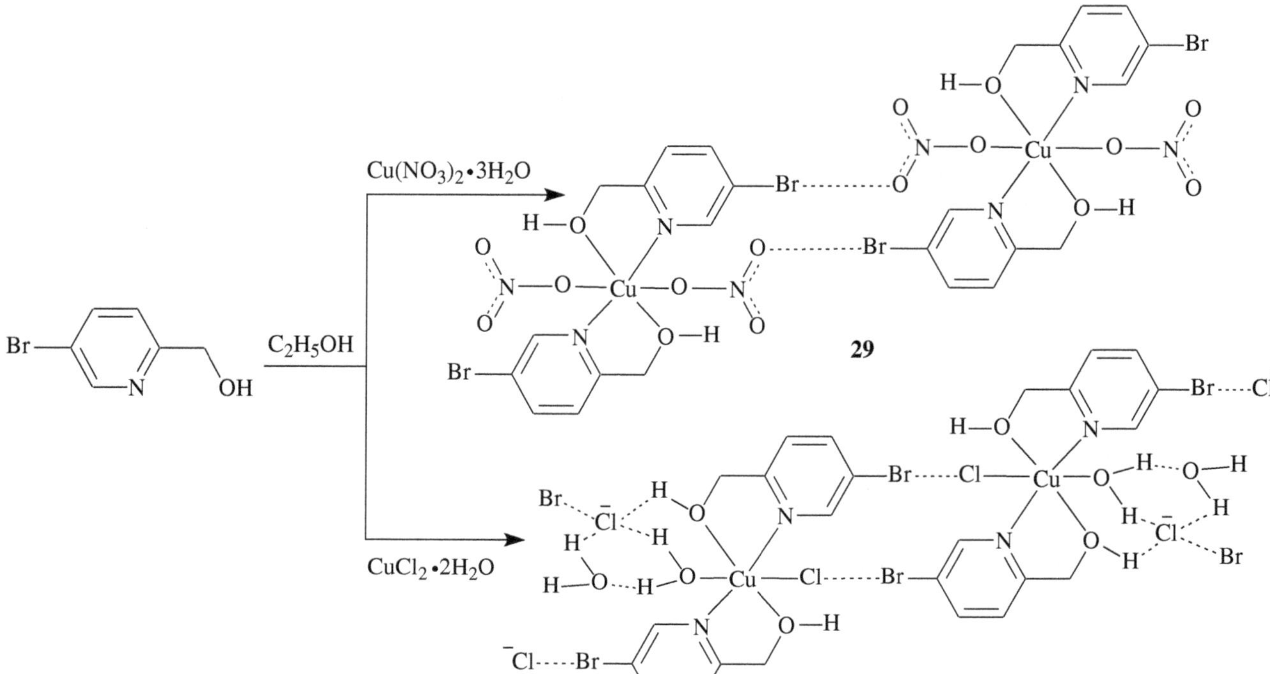

SCHEME 8.17 Synthesis of hydrogen and halogen bond-assisted supramolecular networks in **29** and **30** [31].

Cooperative coordination, hydrogen, and halogen bond-assisted reaction of 5-bromo-2-hydroxymethylpyridine with $Cu(NO_3)_2{\cdot}3H_2O$ and $CuCl_2{\cdot}2H_2O$ in ethanol gives [Cu(5-bromo-2-hydroxymethylpyridine)$_2$(NO_3)$_2$] (**29**) and [CuCl(H_2O)(5-bromo-2-hydroxymethylpyridine)$_2$]Cl·H_2O (**30**), respectively (Scheme 8.17) [31]. In the structure of **29**, the complex moieties aggregate through pairs of C—Br···ONO_2 halogen bonds (Br···O distance of 3.00Å) formed between the 5-bromo-2-hydroxymethylpyridine ligands and the coordinated nitrate anions. The thus formed parallel chains are then further assembled through O—H···ONO_2 hydrogen bonds. The structure of **30** contains coordinating and noncoordinating chloride atoms, both of them being engaged in C—Br···Cl—Cu (3.26Å) and C—Br···Cl (3.30Å) halogen bonding (Scheme 8.17).

Aquabis(dimethylglyoximato)nitrocobalt(III) reacts with 3-chloropyridine, 3,4,5-trichloro-pyridine, 3-bromopyridine, and 3,5-dibromopyridine to give halogen bond-assisted supramolecular architectures **31–34**, respectively, in which the pyridine donor replaces the aqua ligand (Scheme 8.18) [32]. The derivatives possess a cobalt oxime core with one apical position occupied by a nitro ligand, whereas a halopyridine is attached at the other apical position. Strong halogen bonds of the C—Cl/Br···O(oxime) type are observed; the most relevant intermolecular interactions in these compounds are due to very short oxygen···halogen contacts of *ca.* 2.9Å. Halogen bond distances involving bromine are shorter than those associated with chlorine.

Reaction of copper(II) acetate monohydrate with 3-iodobenzoic acid in dioxane gives the dioxane adduct/solvate [Cu_2(3-iodobenzoate)$_4$(1,4-dioxane)$_2$]·2(1,4-dioxane) **35** (Scheme 8.19) [33]. In the structure of **35**, an oxygen atom of the coordinated dioxane behaves as an acceptor (I···O 3.138Å, C—I···O 170.90°) in one direction, while in the second direction it is a carboxylate oxygen atom, which accepts the halogen (I···O 3.376Å, C—I···O 159.80°). The halogen bond to the dioxane oxygen is stronger; two additional dioxane molecules per paddlewheel unit are included in the crystal structure and located between the halogen-bonded sheets. The main driving force of the self-assembly is

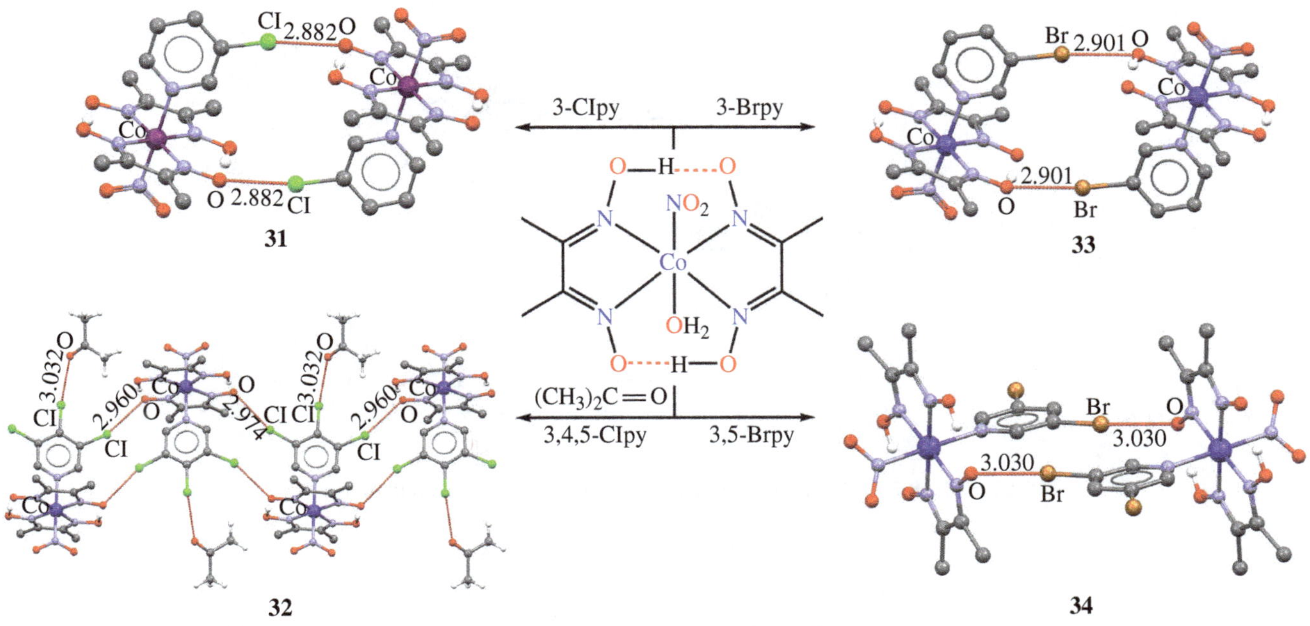

SCHEME 8.18 Synthesis of **31–34**, and halogen bonding pairs in their supramolecular structures. A Cl···O═C halogen bond with a molecule of acetone is also shown in **32** [32].

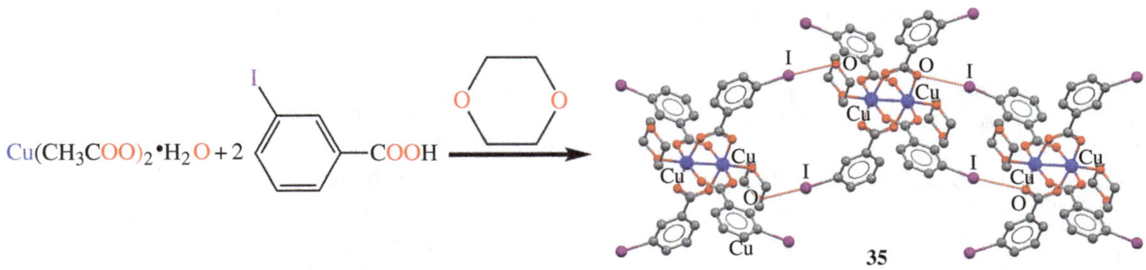

SCHEME 8.19 Halogen bonding in the structure of **35**. Both C—I···O (carboxylato) and C—I···O (ether) act in pairs. The noncoordinated molecules of 1,4-dioxane as well as hydrogen atoms have been omitted for clarity [33].

believed to be C—I···O(ether) and C—I···O(carboxylato) halogen bond interactions that act in centrosymmetric pairs, each of which linking the complex moieties together in different directions and thus forming two-dimensional sheets, which then assemble through C—H···O(ether) and C—H···π interactions.

$CuCl_2{\cdot}2H_2O$-catalyzed C—H oxidative substitution in pyrazole-3,5-dicarboxylic acid (H_3pdc) by bromo/iodo nucleophiles under hydrothermal conditions leads, by a self-assembly process, to the halogenated pyrazole carboxylato-based linear trinuclear copper(II) complexes $[Cu_3(X\text{-}pdc)_2{\cdot}(H_2O)_6]$ (X = Br (**36**), I (**37**), Scheme 8.20) [34]. Both complexes are engaged in extensive hydrogen bonding and centrosymmetric C—Br/I···O pairing (Br···O 2.97 Å and I···O 3.00 Å). A small difference in the length and directionality of the equivalent halogen bonds in these compounds suggests that the short Br/I···O distances are caused not by strong interactions but by the extensive hydrogen bonding network, which brings the moieties into close contact.

Various polyoxometallate anions (POMs) have also been utilized as halogen bond acceptors used in combination with chloro- and bromo-substituted bipyridyl cations to obtain systems in which the crystal structures were governed by hydrogen bonding and, to a lesser extent, halogen bonding of the type C—Cl/Br···O(POM) [35, 36]. The structure of thus formed [5-(4-bromophenyl)-2,4′-bipyridinium]$_4$[Mo_8O_{26}] (**38**, Scheme 8.21) is driven by the C—Br···O(POM)-type halogen bonding (Br···O 3.13 Å) as well as hydrogen bonding between the POM anion and the cations [36].

Another example of a structure governed by Cl···N halogen bonds with Cl as a donor is the 1 : 1 adduct **39** of $AlCl_3$ and cyanuric chloride (Scheme 8.22), in which each of the two crystallographically independent molecules forms chains through Cl···N bonds (Cl···N 3.05 Å) [37]. The chains are then joined together by Cl···N halogen bonding (Cl···N 3.12 and 3.13 Å) to form supramolecular tapes.

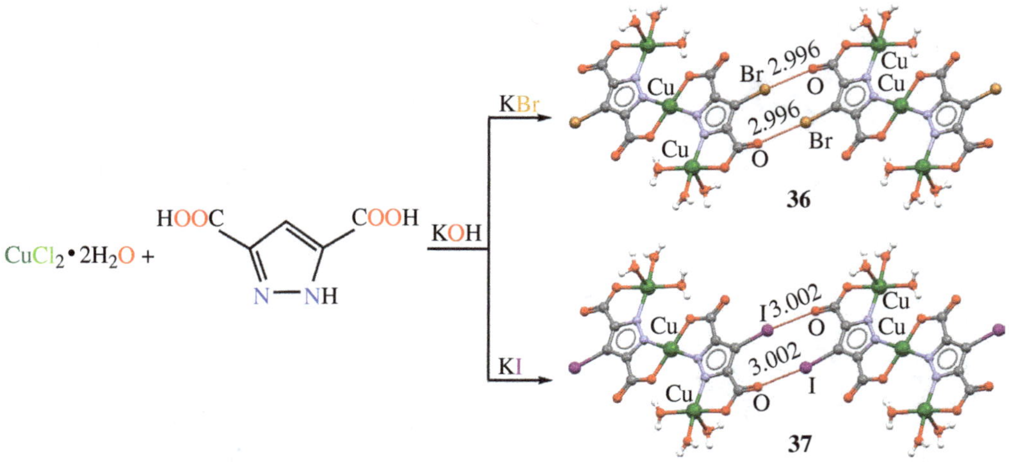

SCHEME 8.20 Synthesis of **36** and **37** under hydrothermal conditions [34].

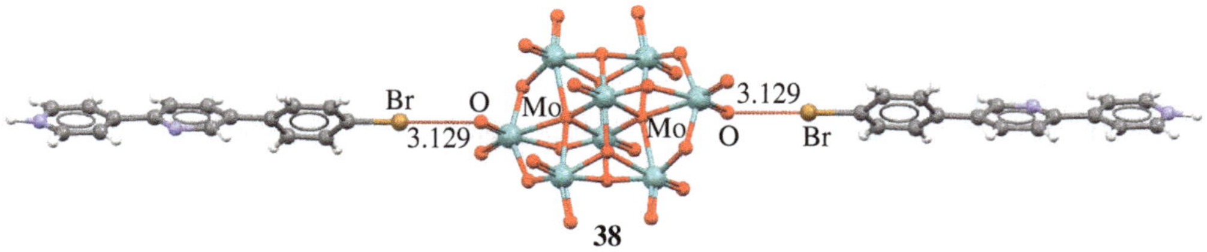

38

SCHEME 8.21 Halogen and hydrogen bonding in the structure of **38** [36].

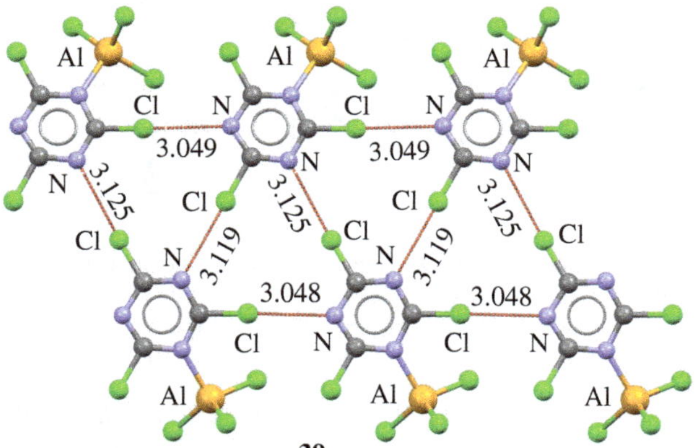

39

SCHEME 8.22 Cl···N halogen bonding in the crystal structure of $[AlCl_3(C_3N_3Cl_3)]$, **39** [37].

8.5 HALOGEN BONDING IN ORGANOMETALLIC COMPOUNDS

Different organometallic derivatives with attached halogen bond donor or acceptor sites can give rise to networks with quite different topologies related to the geometries around metal centers [15, 38–41]. In some cases the metal ion sits at the nodes of the network while its halide or cyanide ligands are involved in transmitting the linear, square planar, or tetrahedral coordination geometry of the metal to the network through halogen bonds [15, 38–41]. If a metal bears ligands where a spacer separates the metal center and the halogen bond acceptors or donors, extra flexibility can be added to the geometry of a supramolecular network. Thus, treatment of Ag_2O with 4,5-dichloro-1,3-dimethylimidazolium nitrate gives the *N*-heterocyclic carbene complex $[Ag(C_5H_6N_2Cl_2)_2]NO_3$ (**40**), which, upon reaction with $AuCl(SMe_2)$, leads to $[Au(C_5H_6N_2Cl_2)_2]NO_3$ (**41**, Scheme 8.23) [38]. Both complexes demonstrate the role of Cl as a donor to construct systems with strong halogen bonds: in both the crystal structures, extensive halogen bonding is observed between the chlorinated organometallic cation and the nitrate anion with a number of Cl···ONO_2 halogen bonds (Cl···O 2.79–3.23 Å for **40**), including an asymmetrically bifurcated Cl···O_2NO halogen bond (Cl···O 2.98 and 3.24 Å for **40**). Similar values were also observed in the case of **41**.

In the crystal structures of the iodocyclopentadienyl complexes $[M(C_5H_4I)(CO)_3]$, where M = Mn (**42**) and Re (**43**), moderately strong C—I···O≡C halogen bonds (I···O 3.23 Å in **42** and I···O 3.23 Å in **43**) are observed (Scheme 8.24) [39]. The molecules in both structures form zigzag chains due to the formation of strong attractive interactions. The attractive nature of halogen–oxygen interactions is caused by electrostatic effects, polarization, charge transfer, and dispersion contributions. The directionality of that type of interaction has been interpreted in terms of charge transfer between the highest occupied molecular orbital of CO and the lowest unoccupied orbital of iodine. The strength of halogen–carbonyl interactions has been characterized as a function of two geometric parameters, the halogen–oxygen distance (I···O) and the halogen–oxygen–carbon angle (I···O—C). A carbonyl group in **42** and **43** is in an eclipsed position relative to the substituted C atom of the η^5-C_5H_4I ring.

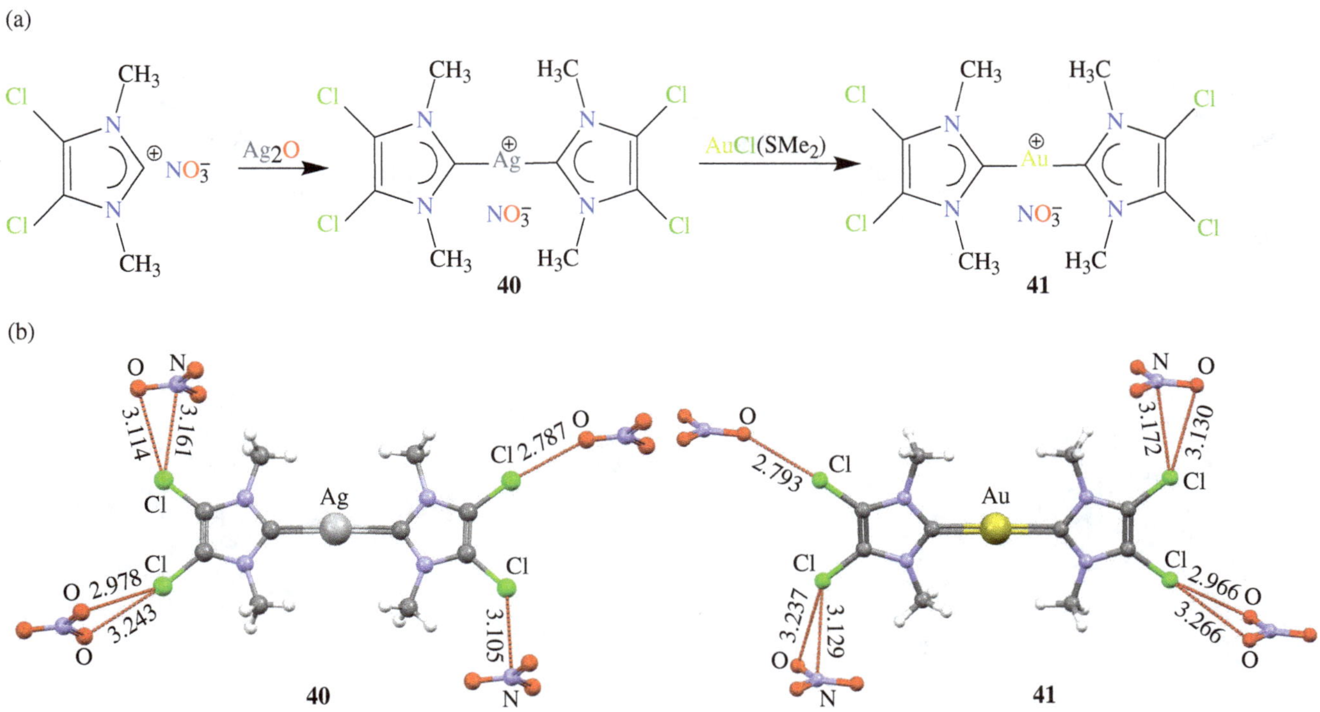

SCHEME 8.23 Synthesis (a) and halogen bonding in the structures of **40** and **41** (b) [38].

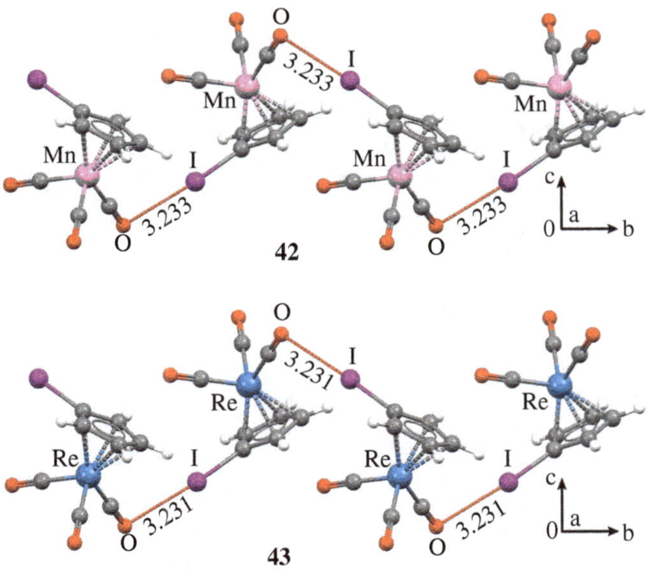

SCHEME 8.24 Halogen-bonded zigzag chains in the structures of **42** and **43** [39].

Two halogen bond-assembled organoruthenium compounds were prepared by addition of $[I(py)_2]BF_4$ to $[Ru(C{\equiv}CH)(dppe)(Cp^*)]$ (dppe = 1,2-bis(diphenylphosphino)ethane; Cp* = permethylcyclopentadienyl) to give the iodovinylidene complex $[Ru({=}C{=}CHI)(dppe)(Cp^*)]BF_4$ (**44**), which can be deprotonated affording compound $[Ru(C{\equiv}CI)(dppe)(Cp^*)]$ (**45**) (Scheme 8.25a) [40]. Remarkably short halogen bonds C—I···FBF_3 (I···F 3.03 Å) are observed between a fluorine atom of the tetrafluoroborate anion and the iodine atom of the iodovinylidene ligand in **44** (Scheme 8.25b) [40]. Complex **45** is the first structurally characterized complex containing an iodoethynyl group.

A new functionalized aryl-imido-bridged cyclopentadienyl titanium compound $[TiCl(Cp^*)(\mu\text{-}4\text{-iodoimidobenzene})]_2{\cdot}[MgCl_2{\cdot}(THF)_4]$ (**46**) was prepared by reaction of $[Ti(Cp^*)Cl_3]$ with 4-iodoaniline and Mg^nBu_2 in THF (Scheme 8.26a) [41]. This imido titanium complex forms a supramolecular chain with a $[MgCl_2(THF)_4]$ unit generated in the synthesis where the metallic fragments are connected by C—I···Cl—Mg halogen bonding interactions (Scheme 8.26b). The compound **46** with this Mg^{2+} unit represents the first example within group 2 metals where the halogen bond motif M—Cl···I—C appears. In the structure of **46**, the I···Cl distance of 3.501 Å is shorter than the sum of the van der Waals radii (3.73 Å), with the C—I···Cl angle of 167.13°. The interaction at the inorganic halogen (Mg—Cl) is markedly bent (I···Cl—Mg angle is 126.95°) and is closer to linear at the organic halogen (C—I). The orientation/interaction is established between the negative charge on the inorganic chlorine, which is accumulated with a toric geometry perpendicular to the axis of the Mg—Cl bond and the area of positive electrostatic potential, or "σ-hole," in the iodine atom that is located on the axis of the I···C bond. However, no halogen bonding is observed in the compound that corresponds to **46** when iodine is replaced by chlorine in the aniline aromatic ring, since the later halogen does not show enough positive potential to be able to form this interaction with $MgCl_2$.

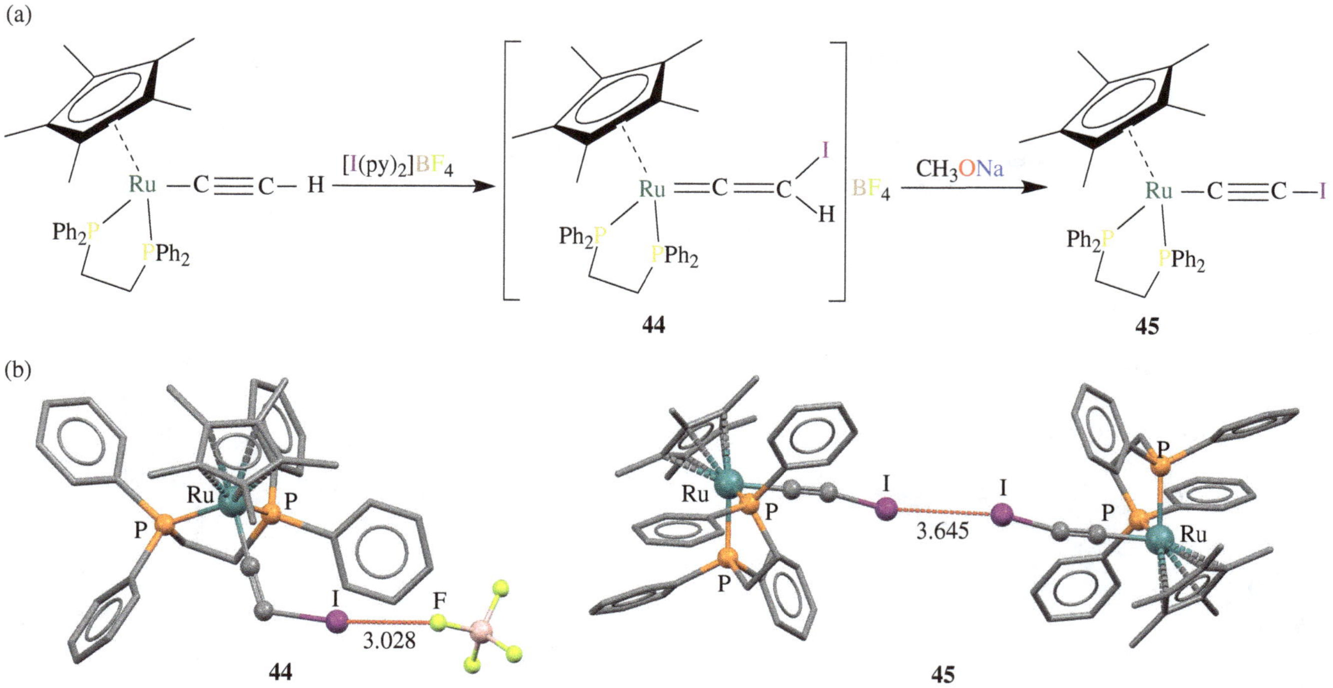

SCHEME 8.25 Synthesis of **44** and **45** (a); C—I···F halogen bonding between the BF_4^- anion and $[Ru(=C=CHI)(Ph_2PCH_2CH_2PPh_2)(Cp^*)]^+$ cation (b) [40].

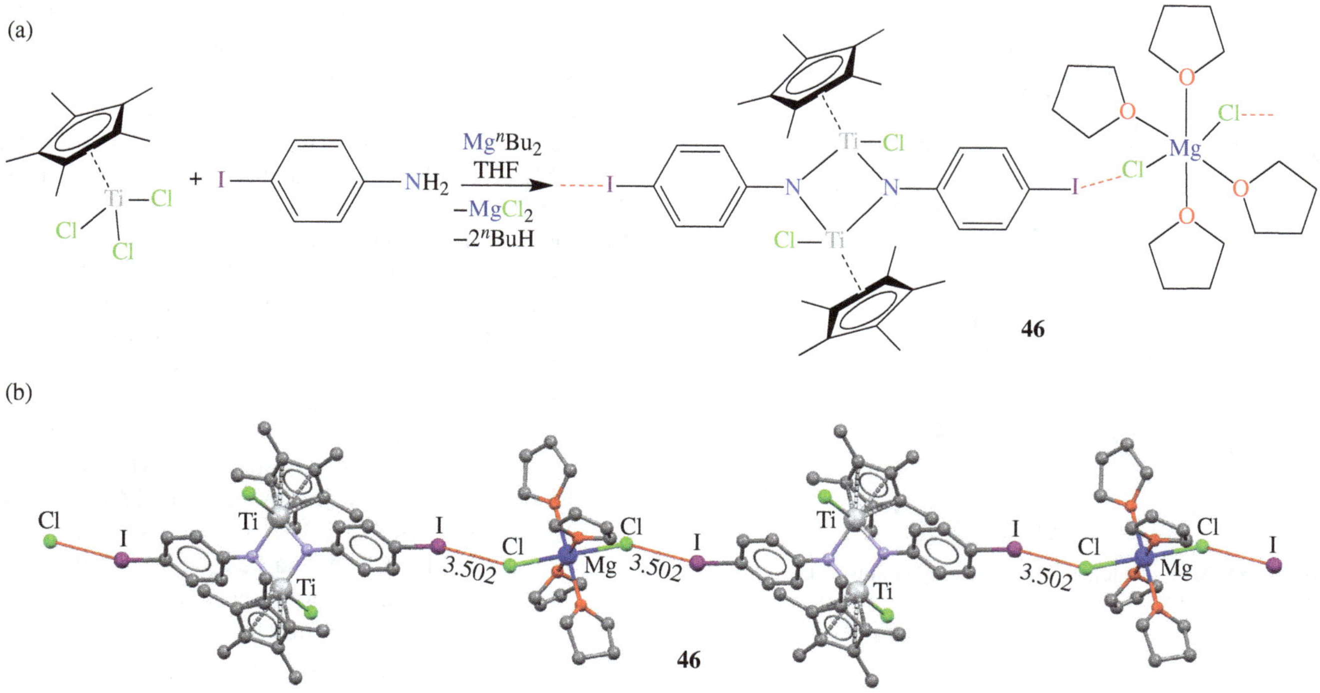

SCHEME 8.26 Synthesis of **46** (a); 1D arrangement in **46** through halogen bonding Mg—Cl···I—C interactions (b) [41].

8.6 FINAL REMARKS

In this chapter we have illustrated the significance of the halogen bonds as a driven force of reaction in the synthesis and design of coordination compounds. Additionally, cooperation of halogen, hydrogen, and coordination bonds in the halogen bond-assisted syntheses, some parameters of halogen bonds and their relations to properties of the metal ions were discussed.

As it can be seen from these examples, the application of halogen bonding in the synthesis of coordination and organometallic compounds is an intensively growing area of research covering also the fields of crystal engineering, supramolecular, materials, and bioinorganic chemistries. It was demonstrated that the halogen atoms bound to metal ions serve good halogen bond acceptors, while the metal centers can be conveniently used as platforms to assemble ligands of various geometries and bearing suitable electron donor groups that can be involved in the formation of halogen bonds at the outer sphere of metal complexes.

The cooperation of coordination, hydrogen, and halogen bonding can be widely used for crystal engineering and material design. The structure of molecular crystals can be controlled by directional bonding, symmetry interactions, and weak-link approaches. Thus, halogen bond-assisted supramolecular metal–organic networks will be increasingly employed by synthetic chemists to engender novel properties and behavior in coordination compounds and extend the fields of applications in biomedicine [4], molecular recognition [6], catalysis [12, 42], etc.

Generally, the halogen bonds with organic donors tend to be stronger than those with donors based on metal complexes. However, in coordination and organometallic compounds, more factors can be applied to regulate the strength and directionality of the halogen bonding and hence the overall supramolecular architecture can be more precisely and subtly tuned. Additionally, coordination can profoundly influence both the halogen bond donor and acceptor properties of the ligands, thus offering an attractive route to self-assembly and halogen bond-driven synthesis in metal–organic systems.

ACKNOWLEDGMENTS

This work has been partially supported by the Baku State University, the Russian Science Foundation (grant no. 14-13-00083), and Foundation for Science and Technology (FCT), Portugal (UID/QUI/00100/2013 and "Investigador 2013" (IF/01270/2013/CP1163/CT0007) programs). M.N.K. thanks FCT for the working contract.

REFERENCES

1. P. Metrangolo, G. Resnati, Halogen Bonding: Fundamentals and Applications (Structure and Bonding); Springer: Heidelberg, 2010.
2. G. Cavallo, P. Metrangolo, T. Pilati, G. Resnati, M. Sansotera, G. Terraneo, *Chem. Soc. Rev.* 2010, *39*, 3772.
3. E. Parisini, P. Metrangolo, T. Pilati, G. Resnati, G. Terraneo, *Chem. Soc. Rev.* 2011, *40*, 2267.
4. R. Berger, G. Resnati, P. Metrangolo, E. Weber, J. Hulliger, *Chem. Soc. Rev.* 2011, *40*, 3496.
5. M. Cametti, B. Crousse, P. Metrangolo, R. Milanic, G. Resnati, *Chem. Soc. Rev.* 2012, *41*, 31.
6. M. Erdélyi, *Chem. Soc. Rev.* 2012, *41*, 3547.
7. A. Priimagi, G. Cavallo, P. Metrangolo, G. Resnati, *Acc. Chem. Res.* 2013, *46*, 2686.
8. P. Politzer, J. S. Murray, M. C. Concha, *J. Mol. Model.* 2007, *13*, 643.
9. L. G. J. Wade, Organic Chemistry; Pearson Education Inc.: Upper Saddle River, 7th edn, 2010.
10. S. Zhu, C. Xing, W. Xu, Z. Li, *Tetrahedron Lett.* 2004, *45*, 777.
11. D. D. Burton, F. Fontana, P. Metrangolo, T. Pilati, G. Resnati, *Tetrahedron Lett.* 2003, *44*, 645.
12. N. Q. Shixaliyev, A. V. Gurbanov, A. M. Maharramov, K. T. Mahmudov, M. N. Kopylovich, L. M. D. R. S. Martins, V. G. Nenajdenko, A. J. L. Pombeiro, *New J. Chem.* 2014, *38*, 4807.
13. R. W. Troff, T. Mäkelä, F. Topić, A. Valkonen, K. Raatikainen, K. Rissanen, *Eur. J. Org. Chem. 2013*, 1617.
14. L. Brammer, G. Mínguez Espallargas, S. Libri, *CrystEngComm.* 2008, *10*, 1712.
15. R. Bertani, P. Sgarbossa, A. Venzo, F. Lelj, M. Amati, G. Resnati, T. Pilati, P. Metrangolo, G. Terraneo, *Coord. Chem. Rev.* 2010, *254*, 677.
16. S. Libri, N. A. Jasim, R. N. Perutz, L. Brammer, *J. Am. Chem. Soc.* 2008, *130*, 7842.
17. S. Das, G. W. Brudvig, R. H. Crabtree, *Chem. Commun.* 2008, 413.
18. R. D. Willett, F. Awwadi, R. Butcher, S. Haddad, B. Twamley, *Cryst. Growth Des.* 2003, *3*, 301.
19. J. P. M. Lommerse, A. J. Stone, R. Taylor, F. H. Allen, *J. Am. Chem. Soc.* 1996, *118*, 3108.
20. L. Brammer, G. Mínguez Espallargas, H. Adams, *CrystEngComm* 2003, *5*, 343.
21. N. Q. Shixaliyev, A. M. Maharramov, A. V. Gurbanov, N. V. Gurbanova, V. G. Nenajdenko, V. M. Muzalevskiy, K. T. Mahmudov, M. N. Kopylovich, *J. Mol. Struc.* 2013, *1041*, 213.
22. A. V. Virovets, D. A. Piryazev, E. V. Lider, A. I. Smolentsev, S. F. Vasilevskii, L. G. Lavrenova, *J. Struct. Chem.* 2010, *51*, 92–98.
23. T. A. Logothetis, F. Meyer, P. Metrangolo, T. Pilati, G. Resnati, *New J. Chem.* 2004, *28*, 760.
24. S. Derossi, L. Brammer, C. A. Hunter, M. D. Ward, *Inorg. Chem.* 2009, *48*, 1666.
25. J. Ormond-Prout, P. Smart, L. Brammer, *Cryst. Growth Des.* 2012, *12*, 205.
26. M. A. M. Abu-Youssef, A. Escuer, F. A. Mautner, L. Öhrström, *Dalton Trans.* 2008, 3553.
27. M. A. S. Goher, B. Bitschnau, B. Sodin, C. Gspan, F. A. Mautner, *J. Mol. Struct.* 2008, *886*, 32.
28. S. Muniappan, S. Lipstman, I. Goldberg, *Chem. Commun. 2008*, 1777.
29. C. Merkens, F. Pan, U. Englert, *CrystEngComm* 2013, *15*, 8153.
30. A. D. Naik, B. Tinant, K. Muffler, J. A. Wolny, V. Schünemann, Y. Garcia, *J. Solid State Chem.* 2009, *182*, 1365.
31. M. Hubrich, M. Peukert, W. Seichter, E. Weber, *Polyhedron* 2010, *29*, 1854.
32. J. M. Rubin-Preminger, U. Englert, *Inorg. Chim. Acta* 2009, *362*, 1135.
33. P. Smart, G. M. Espallargas, L. Brammer, *CrystEngComm.* 2008, *10*, 1335.

34. L. Chen, X. Cao, Z. Li, J. Cheng, Q. Lin, Y. Yao, *Inorg. Chem. Commun.* 2008, *11*, 961.
35. Z. Han, Y. Zhao, J. Peng, A. Tian, Q. Liu, J. Ma, E. Wang, N. Hu, *CrystEngComm.* 2005, *7*, 380.
36. Z. Han, Y. Gao, X. Zhai, J. Peng, A. Tian, Y. Zhao, C. Hu, *Cryst. Growth Des.* 2009, *9*, 1225.
37. S. Tragl, K. Gibson, H.-J. Meyer, *Z. Anorg. Allg. Chem.* 2007, *633*, 470.
38. T. J. Siciliano, M. C. Deblock, K. M. Hindi, S. Durmus, M. J. Panzner, C. A. Tessier, W. J. Youngs, *J. Organomet. Chem.* 2011, *696*, 1066.
39. A. S. Romanov, J. M. Mulroy, M. Y. Antipin, T. V. Timofeeva, *Acta Crystallogr. Sect. C Cryst. Struct. Commun.* 2009, *65*, m431.
40. M. I. Bruce, M. Jevric, C. R. Parker, W. Patalinghug, B. W. Skelton, A. H. White, N. N. Zaitseva, *J. Organomet. Chem.* 2008, *693*, 2915.
41. F. Vidal, M. A. Dávila, A. S. Torcuato, P. Gómez-Sal, M. E. G. Mosquera, *Dalton Trans.* 2013, *42*, 7074.
42. N. Q. Shixaliyev, A. M. Maharramov, A. V. Gurbanov, V. G. Nenajdenko, V. M. Muzalevskiy, K. T. Mahmudov, M. N. Kopylovich, *Cat. Today* 2013, *217*, 76.

9

THE INFLUENCE OF NON-COVALENT INTERACTIONS IN THE STRUCTURE AND DIMENSIONALITY OF HYBRID COMPOUNDS AND COORDINATION POLYMERS

FERDINANDO COSTANTINO[1,2], ANDREA IENCO[2], AND MARCO TADDEI[3]

[1] *Dipartimento di Chimica, Biologia e Biotecnologie, University of Perugia, Perugia, Italy*
[2] *Department at Consiglio Nazionale delle Ricerche, Institute of Chemistry of Organometallic Compounds (CNR–ICCOM), Florence, Italy*
[3] *Laboratory for Catalysis and Sustainable Chemistry, Paul Scherrer Institut (PSI), Villigen, Switzerland*

9.1 INTRODUCTION

Crystalline inorganic–organic hybrid materials are a wide class of compounds extensively studied over the last decades for several reasons, including their high structural and chemical variability and the large number of application fields (catalysis, gas sorption and separation, electro- and photochemical devices, etc.). Among them, coordination polymers (CPs) and metal–organic frameworks (MOFs) are generally defined as compounds containing backbones constructed by metal ions as "nodes" and organic ligands as "linkers" [1]. The main difference between MOFs and CPs is related to the presence, in MOFs, of potential void; hence, CPs represent a wider family of compounds classified in different manners depending on dimensionality and flexibility. An exhaustive review from Kitagawa gave the definition of the classes of CPs based on dimensionality of the covalent (or ionic) bonds or flexibility of framework upon removal of the guest molecules [2]. The six classes reported by Kitagawa are 1D chains, 2D stacked layers, 2D interdigitated layers, 3D pillared layers, 3D expanding and shrinking grids and 3D interpenetrated grids (Scheme 9.1).

These definitions allow to include 1D and 2D compounds into the family of CPs because they can be generally assembled in such a way to recall the 3D compounds in terms of porosity and reactivity (ion exchange, gas sorption, etc.). In these arrangements weak interactions can play a crucial role by extending the dimensionality through the creation of supramolecular bonds between the motifs. The concept of "supramolecular synthons" is then referred to 0D or 1D units linked to each other via H-bond to form macro-units of higher dimensionality. The role of hydrogen bonding in CPs is related to the ability of ligands to form stable hydrogen bonds, thus acting as a "supramolecular glue," as properly defined by Kitagawa in its review. However, the influence of other non-covalent interactions (hydrophobic interactions, $\pi\cdots\pi$ or $CH\cdots\pi$ stacking, halogen bonding interactions, etc.), on dimensionality, porosity, stability, and reactivity of CPs and MOFs, has not been reviewed yet. In this chapter the role of the non-covalent interactions in the assembly of some CPs based on phosphonates, phosphinates, and other auxiliary ligands will be discussed. In particular, some selected examples on layered–pillared compounds based on zirconium and copper aminophosphonates and phosphinates will be presented. Other examples discuss the size of the coordinating metal ions in CPs, the relationships between the covalent interactions, and the coordination environment for same coordination metals. In the provided examples weak interactions are responsible for several structural features: they direct the dimensionality of 2D and 3D CPs upon the synthesis, sensibly affect the water stability, influence the flexibility of the framework, and induce interesting phase transformations.

Non-Covalent Interactions in the Synthesis and Design of New Compounds, First Edition.
Edited by Abel M. Maharramov, Kamran T. Mahmudov, Maximilian N. Kopylovich, and Armando J. L. Pombeiro.

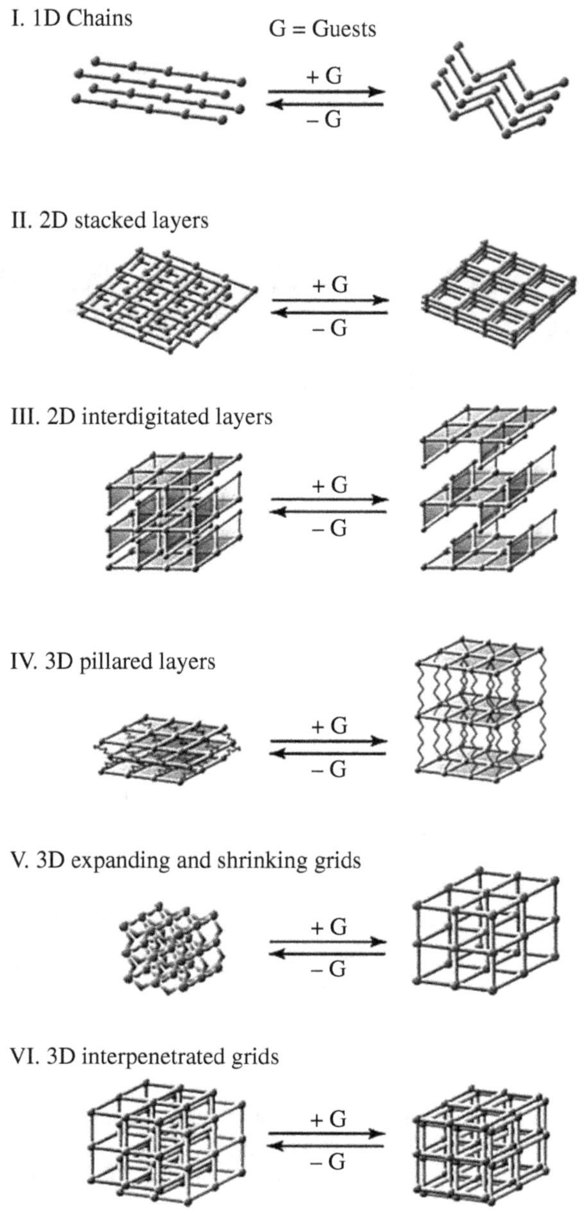

SCHEME 9.1 The six classes of dynamic coordination polymers.

9.2 NON-COVALENT INTERACTIONS IN LAYERED AND 3D OPEN FRAMEWORK ZIRCONIUM AMINOPHOSPHONATES

The chemistry of zirconium phosphonates, hybrid compounds known for their high thermal and chemical stability, has a long history, which started more than 40 years ago [3]. However, it continues to raise a remarkable interest in the scientific community from a fundamental point of view and for the extent of applications in which these materials are still employed. Phosphonic ligands are versatile building blocks for the synthesis of robust CPs owing to the high variation in their molecular structure. Polyphosphonic ligands with variable geometry and coordination capability can be obtained with easy synthetic procedures. Simple monophosphonic or bisphosphonic ligands with formula R—PO_3H_2 or H_2PO_3—R—PO_3H_2 were extensively used in the past as linkers to obtain layered or layered–pillared zirconium phosphonates with formula $Zr(O_3P\text{—}R)_2$ or $Zr(PO_3\text{—}R\text{—}PO_3)$ (R = *n*-alkyl, carboxyalkyl, aminoalkyl, hydroxyalkyl, benzyl, etc.) [4]. In these compounds the structure of layers is similar to that of the parent compound α-ZrP, and two organic R groups per zirconium atom point toward the interlayer region. The free area associated to each R group is 24 $Å^2$, and in general, the interlayer space is filled by these organic groups that are arranged in a compact double (or pillared) layer. Figure 9.1 shows the structure of zirconium benzene phosphonate and a scheme of the layer surface.

R-amino bis(methylenephosphonic) acids were considered good candidates to be employed for the attainment of compounds with different structures and dimensionalities. As a matter of fact, these acids possess two R—PO_3H_2 phosphonic tetrahedra directly linked to a tertiary amino group with pendant R— residues. The higher flexibility degree compared to the simple organic phosphonate (with P—C bond) allowed to increase the variety of coordination with zirconium atom. The molecular structure of these ligands has a candlestick-like conformation in which the two CPO_3H_2 tetrahedra are close to each other representing the "candles," and the organic residue represents the "column" (Fig. 9.2). The reaction of *n*-alkyl-amino-*N*,*N*-bis (methylenephosphonic) acids with zirconium was carried out in mild conditions and in the presence of HF as mineralizing agent. A family of isostructural layered Zr derivatives was thus prepared. In these compounds the phosphonic tetrahedra were linked to zirconium atoms thus designing inorganic layer with the common structure, whereas the alkyl chains occupied the interlayer region with an interdigitated disposition (Fig. 9.2).

This disposition seemed to be confirmed for the *n*-alkyalmino chains with more than three carbon atoms up to 16 because there is a linear correlation between the number of carbon atoms and the interlayer distance. The derivative with five carbon atoms (**C5**) had a good crystallinity degree, and its structure was solved *ab initio* from X-ray powder diffraction data (Fig. 9.3) [5].

The structure of the layer is different from that of α-type zirconium phosphonates. In fact, each layer is constituted by ZrO_5F octahedra and O_3PC tetrahedra connected to each other in different ways. Zr octahedra are placed in two parallel planes, while phosphonate tetrahedra are linked to the same side of the layers and are bridged together by the alkyl chains, via the nitrogen atoms. The two phosphonate groups of each diphosphonic unit are crystallographically and chemically nonequivalent: one of them is triply connected to three zirconium atoms, while the second one is doubly connected to two different zirconium atoms. The third P—OH group points toward the nitrogen atom of an adjacent diphosphonic group. The PO···N distance is only 2.48 Å indicating that a strong hydrogen bond is present between them. Very likely the amino group is protonated by the adjacent phosphonate, and therefore the system has a zwitterionic character. This kind of interactions seems to be very important in building up the connectivity of the layer.

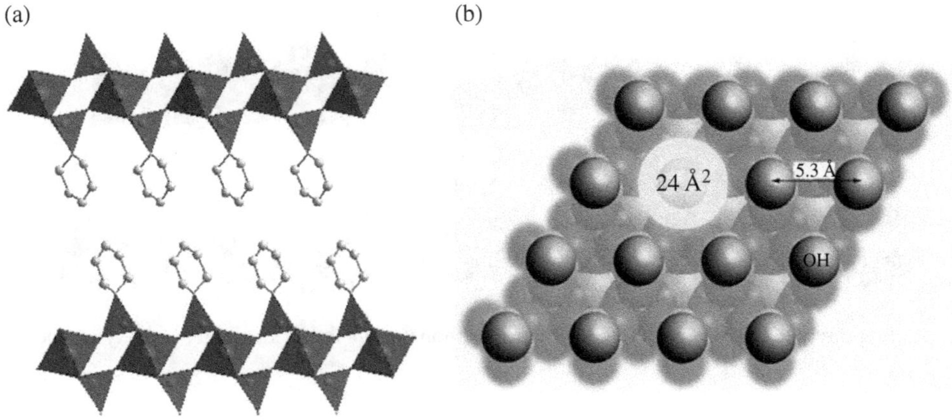

FIGURE 9.1 Polyhedral representation of α-zirconium phenylphosphonate (a) and representation of the layer surface in α-type zirconium phosphonates (b).

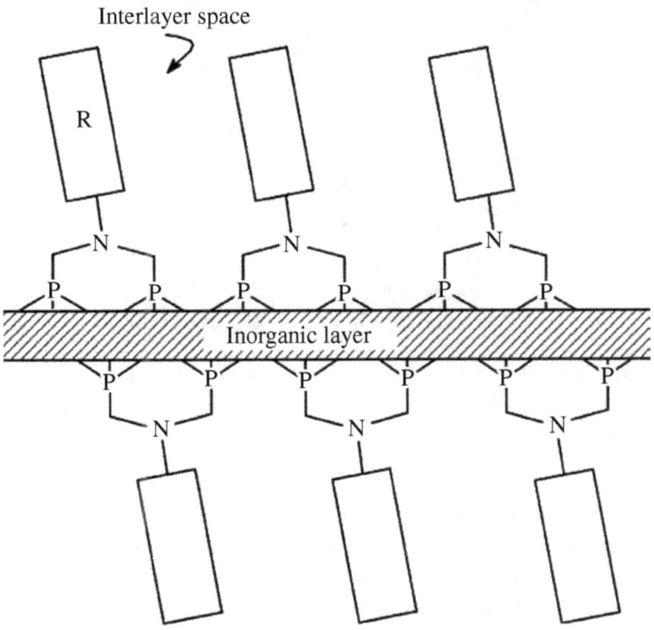

FIGURE 9.2 Scheme of a layered diphosphonate in which both phosphonic groups are bonded to the same face of the inorganic layer and the organic pendants occupy the interlayer space.

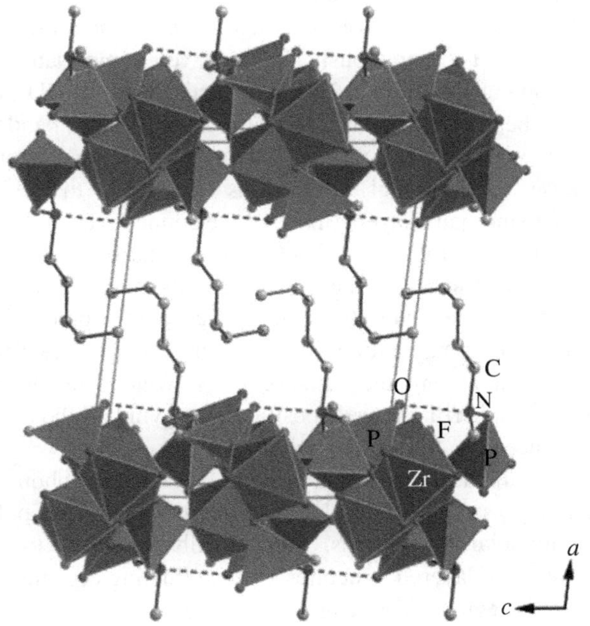

FIGURE 9.3 Polyhedral structure of **C5** viewed along the *b*-axis.

Nitrogen atom has a strong basic character that forces the protonation and consequently the orientation of the neighboring P—O groups. This interaction forces the phosphonate group to be external to the inorganic layer and tends to exclude one P—O group from the zirconium coordination, with the consequent insertion of an additional anionic ligand, F^- in this case, to complete zirconium coordination sphere and counterbalance the charge. Details of the layer structure and of the main weak interactions are shown in Figure 9.4.

Analogous interactions have been also found in similar zirconium derivatives of amino-bis(methylenephosphonic) acids containing benzyl, carboxyalkyl, or hydroxyalkyl groups (Fig. 9.5). These structures have the same layer connectivity and a similar disposition of the organic residues attached to the central N atom.

For further details on their crystal structure, we refer to a review (2008) devoted to these compounds [6], while in this

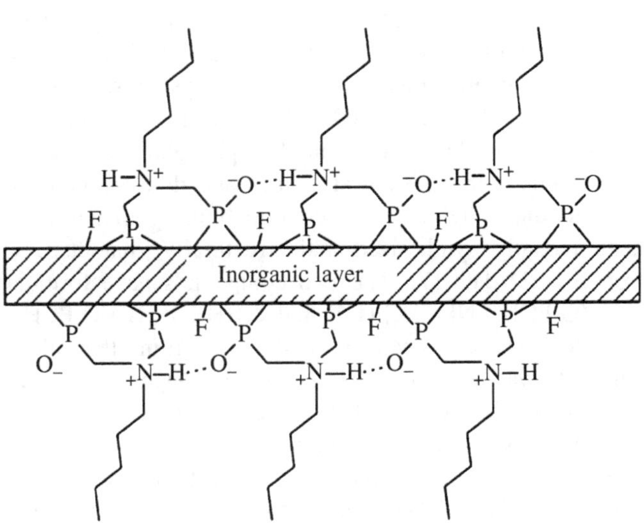

FIGURE 9.4 Layer framework found for layered zirconium biphosphonates, in which the zwitterionic character of the diphosphonate moiety and the non-covalent interactions are shown.

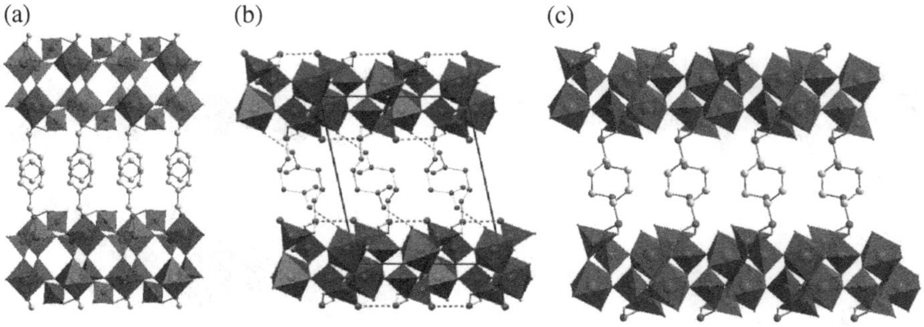

FIGURE 9.5 Polyhedral representation of the structure of Zr bisphosphonate containing alcoholic, carboxylic, and benzylic groups in the interlayer region.

chapter we want to focus on the specific role of the non-covalent interactions in designing the connectivity and the dimensionality of these materials.

Non-covalent intermolecular acid–base interactions, such as the P—OH···N previously shown, have an important role in designing the internal layer connectivity of the 2D systems. The length and the steric hindrance of the pendant organic groups also act as "structure-orienting factors" because their mutual interactions contribute to design alternate inorganic layers (made of the connections of zirconium atom with the polar PO_3C groups) and organic interlayer regions in which the organic groups strongly interact with each other via different non-covalent forces. For example, in benzylamino groups strong $\pi\cdots\pi$ interactions are present among the benzene rings, which are closely packed in a herringbone disposition occupying all the available space (Fig. 9.6).

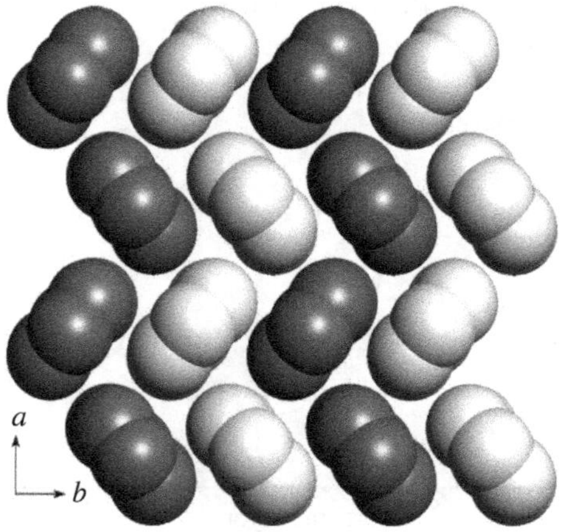

FIGURE 9.6 Close-packed arrangement of benzyl groups, viewed along the *c*-axis, in the structure of zirconium derivative containing benzyl groups.

In all the above reported systems, the 2D phosphonates are likely assembled in solution through equilibrated polar and nonpolar interactions. This equilibrium leads to the formation of layered structures with alternating organic and inorganic regions. The length of the organic residue (alkyl, carboxy- or hydroxyl-alkyl, benzyl, etc.) and the mutual interactions among the pendant groups were found to be the most important factor influencing the dimensionality and the connectivity of this family of compounds together with the P—OH···N interactions, which are always present in all compounds above described. As a confirmation, the synthesis of short-chain amino-alkyl bisphosphonate leads to compounds with different structures and dimensionalities. Methyl- and ethyl-amino zirconium biphosphonates have been recently synthesized and structurally characterized from PXRD data [7]. The two compounds have formula $ZrF_2[(O_3PCH_2)_2NHCH_3]$ (**C1**) and $Zr[(HO_3PCH_2)(O_3PCH_2)NHC2H_5]_2$ (**C2**), respectively. Differently from other alkyl-amino derivatives, which display the same layered structure and composition, **C1** and **C2** showed different stoichiometries and crystal structures (Fig. 9.7).

C1 has a pillared–layered 3D structure with small cavities between the layers. The inorganic layers are formed by the connection of two different types of chains, both running along the *c*-axis with different orientations. A-type chains are built from the connectivity of Zr1 atoms via the tridentate PO_3C tetrahedron: the zirconium atom is coordinated by four oxygen atoms and two fluorine atoms placed in *trans* position. B-type chains are constituted by connecting on Zr2 atoms via a bidentate PO_3C tetrahedron. In this case the zirconium atom also occupies a special position but is coordinated by six oxygen atoms, four of which belong to the doubly connected PO_3C tetrahedra and allow the development of the chain along the *c*-axis. The connection of these two inorganic chains gives 2D planes linked in the third dimension by other biphosphonic groups. Methylamine groups point inside the small cavities formed between adjacent layers. Carbon atom of the methyl group is oriented toward the adjacent F atom with a C···F distance of about 3.2 Å, thus suggesting that a weak non-covalent bond could exist between these two atoms.

The structure of the ethylamino zirconium derivative (**C2**) is also constituted by a three-dimensional network, but the connectivity is different from that of **C1**. **C2** does not possess fluorine and the coordination sphere of all zirconium atoms is filled by oxygen from the phosphonic groups. The structure is based on the connection of discrete inorganic secondary building units (SBUs), via the bridging organic ligands, so an extended inorganic connectivity is missing.

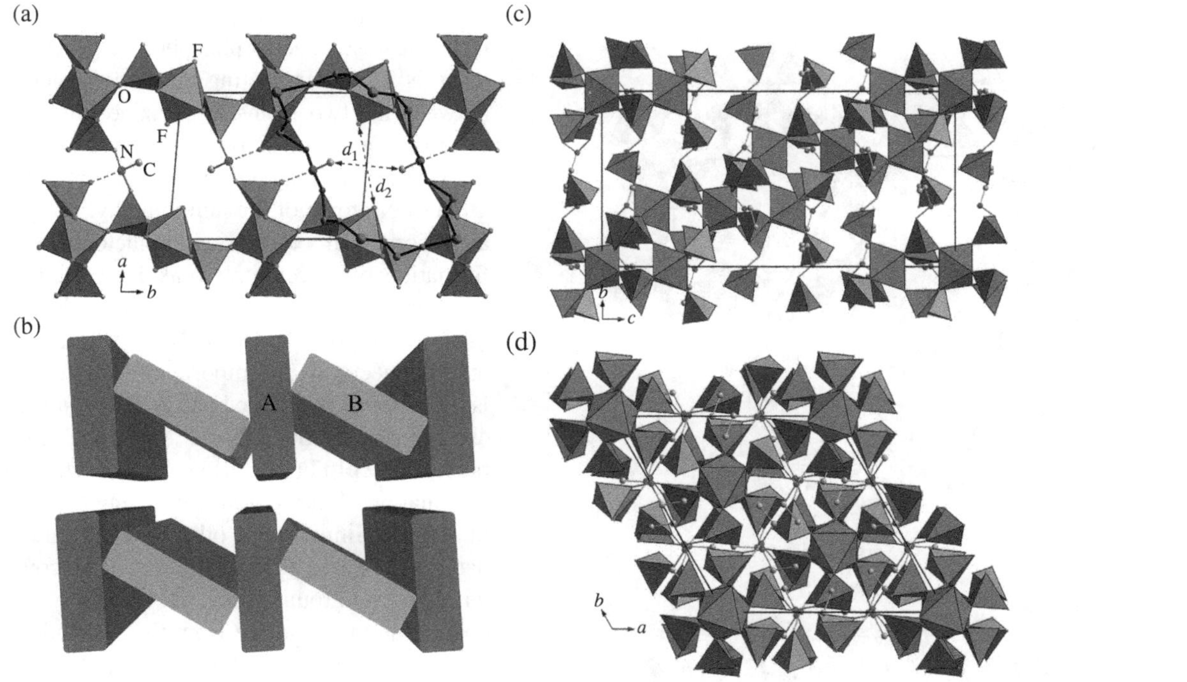

FIGURE 9.7 Representation of **C1** ((a) polyhedral and (b) schematic) and **C2** ((c) viewed along the *a*-axis and (d) viewed along the *c*-axis) structures.

These inorganic SBUs are bridged to each other via the ethylamino groups in a complicated pattern. The structure contains discrete cavities of ellipsoidal shape, where the organic pendant groups are accommodated (Fig. 9.8a). The cavities are not accessible as they are distributed in two different planes, shifted on 1/3 *a* and 1/3 *b* along the *c*-axis in such a way that on the top of each cavity, there is another SBU placed above the center (Fig. 9.8b).

The common layered character of the zirconium aminophosphonates with long tiles was attributed to the structure-orienting effect of the interactions among the pendant moieties. As a matter of fact, we found that derivatives containing different pendant groups adopted the same arrangement. The differences between compounds **C1** and **C2**, with respect to longer-chain phosphonates, arise from the balancing of polar and nonpolar non-covalent interactions that act as structure-directing factors. In **C1** and **C2**, the organic groups attached to the nitrogen atoms are very small, and so the contribution of nonpolar interactions should be smaller than in the longer-chain compounds. Also in these structures the organic nonpolar groups tend to be confined in shared spaces, whose extent depends on the dimension of these groups and therefore on the amount of nonpolar interactions. In **C1**, all methyl groups point toward small interlayer tunnels, interacting with the fluorine atoms exposed, thus completely filling the available space and resembling the arrangement of a rodlike micelle. **C2** displays a different structure, which is formed by discrete cavities where the organic groups are confined, in an arrangement similar to that of spherical micellar systems, whereas the inorganic framework seems to be modeled around these nonpolar regions. The homologous compounds with longer R assemble in layered arrangement and can be then compared to Langmuir–Blodgett films (Fig. 9.9).

Concerning 3D zirconium phosphonates, other two zirconium derivatives with the piperazine *N,N′*-bis(methylenediphosphonic) acid $Zr_2H_4[(O_3PCH_2)_2N_2C_4H_8]_3 \cdot 9H_2O$ (**Zpip1**) and $ZrF_2(O_3PCH_2)_2(NH)_2C_4H_8$ (**Zpip2**) have been obtained under similar synthetic conditions [8]. The two compounds have different compositions and crystal structures (Fig. 9.10). **Zpip1** has a three-dimensional structure, made of infinite inorganic chains of ZrO_6 octahedra and PO_3C tetrahedra, running along the *c*-axis direction and connected by piperazine groups in the *ab* plane. Each zirconium atom lies in a −3 axis, and its coordination sphere is filled by the oxygen belonging to three different phosphonic groups. The zirconium octahedra are bridged by the phosphonic groups thus forming the infinite inorganic chains. For the electroneutrality requirement, four of the six terminal P—O groups per formula unit should be protonated, while the remaining two must be P—O groups. Inorganic chains are packed in a hexagonal pattern and form quasicylindrical channels of about 7 Å centered on the threefold inversion axis. These channels are filled with water molecules, and the polar P—OH and P—O groups point toward the center of the cavities making H-bonds with them.

Compound **Zpip2** has a pillared layered structure made of inorganic planes of ZrO_4F_2 octahedra internally linked by bidentate phosphonic PO_3C tetrahedra. This connectivity creates a quasi-square pattern based on 16-atom (8 polyhedra) rings, which constitute the inorganic planes. The P—O groups not involved in bonds with Zr are directed

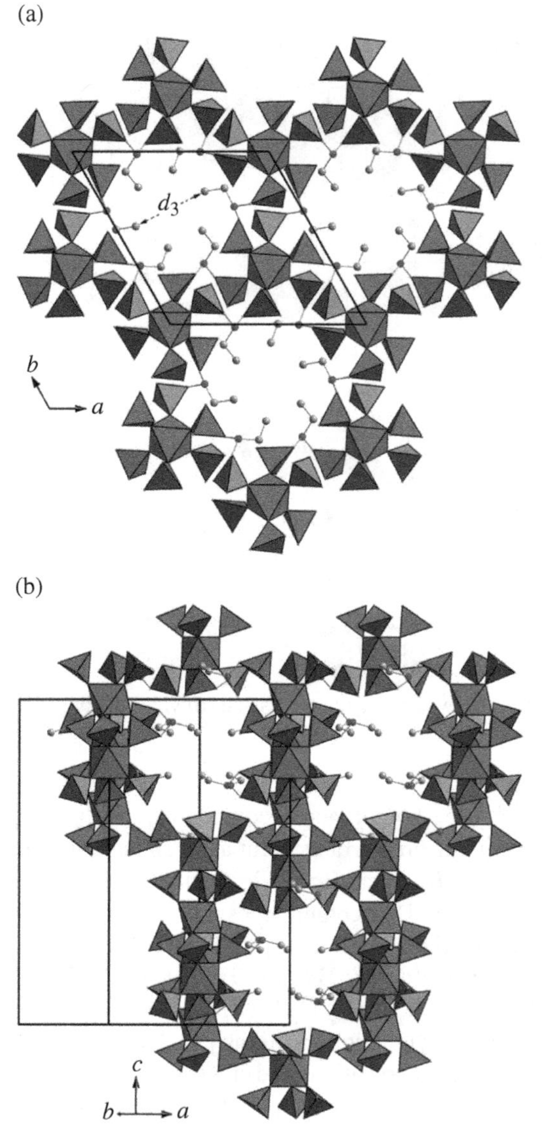

FIGURE 9.8 Polyhedral representation of the hydrophobic cavities in **C2**, viewed along the *c*-axis (a) and along the [210] vector (b).

toward the nitrogen atom of the piperazine moiety belonging to the nearest adjacent phosphonate group. The PO···N distance is 2.74Å, indicating that a hydrogen bond is present between the two atoms. Piperazine rings occupy the interlayer region in a close-packed herringbone arrangement [8].

As previously shown for similar compounds, the protonation of N atoms of the aminomethyl phosphonic groups is an important structure-orienting factor because it induces the formation of directional weak interactions with the polar P—O groups. Also in the two compounds **Zpip1** and **Zpip2**, the presence of protonated N atom is crucial in designing their connectivity. An important feature of the FT-IR spectra is the absence of a large band in the region 2600–2200 cm^{-1}, characteristic of the $R_3N—H^+$ stretching vibrations, which is present in **Zpip2** (Fig. 9.11). This signal is indicative of the protonation of the N atom, according to the crystal structure of **Zpip2**. Similarly to other layered zirconium phosphonates, protonated N atoms induce a non covalent interaction with the P-O groups which are therefore non bonded to zirconium atoms. Fluorine from the synthetic solution replace this missing ligand; in compound **Zpip2** this effect is clearly visible. On the contrary, in **Zpip1** the absence of protonated N atom implies the P—O groups to be linked to zirconium and no F atoms are present.

The two compounds are obtained at different pH (2 for **Zpip2** and 7 for **Zpip1**). The acid pH value for **Zpip2** probably favors the protonation of the basic N atom and provides a larger fraction of positively charged building blocks. It may induce the formation of a compound containing fluorine and strong PO···HN interactions. On the contrary, a higher starting pH value, associated with a larger fraction of unprotonated piperazine diphosphonate molecules, may favor the formation of a solid phase free from fluorine (**Zpip1**) with an open-framework structure.

Non-covalent interactions can be also responsible for unusual solid crystal-to-crystal transformations at high temperature. In this context, two zirconium fluoride

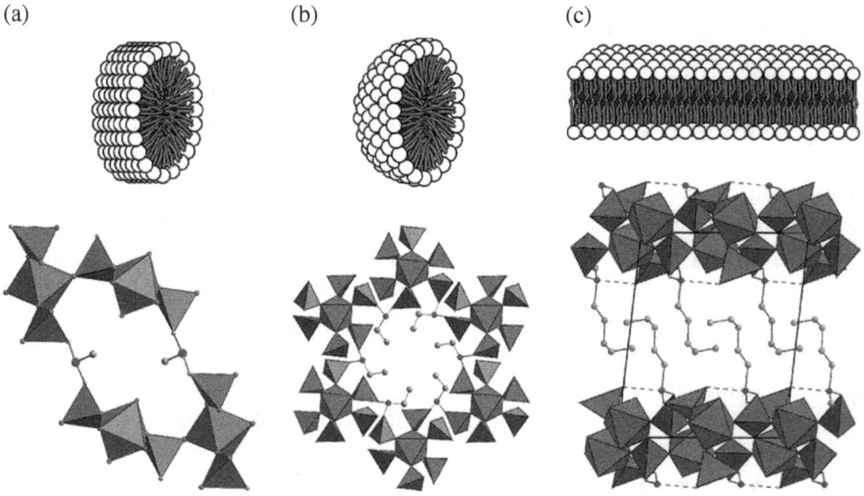

FIGURE 9.9 Analogies among amphiphilic systems and zirconium phosphonates with different alkyl chain lengths: (a) **C1** and nonaggregated units, (b) **C2** and direct micelle arrangement, (c) **C5** and bilayer arrangement.

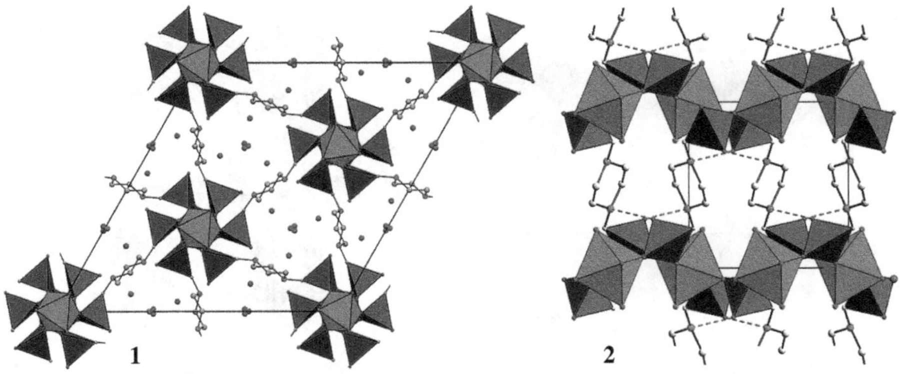

FIGURE 9.10 Polyhedral representation of the structures of **Zpip1** (1) and **Zpip2** (2).

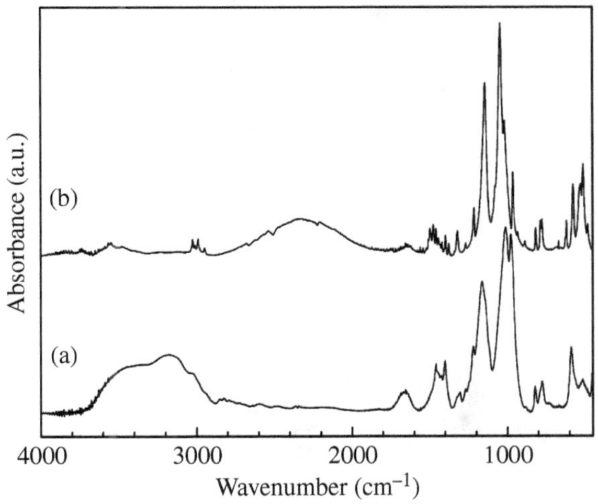

FIGURE 9.11 FT-IR spectra of **Zpip1** (a) and **Zpip2** (b).

carboxypyridine-phosphonates have been recently reported: $ZrF_2(HF)(PO_3CH_2CH_2NC_5H_4CO_2)$ (**Zison**) and $ZrF_2(PO_3CH_2CH_2NC_5H_4CO_2)$ (**Znic**) (Fig. 9.12) [9]. The phosphonic acids were prepared by nucleophilic substitution of Br-ethylphosphonic ester with the heterocyclic N atom of isonicotinic and nicotinic acids.

Zison has a 1D structure constituted of infinite composite building units running along the *c*-axis. They are built from linked PO_3C tetrahedra and ZrO_3F_3 octahedra, constructed by the repetition of a four-membered ring of P—O—Zr—O—P atoms. Within the zirconium coordination sphere, two fluorine atoms are placed in axial position, with the third one occupying the equatorial plane in a *mer* (or *trans*) reciprocal configuration. The isonicotinic groups are placed in the external part of the chain through the linking phosphonoethyl groups, and because of the disposition of the phosphonic tetrahedra, they are also alternatively placed up and down with respect to the plane defined by the zirconium atoms. Each COOH group, belonging to the carboxypyridine ligands, points, with one of the two oxygen atoms, toward the F atom of the adjacent chains with a F···O—C distance of 2.67(2) Å. This short distance strongly suggests that a non-covalent interaction is present between these two atoms or, more precisely, that F is protonated by the COOH group.

The compound **Zison** presents an unusual thermal behavior: TGA analysis reveals a weight loss of about 5% around 280°C. Above this temperature a short plateau indicates the formation of a stable new phase up to about 350°C. The weight loss is compatible with the loss of one molecule of HF per formula unit. The diffraction pattern of the heated compound was then collected, and the structure of the new phase (hereafter **ZisonA**) was solved ab initio from PXRD data. The new structure is layered and arises from the condensation of 1D chains into 2D layers after the replacement of fluorine atoms from the Zr coordination sphere with the carboxylic groups belonging to the organic ligands. The loss of the HF molecule leaves a coordination vacancy in the zirconium octahedron, and therefore it is saturated by the neighboring oxygen atom O4 belonging to a carboxylate group coming from the adjacent chain. The mechanism of phase transformation and the structure of **ZisonA** are shown in Figure 9.13.

Accessory spectroscopic techniques have been also used to confirm this unusual behavior: FT-IR spectra of compound **Zison** and **ZisonA** and **Znic** are shown in Figure 9.14. Despite the variety of signals, some clear indications are derived for the coordination of the zirconium atom. Together with aromatic and aliphatic CH stretching absorptions, the 3100–2800 cm^{-1} spectral region highlights the presence of the intense signal of HF for **Zison** (absorption at 2942 cm^{-1} in Fig. 9.14a): this band is not revealed for **ZisonA** and **Znic**, thus confirming the loss of HF. The C═O and C—O stretching modes at 1736 and 1270 cm^{-1}, characteristic of carboxylic acids, are observed in the IR spectrum of **Zison**. In **ZisonA**, a clear indication of unidentate carboxylate binding to Zr atom comes from the shift of both C═O and C—O vibrations to 1671 and 1341 cm^{-1}, respectively.

A similar effect is also evidenced by the Raman spectra of these compounds thus confirming the results of IR detection. ^{1}H and ^{19}F solid-state MAS-NMR spectra have been also collected for the two compounds; however, for the sake of brevity, we only report the ^{1}H spectrum (Fig. 9.15). ^{1}H MAS-NMR spectra of compounds **Zison** and **ZisonA** clearly show the absence, in the latter compound, of the signal at 12 ppm, which can be assigned to the acidic proton of the HF group,

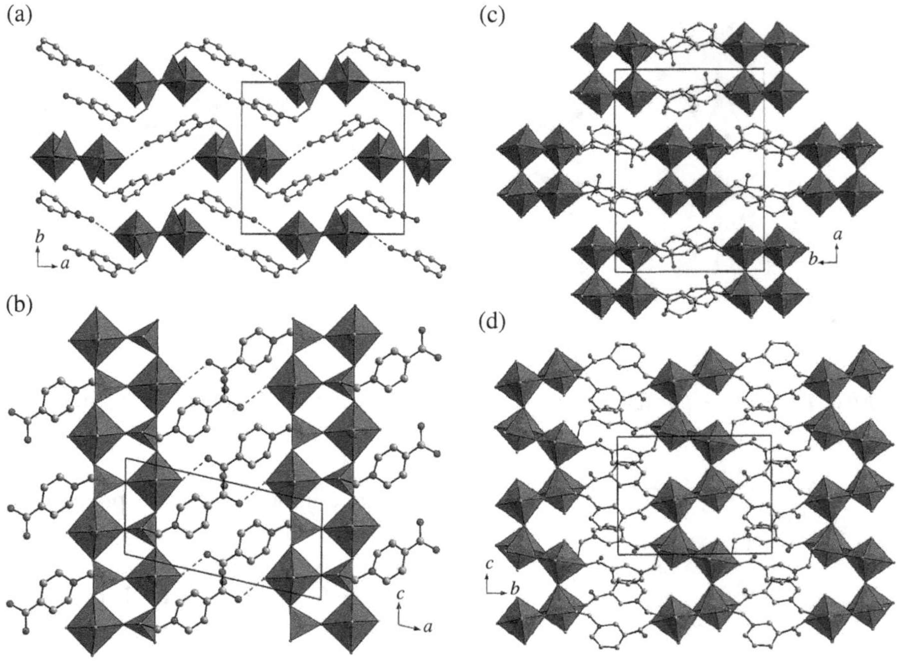

FIGURE 9.12 Polyhedral representation of the structure of (a, b) and Znic (c, d).

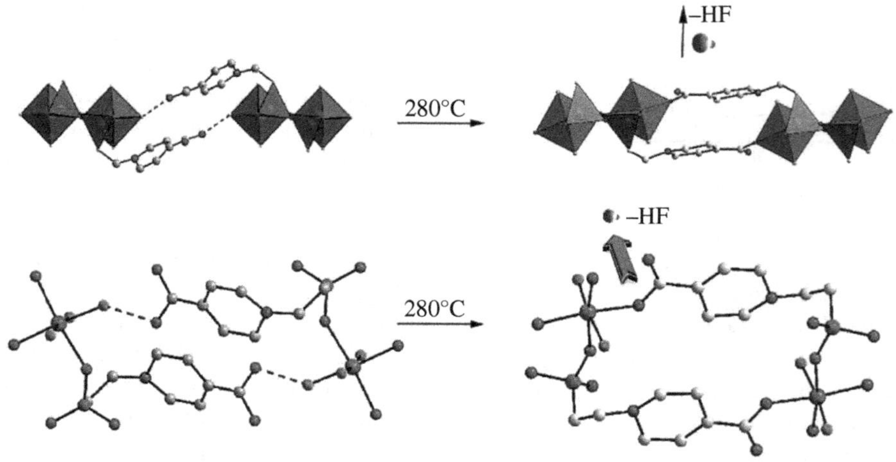

FIGURE 9.13 Temperature-induced transformation of the structure of **Zison** into **ZisonA**.

thus confirming the loss of hydrofluoric acid upon heating. The FT-IR spectrum of compound **Znic** is also shown in Figure 9.14c for comparison as it displays a similar connectivity, without protonated F atoms. Also the ^{19}F MAS-NMR spectra lead to similar results.

9.3 THE ROLE OF π···π STACKING IN THE ASSEMBLY, DIMENSIONALITY, AND STABILITY OF CPs

The number of CPs and MOFs containing aromatic linkers is extremely high; many of them have been constructed by using some of the most common oxygenated and nitrogenated ligands such as terephthalic acids, pyridine, carboxypyridine, 4,4′-, and related molecules. It is well known that the π···π interactions act as structure-orienting factors in building up the framework architecture. As a matter of fact, during synthesis the ligands are self-assembled in supramolecular synthons facing each other by stacking interactions, H-bonds, and other van der Waals forces. This kind of weak interactions is generally competitive with the stronger bonds of the donor atoms with the metal ions and the two driving forces together lead toward the final crystalline structure.

The non-covalent interactions were applied to construct large entities from molecular components [10]; these constructs, held together by intramolecular hydrogen bonding, π···π interactions, van der Waals forces, and other weak interactions, were called "supermolecules." The approach of

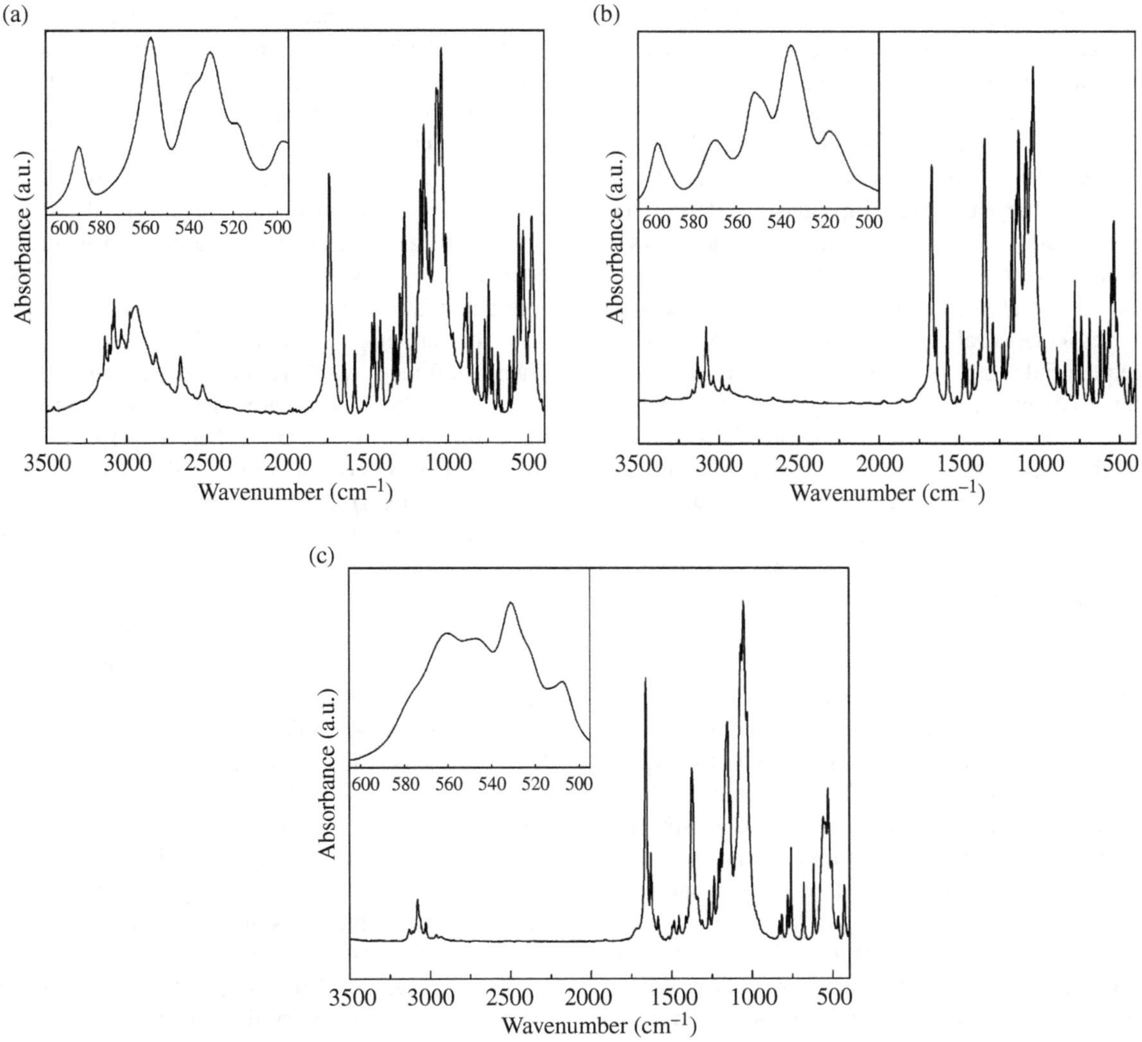

FIGURE 9.14 FT-IR spectra of compounds **Zison** (a), **ZisonA** (b), and **Znic** (c).

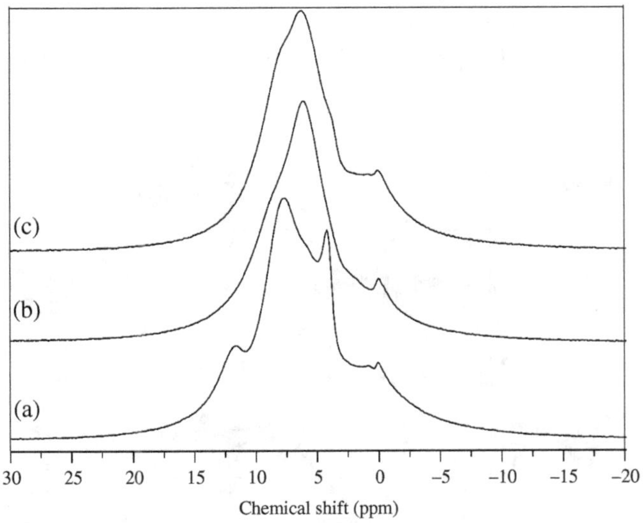

FIGURE 9.15 The ^{1}H MAS spectra of **Zison** (a), **ZisonA** (b), and **Znic** (c). The spectra were recorded at a MAS spinning frequency of 22 kHz.

supermolecules self-assembly could be used in several fields ranging from biochemistry of enzymes, biomimetic, sensors, liquid crystals, crystalline polymers, and so on. The role of the stacking interactions of aromatic moieties cannot be easily schematized, and all the examples that can be found in literature are too numerous to be included in an exhaustive way in this chapter. Thus we provide here just some selected examples, mainly taken from the recent research of our group, to give insights on how stacking interactions can influence the connectivity of CPs, the dimensionality, the metal coordination, and the chemical stability.

For instance, recently copper phosphonates and phosphinates containing 4,4′-bipyridine as auxiliary ligands have been reported [11]. Three copper derivatives of α,α′-p-xylylendiphosphonic acid and α,α′-m-xylylendiphosphonic acids have formula $Cu(HO_3PCH_2)_2C_6H_4{\cdot}3H_2O$ (**Cu-mxyl**), $Cu(HO_3PCH_2)_2C_6H_4[N(CH)_4CC\text{-}(CH)_4N]{\cdot}2H_2O$ (**Cu-mxyl-bipy**), and $Cu(HO_3PCH_2)_2C_6H_4[N(CH)_4CC\text{—}(CH)_4N]{\cdot}2H_2O$ (**Cu-pxyl-bipy**).

Cu-mxyl has a 2D structure made of infinite hybrid organic–inorganic layers stacked along the *a*-axis with an interlayer distance of about 9.9 Å. This arrangement results in infinite 1D chains of edge-sharing copper octahedra connected by phosphonate tetrahedra and linked to each other by the xylene groups along the [010] direction. The structure of compound **Cu-mxyl-bipy** has a different connectivity compared to the previous compound: the layers lie in the [110] plane (Fig. 9.16). The asymmetric unit is constituted by two copper atoms in special positions; they have the same distorted octahedral coordination geometry in which two *trans* equatorial positions are occupied by two equivalent phosphonate oxygen atoms belonging to a monodentate phosphonate group and the other two are filled by two nitrogen atoms belonging to the bipyridine ligands. The coordination sphere of the metal ions is completed by two water molecules in the elongated axial positions that point toward the interlayer region. In this compound the $\pi\cdots\pi$ stacking interactions play a fundamental role in designing the framework architecture. As a matter of fact, the phenyl ring of xylene containing phosphonate groups is sandwiched between two bipyridine ligands so that groups of three $\pi\cdots\pi$ stacked rings can be recognized. The layers are then constituted by a large part of organic fraction (made of the stacked rings) and the copper octahedra and phosphonic tetrahedra chains. In addition, the free oxygen atoms of phosphonate groups are involved in intralayer hydrogen bonds with both coordinating water molecules and free oxygen atoms of phosphonate (Fig. 9.16).

The structure of **Cu-pxyl-bipy** (Fig. 9.17) resembles that of **Cu-mxyl-bipy**, but the different arrangement for the ligand brings to a substantial difference within the layer. As a matter of fact, infinite arrays of stacked aromatic rings can be recognized, involving both the phenyl rings of the phosphonate and one of the two rings of bipyridine. Every bipyridine unit has then a nonstacked ring held between two PO_3C tetrahedra. The free oxygen atoms of the phosphonate groups are involved in intralayer hydrogen bonds both with water molecules coordinating copper and with opposite free phosphonate oxygen atoms, as seen for **Cu-mxyl-bipy**. Hydrogen bonds between coordinating oxygen atoms and water molecules from adjacent layers are also present.

In 2013 Zubieta *et al.* reported the structure of several copper α,α′-*p*-xylylendiphosphonic acid CPs with phenanthroline and 2,2′-bipyridine as auxiliary ligands [12]. For the sake of brevity, we will only discuss three copper-phenanthroline-α,α′-*p*-xylylendiphosphonic acid CPs and focus on the role of the auxiliary ligands. The compounds have formulas $[Cu(H_2O)(o\text{-phen})(pxyl)]\cdot 3H_2O$ (**Cu-pxyl-phen1**), $[Cu_2(H_2O)_2(o\text{-phen})(pxyl)]$ (**Cu-pxyl-phen2**), and $[Cu(o\text{-phen})(pxyl)]\cdot 3H_2O$ (**Cu-pxyl-phen3**) (Fig. 9.18).

Cu-pxyl-phen1 is the 0D compound, and it is constituted by a dinuclear copper(II) cluster made of two square pyramidal copper(II) ions bridged by two bidentate phosphonic groups. The terminal phosphonic groups of each diphosphonic moieties are protonated. Each copper ion possesses an apical water molecule. The one-dimensional structure of **Cu-pxyl-phen2** is constructed from the similar four-membered rings observed in the previous compounds linked through the organic tether of the diphosphonate ligand. The copper ions display the same square pyramidal geometry with the basal plane occupied by the phen nitrogen donors and two phosphonate oxygen donors and a water molecule in the apical position. The aqua ligands assume an *anti*-orientation with respect to the $\{\text{—Cu—O—P—O—}\}_2$ ring.

The structure of **Cu-pxyl-phen3** is 2D, and the layers arise from fragments similar to those observed in the previous compounds. The chain substructure of **Cu-pxyl-phen3** is clearly derived from that of **Cu-pxyl-phen2** with displacement of the aqua ligands by additional diphosphonate linkers providing the expansion into two dimensions. This observation suggests that the chain of **Cu-pxyl-phen2** may be a precursor to the formation of the two-dimensional structure of **Cu-pxyl-phen3** and consequently represents a

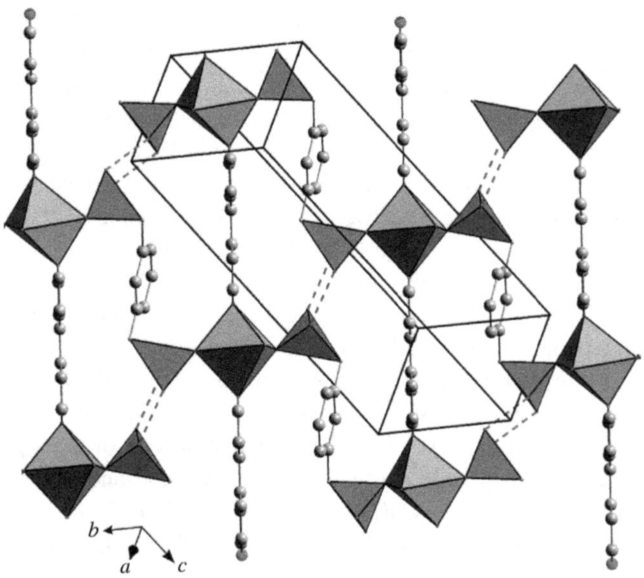

FIGURE 9.16 View of a single layer of **Cu-mxyl-bypi**.

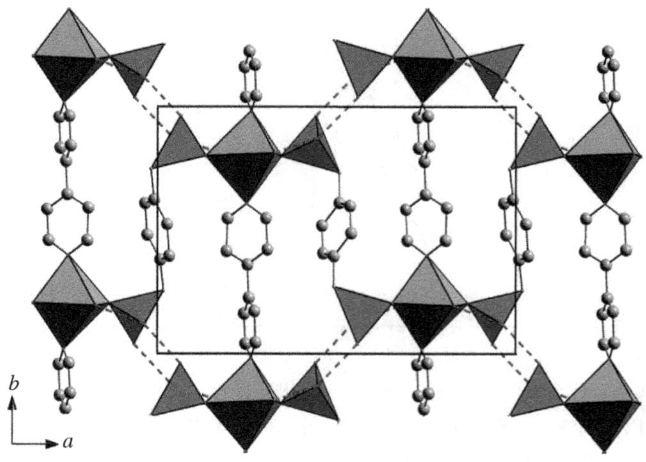

FIGURE 9.17 View of a single layer of **Cu-pxyl-bypi**.

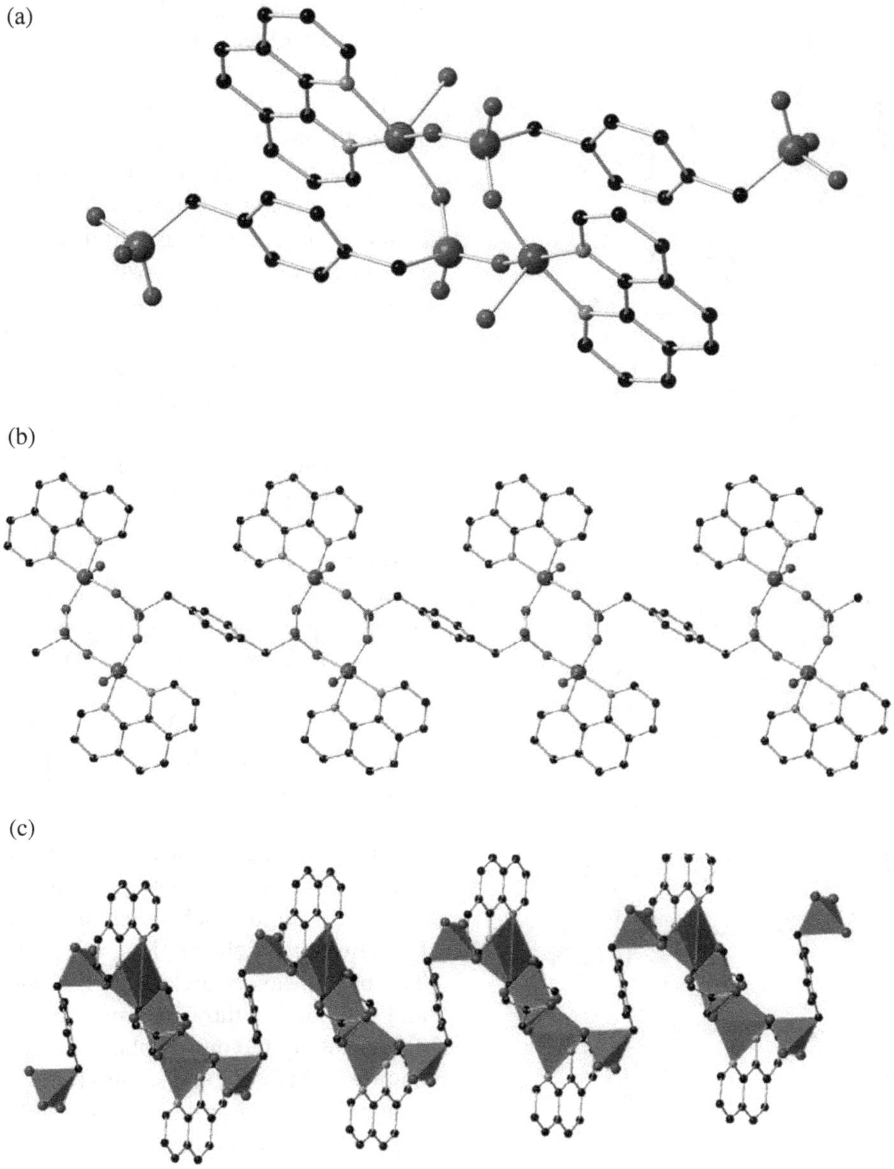

FIGURE 9.18 Ball-and-stick representation of the **Cu-pxyl-phen1** (a), **Cu-pxyl-phen2** (b), and **Cu-pxyl-phen3** (c) structures.

kinetic product. Also in this case the terminal phenanthroline ligands preclude the expansion of the structure in the third dimension, but they are involved in strong π···π interactions among them, designing interdigitated layers.

A connectivity similar to the copper phosphonates above described has been also observed in a homologous copper CP containing an analogous phosphinic ligand. Its molecular structure resembles that of *p*-xylene diphosphonic acid: the ligand is P,P′-diphenyl-*p*-xylenediphosphinate (hereafter pxilp), which possess a phenyl ring in place of one of the two phosphonic P—OH groups [13]. The reaction of this ligand with copper and 4,4′-bipy afforded a copper compound of formula $[Cu(4,4'\text{-bipy})(p(xyl)p)(H_2O)_2]\cdot 2(H_2O)$ (**Cu-pxylp-bypi**). The structure of **Cu-pxylp-bypi** is shown in Figure 9.19a in comparison with those of **Cu-pxyl-bypi** and **Cu-mxyl-bypi** (Fig. 9.19a and b).

The structure is constituted of hybrid layers containing copper octahedra bridged by the phosphinic groups through the xylene rings. 4,4′-Bipyridine connects the copper atoms along the [010] direction. The terminal phenyl rings belonging to the phosphonic groups form infinite rows of π···π stacked aromatic rings, running along the [100] direction, with one of the two rings of the bipyridine (Fig. 9.19). In these examples the presence of π···π stacking interactions turns out to play a dominant role in the building of the crystal structures, 4,4′-bipy, being linked to the copper atom through the basic N atom, preventing the formation of copper clusters thus forming infinite arrays. The presence of the aromatic rings linked to the phosphonic (or phosphinic ligands) makes possible the supramolecular stacking interactions along a plane, preventing the formation of three-dimensional structures.

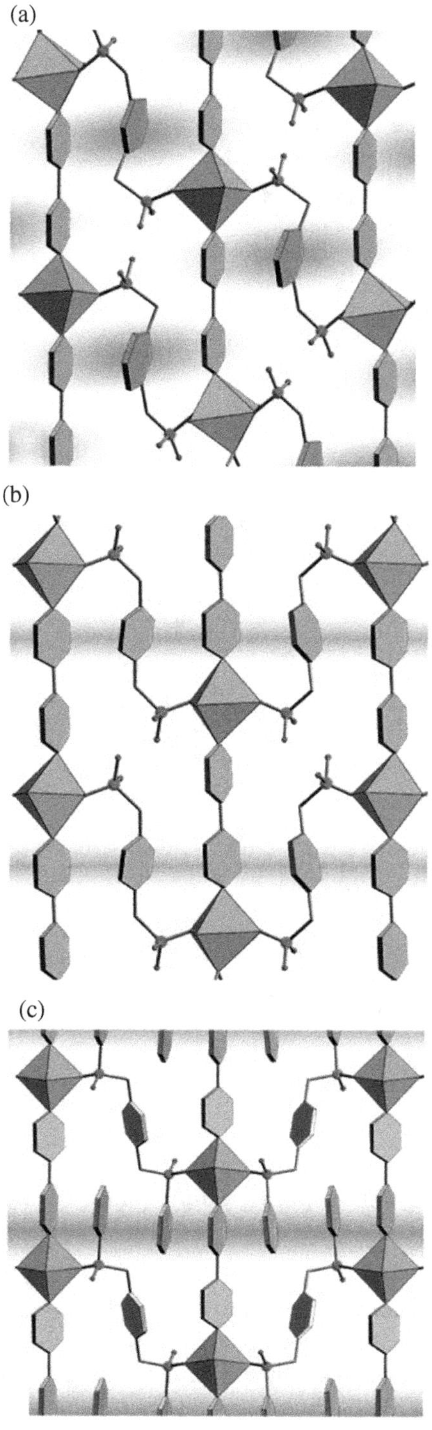

FIGURE 9.19 Representation of the structure of **Cu-pxylp-bypi** (c) in comparison with those of **Cu-mxyl-bypi** (a) and **Cu-pxyl-bypi** (b).

The absence of a phenyl ring between the phosphinic groups drastically changes the coordination capability, thus influencing the connectivity of the metal derivatives. When P,P′-methylenephenyldiphosphonic (pcp) acid is used in the synthesis with copper and 4,4′-bipyridine or 4,4′-bipyridylethane (bipy and bpye), two different CPs of formula $[Cu_2(pcp)_2bipy]\cdot 5H_2O$(**MONT1**)and$[Cu_2(pcp)_2bpye]\cdot 2.5H_2O$ (**MONT2**) are obtained, and they display similar 1D tubular structures (Fig. 9.20) [14].

The presence of a methylene group between the two phosphinates precludes their *trans* configuration in respect to the central group and forces their oxygen to be coordinated to a single copper ion. This coordination mode induces the formation of infinite rows of pentacoordinated copper ions bridged by two oxygen atoms from one pcp group and by the N atom of the bipy or bpye coligands. These infinite rows run along the [001] directions and are bridged by the nitrogenated coligands forming tubular structures. The resultant tubes pack into the crystalline framework through the C—H···π interactions that take place among the phenyl rings of the pcp groups lying in the external part of each tube. The two compounds showed good thermal and chemical resistance and a good absorption selectivity toward CO_2 and polar solvents [15]. However, one of the two CPs, namely, **MONT2**, if left in mother liquors, or if placed in water at reaction conditions, slowly transforms in a 1D phase with formula $Cu_2(H_2O)_2(pcp)_2bpye$ (hereafter **STRIP**) that is a pseudo-polymorphic phase of **MONT2** (Fig. 9.21).

In **STRIP** phase the Cu atoms are pentacoordinated by three oxygen atoms from two different pcp ligands, one water molecule and one nitrogen atom from one bpye ligand. Bpye ligand lies in an inversion center, whereas the phenyl rings of phenylphosphinate groups are placed perpendicular to each other, in such a way that one of them lies into the layer planes and the other points outside. The hybrid layers are stacked along the [110] direction and are packed through strong π···π interactions (about 3.6Å) among the out-of-plane phenyl rings. The transformation of **MONT2** phase into the **STRIP** phase is driven by the hydrolysis of the Cu—O bonds and a rotation of the copper SBU around the [001] direction. At the same time, half of the bpye ligands, after the breaking of one Cu—N bond for each tube side, rotate at 90°, connecting the unsaturated Cu atoms of the opposite side of the tubes and forming two **STRIP** phases for every broken tube (Fig. 9.22) [15].

The other CP, **MONT1**, does not undergo any phase transformation, and it is perfectly stable in hot water, even for a long time. This effect is thought to be due to concomitant effects arising from the metal coordination, the presence of H-bonds between the oxygen atoms connected to the copper ions, and the unfavorable predicted packing if bipy is replaced by bpye. For a better understanding of these effects, other two CPs built with the same ligands and containing Co ions have been used for comparison. Thus, two homologous compounds, **Co–pcp–bipy** and **Co–pcp–bpye**, were reported, and their structure resembles that of **STRIP** phase. In **Co–pcp–bipy**, the shorter length of the bipy ligand induces the reciprocal sliding of adjacent strips so that metal atoms belonging to adjacent strips are not facing each other, and two water molecules per metal atom (placed in *cis*

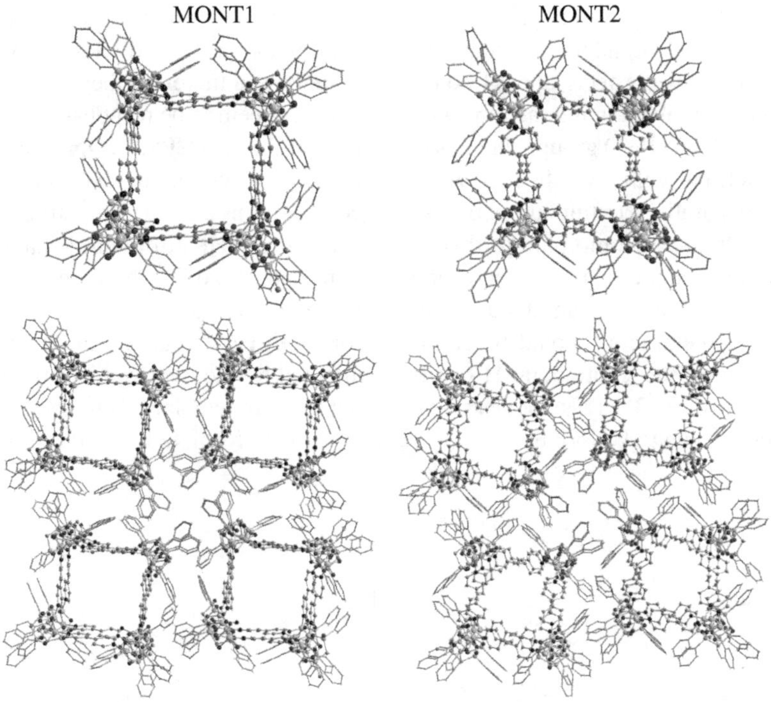

FIGURE 9.20 Structural representation packing of **MONT1** and **MONT2**.

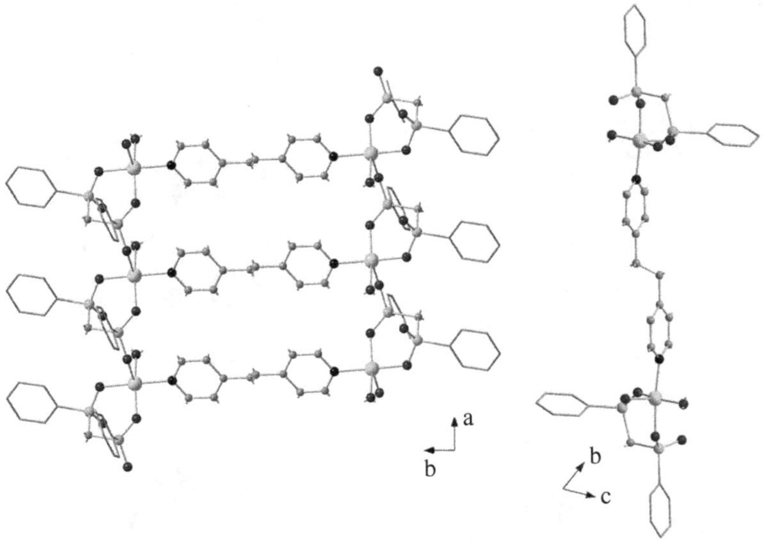

FIGURE 9.21 Ball-and-stick representation of the **STRIP** phase viewed perpendicularly to the *c* (a) and the *a* (b) axes.

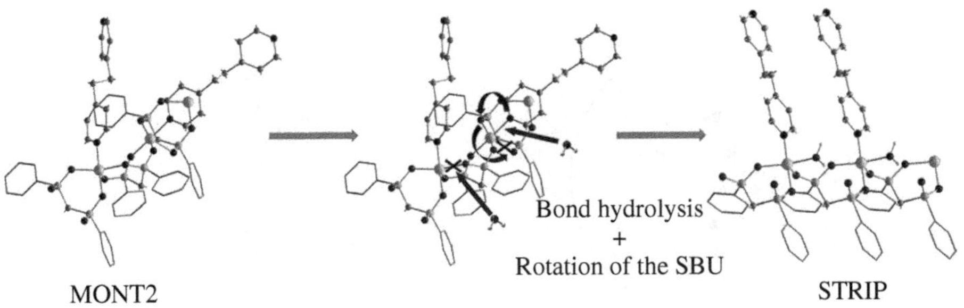

FIGURE 9.22 Detail of the transformation of the **MONT2** phase into the **STRIP** phase.

position) are found. The additional water molecule is required for fulfilling the need of Co to adopt an octahedral geometry. This rearrangement also affects the H-bonds pattern: each water molecule is involved in interactions with another water molecule and with one P—O group, which is in turn involved in two interactions with two distinct water molecules. In this way, two parallel continuous networks are created. If a hypothetic structure isomorphic to that of **Co–pcp–bipy** but containing copper is designed and assuming that Cu would prefer a square pyramidal arrangement, one water molecule per metal atom would be eliminated with the consequence that the continuous H-bonds pattern would be destroyed (Fig. 9.23). This suggests that such a structure is not enough stable to be formed.

It is also worth to be noted that other non-covalent interactions like π···π and CH···π are similar in the three compounds, and therefore they seem to be less important. These features seem to be responsible for the different water stabilities of the two CPS, **MONT1** and **MONT2**. As a matter of fact, the two compounds are very similar in terms of chemical composition and crystal structure. The only reason for expecting such a different behavior should be ascribed to a nonexistence of a thermodynamic stable phase deriving from the structural rearrangement of the tubes, as observed for **MONT2**. Presumably, also **MONT1** compound could be affected by the hydrolysis of the PO—Cu and N—Cu bonds, but the original phase is reformed because it cannot proceed toward the formation of a more stable one.

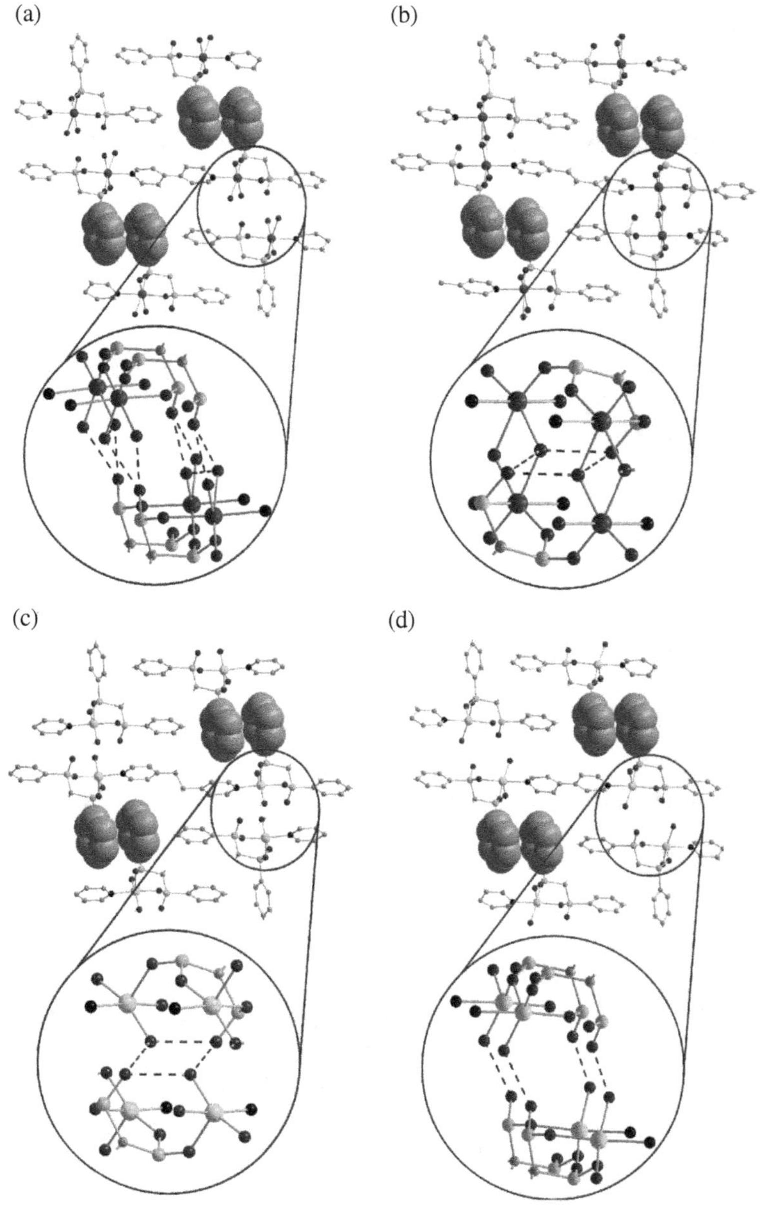

FIGURE 9.23 Comparison of the structure of **STRIP** phase (a) with those of **Co–pcp–bipy** (b) and **Co–pcp–bpye** (c) and a hypothetic **STRIP** phase without a water molecule (d).

9.4 SUPRAMOLECULAR INTERACTIONS AFFECTING THE METAL COORDINATION

The most common coordination mode in transition metals is the octahedral one. Almost all other coordination modes can be obtained and rationalized starting from the octahedral by removing or adding one or more ligands [16]. In supramolecular chemistry metallic building block is used to generate specific directional patterns in order to build a desired molecular shape [17]. In this chapter, we will show some examples where the supramolecular interactions force the metal coordination in CPs to be in certain mode.

In this context, the thiodiacetate (tda) molecule of formula $S(CH_2CO_2)_2^{2-}$ can be used as flexible and versatile ligand [18]. In particular some of us have shown the surprising different structural features of the reaction of $[Mn(tda)(H_2O)]_n$, **1**, with 2,2′-bipyridine (2,2′-bipy) and different substituted 2,2′-bipyridine ligands, namely, 4,4′-dimethyl-2,2′-bipyridine (4,4′-Me_2bipy), 5,5′-dimethyl-2,2′-bipyridine (5,5′-Me_2bipy), and 4,4′-dimethoxy-2,2′-bipyridine (($MeO)_2$ bipy) (Scheme 9.2) [19–21].

All the products are 1D CPs with the same Mn/tda/bipy ratio of 1 : 1 : 1. $[Mn(tda)(2,2'\text{-bipy})]_n$ was the first structurally characterized tetra-carboxylate dimanganese antiferromagnetic system with "copper acetate" core (Fig. 9.24).

Each Mn ion has an approximate trigonal prismatic coordination instead of the expected octahedral core and is surrounded by four oxygen and two nitrogen atoms of the 2,2′-bipy. The bimetallic units are glued together by tda ligands with a very particular μ,μ'–$\kappa^1(O),\kappa^1(O'),\kappa^1(O''),\kappa^1(O''')$ binding mode of tda forming the main 1D polymeric structure. The large separation between the metals (3.50 Å)

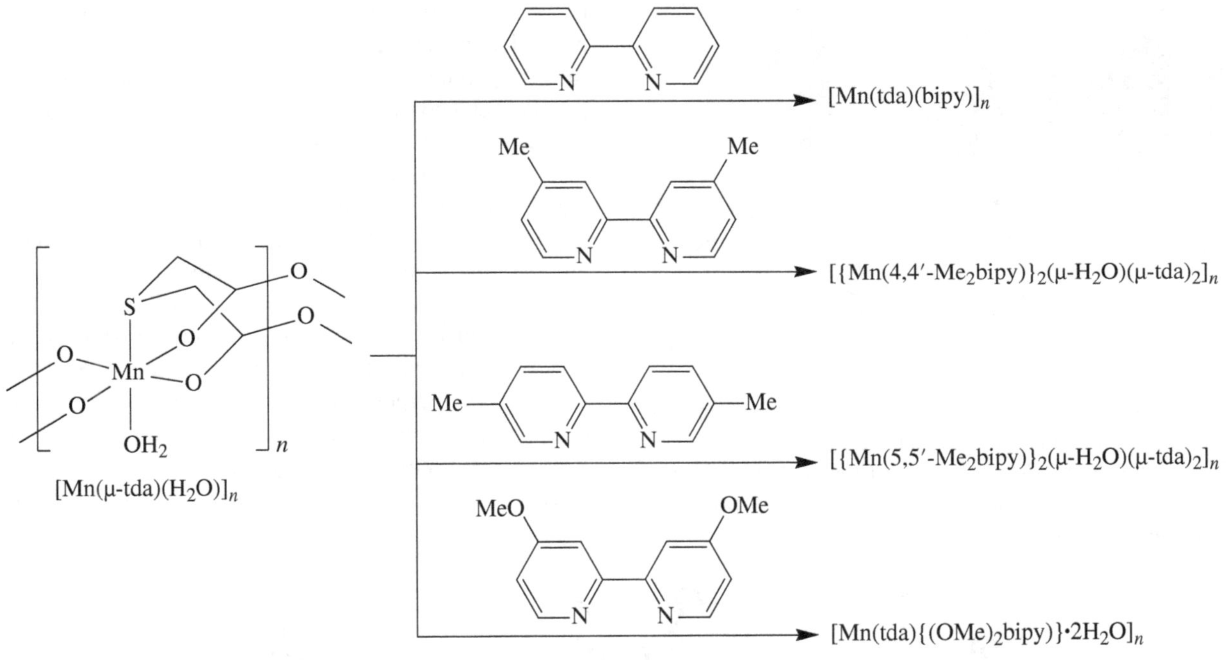

SCHEME 9.2 Reactions of $[Mn(tda)(H_2O)]_n$ with different bipyridines.

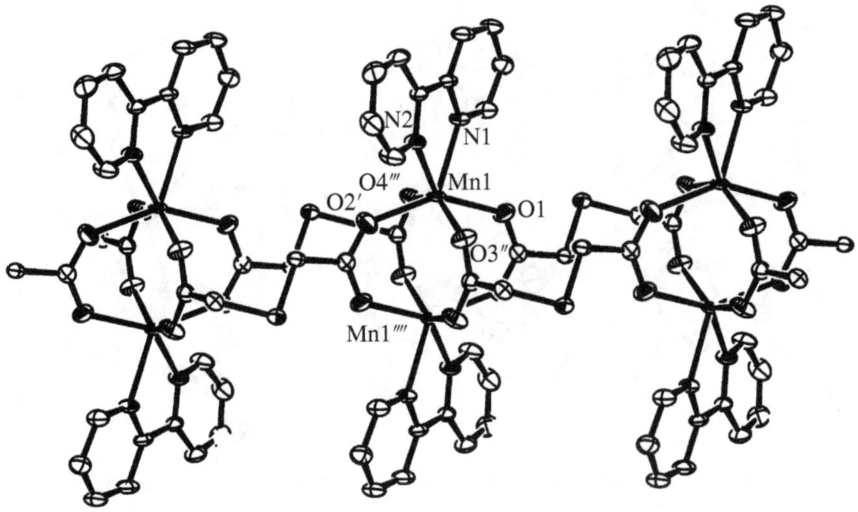

FIGURE 9.24 Drawing of the polymeric chain of $[Mn(tda)(2,2'\text{-bipy})]_n$ along the a-axis. The hydrogen atoms have been omitted for clarity. The view was chosen in order to highlight the local trigonal prismatic coordination of the Mn atoms.

excludes a possible metal—metal bond. Eventually, the 1D ribbons form a 2D slabs due to supramolecular non-covalent interactions. The bipys of consecutive Mn2 units are perpendicularly oriented in respect to the polymeric chain. The distances between two consecutive 2,2′-bipys are as large as 4.0 Å; it allows the 2,2′-bipy of adjacent 1D ribbons be inserted in a gear-like compact arrangement (Fig. 9.25).

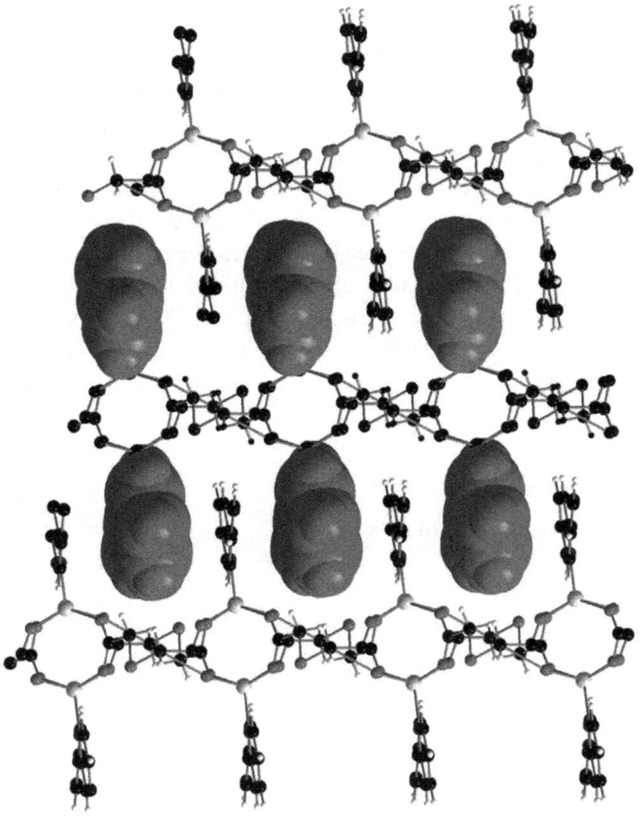

FIGURE 9.25 The 2D slabs formed by 1D polymeric chains of $[Mn(tda)(2,2'\text{-bipy})]_n$, thanks to the perfect interpenetration of the 2,2-bipy ligands. The 2,2-bipy ligands of the central chain were drawn using the van der Waals surfaces in order to highlight the supramolecular non-covalent interactions.

The large number of interactions provides a stabilization to the final crystal structure. In particular they might be the responsible for the local trigonal prismatic coordination of the metal atoms. The latter point was established analyzing the structures with substituted 2,2′-bipys. In particular, some hints come from the two topologically and essentially similar but not isomorphic CPs with 4,4′-Me_2 bipy and 5,5′-Me_2 bipy. Here, only $[\{Mn(4,4'\text{-}Me_2bipy)\}_2(\mu\text{-}H_2O)(\mu\text{-tda})_2]$ will be discussed (Fig. 9.26). The 1D polymeric structure of this compound is maintained as well as 1 : 1 : 1 ratio between Mn, tda, and bipy. The tda chains connect Mn2 units, but the copper acetate core is no longer adopted by the Mn2 units.

While the sulfur atom of tda ligand is still uncoordinated to the metal, only three of the four oxygen atoms are coordinated, having a μ_3-$\kappa^1(O),\kappa^1(O'),\kappa^1(O'')$ coordination mode of the tda ligand. In this case, the metal atoms adopt an expected pseudo octahedral coordination, thanks to the coordination of water molecule in bridging mode between the two metals of the Mn2 unit. A comparison of the coordination mode in 1 and 2 is highlighted in Figure 9.27.

Strong hydrogen bonds are present between the hydrogen atoms of the bridging water molecule and the uncoordinated oxygen atom of tda ligands. As consequence of the new environment of the Mn2 unit, the bipy ligands are almost perpendicular with the dihedral angle around 70°. They are not any more perpendicular in respect to the elongation axis of the polymer, but they are skewed of about 45°. Figure 9.28 shows the supramolecular arrangement of consecutive slabs. There are still a good degree of interpenetration between any

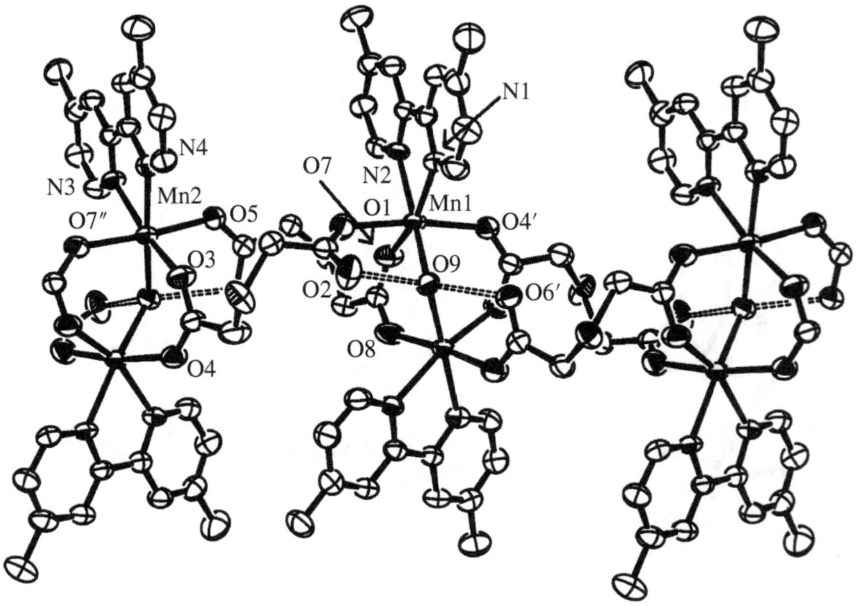

FIGURE 9.26 Drawing of the polymeric chain of $[\{Mn(4,4'Me_2bipy)\}_2(\mu\text{-}H_2O)(\mu\text{-tda})_2]$ along the crystallographic axis *a*. The hydrogen atoms have been omitted for clarity, and dashed lines represent the hydrogen bonds between the water molecule and the uncoordinated carboxylate oxygen atoms.

pair of consecutive ligands, but in this case the separation between the planes of bipyridine is not all equivalent.

The difference between the π···π stacking in **1** and **2** is highlighted in Figure 9.29 where three consecutive bipy ligands are viewed. In **1**, the cooperative effect is evident (the bipys almost perfecting eclipsed), while in **2** only two bipy are eclipsed and the third bipy is tilted of about 10°. Evidently in **2** the overall cumulative effect of the π···π stacking is lost. Comparative DFT study has demonstrated that the configuration of the Mn2 unit in **2** is lower in energy of around 20 kcal mol^{-1} than the corresponding unit in **1**. This means that the staking interactions in **1** are sufficient to force the coordination environment in the Mn2 unit.

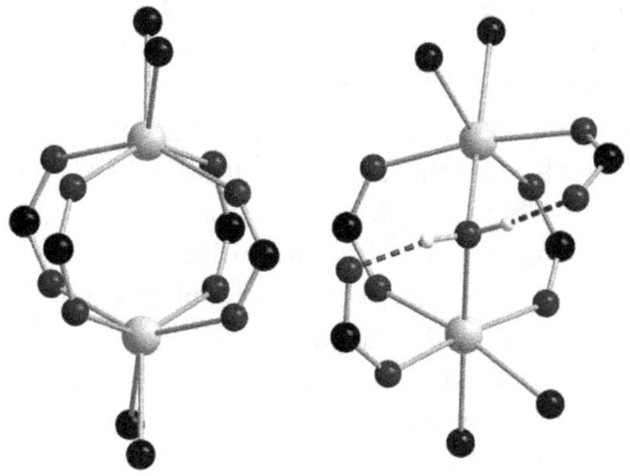

FIGURE 9.27 Comparison of the coordination modes of [Mn(tda)(2,2′-bipy)]$_n$ and [{Mn(4,4′Me_2bipy)}$_2$(μ-H_2O)(μ-tda)$_2$].

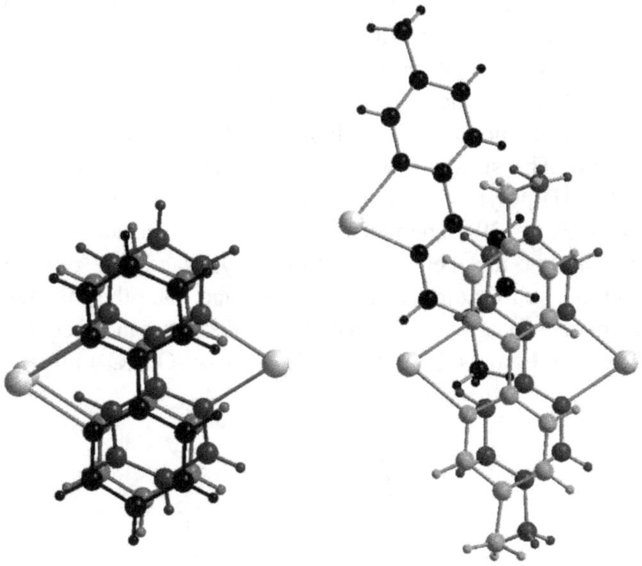

FIGURE 9.29 The different stacking of three consecutive bipy ligands in [Mn(tda)(2,2′-bipy)]$_n$ (left) and [{Mn(4,4′Me_2bipy)}$_2$(μ-H_2O)(μ-tda)$_2$] (right).

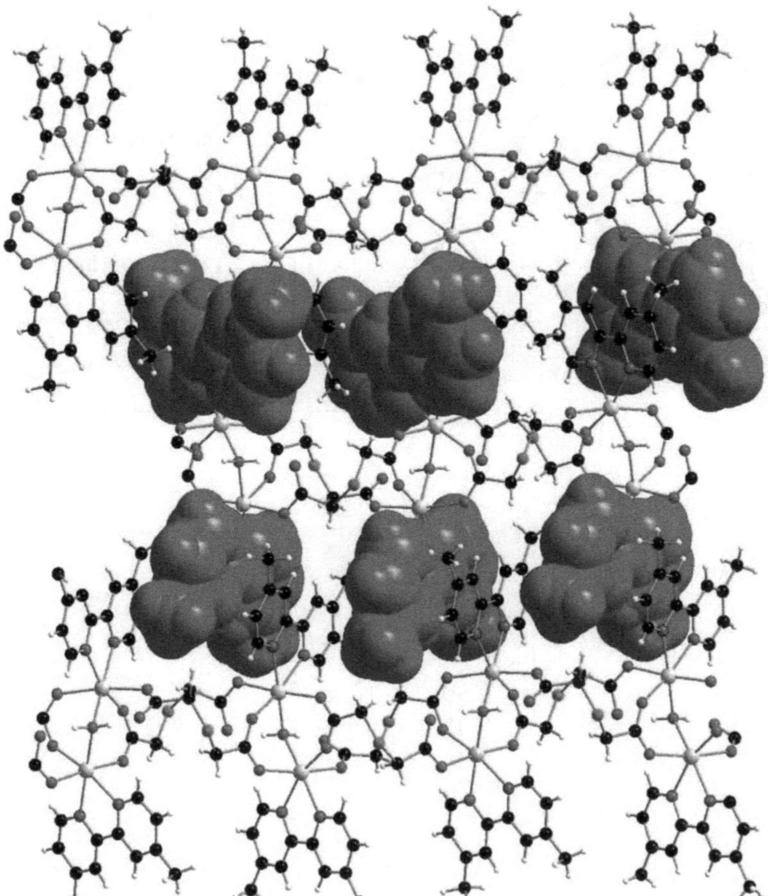

FIGURE 9.28 The 2D slabs formed by 1D polymeric chains of [{Mn(4,4′Me_2bipy)}$_2$(μ-H_2O)(μ-tda)$_2$]. The 4,4′Me_2bipy ligands of the central chain were drawn using the van der Waals surfaces.

The crystal density of **1** is 1.705 g cm^{-3} larger than the density of **2** and **3** (1.540 and 1.520 g cm^{-3}), and this is somehow surprising because the molecular weight of **2** and **3** is about 10% larger than **1**. One may conclude that the larger compactness is the driving force that forces the coordination environment of the metal center.

Finally, a small change in the substituents of the 2,2′-bipyridine like the introduction of the methoxy groups in position 4 and 4′ have a dramatic effect in the resulting crystal. The structure of $[Mn(tda)\{(MeO)_2bipy\}\cdot 2H_2O]_n$ is still a 1D polymer but not formed by the repetition of Mn2 units (Fig. 9.30). The metal coordination is pseudo octahedral by two nitrogen atoms of the bipy ligand, two oxygen and one sulfur atoms of a tda ligand, and one other oxygen atom of a tda ligand coordinated around another metal center. The tda ligands binding mode is μ_2-κ^3(O,S,O′),κ^1(O″), and one oxygen atom is uncoordinated. The latter is used together with the solvent water molecule to pack the 1D ribbons in 2D slabs instead of using the π···π stacking. In conclusion, these examples have shown that the geometry of the building block can be modified in order to attain the most favorable supramolecular architecture.

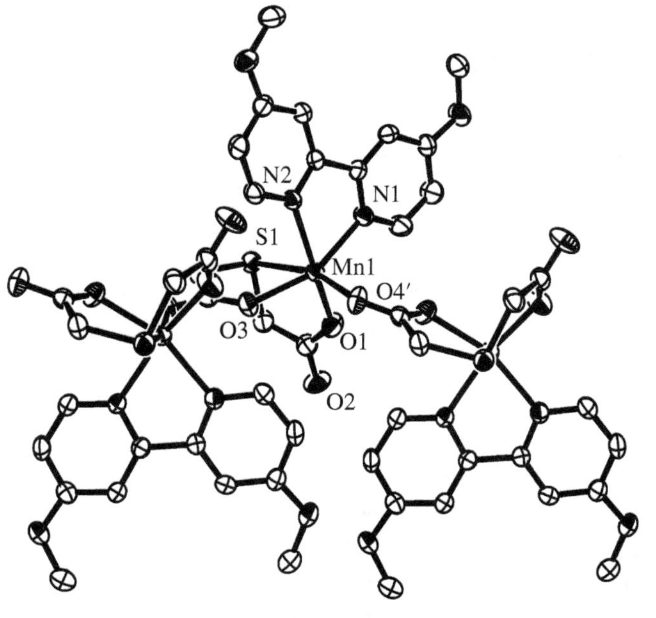

FIGURE 9.30 Drawing of three units of $[Mn(tda)\{(MeO)_2bipy\}\cdot 2H_2O]_n$.

9.5 THE RELATIONSHIP BETWEEN NETWORK DIMENSIONALITY AND CATION SIZE

The alkaline earth metals are suitable to explore the correlations between the dimensionality of the networks and the size of the cations. As known the atomic radius [22] increases from 1.05 to 2.15 Å from beryllium to barium atom, respectively (Table 9.1). In this context, in the work [23] interesting and didactic, Côté and Shimizu used the disulfonated catechol 4,5-dihydroxybenzene-1,3-disulfonate (dSC) (Scheme 9.3) to study a system in which either the s-block metal or the ligand is coordinatively flexible and adaptable. It was demonstrated that the dSC ligand does not coordinate magnesium and results to a solvent-separated crystal structure of formula $[Mg(H_2O)_6]\cdot(dSC)\cdot 3H_2O$ (Fig. 9.31). The magnesium hexaquo species form a charge-assisted hydrogen bonds with the ligand and the other solvent water molecules.

Further, the dSC ligand chelates the Ca^{2+} cations using the catechol unit as well as two oxygen atoms of one sulfonate group. In total, each dSC ligand connects three different calcium ions, and on the other side, each calcium atom is coordinated by three different ligands; the resulting net is a 1D column running parallel to the a axis (Fig. 9.32).

TABLE 9.1 Empirical Atomic Radius in Å for Alkaline Earth Metals [22]

Be	Mg	Ca	Sr	Ba
1.05	1.25	1.80	2.00	2.15

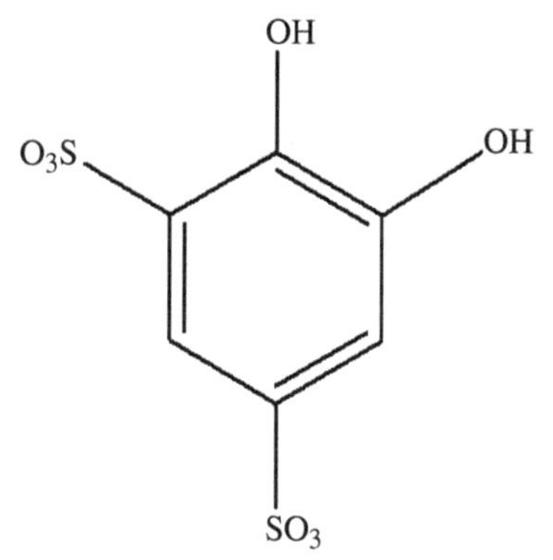

SCHEME 9.3 Schematic representation of 4,5-dihydroxybenzene-1,3-disulfonate (dSC).

In the case of the strontium atom, a 2D polymer of formula $[[Sr(dSC)(H_2O)_4]H_2O]_n$ was obtained. The columnar motif found for the calcium polymer is preserved, and the various columns are fused to a 2D layered structure (Fig. 9.33). Each strontium atom is surrounded by four dSC ligands; one of them chelates using the catechol unit, while the other ligands are bonded through the oxygen atoms of the sulfonate groups. Again, each dSC ligand is bonded to four different strontium ions.

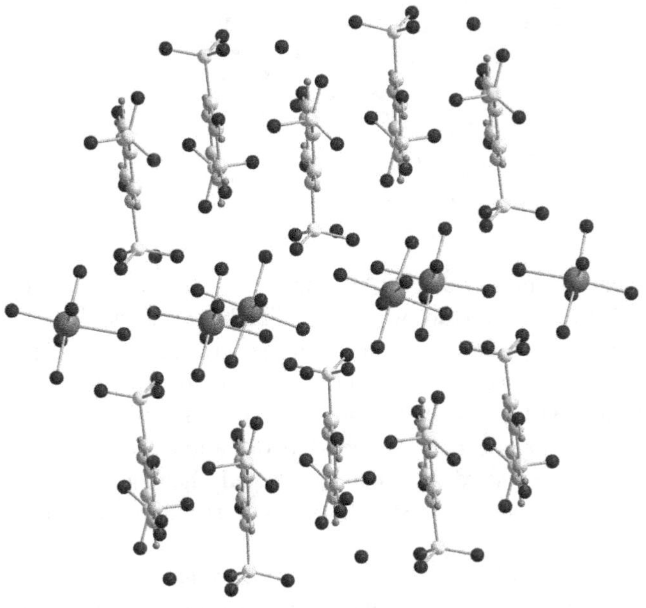

FIGURE 9.31 Packing diagram of $[Mg(H_2O)_6]$ (dSC) $3H_2O$.

Eventually, the $[[Ba(dSC)(H_2O)_4]H_2O]_n$ is a 3D CP (Fig. 9.34). Each barium atom is coordinated by five different dSC ligands. One of the latter bonds the metal ion either with the catechol oxygen atoms or one oxygen atom of the sulfonate group. Each molecule of the ligand connects six different metal atoms. The barium network is porous with channels of around 4Å in diameter filled by water solvent molecules.

The coordination of the alkaline earth metals goes from six in the case of Mg to seven for Ca, eight for Sr, and nine for Ba.

A different behavior was found in the case of the structures built using the 2,5 thiazole [5,4-d] thiazoledicarboxylate (Thz^{2-}; Scheme 9.4) [24]. They form different 1D CPs for magnesium, calcium, and strontium and a 2D CP in the case of Ba.

The different coordination of the Thz ligand in the 1D polymers with Mg, Ca, and Sr polymer is shown in Figure 9.35. The Thz ligand is bridging between two magnesium atoms forming in this way the 1D chain. The metal is surrounded by four oxygen atoms of four water molecules in the equatorial plane and by two oxygen atoms of the Thz ligands in the axial position. The uncoordinated oxygen atoms of the Thz ligands are engaged in strong hydrogen bonds with the coordinated water molecules.

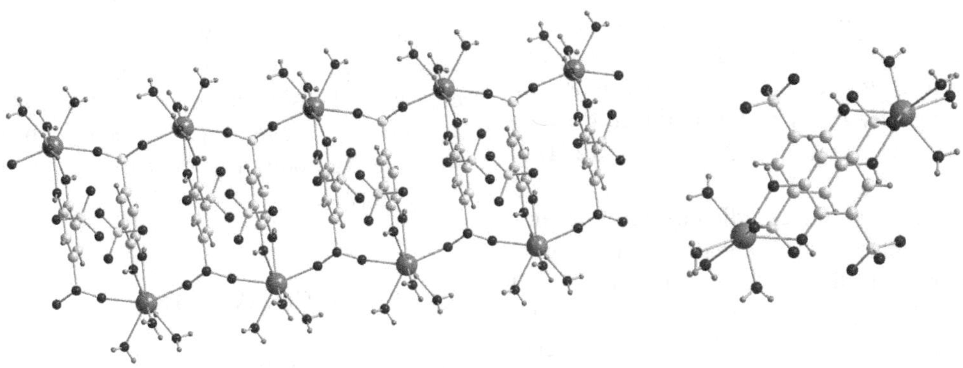

FIGURE 9.32 View perpendicular (left) and along (right) elongation axis of the Ca dSC 1D polymer.

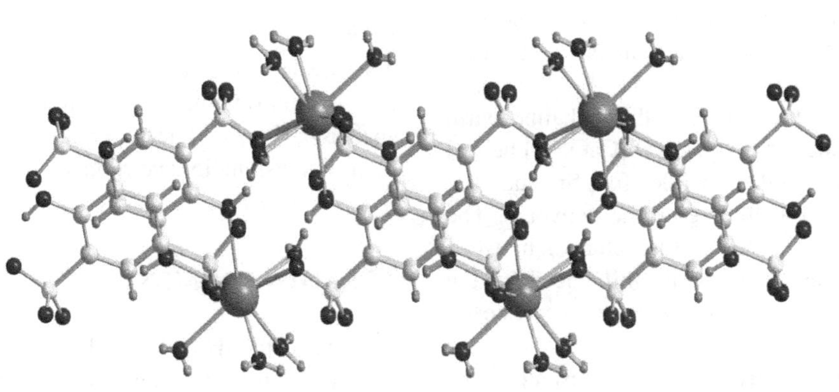

FIGURE 9.33 View of the 2D $[[Sr(dSC)(H_2O)_4]H_2O]_n$ layered structure.

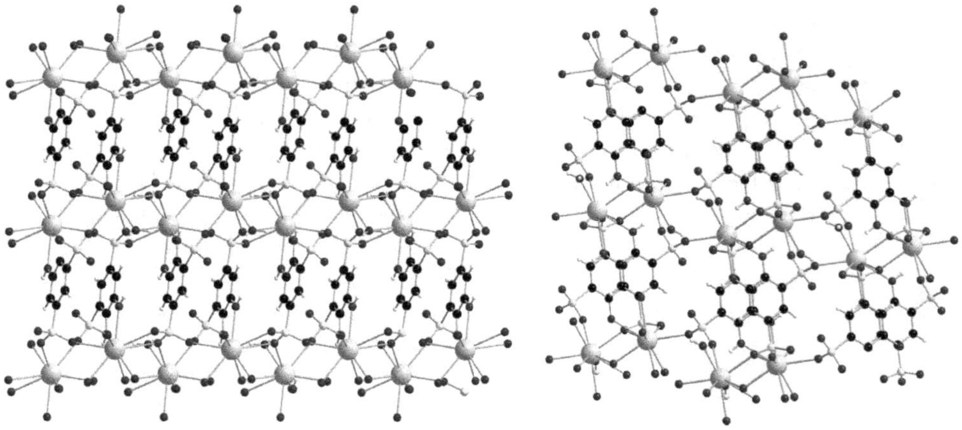

FIGURE 9.34 Two views of the 3D $[[Ba(dSC)(H_2O)_4]\ H_2O]_n$ network.

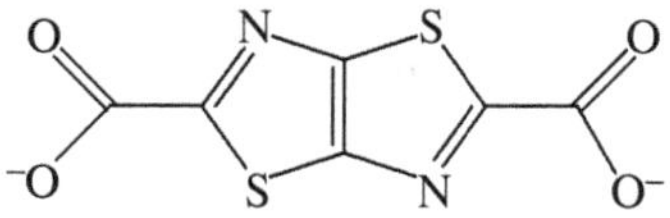

SCHEME 9.4 Schematic representation of 2,5 thiazole [5,4-d] thiazoledicarboxylate (Thz^{2-}).

The 1D CP with Ca^{2+} is built by the repetition of binuclear $Ca_2(OH_2)_8$ and Thz units. The latter ligand connects four different Ca atoms. To provide the charge balance, another Thz anion lies between the 1D chains.

With Sr as well as Ba ions, one of the Thz carboxylate groups is uncoordinated, while the nitrogen atom participates in coordination of the metal atoms. In the polymer backbone, the organic part of the ligand is not present, and the chain is exclusively formed by an extended inorganic connectivity with several water molecules bridged between the metal atoms (Fig. 9.36). As result, a 2D polymer $[Ba(Thz)(H_2O)_7]_n$ is formed.

As can be seen, the increasing coordination requirements of the metal ions from Mg to Ba lead to an apparent increasing of the network dimensionality and to different coordination and chelating modes of the ligand. A similar observation was made by Duan *et al.* [25] using methylene-diisophthalic acid (H_4MDIP) (Scheme 9.5); in this case, a 1D chiral chain $[[Mg(H_4MDIP)(H_2O)]_3H_2O]_n$ is formed. The Mg atom is bonded by four oxygen atoms of water molecules and two oxygen atoms of two different ligands, and the thus formed chains are linked by hydrogen bonds.

2D networks were found for the other alkaline earth metals. In the case of calcium cations, 1D $[CaO_2Ca]$ helical chains are connected together, while for Sr and Ba, [Sr—O—Sr] and [Ba—O—Ba] units glue the networks. The authors notice that the difference in the behavior between Mg and the other cations seems to be influenced by the ability of Mg ion to coordinate strong water molecules even in the presence of strong ligands.

A series of fourteen 3D structures was presented [26] using various aromatic dicarboxylates such as benzene-1,4 dicarboxylic acid, naphthalene-2,6 dicarboxylic acid, 4,5,9,10 tetrahydropyrene-2-7-dicarboxylic acid, and pyrene-2-7-dicarboxylic acid. In the series, the C—C separation between the carboxylate carbon atoms goes from 5.81 to 10.42 Å. In this case, the formation of 3D network is independent of the kind of cation used. Most of the structures exhibit diamond-shaped channels, while in the two cases, the 3D structures present an α-Po-type architectures.

Finally, a Cambridge Structural Database (CSD) search [27] gave 47 structures, which matched the skeleton structure of diphosphonate or diphosphinate ligand type (Scheme 9.6). The results (Table 9.2) show that the structure of Scheme 9.6 is not suitable for the building of 3D networks with alkaline earth metals, but the network formation seems to be the favorite with ions of large radius. In conclusion, the correlation between network dimensionality of the obtained networks and the metal radius seems to be one very difficult to prove, and only in some cases, a direct correlation was reported [23].

9.6 FINAL COMMENTS

The influence of non-covalent interactions, such as hydrophobic, π···π or CH···π stacking, halogen bonding interactions, etc., on dimensionality, porosity, stability, and reactivity of CPs and MOFs was discussed. It was found that non-covalent interactions in the assembly of selected COs based on phosphonates, phosphinates, and auxiliary ligands, such as 4,4′-bipyridine, piperazine, etc., play a crucial role in their dimensionality and properties.

ACKNOWLEDGMENTS

We thank the HYDROLAB2 project of Ente Cassa di Risparmio di Firenze and MIUR—Project FIRB 2010 No. RBFR10CWDA_003 for the financial support.

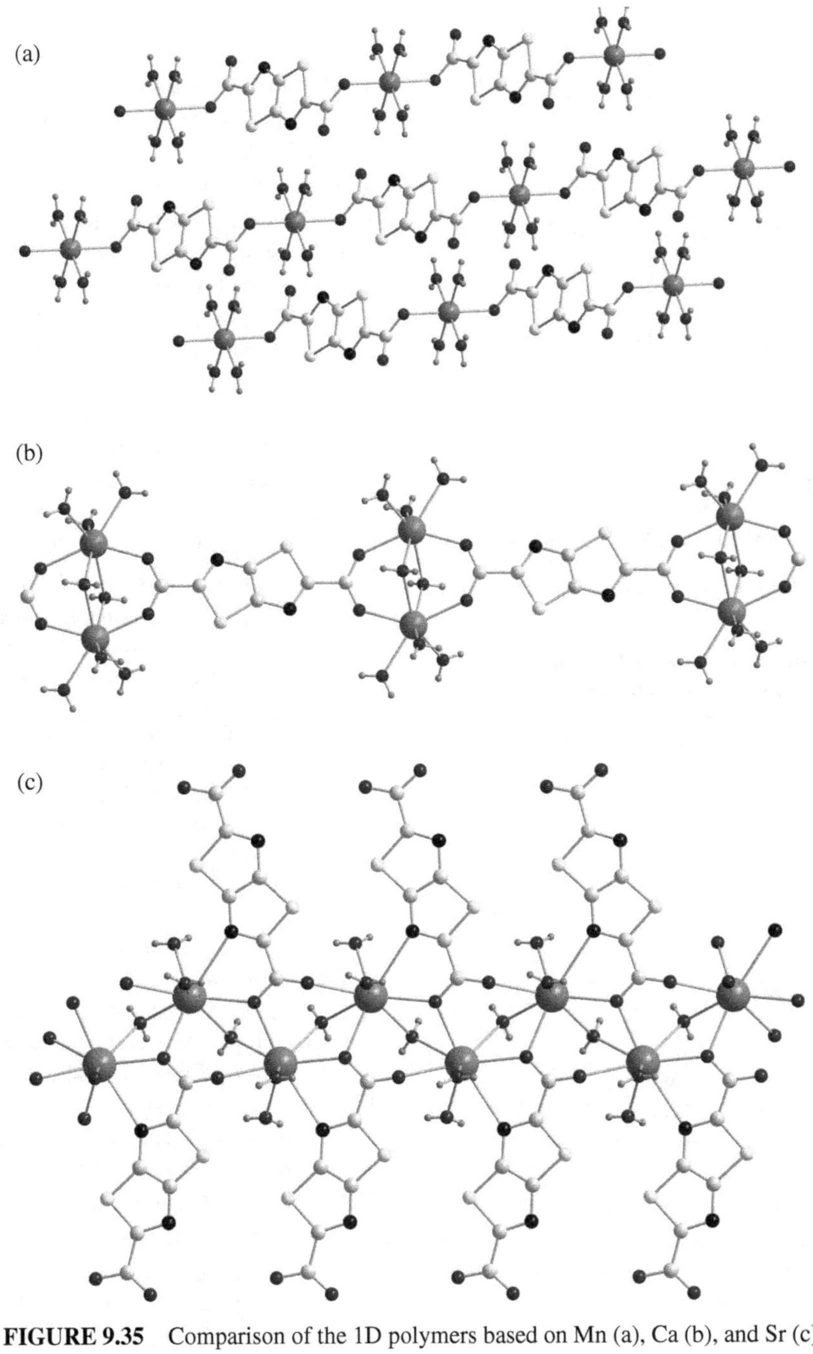

FIGURE 9.35 Comparison of the 1D polymers based on Mn (a), Ca (b), and Sr (c).

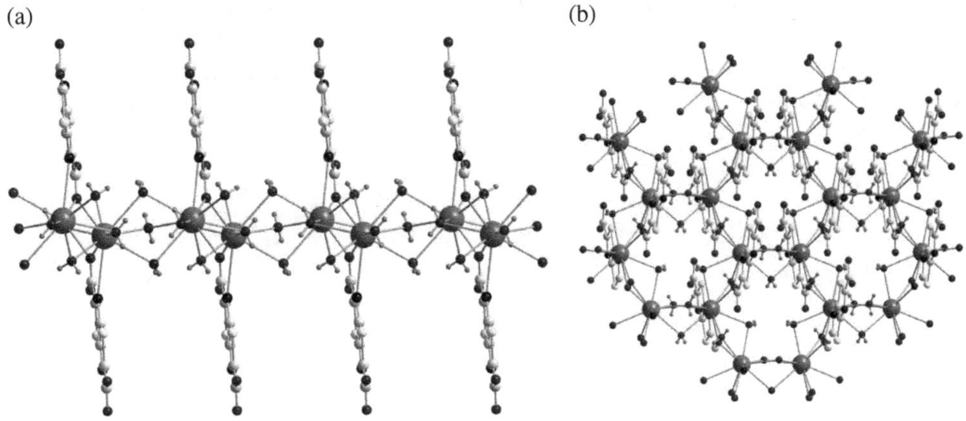

FIGURE 9.36 Two views of the 2D $[Ba(Thz)(H_2O)_7]_n$ coordination polymer.

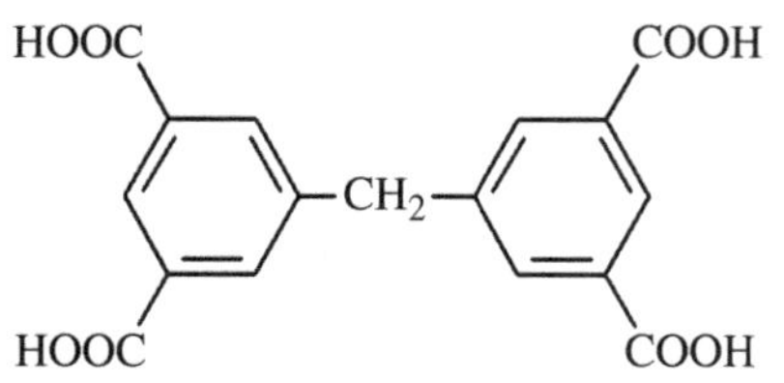

SCHEME 9.5 Schematic representation of methylenediisophthalic acid (H_4MDIP).

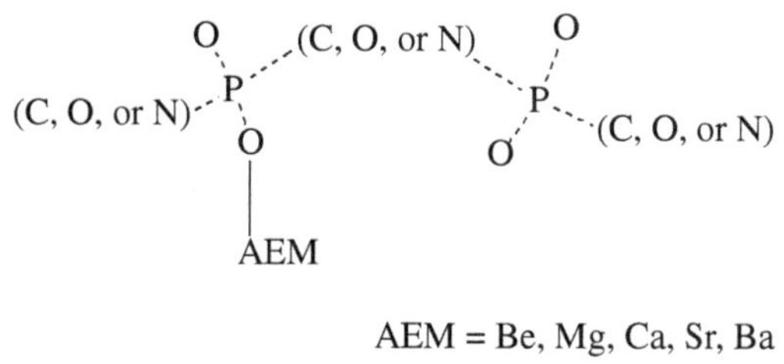

SCHEME 9.6 Skeleton of the structure used as query in the CSD.

TABLE 9.2 Overview of the Results of the CSD Search Using as Query the Fragment Shown in Scheme 9.6

	Hits	0D	1D	2D	3D
Be	1	1	0	0	0
Mg	15	5	8	1	1
Ca	14	2	7	4	0
Sr	10	2	4	4	0
Ba	8	0	4	4	0
Total	47	10	23	13	1

REFERENCES

1. (a) J. R. Long, O. M. Yaghi, *Chem. Soc. Rev.* 2009, *38*, 1213; (b) S. Kitagawa, R. Kitaura, S. Noro, *Angew. Chem. Int. Ed.* 2004, *43*, 2334.
2. S. Kitagawa, K. Uemura, *Chem. Soc. Rev.* 2005, *34*, 109.
3. R. Vivani, F. Costantino, M. Taddei, Zirconium Phosphonates, In *Metal Phosphonate Chemistry: From Synthesis to Applications*, Eds., A. Clearfield, K. Demadis; RSC: Oxford, 2011.
4. G. Alberti, U. Costantino, F. Marmottini, R. Vivani, P. Zappelli, *Angew. Chem. Int. Ed.* 1993, *32*, 1357.
5. U. Costantino, M. Nocchetti, R. Vivani, *J. Am. Chem. Soc.* 2002, *124*, 8428.
6. R. Vivani, G. Alberti, F. Costantino, M. Nocchetti, *Microporous Mesoporous Mater.* 2008, *107*, 58.
7. M. Taddei, R. Vivani, F. Costantino, *Dalton Trans.* 2013, *42*, 9671.
8. M. Taddei, F. Costantino, R. Vivani, *Inorg. Chem.* 2010, *49*, 9664.
9. F. Costantino, M. Taddei, P. Sassi, M. Geppi, *Cryst. Growth Des.* 2012, *12*, 5462.
10. T. R. Cook, Y.-R. Zheng, P. J. Stang, *Chem. Rev.* 2013, *113*, 734.
11. M. Taddei, F. Costantino, R. Vivani, C. Sangregorio, L. Sorace, L. Castelli, *Cryst. Growth Des.* 2012, *12*, 2327.
12. T. M. Smith, J. Vargas, D. Symester, M. Tichenor, C. J. O'Connor, J. Zubieta, *Inorg. Chim. Acta* 2013, *403*, 63.
13. F. Costantino, A. Ienco, S. Midollini, *Cryst. Growth Des.* 2010, *10*, 7.
14. T. Bataille, S. Bracco, A. Comotti, F. Costantino, A. Guerri, A. Ienco, F. Marmottini, *CrystEngComm* 2012, *14*, 7170.
15. M. Taddei, A. Ienco, F. Costantino, A. Guerri, *RSC Adv.* 2013, *3*, 26177.
16. R. Hoffmann, *Angew. Chem. Int. Ed.* 1982, *21*, 711.
17. J.-M. Lehn, *Supramolecular Chemistry*; VCH: Weinheim, 1995.
18. (a) L. Álvarez, A. Grirrane, R. Moyano, E. Álvarez, A. Pastor, A. Galindo, *Polyhedron* 2010, *29*, 3028; (b) A. Grirrane, A. Pastor, E. Álvarez, C. Mealli, A. Ienco, P. Rosa, A. Galindo, *Eur. J. Inorg. Chem.* 2007, 3543; (c) D. del Río, A. Galindo, R. Vicente, C. Mealli, A. Ienco, D. Masi, *Dalton Trans.* 2003, 1813; (d) A. Grirrane, A. Pastor, E. Álvarez, C. Mealli, A. Ienco, D. Masi, A. Galindo, *Inorg. Chem. Commun.* 2005, *8*, 463; (e) A. Grirrane, A. Pastor, E. Álvarez, R. Moyano, A. Galindo, *Inorg. Chem. Commun.* 2007, *10*, 1125; (f) A. Grirrane, A. Pastor, A. Ienco, C. Mealli, A. Galindo, *J. Chem. Soc. Dalton Trans.* 2002, 3771.
19. A. Grirrane, A. Pastor, A. Galindo, A. Ienco, C. Mealli, P. Rosa, *Chem. Commun.* 2003, 512.
20. A. Grirrane, A. Pastor, A. Galindo, D. del Río, A. Orlandini, C. Mealli, A. Ienco, A. Caneschi, J. Fernández Sanz, *Angew. Chem. Int. Ed.* 2005, *44*, 3429.
21. A. Grirrane, A. Pastor, A. Galindo, D. del Río, E. Alvarez, C. Mealli, A. Ienco, A. Orlandini, P. Rosa, A. Caneschi, A.-L. Barra, J. Fernández Sanz, *Chem. Eur. J.* 2011, *17*, 10600.
22. J. C. Slater, *J. Chem. Phys.* 1964, *39*, 3199.
23. A. P. Côté, G. K. H. Shimizu, *Chem. Eur. J.* 2005, *9*, 5361.
24. E. H. L. Falcao, Naraso, R. K. Feller, G. Wu, F. Wudl, A. K. Cheetham, *Inorg. Chem.* 2008, *47*, 8336.
25. X. Duan, J. Lin, Y. Li, C. Zhu, Q. Meng, *CrystEngComm* 2008, *10*, 207.
26. C. A. Williams, A. J. Blake, C. Wilson, P. Hubberstey, M. Schroder, *Cryst. Growth Des.* 2008, *8*, 911.
27. F. Costantino, P. L. Gentili, A. Guerri, A. Ienco, S. Midollini, W. Oberhauser, *Inorg. Chim. Acta* 2012, *391*, 150.

PART III

CRYSTAL DESIGN AND HOST–GUEST COMPOUNDS

10

DIHALOGENS AS HALOGEN BOND DONORS

MATTI HAUKKA[1], PIPSA HIRVA[2], AND KARI RISSANEN[1,3]
[1] *Department of Chemistry, University of Jyvaskyla, Jyvaskyla, Finland*
[2] *Department of Chemistry, University of Eastern Finland, Joensuu, Finland*
[3] *Nanoscience Center, University of Jyvaskyla, Jyvaskyla, Finland*

10.1 INTRODUCTION

According to its definition, "a halogen bond occurs when there is evidence of a net attractive interaction between an electrophilic region associated with a halogen atom in a molecular entity and a nucleophilic region in another, or the same, molecular entity" [1]. This means that a halogen atom is acting as an electrophile. As such, it can form an attractive interaction with a nucleophilic counterpart, that is, with a Lewis base. The nucleophilic halogen-containing molecule, R_1—X, is called halogen bond donor and the Lewis acid, Y—R_2, halogen bond acceptor. The terminology is analogous with the terminology of hydrogen bonds [2]. In hydrogen bonds the hydrogen atom is the electrophile interacting with an electron donor. In a halogen bond donor, R_1—X, the R_1 group can be an organic moiety or another halogen atom. Typically, the acceptor Y is a good electron donor such as nitrogen, sulfur, or π-electron-rich aromatic ring, but it can also be another halogen atom [3]. A true halogen bond (XB) is highly directional (Fig. 10.1) [3]. However, because of the vast number of crystal structures with relatively short halogen–halogen distances and somewhat different contact geometries, the halogen–halogen contacts have been divided into two classes, type I and type II (Fig. 10.1) [3, 4, 7].

The IUPAC definition sets limitations to the geometry of a halogen bond. The R_1—X···A angle should be *ca.* 180° [1]. This means that only the type II halogen–halogen contact should be defined as halogen bond (Fig. 10.1) [4]. The geometric restriction arises strictly from the definition of halogen bond, that is, there should be a contact between the electrophilic region of the halogen on R_1—X and the electrophile. This type of interaction is most commonly rationalized by the σ-hole model. In order to act as an electrophile, the halogen atom must be polarized by the R_1 moiety. Due to polarization, the electron density around the halogen atom is redistributed in such a way that an electron-deficient area is formed opposite to the covalent R—X bond. In the center of this area a small positive region called a σ-hole (Fig. 10.2) appears [3]. The appearance and magnitude of the σ-hole depend essentially on the polarizability of the halogen atom and the electron-withdrawing ability of the R_1 group [3].

In the case of halogen–halogen bond, the electron-deficient area of the halogen bond donor is interacting with the electron-rich area of another halogen leading to type II geometry (Fig. 10.2).

Even if type I halogen interactions and other short halogen–halogen contacts should not be called halogen bonds, this does not mean that they are insignificant. Such interactions may still play an important role in crystal packing and in self-assembly of molecules. Over the last decades a number of papers have been devoted to analysis of these short halogen–halogen contacts [5, 6, 8–20]. However, often other than type II contacts are simply related to the attempts to minimize electronic repulsion between the closely packed atoms by interfacing neutral regions of the polarized atoms [7]. In its extreme form this would mean head-to-head approach of the σ-holes of the halogen atoms.

10.2 TYPICAL HALOGEN BOND DONORS

As stated above, the capability of a halogen atom to act as halogen bond donor is dependent on its polarizability. The halogen–halogen interaction strength increases in the order

Non-Covalent Interactions in the Synthesis and Design of New Compounds, First Edition.
Edited by Abel M. Maharramov, Kamran T. Mahmudov, Maximilian N. Kopylovich, and Armando J. L. Pombeiro.

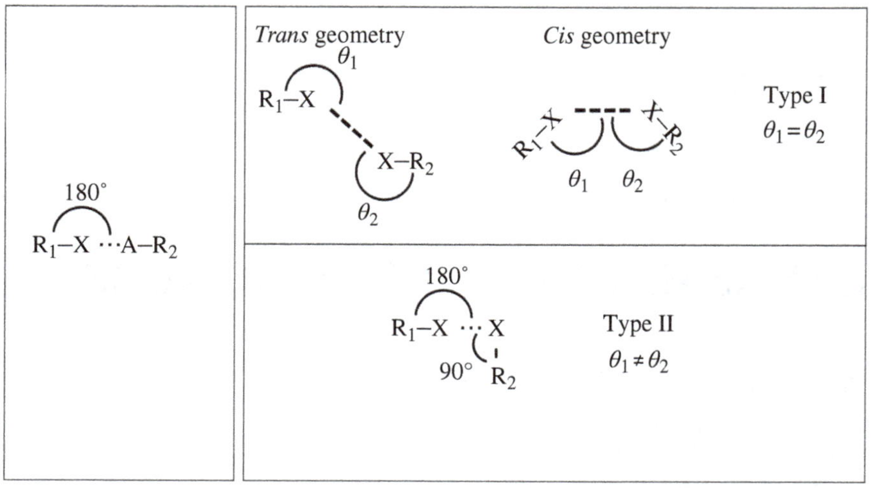

FIGURE 10.1 Left: a true halogen bond involving halogen as XB donor and a heteroatom as halogen bond acceptor. Up right: type I halogen contact between two halogen atoms. Down right: type II halogen bond between two halogen atoms. Only type II is considered as a true halogen bond [4–6].

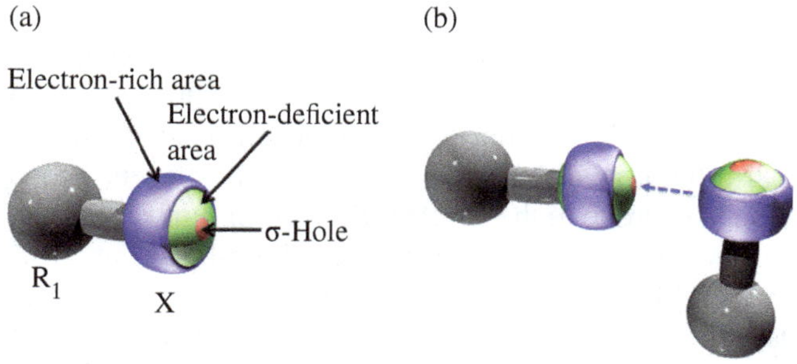

FIGURE 10.2 (a) A schematic representation of a polarized halogen atom in a R_1—X molecule. (b) Type II halogen bond between two halogen atoms.

F···F<Cl···Cl<Br···Br<I···I following the order of the electric dipole polarizabilities [5, 6] and ionic sizes of F, Cl, Br, and I. The fluorine atom is least likely to be strongly polarized, and therefore it is less likely to act as halogen bond donor. In R_1—X, the R_1 group can be another halogen or an organic group but usually not a metal ion. Metal ions are rarely able to polarize halides strongly enough to generate a true σ-hole. It has been suggested that one of the reasons for this is the back-bonding pushing electrons back from the d-orbitals of a metal atom to the p-orbitals of the halogen [21]. When the electron-attracting properties of the R_1 group are increasing, the polarization of X is also enhanced. It has been suggested that eventually even fluorine can be polarized with strongly electron-withdrawing groups [20, 22, 23]. Because the R_1 group should be electron withdrawing, typical halogen bond donors are often fluorinated organic molecules such as CF_3X or fluorinated iodobenzenes. As mentioned above, dihalogens, both homonuclear and heteronuclear, can also act as XB donors [24–31]. Similarly, polyhalogens can also be involved in halogen bonds as XB donors [32]. In the case of homonuclear dihalogens, usually only I_2 and Br_2 are suitable for halogen bonds since both Cl_2 and F_2 are most often far too reactive to be used for this purpose. Based on the crystallographic database analysis and the interaction energies, the halogen bond donor strength of dihalogens follows the order $F_2<Cl_2<Br_2<I_2<IBr<ICl$ independently of the nature of the XB acceptor [31, 33, 34].

In principle, homonuclear X_2 molecules can act as halogen-bonded bridge through both of its halogen atoms. Whether this actually happens depends on the strength of the first XB contact. The strong interaction with the first halogen changes the contact preference of the second halogen due to the charge transfer effect discussed later. Heteronuclear XY dihalogens typically form bonds only through the less electronegative halogen atom. Since the electron density is accumulated on the more electronegative halogen atom, its capability to act as electrophile is strongly reduced. It may even be a nucleophile seeking electrophilic counterparts, that is, acting as XB acceptor.

10.3 TYPICAL HALOGEN BOND ACCEPTORS

The most common XB acceptor atoms are N, S, and O. All of these contain free electron pairs that can easily interact with the electrophiles. Simple Lewis base molecules such as NH_3 or pyridine are examples of excellent halogen bond donors. The acceptors can either be free organic molecules, anions, or ligands coordinated to a metal center as long as the acceptor atom is available for XB. For example, carbon-coordinated cyanide ligands [35] or nitrogen-coordinated thiocyanates can be involved in halogen bonding networks as acceptors (Fig. 10.3) [36, 37]. Both these ligands can use not only the free electron pairs of S, C, or N atoms but also the π-electrons of the triple bonds to form halogen bonds. This is reflected in the geometry of the formed adduct. The *N*-coordinated thiocyanate is able to use both of its free electron pairs on sulfur to form halogen bonds (Fig. 10.3). In the case of metal cyanates of Cr, Fe, and Co with *N*-methyl-3-halopyridinium and *N*-methyl-3,5-dihalopyridinium cations as XB donors, interaction through nitrogen lone pair results in nearly linear system with C—N···X angle >165°, while interaction through π-electrons gives angle closer to 90° (<105°) [35].

10.4 BASIC FEATURES OF DIHALOGENS

The properties of dihalogens have been thoroughly studied over the years both experimentally and computationally. However, for this contribution we carried out a topological analysis of the charge density in homonuclear dihalogens Cl_2, Br_2, and I_2 as well as heteronuclear dihalogens ICl and IBr to illustrate the key features behind the XB behavior of dihalogens. The results are summarized in Figures 10.4 and 10.5. The covalent nature of a bond can be estimated by analyzing the $|V|/G$ ratio of the bond (V is the potential energy density and G the kinetic energy density at the bond critical point (BCP)) [41]. In pure electrostatic interactions the ratio is typically <1 and in pure covalent bonds two or more [42]. The values between one and two are indicative to bonds with both electrostatic and covalent nature. Although the halogen–halogen bonds are dominatingly covalent in all analyzed dihalogens, the interaction energies (E_{int}) vary. Among the homonuclear dihalogens the bond strength decreases in the order of decreasing electronegativity of the halogens (Fig. 10.4). This is mainly due to reduced electron density at the BCP, which are located in the middle of the molecular axis.

The impact of the electronegativity difference can be seen in the Laplacian maps of the electron density for interhalogen

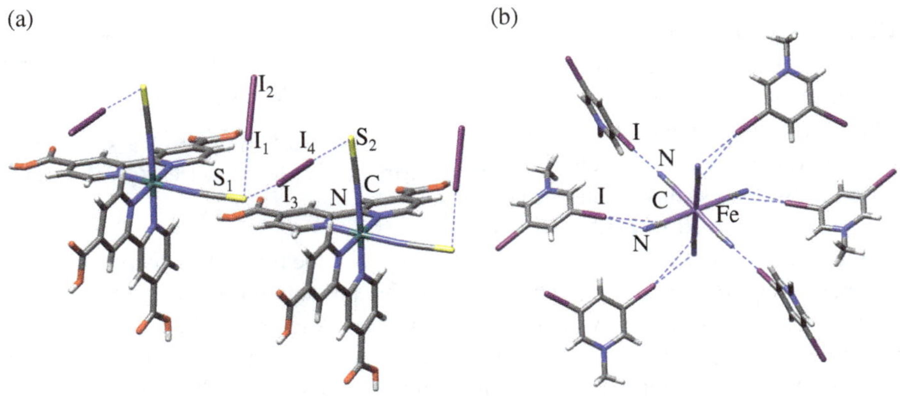

FIGURE 10.3 (a) Halogen bonds involving metal-coordinated thiocyanate. $S_1\cdots I_1$: 2.836(3)Å, $S_1\cdots I_3$: 3.531(2)Å, $S_2\cdots I_4$: 2.954(3)Å, $I_1\cdots I_2$: 2.777(1)Å, $I_3\cdots I_4$: 2.7669(9)Å, $I_1\cdots S_1\cdots I_3$ 78.31(6)° [36]. (b) Halogen bonds between $[Fe(CN)_6]^{3-}$ and *N*-methyl-3,5-diiodopyridinium cation [35].

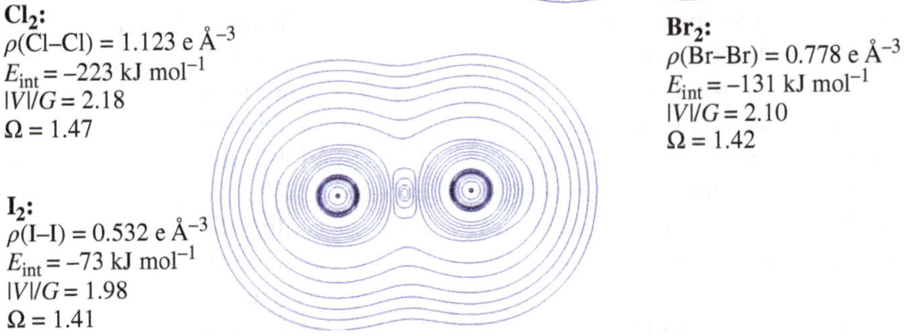

FIGURE 10.4 Laplacian of the electron density in dihalogens Cl_2, Br_2, and I_2 according to the Quantum Theory of Atoms in Molecules (QTAIM) [38]. The wavefunctions were calculated at the PBE0/def2-TZVPPD level of theory [39, 40]. Calculated X—X distances are 1.981, 2.279, and 2.654 Å for Cl_2, Br_2, and I_2, respectively.

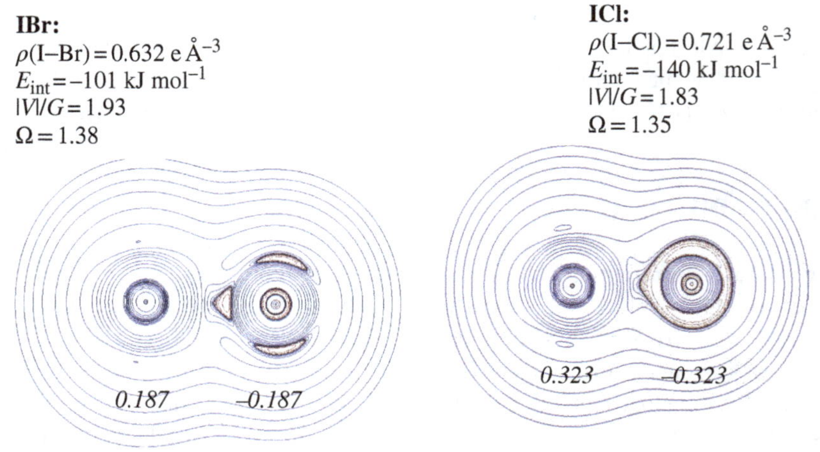

FIGURE 10.5 Laplacian of the electron density in interhalogens IBr and ICl according to the Quantum Theory of Atoms in Molecules (QTAIM) [38]. Wavefunctions were calculated at the PBE0/def2-TZVPPD level of theory [39, 40].

compounds IBr and ICl (Fig. 10.5). First of all, the interaction energy is dependent on the electronegativity difference. As expected, the electron density is concentrated on the more electronegative atom both in IBr and ICl, leaving positive net charge on iodine. Because of the positive charge, iodine is more willing to interact with electron donors; this means stronger halogen bonding interactions compared to I_2 [26, 43, 44]. Simultaneously with the increasing polarization, the electron density at the BCP is increasing (ρ_{BCP} = *c.a.* 0.5, 0.6, and 0.7 e$Å^{-3}$ in I_2, IBr, and ICl, respectively), which also contributes to the strength of the interaction. At the same time, the covalency of the X—Y bond is slightly decreasing in the order I_2 > IBr > ICl, as can be seen in the reducing values of $|V|/G$ and the delocalization index Ω that describes the average number of shared electrons between the interacting atoms at the BCP.

10.5 CHARGE TRANSFER AND DIHALOGENS

The halogen bonds are essentially electrostatic contacts, but other factors such as polarization, dispersion forces, and charge transfer may also play an important role. The relative impact of these forces varies from case to case depending on the nature of the XB donor and acceptor [1]. The effect of charge transfer on halogen bonding has been studied widely both theoretically and experimentally, and in many cases it is the key to strong halogen bonds [37, 45–49]. In the case of dihalogens the impact of charge transfer can be quite drastic. The charge transfer interaction can be rationalized as mixing of the HOMO orbital of the electron donor with the empty σ^*-orbital of X_2 [50, 51]. This results in a weakening and lengthening of the X—X bond. If the charge transfer effect is strong, it can also change the internal charge distribution of the dihalogen molecule. When a homonuclear dihalogen, X_A—X_B, is approaching to an electrophilic region of a halogen bond acceptor with its X_A end, the negative charge accumulates on the X_B end. Simultaneously, the partial

Increasing charge transfer →

$$R_2{-}A\cdots X_A^{+\delta}{-}X_B^{-\delta} \qquad R_2{-}A\cdots X_A^{+} + X_B^{-}$$

SCHEME 10.1 Impact of a strong charge transfer on charge distribution of homonuclear X_2 molecule (see the text for details).

charge of the interacting halogen, X_A, becomes less negative. If the interaction is strong enough, X_A can become positively charged, resembling the charge distribution in heteronuclear dihalogens. In the extreme case, a very intense charge transfer can eventually cause a heterolytic cleavage of the X_A—X_B bond, releasing of X_B^- and leaving, at least formally, a halonium ion, X_A^+, attached to the halogen bond acceptor, A (Scheme 10.1) [51].

Usually the R_2—X_A^+ complexes are not stable. It is more common that heterolytic cleavage leads to formation of R_2—X_A^+–R_2 systems, where halonium cation is acting as a bridging atom. Formation of $[X(py)_2]^+$ (X = I, Br) is one of the classic examples of this type of process (Fig. 10.6) [52–58]. Because of the strong interaction between the pyridine molecule and I_2, the process easily goes all the way to the halonium stage and even isolation of py-I_2 can be challenging (Fig. 10.6).

The N—I distance in py···I_2 is short, only 2.425(8) Å [52]. This is considerably shorter than the sum of van de Waals radii of nitrogen and iodine (3.53 Å). The lengthening of the I—I bond is also noticeable. The I—I distance of a free I_2 molecule in gas phase is 2.654 Å (Fig. 10.4) and in solid state 2.720(8) Å [59], while in py···I_2 it is extended to 2.8043(9) Å.

Even if the X^+ cations are described as halogen bond donors according to the IUPAC definition [1], halonium ions have quite unique properties. First of all, they are capable of forming unusually strong halogen bonds. Because of this, Lewis acid–I^+ systems have often been interpreted as covalent bonds [51]. On the other hand, Erdélyi *et al.* have analyzed N—I^+—N systems both experimentally and computationally and described such contacts as 4-electron 3-center

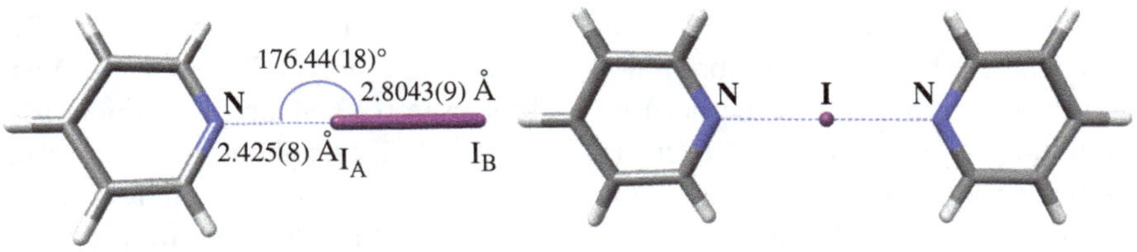

FIGURE 10.6 Crystal structure of Py···I_2 adduct (a) [52] and the cationic $[I(py)_2]^+$ (b). The structure of $[I(py)_2]^+$ was taken from Ref. [53]. The N···I distances in this crystal structure range from 2.255(3) Å to 2.261(3) Å and the N···I···N angle from 177.66(12)° to 180°. The counteranion BF_4^- is omitted for clarity.

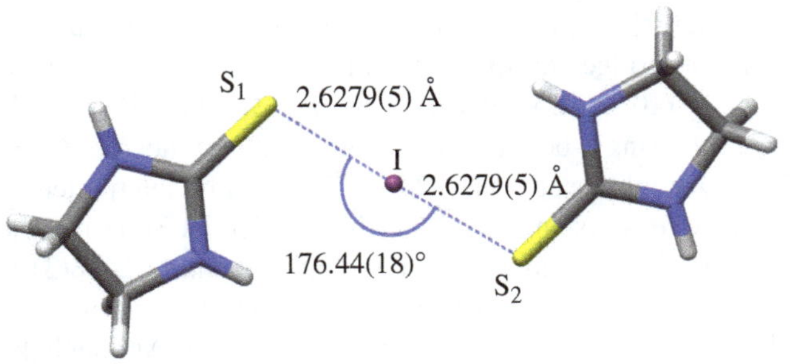

FIGURE 10.7 I^+ bridging two 2-imidazolidinethione units [41]. The counteranions (I^- and I_3^-) have been omitted for clarity.

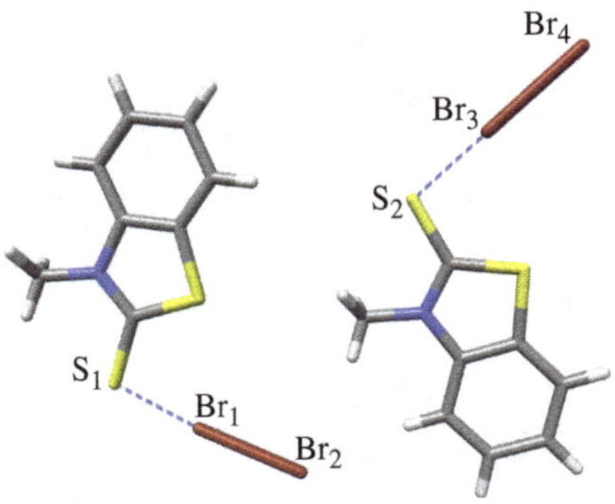

FIGURE 10.8 Structure of [(*N*-methylbenzothiazole-2-thione)Br_2]. The asymmetric unit contains two independent [(*N*-methylbenzothiazole-2-thione)Br_2] systems. The slight differences are due to the packing. The chloroform of crystallization has been omitted for clarity [75]. S_1···Br_1: 2.3244(8) Å, Br_1—Br_2: 2.6673(4) Å, S_1···Br_1—Br_2 174.54(2)°, S_2···Br_3: 2.2931(7) Å, Br_3—Br_4: 2.7504(4) Å, S_2···Br_3—Br_4 178.39(2)°.

bonds [60–64]. This description emphasizes the fact that there is interplay between the XB donors over the I^+ bridge and the positive charge tends to be distributed over the whole N—I^+—N system.

In addition to nitrogen, sulfur is also a powerful charge transfer donor/XB acceptor for diiodine [50]. Systems with bridging I^+ have been synthesized by using various sulfur acceptors [65–71]. A bridging S—I^+—S system with 2-imidazolidinethione provide an example of this type of XBs (Fig. 10.7) [41].

The computational QTAIM [38] analysis of the $[I(2\text{-imidazolidinethione})_2]^+$ system indicates that the S—I^+ contact is not purely electrostatic but have also considerable covalent nature. This is typical for strong halogen bonds involving strong charge transfer effect. The $|V|/G$ ratio is 1.58, indicating the dual electrostatic/covalent nature of the contact. The delocalization index Ω of 0.79 also indicates notable covalency for the S—I^+ contact. The geometry of $[I(2\text{-imidazolidinethione})_2]^+$ resembles closely the geometries of similar metal-bridged complexes of Ag^+, Au^+, Cu^+, and Hg^{2+} [72–74]. Because of the structural similarities, it has been suggested that the S—I^+—S type of halogen bonds could also be called as "halogen bonds with coordinative nature" [41].

10.6 CASE STUDIES

10.6.1 Manipulating Homonuclear X—X Bonds by Thione XB Acceptors

Charge transfer plays a key role in halogen bonds involving X_2 donors. Therefore, by adjusting the electron-donating nature of the XB acceptor, it is possible to control the strength/length of the X—X bond and the charge distribution within the X_2 molecule. This can be clearly seen in a series of three thione-Br_2 adducts, where the Br—Br bond is systematically tuned by altering the electron-donating nature of the thione [75]. The electron-donating capabilities among the studied thione acceptors increase in the order *N*-methylbenzothiazole-2-thione < 2(3*H*)-benzothiazolethione < 2-benzimidazolethione (Figs. 10.8 and 10.9).

The impact of the charge transfer can already be seen with the weakest XB acceptor *N*-methylbenzothiazole-2-thione. The computational Br—Br distance in a free gaseous Br_2 is 2.279 Å (Fig. 10.4), while in [(*N*-methylbenzothiazole-2-thione)Br_2] (Fig. 10.8) the Br—Br distance has been extended to 2.6673(4)–2.7504(4) Å. With stronger acceptor [(2(3*H*)-benzothiazolethione)Br_2] (Fig. 10.9a), the Br_A—Br_B distance is extended further to 2.8141(7) Å, and finally with 2-benzimidazolethione (Fig. 10.9b) the Br—Br bond is broken and the covalent Br—Br bond has been replaced with a halogen bond between S—Br_A^+ halogen bond donor and the anionic bromide acceptor. The bromide is attached to another Br_2 forming Br_3^- anion (Fig. 10.9b). Simultaneously, with the extending Br—Br bond, the S—Br_A distance is shortening from 2.3244(8) Å to 2.2827(13) Å and finally to 2.1824(7) Å [75]. In the first two cases the negative charge of the noninteracting bromine, Br_B, is increasing slightly from −0.46 to −0.50, while the charge of the interacting Br_A remains almost unchanged and close to 0 until the breakage of the Br—Br bond. The bond cleavage leads to formation of a bromium cation with small positive charge. In practice, the positive charge of the formal bromium cation is distributed over the whole [(2-benzimidazolethione)Br]$^+$ molecule.

The S···Br_2 interactions are strong and partially covalent: the $|V|/G$ ratios for S···Br_A (S_1···Br_1 in Fig. 10.8) contacts are 1.69 and 1.76 for [(*N*-methylbenzothiazole-2-thione)Br_2] and [(2(3*H*)-benzothiazolethione)Br_2], respectively. For comparison, the $|V|/G$ ratios for Br—Br bonds in these adducts are only 1.37 and 1.24. When the Br—Br bond is broken in [(2-benzimidazolethione)Br][Br_3], the $|V|/G$ ratio for the closest neighboring bromines reduces to 0.9, which is already a typical value for purely electrostatic contact. At the same time, the $|V|/G$ ratio for the S—Br is increased to 2.19, which could be interpreted as a covalent bond.

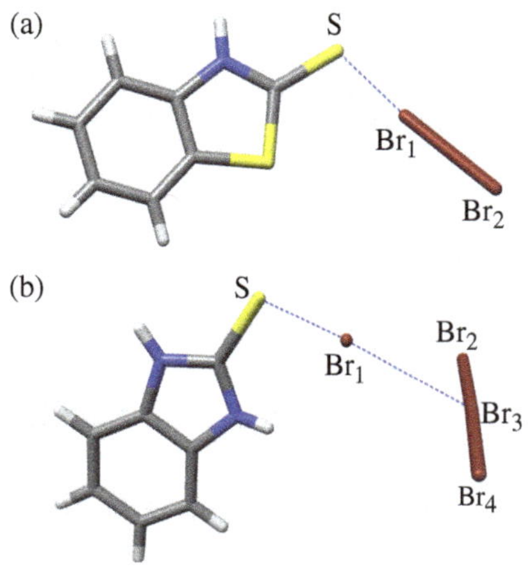

FIGURE 10.9 The halogen bond contacts in [(2(3*H*)-benzothiazolethione)Br_2] (a) and [(2-benzimidazolethione)Br][I_3] (b). (a) S···Br_1: 2.2827(13)Å, Br_1···Br_2: 2.8141(7)Å. (b) S···Br_1: 2.1824(7)Å, Br_1···Br_3: 3.2395(4)Å [75].

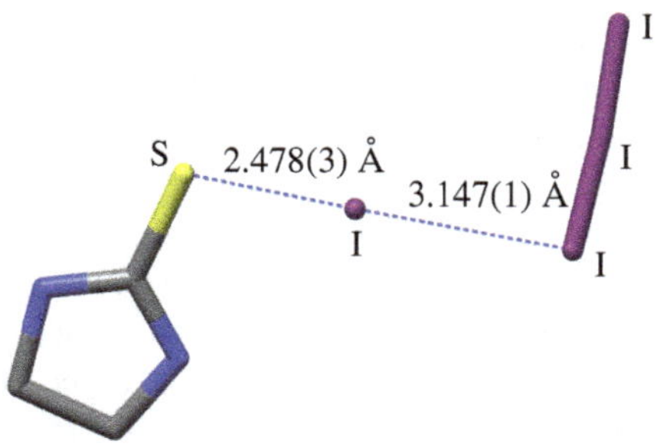

FIGURE 10.10 Ethylenethiourea with I$^+$ and I_3^- [76].

The breakage of the Br—Br bond in [(2-benzimidazolethione)Br][Br_3] is even further supported by the minimal electron sharing (Ω=0.21) and small interaction energy (−11 kJ mol^{-1}) between the neighboring bromines [75]. In [(2-benzimidazolethione)Br]$^+$, the formal halonium cation is not part of a symmetrical bridge as the iodonium in the case of [py_2I]$^+$ (Fig. 10.6b) or [2-imidazolidinethione)$_2$I]$^+$ (Fig. 10.7). This type of "single-bonded" halonium ions are not common. However, ethylenethiourea-2(I_2) [76] could be interpreted as an analogous system for [(2-benzimidazolethione)Br][I_3]. The S···I in ethylenethiourea-2(I_2) is relatively short, 2.478(3)Å. The I···I distance between the sulfur bound iodine and the closest neighboring iodine is already 3.147(1)Å. Although the system has been reported to consist of two interacting I_2 molecules, it is more likely that it is closer to S···I$^+$···I_3^- system (Fig. 10.10). This seems plausible despite the fact that in such a case the I_3^- would not be completely symmetrical. The longer I···I distance is 3.004Å and the shorter 2.850Å.

In thione-Br_2 systems, the increased negative charge on the noninteracting Br_B atom makes it nucleophilic. This can be seen when the packing of [(*N*-methylbenzothiazole-2-thione)Br_2] and [(2(3*H*)-benzothiazolethione)Br_2] is analyzed (Fig. 10.11). In both systems the negatively charged bromine is no longer eager to act as a halogen bond donor; instead, it is taking part in weak halogen bonds and weak hydrogen bonds as halogen bond or hydrogen bond acceptor.

10.6.2 Interactions of Heteronuclear Dihalogens, XY, with Thione XB Acceptors

Heteroatomic dihalogens, XY, are polar by nature (Fig. 10.5) and this has an impact on their XB donor properties. Concentration of the negative charge on the more

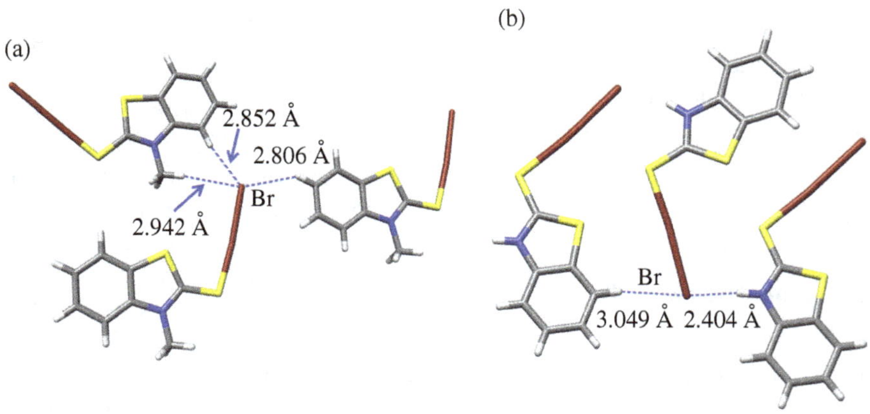

FIGURE 10.11 Packing of [(*N*-methylbenzothiazole-2-thione)Br_2] (a) and [(2(3H)-benzothiazolethione)Br_2] (b) [75].

electronegative halogen, Y, reduces its σ-hole. The Y is then less prone to act as XB donor and more likely to act as a nucleophile. Therefore, only the less electronegative halogen X is typically the XB donor atom. The best choice for XB donor would be heteronuclear dihalogen with iodine as the more electropositive atom.

Only few papers have been published on thione or selone acceptors [77–81] and iodine-containing heteronuclear dihalogens. However, it has been suggested that change in the XB donor character, when the Y is changed, does have impact on both reaction mechanisms and the structures of the final products [78–80].

The crystal structure of [(mbtt)ICl] (mbtt = *N*-methylbenzothiazole-2-thione) has been determined both at room temperature [80] and at 120 K [77]. However, the temperature-dependent structural differences are only minor. The corresponding mbtt structures with heteronuclear IBr and homonuclear I_2, that is, [(mbtt)IBr] [77] and [(mbtt)I_2] [67], have also been determined. Both the increase in the S···I distances when moving from [(mbtt)ICl] to [(mbtt)IBr] and [(mbtt)I_2] (2.5485(6), 2.706(2), and 2.808(3) Å, respectively) and the decrease in the interaction energies (−75, −50, and −39 kJ mol^{-1}, respectively) are indicative to the strengthening of the XB donor properties of the dihalogen in order I_2 < IBr < ICl [77]. The partial charges of the halogens are also changed accordingly. In all cases the iodine that is involved in halogen bonding has a small positive charge that increases when moving from I_2 to IBr and ICl. Simultaneously, the negative charge of the noninteracting halogen increases (Fig. 10.12).

The same trends can be found in the series of [(btt)I_2], [(btt)IBr], and [(btt)ICl] (btt = 2(3)*H*-benzothiazolethione). Just like in homonuclear Br_2, the halogen bonds between the iodine in a heteronuclear dihalogen and the thione acceptors are partially covalent. The |*V*|/*G* ratio ranges from 1.41 in [(mbtt)IBr] to 1.60 in [(btt)ICl] [77].

10.6.3 I_2 in Halogen-Containing Metal Complexes as Weak Halogen Bond Acceptor

Typically, metal ions polarize the halogen atoms only partially. No σ-holes are formed but some polarization does occur. Despite the lack of true σ-hole, rearrangement of the electron density generates electron-rich area perpendicular to the M—X bond, just like in Figure 10.2. A coordinated halogen can then use this electron density and act as halogen bond acceptor. Because of the partial polarization, the angle between the M—X and the halogen bond donor M—X···X is approaching to 90°. The stronger the polarization is, the closer the angle is to 90°. However, steric factors may also have considerable impact on the angle. The direction perpendicular to the M—X bond is relatively free. The M—X halogenido ligand can thus be involved in multiple halogen bonds simultaneously [82]. A number of metal halide complexes involved in XBs have been reported [83–92].

Since the iodido ligand is the most easily polarizable halide, it is also the strongest candidate as acceptor. Just like with organic halogen bond acceptors, charge transfer is involved in contacts between metal halides and dihalogens. Only one end of X_2 is usually involved in strong halogen bond. If a X_2 molecule is bridging two metal halides as a symmetrical bridge, it indicates only weak charge transfer effects. For example, [$RuI_2(H_2dcbpy)(CO)_2$] (H_2dcbpy = 4,4′-dicarboxylic acid 2,2′-bipyridine) forms a relatively strong halogen bond between the iodido ligand and I_2 [93]. The M—I···I distance ranges from 3.1442(2) Å to 3.2376(2) Å (I_1···I_2 and I_5···I_6 in Fig. 10.13).

The I—I distances of I_2 molecules are slightly elongated (2.7672(2) and 2.7401(2) Å for I_2—I_3 and I_4—I_5, respectively) compared to the free I_2. The Ru—I_1···I_2 halogen bond is somewhat stronger than the Ru—I_6···I_5 bond, and therefore the I_2—I_3 bond extends more than the I_4—I_5 bond. Because of the other ends of iodine molecules (I_3, I_4) are weaker donors and are not involved in pure halogen bonds. Instead, they are involved in bifurcated halogen–halogen/halogen–hydrogen interactions. Because of weaker charge transfer, the I_5—I_4···I_3 angle is still nearly linear [177.370(7)°], while the I_2—I_3···I_4 angle is only 137.789(7)° due to the increased negative charge on I_3. According to this, the I_3 in Figure 10.13 is acting dominantly as XB acceptor and I_4 as donor. However, the I_3···I_4 distance is relatively long (3.5371(2) Å), which indicates that the interaction between the I_2 molecules is

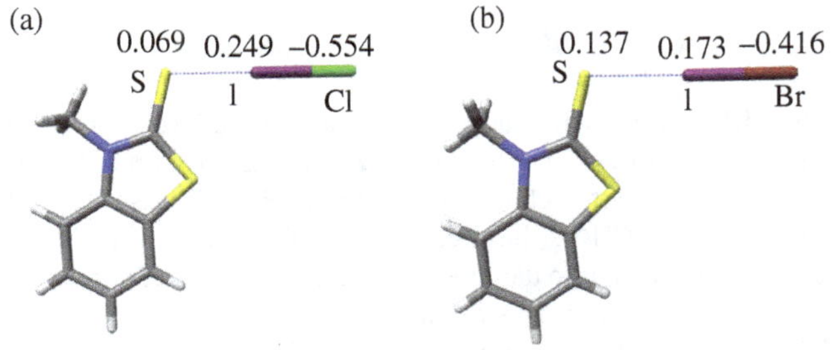

FIGURE 10.12 Partial charges on [(mbtt)ICl] and [(mbtt)IBr] [77].

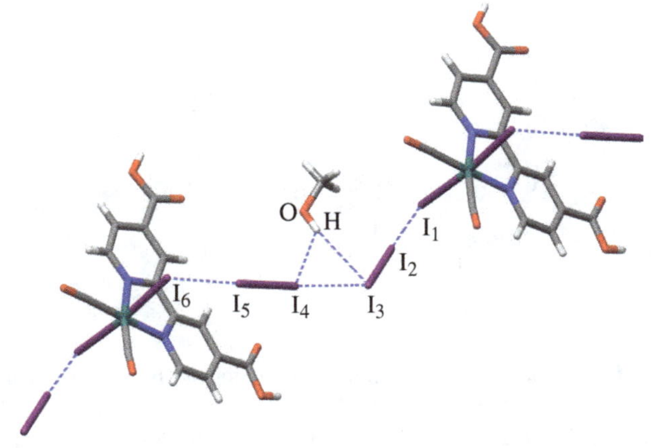

FIGURE 10.13 Halogen bonds between [$RuI_2(H_2dcbpy)(CO)_2$] and I_2 [93].

weaker than the interaction between M—I and I_2. The topological analysis of the systems [93] suggests that the contact between the diiodines may have some amphoteric nature, which means that there may be donor and acceptor features on both interacting iodines [9]. The involvement of hydrogen bond is an indication of changes in charge distribution on I_2 molecules. Concentration of negative charge on I_4 and especially I_3 (Fig. 10.13) makes them more prone to act as hydrogen bond acceptors.

The capabilities of different halide ligands to act as halogen bond acceptors can be seen when comparing structures of $[RuX_2(bpy)(CO)_2]\cdots I_2$ [94]. With $[RuCl_2(bpy)(CO)_2]$, I_2 is bridging two metal complexes symmetrically (Fig. 10.14). As can be seen, the I—I···Cl is nearly linear (174.566(8)°), and the Ru—Cl···I angle 115.76(1)°. The latter is quite far from the 90° angle due to the lower level of polarization on Cl ligand. This means that the negative charge is less concentrated on the chlorido ligand perpendicular to the M—Cl bond. The partial charges of individual iodines on I_2 molecule are equal and only slightly negative, which also indicates low level of charge transfer. This is also the reason why there is no considerable redistribution of charges on diiodine, and therefore, it is capable of using both iodines as weak halogen bond donors. Identical Cl-I distances (3.0421(3) Å) also reflect similar donor properties of both iodines.

Let's again assume the diiodine I_A—I_B is initially approaching the metal halide through I_A iodine. When the chlorido ligand is replaced by a bromido ligand, the I_2 is still bridging two metal complexes as halogen bond donor, but the bridge is no longer symmetrical, indicating increased charge transfer from the metal complex to the I_2 molecule.

The $[RuBr_2(bpy)(CO)_2]\cdots I_2$ adduct has been crystallized together with chloroform of crystallization in two different space groups: centrosymmetric $P2_1/n$ and chiral $P2_1$. In both structures the metal complexes are linked together by I_2 molecules giving chains of complexes. The main difference between these structures is in the orientation of the neighboring chains [94].

From the halogen bonding point of view, both the structures are only slightly different. The shorter Ru—Br···I_A distances range from 3.2381(3) to 3.2938(4) Å, and the longer Ru—Br···I_B distances are between 3.3001(3) and 3.3627(3) Å in the centrosymmetric and chiral structures, respectively. Although the differences in Br···I_A and Br···I_B distances are relatively small even in the chiral structure, there are already some changes in the partial charges of the iodines on I_2 (−0.064 and −0.073 for iodines I_A and I_B, respectively) (Fig. 10.15). Again, both the charges and the structural parameters indicate only weak charge transfer. However, some rearrangement of electron density takes places on both iodines I_A and I_B. Both of them are able to form weak hydrogen bonds with the aromatic C—H groups of the neighboring complex. Even if the hydrogen bonds in this structure are extremely weak, they are still in agreement with the charge differences caused by stronger charge transfer from Br to iodine I_A ($Br_1\cdots I_1$ in Fig. 10.15).

The iodine-containing complex $[RuI_2(bpy)(CO)_2]$ can be cocrystallized with I_2 in two different space groups, centrosymmetric $P\bar{1}$ and chiral $P2_12_12$. In the centrosymmetric structure the I_2 is bridging two metal complexes giving a dimeric overall structure (Fig. 10.16). The iodine bridges are asymmetrical. The shorter M—I···I_A is 3.2553(13) Å and the longer M—I···I_B 3.4108(15) Å. The I_B—I_A···I angle is nearly linear 172.75(2)° and the I_A—I_B···I at the other end of I_2 166.50(2)°. These values are in accordance with the strength of the halogen bonds at different ends of I_2. The stronger M—I···I_A weakens the halogen bond donor properties of I_B. The M—I···I_A and M—I···I_B angles are 97.81(2)° and 98.90(2)°, respectively. Angles close to 90° indicate relatively strong polarization of the metal-coordinated iodines.

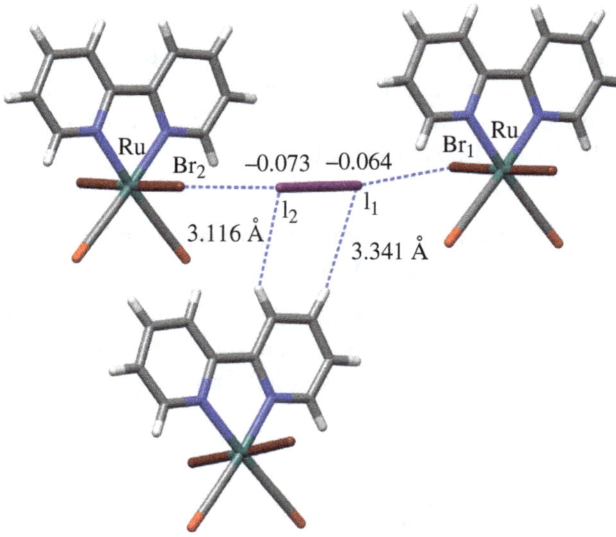

FIGURE 10.15 The halogen bonds and weak hydrogen bond interactions between $[RuBr_2(bpy)(CO)_2]$ and I_2. The structure has been solved in the chiral space group $P2_1$. The solvent of crystallization ($CHCl_3$) has been omitted for clarity. The shorter $Br_1\cdots I_1$ (corresponds to the Ru—Br···I_A distance) is 3.2938(4) Å, the longer $Br_2\cdots I_2$ (corresponds to the Ru—Br···I_B distance) is 3.3627(3) Å, and the I—I bond is 2.7212(3) Å. The I_2—I_1—Br_1, I_1—$I_2\cdots Br_2$ and Ru—$Br_1\cdots I_1$ and Ru—$Br_2\cdots I_2$ angles are 170.28(1)°, 173.80(1)°, 101.30(1)°, and 102.27(1)° [94].

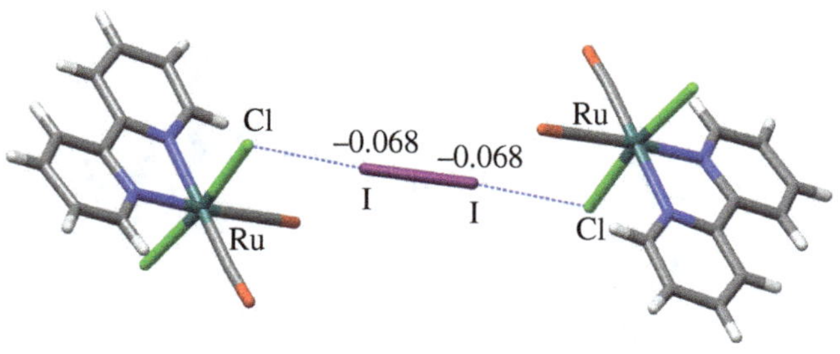

FIGURE 10.14 The symmetrical halogen bonds between $[RuCl_2(bpy)(CO)_2]$ and I_2. The Cl···I distance is 3.0421(3) Å and the I—I bond 2.7317(2) Å. I—I···Cl and Ru—Cl···I angles are 174.566(8)° and 115.76(1)°, respectively [94]. The partial charges are estimated by topological QTAIM analysis.

The differences of I_A and I_B are clearer in the chiral crystals of $[RuI_2(bpy)(CO)_2]\cdots I_2$ (Fig. 10.17).

Here the metal complexes are arranged into a linear chains, linked together by asymmetrical I_2 bridges. In the chiral structure, the shorter M—I···I_A ($I_1\cdots I_2$ in Fig. 10.17) is only 3.1984(2)Å and the longer M—I···I_B ($I_3\cdots I_4$ in Fig. 10.17) 3.7984(3)Å. Similarly, the I_B—I_A···I angle is nearly linear 177.941(7)°, while the I_A—I_B···I angle is 152.083(6)°. This shows that I_A is acting as halogen bond donor forming rather strong halogen bond with the metal-coordinated iodine. The other end of I_2 is no longer able to act as a proper XB donor. The contact with the metal-coordinated iodine is weak and the contact angle quite far from the ideal 180°. The relative charges of I_A and I_B confirm this observations.

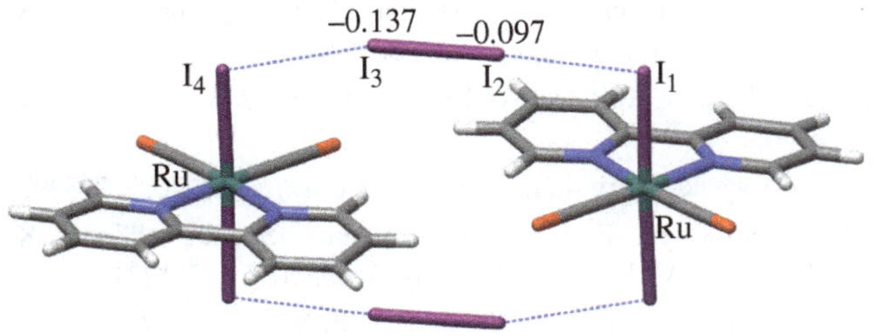

FIGURE 10.16 The halogen bond contacts in the crystal structures of $[RuI_2(bpy)(CO)_2]\cdots I_2$ solved in space group P-1. The shorter $I_1\cdots I_2$ (corresponds to the Ru—I···I_A distance) is 3.2553(13)Å, the longer $I_3\cdots I_4$ (corresponds to the Ru—I···I_B distance) 3.4108(15) and the I—I bond 2.7572(12) Å. The I_3—$I_2\cdots I_1$, I_2—$I_3\cdots I_1$ and Ru—$I_1\cdots I_2$ and Ru—$I_4\cdots I_3$ angles are 172.75(2)°, 166.50(2)°, 97.81(2)°, and 98.90(2)°. The charges have been calculated by computational QTAIM analysis [94].

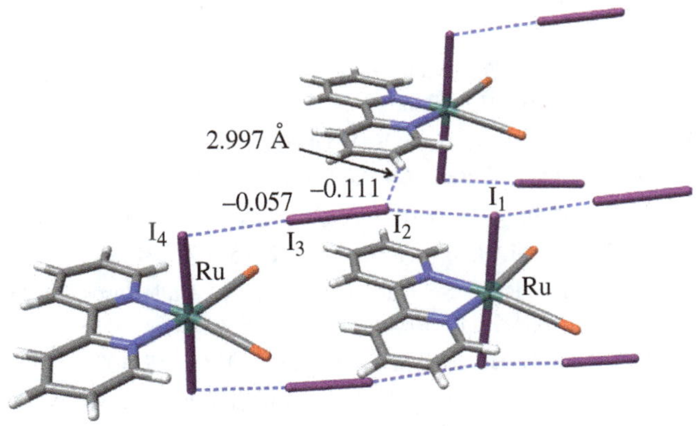

FIGURE 10.17 Crystal structures of $[RuI_2(bpy)(CO)_2]\cdots I_2$ adducts solved in space group $P2_12_12$. The shorter $I_1\cdots I_2$ (corresponds to the Ru—I···I_A distance) is 3.1984(2)Å, the longer $I_3\cdots I_4$ (corresponds to the Ru—I···I_B distance) is 3.7984(3), and the I—I bond is 2.7554(2)Å. The I_3—$I_2\cdots I_1$, I_2—$I_3\cdots I_1$ and Ru—$I_1\cdots I_2$ and Ru—$I_4\cdots I_3$ angles are 177.941(7)°, 152.083(6)°, 97.91(1)°, and 104.26(7)°. The charges have been calculated by computational QTAIM analysis [94].

In both the centrosymmetric and chiral structures, the I_B atom has higher negative charge (Figs. 10.16 and 10.17). Accumulation of the negative charge on I_B converts it to a poor halogen bond donor. In the chiral structure, I_B is already involved in hydrogen bond as hydrogen bond acceptor. The M—I···I_A angle is 97.91(1)°, while the M—I···I_B angle is 104.26(1)°. This also shows that the M—I···I_B contact is no longer a clear electron donor–acceptor contact. It is more likely being an amphoteric halogen–halogen contact rather than a pure halogen bond.

It is quite interesting that the I—I distance does not necessarily change much among the weakly halogen bonded systems. This suggest that the distance is not always particularly indicative for the level of charge transfer in the case of weak XB acceptors. In such cases, the charge transfer effect can be observed more clearly by analyzing packing and by following the partial charges of iodines. All in all, with weak acceptors such as $[RuX_2(bpy)(CO)_2]$, I_2 is capable of acting as double halogen bond donor if the charge transfer is at most moderate. This has been observed in other halogen bonding systems involving metal complexes and I_2 as well. The diiodine molecule can either bridge two similar molecules as in the above examples or different units, as in the case of PCPPdI (2,6-bis[(di-*t*-butylphosphino)methyl]-phenyl (PCP)) and I_2 [82]. In this structure, the I_A end of the diiodine is forming a halogen bond with the Pd coordinated iodine and the I_B end with the π-cloud of an aromatic ring.

10.7 FINAL COMMENTS

Charge transfer has a pivotal role in the halogen bond behavior of dihalogens, especially in the case of homonuclear I_2 and Br_2. When the X_A atom of the homonuclear X_A—X_B dihalogen forms the halogen bond with a Lewis base, the level of charge transfer determines the role of the second halogen, X_B. If the charge transfer is strong, X_B is no longer prone to act as a halogen bond donor. This is due to the increased negative charge on it. Therefore, X_B is more likely to act as a halogen or hydrogen bond acceptor. In the case of heteronuclear dihalogens, only the less electronegative halogen atom is likely to act as halogen bond donor, while the more electronegative atom is more likely to act as acceptor for halogen or hydrogen bonds. This type of behavior makes dihalogens useful tools for crystal engineering and for assembling molecular entities into supramolecular systems.

ACKNOWLEDGMENTS

This work has been partially supported by the Academy of Finland (MH: proj. no. 139571 and KR: 263256, 265328). The COST Action CM1302 is also gratefully acknowledged.

REFERENCES

1. G. R. Desiraju, P. S. Ho, L. Kloo, A. C. Legon, R. Marquardt, P. Metrangolo, P. A. Politzer, G. Resnati, K. Rissanen, (IUPAC Recommendations 2013), *Pure Appl. Chem. 2013*, 85, 1711.
2. P. Metrangolo, H. Neukirch, T. Pilati, G. Resnati, *Acc. Chem. Res.* 2005, *38*, 386.
3. P. Politzer, J. S. Murray, T. Clark, *Phys. Chem. Chem. Phys.* 2010, *12*, 7748.
4. A. Mukherjee, S. Tothadi, G. R. Desiraju, *Acc. Chem. Res.* 2014, *47*, 2514.
5. W. Wang, Y. Zhao, Y. Zhang, Y. B. Wang, *Comput. Theor. Chem.* 2014, *1030*, 1.
6. F. F. Awwadi, R. D. Willett, K. A. Peterson, B. Twamley, *Chem. Eur. J.* 2006, *12*, 8952.
7. P. Metrangolo, G. Resnati, *IUCrJ* 2013, *1*, 5.
8. W. Wang, Y. Zhang, Y. Wang, *Comput. Theor. Chem.* 2013, *1023*, 88.
9. Y. V. Nelyubina, M. Y. Antipin, D. S. Dunin, V. Y. Kotov, K. A. Lyssenko, *Chem. Commun.* 2010, *46*, 5325.
10. H. Wang, X. R. Zhao, W. J. Jin, *Phys. Chem. Chem. Phys.* 2013, *15*, 4320.
11. S. Tothadi, S. Joseph, G. R. Desiraju, *Cryst. Growth Des.* 2013, *13*, 3242.
12. R. Bhaskar, K. Siram, D. P. Karothu, T. N. Guru Row, *Cryst. Growth Des.* 2013, *13*, 1045.
13. S. L. Price, A. J. Stone, J. Lucas, R. S. Rowland, A. E. Thornley, *J. Am. Chem. Soc.* 1994, *116*, 4910.
14. G. R. Desiraju, R. Parthasarathy, *J. Am. Chem. Soc.* 1989, *111*, 8725.
15. N. Ramasubbu, R. Parthasarathy, P. Murray-Rust, *J. Am. Chem. Soc.* 1986, *108*, 4308.
16. S. W. Robinson, C. L. Mustoe, N. G. White, A. Brown, A. L. Thompson, P. Kennepohl, P. D. Beer, *J. Am. Chem. Soc.* 2015, *137*, 499.
17. W. Wang, Y. Zhang, Y. Wang, *J. Phys. Chem. A* 2012, *116*, 12486.
18. M. V. Vener, A. V. Shishkina, A. A. Rykounov, V. G. Tsirelson, *J. Phys. Chem. A* 2013, *117*, 8459.
19. C. M. Reddy, M. T. Kirchner, R. C. Gundakaram, K. A. Padmanabhan, G. R. Desiraju, *Chem. Eur. J.* 2006, *12*, 2222.
20. A. Varadwaj, P. R. Varadwaj, B.-Y. Jin, *Int. J. Quantum Chem.* 2015, *115*, 453.
21. R. Bertani, P. Sgarbossa, A. Venzo, F. Lelj, M. Amati, G. Resnati, T. Pilati, P. Metrangolo, G. Terranec, *Chem. Rev.* 2010, *254*, 677.
22. P. Metrangolo, J. S. Murray, T. Pilati, P. Politzer, G. Resnati, G. Terraneo, *Cryst. Growth Des.* 2011, *11*, 4238.
23. D. Chopra, T. N. G. Row, *CrystEngComm* 2011, *13*, 2175.
24. A. C. Legon, *Phys. Chem. Chem. Phys.* 2010, *12*, 7736.
25. E. V. Bartashevich, V. G. Tsirelson, *Phys. Chem. Chem. Phys.* 2013, *15*, 2530.
26. A. Legon, *Angew. Chem. Int. Ed.* 1999, *38*, 2686.
27. A. C. Legon, J. M. Thumwood, E. R. Waclawik, *Chem. Eur. J.* 2002, *8*, 940.
28. C. A. Legon, *Chem. Eur. J.* 1998, *4*, 1890.
29. S. A. Cooke, G. Cotti, C. M. Evans, J. H. Holloway, Z. Kisiel, A. C. Legon, J. M. A. Thumwood, *Chem. Eur. J.* 2001, *7*, 2295.
30. S. D. Springer, L. A. Rivera-Rivera, B. A. Mcelmurry, Z. Wang, I. I. Leonov, R. R. Lucchese, A. C. Legon, J. W. Bevan, *J. Phys. Chem. A* 2012, *116*, 1213.
31. C. Ouvrard, J. Y. Le Questel, M. Berthelot, C. Laurence, *Acta Crystallogr. Sect. B: Struct. Sci.* 2003, *59*, 512.
32. P. H. Svensson, L. Kloo, *Chem. Rev.* 2003, *103*, 1649.
33. A. C. Legon, *Chem. Commun.* 1998, *24*, 2737.
34. A. C. Legon, *Chem. Commun.* 1998, *23*, 2585.
35. J. E. Ormond-Prout, P. Smart, L. Brammer, *Cryst. Growth Des.* 2012, *12*, 205.
36. M. Tuikka, P. Hirva, K. Rissanen, J. Korppi-Tommola, M. Haukka, *Chem. Commun.* 2011, *47*, 4499.
37. S. V. Rosokha, C. L. Stern, A. Swartz, R. Stewart, *Phys. Chem. Chem. Phys.* 2014, *16*, 12968.
38. R. F. W. Bader, Atoms in Molecules: A Quantum Theory, Oxford University Press: Oxford, 1990.
39. J. P. Perdew, K. Burke, M. Ernzerhof, *Phys. Rev. Lett.* 1996, *77*, 3865.
40. F. Weigend, R. Ahlrichs, *Phys. Chem. Chem. Phys.* 2005, *7*, 3297.
41. L. Koskinen, P. Hirva, E. Kalenius, S. Jääskeläinen, K. Rissanen, M. Haukka, *CrystEngComm* 2015, *17*, 1231.
42. E. Espinosa, I. Alkorta, J. Elguero, E. Molins, *J. Chem. Phys.* 2002, *117*, 5529.
43. P. Politzer, J. S. Murray, *ChemPhysChem* 2013, *14*, 278.
44. A. Karpfen, Theoretical Characterization of the Trends in Halogen Bonding, In Structure and Bonding, (Eds. P. Metrangolo, G. Resnati, Springer, New York, 2008), *126*, 1–15.
45. S. V. Rosokha, A. Traversa, *Phys. Chem. Chem. Phys.* 2015, *17*, 4989.
46. J. W. Zou, Y. J. Jiang, M. Guo, G. X. Hu, B. Zhang, H. C. Liu, Q. S. Yu, *Chem. Eur. J.* 2005, *11*, 740.
47. L. P. Wolters, F. M. Bickelhaupt, *Chemistry Open* 2012, *1*, 96.
48. C. C. Robertson, R. N. Perutz, C. A. Hunter, *Chem. Sci.* 2014, *5*, 4179.
49. C. Wang, D. Danovich, Y. Mo, S. Shaik, *J. Chem. Theory Comput.* 2014, *10*, 3726.
50. M. C. Aragoni, M. Arca, F. A. Devillanova, A. Garau, F. Isaia, V. Lippolis, G. Verani, *Coord. Chem. Rev.* 1999, *184*, 271.
51. K. Rissanen, M. Haukka, *Top. Curr. Chem.* 2015, *359*, 77–90.
52. M. Tuikka, M. Haukka, *Acta Cryst.* 2015, *E71*, o463.
53. C. Álvarez-Rúa, S. García-Granda, A. Ballesteros, *Acta Crystallogr. Sect. E: Struct. Rep. Online* 2002, *58*, o1381.
54. O. Hassel, H. Hope, *Acta Chem. Scand.* 1961, *15*, 407.
55. Y. Kim, E. J. Mckinley, K. E. Christensen, N. H. Rees, A. L. Thompson, *Cryst. Growth Des.* 2014, *14*, 6294.
56. T. Okitsu, S. Yumitate, K. Sato, Y. In, A. Wada, *Chem. Eur. J.* 2013, *19*, 4992.
57. J. M. Chalker, A. L. Thompson, B. G. Davis, *Org. Synth.* 2010, *87*, 288.
58. A. A. Neverov, H. X. Feng, K. Hamilton, R. S. Brown, *J. Org. Chem.* 2003, *68*, 3802.

59. U. Buontempo, A. Di Cicco, A. Filipponi, M. Nardone, P. Postorino, *J. Chem. Phys.* 1997, *107*, 5720.
60. S. B. Hakkert, M. Erdélyi, *J. Phys. Org. Chem.* 2015, *28*, 226.
61. A. C. C. Carlsson, J. Gräfenstein, A. Budnjo, J. L. Laurila, J. Bergquist, A. Karim, R. Kleinmaier, U. Brath, M. Erdélyi, *J. Am. Chem. Soc.* 2012, *134*, 5706.
62. A. C. C. Carlsson, M. Uhrbom, A. Karim, U. Brath, J. Gräfenstein, M. Erdelyi, *CrystEngComm* 2013, *15*, 3087.
63. M. Erdélyi, *Chem. Soc. Rev.* 2012, *41*, 3547.
64. M. Bedin, A. Karim, M. Reitti, A.-C. C. Carlsson, F. Topic, M. Cetina, F. Pan, V. Havel, F. Al-Ameri, V. Sindelar, K. Rissanen, J. Gräfenstein, M. Erdélyi, *Chem. Sci.* 2015, *6*, 3746.
65. M. S. Chernov'yants, I. V. Burykin, Z. A. Starikova, N. E. Erofeev, *J. Mol. Struct.* 2011, *1006*, 379.
66. G. J. Corban, S. K. Hadjikakou, A. C. Tsipis, M. Kubicki, T. Bakas, N. Hadjiliadis, *New J. Chem.* 2011, *35*, 213.
67. G. J. Corban, S. K. Hadjikakou, N. Hadjiliadis, M. Kubicki, E. R. T. Tiekink, I. S. Butler, E. Drougas, A. M. Kosmas, *Inorg. Chem.* 2005, *44*, 8617.
68. A. Tamilselvi, G. Mugesh, *Bioorg. Med. Chem. Lett.* 2010, *20*, 3692.
69. V. Daga, S. K. Hadjikakou, N. Hadjiliadis, M. Kubicki, J. H. Z. dos Santos, I. S. Butler, *Eur. J. Inorg. Chem.* *2002*, 1718.
70. F. Demartin, P. Deplano, F. A. Devillanova, F. Isaia, V. Lippolis, G. Verani, *Inorg. Chem.* 1993, *32*, 3694.
71. G. H.-Y. Lin, H. Hope, *Acta Crystallogr. Sect. B: Struct. Crystallogr. Cryst. Chem.* 1972, *28*, 643.
72. J. Sola, A. López, R. A. Coxall, W. Clegg, *Eur. J. Inorg. Chem.* 2004, 4871.
73. E. S. Raper, *Coord. Chem. Rev.* 1994, *129*, 91.
74. P. D. Akrivos, *Coord. Chem. Rev.* 2001, *213*, 181.
75. L. Koskinen, S. Jääskeläinen, P. Hirva, M. Haukka, *Cryst. Growth Des.* 2015, *15*, 1160.
76. F. H. Herbstein, W. Schwotzert, *J. Am. Chem. Soc.* 1984, *106*, 2367.
77. L. Koskinen, P. Hirva, A. Hasu, S. Jääskeläinen, J. Koivistoinen, M. Pettersson, M. Haukka, *CrystEngComm* 2015, *17*, 2718.
78. M. C. Aragoni, M. Arca, F. Demartin, F. A. Devillanova, A. Garau, F. Isaia, V. Lippolis, G. Verani, *Dalton Trans.* 2005, 2252.
79. M. C. Aragoni, M. Arca, F. Demartin, *Cryst. Growth Des.* 2007, *7*, 1284.
80. F. Demartin, F. A. Devillanova, A. Garau, F. Isaia, V. Lippolis, G. Verani, *Polyhedron* 1999, *18*, 3107.
81. P. D. Boyle, W. I. Cross, S. M. Godfrey, C. A. McAuliffe, R. G. Pritchard, S. Teat, *J. Chem. Soc. Dalton Trans.* 1999, 2219.
82. M. T. Johnson, Z. Džolić, M. Cetina, O. F. Wendt, L. Öhrström, K. Rissanen, *Cryst. Growth Des.* 2012, *12*, 362.
83. J. Le Bras, H. Amouri, J. Vaisserman, *Inorg. Chem.* 1998, *37*, 5056.
84. K. Tebbe, B. Freckmann, *Z. Naturforsch. B Chem. Sci.* 1996, *51*, 999.
85. C. Wieczorrek, *Acta Crystallogr. Sect. C: Cryst. Struct. Commun.* 2000, *56*, 1085.
86. R. Ambrosetti, W. Baratta, D. B. Dell'Amico, F. Calderazzo, F. Marchetti, *Gazz. Chim. Ital.* 1990, *120*, 511.
87. N. Masuhara, S. Nakashima, K. Yamada, *Chem. Lett.* 2005, *34*, 1352.
88. M. F. Belicchi, G. G. Fava, C. Pelizzi, *Acta Cryst.* 1981, *B37*, 924.
89. S. Blanchard, F. Neese, E. Bothe, E. Bill, T. Weyhermüller, K. Wieghardt, *Inorg. Chem.* 2005, *44*, 3636.
90. L. R. Gray, D. J. Gulliver, W. Levason, M. Webster, *Inorg. Chem.* 1983, *7*, 2362.
91. R. A. Gossage, A. D. Ryabov, A. L. Spek, D. J. Stufkens, J. A. M. van Beek, R. van Eldik, G. van Koten, *J. Am. Chem. Soc.* 1999, *121*, 2488.
92. S. Zhao, R. Wang, S. Wang, *Organometallics* 2009, *28*, 2572.
93. M. Tuikka, M. Niskanen, P. Hirva, K. Rissanen, A. Valkonen, M. Haukka, *Chem. Commun.* 2011, *47*, 3427.
94. X. Ding, M. J. Tuikka, P. Hirva, V. Yu. Kukushkin, A. S. Novikov, M. Haukka, Submitted for publication.

11

CONSTRUCTION OF SUPRAMOLECULAR ASSEMBLIES BASED ON ANION–π INTERACTIONS

ANTONIO BAUZÁ AND ANTONIO FRONTERA

Departament de Química, Universitat de les Illes Balears, Palma de Mallorca (Baleares), Spain

11.1 INTRODUCTION

Supramolecular chemistry is one of the most multidisciplinary fields of research. Its rapid development has a profound influence on the increasing effectiveness and success in the synthesis of assemblies of different sizes, shapes, and functionalities. Supramolecular chemists depend on the understanding of the non-covalent forces that are the basis of highly specific recognition, transport, and regulation mechanisms. The orchestration of many biological and chemical processes is often organized by a complicated combination of non-covalent interactions [1, 2]. This is exemplified by the foundation of the life process itself, the ultimate expression of function.

Usually in host–guest chemistry, interactions between targeted guests and rationally designed receptors propel the formation of assemblies of different shapes, sizes, and affinities [3–6]. The correct description and good comprehension of interactions between molecules is necessary for the understanding and progress of the supramolecular chemistry, exemplified by the rational design of new drugs and the developing of synthetic receptors capable to function in competitive media. It relies on strong directional interactions such as hydrogen bonding and σ-hole bonding [7–15] and less directional forces like ion pairing. Moreover, interactions involving π-systems (usually aromatic rings) are very relevant in supramolecular chemistry, exemplified by the cation–π interaction and its importance in enzyme catalysis and protein structure [16–19].

From a traditional point of view, the π-system of an aromatic ring is considered as electron rich (π-basic). The naissance of its counterpart, the anion–π interaction [20–24] (i.e., the attractive interaction between an anion and an electron-poor π-system), was somewhat controversial. Nevertheless, in the last two decades a large number of theoretical [25–28] and experimental [29–33] investigations have been published and confirmed that the anion–π interaction is an important supramolecular bond. Recently, this interaction has been defined as a subclass of a relatively unexplored group of non-covalent interactions named as π-hole bonding [34].

A π-hole is a region of positive electrostatic potential that is perpendicular to a portion of a molecular framework. In case that this portion is an aromatic π-system with electron-withdrawing groups (characterized by a positive electrostatic potential region perpendicular to the aromatic molecular plane), the interaction is known either lp–π [35, 36] or anion–π depending on the nature of the electron donor. Remarkably, anion–π interactions are visible players in fields as diverse as medicine, environmental chemistry, and biochemical processes [20, 37–39]. Moreover, their use in the design of highly selective anion receptors and transport channels confirms their significance in the field of supramolecular chemistry [40, 41]. As a matter of fact, the synthesis of selective receptors for anions that function in competitive media is a topic of interest [42–47] due to the vital function played by some anions. For example, they are ubiquitous in biological systems [48, 49], and some of them are problematic environmental pollutants [50–52]. Therefore, continuous efforts developing a new class of anion receptors based on the anion–π interaction are carried out [20, 53]. For instance, an interesting receptor, which combines hydrogen bonding and anion–π interaction for the binding of anions with neutral π-acceptors, has been recently published by Albrecht and collaborators [54], and its ability

Non-Covalent Interactions in the Synthesis and Design of New Compounds, First Edition.
Edited by Abel M. Maharramov, Kamran T. Mahmudov, Maximilian N. Kopylovich, and Armando J. L. Pombeiro.

to trap anions has been demonstrated both in solution and in the solid state. Similarly, Johnson and collaborators have developed receptors for selective nitrate binding in competitive hydrogen bonding solvents where anion–π interactions facilitate the selectivity [55].

Some early publications on this topic should be placed into perspective. Several initiating reports on weak attractive interactions involving negatively charged species and polarizable aryl groups in host–guest systems [56–60] were reported before the exponential growth of manuscripts studying the anion–π interaction that started at the beginning of this millennium. Moreover, Hiraoka *et al.* [59] demonstrated the formation of highly symmetric $X^{-}\cdots C_6F_6$ anion–π complexes in the gas phase (X=CI, Br, and I) in 1987. In addition, they studied the gas-phase clustering reactions $X^{-}@(C_6F_6)_{n-1}+C_6F_6=X^{-}@(C_6F_6)_n$ for X=F, C1, Br, and I by means of pulsed electron-beam high-pressure mass spectrometry [59]. For fluoride a nucleophilic attack and covalent bond formation were observed, and, conversely, noncovalent $X^{-}\cdots C_6F_6$ complexes were detected for X=CI, Br, and I where the anion was located along the C_6 main symmetry axis of C_6F_6. In 1993 Schneider and coworkers demonstrated attractive interactions between simple organic host and guest molecules by means of NMR measurements [56]. Another initial progress in this direction [61] was published by Woollins and collaborators in 1996. They named the interaction as "π-facial" [61] in their description of the close contact between chloride and the seven-membered aromatic $[S_4N_3]^+$ ring that they observed in the solid-state architecture of thiotrithiazylium chloride.

This chapter is not a bibliographic review of the literature related to anion–π interactions since several surveys have been written for that purpose [20, 23, 24, 62, 63]. Instead, we intend to emphasize the importance of this interaction for the construction of new supramolecular assemblies in the solid state. Moreover, we aimed to point out that the comprehension of anion–π interactions is fundamental in crystal engineering. Indeed, we believe that harvesting the full potential of anion–π interactions is merely in its naissance. The chapter is developed under two main headings. First, the physical nature of the interaction is explained for assimilation by a wide readership. Second, we highlight selected examples from the literature where molecular recognition driven by anion–π bonding is very relevant.

11.2 PHYSICAL NATURE

The physical nature of the anion–π interaction has been widely studied using high-level ab initio and density functional theory (DFT) calculations. Moreover, several partition energy schemes have been used in order to decompose the total interaction energy into individual components [64, 65]. The general conclusion is that electrostatic forces and ion-induced polarization are the main contributions to the anion–π interaction [66–68]. The electrostatic term is explained by means of the permanent quadrupole moment (Q) of the arene, which is the first nonzero multipole moment in symmetric arenes. Moreover, in most asymmetric arenes where the dipole moment (μ) is nonzero, the μ_z component is approximately 0, and, likewise symmetric arenes, the electrostatic attraction is basically due to the existence of the adequate component (perpendicular to the ring plane) of the quadrupole moment (Q_{zz}), which describes the charge distribution on both sides of the aromatic plane. The Q_{zz} of benzene is negative but can be turned into positive by attaching electron-withdrawing substituents to the ring (see Fig. 11.1a). Similarly, the Q_{zz} of pyridine is negative but can be turned into a positive value by attaching a coordination metal to the nitrogen atom [63]. Therefore, the electrostatic charge–quadrupole interaction between an anion and an aromatic ring can become attractive either attaching electron-withdrawing substituents or coordinating metal ions in case of heteroaromatic rings.

The ion-induced polarization of the π-electron system by the anion is significant, inducing a dipole (see Fig. 11.1b). Therefore, a polarization contribution to the total interaction energy is derived from the interaction of the anion with the induced dipole [64–67]. An alternative explanation on the nature of anion–π interactions involving benzene rings has been proposed by Wheeler and Houk [69], who examined substituent effects in $Cl^{-}\cdots C_6H_{6-n}X_n$ complexes. In contrast to the intuitive view where the substituent induces changes in the aryl π-system, Wheeler and Houk propose a model where substituent effects in these systems can be attributed mainly to direct interactions between the anion and local C—X dipoles.

The additivity of the anion–π interaction is another important aspect that should be considered for the use of electron-deficient rings as efficient building blocks. As a matter of fact the additivity in halide complexes has been examined and demonstrated theoretically [70] using several *s*-triazines as examples. That is, the interaction energy of a ternary "sandwich" complex (π···X⁻···π) is essentially twice the interaction energies of the corresponding binary complexes (X⁻···π) for X=Cl and Br. Regarding charge transfer effects of anion–π interactions, the values strongly depend on the theoretical method used to derive the atomic charges. In general, NPA and AIM charges predict negligible charge transfer effects, and Merz–Kollman and CHelpG charges predict values ranging within 0.1–0.25e. A molecular orbital analysis of the anion–π interaction has also been reported [71, 72] with results compared to the cation–π interaction, and a totally different picture emerges. While the atomic orbitals of the cation do not participate in the molecular orbitals of the cation–π complex, the atomic orbitals of the anion have an active participation in the molecular orbitals of the anion–π complex.

A proper understanding of the physical nature (*vide supra*) is necessary to explain, on one hand, the dual binding mode exhibited by arenes with negligible quadrupole moments [71, 73]. That is, since electrostatic and polarization effects

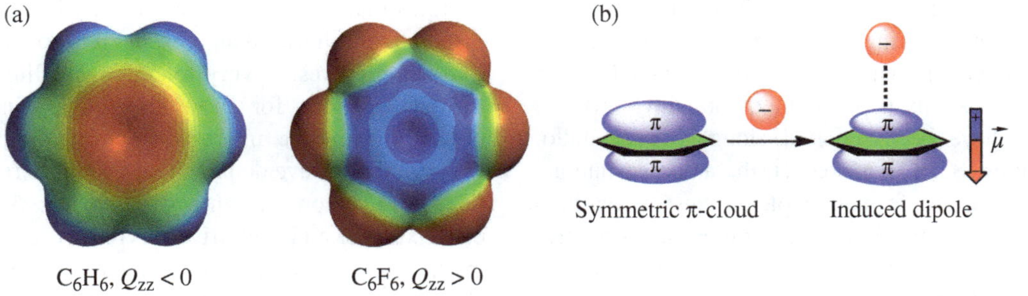

FIGURE 11.1 (a) Molecular electrostatic potential of benzene and hexafluorobenzene. (b) Schematic representation of the ion-induced dipole.

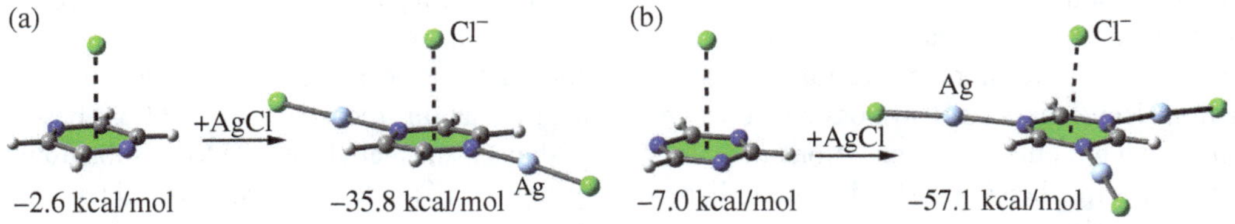

FIGURE 11.2 Interaction energies of pyrazine (a) and triazine (b) anion–π complexes from Refs. [79, 80].

are the main factors contributing to both anion–π and cation–π interactions, molecules with very small Q_{zz} values such as 1,3,5-trifluorobenzene (Q_{zz}=0.57 B) and *s*-triazine (Q_{zz}=0.90 B) are able to interact with both anions and cations. The interaction is thus dominated by the polarization term, which is always attractive, and the electrostatic term is negligible. On the other hand, it is useful to rationalize the interaction of anions with electron-rich aromatic rings, such as benzene, which are not strongly repulsive. This is due to a compensating effect between the electrostatic (unfavorable) and ion-induced polarization (favorable) forces that roughly cancel each other out. For this reason, the anion–π interaction energy of benzene with chloride is very small but favorable [74, 75].

An interesting example of this compensating effect has been described in complexes of anions with isocyanuric acid [76]. When the oxygen atoms of isocyanuric acid are progressively replaced by sulfur atoms, the four possible (thio) cyanuric acids unexpectedly present similar binding energies with chloride (~15 kcal/mol). In this particular system, the quadrupole moment progressively decreases on going from cyanuric acid to trithiocyanuric acid, and, simultaneously, the molecular polarizability progressively increases resulting in invariant interaction energy values.

The electrostatic term depends on the magnitude of the Q_{zz} and the polarization term on the magnitude of the molecular polarizability parallel to the main symmetry axis (denoted as $\alpha_{\parallel}$) in symmetric arenes (or perpendicular to the ring plane in asymmetric arenes, denoted as α_{zz}), which are intrinsic properties of the π-system. Therefore, it is clear that in order to design an efficient anion receptor based on the anion–π interaction, the π-binding units should have a large and positive quadrupole moment and a large molecular polarizability. However, there is a limitation for the first condition due to the reduced number of strong electron-withdrawing groups available for constructing the binding blocks. To have a large value of Q_{zz}, the use of $-NO_2$ and $-CN$ groups is required. However, it is synthetically complicated to attach more than three strong electron-withdrawing groups and the spacer to the aromatic ring to build the receptor. An intelligent solution to this limitation is the utilization of heteroaromatic rings, especially di-, tri-, and tetrazines. The metal coordination to heteroaromatic rings strongly increases the π-acidity of the ring, thus favoring the anion–π interactions [77, 78]. It has been demonstrated that the coordination of both pyrazine (Fig. 11.2a) and *s*-triazine (Fig. 11.2b) to Ag^I dramatically enhances their ability to establish anion–π binding [77, 81].

The same behavior has been demonstrated [78] for *s*-tetrazine using both theory and experiment. As a matter of fact, coordinated to four Ag^I atoms *s*-tetrazine is a very powerful anion–π acceptor. Theoretically, the interaction energy of *s*-tetrazine with nitrate ion strengthens from −9.6 to −62.4 kcal/mol when the arene is tetracoordinated to Ag^I. Experimentally, several X-ray crystal structures of *s*-tetrazine μ_4 coordinated to Ag^I have been reported, exhibiting very close contacts between the anion and the *s*-tetrazine ring (Fig. 11.3) indicating strong anion–π interactions, in agreement with theoretical predictions. Interestingly the anion–π distance for the perchlorate anion in the X-ray structure shown in Figure 11.3a (2.61 Å) is the shortest reported to date.

As aforesaid, the coordination of the heteroaryl ring to a transition metal largely increases the anion binding capability of the arene. In general, the anion–π binding ability of the receptor increases as the oxidation state of the metal center grows. Moreover, the electronic nature of the other ligands coordinated to the metal also influences the interaction strength [20, 53]. The coordination of the heteroaromatic ring is somehow similar to the effect of protonation. For example, pyridine and related azines can be easily protonated by adjusting the pH of the medium, thus increasing the anion binding ability of the ring (anion–π^+ interactions). The geometric and energetic features of anion–π^+ complexes of several aromatic cations (tropylium, quinolizinylium) and various anions have been reported along with crystallographic structures [82, 83]. Moreover, the anion–π^+ interaction participates in the formation of a robust recognition motif in the transition metal malonate complexes using protonated 2-amino-4-picoline and 2-aminopyridine as the auxiliary ligands [79, 80]. Lastly, the importance of this interaction in protonated purine and pyrimidine bases has been recently reviewed [84]. As expected, these complexes exhibit very large (>80 kcal/mol) interaction energies due to strong electrostatic effects that dominate the interaction.

The utilization of a very polarizable π-system with a large and positive quadrupole moment warrants a strong anion–π interaction. However, it is also possible to establish a strong anion–π interaction between anions and electron-rich aromatic rings like benzene. This is accomplished whenever the aromatic ring is simultaneously interacting with a cation by the opposite side of the ring. The simultaneous interaction of anions and cations on different faces of the same π-system has been studied [85–88]. The π-system can act as charge insulator or as intermediate in the transfer of *information* between the charged systems. These ternary anion–π–cation complexes present interaction energies that are large and negative, while the equilibrium distances are shorter than those in the corresponding binary ion–π complexes. A very relevant and illustrative experimental evidence for these theoretical findings can be obtained from the interesting work of Atwood's group [89–92]. They have demonstrated using X-ray analysis and ^{1}H NMR titration experiments that the host–guest behavior of calixarenes and cyclotriveratrylenes is drastically altered in the presence of transition metal centers (Ru, Ir, Rh) allowing anionic guest species to be included within the molecular cavity (Fig. 11.4).

Another strategy to generate π-systems with strong anion binding ability is the annelation of electron-deficient heteroaryl rings. Energetically, it has a comparable effect to the introduction of strong electron-withdrawing groups. This approach has been used by Kim Dunbar *et al.* to construct a neutral anion receptor (Fig. 11.5) based on 1,4,5,8,9,12-hexaazatriphenylene (HAT) substituted with cyano groups that is able to interact with three anions by one side of the polycyclic ring and with one anion by the opposite site [93]. In addition, Ballester and coworkers [94] have studied the anion–π binding ability of $HAT(CN)_6$ both theoretically and experimentally. It was concluded that the formation of typical anion–π non-covalent complexes, charge transfer complexes, or electron transfer adducts strongly depends on the type of anion.

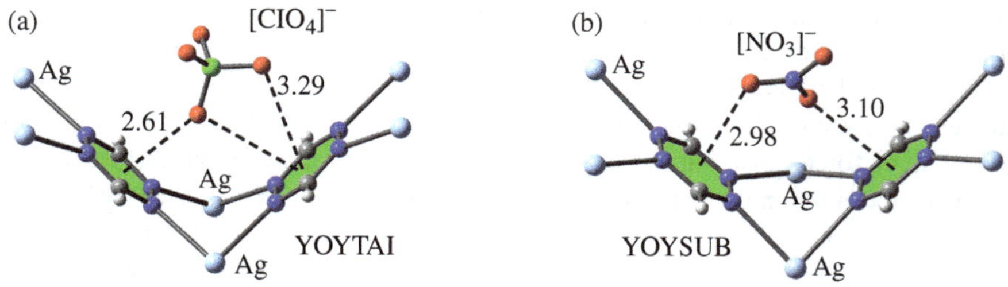

FIGURE 11.3 Fragments of the X-ray crystal structures containing μ_4-coordination of 1,2,4,5-tetrazine [79]. The relevant anion–π interactions are indicated by dashed lines (distances in Å). The CCDC reference codes are indicated.

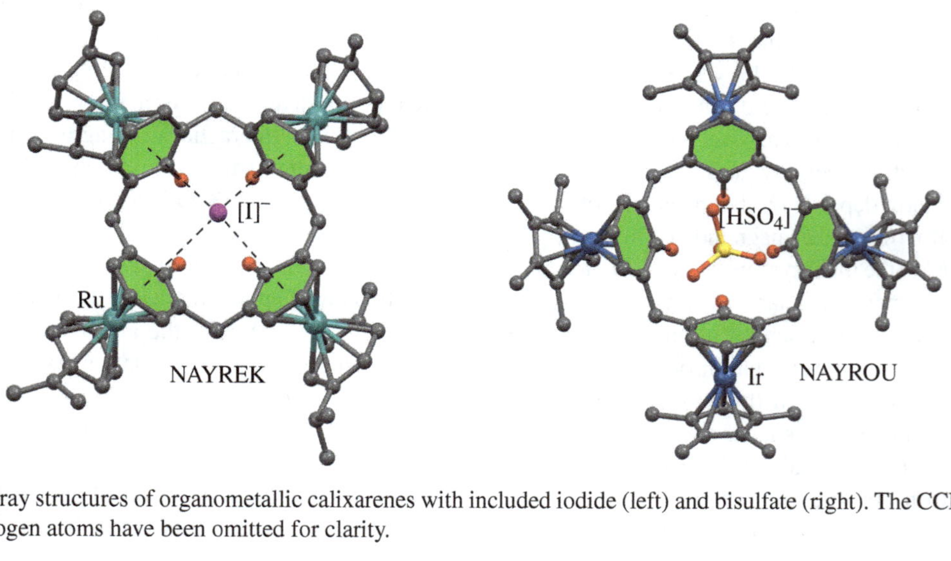

FIGURE 11.4 X-ray structures of organometallic calixarenes with included iodide (left) and bisulfate (right). The CCDC reference codes are indicated. Hydrogen atoms have been omitted for clarity.

A remarkable example of this particular strategy is the utilization of the bicyclic aromatic arene pyridazino[4,5-*d*]pyridazine and related polycyclic rings. The pyridazino[4,5-*d*]pyridazine system preserves the N-donor ability of pyridazine to form transition metal complexes with strong π-acidic character. Moreover, it is easily available by a one-pot synthesis involving inverse electron demanding Diels–Alder cycloaddition [95]. The ability to establish anion–π interactions of this ligand has been analyzed theoretically [96, 97], and it has been demonstrated that the interaction energies with several anions are more favorable than those of pyrazine. The theoretical work also included the study of larger polycyclic arenes based on pyridazino[4,5-*d*]pyridazine, and it is observed that the interaction energy increases with the number of fused rings (Fig. 11.6).

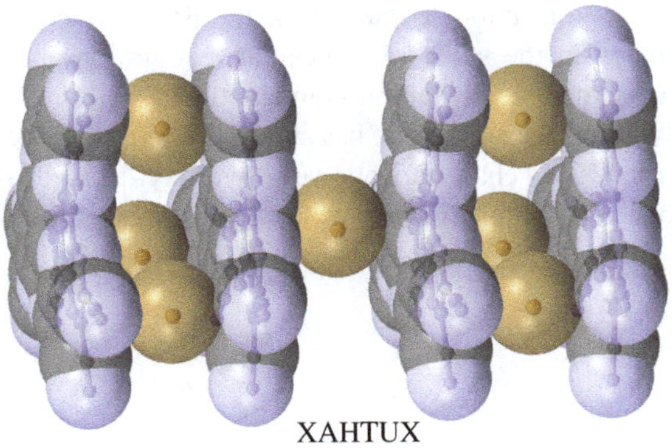

FIGURE 11.5 X-ray structure of $[HAT(CN)_6]_2[Br^-]_3$. The CCDC reference code is indicated.

The pyridazino[4,5-*d*]pyridazine ligand (L_{pp}) has been utilized by Gural'skiy *et al.* [95] to investigate the ability of this ring to establish anion–π interactions providing an attractive structural synthon and an unique π-bifunctional building block for novel polymeric and molecular anion receptors. Two examples are represented in Figure 11.7. In one of them, namely, $[Cu(H_2O)_2(L_{pp})_2](ClO_4)_2$, the metal-organic portion adopts a 2D square grid structure where a pair of anions is encapsulated into each square, establishing close anion–π interactions. The other one is a hydroxopolymer $\{[Cu(\mu\text{-}OH)(L_{pp})](H_2NSO_3)\cdot H_2O\}_n$, which possesses a very remarkable anion–π interaction where the ligand is a double receptor for oxygen and nitrogen sites of anion. The dimensions of the sulfamate anion and pyridazino[4,5-*d*]pyridazine ligand perfectly match facilitating an effective double binding (anion–π and lp–π), in which the corresponding distances are 3.14 and 3.10 Å.

11.3 EXAMPLES OF SUPRAMOLECULAR ASSEMBLIES

The first example we wish to emphasize is devoted to coordination-driven self-assembly, probably one of the main topics of supramolecular chemistry. The structural versatility of metal ions and the directionality of metal–ligand interactions have been used as strategy to promote the spontaneous

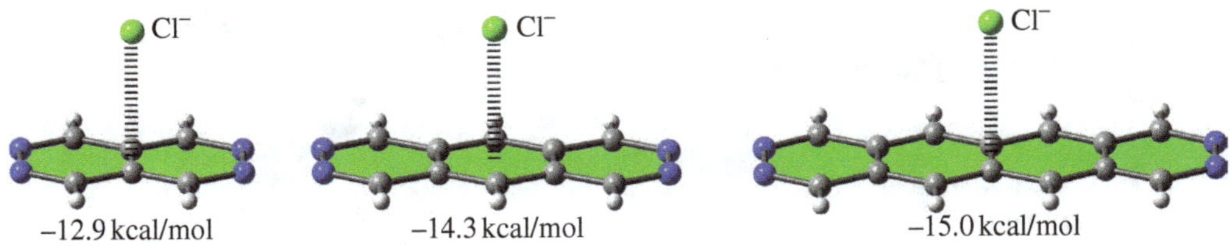

FIGURE 11.6 Interaction energies of Cl^-–π complexes of pyridazino[4,5-*d*]pyridazine and related rings [96].

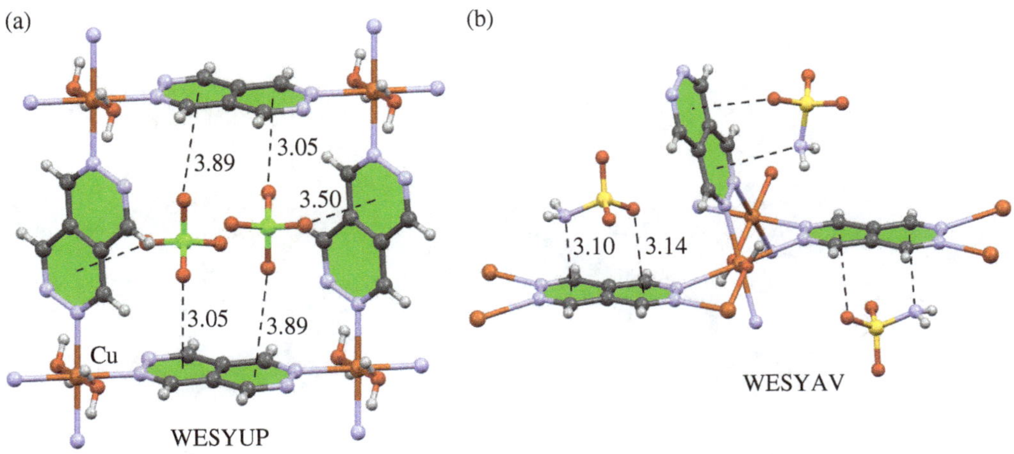

FIGURE 11.7 Fragments of the X-ray crystal structures containing μ^2- and μ^4-coordination of pyridazino[4,5-*d*]pyridazine [95]. The relevant anion–π interactions are indicated by dashed lines (distances in Å). The CSD reference codes are indicated.

assembly of elegant supramolecular architectures [98–100] with unusual properties and applications or intriguing host–guest behavior [101–103]. In 2013 Dunbar and collaborators [104] provided unambiguous evidence that anion–π interactions are the main driving force in the templation process leading to the formation of Fe(II) metallacycles with π-acidic cavities using 3,6-bis(2-pyridyl)-1,2,4,5-tetrazine as chelating ligand. Experimental evidences obtained in the solid state and in solution demonstrate that the inclusion of anions play a decisive role in the formation and size of the supramolecular polygons. It was demonstrated by X-ray crystallography, NMR spectroscopies, and CV and mass spectrometry studies that $[BF_4]^-$ and $[ClO_4]^-$ anions template the molecular squares, whereas the larger anions $[SbF_6]^-$, $[AsF_6]^-$, and $[PF_6]^-$ induce formation of molecular pentagons in high yields (Fig. 11.8). Additional evidence for the role of the anions as templates and stabilizing Fe(II) metallacycles of specific size was obtained by the fact that random Fe(II)/ligand oligomers are formed in the presence of nontemplating anions like $[CF_3SO_3]^-$ and the self-assembly of the respective closed polygons is not observed unless the appropriate templating anions are added. Furthermore, the interconversion between molecular squares and pentagons is possible in solution by adding an excess of the appropriate anion.

Anion–π interactions have been systematically studied by Wang *et al.* [105] using a well-designed and neutral tetraoxacalix[2]arene[2]triazine as an electron-deficient macrocyclic host. By means of several experimental techniques like electrospray ionization mass spectrometry, fluorescence titration, and X-ray crystallography, they demonstrated that tetraoxacalix[2]arene[2]triazine form 1 : 1 complexes with several polyatomic anions of different geometries and shapes. X-ray molecular structures of the complexes showed that two opposing triazine rings of tetraoxacalix[2]arene[2] triazine act as a pair of tweezers to interact with the included anions through cooperative anion–π interactions (Fig. 11.9). This study on anion–π interaction motifs provided a new dimension in the study of molecular recognition and self-assembly mediated by anion–π interactions.

The same research group has also published [106] the synthesis and X-ray characterization of a hydroxyl-functionalized tetraoxacalix[2]arene[2]triazine host molecule through a

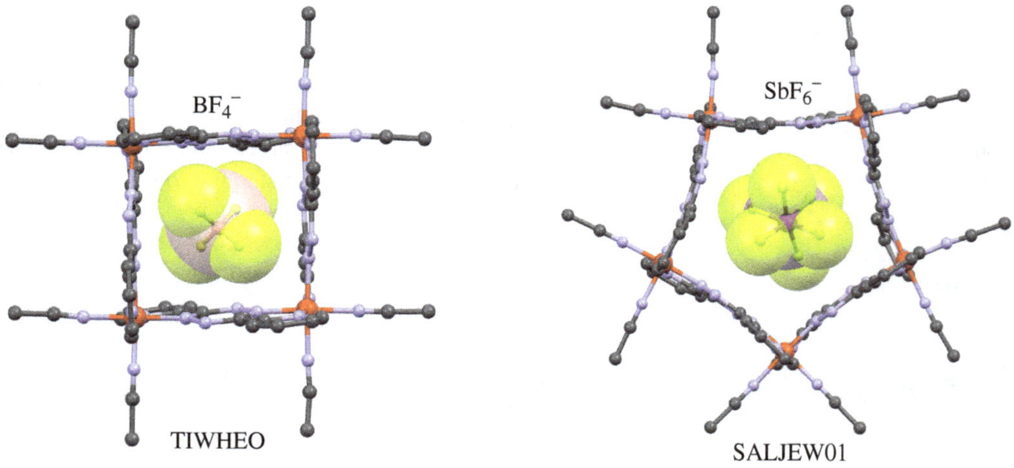

FIGURE 11.8 X-ray crystal structures of the molecular squares and pentagons constructed by Fe^{II} and 3,6-bis(2-pyridyl)-1,2,4,5-tetrazine [104]. The CSD reference codes are indicated.

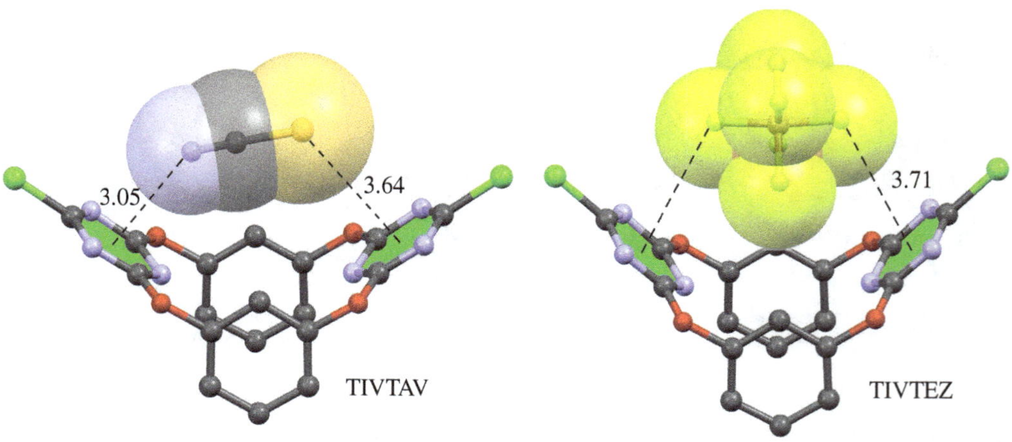

FIGURE 11.9 X-ray crystal structures of TIVTAV and TIVTEZ [105]. The CSD reference codes are indicated. Distances in Å.

fragment coupling followed by $AlCl_3$-mediated deprotection protocol. Interestingly, the simple introduction of hydroxyl groups on the larger rim rendered the macrocyclic molecule as a unique donor–acceptor functional building block. Consequently, infinite self-assemblies were obtained under the direction of cooperative anion–π and lone-pair electron–π interactions and intermolecular hydrogen bonding. This study is a relevant example where the anion–π interaction directed the self-assembly previously designed through the rational fabrication of building blocks.

A rare noninterpenetrated three-dimensional (3D) hydrogen-bonded supramolecular framework containing polyoxometalate (POM) guest molecules has been reported by Liao *et al.* [107]. Interestingly, the trapped POM anions effectively control the geometry and the interpenetration of the hydrogen-bonded supramolecular framework via directional anion–π interactions (Fig. 11.10). This result indicates that anion–π interactions are, on one hand, applicable to the stabilization and immobilization of functional POM anions and, on the other hand, directly affect the final structure of host–guest supramolecular materials. A combination of hydrogen bonds and anion–π interactions with *N,N′*-di(4-pyridyl)-1,4,5,8-naphthalenediimide tectons facilitates the self-assembly process and controls the structures of POM-based solid host–guest materials and their supramolecular architectures.

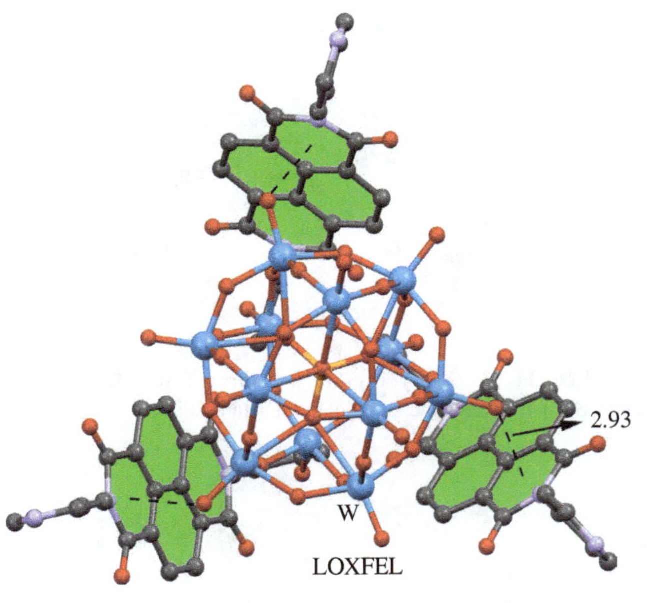

FIGURE 11.10 X-ray crystal structure of LOXFEL [107]. The CSD reference code is indicated. Distance in Å.

Two pH-dependent 4′-(4-pyridyl)-2,2′:6′,2″-terpyridine (PTP) salts show different structural architecture depending on the anion [108]. The solid-state structures exhibit the participation of the pyridine rings in hydrogen bonding (N–H···Br^-) and anion–π interactions (Fig. 11.11). In both compounds, the anions are in close proximity to the electron-deficient PTP cores, forming distinctive anion–π interactions, thus generating interesting supramolecular anion–π/π–anion and anion–π/π–π/π–anion assemblies. In this experimental work, the role of the anion–π interaction in solid state has been investigated, giving credence to its role as a valuable supramolecular force in building multidimensional structures. In addition, this investigation includes a computational study, highlighting the impact of the anion–π interaction on the final self-assembly structure. Anion–π interactions, in concert with π–π stacking and hydrogen bonding interactions, are crucial for the crystal packing phenomena of both compounds.

The same group extended their study to a similar ligand, that is, the triply protonated 4′-(4-pyridyl)-3,2′:6′,3″-terpyridine) molecule and perchlorate as counterion [109]. Crystallography reveals that π^+–π^+, π^+–π, and various anion–π interactions are the major driving forces in stabilization of the several assemblies observed in the solid-state structure. Furthermore, using a similar strategy, Orvay *et al.* [110] have synthesized and X-ray characterized five proton

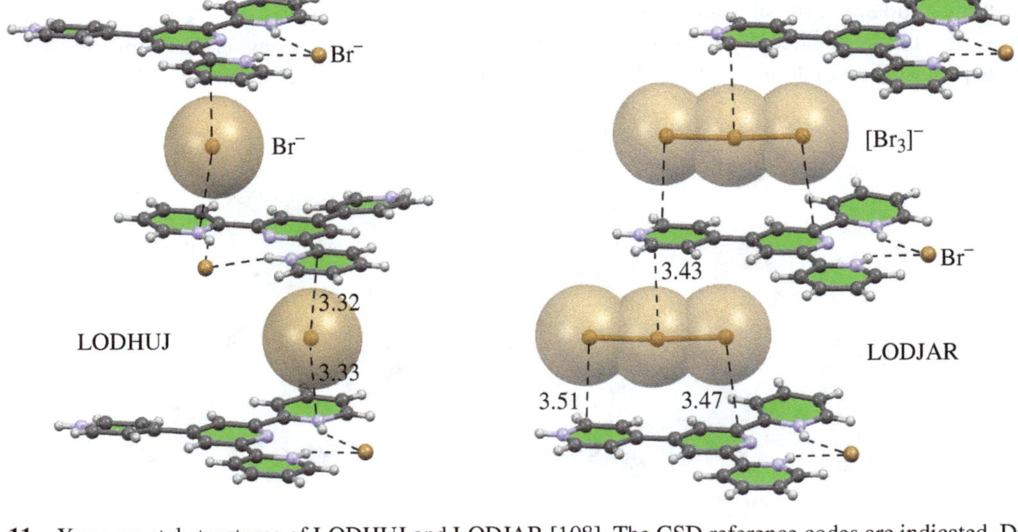

FIGURE 11.11 X-ray crystal structures of LODHUJ and LODJAR [108]. The CSD reference codes are indicated. Distances in Å.

transfer compounds using 4,6-di(1*H*-imidazol-1-yl)pyrimidine (*bimipyr*) and different counterions. In the crystal structures of the five salts, anion–π interactions involving the aromatic rings are crucial for the generation of 3D supramolecular frameworks in the solid state.

Two interesting examples are included in Figure 11.12. In the first one (YOGJIP), a remarkable supramolecular architecture through two anion–π and one lp–π interactions is observed involving the three rings of the *bimipyr* moiety, the chloride ions, and the water molecule. The oxygen atom of the water molecule is oriented toward the π-cloud of the pyrimidine ring forming a strong lp–π interaction. In addition, two chloride ions interact with the imidazolium rings. In the second one (YOGKAI), the $[ZnCl_4]^{2-}$ anion establishes a double anion–π interaction where two Cl atoms interact with the imidazolium rings.

Podgajny *et al.* [111] have combined the π-acidity of pyrazine-*N,N*′-dioxide molecule with $[W(CN)_8]^{3-}$ to generate extended networks governed by supramolecular anion–π interactions between $[W(CN)_8]^{3-}$ and pyrazine-*N,N*′-dioxide (Fig. 11.13). Moreover, the electronic interactions between pyrazine-*N,N*′-dioxide and $[W(CN)_8]^{3-}$ were manifested by the visible light absorption bands of the solid state. This interesting investigation is the first report recognizing and describing the anion–π interactions in octacyanometalate-based coordination networks.

A very significant work in this field has been recently reported by Stoddart's group [112] where the template-directed synthesis of a macrobicyclic cyclophane, namely, BlueCage^{6+}, was investigated. It is composed of six pyridinium rings fused with two central triazines and bridged by three *para* xylylene units (Fig. 11.14). These moieties provide the cage with a remarkably electron-poor cavity, which makes it a powerful receptor for polycyclic aromatic hydrocarbons (e.g., pyrene). The X-ray structure reveals that one of the six PF_6^- counterions of BlueCage^{6+} occupies the cavity as a consequence of multiple anion–π interactions.

Giese *et al.* [113] have demonstrated that the anion–π interaction is an influential force that determines molecular microstructures in the solid state. In particular, the obtained crystal structures of simple triphenyl(pentafluorobenzyl) phosphonium salts provided crucial data on the influence of anion size on the molecular structure of bis(pentafluorobenzyl) phosphonium cations containing two adjacent electron-deficient rings. Whereas the bromide anions interact with the pentafluorobenzene unit by anion–π interactions in a 1 : 1 mode, *Z*-configuration is observed (Fig. 11.15). The bulkier anions iodide, tetrafluoroborate, and hexafluorophosphate result in a 1 : 2 tweezer-like anticonfiguration, in which one anion interacts simultaneously with two pentafluorobenzene units (Fig. 11.15). When the spatial separation of the two electron-deficient rings matches the size of the anion, anion–π interactions induce a conformational change from the *anti*-form observed for the smaller anion to the tweezer-like *syn*-form for the bulkier one.

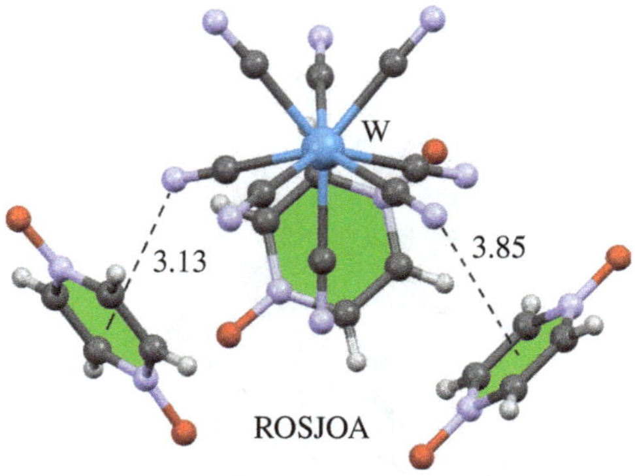

FIGURE 11.13 X-ray crystal structure of ROSJOA [111]. The CSD reference code is indicated. Distances in Å.

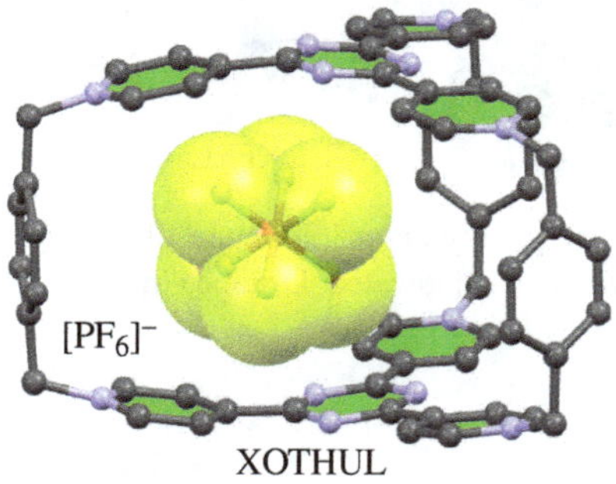

FIGURE 11.14 X-ray crystal structure of XOTHUL [112]. The CSD reference code is indicated.

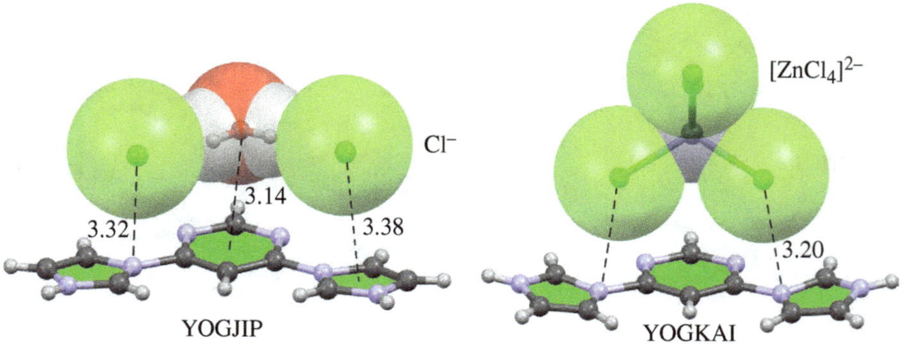

FIGURE 11.12 X-ray crystal structures of YOGJIP and YOGKAI [110]. The CSD reference codes are indicated. Distances in Å.

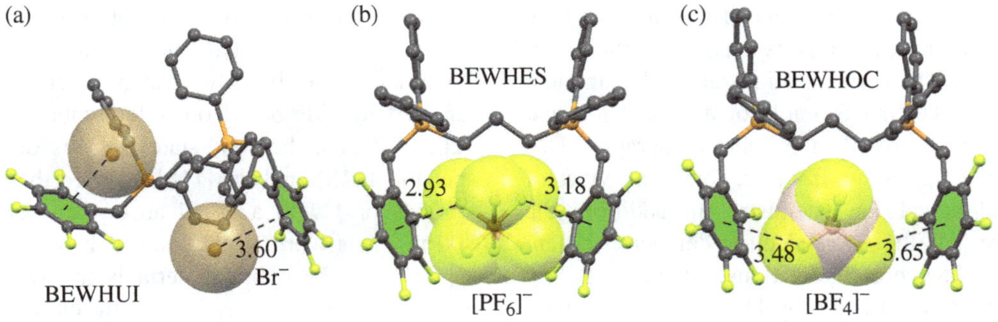

FIGURE 11.15 X-ray crystal structures of BEWHUI, BEWHES, and BEWHOC [113]. The CSD reference codes are indicated. Distances in Å.

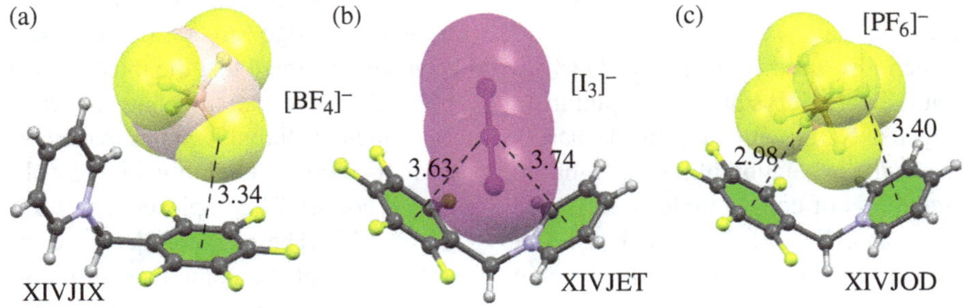

FIGURE 11.16 X-ray crystal structures of XIVJIX, XIVJET, and XIVJOD [114]. The CSD reference codes are indicated. Distances in Å.

The same group has also used *N*-(pentafluorobenzyl)pyridinium salts to investigate the anion size dependence in solid-state structures exhibiting anion–π interactions [114]. In this example, both C_6F_5 and pyridinium moieties are involved in the anion–π contacts (Fig. 11.16). In the solid-state structures, the anion is embedded in a number of hydrogen bonds (C—H···X) and anion–π contacts. The supramolecular complexes are different depending on the size of the anion. That is, the crystal structures of *N*-(pentafluorobenzyl)pyridinium salts of I_3^- and PF_6^- show a cooperative anion–π interaction between the two electron-deficient arenes and the anion located between them. In contrast, this cooperativity is not observed for other anions like bromide of BF_4^- (Fig. 11.16a).

The ability of the pentafluorophenyl group to act as π-acceptor for anions has been also analyzed in depth [115], providing comprehensive study on the importance of anion–π interactions in the solid state. The study has been focused on purely organic and charge-neutral fluorophenyl groups bearing a positive charge located at a side chain. The detailed statistical analysis of a series of structural data sets demonstrates the geometrical variability of anion–π bonding in the solid state. It also showed the directing role of substituents at the arene for the final position of the anions above π-systems. The structural variety of the interactions between anions and electron-deficient arenes was analyzed using the hapticity concept. Then results showed that anions prefer (nearly 90% probability) the position above the electron-deficient π-system, while the high variability of the anion position is caused by interplay of various non-covalent forces. Moreover, the importance of the directing substituents at fluoroarene was demonstrated. The authors proposed that the hapticity is helpful to describe anion–π interactions in crystal structures [115].

Stoddart's group [116] has reported interesting anion–π recognition in molecular triangular prisms formed by naphthalenediimide (NDI) binding blocks. They have used both theory and experiment to demonstrate through-space orbital interactions and electron sharing phenomena between the anion and the π-acidic surface of the NDI in their synthesized triangular, redox-active prisms (Fig. 11.17). The resulting electronic communication among the three NDI moieties leads to an unusually large number of individually accessible redox states in the triangular prisms, opening the door to potential applications in the molecular electronics field. The electron-deficient cavities of the molecular prisms are ideal for studying anion–π interactions. This ability is evidenced by the trapping of linear I_3^- anions inside the prismatic cavities, causing a significant change in the packing of the prisms in the extended solid-state architecture. The inclusion of I_3^- anions induces π–π stacking of the chiral prisms into supramolecular helices, providing a remarkable example of anion-induced self-assembly with potential applications as ion channels. In addition, the chirality endowed by the six stereogenic centers of the cyclohexane corners in the occupied prisms dictates the either right- or left-handedness associated with their packing in the solid state.

In a fascinating work, Kumar *et al.* [117] have demonstrated the extraordinary stability of NDI radical ion and its "ultra-electron-deficient" precursor. They have established new elements of stabilization utilizing the phosphonium group, enabling the isolation of the first NDI-based radical ion $[(NDI^{\bullet+})BPh_4^-]$ (Fig. 11.18). One of the most striking findings is its extraordinary stability leading to their synthesis, purification, and crystallization, all under ambient conditions. This is the first example of a radical ion that exhibits endurance toward silica gel chromatography. This research is highly appealing from the point of view of synthetic radical/radical ion chemistry. In addition, the crystal structure of the ultra-electron-deficient system provided direct evidence of interactions between the BF_4^- and the NDI core $[(NDI^{2+})BF_4^-]$ (Fig. 11.18). The facile electron transfer even in nonpolar solvents highly modulates the optical properties of the radical ion, providing new attractive scenarios in several fields like switchable panchromatic materials and phosphorus-based stable radical ions and, in general, spin-based research.

The NDI core has been also used by Liu *et al.* [118] to generate a panchromatic hybrid crystal of iodoplumbate nanowires and J-aggregated NDIs with long-lived charge-separated states even upon the irradiation of indoor lighting. Therefore, this investigation is of considerable significance in research fields like solar cells and photocatalysis. The organic–inorganic photoactive hybrid is composed of anionic iodoplumbate nanowires that interact with the NDIs by means of charge-assisted anion–π interactions (Fig. 11.19). Remarkably, the black crystal is much darker than other reported hybrids based on iodoplumbate nanowires [119, 120]. Actually, the solid-state UV–Vis diffuse reflectance spectral (DRS) analysis [118] revealed that it is a panchromatic hybrid with a broad absorption band from 200 to 800 nm. It is worth mentioning that the long-lived charge separation in this hybrid material is mainly due to the intense and panchromatic absorption and the close contacts between organic and inorganic units through anion–π interactions that are beneficial to the electron transfer.

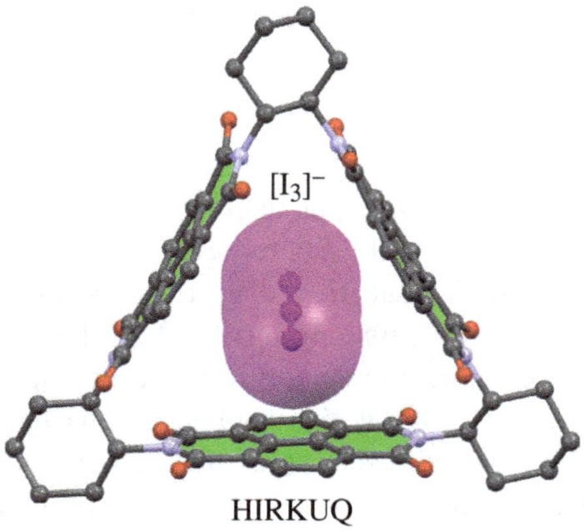

FIGURE 11.17 The X-ray crystal structure of HIRKUQ [116]. The CSD reference code is indicated. H-atoms are omitted for clarity.

The interplay between π–π and anion–π interactions has been investigated combining theoretical studies and search in the Cambridge Structural Database [75, 121, 122]. It has been demonstrated that the interplay between ion–π and π–π interactions can lead to strong cooperativity effects. A recent experimental study [123] has evidenced the formation of an anion/π–π/anion assembly in the solid-state structures of several copper(II) complexes with a tetradentate Schiff base (Fig. 11.20). These assemblies are crucial to explain the 3D architecture of the complexes in the solid state. This work also includes a theoretical study where the contribution of each interaction to the formation of the assembly and the interplay between the anion–π and π–π interactions is investigated and the mutual reinforcement of both interactions has been demonstrated computationally. It was also found that the systematic change of the halide coligands in these compounds (from Cl to I) has a very small influence upon the geometrical and energetic features of the anion/π–π/anion assemblies in the solid state.

A significant investigation that highlights the importance of the anion–π interactions in emerging fields of research has been reported by Sabek *et al.* [124] They have demonstrated the performance of chloride-selective electrodes based on "two-wall" aryl-extended calix[4]pyrroles and multiwall carbon nanotubes. When the meso-aryl substituents of the calix[4]pyrrole core are electron deficient, attractive anion–π interactions exist between the aromatic walls of the receptor

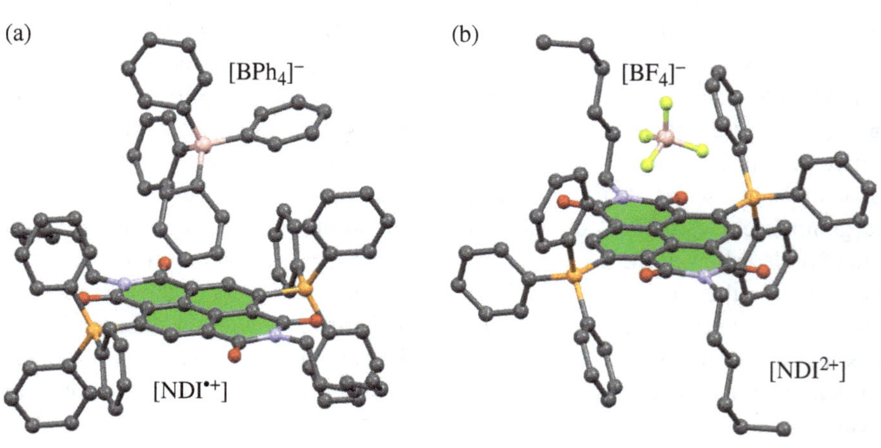

FIGURE 11.18 (a) Crystal structure of $[(NDI^{\bullet+})BPh_4^-]$. H-atoms are omitted for clarity. (b) Partial view of the crystal structure of $[(NDI^{2+})2BF_4^-]$; only one anion is shown. H-atoms are omitted for clarity.

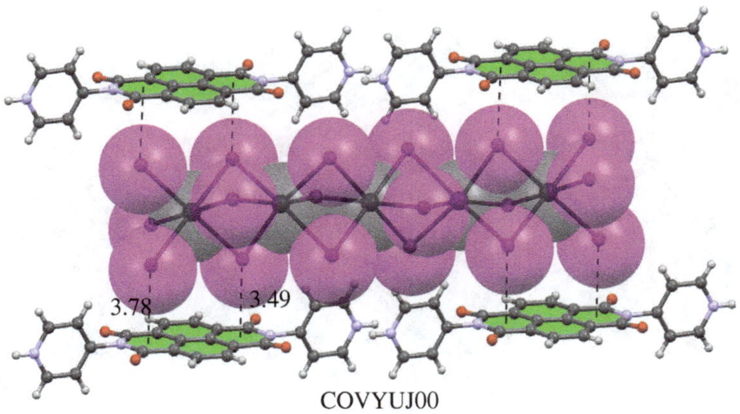

FIGURE 11.19 The X-ray crystal structure of COVYUJ00 [118]. The CSD reference code is indicated. Distances in Å.

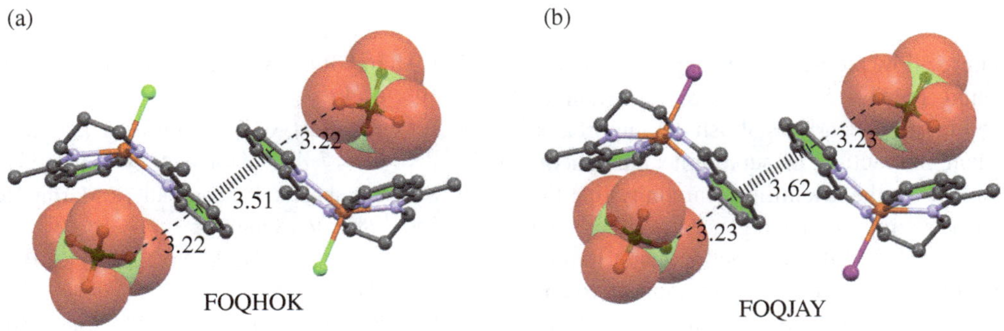

FIGURE 11.20 X-ray crystal structures of FOQHOK and FOQJAY [110]. The CSD reference codes are indicated. Distances in Å. H-atoms omitted for clarity.

and the sandwiched anion. These anion–π interactions are shown to remarkably control the selectivity of the electrodes. Calix[4]pyrrole, bearing a *p*-nitro withdrawing group on each of the meso-phenyl rings, afforded sensors that display anti-Hofmeister behavior against the lipophilic salicylate and nitrate anions. Based on the experimental data, the authors proposed a series of principles that are helpful to predict the suitability of synthetic receptors for use as anion-specific ionophores. As a matter of fact, the synthesized sensors delivered excellent results in the direct detection of chloride ions in bodily fluids due to the crucial role of anion–π interactions.

An interesting research that proposed the utilization of neutral chiral supramolecular assemblies to recognize anions has been reported by Kumar *et al.* [125]. They have demonstrated, using a chiral ligand, that two different supramolecular assemblies of Cu(II) can be obtained depending on the solvent of crystallization. The solvent (MeOH vs. DMSO) strongly influences the supramolecular network formation due to different hydrogen bonding interactions and orientation of the ligands around the metal center. The different binding ability toward various anions was demonstrated chromogenically and by fluorescence spectroscopy. The receptor–anion complex formation is established based on the anion–π and hydrogen bonding interactions. This research paves new avenues for the application of such neutral supramolecular assemblies as new receptors for anions.

Finally, we would like to emphasize the importance of anion–π interactions in hybrid inorganic–organic materials formed by POM and organic ligands. In particular [126] the synthesis and X-ray characterization of several hybrid inorganic–organic assemblies generated from $[SiW_{12}O_{40}]^{4-}$ as Keggin-type POM and trinuclear lanthanide clusters of formulas $\{Na(H_2O)_3[Ln(L)(H_2O)_3]_3\}^{4+}$ have been reported, where Ln metal was La, Ce, or Eu and the ligand (L) was either chelidamic acid or 2,6-dicarboxy-4-hydroxypyridine. In the solid-state structure of all three compounds, unprecedented anion–π interactions were observed between the POM, which is a tetraanion, and the aromatic rings that played a crucial role in the crystal packing formation (Fig. 11.21 for the Eu structure). This interesting investigation is the first report that describes and analyzes this interaction in Keggin-type POM-based inorganic–organic frameworks.

11.4 FINAL COMMENTS

The purpose of this chapter is to explain the bonding relationship between π-acidic systems and anions from two perspectives. First, the theoretical work and the understanding

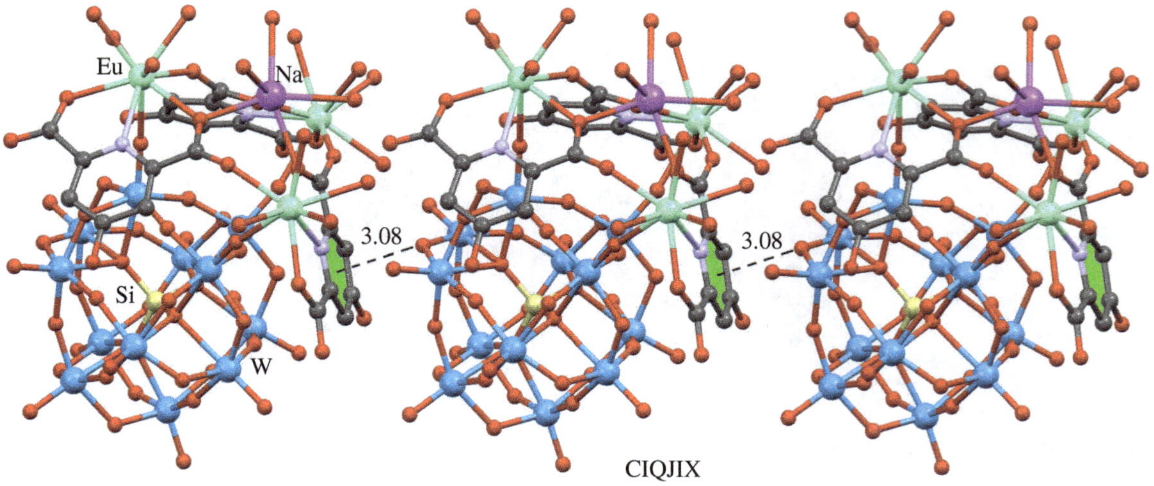

FIGURE 11.21 X-ray crystal structure of CIQJIX [126]. The CSD reference code is indicated. Distances in Å. H-atoms omitted for clarity.

of the physical nature allow predicting and explaining unexpected experimental results. Second, selected examples retrieved from the literature clearly establish the importance of this non-covalent interaction in supramolecular chemistry. It is now evident for the scientific community that the anion–π interaction is an important and general non-covalent binding force. The potential of this interaction is not limited to the design of novel hosts and sensors. Its important role in RNA recognition motifs, enzymatic chemistry, catalysis, and crystal engineering has provided a new dimension to this interaction.

ACKNOWLEDGMENTS

We are grateful to Carol Garau, Xavier Lucas, Daniel Escudero, David Quiñonero, and Pere M. Deyà with whom we have had the good fortune to work, and their names are contained within the pertinent references. We thank CONSOLIDER-Ingenio 2010 (project CSD2010-0065 FEDER funds) for financial support.

REFERENCES

1. H. J. Schneider, *Angew. Chem. Int. Ed.* 2009, *48*, 3924.
2. H. J. Schneider, A. Yatsimirski, Principles and methods in supramolecular chemistry, John Wiley & Sons, Ltd, Chichester, 2000.
3. J. M. Lehn, Supramolecular chemistry concepts and perspectives, Wiley–VCH, Weinheim, 1995.
4. F. Vögtle, Supramolecular chemistry: an introduction, John Wiley & Sons, Inc., New York, 1993.
5. P. D. Beer, P. A. Gale, D. K. Smith, Supramolecular chemistry, Oxford University Press, Oxford, 1999.
6. J. W. Steed, J. L. Atwood, Supramolecular chemistry, John Wiley & Sons, Ltd, Chichester, 2000.
7. P. Murray-Rust, W. D. S. Motherwell, *J. Am. Chem. Soc.* 1979, *101*, 4374.
8. P. Murray-Rust, W. C. Stallings, C. T. Monti, R. K. Preston, J. P. Glusker, *J. Am. Chem. Soc.* 1983, *105*, 3206.
9. N. Ramasubbu, R. Parthasarathy, P. Murray-Rust, *J. Am. Chem. Soc.* 1986, *108*, 4308.
10. P. Metrangolo, H. Neukirch, T. Pilati, G. Resnati, *Acc. Chem. Res.* 2005, *38*, 386.
11. P. Politzer, J. S. Murray, *ChemPhysChem* 2013, *14*, 278.
12. P. Politzer, J. S. Murray, T. Clark, *Phys. Chem. Chem. Phys.* 2013, *15*, 11178.
13. P. Politzer, J. S. Murray, T. Clark, *Phys. Chem. Chem. Phys.* 2010, *12*, 7748.
14. A. Bundhun, P. Ramasami, J. S. Murray, P. Politzer, *J. Mol. Model.* 2013, *19*, 2739.
15. A. Bauzá, T. J. Mooibroek, A. Frontera, *Angew. Chem. Int. Ed.* 2013, *52*, 12317.
16. J. C. Ma, D. A. Dougherty, *Chem. Rev.* 1997, *97*, 1303.
17. N. Zacharias, D. A. Dougherty, *Trends Pharmacol. Sci.* 2002, *23*, 281.
18. N. S. Scrutton, A. R. C. Raine, *Biochem. J.* 1996, *319*, 1.
19. J. P. Gallivan, D. A. Dougherty, *Proc. Natl. Acad. Sci. U. S. A.* 1999, *96*, 9459.
20. A. Frontera, P. Gamez, M. Mascal, T. J. Mooibroek, J. Reedijk, *Angew. Chem. Int. Ed.* 2011, *50*, 9564.
21. B. L. Schottel, H. T. Chifotides, K. R. Dunbar, *Chem. Soc. Rev.* 2008, *37*, 68.
22. C. Caltagirone, P. A. Gale, *Chem. Soc. Rev.* 2009, *38*, 520.
23. P. Gamez, T. J. Mooibroek, S. J. Teat, J. Reedijk, *Acc. Chem. Res.* 2007, *40*, 435.
24. B. P. Hay, V. S. Bryantsev, *Chem. Commun.* 2008, *44*, 2417.
25. D. Quiñonero, C. Garau, C. Rotger, A. Frontera, P. Ballester, A. Costa, P. M. Deyà, *Angew. Chem. Int. Ed.* 2002, *41*, 3389.
26. M. Mascal, A. Armstrong, M. D. Bartberger, *J. Am. Chem. Soc.* 2002, *124*, 6274.
27. I. Alkorta, I. Rozas, J. Elguero, *J. Am. Chem. Soc.* 2002, *124*, 8593.

28. D. Y. Kim, N. J. Singh, K. S. Kim, *J. Chem. Theory Comput.* 2008, *4*, 1401.
29. B. L. Schottel, J. Bacsa, K. R. Dunbar, *Chem. Commun.* 2005, *41*, 46.
30. B. Han, J. J. Lu, J. K. Kochi, *Cryst. Growth Des.* 2008, *8*, 1327.
31. M. Mascal, I. Yakovlev, E. B. Nikitin, J. C. Fettinger, *Angew. Chem. Int. Ed.* 2007, *46*, 8782.
32. R. J. Götz, A. Robertazzi, I. Mutikainen, U. Turpeinen, P. Gamez, J. Reedijk, *Chem. Commun.* 2008, *44*, 3384.
33. M. Albrecht, M. Müller, O. Mergel, K. Rissanen, A. Valkonen, *Chem. Eur. J.* 2010, *16*, 5062.
34. J. S. Murray, P. Lane, T. Clark, K. E. Riley, P. Politzer, *J. Mol. Model.* 2012, *18*, 541.
35. M. Egli, S. Sarkhel, *Acc. Chem. Res.* 2007, *40*, 197.
36. T. J. Mooibroek, P. Gamez, J. Reedijk, *CrystEngComm* 2008, *10*, 1501.
37. C. Estarellas, A. Frontera, D. Quiñonero, P. M. Deyà, *Angew. Chem. Int. Ed.* 2011, *50*, 415.
38. S. Chakravarty, Z. Z. Sheng, B. Iverson, B. Moore, *FEBS Lett.* 2012, *586*, 4180.
39. D. D. Jenkins, J. B. Harris, E. F. Howell, R. J. Hinde, J. Baudry, *J. Comput. Chem.* 2013, *34*, 518.
40. R. E. Dawson, A. Hennig, D. P. Weimann, D. Emery, V. Ravikumar, J. Montenegro, T. Takeuchi, S. Gabutti, M. Mayor, J. Mareda, C. A. Schalley, S. Matile, *Nat. Chem.* 2010, *2*, 533.
41. N. Sakai, J. Mareda, E. Vauthey, S. Matile, *Chem. Commun.* 2010, *46*, 4225.
42. P. A. Gale, C. Caltagirone, *Chem. Soc. Rev.* 2015, *44*, 4212.
43. A. Caballero, F. Zapata, P. D. Beer, *Coord. Chem. Rev.* 2013, *257*, 2434.
44. P. A. Gale, *Chem. Commun.* 2011, *47*, 82.
45. M. Wenzel, J. R. Hiscock, P. A. Gale, *Chem. Soc. Rev.* 2012, *41*, 480.
46. P. A. Gale, R. Quesada, *Coord. Chem. Rev.* 2006, *250*, 3219.
47. K. Bowman-James, A. Bianchi, E. García-España, Anion coordination chemistry, Wiley–VCH, Weinheim, 2011.
48. J. M. Berg, *Acc. Chem. Res.* 1995, *28*, 14.
49. F. M. Ashcroft, Ion channels and disease, Academic Press, San Diego and London, 2000.
50. B. Moss, *Chem. Ind.* 1996, *11*, 407.
51. T. Asakura, S. Y. Kim, Y. Morita, M. Ozawa, *J. Nucl. Radiochem. Sci.* 2005, *6*, 267.
52. S. Benner, *Nat. Geosci.* 2010, *3*, 5.
53. A. Frontera, D. Quiñonero, P. M. Deyà, *WIREs Comput. Mol. Sci.* 2011, *1*, 440.
54. M. Giese, M. Albrecht, T. Krappitz, M. Peters, V. Gossen, G. Raabe, A. Valkonen, K. Rissanen, *Chem. Commun.* 2012, *48*, 9983.
55. M. M. Watt, L. N. Zakharov, M. M. Haley, D. W. Johnson, *Angew. Chem. Int. Ed.* 2013, *52*, 10275.
56. H. J. Schneider, F. Werner, T. Blatter, *J. Phys. Org. Chem.* 1993, *6*, 590.
57. H. J. Schneider, *Angew. Chem. Int. Ed. Engl.* 1991, *30*, 1417.
58. K. Hiraoka, S. Mizuse, S. Yamabe, *J. Phys. Chem.* 1987, *91*, 5294.
59. K. Hiraoka, S. Mizuse, S. Yamabe, *J. Chem. Phys.* 1987, *86*, 4102.
60. A. Bauzá, D. Quiñonero, P. M. Deyà, A. Frontera, *New J. Chem.* 2013, *37*, 2636.
61. J. R. Galán-Mascarós, A. M. Z. Slawin, J. D. Woollins, D. J. Williams, *Polyhedron* 1996, *15*, 4603.
62. P. Ballester, Recognition of anions, Springer, Berlin and Heidelberg, 2008.
63. A. Frontera, *Coord. Chem. Rev.* 2013, *257*, 1716.
64. D. Kim, P. Tarakeshwar, K. S. Kim, *J. Phys. Chem. A* 2004, *108*, 1250.
65. D. Y. Kim, N. J. Singh, J. W. Lee, K. S. Kim, *J. Chem. Theory Comput.* 2008, *4*, 1162.
66. C. Garau, A. Frontera, D. Quiñonero, P. Ballester, A. Costa, P. M. Deyà, *ChemPhysChem* 2003, *4*, 1344.
67. D. Quiñonero, C. Garau, A. Frontera, P. Ballester, A. Costa, P. M. Deyà, *Chem. Phys. Lett.* 2002, *359*, 486.
68. A. Bauzá, P. M. Deyà, A. Frontera, D. Quiñonero, *Phys. Chem. Chem. Phys.* 2014, *16*, 1322.
69. S. E. Wheeler, K. N. Houk, *J. Phys. Chem. A* 2010, *114*, 8658.
70. C. Garau, D. Quiñonero, A. Frontera, P. Ballester, A. Costa, P. M. Deyà, *J. Phys. Chem. A* 2005, *109*, 9341.
71. C. Garau, A. Frontera, D. Quiñonero, P. Ballester, A. Costa, P. M. Deyà, *J. Phys. Chem. A* 2004, *108*, 9423.
72. C. Garau, A. Frontera, D. Quiñonero, P. Ballester, A. Costa, P. M. Deyà, *Chem. Phys. Lett.* 2004, *399*, 220.
73. C. Garau, D. Quiñonero, A. Frontera, P. Ballester, A. Costa, P. M. Deyà, *Org. Lett.* 2003, *5*, 2227.
74. A. Bauzá, D. Quiñonero, A. Frontera, P. Ballester, *Int. J. Mol. Sci.* 2015, *16*, 8934.
75. D. Quiñonero, A. Frontera, C. Garau, P. Ballester, A. Costa, P. M. Deyà, *ChemPhysChem* 2006, *7*, 2487.
76. A. Frontera, F. Saczewski, M. Gdaniec, M. Dziemidowicz-Borys, A. Kurland, P. M. Deyà, D. Quiñonero, *Chem. Eur. J.* 2005, *11*, 6560.
77. D. Quiñonero, A. Frontera, P. M. Deyà, *ChemPhysChem* 2008, *9*, 397.
78. I. A. Gural'skiy, D. Escudero, A. Frontera, P. V. Solntsev, E. B. Rusanov, A. N. Chernega, H. Krautscheid, K. V. Domasevitch, *Dalton Trans.* 2009, *38*, 2856.
79. A. Das, S. R. Choudhury, B. Dey, S. K. Yalamanchili, M. Helliwell, P. Gamez, S. Mukhopadhyay, C. Estarellas, A. Frontera, *J. Phys. Chem. B* 2010, *114*, 4998.
80. A. Das, S. Ray Choudhury, C. Estarellas, B. Dey, A. Frontera, J. Hemming, M. Helliwell, P. Gamez, S. Mukhopadhyay, *CrystEngComm* 2011, *13*, 4519.
81. A. Bauzá, D. Quiñonero, P. M. Deyà, A. Frontera, *Theor. Chem. Accounts* 2012, *131*, 1219.
82. C. Estarellas, A. Frontera, D. Quiñonero, P.M. Deyà, *J. Chem. Theory Comput.* 2008, *4*, 1981.
83. D. Quiñonero, A. Frontera, D. Escudero, P. Ballester, A. Costa, P. M. Deyà, *ChemPhysChem* 2007, *8*, 1182.
84. J. J. Fiol, M. Barceló-Oliver, A. Tasada, A. Frontera, À. Terrón, Á. García-Raso, *Coord. Chem. Rev.* 2013, *257*, 2705.
85. C. Garau, D. Quiñonero, A. Frontera, P. Ballester, A. Costa, P. M. Deyà, *New J. Chem.* 2003, *27*, 211.

86. I. Alkorta, J. Elguero, *J. Phys. Chem. A* 2003, *107*, 9428.
87. I. Alkorta, D. Quiñonero, C. Garau, A. Frontera, J. Elguero, P. M. Deyà, *J. Phys. Chem. A* 2007, *111*, 3137.
88. D. Quiñonero, A. Frontera, P. M. Deyà, I. Alkorta, J. Elguero, *Chem. Phys. Lett.* 2008, *460*, 406.
89. J. W. Steed, R. K. Juneja, J. L. Atwood, *Angew. Chem. Int. Ed.* 1995, *33*, 2456.
90. M. Staffilani, K. S. B. Hancock, J. W. Steed, K. T. Holman, J. L. Atwood, R. K. Juneja, R. S. Burkhalter, *J. Am. Chem. Soc.* 1997, *119*, 6324.
91. K. T. Holman, M. M. Halihan, S. S. Jurisson, J. L. Atwood, R. S. Burkhalter, A. R. Mitchell, J. W. Steed, *J. Am. Chem. Soc.* 1996, *118*, 9567.
92. M. Staffilani, G. Bonvicini, J. W. Steed, K. T. Holman, J. L. Atwood, M. R. J. Elsegood, *Organometallics* 1998, *17*, 1732.
93. H. T. Chifotides, B. L. Schottel, K. R. Dunbar, *Angew. Chem. Int. Ed.* 2010, *49* 7202.
94. G. Aragay, A. Frontera, V. Lloveras, J. Vidal-Gancedo, P. Ballester, *J. Am. Chem. Soc.* 2013, *135*, 2620.
95. I. A. Gural'skiy, P. V. Solntsev, H. Krautscheid, K. V. Domasevitch *Chem. Commun.* 2006, *42*, 4808.
96. D. Escudero, A. Frontera, D. Quiñonero, P.M. Deyà, *J. Comput. Chem.* 2009, *30* 75.
97. X. Lucas, C. Estarellas, D. Escudero, A. Frontera, D. Quiñonero, P.M. Deyà, *ChemPhysChem* 2009, *10*, 2256.
98. R. Chakrabarty, P. S. Mukherjee, P. J. Stang, *Chem. Rev.* 2011, *111*, 6810.
99. M. Fujita, M. Tominaga, A. Hori, B. Therrien, *Acc. Chem. Res.* 2005, *38*, 371.
100. M. Fujita, K. Umemoto, M. Yoshizawa, N. Fujita, T. Kusukawa, K. Biradha, *Chem. Commun.* 2001, 509.
101. M. D. Pluth, R. G. Bergman, K. N. Raymond, *Acc. Chem. Res.* 2009, *42*, 1650.
102. R. A. Bilbeisi, J. K. Clegg, N. Elgrishi, X. de Hatten, M. Devillard, B. Breiner, P. Mal, J. R. Nitschke, *J. Am. Chem. Soc.* 2012, *134*, 5110.
103. J. Rebek, Jr. *Acc. Chem. Res.* 2009, *42*, 1660.
104. H. T. Chifotides, I. D. Giles, K. R. Dunbar, *J. Am. Chem. Soc.* 2013, *135*, 3039.
105. D.-X. Wang, M.-X. Wang, *J. Am. Chem. Soc.* 2013, *135*, 892.
106. W. Liu, Q.-Q. Wang, Y. Wang, Z.-T. Huang, D.-X. Wang, *RSC Adv.* 2014, *4*, 9339.
107. J.-Z. Liao, X.-J. Dui, H.-L. Zhang, X.-Y. Wu, C.-Z. Lu, *CrystEngComm* 2014, *16*, 10530.
108. P. Manna, S. K. Seth, A. Bauzá, M. Mitra, S. R. Choudhury, A. Frontera, S. Mukhopadhyay, *Cryst. Growth Des.* 2014, *14*, 747.
109. P. Manna, S. K. Seth, M. Mitra, S. R. Choudhury, A. Bauzá, A. Frontera, S. Mukhopadhyay, *Cryst. Growth Des.* 2014, *14*, 5812.
110. F. Orvay, A. Bauzá, M. Barceló-Oliver, A. García-Raso, J. J. Fiol, A. Costa, E. Molins, I. Mata, A. Frontera, *CrystEngComm* 2014, *16*, 9043.
111. R. Podgajny, D. Pinkowicz, B. Czarnecki, M. Kozieł, S. Chorąży, M. Wis, W. Nitek, M. Rams, B. Sieklucka, *Cryst. Growth Des.* 2014, *14*, 4030.
112. N. Hafezi, J. M. Holcroft, K. J. Hartlieb, E. J. Dale, N. A. Vermeulen, C. L. Stern, A. A. Sarjeant, J. F. Stoddart, *Angew. Chem. Int. Ed.* 2015, *54*, 456.
113. M. Giese, M. Albrecht, A. Valkonen, K. Rissanen, *Eur. J. Org. Chem.* 2013, 3247.
114. M. Giese, M. Albrecht, T. Repenko, J. Sackmann, A. Valkonen, K. Rissanen, *Eur. J. Org. Chem.* 2014, 2435
115. M. Giese, M. Albrecht, A. Valkonen, K. Rissanen, *Chem. Sci.* 2015, *6*, 354.
116. S. T. Schneebeli, M. Frasconi, Z. Liu, Y. Wu, D. M. Gardner, N. L. Strutt, C. Cheng, R. Carmieli, M. R. Wasielewski, J. R. Stoddart, *Angew. Chem. Int. Ed.* 2013, *52*, 13100.
117. S. Kumar, M. R. Ajayakumar, G. Hundal, P. Mukhopadhyay, *J. Am. Chem. Soc.* 2014, *136*, 12004.
118. J.-J. Liu, Y.-F. Guan, C. Jiao, M.-J. Lin, C.-C. Huang, W.-X. Dai, *Dalton Trans.* 2015, *44*, 5957.
119. J. Fujisawa, N. Tajima, K. Tamaki, M. Shimomura, T. Ishihara, *J. Phys. Chem. C* 2007, *111*, 1146.
120. T. Fukumoto, M. Hirasawa, T. Ishihara, *J. Lumin.* 2000, *497*, 87.
121. A. Frontera, D. Quiñonero, C. Garau, A. Costa, P. Ballester, P. M. Deyà, *J. Phys. Chem. A* 2006, *110*, 9307.
122. A. Frontera, D. Quiñonero, A. Costa, P. Ballester, P. M. Deyà, *New J. Chem.* 2007, *31*, 556.
123. A. Bhattacharyya, P. K. Bhaumik, A. Bauzá, P. P. Jana, A. Frontera, M. G. B. Drew, S. Chattopadhyay, *RSC Adv.* 2014, *4*, 58643.
124. J. Sabek, L. Adriaenssens, T. Guinovart, E. J. Parra, F. X. Rius, P. Ballester, P. Blondeau, *Chem. Eur. J.* 2015, *21*, 448.
125. N. Kumar, S. Khullar, S. K. Mandal, *Dalton Trans.* 2015, *44*, 1520.
126. M. Mirzaei, H. Eshtiagh-Hosseini, N. Lotfian, A. Salimi, A. Bauzá, R. Van Deun, R. Decadt, M. Barceló-Oliver, A. Frontera, *Dalton Trans.* 2014, *43*, 1906.

12

ASYMMETRIC AZAMACROCYCLES AS CHIRAL SOLVATING AGENTS

KOICHI TANAKA

Department of Chemistry and Materials Engineering, Faculty of Chemistry, Materials and Bioengineering, Kansai University, Osaka, Japan

12.1 INTRODUCTION

There is increasing demand for rapid and accurate methods for determining enantiomeric purity of chiral compounds in biological and pharmaceutical chemistry [1]. Among various chromatographic and spectroscopic methods, ^{1}H-NMR spectroscopy is useful because the enantiomeric purity can be determined quickly and economically, using only a small amount of deuterated solvent and a few milligrams of chiral compounds. Enantiomer discrimination in NMR spectroscopy is achieved either (i) by forming a covalent bond to the analyte with "chiral derivatizing agent" (CDA) or (ii) by forming a diastereomeric complex with a "chiral solvating agent" (CSA) [2–4]. The diastereomeric species show chemical shift nonequivalence of their ^{1}H-NMR signals, allowing enantiomeric excess determination of the analyte by direct integration of ^{1}H-NMR signals. The use of a CDA may involve the possibility of kinetic resolution in the covalent bond-forming step, while that of a CSA does not possess such drawbacks. The use of a CSA is also advantageous because both the starting chiral compounds and the CSA can be easily recovered after the measurement.

The wide variety of CSAs, such as pincer–diamines [5], crown ethers [6], calixarenes [7], cyclodextrins, [8] porphyrins [9], and macrocycles [10] have already been reported. Among these CSAs, chiral azamacrocyclic compounds have been recognized as successful and promising chiral selectors for molecular recognition mainly because of their inherent reduced flexibility and complexation ability. Here we give an overview of various types of artificial macrocyclic receptors, such as BINOL-based macrocyclic amines, macrocyclic amides, macrocyclic amines containing a *trans*-1,2-diaminocyclohexyl moiety, and aza-crown macrocycles, which have been developed mainly for chiral analysis and separation. Excellent books and reviews on related topic have already been published some time ago (e.g., see [1]); however, here we would like to introduce some of more recent achievances.

12.2 BINOL-BASED CHIRAL MACROCYCLIC AMINES

Chiral macrocycle **1** (Scheme 12.1) with a C_2 symmetric BINOL unit was found to be an extremely versatile CSA for a wide range of chiral compounds such as carboxylic acids, oxazolidinones, carbonates, lactones, alcohols, sulfoxides, sulfoximides, isocyanates, and epoxides [11]. ^{1}H-NMR signals for the enantiomers of **3**, **4**, and **5–13** (Scheme 12.2) were well split upon complexation with the chiral macrocycle **1a**. Good enantiomeric discrimination was achieved in many cases ($\Delta\Delta\delta > 0.15$ ppm). In the case of sulfoxide **10**, only 5 mol% of **1a** was enough for the complete splitting ($\Delta\Delta\delta = 0.55$ ppm) of the ^{1}H-NMR signals of **10**. Host **1b** with a 3,5-bis(trifluoromethyl)phenyl group at the 3,3′-positions showed improved enantioselectivity for small guest molecules such as 2-chloropropionic acid and methyl lactate as shown in Table 12.1 [12]. Chiral macrocycle **2** is also an effective CSA for the carboxylic acids, and in many cases baseline resolution of signals was observed [13].

Non-Covalent Interactions in the Synthesis and Design of New Compounds, First Edition.
Edited by Abel M. Maharramov, Kamran T. Mahmudov, Maximilian N. Kopylovich, and Armando J. L. Pombeiro.

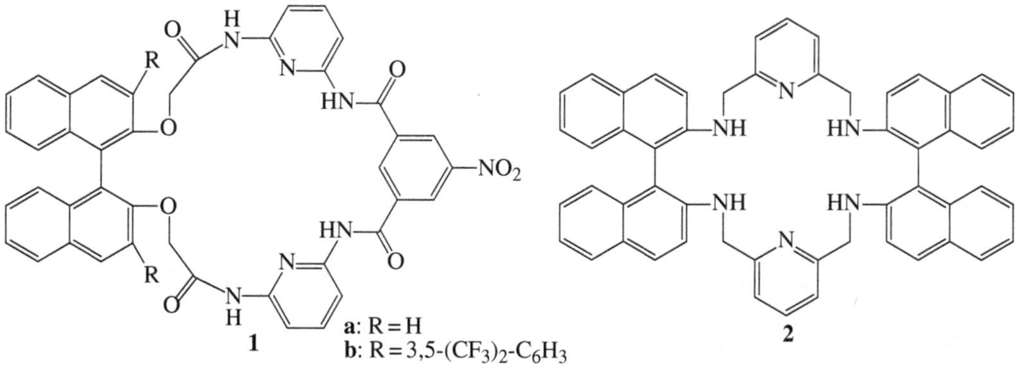

1 a: R = H
b: R = 3,5-$(CF_3)_2$-C_6H_3

2

SCHEME 12.1 BINOL-based chiral macrocyclic amines.

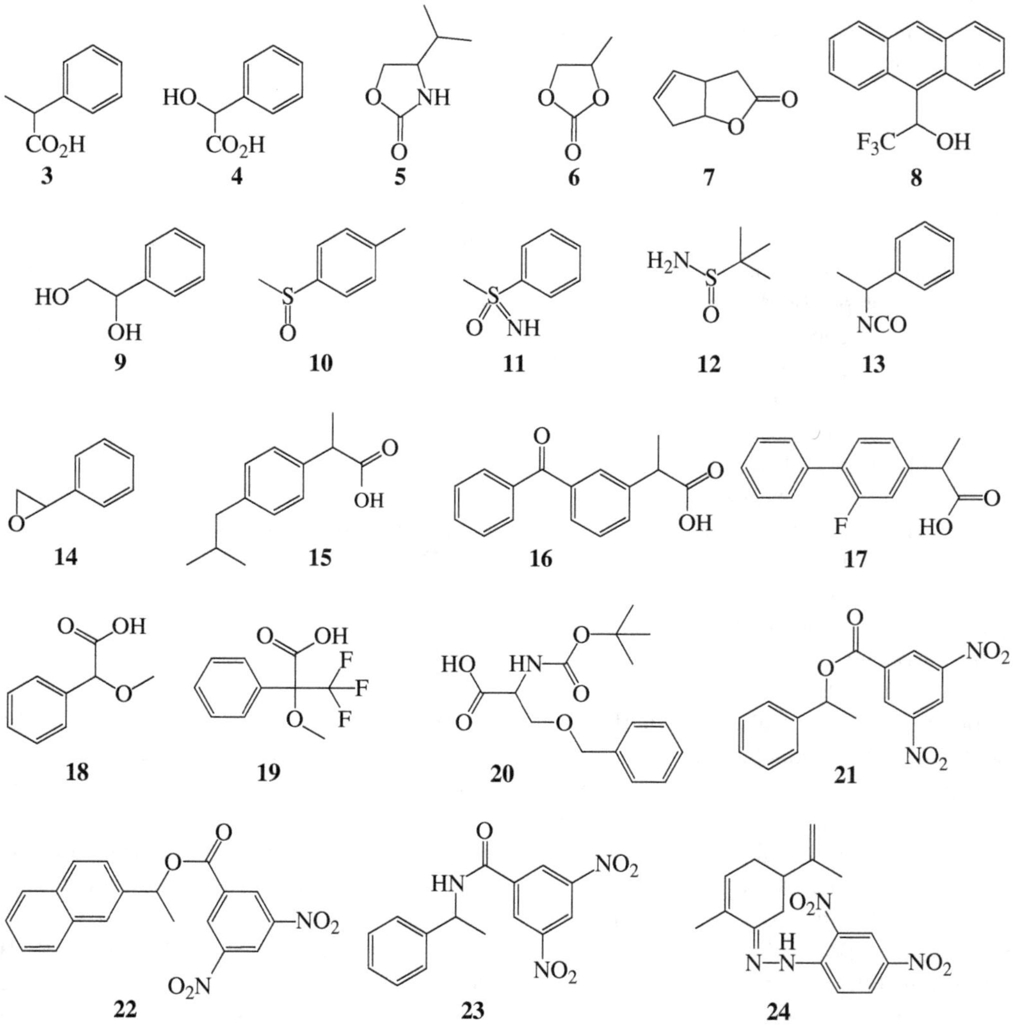

SCHEME 12.2 Guest compounds.

TABLE 12.1 Binding Constants of Host (*R*)-1a and (*R*)-1b for Selected Guests [12]

Guest	(*R*)-**1a**	(*R*)-**1b**
(*R*)-**3**	1670	550
(*S*)-**3**	3050	760
(*R*)-**10**	610	66
(*S*)-**10**	2600	93
(*R*)-**51b**	3100	1470
(*S*)-**51b**	3080	8760

12.3 CHIRAL MACROCYCLIC AMIDES

Tetraamidic CSAs **25** and **26** (Scheme 12.3) were prepared and enantiodiscrimination experiments were carried out by ^{1}H-NMR of guests **29–33** (Scheme 12.4) [14]. In the presence of **26**, the proton resonances of guest **29a** undergo splitting, ranging from about 0.01 ppm for the para (*p*)-proton of the 3,5-dinitrophenyl moiety to 0.13 ppm for the amide proton. Interestingly, these nonequivalences are significantly

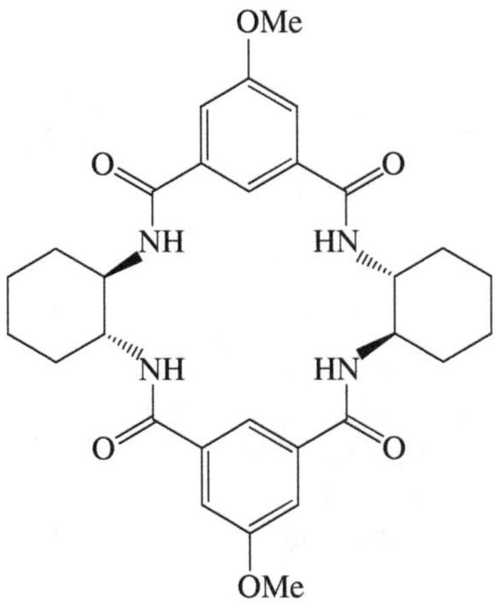

25

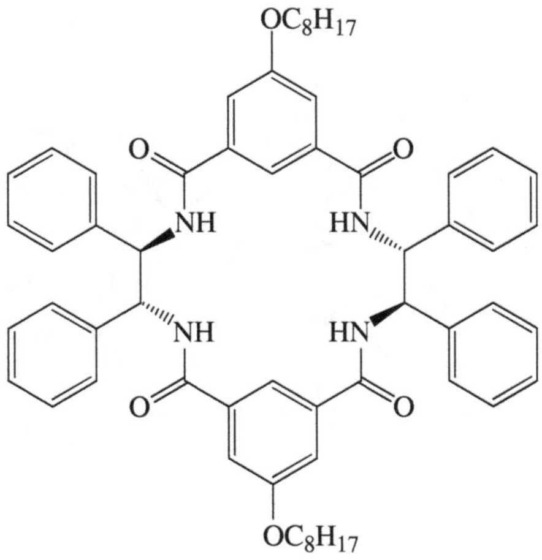

26

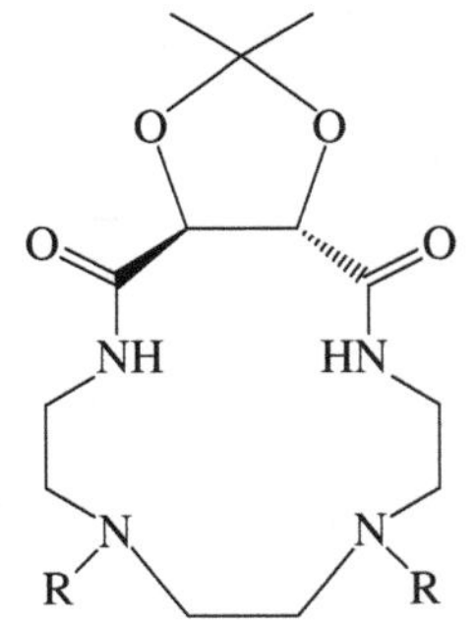

27
a: R = H
b: R = CH_2Ph

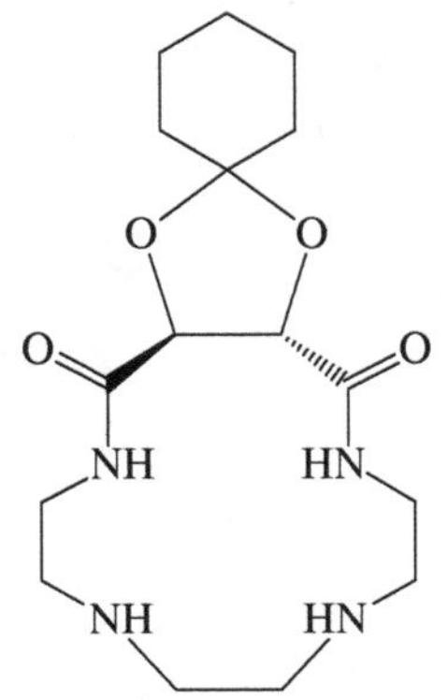

28

SCHEME 12.3 Chiral macrocyclic amides.

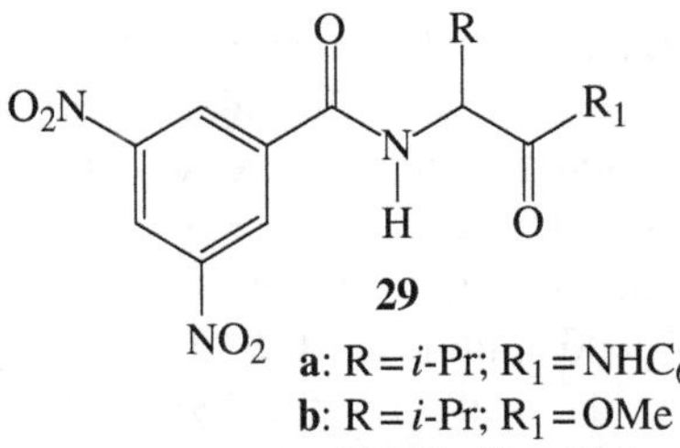

29
a: R = *i*-Pr; R_1 = NHC_6H_{13}
b: R = *i*-Pr; R_1 = OMe
c: R = Ph; R_1 = OMe
d: R = Me; R_1 = OMe

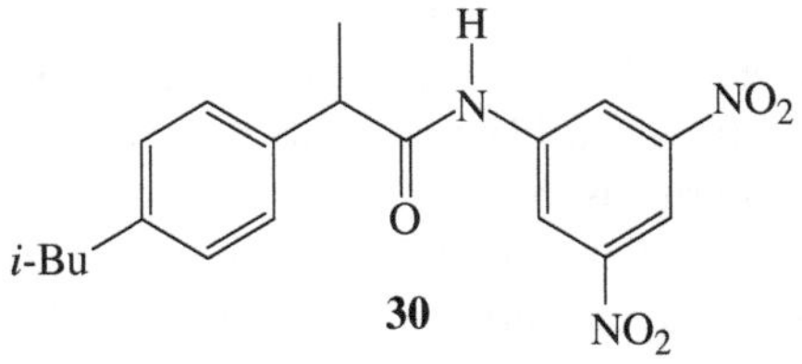

30

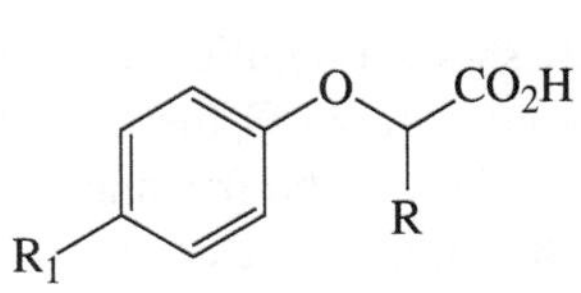

31
a: R = C_6H_{13} ; R_1 = Cl
b: R = C_2H_5 ; R_1 = Me

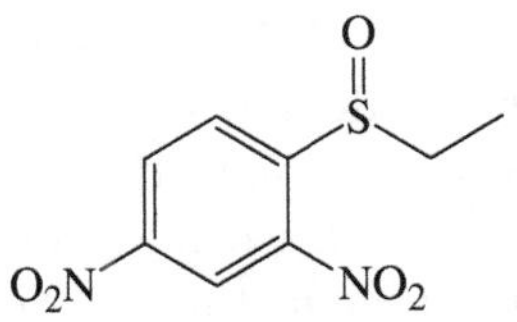

32

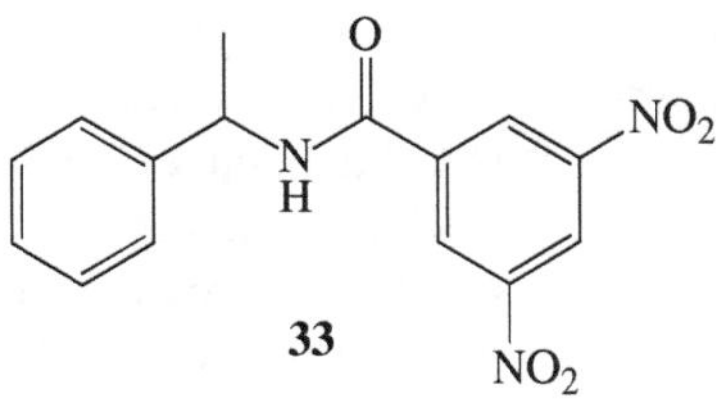

33

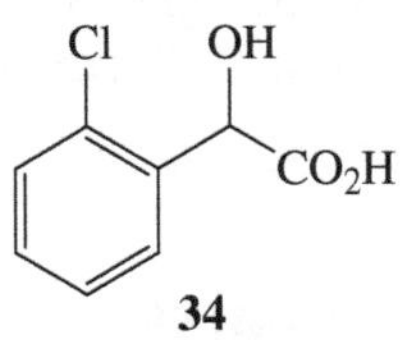

34

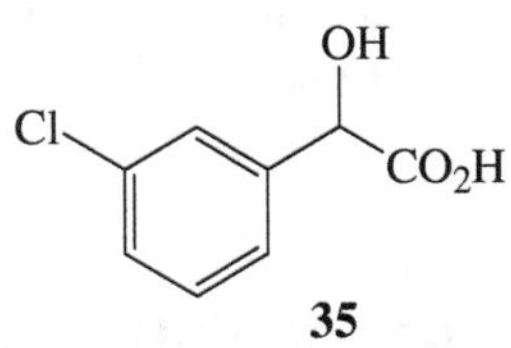

35

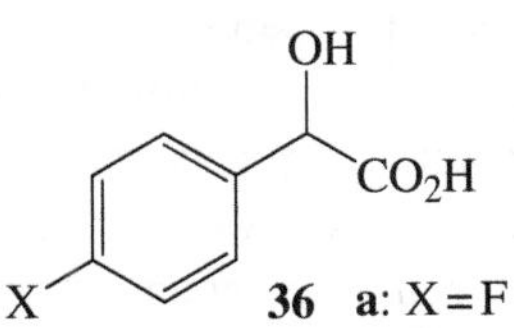

36 **a**: X = F
b: X = Cl
c: X = Br

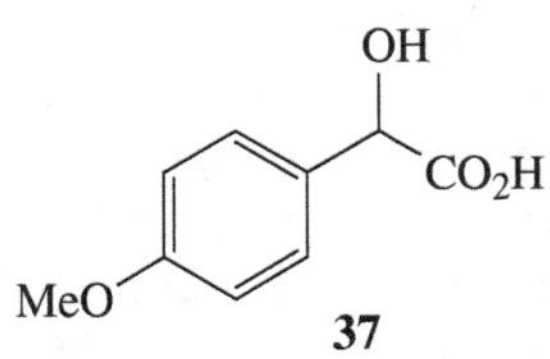

37

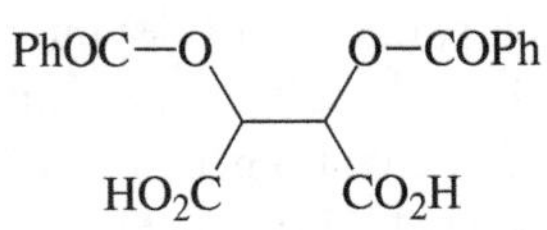

38

SCHEME 12.4 Guest compounds.

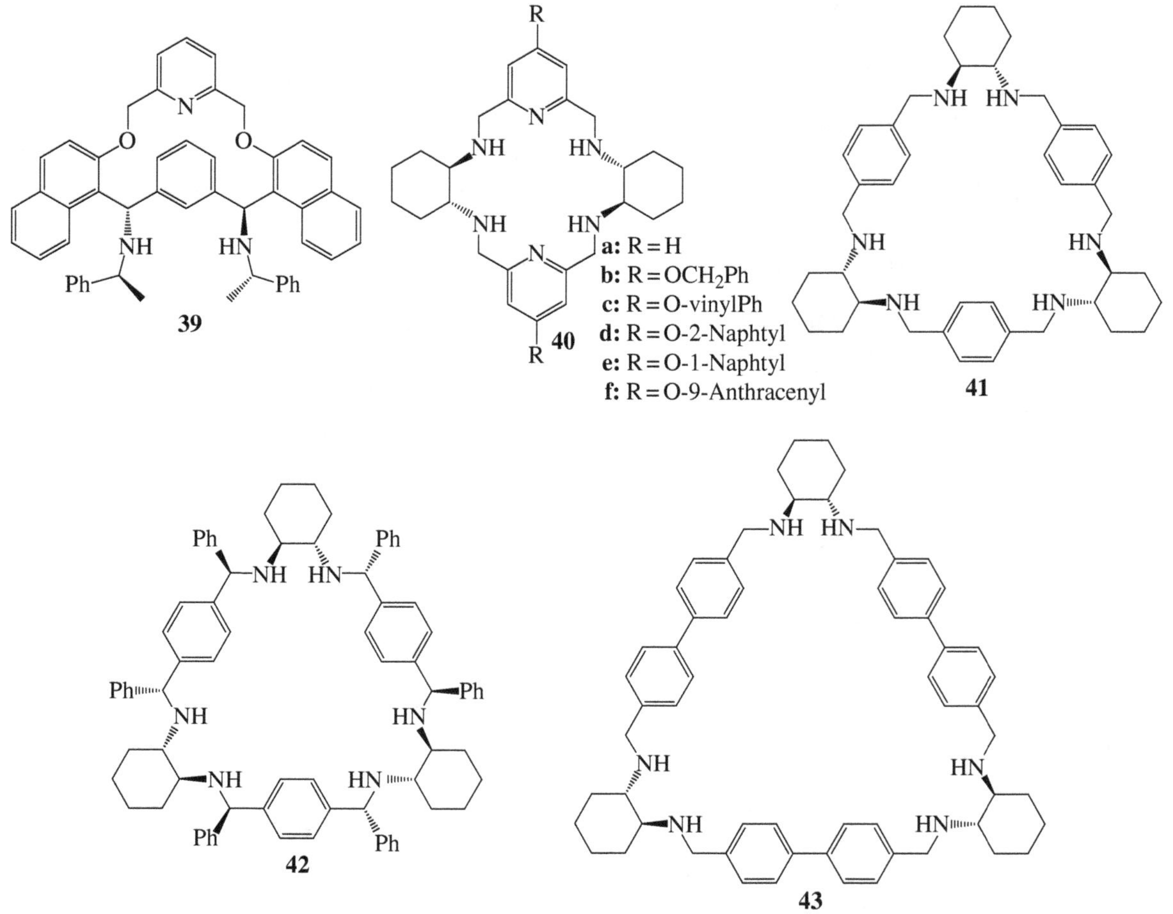

SCHEME 12.5 Chiral macrocyclic amines.

increased on lowering the temperature. For example, chiral sulfoxide **32** shows a significantly high nonequivalence of the aromatic protons. At 25°C, these are 0.03, 0.05, and 0.12 ppm for the meta' (m'), ortho (o), and meta (m) protons, respectively. These values increase at −40°C to 0.1 ppm for m', 0.14 ppm for o, and 0.43 ppm for m. However, the enantiodiscriminating efficiency of **25** is lower than that of **26**.

The enantioselectivity of chiral macrocyclic polyamides **27** and **28**, derived from L- or D-tartaric acid (Scheme 12.3), was also investigated [15]. As a CSA, the macrocycle **27b** has an excellent enantiomeric discriminating ability for mandelic acid and its derivatives **34–37**, while macrocycle **28** has the best enantioselectivity toward dibenzoyltartaric acid **38**.

12.4 CHIRAL MACROCYCLIC AMINES

Chiral macrocyclic compound **39** (Scheme 12.5) with a 16-member ring was prepared from C_2-symmetric aminonaphthol [16]. Host **39** shows excellent discrimination ability (up to 0.80 ppm) for the enantiomers of a wide variety of carboxylic acids **3, 4, 18, 45, 51a**, and **50a–50d** (Scheme 12.6) in ^{1}H-NMR spectroscopy. In the presence of **39**, the ^{1}H-NMR N*H*COCH$_3$ signals of **50a–50d** move downfield while all the other observable proton signals of the carboxylic acids **3**, **4**, **18**, **45**, **51a**, and **50a–50d** move upfield. The efficient use of polyazamacrocycle **40** as CSA has been studied for determining the enantiomeric excess of different carboxylic acids **3**, **4**, **18**, **36b**, **37**, **47**, **51a**, and **52** [17, 18]. Formation of multimolecular diastereomeric complexes in solution gives good splitting of the ^{1}H-NMR signals for the enantiomers of the acids (up to $\Delta\Delta\delta = 0.20$ ppm) using a small amount (as low as 0.125 eq.) of the polyazamacrocycles **40**.

Trianglamines (*S,S,S,S,S,S*)-**41** and (*S,S,S,S,S,S*)-**43** can be readily synthesized through [3+3] cyclocondensation of enantiomerically pure 1,2-diaminocyclohexane with terephthaldehyde or 4,4′-diformylbiphenyl, followed by $NaBH_4$ reduction. The D_3 symmetrical chiral structures of **41** and **43** provide a highly asymmetric environment for the chiral guest molecules through various interactions such as hydrogen bonding, π–π stacking, and CH–π interactions. Thus, chiral trianglamine macrocycles act as ^{1}H-NMR chiral shift agents for determining absolute configuration and enantiomeric purity of neutral guest compounds [19]. Table 12.2 shows the chemical shift differences ($\Delta\Delta\delta$) between the enantiomers of *rac*-alcohols **56–60** in the presence of 1 or 2 eq. of (*S,S,S,S,S,S*)-**41** or (*S,S,S,S,S,S*)-**43** in $CDCl_3$.

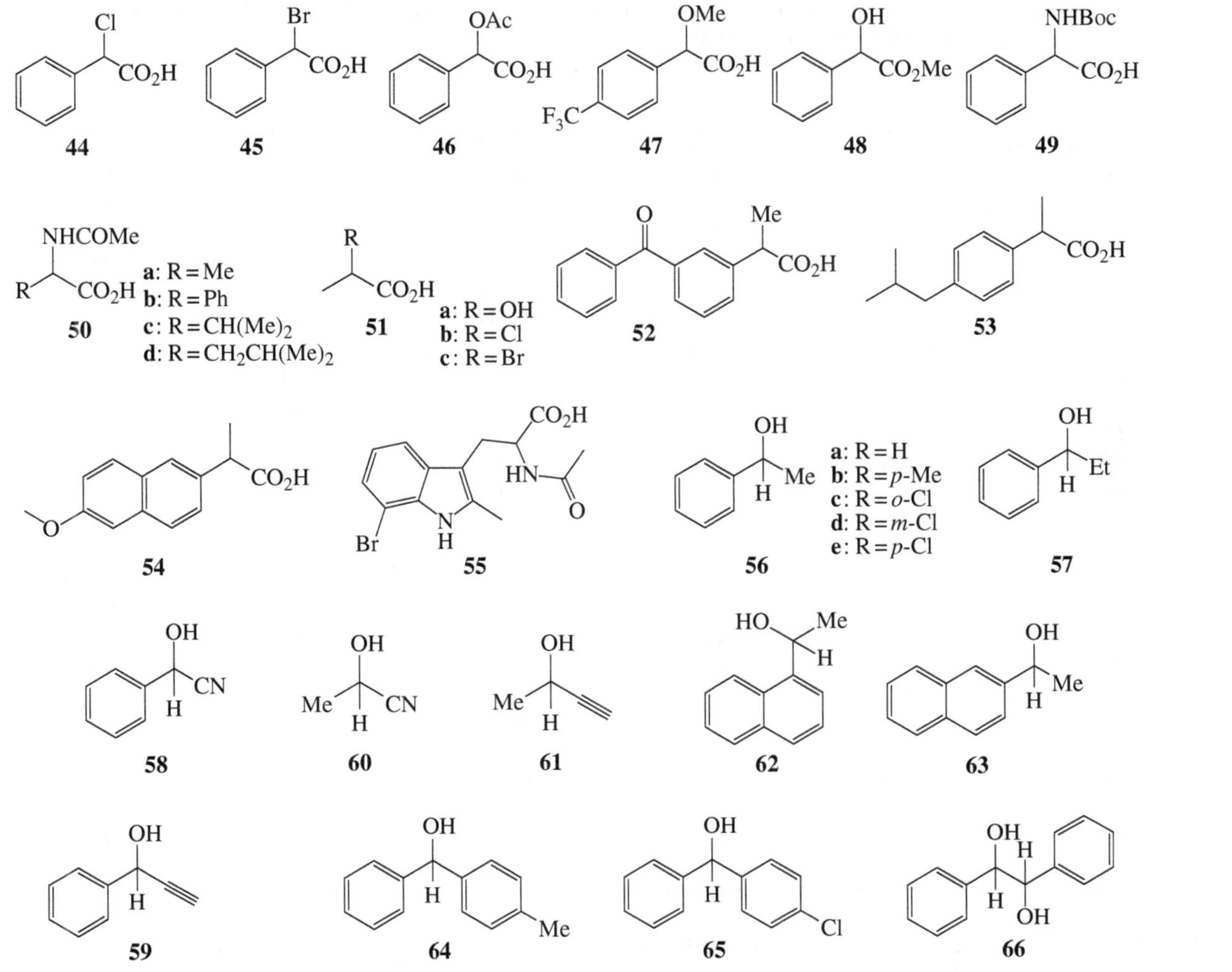

SCHEME 12.6 Guest compounds.

As can be seen the methine proton signal (CαH) of 1-phenylethanol **56a** in the presence of (*S,S,S,S,S,S*)-**41** is split into two quartets with a $\Delta\Delta\delta$ value of 0.064 ppm due to the diastereomeric host–guest interactions of the enantiomers of **56a** with (*S,S,S,S,S,S*)-**41**. The methyl proton signal is also split into two doublets ($\Delta\Delta\delta = 0.040$ ppm) in the presence of (*S,S,S,S,S,S*)-**41**. For the chloro derivatives **56c–56e**, the chemical shift of the CαH proton is in the order *o*-Cl-**56c** > *m*-Cl-**56d** > *p*-Cl-**56e**. For compounds **57** and **58**, the signals for the protons attached to the stereogenic center are split. 1-Phenyl-2-propyn-1-ol **59** shows smaller nonequivalences of both CαH and C≡CH protons, whereas 3-butyn-2-ol **61**, without an aromatic ring, exhibits no signal splitting. Conversely, 2-hydroxypropanenitrile **60**, possessing an electronegative CN group, shows baseline resolution, demonstrating that an electronegative group attached to the stereogenic center is more important than an aromatic ring for better signal splitting. In α- and β-1-naphthyl-1-ethanols **62** and **63**, both CαH and CH_3 proton signals are resolved successfully. The diphenylmethanol derivatives **64** and **65** and 1,2-diphenylethan-1,2-diol **66** show smaller nonequivalences of the signal than that observed in 1-phenylethanol derivatives **56a–56d**. As can be seen from Table 12.3, (*S,S,S,S,S,S*)-**41** exhibits a better discriminating ability than (*S,S,S,S,S,S*)-**43** except for compounds **58** and **66**.

In the case of compounds **56a**, **57**, **59**, and **60**, the α-methine proton (CαH) of the (*R*)-enantiomer appears on ^{1}H NMR spectra at a higher field than that of the (*S*)-enantiomer in the presence of (*S,S,S,S,S,S*)-**41** or (*S,S,S,S,S,S*)-**43** (Table 12.2). In the case of relatively bulky guest compounds **64** and **65** with two benzene rings, the α-methine proton signal of the (*S*)-enantiomer appears at a higher field than that of the (*R*)-enantiomer in the presence of (*S,S,S,S,S,S*)-**41**. In the NMR spectra of compound **66**, the α-methine proton signal of the (*S*)-enantiomer appears at a higher field in the presence of (*S,S,S,S,S,S*)-**41** while that of the (*R*)-enantiomer can be seen at a higher field in the presence of (*S,S,S,S,S,S*)-**43**. The enantiomeric excess (% ee) of the guest alcohol can be determined by integrating the corresponding CαH ^{1}H-NMR signal in the presence of the chiral trianglamine. For example, samples containing different % ee's of **66** were prepared and their NMR spectra in the presence of (*S,S,S,S,S,S*)-**41** were measured. An excellent linear correlation ($R^2 = 0.999$) between theoretical and observed % ee values was observed.

TABLE 12.2 Chemical Shift Difference (ΔΔδ) Between Enantiomers of *rac*-Alcohols (56–66) in the Presence of (*S,S,S,S,S,S*)-41 or (*S,S,S,S,S,S*)-43 in $CDCl_3$

		(*S,S,S,S,S,S*)-**41**		(*S,S,S,S,S,S*)-**43**	
Alcohol	Signal	Ratio	ΔΔδ (ppm)	Ratio	ΔΔδ (ppm)
56a	CαH	1 : 1	0.064	2 : 1	0.011
	CH_3	1 : 1	0.040	2 : 1	0.016
56b	CαH	2 : 1	0.061	2 : 1	0.010
56c	CαH	2 : 1	0.078	1 : 1	0.012
	CH_3	2 : 1	0.027	1 : 1	0.016
56d	CαH	1 : 1	0.047	2 : 1	0.007
	CH_3	1 : 1	0.014	2 : 1	0
56e	CαH	1 : 1	0.044	2 : 1	0.005
	CH_3	1 : 1	0.026	2 : 1	0
57	CαH	2 : 1	0.047	1 : 1	0.012
58	CαH	1 : 1	0.040	2 : 1	0.064
59	CαH	1 : 1	0.014	1 : 1	0.006
	C≡CH	1 : 1	0.014	1 : 1	0.006
60	CαH	1 : 1	0.099	1 : 1	0.035
	CH_3	1 : 1	0.016	1 : 1	0.008
61	CαH	1 : 1	0	1 : 1	0
	C≡CH	1 : 1	0	1 : 1	0
62	CαH	1 : 1	0.059	1 : 1	0.009
	CH_3	1 : 1	0.016	1 : 1	0
63	CαH	1 : 1	0.032	1 : 1	0
	CH_3	1 : 1	0.017	1 : 1	0
64	CαH	1 : 1	0.010	1 : 1	0
	CH_3	1 : 1	0.013	1 : 1	0
65	CαH	2 : 1	0.012	2 : 1	0
66	CαH	2 : 1	0.016	2 : 1	0.020

TABLE 12.3 Absolute Configuration of Enantiomer Showing More Up-field Shift in the ¹H-NMR Spectra in the Presence of (*S,S,S,S,S,S*)-41 or (*S,S,S,S,S,S*)-43

Alcohol	Host	Enantiomer Showing More Upfield Shift
56a	(*S,S,S,S,S,S*)-**41**	*R*
58	(*S,S,S,S,S,S*)-**41**	*R*
58	(*S,S,S,S,S,S*)-**43**	*R*
59	(*S,S,S,S,S,S*)-**41**	*R*
60	(*S,S,S,S,S,S*)-**41**	*R*
60	(*S,S,S,S,S,S*)-**43**	*R*
64	(*S,S,S,S,S,S*)-**41**	*S*
65	(*S,S,S,S,S,S*)-**41**	*S*
66	(*S,S,S,S,S,S*)-**41**	*S*
66	(*S,S,S,S,S,S*)-**43**	*R*

In comparison with (*S,S,S,S,S,S*)-**41**, the macrocycle **42** with phenyl substituents in the benzylic positions allows use of a lower amount of the chiral shift reagent for observing good enantiodiscrimination for racemic acids **3**, **4**, **44**, **45**, **46**, **53**, **54**, **55**, and **56** [20].

The chiral rhombamine macrocycles **67** and **68** (Scheme 12.7) were prepared by treating enantiomerically pure (*R,R*)-1,2-cyclohexanediamine with a suitable dialdehyde, followed by $NaBH_4$ reduction of the intermediate macrocyclic imine [21]. The structure of (*R,R,R,R*)-**67a** was determined by X-ray crystallography. Intramolecular hydrogen bonds were observed between N1—H11···N4 (2.301 Å) and N3—H37···N2 (2.231 Å) as shown in Figure 12.1.

Ring-expanded novel chiral rhombamine macrocycles **68** were also synthesized according to the routes analogous to those reported for **67** [22]. The structure of (*R,R,R,R*)-**68a** was also determined by X-ray crystallography; this compound crystallizes in the triclinic space group P_1. The asymmetric unit contains one (*R,R,R,R*)-**68a** molecule and four 1,4-dioxane molecules, which are located in the cavity of (*R,R,R,R*)-**68a** as shown in Figure 12.2.

Chiral shift experiments were carried out by obtaining ¹H-NMR spectra (400 MHz) of mixtures of (*R,R,R,R*)-rhombamine macrocycles **67** or **68** racemic carboxylic acids **3**, **4**, **18**, and **45** in $CDCl_3$ at room temperature. Upon addition of 0.25 eq. (*R,R,R,R*)-**68a**, the chemical shift values of the methine protons (CαH) attached to the asymmetric carbons of (*S*)- and (*R*)-mandelic acid **4** exhibited 0.437 and 0.346 ppm upfield shifts, respectively. Furthermore, upon gradual addition of (*R,R,R,R*)-**68a**, the ¹H-NMR signals for the CαH protons of **4** moved upfield, and the difference in the chemical shifts of the two enantiomers increased gradually. The difference of 0.091 ppm for the two enantiomers at 0.25 eq. was found to be the best result for chiral recognition (Fig. 12.3).

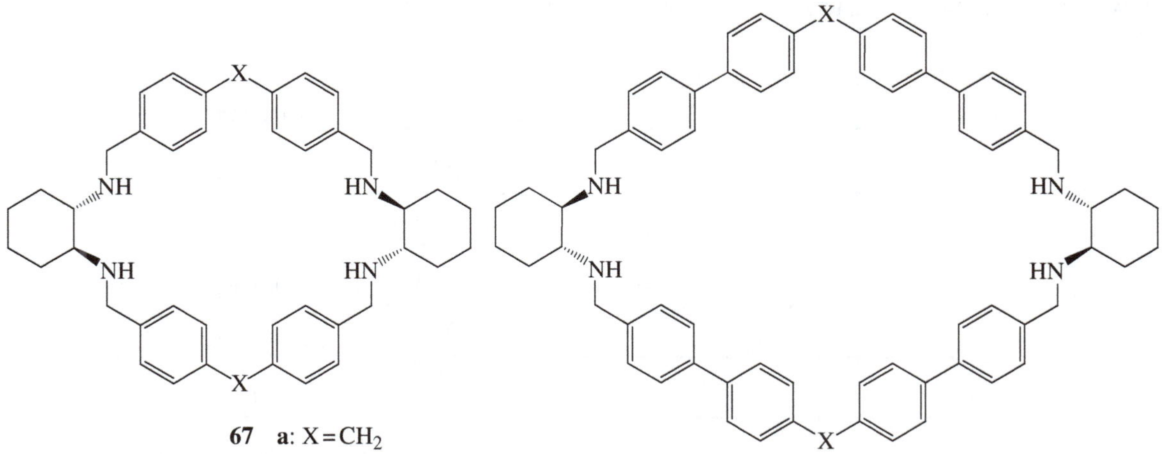

67 **a**: $X=CH_2$
b: $X=O$

68 **a**: $X=CH_2$
b: $X=O$

SCHEME 12.7 Chiral rhombamine macrocycles.

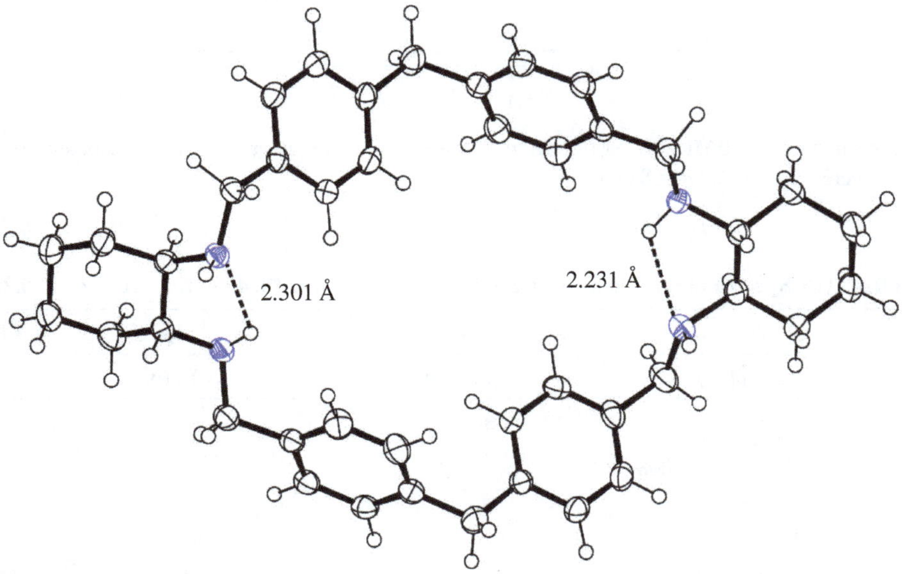

FIGURE 12.1 Ortep drawing of (*R,R,R,R*)-**67a** molecule. Thermal displacement ellipsoids are drawn at the 50% probability level. H atoms are represented by spheres. Intramolecular hydrogen bonds are represented as broken lines.

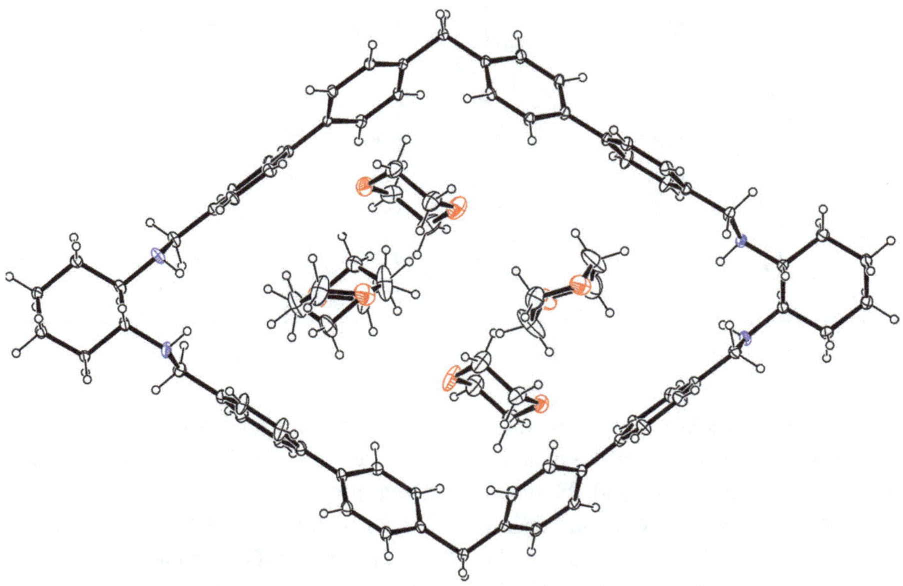

FIGURE 12.2 ORTEP drawing of 1 : 4 complex of (*R,R,R,R*)-**68a** with 1,4-dioxane.

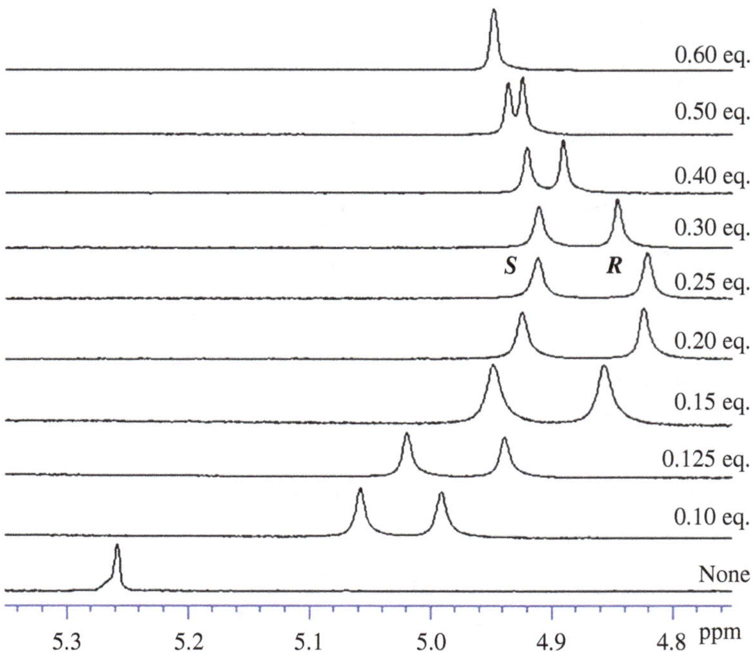

FIGURE 12.3 Variations in part of 400 MHz ^{1}H-NMR spectrum corresponding to the benzylic C—H resonance of *rac*- mandelic acid (**4**, 20.0 mM in $CDCl_3$) upon increasing (*R,R,R,R*)-**68a** content.

TABLE 12.4 Partial ^{1}H-NMR Spectra of *rac*-Phenylacetic Acid Derivatives 3, 4, 18, 45 in the Presence of 0.25 eq. of 67 and 68

Racemate	67a ΔΔδ (ppm)		67b ΔΔδ (ppm)		68a ΔΔδ (ppm)		68b ΔΔδ (ppm)	
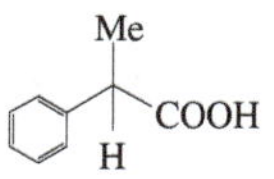 **3**	CH_3 (R S R S; 1.45, 1.40)	0.008	CH_3 (1.45)	0	CH_3 (1.5)	0	CH_3 (1.50, 1.45)	0
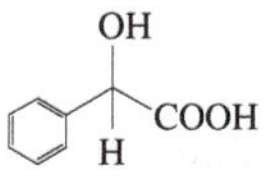 **4**	CαH (S R; 4.8)	0.058	CαH (S R; 4.8)	0.052	CαH (S R; 4.8)	0.091	CαH (S R; 5.0)	0.097
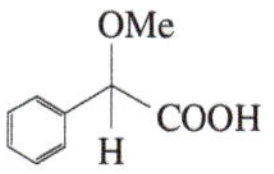 **18**	CαH (R S; 4.6)	0.024	CαH (R S; 4.6)	0.017	CαH (R S; 4.6)	0.037	CαH (R S; 4.6)	0.049
	CH_3 (R S; 3.4)	0.017	CH_3 (R S; 3.4)	0.003	CH_3 (R S; 3.4)	0.022	CH_3 (R S; 3.4)	0.037
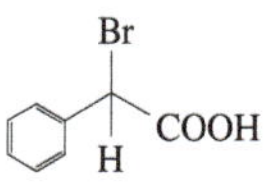 **45**	CαH (R S; 5.2)	0.034	CαH (R S; 5.2)	0.024	CαH (5.3)	0	CαH (R S; 5.4, 5.2)	0.006

The chemical shift differences (ΔΔδ) for the methine protons attached to the asymmetric carbons of the enantiomers of racemic carboxylic acids **3**, **4**, **18**, and **45** in the presence of 0.25 eq. of (*R,R,R,R*)-**67a**, **67b**, **68a**, and **68b** in $CDCl_3$ are listed in Table 12.4. For all of the tested carboxylic acids **3**, **4**, **18**, and **45**, the signals for the protons (CαH) attached to the stereogenic centers of the enantiomers were split, and the baseline separation was sufficient for accurate integration.

For mandelic acid **4**, the ΔΔδ values (0.091 and 0.097) in the presence of **68a** and **68b**, respectively, were much larger than those (0.058 and 0.052) in the presence of **67a** and **67b**,

TABLE 12.5 Partial ¹H NMR Spectra of *rac*-Propionic Acid Derivatives 51a–c in the Presence of 0.25 eq. of 67 and 68

	67a	67b	68a	68b
Racemate	ΔΔδ (ppm)	ΔΔδ (ppm)	ΔΔδ (ppm)	ΔΔδ (ppm)
51a (OH)	CH_3 0.008	CH_3 0	CH_3 0.017	CH_3 0.006
51b (Cl)	CαH 0.007	CαH 0.002	CH_3 0.018	CH_3 0.023
51c (Br)	CH_3 0.013	CH_3 0.006	CH_3 0.006	CH_3 0.011

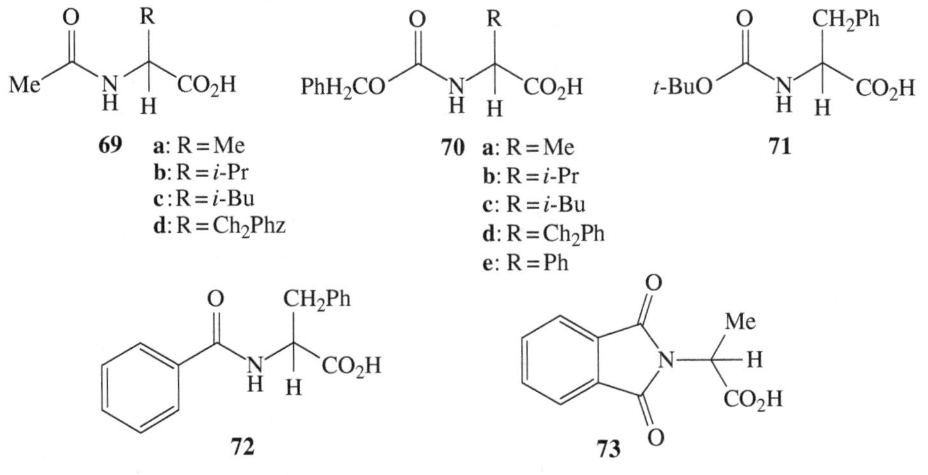

SCHEME 12.8 Amino acid derivatives.

respectively. Similarly, efficient chiral discrimination was observed for the methine proton (CαH) signals of α-methoxy phenylacetic acid derivative **18** in the presence of **68a** and **68b**, while α-bromo- and α-methylphenyl propionic acids **45** and **3** gave poor results. It is also interesting to note that chiral discrimination was observed for the OCH_3 signals of **18**, and the nonequivalence was much larger in the presence of **68a** and **68b** than in the presence of **67a** and **67b**. For the propionic acid derivatives **51a–c**, chiral discrimination was observed for the CH_3 and CαH signals (Table 12.5). Generally, better separation (higher ΔΔδ values) was observed in the presence of **68a** and **68b**.

Macrocycles **68a** and **68b** show excellent enantiodiscriminating ability for several amino acid derivatives including *N*-carbobenzoxy amino acids **70a–e** (Scheme 12.8), *N*-Boc amino acid **71**, *N*-benzoyl amino acid **72**, and *N*-phthaloyl alanine **73**, while macrocycles **67a** and **67b** show good enantiodiscriminating ability for sterically less bulky *N*-acetyl amino acids (Table 12.6). Generally, macrocycle **68** provides larger ΔΔδ values for the α-CH_3 and NH protons of these guest molecules than macrocycle **67**. In particular, upon addition of 0.25 eq. of (*R,R,R,R*)-**68a**, the chemical shift values of the NH proton signals for *N*-carbobenzoxy-phenylalanine **70d** give the largest chemical shift nonequivalence (ΔΔδ = 0.214 ppm). Similarly, (*R,R,R,R*)-**68** provides clear baseline separation of the NH and α-Me proton signals for *N*-carbobenzoxy amino acids **70e**, *N*-Boc amino acid **71**, *N*-benzoyl amino acid **72**, and *N*-phthaloyl alanine **73**. These results imply that the molecular structure of macrocycle **68** fits these guest molecules better than that of macrocycle **67**.

The ee of mandelic acid **4** was determined by integration of the corresponding ¹H-NMR signals in the presence of 0.25 eq. of (*R,R,R,R*)-**67a**. Samples containing different ee's of **36b** were also prepared, and their ¹H-NMR spectra in the presence of (*R,R,R,R*)-**67a** were measured (Fig. 12.4). An excellent linear correlation ($R^2 = 0.999$) between theoretical and observed % ee values was observed (Fig. 12.5). The Job plots of (*R,R,R,R*)-**67a** with (*R*)- and (*S*)-**4** with a total

TABLE 12.6 Partial ^{1}H NMR Spectra of *rac*-α-Amino Acid Derivatives 69a–d, 70e–d, 71, 72, 73 in the Presence of 0.25 eq. of 67 and 68

Racemate	67a		67b		68a		68b	
	ΔΔ*δ* (ppm)		ΔΔ*δ* (ppm)		ΔΔ*δ* (ppm)		ΔΔ*δ* (ppm)	
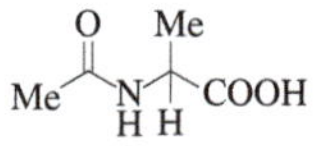 **69a**	$COCH_3$ *S R* 1.8	0.104	$COCH_3$ 2.0	0	$COCH_3$ 2.00	0	$COCH_3$ 2.0	0
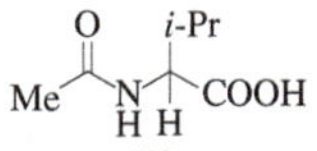 **69b**	$COCH_3$ *RS* 2.0	0.008	$COCH_3$ *RS* 2.0	0.008	$COCH_3$ 2.0	0	$COCH_3$ 2.0	0
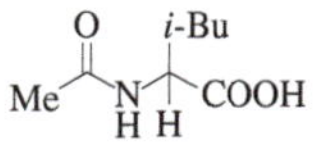 **69c**	$COCH_3$ *S R* 1.9	0.040	$COCH_3$ *RS* 2.0	0.004	$COCH_3$ 2.0	0	$COCH_3$ 2.0	0
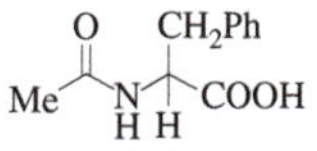 **69d**	$COCH_3$ *S R* 1.9	0.040	$COCH_3$ *R S* 1.9	0.041	$COCH_3$ 1.9	0	$COCH_3$ 2.0 1.9	0
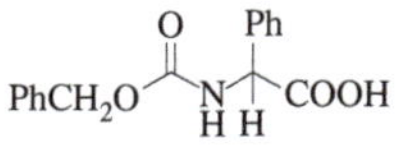 **70e**	NH 6.3 6.2	0	NH 6.2	0	NH *R S* 6.2	0.052	NH *R S* 6.1 6.0	0.070
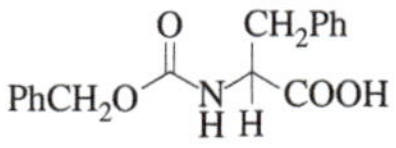 **70d**	NH *RS* 5.4	0.032	NH *R S* 5.6	0.061	NH *R S* 5.6 5.4	0.214	NH *R S* 5.6 5.4	0.208
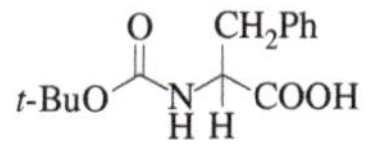 **71**	NH *RS* 5.2	0.042	NH *R S* 5.2	0.080	NH *R S* 5.4 5.2	0.181	NH *R S* 5.4 5.2	0.189
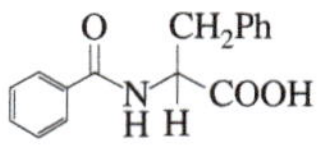 **72**	NH *RS* 7.0	0.022	NH *R S* 7.0	0.034	NH *R S* 7.0 6.9 6.8	0.166	NH *R S* 6.9	0.020
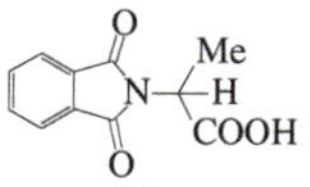 **73**	CH_3 *R S* 1.6 1.5	0.085	CH_3 *R S* 1.6 1.5	0.039	CH_3 *R S* 1.6 1.5	0.084	CH_3 *R S* 1.6	0.055

concentration of 40 mM are shown. A maximum was observed when the ratio of (*R,R,R,R*)-**67a** versus (*R*)- or (*S*)-**4** was 1 : 4 ($X = 0.2$), which suggests that the host (*R,R,R,R*)-**67a** forms a 1 : 4 complex with (*R*)- or (*S*)-**4**.

In the formation of a 1 : 4 diastereomeric complex of (*R,R,R,R*)-**67a** with a chiral carboxylic acid, hydrogen bonding, CH–π, and π–π interactions may be responsible for the chiral recognition. The analysis of 2D NOESY spectra of (*R*)-**4** in the presence of 0.25 eq. of (*R,R,R,R*)-**67a** showed a cross peak between the CαH proton of (*R*)-**4** and aromatic protons of (*R,R,R,R*)-**67a** probably due to the CH–π interaction between (*R,R,R,R*)-**67a** and guest in addition to the major acid–base attractive interaction (Fig. 12.6).

In many cases, the protons distant from the stereogenic center, such as the CH_3 group in *N*-acetyl alanine **69a**, can be split completely in the presence of chiral host (*R,R,R,R*)-**67a**. This may be due to the CH–π interaction between aromatic groups in the hosts and the CH_3 protons in guests.

Calixarene-like chiral macrocyclic amine (*S,S,S,S,S,S*)-**74** (Scheme 12.9) was prepared by treating enantiomerically pure (*S,S*)-1,2-cyclohexanediamine with 2-hydroxy-5-methylbenzene-1,3-dialdehyde followed by $NaBH_4$ reduction of the intermediate macrocyclic imine. The chiral shift experiments were carried out by measuring ^{1}H-NMR spectra (400 MHz) of a mixture of (*S,S,S,S,S,S*)-**74** and racemic carboxylic acids in $CDCl_3$ [23].

Upon addition of 0.1 eq. of (*S,S,S,S,S,S*)-**74**, the chemical shift values of the methine CαH proton signal of (*S*)- and (*R*)-*o*-chloromandelic acid **34** exhibited 0.303 and 0.368 ppm upfield shifts, respectively. Upon gradual addition of (*S,S,S,S,S,S*)-**74**, the ^{1}H NMR signal of the CαH proton of **34** moved upfield and the chemical shift difference between the

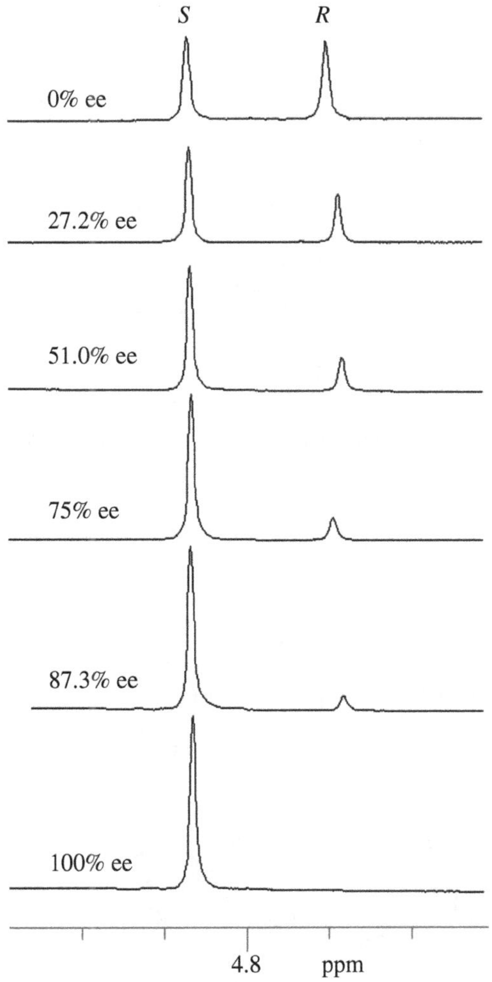

FIGURE 12.4 Selected region of the 400 MHz NMR spectra of *rac*-**4** of various enantiomeric purities in the presence of 0.25 eq. of (*R*,*R*,*R*,*R*)-**67a**.

two enantiomers increased gradually, until the addition of 0.4 eq. of (*S*,*S*,*S*,*S*,*S*,*S*)-**74**. However, 0.2 eq. proved to be the best for the chiral recognition showing a 0.204 ppm difference, since the more upfield signal of the methine proton of (*R*)-**34** was broadened by adding over 0.2 eq. of (*S*,*S*,*S*,*S*,*S*,*S*)-**74**. Figure 12.7 shows the Job plots of (*S*,*S*,*S*,*S*,*S*,*S*)-**74** with (*R*)- and (*S*)-**34** with a total concentration of 40 mM. A maximum was observed when the ratio of (*S*,*S*,*S*,*S*,*S*,*S*)-**74** versus (*R*)- or (*S*)-**34** was 1 : 4 (X=0.2), which indicates that the host forms a 1 : 4 complex with (*R*)- or (*S*)-**34**.

Table 12.7 summarizes the chemical shift differences ($\Delta\Delta\delta$) between the enantiomers of various types of *rac*-carboxylic acids **3**, **4**, **18**, **34**, **35**, **36a–c**, **45**, **51a–c**, and **69a–d** in the presence of 0.25 or 0.5 eq. of (*S*,*S*,*S*,*S*,*S*,*S*)-**74** in $CDCl_3$. For all the carboxylic acids tested, the signals for the protons attached to the stereogenic center were split. Acids **4**, **34**, **36a**, **36c**, and **18** in particular presented baseline separation enough for accurate integration. In contrast, lactic acid **51a** shows smaller nonequivalencies than the mandelic acids, suggesting that the aromatic ring is necessary for good signal separation. For the amino acid derivatives **69a–d**, chiral discrimination was observed for the $COCH_3$ signals. The relationship between the absolute configuration and the upfield shift of the proton signals of the carboxylic acid in the presence of (*S*,*S*,*S*,*S*,*S*,*S*)-**74** was examined, and the signals of the (*R*)-enantiomer appeared at a higher magnetic field than that of the (*S*)-enantiomer in the presence of (*S*,*S*,*S*,*S*,*S*,*S*)-**74** (Table 12.7).

Fluorescence-based enantioselective sensors have attracted considerable attention because of their potential applications in real-time analysis of chiral organic compounds. The secondary nitrogen atoms adjacent to the phenol

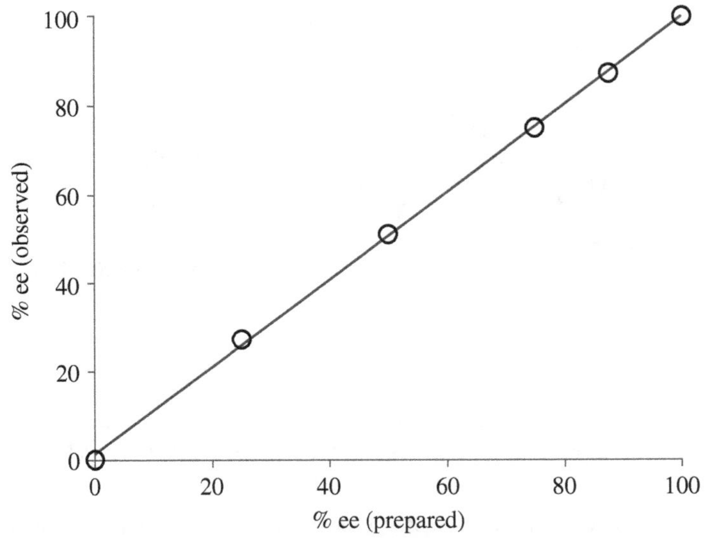

FIGURE 12.5 Correlation between theoretical and observed % ee values of **4**.

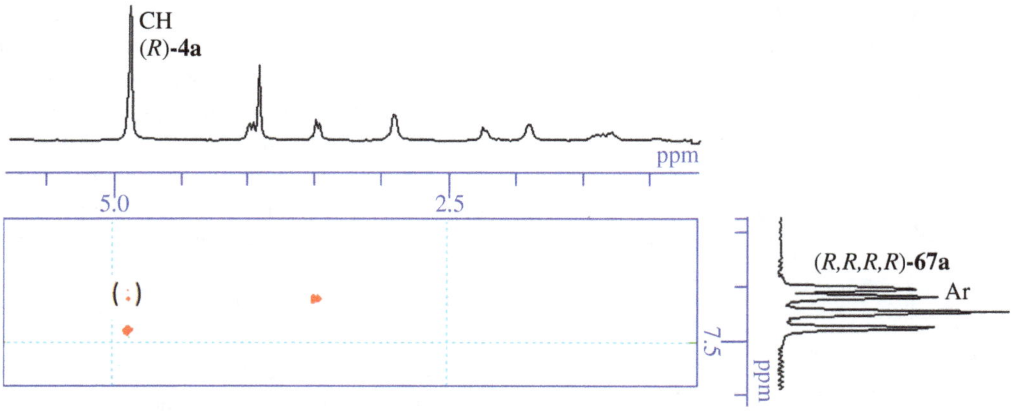

FIGURE 12.6 Selected region of the 2D NOESY spectra (400 MHz in $CDCl_3$) of (*R*)-**4a** in the presence of 0.25 eq. of (*R,R,R,R*)-**67a**.

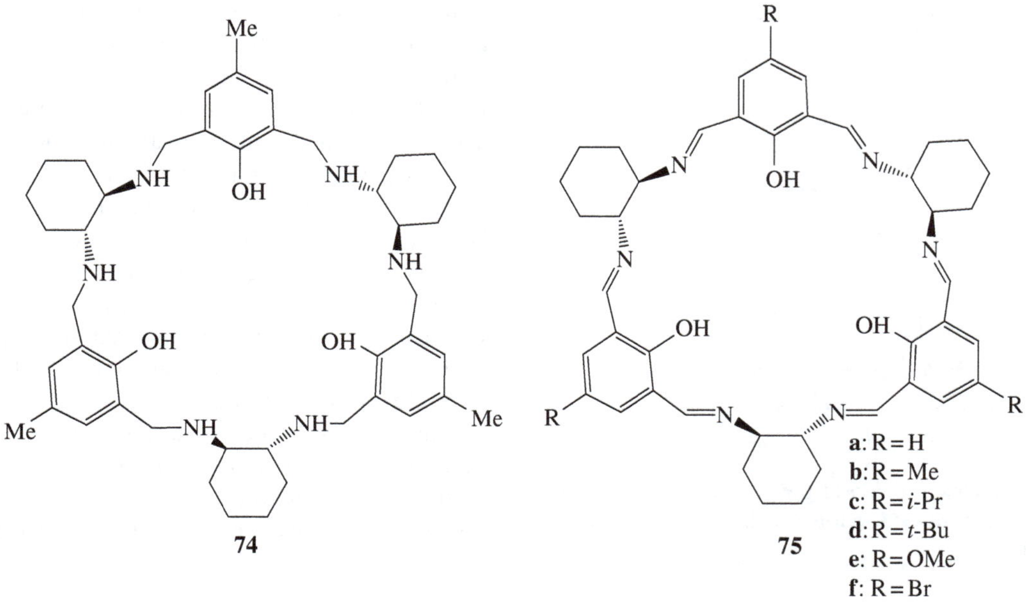

SCHEME 12.9 Calixarene-like chiral macrocyclic amines.

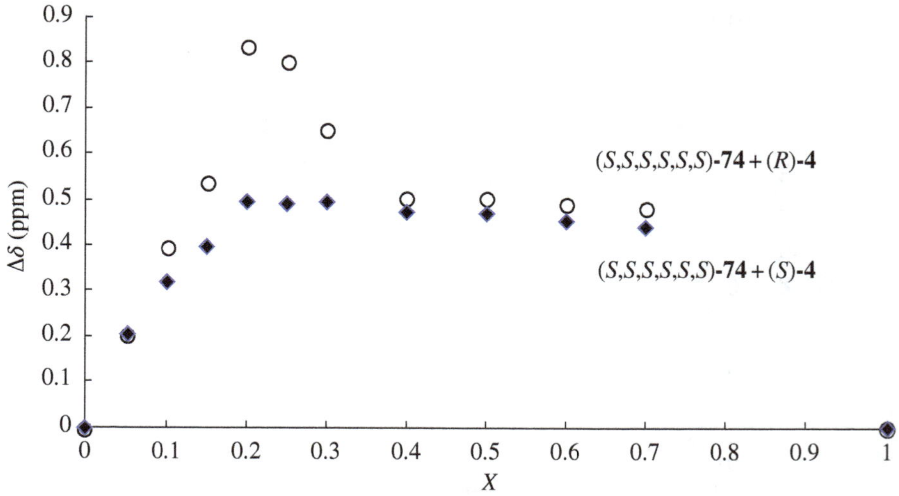

FIGURE 12.7 Job plot of (*S,S,S,S,S,S*)-**74** with (*R*)- and (*S*)-**34**. $\Delta\delta$ stands for chemical shift change of the CH proton of **34** in the presence of (*S,S,S,S,S,S*)-**74**. *X* stands for mole fraction of host, (*X* = [(*S,S,S,S,S,S*)-**74**]/[(*S,S,S,S,S,S*)-**74**] + [**34**]). Total concentration is 40 mM.

TABLE 12.7 Measurement of [1]H Chemical Non-equivalencies (ΔΔ*δ*) of the Guests in the Presence of (*S*,*S*,*S*,*S*,*S*,*S*)-74 by ^{1}H NMR Spectroscopy (400 MHz) in $CDCl_3$ at 25°C

Guest	Proton	Ratio	ΔΔ*δ* (ppm)	Enantiomer Showing More Upfield Shift
4	CαH	1 : 4	0.234	*R*
34	CαH	1 : 4	0.204	*R*
35	CαH	1 : 2	0.043	*R*
36b	CαH	1 : 2	0.094	
36a	CαH	1 : 4	0.323	
36c	CαH	1 : 4	0.296	
45	CαH	1 : 4	0.061	
18	CαH	1 : 4	0.202	*R*
18	OCH_3	1 : 2	0.023	*R*
3	CαH	1 : 4	0.022	*R*
3	CH_3	1 : 4	0.003	*R*
51a	CαH	1 : 2	0.042	
51b	CαH	1 : 4	0.018	*R*
51b	CH_3	1 : 4	0.028	*R*
51c	CαH	1 : 4	0.020	
51c	CH_3	1 : 4	0.033	
69a	$COCH_3$	1 : 4	0.025	*R*
69b	$COCH_3$	1 : 2	0.030	*R*
69c	$COCH_3$	1 : 4	0.024	*R*
69d	$COCH_3$	1 : 4	0.025	*R*

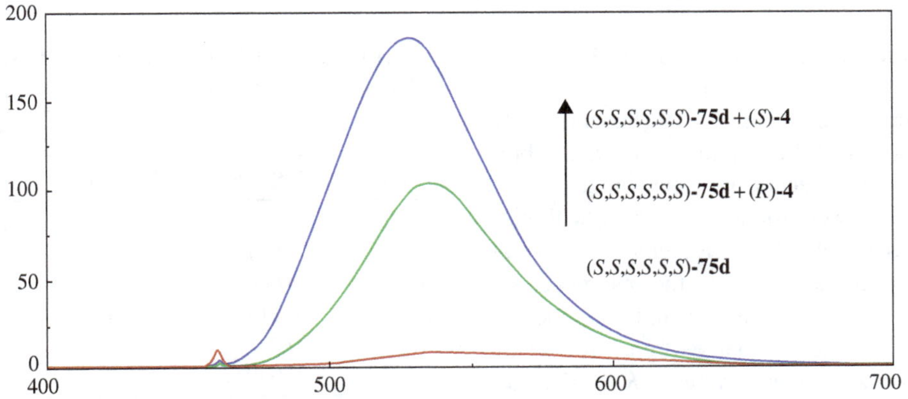

FIGURE 12.8 Fluorescence spectra of (*S*,*S*,*S*,*S*,*S*,*S*)-**75d** (1.0×10^{-4} M in $CHCl_3$, $\lambda_{ex} = 460$ nm) with and without (*R*)- and (*S*)-mandelic acid **4** (3.0×10^{-3} M).

rings of the chiral salen macrocycle (*S*,*S*,*S*,*S*,*S*,*S*)-**75** (Scheme 12.9) may interact with acids and can turn on the fluorescence of the phenol by inhibiting the fluorescence quenching of the photoinduced electron transfer (PET) of the nitrogen lone pair electrons [24]. Because the interaction of (*S*,*S*,*S*,*S*,*S*,*S*)-**75** with two enantiomers of mandelic acid should generate diastereomers, enantioselective fluorescence enhancements may occur. Figure 12.8 shows the fluorescence spectra of (*S*,*S*,*S*,*S*,*S*,*S*)-**75** in the presence of (*R*)- and (*S*)-**4**. As shown in the spectra, when (*S*,*S*,*S*,*S*,*S*,*S*)-**75d** was treated with (*R*)- or (*S*)-**4**, significant fluorescence enhancement was observed at 527 nm. The (*S*)-**4** enantiomer enhanced the emission twice as much as (*R*)-**4**.

To confirm the type of molecular recognition, the ^{1}H-NMR spectra using various ratios of (*S*,*S*,*S*,*S*,*S*,*S*)-**75d** and **4** in $CDCl_3$ were recorded. When the ratio of (*S*,*S*,*S*,*S*,*S*,*S*)-**75d** to *rac*-**4** was 1 : 2, a methine proton signal appeared and was split into two singlets at *δ* 4.84 and 4.88 ppm for the (*R*)- and (*S*)-enantiomers, respectively. This suggests that in the host mandelic acid complex, (*S*)-mandelic acid is probably located much deeper inside the cavity of (*S*,*S*,*S*,*S*,*S*,*S*)-**75d** than (*R*)-mandelic acid, which significantly shields the methine proton of (*S*)-**4**. This could be the origin of the dramatic difference in the fluorescence response of (*S*,*S*,*S*,*S*,*S*,*S*)-**75d** toward the two enantiomers of mandelic acid.

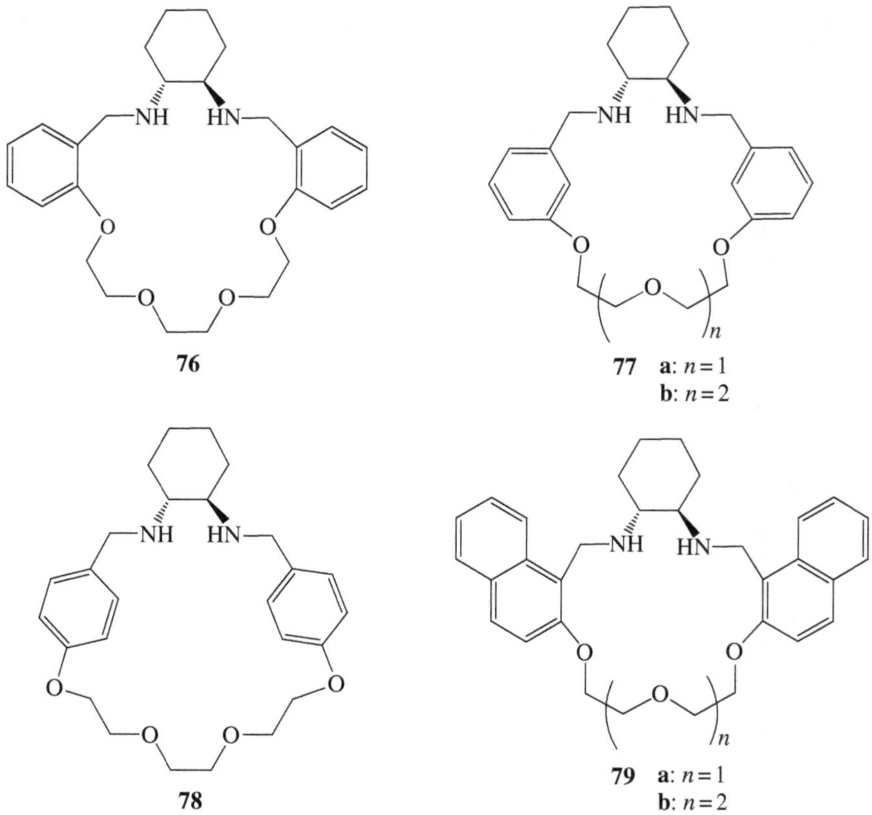

SCHEME 12.10 Chiral aza-crown macrocycles.

12.5 CHIRAL AZA-CROWN MACROCYCLES

A series of new chiral macrocycles **76–79** (Scheme 12.10) containing the (*R*,*R*)-*trans*-1,2-diaminocyclohexane subunit and oligoethylene glycol spacers has been prepared [25]. Most of the macrocycles were found to associate with mandelic acid **4** enantioselectively. ^{1}H-NMR studies demonstrate that in most cases molecule **4** binds to the macrocycles to form complexes of 2 : 1 stoichiometry and that the 2 : 1 complexes generally provide the source of enantiodiscrimination. The *meta*-linked macrocycles (*R*,*R*)-**77a** and (*R*,*R*)-**77b** as well as the naphthyl-containing *ortho*-linked macrocycle (*R*,*R*)-**79** are shown to be effective CSAs for mandelic acid derivatives and were used to accurately determine the % ee of a number of mandelic acid derivatives (Table 12.8).

TABLE 12.8 ^{1}H-NMR Chemical Shift Nonequivalences for Mandelic Acid Derivatives[a]

	Macrocycle			
Guest	(*R*,*R*)-**76**	(*R*,*R*)-**77**	(*R*,*R*)-**78**	(*R*,*R*)-**79**
4	0.077	0.102	0.043	0.054
34	0.077	0.079	0.047	0.033
35	0.060	0.074	0.046	0.052
36b	0.114	0.130	0.034	0.048
36c	0.078	0.128	0.042	0.038
37	0.090	0.105	0.022	0.032

[a] ^{1}H-NMR chemical shift differences ($\Delta\delta$, ppm) between the benzylic C—H resonances of enantiomers of mandelic acid derivatives (ca. 10 mM) observed upon titration of macrocycles in $CDCl_3$ at 23°C.

12.6 FINAL COMMENTS

In summary, the chiral macrocycles with that are fixed in appropriate positions described above can be effectively used for the chiral recognition and discrimination in NMR spectroscopy. Various intermolecular interactions between chiral macrocycles and racemic guest molecules take place during chiral discriminating steps, such as CH–π and π–π interactions, hydrogen bonding interactions, and salt formation. The design of new macrocyclic receptors will be further perfected to perform higher level chiral recognition and discrimination in NMR spectroscopy.

REFERENCES

1. L. Pu, *Chem. Rev.* 2004, *104*, 1687.
2. T. J. Wenzel, J. D. Wilcox, *Chirality* 2003, *15*. 256.
3. J. M. Seco, E. Quiñoá, R. Riguera, *Chem. Rev.* 2004, *104*, 17.
4. T. J. Wenzel, C. D. Chisholm, *Chirality* 2011, *23*. 190.

5. (a) D. Yang, X. Li, Y. E. Fan, D. W. Zhang, *J. Am. Chem. Soc.* 2005, *127*, 7996; (b) C. Pena, J. Gonzalez-Sabin, I. Alfonso, F. Rebolledo, V. Gotor, *Tetrahedron Asymmetry* 2007, *18*, 1981; (c) C. Pena, J. Gonzalez-Sabin, I. Alfonso, F. Rebolledo, V. Gotor, *Tetrahedron* 2008, *64*, 7709.

6. (a) T. J. Wenzel, J. E. Thurston, *J. Org. Chem.* 2000, *65*, 1243; (b) A. E. Lovely, T. J. Wenzel, *Org. Lett.* 2006, *8*, 2823; (c) A. E. Lovely, T. J. Wenzel, *J. Org. Chem.* 2006, *71*, 9178; (d) C. D. Chisholm, F. Fülöp, E. Forró, T. J. Wenzel, *Tetrehedron Asymmetry* 2010, *21*, 2289.

7. (a) C. M. O'Farrell, J. M. Chudomei, J. M. Collins, C. F. Dignam, T. J. Wenzel, *J. Org. Chem.* 2008, *73*, 2843; (b) K. A. Hagan, C. M. O'Farrell, T. J. Wenzel, *Eur. J. Org. Chem.* 2009, 4825; (c) N. H. Pham, T. J. Wenzel, *J. Org. Chem.* 2011, *76*, 986.

8. (a) T. J. Wenzel, E. P. Amonoo, S. S. Shariff, S. E. Aniagyei, *Tetrahedron Asymmetry* 2003, *14*, 3099; (b) K. A. Provencher, M. A. Weber, L. A. Randall, P. R. Cunningham, C. F. Dignam, T. J. Wenzel, *Chirality* 2010, *22*, 336; (c) E. Rudzinska, G. Dziedziola, L. Berlicki, P. Kafarski, *Chirality* 2010, *22*, 63.

9. (a) R. Schwenniger, J. Schlogl, J. Maynollo, K. Gruber, P. Ochsenbein, H. B. Burgi, R. Konrat, B. Krautler, *Chem. Eur. J.* 2001, *7*, 2676; (b) M. Claeys-Bruno, D. Toronto, J. Pecaut, M. Bardet, J.-C. Marchon, *J. Am. Chem. Soc.* 2001, *123*, 11067; (c) T. Ema, N. Ouchi, T. Doi, T. Korenaga, T. Sakai, *Org. Lett.* 2005, *7*, 3985.

10. (a) T. J. Wenzel, J. E. Thurston, D. C. Sek, J.-P. Joly, *Tetrahedron Asymmetry* 2001, *12*, 1125; (b) G. Heinrichs, L. Vial, J. Lacour, S. Kubik, *Chem. Commun.* 2003, 1252; (c) X. M. Yang, X. J. Wu, M. H. Fang, Q. Yuan, E. Q. Fu, *Tetrahedron Asymmetry* 2004, *15*, 2491; (d) Y.-S. Zheng, C. Zhang, *Org. Lett.* 2004, *6*, 1189; (e) C. F. Dignam, C. J. Richards, J. J. Zopf, L. S. Wacker, T. J. Wenzel, *Org. Lett.* 2005, *7*, 1773; (f) X. X. Liu, Y. S. Zheng, *Tetrahedron Lett.* 2006, *47*, 6357; (g) A. E. Lovely, T. J. Wenzel, *Chirality* 2008, *20*, 370; (h) T. Ema, *J. Incl. Phenom. Macrocycl. Chem.* 2012, *74*, 41.

11. T. Ema, D. Tanida, T. Sakai, *Org. Lett.* 2006, *8*, 3773.

12. T. Ema, D. Tanida, K. Hamada, T. Sakai, *J. Org. Chem.* 2008, *73*, 9129.

13. K. Gospodarowicz, M. Holynska, M. Paluch, J. Lisowski, *Tetrahedron* 2012, *68*, 9930.

14. G. U. Barretta, F. Balzano, J. Martinelli, M. Berni, C. Villani, F. Gasparrini, *Tetrahedron Asymmetry* 2005, *16*, 3746.

15. B. Li, X. Yang, X. Wu, Z. Luo, C. Zhong, E. Fu, *Supramol. Chem.* 2006, *18*, 507.

16. F. Ma, L. Ai, X. Shen, C. Zhang, *Org. Lett.* 2007, *9*, 125.

17. A. Gonzalez-Alvarez, I. Alfonso, V. Gotor, *Tetrahedron Lett.* 2006, *47*, 6397.

18. E. Busto, A. Gonzalez-Alvarez, V. Gotor-Fernandez, I. Alfonso, *Tetrahedron* 2010, *66*, 6070.

19. K. Tanaka, N. Fukuda, T. Fujiwara, *Tetrahedron Asymmetry* 2007, *18*, 2657.

20. A. Gualandi, S. Grilli, D. Savoia, M. Kwit, J. Gawronski, *Org. Biomol. Chem.* 2011, *9*, 4234.

21. K. Tanaka, Y. Nakai, H. Takahashi, *Tetrahedron Asymmetry* 2011, *22*, 178.

22. K. Tanaka, T. Iwashita, C. Sasaki, H. Takahashi, *Tetrahedron Asymmetry* 2014, *25*, 602.

23. K. Tanaka, N. Fukuda, *Tetrahedron Asymmetry* 2009, *20*, 111.

24. K. Tanaka, T. Tsuchitani, N. Fukuda, A. Masumoto, R. Arakawa, *Tetrahedron Asymmetry* 2012, *23*, 205.

25. T. P. Quinn, P. D. Atwood, J. M. Tanski, T. F. Moore, J. F. Folmer-Andersen, *J. Org. Chem.* 2011, *76*, 10020.

13

NEW STRATEGIES FOR THE DESIGN OF INCLUSION COMPOUNDS WITH CUCURBITURIL HOSTS

Na'il Saleh

Department of Chemistry, College of Science, United Arab Emirates University, Al-Ain, United Arab Emirates

13.1 INTRODUCTION

The development of nanostructured materials with stimuli-responsive, switchable functions is at the frontier of supramolecular chemistry [1]. One of the approaches toward that end is to utilize non-covalent interactions to convert cheap, naturally abundant materials into useful molecular tools by virtue of simple and straightforward procedures [1]. Within the host–guest supramolecular approach, one or several small organic molecules (guests) are sequestered and held in the cavity of a molecular container (host) by means of non-covalent intermolecular interactions such as van der Waals, dipole–dipole, metal–ligand, electrostatic interactions, solvophobic effects and hydrogen bonding. The host–guest interactions endow the traditional nanostructures with unique stimuli-responsive abilities. In particular, the application of host–guest concepts in surface technology has advanced the field toward the complicated molecular recognition systems in nature.

Stimuli-responsive supramolecular assembly formed by host–guest interactions has been well established in solution [2], and this chapter will focus on the stimuli-responsive supramolecular nanostructures based on cucurbit[*n*]uril (CB*n*) ($n = 6, 7, 8$, and 10) host–guest chemistry (Scheme 13.1 for CB7 and CB8) [3]. The host–guest interaction of CB*n* is based on two main intermolecular forces: a hydrophobic effect and ion dipole interactions at the carbonyl portals. Ostensibly, the utility of CB*n* is endowed with size selectivity, high binding affinity, unique recognition properties toward cations, and water solubility. There are numerous studies on the host–guest chemistry of CB homologues and their recognition properties, as well as a number of excellent review papers outlining the synthesis, properties, and applications of CB homologues [4].

This chapter mainly focuses on stimuli-responsive systems using CB*n* molecular containers. In Section 13.2, it highlights the unique hosting properties of CB*n* that inspired researchers to design inclusion compounds on surfaces with distinct stimuli-responsive functions. These CB*n*-based stimuli-responsive systems were geared toward some real applications, as tackled by Section 13.3. The chapter does not include systems that are based on functionalized CB*n* homologues and limits the discussion to switchable host–guest nanostructured materials in solid state.

13.2 PROMINENT FEATURES OF CB*n* FOR HOST–GUEST CHEMISTRY

The use of supramolecular approach can offer many advantages, particularly when reversibly controlled structures are demanded. Thus, new strategies were adopted for the design of different stimuli-responsive nanostructures by taking advantage of the interesting features of CB*n*, such as the polar carbonyl portal, their ability to accommodate more than one guest in their cavity, and selective and high affinity of the molecular containers to some guests such as azo dyes, viologen, ADA derivatives, etc. In this section, we mainly review the host–guest chemistry of CB*n* in relation to their stimuli-responsive features in CB*n*-containing nanostructures.

Non-Covalent Interactions in the Synthesis and Design of New Compounds, First Edition.
Edited by Abel M. Maharramov, Kamran T. Mahmudov, Maximilian N. Kopylovich, and Armando J. L. Pombeiro.

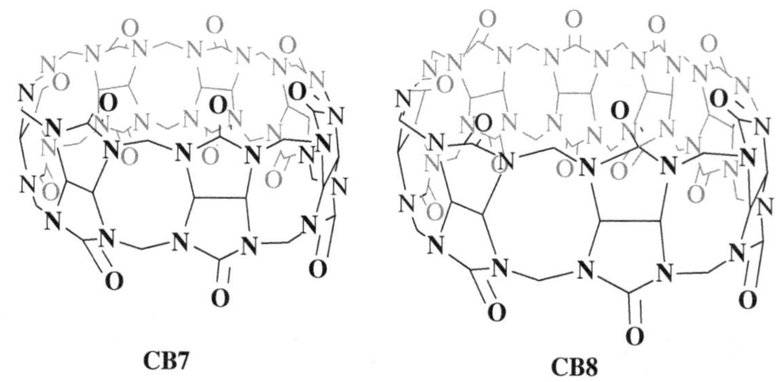

SCHEME 13.1 Chemical structures for cucurbit[*n*]uril macrocycles (CB7 and CB8).

13.2.1 Ultrahigh Binding and Selective Affinity Toward Guest Molecules

The group of Isaacs reported ultrahigh binding affinity (K_a up to $10^{17}\,M^{-1}$) of CB7/8 with selective guests [5]. This property can be used to construct guest competition responsive systems. For example, in collaboration with Rotello research group, Isaacs *et al.* showed that 1-ADA can be used to displace terminal dihexaneamine from the cavity of CB7, which was first functionalized on gold nanoparticles leading to an increase in cytotoxicity of the resulted gold nanoparticles (see below).

13.2.2 Rigid, Symmetrical Structures with Varying Cavity Sizes

CB*n* homologues can exist in different sizes. Researchers, in particular, took advantage of the ability of CB8 to accommodate more than one guest in their cavity, allowing the preparation of redox-stimuli-responsive nanostructures (see below). The size of cavity of CB7 was also suitable to selectively bind *trans* configuration over *cis* forms of some of the sequestered guest molecules leading to light-stimuli-responsive systems (see below).

13.2.3 Preferential Binding Toward Cations Over Neutral Species

Several reports have demonstrated the ability of CB*n* hosts to modulate the pK_as (protonation states) of embedded guests [6]. This comes from the preferential binding of cationic guests over the neutral forms, which was utilized to provide pH control over the sequestration and release of guest molecules by some nanoparticles (see below).

13.2.4 Polar Carbonyl Group

CB*n* facilitates the formation of organic–inorganic composite materials. Although CB lacks anchoring groups that can form strong ionic and covalent bonds with metal and metal-oxide surfaces, its carbonyl groups, which are significantly polar, are capable of hydrogen bonding in a multivalent fashion to surface hydroxyl groups or, alternatively, of forming electrostatic and van der Waals interactions on the surface of the nanoparticles [7]. Moreover, the characteristics of two identical portals decorated with partially negatively charged carbonyl groups and their rigid, symmetrical structures were advantageous for the preparation of metal nanoparticles; CB*n* acted as a capping agent to stabilize the gold or silver nanoparticles in the aqueous solutions [8].

13.3 STIMULI-RESPONSIVE SUPRAMOLECULAR NANOSTRUCTURES

The incorporation of macrocyclic hosts such as crown ethers, cyclodextrins, calixarenes, cucurbiturils, and pillararenes as building blocks into the traditional nanostructures leads to the fabrication of supramolecular nanostructures with reversible stimuli-responsive nature. The following section summarizes reports that utilized the rich host–guest chemistry of CB*n* to construct a number of different nanostructures. These new strategies in designing dynamic systems were based on the unique hosting properties of CB*n*, such as high and selective binding affinity toward some guest molecules, suitable cavity structures for distinguishing *cis* and *trans* isomers, the large size of CB8 enabling the formation of ternary complexes with electron-rich (e.g., naphthol) and electron-deficient (e.g., viologen cation) moieties, and higher affinity toward cationic guest over neutral species.

13.3.1 Guest Competition Supramolecular Gold Nanoparticles for *In Vitro* Therapeutics

Isaacs group in collaboration with the Rotello group utilized the *high affinity and highly selective binding interactions of CB7* with terminal hexanediamine and 1-ADA to switch the cytotoxicity of gold nanoparticles against MCF-7 cells in a

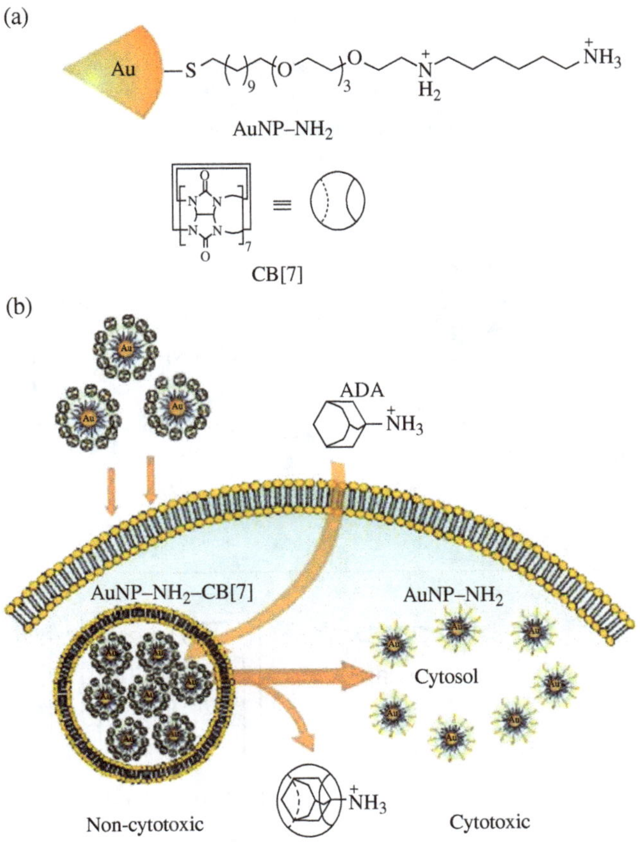

FIGURE 13.1 Guest competition gold nanoparticles triggered by 1-adamantylamine (ADA) (Reprinted with permission from Ref. [9]. Copyright 2010 Nature Publishing Group).

stimuli-responsive manner [9]. Before the addition of CB7 (Fig. 13.1a), the nanoparticles decorated with diaminohexane (AuNP–NH_2) were cytotoxic, but upon complexation with CB7, the complex-decorated nanoparticles become entrapped in endosomes leading to nanoparticles with reduced cytotoxicity (Fig. 13.1b). The toxicity was then restored by the addition of 1-ADA competitor, which triggers the release of the cytotoxic nanoparticle AuNP–NH_2 from the endosome. This competitive displacement happens because ADA has a higher affinity toward CB7 compared to diaminohexane group.

13.3.2 Guest Competition Supramolecular Beads for Protein Isolation

The high affinity and highly selective binding interactions of CB7 with ferrocene derivatives (AFc and BAFc) were used by Kim group for isolating plasma membrane proteins as illustrated in Figure 13.2 [10]. The plasma membrane proteins could be separated from protein mixtures using CB7 host–guest functionalized beads. The stimuli-responsive method was built on the idea that the captured proteins were readily removed from the beads via the addition of a second ferrocene derivative (BAFc), which had a higher binding affinity for CB7 than AFc.

13.3.3 Light-Responsive Supramolecular Hybrid Colloids

Light is one of the most widely used stimuli, since it can be easily controlled simply by adjusting its wavelength and intensity. The group of Scherman prepared photoresponsive hybrid raspberry-like colloids (HRCs) by taking advantage of *the ability of CB8 to bind two guest molecules*: methyl viologen (MV) and 4-hydroxyazobenzene (Azo) (Fig. 13.3) [11]. Ternary complex has formed between azo-functionalized silica microspheres (the core), MV-decorated polymeric nanoparticles (the corona), and CB8, leading to the generation of HRCs in water. Upon light irradiation, the conformational equilibrium of *trans* Azo/*cis* Azo made the complexation process reversible leading to light-controlled assembly/disassembly systems.

13.3.4 Heat-Responsive Supramolecular Hydrogel for the Preparation of Self-Healing Materials

The non-covalent interactions with CB*n* can offer reversibility to the structure of hydrogels, which are prepared by cross-linking of appropriate polymers through covalent bonds and non-covalent bonds. Within this approach, the group of Scherman *utilized the two-guest binding of CB8* in preparing supramolecular hydrogel through a ternary complex formation between the host CB8 and hydrophilic copolymers bearing either pendant MV or naphthoxy derivatives (Fig. 13.4) [12]. Heating the resulting supramolecular hydrogel has reversed its mechanical properties (self-healing) after the damage caused by deformation, owing to the cross-links of non-covalent bonding.

13.3.5 pH-Responsive Supramolecular Micelles for *In Vitro* Drug Delivery

Most biological functions can be associated to the pH stimuli-responsive host–guest systems as different tissues and cellular compartments exhibit different pH values. For instance, the group of Wang benefited *from the ability of CB8 to form ternary complex with two guest units* to generate pH-responsive supramolecular prodrug micelles for intracellular drug delivery (Fig. 13.5) [13]. The self-assembled micelles formed in water in the presence of CB8 host contain two guest units: naphthalene-terminated poly(ethylene glycol) (PEG-Np) and MV-conjugated doxorubicin (MV-DOX). The hydrophobic DOX drugs

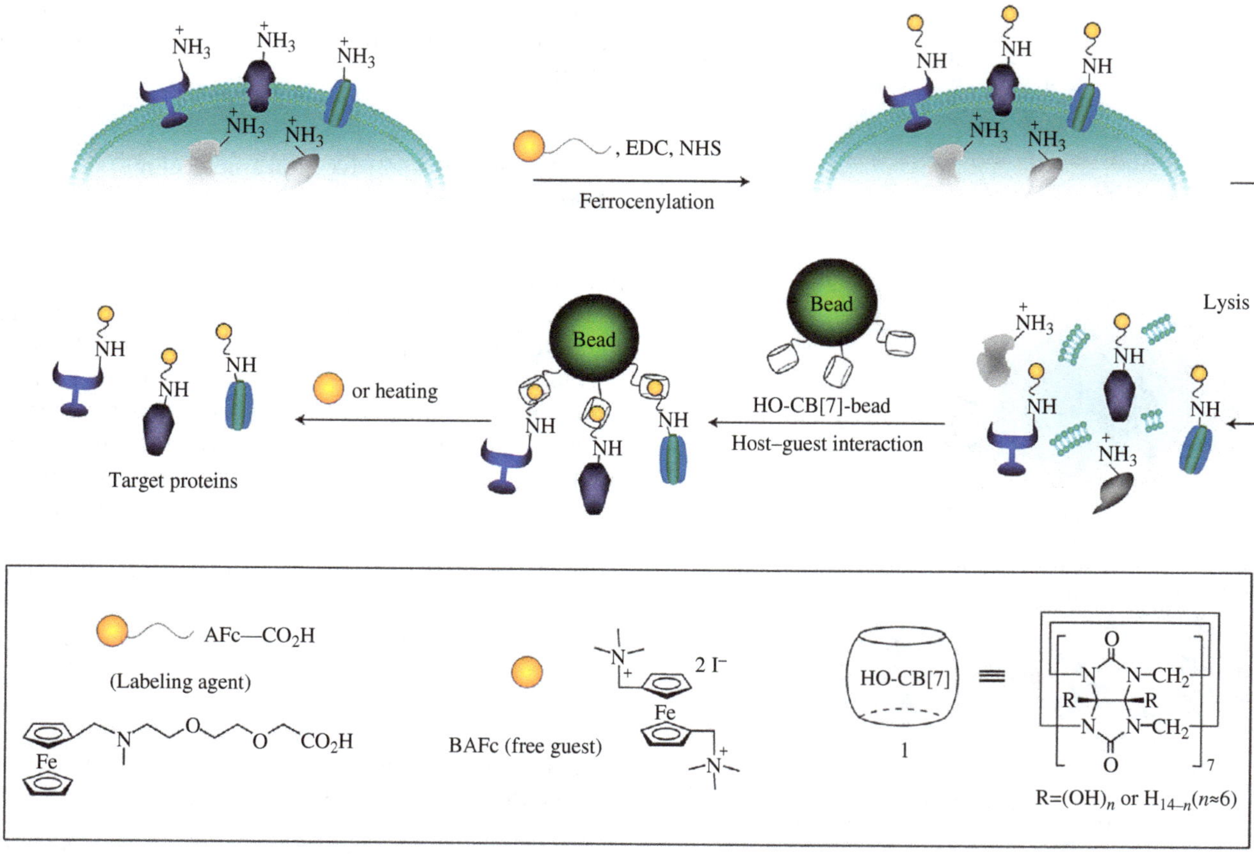

FIGURE 13.2 Isolation of plasma membrane proteins triggered by ferrocene derivatives. 1-Ethyl-3-(3-dimethylaminopropyl) carbodiimide (EDC), *N*-hydroxysuccimidyl Sepharose (NHS). (Adapted with permission from Ref. [10]. Copyright 2011 Nature Publishing Group).

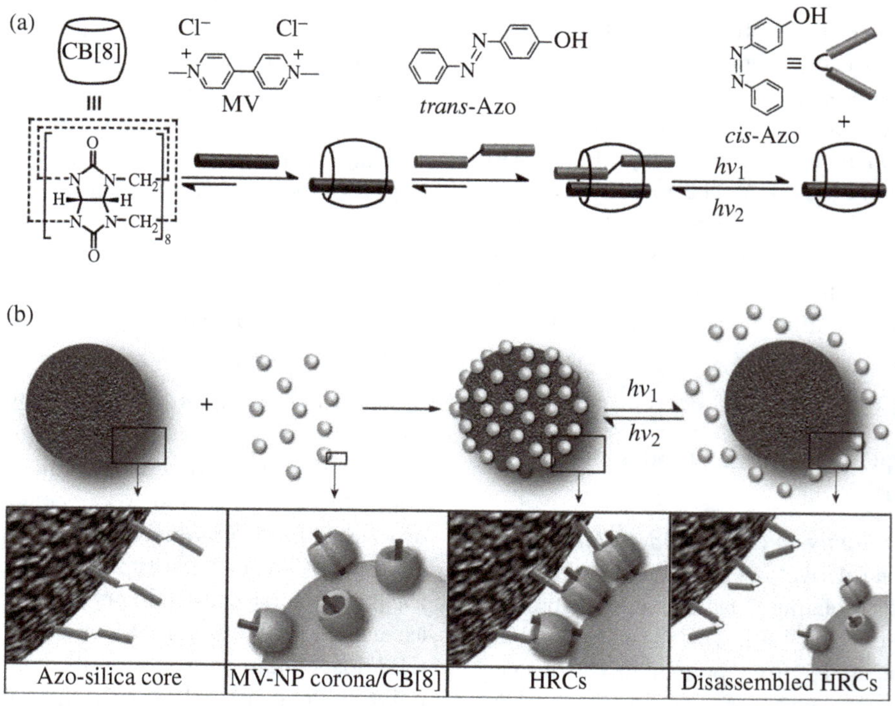

FIGURE 13.3 (a) Light-stimulated ternary complex formation between MV, *trans* azo, and CB8; (b) Photoresponsive hybrid raspberry-like colloids (HRCs) prepared by the host–guest chemistry of CB8. (Reprinted with permission from Ref. [11]. Copyright 2014 Wiley-VCH Verlag GmbH & Co. KGaA).

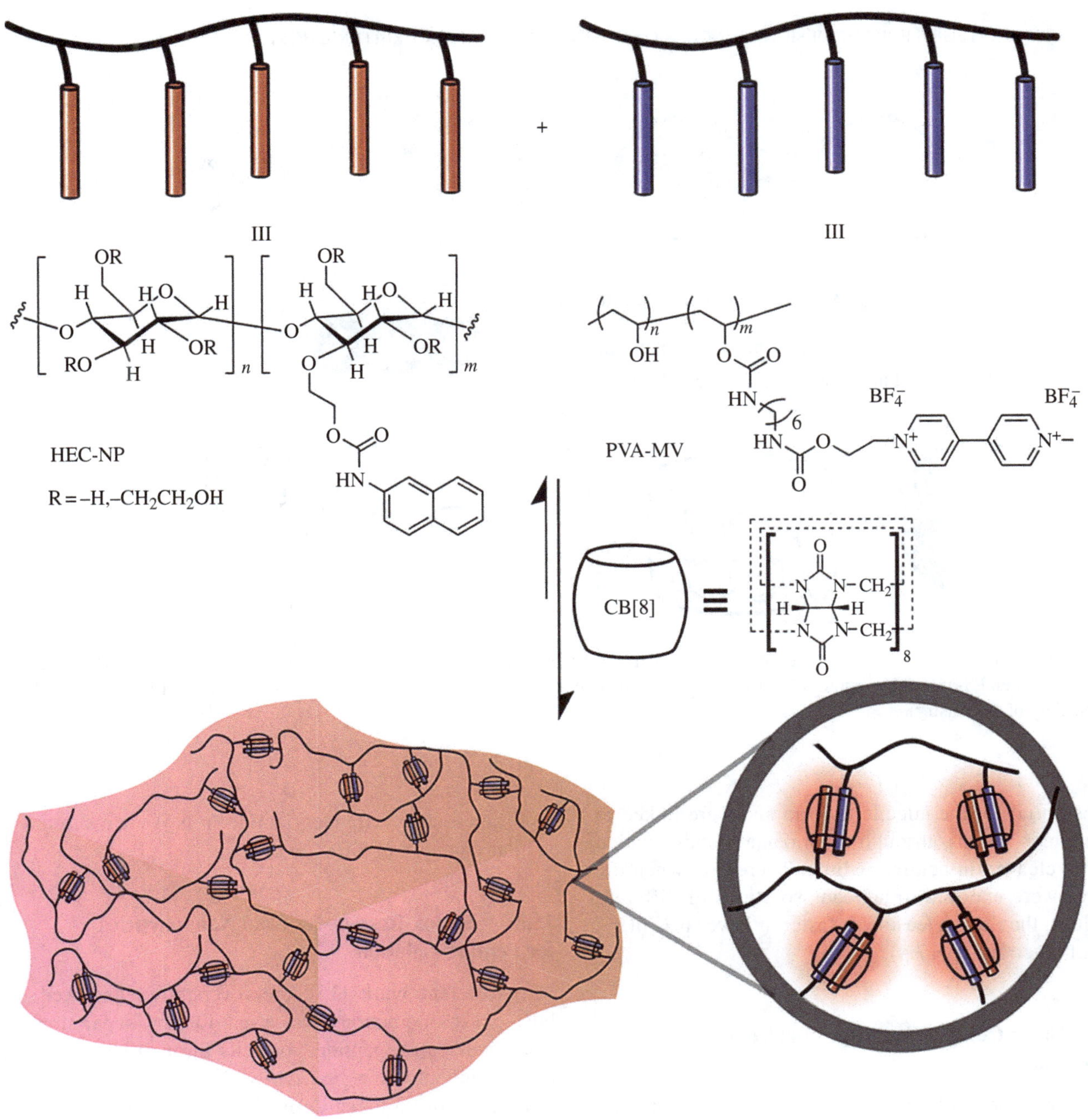

FIGURE 13.4 Heat-responsive cellulosic hydrogel in the presence of CB8 (Reprinted with permission from Ref. [12]. Copyright 2012 American Chemical Society).

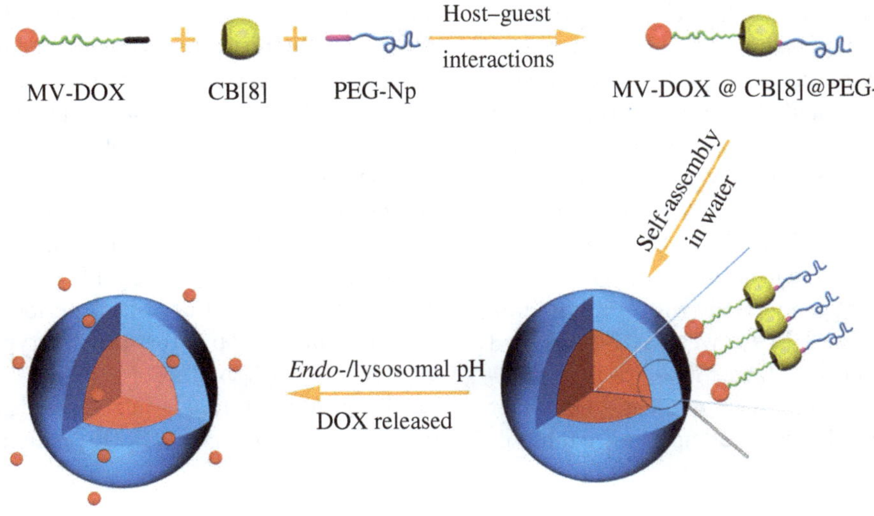

FIGURE 13.5 Controlled release of DOX from supramolecular prodrug micelles triggered by endo-/lysosomal pH (Reprinted with permission from Ref. [13]. Copyright 2014 Royal Society of Chemistry).

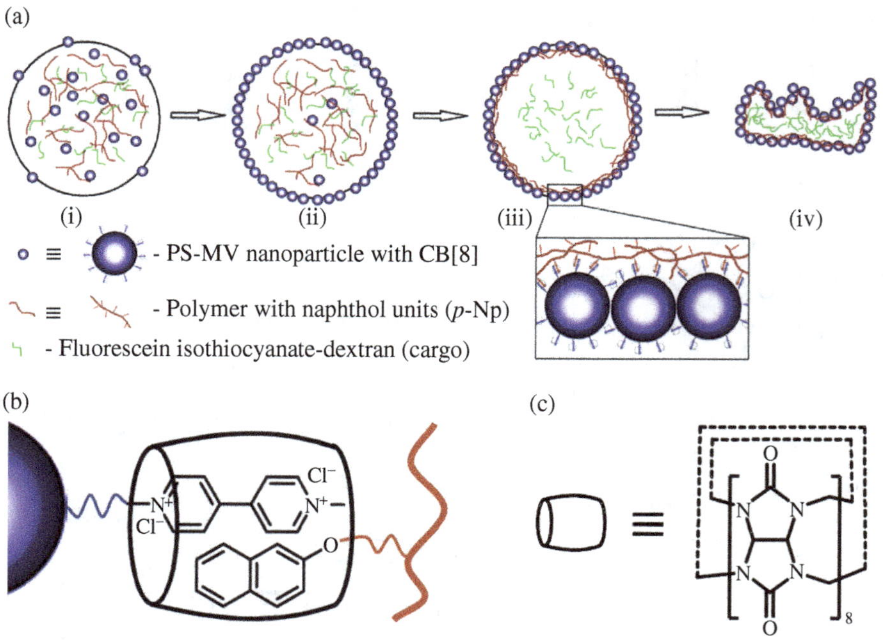

FIGURE 13.6 (a) Colloidosome formation/dissociation triggered by adamantylamine (ADA); (b) ternary supramolecular complex formation between PS-MV, *p*-Np, and CB8. (c) The molecular structure of CB8 (Reprinted with permission from Ref. [14]. Copyright 2014 Royal Society of Chemistry).

were essential for the micelle to form and were linked to the polymer chains through hydrazone bonds, which could be cleaved in acidic medium. Therefore, when the micelles were exposed to aqueous solutions of different pH values, the drug release was faster at lower pH (pH 5) than at higher one (pH 7.4).

13.3.6 Guest Competition Supramolecular Colloidosomes

In a recent work [14], the group of Scherman utilized *the ability of CB8 to form ternary complex* in preparing stimuli-responsive supramolecular colloidosomes. Colloidosomes are microcapsules whose shells are composed of self-assembled colloidal nanoparticles. These capsules were demonstrated to be capable to retain and subsequently release a cargo upon responding to 1-ADA as a trigger. The addition of ADA disassembles the ternary supramolecular complex, because it forms stronger complex with CB8 than MV and naphthol (Np) moieties (Fig. 13.6) [14]. The system was first prepared by isolating monodisperse oil droplets using a microfluidic technique, and then the polystyrene (PS)-based nanoparticles, which are decorated with the binding motifs of MV for CB8 (PS-MV), were self-assembled at the water–oil droplet interface. The PS-MV particles were reversibly cross-linked with a copolymer of poly(*N,N*-dimethylacrylamide), comprising naphthol-modified hydroxyethylacrylamide (p-Np) via a supramolecular complex of CB8.

13.3.7 Redox-Responsive Gold Nanoparticles for Peptide Isolation

In a different work [15], the reversibility endowed by host–guest interactions was used for the separation of small molecules, peptides, or cells through electrochemical stimuli utilizing *the ability of CB8 to form ternary complex* with viologen derivatives and the targeted peptides. In this case, the CB8 selectively forms ternary complex with N-tryptophan-containing peptide in the peptide mixture and the terminal viologen attached to the gold surfaces (μCP) (Fig. 13.7a). Only green fluorescent periodic dots of the fluorescent tags conjugated to the isolated peptides were observed by fluorescence microscopy, indicating that N-tryptophan-containing peptides were successfully immobilized onto the surfaces as uniform peptide arrays (Fig. 13.7b). Upon reduction of the viologen derivatives linked to the surface, the N-tryptophan-containing peptides were released (Fig. 13.7c); the retention and release cycle could be repeated as monitored by observing green fluorescent periodic dots using fluorescence microscopy for the N-tryptophan-containing peptides (Fig. 13.7b–d).

The groups of Jonkheijm and Huskens have *utilized the CB8 ternary complex* to demonstrate controlled

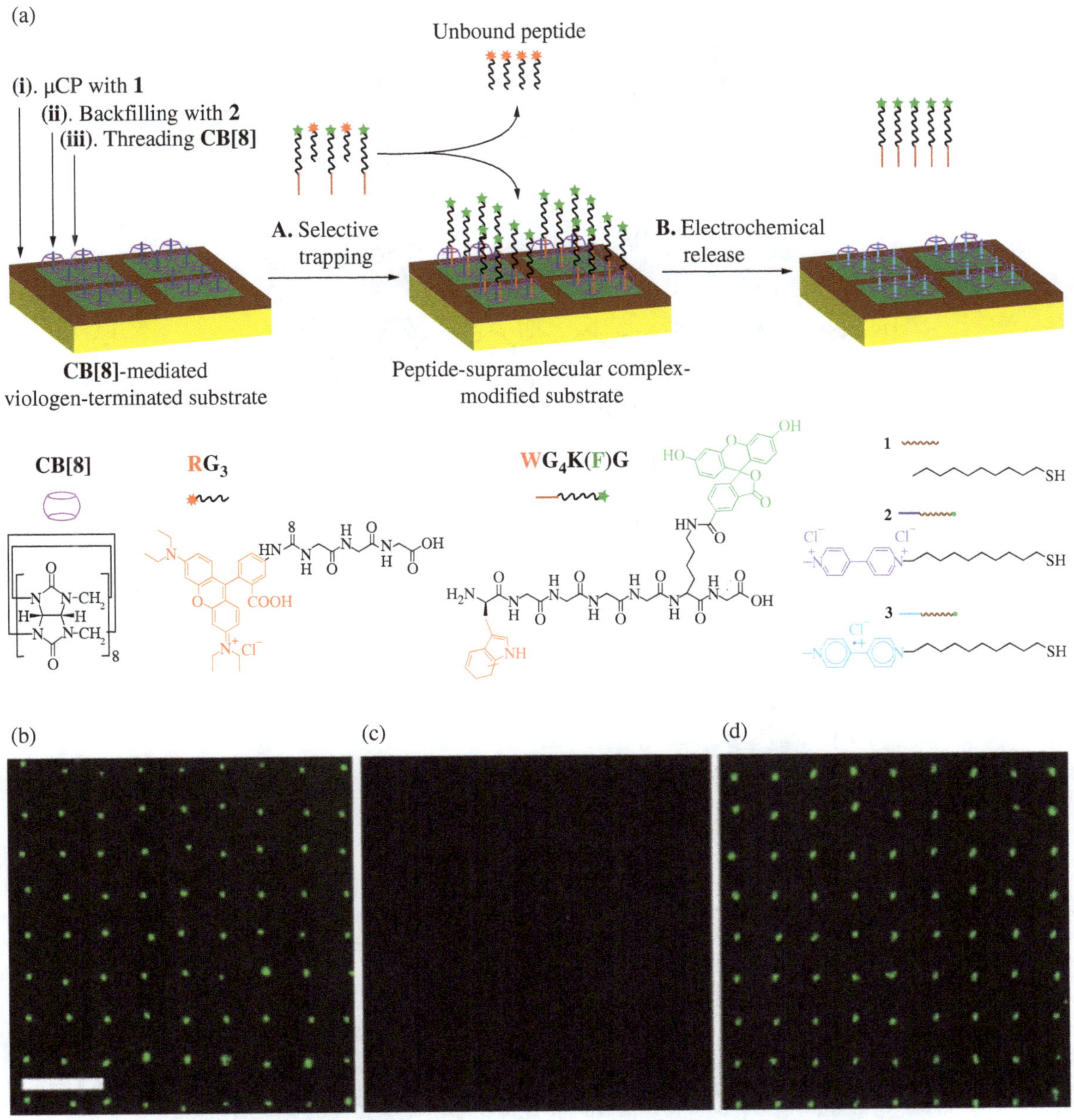

FIGURE 13.7 (a) CB8-based ternary complex for peptide trapping and electrochemical release. Fluorescence microscopy images of (b) the original peptide array, (c) no pattern after the reduction and washing of the substrate, and (d) the recovered peptide array (Reprinted with permission from Ref. [15]. Copyright 2011 American Chemical Society).

peptides adhesion and release by electrochemical activation [16]. The gold surfaces were functionalized by viologen derivatives that could form stable ternary complex with CB8 together with tripeptide arginine–glycine–aspartic acid (RGD) ligands (Fig. 13.8a). Applying an electrical potential dissociated such host–guest complex by reducing the viologen unit, thus resulting in the release of RGD ligands from the surface. In another experiment, mouse myoblast cells were bound to the RGD, and the results also demonstrated electrochemical-responsive cellular adhesion and release from the surface utilizing CB host–guest chemistry.

13.3.8 Redox-Responsive Supramolecular Polymer

The strategy in designing stimuli-responsive systems was expanded to supramolecular polymerization. *The host–guest interactions between CB8, calix*[4]*arenes, and viologen derivatives* were used to prepare a ternary linear polymer with reduction/oxidation responsiveness (Fig. 13.9) [17].

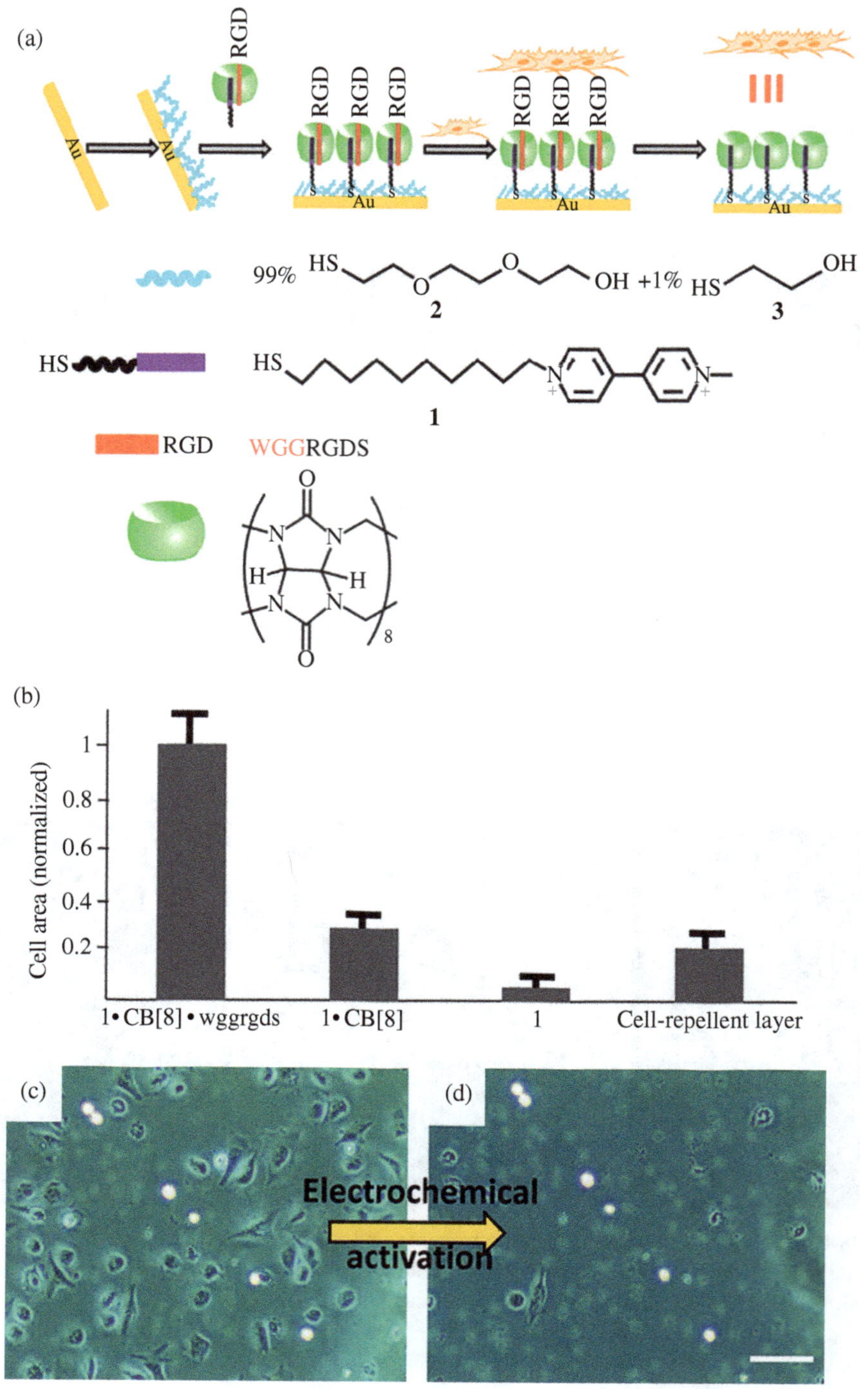

FIGURE 13.8 (a) Redox-responsive gold nanoparticles; (b) normalized projected cell area for different functionalized slides. Bright field images of cells before (c) and after (d) electrochemical activation (Reprinted with permission from Ref. [16]. Copyright 2012 John Wiley & Sons, Inc.).

The ternary polymer is composed of two monomers: **11** (possessing two calix[4]arene units) and **12** (containing two viologen groups) and CB8 leading to the formation of a high molecular weight cyclic supramolecular polymer with reversible assembly/disassembly ability induced by electrochemical or chemical stimuli.

Another group prepared redox-responsive layer-by-layer assemblies of two polymer matrixes poly(methylviologen) ($PMVC_nH_{2n+1}$)/CB8 complex and azide-functionalized poly(acrylic acid) ($PAA-N_3$) [18]. Such multilayer assemblies formed *a stable selective ternary complex* with 1-(anthracen-2-ylmethyl)pyridinium bromide (AnPy) in the

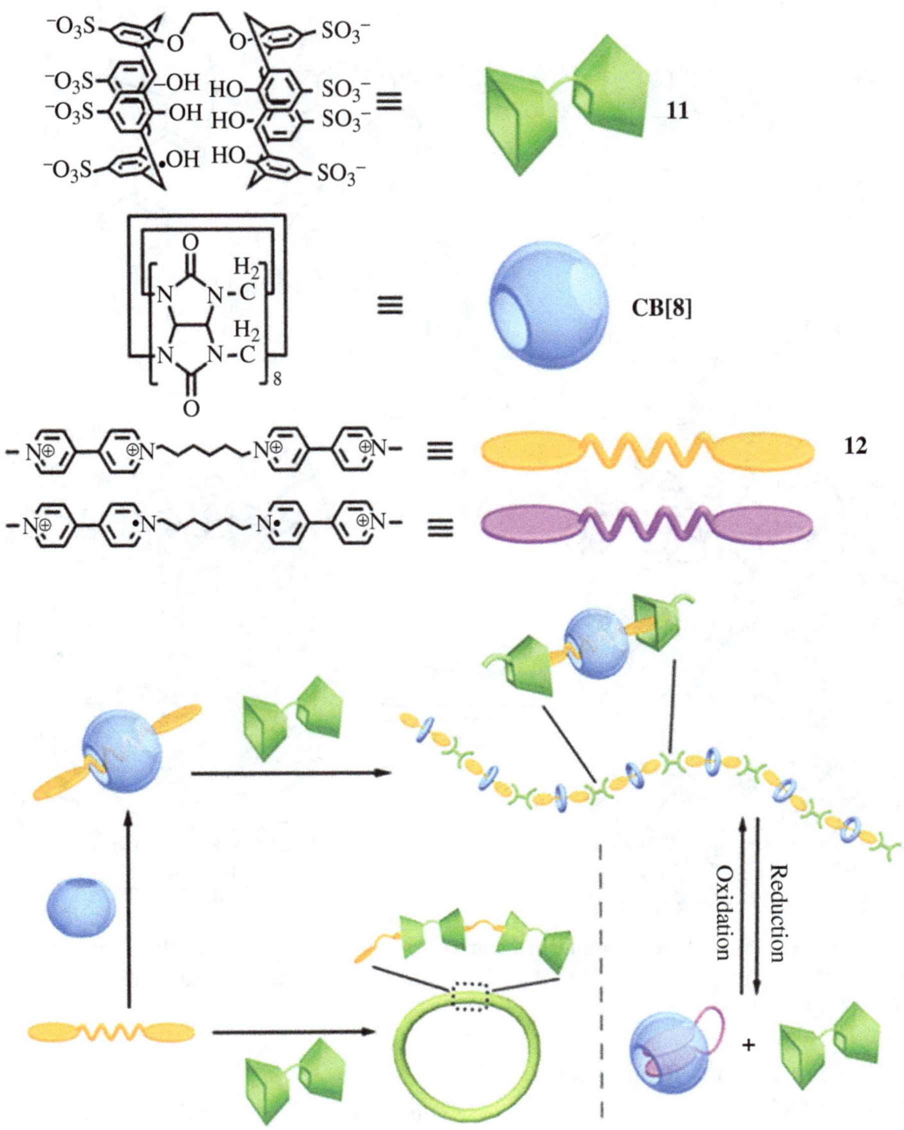

FIGURE 13.9 Calixarene-based supramolecular polymers controlled by the addition of CB[8] (Adapted with permission from Ref. [17]. Copyright 2012 John Wiley & Sons, Inc.).

presence of another anthracene derivative (9An-Py) due to steric congestion (Fig. 13.10). The addition of a reducing agent like $NaBH_4$ expelled the AnPy guest out from the surface as the viologen units were reduced to their radical cation form, inducing the disassociation of the heteroternary host–guest complex.

13.3.9 Light-Responsive Mesoporous Silica Nanoparticles

The light as stimuli was employed by the group of Yang for the controlled release of cargos from mechanized mesoporous nanoparticles (MSNs, Fig. 13.11) by utilizing the host–guest chemistry of CB7 [19]. They took advantage of the *suitable cavity and reasonable water solubility of CB7, as well as their ability to form stable host–guest complexes with trans cinnamamide derivatives*, which undergo a *trans*-to-*cis* conformational change upon UV light irradiation (300 nm), leading to complex dissociation due to steric hindrance effect. In more details, the surfaces of MSNs were decorated with cinnamamide-containing stalks encircled by CB7 rings sealing the nanopores of the MSNs and preventing the release of the preloaded cargos. The cargo was then released upon exposure to light since the isomerization of *trans*-to-*cis* cinnamamide units results in the dissociation of CB7 rings from the stalks, opening the gates and releasing the cargos.

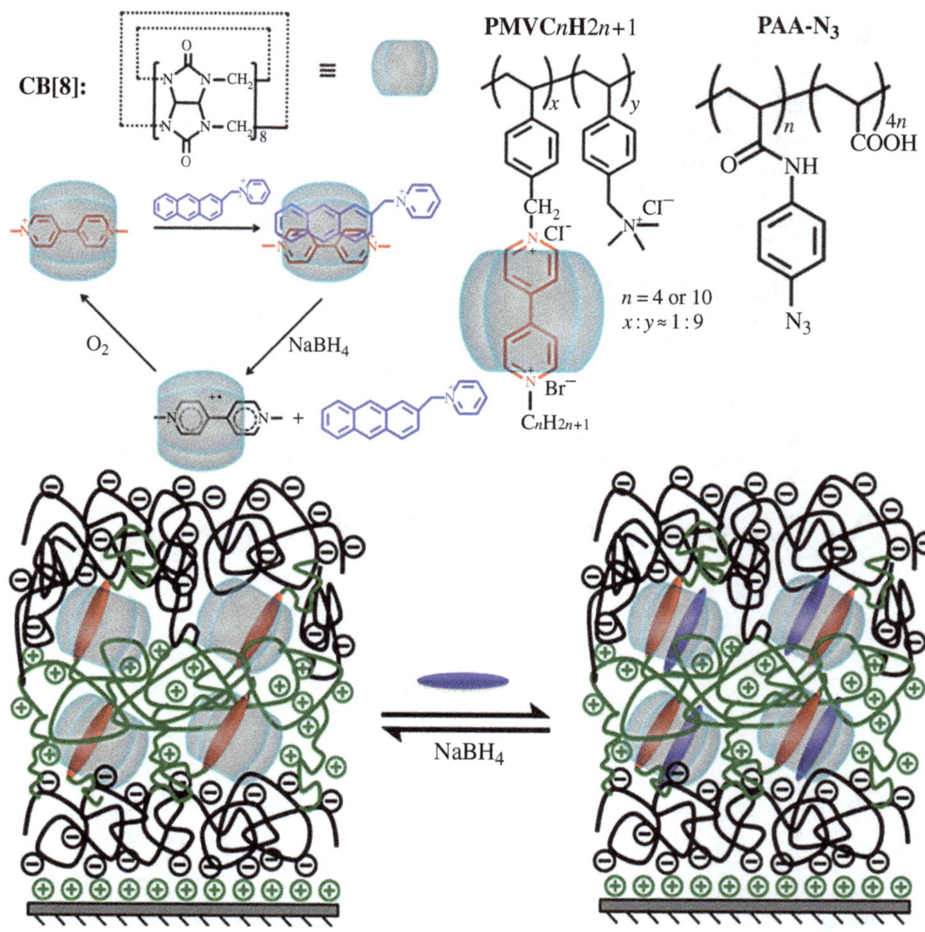

FIGURE 13.10 Redox-controlled reversible sequestering of 1-(anthracen-2-ylmethyl)pyridinium bromide (AnPy) in CB8, and schematic representation of layer-by-layer assembly of the PMVC$_n$H$_{2n+1}$-CB8 complex and PAA-N$_3$ (Adapted with permission from Ref. [18]. Copyright 2012 John Wiley & Sons, Inc.).

13.3.10 Dual Stimuli (pH and Light)-Responsive Mesoporous Silica Nanoparticles

The same group of Yang constructed a system based on MSNs yet with dual stimuli-responsive behaviors for applications in drug delivery and molecular electronics (Fig. 13.12) [19]. The idea was built on similar azobenzene–CB–pseudorotaxane supramolecular switching in response to pH and light stimuli. The azobenzene stalks were attached at the inner wall of MSNs pores. *The binding strength between the stalks and the CB circles was suppressed when pH was increased or upon the UV light-induced trans-to-cis isomerization*, thus releasing the cargo from MSNs into solution (Fig. 13.12a). The process was restored and the cargo was loaded back upon irradiation with visible light, which converts *cis* back to the *trans* configuration or upon decreasing pH (Fig. 13.12b). Overall, the authors claimed the establishment of AND logic type of systems.

13.4 CONCLUDING REMARKS

This chapter summarizes some reports that have utilized host–guest complexation of CBn in constructing various stimuli-responsive supramolecular nanostructures such as supramolecular metal nanoparticles, beads, colloids, hydrogels, micelles, colloidosomes, and polymers. It is concluded that certain strategies were taken into account to rationalize the CBn host–guest chemistry in order to construct special stimuli-responsive nanostructured systems. The high and selective binding affinity to certain guests has led to guest competition-responsive systems. The ability of CB8 to form selective ternary complexes has endowed the new nanostructures with redox-controlled switching. The suitable cavity of CB7 has allowed to construct light-controlled *cis/trans* switching systems, while the preferential affinity toward cationic guest over neutral molecular has facilitated the preparation of systems with pH-controlled functions. On the top of that, CBn with its negative carbonyl portals helped in stabilizing the metal nanoparticles during the functionalization process. Such unique hosting properties of CB have inspired many researchers to rationally design novel stimuli-responsive systems, which can be used for the fabrication of self-healing materials, *in vitro* therapeutics, peptide and protein isolations, and cellular transportation. Future applications in cancer therapy and electronic materials also remain an option.

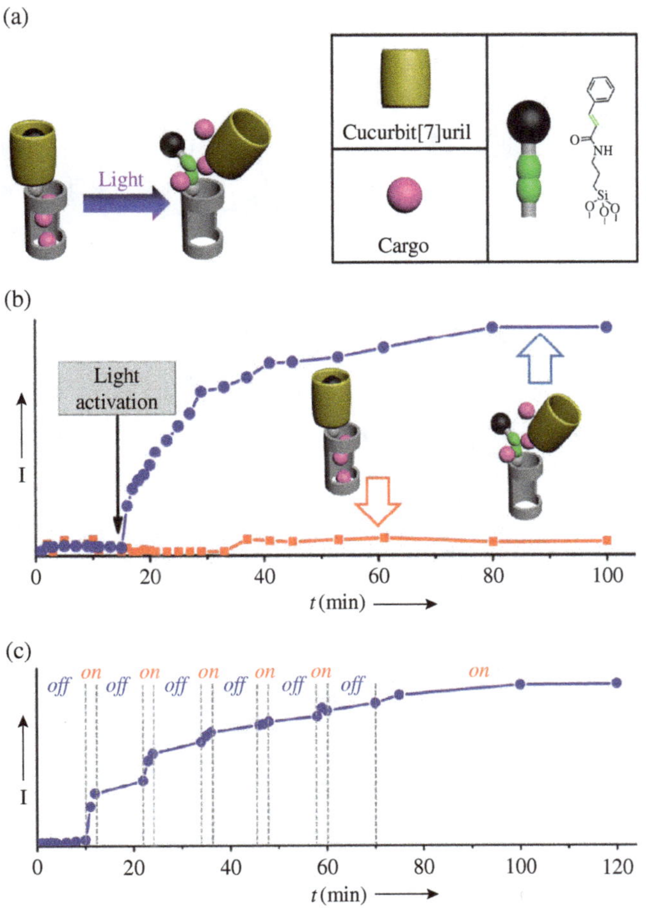

FIGURE 13.11 (a) Light-switchable mechanized MSNs using CB7 and their release profiles upon (b) continuous light and (c) pulsed light irradiation (Reprinted with permission from Ref. [19]. Copyright 2014 American Chemical Society).

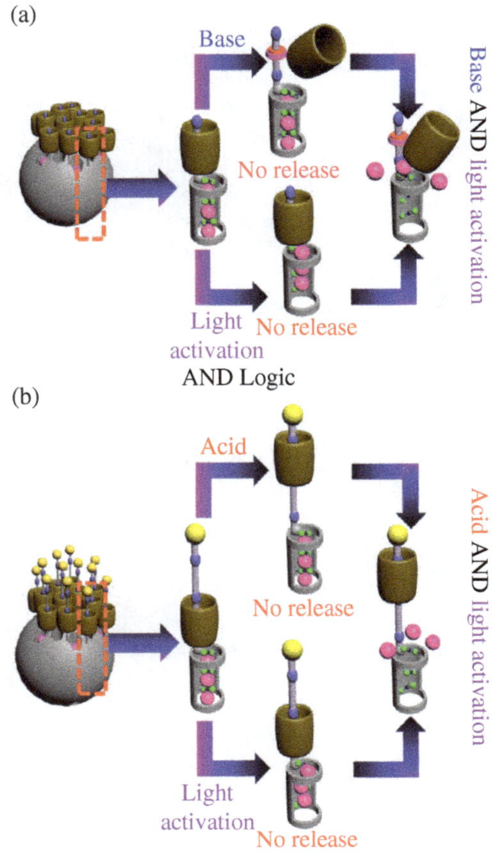

FIGURE 13.12 Dual-controlled systems consisting of (a) base-responsive nanovalves (raising pH) and azobenzene stalks and (b) acid-responsive nanovalves (lowing pH) and azobenzene stalks (Reprinted with permission from Ref. [19]. Copyright 2014 American Chemical Society).

ACKNOWLEDGMENTS

The author thanks the research program in United Arab Emirates University for financial grant number **UPAR (9) 2014.**

REFERENCES

1. P. A. Gale, J. W. Steed (Eds.), Supramolecular Chemistry: From Molecules to Nanomaterials, John Wiley & Sons, Ltd, Chichester, 2012.
2. L. Isaacs, *Acc. Chem. Res.* 2014, *47*, 2052.
3. W. A. Freeman, W. L. Mock, N. Y. Shih, *J. Am. Chem. Soc.* 1981, *103*, 7367.
4. K. I. Assaf, W. M. Nau, *Chem. Soc. Rev.* 2015, *44*, 394.
5. L. Cao, M. Šekutor, P. Y. Zavalij, K. Mlinarić-Majerski, R. Glaser, L. Isaacs, *Angew. Chem. Int. Ed.* 2014, *53*, 988.
6. A. L. Koner, I. Ghosh, N. Saleh, W. M. Nau, *Can. J. Chem.* 2011, *89*, 139.
7. F. Benyettou, I. Milosevic, Y. Lalatonne, F. Warmont, R. Assah, J.-C. Olsen, M. Jouaid, L. Motte, C. Platas-Iglesias, A. Trabolsi, *J. Mat. Chem. B* 2013, *1*, 5076.
8. S. Gürbüz, M. Idrisa, D. Tuncel, *Org. Biomol. Chem.* 2015, *13*, 330.
9. C. Kim, S. S. Agastil, Z. Zhu, L. Isaacs, V. M. Rotello, *Nat. Chem.* 2010, *2*, 962.
10. D. W. Lee, K. M. Park, M. Banerjee, S. H. Ha, T. Lee, K. Suh, S. Paul, H. Jung, J. Kim, N. Selvapalam, S. H. Ryu, K. Kim, *Nat. Chem.* 2011, *3*, 154.
11. Y. Lan, Y. Wu, A. Karas, O. A. Scherman, *Angew. Chem. Int. Ed.* 2014, *53*, 2166.
12. E. A. Appel, X. J. Loh, S. T. Jones, F. Biedermann, C. A. Dreiss, O. A. Scherman, *J. Am. Chem. Soc.* 2012, *134*, 11767.
13. Y. Wang, D. Li, H. Wang, Y. Chen, H. Han, Q. Jin, J. Ji, *Chem. Commun.* 2014, *50*, 9390.
14. G. Stephenson, R. M. Parker, Y. Lan, Z. Yu, O. A. Scherman, C. Abell, *Chem. Commun.* 2014, *50*, 7048.
15. F. Tian, M. Cziferszky, D. Z. Jiao, K. Wahlstrom, J. Geng, O. A. Scherman, *Langmuir* 2011, *27*, 1387.
16. Q. An, J. Brinkmann, J. Huskens, S. Krabbenborg, J. de Boer, P. A. Jonkheijm, *Angew. Chem. Int. Ed.* 2012, *51*, 12233.
17. H. Qian, D.-S. Guo, Y. Liu, *Chem. Eur. J.* 2012, *18*, 5087.
18. J. W. Zhang, Y. L. Liu, B. Yuan, Z. Q. Wang, M. Schönhoff, X. Zhang, *Chem. Eur. J.* 2012, *18*, 14968.
19. Y.-W. Yang, Y.-L. Sun, N. Song, *Acc. Chem. Res.* 2014, *47*, 1950.

14

PARALLEL INTERACTIONS OF AROMATIC MOLECULES AT LARGE HORIZONTAL DISPLACEMENTS

DUŠAN P. MALENOV[1] AND SNEŽANA D. ZARIĆ[2,3]

[1] *Innovation Center of the Faculty of Chemistry, University of Belgrade, Belgrade, Serbia*
[2] *Department of Chemistry, University of Belgrade, Belgrade, Serbia*
[3] *Department of Chemistry, Texas A&M University at Qatar, Doha, Qatar*

14.1 INTRODUCTION

Aromatic molecules are of essential role in nature. They can be encountered in large number of molecular systems, including biological systems, materials, drugs, crystals, and nanosystems [1–8]. Aromatic molecules can form various types of interactions, mostly through their π-systems [8]. Notable examples of these interactions include stacking interactions, which are essential for the structure of DNA [9], as well as for the stabilization of protein tertiary structure [10]. Also, XH/π (where X can be O, N, C, S), cation/π, and anion/π interactions have been recognized as important stabilizing forces in many systems [11–13]. Aromatic molecules can also form weak but very important CH—O interactions [14, 15].

Benzene molecule is the prototype molecule for studying non-covalent interactions between aromatic molecules. Interactions formed between two benzenes can be parallel (stacking) interactions (interaction energy of −2.73 kcal/mol) and aromatic CH/π interactions (interaction energy −2.84 kcal/mol) [8].

Until recent years, most of the studies of aromatic interactions through π-system focused on interactions with relatively small horizontal displacements [8, 16]. It should be noted that, relative to benzene ring center, carbon atoms are at horizontal displacement of approximately 1.4 Å and hydrogen atoms at horizontal displacement of approximately 2.5 Å (Fig. 14.1). Interacting molecules can be considered at large horizontal displacement relative to benzene if all of its atoms are out of the ring plane and beyond the C—H bond region, that is, with $r > 2.5$ Å. During the last few years, several studies were published that showed that substantial parallel interactions are formed at large horizontal displacements [17–24]. This review will summarize the most important findings of these studies and introduce these important interactions to wider audience.

14.2 PARALLEL WATER–AROMATIC INTERACTIONS AT LARGE HORIZONTAL DISPLACEMENTS

14.2.1 Parallel Interactions of Aromatic Molecules with Uncoordinated Water

Interactions of aromatic systems with water molecules are very important for many chemical and biological systems [25–30]. Contacts between aromatic systems and water are of particular importance in nanotubes [28], nanoporous materials [29], and water-transporting proteins (aquaporins) [30]. Therefore, water–aromatic interactions have been a topic of many extensive studies, both experimental and theoretical [31–34].

Prototypes for water–aromatic interactions are water–benzene interactions, which have been extensively studied. It was noted that water and benzene can form two common types of interactions, namely, OH/π (where one water O—H bond is pointed toward benzene plane) (Fig. 14.2) [31] and CH—O interactions (where C—H bond of benzene is proton donor to water oxygen) [14, 15]. Additionally, lone-pair/π interactions can also be encountered [35], but they were

Non-Covalent Interactions in the Synthesis and Design of New Compounds, First Edition.
Edited by Abel M. Maharramov, Kamran T. Mahmudov, Maximilian N. Kopylovich, and Armando J. L. Pombeiro.

shown to be repulsive in nature [18]. In the past few years, another type of water–aromatic interactions was discovered and thoroughly investigated [17–21]. These interactions are called parallel alignment interactions, and they will be discussed in this chapter.

Preliminary search of the Cambridge Structural Database (CSD) within the structures that contain water–phenyl interactions found a large number of contacts with d_O distance shorter than 6.0Å [20]. Plot of normal R_O distances versus offset r_O values (Fig. 14.3) revealed that the majority of those contacts can be found in the ellipsoid with vertical axis of 4.0Å and horizontal axis of 5.5Å, that is, with normal distances shorter than 4.0Å and offset values shorter than 5.5Å. Further searches were performed by using the contacts within this ellipsoid.

Crystal structures in the CSD were studied in detail by analyzing more parameters (Fig. 14.2), in order to determine the types of interactions and preferred geometries. Table 14.1 enlists the parameters that define four types of water–aromatic interactions. By using the parameters listed in Table 14.1, a total of 5234 contacts found in CSD search were separated into five groups. It was found that CH—O interactions are the most frequently present in crystal structures, rather than OH/π interactions. It is very interesting that

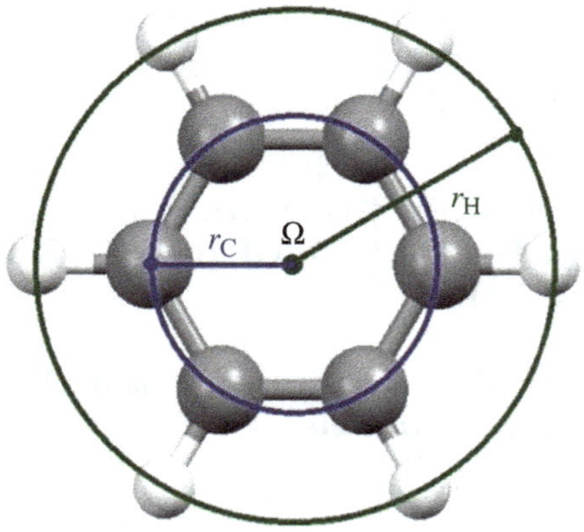

FIGURE 14.1 Offsets of benzene carbon atoms and benzene hydrogen atoms relative to center of benzene ring, since $r_C \approx 1.4$Å and $r_H \approx 2.5$Å.

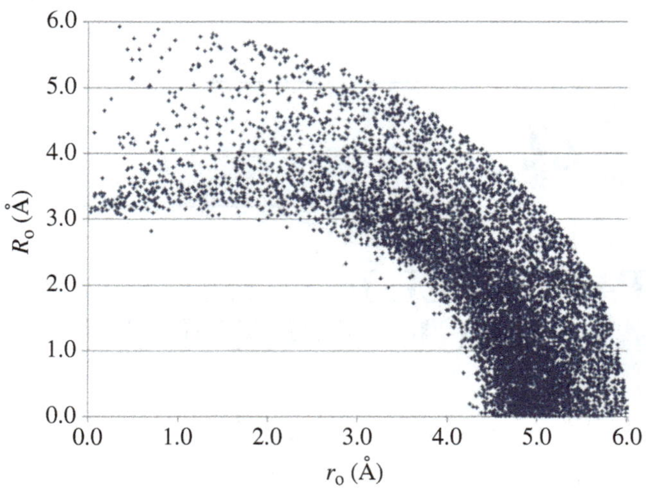

FIGURE 14.3 Oxygen normal distance (R_O) versus offset (r_O) plot for water–phenyl contacts found in the CSD with d_O distance shorter than 6.0Å.

TABLE 14.1 Geometrical Parameters for OH/π, CH—O, Parallel Alignment, and Lone-Pair/π Interactions and Their Presence in Crystal Structures Found in the CSD

Type	Parameters	Number of Contacts	Percentage
OH/π	$d(H_1\cdots\Omega)<3.5$Å $\measuredangle(O\text{—}H_1\cdots\Omega)>110°$ $\beta<30°$	141	2.7
CH-O	$d(H\cdots O)<3.2$Å $(C\text{—}H\cdots O)>110°$ $d(H\cdots O)<d(H\cdots H_1)$	2333	44.6
Parallel alignment	$80°<\Theta<100°$	837	16.0
Lone-pair/π	$d(O\cdots\Omega)<3.5$Å $d(O\cdots\Omega)<d(H_1\cdots\Omega)$	11	0.2
Others	Not satisfying any group of parameters of the above	1912	36.5

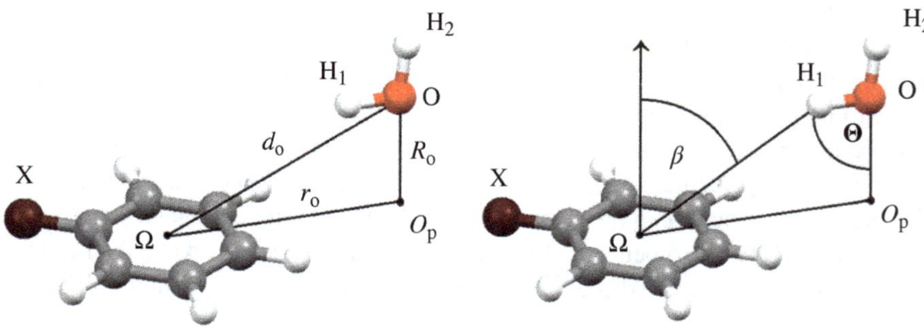

FIGURE 14.2 Geometrical parameters that define water–phenyl interactions: Ω is the center of phenyl ring, H_1 is water hydrogen atom closer to this center, O is water oxygen and O_p is the projection of water oxygen on the average phenyl ring plane, and X is any atom or group; β is the angle between $\Omega\cdots H_1$ and normal of the ring plane, d_O is the distance between water oxygen and phenyl ring center, r_O is the displacement (offset) of O_p, and R_O is the shortest (normal) distance between water oxygen and average ring plane; Θ is O_p—O—H_1 angle.

parallel alignment interactions are significantly more abundant than OH/π interactions (Table 14.1).

Similar trends were noted in protein crystal structures. Although hydrogen atom positions cannot be determined in crystal structures of such large molecules, some conclusions can be drawn from the positions of water oxygen atoms. The density plot of normal distance R_O versus offset value r_O showed that most of water–aromatic contacts are with water molecule with large offset values r_O, outside the ring and very close to the ring plane (Fig. 14.4a), indicating CH–O and parallel alignment interactions. If the plot is area corrected,[1] it can be noted that there is a preference for OH/π interactions, since water oxygen prefers position near the center of the phenyl ring (Fig. 14.4b). These density plots are very similar to plots for crystal structures from the CSD (Fig. 14.4c and d), implying that it is justified to assume that interactions in proteins are similar to interactions in crystal structures of small molecules.

The plot of normal distances R_O versus offset values r_O for parallel interactions only (Fig. 14.5) shows that water prefers the position in which its oxygen atom is out of the ring plane and beyond the region of C–H bonds (r_O>2.5Å). This plot points toward the preference for parallel alignment interactions at large horizontal displacements, which were examined in greater detail by analysis of crystal structures deposited in the CSD and by performing calculations of interaction energies.

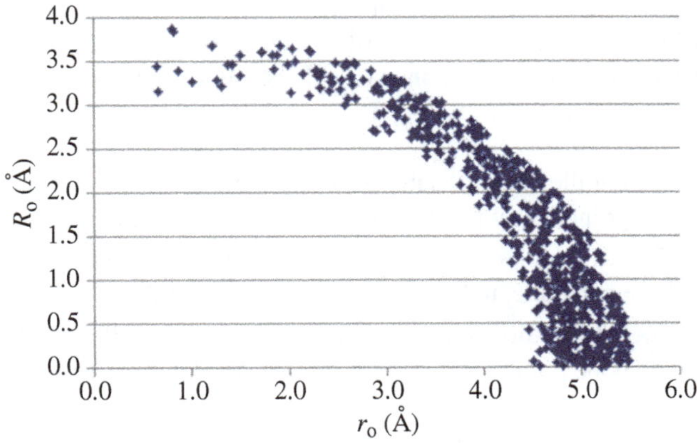

FIGURE 14.5 Oxygen normal distance (R_O) versus oxygen offset (r_O) plot for parallel alignment interactions between water and phenyl ring in the CSD crystal structures.

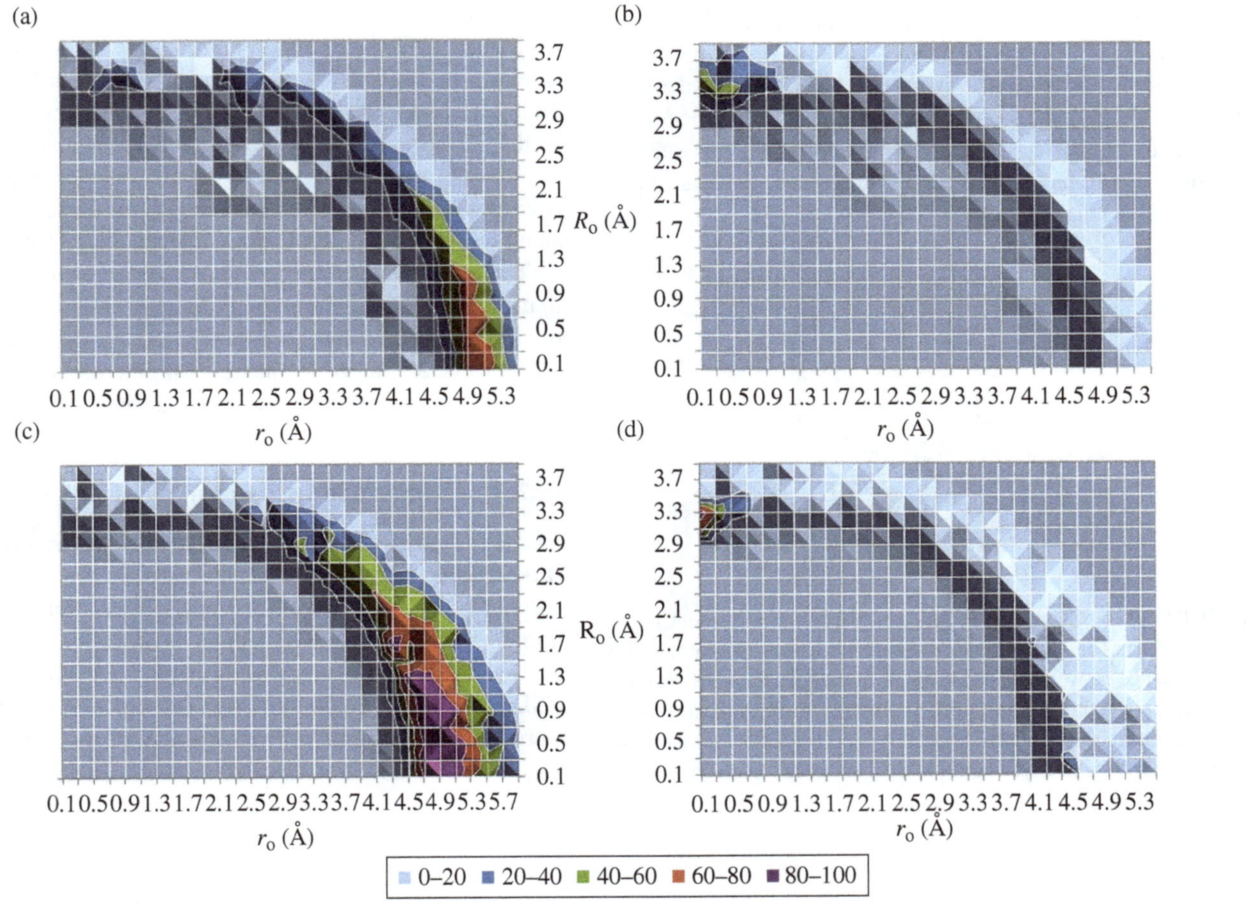

FIGURE 14.4 Density plots of oxygen normal distance (R_O) versus oxygen offset (r_O) for water–phenyl interactions found in PDB (uncorrected (a), area corrected (b)) and CSD crystal structures (uncorrected (c), area corrected (d)).

[1] The area correction assumes that the acceptor is of multiatom type and has the finite area, so frequency distributions and density plots must be corrected. The number of interactions with offset r is proportional to $2\pi r$; if N is the number of interactions in the range from r to $(r+\Delta r)$, N/r values must be used in distributions and density plots instead of N.

14.2.2 Cambridge Structural Database Search

It can be considered that C_6-aromatic rings form parallel interaction with water molecule if the following conditions are fulfilled: normal distance R_{H1} is shorter than 4.0 Å, offset r_{H1} is shorter than 3.5 Å, absolute ΔR_1 value is shorter than 0.1 Å ($\Delta R_1 = R_{H1} - R_O$), and Δr_1 values are negative ($\Delta r_1 = r_{H1} - r_O$) (Fig. 14.6) [20]. By using the parameter ΔR_2 ($\Delta R_2 = R_{H2} - R_O$), the contacts can be separated into three groups (Fig. 14.7):

1. Parallel-up interactions, where nonparallel O—H bond points away from the benzene ring plane ($\Delta R_2 > 0.1$ Å)
2. All-parallel interactions, where both water O—H bonds are parallel to benzene ring plane (0.1 Å > ΔR_2 > −0.1 Å)
3. Parallel-down interactions, where nonparallel O—H bond points toward the benzene ring plane (−0.1 > ΔR_2)

Parallel-up and all-parallel interactions were analyzed separately from parallel-up interactions.

14.2.3 Parallel-Up and All-Parallel Water–Aromatic Interactions

Parallel-up and all-parallel water–aromatic interactions were analyzed in crystal structures deposited in the CSD November 2009 release (version 5.31) [17]. A total of 71 all-parallel interactions, where the entire water molecule is parallel to aromatic ring plane, were found in this search. These interactions showed no preference for any particular offset values (Fig. 14.8a).

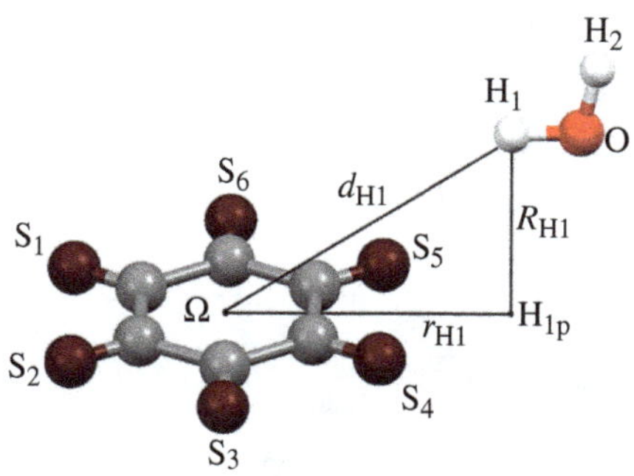

FIGURE 14.6 Geometric parameters of parallel alignment interactions of water and C_6-aromatic rings. Ω is aromatic ring center; d_{H1} is the distance between Ω and H_1; R_{H1} is normal distance between H_1 and mean plane of the phenyl ring; horizontal displacement (offset) r_{H1} is the distance from Ω to the projection of H_{1p} of H_1 atom on the mean plane of the phenyl ring; S_1 to S_6 is any atom or group.

Parallel-up water–aromatic interactions, where one water O—H bond is parallel to aromatic ring plane and the other one points away from the ring plane, were more frequent in the CSD crystal structures. The search yielded 716 of these interactions. The normal distance R_{H1} versus offset r_{H1} plot for these interactions showed that in most of the contacts normal distances are longer than 3.0 Å (Fig. 14.8b), similar to normal distances for two stacked benzene molecules (3.3–3.8 Å) [36–38]. In contacts with water molecules above the aromatic ring ($r_{H1} < 1.5$ Å), all normal distances are longer than 3.0 Å. However, in the C—H bond region (1.5 Å < r_{H1} < 2.5 Å), there are contacts with normal distances shorter than 3.0 Å. Outside of the C—H bond region starts the large horizontal displacement area, which is the area with the largest portion of parallel-up water–aromatic interactions (Fig. 14.8b). Also, the number of contacts with short R_{H1} values is very high.

14.2.4 Parallel-Down Water–Aromatic Interactions

Parallel-down water–aromatic interactions were studied within crystal structures from the CSD November 2011 release (version 5.33) [19]. A total of 707 parallel-down interactions were found in this search.

The normal distance R_{H1} versus offset r_{H1} plot for these interactions (Fig. 14.9) shows that the number of contacts increases with the increase of offset values, ending with more than 50% of the contacts at large horizontal displacements (r_{H1} > 2.5 Å). Similar to parallel-up water–aromatic interactions, normal distances are shortest for interactions at large offsets—a significant number of interactions are with R_{H1} < 3.0 Å.

The CSD searches can be summarized into the conclusion that parallel alignment water–aromatic interactions (parallel-up, parallel-down, and all-parallel) at large horizontal displacements are more frequent than interactions at small offsets, that is, with water above the aromatic ring.

14.2.5 *Ab Initio* and DFT Calculations

14.2.5.1 Parallel-Up and All-Parallel Water–Aromatic Interactions In order to obtain data on the strength of parallel alignment water–benzene interactions at large horizontal displacements, calculations of interaction energies on several levels of theory were performed. Similar to the CSD searches, parallel-up and all-parallel interactions were studied separately from parallel-down interactions. In the first study, five model systems divided into two

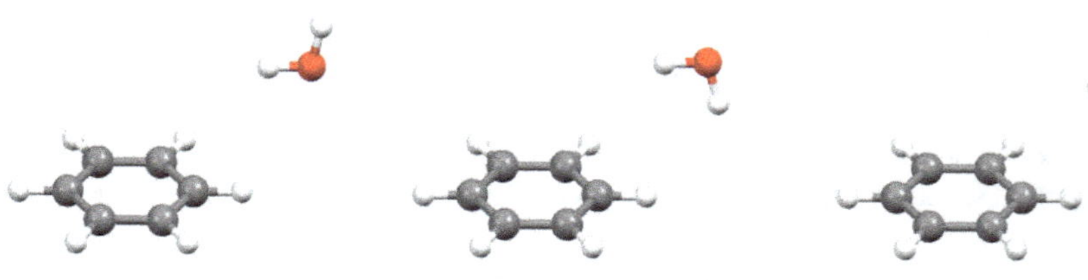

FIGURE 14.7 Parallel-up, parallel-down, and all-parallel interactions between water and C_6-aromatic ring.

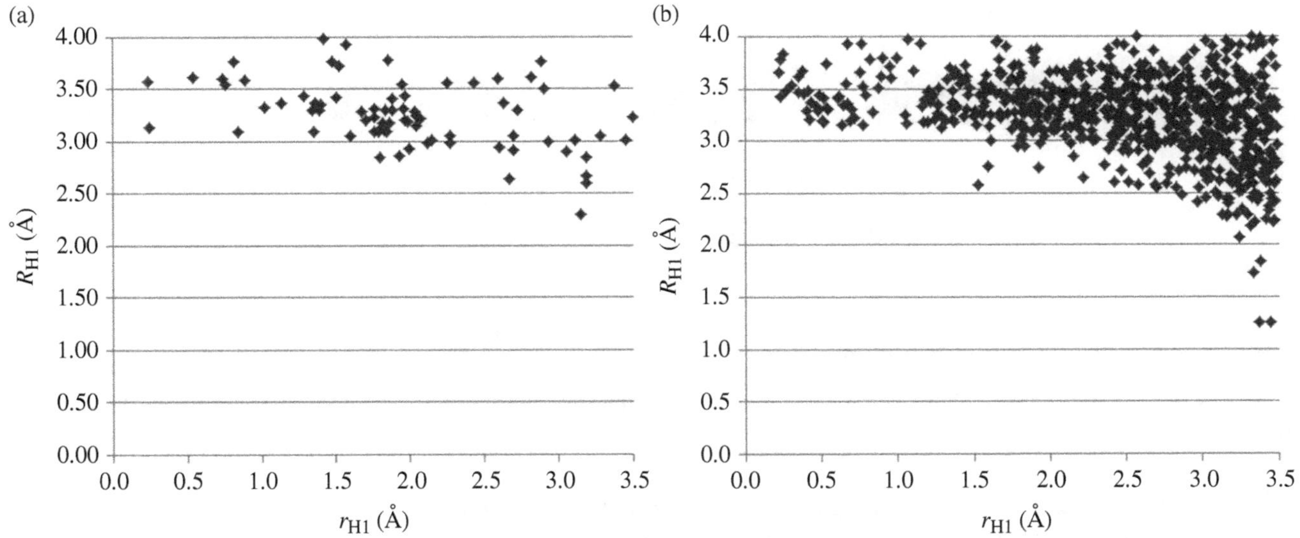

FIGURE 14.8 Correlation of normal distance R_{H1} and offset r_{H1} for all-parallel (a) and parallel-up water–aromatic interactions (b).

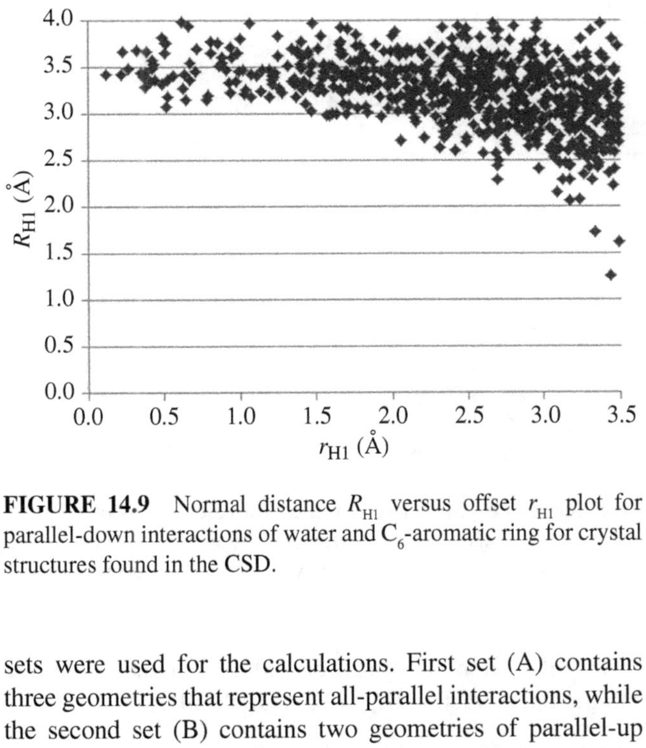

FIGURE 14.9 Normal distance R_{H1} versus offset r_{H1} plot for parallel-down interactions of water and C_6-aromatic ring for crystal structures found in the CSD.

sets were used for the calculations. First set (A) contains three geometries that represent all-parallel interactions, while the second set (B) contains two geometries of parallel-up interactions (Fig. 14.10) [21]. The calculations were performed by using supramolecular method, including MP2 and "gold standard" CCSD(T) calculations [39], and also by using symmetry-adapted perturbation theory (SAPT) [40] and extended transition state (ETS) [41] approaches, which provide the data on various components of energy.

All levels of theory agree that parallel-up interactions are stronger than all-parallel interactions (Table 14.2). The most stable geometry is B_2, with CCSD(T) energy of −2.45 kcal/mol at complete basis set, making parallel alignment water–benzene interactions at large horizontal displacements stronger than CH—O interactions (−1.41 kcal/mol) [42] and less strong than OH/π interactions (−3.19 kcal/mol) [42]. Also, parallel alignment interactions at large offsets are comparable in energy with benzene–benzene parallel interactions (the most stable being with energy of −2.73 kcal/mol) [43]. It is also interesting to note that water–benzene parallel alignment interactions at large horizontal displacements are significantly stronger than interactions at small displacements. The strongest interaction with water molecule positioned above the benzene ring is with interaction energy of −0.78 kcal/mol [21, 42].

Both SAPT and ETS energy decomposition methods point on electrostatics as the dominant energy component responsible for substantial attraction (Tables 14.3 and 14.4). Dispersion is also recognized as important force, but it is in all cases canceled by the exchange term, as it was observed in SAPT approach (Table 14.3); this cancelation is also present in most of the systems observed by the ETS approach (Table 14.4). Additional calculations with ETS–natural orbitals for chemical valence (NOCV) [44] method revealed $\pi \rightarrow \sigma^*$(O—H) charge transfer between the monomers in three model systems with strongest interactions (A_3, B_1, and B_2), leading to charge accumulation in water–benzene region and partial covalency of these interactions.

14.2.5.2 Parallel-Down Water–Aromatic Interactions

Interaction energies of parallel-down interactions were calculated at MP2/def2-QZVP level of theory [19]. Water molecule was moved between two C—H bonds, starting from offset r_{H1} = 0.0 Å to offset r_{H1} = 3.5 Å (Fig. 14.11). However, geometry for r_{H1} = 0.0 Å satisfies criteria for OH/π interactions and therefore its substantial interaction energy of −3.14 kcal/mol is not surprising. Normal distance R_{H1} for this interaction is at 3.5 Å, which is similar to normal distances for stacking interactions between aromatic molecules. At large horizontal displacements (r_{H1} = 3.0 Å), the parallel-down interaction is not as strong as parallel-up and all-parallel interactions. Interaction energy is only

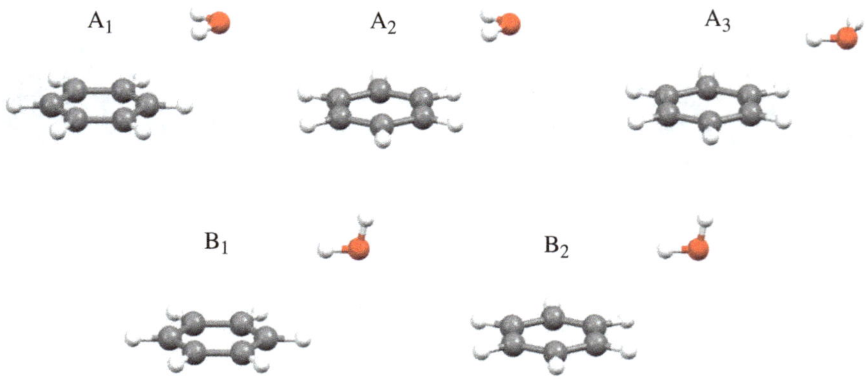

FIGURE 14.10 Model systems used for calculations of interactions energies for all-parallel (A_1, A_2, A_3) and parallel-up (B_1, B_2) interactions of water and benzene.

TABLE 14.2 Interaction Energies and Geometrical Parameters Calculated at Various Levels of Theory for Water–Benzene Parallel Alignment Interactions

	A_1	A_2	A_3	B_1	B_2
r_{H1} (Å)	2.9	2.9	2.8	2.5	2.6
R_{H1} (Å)	2.9	2.8	2.5	2.5	2.3
ΔE (MP2/cc-pVQZ) (kcal/mol)	−1.27	−1.31	−1.58	−2.08	−2.32
ΔE (CCSD(T)/CBS) (kcal/mol)	−1.25	−1.46	−1.73	−2.18	−2.45
E (SAPT/aug-cc-pVDZ) (kcal/mol)	−1.24	−1.10	−1.53	−1.94	−2.13
E (ETS/PBE-D3) (kcal/mol)	−1.83	−1.79	−2.13	−2.42	−2.72

Model systems presented in Figure 14.10.

TABLE 14.3 SAPT Interaction Energy Decomposition for Water–Benzene Parallel Alignment Interactions

	A_1	A_2	A_3	B_1	B_2
Electrostatics	−1.37	−2.14	−1.45	−2.33	−2.36
Dispersion	−1.21	−1.94	−1.27	−1.65	−1.82
Exchange	+1.53	+3.31	+1.45	+2.36	+2.51
Induction	−0.19	−0.32	−0.27	−0.42	−0.47
Exchange + dispersion	+0.32	+1.37	+0.18	+0.71	+0.69
Total	*−1.24*	*−1.10*	*−1.53*	*−1.94*	*−2.13*

Model systems presented in Figure 14.10.

TABLE 14.4 ETS/PBE-D3 Interaction Energy Decomposition Water–Benzene Parallel Alignment Interactions

	A_1	A_2	A_3	B_1	B_2
Electrostatics	−1.25	−1.96	−1.33	−2.07	−2.31
Dispersion	−0.89	−1.19	−0.90	−0.94	−1.03
Orbital	−0.45	−0.67	−0.60	−0.85	−0.89
Pauli (exchange)	+0.76	+2.03	+0.70	+1.43	+1.44
Total	*−1.83*	*−1.79*	*−2.13*	*−2.43*	*−2.79*

Model systems presented in Figure 14.10.

−1.11 kcal/mol, while parallel-up interaction is more than twice as strong (−2.45 kcal/mol), and all-parallel interaction (model system A_3, Fig. 14.10) is moderately stronger (−1.73 kcal/mol, Table 14.2). Normal distance for parallel-

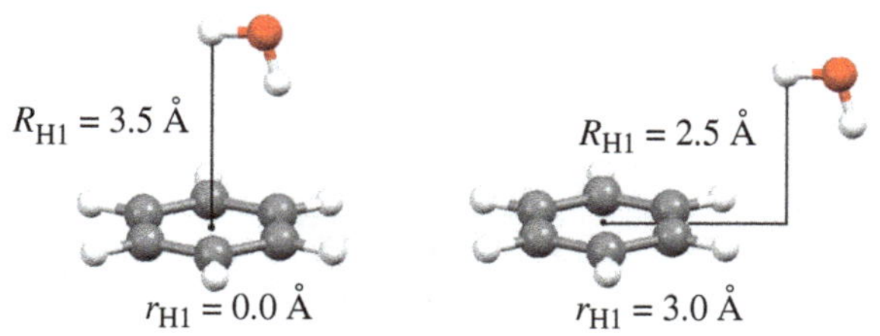

FIGURE 14.11 Model system used for calculations of parallel-down interactions between water and benzene; the most stable geometries with offsets of r_{H1} = 0.0 Å and r_{H1} = 3.0 Å are presented.

down interaction at offset of 3.0 Å is higher than for parallel-up interaction (R_{H1} = 2.3 Å, Table 14.2).

14.2.5.3 Influence of Parallel-Up Interactions on Other Water–Aromatic Interactions The influence of parallel-up interactions on other water-benzene interactions was studied at MP2/cc-pVQZ level of theory [18]. This study was evaluating two different energies: cooperativity energy (ΔE_{COOP}), which encompasses the contribution of all possible interacting pairs [45], and $\Delta\Delta E$, which takes into account only the pairs that are close in space, that is, effectively interacting [18].

Four model systems used in this study all contain two water–benzene interactions; one of them is always parallel-up interaction (B/W2) and the other one being OH/π, CH—O, parallel-up, or lone-pair/π interaction (B/W1) (Fig. 14.12).

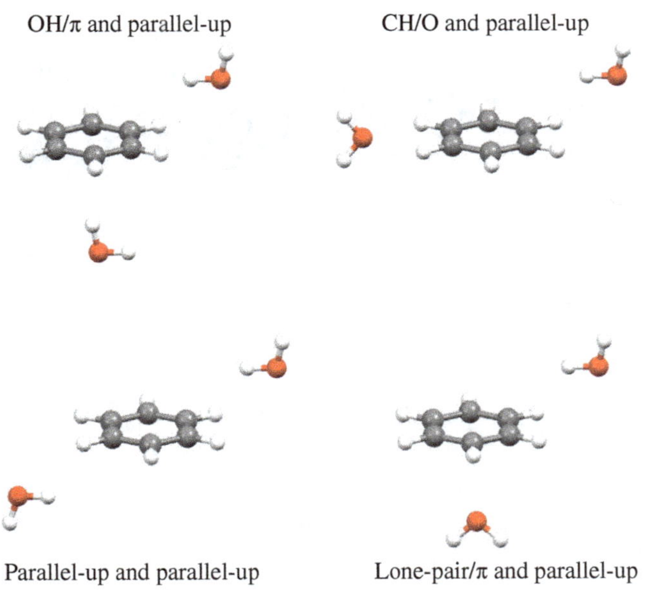

FIGURE 14.12 Model systems used for the evaluation of influence of parallel-up interactions on other attractive water–benzene interactions; parallel-up interaction is B/W2; the other interactions, OH/π, CH—O, parallel-up, or lone-pair/π interaction are B/W1.

The geometry of parallel-up interaction B/W2 was kept rigid, while B/W1 interactions were varied in order to find the position with the lowest energy of the system.

Both ΔE_{COOP} and $\Delta\Delta E$ values (Table 14.5) agree that parallel-up interactions have no significant influence on OH/π, CH—O, and other parallel-up interactions. This can be explained by very small electron transfer caused by parallel-up interactions (only 0.6 me transferred from benzene to water) in comparison with electron transfer caused by OH/π and CH—O interactions (up to 2.1 me). However, this small electron transfer is sufficient to influence lone-pair/π interactions, which become less repulsive in the presence of parallel-up interactions, as witnessed by $\Delta\Delta E$ value of −0.16 kcal/mol.

$\Delta\Delta E$ values for systems in which influence of other interactions on parallel-up interactions was studied are almost identical to the values presented here, implying that parallel-up interactions do not influence other water–aromatic interactions, nor other water–aromatic interactions influence parallel-up interactions.

14.2.6 Parallel Water–Aromatic Interactions at Large Horizontal Displacements in Supramolecular Structures

Comparing the abundance and strength of attractive water–benzene interactions, one can observe strong OH/π interactions and abundant CH—O interactions (Table 14.1) (Fig. 14.13). The reason of this great abundance of CH—O interactions in comparison with other types might be the fact that they have the largest area at their disposal, which can be rationalized by comparing uncorrected and corrected density plots of water–phenyl interactions. OH/π interactions are

TABLE 14.5 Influence of Parallel-Up Interaction on Other Water–Benzene Interactions

	ΔE(W1/B)	ΔE(W1/BW2)	$\Delta\Delta E^{a}$	ΔE_{COOP}^{b}
OH/π	−3.22	−3.18	+0.04	+0.07
CH/O	−1.51	−1.57	−0.06	+0.01
Parallel-up	−2.32	−2.35	−0.03	+0.04
LP/π	+0.55	+0.39	−0.16	−0.07

All energies are given in kilocalories per mole.

[a] $\Delta\Delta E$ is the measure of strengthening (if it is negative) or weakening (if it is positive) of other interactions in the presence of parallel-up interaction; $\Delta\Delta E = \Delta E(W1/BW2) - \Delta E(W1/B)$.

[b] ΔE_{COOP} is cooperativity energy, calculated as the difference between the energy of the whole system and the sum of all dimer interaction energies $\Delta E_{COOP} = (\Delta E(W1/B/W2)) - (\Delta E(W1/B) + \Delta E(B/W2) + E(W1/W2))$.

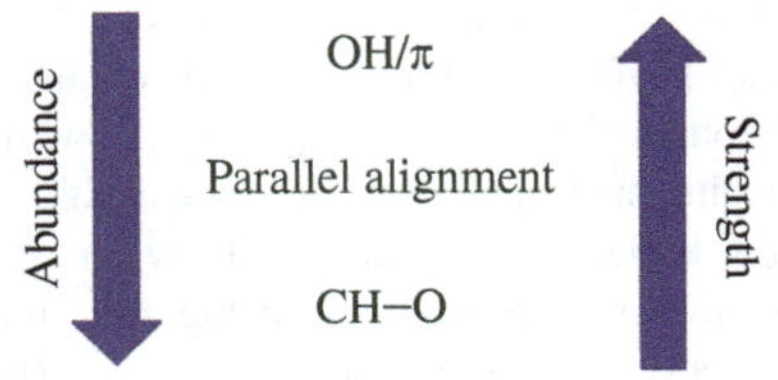

FIGURE 14.13 Comparison of strength and abundance in crystal structures for attractive water–benzene interactions.

stronger than parallel alignment interactions (−3.19 and −2.45 kcal/mol, respectively), but they occur significantly less in crystal structures.

Lower abundance of OH/π interactions, in spite of their greater strength, can be explained by additional simultaneous interactions that stabilize supramolecular structure. Namely, in OH/π interactions, one of water O—H bonds is occupied by the aromatic π-system, and it is disabled in forming simultaneous interactions. On the other hand, both O—H bonds of water molecule forming parallel alignment interaction at large horizontal displacement are available for surrounding molecules and possess the ability to form simultaneous interactions (Fig. 14.14), leading to additional stabilization of supramolecular structure [17]. Additionally, π-system of aromatic ring is also unoccupied by the water molecule and can also participate in additional interactions (Fig. 14.14).

14.2.7 Coordinated Water–Aromatic Interactions at Large Horizontal Displacements

It is known that the properties of water can be substantially influenced by the presence of metal ions [47]. Recent theoretical studies showed that water molecules coordinated to metal ions within neutral and positively charged complex form stronger OH/π interactions with benzene than uncoordinated water molecules (interaction energies are −6.50 and −14.85 kcal/mol for neutral charged complex and positively charged complex, respectively, and −3.36 kcal/mol for uncoordinated water) [48].

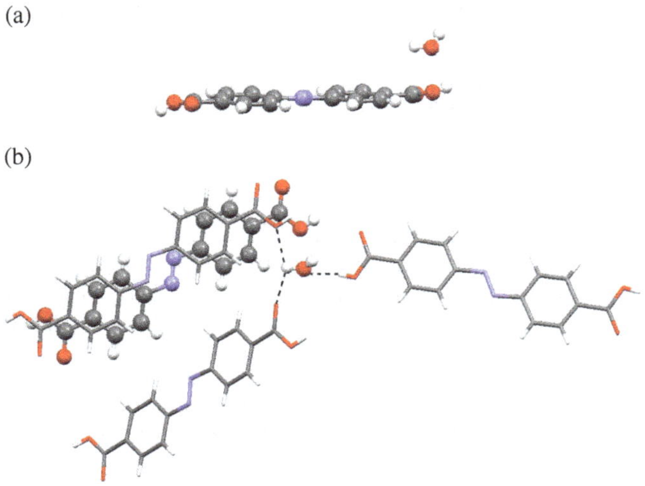

FIGURE 14.14 In crystal structure SENRUZ (4,4-azodibenzoic acid dihydrate) [46], water molecule forms parallel-up interaction at large horizontal displacement (r_{H1} = 2.83 Å) with aromatic ring with two substituents (a); as a consequence of this parallel interaction at large horizontal displacement, water molecule forms additional hydrogen bonds and aromatic ring forms additional stacking interactions with surrounding molecules (b).

In order to determine the influence of coordination on water–benzene parallel alignment interactions at large horizontal displacements, the CSD search and quantum chemical calculations were performed [19].

14.2.7.1 Cambridge Structural Database Search Contacts of coordinated water with C_6-aromatic ring in crystal structures from the CSD November 2011 release (version 5.33) were studied by analyzing the same parameters used in searches of interactions of uncoordinated water (Fig. 14.6) [19]. The search yielded 333 parallel alignment interactions of coordinated water and C_6-aromatic ring, which were then separated into three groups, similar to interactions of uncoordinated water:

1. Parallel-up interactions (71 contacts)
2. All-parallel interactions (35 contacts)
3. Parallel-down interactions (227 contacts)

All-parallel interactions were not particularly analyzed in this study.

In parallel-up interactions of coordinated water, the ligand in *cis* position to the water often interacts with aromatic group; however, if the ligand is voluminous, the parallel interaction of aqua ligand and aromatic ring is disabled. Therefore, the number of parallel-up interactions is much smaller than number of parallel-down interactions. However, it can be noted that both types of interactions prefer large horizontal displacements. The normal distance R_{H1} versus offset r_{H1} plot for both parallel-up (Fig. 14.15a) and parallel-down interactions (Fig. 14.15b) shows high tendency for large horizontal displacements. For parallel-up

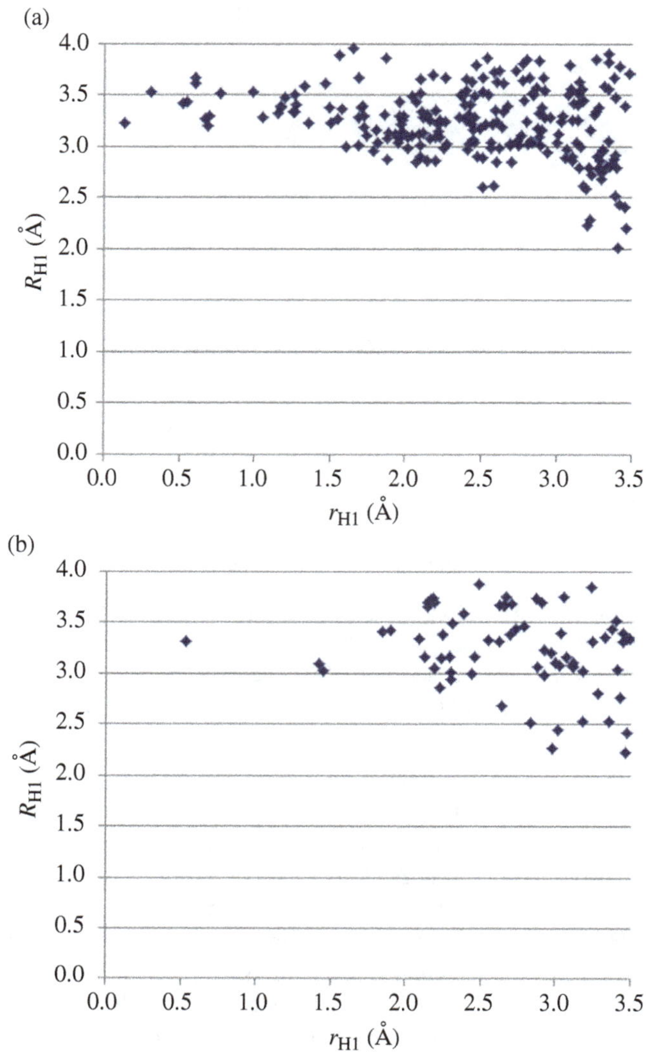

FIGURE 14.15 The normal distance R_{H1} versus offset r_{H1} plot for parallel-up (a) and parallel-down (b) interactions of coordinated water with C_6-aromatic ring in crystal structures found in the CSD.

interactions, only three contacts are in aromatic ring region and only 21 in C—H bond region (Fig. 14.15a). For parallel-down interactions, small number of interactions is in aromatic ring region and larger number of interactions outside that region, particularly with offsets larger than 2.5 Å (Fig. 14.15b).

14.2.7.2 Interaction Energy Calculations for Parallel-Down Interactions of Coordinated Water with Benzene Interaction energies of parallel-down interactions between benzene and water coordinated to metal were calculated by using three aqua complexes of different charges. For neutral aqua complex, $[ZnCl_2(H_2O)_4]$ was used, while the calculations for positively charged complexes were done with $[ZnCl(H_2O)_5]^+$ and $[Zn(H_2O)_6]^{2+}$ (Fig. 14.16) [19]. The calculations were performed at MP2/def2-QZVP level of theory, which is in good agreement with CCSD(T)/CBS level for parallel interactions of uncoordinated water and benzene [48].

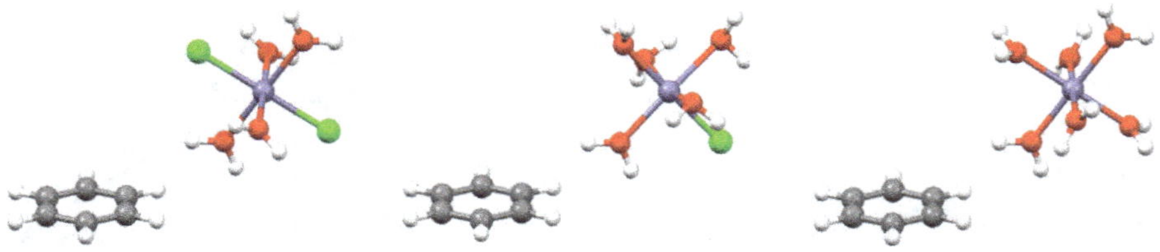

FIGURE 14.16 Model systems used for calculations of interaction energies for parallel-down interactions of coordinated water with benzene; aqua complexes with different charges were neutral $[ZnCl_2(H_2O)_4]$, positive $[ZnCl(H_2O)_5]^+$, and double positive $[Zn(H_2O)_6]^{2+}$; geometries in this figure are with $r_{H1} = 3.0$ Å.

TABLE 14.6 Interaction Energies and Geometric Parameters of Parallel-Down Interactions of Benzene with Uncoordinated Water and Coordinated Water (Aqua Complexes of Different Charges)

	$r_{H1} = 0.0$ Å[a]		$r_{H1} = 3.0$ Å	
Interacting Water	ΔE (kcal/mol)	R_{H1} (Å)	ΔE (kcal/mol)	R_{H1} (Å)
H_2O	−3.14	3.5	−1.11	2.5
$[ZnCl_2(H_2O)_4]$	−4.70	3.3	−1.25	2.5
$[ZnCl(H_2O)_5]^+$	−9.82	3.1	−3.71	2.1
$[Zn(H_2O)_6]^{2+}$	−14.89	3.1	−7.51	1.7

[a] When $r_{H1} = 0.0$ Å, water hydrogen that belongs to parallel O—H bond is above benzene ring center, and this geometry satisfies criteria for both parallel-down and OH/π interactions.

Energies of hydrogen bonds between coordinated and uncoordinated water were shown to be higher than energies of hydrogen bonds between two uncoordinated water molecules [49]. Parallel-down interactions of coordinated water are also stronger than interactions of uncoordinated water. For small horizontal displacements, that is, for geometries with water molecule above the benzene, interaction can be considered both OH/π and parallel-down, which can explain high interaction energies. Interaction of neutral complex is stronger than interaction of uncoordinated water (−4.70 and −3.14 kcal/mol, respectively, Table 14.6). The energies are highly dependent on the charge of the complex, and therefore, interactions of double positively charged $[Zn(H_2O)_6]^{2+}$ complex (−14.89 kcal/mol) are significantly stronger than interactions of neutral complex and single positively charged complex (−4.70 and −9.82 kcal/mol, respectively). Additionally, normal distances for interactions of coordinated water are shorter than those for uncoordinated water (3.5 Å), and they are also dependent on the charge of the complex (3.3 Å for $[ZnCl_2(H_2O)_4]$ and 3.1 Å for $[ZnCl(H_2O)_5]^+$ and $[Zn(H_2O)_6]^{2+}$, Table 14.6). Due to the fact that these complexes are quite large and these interactions are of similar geometries to OH/π interactions, their strength is very similar to the strength of OH/π interactions of coordinated water molecules with benzenes [48].

At large horizontal displacements, interaction energies are also stronger and normal distances shorter in comparison with uncoordinated water molecule [19]. Parallel-down interaction between $[ZnCl_2(H_2O)_4]$ complex and benzene is only slightly stronger than parallel-down interaction of uncoordinated water and benzene (−1.25 and −1.11 kcal/mol, respectively), but as the charge of the complex grows, the energy goes up to −7.51 kcal/mol for $[Zn(H_2O)_6]^{2+}$ (Table 14.6). Normal distance for this complex is very short, only 1.7 Å, which is much shorter than for uncoordinated water (2.3 Å, Table 14.2). However, it is interesting to note that parallel-up interaction of uncoordinated water is almost as twice as strong as parallel-down interaction of water within neutral complex (−2.45 and −1.25 kcal/mol) for similar offsets (2.6 and 2.5 Å, respectively).

14.2.7.3 Parallel Interactions of Aromatic Molecules with Coordinated Water in Supramolecular Structures It was shown that parallel-up interactions of uncoordinated water at large horizontal displacements are stronger than interactions at small displacements (Table 14.2) [19]. This is in good agreement with the preference for large horizontal displacements found in crystal structures from the CSD (Fig. 14.7a). However, parallel-down interactions of coordinated water at large horizontal displacements are less strong than interactions above the benzene ring (Table 14.6), but they are more frequent in crystal structures (Fig. 14.17b). These dissimilarities are the consequence of supramolecular structures, which have very strong influence on water–aromatic interactions.

14.3 PARALLEL INTERACTIONS BETWEEN AROMATIC RINGS AT LARGE HORIZONTAL DISPLACEMENTS

14.3.1 Parallel Aromatic–Aromatic Interactions at Large Horizontal Displacements

Aromatic–aromatic interactions are essential interactions for important biological systems, most notably DNA and proteins [9, 10]. Benzene–benzene interactions are prototype for studying interactions between aromatic molecules. High

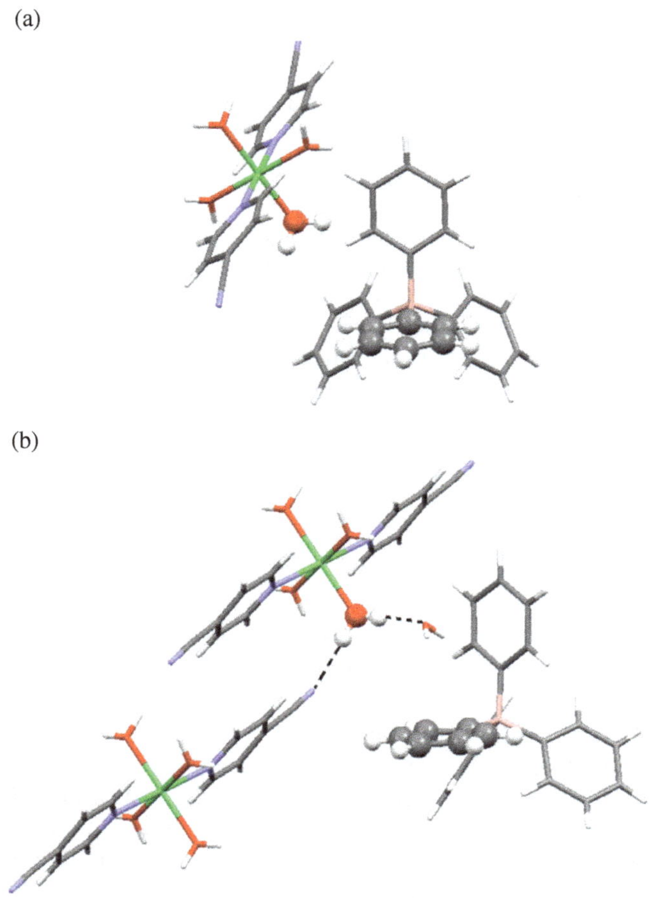

FIGURE 14.17 In crystal structure CAGHUO (tetraaqua-bis (4-cyanopyridine)nickel(II)-bis(tetraphenylborate)-4-cyanopyridine solvate tetrahydrate) [50], coordinated water molecule forms parallel-down interaction at large horizontal displacement (r_{HI} = 2.62 Å) with phenyl ring (a); this large horizontal displacement enables coordinated water to form hydrogen bonds with cyano group of another aromatic ring and uncoordinated water molecule, while phenyl ring can form OH/π interaction with the same uncoordinated water (b).

level *ab initio* calculations of interaction energies between benzene molecules showed two minima at potential energy surface for benzene dimer, which are nearly isoenergetic [43]. In first of these minima, two benzenes form aromatic CH/π interactions when C—H bonds of one benzene interact with π-system of another benzene; interaction energy of this (global minimum) is −2.84 kcal/mol, as determined by CCSD(T)/CBS calculations [43]. Second minimum has the energy of −2.73 kcal/mol (also at CCSD(T)/CBS level) [43], with geometry of stacking interaction, where two benzenes are mutually parallel, with horizontal displacement of approximately 1.5 Å. Huge importance of these interactions prompted many new studies, which resulted in the discovery of some novel parallel interactions of aromatic molecules, which will be discussed below.

14.3.1.1 Cambridge Structural Database Search of Parallel Benzene/Benzene Interactions The CSD November 2010 release (version 5.32) was subjected to search for parallel

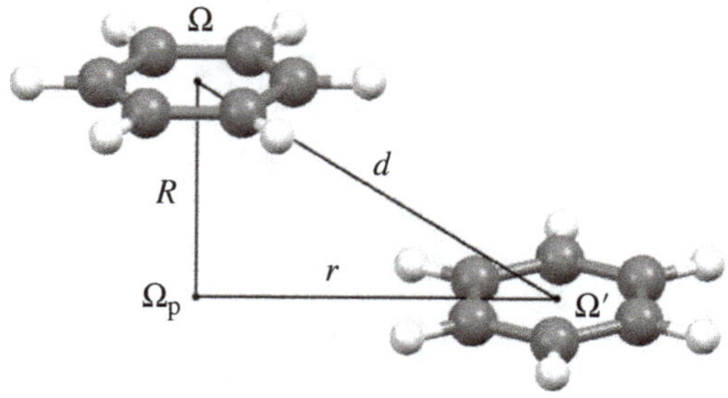

FIGURE 14.18 Geometrical parameters of parallel benzene–benzene interactions; d is the distance between the centers (Ω and Ω′) of benzene molecules; R is the normal distance between the planes of interacting rings; Ω_p is the projection of the center of one ring to the plane of the other ring; r is horizontal displacement (offset) the distance from Ω′ to Ω_p; the angle between benzene planes is smaller than 10°; in this figure, geometry with r = 5.0 Å is presented.

benzene–benzene contacts [22]. It was considered that two benzene molecules form parallel interaction if the angle between their mean planes is less than 10°, the distance d between their centers is less than 6.0 Å, and normal distance between their planes is less than 4.0 Å (Fig. 14.18).

The search yielded 1824 parallel benzene–benzene contacts. The distribution of offset values for these contacts shows that the molecules prefer interactions at large horizontal displacements (Fig. 14.19), since 64% of all contacts (1173 of 1824) are with offsets values larger than 4.5 Å. The peak of this distribution is for offsets between 5.0 and 5.5 Å; since in benzene molecule hydrogen atoms are at 2.5 Å of benzene center, it can be concluded that at these offset values only hydrogen atoms of benzene molecules overlap (Fig. 14.18).

Normal distances versus offset values plot (Fig. 14.20) shows that for contacts with offset values below 3.5 Å all normal distances are in range from 3.0 to 4.0 Å, with most of them being around 3.5 Å, which is typical for stacking interactions [36–38]. For horizontal displacements larger than 3.5 Å, and particularly over 4.5 Å, normal distances are mostly lower than 3.0 Å (Fig. 14.20).

14.3.1.2 Protein Data Bank Search of Parallel Interactions Between Phenylalanine Residues December 2013 release of Protein Data Bank was used for the search of parallel interactions between two phenylalanine residues. The full protein set was reduced by using PDBselect list of nonredundant protein chains (November 2012 release), the threshold being 25% and with resolution of 3.0 Å or better. The search parameters were similar to those for the CSD search (Fig. 14.18), and the contacts were included if they are within the area corresponding to the ellipsoid defined by offset r of 7.0 Å and normal distance R of 6.0 Å [24]. Search results show that in protein crystal structures, there is a slightly larger preference for offsets above 3.0 Å over offsets below 3.0 Å (Fig. 14.21). It is interesting to notice that even though the available area for these interactions corresponds to the ellipsoid with R = 6.0 Å and r = 7.0 Å, the vast majority of the interactions can be placed in the area

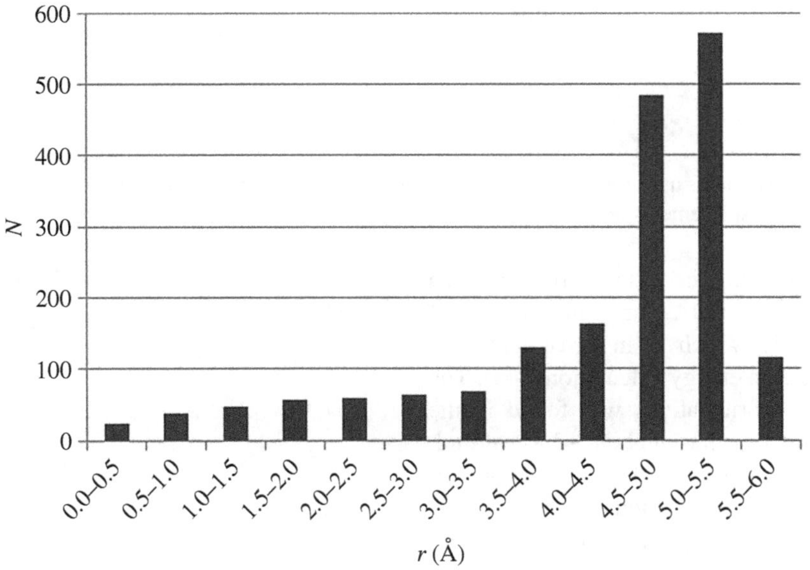

FIGURE 14.19 The distribution of offset values (r) for parallel interactions of benzene molecules in CSD crystal structures.

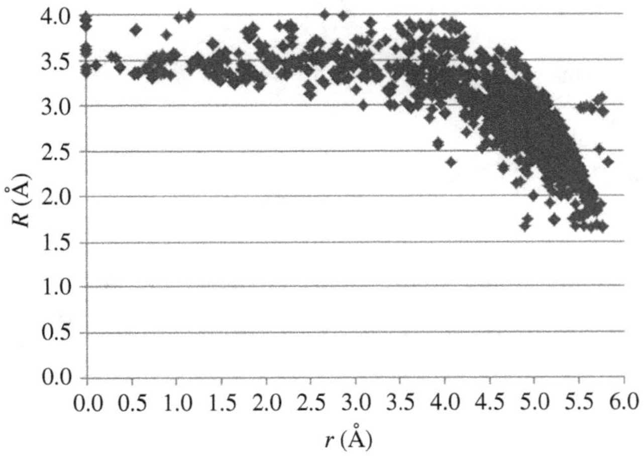

FIGURE 14.20 Normal distances (R) versus offset values (r) plot for parallel interactions between benzene molecules in CSD crystal structures.

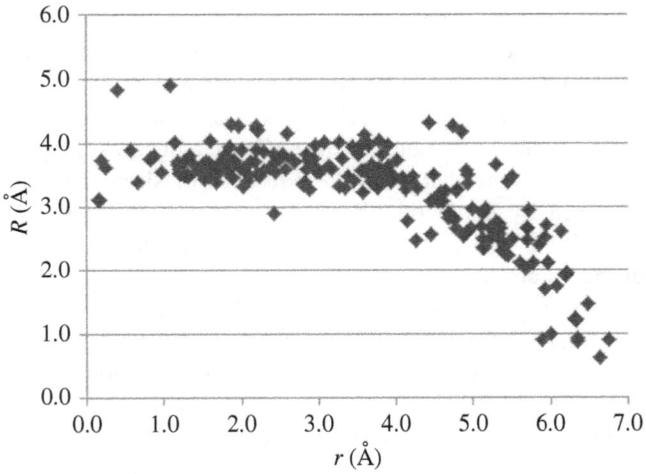

FIGURE 14.22 Normal distances (R) versus offset values (r) plot for parallel interactions between phenylalanine side chains in PDB crystal structures.

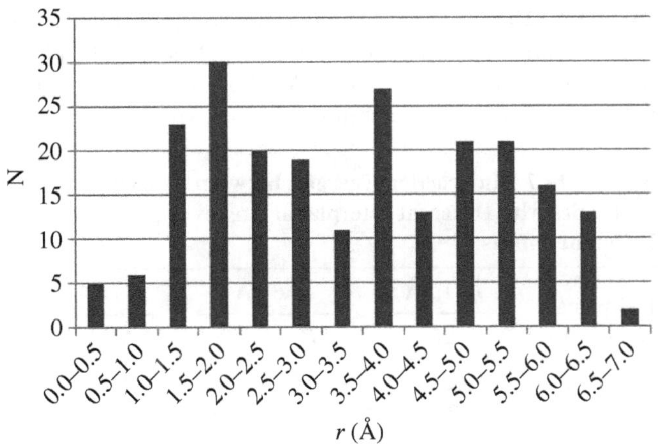

FIGURE 14.21 The distribution of offset values (r) for parallel interactions between phenylalanine side chains in PDB crystal structures.

within $R = 4.0$ Å and $r = 6.0$ Å (Fig. 14.22), which is in excellent agreement with the findings of the CSD search (Fig. 14.20) [22].

14.3.1.3 DFT Calculations of Interaction Energies of Aromatic–Aromatic Interactions Interaction energies between parallel benzene molecules were calculated with B2PLYP density functional [51] by using the second-generation Grimme dispersion correction (D2) [52] and def2-TZVP basis set [53]. Three orientations were considered (Fig. 14.23) by keeping the monomer geometries unchanged and by changing the offset values in the range 0.0–6.0 Å with 0.5 Å step and finding the optimal normal distance for each horizontal displacement [22]. Basis set superposition error (BSSE) was not corrected in these calculations, since this functional is parameterized to give good results for aromatic–aromatic interactions without this correction [54].

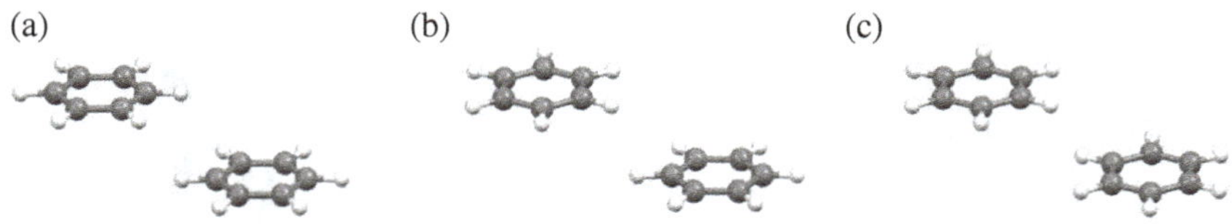

FIGURE 14.23 Three model systems used for calculations of energies of parallel interactions between benzene molecules; the presented geometries are with horizontal displacements of 5.0Å.

The first stationary point, face-to-face, or sandwich geometry, with offset of r=0.0Å, has the energy of −1.60 kcal/mol (Fig. 14.24), which is in agreement with previous results of interaction energy calculations [43]. The strongest interaction in all orientations was found in the offset range 1.5–2.0Å, with energies of about −2.8 kcal/mol, as established by earlier studies [43]. Further increase of offset values produces weaker interactions. However, only for orientation A interaction energy change is relatively steep. For orientations B and C, interactions are surprisingly strong at large horizontal displacements. Interaction energy for orientation C at r=4.5Å is −2.0 kcal/mol, which is more than 70% of the strongest interaction at r= 1.5Å (Table 14.7). The used method, B2PLYP-D2, is very reliable, since the calculated CCSD(T)/CBS value for orientation B at this offset is −1.98 kcal/mol. Substantial interaction energies at these offsets are the consequence of large reduction of repulsion and somewhat smaller reduction of dispersion, as observed by comparing B3LYP (without dispersion) and B3LYP-D2 interaction energies. The calculated normal distances are in excellent agreement with normal distances in CSD crystal structures (Fig. 14.20). At horizontal displacements larger than 4.0Å, normal distances drop to values lower than 3.0Å, and for r>5.0Å, they are even lower than 2.0Å (Fig. 14.24b).

Detailed study on benzene–benzene potential energy surface showed that interactions between benzene molecules with interplanar angles from 10° to 40° (Fig. 14.25) are similar to interactions between two parallel benzene molecules [24]. Minimum of energy for interplanar angles of 0°, 20°, and 40° was found to be at r= 1.5Å, and interaction energies are very similar (Table 14.7). At horizontal displacements of 3.5 and 4.5Å, interaction energies for interplanar angles of 20° and 40° are substantial (from −2.01 to −2.13 kcal/mol, Table 14.7), which is more than 70% of the strongest interaction energy, as it was previously determined for parallel benzene–benzene contacts [22]. For interplanar angles over 50°, potential energy curves are different, indicating aromatic C—H/π nature of interactions (Fig. 14.26) [24].

14.3.1.4 Aromatic–Aromatic Interactions at Large Horizontal Displacements in Supramolecular Structures

Similar to interactions of benzene with water molecules, which were discussed in Section 1.2, there is a disagreement between results of the CSD search and interaction energy calculations for parallel benzene–benzene interactions. The CSD data show that the most frequent geometries in the CSD are the ones with large horizontal displacements, while

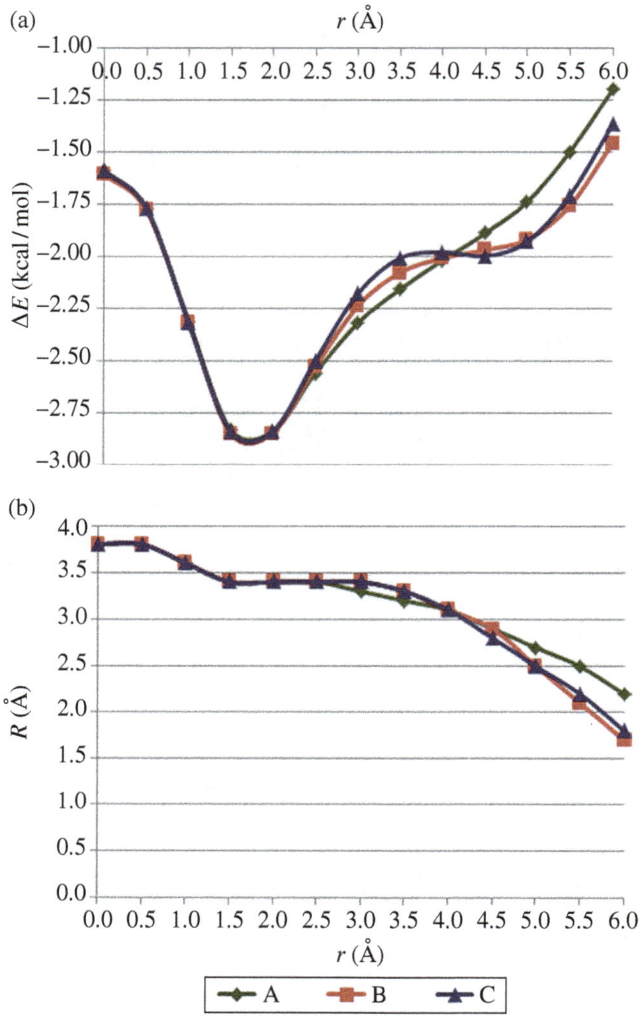

FIGURE 14.24 Interaction energies (ΔE) for parallel interactions of benzene molecules (model systems A, B, and C, Fig. 14.23) for offset values from 0.0 to 6.0Å (a) and plot of normal distances (R) versus offsets (r) for geometries with these energies (b).

TABLE 14.7 Interaction Energies Between Benzene Molecules with Different Interplanar Angles at Different Offsets

Interplanar Angle (°)	r=1.5Å	r=3.5Å		r=4.5Å	
	ΔE	ΔE	%	ΔE	%
0	−2.85	−2.16	76	−2.00	70
20	−2.73	−2.06	75	−2.01	74
40	−2.90	−2.13	73	−2.13	73

"%" Denotes percentage of the strongest interaction energy (at r=1.5Å, 100%); all energies (ΔE) are given in kilocalories per mole.

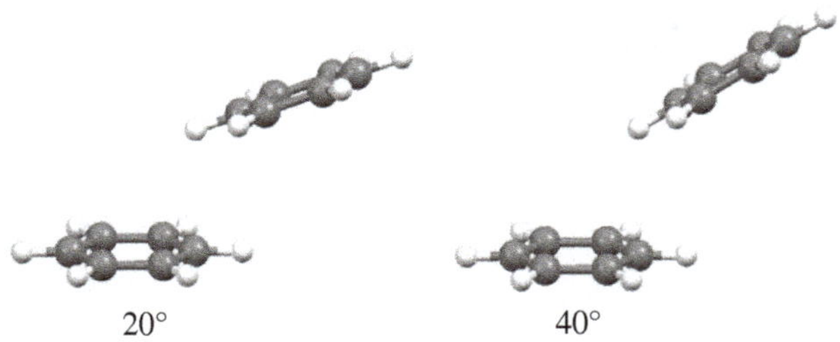

FIGURE 14.25 Model system for calculations of benzene–benzene interaction energies for interplanar angles of 20° and 40°; geometries with $r = 3.5$ Å are presented.

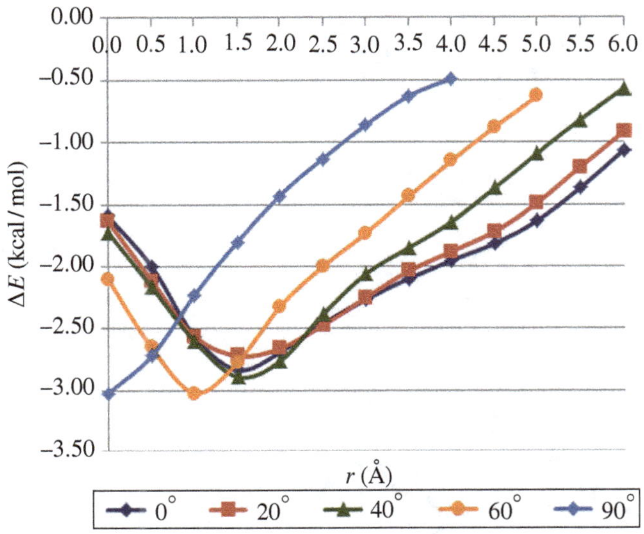

FIGURE 14.26 Curves of interaction energies between benzene molecules with interplanar angles of 0°, 20°, 40°, 60°, and 90°, calculated at B2PLYP-D2/def2-TZVP level.

DFT calculations demonstrate the strongest interactions in stacking area ($r = 1.5$–2.0 Å) [22]. Stabilization of supramolecular structures proves to be essential for benzene interactions as well.

Mutually parallel benzene molecules have the tendency to form simultaneous interactions with surrounding molecules if they are at large horizontal displacements (Fig. 14.27). For interactions at large horizontal displacements, both faces of benzene rings can simultaneously interact with surrounding molecules. However, for interactions at smaller horizontal displacement, one face of each benzene is prevented from forming simultaneous interactions. The additional simultaneous interactions of faces, along with the additional interactions of edges (e.g., C—H can form CH—O interactions; Fig. 14.27), provide further stabilization of supramolecular structures in crystals [22].

14.3.2 Heteroaromatic/Heteroaromatic Interactions at Large Horizontal Displacements

One of the simplest heteroaromatic molecules, pyridine, is considered the model molecule for studying interactions of aromatic rings with nitrogen atoms [8, 55]. Nitrogen heteroaromatic rings are common feature of nucleic acids [56, 57], as well as many drugs and materials [8]. Interactions between two pyridine molecules were calculated at high levels of quantum chemistry, giving two energy minima, similar to ones found for benzene–benzene dimer [43]. Stacking interactions of two pyridines are, however, somewhat stronger than CH/π interactions (interaction energies are −3.80 and −3.56 kcal/mol, respectively) [58]. Recently, study on pyridine–pyridine interactions at large horizontal displacement was performed [23], and it will be discussed in this section.

14.3.3 Cambridge Structural Database Search for Pyridine–Pyridine Interactions

Parallel pyridine–pyridine interactions were studied by performing the CSD search on November 2011 release (version 5.33) [23]. Similar parameters to those for benzene–benzene parallel interactions were set (Fig. 14.28). However, since previous investigations showed huge influence of hydrogen bonding on pyridine–pyridine stacking [59], the contacts that included hydrogen bonds between pyridine molecules and X—H species (where X can be O, N, S, F, and Cl) were excluded from this search data set.

The search gave 166 parallel pyridine–pyridine contacts. The parameter T (torsion angle N-Ω-Ω′-N′, Fig. 14.28) showed clear preference for head-to-tail orientation, which is the most stable orientation found by performing quantum chemical calculations [58].

The plot of normal distances versus offset values (Fig. 14.29a) and distribution of offset values (Fig. 14.29b) showed large preference for horizontal displacements higher than 4.0 Å (74% of all the contacts), similar to benzene–benzene interactions. All of the contacts at horizontal displacements lower than 4.0 Å have normal distances in the range from 3.0 to 4.0 Å, while for larger offset values, normal distances can be lower than 3.0 Å, and even lower than 2.0 Å (Fig. 14.29b).

14.3.4 DFT Calculations of Energies of Pyridine–Pyridine Interactions

Calculations of interaction energies between two parallel pyridine molecules were performed at B2PLYP-D2/def2-TZVP level of theory [51–53], without BSSE correction, as was previously done for benzene–benzene interactions [22]. Due to almost unanimous preference for torsion angles T close to 180°, only model systems with head-to-tail orientations were considered; three model systems used in this study are presented in Figure 14.30. Due to lower symmetry of pyridine in comparison to benzene, all model systems had two subsystems, labeled as (+) and (−), where (+) denotes that rigid monomers displaced in the way nitrogen atom of one ring moves away from the other ring, while (−) means that at small horizontal displacements

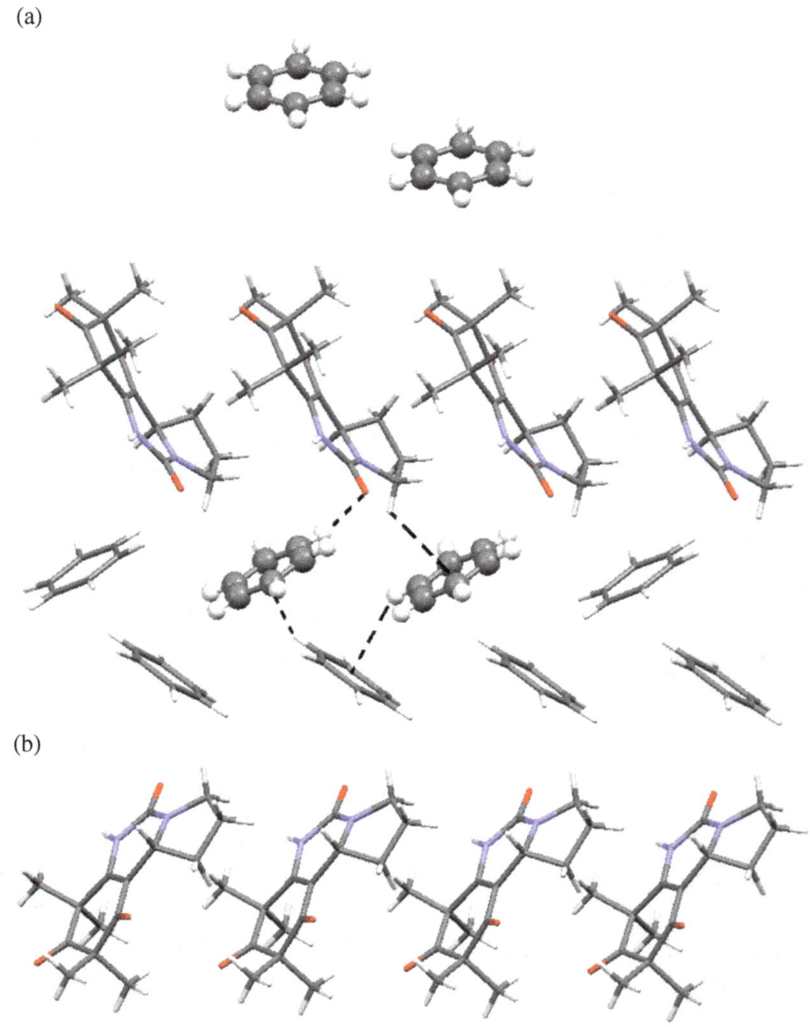

FIGURE 14.27 Parallel benzene–benzene interaction at large horizontal displacement (r = 5.10 Å) in crystal structure CENNUE (syncarpurea benzene solvate) (a); both benzene molecules form additional aromatic CH/π interactions with surrounding benzene molecules, CH/π interactions with syncarpurea molecules, and CH/O interactions with O atoms of syncarpurea molecules (b).

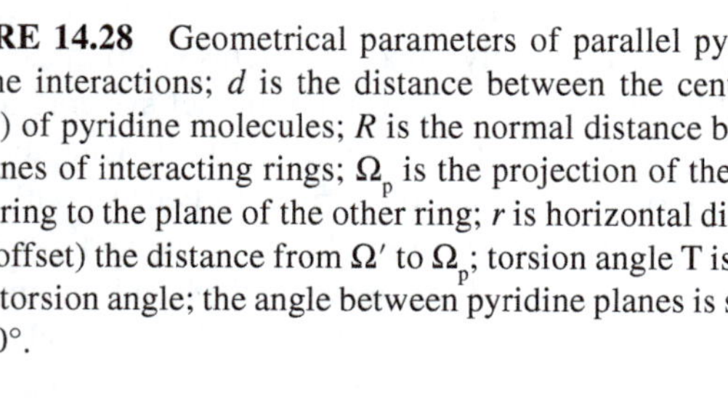

FIGURE 14.28 Geometrical parameters of parallel pyridine–pyridine interactions; d is the distance between the centers (Ω and Ω′) of pyridine molecules; R is the normal distance between the planes of interacting rings; Ω_p is the projection of the center of one ring to the plane of the other ring; r is horizontal displacement (offset) the distance from Ω′ to Ω_p; torsion angle T is N–Ω–Ω′–N′ torsion angle; the angle between pyridine planes is smaller than 10°.

nitrogen of one ring is placed above the other ring (Fig. 14.30) [23]. For model system C, these subsystems are equivalent.

Interaction energies are similar for positive offsets in all orientations (Fig. 14.31a). The minimum of potential energy curve was found at r = +1.5 Å for orientation C, with energy of −4.12 kcal/mol, which is stronger than previous quantum chemical calculations gave [5]. However, this energy can be accounted as accurate, since it is in good agreement with CCSD(T)/CBS value of −3.99 kcal/mol [59].

For large positive offset values, interactions are also substantially strong, being the strongest also in model system C, with interaction energies of −2.5 kcal/mol at r = 4.0 Å and −2.0 kcal/mol for r = 5.0 Å. This means that 60% of the strongest interaction energy is preserved at large offset of r = 4.0 Å and 50% at r = 5.0 Å, which are still large portions of energies, even though not as large as in benzene–benzene system (Table 14.8). It can also be noted that there is large influence

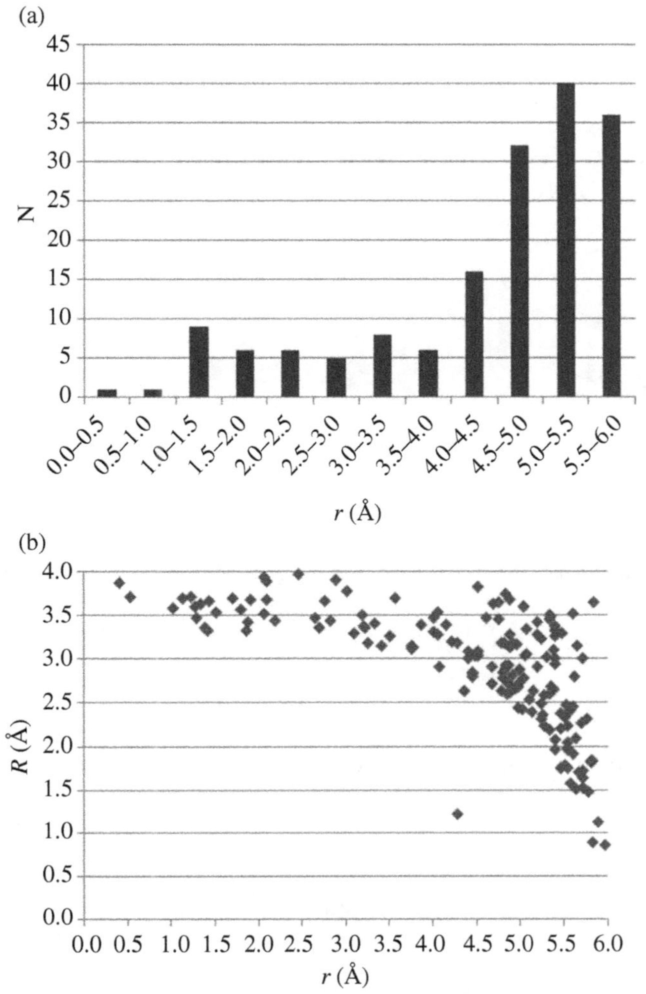

FIGURE 14.29 The distribution of offset values (a) and normal distances versus offset values plot (b) for parallel interactions of pyridine molecules in CSD crystal structures.

of heteroatom, since face-to-face geometry (r=0.0Å) is more stable than geometries at large offsets (−2.9 and −2.5 kcal/mol, respectively).

For negative offsets, interactions that involve nitrogen atoms are pronounced (Fig. 14.30). Since C(−) is equivalent to C(+), the strongest interaction at negative offsets is also with energy of −4.12 kcal/mol. Other orientations are less stable, with minima at potential energy curves of −3.0 and −3.4 kcal/mol for A(−) and B(−), respectively. In orientation A(−), there are unfavorable interactions of nitrogen of one ring and nitrogen atom and center of the other ring. The interaction in this model system even becomes repulsive at r=−4.5Å, since N—N interactions become more pronounced. This is also witnessed by the fact that normal distances in this orientation never drop to values lower than 3.0Å (Fig. 14.31b). For orientation B(−), there is a new minimum at r=−5.5Å, with interaction energy of −2.7 kcal/mol, due to very favorable CH—N hydrogen bonds; at these offsets normal distances reach minimal values below 1.5Å.

As observed in benzene–benzene dimers, substantial interactions at large horizontal displacements are the consequence of significant decrease in repulsion and less pronounced decrease in dispersion.

14.3.5 Pyridine–Pyridine Interactions at Large Horizontal Displacements in Supramolecular Structures

Supramolecular structures play crucial role in pyridine–pyridine interactions. The loss in energy by moving to larger offsets is easily compensated by forming additional interactions through faces of aromatic rings, which would be prevented from interacting if the offsets were smaller. In crystal structure KINLIC (Fig. 14.32), both parallel pyridines at large horizontal displacements form additional aromatic and aliphatic CH/π interactions surrounding molecules [23].

14.3.6 DFT Study on Parallel Aromatic/Heteroaromatic Interactions: Benzene/Pyridine System

The CSD does not contain any parallel benzene–pyridine contact that would satisfy the interaction criteria described in previous searches [23]. Therefore, only DFT study of these interactions was possible, and it was performed by using the same methodology as for pyridine–pyridine interactions [23]. Three model systems were constructed, with (+) and (−) subsystems (Fig. 14.33), similar to pyridine–pyridine parallel interactions.

Minimum at potential energy curve was calculated at r=1.5Å for orientation B(+), with energy of −3.54 kcal/mol, while minima for A(+) and C(+) are almost equal in energy (Fig. 14.34). Interactions at large horizontal displacements are also substantially strong; the strongest interaction for r=4.0Å was in system A(+), with energy of −2.20 kcal/mol, while at r=5.0Å interaction in system C(+) is −1.95 kcal/mol. This means that 62% of interaction energy can be preserved at r=4.0Å, while for r=5.0Å the percentage is only a bit lower (55%), which is less than in benzene–benzene system but more than in pyridine–pyridine system (Table 14.8). Contrary to pyridine–pyridine interactions, these benzene–pyridine interactions are similarly strong at large horizontal displacements as for face-to-face geometry (Fig. 14.34 and Table 14.8).

At negative offsets, there is a similar behavior as in pyridine–pyridine interactions, with nitrogen atom of pyridine facing some repulsion with benzene ring center. However, potential energy curve for A(−) reaches maximum for r=−4.5Å (Fig. 14.34), but it is never repulsive. As a consequence of only one CH—N hydrogen bond, the minimum for benzene–pyridine B(−) orientation is very shallow and weaker than the same minimum for pyridine–pyridine B(−) orientation by 0.8 kcal/mol.

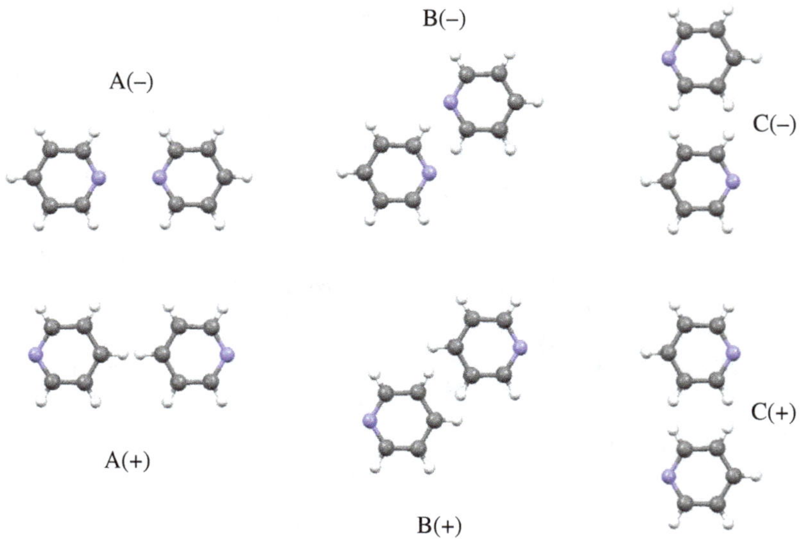

FIGURE 14.30 Top view of parallel pyridine–pyridine orientations used for calculations of interaction energies; geometries with positive (+) and negative (−) offset values of 5.0Å are presented.

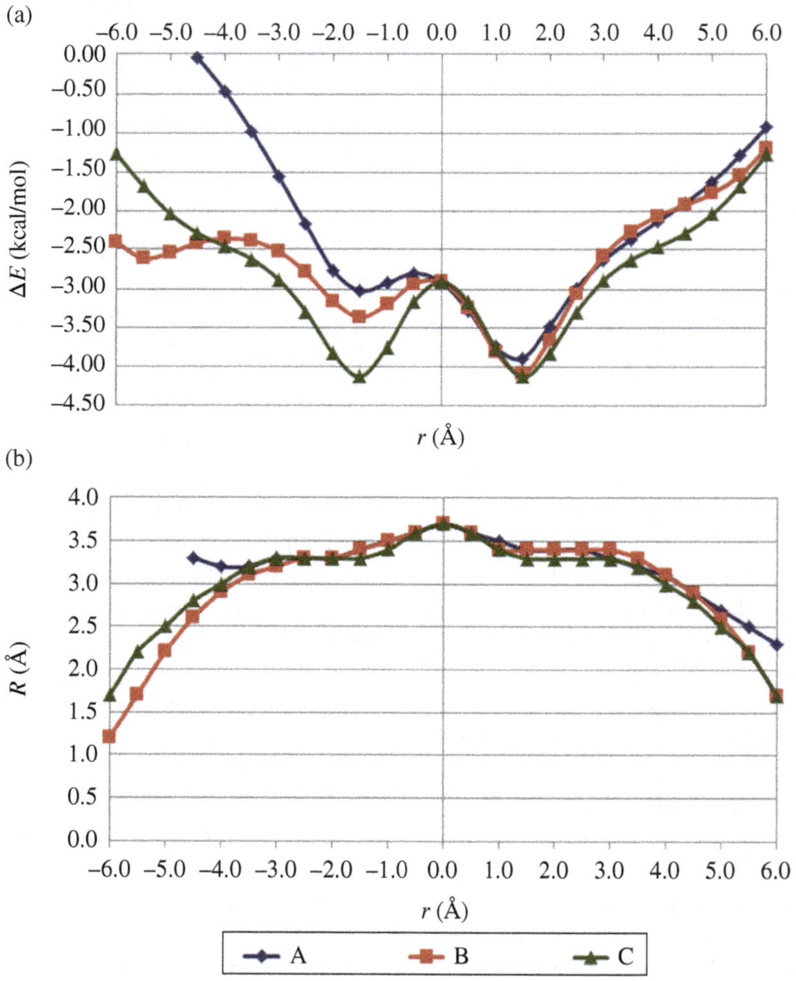

FIGURE 14.31 Interaction energies (ΔE) for parallel interactions of pyridine molecules (model systems A, B, and C, Fig. 14.29) for offset values from −6.0 to 0.0Å and from 0.0 to 6.0Å (a) and plot of normal distances (R) versus offsets (r) for geometries with these energies (b).

14.4 CONCLUSIONS

This review presented recent findings about parallel interactions formed by aromatic molecules. Studies of aromatic–aromatic and water–aromatic interactions showed important role played by them at large horizontal displacements, that is, out of aromatic ring and beyond the C—H bond region.

Parallel alignment interactions of water and aromatic rings are frequent in crystal structures, as evidenced both from the CSD and from the Protein Data Bank. These interactions are more frequent than OH/π interactions, which have stronger energy. There is a preference for large horizontal displacements of water molecules, with offsets larger than 2.5 Å. These interactions are stronger than interactions at small horizontal displacements, with energies up to −2.45 kcal/mol for water–benzene system. The dominant force responsible for this substantial attraction is electrostatics, with dispersion being canceled by exchange repulsion. Parallel alignment interactions of water with benzene do not have significant influence on other attractive water–benzene interactions. There is a significant influence of water coordination to metal ions on interaction energies, since interactions of double positively charged aqua complexes have the energies up to −14.88 kcal/mol.

The preference for large horizontal displacements was also observed for parallel aromatic–aromatic contacts in crystal structures from CSD and PDB structures. There is a huge preference for horizontal displacements over 4.5 Å for benzene–benzene contacts and a similar one for pyridine–pyridine contacts. Even though these interactions are weaker than classical stacking interactions, their interaction energies are surprisingly strong, with values of −2.0 kcal/mol for benzene–benzene and −2.5 kcal/mol for pyridine–pyridine system, which is a substantial portion of the strongest stacking interaction energies (71% for benzene–benzene and 60% for pyridine–pyridine). These interactions are the result of significant reduction of repulsion at large horizontal displacements, followed by smaller reduction of dispersion.

Parallel interactions of aromatic molecules at large horizontal displacements are at significant importance, since they are substantially strong and, at the same time, enable aromatic rings to form additional interactions through both of their faces, therefore forming stable and versatile supramolecular structures. They are essential for all molecular systems that contain aromatic rings and can be exploited in the design of new materials and other compounds.

TABLE 14.8 Comparison of Energies of Benzene–Benzene, Pyridine–Pyridine, and Benzene–Pyridine Interactions at Different Horizontal Displacements

	r=0.0Å		r=1.5Å	r=4.0Å		r=5.0Å	
	ΔE	%	ΔE	ΔE	%	ΔE	%
Benzene–benzene	−1.61	56	−2.85	−2.02	71	−1.93	68
Pyridine–pyridine	−2.91	71	−4.12	−2.46	60	−2.04	50
Benzene–pyridine	−2.11	60	−3.54	−2.20	62	−1.95	55

"%" Denotes percentage of the strongest interaction energy, which is for r=1.5Å in all investigated systems; all energies (ΔE) are given in kilocalories per mole.

(a)

(b)

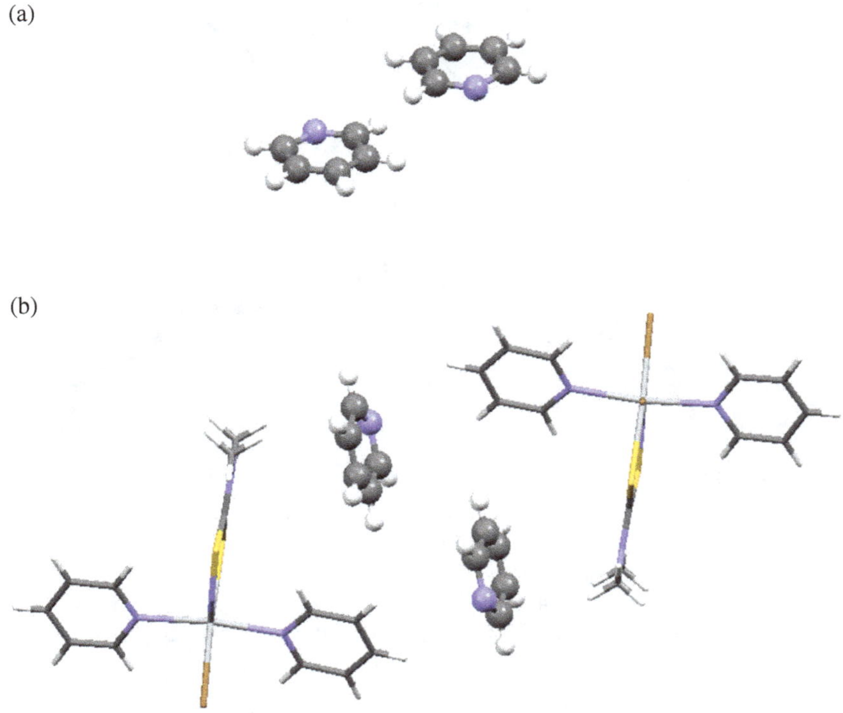

FIGURE 14.32 Parallel pyridine–pyridine interaction in crystal structure KINLIC (dibromo-(dimethylamino(thiocarbonyl)thiamin,S)-dipyridyl-titanium(IV) pyridine solvate) [60] with large horizontal displacement (r=4.29Å) (a); as a consequence of these pyridines being at large offsets, both pyridine molecules can form two additional CH/π interactions with ligands of surrounding titanium(IV) complex (b).

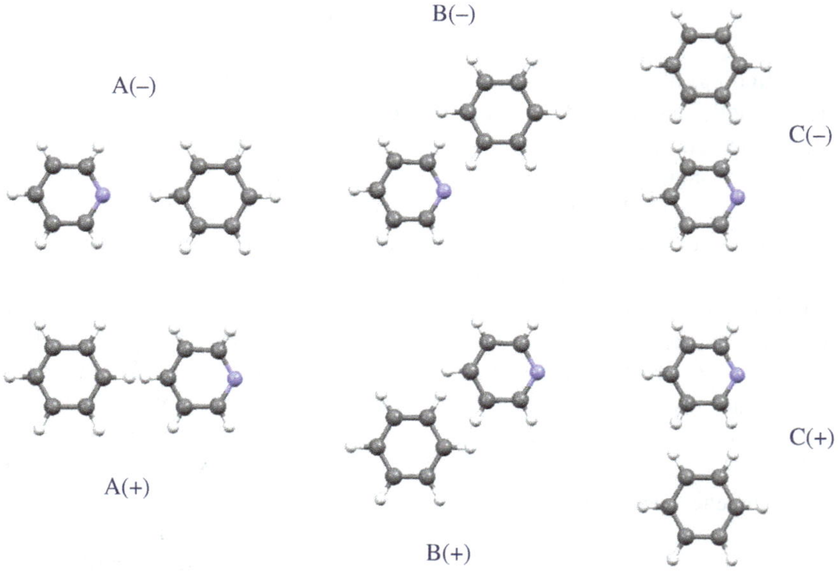

FIGURE 14.33 Top view of parallel benzene–pyridine orientations used for calculations of interaction energies; geometries with positive (+) and negative (–) offset values of 5.0 Å are presented.

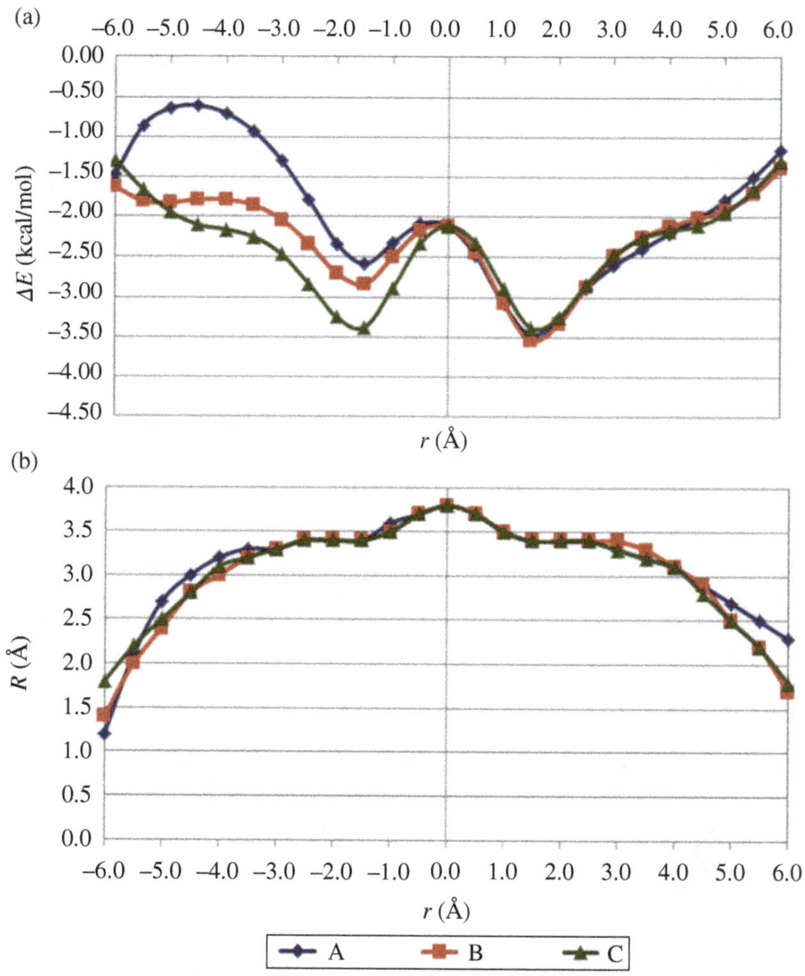

FIGURE 14.34 Interaction energies (ΔE) for parallel benzene–pyridine molecules (model systems A, B, and C, Fig. 14.33) for offset values from –6.0 to 0.0 Å and from 0.0 to 6.0 Å (a) and plot of normal distances (R) versus offsets (r) for geometries with these energies (b).

ACKNOWLEDGMENTS

This work was supported by the Ministry of Education, Science and Technological Development of the Republic of Serbia (Grant No. 172065). High-performance computing resources and services used in this work were partially provided by the IT Research Computing group in Texas A&M University at Qatar, which is funded by the Qatar Foundation for Education, Science and Community Development. The authors are grateful to Dr. Horst Borrman of the Max Planck Institute for Chemical Physics of Solids in Dresden, Germany, for his support.

REFERENCES

1. C. Bissantz, B. Kuhn, M. Stahl, *J. Med. Chem.* 2010, *53*, 5061.
2. M. L. Ma, Y. Kuang,Y. Gao, Y. Zhang, P. Gao, B. Xu, *J. Am. Chem. Soc.* 2010, *132*, 2719.
3. H. Robson Marsden, J. G. E. M. Fraaije, A. Kros, *Angew. Chem. Int. Ed.* 2010, *49*, 8570.
4. H. J. Schneider, *Angew. Chem. Int. Ed.* 2009, *48*, 3924.
5. B. Meszaros, P. Tompa, I. Simon, Z. Dosztanyi, *J. Mol. Biol.* 2007, *372*, 549.
6. A. Lattanzi, C. De Fusco, A. Russo, A. Poater, L. Cavallo, *Chem. Commun.* 2012, *48*, 1650.
7. M. J. M. Munoz, G. Fernandez, *Chem. Sci.* 2012, *3*, 1395.
8. L. M. Salonen, M. Ellermann, F. Diederich, *Angew. Chem. Int. Ed.* 2011, *50*, 4808.
9. V. L. Malinovskii, F. Samain, R. Haener, *Angew. Chem. Int. Ed.* 2007, *46*, 4464.
10. R. Fasan, R. L. A. Dias, K. Moehle, O. Zerbe, D. Obrecht, P. R. E. Mittl, M. G. Gruetter, J. A. Robinson, *ChemBioChem* 2006, *7*, 515.
11. O. Takahashi, Y. Kohno, M. Nishio, *Chem. Rev.* 2010, *110*, 6049.
12. J. C. Ma, D. A. Dougherty, *Chem. Rev.* 1997, *97*, 1303.
13. P. Gamez, T. J. Mooibroek, S. J. Teat, J. Reedijk, *Acc. Chem. Res.* 2007, *40*, 435.
14. D. Ž. Veljković, V. B. Medaković, J. M. Andrić, S. D. Zarić, *CrystEngComm* 2014, *16*, 10089.
15. D. Ž. Veljković, G. V. Janjić, S. D. Zarić, *CrystEngComm* 2011, *13*, 5005.
16. E. A. Meyer, R. K. Castellano, F. Diederich, *Angew. Chem. Int. Ed.* 2003, *42*, 1210.
17. G. V. Janjić, D. Ž. Veljković, S. D. Zarić, *Cryst. Growth Des.* 2011, *11*, 2680.
18. D. P. Malenov, G. V. Janjić, D. Ž. Veljković, S. D. Zarić, *Comput. Theor. Chem.* 2013, *1018*, 59.
19. D. Z. Vojislavljević-Vasilev, G. V. Janjić, V. B. Medaković, J. P. Blagojević, S. D. Zarić, *ChemPhysChem* 2014, *15*, 2386.
20. G. V. Janjić, S. N. Malkov, M. V. Živkovic, S. D. Zarić, *Phys. Chem. Chem. Phys.* 2014, *16*, 23549.
21. M. P. Mitoraj, G. V. Janjić, V. B. Medaković, D. Ž. Veljković, A. Michalak, S. D. Zarić, M. K. Milčić, *J. Comput. Chem.* 2015, *36*, 171.
22. D. B. Ninković, G. V. Janjić, D. Ž. Veljković, D. N. Sredojević, S. D. Zarić, *ChemPhysChem* 2011, *12*, 3511.
23. D. B. Ninković, J. M. Andrić, S. D. Zarić, *ChemPhysChem* 2013, *14*, 237.
24. D. B. Ninković, J. M. Andrić, S. N. Malkov, S. D. Zarić, *Phys. Chem. Chem. Phys.* 2014, *16*, 11173.
25. T. A. Pascala, W. A. Goddarda, Y. Junga, *Proc. Natl. Acad. Sci. U. S. A.* 2011, *108*, 11794.
26. A. Simon, M. Rapacioli, J. Mascetti, F. Spiegelman, *Phys. Chem. Chem. Phys.* 2012, *14*, 6771.
27. T. Steiner, G. Koellner, *J. Mol. Biol.* 2001, *305*, 535.
28. G. Hummer, J. C. Rasaiah, J. P. Noworyta, *Nature* 2001, *414*, 188.
29. R. Natarajan, J. P. H. Charmant, A. G. Orpen, A. P. Davis, *Angew. Chem. Int. Ed.* 2010, *49*, 5125.
30. P. Agre, *Angew. Chem. Int. Ed.* 2004, *43*, 4278.
31. K. P. Gierszal, J. G. Davis, M. D. Hands, D. S. Wilcox, L. V. Slipchenko, D. Ben-Amotz, *J. Phys. Chem. Lett.* 2011, *2*, 2930.
32. L. V. Slipchenko, M. S. Gordon, *J. Phys. Chem. A* 2009, *113*, 2092.
33. S. Tsuzuki, *Struct. Bond.* 2005, *115*, 149.
34. M. V. Sigalov, E. P. Doronina, V. F. Sidorkin, *J. Phys. Chem. A* 2012, *116*, 7718.
35. A. Jain, V. Ramanathan, R. Sankararamakrishnan, *Protein Sci.* 2009, *18*, 595.
36. C. Janiak, *J. Chem. Soc. Dalton Trans.* 2000, 3885.
37. T. Steiner, *Biophys. Chem.* 2002, *95*, 195.
38. T. Janowski, P. Pulay, *Chem. Phys. Lett.* 2007, *447*, 27.
39. M. O. Sinnokrot, C. D. Sherrill, *J. Phys. Chem. A* 2004, *108*, 10200.
40. B. Jeziorski, R. Moszynski, K. Szalewicz, *Chem. Rev.* 1994, *94*, 1887.
41. T. Ziegler, A. Rauk, *Inorg. Chem.* 1979, *18*, 1755.
42. B. D. Ostojić, G. V. Janjić, S. D. Zarić, *Chem. Commun.* 2008, 6546.
43. E. C. Lee, D. Kim, P. Jurecka, P. Tarakeshwar, P. Hobza, K. S. Kim, *J. Phys. Chem. A* 2007, *111*, 3446.
44. A. Michalak, M. Mitoraj, T. Ziegler, *J. Phys. Chem. A* 2008, *112*, 1933.
45. O. Mo, M. Yañez, J.E. Del Bene, I. Alkorta, J. Elguero, *ChemPhysChem* 2005, *6*, 1411.
46. X.-W. Wang, Z.-F. Chen, H. Liang, H.-K. Fun, *J. Guangxi Normal Univ.* 2005, *23*, 53
47. D. Paschek, R. Ludwig, *Angew. Chem. Int. Ed.* 2011, *50*, 352.
48. D. Z. Vojislavljević, G. V. Janjić, D. B. Ninković, A. Kapor, S. D. Zarić, *CrystEngComm* 2013, *15*, 2099.
49. J. M. Andrić, G. V. Janjić, D. B. Ninković, S. D. Zarić, *Phys. Chem. Chem. Phys.* 2012, *14*, 10896.
50. R. K. Barman, B. K. Das, *CrystEngComm* 2002, *4*, 80.
51. S. Grimme, *J. Chem. Phys.* 2006, *124*, 034108.
52. S. Grimme, *J. Comput. Chem.* 2006, *27*, 1787.
53. F. Weigend, R. Ahlrichs, *Phys. Chem. Chem. Phys.* 2005, *7*, 3297.
54. J. C. Sancho-Garcia, A. J. Perez-Jimenez, *J. Chem. Phys.* 2009, *131*, 084108.

55. E. G. Hohenstein, C. D. Sherrill, *J. Phys. Chem. A* 2009, *113*, 878.
56. P. Mignon, S. Loverix, F. D. Proft, P. Geerlings, *J. Phys. Chem. A* 2004, *108*, 6038.
57. B. K. Mishra, J. S. Arey, N. Sathyamurthy, *J. Phys. Chem. A* 2010, *114*, 9606.
58. I. Geronimo, E. C. Lee, N. J. Singh, K. S. Kim, *J. Chem. Theory Comput.* 2010, *6*, 1931.
59. D. B. Ninković, G. V. Janjić, S. D. Zarić, *Cryst. Growth Des.* 2012, *12*, 1060.
60. B. Meller-Rehbein, H. W. Roesky, M. Noltemeyer, *Chem. Ber.* 1991, *124*, 523.

15

SELECTIVE MOLECULAR BINDING AND NANOSUPRAMOLECULAR ASSEMBLY OF *p*-SULFONATOCALIX[*n*]ARENES

Yu Liu and Yi-Xuan Wang

Department of Chemistry, State Key Laboratory of Elemento-Organic Chemistry, Collaborative Innovation Center of Chemical Science and Engineering (Tianjin), Nankai University, Tianjin, China

15.1 INTRODUCTION

Discrete host–guest complexation represents an advantageous approach to synthesize versatile inclusions and further complicated assemblies. Compared with modulating the physicochemical property of the substrates via covalent derivation, host–guest complexation not only involves facile synthesis but also appears as a modular method, which allows the functional component to be easily adjusted, replaced, and even combined.

Calixarenes [1], composed of phenolic units linked by methylene groups at the 2,6-positions, represent one of the most widely studied classes of organic supramolecular hosts and are described as "macrocycles with (almost) unlimited possibilities" due to their simple modification. In most cases, they can serve as simple scaffolds to build podand-like receptors where the calixarene cavity very often remains unexploited [2].

The cavity complexation is considered as the most iconic feature of macrocyclic hosts and therefore an appealing avenue that leads to functional supramolecular inclusions and further nanoarchitectures. However, the cavity inclusion property of maternal calixarene is not as good as other common macrocycles such as crown ether, cyclodextrin, and cucurbituril. Extensive chemical modification of maternal calixarene is always demanded to achieve the desired *endo* complexation [3].

Among various calixarene derivatives, *p*-sulfonatocalix[*n*]arenes (SC*n*As, $n=4$–8) (Scheme 15.1), firstly reported by Shinkai in 1984 [4], represent a prominent family of water-soluble calixarene derivatives with robust cavity binding properties in aqueous media. Firstly, SC*n*As are easily sulfonated directly on the upper rim of calixarene with satisfied yields. Secondly, SC*n*As have high water solubility, and the driving forces of cavity inclusion, such as hydrophobic and π-stacking interactions, are more effective in aqueous media. Thirdly, the upper-rim sulfonate groups provide additional anchoring points besides the intrinsic π-electron-rich cavities, which enable SC*n*As to display especially strong binding ability and high molecular selectivity toward a variety of organic ions [5–8].

Finally, SC*n*As are biocompatible. *In vitro*, SC4A shows zero hemolytic toxicity for concentration up to 5 mM and a lack of nonspecific immune response [9]. *In vivo*, SC4A shows no acute toxicity for single injected doses equivalent to 2–5 g in humans and is rapidly cleared *via* elimination in urine without accumulation in liver [10]. Such intrinsic biocompatibility endows SC*n*As with highly diverse biological and pharmaceutical applications [11, 12].

In this chapter, we discuss some investigations concerning the binding ability and structures of SC*n*As and the construction of their supramolecular assemblies.

15.2 SELECTIVE MOLECULAR BINDING

15.2.1 Binding with Organic Ammonium Cations

Benefiting from the π-electron-rich cavities and the additional anchoring points offered by sulfonate groups, SC*n*As possess strong binding abilities and high selectivity toward

Non-Covalent Interactions in the Synthesis and Design of New Compounds, First Edition.
Edited by Abel M. Maharramov, Kamran T. Mahmudov, Maximilian N. Kopylovich, and Armando J. L. Pombeiro.

various organic cations, which is mainly due to the cooperation of π-stacking and charge interactions. Organic ammonium cations are a typical class of cationic guests, including primary ammonium, secondary ammonium, and quaternary ammonium cations. Herein we select 19 guests that have been previously investigated upon complexation with SC*n*As (Scheme 15.2) and list their complex stability constants (K_S) and other thermodynamic parameters for the inclusion complexation in Table 15.1.

As a result of ease of synthesis and steady conformation, SC4A has received much more attention than other SC*n*As. The complexation of SC4A with organic ammonium cations is prominently driven by enthalpy factor, accompanied by some either positive or negative entropy changes. In the case of binding with linear alkylammonium ions (G1–G7) (Scheme 15.2), the dominant enthalpy changes are mainly ascribable to the stability constants, van der Waals interactions associated with the inclusion of the alkyl chains, and favorable entropy changes originated from the major positive contribution of desolvation effect as well as the minor negative contribution of loss of conformational freedom [14–16]. Since the phenolic hydroxyls do not play a major role in association, the binding geometries between pH 7.1 and pH 2 are similar.

Liu and coworkers studied the binding behaviors of SC4A with cyclic alkylammonium ions (G8–G10) (Scheme 15.2) [17]. The complexation is in good agreement with the 1 : 1 binding stoichiometry, which is also enthalpy stabilized [17].

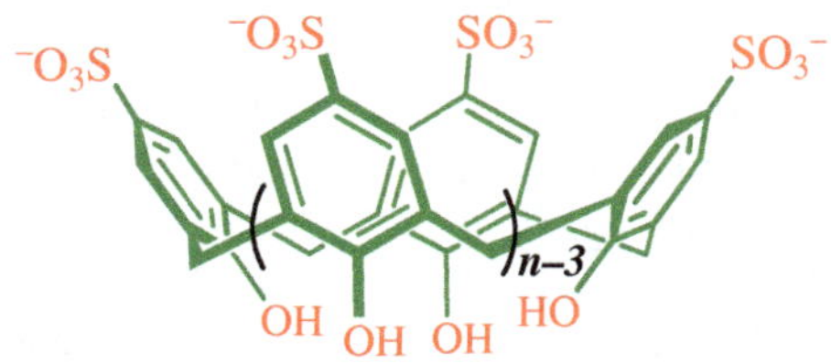

SCHEME 15.1 Structures of four common *p*-sulfonatocalixarenes.

However, pH value affects the binding ability significantly, because the guests could only be completely protonated in an acidic solution.

The interactions of quaternary ammonium cations G11 and G12 (Scheme 15.2) with SC*n*As were firstly investigated by Shinkai *et al.* via NMR spectroscopy [18, 19]. It was found that SC4A and SC6A could form stoichiometric 1 : 1 complexes with the guests, and the binding abilities of SC6A were weaker than that of SC4A and SC8A. Interestingly, SC8A formed stoichiometric 1 : 2 complexes with the guests, which is primarily attributed to the large and flexible ring of SC8As and the different driving forces for complexation. Compared with SC4A, SC5A possesses a wider cavity, leading to different inclusion geometries. For the case of complexation of SC4A and SC5A with quaternary ammonium guests (G12–G15) (Scheme 15.2) [20], the driving forces are the same, namely, the synergy of electrostatic and C—H···π interactions. Unlike that, G12 [18] and G13 are included into the cavity of SC4A without regioselectivity; SC5A is able to accommodate simultaneously both the charged group and the aromatic moiety of ditopic G12 and G13 (Scheme 15.2).

Liu and coworkers have also studied the interactions of some quaternary ammonium cations (G13 and G16–G18) (Scheme 15.2) with SC4A and SC5A in pH 7.2 phosphate buffer solutions via ITC and NMR measurements [21]. Owing to its smaller cavity with relatively high π-electron density, which contributes to the good size fit and strong C—H···π/π···π interactions between host and guest, SC4A affords the strongest binding abilities toward all the guests. Note that the guests bearing electronegative carboxyl groups show much weaker binding abilities toward all the hosts than the nonbearing ones. Recently, Liu and coworkers reported a spherical quaternary ammonium guest (G19) toward SC4A, which shows K_S value up to $10^7\,M^{-1}$ with favorable entropy and enthalpy components [22]. More detailed information in this field can be found in the previous review [13].

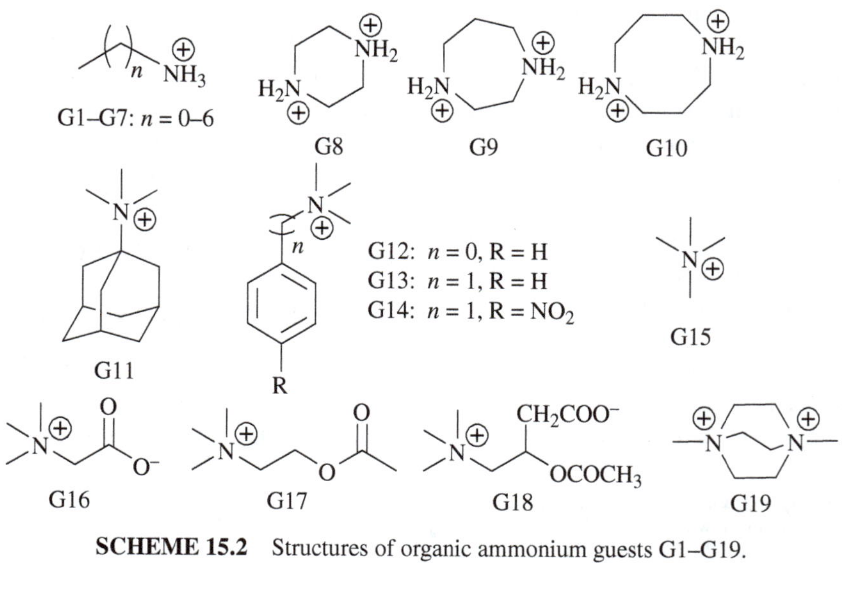

SCHEME 15.2 Structures of organic ammonium guests G1–G19.

TABLE 15.1 Complex Stability Constants (K_S/M), Standard Enthalpy ($\Delta H°$/(kJ/mol)), and Entropy Changes ($T\Delta S°$/(kJ mol)) for Intermolecular Complexation of Organic Ammonium Cations with SC*n*As in Aqueous Solution at 298.15 K [13]

Host	Guest	pH	$\lg K_S$	$\Delta H°$	$T\Delta S°$	Method
SC4A	G1	2	2.65	−11.5	3.6	ITC
	G2	2	3.58	−16.5	3.9	ITC
	G3	7.1	4.12	−16.89	6.61	ITC
	G4	7.1	4.01	−17.94	4.94	ITC
	G5	7.1	3.81	−20.24	1.48	ITC
	G6	7.1	3.6	−20.42	0.15	ITC
	G7	7.1	3.39	−20.86	−1.51	ITC
	G8	7.2	3.02	−20.05	−2.85	ITC
		2	2.95	−9.13	7.71	ITC
	G9	7.2	3.93	−26.4	−3.96	ITC
		2	4.07	−20.32	2.91	ITC
	G10	7.2	4.02	−13.8	9.15	ITC
		2	4.14	−17.48	6.15	ITC
	G11	7.3	4.32	−23.9	0.8	NMR
	G12	7.3	4.6	—	—	NMR
		7.3	3.75	−25.9	−4.5	NMR
	G13	7.3	4.1	—	—	NMR
		7.2	4.08	−32.34	−9.00	ITC
	G14	7.3	4.2	—	—	NMR
	G15	2	4.40	−26.0	−0.9	ITC
		7.3	4.9	—	—	NMR
	G16	7.2	2.62	−26.23	−11.25	ITC
	G17	7.2	4.14	−30.12	−6.49	ITC
	G18	7.2	2.81	−25.76	−9.71	ITC
	G19	7.0	7.2	−22.9	18.0	ITC
SC5A	G12	7.3	4.2	—	—	NMR
	G13	7.3	4.0	—	—	NMR
	G14	7.3	5.1	—	—	NMR
	G15	7.3	3.6	—	—	NMR
SC6A	G11	7.3	3.0	0.63	16.6	NMR
	G12	7.3	2.74	−1.0	14.6	NMR
	G13	7.2	3.64	−34.98	−14.23	ITC
	G16	7.2	2.42	−31.71	−17.87	ITC
	G17	7.2	3.73	−35.06	−13.81	ITC
	G18	7.2	2.75	−31.21	−15.48	ITC
SC8A[a]	G11	7.3	3.72	0.0	21.2	NMR
			3.66	0.0	20.8	NMR
	G12	7.3	4.28	0.0	24.5	NMR
			4.23	0.0	24.1	NMR

[a] In the case of SC8A-involved complexation, there is a two-step binding process, thus giving two independent stability constants.

15.2.2 Binding with Aromatic Cations

Aromatic cations are another typical class of cationic guests, including imidazolium, pyridinium, and viologens. Liu and coworkers systematically investigated the inclusion phenomena of SC4As with these aromatic cations utilizing ITC measurement and NMR spectroscopy [22]. SC4A and SC5A are selectively employed due to their preferred cone shape. The selected guest molecules (G20–G32) at this tube are illustrated in Scheme 15.3. The binding behaviors of SC4A with G20 and G21 are quite similar. SC4A forms stoichiometric 1 : 1 complexes with both G20 and G21, and the aromatic portion of the guests is encapsulated into the SC4A cavity, while the positive charged N atoms are fixed at the upper rim composed of sulfonate groups. It is also found that the ditopic derivatives G22–G27 form 1 : 1 complexes with SC4A, although two identical binding sites have been grafted on one gemini guest [22]. One rational explanation is that the complexation of the second SC4A is prevented by the charge repulsion. The assumed 1 : 2 complexes are negatively six charged (the ionization of phenolic hydroxyls is not considered), whereas the electrostatic repulsion between sulfonate groups of two SC4As is so remarkable that the 1 : 2 complexation cannot occur, although the binding affinities of SC4A with the terminal imidazolium and pyridinium groups are strong (up to the magnitude of $10^5\,M^{-1}$) (Table 15.2).

Viologens have a number of applications, for example, in the design of herbicides (paraquat and diquat) [24] and electrochromic display devices [25]. Hence, studies at complexation of viologens with SC*n*As are promised to be quite meaningful. Liu and coworkers investigated the binding behavior of SC4A and SC5A with G28 (Scheme 15.3) and its radical cation in different pH conditions (Table 15.2) [23]. It can be deduced from the ^{1}H NMR and 2D ROESY experiments that the binding modes of SC4A and SC5A with G28 are different. G28 is included into the cavity of SC4A in its axial orientation with the methyl group being preferential, while it lies at the upper-rim midsection of SC5A in the latitudinal orientation (Scheme 15.4).

The ITC experiments show that both SC4A and SC5A form stoichiometric 1 : 1 complexes with G28, and the binding abilities become stronger when the pH value increases, which is driven by enthalpy term. The host selectivity for SC4A/SC5A pairs is reversed with increasing pH value, mainly due to the difference in π-electron density of the calixarene's cavity and the different binding modes of the two hosts with G28.

The effective inclusion of the radical cation G28 into SC4A and SC5A was also proved by cyclic voltammetry (CV). However, due to the weaker π-electron acceptor and hydrogen-bonding donor abilities of the radical cation, the binding affinities of SC4A or SC5A with the radical cation are weaker than those with G28.

The complexation of SC4A with viologen-type gemini guests (G29–G32) (Scheme 15.3) was also investigated by NMR measurements [22]. Distinguished from imidazolium and pyridinium, all viologen-based guests G29–G32 form the desired 1 : 2 complexes with SC4A, which is ascribed to decreased negative charge of the complexes and increased host–guest binding affinity.

Since high toxicity of viologens poses considerable risks to human health and environment, considerable endeavors have been made to neutralize the viologen poisoning [23]. A new therapeutic protocol based on the concept of host–guest chemistry was put forward by Liu and coworkers [23],

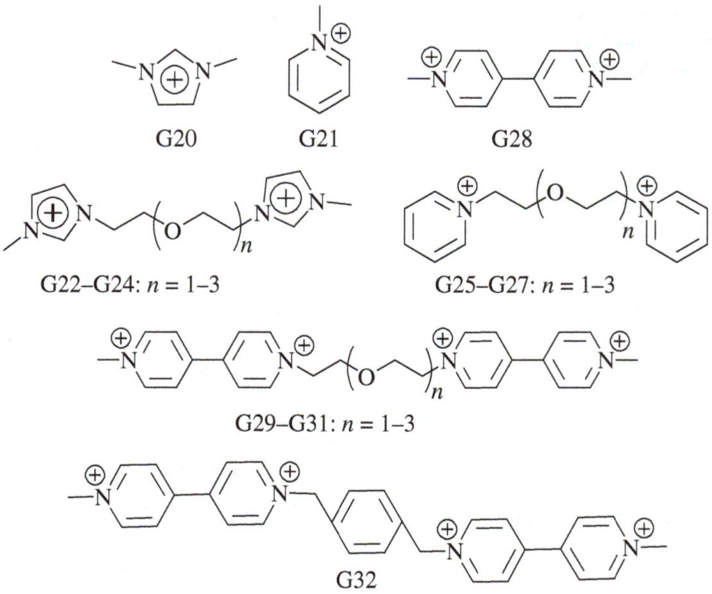

SCHEME 15.3 Structures of aromatic cationic guests G20–G32.

TABLE 15.2 Complex Stability Constants (K_S/M), Standard Enthalpy ($\Delta H°$/(kJ/mol)), and Entropy Changes ($T\Delta S°$/(kJ/mol)) for Intermolecular Complexation of Aromatic Cationic Guests with SC*n*As in Aqueous Solution at 298.15 K [22, 23]

Host	Guest	Medium	$\lg K_S$	$\Delta H°$	$T\Delta S°$
SC4A	G20	Water (pH 7.0)	5.5	−28.0	3.2
	G21	Water (pH 7.0)	5.8	−31.2	1.9
	G28	Water (pH 7.0)	6.0	−27.2	6.8
		PBS (0.1 M, pH 2.0)	4.49	−28.18	−2.53
		PBS (0.1 M, pH 7.2)	4.97	−31.98	−3.62
		PBS (0.1 M, pH 12.0)	4.97	−32.83	−4.46
SC5A	G28	PBS (0.1 M, pH 2.0)	3.74	−20.58	0.79
		PBS (0.1 M, pH 7.2)	5.40	−31.52	−0.67
		PBS (0.1 M, pH 12.0)	4.53	−33.11	−1.53

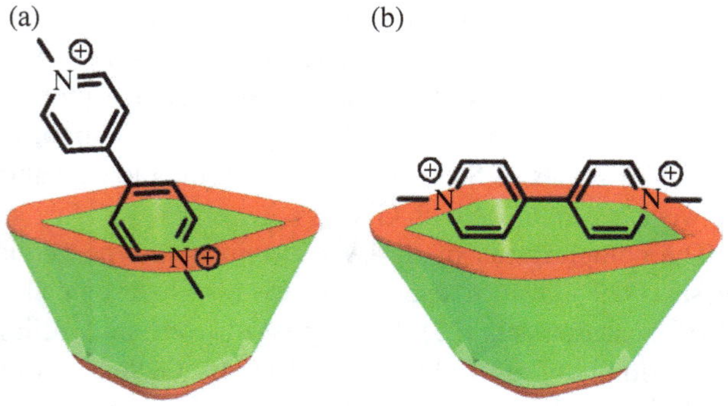

SCHEME 15.4 Deduced binding manners of (a) SC4A and (b) SC5A with G28 [23].

where the toxicity of viologens can be effectively inhibited by the host–guest complexation of SC*n*As; several factors contribute to this protocol, namely, high binding abilities to viologens [26], inducing negative shifts in the reduction potentials of viologens, hydrogen transfer to (toxic) radicals, and coordination of transition metal ions (Scheme 15.5) [27].

The therapeutic effect was actually validated by mice tests. The mortality rate was significantly decreased when the viologen-poisoned mice were administered with SC5A even two hours later, and the ingestion of SC5A can also effectively prevent viologen-induced destruction of structures of the tissues. Subsequently, Qi *et al.* reported the paraquat detoxification with SC4A by a pharmacokinetic study *in vivo* [28]. It was found that the absorption of paraquat was efficiently impeded after SC4A intervention by forming stable host–guest complex.

15.2.3 Binding with Biofunctional Molecules

15.2.3.1 Binding with Amino Acids Due to their excellent biocompatibility, SC*n*As are expected to serve as efficient artificial receptors for biological applications. Therefore, the ability of SC*n*As to form complexes with compounds of biological interest has been the subject of considerable research. The interactions of SC*n*As with amino acids were systematically studied by Coleman *et al.* [29–33] and Ungaro *et al.* [34] via NMR, microcalorimetry, and reversed-phase high-performance liquid chromatography (Table 15.3).

It is found that the SC4A can form stoichiometric 1 : 1 complexes with the basic amino acids lysine and arginine, with the highest K_S values up to $10^3\,M^{-1}$ [31].The binding mode was confirmed by NMR experiments and suggests that the lateral aliphatic chains or aromatic parts of the amino acids are immersed into the cavity of the calixarenes. The binding process is driven by favorable enthalpy changes, which are ascribed to the van der Waals interactions between the included apolar part of the guest and the hydrophobic cavity of the calixarenes, and also favorable entropy factor, derived from the desolvation of the charged groups.

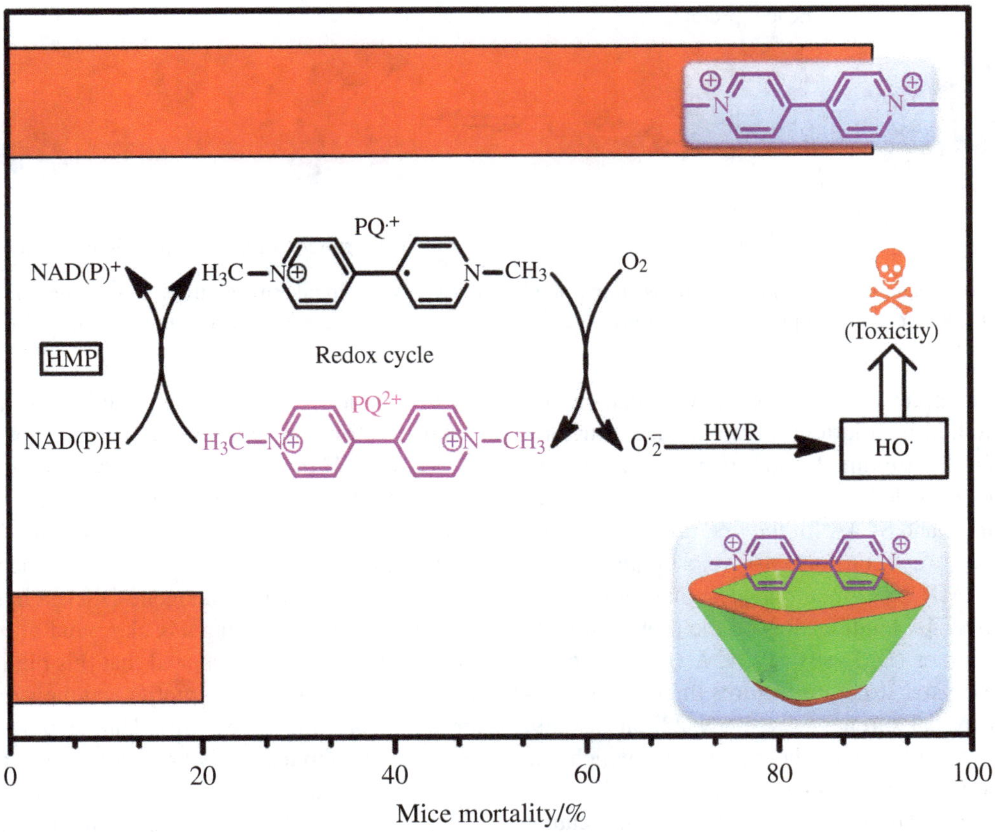

SCHEME 15.5 Structures and biochemical mechanism of viologen toxicity and detoxification by complexation with SC5A (HWR, Haber–Weiss reaction; HMP, hexose monophosphate pathway) [27].

TABLE 15.3 Complex Stability Constants (K_s/M) for Intermolecular Complexation of Amino Acids with SC*n*As in Aqueous Solution at 298.15 K [11]

	Lys	Arg	Asp	Trp	Ser
SC4A	1356 ± 349	1,546 ± 721	512 ± 24	263 ± 30	120 ± 14
SC6A	2200 ± 276	3,090 ± 1,024	354 ± 38	1448 ± 293	112 ± 38
SC8A	4288 ± 1543	10,083 ± 6,705	614 ± 68	3465 ± 1156	320 ± 32

For SC8As, 1 : 2 stoichiometric complexes are also obtained with arginine and lysine. Furthermore, Coleman *et al.* [32] investigated the binding behavior of SC*n*As with several oligopeptides composed of these basic amino acids lysine and arginine (lysyl–lysine, arginyl–arginine, lysyl–arginine, arginyl–lysine, lysyl–lysyl–lysine, and arginyl–arginyl–arginine). Unlike SC4A that only forms 1 : 1 complexes, both 1 : 1 and 1 : 2 complexes with SC6A were observed. It is because SC6A adopts in solution a conformation of the 1,2,3-alternate type upon binding with two guests, as observed in the solid state. For SC8A, the binding behavior is more complicated, and highly ordered assembly may occur.

The binding modes of SC*n*As with amino acids in the solid state have been comparatively investigated by Selkti [35], Raston [36, 37], Coleman [38], and Atwood *et al.* [39], and the detailed information is reviewed in the recent publication [11].

*15.2.3.2 **Binding with Proteins*** Owing to the high binding affinity of SC*n*As to basic amino acids, the protein, which is rich in arginines and lysines, is the desired biofunctional guest for complexation with calixarenes. Coleman *et al.* investigated the interaction of bovine serum albumin (BSA) with SC*n*As by electrospray ionization mass spectrometry (ESI-MS), dynamic light scattering (DLS), and atomic force microscopy [40]. It is found that BSA is endowed with one strong and two weaker binding sites with SC*n*As. The effects on the structure of thin films formed by surface deposition of BSA show that SC*n*As reticulate the films and produce essentially planar systems. ESI-MS was further used to determine K_S and binding stoichiometries for SC*n*As and their derivatives with BSA [41]. The titration experiments show that SC4A exhibits strong binding capacities at three nonequivalent binding sites, and the K_S values decrease with increasing cavity size.

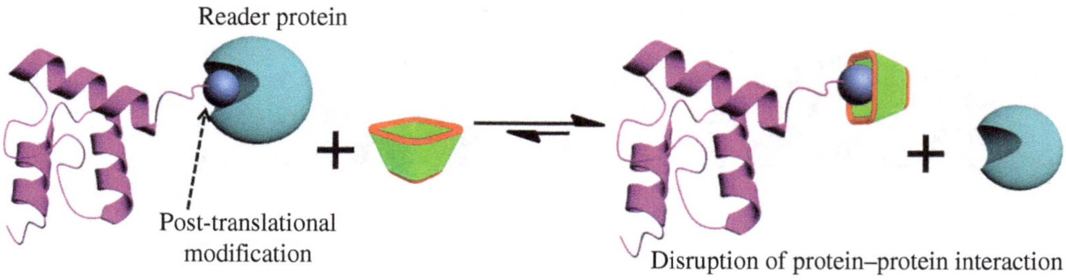

SCHEME 15.6 Schematic illustration of post-translational modifications (e.g., lysine methylation) on unstructured protein tails, which serve as recruitment sites for protein–protein interactions (PPIs), and the disruption of the PPI via a competition from SC4A [46].

Although the complexation of proteins with calixarenes has been well studied in solution, the solid-state structural investigations on the supramolecular complexes are still rare. Gratifyingly, Crowley *et al.* realized the cocrystallization of cytochrome c and SC4A for the first time [42], which is commended as "a milestone for biosupramolecular chemistry" [43]. NMR experiments suggest that cytochrome c has at least two SC4A binding sites on the protein surface with different affinities (K_S(Lys4) = 1220, K_S(Lys87) = 633). Interestingly, the crystal structure features three lysine residues complexed by calixarenes—Lys4, Lys22, and Lys89. All of them are complexed in the same way, which the methylene chain is deeply immersed into the calixarene cavity, with the amino group pointing outward and locating between two sulfonate groups. The cytochrome c-SC4A crystal structure clearly shows that SC4A can camouflage and modify the protein surface via host–guest chemistry, which significantly promotes crystallization by facilitating the buildup of protein assemblies. Therefore, Crowley *et al.* proposed this supramolecular strategy as a rapid and simple approach to drive protein assembly and crystallization.

Recently, Crowley *et al.* prepared a linear assembly of lysozyme mediated by protein–calixarene interactions, including encapsulation of the highly exposed *C*-terminal Arg128 and a complicated protein-bound complex of SC4A, Mg^{2+}, and a fragment of polyethylene glycol [44]. The crystal structures of protein–SC4A complexes provide useful information on binding behavior between macrocyclic hosts and cationic proteins. Additionally, SC4A promises potential application in regulating protein–protein interactions via complexation with the reference point on protein surface.

Post-translational modifications of lysine, involving methylation to the mono-, di-, or trimethylated amine, lead to chemical and structural alterations that have pronounced influences on protein interactions and signaling pathways. In order to explore the detailed structural information of ligands in complex with methylated lysine, Crowley *et al.* studied a crystal structure of SC4A bound to a lysozyme-containing dimethyllysine [45]. SC4A not only selectively encapsulated the side chain of Lys116-Me_2 but also acted as "molecular glue" at protein–protein interfaces in the crystal. This complex proves the tendency for Lys-Me_2 residues to form cation–π interactions, which are important in the protein recognition of histone tails bearing methylated lysines.

In addition to facilitating protein crystallization to get detailed structure information, small-molecule ligand is expected to further modulate the physiological behavior and biological function of proteins by interacting with protein surfaces. Hof *et al.* reported that SC*n*As bind selectively and with high affinity to histone trimethyllysine motifs involved in gene regulation and oncogenesis, even in highly complex protein substrates bearing a background of many unmethylated lysines and arginines, and that this binding disrupts the interaction between these motifs and their epigenetic reader proteins (Scheme 15.6) [46]. Unlike other typical small-molecule antagonists, SC*n*As do not have extended structures beyond the Kme3-binding motifs, resulting in broad selectivity. While small molecules are limited to biological targets that contain well-defined binding pockets, supramolecular hosts are uniquely able to address and bind to protein-based signaling motifs that are unstructured or that are of unknown structure. This supramolecular inhibitor is a very valuable model for future design of artificial receptor molecules regulating protein bioactivities.

15.2.3.3 Binding with Pharmic Molecules Discrete host–guest complexation represents an approach to supramolecular drug delivery, complementing existing supramolecular drug formulation strategies based on liposomes, polymersomes, nanoparticles, hydrogels, and other related microheterogeneous systems [47]. As new drug formulations, complexation of drugs with macrocyclic compounds is desired to enhance the bioavailability and solubilization of drugs to protect drugs from photochemical and thermal degradation and to alter the chemical reactivity of pharmic molecules. Yang *et al.* have explored the solubility enhancement of three practically insoluble drugs by SC*n*As in acidic aqueous solution, including nifedipine (an important calcium channel blocker that is used extensively for the clinical management of a number of cardiovascular diseases) [48], furosemide (a high ceiling diuretic [49]), and niclosamide (an anthelmintic drug that is active against most tapeworms and also used as a molluscicide for the treatment of water in schistosomiasis control programs) [50] (Scheme 15.7). They validated that molecular size and concentration of the calixarenes significantly affect the enhancement in the solubility of the drugs. The possible mechanisms at the complexation may be a combination of hydrogen bonding, hydrophobic bonding,

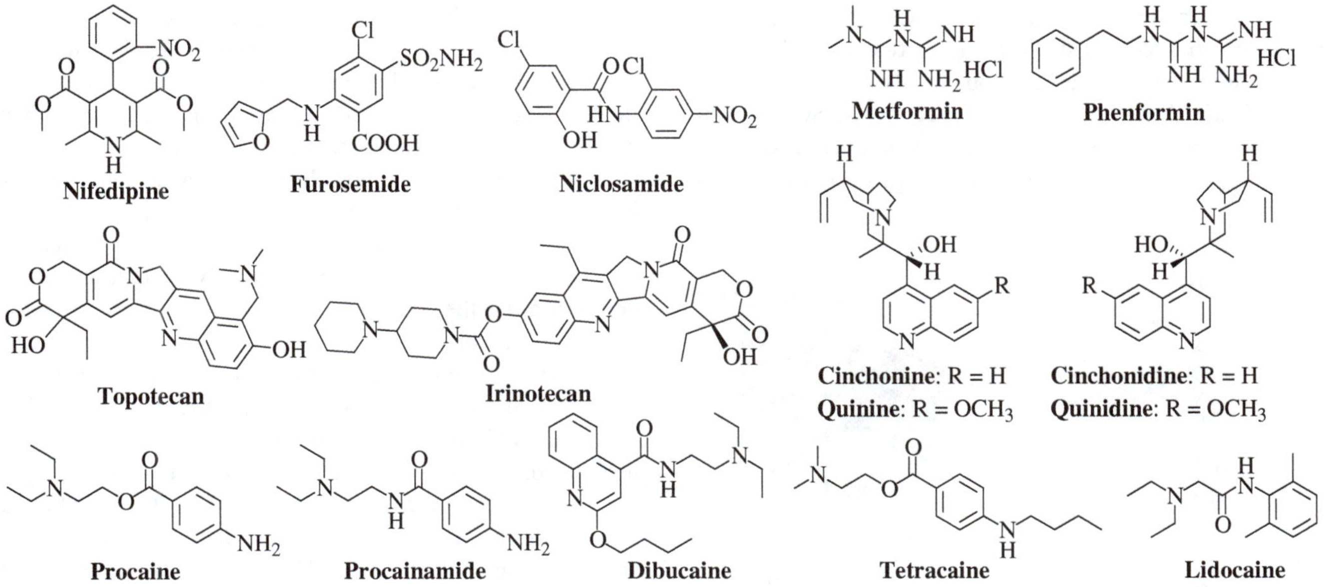

SCHEME 15.7 Structures of some pharmic guests.

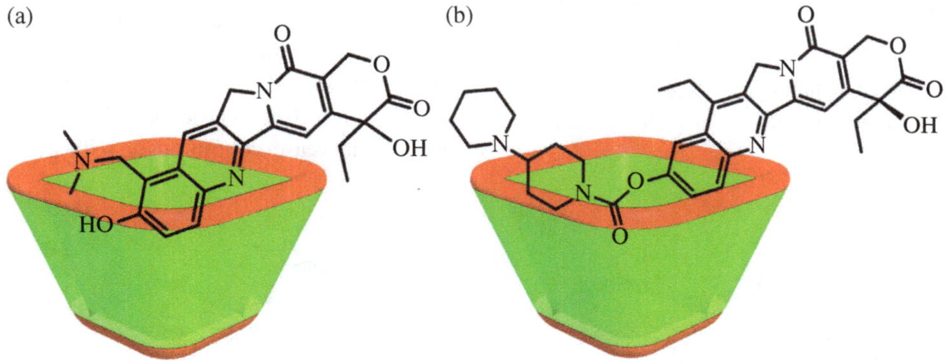

SCHEME 15.8 Possible binding modes of (a) SC4A–TPT [51] and (b) SC4A–CPT-11 [52] complexes.

and electron donor–acceptor interactions. As most of synthetic drugs are insoluble in water, these results are interesting for their solubilization in aqueous media.

Liu and coworkers studied the binding behavior of SC4A with two derivatives of the anticancer alkaloids camptothecin, topotecan (TPT) [51], and irinotecan (CPT-11) [52] (Scheme 15.7), both of which are used clinically to treat cancers. The binding mechanics and the binding modes (Scheme 15.8) were investigated by UV, NMR, and differential scanning calorimetry (DSC) analyses. Both two drugs formed 1 : 1 complex with SC4A, and the K_S values were determined as $5.6\times10^3\,M^{-1}$ for SC4A–TPT complex and $7.2\times10^3\,M^{-1}$ for SC4A–CPT-11 complex. Aqueous solubility study shows that SC4A could enhance the water solubility of the drugs. In addition, the antiproliferative activity of the SC4A–CPT-11 complex was evaluated *in vitro*. The obtained results indicate that the complexation of CPT-11 and SC4A can increase the antiproliferative activity of CPT-11, which has potential application in the medical fields.

The binding geometries, abilities, and thermodynamic parameters for the intermolecular complexation of two water-soluble calixarenes, SC4A and SC5A, with biguanidinium guests, metformin and phenformin, were explored by Liu and coworkers [53]. These guests are important for the treatment of hyperglycemia in patients with noninsulin-dependent diabetes mellitus. The obtained results show that biguanidinium guests are captured by calixarenes with the alkyl or aromatic moieties immersed into the cavities and the guanidinium group fixed at the upper rims (Scheme 15.9). Thermodynamically, the complexation of SC4A with biguanidiniums is stronger and dominantly driven by the enthalpy changes, while the complexation of SC5A is driven almost equally by the enthalpy and entropy changes.

Recently, Liu and coworkers examined the binding behavior of SC*n*As with four cinchona alkaloids (quinidine, quinine, cinchonine, and cinchonidine) (Scheme 15.7) in aqueous and phosphate buffer solutions (pH 7.2 and 2.0, respectively) [54]. Cinchona alkaloids, natural products typically extracted from the bark of *Cinchona ledgeriana trees*, are used to treat malaria and cardiac arrhythmias, as chiral modifiers in heterogeneous catalysis and as flavors in food and drinks. Among the used hosts, SC4A forms the most stable complexes with cinchona alkaloids, due to the preorganized bowl shape and high π-electron density.

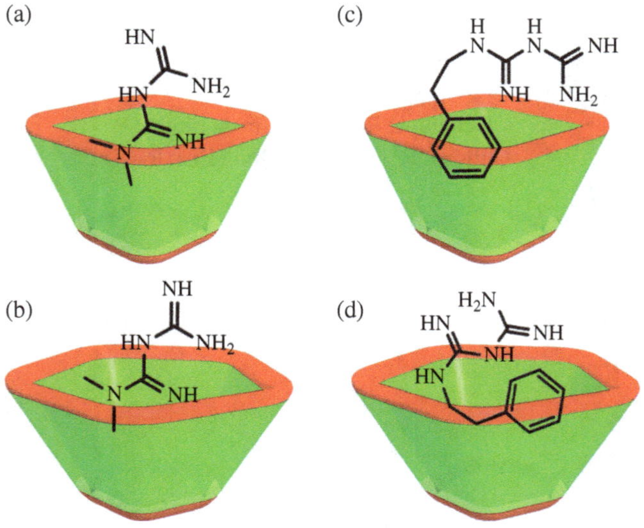

SCHEME 15.9 The deduced binding geometries of (a, b) SC4A and (c, d) SC5A with metformin and phenformin according to NMR spectral analyses [53].

The K_S values obtained at pH 2.0 are higher than those at pH 7.2 as a result of charge interaction. The K_S values in phosphate buffer are 1–2 orders of magnitude lower than those in aqueous solution, arising from the competitive binding of Na^+ with calixarenes as well as the effect of ionic strength. Moreover, desired solubility enhancement of quinidine was achieved by the complexation of SC4A. The included drugs were successfully released from the calixarene cavity in the presence of negatively charged micelles, demonstrating the potential of SC4A as a drug carrier for cinchona alkaloids.

In order to improve the applications of local anesthetics (LA), which alleviate the unpleasant sensations by causing reversible inhibition of axonal voltage-gated sodium channels responsible for depolarization, Liu and coworkers systematically investigated the complexation of SC*n*As with five LA guests, procaine, procainamide, tetracaine, dibucaine, and lidocaine (Scheme 15.7) [55]. ^{1}H NMR spectra show that, at pD 7.2, LA guests are encapsulated by SC*n*As with the triethylammonium moieties immersed into the cavities of SC*n*As. At pD 2.0, SC*n*As bind the triethylammonium moieties of procaine and procainamide but unselectively bind the tertiary ammonium and butylanilinium/butyloxy moieties of tetracaine/dibucaine (Scheme 15.10). Thermodynamically, the complexation of SC*n*As with the LA guests is predominately controlled by the enthalpy term, and the binding abilities increase continuously with decreasing cavity size.

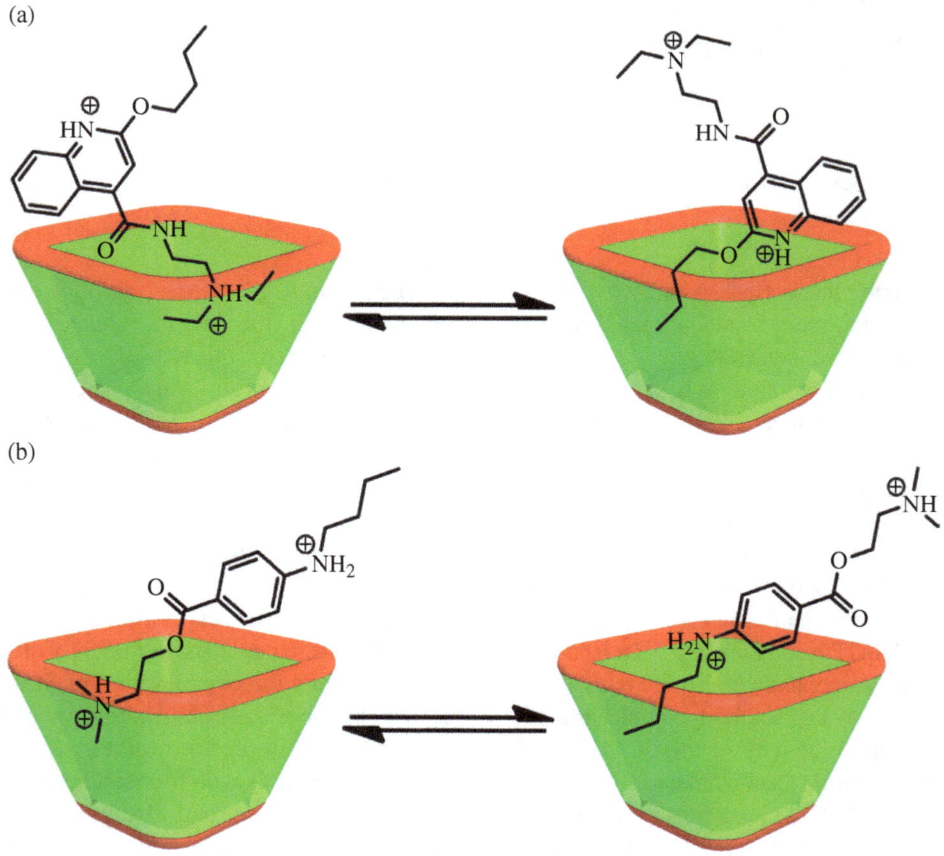

SCHEME 15.10 The deduced binding equilibriums of SC4A with (a) dibucaine and (b) SC4A with tetracaine under acidic conditions [55].

As a result, SC4A exhibits the higher complex constant to LA guests, 3–7 times larger than SC5A. These results may provide a potential clinical application in anesthetic studies.

15.3 NANOSUPRAMOLECULAR ASSEMBLY

15.3.1 Supramolecular Polymers

Supramolecular polymers, emerging from the fusion of supramolecular chemistry and polymer science, have now blossomed into an important and independent research area. Different from conventional polymers, which are covalently synthesized, the monomeric units of supramolecular polymers are held together with directional and reversible non-covalent interactions [56, 57]. As well known, covalent bond formation is generally irreversible and occurs under kinetic control, while supramolecular non-covalent chemistry is reversible and proceeds under thermodynamic control [58]. These features result in intrinsic differences between the supramolecular and conventional polymers.

There are three major advantages of supramolecular polymers: (i) the reversibility of non-covalent interactions facilitates their processing and recycling and also endows them with responsiveness to external stimuli; (ii) supramolecular polymers have the capability of self-healing and adaptation; (iii) they can act as shape-memory materials if combined with covalent cross-linking [59]. Among the non-covalent interactions, hydrogen and metal bonding have been widely used so far to drive the self-assembling of supramolecular polymers [60, 61]. The host–guest interactions have also attracted considerable attention and occupy a special position in supramolecular polymerization with crown ethers, cryptands, cyclodextrins, cucurbiturils, calixarenes, and resorcinarenes as host monomer units [62–65]. Herein, we focus solely on supramolecular polymers based on SC*n*As.

Inspired by the desired inclusion properties of SC*n*As, Liu and coworkers designed a series of water-soluble calixarene-based supramolecular polymers using the "cleanest" solvent (i.e., water). A minimum of two interacting sites is needed to form supramolecular polymers from two complementary monomers (host and guest), and therefore, the homoditopic bis(*p*-sulfonatocalix[5]arenes) (bis-SC5A) and bis(*p*-sulfonatocalix[4]arenes) (bis-SC4A) (Scheme 15.11) bridged at their lower rims were prepared, where the divergent tail-to-tail arrangement of two cavities favors iterative intermolecular inclusion that leads to supramolecular polymers. Two different supramolecular polymers with 2D netlike and 1D linear topological structures were constructed by employing tetracationic and dicationic porphyrins as model guests, respectively, upon complexation with bis-SC5A (Scheme 15.12) [66].

Corresponding to the molecular design, the assembly morphologies are well manipulated by the inherent binding sites of the molecular building blocks through strong host–guest interactions. Upon determining the binding stoichiometries between calixarene host and porphyrin guests by the Job method, it was found that the complexation exceeds the conventional 1 : 1 (one cavity to one binding site) molar ratio, and bis-SC5A forms 2 : 5 and 1 : 5 complexes with tetracationic and dicationic porphyrins. It is noticed that the host–guest stoichiometries satisfy the charge matching: the

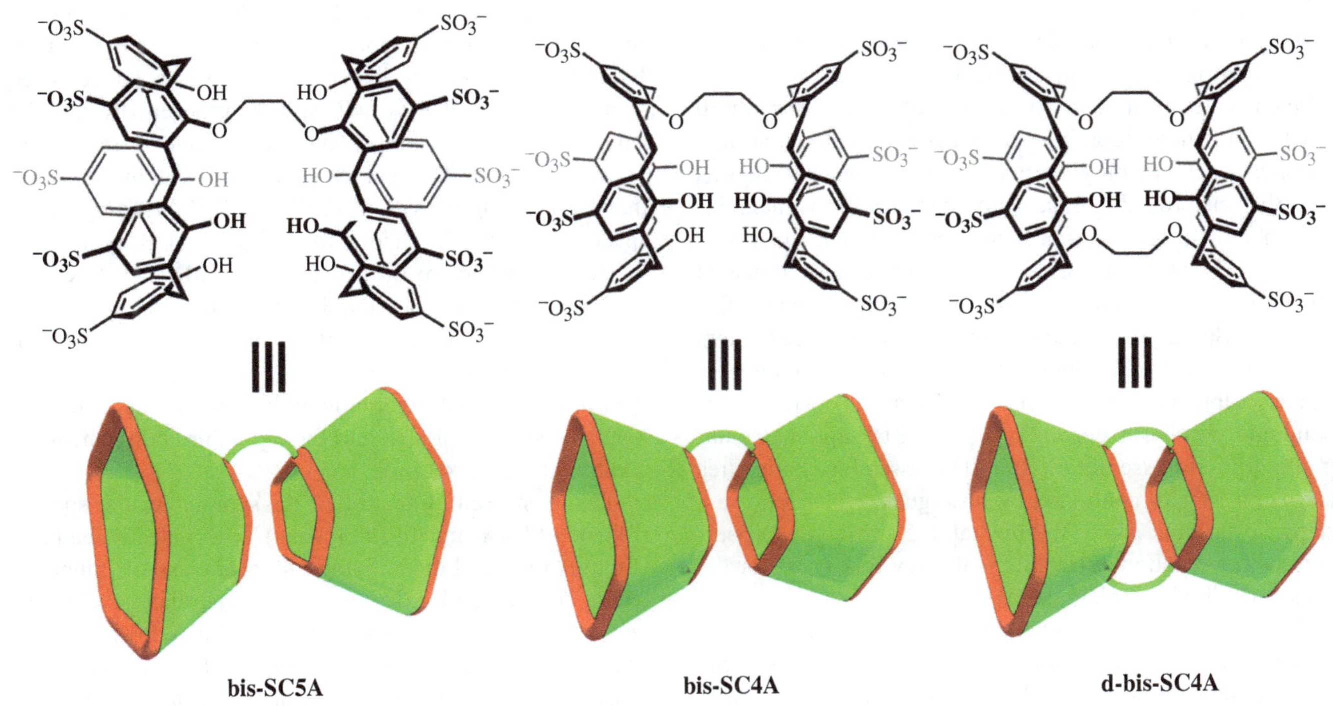

SCHEME 15.11 Structures of three bis-SC*n*As.

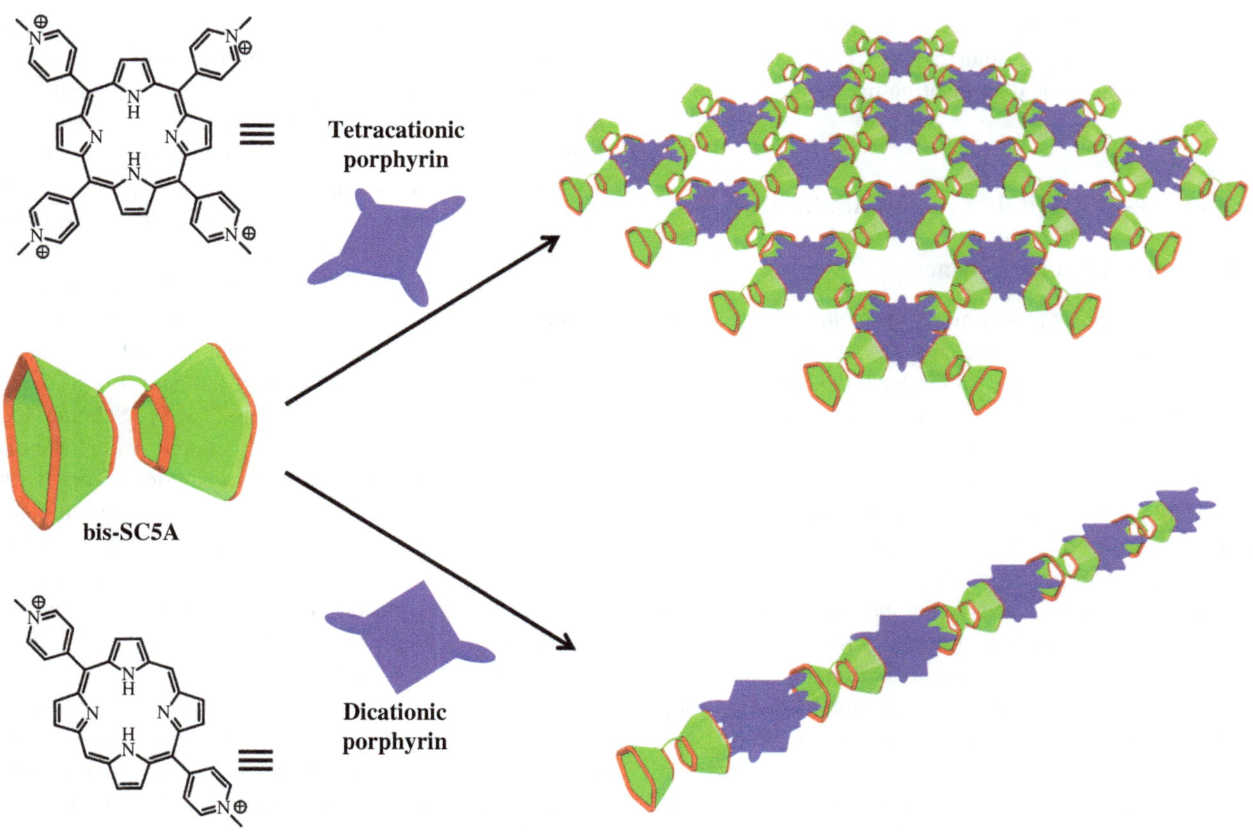

SCHEME 15.12 Schematic representation of the construction of supramolecular 2D network and 1D linear polymers based on the complexation of the ditopic host bis-SC5A with tetracationic and dicationic porphyrin guests [66].

number of negative sulfonate groups on the calixarenes is equal to the number of positive charges on the porphyrins. Hence, besides the host–guest interactions between calixarene and pyridinium, charge interactions also play a pivotal role in self-assembly.

Porphyrins are fascinating antenna molecules possessing solar light-harvesting capability, whereas calixarenes are efficient electron donors on account of their electron-rich cavities. In these supramolecular systems, calixarene and porphyrin act not only as interacting tectons but also as effective electron donor–acceptor pairs, and an unambiguous photoinduced electron transfer is observed, which was validated by (time-resolved) fluorescence and transient absorption spectroscopy, as well as electrochemical CV method. Subsequent studies by Micali and coworkers reported an alternative method for the complexation of a water-soluble octacationic bis(calix[4]arenes) with a tetraanionic porphyrin generating 2D or 3D supramolecular multiporphyrin assemblies [67]. The assembled structures are hierarchically controlled by host–guest molar ratio in a stepwise fashion, where the pivotal 4:5 (host–guest) complex with a cruciform structure is the key fork point for the subsequent higher-order assembly.

Further, Liu *et al.* reported the construction of supramolecular polymers based on bis-SC4A and homoditopic bis(viologens) guests [68, 69]. The iterative complexation of bis-SC4A with ethylene-bridged bis(viologens) (EBV^{4+}) generates a linear polymer, whereas the iterative complexation of bis-SC4A with hexamethylene-bridged bis(viologens) (HBV^{4+}) generates merely a cyclic oligomer (Scheme 15.13).

The topological difference between EBV^{4+} and HBV^{4+} is ascribed to the fact that longer flexible spacer favors formation of cyclic oligomer, rather than linear polymer. Interestingly, a ternary linear polymer was obtained when cucurbit[*n*]uril (CB[*n*]) ($n=7, 8$) was further involved in the calixarene–viologen binary system. As can be seen from Scheme 15.13, HBV^{4+} possesses a flexible aliphatic spacer that can be captured by CB[*n*]. With CB[*n*] threading on the spacer of HBV^{4+}, the rigidity of the $HBV^{4+}\subset CB[n]$ complex increases much more than that of free HBV^{4+}. It has been well established that building blocks with more rigid spacer groups favor the formation of linear polymer, rather than cyclic oligomer [70]. Consequently, in this design, one can control the assembly morphology by not only adjusting the length of spacer but also introducing a third component for additional complexation.

The most exciting feature of the supramolecular species from bis-SC4A and bis(viologens) is the electrochemical stimulus responsibility. Viologens are a class of important redox couples and exist in dicationic, radical cation, or neutral form intertransformed via reversible redox processes. The complex stability of *p*-sulfonatocalixarene is highly sensitive to the redox properties of viologen, decreasing in magnitude from dication to radical cation to neutral form,

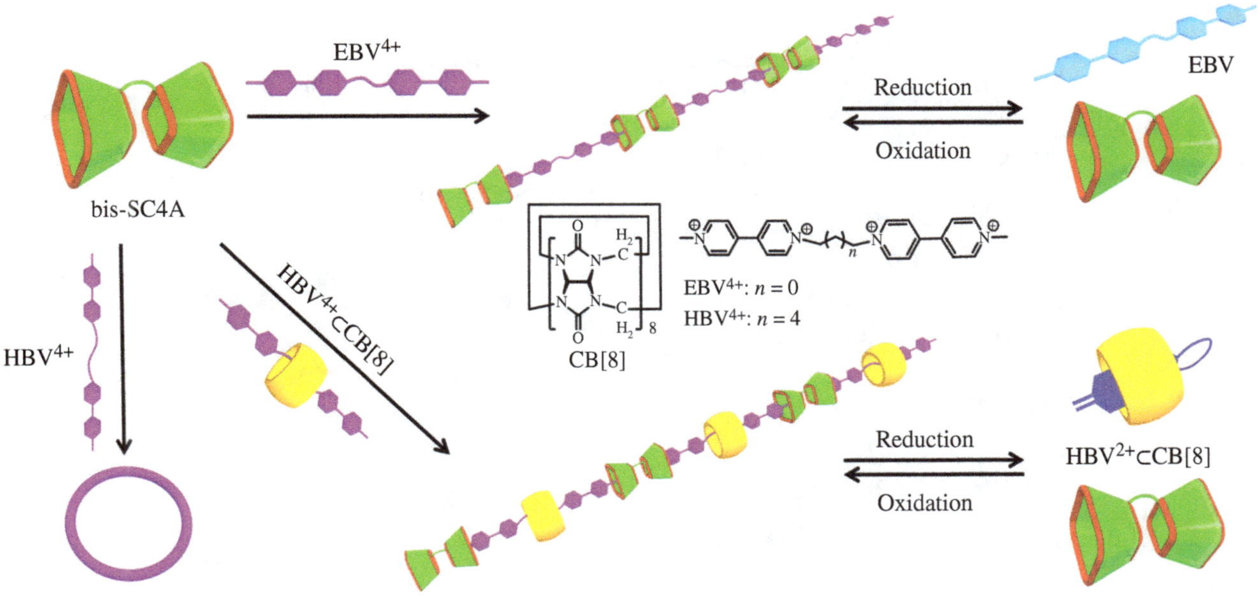

SCHEME 15.13 Chemical structures of CB[8], EBV^{4+}, and HBV^{4+} and cartoon illustration of supramolecular binary polymer, cyclic oligomer, and ternary polymer formed by bis-SC4A upon complexation with EBV^{4+}, HBV^{4+}, and $HBV^{4+}\subset CB[8]$ [2]pseudorotaxane [68, 69].

which means that assembly/disassembly of the polymer can be manipulated by electrochemical switching and redox control [23]. CV experiments show that the assembly of bis-SC4A with EBV^{4+} is dissipated upon consecutive two-step reduction processes of EBV^{4+} to its neutral form but not upon one-step reduction to its radical cation form. It means that viologen cannot be released from the cavity of calixarenes until reduced to water-insoluble neutral form, and therefore, reversible control of the assembly/disassembly processes seems to be difficult to practical operation. As a subtle evolution, the ternary polymer from bis-SC4A, HBV^{4+}, and CB[8] disassembles to free bis-SC4A and molecular loop $HBV^{2+}\subset CB[8]$ once HBV^{4+} is reduced to its radical cation form HBV^{2+} ([2]pseudorotaxane $HBV^{4+}\subset CB[8]$ transfers to molecular loop $HBV^{2+}\subset CB[8]$ undergoing one-step reduction) [71]. The reversible assembly/disassembly processes were conveniently followed by UV–Vis and DLS measurements. In the ternary system, CB[8] not only modulates the topology of supramolecular assemblies but also makes the switching between assembly and disassembly more practically operational.

Recently, Liu *et al.* prepared a novel bis(*p*-sulfonatocalix[4]arene), d-bis-SC4A (Scheme 15.11), with two ethyl bridges at the lower rim, with an improved molecular conformation and rigidity for supramolecular polymerization in comparison with bis-SC4A, with a single ethyl bridge [72]. A highly soluble linear polymer was then fabricated by the complexation of d-bis-SC4A with homoditopic triglycol-bridged bis(viologens) (G30, Scheme 15.3) in water. The resulting d-bis-SC4A–G30 polymer exhibits a high degree of polymerization and high viscosity. It is inferred that d-bis-SC4A presents a more rigid and fixed conformation, and thereby the d-bis-SC4A–G30 assembly tends to stretch in solution, which leads to the enhancement in the viscosity.

Combining the same bis-SC4A as the homoditopic host, α-cyclodextrin (α-CD) as the threading wheel, and suberyl dicholine (DiCh) as the enzyme-responsive axle, Liu *et al.* prepared an enzyme-responsive supramolecular polymer by complexation of bis-SC4A with DiCh-based pseudorotaxane (Scheme 15.14) [73]. Compared with binary polymer DiCh–bis-SC4A, ternary polymer DiCh@α-CD–bis-SC4A presents not only a larger polymeric size but also better size distribution and topology. Threading α-CD on DiCh effectively regulates the spacer flexibility and therefore promotes the supramolecular polymerization. In addition, the obtained supramolecular polymer is capable of dispersal, on account of the hydrolysis of DiCh by cholinesterases. This novel enzyme-responsive approach promises potential application in the design of biodegradable materials.

The formation of the supramolecular orthogonal polymer based on self-complementary host–guest interactions was also studied by the same group [74]. The polymer consists of two homoditopic hosts bis-SC4A and bis(β-cyclodextrins) (bis-CD), with one heteroditopic adamantane–viologen guest (AdMV). Two independent host–guest, SC4A–viologen, and CD–adamantane interactions lead to stepwise polymerization in an orthogonal way. Such an orthogonal system represents a proof that the supramolecular heteropolymers could be applied in biological process mimicry.

Based on the homoditopic host bis-SC4A, Tian *et al.* reported an electrochemical and pH dual stimuli-responsive supramolecular polymer by employing a viologen and dimethylamino unit containing connector decorated with a fluorophore as guest [75]. There are two diverse binding

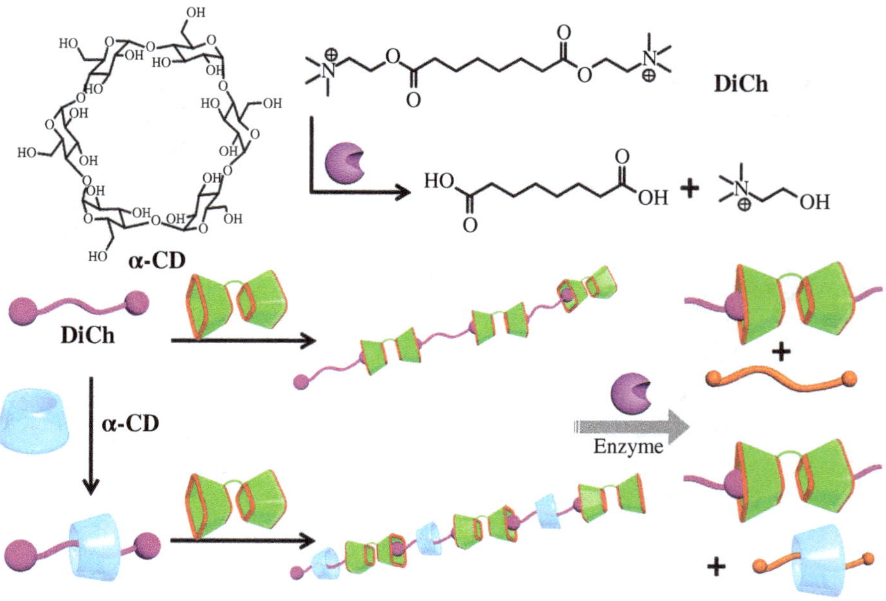

SCHEME 15.14 Structural illustration of the building blocks and schematic representation of the enzyme-responsive supramolecular polymers [73].

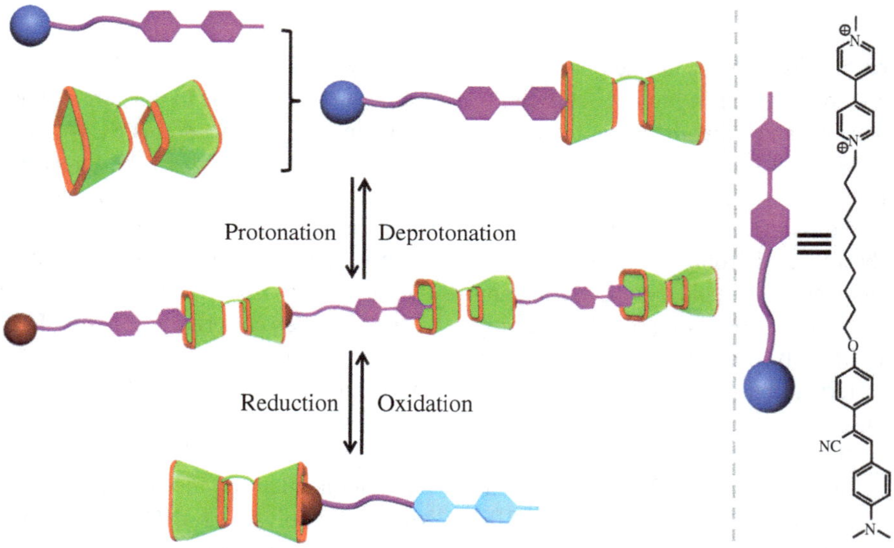

SCHEME 15.15 Structural illustration of the reversible assembly/disassembly process of supramolecular polymer based on bis-SC4A and heteroditopic guest induced by the respective stimuli of protonation/deprotonation and electrochemical redox [75].

sites on the guest, pH-sensitive dimethylamino and redox-sensitive viologen portions, and thereby, the corresponding system can be reversibly controlled via two pathways. The polymer is formed when two complexation processes occur simultaneously, that is, the viologen moiety is included into one cavity of SC4A and the protonated dimethylamino moiety is in the other cavity (Scheme 15.15). For the pH route, deprotonation would result in the release of dimethylamino moiety from the calixarene cavity, and the polymer disassembles to a viologen–calixarene complex. For the redox route, reduction would result in the release of neutral form of viologen from the calixarene cavity, and the polymer disassembles to a dimethylamino–calixarene complex.

Benefiting from the cleanness, high efficiency, and low cost, light has become one of the most desirable stimuli for regulating molecular structures and aggregate behaviors of molecular assembly [76]. Liu *et al.* reported a linear supramolecular polymer based on the intrinsically rigid guest bis-4-(*N*-methylpyridinium)azobenzene (*trans*-AzobPy) and the ditopic host bis-SC4A [77]. Upon irradiation with UV and visible light, the configuration of AzobPy switches between the *trans* and the *cis* forms, accompanied by the reversible bending of the straight spacer, which results in the morphological conversion of the supramolecular assembly between linear assembly and globular nanoparticles (Scheme 15.16). Recently, Tian *et al.* also reported a

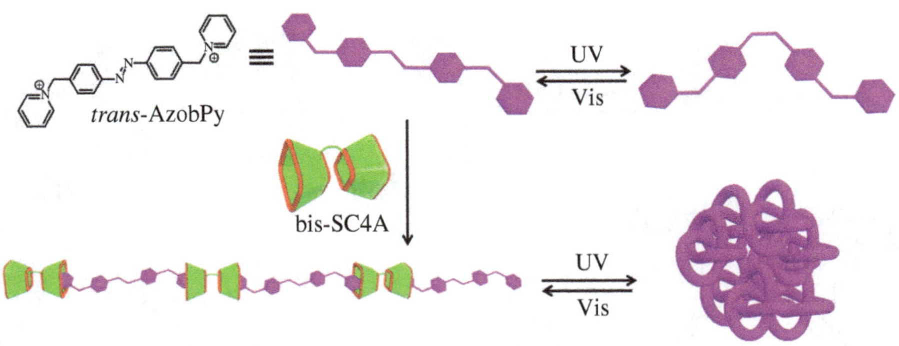

SCHEME 15.16 Structural illustration of bis-SC4A and AzobPy and schematic of the morphological conversion between linear supramolecular polymer and spherical nanoparticle based on the *trans–cis* photoisomerization of AzobPy induced by light stimulus [77].

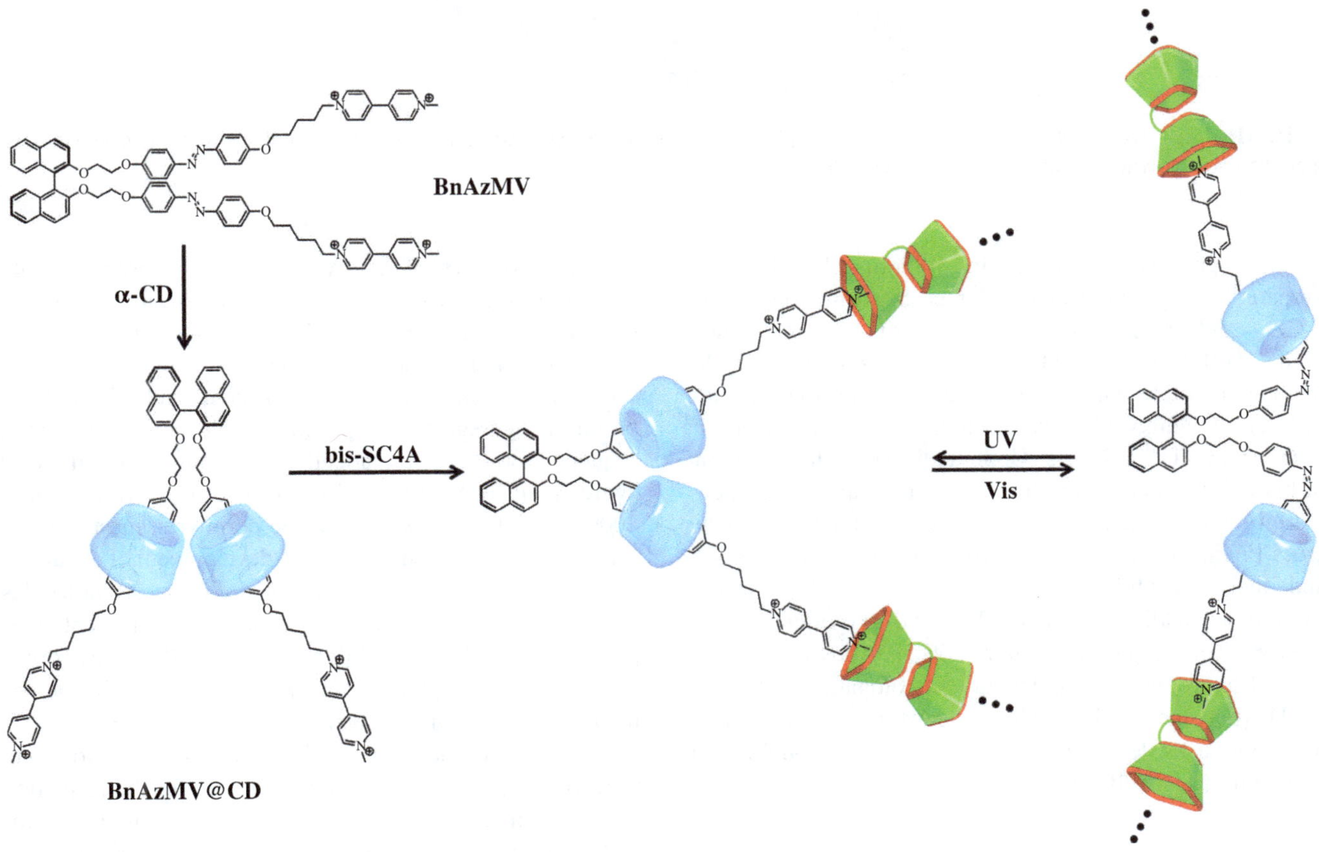

SCHEME 15.17 Schematic description of the self-assembly process for forming α-CD-based pseudo[3]rotaxane (BnAzMV@CD), the supramolecular polymer with azobenzene in the *trans* configuration, and the corresponding UV-irradiated polymer containing *cis*-azobenzene [79].

light-responsive supramolecular polymer, consisting of bis-SC4A and a water-soluble dithienylethene derivative 2-bis[2-methyl-5-(4-methylpyridyl)-3-thienyl] cyclopentene [78]. The resulting polymer exhibits a good photochromic property and morphology change under alternative UV–Vis light stimuli, owing to the photoisomerization of the dithienylethene guest.

Moreover, Tian and coworkers constructed a novel light-driven chiral supramolecular polymer via host–guest interaction between bis-SC4A and an α-CD-based pseudo[3] rotaxane (BnAzMV@CD), involving an axially binaphthyl derivative as the chiral center, two α-CD-included azobenzene moieties as photoresponsive tectons, and viologens as binding sites to SC4As (Scheme 15.17) [79]. The initial supramolecular polymer shows an obvious starlike morphology with multiple branches while appearing as linear belts after UV irradiation. Interestingly, such helical supramolecular polymer molecules with lengths of hundreds of nanometers to micrometers in water were observed in the native state for the first time.

Unlike α- and β-CDs, γ-CD possesses the capacity of simultaneously including two small aromatic guests because

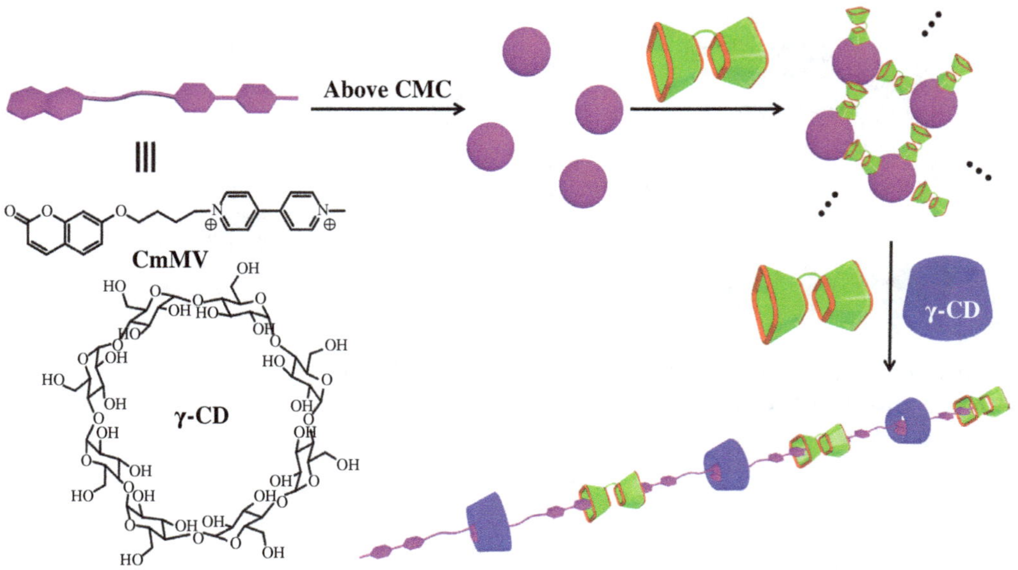

SCHEME 15.18 Schematic representation of the morphology transition from spherical micelles to the amorphous worm-like network and then the linear polymer, regulated by host–guest interactions [80].

of the larger hydrophobic cavity. Accordingly, Ma *et al.* prepared a supramolecular polymer by using bis-SC4A, γ-CD, and a heteroditopic coumarin–viologen guest (CmMV) [80]. Free CmMV firstly forms spherical micelle in aqueous solution above the critical micelle concentration (CMC), showing visible spheres in its transmission electron microscopy (TEM) image. Upon addition of small amount of bis-SC4A, micelles are transferred into large amorphous wormlike network, which is ascribable to the non-covalent cross-linking via the host–guest interaction between SC4A and the hydrophilic viologen on the micelle surface. Continuous addition of bis-SC4A and subsequent γ-CD finally generates the linear supramolecular polymer, due to the 2 : 1 hydrophobic complexation of coumarin moiety and γ-CD (Scheme 15.18). This interesting transformation represents a facile approach to control the morphology of multicomponent self-assembly systems.

15.3.2 Calixarene-Induced Aggregation

CDs, CBs, and SC*n*As are three kinds of widely studied water-soluble macrocycles. CDs and CBs are apt to form threading structures with guest molecules, which are always employed to fabricate various rotaxanes and catenanes. One of the representative outcomes afforded by such inclusion is preventing self-aggregation of guest molecules [81, 82]. Opposite to CDs and CBs, SC*n*As are capable of promoting the self-aggregation of aromatic or amphiphilic molecules by lowering the critical aggregation concentration (CAC), enhancing the aggregate stability, and regulating the degree of order in the aggregates [83]. Liu *et al.* coined this unique self-assembling strategy as *calixarene-induced aggregation* (*CIA*) [84].

In fact, the prime application of CIA can be traced back to the complexation between the tetrakis(hydroxycarbonylm ethoxy) derivative of SC4A and cationic porphyrin with tunable host–guest stoichiometry ranging from 4 : 1 to 4 : 7, reported by Randaccio, Purrello, and Sciotto early in 2001 [85]. Following this idea, they performed systematically a series of excellent works on non-covalent synthesis of multi-porphyrin aggregates with programmable stoichiometry and sequence [86–88]. Dramatically, the complexation-induced aggregation by SC*n*As did not receive due attention until 2009, possibly because of the sole porphyrin guest species. Since 2009, it has sprung from the development of expanded guest species, and now more than 20 guest molecules (Scheme 15.19), applicable as aromatic fluorescent dyes [66, 85, 89–96], amphiphilic surfactants [97–105], drugs [106], and nonamphiphilic polycations [72, 107], can be induced to self-aggregate using CIA strategy [108].

For these aromatic or amphiphilic guests employed in CIA, two opposing forces are responsible for the formation of highly ordered assemblies. The hydrophobic effect (or π-stacking interaction) drives the segregation of the alkyl chains (or aromatic portions) in water, thus providing the impetus for self-organization. On the other hand, repulsive electrostatic forces between the polar heads, positively charged organic ammonium cations, prevent the formation of large three-dimensional assemblies. Upon complexation with SC*n*As, the intrinsic electrostatic repulsion between the polar heads was replaced by electrostatic attraction between the polar heads and the sulfonate groups of SC*n*As, therefore boosting the desired aggregation of guests.

Liu *et al.* postulated that the CIA of guests observed upon addition of SC*n*As occurred in two steps (Scheme 15.20; taking the dicationic perylene bisimide guest as example [89]). First, the host and guest molecules instantaneously form a complex in which the two head groups of the guest are captured by the cavities of two host molecules, forming a 2 : 1 capsule-like complex driven by the host–guest

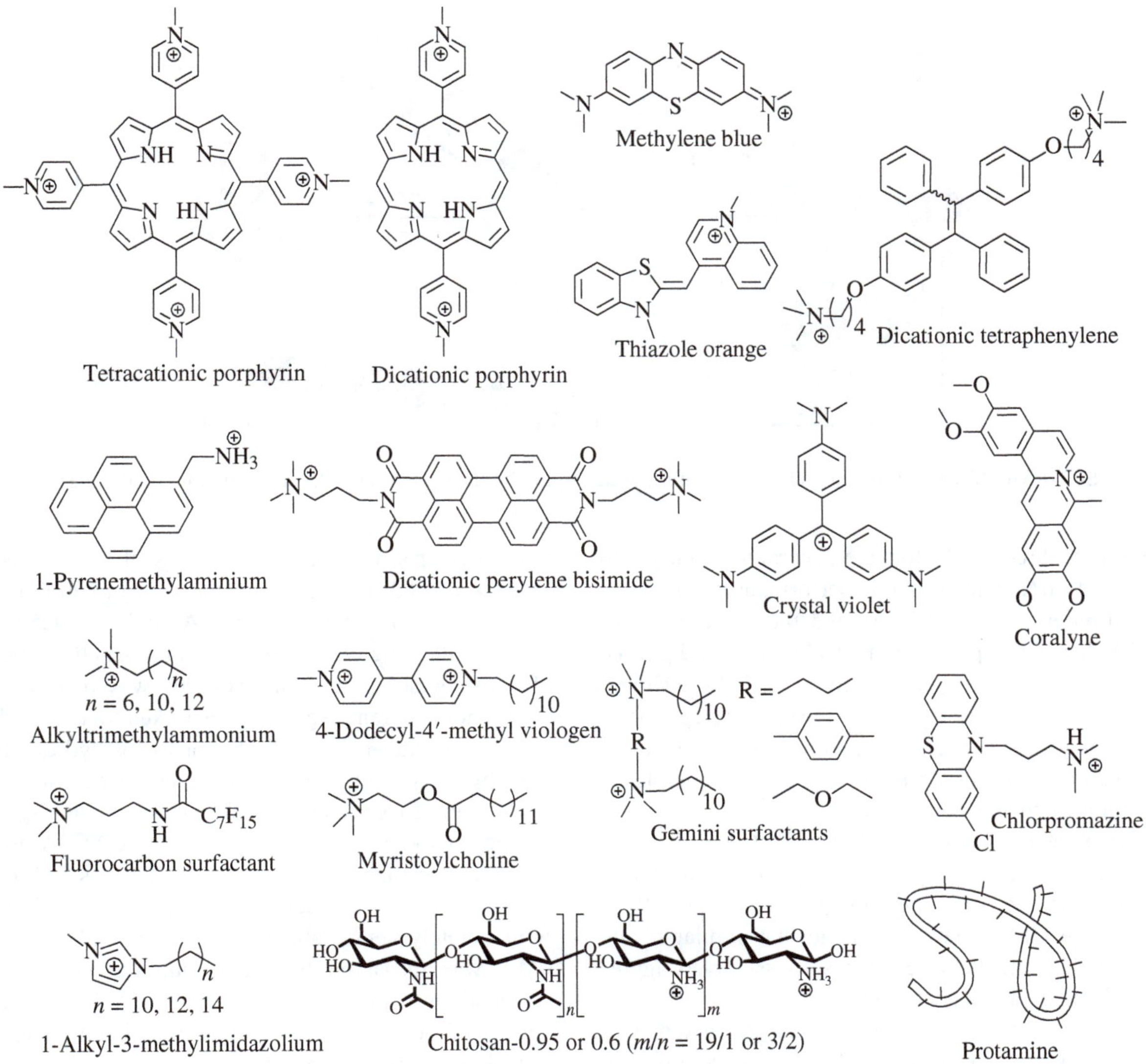

SCHEME 15.19 Chemical structures of guest molecules used for CIA.

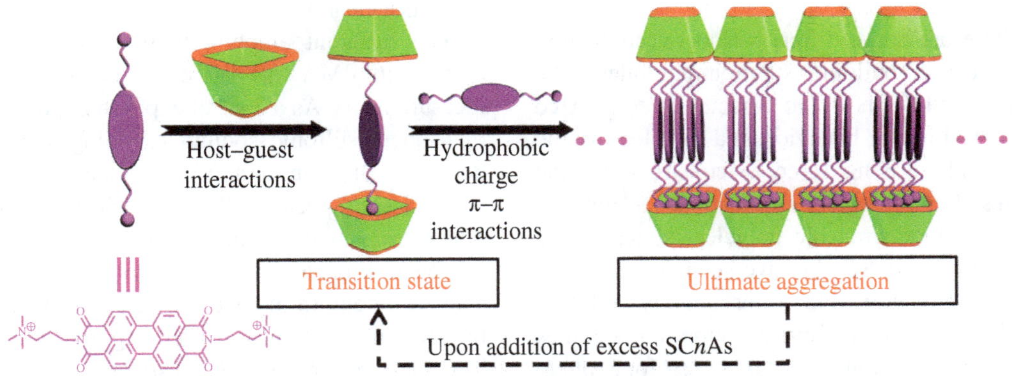

SCHEME 15.20 Schematic illustration of CIA with dicationic perylene bisimide as the guest [89].

interactions. Subsequently, additional guest molecules are readily integrated into the 2 : 1 complexes, resulting in the formation of the 2 : n complexes and then large three-dimensional aggregates. Several non-covalent interactions, including host–guest, charge, π-stacking, and hydrophobic effects, contribute to the stabilization of the ultimate aggregates, where strong host–guest interactions are an indispensable prerequisite for CIA. The further addition of excess SCnAs is prone to separate the aggregation of guests, indicating that the host–guest interactions are more dominant than the charge ones. The preorganized cyclic scaffold of SCnAs also plays a crucial role in CIA.

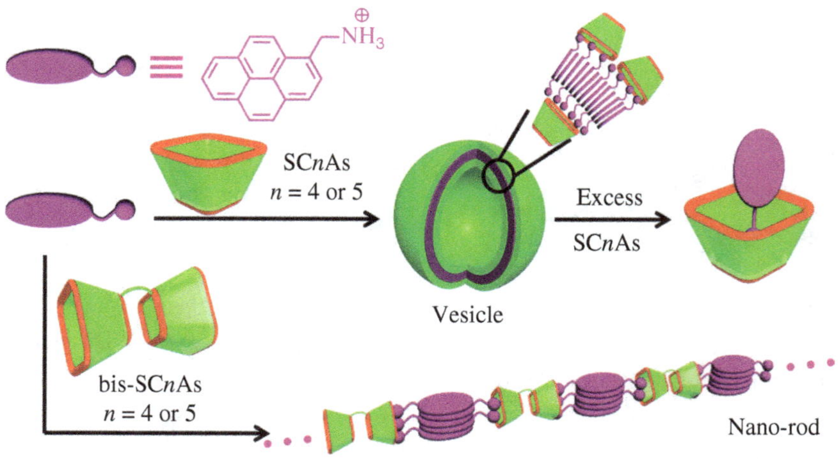

SCHEME 15.21 Different architectures of PMA aggregates in the presence of SC*n*As or bis-SC*n*As [94, 95].

In summary, three key factors are generally demanded to achieve the desired CIA of high performance: (i) strong binding affinities between SC*n*As and the polar heads of guests, (ii) charge compensation between hosts and guests, and (iii) the preorganized cyclic scaffold of SC*n*As. It should be noticed that CIA means prompting the self-aggregation of special model guests. For the guests without any tendency of self-aggregation, SC*n*As form merely simple 1 : 1 complexes but not complicated 1 : *n* complexes.

The CIA strategy cannot only lower the CAC of guests but also enhance the aggregate stability/compactness and regulate the degree of order in the aggregates [89]. Admittedly, the CIA behavior is somewhat similar to the polyion complex micelle where the self-assembled aggregates are formed at concentrations well below the CAC when flexible and strong polyelectrolytes are mixed with oppositely charged surfactants [99]. In this way, SC*n*As can serve as simple models to mimic polyanions and to get insight into the complex behavior displayed by these polyelectrolytes.

Definitely, there are some differences between SC*n*As and polyelectrolytes in building supra-amphiphiles. The flexible polyelectrolytes show no special preorganized structure and cavity binding behavior, and the electrostatic attraction is the only driving force to construct polyion complex micelles. For biological applications, a constant challenge is finding biodegradable polyelectrolytes to use for controlled drug release. SC*n*As have been demonstrated to be significantly biocompatible. Consequently, a lot of endeavors have been donated to fabricate supra-amphiphiles based on CIA, which are of fundamental interest for application in drug delivery.

Employing SC5A as the host and 1-pyrenemethylaminium (PMA) as the guest, respectively, self-assembled binary supramolecular vesicles were fabricated (Scheme 15.21) [94]. DLS was employed to identify the self-assembled size of SC5A–PMA amphiphiles, giving an average diameter of 99 nm with a narrow size distribution. TEM images show the hollow spherical morphology, indicating convincingly the vesicular structure. The thickness of the bilayer membrane is about 3 nm, which is in the same order of magnitude as the sum of one PMA length (7 Å) and two SC5A heights (14 Å), indicating that the vesicle is unilamellar. The model of forming supramolecular vesicles was deduced as the hydrophobic pyrene segments are packed together, while the inner- and outer-layer surfaces consist of hydrophilic phenolic hydroxyl groups of SC5A, which are exposed to water. SC5A and PMA are connected together by host–guest and charge interactions. Notably, neither free SC5A nor PMA can form nanosized aggregates by themselves. SC5A has no tendency to self-aggregation in aqueous solution [109]. A moderately appreciable CAC value of 0.27 mM for free PMA was measured by monitoring the polarity of the surrounding medium of pyrene. However, free PMA shows no Tyndall effect, revealing that free PMA cannot form nanosized aggregates but may just form small dimers and oligomers. In the presence of SC5A, the CAC values decrease more than 3 times. A solution of SC5A with PMA exhibits a clear Tyndall effect.

The different amphiphilicity between free PMA and SC5A with PMA was also evaluated by the surface tension measurements. As a control experiment, the excess addition of 4-phenolsulfonic sodium (building subunit of SC*n*As) does not cause any appreciable change in both absorption and emission spectra of PMA, which proves undoubtedly the crucial role of host–guest complexation. Interestingly, the aggregation behavior of PMA with bis-SC*n*As is dramatically different from that with SC*n*As, which showed the fabrication of 1D linear nanorod aggregates (Scheme 15.21) [95]. The reasons for the distinctly different aggregation behavior of PMA with bis-SC*n*As may be the distinguishable number and rigidity of their cavity compared with that of SC*n*As. The most fascinating aspect of supramolecular chemistry is that two or more components can self-assemble into higher-order structures, which exhibit particular properties and functions that the individual component cannot achieve. The present case highlights the fascinating feature of self-assembly that neither of the two individual

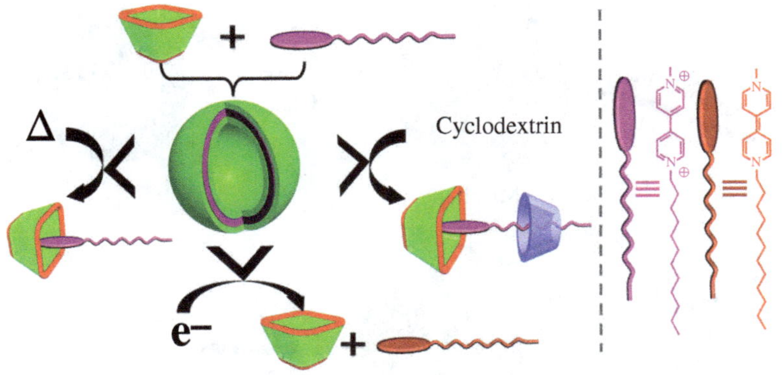

SCHEME 15.22 Formation of a multistimulus-responsive supramolecular binary vesicle composed of SC4A and an asymmetric viologen [103].

components possess typical amphiphilicity but can form nanoscaled aggregates upon exquisitely mixing with each other at a specific host–guest ratio.

The obtained vesicle is stable at room temperature, and the vesicular structure can be maintained at least over one week. Moreover, the assembly/disassembly process can be accomplished in the timescale of minutes with changing temperature. The temperature response is reasonably acceptable because both the complexation of SC*n*As with organic guests and the π···π-stacking of aromatic pyrenes are enthalpy driven, which would be weakened upon warming. Consequently, the binary vesicle exhibits not only benign water solubility and self-labeled fluorescence but also thermal reversibility. It can disassemble when the temperature increases up to 35–40°C, which can act as potential delivery vehicle for drugs.

In nature, the change in behavior of biomacromolecules, such as proteins and nucleic acids, is often a result of the response not to a single factor but to a combination of environmental changes. To mimic this feature, construction of materials with multistimuli responses in a predictable manner would be of great interest. Liu *et al.* further constructed supramolecular binary vesicles based on host–guest complexation of SC4A with asymmetric viologen (Scheme 15.22) [103]. The CAC value of asymmetric viologen decreases pronouncedly by a factor of ca. 1000 due to the complexation of SC4A. As expected, the obtained vesicle is multistimuli responsive, benefiting from the intrinsic advantages of supramolecular chemistry. Reduction of asymmetric viologen to radical cation state can adjust the vesicle to smaller species and *vice versa*. Reduction to neutral form, increasing temperature, and inclusion of cyclodextrins can all disrupt the vesicle architecture to trigger the efficient release of the entrapped doxorubicin from the interior of the vesicle. Cell experiments show that the loading of doxorubicin by a vesicle does not affect the therapeutic effect of doxorubicin for cancer cells, whereas its damage for normal cells is reduced.

Enzyme-responsive, amphiphilic self-assembly represents one of the increasingly significant topics in biomaterials research and finds feasible applications to the controlled release of therapeutic agents at specific sites where the target enzyme is located. Based on the concept of supra-amphiphiles [110], Liu *et al.* constructed an enzyme-responsive supramolecular vesicle as an operational targeted drug delivery system, using SC4A as the macrocyclic host and natural enzyme-cleavable myristoylcholine as the guest molecule (Scheme 15.23) [100]. Although myristoylcholine is a natural enzyme-active molecule, it is defective in fabricating an enzyme-responsive assembly independently as the CACs of the substrate and product are similar. Both myristoylcholine and myristic acid form micelles with CACs of 2.5 and 4.5 mM, respectively. The complexation of SC4A with myristoylcholine directs the formation of a supramolecular binary vesicle, where the CAC of myristoylcholine decreases significantly by a factor of ca. 100. By taking the advantage of the dynamic equilibrium characteristics of non-covalent interactions, enzyme is active to disintegrate the self-assembled vesicle. Although the cleavage rate may be reduced, it is certainly advantageous to covalent modification of the substrates, which often results in a loss of enzyme recognition. The binary vesicle from SC4A and myristoylcholine exhibits highly specific and efficient responsiveness to cholinesterase. An enzyme-induced cleavage of myristoylcholine triggers a cascade of events, loss of hydrophilic–hydrophobic balance of binary superamphiphile, disassembly of vesicle, and tandem release of entrapped drugs. Cholinesterase is overexpressed in Alzheimer's disease, and therefore, the present system may have potential for the delivery of Alzheimer's drugs.

A trypsin-responsive supramolecular vesicle was recently constructed from SC4A as the macrocyclic host and protamine as the enzyme-degradable guest (Scheme 15.24) [72]. Different from the guests used previously for CIA, protamine is not a small molecule but rather a nonamphiphilic cationic protein, and its use represents a major extension in the range of substrates that can be induced to form supramolecular assemblies by SC*n*As. Subsequently, supramolecular polymeric vesicles were fabricated by the complexation of SC4A and chitosan (Scheme 15.24) [107]. As a result of high deacetylation degree, chitosan possesses multivalent electrostatic interactions with SC4A, which leads to the final

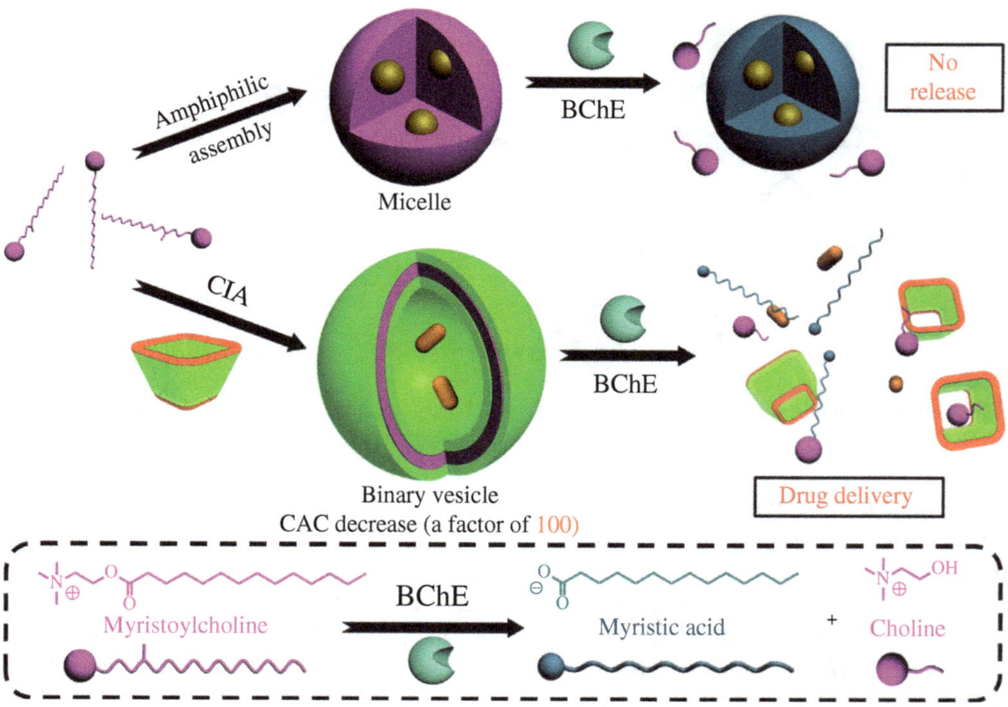

SCHEME 15.23 Enzymatic responsiveness of amphiphilic assemblies of myristoylcholine fabricated in the absence or presence of SC4A [100].

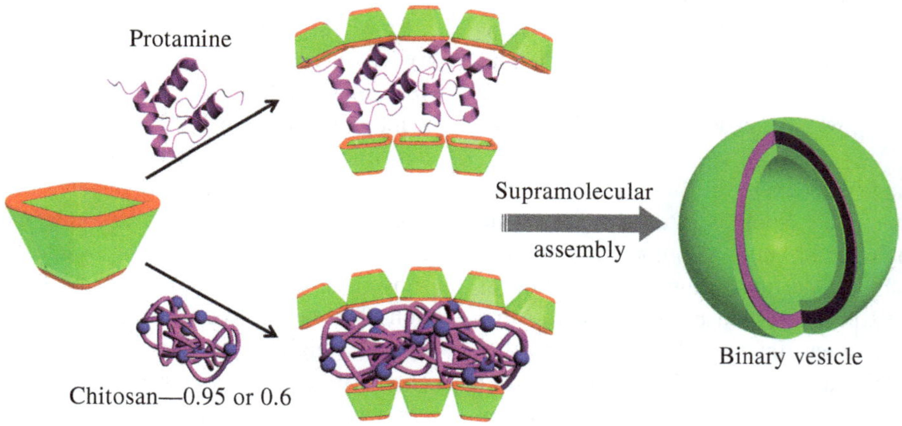

SCHEME 15.24 Formation of supramolecular binary vesicle composed of SC4A and nonamphiphilic polycations [72, 107].

formation of stable supramolecular binary assembly. Unlike the CIA assemblies, which are driven by enthalpy term and therefore disassemble upon increasing temperature, this supramolecular polymeric vesicle fuses into larger nanoparticles with increasing temperature. This unusual temperature response broadens the application of CIA assemblies in resembling the membrane fusion process.

Upon complexation with reaction products, SC*n*As are found to be capable to catalyze the alcoholysis of *N*-acetyl-L-amino acids in methanol (particularly histidine, lysine, and arginine) [111]. Recently, Liu *et al.* reported a self-assembled fluorescent organic nanoparticle that exhibited CIA and aggregation-induced emission (AIE) in aqueous media, by employing SC4A and bis-SC4A as the hosts and dicationic tetraphenylene (QA-TPE) as the guest [96]. Compared with free QA-TPE, the π-aggregation of TPE by CIA becomes more compact, leading to the desired AIE fluorescence. More importantly, the photoreactivity of TPE by CIA was pronouncedly enhanced, while no acceleration of photoreaction was observed for the aggregates of QA-TPE above its CAC. One reasonable explanation is that the complexed QA-TPE existed in a compact aggregate, and the intramolecular rotations of the phenyl groups in the TPE unit were effectively restricted by π···π-stacking of the neighboring TPE units, which favored photocyclization of QA-TPE (Scheme 15.25). This principle is expected to be adaptable for the development of various stimulus-triggered and self-assembled materials for supramolecular catalysis.

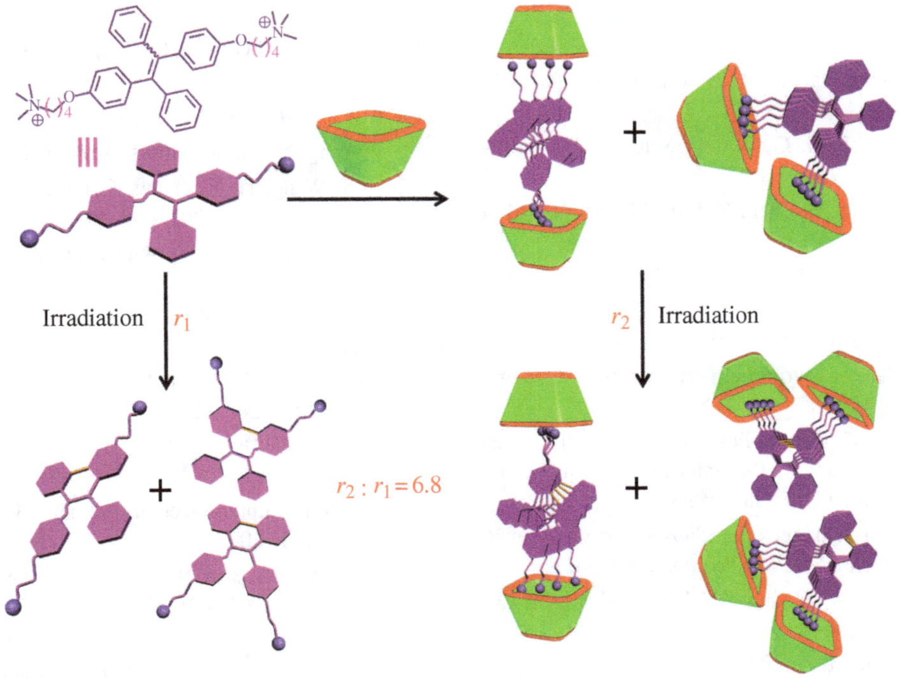

SCHEME 15.25 Schematic illustration of the photocyclization of free QA-TPE and the SC4A–QA-TPE nanoparticles [96].

15.4 FINAL COMMENTS

In summary, in this chapter we have summarized the binding structures and properties of SC*n*As with various guest molecules, their thermodynamic origins, and the construction of their highly ordered supramolecular assembly. Benefiting from the unique structures of 3D cavity and additional binding sites of sulfonate groups, SC*n*As exhibit strong binding abilities to organic cations, leading to the formation of versatile complexes with different functions. Particularly, the protein camouflage via host–guest complexation of SC4A can be mentioned as a novel approach not only to facilitate protein crystallization but also to regulate the physiological behavior and biological function of proteins. Moreover, the exquisite nanosupramolecular assemblies based on SC*n*As show diverse applications in many fields, including biomedicine, sensor/probe, materials science, catalysis, and so on. Although there have been so many fascinating non-covalent SC*n*A-based complexes, the covalently modified derivatives of SC*n*A with excellent binding capacity, especially modified with functional groups, are still limited. Therefore, we believe that delicate non-covalent interaction-driven complexes based on functionalized SC*n*A will be much more prosperous and valuable in the near future.

ACKNOWLEDGMENTS

We thank "973" Program (no. 2011CB932502) and NNSFC (nos. 91227107 and 21432004) for the financial support.

REFERENCES

1. V. Böhmer, *Angew. Chem. Int. Ed.* 1995, *34*, 713.
2. C. Gaeta, F. Troisi, P. Neri, *Org. Lett.* 2010, *12*, 2092.
3. D. Coquière, A. de la Lande, S. Martì, O. Parisel, T. Prangé, O. Reinaud, *Proc. Natl. Acad. Sci. U. S. A.* 2009, *106*, 10449.
4. S. Shinkai, S. Mori, T. Tsubaki, T. Sone, O. Manabe, *Tetrahedron Lett.* 1984, *25*, 5315.
5. S. Shinkai, K. Araki, T. Matsuda, N. Nishiyama, H. Ikeda, I. Takasu, M. Iwamoto, *J. Am. Chem. Soc.* 1990, *112*, 9053.
6. G. Arena, A. Casnati, A. Contino, G. G. Lombardo, D. Sciotto, R. Ungaro, *Chem. Eur. J.* 1999, *5*, 738.
7. Y. Liu, D.-S. Guo, H.-Y. Zhang, Y.-H. Ma, E.-C. Yang, *J. Phys. Chem. B* 2006, *110*, 3428.
8. J. Cui, V. D. Uzunova, D.-S. Guo, K. Wang, W. M. Nau, Y. Liu, *Eur. J. Org. Chem.* 2010, 1704.
9. M. H. Paclet, C. F. Rousseau, C. Yannick, F. Morel, A. W. Coleman, *J. Incl. Phenom. Macrocycl. Chem.* 2006, *55*, 353.
10. A. W. Coleman, S. Jebors, S. Cecillon, P. Perret, D. Garin, D. Marti-Battle, M. Moulin, *New J. Chem.* 2008, *32*, 780.
11. F. Perret, A. N. Lazar, A. W. Coleman, *Chem. Commun.* 2006, 2425.
12. F. Perret, A. W. Coleman, *Chem. Commun.* 2011, *47*, 7303.
13. D.-S. Guo, K. Wang, Y. Liu, *J. Incl. Phenom. Macrocycl. Chem.* 2008, *62*, 1.
14. C. Bonal, Y. Israëli, J.-P. Morel, N. Morel-Desrosiers, *J. Chem. Soc., Perkin Trans. 2* 2001, 1075.
15. M. Stödeman, N. Dhar, *J. Chem. Soc., Faraday Trans.* 1998, *94*, 899.
16. M. Stödeman, N. Dhar, *Thermochim. Acta* 1998, *320*, 33.
17. Y. Liu, E.-C. Yang, Y. Chen, *Thermochim. Acta* 2005, *429*, 163.

18. S. Shinkai, K. Araki, T. Matsuda, O. Manabe, *Bull. Chem. Soc. Jpn.* 1989, *62*, 3856.
19. S. Shinkai, K. Araki, O. Manabe, *J. Am. Chem. Soc.* 1988, *110*, 7214.
20. G. Arena, S. Gentile, F. G. Gulino, D. Sciotto, C. Sgarlata, *Tetrahedron Lett.* 2004, *45*, 7091.
21. L.-H. Wang, D.-S. Guo, Y. Chen, Y. Liu, *Thermochim. Acta* 2006, *443*, 132.
22. H.-X. Zhao, D.-S. Guo, Y. Liu, *J. Phys. Chem. B* 2013, *117*, 1978.
23. D.-S. Guo, L.-H. Wang, Y. Liu, *J. Org. Chem.* 2007, *72*, 7775.
24. L. A. Summers, The Bipyridinium Herbicides, Academic Press, New York, 1980.
25. P. M. S. Monk, The Viologens. Physicochemical Properties, Synthesis and Applications of the Salts of 4, 4′-Bipyridine, John Wiley & Sons, Ltd, Chichester, 1998.
26. R. J. Dinis-Oliveira, J. A. Duarte, A. Sánchez-Navarro, F. Remião, M. L. Bastos, F. Carvalho, *Crit. Rev. Toxicol.* 2008, *38*, 13.
27. K. Wang, D.-S. Guo, H.-Q. Zhang, D. Li, X.-L. Zheng, Y. Liu, *J. Med. Chem.* 2009, *52*, 6402.
28. G.-F. Wang, X.-L. Ren, M. Zhao, X.-L. Qiu, A.-D. Qi, *J. Agric. Food Chem.* 2011, *59*, 4294.
29. E. Da Silva, A. W. Coleman, *Tetrahedron* 2003, *59*, 7357.
30. N. Douteau-Guevel, A. W. Coleman, J.-P. Morel, N. Morel-Desrosiers, *J. Phys. Org. Chem.* 1998, *11*, 693.
31. N. Douteau-Guevel, A. W. Coleman, J.-P. Morel, N. MorelDesrosiers, *J. Chem. Soc., Perkin Trans. 2* 1999, 629.
32. N. Douteau-Guevel, F. Perret, A. W. Coleman, J.-P. Morel, N. Morel-Desrosiers, *J. Chem. Soc., Perkin Trans. 2* 2002, 524.
33. O. I. Kalchenko, F. Perret, N. Morel-Desrosiers, A. W. Coleman, *J. Chem. Soc., Perkin Trans. 2* 2001, 258.
34. G. Arena, A. Contino, F. G. Gulino, A. Magri, F. Sansone, D. Sciotto, R. Ungaro, *Tetrahedron Lett.* 1999, *40*, 1597.
35. M. Selkti, A. Tomas, A. W. Coleman, N. Douteau-Guevel, I. Nicolis, F. Villain, C. de Rango, *Chem. Commun.* 2000, 161.
36. J. L. Atwood, T. Ness, P. J. Nichols, C. L. Raston, *Cryst. Growth Des.* 2002, *2*, 171.
37. P. J. Nichols, C. L. Raston, *Dalton Trans.* 2003, 2923.
38. A. Lazar, E. Da Silva, A. Navaza, C. Barbey, A. W. Coleman, *Chem. Commun.* 2004, 2162.
39. J. L. Atwood, S. J. Dalgarno, M. J. Hardie, C. L. Raston, *Chem. Commun.* 2005, 337.
40. L. Memmi, A. Lazar, A. Brioude, V. Ball, A. W. Coleman, *Chem. Commun.* 2001, 2474.
41. E. Da Silva, C. F. Rousseau, I. Zanella-Cléon, M. Becchi, A. W. Coleman, *J. Incl. Phenom. Macrocycl. Chem.* 2006, *54*, 53.
42. R. E. McGovern, H. Fernandes, A. R. Khan, N. P. Power, P. B. Crowley, *Nat. Chem.* 2012, *4*, 527.
43. T. Schrader, *Nat. Chem.* 2012, *4*, 519.
44. R. E. McGovern, A. A. McCarthy, P. B. Crowley, *Chem. Commun.* 2014, *50*, 10412.
45. R. E. McGovern, B. D. Snarr, J. A. Lyons, J. McFarlane, A. L. Whiting, I. Paci, F. Hof, P. B. Crowley, *Chem. Sci.* 2015, *6*, 442.
46. K. D. Daze, T. Pinter, C. S. Beshara, A. Ibraheem, S. A. Minaker, M. C. F. Ma, R. J. M. Courtemanche, R. E. Campbell, F. Hof, *Chem. Sci.* 2012, *3*, 2695.
47. I. Ghosh, W. M. Nau, *Adv. Drug Deliv. Rev.* 2012, *64*, 764.
48. W. Yang, M. M. De Villiers, *Eur. J. Pharm. Biopharm.* 2004, *58*, 629.
49. W. Yang, M. M. De Villiers, *J. Pharm. Pharmacol.* 2004, *56*, 703.
50. W. Yang, M. M. de Villiers, *AAPS J.* 2005, *7*, E241.
51. G.-S. Wang, H.-Y. Zhang, F. Ding, Y. Liu, *J. Incl. Phenom. Macrocycl. Chem.* 2011, *69*, 85.
52. G.-S. Wang, H.-Y. Zhang, D. Li, P.-Y. Wang, Y. Liu, *Supramol. Chem.* 2011, *23*, 441.
53. D.-S. Guo, H.-Q. Zhang, F. Ding, Y. Liu, *Org. Biomol. Chem.* 2012, *10*, 1527.
54. L.-H. Wang, P. Du, J. Yang, D.-S. Guo, Y. Liu, *Supramol. Chem.* 2014, *26*, 10.
55. Q. Li, D.-S. Guo, H. Qian, Y. Liu, *Eur. J. Org. Chem.* 2012, 3962.
56. A. Ciferri, Supramolecular Polymers, Taylor & Francis Group, Boca Raton, 2005.
57. L. Brunsveld, B. J. B. Folmer, E. W. Meijer, R. P. Sijbesma, *Chem. Rev.* 2001, *101*, 4071.
58. S. J. Rowan, S. J. Cantrill, G. R. L. Cousins, J. K. M. Sanders, J. F. Stoddart, *Angew. Chem. Int. Ed.* 2002, *41*, 898.
59. S. Seiffert, J. Sprakel, *Chem. Soc. Rev.* 2012, *41*, 909.
60. R. J. Wojtecki, M. A. Meador, S. J. Rowan, *Nat. Mater.* 2011, *10*, 14.
61. T. F. A. De Greef, M. M. J. Smulders, M. Wolffs, A. P. H. J. Schenning, R. P. Sijbesma, E. W. Meijer, *Chem. Rev.* 2009, *109*, 5687.
62. B. Zheng, F. Wang, S. Dong, F. Huang, *Chem. Soc. Rev.* 2012, *41*, 1621.
63. A. Harada, Y. Takashima, H. Yamaguchi, *Chem. Soc. Rev.* 2009, *38*, 875.
64. J. M. Zayed, N. Nouvel, U. Rauwald, O. A. Scherman, *Chem. Soc. Rev.* 2010, *39*, 2806.
65. Y. Liu, Y. Yu, J. Gao, Z. Wang, X. Zhang, *Angew. Chem. Int. Ed.* 2010, *49*, 6576.
66. D.-S. Guo, K. Chen, H.-Q. Zhang, Y. Liu, *Chem. Asian J.* 2009, *4*, 436.
67. A. D'Urso, D. A. Cristaldi, M. E. Fragalà, G. Gattuso, A. Pappalardo, V. Villari, N. Micali, S. Pappalardo, M. F. Parisi, R. Purrello, *Chem. Eur. J.* 2010, *16*, 10439.
68. D.-S. Guo, S. Chen, H. Qian, H.-Q. Zhang, Y. Liu, *Chem. Commun.* 2010, *46*, 2620.
69. H. Qian, D.-S. Guo, Y. Liu, *Chem. Eur. J.* 2012, *18*, 5087.
70. K. Ohga, Y. Takashima, H. Takahashi, Y. Kawaguchi, H. Yamaguchi, A. Harada, *Macromolecules* 2005, *38*, 5897.
71. W. S. Jeon, A. Y. Ziganshina, J. W. Lee, Y. H. Ko, J.-K. Kang, C. Lee, K. Kim, *Angew. Chem. Int. Ed.* 2003, *42*, 4097.
72. K.-P. Wang, D.-S. Guo, H.-X. Zhao, Y. Liu, *Chem. Eur. J.* 2014, *20*, 4023.
73. D.-S. Guo, T.-X. Zhang, Y.-X. Wang, Y. Liu, *Chem. Commun.* 2013, *49*, 6779.
74. H.-X. Zhao, D.-S. Guo, L.-H. Wang, H. Qian, Y. Liu, *Chem. Commun.* 2012, *48*, 11319.
75. X. Ma, R. Sun, W. Li, H. Tian, *Polym. Chem.* 2011, *2*, 1068.
76. M. Burnworth, L. Tang, J. R. Kumpfer, A. J. Duncan, F. L. Beyer, G. L. Fiore, S. J. Rowan, C. Weder, *Nature* 2011, *472*, 334.

77. H. Qian, D.-S. Guo, Y. Liu, *Asian J. Org. Chem.* 2012, *1*, 155.
78. X. Yao, T. Li, S. Wang, X. Ma, H. Tian, *Chem. Commun.* 2014, *50*, 7166.
79. R. Sun, C. Xue, X. Ma, M. Gao, H. Tian, Q. Li, *J. Am. Chem. Soc.* 2013, *135*, 5990.
80. Q. Zhang, X. Yao, D.-H. Qu, X. Ma, *Chem. Commun.* 2014, *50*, 1567.
81. F. Biedermann, E. Elmalem, I. Ghosh, W. M. Nau, O. A. Scherman, *Angew. Chem. Int. Ed.* 2012, *51*, 7739.
82. R. N. Dsouza, U. Pischel, W. M. Nau, *Chem. Rev.* 2011, *111*, 7941.
83. D.-S. Guo, Y. Liu, *Chem. Soc. Rev.* 2012, *41*, 5907.
84. D.-S. Guo, Y. Liu, *Acc. Chem. Res.* 2014, *47*, 1925.
85. L. D. Costanzo, S. Geremia, L. Randaccio, R. Purrello, R. Lauceri, D. Sciotto, F. G. Gulino, V. Pavone, *Angew. Chem. Int. Ed.* 2001, *40*, 4245.
86. G. Moschetto, R. Lauceri, F. G. Gulino, D. Sciotto, R. Purrello, *J. Am. Chem. Soc.* 2002, *124*, 14536.
87. F. G. Gulino, R. Lauceri, L. Frish, T. Evan-Salem, Y. Cohen, R. D. Zorzi, S. Geremia, L. D. Costanzo, L. Randaccio, D. Sciotto, R. Purrello, *Chem. Eur. J.* 2006, *12*, 2722.
88. A. D'Urso, M. E. Fragalà, R. Purrello, *Chem. Commun.* 2012, *48*, 8165.
89. D.-S. Guo, B.-P. Jiang, X. Wang, Y. Liu, *Org. Biomol. Chem.* 2012, *10*, 720.
90. N. Basílio, Á. Piñeiro, J. P. D. Silva, L. García-Río, *J. Org. Chem.* 2013, *78*, 9113.
91. V. Lau, B. Heyne, *Chem. Commun.* 2010, *46*, 3595.
92. O. Varga, M. Kubinyi, T. Vidóczy, P. Baranyai, I. Bitter, M. Kállay, *J. Photochem. Photobiol. A Chem.* 2009, *207*, 167.
93. M. Megyesi, L. Biczók, *J. Phys. Chem. B* 2010, *114*, 2814.
94. K. Wang, D.-S. Guo, Y. Liu, *Chem. Eur. J.* 2010, *16*, 8006.
95. K. Wang, D.-S. Guo, Y. Liu, *Chem. Eur. J.* 2012, *18*, 8758.
96. B.-P. Jiang, D.-S. Guo, Y.-C. Liu, K.-P. Wang, Y. Liu, *ACS Nano* 2014, *8*, 1609.
97. N. Basilio, L. García-Río, *Chem. Eur. J.* 2009, *15*, 9315.
98. V. Francisco, N. Basilio, L. García-Río, J. R. Leis, E. F. Maques, C. Vázquez-Vázquez, *Chem. Commun.* 2010, *46*, 6551.
99. N. Basilio, B. Gómez, L. García-Río, V. Francisco, *Chem. Eur. J.* 2013, *19*, 4570.
100. D.-S. Guo, K. Wang, Y.-X. Wang, Y. Liu, *J. Am. Chem. Soc.* 2012, *134*, 10244.
101. N. Basilio, M. Martín-Pastor, L. García-Río, *Langmuir* 2012, *28*, 6561.
102. V. Wintgens, C. L. Coeur, C. Amiel, J.-M. Guigner, J. G. Harangozó, Z. Miskolczy, L. Biczók, *Langmuir* 2013, *29*, 7682.
103. K. Wang, D.-S. Guo, X. Wang, Y. Liu, *ACS Nano* 2011, *5*, 2880.
104. Z. Q. Li, C. X. Hu, Y. Q. Chen, H. Xu, X. L. Cao, X. W. Song, H. Y. Zhang, Y. Liu, *Sci. China Chem.* 2012, *55*, 2063.
105. Y. Cao, Y.-X. Wang, D.-S. Guo, Y. Liu, *Sci. China Chem.* 2014, *57*, 371.
106. Z. Qin, D.-S. Guo, X.-N. Gao, Y. Liu, *Soft Matter* 2014, *10*, 2253–2263.
107. S. Peng, K. Wang, D.-S. Guo, Y. Liu, *Soft Matter* 2015, *11*, 290.
108. N. Basilio, V. Francisco, L. García-Río, *Int. J. Mol. Sci.* 2013, *14*, 3140.
109. M. Rehm, M. Frank, J. Schatz, *Tetrahedron Lett.* 2009, *50*, 93.
110. C. Wang, Z. Wang, X. Zhang, *Acc. Chem. Res.* 2012, *45*, 608.
111. K. Goto, Y. Yano, E. Okada, C.-W. Liu, K. Yamamoto, R. Ueoka, *J. Org. Chem.* 2003, *68*, 865.

16

SYNTHESIS, DESIGN, CHARACTERIZATION, AND APPLICATION OF METALLO-SUPRAMOLECULAR POLYMERS

LIPENG HE AND WEIFENG BU

Key Laboratory of Nonferrous Metals Chemistry and Resources Utilization of Gansu Province, State Key Laboratory of Applied Organic Chemistry, College of Chemistry and Chemical Engineering, Lanzhou University, Lanzhou, China

16.1 INTRODUCTION

Through non-covalent forces, such as host–guest recognition, hydrogen bonding, metal–ligand coordination, π–π stacking, electrostatic, and charge transfer interactions, small building blocks can rapidly and easily self-assemble into supramolecular polymers (SPs) [1–6]. Due to their multiple responsiveness to external stimuli such as pH values, temperature, ions, photoirradiation, and solvent compositions, the assembly and disassembly of SPs occur reversibly, which endow SP materials with superb self-healing, shape memory, and degradability [1–6].

Metallo-supramolecular chemistry was introduced by Constable in 1994 and has been reviewed recently by Ward and Raithby in the context of self-assembly techniques to meet functionalization/utilization criteria [7, 8]. Metal ions possess a variety of properties useful in electronics, optics, magnetics, and catalysis. The incorporation of metal centers into polymeric structures can form diverse metallo-supramolecular polymers (MSPs). Metal ions have continuously varied from transition group and main group metals to lanthanides and a variety of ligands have been synthesized. Accordingly, the coordination interactions vary from strong and essentially covalent bonds that lead to irreversible or static binding between the metal ions and ligands to weak, labile, and non-covalent bonds that endow MSPs with reversible and dynamic properties. In the former case, MSPs are as robust as covalent polymers with really high molecular weights from 10 to 100 thousands. This is very different from other kinds of SPs. In the latter case, however, the dynamic nature leads to the formation of self-healable and degradable MSPs under some specific conditions. Furthermore, these metal-based SPs own both versatile functions of metal components and good processing ability of polymers [9–15]. Fascinating functions arise from metal ion centers such as luminescence, electro or chemical redox, catalytic, and magnetic properties. The metal-based properties together with the SP features will benefit their wide applications in the fields of optoelectronics, self-healing materials, and biomedicine. Therefore, MSPs have attracted more and more widespread and expanding interests in academic and industrial circles. Recent studies focus more on new topological structures and properties. Now, MSPs have been developed from one-dimension to two- and three-dimensions. Moreover, the correlation between coordination and other non-covalent interactions has endowed MSPs with new features like hierarchical self-assembly and multiresponsiveness [16–19].

In this chapter, we will introduce the basic synthetic strategies toward MSPs. New topological structures of MSPs and orthogonal MSPs will be discussed in detail. In addition, we highlight the potential applications of MSPs in the fields of materials science and biomedicine.

16.2 SYNTHESIS OF MSPs

MSPs have attracted great interest due to their combination of polymer features together with optical, electronic, magnetic, and catalytic potentials of metal ions. Usually,

Non-Covalent Interactions in the Synthesis and Design of New Compounds, First Edition.
Edited by Abel M. Maharramov, Kamran T. Mahmudov, Maximilian N. Kopylovich, and Armando J. L. Pombeiro.

TABLE 16.1 Binding Constants of Zn^{II}, Fe^{II}, and Co^{II} Complexes with Different Types of Polypyridine Ligands

Ligand	Name[a]	Metal Ion	K	Solvent
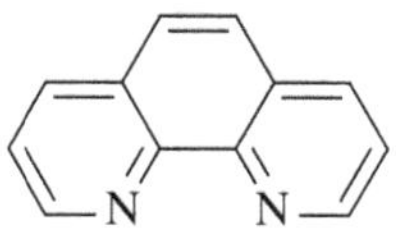	Phen	Zn^{II}	$\sim 10^{13}\,M^{-3}$	H_2O
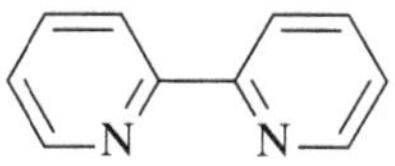	BPY	Zn^{II}	$\sim 10^{17}\,M^{-3}$	H_2O
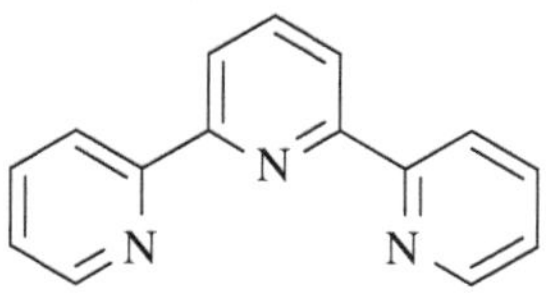	TPY	Zn^{II} Fe^{II}	$2.0 \times 10^{14}\,M^{-2}$ $\sim 10^{21}\,M^{-2}$	CH_3CN
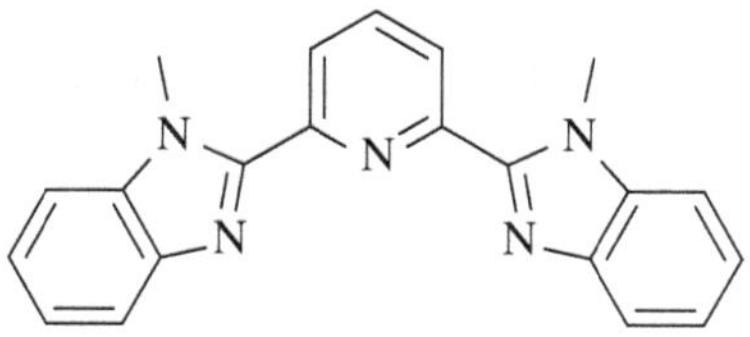	BIP	Zn^{II} Fe^{II}	$\sim 10^{6}\,M^{-2}$ $\sim 10^{14}\,M^{-2}$	$CHCl_3/CH_3CN$

[a] The abbreviations of the ligands.

MSPs can be classified into two main species: (1) one-dimensional linear polymers and (2) multidimensional polymers such as star-shaped, hyperbranched, and cross-linked polymers. In this section, we primarily focus on the basic synthetic strategies toward one-dimensional MSPs.

Due to the high diversity of ligands and metal ions leading to various binding strength and coordination mode, the structures and properties of MSPs can be adjusted widely. For example, the coordination of Ru^{II} and 2,2′:6′,2″-terpyridine (TPY) ligand, as a result of the formation of almost covalent bonds between them, is irreversible and nondynamic in nature. Nevertheless, lanthanide complexes with 2,6-bis(*N*-methylbenzimidazolyl)pyridine (BIP) are really reversible and dynamic, which can readily respond to temperature, pH, and ions. With the development of MSPs, a great variety of ligands are synthesized. Among these ligands, pyridine derivatives (monodentate and multidentate ligands) are especially popular for their high binding affinity with transition metals and lanthanide cations. Here, we pay more attention to this type of ligands.

The main difference between covalent polymers and SPs is that SPs constructed by weak non-covalent interactions are dynamic and reversible. Therewith, experimental conditions such as solvents, temperature, and concentrations have great influence on the polymerization of SPs. For the reversible SPs, Equation 16.1 can be used to estimate the value of degree of polymerization (DP) [2, 20, 21]:

$$\mathrm{DP} \approx \left(K\left[M\right]\right)^{1/2} \tag{16.1}$$

where K is the binding constant and M is the monomer concentration.

According to this equation, to obtain SPs with a high DP, large binding constants and high concentrations are of the essence. As shown in Table 16.1, Zhang and coworkers summarize binding constants of Zn^{II}, Fe^{II}, and Co^{II} complexes with different types of pyridine ligands in specific solvents [19]. According to different pyridine derivatives and coordination ions, K varies from $10^{6}\,M^{-1}$ to $10^{21}\,M^{-2}$. Generally, the higher binding constant is, the larger the DP is under the specific concentrations. Through designing the rational ligands and choosing the reasonable metal ions, scientists can construct their expected MSPs. In addition, the DP of MSPs can be adjusted by the addition of monofunctional chain stoppers. Usually, the higher concentration of chain stoppers leads to a lower DP of the resulting MSPs.

16.2.1 Basic Synthetic Strategies Toward Linear MSPs

Generally, metal complex moieties can be attached to polymers as a part of main or side chains or end groups (Scheme 16.1) [22]. The spacer blocks in the main and side chains are either conjugated or nonconjugated groups, which are small molecule or polymer groups. Type **I** possesses the metal ions as a part of the main chains by (a) metal–ligand coordination or (b) covalent interaction, which is the most facile and classical MSPs. In detail, type **I** of MSPs is divided into two classes as follows. In type **Ia** of MSPs, ligands alternate with metal ions to form a main chain. Of difference is that in type **Ib**, metal ions are directly attached to a polymer main chain without any other side groups. In type **II**, metal ions or their complexes are attached to a polymer main chain at the sides by coordination or electrostatic interactions. Through similar coordination or electrostatic interactions, metallo-supramolecular complexes are linked to the ends of a polymer chain, leading to the formation of type **III** of MSPs. Depending on the designed structures and properties and the final applications, one can choose a suitable strategy to synthesize the fascinating MSPs.

Three main routes are widely applied to synthesize type **I** MSPs (Scheme 16.2). In the route **Ia**1, a ditopic ligand is synthesized at first and then the coordination self-assembly of the ligand and metal ion leads to the formation of type **Ia** of MSPs. This method is very simple and facile and thus is applied frequently to fabricate linear MSPs. In the route **Ia**2, metallo-supramolecular complexes are polymerized by covalent interactions, such as C–C cross-coupling and/or click reactions. Therefore, it is very easy to obtain the MSPs bearing different metal ions through the route **Ia**2. However, the metal complex must be stable under the reaction conditions in this method. The molecular weight of the type **Ia** of MSPs synthesized through the route **Ia**2 can be estimated by the Carothers equation [23]. Type **Ib** MSPs are synthesized by direct coordination reactions of metal ions with covalent polymers containing coordination sites in the main chains. By controlling the ratio of metal ions with coordination sites, scientists can regulate facilely the photophysical and redox properties of the resulting MSPs.

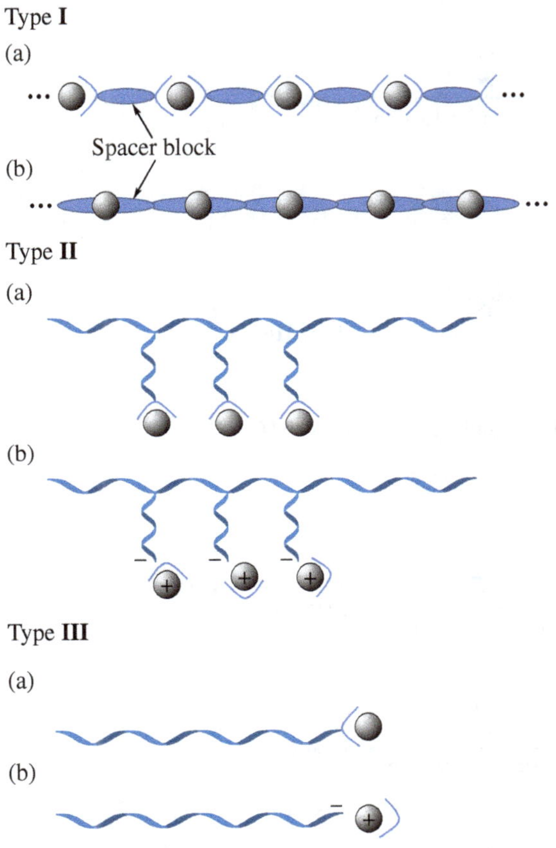

SCHEME 16.1 Schematic representation of the topological structures of one-dimensional linear MSPs.

In order to synthesize type **II** of MSPs, four routes are widely applied (Scheme 16.3). Metal complex monomers can be polymerized to form MSPs with the metal complex in side chain (route **IIa**1). In this way, the monomer must be stable under the polymerization conditions and neither uncoordinated monomers nor free metal ions exist after the polymerization. For the route **IIa**2, type **II** of MSPs was synthesized by covalently grafting metal complexes onto a polymer. Similarly, this procedure also requires the stability of the metal complexes under the reaction condition. The above grafted reaction is usually incomplete. However, the unreacted sites of the polymer can be further used to synthesize MSPs bearing different metal complexes. Compared with the MSPs with single metal ions, the MSPs containing various metal ions have some extra properties and potential applications as emissive and photovoltaic materials. In the route **IIa**3, ligands, as the side chains, are attached to a polymer. The resulting macromolecular ligands can coordinate with metal ions, leading to the formation of type **II** MSPs. Similar to the route **IIa**2, the unreacted ligands are still present in the final MSPs and the MSPs containing various metal ions can be constructed expectedly. Type **IIb** of MSPs is synthesized through facile ion exchange of an ionic polymer with oppositely charged metal complexes, usually at room temperature.

Through coordination, covalent, or electrostatic interactions, type **III** of MSPs can be similarly synthesized by the

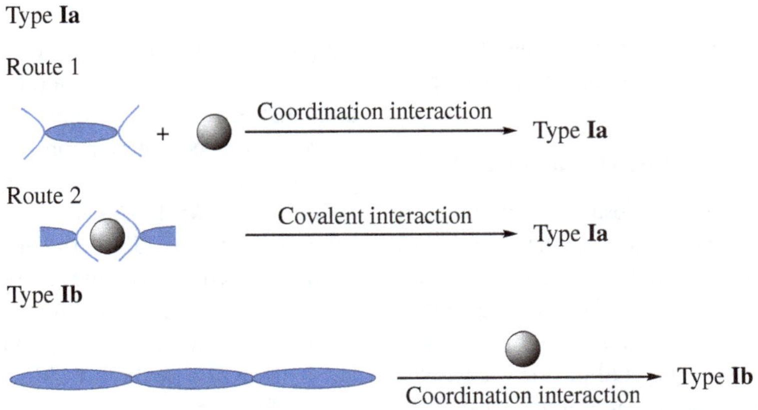

SCHEME 16.2 Schematic representation of the synthesis routes of type **I**.

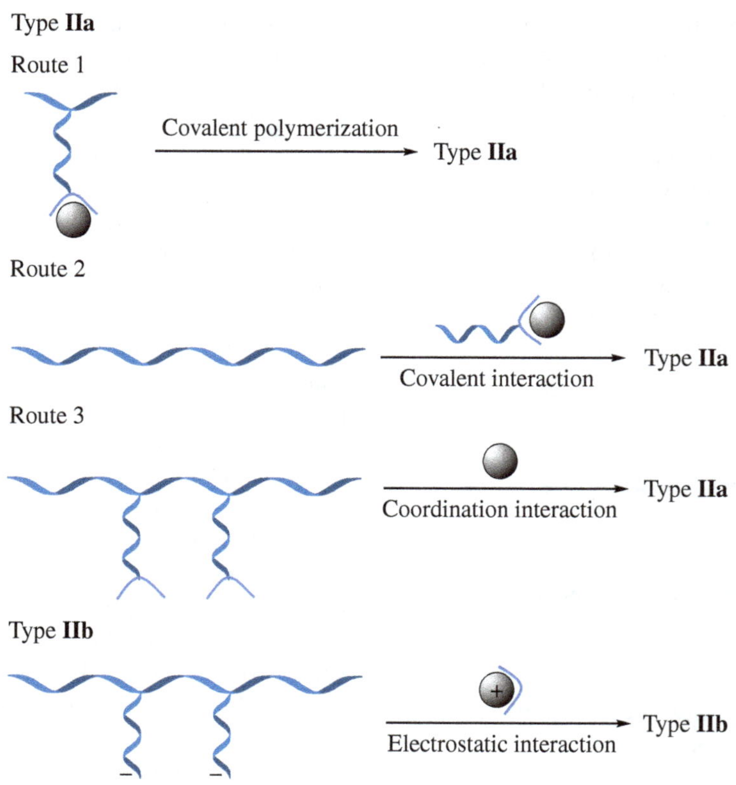

SCHEME 16.3 Schematic representation of the synthesis routes of type **II**.

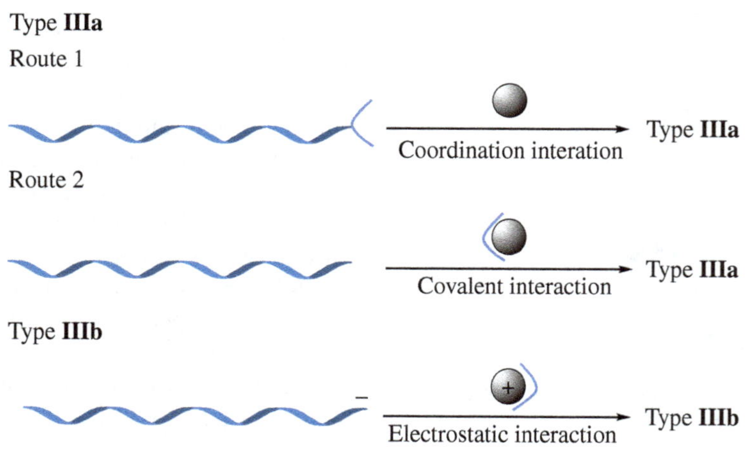

SCHEME 16.4 Schematic representation of the synthesis routes of type **III**.

methods as described for types **I** and **II** (Scheme 16.4). In this section, we will not discuss the routes any more in detail.

16.2.2 Linear MSPs Based on Polypyridine Ligands

At the early stage, MSPs were decomposable or oligomeric in solution. The original attempt starts from the complexation of monodentate pyridine and metal ion. However, the interaction between them is so weak that the resulting MSPs do not process typical polymer features. In 1990, Serrano *et al.* reported a series of liquid crystalline MSPs, which were synthesized by the condensation of a Schiff base–Cu^{II} complex and acid dichlorides [24]. The resulting MSPs were soluble in chlorinated solvents. However, the molecular weight obtained from vapor pressure osmometry (VPO) was very low, corresponding to a relatively low DP (5 ~ 6). Soon after that, Eisenbach and Schubert fabricated a MSP by the coordination polymerization of bipyridine (BPY)-ended polyether and Cu^{I} ions [25]. Compared with the pure ligand solution, the MSP showed a significant increase in the solution viscosity. Gel permeation chromatography (GPC) results showed a three times increase of the molar mass after the complexation. Although these early reports had some defects, they initiated a new polymer system and brought up the construction methods of MSPs.

The first well-defined example of MSPs was reported by Rehahn *et al.* (Scheme 16.5) [26]. A ligand monomer, ditopic phenanthroline (**1**), coordinated with 1.0 equivalent of metal

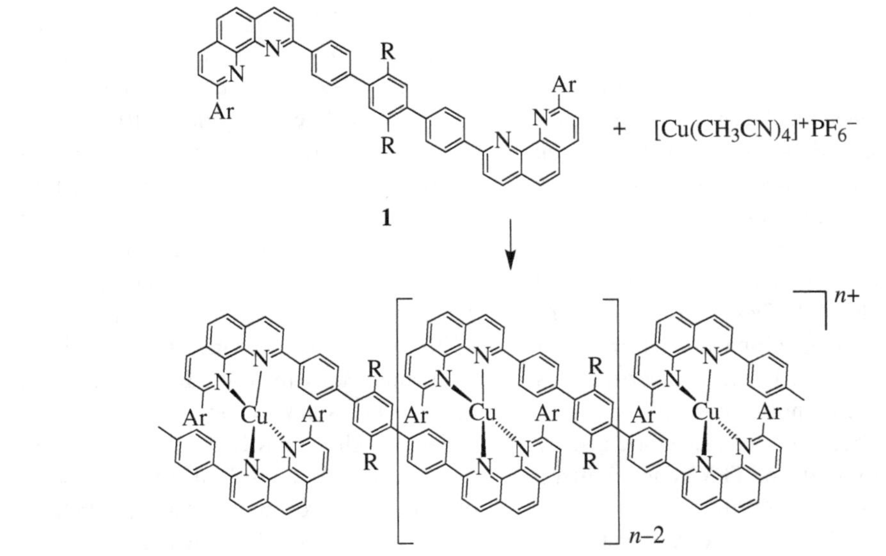

SCHEME 16.5 Schematic representation of the formation of MSP based on a ditopic phenanthroline ligand [26].

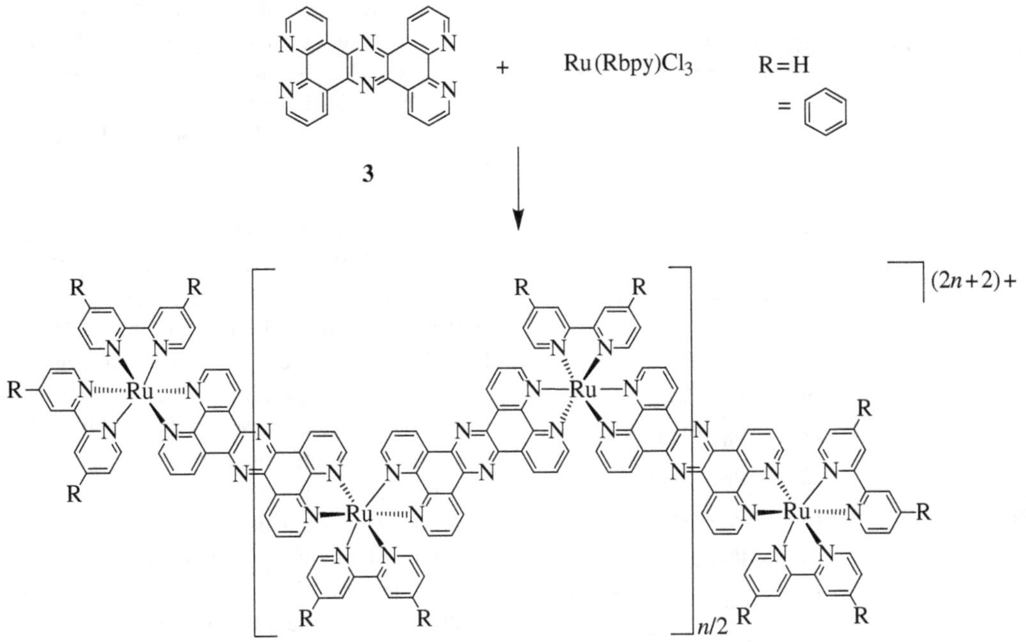

SCHEME 16.6 Schematic representation of the formation of MSPs based on a tetrapyridophenazine and BPY [29].

ion $[Cu(MeCN)_4]PF_6$, leading to the formation of a MSP (**2**). The broadened ^1H NMR signals confirmed the successful polymerization. The lack of end group absorptions hinted that the MSP had a high DP of greater than 30. The resulting MSP **2** was inert and soluble in noncoordinating solvents such as tetrachloroethane. However, due to the exchange reaction with other competitive ligands or solvents, the MSP readily decomposed to oligomers. Besides this, Rehahn and coworkers also reported a series of other MSPs containing different metal ions, such as Cu^I and Ag^I, and different phenanthroline-based ligands [27, 28]. Similarly, these MSPs also displayed dynamic and reversible features in coordinating solvents and are stable in noncoordinating solvents.

To fabricate more stable and completely irreversible MSPs, Rehahn *et al.* introduced Ru^{II} ions into MSP systems for its high binding constants (Scheme 16.6) [29, 30]. They combined a bitopic ligand tetrapyridophenazine (**3**) with a 2,2′-bipyridine Ru^{II} complex ($Ru(BPY)Cl_3$) to construct the MSPs (**4**). The formed MSPs had a rigid main chain and high overall charges. Moreover, they were soluble both in polar organic solvents and water. When the substituted group R of

BPY moiety was phenyl group, the molecular weight could reach a fairly high value of 47 kDa.

2,2′:6′2″-Terpyridine (TPY) and its derivatives have always received wide attention since the MSP system is presented. A variety of transition and lanthanide metal ions can coordinate with them to fabricate MSPs. The difference is that the complexation of TPY-based ligands with transition metal ions can form more stable MSPs. Furthermore, the 4′-position of TPY (**5**) is facile to be modified and the usual substitutions are listed in Table 16.2. These substituted TPYs link with flexible or rigid π-conjugated spacers to form di- or polytopic ligands, which can further fabricate diverse MSPs after the complexation with different metal ions.

In 1995, Constable *et al.* reported a MSP fabricated by adding metal ions into the solutions of TPY-functionalized monomers [31]. Soon after that, Kurth and coworkers prepared this kind of MSP (**7**) through the complexation of 1,4-bis-(2,2′:6′,2″-terpyridine-4′-yl)benzene (**6**) and Fe^{II} ion (Fig. 16.1) [32, 33]. The molecular weight was determined by analytical ultracentrifugation. According to the Swedberg formula, the molecular weight was roughly estimated at 14.9 kDa, which was consistent with 25 repeating units. Through electron diffraction in combination with molecular modeling, they obtained a crystal structure of **7**, which evidenced the linear polymeric structure [32].

TABLE 16.2 Important 4′-Substituted TPY

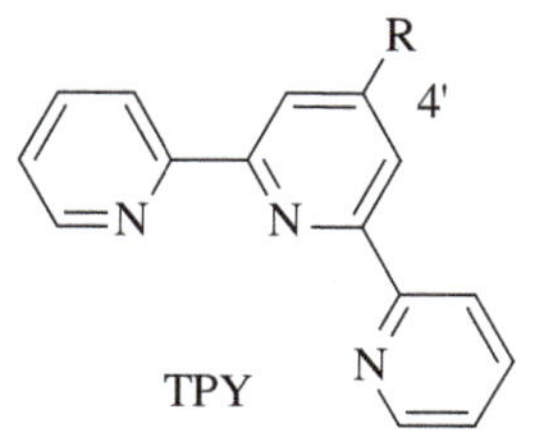

	R		R
1	–OH	**9**	–Ph–NH_2[a]
2	–Cl	**10**	–Ph–CHO[a]
3	–OTf	**11**	–Ph–CH_2CN[a]
4	4-N_3	**12**	–Ph–CH=CH_2[a]
5	–Ph–Br[a]	**13**	–Ph–CH_2Br[a]
6	–Ph–$B(OH)_2$[a]	**14**	–Ph–$CH_2PPh_3^+Br^-$[a]
7	–Ph–I[a]	**15**	–Ph–$CH_2PO(OEt)_2$[a]
8	–Ph–C≡CH[a]	**16**	–Ph–$Sn(n\text{-}Bu)_3$[a]

[a] Ph is the abbreviation of phenyl group.

Rehahn and coworkers fabricated a Ru^{II}–TPY MSP (**10**) through two routes: (**Ia**1) the coordination assembly of a bitopic TPY-functionalized ligand (**8**) and a Ru^{II} ion and (**Ia**2) the Suzuki cross-coupling polymerization of a bitopic bromine-substituted $Ru(TPY)_2$ complex (**9**) and a 1,4-phenylenediboronic acid derivative (Scheme 16.7) [34]. ^{1}H NMR, ^{13}C NMR, and viscosity measurements indicated that the DP of **10** obtained by the route **Ia**1 exceeded 30 but obtained by the route **Ia**2 was very low.

Early studies of MSPs paid much attention to the rigid conformation. As a result, these MSPs had poor solubility in solvents. To overcome this defect, Schubert *et al.* introduced flexible chains into TPY ligands (Scheme 16.8, **11–13**) to construct a series of MSPs [35–38]. Such MSPs were not only soluble in water but also owned typical polymer properties. These MSPs could decompose into oligomers or monomers upon acid, base, electrochemical, or thermal treatments. The reversibility of metal–ligand bonds was further studied by the addition of the strong ligand (*N*-(2-hydroxyethyl)-ethylenediamine-*N*,*N*′,*N*′-triacetic acid, HEDTA) in aqueous solution. Upon addition of $FeCl_2$, the MSPs formed again. The reversibility of MSPs based on the TPY–Fe^{II} complex system was also demonstrated by the groups of Abruña, Kimura, Craig, and Meijer [39–42].

Recently, BIP-based ligands were also used to successfully construct MSPs. BIP was attached onto the ends of different flexible or rigid macromolecular spacers to obtain a series of BIP-based ligands **14** and **15** (Scheme 16.9) [43–45]. These ligands could coordinate with metal ions such as Fe^{II}, Co^{II}, Zn^{II}, and Cd^{II} ions, leading to the formation of MSPs with typical polymer features. When the BIP ligand was modified at the ends of a conjugated spacer (**14**), like poly(*p*-phenylene ethynylene), the coordination self-assembly of **14** with Zn^{II} and Cd^{II} ions could be used to develop light-emitting MSP materials. Utilizing poly(tetrahydrofuran) (M_n=2900) as a spacer (**15**), the formed MSP could be easily prepared as mechanically stable films. The films exhibited a thermoplastic elastomeric behavior, in which the metal complexes and polymer segments were microphase separated.

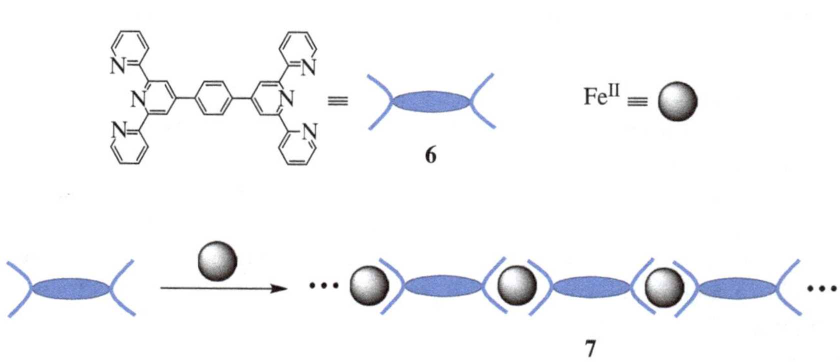

FIGURE 16.1 Metal ion-induced self-assembly of Fe^{II} and 1,4-bis(2,2′:6′,2″-terpyridine-4-yl)benzene results in the formation of a MSP [32].

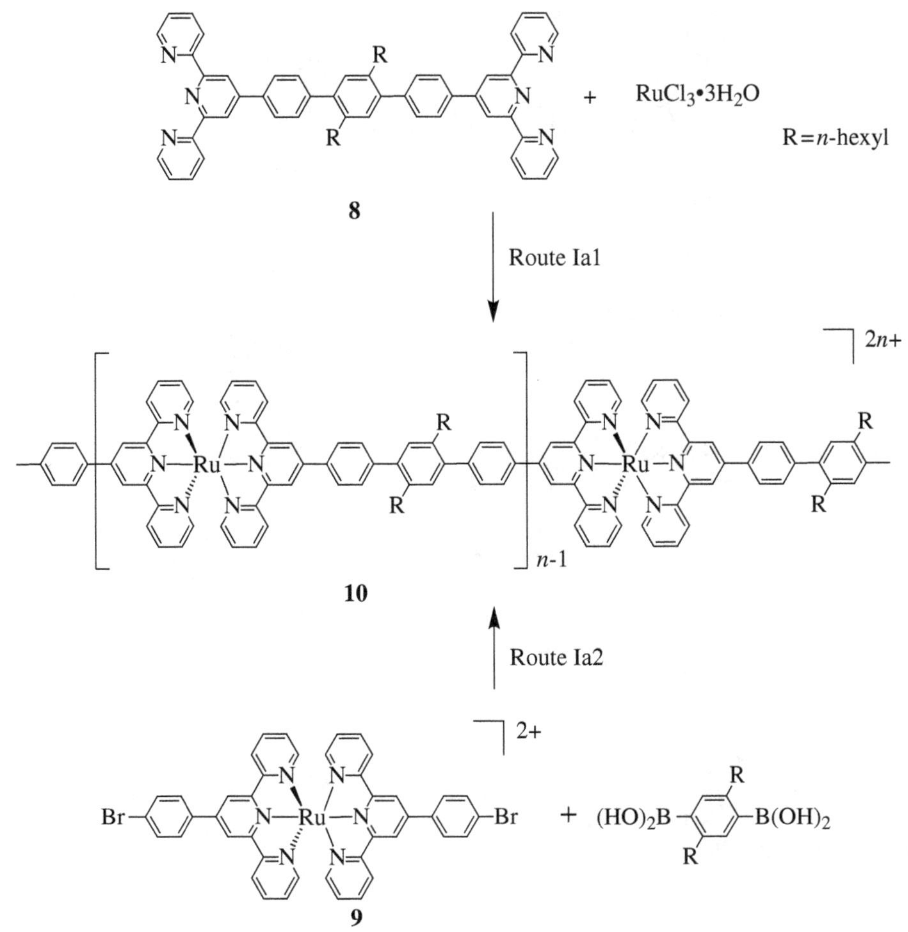

SCHEME 16.7 A linear Ru^{II} MSP fabricated through different routes [34].

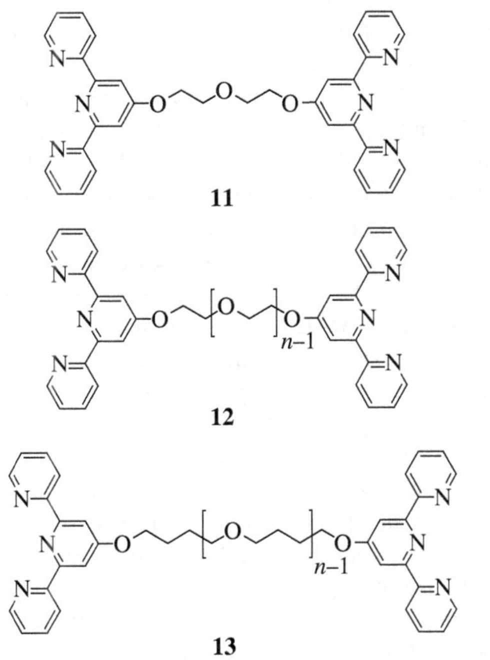

SCHEME 16.8 Schematic representation of water-soluble bis-terpyridine. The two ligands were linked by ethylene oxide, poly(ethylene oxide), and poly(oxytetramethylene), respectively.

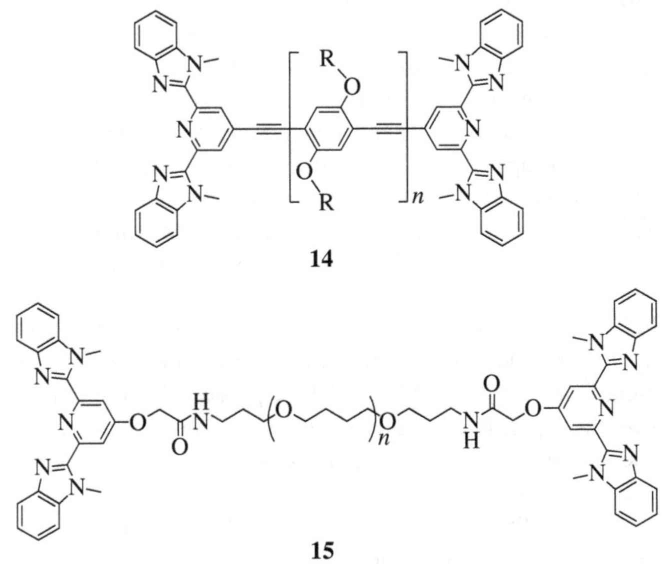

SCHEME 16.9 Schematic representation of rigid and flexible bis-BIP. The two BIPs were linked by poly(*p*-phenylene ethynylene) and poly(tetrahydrofuran), respectively.

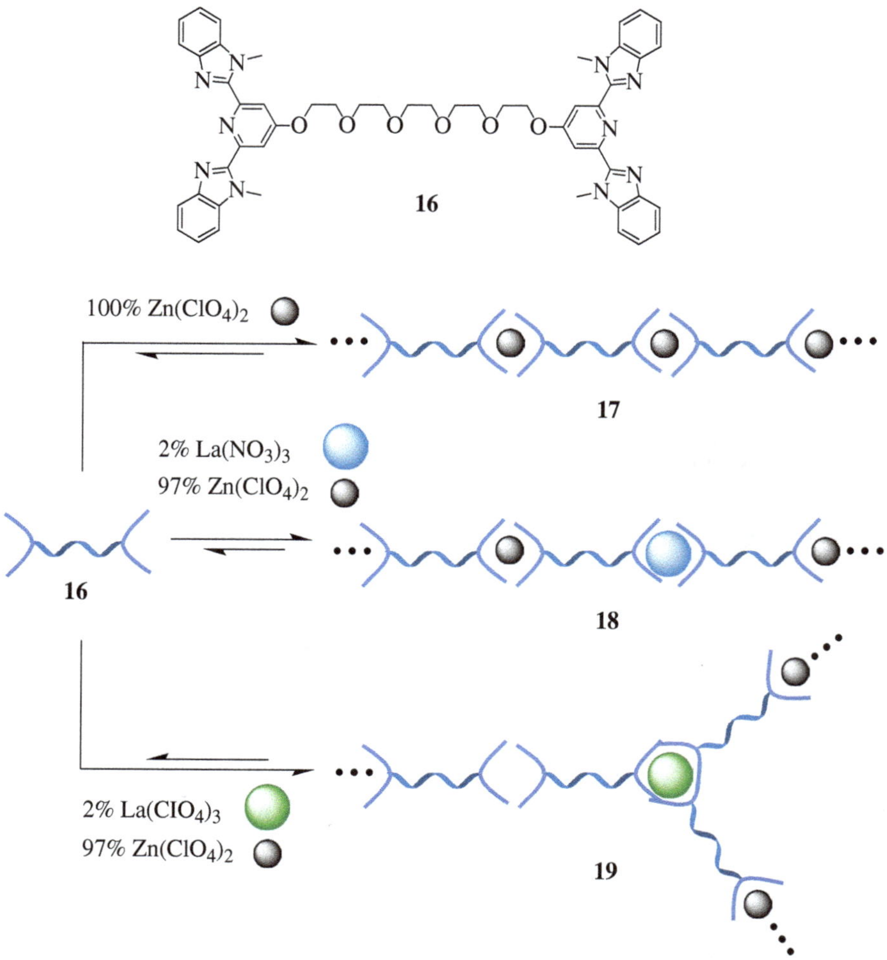

FIGURE 16.2 Formation of MSPs based on the coordination of a ditopic BIP ligand with metal salts in acetonitrile [48].

Treating the solution of oligo(ethylene glycol)-linked bis-BIP monomer with Zn^{II} ion, the reduced viscosity exhibited a steady increase, indicating that a linear MSP is formed [46]. When the Zn^{II}/ligand ratio exceeded 1 : 1, the reduced viscosity started to decrease, suggesting the depolymerization of the MSP occurred. Other transition metal ions, like Fe^{II}, Co^{II}, and Cd^{II}, were also used to fabricate BIP-based MSPs [47]. Different metal ions endowed the MSPs with different properties. Thermal measurements suggested that MSPs with weaker binding, more labile metal ions (e.g., Zn^{II}, Cd^{II}) showed less thermal stability than those with stronger binding, less labile metal ions (e.g., Fe^{II}, Co^{II}). Rowan *et al.* demonstrated that a penta(ethylene glycol)-linked bis-BIP (**16**) could form metallo-supramolecular gels upon the addition of transition metal and lanthanide metal ions (Fig. 16.2, **17–19**). These gels exhibited multiple stimuli-responsive behaviors, including thermo-, optico-, chemo-, and mechanical responses, as well as light-emitting and shape-memory properties [48].

Besides the coordinated self-assembly of ligands and metal ions, the facile electrostatic combination of ionic polymers with oppositely charged metal complexes also can result in the formation of MSPs. Such MSPs not only inherit the polymer nature but also own the photophysical and redox properties of the metal complexes. In 2011, Bu *et al.* reported a series of MSPs fabricated by the electrostatic complexation of block copolymers poly(ethylene oxide-*b*-acrylate sodium) (E_{251}-*b*-A_{51}, **22**) and poly(styrene-*b*-acrylate sodium) (S_{586}-*b*-A_{55}, **23**) and the platinum(II) complexes (**20** and **21**) with N^N^N ligands (BIP and TPY) (Fig. 16.3) [49]. UV–Vis absorption and luminescence titration experiments were performed to monitor the electrostatic self-assembly process. When the mixing fraction between **20** and the carboxylate group of E_{250}-*b*-A_{51} was 1 : 1, the metal–metal-to-ligand charge transfer (MMLCT) absorbance at 545 nm and ^{3}MMLCT luminescence intensity at 605 nm reached their maximum values, respectively, because of an intermolecular association through Pt···Pt and π–π stacking interactions. However, when E_{250}-*b*-A_{51} was titrated with **21**, the MMLCT absorbance at 470 nm and ^{3}MMLCT luminescence intensity at 615 nm reached their maximum values at a molar ratio of close to 2 : 1 between the complex and carboxylate group. In the latter case, it is proposed that an acrylate ligand of A_{51} would replace the coordinated Cl^- in $[Pt(TPY)Cl]^+$ **21** and another acrylate acted as a counteranion. Dynamic light scattering (DLS) measurement indicated that the aggregates were present after the electrostatic self-assembly. Further evidence for the aggregates was revealed by bright-field

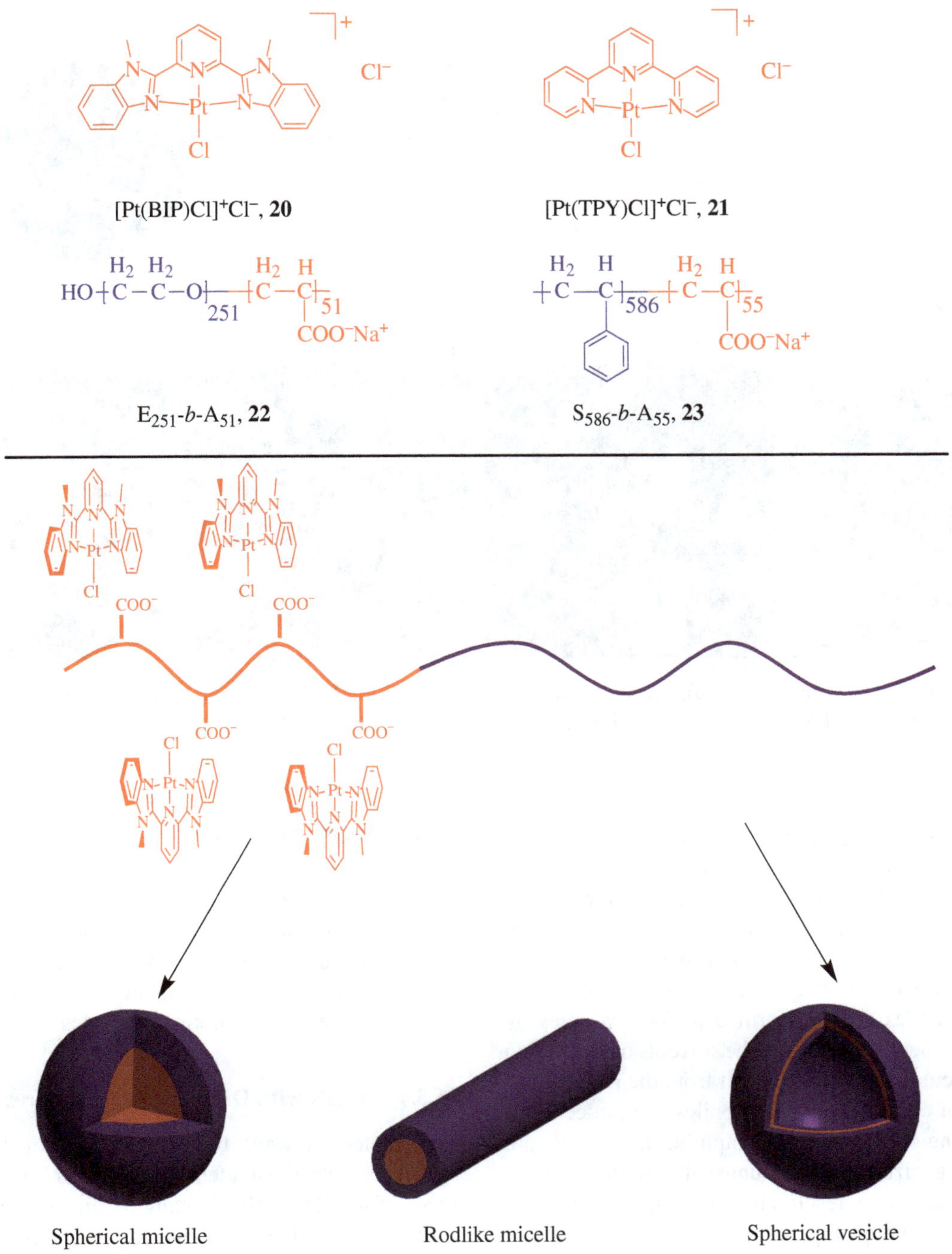

FIGURE 16.3 Structural formulas of block copolymers and platinum(II) complexes and schematic representation of their self-assembling morphologies in solvents [49].

transmission electron microscopy (BF-TEM) and high-angle annular dark-field scanning TEM (HAADF-STEM). The MSPs self-assembled into spherical and rodlike micelles and vesicles in solvents.

Afterward, utilizing the same strategy, the same group fabricated another MSP (**24**) (Fig. 16.4a) [50]. The MSP **24** aggregates into star micelles in chloroform but wormlike micelles together with a few spherical micelles in toluene. The star micelle has a small ionic core and an expanded polystyrene corona, similar to star polymers. This easily established model of star polymers was used to study their intra- and interstar interactions and self-assembled behavior in solvents of variable quality. The MSP was first dispersed in chloroform. When methanol, as a poor solvent, was directly added into the chloroform solution with a volume ratio of 33%, BF-TEM, HAADF-STEM, and scanning electron microscopy (SEM) images revealed the MSP formed vesicular aggregates (Fig. 16.4b–e). In these vesicles, the star micelles retained their core–shell nanostructures. The polystyrene corona shrank significantly, while the ionic core showed a slight compaction. Such compactness was assigned to the weakened solvent quality in the chloroform/methanol

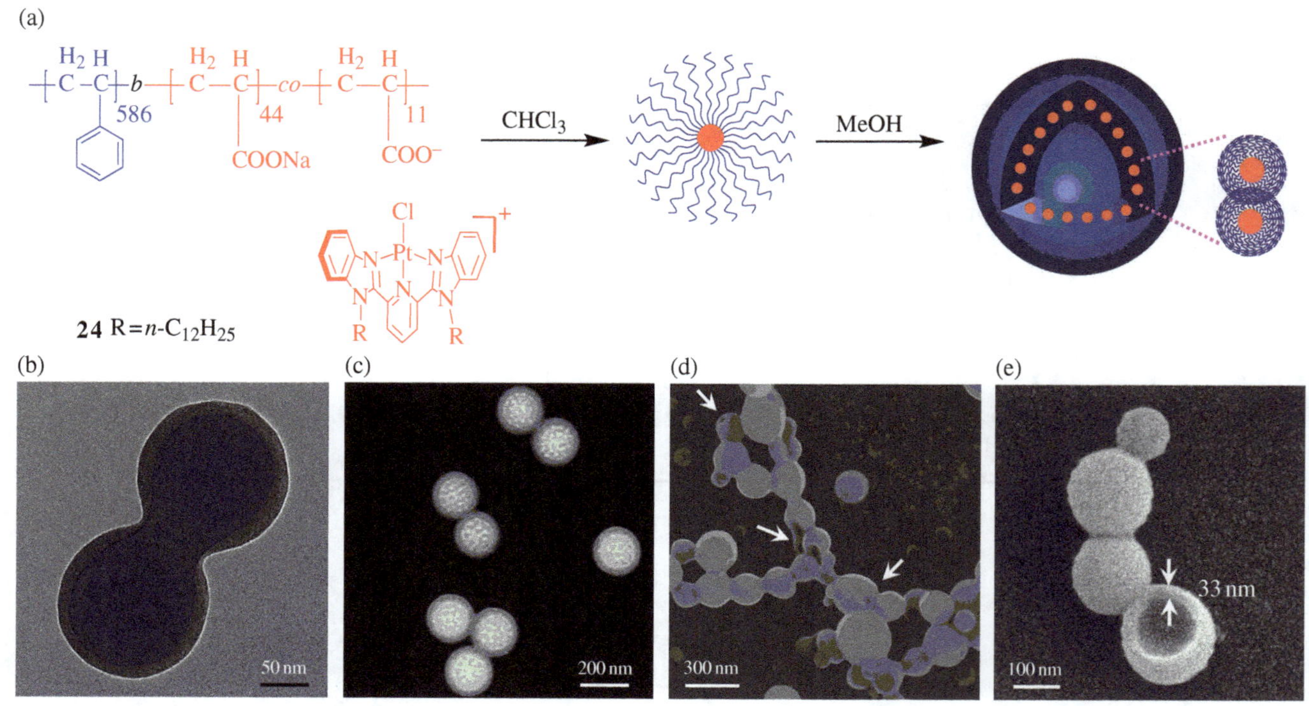

FIGURE 16.4 (a) Structural formula of MSP and schematic representation of its self-assembling morphologies from micellar to vesicular. (b) BF-TEM, (c) HAADF-STEM, and (d and e) SEM images of the vesicles [50]. Reprinted with permission. Copyright 2015 American Chemical Society.

mixture leading to the presence of intrastar micelle van der Waals attractive interactions between the segments of the polystyrene chains. The driving forces for the formation of vesicles were proposed to be a delicate balance of van der Waals attractions and entropic repulsions between MSP stars under the present solvent condition. This was the first time that the star micelles are transformed to form vesicles by packing the contractive micelles in the solvents of worsening quality. This picture is totally different from the morphological evolution of conventional amphiphiles from micellar to vesicular, during which the amphiphilic molecules are commonly reorganized. Such an unusual system offered a considerable potential to develop hierarchical self-assembly in solvents of variable quality.

In addition, Bu and coworkers reported a series of platinum(II) complex-terminated polystyrenes (**25–27**), which were obtained through the electrostatic self-assembly of **20** with various sulfonate-terminated polystyrenes (Fig. 16.5) [51]. The resulting MSPs self-assembled into vesicular structures in the chloroform/methanol (v/v = 1) mixed solvents, which further fused into large vesicles due to Pt···Pt and π–π stacking interactions between the platinum(II) complexes.

Besides the above-mentioned ligands, 2,6-di(pyrazol-1-yl) pyridine [52, 53] and pyridine-2,6-dicarboxylate [54, 55] have also been well documented to construct various MSPs. In the future, with the development of chemical science, more and more coordination systems will be introduced to enrich the types of MSPs.

16.3 DESIGN OF MSPs

In decades, MSPs have experienced a large development in terms of their syntheses; however, science will not stop and move forward. The design of MSPs bearing novel architectures and multiple properties is imperative. According to the different targets, two different strategies are brought up.

16.3.1 MSPs with Different Architectures

The strategy toward the architecture-directed design of ligands is the earliest attempt, which has been demonstrated successfully. Through adjusting the ligand structures and/or choosing metal ions, star-shaped and hyperbranched MSPs and MSP networks can be constructed.

16.3.1.1 Star-Shaped MSPs Compared with linear polymers with the same molecular weights, star-shaped polymers have lower viscosity in solution and melt states due to their topologically branched structures. Usually, they are synthesized by controlled/living ionic and radical polymerization techniques, where the linear polymer arms are attached to the organic core by covalent bonds. The replacement of the covalent connection with non-covalent interactions can form supramolecular star-shaped polymers, which is an efficient approach to developing polymer materials with dynamic nature, controlled hierarchical nanostructures, and functional properties. The involved non-covalent interactions include hydrogen bonds, host–guest

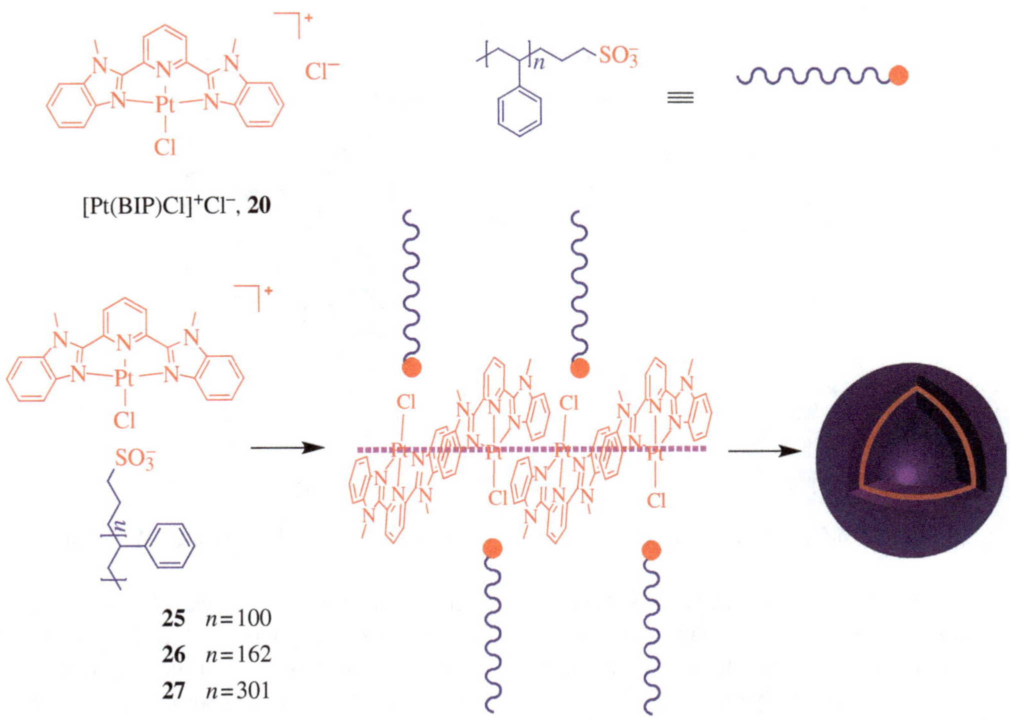

FIGURE 16.5 Structural formulas of MSPs and their self-assembly into vesicles in chloroform/methanol mixed solvents [51].

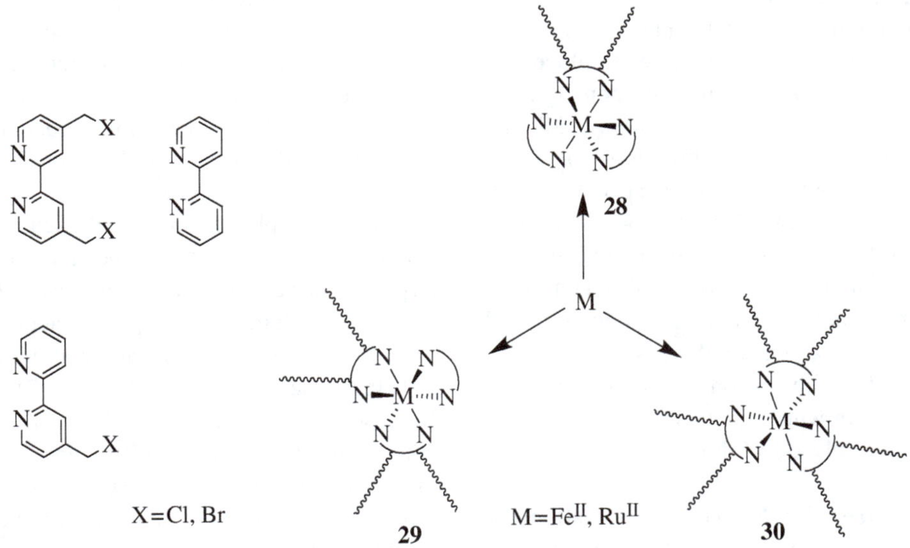

SCHEME 16.10 Schematic representation of metal-centered star-shaped MSPs based on the BPY–metal complexes.

recognition, metal–ligand coordination, and electrostatic forces. Metal-centered star-shaped polymer is always one of current research hotspots, in which the metal complex acts as the core and the polymer chains act as the arms grafted on the complex.

BPY can form a tris-BPY complex with various transition metal ions and its 4- and/or 4′-position can be functionalized easily. So it is widely used as a core of star-shaped MSPs. Through reasonable choice of BPY-based ligands, the arm numbers of the star-shaped polymers can be precisely adjusted (Scheme 16.10, **28–30**) [56–62]. In 1997, Fraser *et al.* presented a pioneering route to synthesize metal-centered starlike polymer. The 4- and 4′-positions of BPY were prefunctionalized by halomethyl groups and then coordinated with Fe^{II} and Ru^{II} to form the metalloinitiators, which could initiate living oxazoline polymerizations. Well-defined star-shaped MSPs with low polydispersities were obtained. Interestingly, Fe^{II} and Ru^{II} initiators exhibited obviously different polymerization chemistry. Using the hexafunctional Fe^{II}-centered initiator $[Fe\{BPY(CH_2Cl)_2\}_3](PF_6)_2$, star-shaped polymers with high molecular weights (M_w=~120kDa, DP>200) and narrow polydispersities were obtained. Even higher M_w polymers but with broadening polydispersities could be achieved. In contrast, the

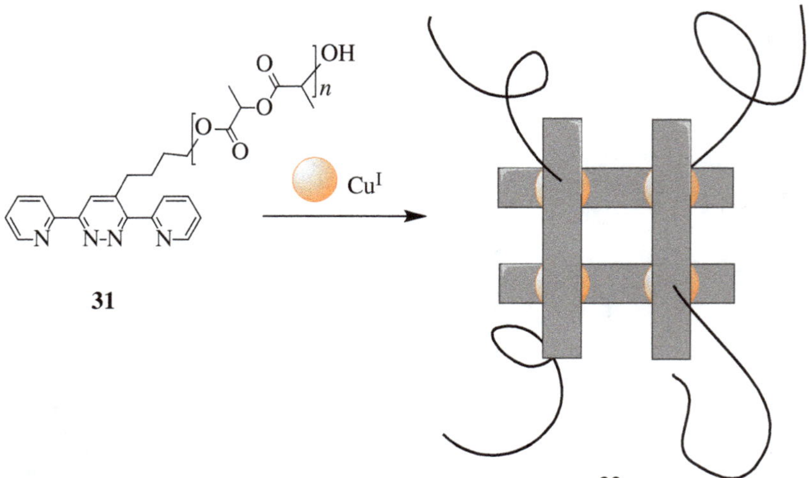

FIGURE 16.6 Schematic representation of formation of a grid-cored star-shaped MSP [63].

Ru^{II}-centered polymers had an upper M_w limit (~25 kDa) with narrow polydispersities. When using di-, tetra-, and hexafunctional Ru^{II}-centered initiators, star-shaped MSPs with different arm numbers were successfully obtained. Moreover, upon sequential addition of two different oxazoline monomers into the hexafunctional Fe^{II} initiators, amphiphilic Fe^{II}-centered star-shaped block copolymers with six arms were also prepared. In addition, the metal-centered complexes modified with functional groups could be employed to initiate other living polymerization such as atom transfer radical polymerization (ATRP) and reversible addition–fragmentation chain transfer polymerization (RAFT).

In 2003, Schubert *et al.* prepared a star-shaped MSP (**32**) through a different method (Fig. 16.6) [63, 64]. They first synthesized a hydroxy-functionalized 3,6-bis(2-pyridyl)pyridazine ligand. Subsequently, utilizing a controlled aluminum alkoxide-based polymerization, the ligand was polymerized with L-lactide. The complexation of the resulting poly(L-lactide) macroligands (**31**) with Cu^I ions led to the formation of a starlike polymer **32**, in which the metallo-grid was surrounded by four arms of poly(L-lactide).

16.3.1.2 Hyperbranched MSPs The combination of dendritic ligands with metal ions leads to the formation of hyperbranched MSPs with interesting redox, catalytic, and luminescence properties. Therefore, the hyperbranched MSPs have attracted an increasing interest. Through adjusting the ligand structures from the ditopic to polytopic and/or choosing metal ions, hyperbranched MSPs can be prepared purposively (Scheme 16.11).

In most cases, the polytopic ligands were built up from single ligands through covalent bonds. 4′-Substituted 2,2′:6′,2″-terpyridine is one of the most classical ligands for the construction of polytopic ligands. Many rigid, fully π-conjugated polytopic TPY ligands are utilized as the building blocks of hyperbranched MSPs. In 1992, Thompson *et al.* synthesized a rigid tris-TPY **33** with a phenyl moiety as a core [65]. Tris-TPYs **34** and **35**, which were analogues of **33**, were prepared by Suzuki cross-coupling reactions of functionalized 4′-phenyl-2,2′:6′,2″-terpyridine [66]. A structurally related derivative **36** bearing three U-shaped phenyl-substituted TPYs was reported later by Winter *et al.* [67]. Due to the rigid structures and the absence of any solubilizing side chains, the hyperbranched MSPs fabricated by ligands **33–36** with metal ions had a very low solubility in most organic solvents. He *et al.* reported a tris-TPY **37**, which could self-assemble into chiral architectures with Ag^I ions [68]. The symmetric tris-TPY **38** was obtained by Nishimori *et al.* by a Stille cross-coupling reaction [69]. Newkome's group reported a tetrakis-TPY-porphyrin derivative **39** as a potential core for the construction of hyperbranched MSPs containing porphyrin units [70]. Schmittel and coworkers synthesized a D_{6h} symmetric hexakis-TPY **40** [71, 72]. The terpyridyl arms were attached to a hexaethynylbenzene core via a Sonogashira cross-coupling reaction.

With respect to potential applications as functional materials in the fields of photovoltaics or OLEDs, several highly fluorescent tris- and tetrakis-TPY (**41–47**, Scheme 16.11) containing rigid hexaalkyl-truxene or dialkoxybenzene moieties were prepared by Yuan [73], Wang [74], Winter [75, 76], and coworkers. In contrast to the above-mentioned unsubstituted polytopic TPYs, the incorporation of solubilizing alkyl or alkoxy groups in **41–47** resulted in the fabrication of hyperbranched MSPs with good solubility in most organic solvents.

16.3.1.3 MSP Networks Except for linear, star, and hyperbranched MSPs, supramolecular polymer networks have also drawn much attention. Supramolecular polymer networks are usually obtained by cross-linking polymeric backbones through non-covalent interactions. For MSP networks, the cross-linkers could be polytopic ligands or lanthanide ions because they can provide more than two coordination sites.

In 2006, Rowan *et al.* demonstrated that ditopic ligands (**48**) could bind with large lanthanide ions such as La^{III} and

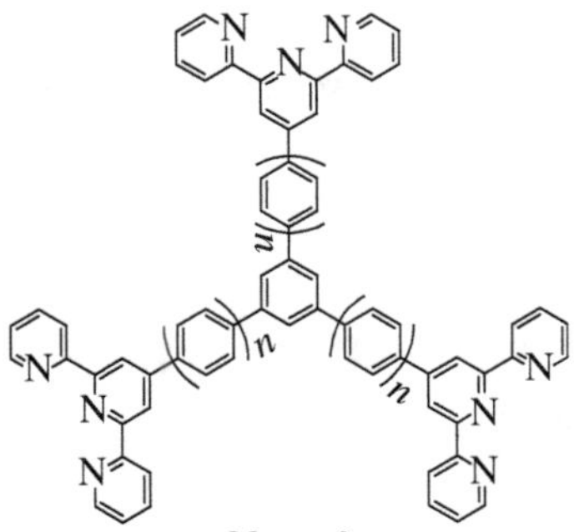

33 n=0
34 n=1
35 n=2

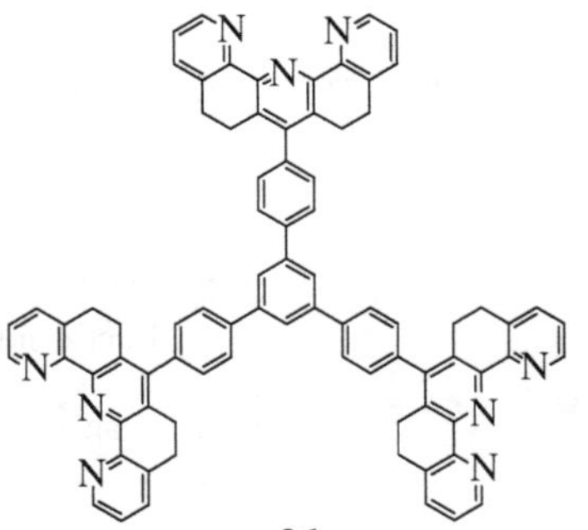

36

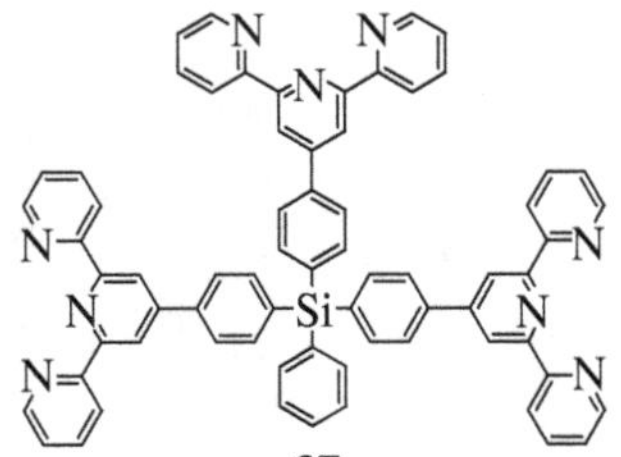

37

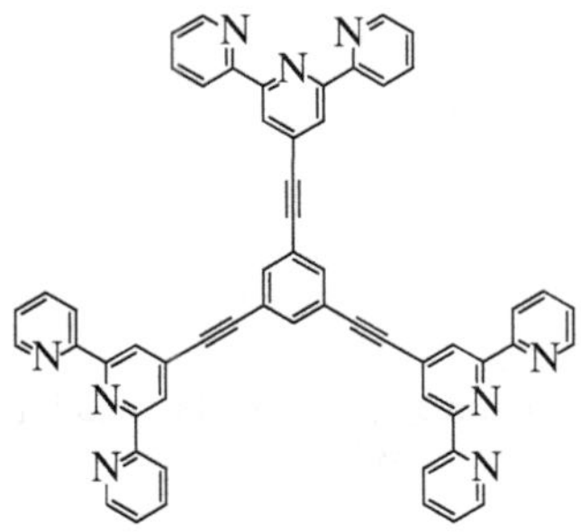

38

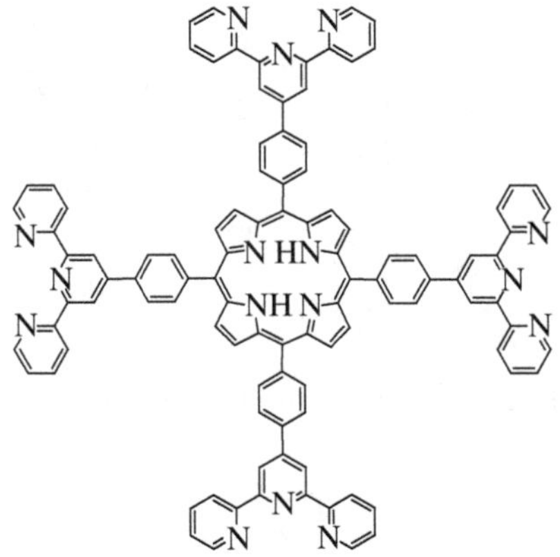

39

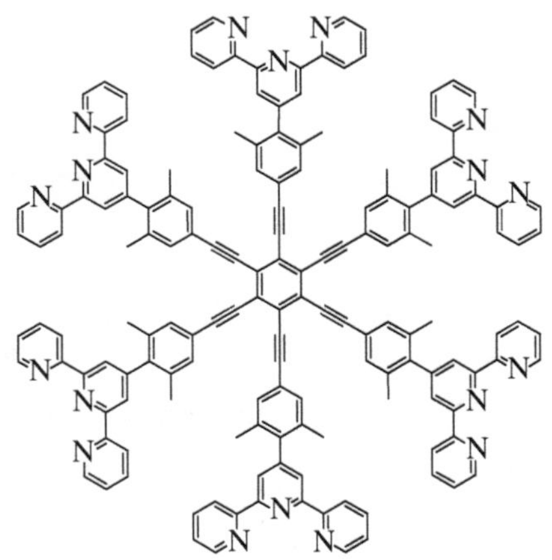

40

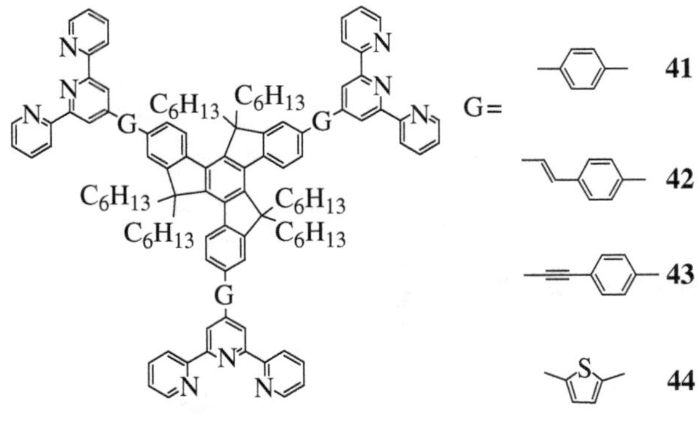

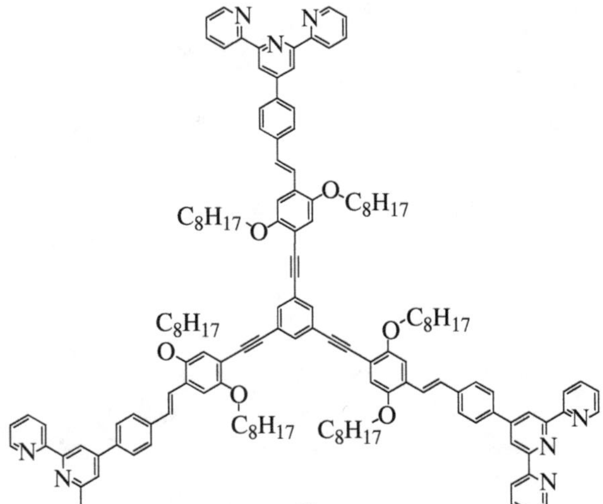

45

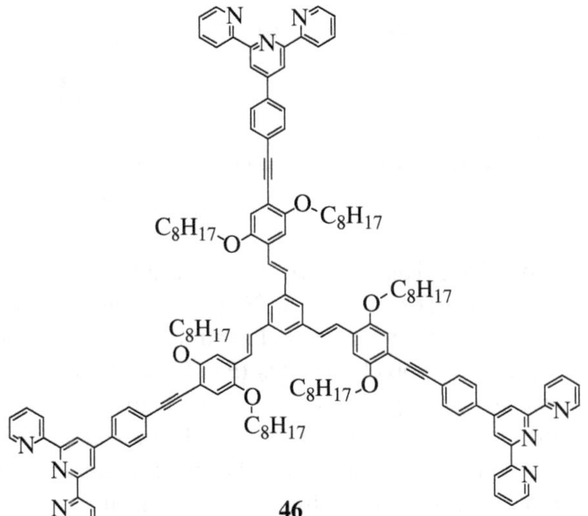

46

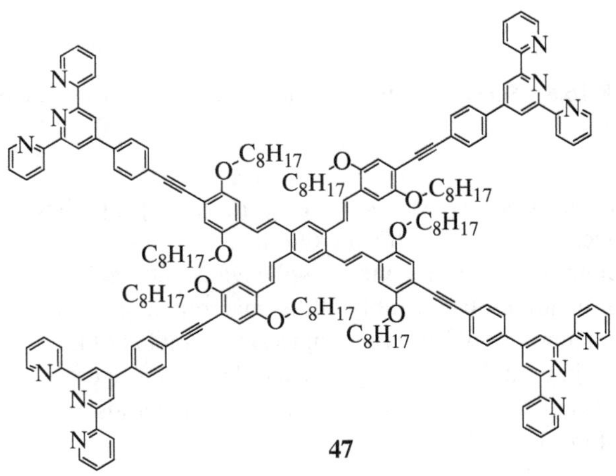

47

SCHEME 16.11 Schematic representation of various π-conjugated TPY derivatives.

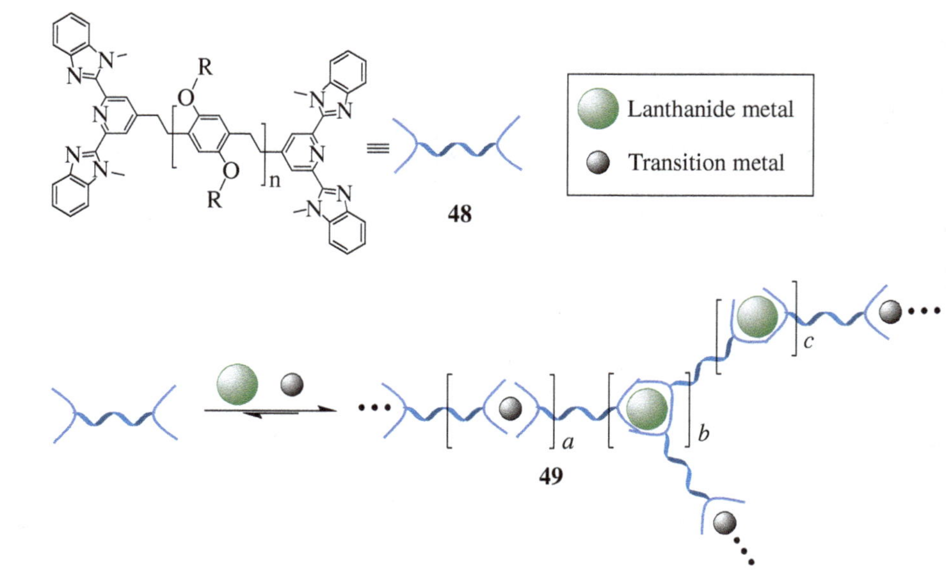

FIGURE 16.7 Schematic representation of a MSP network through the coordination of a bis-BIP building block with Fe^{II} and La^{III} [77].

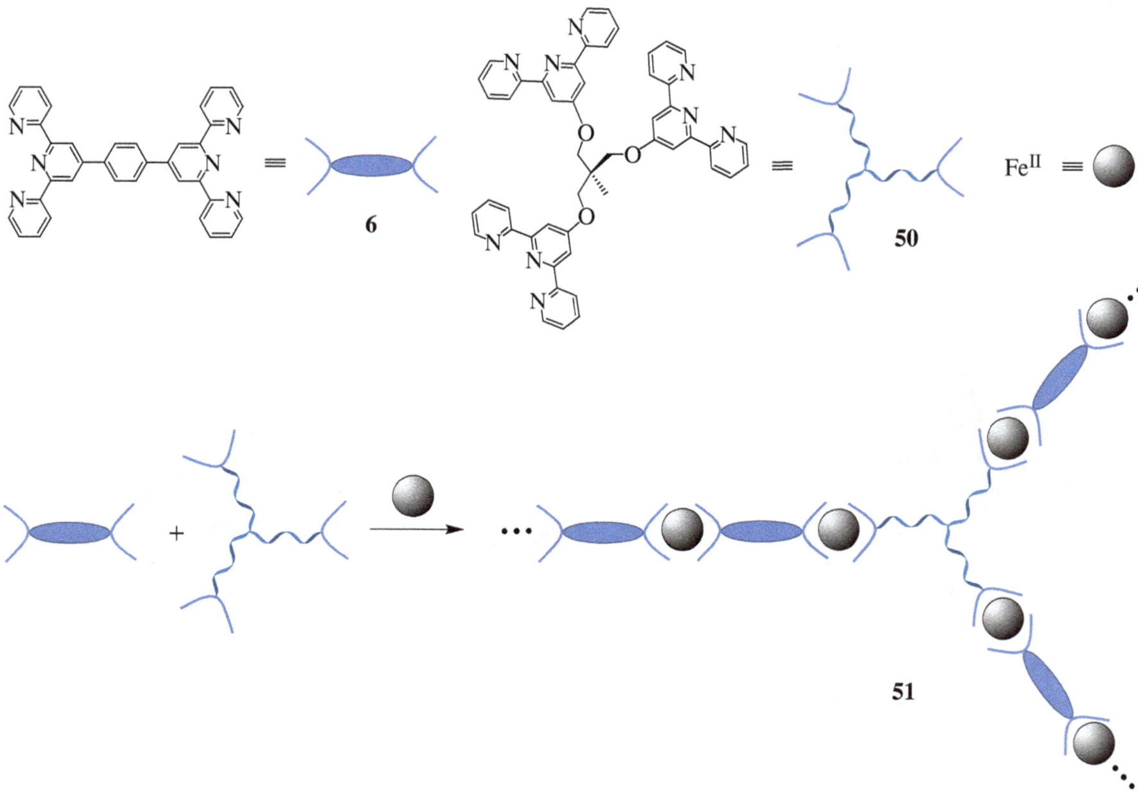

FIGURE 16.8 Schematic representation of a MSP network fabricated by the self-assembly of a bis-TPY and a tris-TPY with metal ions [78].

Eu^{III} to form cross-linked networks (Fig. 16.7, **49**) [77]. They synthesized a ditopic BIP ligand **48** linked by a low molecular weight poly(2,5-dioctyloxy-*p*-phenylene ethynylene). The coordination self-assembly of this macromonomer with equimolar amounts of Zn^{II} or Fe^{II} and a small amount of La^{III}, which could bind three BIP ligands and thus acted as cross-linkers, resulted in the formation of MSP materials with excellent mechanical properties.

In 2007, Kurth *et al.* synthesized a new tris-TPY ligand (**50**). The mixture of **50** and **6** together with Fe^{II} led to the formation of cross-linked metallo-supramolecular coordination polyelectrolytes (**51**) (Fig. 16.8) [78]. Due to low cross-linking degrees (9.5% and below), soluble homogeneous networks were obtained. The molecular weights of the networks were large and depended on the cross-linking degree. Through layer-by-layer self-assembly, the soluble networks could fabricate smooth hybrid multilayer films with poly(sodium styrenesulfonate).

These examples highlight the potentials of the architecture-directed design of ligands as building blocks for star-shaped and hyperbranched MSPs and MSP networks. However, the tedious synthesis and low yields of polytopic TPYs still limited their structural diversity and applications to fabricate functional MSPs.

16.3.2 MSPs Fabricated by Orthogonal Self-Assembly

Recently, an alternative strategy is attracting considerable attention. In this strategy, MSPs are fabricated by two or more non-covalent interactions in an efficient manner. Generally, the multiple non-covalent interactions involved in one MSP system should be orthogonal to each other. In other words, the non-covalent interactions are independent and have no intention of intervening in each other [16–19].

The MSPs constructed by orthogonal self-assembly can generate more ordered and hierarchical structures with novel properties. Moreover, since their self-assemblies are dynamic, they are responsive to various external stimuli such as pH, temperature, light, and ions, which open up a chance to explore novel materials with tunability and reversibility. Because the metal–ligand coordination bonds have a high direction, tunable binding strengths, and large association constants, they have a huge potential for orthogonal self-assembly with other non-covalent interactions, such as metal–ligand coordination, hydrogen bonding, electrostatic, and host–guest interactions.

16.3.2.1 Combination of Two Different Metal–Ligand Coordination Interactions Due to various binding strengths and coordination modes, different metals can bind different ligands and thus form different coordinated geometries. The combination of two or more different metal–ligand coordination building blocks in one ligand will be a convenient way to self-assemble hierarchically with different metal ions.

In 2002, Sanders *et al.* reported a MSP (**55**) fabricated by two orthogonal coordination interactions (Fig. 16.9) [79]. The mixture of bis-diphenylphosphine-substituted porphyrin–Zn (**52**) and 1.0 equimolar porphyrin–Rh (**53**) led to the formation of a cyclic oligomer (**54**) and linear polymers (**55**) in $CHCl_3$. After addition of 0.5 equimolar 4,4′-bipyridine, which could selectively coordinate the metal ion Zn^{II}, all linear assemblies **55** transformed into cyclic porphyrin tetramers **54**. This early phenomenon provided chemists an approach using orthogonal combination of two coordination interactions to construct MSPs hierarchically.

Recently, Wang and coworkers used a heterometallic coordination-driven hierarchical self-assembly to construct a MSP network (Fig. 16.10, **58**) [80]. Two TPY ligands were linked covalently by click reactions to synthesize a multifunctional ligand **56**. Upon addition of $FeCl_2$, a linear

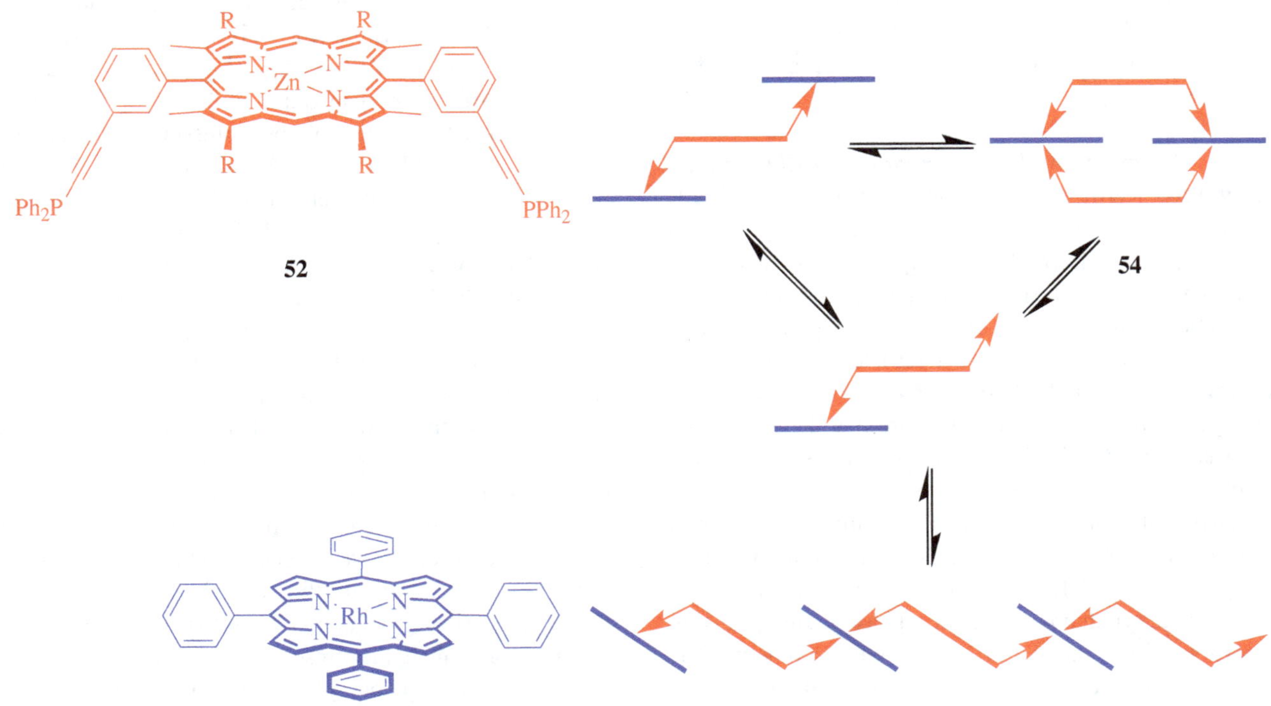

52 **54** **53** **55**

FIGURE 16.9 Cyclic oligomer and linear polymers fabricated by the combination of two different metal–ligand coordination interactions [79].

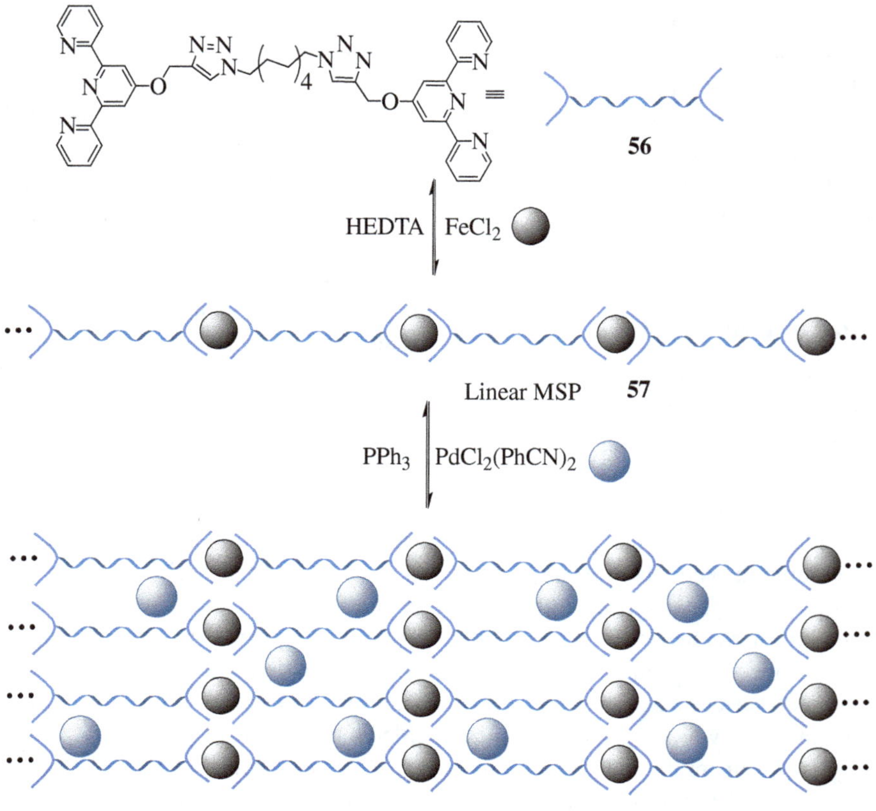

FIGURE 16.10 MSP network fabricated by the combination of two different metal–ligand coordination interactions [80].

supramolecular polymer **57** was initially constructed. And then $PdCl_2(PhCN)_2$ was added to coordinate with 1,2,3-triazole units, resulting in the formation of a MSP network **58**.

16.3.2.2 Combination of Coordination and Electrostatic Interactions Kurth *et al.* developed the hierarchical self-assembly based on ditopic TPY-containing MSPs (Fig. 16.11) [81–85]. Due to the high charged density of MSPs, they were well soluble in aqueous solution. The electrostatic self-assembly of **7** with an amphiphilic molecule, dihexadecyl phosphate (DHP), led to the formation of polyelectrolyte–amphiphile complexes (PACs, **59**). The resulting **59** had a stratified architecture that the MSP main chain was embedded in the amphiphile layers. Due to the hydrophobic surface, the PACs were soluble in common organic solvents and could form Langmuir monolayers at the air–water interface, which could be transferred to prepare highly ordered LB films. Upon heating the multilayer LB film, the alkyl chains of the mesophase melted, causing a distortion of the metal ion coordination geometry. The unfavorable coordination of the TPYs around the Fe^{II} cation gave rise to a lower energy gap between the d-orbital subsets, resulting in a reversible transition from a diamagnetic low-spin state to a paramagnetic high-spin one.

In 2012, Bu *et al.* replaced the acetate counterion of **7** with sulfonate-terminated polystyrenes (ST-S_n) and fabricated a series of MSP brushes (Fig. 16.12, **60–62**) [86]. Due to the difficult reorganization of the ST-S_n chains at the air–water interface, the collapse pressures of Langmuir monolayers of MSP brushes (**60–62**) were *ca.* 30 mN/m, which were much smaller than the collapse pressure (62 mN/m) of DHP-enwrapped **7**, **59**. The lengths of ST-S_n and **7** chains were within a similar order. Scanning force microscopy (SFM) measurement was employed to image the individual molecules of **60–62**, revealing that the short worms were dominant in **60** and **61** along with a few spheres, while **62** was basically spherical. When the length of the backbones is significantly longer than that of the side chains, the molecular brushes prefer to adopt a cylindrical shape because of intramolecular excluded volume effects. Due to a delicate balance between the repulsive entropic effects and van der Waals attractive forces in chloroform/methanol (v/v = 1:1) mixture solvents, the MSP brushes self-assembled into vesicles. The rigid chain of **7** had to bend in the vesicles, which accompanied the presence of a lateral tension and thus led to a spontaneous vesicle fusion with an hour-scale fusion time. For this much longer fusion process, various intermediates, such as the arrow-like and stalklike intermediate and hemifusion diaphragm, were clearly

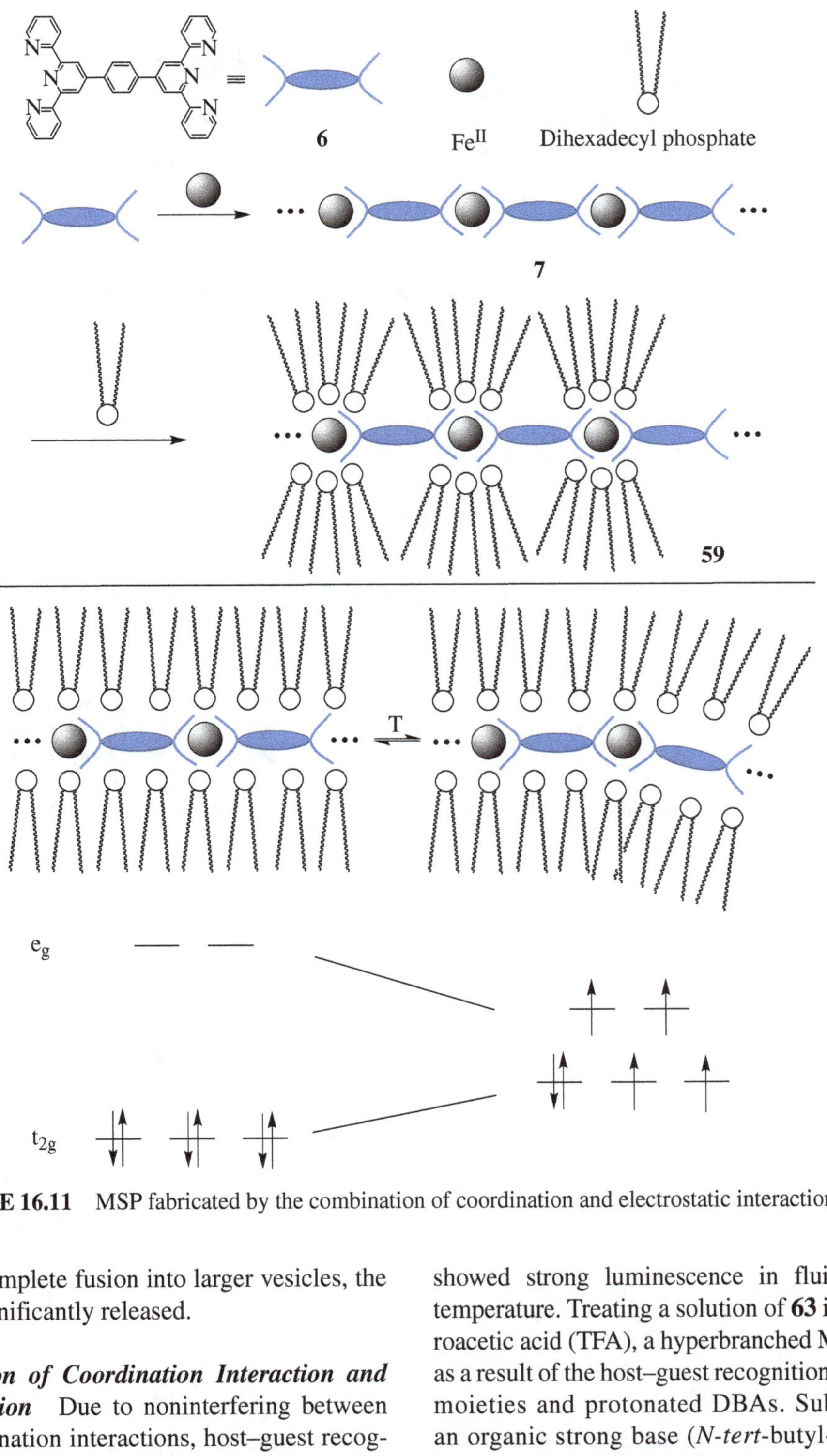

FIGURE 16.11 MSP fabricated by the combination of coordination and electrostatic interactions [81].

observed. After the complete fusion into larger vesicles, the lateral tension was significantly released.

16.3.2.3 Combination of Coordination Interaction and Host–Guest Recognition Due to noninterfering between host–guest and coordination interactions, host–guest recognition is one of the likeliest candidates for the orthogonal self-assembly with MSPs. Through cucurbit[*n*]uril [87–90], crown ether [91–117], cyclodextrin [118–123], and pillar[*n*] arene-based host–guest recognition [124, 125], many linear and hyperbranched MSPs and MSP networks have been developed powerfully.

In 2013, Bu *et al.* synthesized a AB_2 monomers [Pt(diimine)(C≡CR)$_2$] (**63**) with a dibenzo[24]crown-8 (DB24C8) ring and two secondary dibenzylamine centers (DBAs) (Fig. 16.13) [105]. The platinum(II) complex **63** showed strong luminescence in fluid solutions at room temperature. Treating a solution of **63** in CH_2Cl_2 with trifluoroacetic acid (TFA), a hyperbranched MSP **64** was fabricated as a result of the host–guest recognition between the DB24C8 moieties and protonated DBAs. Subsequent addition of an organic strong base (*N-tert*-butyl-*N′*,*N′*,*N″*,*N″*,*N‴*,*N‴*-hexamethylphosphorimidic triamide, P_1-*t*-Bu) led to the complete dissociation of the hyperbranched MSP **64**.

The acid–base-controlled process also induced a controllable absorption and luminescence of the platinum(II) complex. The groups of Huang [102] and Yang [125] reported a series of organoplatinum(II) metallacycle-containing hyperbranched MSPs fabricated by hierarchical self-assembly. They first synthesized host-functionalized organoplatinum(II) metallacycles (A_m) by coordination-driven self-assembly and then the ditopic guest molecules

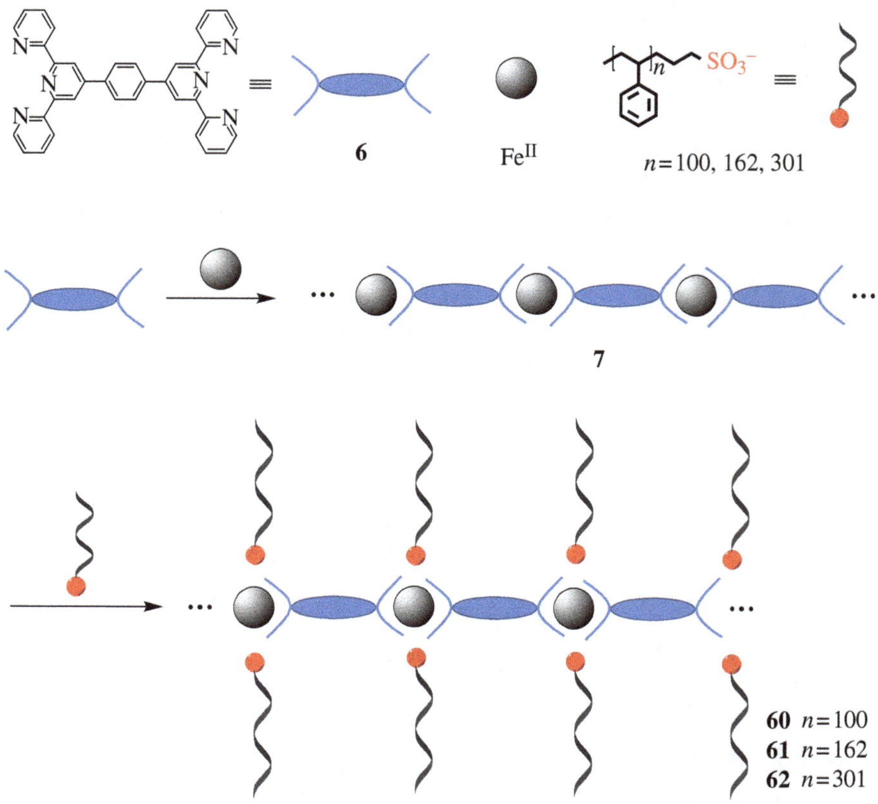

FIGURE 16.12 MSP brushes fabricated by the combination of coordination and electrostatic interactions [86].

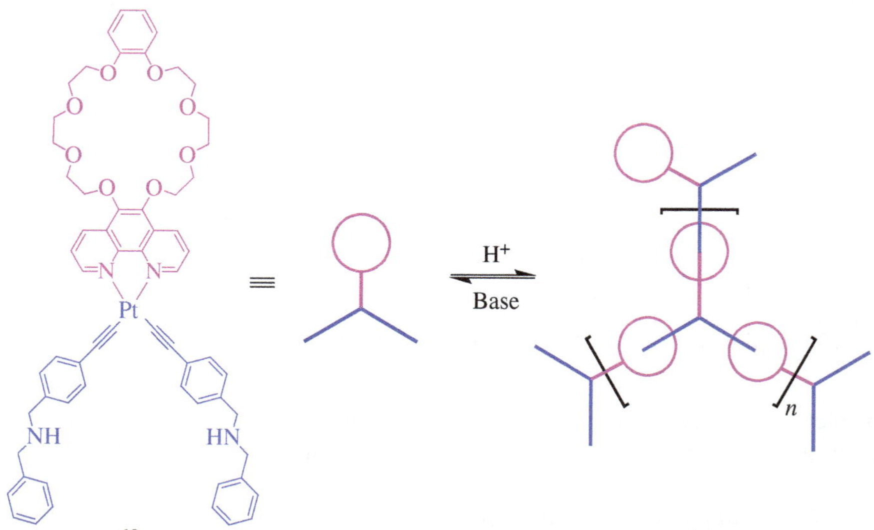

63 64

FIGURE 16.13 Schematic representation of the reversible formation of a hyperbranched MSP [105].

(B_2) were added. As a result of the corresponding host–guest interactions, hyperbranched MSPs of ($A_m + B_2$) were formed. At high concentrations, these hyperbranched MSPs formed gels. The resulting metallogels were responsive to temperature, pH, ions, and extra host and/or guest molecules.

In 2014, Bu *et al.* reported the fabrication of a light-emitting MSP (**66**) by complexing Zn^{II} with conjugated bis-TPY ligand (**65**) bearing two DB24C8 groups (Fig. 16.14) [106].

MP-Zn formed linear oligomers at low concentrations and linear polymers with a high DP at high concentrations. As a result, MP-Zn showed a concentration controllable fluorescence. This is the first report for concentration and thus DP controllable fluorescence for MSPs. Treating a solution of MP-Zn with ditopic guest **68** at a low concentration, the resulting MSP networks showed a decreased fluorescence. Importantly, the process was reversible. Upon addition of

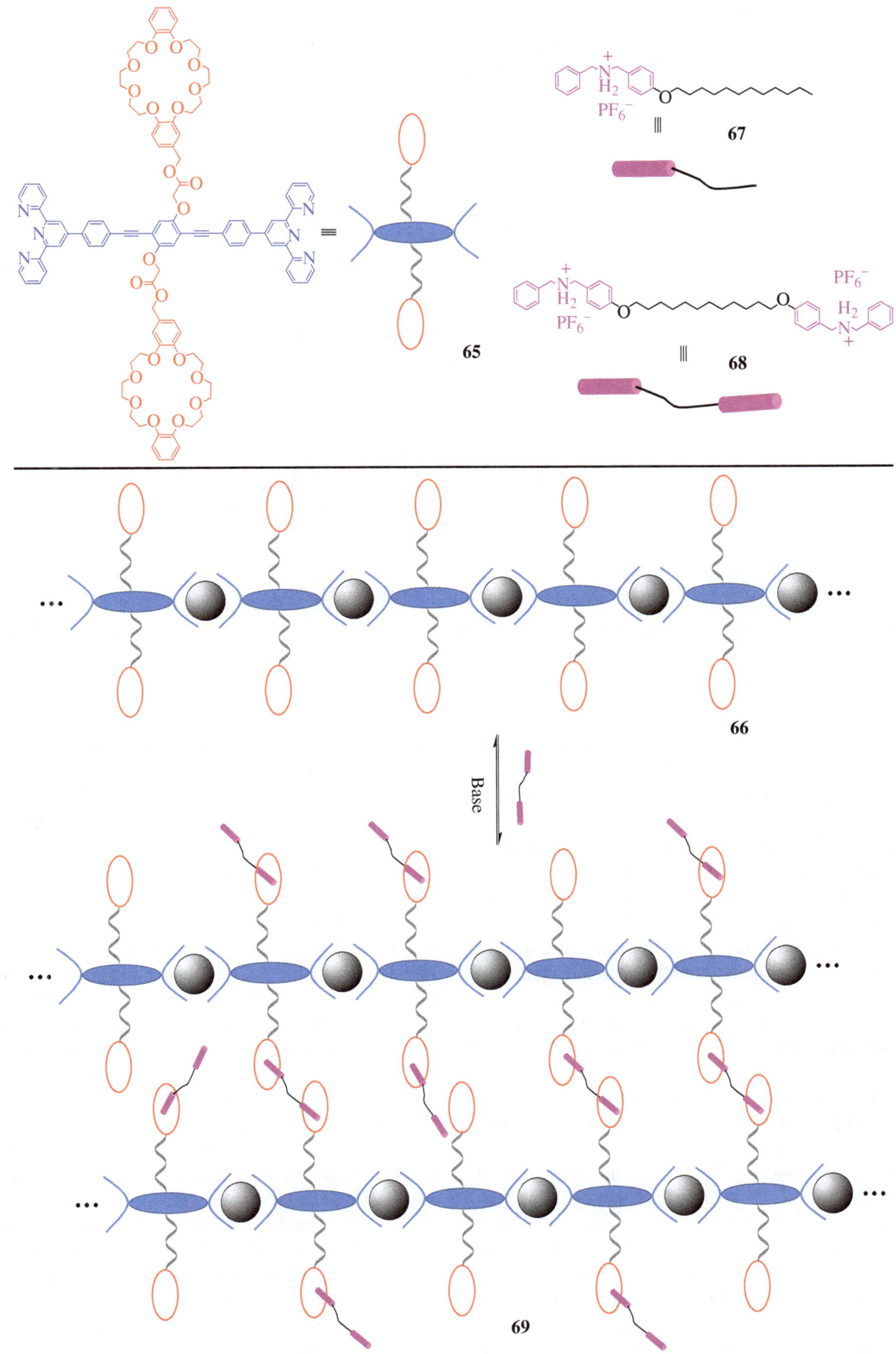

FIGURE 16.14 Schematic representation of the reversible formation of MSP network [106].

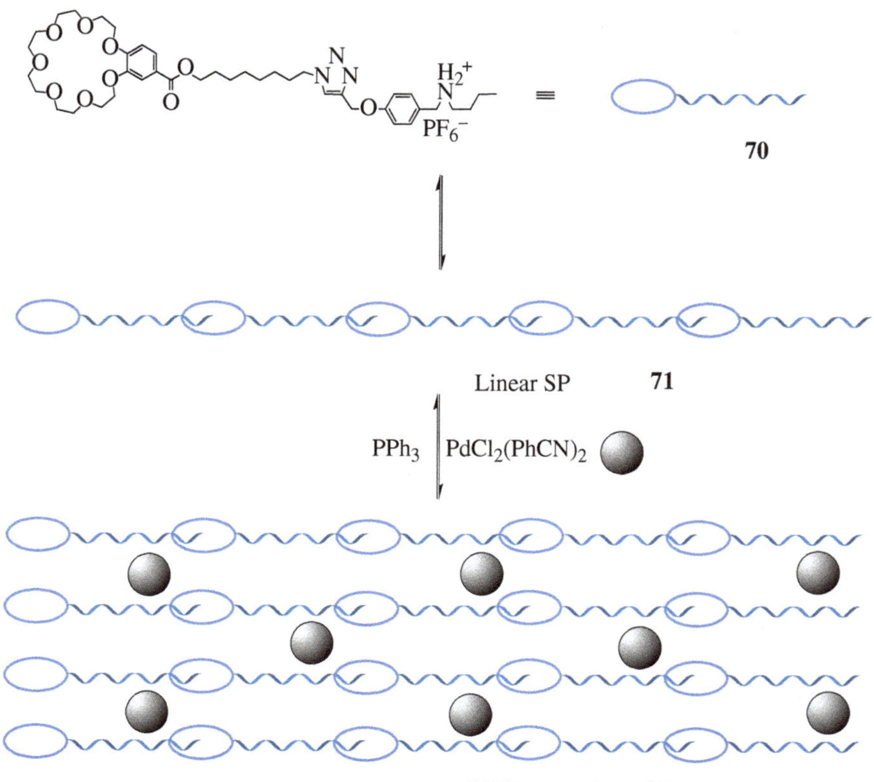

FIGURE 16.15 MSP network fabricated by the combination of coordination interaction and host–guest recognition [96].

one-end guest **67**, the fluorescence only showed a slight decrease. However, upon titration of **66** with **68** at a high concentration, the fluorescence band showed a significant increase in its intensity.

Yin and coworkers reported a flexible ditopic ligand that two TPY moieties were linked together with DB24C8 group [107]. Through orthogonal self-assembly driven by coordination and host–guest interactions, a flexible MSP network was prepared. On the other hand, Huang *et al.* developed a series of MSP network gels through the orthogonal self-assembly of host–guest and coordination interactions, all of which were multiresponsive (pH, metallo-, thermo-, and cation responsive), shape persistent, and elastic (Fig. 16.15) [96, 116].

The ditopic monomers had similar structures: the terminal functional groups were crown ethers and their complementary guests, and 1,2,3-triazole groups were embedded in the middle of the monomers (**70**). The ditopic monomer **70** could self-assemble to form linear polymers (**71**) at high concentrations. The 1,2,3-triazole moieties were linked together upon subsequent addition of Pd^{II}, leading to the formation of MSP networks (**72**). The deconstruction of MSP networks was triggered by the addition of competitive ligand, PPh_3. The conversion from linear to networked structures gave rise to a reversible sol–gel transition at high concentrations.

16.3.2.4 Combination of Coordination Interaction and Hydrogen Bonding Hydrogen bonding also can be combined with coordination interactions in an orthogonal way. Schubert and coworkers attached a quadruple hydrogen bonding ureidopyrimidinone (UPy) unit and a TPY moiety to two ends of linear spacer blocks (molecule or polymer, **73**) (Fig. 16.16) [126–128].

^1H NMR spectrum indicated the dimerization of the hydrogen bonding units in nonpolar solvents (**74**). The addition of metal ions led to the formation of a linear MSP (**75**). Viscometry measurements proved that polymeric species were formed in solution. When the spacers were small molecules, the resulting MSPs had a poor solubility, which made them difficult to investigate the DPs and ring-chain equilibrium. However, when the spacer blocks were combined into polymers, like poly(ε-caprolactone) and polystyrene, the resulting MSPs showed good solubility in nonpolar solvents. In the case of poly(ε-caprolactone), the coordinated polymer showed a film-forming property and its film was transparent, greatly differing from the original polymer. The successive addition of a competitive ligand, metal ion, or quadruple hydrogen bonding unit could deconstruct the MSPs. Hirsch *et al.* developed a facile strategy for the self-assembly of hyperbranched MSPs, which was based on the orthogonal combination of Hamilton receptor–cyanuric acid and $[Ru(BPY)_3]^{2+}$ coordination interactions [129].

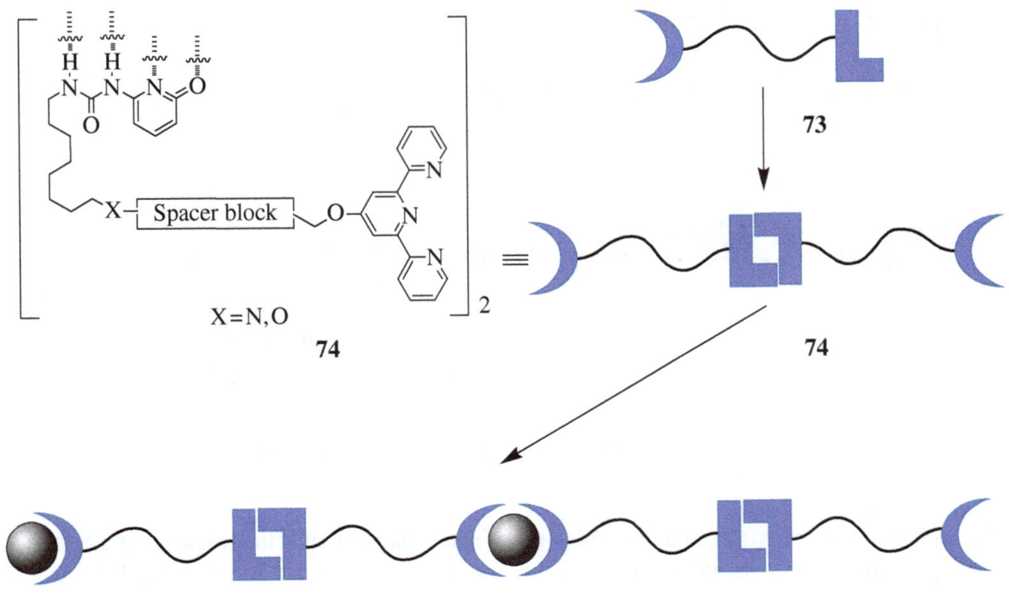

FIGURE 16.16 Linear MSP fabricated by the combination of coordination interaction and hydrogen bonding.

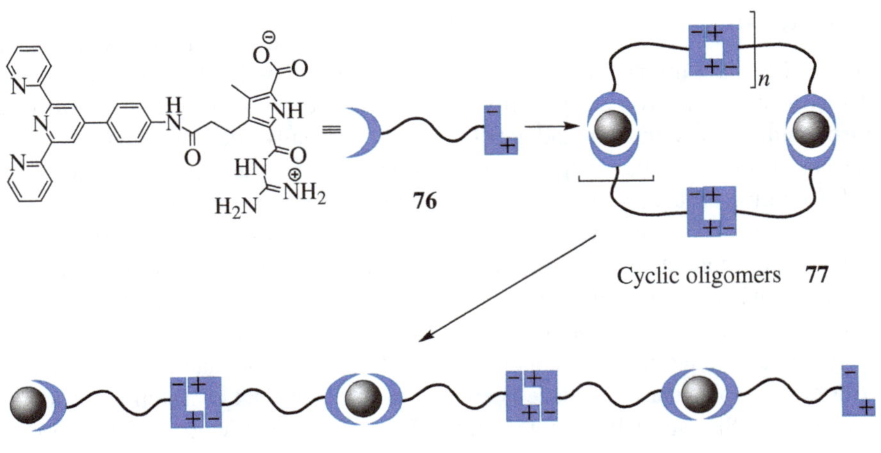

FIGURE 16.17 Cyclic and linear MSP fabricated by the combination of coordination interaction and hydrogen bonding interaction [130].

Hydrogen bonds are only stable in nonpolar solvents. In other words, hydrogen bonds are very weak in polar solvents and especially protic solvents. In 2011, Schmuck *et al.* reported a switchable MSP (**77** and **78**) from a small heteroditopic monomer (**76**) with two orthogonal binding interactions in very polar solvents (Fig. 16.17) [130].

One terminal group of the monomer **76** was a guanidinio-carbonyl pyrrole carboxylate zwitterion motif and another was a TPY unit. The self-assembly of **76** was based on two orthogonal binding interactions, the formation of a Fe^{II}–TPY complex and the dimerization of a self-complementary zwitterion. Both the interactions had a sufficient stability in DMSO and even aqueous solutions, which ensured the formation of a linear MSP **78** with a considerable length (up to 100 nm). The MSP formed rings of **77** at low concentrations and linear chains at millimolar concentrations and above. The linear MSP of **78** then further aggregated into larger globular aggregates. The MSPs could be furthermore switched on and off either upon addition of a competitive ligand and Fe^{II} or addition of acid or base.

These studies have demonstrated that the orthogonal combination of coordination interaction with other non-covalent interactions can pave a new route to develop smart MSPs with controllable topologies and properties.

16.4 APPLICATIONS OF MSPs

SPs, fabricated from building blocks through non-covalent interactions, not only have the classical properties of polymer materials but also own the stimuli responsiveness, self-healing, shape memory, and degradability. Moreover, in the

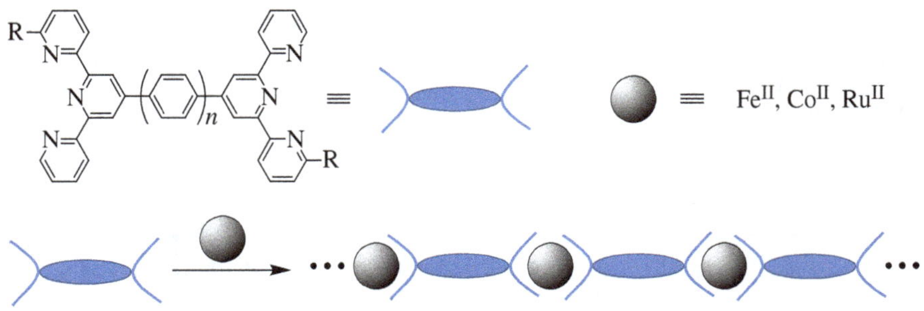

79 R=H, n=1 **80** R=H, n=2
81 R=OMe, n=1 **82** R=OMe, n=2
83 R=Br, n=1

FIGURE 16.18 MSPs with different TPY derivatives and metal ions [131].

case of MSPs, the incorporation of metal centers endows them with a variety of properties useful in optics, electronics, magnetics, and catalysis. In this section, we will summarize the applications of MSPs in the fields of optoelectronic, self-healing, and biomedical materials.

16.4.1 Optoelectronic Materials

Electrochromic materials (ECMs) have drawn great attention as molecular switches for optical and electronic applications including antiglare mirror and glass, indicator and label, smart window, and information storage. These applications require ECMs with good coloration efficiency, high contrast ratio, long-term stability, and write–erase efficiency. Some other properties are application dependent; for example, displays need fast switching rates, whereas smart windows can operate with longer response times of up to several minutes. For these purposes, a vast number of materials, like conducting polymers, metal oxides, and molecular dyes, have been studied. Among these materials, MSP materials are potential candidates for the ECMs due to their high performances.

Kurth *et al.* reported a series of MSPs constructed from rigid, π-conjugated bis-TPYs (**79–83**) and metal ions (Fig. 16.18) [131]. According to the different substituents at 6-position of TPY and/or metal ion, the resulting MSPs displayed various colors spanning the entire visible range. Electrochemical and electrochromic properties of the MSPs also were affected by the substituents. Commonly, the MSPs assembled from the ligands modified by electron-rich group (OMe) exhibited higher switching stability, reversibility, and lower switching potentials than the electron-deficient analogues with H- and Br-substitutions. The response times could be tuned by the design of the ligands or by the choice of the metal ions, which underwent a broad timescale varying from 1 s to several minutes. The optical memory also was enhanced from 30 s to more than 15 min.

Photoluminescent materials have also received significant interest especially in the fields of organic light-emitting diodes (OLEDs). Conjugated MSPs have been investigated for their higher electroluminescence quantum yields. The groups of Würthner, Kurth, and Schubert synthesized a series of conjugated MSPs [132–135]. It was found that the fluorescence properties of MSPs could be tuned either by the metal ions or functional groups at the TPY and spacer blocks. The electron-rich group like OMe or the conjugation of the spacer blocks led to a high quantum yield, while electron-deficient groups, such as Br and CN, gave rise to a low quantum yield. A profound example of the substitution effect was a Fe^{II}-based MSP constructed by the self-assembly of triethylene glycol (TEG)-substituted BPY ligand with Fe^{II} [133]. Usually, the fluorescence of bis-TPY–Fe^{II} MSPs was quenched, but the TEG-substituted Fe^{II} MSP showed a three-fold higher quantum yield than the unsubstituted analogue. A series of bis-TPY–Zn^{II} MSPs with diverse rigid linear π-conjugated moieties were also investigated [134]. The emissions, strongly depending on the nature of the conjugated groups, varied from bright purple to green.

In 2006, Hagemann *et al.* prepared a so-called "all-in-one molecule" **84** by a stepwise and unidirectional synthesis for solar cell applications (Scheme 16.12) [136]. The central part of the MSP **84** was consisted of porphyrin–Zn^{II} and bis-TPY–Ru^{II} complex. The pendant arms were oligohexylthiophene and oligo(phenylene-vinylidene) chains, as antennae for light harvesting and donor and acceptor groups for photoinduced charges, respectively. The MSP **84** was introduced into a bulk heterojunction photovoltaic device. The performance of light-to-electricity conversion was very poor with a short circuit current (4.4 μA cm^{-2}) and a low efficiency (0.000081%). Duprez *et al.* utilized the donor- and acceptor-type TPY macroligands to prepare two homopolymers (**85** and **86**, Scheme 16.12) and a copolymer (**87**, Scheme 16.12) [137]. These MSPs were fabricated into two different types of solar cells (a polymer solar cell and a dye-sensitized solar cell (DSSC)). Very low efficiencies were found in the resulting polymer solar cells (<0.01%). The poor film-forming ability led to low device resistance, short circuit current (I_{sc}), and open circuit voltage (V_{oc}). When sublimating C_{60} onto the film of **85**, the photovoltaic performance was slightly improved. When they were applied in DSSCs, much better performances were achieved.

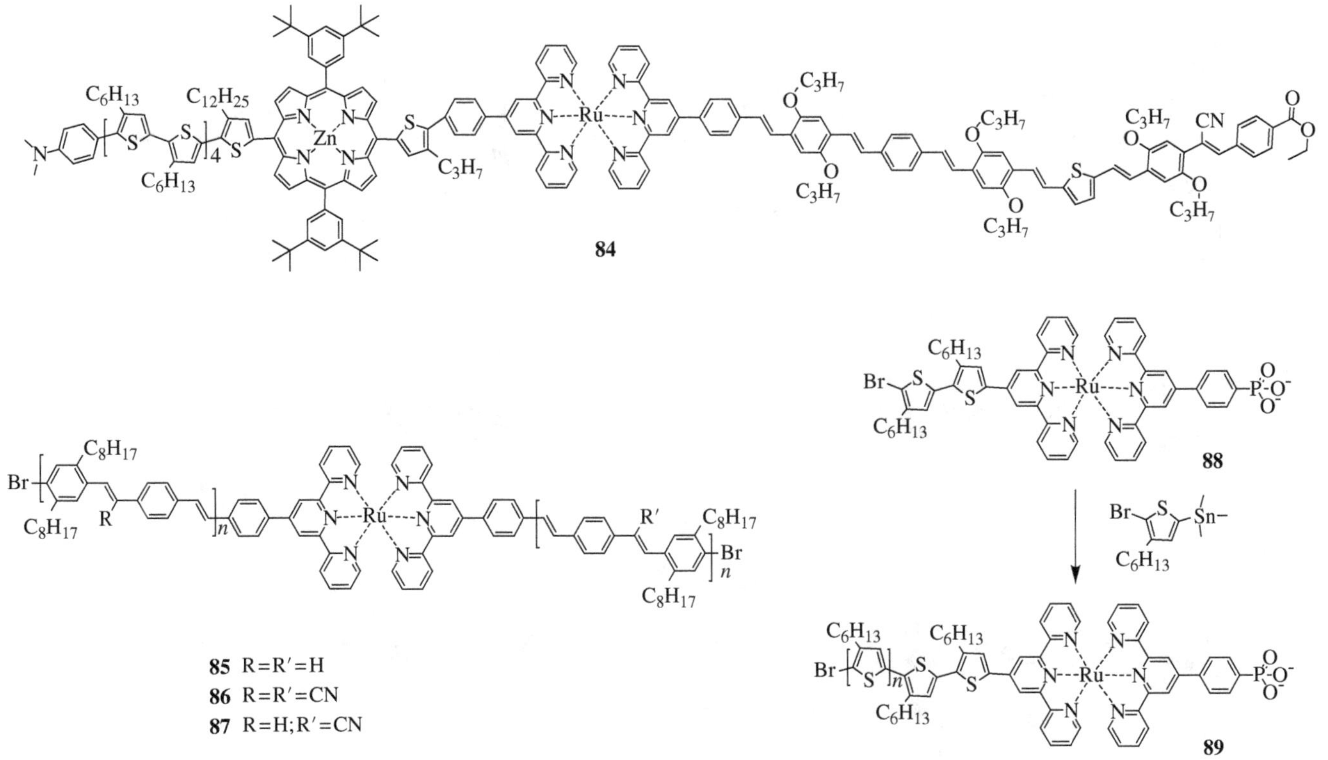

84

85 R=R′=H
86 R=R′=CN
87 R=H;R′=CN

88

89

SCHEME 16.12 Schematic representation of the Ru^{II}-containing MSPs used for solar cell applications.

For **87**, *I*/*V*-promising characteristics (I_{sc} = 0.71 mA cm^{-2}, V_{oc} = 315 mV, and P_{max} (maximum power) = 0.22 mW) were achieved. Krebs and coworkers attached a poly(3-hexylthiophene) chain to the Ru^{II} complex **88** through a Stille cross-coupling reaction, leading to the formation of a MSP **89** (Scheme 16.12) [138]. When fabricated into a polymer photovoltaic device, very poor efficiencies were obtained. However, the incorporation of **88** or **89** into photoelectrochemical solar cells revealed better performances. For **89**, the *I*/*V* characteristics were as follows: I_{sc} = 0.456 mA cm^{-2}, V_{oc} = 555 mV, and P_{max} = 0.055 mW.

16.4.2 Self-Healing Materials

Self-healing materials are able to heal inflicted mechanical damages and thus to recover their mechanical properties. Intrinsically, the self-healing feature originates from the reversible changes of the specific materials. As a consequence, various dynamic covalent bonds and noncovalent bonds are utilized for the self-healing process. In the last decade, several self-healing materials have been developed. However, only few MSP examples own the self-healing feature.

In 2011, Rowan and coworkers presented a self-healable MSP (**91**) (Fig. 16.19). They linked two BIP groups with poly(ethylene-*co*-butylene) to form a macromolecular ligand **90**. The MSP **91** was formed through the coordination self-assembly of the resulting ligand and Zn^{II} ions [139]. Upon the irradiation of UV light, the MSP **91** was excited and the adsorbed energy was converted into heat, which resulted in a temporary dissociation and a decrease of the viscosity, thereby causing a quick defect healing (Fig. 16.19). The light–heat conversion for self-healing **91** was innovative and may be extended to other SP materials.

Holten-Andersen and coworkers reported MSP networks fabricated by coordination between Fe^{III} and catechol-based macromolecular ligands (**92**, Fig. 16.20) [140]. The coordination has been correlated to the hardness and high extensibility of the cuticle of mussel byssal threads and proposed to endow self-healing properties. Inspired by this biological material, they developed a simple method to control catechol–Fe^{III} interpolymer cross-linking. According to the different pH values, the mono-, bis-, and triscatechol–Fe^{III} complexes **93–95** can be transformed into each other, thus providing the ability of cross-linking control. The catechol–Fe^{III} cross-linked polymer gels at high pH displayed high elastic moduli and self-healing properties.

Recently, Schubert *et al.* reported a series of TPY-containing polymers **97–99** synthesized by the copolymerization of a TPY monomer **96** and other monomers like methyl methacrylate, butyl methacrylate, and lauryl methacrylate (Fig. 16.21) [141]. Subsequently, the copolymers were cross-linked by the addition of Fe^{II} ion, leading to the formation of MSP networks **100–102**. The films fabricated from the MSP networks were mechanically stable with no melting point in DSC curves (>250°C) as well as optical microscope observations (>300°C). Upon heating, the MSP films with small scratch showed self-healing capabilities.

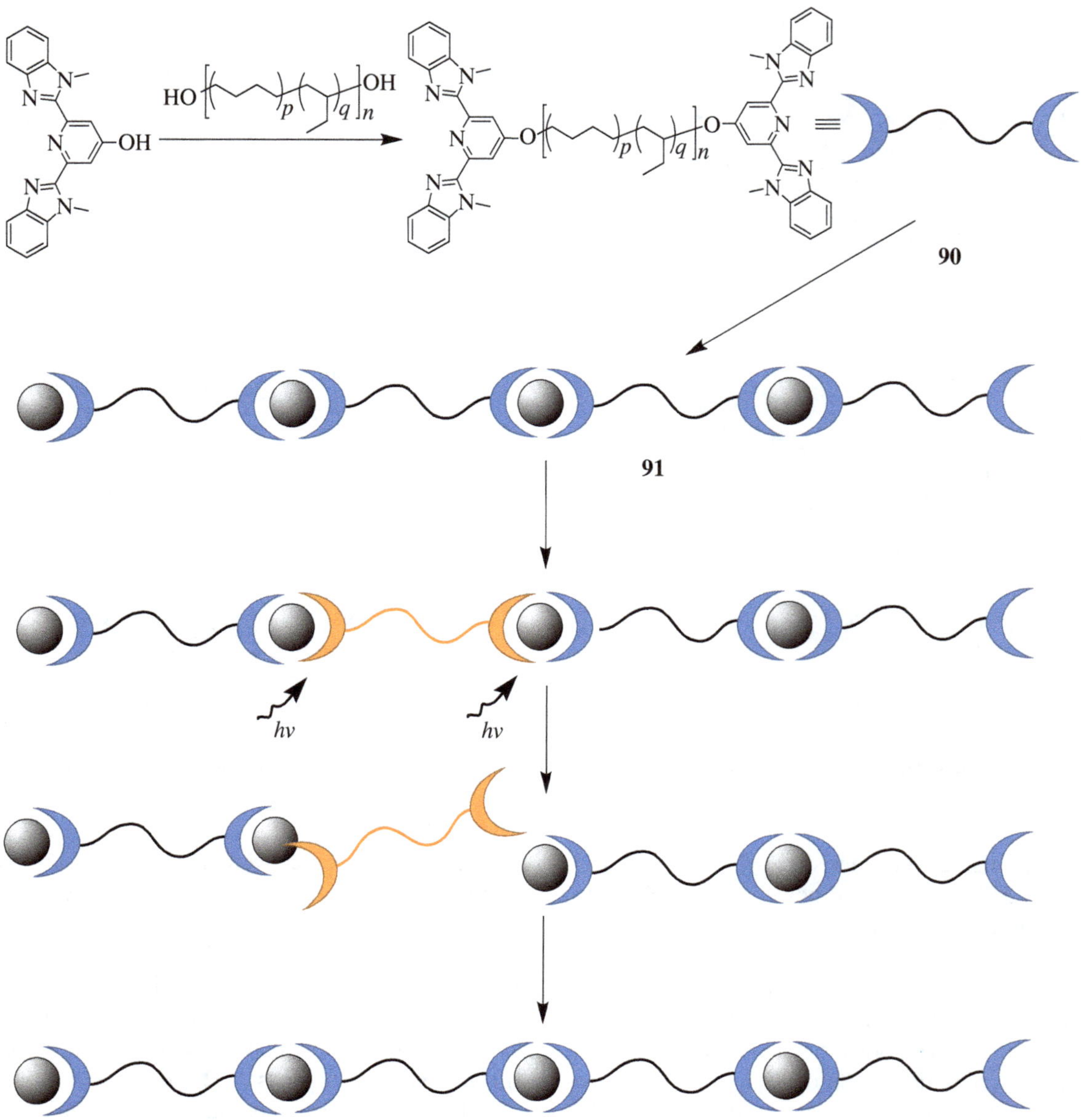

90

91

FIGURE 16.19 Synthesis of a BIP-containing MSP and its photohealable feature [139].

16.4.3 Biomedical Materials

The incorporation of bioactive components can endow MSPs with unique properties, such as biodegradation and responsiveness to biological stimuli. Such MSPs, especially orthogonal MSPs, show potential applications in biomedical fields, such as protein and drug delivery, tissue engineering, gene transfection, and bioimaging.

Zhu *et al.* reported an AA–BB-type MSP (**105**, Fig. 16.22) [142]. This cationic MSP was fabricated by host–guest recognition of a ferrocene dimer (**103**) and a β-CD dimer **104**, which showed redox-induced reversible polymerization and depolymerization upon alternating treatment with hydrogen peroxide (H_2O_2) and glutathione. In contrast to the branched poly(ethylene imine), the MSP **105** exhibited very low cytotoxicity and moderate gene transfection efficiency in COS-7 cells. The addition of H_2O_2 induced the *p*DNA release from the MSP-*p*DNA polyplexes as a result of H_2O_2-triggered depolymerization of the MSP. However, the MSP displayed much lower transfection efficiency in HeLa and MCF-7 cells, which resulted from the dissociation of the polyplexes triggered by the excess H_2O_2 in cancer cells.

Supramolecular polymers also have shown their potential applications in bioimaging and diagnosis. Liu *et al.* reported a MSP **108** based on host–guest interactions between a β-CD dimer **106** and a porphyrin–Mn^{III} star-shaped polymer **107** (Fig. 16.23) [143]. Porphyrin–Mn^{III} in **108** would be reduced effectively to porphyrin–Mn^{II} by sodium ascorbate. However, the free porphyrin–Mn^{III} could not be reduced under the same conditions. The toxicity of **108** *in vitro* was estimated and the results revealed the MSP had no cellular toxicity. Compared with commercially available Gd contrast agents, the MR imaging characterization revealed the MSP exhibited considerably enhanced T_1 relaxivity *in vitro*. Further MR imaging *in vivo* demonstrated that the MSP **108** exhibited remarkably positive contrast enhancement within the blood, kidney, and urinary bladder.

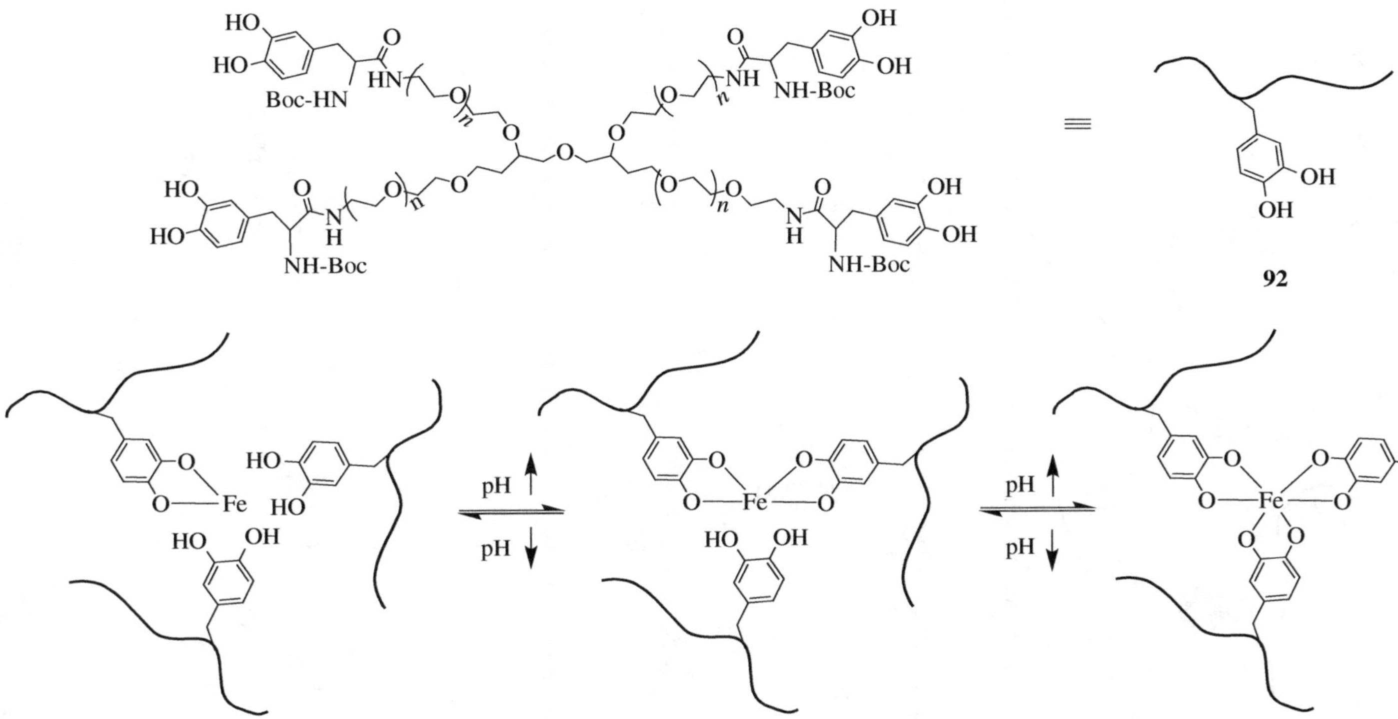

FIGURE 16.20 pH-dependent self-healing of catechol–Fe^{II} complexes [140].

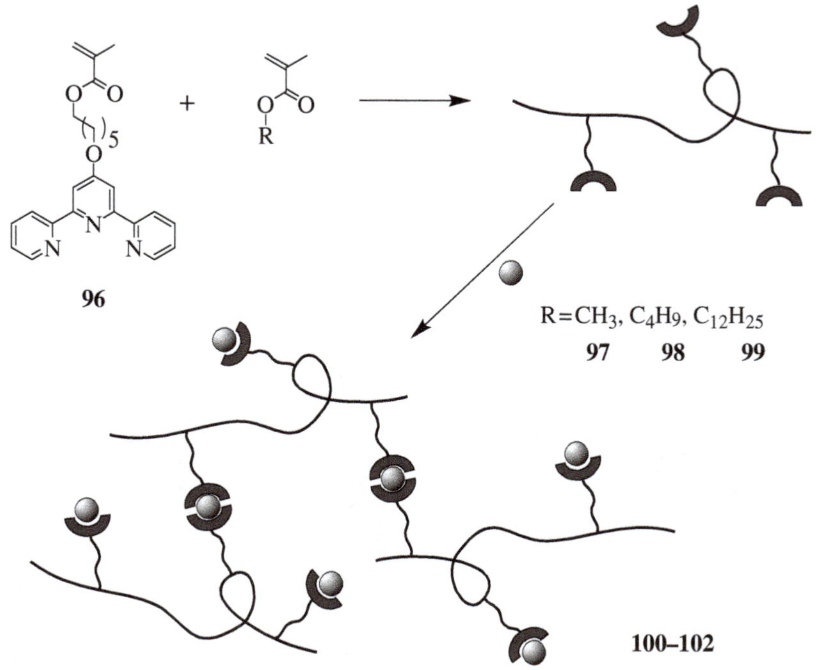

FIGURE 16.21 Schematic representation of the synthesis of a MSP network [141].

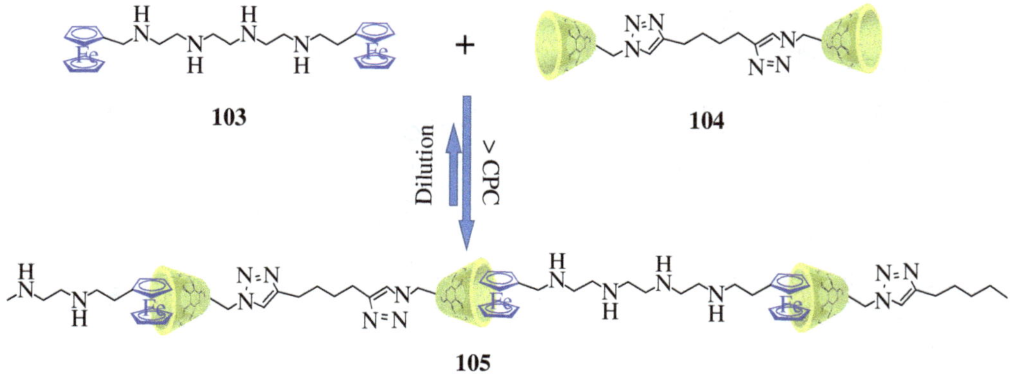

FIGURE 16.22 Schematic representation of a cationic supramolecular polymer constructed via host–guest interactions between a ferrocene dimer and a β-CD dimer [142].

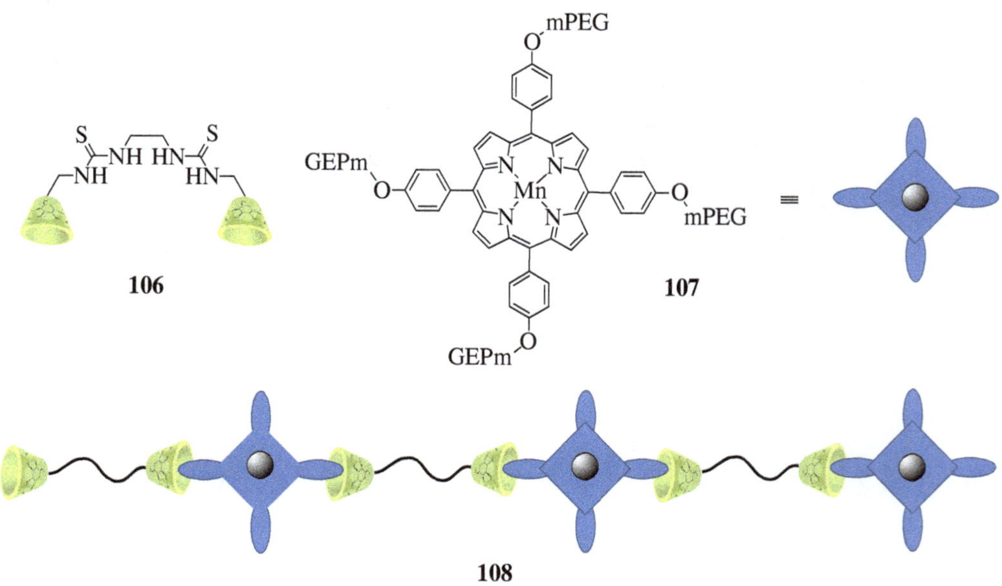

FIGURE 16.23 Schematic representation of supramolecular polymers based on host–guest interactions between a β-CD dimer and a porphyrin–Mn^{III} derivative [143].

16.5 CONCLUSIONS

The incorporation of metal centers into polymeric structures can form diverse MSPs. They not only own the versatile properties of metal components useful in electronics, optics, magnetics, and catalysis but also have good processing ability, a typical feature of polymers. This chapter provides an overview of recent developments in MSPs. In the section about the basic synthetic strategies toward MSPs, we have intended to give a classification in accordance with the position of metal complex in MSPs. This chapter also deals with the design and characterization of MSPs. Through two different strategies—(1) architecture-directed design of ligands and/or the choice of metal ions and (2) orthogonal self-assembly of coordination interaction with other non-covalent interactions—linear MSPs, star-shaped MSPs, hyperbranched MSPs, and MSP networks can be constructed. In addition, we highlight the potential applications of MSPs in the fields of optoelectronic, self-healing, and biomedical materials.

ACKNOWLEDGMENTS

This work was supported by the National Natural Science Foundation of China (21474044).

REFERENCES

1. L. Brunsveld, B. J. B. Folmer, E. W. Meijer, R. P. Sijbesma, *Chem. Rev.* 2001, *101*, 4071.
2. T. F. A. D. Greef, M. M. J. Smulders, M. Wolffs, A. P. H. J. Schenning, R. P. Sijbesma, E. W. Meijer, *Chem. Rev.* 2009, *109*, 5687.
3. T. Aida, E. W. Meijer, S. I. Stupp, *Science*, 2012, *335*, 813.
4. X. Yan, F. Wang, B. Zheng, F. Huang, *Chem. Soc. Rev.* 2012, *41*, 6042.
5. E. A. Appel, J. Del Barrio, X. J. Loh, O. A. Scherman, *Chem. Soc. Rev.* 2012, *41*, 6195.
6. G. R. Whittell, M. D. Hager, U. S. Schubert, I. Manners, *Nat. Mater.* 2011, *10*, 176.
7. D. Michael, *J. Chem. Soc. Dalton Trans.* 1990, *29*, 1405.
8. M. D. Ward, P. R. Raithby, *Chem. Soc. Rev.* 2013, *42*, 1619.
9. U. Schubert, *Chem. Mater.* 2001, *13*, 3487.
10. U. Schubert, *Chem. Soc. Rev.* 2011, *40*, 575.
11. U. S. Schubert, C. Eschbaumer, *Angew. Chem., Int. Ed.* 2002, *41*, 2892.
12. R. Hoogenboom, U. S. Schubert, *Chem. Soc. Rev.* 2006, *35*, 622.
13. C. Fustin, P. Guillet, U. S. Schubert, J. Gohy, *Adv. Mater.* 2007, *19*, 1665.
14. M. Chiper, R. Hoogenboom, U. S. Schubert, *Macromol. Rapid Commun.* 2009, *30*, 561.
15. U. S. Schubert, G. R. Newkome, I. Manners, Ed. Metal-Containing and Metallosupramolecular Polymers and Materials, American Chemical Society, Washington, DC, 2006.
16. S.-L. Li, T. Xiao, C. Lin, L. Wang, *Chem. Soc. Rev.* 2012, *41*, 5950.
17. X.-Y. Hu, T. Xiao, C. Lin, F. Huang, L. Wang, *Acc. Chem. Res.* 2014, *47*, 2041.
18. P. Wei, X. Yan, F. Huang, *Chem. Soc. Rev.* 2015, *44*, 815.
19. L. Yang, X. Tan, Z. Wang, X. Zhang, *Chem. Rev.* 2015, *115*, 7196.
20. L. Brunsveld, B. J. B. Folmer, E. W. Meijer, R. P. Sijbesma, *Chem. Rev.* 2001, *21, 101*, 4071.
21. A. Cifferi, *Macromol. Rapid Commun.* 2002, *23*, 511.
22. A. Wild, A. Winter, F. Schlütter, U. S. Schubert, *Chem. Soc. Rev.* 2011, *40*, 1459.
23. W. H. Carothers, *Trans. Faraday Soc.* 1936, *32*, 39.
24. M. Marcos, L. Oriol, J. L. Serrano, P. J. Alonso, J. A. Puertolas, *Macromolecules*, 1990, *23*, 5187.
25. C. D. Eisenbach, U. S. Schubert, *Macromolecules*, 1993, *26*, 7372.
26. U. Velten, M. Rehahn, *Chem. Commun.* 1996, 2639.
27. U. Velten, B. Lahn, M. Rehahn, *Macromol. Chem. Phys.* 1997, *198*, 2789.
28. B. Lahn, M. Rehahn, *Macromol. Symp.* 2001, *163*, 157.
29. R. Knapp, A. Schott, M. Rehahn, *Macromolecules*, 1996, *29*, 478.
30. S. Kelch, M. Rehahn, *Macromolecules*, 1997, *30*, 6185.
31. E. C. Constable, *Macromol. Symp.* 1995, *98*, 503.
32. U. Kolb, K. Buescher, C. A. Helm, A. Lindner, A. F. Thuenemann, M. Menzel, M. Higuchi, D. G. Kurth, *Proc. Natl. Acad. Sci. U.S.A.* 2006, *103*, 10202.
33. M. Schütte, D. G. Kurth, M. R. Linford, H. Cölfen, H. Möhwald, *Angew. Chem., Int. Ed.* 1998, *37*, 2891.
34. S. Kelch, M. Rehahn, *Macromolecules*, 1999, *32*, 5818.
35. U. S. Schubert, O. Hien, C. Eschbaumer, *Macromol. Rapid Commun.* 2000, *21*, 1156.
36. S. Schmatloch, M. F. Gonźalez, U. S. Schubert, *Macromol. Rapid Commun.* 2002, *23*, 957.
37. S. Schmatloch, A. M. J. van den Berg, A. S. Alexeev, H. Hofmeier, U. S. Schubert, *Macromolecules*, 2003, *36*, 9943.
38. U. Mansfeld, A. Winter, M. D. Hager, W. Günther, E. Altuntas, U. S. Schubert, *J. Polym. Sci., Part A: Polym. Chem.* 2013, *51*, 2006.
39. S. Bernhard, K. Takada, D. J. Díaz, H. D. Abruña, H. Mürner, *J. Am. Chem. Soc.* 2001, *123*, 10265.
40. M. Kimura, M. Sano, T. Muto, K. Hanabusa, H. Shirai, N. Kobayashi, *Macromolecules*, 1999, *32*, 7951.
41. G. D. Storrier, S. B. Colbran, D. J. C. Craig, *J. Chem. Soc., Dalton Trans.* 1997, 3011.
42. A. El-Ghayoury, A. P. H. J. Schenning, E. W. Meijer, *J. Polym. Sci., Part A: Polym. Chem.* 2002, *40*, 4020.
43. M. Burnworth, S. J. Rowan, C. Weder, *Macromolecules*, 2012, *45*, 126.
44. D. Knapton, S. J. Rowan, C. Weder, *Macromolecules*, 2006, *39*, 651.
45. M. Burnworth, J. D. Mendez, M. Schroeter, S. J. Rowan, C. Weder, *Macromolecules*, 2008, *41*, 2157.
46. P. K. Iyer, J. B. Beck, C. Weder, S. J. Rowan, *Chem. Commun.* 2005, 319.
47. J. B. Beck, J. M. Ineman, S. J. Rowan, *Macromolecules*, 2005, *38*, 5060.

48. W. Weng, J. B. Beck, A. M. Jamieson, S. J. Rowan, *J. Am. Chem. Soc.* 2006, *128*, 11663.
49. N. Liu, B. Wang, W. Liu, W. Bu, *Chem. Commun.* 2011, *47*, 9336.
50. N. Liu, Q. He, W. Bu, *Langmuir*, 2015, *31*, 2262.
51. F. Qu, N. Liu, W. Bu, *RSC Adv.* 2014, *4*, 9750.
52. C. Rajadurai, O. Fuhr, R. Kruk, M. Ghafari, H. Hahn, M. Ruben, *Chem. Commun.* 2007, 2636.
53. S. Basak, P. Hui, S. Boodida, R. Chandrasekar, *J. Org. Chem.* 2012, *77*, 3620.
54. T. Vermonden, W. M. de Vos, A. T. M. Marcelis, E. J. R. Sudhölter, *Eur. J. Inorg. Chem.* 2004, 2847.
55. T. Vermonden, J. van der Gucht, P. de Waard, A. T. M. Marcelis, N. A. M. Besseling, E. J. R. Sudhölter, G. J. Fleer, M. A. Cohen Stuart, *Macromolecules*, 2003, *36*, 7035.
56. J. J. S. Lamba, C. L. Fraser, *J. Am. Chem. Soc.* 1997, *119*, 1801.
57. J. E. Collins, C. L. Fraser, *Macromolecules*, 1998, *31*, 6715.
58. J. E. McAlvin, C. L. Fraser, *Macromolecules*, 1999, *32*, 6925.
59. J. E. McAlvin, C. L. Fraser, *Macromolecules*, 1999, *32*, 134.
60. R. M. Johnson, P. S. Corbin, C. Ng, C. L. Fraser, *Macromolecules*, 2000, *33*, 7404.
61. L. Viau, M. Even, O. Maury, D. M. Haddleton, H. Le Bozec, *Macromol. Rapid Commun.* 2003, *24*, 630.
62. M. Chen, K. P. Ghiggino, A. Launikonis, A. W. H. Mau, E. Rizzardo, W. H. F. Sasse, S. H. Thang, G. J. Wilson, *J. Mater. Chem.* 2003, *13*, 2696.
63. R. Hoogenboom, D. Wouters, U. S. Schubert, *Macromolecules*, 2003, *36*, 4743.
64. R. Hoogenboom, B. C. Moore, U. S. Schubert, *Macromol. Rapid Commun.* 2010, *31*, 840.
65. E. C. Constable, A. M. W. Cargill Thompson, *J. Chem. Soc., Dalton Trans.* 1992, 3467.
66. E. C. Constable, A. M. W. Cargill Thompson, *J. Chem. Soc., Chem. Commun.* 1992, 617.
67. A. Winter, J. Hummel, N. Risch, *J. Org. Chem.* 2006, *71*, 4862.
68. X.-R. He, Q.-K. Li, Y.-L. Li, N. Wang, Y. Song, X.-F. Liu, M.-J. Yuan, W. Xu, H.-B. Liu, S. Wang, Z.-G. Shuai, D.-B. Zhu, *J. Phys. Chem. B* 2007, *111*, 8063.
69. Y. Nishimori, K. Kanaizuka, M. Murata, H. Nishihara, *Chem. Asian J.* 2007, *2*, 367.
70. T.-J. Cho, C. D. Shriner, S.-H. Hwang, C. N. Moorefield, B. Courneya, L. A. Godinez, J. Manriquez, K.-U. Jeong, S. Z.-D. Cheng, G. R. Newkome, *Chem. Commun.* 2007, 4456.
71. M. Schmittel, P. Mal, *Chem. Commun.* 2008, 960.
72. M. Schmittel, B. He, *Chem. Commun.* 2008, 4723.
73. S.-C. Yuan, H.-B. Chen, Y. Zhang, J. Pei, *Org. Lett.* 2006, *8*, 5701.
74. J.-L. Wang, Y.-T. Chan, C. N. Moorefield, J. Pei, D. A. Modarelli, N. C. Romano, G. R. Newkome, *Macromol. Rapid Commun.* 2010, *31*, 850.
75. A. Winter, D. A. M. Egbe, U. S. Schubert, *Org. Lett.* 2007, *9*, 2345.
76. A. Winter, C. Friebe, M. D. Hager, U. S. Schubert, *Eur. J. Org. Chem.* 2009, 801.
77. D. Knapton, P. K. Iyer, S. J. Rowan, C. Weder, *Macromolecules*, 2006, *39*, 4069.
78. T. K. Sievers, A. Vergin, H. Möhwald, D. G. Kurth, *Langmuir*, 2007, *23*, 12179.
79. E. Stulz, Y.-F. Ng, S. M. Scott, J. K. M. Sanders, *Chem. Commun.* 2002, 524.
80. Y.-K. Tian, L. Chen, Y.-J. Tian, X.-Y. Wang, F. Wang, *Polym. Chem.* 2013, *4*, 453.
81. Y. Bodenthin, U. Pietsch, H. Möhwald, D. G. Kurth, *J. Am. Chem. Soc.* 2005, *127*, 3110.
82. D. G. Kurth, P. Lehmann, M. Schütte, *Proc. Natl. Acad. Sci. U.S.A.* 2000, *97*, 5704.
83. D. G. Kurth, N. Severin, J. P. Rabe, *Angew. Chem., Int. Ed.* 2002, *41*, 3681.
84. G. Schwarz, Y. Bodenthin, Z. Tomkowicz, W. Haase, T. Geue, J. Kohlbrecher, U. Pietsch, D. G. Kurth, *J. Am. Chem. Soc.* 2011, *133*, 547.
85. Y. Bodenthin, G. Schwarz, Z. Tomkowicz, T. Geue, W. Haase, U. Pietsch, D. G. Kurth, *J. Am. Chem. Soc.* 2009, *131*, 2934.
86. L. He, S. Bi, H. Wang, B. Ma, W. Liu, W. Bu, *Langmuir*, 2012, *28*, 14164.
87. D. Whang, Y.-M. Jeon, J. Heo, K. Kim, *J. Am. Chem. Soc.* 1996, *118*, 11333.
88. D. Whang, K. Kim, *J. Am. Chem. Soc.* 1997, *119*, 451.
89. E. Lee, J. Heo, K. Kim, *Angew. Chem., Int. Ed.* 2000, *39*, 2699.
90. Y. Liu, Z. Huang, X. Tan, Z. Wang, X. Zhang, *Chem. Commun.* 2013, *49*, 5766.
91. C. Zhang, S. Li, J. Zhang, K. Zhu, N. Li, F. Huang, *Org. Lett.* 2007, *9*, 5553.
92. Y. Ding, P. Wang, Y.-K. Tian, Y.-J. Tian, F. Wang, *Chem. Commun.* 2013, *49*, 5951.
93. Y.-K. Tian, F. Wang, *Macromol. Rapid Commun.* 2014, *35*, 337.
94. Y.-K. Tian, Y.-G. Shi, Z.-S. Yang, F. Wang, *Angew. Chem., Int. Ed.* 2014, *53*, 6090.
95. Y.-K. Tian, Z.-S. Yang, X.-Q. Lv, R.-S. Yao, F. Wang, *Chem. Commun.* 2014, *50*, 9477.
96. X. Yan, D. Xu, X. Chi, J. Chen, S. Dong, X. Ding, Y. Yu, F. Huang, *Adv. Mater.* 2012, *24*, 362.
97. X. Yan, M. Zhou, J. Chen, X. Chi, S. Dong, M. Zhang, X. Ding, Y. Yu, S. Shao, F. Huang, *Chem. Commun.* 2011, *47*, 7086.
98. J. Chen, X. Yan, Q. Zhao, L. Li, F. Huang, *Polym. Chem.* 2012, *3*, 458.
99. M. Zhang, D. Xu, X. Yan, J. Chen, S. Dong, B. Zheng, F. Huang, *Angew. Chem., Int. Ed.* 2012, *51*, 7011.
100. X. Yan, D. Xu, J. Chen, M. Zhang, B. Hu, Y. Yu, F. Huang, *Polym. Chem.* 2013, *4*, 3312.
101. R. Chakrabarty, P. S. Mukherjee, P. J. Stang, *Chem. Rev.* 2011, *111*, 6810.
102. X. Yan, T. R. Cook, J. B. Pollock, P. Wei, Y. Zhang, Y. Yu, F. Huang, P. J. Stang, *J. Am. Chem. Soc.* 2014, *136*, 4460.
103. S. Dong, Y. Luo, X. Yan, B. Zheng, X. Ding, Y. Yu, Z. Ma, Q. Zhao, F. Huang, *Angew. Chem., Int. Ed.* 2011, *50*, 1905.
104. P. R. Ashton, E. J. T. Chrystal, P. T. Glink, S. Menzer, C. Schiavo, N. Spencer, J. F. Stoddart, P. A. Tasker, A. J. P. White, D. J. Williams, *Chem. Eur. J.* 1996, *2*, 709.
105. B. Yu, S. Guo, L. He, W. Bu, *Chem. Commun.* 2013, *49*, 3333.

106. L. He, J. Liang, Y. Cong, X. Chen, W. Bu, *Chem. Commun.* 2014, *50*, 10841.
107. J. Yuan, Q. Li, Q. Hu, Q. Wu, C. Li, H. Qiu, M. Zhang, S. Yin, *Chem. Commun.* 2014, *50*, 722.
108. J. Zhan, M. Zhang, M. Zhou, B. Liu, D. Chen, Y. Liu, Q. Chen, H. Qiu, S. Yin, *Macromol. Rapid Commun.* 2014, *35*, 1424.
109. G. Du, E. Moulin, N. Jouault, E. Buhler, N. Giuseppone, *Angew. Chem., Int. Ed.* 2012, *51*, 12504.
110. X. Yan, B. Zheng, F. Huang, *Polym. Chem.* 2013, *4*, 2395.
111. M. Zhang, X. Yan, F. Huang, Z. Niu, H. W. Gibson, *Acc. Chem. Res.* 2014, *47*, 1995.
112. P. Wei, B. Xia, Y. Zhang, Y. Yu, X. Yan, *Chem. Commun.* 2014, *50*, 3973.
113. P. Wei, S. Li, Y. Zhang, Y. Yu, X. Yan, *Polym. Chem.* 2014, *5*, 3972.
114. F. Huang, D. S. Nagvekar, C. Slebodnick, H. W. Gibson, *J. Am. Chem. Soc.* 2005, *127*, 484.
115. P. Wei, J. Li, X. Yan, Q. Zhou, *Org. Lett.* 2014, *16*, 126.
116. F. Wang, J. Zhang, X. Ding, S. Dong, M. Liu, B. Zheng, S. Li, L. Wu, Y. Yu, H. W. Gibson, F. Huang, *Angew. Chem., Int. Ed.* 2010, *49*, 1090.
117. H. Xing, P. Wei, X. Yan, *Org. Lett.* 2014, *16*, 2850.
118. Y. Liu, Y.-L. Zhao, H.-Y. Zhang, H.-B. Song, *Angew. Chem., Int. Ed.* 2003, *42*, 3260.
119. Y.-W. Yang, Y. Chen, Y. Liu, *Inorg. Chem.* 2006, *45*, 3014.
120. Y. Liu, Y. Song, H. Wang, H.-Y. Zhang, X.-Q. Li, *Macromolecules*, 2004, *37*, 6370.
121. Y. Liu, Z.-X. Yang, Y. Chen, Y. Song, N. Shao, *ACS Nano*, 2008, *2*, 554.
122. L. Zhu, M. Lu, Q. Zhang, D. Qu, H. Tian, *Macromolecules*, 2011, *44*, 4092.
123. H. Masai, J. Terao, S. Seki, S. Nakashima, M. Kiguchi, K. Okoshi, T. Fujihara, Y. Tsuji, *J. Am. Chem. Soc.* 2014, *136*, 1742.
124. Z. Zhang, C. Han, G. Yu, F. Huang, *Chem. Sci.* 2012, *3*, 3026.
125. Z.-Y. Li, Y. Zhang, C.-W. Zhang, L.-J. Chen, C. Wang, H. Tan, Y. Yu, X. Li, H.-B. Yang, *J. Am. Chem. Soc.* 2014, *136*, 8577.
126. H. Hofmeier, A. El-Ghayoury, A. P. H. J. Schenning, U. S. Schubert, *Chem. Commun.* 2004, *40*, 318.
127. H. Hofmeier, R. Hoogenboom, M. E. L. Wouters, U. S. Schubert, *J. Am. Chem. Soc.* 2005, *127*, 2913.
128. U. Mansfeld, M. D. Hager, R. Hoogenboom, C. Ott, A. Winter, U. S. Schubert, *Chem. Commun.* 2009, *45*, 3386.
129. F. Grimm, K. Hartnagel, F. Wessendorf, A. Hirsch, *Chem. Commun.* 2009, *45*, 1331.
130. G. Gröger, W. Meyer-Zaika, C. Böttcher, F. Gröhn, C. Ruthard, C. Schmuck, *J. Am. Chem. Soc.* 2011, *133*, 8961.
131. F. S. Han, M. Higuchi, D. G. Kurth, *J. Am. Chem. Soc.* 2008, *130*, 2073.
132. R. Dobrawa, F. Würthner, *Chem. Commun. 2002*, 1878.
133. R. R. Pal, M. Higuchi, Y. Negishi, T. Tsukuda, D. G. Kurth, *Polym. J.* 2010, *42*, 336.
134. A. Winter, C. Friebe, M. Chiper, M. D. Hager, U. S. Schubert, *J. Polym. Sci., Part A: Polym. Chem.* 2009, *47*, 4083.
135. F. Han, M. Higuchi, D. G. Kurth, *Tetrahedron*, 2008, *64*, 9108.
136. O. Hagemann, M. Jørgensen, F. C. Krebs, *J. Org. Chem.* 2006, *71*, 5546.
137. V. Duprez, M. Biancardo, H. Spangaard, F. C. Krebs, *Macromolecules*, 2005, *38*, 10436.
138. F. C. Krebs, M. Biancardo, *Sol. Energy Mater. Sol. Cells* 2006, *90*, 142.
139. M. Burnworth, L. M. Tang, J. R. Kumpfer, A. J. Duncan, F. L. Beyer, G. L. Fiore, S. J. Rowan, C. Weder, *Nature*, 2011, *472*, 334.
140. N. Holten-Andersen, M. J. Harrington, H. Birkedal, B. P. Lee, P. B. Messersmith, K. Y. C. Lee, J. H. Waite, *Proc. Natl. Acad. Sci. U.S.A.* 2011, *108*, 2651.
141. S. Bode, L. Zedler, F. H. Schacher, B. Dietzek, M. Schmitt, J. Popp, M. D. Hager, U. S. Schubert, *Adv. Mater.* 2013, *25*, 1634.
142. R. Dong, Y. Su, S. Yu, Y. Zhou, Y. Lu, X. Zhu, *Chem. Commun.* 2013, *49*, 9845.
143. M. Sun, H. Zhang, B. Liu, Y. Liu, *Macromolecules*, 2013, *46*, 4268.

PART IV

CATALYSIS

17

CATALYTIC APPLICATIONS OF METAL COMPLEXES IMMOBILIZED BY NON-COVALENT INTERACTIONS ONTO CHEMICALLY DERIVED GRAPHENES AND RELATED MATERIALS

SARA SABATER AND JOSE A. MATA

Departamento de Química Inorgánica y Orgánica, Institute of Advanced Materials (INAM), Universitat Jaume I, Castellón de la Plana, Spain

17.1 INTRODUCTION

Despite the attractive properties of many homogeneous catalysts in terms of activity and selectivity, they suffer from difficulties associated with separating the catalysts from the products, an important drawback that makes industries to be relatively reluctant to use them for the large-scale production. Another important disadvantage for the industrial use of homogeneous catalysts concerns catalyst recycling, for which an enormous effort is currently being carried out in the immobilization of known catalysts onto solid supports.

Catalyst recovery follows the definition of green chemistry proposed by Anastas and Warner as "the utilization of a set of principles that reduces or eliminates the use or generation of hazardous substances in the design, manufacture and application of chemical products" [1] and has encouraged the scientific community to look for efficient ways to separate homogeneous catalysts from the reaction media and subsequent recycling. The main goal in recycling a catalyst refers to achieving consistent high yields in repeated runs that testify the robustness of the supported catalytic system, a necessary requirement for long-term applications.

Undoubtedly, the most widely used method for the immobilization of a homogeneous catalyst is the formation of a covalent bond between the solid support and the ligand. This general method has two important drawbacks. The first one, arising from the need of an additional functionalization of the ligand and solid surface (which, in turn, implies an increase in the catalyst preparation cost), and the second one, arising from the possible changes in chemical reactivity and therefore in catalytic activity, derived from the formation of the covalent bond between the support and the catalyst. As an alternative, non-covalent interactions are merging as a powerful tool that gives rise to supported catalysts without chemical modification of the homogeneous catalyst or the support.

The most common non-covalent interactions are hydrogen bonding, van der Waals forces, and π-stacking. The relative energy of a single non-covalent interaction is much lower than a typical covalent bond, but on account of their multiplicity, the overall influence should not be negligible on the anchoring of the molecular complex on surface of the support. The stability against leaching may become a real problem that needs to be avoided by a careful selection of the reaction conditions, immobilization method, nature of the support, and catalytic complex. The main advantage of immobilizing homogeneous catalyst by non-covalent interactions is the simplicity of catalyst design and ready availability. Further modifications to achieve high selectivity and stability are easily obtained by tuning the electronic and steric properties of the ligands.

Well-defined supported molecular complexes for catalytic applications should facilitate precise characterization of the hybrid materials and study of the reaction mechanisms. The reaction mechanism can be studied in solution at the molecular level with the well-defined homogeneous organometallic

Non-Covalent Interactions in the Synthesis and Design of New Compounds, First Edition.
Edited by Abel M. Maharramov, Kamran T. Mahmudov, Maximilian N. Kopylovich, and Armando J. L. Pombeiro.

techniques. The support may play a role during the catalytic process, not only acting as a simple platform for anchoring the molecular complex but also enhancing the catalytic activity by avoiding aggregation and facilitating the interaction between the substrates and the catalytic active species. Figure 17.1 describes the key points in the rational design of a supported molecular complex for catalytic applications.

Graphene is a two-dimensional carbon sheet of one atom thickness and it has inspired tremendous interests in physics, materials science, chemistry, and biology [2]. Graphene features unique electronic, mechanical, optical, thermal, and chemical properties, which are inherent, from the two-dimensional structure. However, despite the excellent properties and characteristics of pristine graphene, the synthesis of this material in its pure form is still challenging the scientific community. The difficulties found in the synthesis of pristine graphene have promoted the interest of derived graphene nanocomposites and hybrids for real applications. Chemically derived graphenes (CDGs), such as graphene oxide (GO) and reduced graphene oxide (rGO), are ideal candidates for the immobilization of active transition metal complexes for catalytic applications due to its inertness, stability, large-scale availability, and the presence of functional groups [3]. The properties of CDGs strongly depend on the synthetic procedure used and pave the way for the use of related materials based on graphene. Another interesting properties of CDGs as support for catalytic applications arise from the high specific surface area, estimated in the case of pristine graphene of *ca.* 2600 m^2/g, and also from the tailored pore size control.

In general, catalyst immobilization onto rGO has been carried out mostly using the carboxylic or hydroxyl groups present at the edges of rGO sheets. These covalent functionalization strategies modify the carbon hybridization and therefore lead to a partial loss of rGO properties such as conductivity. As an alternative, the immobilization by non-covalent interactions on the surface of rGO assures to maintain the properties of the material. In this context, different synthetic strategies are contemplated for anchoring molecular complexes on the surface of CDGs based on non-covalent interactions. The π-stacking or electrostatic interactions are good approaches for the functionalization of graphene with well-defined molecular catalyst preserving the inherent properties of the material [4].

17.2 PREPARATION OF CDGs

Graphene derivatives are complex materials considering the composition, number and nature of functional groups, and the size and number of sheets. The control of all these factors depends on the methodological approach used for their synthesis and determines the inherent properties of the material [5]. The preparation of CDGs is an open research field, and most efforts persuade industrial-scale production of graphene.

There are different ways to prepare graphene materials. This chapter is focused on the use of graphene materials prepared by chemical procedures. These protocols are based on a sequence of redox reactions combined with exfoliation processes (Scheme 17.1). The materials involved in this procedure are three-dimensional (3D) materials, including graphite and graphite oxide and single- to few-layer materials including GO, rGO, and graphene (Fig. 17.2). The chemical synthesis of graphene is a top-down method for the preparation of CDGs and is considered a scalable procedure for the future industrial applications of graphene materials. Since a well-defined classification of these materials has not been yet established and the nomenclature used in the recent bibliography is still puzzling, we have followed the most recent guidelines according to the models and physicochemical properties to assist in the classification of the graphene materials.

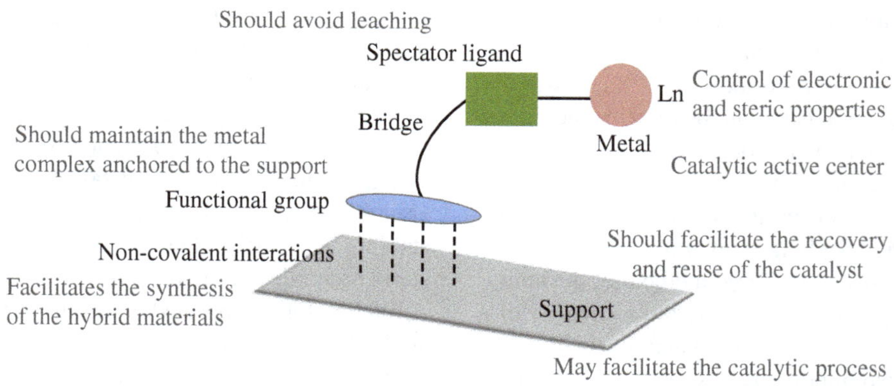

FIGURE 17.1 Rational design of supported molecular complexes for catalytic applications.

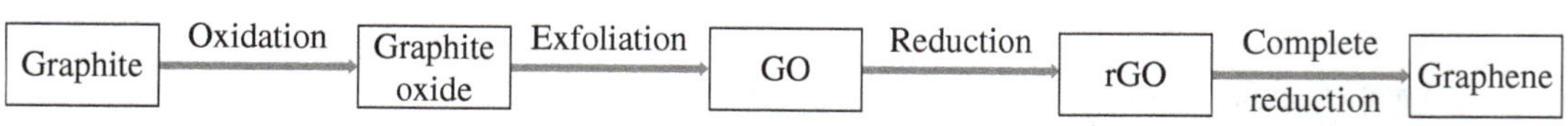

SCHEME 17.1 Chemical synthesis of graphene materials.

The oxidation of graphite using strong oxidants leads to graphite oxide (Fig. 17.2), which is a well-known material that contains epoxides at the basal planes and hydroxyl and carboxylic functional groups at the edges. Graphite oxide was first developed by B. C. Brodie [6] in 1859 using potassium chlorate and fuming nitric acid, and ever since many other variants have been used. Nowadays, the typical procedure for oxidizing graphite is the Hummers method, which uses potassium permanganate, sodium nitrate, and concentrated sulfuric acid [7]. The introduction of oxygen functionalities results in an increase of the interlayer distance compared to graphite (~0.6 nm) and a decrease of π–π interactions between the graphite layers. As a consequence, graphite oxide can be easily exfoliated by ultrasounds rendering single layers of GO.

The exact structure of GO is and has been under intense debate and many models have been proposed. The most accepted model was proposed by Lerf and Klinowski [8] and assumes the presence of epoxides and hydroxyl groups at the basal planes, carboxylic groups at the edges, and sp^3-hybridized carbon atoms leading to a nonstoichiometric amorphous material. The presence of functional groups and imperfections disrupts the electronic structure of the graphene material making it electrically insulating. A characteristic property of GO is the high content of oxygen; the average elemental analysis: C/O ratio is about 2–4 [3b]. Suspensions of GO in polar solvents are stable for weeks due to the hydrophilic character of the sheets.

Reduction of GO leads to a material known as rGO. The complete reduction of GO should give pristine graphene, but in practice the rGO material is quite different. The rGO is an intermediate material whose properties are different of pristine graphene and GO. The most common procedure for the reduction of GO is the chemical reduction using hydrazine monohydrate [9], although other chemicals have been used, such as sodium borohydride, hydroquinone, and vitamin C. The rGO obtained using hydrazine has a surface area of 450 m^2/g and a conductivity of 1200 S/m; those properties are very far from of the calculated ones for the pristine graphene (Table 17.1). The reduction process removes many oxygen functionalities of the GO restoring the π-network of the material and increasing the surface area and conductivity.

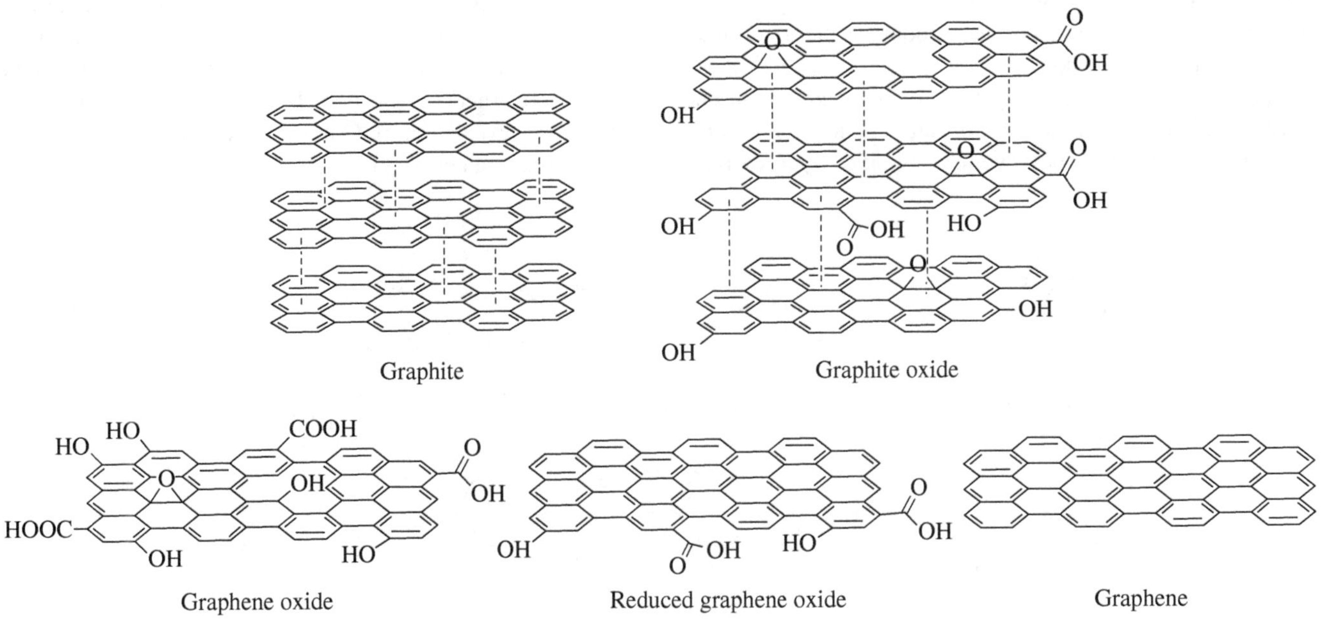

FIGURE 17.2 Representative models for graphene and related materials.

TABLE 17.1 General Properties of Graphene Materials

	Graphite	Graphite Oxide	GO	rGO[a]	Graphene[b]
Surface area[c] (m^2/g)	4–10	4–10	20–30	400–500	2600
Electrical conductivity (S/m)	1×10^5	Insulator	Insulator	600–2400	1×10^8
C/O relation	—	2–4	2–4	10–14	—
Monolayer thickness (nm)	—	—	0.6–1	0.35–0.40	0.34
Dimensionality	3D	3D	2D	2D	2D
Number of layers	Multilayer	Multilayer	<10	<10	1

[a] Values for rGO obtained using hydrazine.
[b] Estimated values for pristine graphene.
[c] Brunauer–Emmett–Teller (BET) surface area measurement.

The chemical synthesis of graphene has a major disadvantage of introducing impurities in the CDGs. For example, the reduction of GO with hydrazine introduces nitrogen impurities covalently bound to graphene.

17.3 FUNCTIONALIZATION BY NON-COVALENT INTERACTIONS

Covalent functionalization of CDGs can be achieved by different methods due to the presence of functional groups such as hydroxyls, carboxyls and epoxides [10]. In contrast, in pristine graphene, due to the lack of functional groups, the non-covalent interactions are the most convenient tool for functionalization. Non-covalent interactions are frequently employed to prepare CDG hybrids and nanocomposites with surfactants, small organic molecules, polymers, or even biomolecules, such as proteins or DNA. These non-covalent interactions between organic molecules and graphene may promote the attachment of organic species onto graphene surfaces by means of π-stacking, electrostatic, or hydrophobic interactions at mild conditions. Under these conditions, the functionalized graphenes maintain the properties of the material and thus serve as a model for the functionalization of pristine graphene.

The π-electron aromatic system found in graphene and derivatives offers a unique opportunity to anchor organic molecules containing polycyclic aromatic hydrocarbons (PAHs) by non-covalent π–π interactions. The energy for a typical π-stacking interaction ranges in the order of 2 kJ/mol, and in principle it should play a role, similar to the hydrogen bonding, in binding and conformations adopted [11]. The strength of π–π interactions depends on the number of aromatic rings in the π-electron system of the adsorbed molecules and on the contact curvature between the aromatic region of the molecule and the surface of graphene materials. In the case of graphene, GO, and rGO materials, the adsorption takes place at the basal planes of the graphitic materials.

The immobilization of polyaromatic hydrocarbons onto graphene derivatives has been developed by analogy with graphite and carbon nanotubes (CNTs). The non-covalent interactions between polyaromatic hydrocarbons (Fig. 17.3) and graphene are based on van der Waals forces (π-stacking) and are controlled by thermodynamics.

A comparative study of the adsorption of different PAHs on CDGs has been carried out using pyrene, anthracene, and naphthalene. The study reveals that the adsorption of PAHs does not depend on pH and that the adsorption capacity decreases in the order naphthalene > pyrene > anthracene. rGO has a larger adsorption than GO and graphite materials. The interaction mechanism of pyrene and anthracene is based on flat surface adsorption and in the case on naphthalene is governed by the pore-filling mechanism [12]. The study also reveals that rGO has a larger adsorption than GO and graphite materials [13].

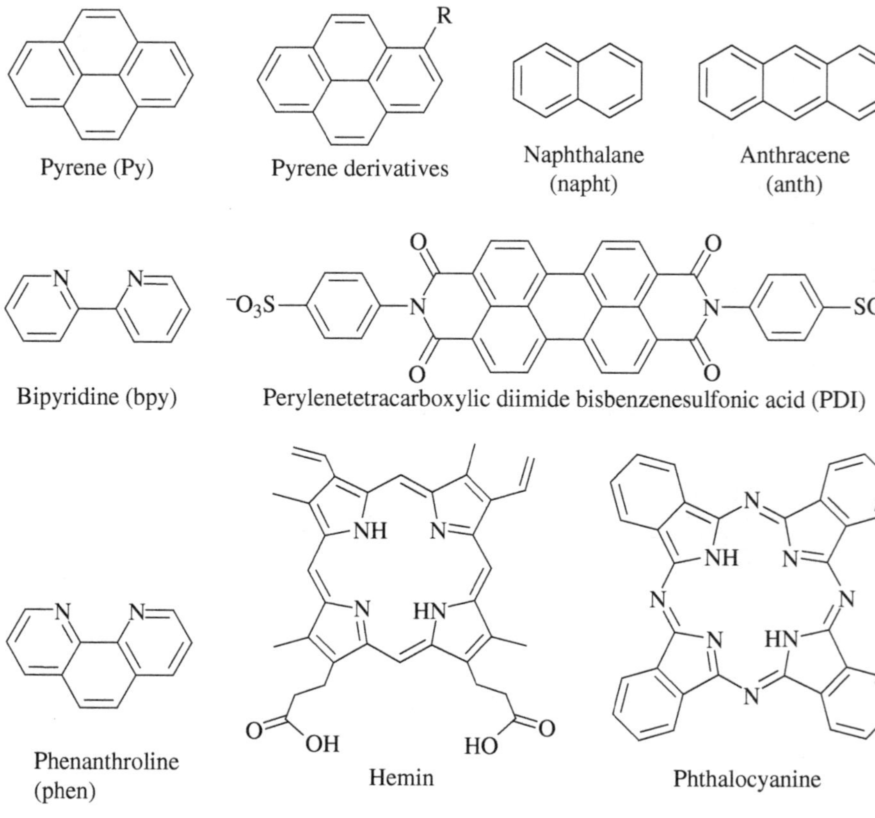

FIGURE 17.3 Polyaromatic molecules that have been immobilized on the surface of chemically derived graphenes.

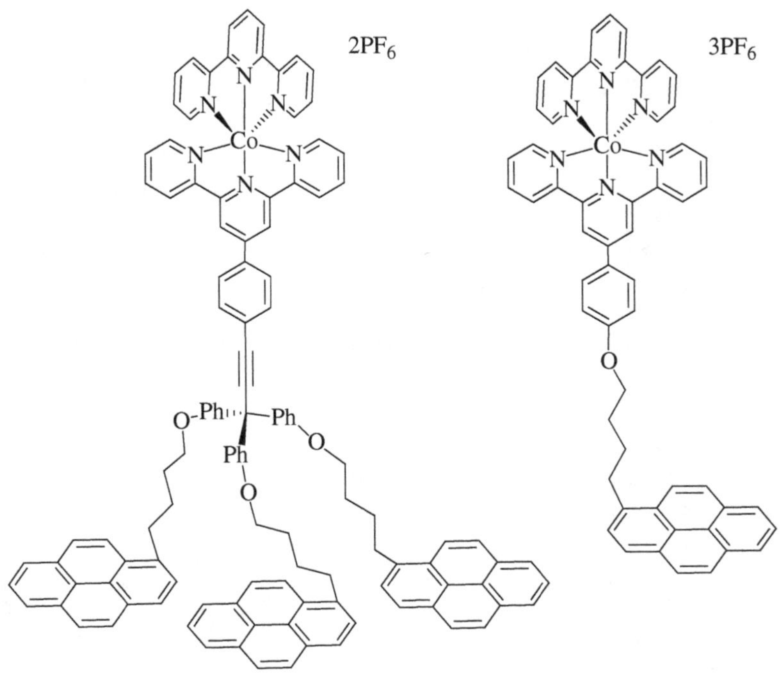

FIGURE 17.4 Model of cobalt complexes used for determination of binding characteristics between pyrene units and graphene.

The kinetic and thermodynamic binding parameters of aromatic systems (Fig. 17.4) on graphene have been evaluated by means of electrochemical studies [14]. Dichtel and coworkers described the immobilization of a tripodal motif that binds multivalently to graphene though three pyrene moieties. The results reveal that the tripodal binding motifs form stable monolayers with a molecular footprint of 2.3 nm^2 and a ΔG_{ads}= −38.8 ± 0.2 kJ/mol, while similar model compounds, bearing a single pyrene units, readily desorb from graphene in organic solvents. Interestingly, the binding energies of the tripodal and monopodal compounds are similar (ΔG_{ads}= −38.8 and −38.3 kJ/mol, respectively), but they differ in their kinetic stability, in which the tripodal motif desorbs 1000 times slower than the monopodal analogue.

17.4 CATALYTIC APPLICATIONS

The immobilization of molecular complexes onto graphene surfaces and CNTs leads to hybrid materials that may be used for catalytic applications. The idea of combining highly active catalysts with the possibility of separating them from products is very attractive from the industrial point of view. In addition, the support may facilitate the catalytic process by synergistic effects, increasing the stability of the catalyst or allowing reagents to come closer and react.

17.4.1 Complexes Immobilized on CNTs

The catalytic applications of materials derived from the immobilization of molecular complexes on the surface of CNTs by non-covalent interactions are a well-established field developed in the last years. In this section, the main properties of such materials focused on pyrene derivatives will be described. The general properties of CNTs and CDGs are similar. The main difference is the curvature nature of the CNTs.

Hybrid materials based on CNTs containing metal complexes with a pyrene-tagged ligand have been described for gold, ruthenium, and rhodium (Fig. 17.5). The gold molecular complex (Py–Au–CNT) supported onto CNTs has recently been reported and represents the first example of a gold complex immobilized on the surface of CNTs [15]. The gold complex containing a pyrene tag is anchored on the surface of CNTs by π–π interactions. The immobilization process is carried out by stirring a solution of the gold complex with a suspension of CNTs. The results show that the incorporation of the molecular complex on the CNTs is dependent on the solvent polarity and temperature. The hybrid material is active in the cyclization of enynes and propargylic carbonates. The catalytic activity of the molecular complex and the hybrid material is identical, indicating that the support does not take part in the catalytic process. The hybrid material is recycled up to five times using dichloromethane as solvent. The catalytic process is homogeneous in nature, in which the boomerang effect is taking place. The supported gold molecular complex is desorbed from the surface of the CNTs during the catalytic process and is reimmobilized at low temperatures.

The pyrene-tagged ruthenium Hoveyda–Grubbs (Py–Ru–CNT)-type catalyst has also been anchored onto CNTs by π–π interactions [16]. The immobilization of the molecular complex is affected by the solvent polarity and temperature. Low temperatures and polar solvents such as acetone prevent the molecular complex desorption. The hybrid material is

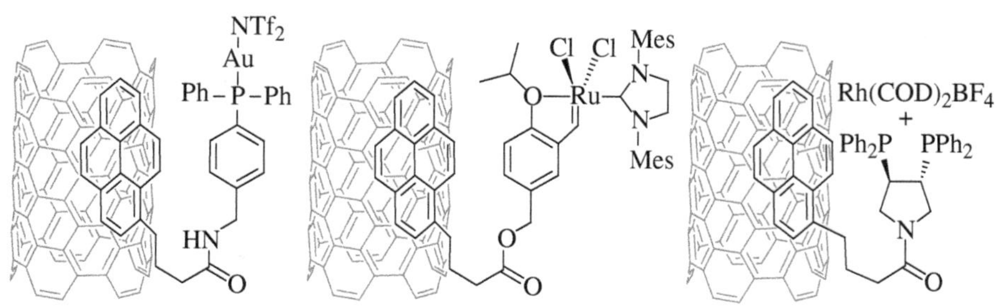

Py-Au–CNT Py-Ru–CNT Py-Rh–CNT

FIGURE 17.5 Hybrid materials based on CNTs containing molecular complexes with a single pyrene-tagged ligand.

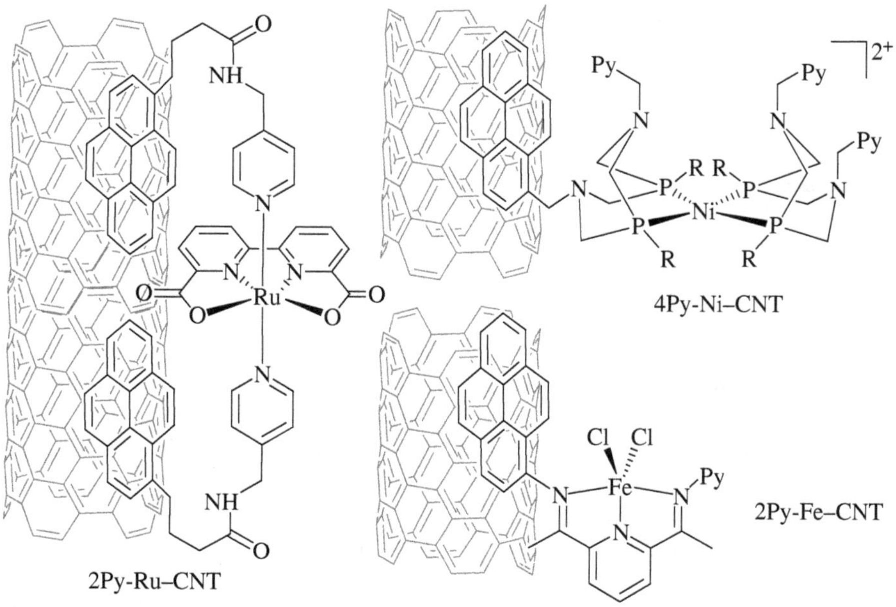

4Py-Ni–CNT

2Py-Fe–CNT

2Py-Ru–CNT

FIGURE 17.6 Hybrid materials based on CNTs with molecular complexes containing ligands with multiple pyrene units.

active in ring-closing metathesis of a variety of substrates. The supported ruthenium catalyst can be recycled up to six or seven times. In the mechanism of the catalytic reactions, the boomerang effect takes place as well.

The rhodium complex (Py–Rh–CNT) containing a chiral pyrene–phosphine-tagged ligand was anchored onto CNTs by π– π interactions. The catalytic activity was tested in the asymmetric hydrogenation of α-dehydroamino esters. The adsorption of the chiral catalyst is carried out in polar solvents, and no adsorption is observed when using dichloromethane. The hybrid chiral material can be recycled up to nine times without any decrease in activity and enantioselectivity. In this case, the boomerang effect also takes place, as previously described for the gold and ruthenium hybrid materials containing a pyrene tag.

In order to improve the binding properties of pyrene-tagged ligands and CNTs, ruthenium, nickel, and iron complexes containing multiple pyrene units have been synthesized and immobilized (Fig. 17.6). The ruthenium molecular complex (2Py–Ru–CNT) with two pyrene units is a highly active electrocatalyst in water oxidation. The electrolysis is carried out in neutral aqueous solution and at relatively low potential. The system is stable for 10 h with an average turnover frequency (TOF) of $0.3\,s^{-1}$ [17]. The nickel molecular complex (4Py–Ni–CNT) with four pyrene units is active in the electrocatalytic hydrogen evolution and uptake. Interestingly, the catalytic activity is maintained in the presence of carbon monoxide, indicating the stability conferred by the support [18]. The iron molecular complex (2Py–Fe–CNT) with two pyrene units displays high activity in ethylene polymerization up to $10^7\,g_{PE}/(mol_{Fe}\,h)$ (PE = polyethylene). The supported iron complex shows better catalytic performance than its homogeneous counterpart, indicating the benefits of the support [19].

17.4.2 Complexes Immobilized on CDGs

With a two-dimensional sheetlike structure, graphene represents an interesting geometrical support for molecular catalysts with a large open surface area that is readily accessible to substrates/products with a small diffusion barrier. The preparation of hybrid materials based on graphene is generally performed by simultaneous reduction of GO and metal salts. The hybrid materials obtained show the presence of

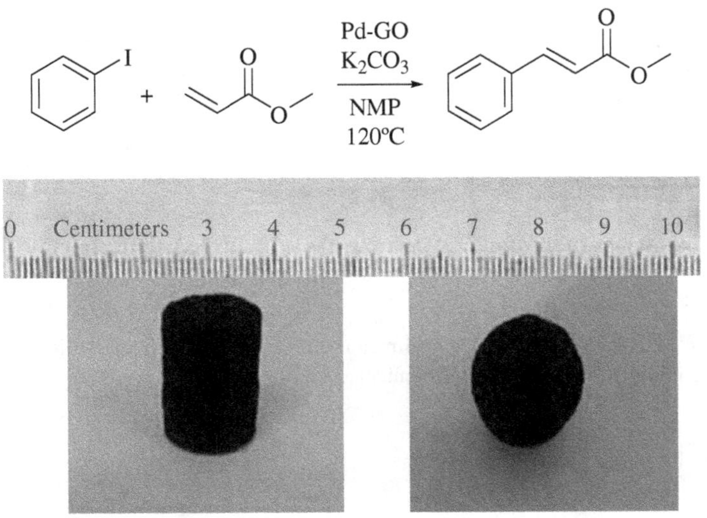

FIGURE 17.7 Heck reaction of phenyl iodide and methyl acrylate using Pd/GO at a catalyst loading of 0.1 mol%. Front and top view of the fixed-bed Pd/GO reactor catalyst. Adapted with permission from reference [20]. Copyright 2010 WILEY-VCH Verlag GmbH & Co.

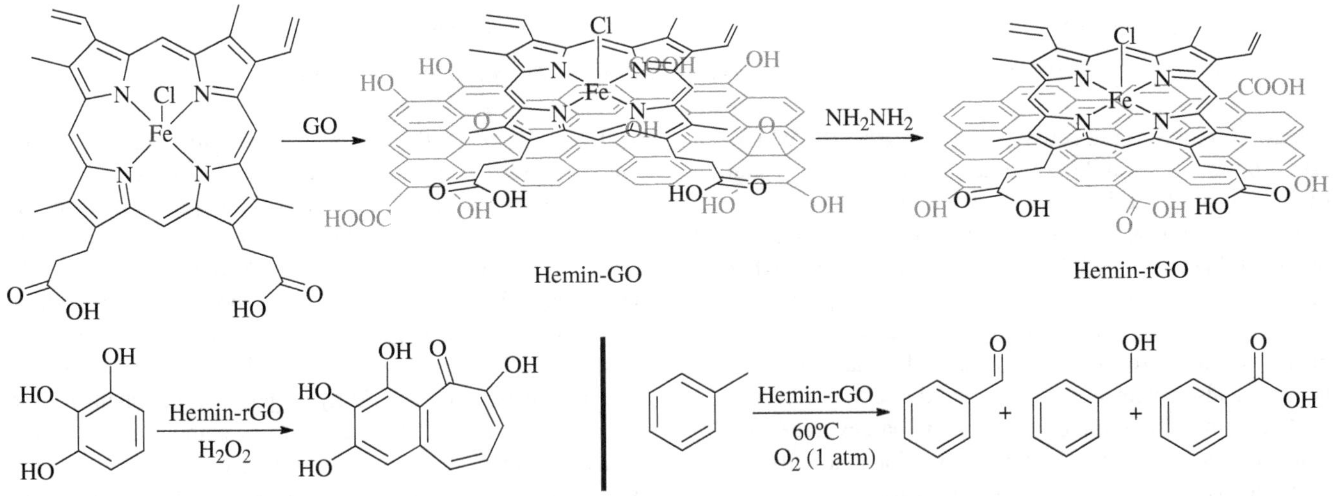

FIGURE 17.8 Synthesis of hemin–rGO hybrid material (top) and catalytic applications in the oxidation reaction of pyrogallol using H_2O_2 and toluene using O_2 (bottom). Adapted with permission from Ref. [21]. Copyright 2012 Wiley-VCH Verlag GmbH & Co.

metal nanoparticles immobilized on the surface of rGO. The nanoparticles formed are retained on the surface of graphene by a combination of electrostatic interactions, overlapping of the extended π-orbital of the graphene with the d-orbitals of the transition metals, and allocation of nanoparticles at the defects of graphene.

Information on the use of immobilization by non-covalent interactions of metal complexes onto graphene derivatives for catalytic applications is scarce. As a first approach, Wang and coworkers described the synthesis of a 3D macroassembly of GO promoted by palladium [20]. The hydrothermal decomposition of palladium chloride, glucose, and GO yields 3D graphene composites (Fig. 17.7). The palladium-based material is active in the Heck reaction of phenyl iodide and methyl acrylate and has a great potential for fixed-bed process in industry.

Duan, Huang, and coworkers described the synthesis of a hemin–graphene hybrid material through π–π interactions (Fig. 17.8) [21]. Hemin, an iron porphyrin, is the active center of heme proteins such as cytochromes, peroxidase, myoglobins, and hemoglobins and can catalyze a variety of oxidation reactions like peroxidase enzymes. The hemin adsorption on the graphene was analyzed by UV–Vis and AFM studies. The UV–Vis spectrum of the hybrid material shows a blue-shifted band compared to free hemin, which suggests the formation of an axial ligation to the iron center of hemin likely because of cation–π interactions between iron centers and graphene.

The hybrid material acts as a biomimetic catalyst in the oxidation of pyrogallol, with exceptionally high catalytic activity and substrate binding affinity (Fig. 17.8). Interestingly, the monomeric hemin–graphene conjugate is

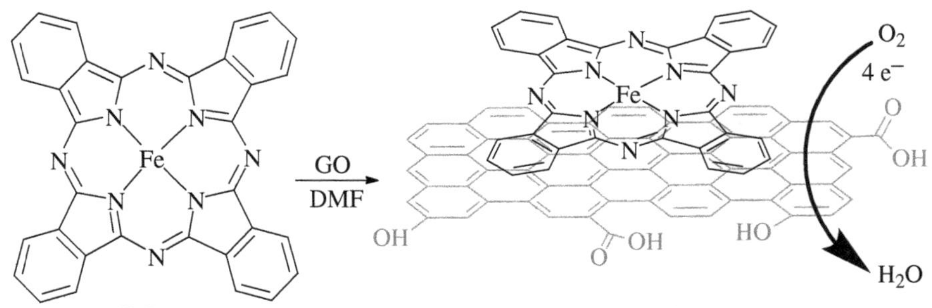

FIGURE 17.9 Synthesis of iron phthalocyanine–rGO hybrid material and catalytic application in oxygen reduction reaction. Adapted with permission from Ref. [23]. Copyright 2013 American Chemical Society.

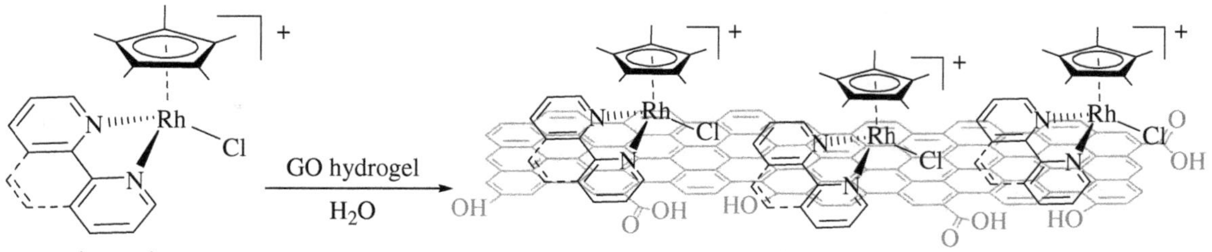

SCHEME 17.2 Synthesis of rhodium–graphene oxide hydrogel hybrid materials.

at least 100 times more active than the free hemin, which did not show a measurable catalytic activity in the oxidation [21]. The applicability of the hemin–graphene conjugate was also demonstrated in the oxidation of primary C–H bond in toluene (Fig. 17.8) [22]. The hemin–graphene hybrid material was proved to be an active catalyst in toluene oxidation under mild conditions. The results reveal that graphene not only prevents the nondesired dimerization of the hemin but also enhances the activity of the central iron core.

The immobilization of an iron phthalocyanine (FePc) onto rGO sheets through π–π interactions has also been described (Fig. 17.9) [23]. This hybrid material is active in the oxygen reduction reaction (ORR). The graphene–iron phthalocyanine exhibits comparable activity, superior stability, methanol tolerance, and anti-CO poisoning effects compared with commercial Pd/C catalyst for the four-electron ORR in alkaline solutions. The improved ORR performance of FePc by graphene support is attributed to increase dispersibility of FePc on graphene sheets where more active sites are exposed and the facilitated electron transfer in the ORR by the graphene support.

A cationic molecular rhodium complex containing 1,4′-phenanthroline (phen) or 2,2′-bipyridine (bpy) has been immobilized on a 3D-structured GO hydrogel [24]. The immobilization process consists in the immersion of the GO hydrogel on a water solution of the rhodium complex (Scheme 17.2). The rhodium complex is retained onto the graphene surface through π-stacking and π–cation interactions. The results reveal that the adsorption rate and the amount of the Rh(phen) complex are higher than that of Rh(bpy); the higher adsorption is attributed to the higher strength of π-interaction of the phenanthroline ligand with graphene.

The 1,4'-phenanthroline ligand contains three aromatic rings, so the strength of the π-interaction of the complex is higher compared to the bipyridine ligand [25]. The catalytic properties of the rhodium– GO hydrogel have been tested in the electrochemical regeneration of NAD(P)H (Fig. 17.10). The results show that the activity of the molecular rhodium complex is almost negligible and the catalytic activity is remarkably increased when the complex is immobilized onto the surface of the GO hydrogel. The Rh(phen) hybrid material was reused up to seven times without any decrease in activity. In contrast, the catalytic activity of the Rh(bpy) hybrid material could be only maintained for three runs (Fig. 17.10). The difference in activity of both complexes is attributed to their adsorption and stability in the GO hydrogel.

The use of an *N*-heterocyclic carbene (NHC) ligand with a pyrene tag allows to anchor ruthenium and palladium complexes onto the surface of rGO by π-stacking interactions (Scheme 17.3) [26]. Complexes with NHC ligands are an important class of transition metal-based catalysts, normally acting as spectator ligands [27]. From the electronic properties, the σ-donor capacity of NHC ligands leads to strong metal–carbon bonds that provide stable complexes suitable to operate under harsh conditions [28]. These important properties and applications make NHC ligands good candidates for the preparation of reusable catalysts. Interestingly, the immobilization of complexes based on NHC ligands without the pyrene tag was unsuccessful, thus confirming that the pyrene tag is responsible for the adsorption onto the graphene surface. The mild reaction conditions used for immobilization imply that the properties of the molecular complex and the graphene material should be preserved.

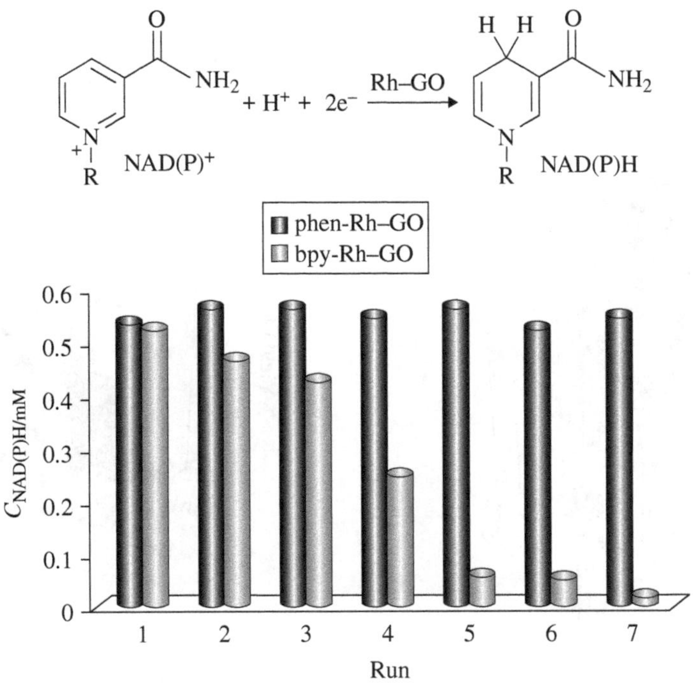

FIGURE 17.10 Recycling experiment for the electrochemical regeneration of NAD(P)H using Rh–GO complexes. The initial concentration of $NAD(P)^+$ was 1 mM, and the regeneration yield of $NAD(P)H^+$ was determined after 120 min reaction for each run. Adapted with permission from Ref. [25]. Copyright 2007 Royal Society of Chemistry.

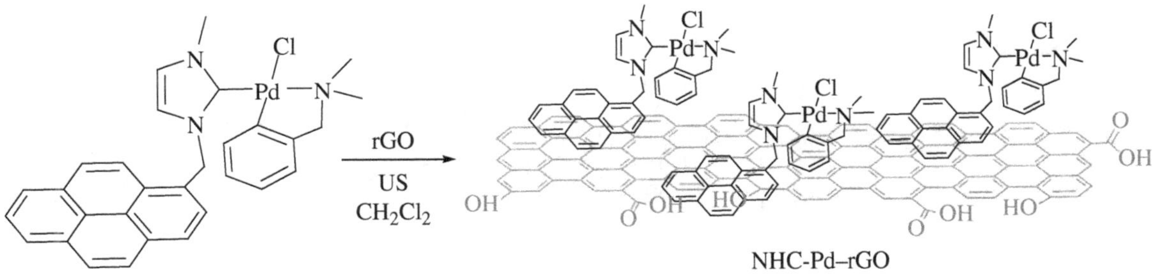

SCHEME 17.3 Synthesis of palladium–rGO hybrid material using ultrasounds.

The graphene decorated with the palladium complex is active in the reduction of alkenes and nitroarenes using molecular hydrogen [26]. The catalytic results are improved in the hybrid materials, compared to the catalytic outcomes provided by the homogeneous palladium complex. The recycling experiments reveal that the catalytic activity is maintained at least for 10 consecutive runs. The time-course recycling experiment reveals that the catalyst activity is maintained at different times for three runs (Fig. 17.11). The ICP-MS analysis of the material after the tenth run indicates a palladium leaching of less than 10%. Morphological analysis by HRTEM showed the formation of well-dispersed palladium nanoparticles. The decomposition of the palladium molecular complex during the catalytic reactions renders palladium nanoparticles anchored on the surface of the material.

The ruthenium–rGO material (Scheme 17.4) is active in the oxidation of alcohols to afford the corresponding ketone or aldehyde derivatives [26].The catalytic results reveal that the immobilization of the ruthenium molecular complex onto rGO provides a benefit in the catalytic process, as has also been observed in the case of the palladium hybrid material. The ruthenium–rGO material was reused up to 10 times without any decrease in activity, and the catalyst activity is maintained at different reaction times as shown by the time-course recycling experiment (Fig. 17.12).

The analysis of the material after the recycling experiments by ICP-MS accounts for a negligible leaching of ruthenium. Morphological analysis by HRTEM of the ruthenium–rGO material before and after the recycling experiments revealed the same well-dispersed ruthenium distribution, without modification of the support morphology and the absence of ruthenium nanoparticles. In the case of the ruthenium material, the molecular character of the complex is preserved after the catalytic experiments without the formation of nanoparticles or metal lost by leaching.

A hot filtration experiment reveals that the catalysis was heterogeneous in nature. This experiment confirmed the absence of catalytic active species in solution due to ruthenium leaching or desorption of the molecular catalyst. No

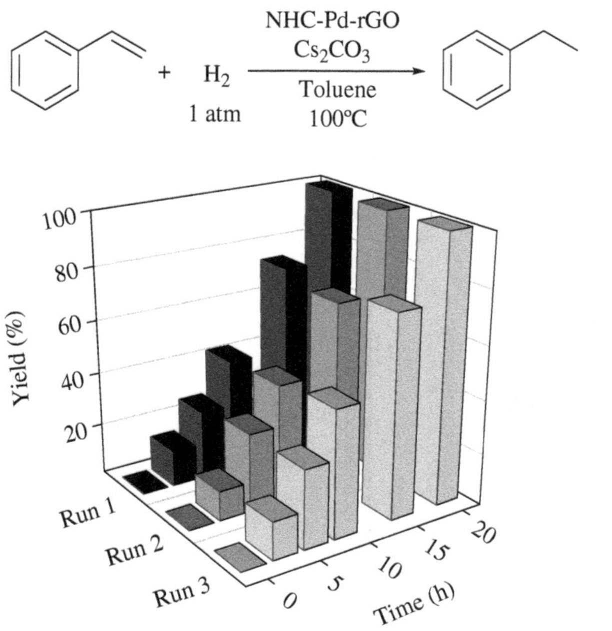

FIGURE 17.11 Time-course recycling experiment of styrene hydrogenation using NHC–Pd–rGO at a catalyst loading of 0.01 mol%. Adapted with permission from Ref. [26]. Copyright 2014 American Chemical Society.

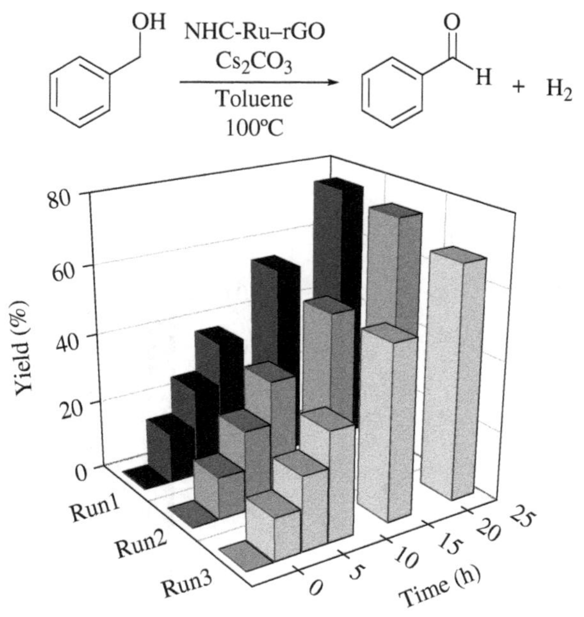

FIGURE 17.12 Time-course recycling experiment in benzyl alcohol oxidation using NHC–Ru–rGO at a catalyst loading of 0.03mol%. Adapted with permission from Ref. [26]. Copyright 2014 American Chemical Society.

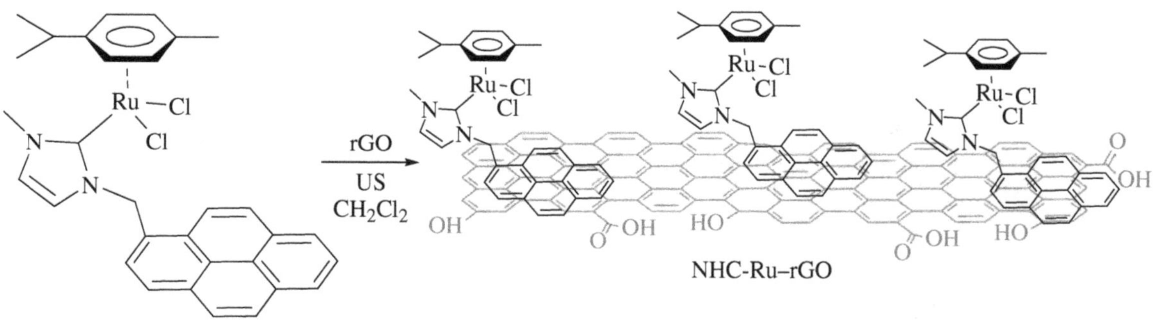

SCHEME 17.4 Synthesis of ruthenium–rGO hybrid material.

catalytic activity was observed by the so-called boomerang effect that consists in the releasing of the molecular complex during the catalytic experiment and subsequent readsorption. The ruthenium molecular complex is strongly anchored on the surface of rGO, and no desorption was observed by increasing the temperature or changing the solvent polarity.

More interestingly, this methodology allows the simultaneous immobilization of ruthenium and palladium metal complexes onto the surface of rGO (Scheme 17.5) [29]. The material obtained shows a homogeneous distribution of both metals along the material. The palladium–ruthenium–rGO hybrid material is active in the hydrodefluorination of fluoroarenes.

Interestingly, the catalyst is only active when it is immobilized onto the graphene surface; the homogeneous hydrodefluorination with the ruthenium and palladium molecular complexes did not work under the same reaction conditions. The catalyst is only active when it is immobilized onto the graphene surface. The enhancement of the activity of the immobilized catalyst is attributed to an activation process of the molecular complexes where the support may play an important role in the stabilization of the catalytic active species (Fig. 17.13). The analysis of the catalyst before and after being used in the catalytic reaction indicates that the active catalytic species are formed by small palladium nanoparticles combined with the ruthenium molecular complex. The time-course reaction profiles show that the curve corresponding to the first run has a sigmoidal form, indicating that the rate of formation of benzene is slowly increased with the reaction time, as a consequence of the gradual activation of the catalyst. Run 2 needs 12 h to completion, and runs 3–8 display *quasi*-identical profiles, achieving completion in about 3 h. The similarities of the reaction profiles displayed for runs 3–8 are a very good indication that the catalyst deactivation is negligible (Fig. 17.13).

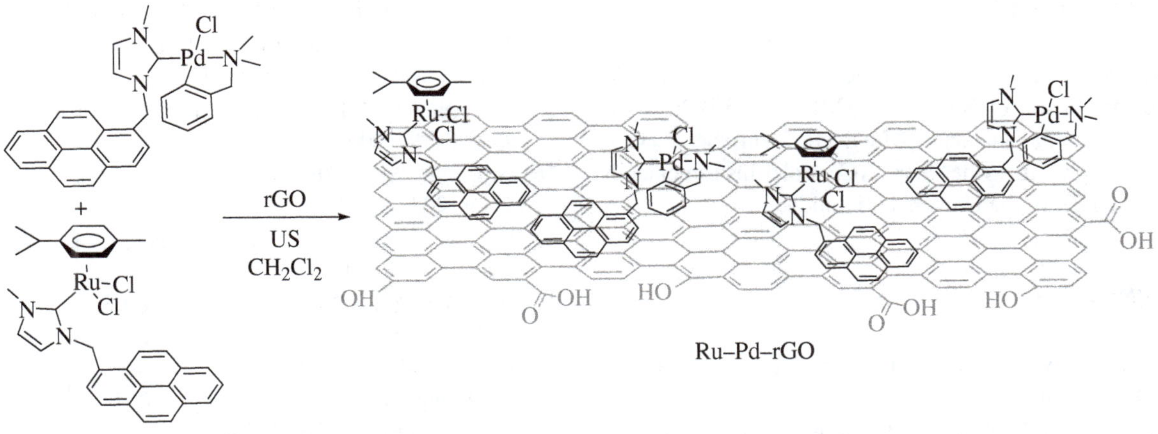

SCHEME 17.5 Synthesis of the heterobimetallic ruthenium–palladium–rGO hybrid material.

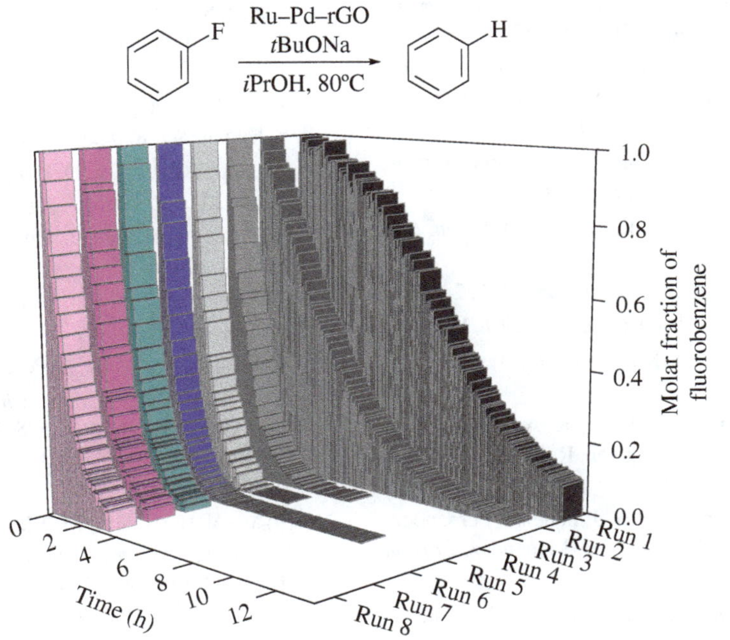

FIGURE 17.13 Reaction profiles for eight successive cycles in the hydrodefluorination of fluorobenzene using the Ru–Pd–rGO hybrid material. The reactions were carried out in *i*PrOH and the presence of *t*BuONa at 80°C. After each run, the catalyst was filtered, washed with CH_2Cl_2, and dried. The monitoring of the reactions was carried out by ^{19}F NMR spectroscopy. The first two runs show an induction period for the formation of the active catalytic species. Adapted with permission from Ref. [29]. Copyright 2015 American Chemical Society.

17.5 PERSPECTIVES

The functionalization of pristine graphene requires strong non-covalent interactions in order to preserve the integrity and properties of the material. The modular methodology by π-stacking allows the incorporation of well-defined molecular complexes onto CDGs and related materials where the properties of the complex are not altered by the effect of immobilization. The catalytic properties of the hybrid graphene materials are improved compared to the catalytic outcomes provided by the homogeneous analogues due to an increase of catalyst stability. Therefore, the immobilization of molecular complexes allows the preparation of well-defined solid catalysts.

The hybrid materials can easily be reused and separated from the reaction media by simple filtration, making them suitable for industrial applications. As a result, the non-covalent interactions can be foreseen as a promising methodology for the preparation of hybrid graphene materials as the catalytically active species are strongly immobilized on the surface. Non-covalent interactions have been shown to produce supported molecular catalysts onto graphene derivatives and demonstrated their great potential in practical applications. The simplicity in the design of these heterogeneous catalytic systems, in combination with its great catalytic performance and recyclability, may inspire future researches in the field of catalyst heterogenization.

ACKNOWLEDGMENTS

This work has been partially supported by the Ministerio de Ciencia e Innovación of Spain (CTQ2011-24055/BQU). We thank the "Generalitat Valenciana" for a fellowship (S. Sabater). The authors are grateful to the "Serveis Central d'Instrumentació Científica (SCIC)" of the Universitat Jaume I.

REFERENCES

1. P. T. Anastas, J. Warner, Green Chemistry: Theory and Practice, Oxford University Press, Oxford, 1998.
2. (a) D. C. Elias, R. R. Nair, T. M. G. Mohiuddin, S. V. Morozov, P. Blake, M. P. Halsall, A. C. Ferrari, D. W. Boukhvalov, M. I. Katsnelson, A. K. Geim, K. S. Novoselov, *Science* 2009, *323*, 610; (b) J. Li, X. Yang, *Prog. Chem.* 2013, *25*, 380; (c) A. K. Geim, K. S. Novoselov, *Nat. Mater.* 2007, *6*, 183.
3. (a) D. R. Dreyer, R. S. Ruoff, C. W. Bielawski, *Angew. Chem. Int. Ed.* 2010, *49*, 9336; (b) D. R. Dreyer, S. Park, C. W. Bielawski, R. S. Ruoff, *Chem. Soc. Rev.* 2010, *39*, 228.
4. G. M. Scheuermann, L. Rumi, P. Steurer, W. Bannwarth, R. Muelhaupt, *J. Am. Chem. Soc.* 2009, *131*, 8262.
5. A. Ambrosi, C. K. Chua, A. Bonanni, M. Pumera, *Chem. Rev.* 2014, *114*, 7150.
6. B. C. Brodie, *Philos. Trans. R. Soc. Lond.* 1859, *149*, 249.
7. W. S. Hummers, R. E. Offeman, *J. Am. Chem. Soc.* 1958, *80*, 1339.
8. A. Lerf, H. Y. He, M. Forster, J. Klinowski, *J. Phys. Chem. B* 1998, *102*, 4477.
9. S. Stankovich, D. A. Dikin, R. D. Piner, K. A. Kohlhaas, A. Kleinhammes, Y. Jia, Y. Wu, S. T. Nguyen, R. S. Ruoff, *Carbon* 2007, *45*, 1558.
10. M. Blanco, P. Alvarez, C. Blanco, M. V. Jimenez, J. Fernandez-Tornos, J. J. Perez-Torrente, L. A. Oro, R. Menendez, *Carbon* 2015, *83*, 21.
11. C. Janiak, *J. Chem. Soc. -Dalton Trans.* 2000, 3885.
12. Y. B. Sun, S. B. Yang, G. X. Zhao, Q. Wang, X. K. Wang, *Chem. Asian J.* 2013, *8*, 2755.
13. A. Hirsch, J. M. Englert, F. Hauke, *Acc. Chem. Res.* 2013, *46*, 87.
14. J. A. Mann, J. Rodriguez-Lopez, H. D. Abruna, W. R. Dichtel, *J. Am. Chem. Soc.* 2011, *133*, 17614.
15. C. Vriamont, M. Devillers, O. Riant, S. Hermans, *Chem. Eur. J.* 2013, *19*, 12009.
16. G. Liu, B. Wu, J. Zhang, X. Wang, M. Shao, J. Wang, *Inorg. Chem.* 2009, *48*, 2383.
17. F. Li, B. Zhang, X. Li, Y. Jiang, L. Chen, Y. Li, L. Sun, *Angew. Chem. Int. Ed.* 2011, *50*, 12276.
18. P. D. Tran, A. Le Goff, J. Heidkamp, B. Jousselme, N. Guillet, S. Palacin, H. Dau, M. Fontecave, V. Artero, *Angew. Chem. Int. Ed.* 2011, *50*, 1371.
19. L. Zhang, W. Zhang, P. Serp, W.-H. Sun, J. Durand, *ChemCatChem* 2014, *6*, 1310.
20. Z. Tang, S. Shen, J. Zhuang, X. Wang, *Angew. Chem. Int. Ed.* 2010, *49*, 4603.
21. T. Xue, S. Jiang, Y. Qu, Q. Su, R. Cheng, S. Dubin, C.-Y. Chiu, R. Kaner, Y. Huang, X. Duan, *Angew. Chem. Int. Ed.* 2012, *51*, 3822.
22. Y. Li, X. Huang, Y. Li, Y. Xu, Y. Wang, E. Zhu, X. Duan, Y. Huang, *Sci. Rep.* 2013, 3,1787.
23. Y. Jiang, Y. Lu, X. Lv, D. Han, Q. Zhang, L. Niu, W. Chen, *ACS Catal.* 2013, *3*, 1263.
24. J. S. Lee, S. H. Lee, J. Kim, C. B. Park, *J. Mater. Chem. A* 2013, *1*, 1040.
25. S. Gotovac, H. Honda, Y. Hattori, K. Takahashi, H. Kanoh, K. Kaneko, *Nano Lett.* 2007, *7*, 583.
26. S. Sabater, J. A. Mata, E. Peris, *ACS Catal.* 2014, *4*, 2038.
27. (a) R. H. Crabtree, *Coord. Chem. Rev.* 2007, *251*, 595; (b) S. Diez-Gonzalez, N. Marion, S. P. Nolan, *Chem. Rev.* 2009, *109*, 3612; (c) M. Poyatos, J. A. Mata, E. Peris, *Chem. Rev.* 2009, *109*, 3677.
28. (a) W. A. Herrmann, M. Elison, J. Fischer, C. Kocher, G. R. J. Artus, *Angew. Chem. Int. Ed.* 1995, *34*, 2371; (b) D. Bourissou, O. Guerret, F. P. Gabbai, G. Bertrand, *Chem. Rev.* 2000, *100*, 39.
29. S. Sabater, J. A. Mata, E. Peris, *Organometallics* 2015, *34*, 1186–1190.

18

COOPERATION OF NON-COVALENT INTERACTIONS AND COORDINATION IN CATALYSIS

ABEL M. MAHARRAMOV[1], KAMRAN T. MAHMUDOV[1,2,*], MAXIMILIAN N. KOPYLOVICH[2], RAFIGA A. ALIYEVA[1], AND ARMANDO J. L. POMBEIRO[2,*]

[1] *Department of Chemistry, Baku State University, Baku, Azerbaijan*

[2] *Centro de Química Estrutural, Instituto Superior Técnico, Universidade de Lisboa, Lisbon, Portugal*

18.1 INTRODUCTION

Non-covalent interactions, such as electrostatic interactions, hydrogen and halogen bonds, van der Waals and hydrophobic forces, π-effects, etc., play an essential role in the action of many natural catalysts, enzymes. In the last decades these interactions have been successfully exploited in organocatalysis and offered new reactivities and promotion of stereospecificity. The structure and properties of organocatalysts can be tuned according to the synthetic needs and design of the desired products. Organocatalysts are among the most useful tools in synthetic chemistry since they promote various organic transformations through unique activation modes as compared to conventional catalytic systems with transition metals, where activation generally occurs at the metal center, with the ligands serving mainly to tailor its steric and electronic properties. A recent approach converges organocatalysis and metal catalysis, trying to combine specificity and flexibility of organocatalysis with high activity and durability of metal catalysts [1–3]. In some of those cases, catalysts activate bonds through the cooperation of coordination and non-covalent interactions, thereby enabling new reactivity and selectivity patterns inaccessible for each type of catalysis alone. The mechanisms by which the two activating patterns cooperate often remain to be understood in detail, and one of main challenges to overcome is to avoid negative interferences between the merging organocatalytic and metal catalytic systems, that is, to create a useful compatibility between them.

Although this cooperative catalysis has evolved only recently as a successful principle for the design of modern catalysts, it is used extensively in biological enzymatic systems. Similarly, through the use of a multifunctional mechanism, many artificial catalysts have been able to activate otherwise unreactive covalent bonds and, in some cases, to achieve exceptional chemo-, diastereo-, and enantioselectivity [1]. The classification of such a type of catalytic systems can be summarized as follows (Scheme 18.1) [2, 3]:

a. *Cooperative catalysis*, where two catalysts are directly involved in the same catalytic cycle, working cooperatively to form a product
b. *Synergistic catalysis*, where two catalysts simultaneously activate substrates A and B in two directly coupled catalytic cycles, leading to the formation of a product P
c. *Sequential or relay catalysis*, where both catalysts perform two distinct catalytic cycles for consecutive reactions, whereby the substrates (A and B) first react to form an intermediate (Int) in the first catalytic cycle, which subsequently is converted to the final product (P) by another catalyst

In this chapter, the cooperation of an organocatalyst and a metal catalyst will be illustrated by several examples where the roles of non-covalent interactions and coordination in the course of the catalytic process are equally important.

*Corresponding authors Kamran T. Mahmudov (kamran_chem@mail.ru) and Armando J. L. Pombeiro (pombeiro@tecnico.ulisboa.pt)

Non-Covalent Interactions in the Synthesis and Design of New Compounds, First Edition.
Edited by Abel M. Maharramov, Kamran T. Mahmudov, Maximilian N. Kopylovich, and Armando J. L. Pombeiro.

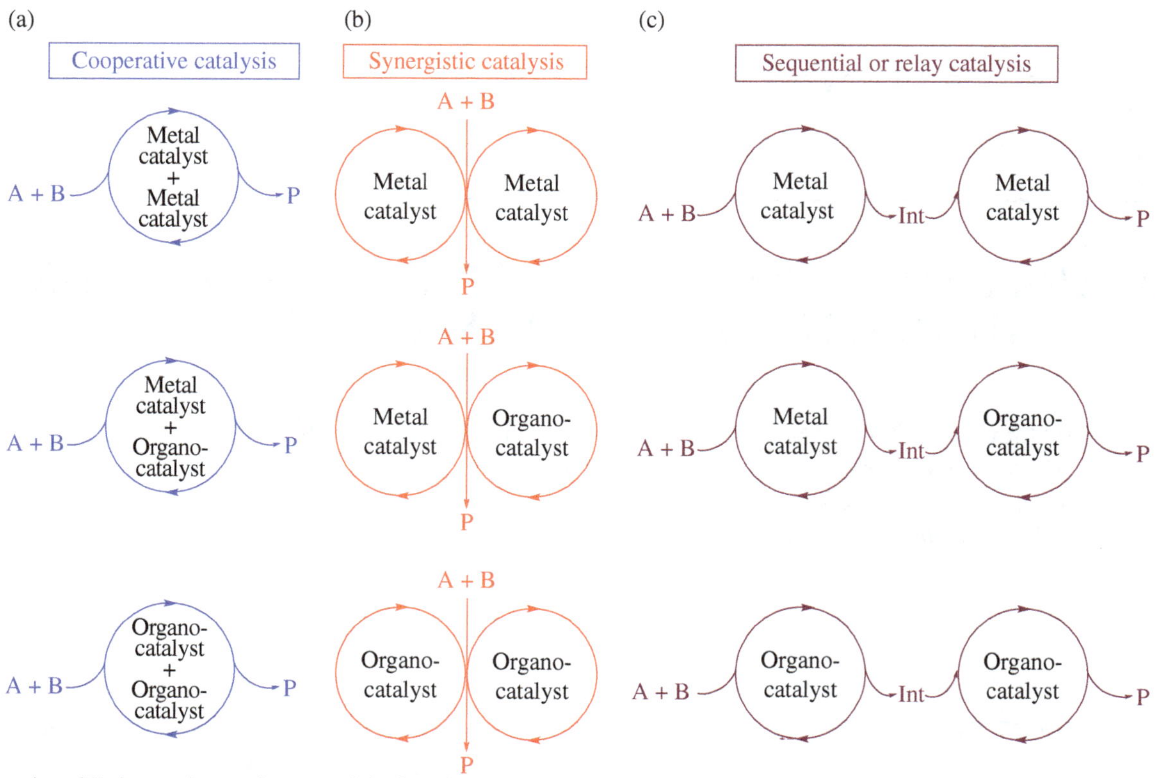

A and B denote the starting materials; P and Int denote the reaction product and an intermediate, respectively.

SCHEME 18.1 The concepts of cooperative (a), synergistic (b), and sequential or relay (c) catalyses and their types [2, 3].

The provided examples will hopefully shed some light on the synthetic utility and potential of the cooperative effects in synthetic chemistry.

18.2 HYDROGEN BONDING COOPERATION WITH COORDINATION

The directionality and versatility of hydrogen bonding, which plays a crucial role in many catalytic processes [4], offer many possibilities for the creation and stabilization of intermediates in these reactions. Combination of several hydrogen bonds in a specific binding motif is a valuable tool to facilitate a desired interaction, while their particular orientation can enhance the specificity of the bond activation.

The strength of single hydrogen bonds is related to the nature of the donor and acceptor, and its magnitude ranges usually from 2 to 20 kcal/mol. Metal–ligand interactions are generally much stronger, their energies being usually in the range of 10–30 kcal/mol per interaction, that is, between those of strong covalent bonding in carbon-based systems and the weak non-covalent interactions. A feature of transition metal–ligand coordination of interest in catalysis is its significant directional freedom in the sense that it can lead to various spatial geometries. Moreover, transition metal complexes provide a pool of different acceptor subunits, which can be linked in various ways to form intermediates of lower energy if compared to starting materials.

A combination of organocatalysts, where a non-covalent interaction can play a major role in the reaction profile, and transition metal catalysts has obvious benefits, if the advantages of both types are added with a possible synergy or other positive effects. Thus, many organocatalysts, such as primary and secondary amines, Bronsted acids (e.g., phosphoric acids based on the chiral 1,1′-bi-2-naphthol scaffold), thioureas, nucleophilic Bronsted and Lewis bases, chiral phase transfer agents, *N*-heterocyclic carbenes, etc., have been successfully combined with metal catalysts (of Pd, Rh, Ru, Cu, Ni, Zn, Fe, Ir, Co, Mn, Ti, Y, In, Nb, Au, Ag, Pt, V, etc.) to give improved yields and/or selectivities if compared with individual catalytic systems. Due to versatile reactivity patterns of transition metals and organocatalysts, new exciting discoveries have been reported, in many cases leading to the generation of previously unattainable compounds. Remarkably, this strategy not only promotes particular reactions but also enables multiple transformations in a one-pot process. In this section, several examples of cooperation of both tools, hydrogen bonding and coordination, are presented.

[Rh(COD)(κC_2-PIN)Br] (COD = 1,5-cyclooctadiene, PIN = 1-isopropyl-3-(5,7-dimethyl-1,8-naphthyrid-2-yl)imidazol-2-ylidene) catalyzes the hydration of a wide variety of organonitriles at ambient temperature, providing the corresponding organoamides (Scheme 18.2a) [5]. An identified 1 : 1 catalyst–water adduct with ligated organonitrile is an essential intermediate, where the pendant naphthyridine

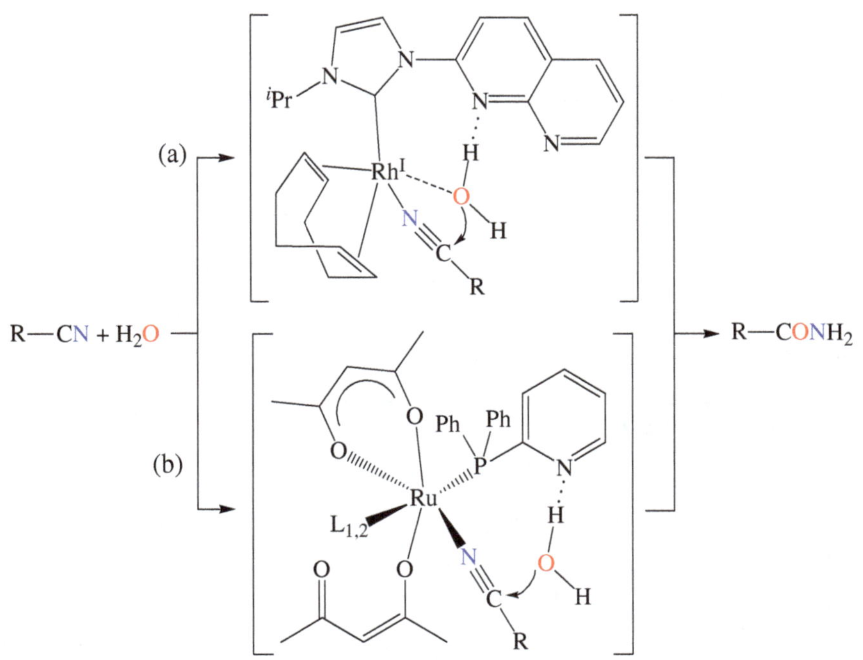

SCHEME 18.2 Rhodium(I) (a) and ruthenium(II) (b) complexes in hydrolysis of nitriles to carboxy amides [5, 6].

paves way for a water molecule to the active metal site using double H-bonded interactions. Calculations reveal a multi-component cooperativity involving proton transfer from the water to the naphthyridine nitrogen and a complementary interaction between the hydroxide and the cyano carbon of the organonitrile ligand. Thus, bifunctional water activation and cooperative proton migration are recognized as key steps in the catalytic cycle. A cooperative action of hydrogen bonding and coordination is also observed in the hydration of organonitriles to organoamides under neutral conditions catalyzed by bis(acetylacetonato)ruthenium(II) complexes bearing 6-diphenylphosphino-*N*-pivaloyl-2-aminopyridine (6-DPPAP), and 3-diphenylphosphinoisoquinolone (3-DPICon) ligands (Scheme 18.2b) [6].

[$Cp_2Zr(H)Cl$] (Schwartz's reagent; Scheme 18.3) is used as a reducing agent in the effective and selective direct transformation of a variety of tertiary amides, including Weinreb amides, to the corresponding aldehydes under mild reaction conditions [8]. Primary and secondary amides also proved to be viable substrates for the reduction, although yields were lower as compared to the corresponding tertiary amides. Experiments with isotopes indicate that the hydrogen atom of the aldehyde group originates from [$Cp_2Zr(H)Cl$], not from water, while the source of the carbonyl oxygen in the aldehyde product is a water added in a following step, rather than the carbonyl oxygen of the reacting amide; this implies that a hydrolysis must occur in the reaction process.

The proposed mechanism accounts for the cooperative role of hydrogen bonding and coordination in the reductive C—N cleavage (Scheme 18.3) [7]. The first step of the reaction is coordination of the C=O moiety to [$Cp_2Zr(H)Cl$] through an "inner" mode of action (intermediate **I**), which leads to the formation of a Zr—O intermediate (**II**), confirmed experimentally [8]. Under the reaction conditions (intermediates **III** and **IV**), the cleavage of the C—O bond of the Zr—O intermediate results in the formation of an iminium cation (**V**); this process occurs with assistance of hydrogen bonding to water. Then a water molecule attacks the iminium cation to produce an amine intermediate (**VI**), and a water-catalyzed elimination reaction occurs to yield the aldehyde product (**VII** and **VIII**). Thus, the combined actions of the coordination of the amide to the metal center and hydrogen bonding to water assist the cleavage of the C—O and the C—N bonds in the course of the reaction.

Transition metal-catalyzed carbene insertion into a N—H bond is an efficient method to form a carbon–nitrogen bond, and asymmetric versions are known. An asymmetric N—H insertion reaction of methyl α-diazo-α-phenylacetate into the N—H bond of *tert*-butyl carbamate ($BocNH_2$) cooperatively catalyzed by dirhodium(II) carboxylates and chiral spiro phosphoric acids (SPAs) allows to reach high yields and enantioselectivities [9]. According to the proposed mechanism (Scheme 18.4), treatment of the dirhodium complex [Rh_2L_4] (L = Ph_3CCOO) with the diazo compound leads to a rhodium carbene intermediate **I**, which reacts with $BocNH_2$ to generate an ylide **II**. SPA assists the proton transfer via a seven-membered H-bonded ring intermediate (**III**) to form the N—H insertion product, and the dirhodium catalyst and chiral SPA are simultaneously liberated. The chiral induction occurs in the proton transfer step. Both chiral SPA and dirhodium catalyst are involved in the proton transfer step and influence the high enantioselectivity of the reaction (*ee* 92%).

Another cooperative action of a chiral H-bond donor (the Bronsted base cupreine) and a transition metal catalyst (the Lewis acid (LA) $AgNO_3$) was observed in the

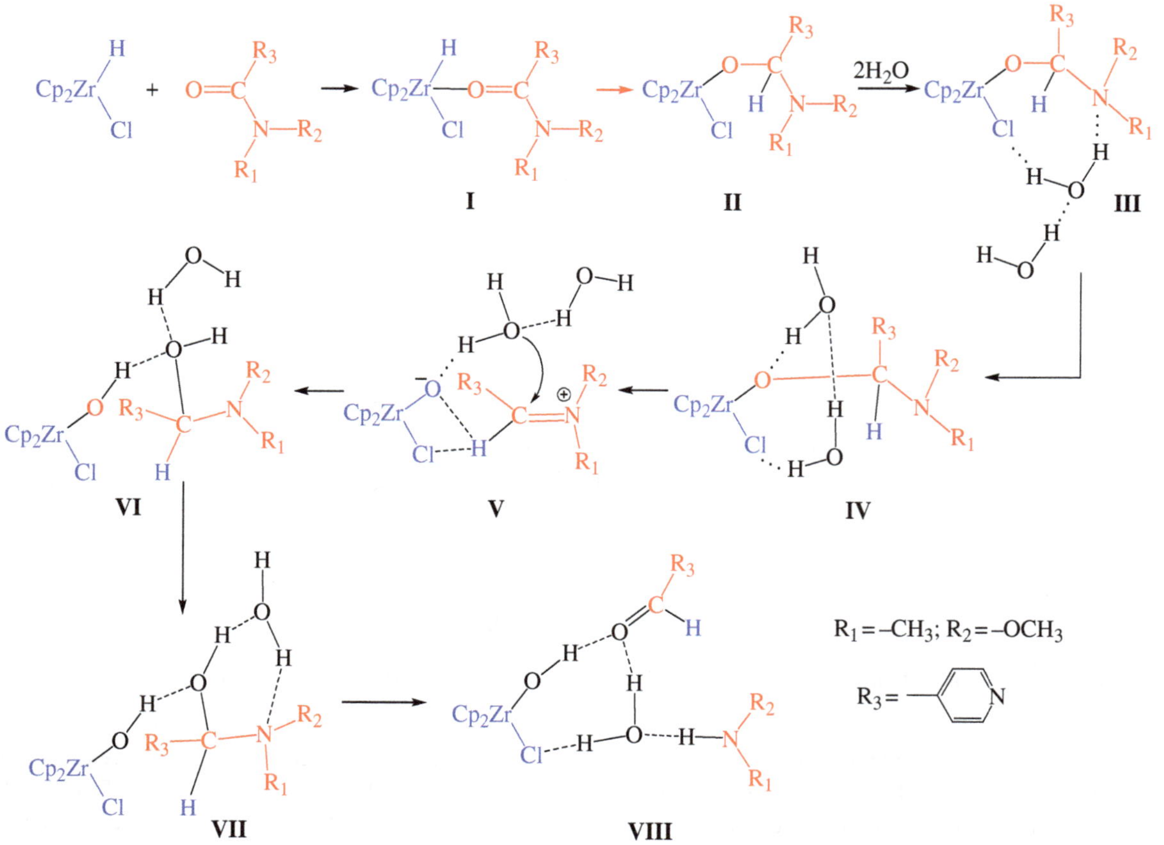

SCHEME 18.3 Proposed mechanism of transformation of amides to aldehydes [7].

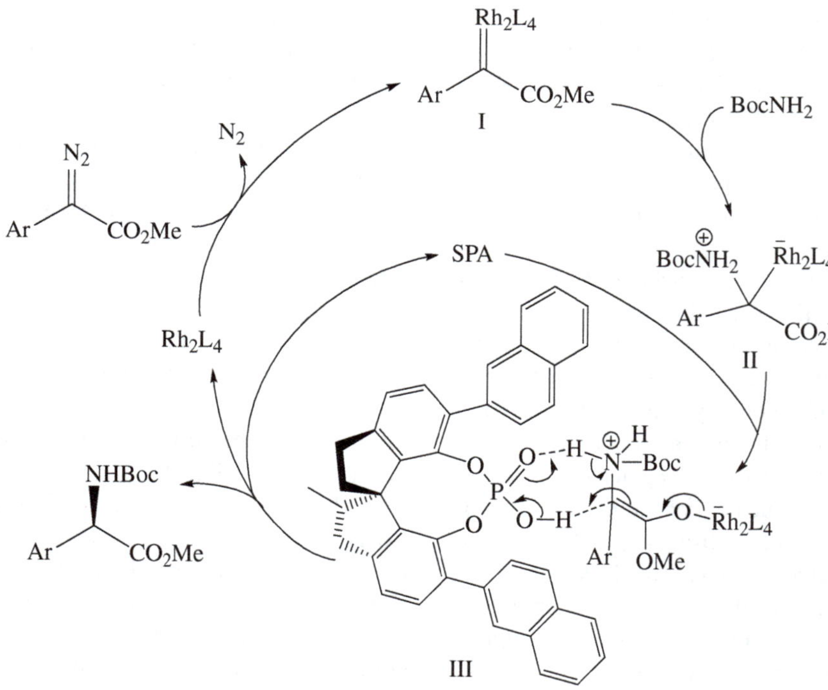

SCHEME 18.4 Proposed mechanism of insertion of carbene from methyl α-diazo-α-phenylacetate into $BocNH_2$ [9].

[3+2] cycloaddition of isocyanoacetates to α,β-unsaturated ketones, providing 2,3-dihydropyrroles in up to 89% *ee* [10]. According to the proposed mechanism (Scheme 18.5), both tools, coordination and hydrogen bond, play equally important roles in the formation of stable enantioenriched 2-pyrrolines. Thus, firstly the isocyano group coordinates to silver(I), thus increasing the acidity of the α-proton and allowing the quinuclidine nitrogen of cupreine to

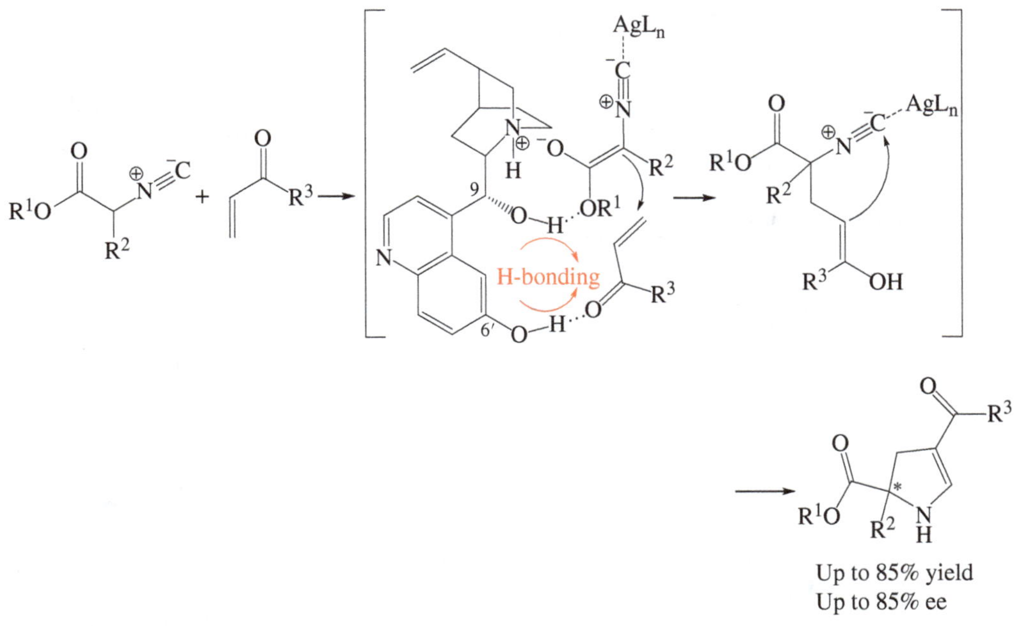

SCHEME 18.5 [3+2] cycloaddition of isocyanoacetates to α,β-unsaturated ketones [10].

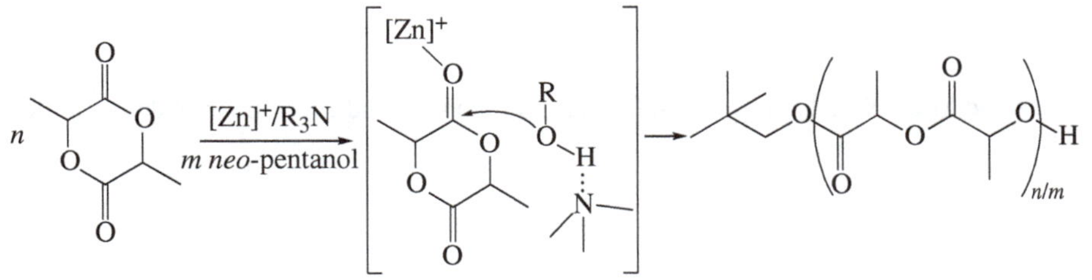

SCHEME 18.6 Polymerization of lactide catalyzed by a cationic zinc complex and a tertiary amine [11].

deprotonate the isocyanoacetate. After deprotonation, a rigid chiral ion pair is formed and stabilized by the hydrogen bonding interaction between the C9-hydroxy group and the substrate. Hence, the C6′-hydroxy group plays a key role in the transfer of stereospecificity through a well-defined hydrogen bonding with vinyl ketone. Then the vinyl ketone undergoes 1,4-addition, generating the stereogenic center of the product. Finally, a 5-*endo-dig* cyclization takes place to afford the final product under the electrophilic silver isocyanide activation [10].

A catalytic system, combining a cationic zinc complex and a tertiary amine, efficiently promotes polymerization of lactide under mild conditions [11]. The two-component catalyst cooperatively activates the lactide monomer and propagating alcohol (Scheme 18.6). The electrophilic zinc cation is coordinated by the monomer through the carbonyl moiety, while the amine orients the alcohol by hydrogen bonding. This example illustrates the possibility to apply the cooperation H-bonding and coordination in polymer chemistry, in particular in ring-opening polymerization reactions.

The fluorinated aminophosphines *rac*-(±)-TF-BiphamPhos (Scheme 18.7) were applied as ligands for the efficient Cu^{I}-catalyzed Michael addition of alkylidene bisphosphonates and azomethine ylides, providing nonproteinogenic α-amino acid derivatives with geminal bisphosphonates in good yields and with high diastereo- (>99 : 1) and enantioselectivities (95% *ee*) [12]. The proposed active species contain a Cu^{I} complex with the TF-BiphamPhos ligand and deprotonated azomethine ylide acting as a (N,O)-chelate. In the catalytic cycle, the amino moiety, coordinated to the copper atom, behaves as a hydrogen bond donor/acceptor. The nucleophilic addition of the active species to the bisphosphonate electrophile generates the carbanionic intermediate **I**, which is stabilized by a hydrogen bond established between the P=O oxygen atom of the phosphonate group and one N—H protons of the chiral ligand [12, 13]. The hydrogen-bound proton then migrates to the carbanion yielding intermediate **II** with the neutral ligand. Coordination of the azomethine ylide to the copper center regenerates the active species.

Acylguanidine-functionalized phosphine (L; Scheme 18.8a) is used with $[Rh(acac)(CO)_2]$ as an efficient catalytic system in hydroformylation of vinylacetic acid and other unsaturated acids yielding preferentially linear aldehydes [14]. The acylguanidinium functional group in the *meta* position with respect to the catalyst-binding phosphine unit supports the concerted coordination and hydrogen bonding, which lead to a specific conformational orientation of the substrate [15]. As a result, a guanidinium–carboxylate ion pair is formed (intermediate **I**) and directs the double bond of the substrate toward the Rh—H bond, thus providing the metalation of the

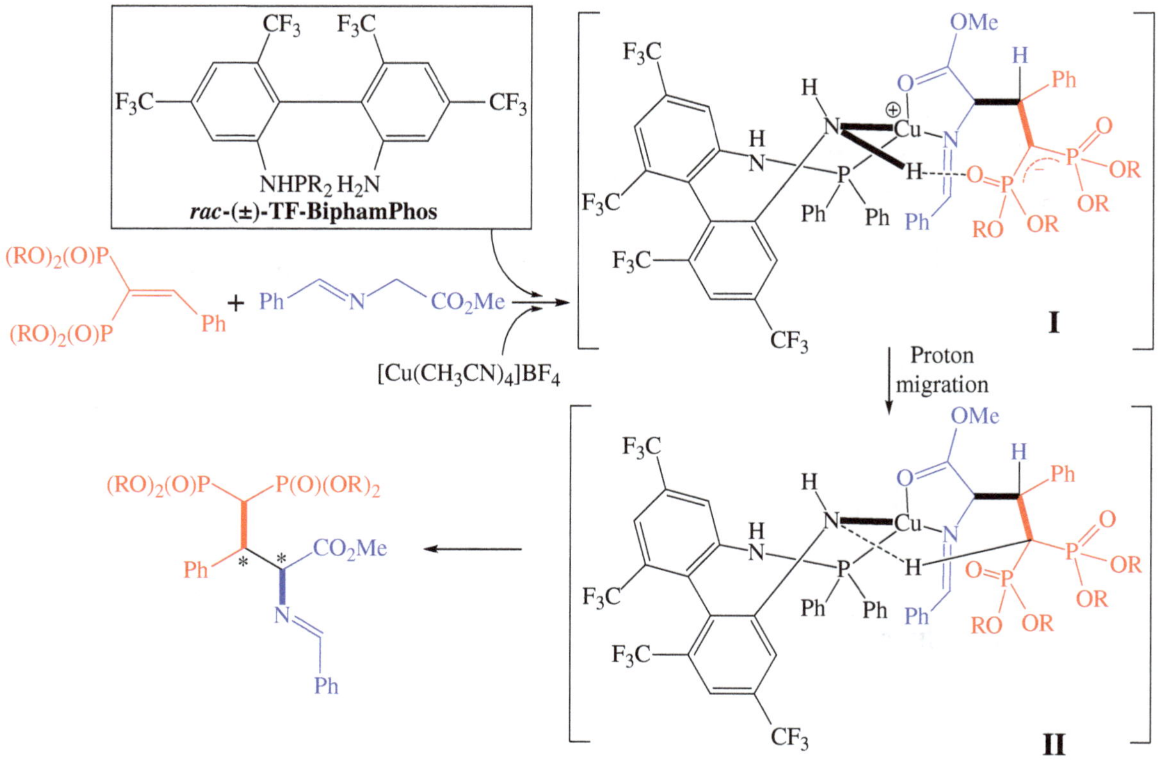

SCHEME 18.7 Stereospecific Michael addition of alkylidene bisphosphonates and azomethine ylides catalyzed by a Cu[I]-*rac*-(±)-TF-BiphamPhos system [12].

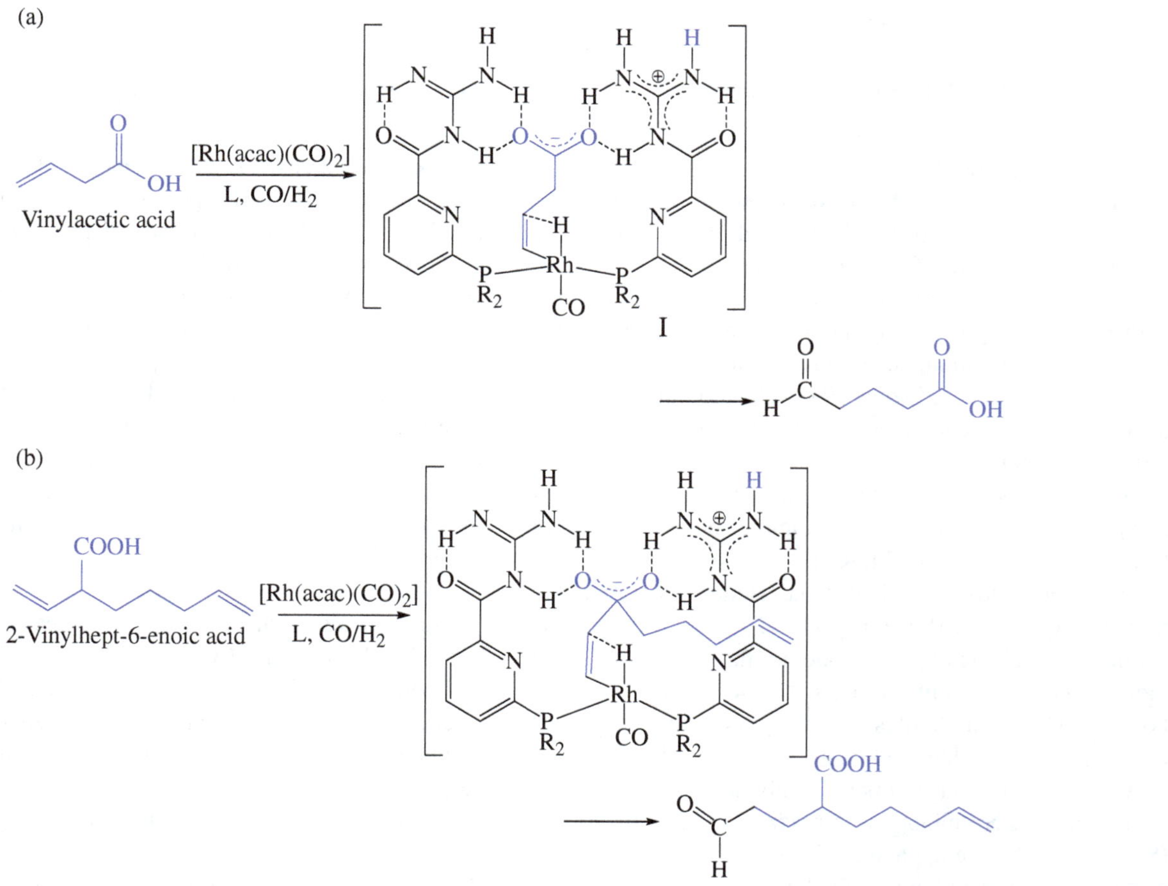

SCHEME 18.8 Hydroformylation of vinylacetic acid and 2-vinylhept-6-enoic acid catalyzed by an acylguanidine-functionalized phosphine-[Rh(acac)(CO)$_2$] system [14, 15].

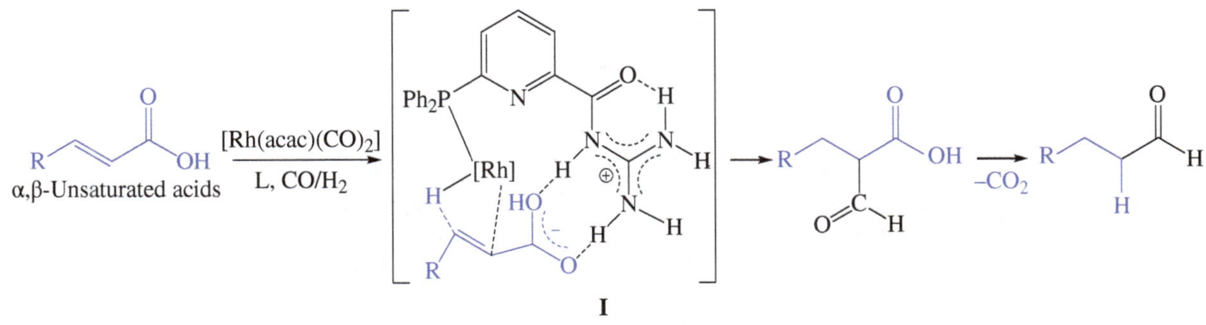

SCHEME 18.9 Synthesis of linear aldehydes from α,β-unsaturated acids catalyzed by an acylguanidine-functionalized phosphine in combination with $[Rh(acac)(CO)_2]$ [16].

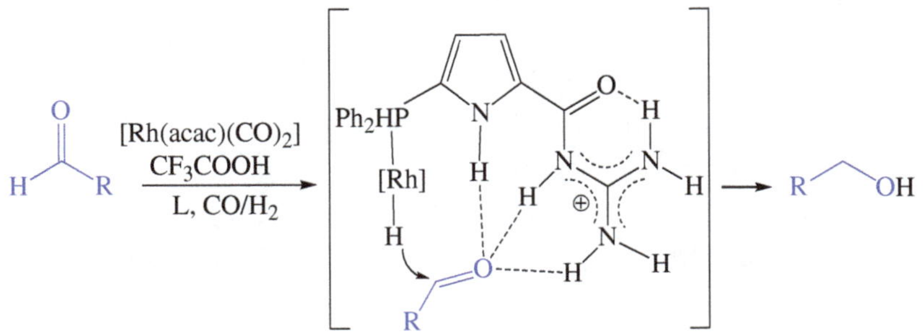

SCHEME 18.10 Hydrogenation of aldehydes using phosphino-1*H*-pyrrole-2-acylguanidine and $[Rh(acac)(CO)_2]$ as a cooperative catalytic system [17].

C═C bond at the γ-position of the carboxylate anion. A proton transfer between the substrate and the guanidine moiety occurs spontaneously in the course of the reaction. The synthetic utility of the catalytic system is reinforced by its ability to transform selectively the double bond located in the β,γ-position to the carboxylic group and not the other double bonds present in the molecule (Scheme 18.8b). Hence, for unsaturated acids, the ligand–substrate interaction brings the β,γ-double bond at an ideal position for the accomplishment of the subsequent rate-determining hydrometalation step [15].

Similarly to the above mentioned system, a catalytic combination of the acylguanidine-functionalized phosphine ligand and $[Rh(acac)(CO)_2]$ was applied to the regioselective hydroformylation of α,β-unsaturated acids to give α-formyl acids, which, after their decarboxylation, produce linear aldehydes (Scheme 18.9) [16]. In this case, the carboxylate–guanidinium hydrogen bond-assisted interaction directs the Rh attack at the α-position of the carboxylate anion (intermediate **I**). One can assume that the proper orientation of the carboxylate can be directed by two guanidine moieties, related to the case described above (Scheme 18.8).

Another related cooperative catalytic system involves phosphino-1*H*-pyrrole-2-acylguanidine with $[Rh(acac)(CO)_2]$ (Scheme 18.10) and was used for hydrogenation of aldehydes with yields ranging between 80 and 99% [17]. In this catalytic reaction, phosphino-1*H*-pyrrole-2-acylguanidine proved to be far more efficient than the acylguanidine-functionalized phosphine ligand, while CF_3COOH has a promoting effect possibly due to its role in protonation of theguanidinium group. The guanidinium cation can also be formed by (i) heterolytic cleavage of H_2 or (ii) oxidative addition of H_2 to the rhodium atom followed by proton transfer to the guanidine moiety. As a result, multicenter hydrogen bonds occur between the guanidinium group of the phosphine ligand and the oxygen atom of the aldehyde. These interactions limit the degree of conformational flexibility of the aldehyde and direct the carbonyl groups toward the Rh─H bond. The catalytic system provides both acidic and hydridic hydrogen atoms from the guanidinium moiety and the rhodium center, respectively. The N─H proton of pyrrole also participates in the binding of aldehyde, accounting for the higher efficiency of the phosphino-1*H*-pyrrole-2-acylguanidine proligand in comparison with that of the acylguanidine-functionalized phosphine.

Unsubstituted and CH_3-substituted aminoferrocenylphosphines (FcPN; Scheme 18.11) promoted the formation of opposite enantiomers in the silver-catalyzed [3+2] cycloaddition of azomethine ylides to dimethyl maleate [18]. When the N─H proton of an unsubstituted aminoferrocenylphosphine interacts with the carbonyl oxygen of the incoming dimethyl maleate substrate, an attack on the top face of the silver-bound azomethine ylide occurs, and the thus formed intermediate **I** is stabilized by coordination and hydrogen bond interactions. If the CH_3-substituted aminoferrocenylphosphine is involved in the reaction, its dimethylamino moiety hampers the formation of hydrogen bonds, and steric repulsion between the $N(CH_3)_2$ group and dimethyl maleate

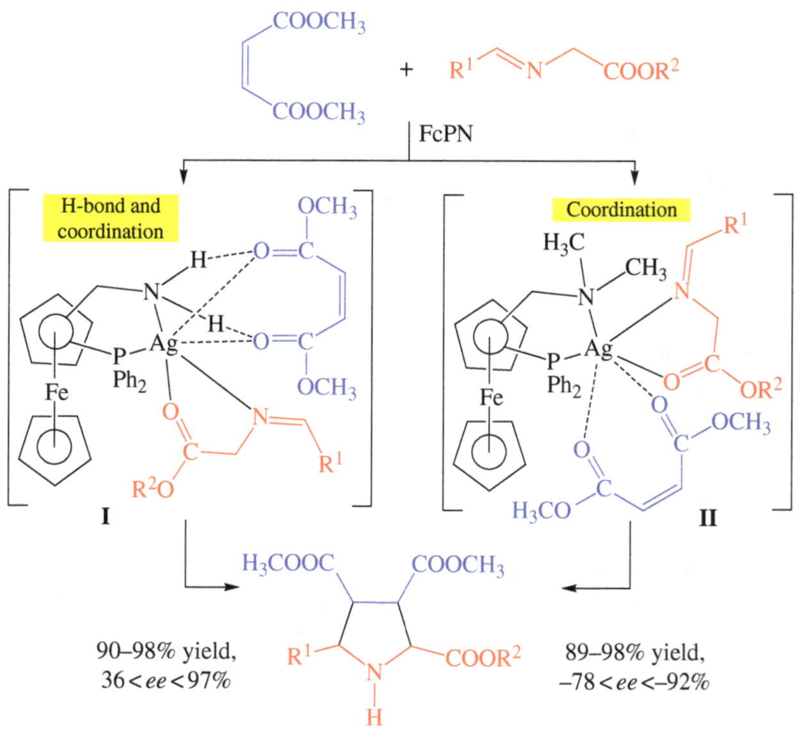

SCHEME 18.11 Hydrogen bond interactions direct the formation of opposite enantiomers in the silver-catalyzed asymmetric [3+2] cycloaddition of azomethine ylides to dimethyl maleate [18].

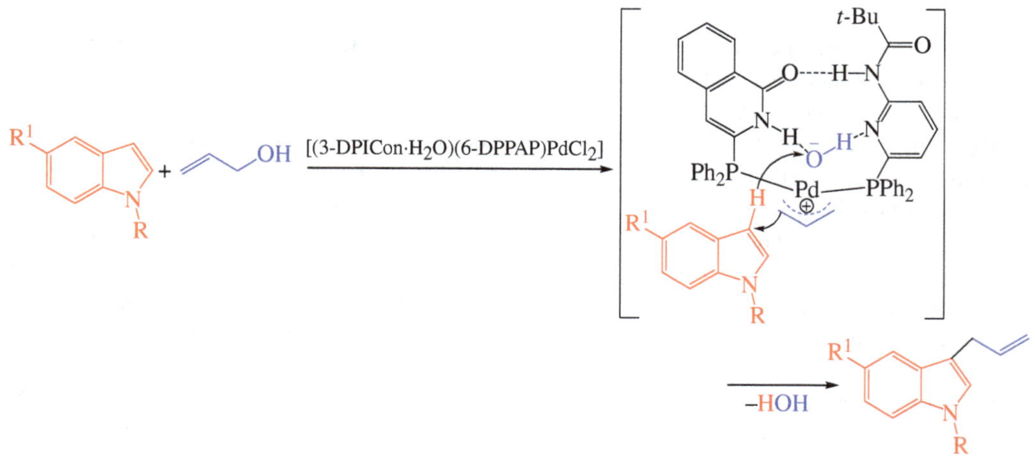

SCHEME 18.12 Allylation of indole derivatives catalyzed by a self-assembled palladium–phosphane catalyst system [19].

causes the attack on the bottom face of the ylide (**II**). In this case, the ligand–substrate interaction does not affect the efficiency of the catalyst but directs the reversion of enantioselectivity.

Self-assembled palladium–phosphane catalysts of the type [(3-DPICon·H_2O)(6-DPPAP)$PdCl_2$], derived from $PdCl_2$, 3-DPICon, and 6-DPPAP, have been successfully employed for the selective allylation of *N*-heterocycles with allylic alcohols (Scheme 18.12) [19]. Upon this reaction, a complementary hydrogen bond system is formed between isoquinolone and aminopyridine groups, which appended to the *cis*-positioned phosphine ligands. Hydrogen bond interactions between the hydroxyl group of the allylic alcohol and the N—H proton of the isoquinolone moiety make the hydroxyl a better leaving group and favor the formation of a η^3-allyl palladium intermediate, which then leads to allylic functionalized *N*-heterocycle.

An unanticipated role of the counteranion in the course of Pd-catalyzed Tsuji–Trost cyclization of β,γ-dihydroxyl allyl acetates to give 2,5-disubstituted 3-hydroxytetrahydrofurans (Scheme 18.13) provided a high diastereoselectivity [20]. In this case, an additional hydrogen bond established between the acetate anion and the allylic C—H proton favors a particular transition state (**I** vs. **II**; Scheme 18.13). A quite strong hydrogen bond with the π-allyl system is established in **I**. This interaction is not geometrically feasible in **II**, the

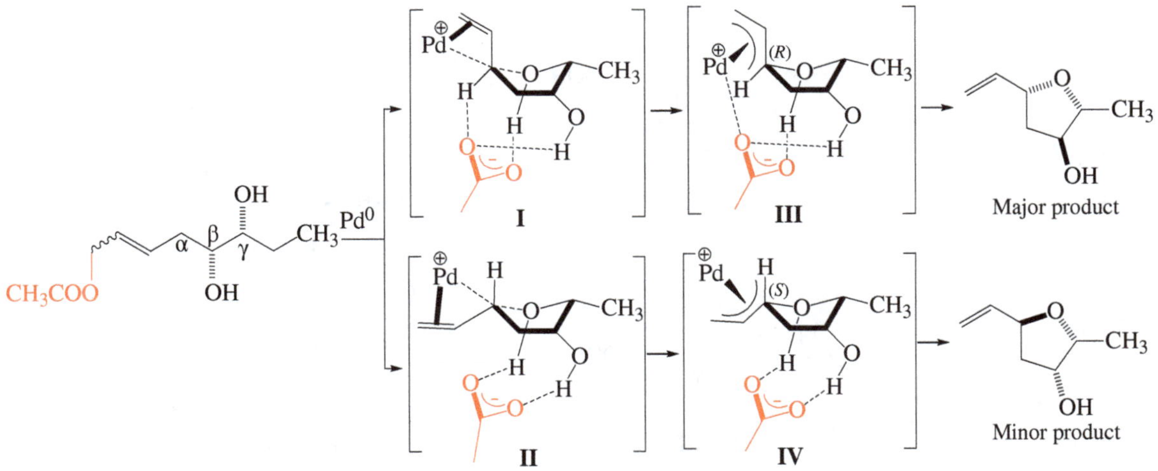

SCHEME 18.13 Hydrogen bond interactions within a metal–anion–substrate intermediate drive diastereoselectivity in cyclization of β,γ-dihydroxyl allyl acetates [20].

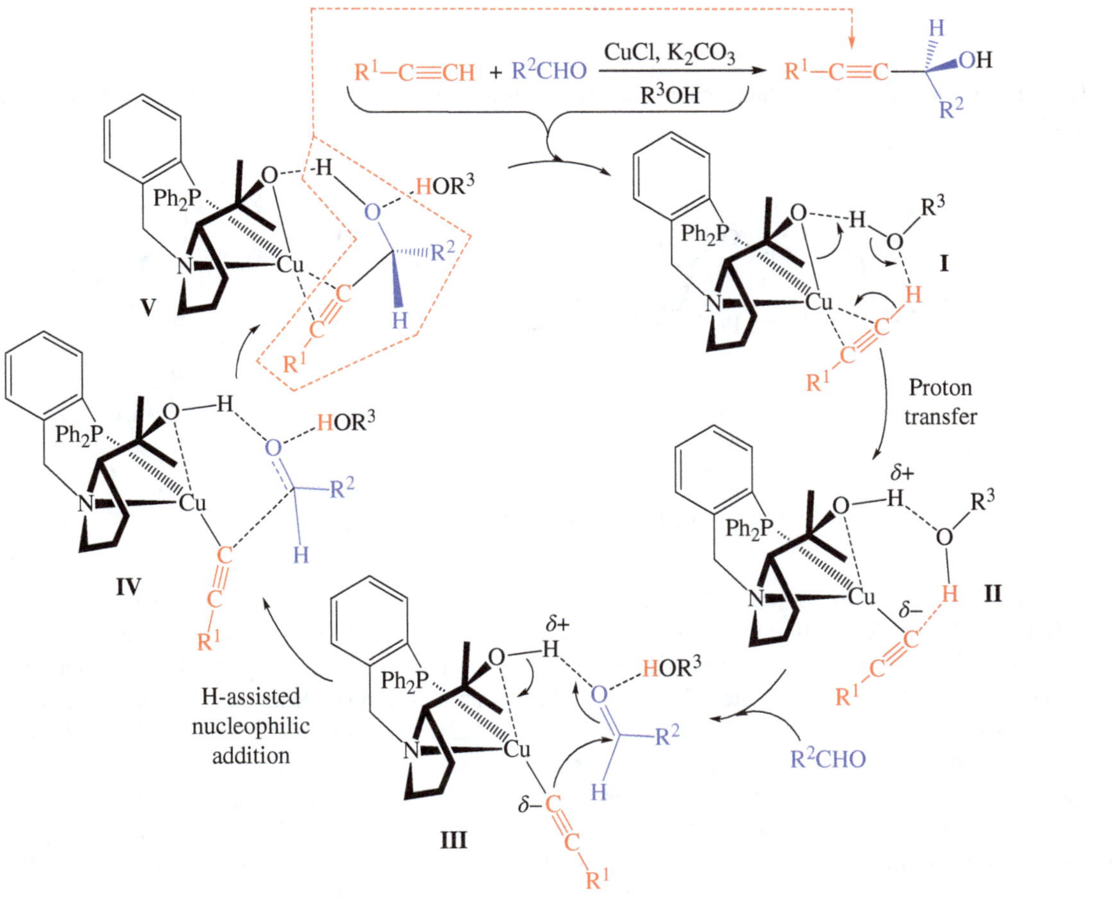

SCHEME 18.14 Enantioselective synthesis of propargylic alcohols by alkynylation of aldehydes with terminal alkynes [21].

free energy of which is higher than that in **I** by 2.1 kcal/mol. Both diastereomers **III** and **IV** exhibit two O···H—O interactions, the former also showing a weak O···Pd interaction, which would not account for the selectivity. Thus, multiple hydrogen bonding interactions stabilize **I** and furnish the diastereoisomer that is observed experimentally as a major product.

Hydrogen bonding interactions play an essential role in the cooper(I)-catalyzed direct alkynylation of aldehydes with terminal alkynes to afford nonracemic secondary propargylic alcohols, versatile synthetic intermediates for nonracemic compounds [21]. In this case, copper(I) ions cooperate with prolinol-based hydroxyamino phosphane chiral proligands (L; Scheme 18.14). The hydroxy group of

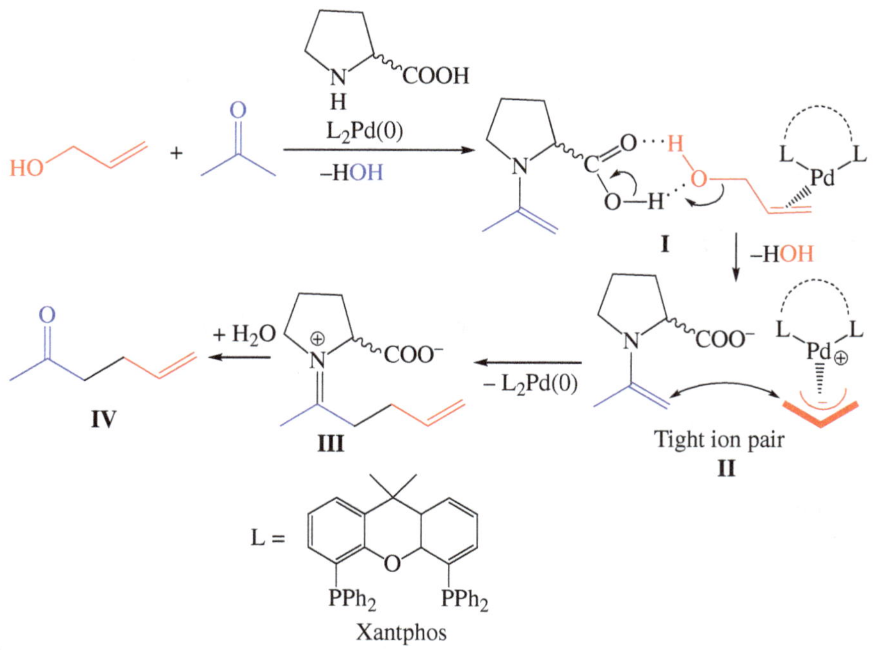

SCHEME 18.15 Cooperation of palladium and proline cocatalysts in allylic alkylation of enolizable ketones and aldehydes with allylic alcohols [22].

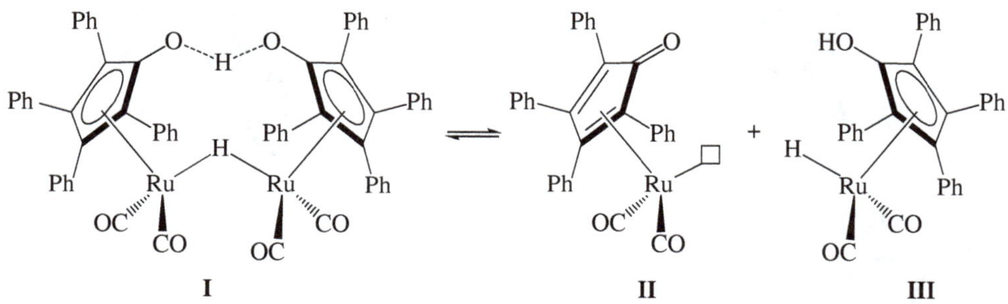

SCHEME 18.16 Shvo's catalyst [23].

the proligand displays an important role, while protic solvents promote a favorable reaction rate and enantioselectivity. A mechanistic study reveals the formation of a Cu^I alkoxide π-complex **I** with a η^2-coordinated alkyne. The alkoxide oxygen atom and the terminal hydrogen atom of the alkyne form a hydrogen bonding with an alcohol molecule. The complex **I** is transformed, through intramolecular proton transfer, to a η^1-acetylide complex **II**, where the OH group of the ligand, bound to the Cu atom, provides an acidic hydrogen bond donor site and the acetylide carbon atom plays the role of a nucleophilic site. The following replacement of the alcohol with an aldehyde leads to **III** where extensive hydrogen bonds provide interaction of the aldehyde group with an alcohol molecule. The following enantiodiscriminating nucleophilic addition proceeds through a six-centered intermediate **IV** to yield the alkoxocopper(I)/propargylic alcohol complex **V** upon a proton-assisted carbon–carbon bond formation. The alkyne exchange between **V** and the substrate releases propargylic alcohol and completes the catalytic cycle.

Cooperation of proline and a palladium–xantphos complex (xantphos = (9,9-dimethyl-9,9*a*-dihydro-4*aH*-xanthene-4,5-diyl)bis(diphenylphosphine)) is essential for the allylic alkylation of enolizable ketones and aldehydes with allylic alcohols [22]. Upon reaction, a π-allyl Pd intermediate is generated from attack to the allylic alcohol of the enamine formed *in situ* from the corresponding enolizable carbonyl substrate and proline (Scheme 18.15). Thus, the carboxylic functionality of proline is involved, through hydrogen bonding, in the ionization step of the palladium complex with allylic alcohol and protonation of the hydroxy leaving group (intermediate **I**). This results in the formation of a tight ion pair between the enamine and the π-allyl ligand (intermediate **II**), which allows a rapid allylation with subsequent liberation of the inorganic catalyst. The obtained zwitterionic azomethine **III** is then hydrolyzed leading to the product **IV**.

A versatile ruthenium Shvo catalyst, $\{[Ph_4(\eta^5\text{-}C_4CO)]_2H]\}Ru_2(CO)_4(\mu\text{-}H)$ (**I**; Scheme 18.16), utilizes noninnocent cyclopentadienone ligands to stabilize the metal in low oxidation states [23]. The catalyst **I** is in equilibrium with its

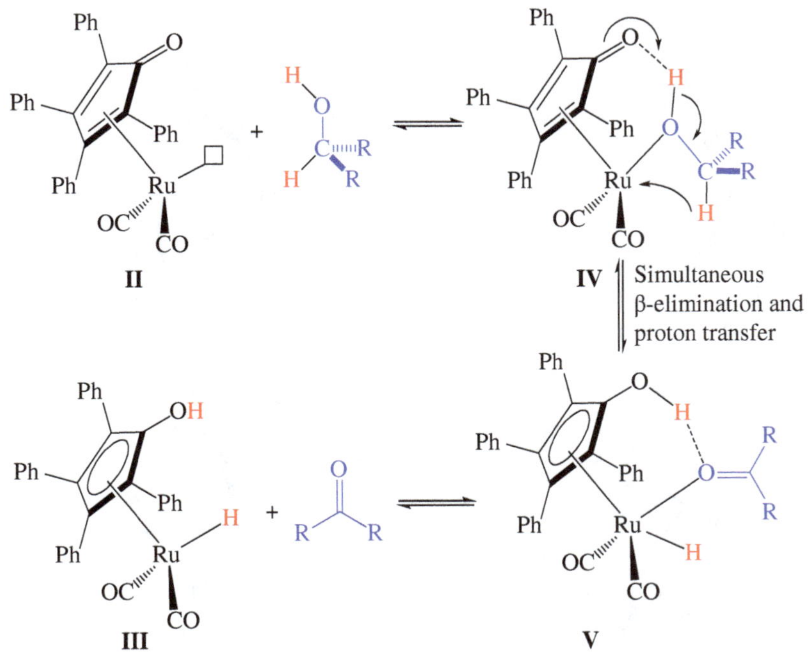

SCHEME 18.17 Shvo's catalyst in dehydrogenation of alcohols or hydrogenation of ketones [24–26].

monomers **II** and **III**. The monomer **III** is able to hydrogenate a hydrogen acceptor, whereas **II** can dehydrogenate a hydrogen donor; these processes interconvert **II** and **III**. The Shvo catalyst is an example of a metal–ligand bifunctional catalyst where the redox function is distributed among the metal center and a cyclopentadienone ligand.

This bifunctional Ox-Red Shvo catalyst and its analogues have been extensively used for hydrogenation of aldehydes, ketones, alkynes, and alkenes; transfer hydrogenation; disproportionation of aldehydes to esters; amine–amine coupling; dynamic kinetic resolution; isomerization of allylic alcohols; and hydroboration reactions. For instance, concerted transfer of OH and RuH hydrogens from **III** to a ketone leads to **II** and alcohol (Scheme 18.17) [24]. The reverse reaction, the dehydrogenation of an alcohol to ketone, is also theoretically possible with **II** as a catalyst. According to the proposed mechanism, an initial coordination of the alcohol to the metal center leads to a H-bond-assisted intermediate **IV** [25, 26], which undergoes simultaneous β-hydride elimination and proton transfer to produce the resulting ketone **V**. The intermediate **IV** is formally an 18-electron complex, and such a species can undergo a $\eta^4 \rightarrow \eta^2$ ring slip to form the open coordination site necessary for subsequent β-hydride elimination and proton transfer to produce **V**. Dissociation of the ketone from **V** leads to the ruthenium hydride **III**.

In parallel to the development of Shvo's catalyst for ketone and aldehyde reduction, its application in catalytic imine hydrogenation has also been extensively explored. For instance, the Shvo's hydroxycyclopentadienyl system was used for the transfer hydrogenation of imines with 2-propanol as a terminal reductant [27]. The related Noyori ruthenium(arene)diamine catalytic system has been also applied for the transfer hydrogenation of imines utilizing a formic acid/triethylamine solvent system; good yields and enantioselectivities were achieved [28]. In these catalytic imine hydrogenations, the acidic OH group of the Shvo catalyst **I** protonates the imine group leading to the iminium cation **II** (Scheme 18.18) [29]. Subsequent hydride addition to the iminium cation (**II** → **III**) followed by coordination of the formed amine affords **IV**. The hydrogen transfer is still concerted in this mechanism, and depending on whether the imine is electron deficient or electron rich, the hydrogen transfer is rate determining or not, respectively. In the latter case the coordination of the formed amine to the unsaturated ruthenium complex (**III** → **IV**) is rate determining. In this case the hydride shift from carbon to ruthenium competes with coordination.

18.3 HALOGEN BONDING COOPERATION WITH COORDINATION OR HYDROGEN BONDING

Over the past few years, halogen bonding between halogen atoms and electron-rich partners has attracted widespread interest in a variety of chemical and biological applications [30–35]. Thus, halogen bonding is currently being exploited extensively to control the crystallization of organic compounds in the design of new materials and in supramolecular chemistry [33, 34].

The halogen bond refers to a non-covalent interaction of general structure R—X···B, in which X is an electrophilic halogen atom (typically chlorine, bromine, or iodine) and B is any negative site, including a Lewis base, a π region of an aromatic, a molecule containing a double bond, and an anion. This type of interactions is typically characterized by

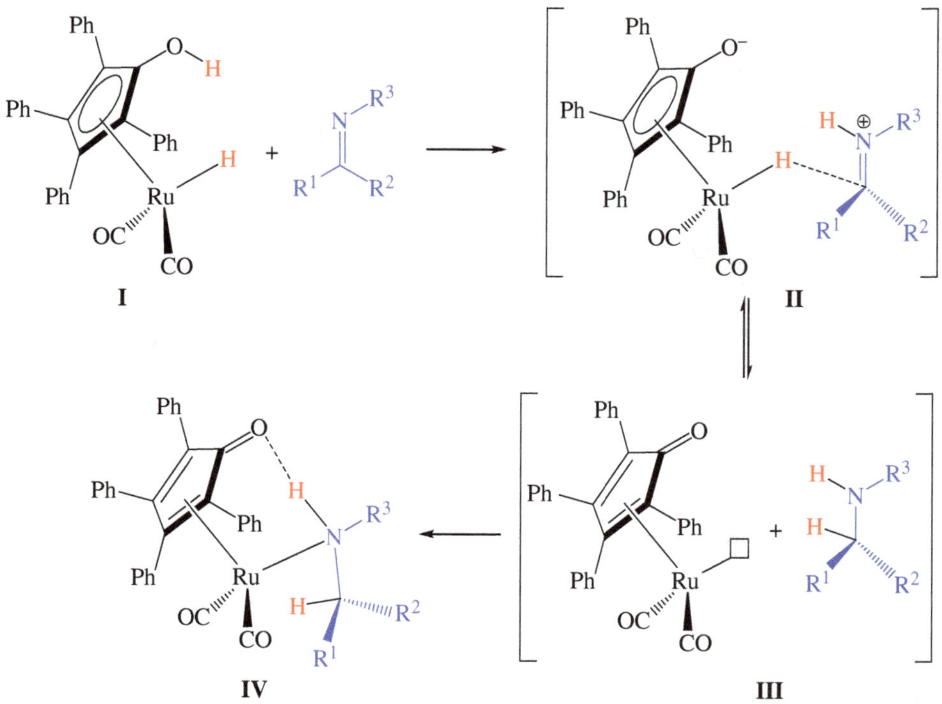

SCHEME 18.18 Shvo's catalyst in hydrogenation of imines to amines [29].

unusually short X···B internuclear separations and a linear R—X···B bond angle. Generally, halogenated compounds are noteworthy for their ability to participate in several distinct types of non-covalent interactions. The positive region on the outermost portion of the halogen's surface, called the σ-hole, can interact attractively with electronegative atoms. The more effective electron withdrawing the group R bound to X is, the stronger is halogen bonding to which it gives rise.

Until now, halogen bonding has been mainly used for molecular recognition to develop receptors with specific selectivity for anions as well as in effective separation processes [30]. Only several examples of application of halogen bonding in catalysis have been reported, and an example concerns the coordination and halogen bond-assisted diketonate-ligated Cu(I)-catalyzed cross-coupling between aryl halides PhX (X = I, Br, Cl) and cyclohexylamine (Scheme 18.19), leading to C—N bond formation [36]. In this case, cyclohexylamine (NH_2Cy) coordinates to (ket)Cu^I (ket = 2-acetylcyclohexanoate) giving the (ket)$Cu^I(NH_2Cy)$ complex, which then reacts with a base (e.g., Cs_2CO_3) to give an anionic intermediate **I**. Phenyl iodide is involved in halogen bond formation with an anionic nitrogen of **I**, thus producing **II**, which undergoes intramolecular oxidative addition to give **III**. Subsequently, reductive elimination leads to the reaction product, PhNHCy, with regeneration of the (ket)Cu^I complex.

The cooperation of coordination, hydrogen, and halogen bonding is observed in asymmetric α-fluorination of aliphatic acyl chlorides [37]; the catalytic system is based on a combination of the chiral nucleophile benzoylquinine and a transition metal LA cocatalyst, such as $[PdCl_2(PPh_3)_2]$; an alkali metal LA salt ($LiClO_4$); *N*-fluorodibenzenesulfonamide (NFSi); and a variety of nucleophiles, for example, methanol and aromatic amines, in the presence of the Hünig base (Scheme 18.20). Under these conditions, aliphatic acyl chlorides are deprotonated by the Hünig base producing 2-substituted ethenones, which react with the chiral nucleophile benzoylquinine and $[PdCl_2(PPh_3)_2]$ leading to dually activated enolates **I**. The parallel activation of NFSi by the LA (Li^+) occurs via the adduct **II**. Interaction of both intermediates, **I** and **II**, creates a system **III**, where coordination, hydrogen, and halogen bonding act cooperatively to promote nucleophilic attack, which leads to the asymmetric α-fluorination products, with good yields and enantioselectivities.

Hydrogen and halogen bonds also assisted the enantioselective bromocycloetherification of 5-arylpentenols using a chiral Brønsted acid and an achiral Lewis base ($Ph_3P{=}S$) (Scheme 18.21) to provide enantiomeric induction [38]. A high selectivity was achieved by a combination of substrate and catalyst control; the enantiomeric ratios of the exo products increased up to 97 : 3 [39].

According to the proposed mechanism [40], the combination of *N*-bromosuccinimide, triphenylphosphine sulfide ($Ph_3P{=}S$), and chiral phosphoric acid leads to the intermediate adduct **I** (Scheme 18.22), which is stabilized by hydrogen and halogen bonds. The following proton transfer from chiral phosphoric acid to *N*-bromosuccinimide allows liberation of succinimide, thus leading to the halogen bond-stabilized intermediate **II**. Reaction of 5-arylpentenol with **II** gives the brominated intermediate **III** with liberation of $Ph_3P{=}S$. This unstable intermediate **III** produces *exo* (5-membered) and *endo* (6-membered) cyclic products.

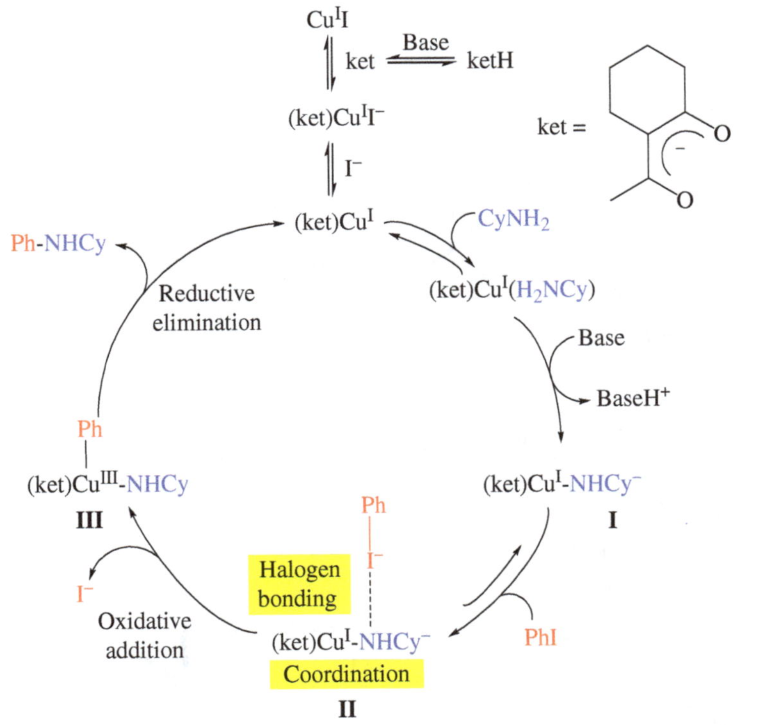

SCHEME 18.19 Cross-coupling reaction between PhI and cyclohexylamine catalyzed by a (2-acetylcyclohexanoate)-Cu(I) system [30].

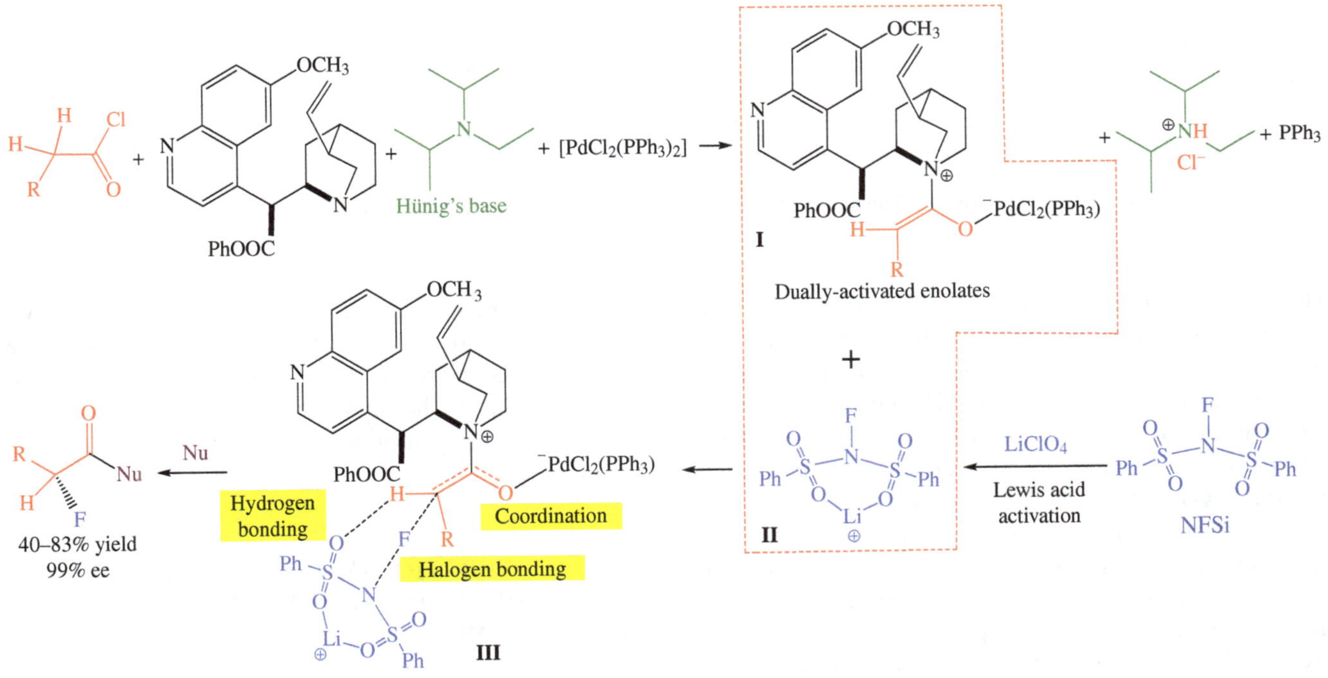

SCHEME 18.20 Enantioselective α-fluorination of aliphatic acyl chlorides [37].

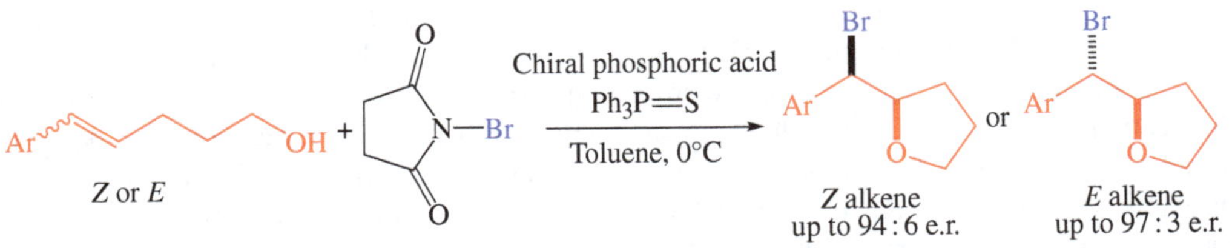

SCHEME 18.21 Enantioselective synthesis of bromomethyltetrahydrofurans upon bromocycloetherification of (*E/Z*) 5-arylpentenols [38, 39].

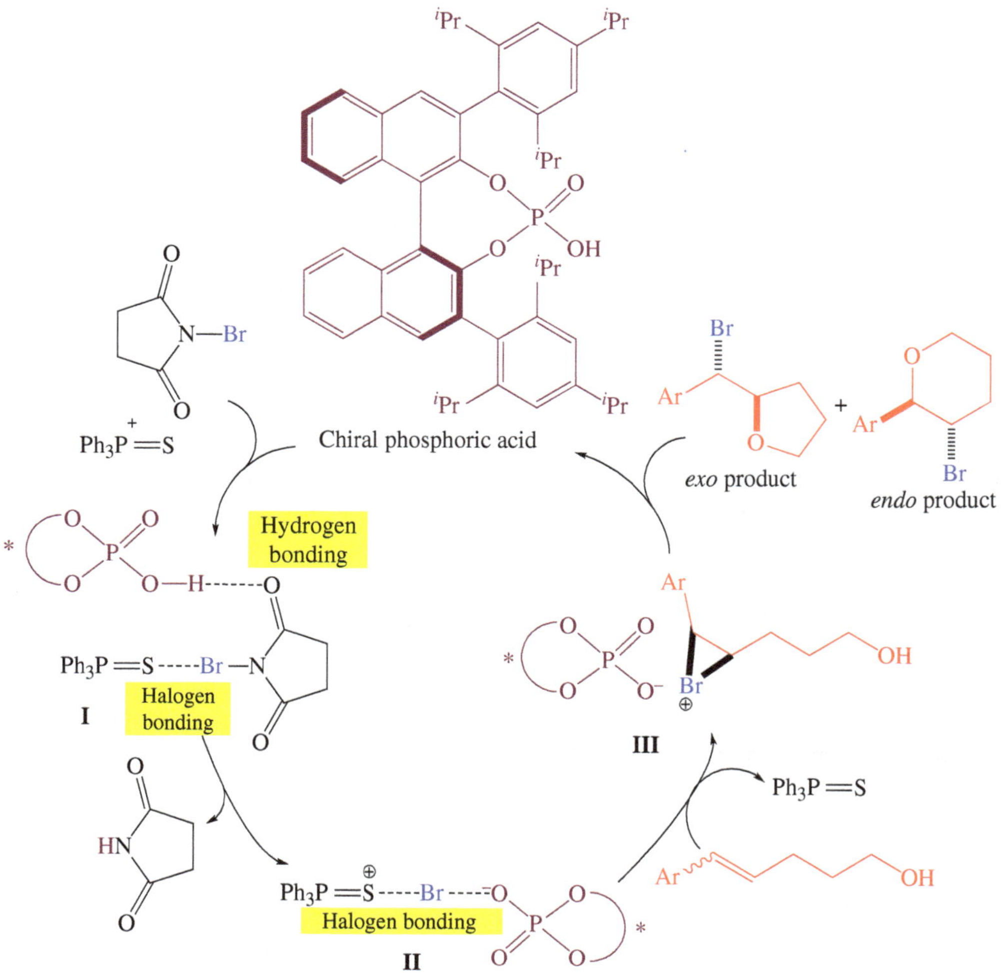

SCHEME 18.22 Proposed mechanism of enantioselective bromocycloetherification of (*E/Z*) 5-arylpentenols [40].

Although the expected product of 6-*endo* cyclization was formed with negligible enantioselectivity, the product of anti-Markovnikov 5-*exo* cyclization is produced with higher enantioselectivity in all cases [38]. The effect of electronic perturbations is strong, particularly among the (*E*)-configured substrates. A moderately stronger electron-rich aryl group unsurprisingly hampers the 5-*exo* cyclization.

The cooperation of hydrogen and halogen bonding also promotes the enantioselective iodolactonization of 4-benzylpent-4-enoic acid with *N*-chlorophthalimide (NCP) in the presence of a chiral phosphate catalyst (CPC) (Scheme 18.23a) to give the corresponding iodolactones in high yields with high enantioselectivity [41]. In this reaction, the Lewis acidic NCP seems to activate I_2 through halogen bonding interactions in toluene at −40°C to form the active iodinating intermediate **I** (X = I; Scheme 18.23b), the presence of which is confirmed by Raman spectroscopy. Subsequently, interaction of **I** with CPC gives the halogen bond adduct **II** and then a charge-stabilized halogen bond system **III**. Since this step generates ICl (X = Cl), which can also act as an iodinating reagent through the activation by NCP, the use of 0.5 equivalents of I_2 was sufficient to complete the iodolactonization. The reaction of benzyl-pent-4-enoic acid with the intermediate **III** leads to the charge-assisted hydrogen-bonded species **IV**. Thus, electrophilic iodination of the double bond of 4-benzylpent-4-enoic acid followed by cyclization gives the iodolactones (**V**) and phthalimide.

18.4 COOPERATION OF COORDINATION WITH OTHER TYPES OF NON-COVALENT INTERACTIONS

Not only hydrogen and halogen bonding but also other types of weak non-covalent interactions, such as cation–π, anion–π, polar–π, C—H–π, π–π, pnicogen bond, or steric repulsions, can play an essential role within the coordination sphere of catalysts allowing in some instances to control effectively their activity and reaction pathways [42–45]. For instance, non-covalent steric hindrance can be crucial for the achievement of a high selectivity by affecting the conformations of the reaction intermediates and/or transition states. The above mentioned non-covalent interactions are rather

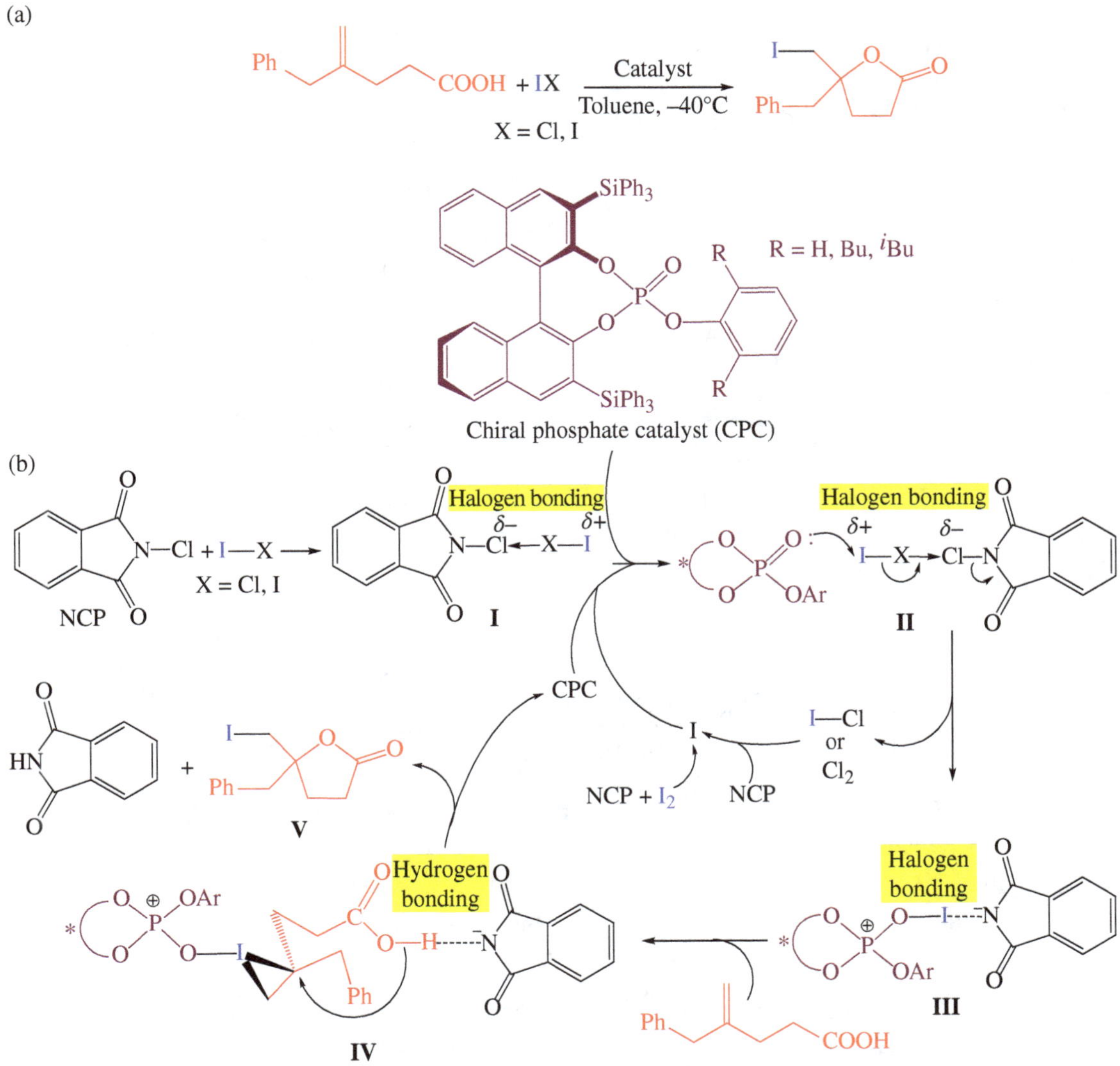

SCHEME 18.23 (a) Enantioselective iodolactonization of 4-benzylpent-4-enoic acid with *N*-chlorophthalimide. (b) Proposed mechanism of the reaction [41].

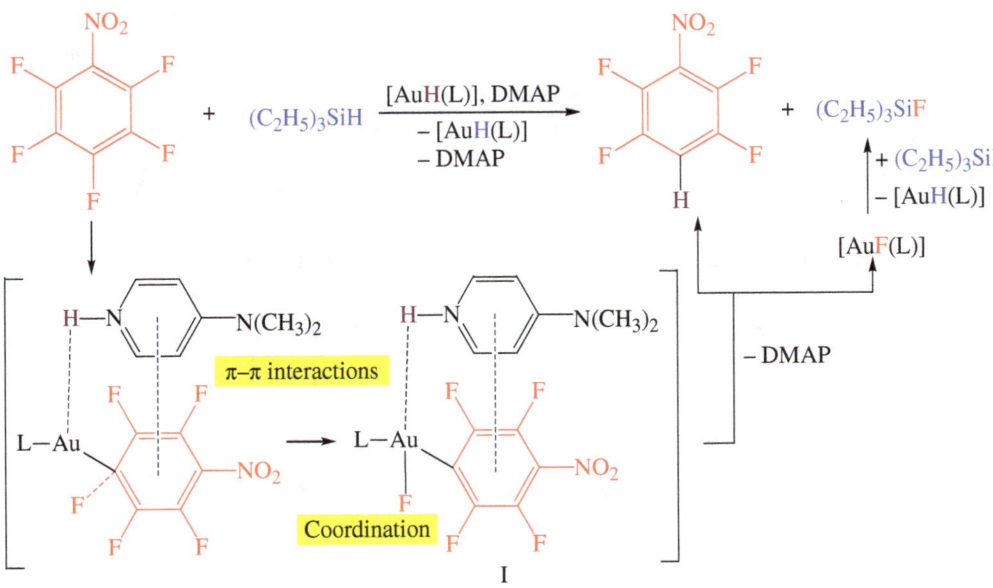

SCHEME 18.24 Coordination and π–π interaction-assisted C–F bond activation [46].

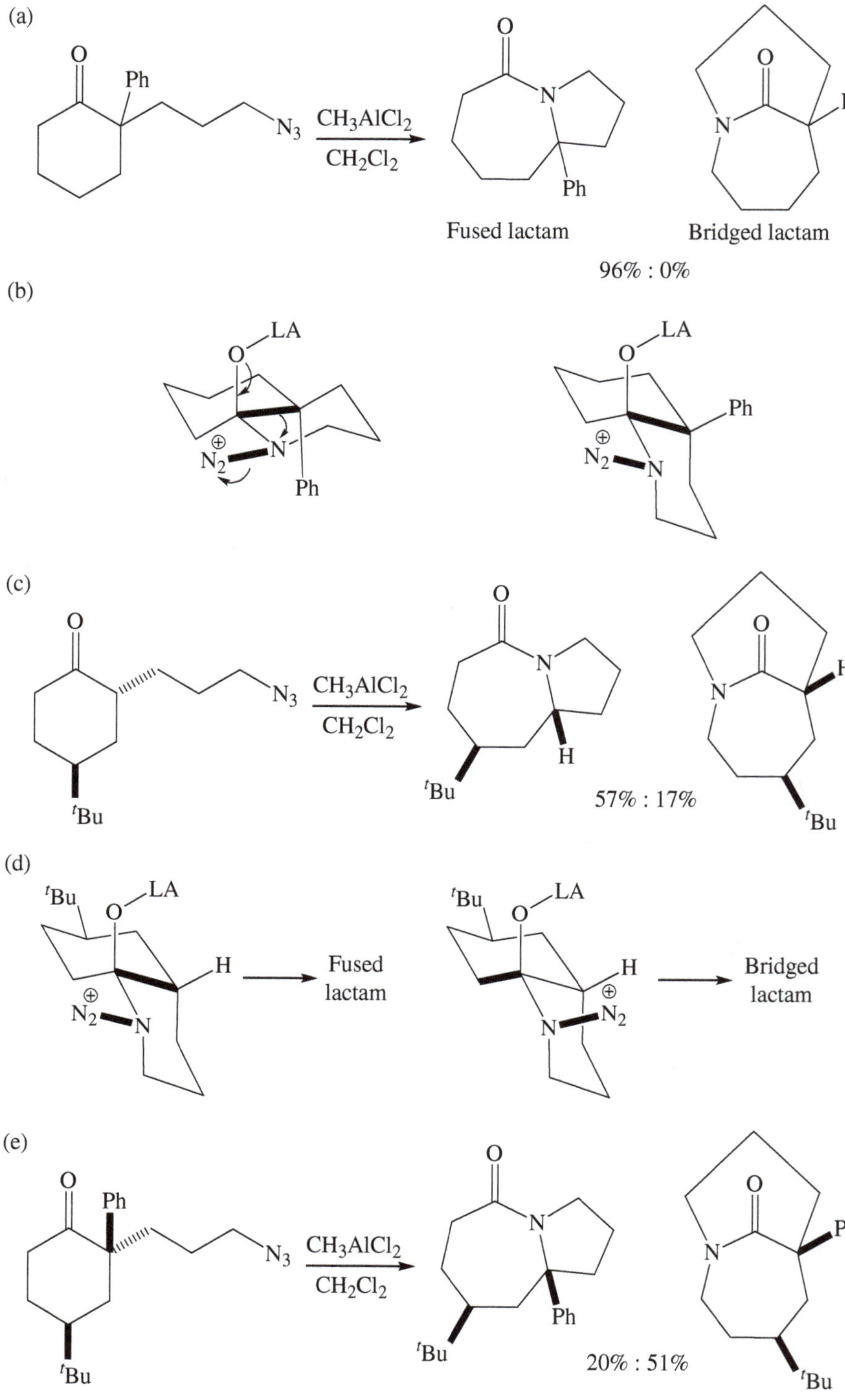

SCHEME 18.25 Intramolecular Schmidt reactions to bridged bicyclic lactams [47].

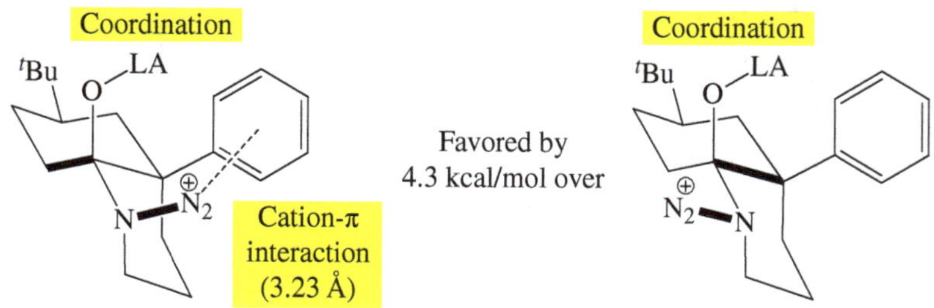

SCHEME 18.26 Cooperation of coordination and cation–π interactions in intramolecular Schmidt reactions [47].

subtle, and therefore it can be difficult to predict the dominant conformers for key intermediates, but information on the factors that control the conformation of the reactive species will be valuable for the design of new cooperative catalysts with a high reactivity and selectivity. Understanding of the nature of the interaction between the substrate(s) and the catalyst helps optimize the activity and selectivity.

It was demonstrated [46] that π–π interactions together with coordination allow effectively to activate the C–F bond in the reaction of perfluoroarenes with silanes (Scheme 18.24). In this reaction an attractive π–π interaction between the organic additive *p-N,N*-dimethylaminopyridine (DMAP), which acts as a electron donor, and the perfluorinated arene (electron acceptor) is catalyzed by gold hydride [AuH(L)] (L = *N,N'*-bis(2,4,6-trimethylphenyl)imidazolin-2-ylidene), resulting mainly in *para*-hydrodefluorinated products with yields up to 90%. According to the proposed mechanism, firstly a reaction of perfluoroarenes with the gold catalyst produces a coordination and π–π interaction-assisted intermediate **I** (Scheme 18.24). Liberation of DMAP leads to *para*-hydrodefluorinated product and the fluoro complex [AuF(L)]. Subsequently, interaction of [AuF(L)] with the silane gives the fluorinated silane and regenerates the hydride catalyst [AuH(L)]. An analysis of the charge distribution and transition state geometries indicates that this hydrodefluorination process is controlled by π–π interactions between DMAP and perfluoroarenes, accompanied by changes of partial atomic charges. Thus, the activation barrier in the presence of DMAP (31.6 kcal/mol) is significantly lower than the corresponding one in its absence (40.8 kcal/mol). Moreover, the yield of the reaction increases with the stronger electron-withdrawing groups and degree of fluorination on the arenes, and the hydrodefluorination reaction also tolerates different functional groups (such as formyl, alkynyl, ketone, ester, and carboxylate groups) without their reduction or hydrogenation [46].

An interesting cooperation of coordination and cation–π effect was observed [47] in the intramolecular Schmidt cyclization where an azide moiety reacts with a carbonyl via a LA (e.g., CH_3AlCl_2) to give two different lactam products (Scheme 18.25a). There are two possible transition states (epimerized for clarity) (Scheme 18.25b), and both of them lead to the observed fused lactam (Scheme 18.25a). However, the situation is different with a conformationally locking *tert*-butyl group (Scheme 18.25c). In this case, the bridged lactam is preferentially formed. One cannot flip the ring and the minor product presumably arises from migration of the less electron-rich carbon–carbon bond (Scheme 18.25d). Since the diazonium intermediate is high in energy, one can expect this reaction to be quite exothermic.

With the introduction of a phenyl group, selectivity increases for the bridged product (Scheme 18.25e). It is proposed that the transition state leading to the bridged lactam is stabilized by a cooperation of coordination and cation–π interaction (Scheme 18.26). Mechanistic studies show a substantial difference in energy between these models (4.3 kcal/mol) [47].

18.5 FINAL COMMENTS

Cooperation of coordination and non-covalent interactions in catalysis is attracting an increasing attention as it can enable the development of unprecedented transformations that are not possible by use of either of the catalytic type systems alone and can improve the reactivity, efficiency, and stereocontrol of known reactions. This cooperative catalysis is a rapidly expanding discipline that has benefited from the development of both homogeneous catalysis and supramolecular chemistry.

In this chapter, such a type of cooperation is illustrated with several recent relevant examples, with the aim of shedding light on the synthetic utilities and potentials of this concept as a novel tool in the synthesis of organic compounds. Thus, various types of coordination and non-covalent interactions, such as hydrogen and halogen bonding, π–π, or cation–π, have been applied successfully for promoting a variety of transformations. In particular, an impressive progress has occurred in the combination of coordination with hydrogen and halogen bonding. However, despite the remarkable developments, the concept of cooperative catalysis so far is still limited in number when compared to transition metal (coordination) catalysis or organocatalysis, Moreover, mechanistic investigations of these dual catalyst reactions are highly desirable to understand their occurrence, optimize the conditions, and search for new catalytic systems. Thus, further highlighting of the potential of this new and exciting research area is required to overcome the challenges that still remain for its development.

ACKNOWLEDGMENTS

This work has been partially supported by the Baku State University as well as Foundation for Science and Technology (FCT), Portugal [UID/QUI/00100/2013 and "Investigador 2013" (IF/01270/2013/CP1163/CT0007) programs]. M.N.K. thanks FCT for the working contract.

REFERENCES

1. Z. Shao, H. Zhang, *Chem. Soc. Rev.* 2009, *38*, 2745.
2. N. T. Patil, V. S. Shinde, B. Gajula, *Org. Biomol. Chem.* 2012, *10*, 211.
3. Z. Du, Z. Shao, *Chem. Soc. Rev.* 2013, *42*, 1337.
4. P. M. Pihko (Ed.), Hydrogen Bonding in Organic Synthesis; Wiley-VCH: Weinheim, 2009.
5. P. Daw, A. Sinha, S.M.W. Rahaman, S. Dinda, J.K. Bera, *Organometallics* 2012, *31*, 3790.
6. T. Šmejkal, B. Breit, *Organometallics* 2007, *26*, 2461.
7. J. Wang, H. Xu, H. Gao, C.-Y. Su, C. Zhao, D. L. Phillips, *Organometallics* 2010, *29*, 42.
8. J. T. Spletstoser, J. M. White, A. R. Tunoori, G. I. Georg, *J. Am. Chem. Soc.* 2007, *129*, 3408.

9. B. Xu, S.-F. Zhu, X.-L. Xie, J.-J. Shen, Q.-L. Zhou, *Angew. Chem. Int. Ed.* 2011, *50*, 11483.
10. C. Arróniz, A. Gil-González, V. Semak, C. Escolano, J. Bosch, M. Amat, *Eur. J. Org. Chem.* 2011, 3755.
11. (a) E. Piedra-Arroni, P. Brignou, A. Amgoune, S. M. Guillaume, J.-F. Carpentierc, D. Bourissou, *Chem. Commun.* 2011, *47*, 9828; (b) P. Brignou, S. M. Guillaume, T. Roisnel, D. Bourissou, J.-F. Carpentier, *Chem. Eur. J.* 2012, *18*, 9360.
12. Z.-Y. Xue, Q.-H. Li, H.-Y. Tao, C.-J. Wang, *J. Am. Chem. Soc.* 2011, *133*, 11757.
13. M. Wang, C.-J. Wang, Z. Lin, *Organometallics* 2012, *31*, 7870.
14. T. Šmejkal, B. Breit, *Angew. Chem. Int. Ed.* 2008, *47*, 311.
15. T. Šmejkal, D. Gribkov, J. Geier, M. Keller, B. Breit, *Chem. Eur. J.* 2010, *16*, 2470.
16. T. Šmejkal, B. Breit, *Angew. Chem. Int. Ed.* 2008, *47*, 3946.
17. L. Diab, T. Šmejkal, J. Geier, B. Breit, *Angew. Chem. Int. Ed.* 2009, *48*, 8022.
18. W. Zeng, G.-Y. Chen, Y.-G. Zhou, Y.-X. Li, *J. Am. Chem. Soc.* 2007, *129*, 750.
19. I. Usui, S. Schmidt, M. Keller, B. Breit, *Org. Lett.* 2008, *10*, 1207.
20. M. Arthuis, R. Beaud, V. Gandon, E. Roulland, *Angew. Chem. Int. Ed.* 2012, *51*, 10510.
21. T. Ishii, R. Watanabe, T. Moriya, H. Ohmiya, S. Mori, M. Sawamura, *Chem. Eur. J.* 2013, *19*, 13547.
22. I. Usui, S. Schmidt, B. Breit, *Org. Lett.* 2009, *11*, 1453.
23. B. L. Conley, M. K. Pennington-Boggio, E. Boz, T. J. Williams, *Chem. Rev.* 2010, *110*, 2294.
24. J. B. Johnson, J.-E. Bäckvall, *J. Org. Chem.* 2003, *68*, 7681.
25. (a) G. Csjernyik, A. H. Éll, L. Fadini, B. Pugin, J. E. Bäckvall, *J. Org. Chem.* 2002, *67*, 1657; (b) Y. R. S. Laxmi, J.-E. Bäckvall, *Chem. Commun.* 2000, 611; (c) M. L. S. Almeida, M. Beller, G.-Z. Wang, J.-E. Bäckvall, *Chem. Eur. J.* 1996, *2*, 1533.
26. (a) N. Menashe, Y. Shvo, *Organometallics* 1991, *10*, 3885; (b) N. Menashe, E. Salant, Y. Shvo, *J. Organomet. Chem.* 1996, *514*, 97.
27. J. S. M. Samec, J.-E. Bäckvall, *Chem. Eur. J.* 2002, *8*, 2955.
28. N. Uematsu, A. Fujii, S. Hashiguchi, T. Ikariya, R. Noyori, *J. Am. Chem. Soc.* 1996, *118*, 4916.
29. C. P. Casey, J. B. Johnson, *J. Am. Chem. Soc.* 2005, *127*, 1883.
30. P. Metrangolo, G. Resnati, Halogen Bonding: Fundamentals and Applications (Structure and Bonding); Springer: Heidelberg, 2010.
31. E. Parisini, P. Metrangolo, T. Pilati, G. Resnati, G. Terraneo, *Chem. Soc. Rev.* 2011, *40*, 2267.
32. M. Erdélyi, *Chem. Soc. Rev.* 2012, *41*, 3547.
33. A. Priimagi, G. Cavallo, A. Forni, M. Gorynsztejn-Leben, M. Kaivola, P. Metrangolo, R. Milani, A. Shishido, T. Pilati, G. Resanti, G. Terraneo, *Adv. Funct. Mater.* 2012, *22*, 2572.
34. A. Priimagi, G. Cavallo, P. Metrangolo, G. Resnati, *Acc. Chem. Res.* 2013, *46*, 2686.
35. T. M. Beale, M. G. Chudzinski, M. G. Sarwar, M. S. Taylor, *Chem. Soc. Rev.* 2013, *42*, 1667.
36. G. Lefèvre, G. Franc, C. Adamo, A. Jutand, I. Ciofini, *Organometallics* 2012, *31*, 914.
37. J. Erb, D. H. Paull, T. Dudding, L. Belding, T. Lectka, *J. Am. Chem. Soc.* 2011, *133*, 7536.
38. S. E. Denmark, M. T. Burk, *Org. Lett.* 2012, *14*, 256.
39. S. E. Denmark, M. T. Burk, *Proc. Natl. Acad. Sci. U. S. A.* 2010, *107*, 20655.
40. F. Lv, S. Liu, W. Hu, *Asian J. Org. Chem.* 2013, *2*, 824.
41. H. Nakatsuji, Y. Sawamura, A. Sakakura, K. Ishihara, *Angew. Chem. Int. Ed.* 2014, *53*, 6974.
42. P. W. N. M. van Leeuwen (Ed.), Supramolecular Catalysis; Wiley-VCH: Weinheim, 2008.
43. M. Raynal, P. Ballester, A. Vidal-Ferran, P. W. N. M. van Leeuwen, *Chem. Soc. Rev.* 2014, *43*, 1660.
44. M. Raynal, P. Ballester, A. Vidal-Ferran, P. W. N. M. van Leeuwen, *Chem. Soc. Rev.* 2014, *43*, 1734.
45. S. Scheiner, *Acc. Chem. Res.* 2013, *46*, 280.
46. H. Lv, J.-H. Zhan, Y.-B. Cai, Y. Yu, B. Wang, J.-L. Zhang, *J. Am. Chem. Soc.* 2012, *134*, 16216.
47. L. Yao, J. Aube, *J. Am. Chem. Soc.* 2007, *129*, 2766.

19

HYBRID ORDERED MESOPOROUS MATERIALS AS SUPPORTS FOR PERMANENT ENZYME IMMOBILIZATION THROUGH NON-COVALENT INTERACTIONS

VICTORIA GASCÓN, CARLOS MÁRQUEZ-ALVAREZ, ISABEL DÍAZ, AND ROSA M. BLANCO

Instituto de Catálisis y Petroleoquímica, CSIC, Madrid, Spain

19.1 INTRODUCTION

The immobilization of enzymes is widely acknowledged as a requirement for their industrial application. These natural catalysts can work in mild conditions of temperature and pH and absence of organic solvents. Besides, enzymes display high regioselectivity, enantioselectivity, and/or substrate specificity. These excellent properties make them optimum catalysts for a number of industrial transformations that can be performed under environmentally nonaggressive conditions while avoiding several reaction steps for activation or protection of reactive groups of the substrates. The lability derived from their proteic nature together with the solubility in the reaction medium may limit the practical application of enzymes, which are also costly to obtain. Immobilization can help to overcome these difficulties: the biocatalysts can be easily separated from reaction media and reused in subsequent reaction cycles, and often immobilization can also contribute to their stability. Additionally, immobilized enzymes permit continuous operation of processes, rapid termination of reactions, and greater variety of reactor designs.

Generally, compared with the free enzymes, immobilized ones undergo decrease in activity and increase in Michaelis constant. This depletion results from the change in the microenvironment where the enzyme works and depends on the nature of the support and the design of the reactor. In addition, mass transfer limitations are unavoidable, but these can be reduced by the proper reactor design and by appropriate characteristics of the support. As result, numerous materials and methods have been studied, and a wide variety of enzyme immobilization systems have been reported using support materials of different origin and chemical nature.

19.1.1 Materials Used as Supports

According to the chemical nature of the support materials, these can be inorganic and organic carriers. Inorganic materials include silica gel, sepiolite, Celite, mesoporous silica, glass beads, montmorillonite, alumina, titania, zirconia, zeolites, and metals (e.g., gold). Organic carriers can be divided into natural and synthetic. Natural carriers comprise chitin, chitosan, cellulose, agarose, or alginate. Synthetic polymers used are poly(vinyl)alcohol (PVA) (commercialized as LentiKats), cross-linked PVA, polypropylene (Accurel), polystyrene foam (porous), poly(acrylic acid-co-acrylamide), poly(hydroxybutyrate) nanoparticles or beads, poly(acrylonitrile) (PAN) (PAN electrospun fibers of high porosity), Polyclar (polyvinylpolypyrrolidone), Mikroperl, epoxy-activated polymer supports for covalent enzyme binding that are commercially available like Eupergit and Sepabeads, and resins like Affi-Gel 10 (couples neutral or basic proteins), Affi-Gel 15 (couples acidic proteins), or IRA-400 (ionic exchange resin) [1–6]. In the last decade new nanomaterials (nanotubes, nanoparticles, or nanofibers) with improved properties (small size, large surface-to-volume ratio, etc.) are being incorporated. Some examples are polymethyl methacrylate (PMMA) cores with

Non-Covalent Interactions in the Synthesis and Design of New Compounds, First Edition.
Edited by Abel M. Maharramov, Kamran T. Mahmudov, Maximilian N. Kopylovich, and Armando J. L. Pombeiro.

cellulase shells, mesoporous silica nanoparticles, and carbon nanotubes [7–9]. Magnetic nanoparticles can be rapidly and easily removed from the reaction mixture by applying a magnetic field; they may be coated with different materials or be used uncoated [10].

As indicated above, a variety of materials used as supports for enzymes exploited covalent or non-covalent bonding for the immobilization. The internal structure of the materials is also diverse: some materials are crossed fibers (Eupergit), and others are nonporous inorganic materials (Celite). The most promising ones are believed to be porous materials: agarose, silica and controlled-pore glass, among others, and especially mesoporous nanoparticles.

19.1.2 Morphology of the Supports

Because non-covalent interactions are weak and reversible forces, the morphology and size of support particles are of relevance in these systems. Hence, external morphology and internal structure of the particle are parameters to study as well as the size, since both may affect the rate and amount of adsorbed protein in the pores [11, 12]. Shorter pore channels result in higher enzyme loading values. Tortuosity of the pore channels also affects both enzyme loading and leaching in opposite ways [13]. Pore size is maybe the parameter that affects more strongly several crucial aspects of the immobilization and stabilization of the enzyme. First and obviously, pore diameter must be larger than molecular dimensions of the enzyme. But how much? It has been demonstrated that too wide pore size permits a fast immobilization process, but the larger the pores, the smaller is the surface area and hence the enzyme loading. On the other hand, non-covalently linked enzymes are less restricted to be released, and consequently leaching is not prevented. Such variety of variables to consider suggests that there is not a single universal enzyme–support system. The nature of the attachment and the properties and characteristics of the support must be carefully studied to get the best choice.

19.1.3 Functionalization of the Support

Unspecific immobilization is not advisable, and so the presence of proper functional groups on the support surface is needed, aiming to drive the immobilization either by covalent or non-covalent interactions. Since most materials used are inert, organic groups must be introduced through different strategies. A quite common one is grafting with a bifunctional reagent. One functionality is required to react with the support, whereas the other one remains untouched to interact with the enzyme. Concerning the reagents, glutaraldehyde is a very common bifunctional compound used for support activation, mainly with covalent attachment purposes. Silanes are also widely used (e.g., trialkoxysilanes bearing different functional groups like aminopropyl, octyl, mercaptopropyl, vinyl, phenyl, etc., all commercially available); reaction with alkyltriethoxysilane is used to functionalize siliceous materials [12, 14].

The introduction of amines on inert supports to induce hydrogen bonds or electrostatic interactions is another common strategy that can be performed using not only aminopropyltriethoxysilane but also diethylamine [15], diethylaminoethyl (DEAE) [16], or monoaminomethyl-*N*-aminoethyl groups [17] as carrier modifiers.

A variation of this strategy is the chemical modification of groups present on the matrix surface (e.g., oxidation of epoxide groups in cross-linked sepharose to generate aldehyde groups for covalent bonding [18]).

Another possibility is the incorporation of organic (functional or active) groups upon the synthesis of the material. This strategy can be applied to obtain hybrid organosiliceous materials by co-condensation pathways [13, 19]. Also, organic polymers can be synthesized starting with monomers bearing the desired groups, for example, polyethyleneimine [20] or poly(styrene sulfonate) [21], where the length of the chain obtained can also be controlled.

19.1.4 Covalent versus Non-covalent Enzyme Immobilization

Immobilization methods are mainly physical entrapment and chemical interactions. Chemical (covalent) immobilization involves enzyme–matrix covalent bonds, enzyme–matrix cross-linking, or cross-linking of enzyme with multifunctional reagents. Non-covalent adsorption methods are the simplest ones. Advances in this area permit one to design the most suitable immobilization system. Cao elevated enzyme immobilization to the status of art [22]. The process starts with the choice of a suitable carrier, condition and enzyme, and a compromise method to satisfy the catalytic and noncatalytic needs required for the application.

Covalent attachment of enzyme to supports bearing reactive groups has attracted attention because of the possibility to increase stability through the formation of multiple bonds [18]. The covalent nature of the enzyme–support link also prevents enzyme release from the support. On the other hand, the supports cannot be reused after enzyme inactivation. The presence of covalently linked enzyme in not sufficiently wide pores prevents the access of new molecules, thus limiting the enzyme loading. Catalytic activity can suffer from severe drops because of the chemical modification and distortion of the tertiary structure of the protein subsequent to the covalent nature of the attachment, especially if it is performed through multiple points [23].

Non-covalent enzyme immobilization through weak forces such as hydrogen bonds or ionic or hydrophobic interactions does not involve severe distortion of the tertiary protein structure, so catalytic activity is better preserved. Besides, reutilization of supports after enzyme inactivation is feasible. However, the weak nature of the adsorption often results in enzyme leaching.

19.2 NON-COVALENT METHODS OF ENZYME IMMOBILIZATION

Immobilization by adsorption is cost efficient, simple, and fast. These are advantages versus covalent attachment, which involves several steps (higher cost and time consuming) and chemical modification of the protein (loss of catalytic activity). The reversible nature of immobilization permits to recover support materials, but on the other hand it also is responsible for enzyme leaching.

Adsorption of proteins can be driven by diverse forces such as (in increasing intensity order) van der Waals forces, hydrogen bonding, hydrophobic interactions, or electrostatic forces. Van der Waals forces are present in any system, but they lack intensity to efficiently link the protein. Hydrogen bonding with support occurs through polar amino acids or sugar molecules from glycoproteins. Hydrophobic interactions result from entropic effects derived from the loss of water molecules from the hydrophobic surfaces of protein and support. Electrostatic interactions involve opposite charges of support and polar amino acids of the enzyme. In enzyme immobilization systems, more than one of these binding interactions occur simultaneously, so it is difficult to point out a specificity regarding binding sites [24].

An interesting non-covalent bonding system is the site-specific affinity attachment, which ensures identical orientation of all enzyme molecules adsorbed [25]. Such specific interactions of antigen/antibody systems can be used not only for enzyme immobilization but also (and preferably) for protein purification, for example, antibodies. The chelation between polyhistidine-tag (6xHis-tag) and divalent metal ions (Fe^{2+}, Cu^{2+}, Co^{2+}, Ni^{2+}, or Zn^{2+}) is a specific non-covalent method [26]. The imidazole ring of histidine can coordinate with the cation to form a stable conjugate with high specificity and affinity. The enzyme must have a 6xHis-tag, and the support must be activated with nitrilotriacetic acid (NTA) or iminodiacetic acid in order to chelate the divalent metal ions.

Another bioaffinity technique is the avidin–biotin system [26, 27]. The avidin–biotin complex possesses the strongest known non-covalent interaction between a protein and a ligand. The bond formation between biotin and avidin is very rapid, and once formed, it is unaffected by extreme pH, temperature, organic solvents, and other denaturing agents. Biotin is a water-soluble B complex vitamin that has a fused ring structure containing a tetrahydroimidizalone ring with a tetrahydrothiophene ring. Avidin is a homotetramer protein. Each one of its subunits can bind one molecule of biotin very specifically, almost as strongly as in a covalent bond. Avidin-coated nanomaterials have been used, including gold, glass, or apoferritin nanoparticles.

There are many commercially available biotinylation reagents for targeting specific functional groups or residues, including primary amines, sulfhydryls, carboxyls, and carbohydrates, which allows the choice of a reagent that does not inactivate the target macromolecule. The avidin–biotin detection system allows an almost unlimited number of primary detection reagents (i.e., antibodies, nucleic acid probes, and ligands) to be easily captured, recovered, immobilized, or detected. The avidin–biotin interaction is widely used in specific biotechnological applications. Regarding enzyme immobilization, these are not common techniques to apply although there are examples reported in the literature [28].

19.3 ORDERED MESOPOROUS MATERIALS

This part of the chapter describes the use of siliceous ordered mesoporous materials for non-covalent immobilization of enzymes. Mesoporous materials offer a high surface area, which allows to obtain high enzyme loadings per unit weight or volume of catalyst. Regular pores with ordered distribution may favor the diffusion of enzyme molecules through the porous network (favoring high enzyme loading) and also the diffusion of substrates and products (reducing diffusional restrictions of reaction). Siliceous materials are robust and easy to handle in reactors. These materials can be easily functionalized with organic groups to increase the chemical affinity of the enzyme for the surfaces of these hybrid materials. Finally, the hybrid mesoporous materials are synthesized with the desired pore sizes, which enable enzyme diffusion while confining it to decrease leaching.

19.3.1 Ordered Mesoporous Silica Materials: Control of Mesostructure and Pore Size

Ordered mesoporous silica materials (OMSM) possess large surface area and pore volume, uniform and tunable pore size, and chemically modifiable surface, which make them promising candidates for enzyme immobilization [29]. Since the synthesis of the so-called M41S family of mesoporous silicas such as MCM-41 and MCM-48, with two-dimensional (2D) hexagonal and cubic bicontinuous arrangement of uniform pore channels, using cationic surfactants as templates, was reported in the early 1990s [30, 31], there have been considerable efforts devoted to the synthesis of mesostructured porous silicas making use of the supramolecular arrangement of surfactants of various types (cationic, anionic, zwitterionic, neutral, and nonionic) to direct the pore structure ordering [32–35]. There are several excellent reviews that summarize synthesis, characterization, and applications of ordered mesoporous silicas [36–41]. In this section, we review briefly the development on the control of structure and pore size of mesoporous silicas of interest for enzyme adsorption.

The synthesis of OMSM is generally carried out by a sol–gel process in which hydrolysis and condensation of the silica precursors take place in a clear, homogeneous aqueous solution of surfactants. In these solutions, surfactant content is higher than the critical micellar concentration, and, therefore, these molecules self-aggregate to form micelles that can exhibit different shapes (e.g., spheroids, discs, or rods) depending on the nature of the surfactant and physicochemical conditions of the solution [42]. Typically, the surfactant concentration used is lower than that required for micelles to organize into liquid crystalline phases. Nonetheless, composite species are formed by non-covalent binding of the polymeric silicate species in solution to the external surface of micelles, which spontaneously assemble generating an ordered mesophase.

It is widely accepted that this process takes place through a cooperative self-assembly mechanism [43], originally proposed by Stucky *et al.* [44], who stated that changes in the silicate oligomers occurring as the polymerization progresses and arrangement of surfactant species mutually influence each other leading to transformations of micelles shape and, eventually, growing of the liquid crystal-like array [45]. After precipitation of the ordered mesophase, longer synthesis time and/or higher temperature allows further condensation and cross-linking of silicate species, rendering a stable silica network resistant to collapse when the entrapped surfactant molecules are subsequently removed by solvent extraction or calcination to obtain the porous material.

Based on this synthesis strategy, mesoporous materials showing a large variety of mesostructures have been obtained, with pore sizes ranging from 1.5 to around 20 nm and different particle morphologies (rod, platelet, sphere, etc.). The ordered pore networks of these materials exhibit mainly hexagonal (both 2D and three-dimensional (3D)) and cubic (3D and bicontinuous) symmetries. Examples of representative materials with periodic pore arrangements in different space groups are shown in Figure 19.1.

Mesostructures with 2D hexagonal symmetry (space group *p6mm*) correspond to a hexagonal close packing of unidirectional cylinder-like pore channels. Characteristic materials having this mesostructure are MCM-41 and SBA-3 (synthesized with cationic surfactants), FSM-16 (so-called folded sheets mesoporous materials, obtained by intercalation of cationic surfactants in a layered silicate), and SBA-15 and CMI-1 (prepared by nonionic surfactant templating). Other materials with channel type of pores are MCM-48 (obtained with cationic surfactants) and KIT-6 (synthesized with nonionic surfactants), which possess two interweaving nonintersecting channels (bicontinuous cubic symmetry; space group *Ia–3d*). Mesostructured materials with interconnecting pore cavities having 3D hexagonal (space group $P6_3/mmc$) or cubic symmetries (space groups *Pm-3m*, *Pm-3n*, *Fd-3m*, *Fm-3m*, and *Im-3m*) are SBA-2 (synthesized with cationic surfactants) and SBA-12, SBA-16, FDU-5, and FDU-12 (obtained with nonionic surfactants), among others.

Due to the large dimensions of biomolecules, only few materials have been used as carriers of proteins or enzymes [29, 46–48]. The OMSM most frequently used for enzyme immobilization are FSM-16, MCM-41, and SBA-15. Other OMSM that have been employed are MCM-48, KIT-6, SBA-16, FDU-5, and FDU-12. Also surfactant-templated mesoporous silicas with uniform pore size but disordered porous structures, like HMS, MSU, and mesocellular foams (MCF), as well as periodic mesoporous organosilicas (PMOs) (which possess a hybrid organic–inorganic framework and ordered mesoporous structures) have been used.

In order to achieve successful immobilization of enzymes, several specific properties of the support play a critical role, like pore size, surface chemistry, and pore structure [49]. OMSM have pores of uniform size, and, most importantly, pore size can be tuned in the nanometer range, within the dimensions of protein molecules, which typically range between 2 and 10 nm [48]. Different synthesis parameters can be controlled in order to tailor the pore size of mesoporous materials [43, 50], including (i) the chain length of the hydrophobic region of the surfactant [30, 32, 33, 51]; (ii) the temperature and time of synthesis (hydrolysis, condensation, and hydrothermal treatment) [52–55]; (iii) addition of organic cosolvents, cosurfactants, and/or micelle swelling agents [56–60]; (iv) addition of salts [55, 61]; and (v) selection of the silica precursor, for example, alkoxides, like tetramethyl and tetraethyl orthosilicate, or sodium metasilicate [62].

19.3.2 Hybrid Organosilica Materials

Regardless of the nature or shape of the material, the enzyme–matrix binding is a surface phenomenon, that is, it happens between the surface groups from both protein and support. Therefore, special attention must be paid to the surface functional groups to drive non-covalent interaction. These functional groups may be naturally present on their surfaces (for instance, amino groups in chitin); otherwise the carrier must be chemically modified to introduce them. There are various ways to incorporate organic functional groups on the surface of siliceous ordered mesoporous materials. This

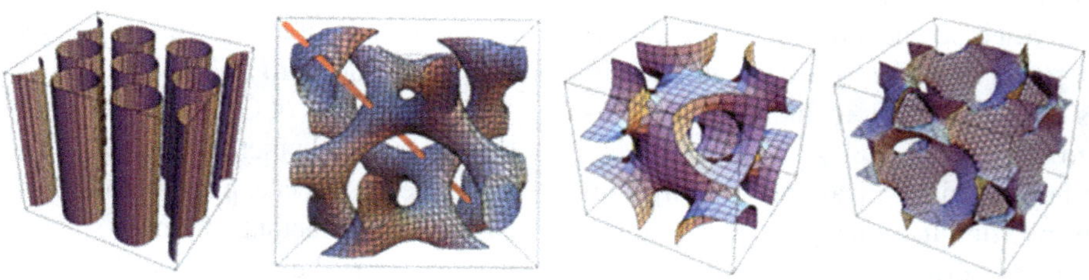

FIGURE 19.1 Pore structures of (from left to right): SBA-15 (*p6mm*), KIT-6 (*Ia-3d*), SBA-16 (*Im-3m*), and FDU-12 (*Fm-3m*). Source: Serra *et al.* [13]. Reproduced with permission from Elsevier.

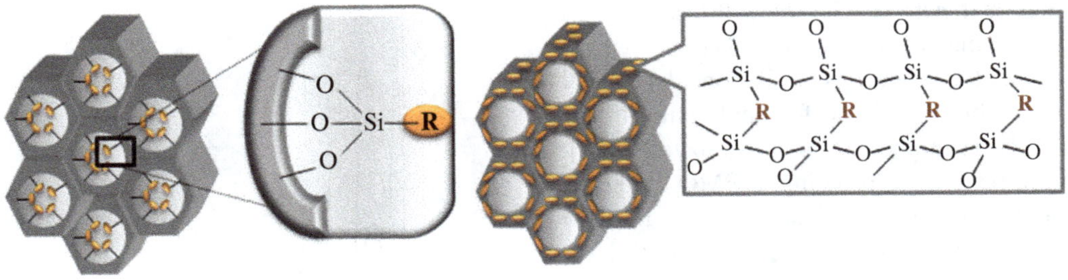

FIGURE 19.2 Hybrid organosilica material with pendant organic groups (left) and periodic mesoporous organosilica with bridging organic groups within the framework (right).

enables to increase the chemical affinity between support and enzyme by modifying surface properties of the material, such as hydrophilicity/hydrophobicity [50, 63]. There are two methods to carry out the organic functionalization of mesoporous materials: grafting after the synthesis of the silica ("two-step" approach) and co-condensation or direct incorporation during the synthesis ("one-pot" approach) [49, 64, 65]. Both methods provide pendant organic groups covalently bonded to silicon atoms of the inorganic framework (Fig. 19.2, left).

In the two-step approach, the OMSM is synthesized and the surfactant is removed by calcination or solvent extraction prior to organic functionalization. OMSM possess high concentration of surface silanol groups that can act as appropriate anchoring points for grafting via silanization reactions with organoalkoxysilanes, allowing to achieve relatively high density of surface organic groups [63, 66, 67]. After the anchoring reaction, the original silica structure is usually kept unchanged, except for a slight decrease of pore volume and pore diameter due to the added groups on pore walls [68, 69]. This pore size reduction has to be taken into account when synthesizing the pure silica material to ensure that the final pore size once anchored the functional groups is suitable for the immobilization of the target enzyme.

To accomplish a high degree of functionalization, it is necessary to retain a high concentration of hydroxyl groups on the silica surface after removal of the surfactant. If this is carried out by solvent extraction, the loss of surface silanol groups is minimized. However, when the surfactant is removed by calcination, a condensation of silanol groups occurs, so the number of anchors is reduced. This situation can be reversed by a subsequent hydration treatment of the support by boiling in water, followed by a dehydration treatment to remove the excess of water incompatible with further functionalization. In this way it is possible to provide a much higher number of silanol groups that would allow to obtain higher surface density of functional organic groups. This method can be used to increase the hydrophobicity of the support surface by incorporating aryl or alkyl groups, introducing reactive functional groups (halides, epoxides, olefins, etc.), which subsequently react with other molecules of interest, or to selectively functionalize the internal and external surfaces.

The direct or single-step synthesis, is based on the co-condensation of a tetraalkoxysilane, $(RO)_4Si$ (where R is Et or Me), and an organoalkoxysilane containing at least one Si—C bond, $(RO)_3Si$—R′ (where R is Et or Me and R′ is a nonhydrolyzable organic moiety), in the presence of surfactant templates [63, 64, 66, 67, 70]. Surfactant removal from the resulting mesostructured functionalized material has to be carried out by extraction, for calcination is not suitable due to the thermal instability of the incorporated organic groups.

The main advantage of this routine over the grafting method is the reduced number of synthesis steps needed [65]. Furthermore, the maximum loading of organic groups is usually higher, and its distribution along the surface of the mesoporous material can be more homogeneous. The main drawbacks are that not all types of organic functionalities can be incorporated in this way, since they may not be compatible with the synthesis conditions, and that control of the pore diameter is more difficult. Also, the final mesostructure is not highly ordered, since the degree of mesoscopic ordering decreases with increasing concentration of the organosilane in the synthesis mixture [63].

In 1999, three independent research groups developed simultaneously a new class of mesostructured siliceous materials called PMOs that incorporate organic species bridging two silicon atoms within the silica framework [71–73]. The synthesis of these materials is similar to that of ordered mesoporous silica described above, but using as silicon source an alkoxysilane of the type $(R'O)_3Si$—R— $Si(OR')_3$, where R is a bridging organic group bonded to two silicon atoms and R′ is a methyl or ethyl group. These materials combine the structural characteristics of ordered mesoporous silica with chemical functionalization of organic polymers [74, 75]. In contrast to mesostructured organosilicas obtained by the grafting or co-condensation methods, which incorporate organic moieties as pendant groups, in PMOs the organic groups are integrated into the 3D structure of the silica matrix via two covalent bonds (Fig. 19.2, right). Thus, the pore spaces are fully accessible for enzyme immobilization. Furthermore, the organic groups are distributed in a perfectly homogeneous way and in high concentration.

Originally, the bridging groups used to synthesize PMOs were limited to relatively simple species, such as short-chain alkanediyls and alkenediyls. However, the number of bridging organic groups employed in the design of PMOs has increased steadily since 1999, including other functionalities like aromatic rings and heterocycles [74, 76], as well as metal complexes and small dendrimers [73, 77].

Due to their specific properties, it is considered that PMO materials have high potential for application in the immobilization of large molecules such as enzymes [78]. Indeed, the immobilization of proteins in PMO materials was reported for the first time by Hudson *et al.* in 2005 [46]. Most of the PMO structures reported to date have pore sizes in the range 2–6 nm, which makes them unsuitable for the adsorption of larger molecules. However, pore size of PMOs can be increased by means of several synthesis strategies, such as the use of swelling agents, addition of salts, and low-temperature synthesis [79–81], which allow to obtain materials able to adsorb enzymes of high molecular weight.

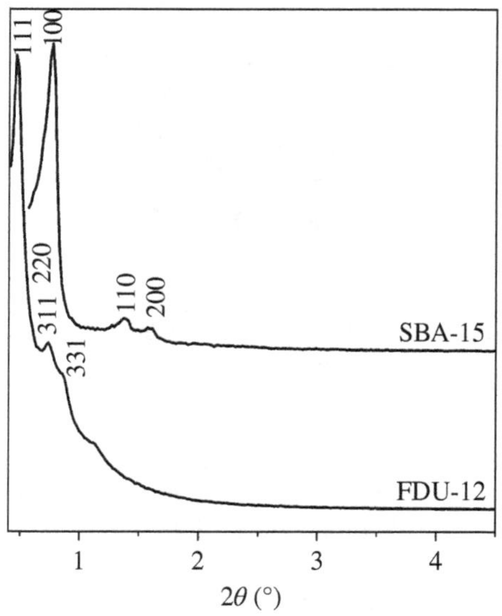

FIGURE 19.3 Characteristic XRD patterns of two mesostructured silica samples indexed according to *p6mm* (SBA-15) and *Fm-3m* space group symmetries (FDU-12). The calculated unit cell parameters (a_0) are 12.6 and 40 nm, respectively.

19.3.3 Characterization of the Pore Structure of Ordered Mesoporous Materials

The presence of a long-range periodic arrangement of uniform channels and/or cavities in ordered mesoporous materials can be identified by X-ray diffraction (XRD). Interplanar spacings of these periodic structures are in the nanometer range, and, hence, diffractograms typically show peaks in the low-angle region (Fig. 19.3). In contrast, diffraction peaks are not observed at high angles due to the lack of crystalline order in the silica framework. XRD is thus a commonly used technique to identify the pore structure topologies and to determine the corresponding unit cell parameters from which pore-to-pore distances (the sum of pore size and pore wall thickness) can be calculated.

However, it is very difficult to elucidate new mesophase structures or to determine the space group from powder XRD patterns. In order to get structure solution, high-resolution transmission electron microscopy (TEM) is required. New electron crystallography methods have been developed that allow to obtain crystal structure factors through Fourier diffractograms corresponding to TEM images from different zone axes and to reconstruct 3D electrostatic potential maps for a given space group symmetry [82]. As an example, Figure 19.4 shows images of two different orientations of the pore structures of two ordered mesoporous materials, revealing the 2D hexagonal arrangement of pore channels of SBA-15 and the cubic packing of spherical pore cavities of FDU-12.

Physisorption measurements are necessary to achieve complete characterization of the pore structure of ordered mesoporous materials. Textural characterization is typically carried out by measuring nitrogen adsorption–desorption isotherms at −196°C, from which specific surface area, pore volume, and pore size can be estimated.

Characteristic nitrogen adsorption–desorption isotherms of OMSM are of type IV according to the IUPAC classification [83]. These isotherms exhibit a steep increase of the adsorbed volume at intermediate relative pressure values due to capillary condensation in mesopores of uniform size. If capillary condensation occurs at a relative pressure higher than 0.4, the isotherms also show hysteresis between the adsorption and desorption branches, as in the case of the isotherm of an SBA-15 sample shown in Figure 19.5 (left). In the case of SBA-15 and other materials obtained using nonionic block copolymers, like poly(ethylene oxide)–poly(propylene oxide)–poly(ethylene oxide) (PEO–PPO–PEO), the isotherm is actually a combination of type I and type IV, showing adsorption at very low relative pressure due to the presence of micropores, which are created due to the insertion of hydrophilic chains (PEO) in the silica framework.

Traditionally, the determination of pore size distribution in the mesopore range is carried out by using the thermodynamics-based approach developed by Barrett, Joyner, and Halenda (BJH) [84]. As a result of the steep increase of the adsorbed volume in the mesopore region of the adsorption isotherms of OMSM, the obtained pore size distributions are very narrow (Fig. 19.5, right), which proves that these materials possess pores of a highly uniform size. Nevertheless, it has to be pointed out that pore sizes determined by the BJH method are underestimated by ca. 10 Å [36] and corrections have been proposed that can be applied with certain limitations [85]. More accurate pore size analysis can be provided by modern microscopic approaches based on nonlocal density functional theory [86].

As a complementary tool, scanning electron microscopy (SEM) can be also a useful technique to characterize OMSM as it provides information about size and morphology of particles (Fig. 19.6), which can be relevant, together with textural characterization data, to evaluate potential diffusion limitations that might affect enzyme adsorption capacity of the supports.

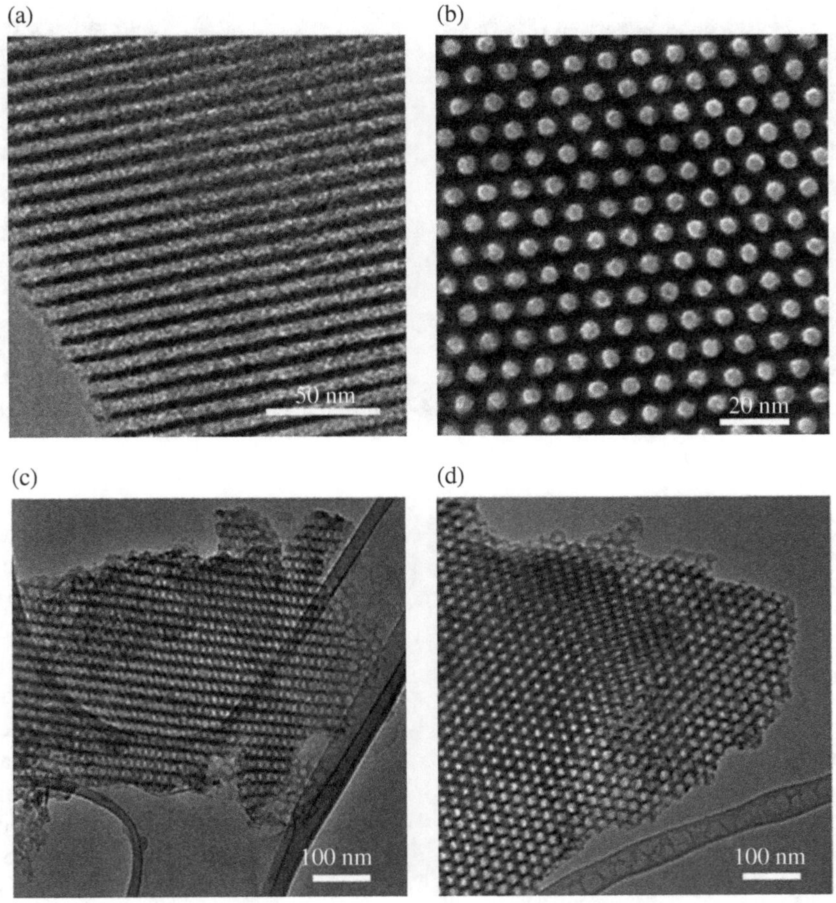

FIGURE 19.4 TEM images of SBA-15 (top) and FDU-12 materials (bottom) with projections along two different zone axes.

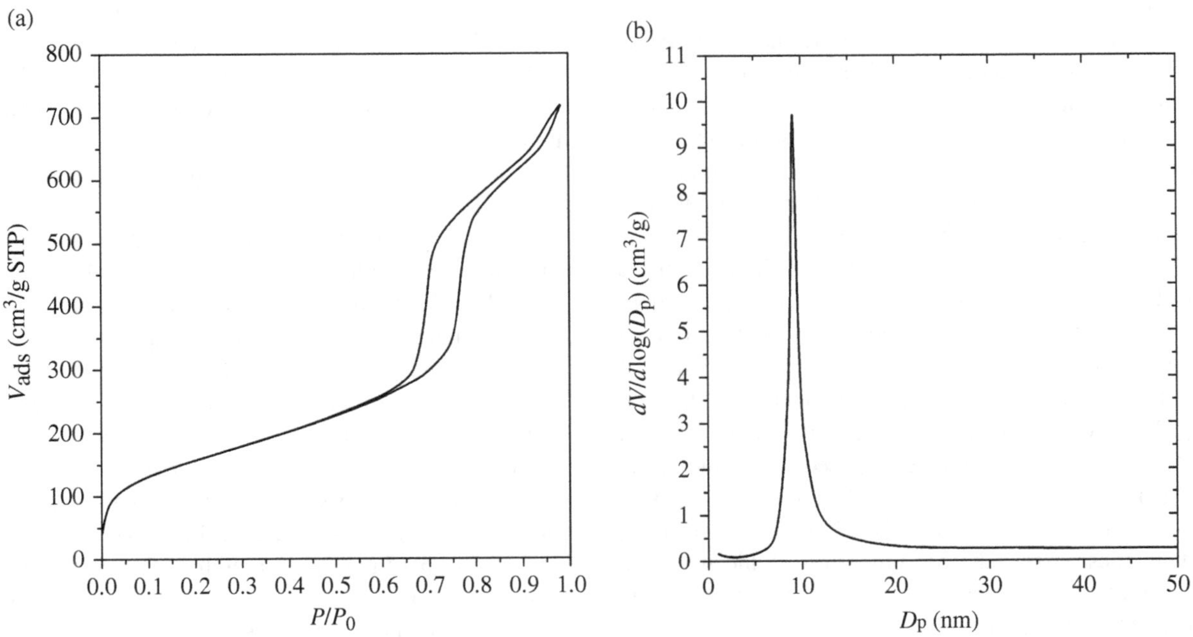

FIGURE 19.5 *Left*: nitrogen adsorption–desorption isotherm at −196 °C of a SBA-15 sample functionalized with aminopropyl groups. The calculated specific surface area (BET) and total pore volume are 576 m^2/g and 1.1 cm^3/g, respectively. *Right*: BJH pore size distribution showing the highly uniform pore diameter of the sample, with a calculated pore diameter of 9.1 nm. Source: Adapted from Ref. [19].

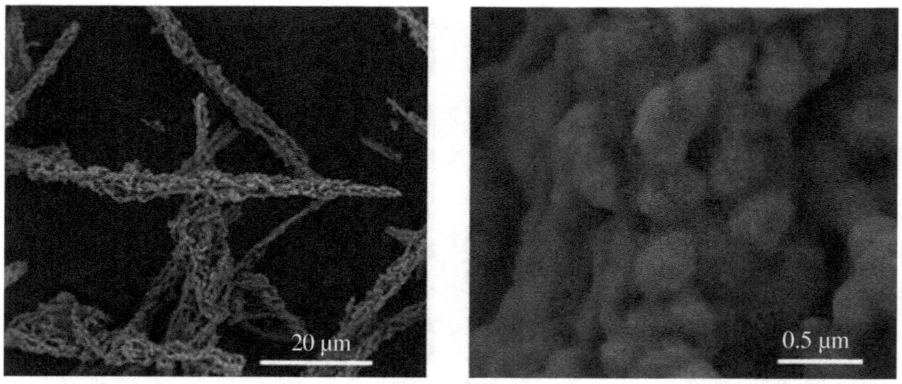

FIGURE 19.6 SEM micrographs showing particle shape of an SBA-15 material. Pore entrances can be observed in the higher magnification image.

19.4 NON-COVALENT IMMOBILIZATION OF ENZYMES ON ORDERED MESOPOROUS MATERIALS

19.4.1 Effect of Structure and Functionalization of the Material Supports on Enzyme Loading and Activity

As commented above, many variables must be considered to define the system. As an example, the enzyme lipase from *Candida antarctica* B shares with all lipases a hydrophobic domain on its surface, which is a part of the activation mechanism of this family of enzymes. So the immobilization can be driven by the interaction with this domain, via introducing hydrophobic groups on the surface of the siliceous matrix. Another example is laccase, an oxidative enzyme with low PI (around 4). In this case, the best choice is to promote electrostatic interactions with, for example, protonated amines on the surface of the support. Ionic and hydrophobic interactions are strong enough to achieve enzyme immobilization and retention under certain conditions. Improvement of the properties of the catalyst depends on some other variables that must come from the ones of the support material.

Pore size or pore shape of the support materials can greatly affect enzyme loading or enzyme activity. The influence of pore shape was extensively studied for lipase using ordered mesoporous silica supports with various pore structures (Fig. 19.1) [13]. The pore connectivity provided by the uniform structure of OMSM favors the diffusion of enzyme inward the channels and consequently the enzyme loading. Pores can be distributed in parallel channels as in SBA-15, or the channels may be disposed in a cubic structure, which increases tortuosity, as in KIT-6. Materials with cage-like pores (SBA-16 and FDU-12) have also been studied. In these materials the dimension of the windows connecting the wide cages may limit the access of the enzyme and the enzyme loading.

A summary of physicochemical properties of supports, lipase immobilization yields, and biocatalyst activities can be found in Table 19.1. Lipase immobilized on purely siliceous SBA-15 and KIT-6 with cylindrical pores of 8.8 and 8.4 nm diameters, respectively, displayed similar lipase loading (44 and 37 mg/g, respectively). The process takes double time for KIT-6 because of the tortuosity of the porous network versus the straight cylinder channels of SBA-15.

Lipase dimensions are approximately $3\times4\times5$ nm [87]. Therefore, only methyl groups were anchored to the pores in order to promote hydrophobic interactions without compromising the pore size. However, interactions were not strong enough to compensate the narrower size of pores and enzyme loadings of the respective methylated materials. PMO with pore structure similar to that of SBA-15 (spacer group *p6mm*) having ethylene bridging groups within the framework succeeded to increase the lipase loading (91 mg/g) even though the pore size was narrower. In the extreme situation, the enzyme loading achieved with an amorphous silica support with almost 30 nm average pore diameter containing highly hydrophobic octyl groups was much higher (200 mg/g) despite the surface area of this material is significantly lower (around 300 m^2/g vs. some 900 m^2/g of SBA-15 and KIT-6). Thus, a stronger interaction favors a higher enzyme loading (octyl > ethyl > methyl). However, enzyme leaching does not follow the same pattern as shown in Fig. 19.7.

Materials with cage-type pores must be discussed separately because the dimension of the window may be the controlling step for successful enzyme immobilization. However, methylation permits to increase the lipase loading in both, FDU-12 and especially SBA-16.

Laccase was immobilized on two similar purely siliceous SBA-15 materials: rodlike particles with longer channels (SBA-15-L) and dish-shaped particles with shorter channels (SBA-15-S), with narrow pore size (7.8 and 6.4 nm, respectively) (Table 19.2). The dimensions of laccase ($6\times5\times5$ nm) [12] match too closely with these pore diameters, and, therefore, anchoring of organic groups to enhance enzyme–support interactions was not feasible. As these supports are purely siliceous materials, interaction with the enzyme is nonspecific. Nevertheless, higher enzyme loading was achieved with the material with the narrower pores because of its particle size.

TABLE 19.1 Immobilization of Lipase on Siliceous and Hybrid OMSM with Channel-Like Pores (Upper Part of Table), Amorphous Silica (Center), and OMSM with Cage-Like Pores (Bottom Part of the Table)

	Material	D_p (nm)	S_{BET} (m^2/g)	V_p (cm^3/g)	mmol C/g SiO_2	t_c (min)	Loading (mg/g)	Catal. Activity (U/g cat)	Catal. Efficiency (U/mgE)
Channel-like pores	SBA-15	8.8	890	1.31		120	44	2,640	60
	Me-SBA-15	7.9	794	1.01	1.0	200	23	2,009	87
	KIT-6	8.4	917	1.20		280	37	2,442	66
	Me-KIT-6	7.7	984	1.29	1.2	50	35	2,870	82
	PMO	7.1	960	1.18	16.1	60	91	18,280	202
Amorphous	AS	27.9	305	2.70		120	45		
	Oc-AS-Me	23.6	294	2.6	4.2	60	200	13,000	83
Cage-like pores	SBA-16	10.2	659	0.73		1,440	5	360	72
	Me-SBA-16	8.8	508	0.85	0.7	1,440	30	2,310	77
	FDU-12	10.4	261	0.38		1,440	28	2,240	80
	Me-FDU-12	9.7	628	0.98	1.2	10	37	4,144	112

D_p, BJH pore diameter calculated from adsorption branch.
S_{BET}, BET surface area.
V_p, pore volume estimated at a relative pressure of 0.97.
t_c, contact time at which the maximum loading is achieved.
Loading, maximum enzymatic loading, expressed in milligrams of enzyme per gram of solid.
Catalytic activity, activity units per gram of catalyst in tributyrin test.
Catalytic efficiency, activity versus loading (units per milligram of immobilized enzyme). Expressed in units of tributyrin converted per milligram of enzyme inside the material.
mmol C/g SiO_2, hydrophobic (aliphatic carbon) groups content determined from TG analysis (weight loss in the temperature range 150–600°C).
AS, amorphous silica.

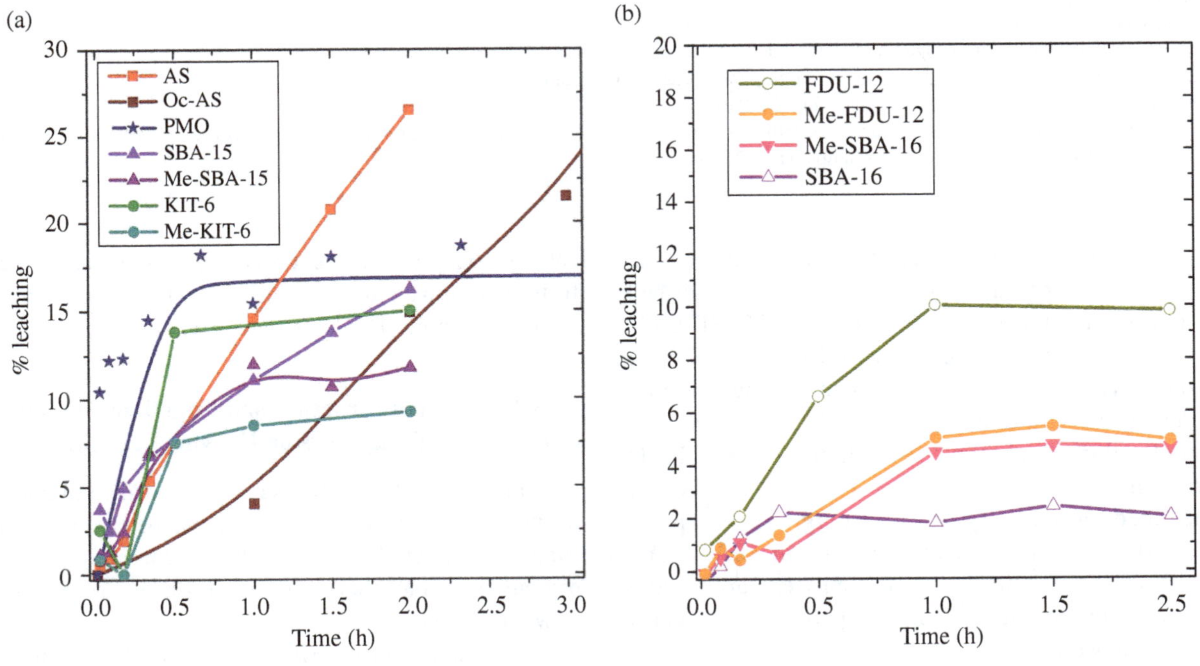

FIGURE 19.7 Leaching of lipase from supports: (a) channel-like pore materials and (b) cage-type pore materials. Source: Adapted from Ref. [13].

Despite diffusion of the enzyme is handicapped, the smaller size of the particle and the shorter length of the channels increase the accessible surface area for enzyme immobilization. On the opposite, most of the surface area is internal and not easily accessible in larger particles with longer channels (SBA-15-L), and thus the enzyme loading is lower. The enzyme loading on amorphous silica (with wide pores) can be increased an order of magnitude through grafting with amine groups. Because of the steric hindrances of SBA-15-L and SBA-15-S, it is hard to think on anchoring of functional groups to promote specific and non-covalent interactions unless the pore size is increased. Therefore SBA-15 materials with expanded pore size were synthesized.

TABLE 19.2 Immobilization of Laccase on SBA-15 and Amorphous Silica Supports

Material	D_p (nm)	Loading (mg/g)	Catal. Efficiency (U/mg Enz)	Particle Size (μm)
SBA-15-L	7.8	31.3	0.134	1×2
SBA-15-S	6.4	38.4	0.159	5×0.2
AS	27.9	15.9	0.14	70×70
Ng-AS	27.9	187.1	0.91	70×70

Initial amount of enzyme solution: 50 mg/g support. Contact time was 3 h in all cases.
D_p, BJH pore diameter calculated from adsorption branch.
Loading, maximum enzymatic loading, expressed in milligrams of enzyme per gram of solid.
Particle size, length × width (μm).
Catalytic efficiency, activity versus loading (units per milligram of immobilized enzyme). Expressed in units of syringaldazine converted per milligram of enzyme inside the material. Specific activity of soluble laccase: 5.68 U/mg.

TABLE 19.3 Laccase Immobilization on OMM Materials Functionalized with Amine Groups

Material	D_p (nm)	V_p (cm³/g)	S_{BET} (cm²/g)	mmol N/g SiO_2	t_c (h)	Loading (mg/g)	Catal. Activity (U/g)	Catal. Efficiency (U/mg)
SBA-15	15.0	1.2	427	—	24	14.2	0.56	0.04
Ng-SBA-15	11.2	1.0	339	1.2	2.0	170.0	50.7	0.30
Nc-SBA-15	17.6	1.2	385	1.5	3.5	173.8	56.0	0.32
AS	~30	2.5	296	—	24	15.8	0.52	0.03
Ng-AS	~30	2.1	236	1.8	24	187.1	169.5	0.91
PMA	10.4	0.6	264	1	90	88	29	0.33

D_p, BJH pore diameter calculated from adsorption branch.
V_p, pore volume estimated at a relative pressure of 0.97.
S_{BET}, BET surface area.
Amine groups incorporated in the silica framework = mmol N/g SiO_2.
t_c, contact time at which the maximum loading is achieved.
Loading, maximum enzymatic loading, expressed in milligrams of enzyme per gram of solid.
Catalytic activity, activity units per gram of catalyst in ABTS test.
Catalytic efficiency, activity versus loading (units per milligram of immobilized enzyme). Expressed in units of ABTS converted per milligram of enzyme inside the material. The activity of enzyme in solution was 1.25 U/mg.

SBA-15 material with wider pore diameter bearing amine groups were prepared by the two methods described above: grafting of siliceous SBA-15 (sample Ng-SBA-15) and co-condensation (sample Nc-SBA-15). Also in these cases the enzyme loading (around 170 mg/g) increased by one order of magnitude compared to the siliceous non-functionalized material (Table 19.3). In the case of a PMO support with two types of bridging groups, the maximal enzyme loading capacity was 88 mg/g. This support (sample PMA) shares the *p6mm* symmetry of SBA-15 and has a pore size of 10.4 nm. It contains ethylene groups and secondary amine groups becoming part of the hybrid siliceous framework.

The catalytic activity of the immobilized enzyme is better preserved when specifically driven non-covalent interactions are responsible for the enzyme immobilization. Actually, SBA-15 and AS, both purely siliceous, display catalytic efficiencies one order of magnitude lower than their aminated counterparts. The biocatalyst prepared on PMA also showed high value of catalytic efficiency although the enzyme loading was not as high as in both amine-functionalized SBA-15 materials.

Thus, the non-covalent interactions with the support permit to obtain high enzyme loading while preserving catalytic activity, and the properties of the catalysts can be further improved by the design of the support material, namely, the pore shape and pore size.

19.4.2 Effect of Structure and Functionalization of the Material Supports on Enzyme Leaching

The main handicap for the use of non-covalent interaction in enzyme immobilization is the release of enzyme from the support material. A close match between enzyme and pore dimensions may help diminish or even prevent this unwanted effect.

Leaching tests results (Fig. 19.7) show that when lipase is nonspecifically immobilized on purely siliceous silica, either amorphous with large pores (AS) or cylinder pore channels (SBA-15), constant enzyme release cannot be prevented. However, the winding pores of KIT-6 were efficient to retain enzyme and prevent leaching. The combination of this effect with hydrophobic interactions promoted by anchoring methyl groups (Me-KIT-6) further reduced the initial leaching step. Even weak hydrophobic interaction with methyl groups was efficient to retain lipase in the cylinder pore channel structure (Me-SBA-15), as leaching was only detected just after

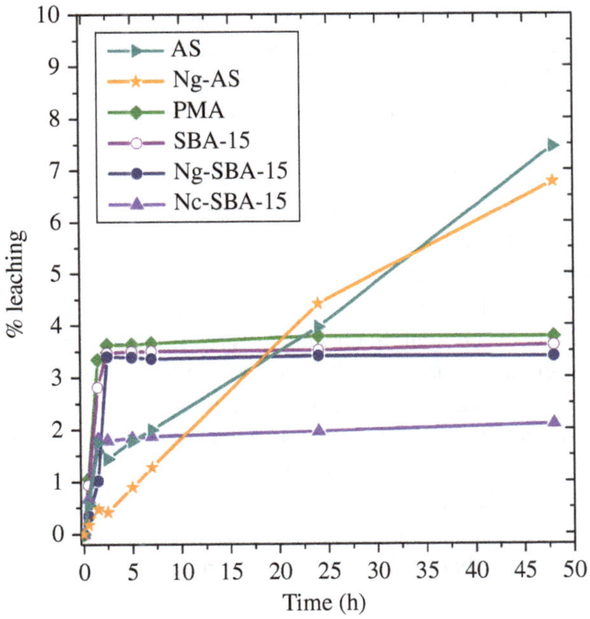

FIGURE 19.8 Leaching of laccase from supports. Source: Adapted from Refs. [19] and [88].

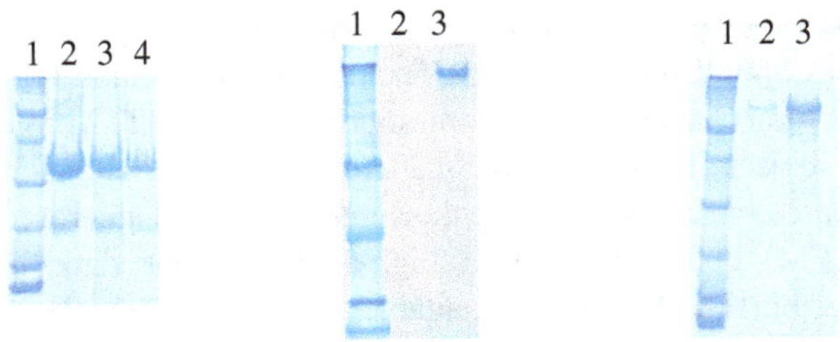

FIGURE 19.9 Electrophoresis of supernatants of biocatalysts suspended in different conditions. Left: protein standard (1), lipase solution (2), PMO (3), and Oc-AS (4). Center: protein standard (1), PMA at pH6.5 (2), and Ng-AS at pH6.5 (3). Right: protein standard (1), PMA at pH14.0 (2), and laccase solution (3). Source: Reproduced with permission from the Royal Society of Chemistry from Ref. [88].

suspension, due to enzyme molecules immobilized on the external surface of the particles or at the entrances of the channels. After this initial step, enzyme was not leached as indicated by the plateaux in the curves.

The same happened in PMO, where ethylene groups (higher hydrophobicity) were engaged in the interaction. But the lipase immobilized on the highly hydrophobic octyl groups of Oc-SA was released despite the stronger interaction: in this material the pore size is some fourfold larger than enzyme dimensions.

These results indicate that the joint effect of pore shape and chemical affinity is responsible for enzyme leaching prevention.

Laccase immobilized on amino-containing materials displayed similar results (Fig. 19.8). All the supports tested shared the same *p6mm* symmetry of SBA-15, with channel-like parallel cylinder pores. Leaching was prevented in all cases, including the purely siliceous SBA-15. Leaching profiles of OMM are identical in lipase and laccase catalysts: there is an initial enzyme release that probably corresponds to enzyme molecules immobilized on external surfaces of the particles or on the terminal edges of the pores. After these loose molecules are leached, the plateaux indicate that no more enzyme is being released

Only the amorphous silica allowed a constant release of laccase, even after functionalization with amine groups. These results corroborate that the non-covalent interactions and the shape of the pores prevent enzyme leaching. Noticeably, lower level of laccase leaching was detected from Nc-SBA-15, with higher content in amine groups (1.5 mmol/g vs. 1.2 mmol/g of Ng-SBA-15). Thus, stronger non-covalent interaction actively contributes to enzyme retention within the porous network.

19.4.3 Enzyme Retention within Pores

The presence of the enzyme in the internal surface of pore channels after suspension of the catalysts under conditions favoring enzyme release can be deduced from the remaining catalytic activity. However, it was demonstrated through indirect and direct evidences.

19.4.3.1 Indirect Evidence of Enzyme Retention inside Hybrid Mesoporous Materials Once the enzyme was no more desorbed from supports, the catalysts were submitted to polyacrylamide gel electrophoresis (PAGE) conditions—sodium dodecyl sulfate (SDS), mercaptoethanol, and 100°C—to obtain totally denatured proteins. The linear peptide chain can then easily exit the pores and the electrophoresis (EF) of the supernatant should show the corresponding protein bands.

Figure 19.9 shows the protein bands of lipase and laccase in the EF gels, corroborating the retention of the enzymes within the hybrid siliceous materials. The bands of lipase from PMO and Oc-AS (left) and the band of laccase from Ng-AS (center) are well resolved. However, laccase was not detected in the EF test at pH6.5 for the PMA support (Fig. 19.9, center). Only when the laccase-PMA catalyst was suspended at pH14.0 after denaturation the band was detected (Fig. 19.9, right). This means that there is a highly intense retention of laccase in the narrow pores of the hybrid secondary amine siliceous support.

Thus, specific non-covalent interactions with the support together with the close match between enzyme and pore sizes contribute to retain the enzyme inside the porous network. This result shows that the main drawback (maybe the only one) of the use of non-covalent interactions for enzyme immobilization can be overcome. Lipase and laccase linked to these materials through hydrophobic or electrostatic interactions, respectively, keep the advantages of non-covalent enzyme immobilization (high enzyme loading, high activity retention) plus the main advantage of covalent immobilization (permanent enzyme retention).

*19.4.3.2 **Direct Evidence of the Presence of Enzyme inside Pores after Leaching Tests*** The most direct method for the structural characterization of mesoporous solids is TEM. When the subject of interest is low scattered guest materials, such as C-based materials, scanning transmission electron microscopy (STEM) combined with a high-angle annular dark field detector (HAADF) is more suitable as the contrast is related to the atomic number Z. With the modern electron microscopes, which incorporate spherical aberration (Cs) correctors, probes down to sub-angstrom resolution with a high current can be achieved, facilitating for the case of ordered mesoporous materials a detailed analysis of the pores. For light elements introduced in the pores, a combination of STEM–HAADF and electron energy loss spectroscopy (EELS), using the spectrum image acquisition mode, is necessary in order to obtain a conclusive confirmation of the presence of enzymes in the pores. Besides, the analyses have to be carried out at an accelerating voltage of 80 kV proving a good stability of the mesoporous materials at this beam energy. Using these combined advanced electron microscopy techniques, lipase and laccase enzymes have been located inside the channels of PMO and PMA mesoporous solids [89].

Figures 19.10 and 19.11 correspond to the [001] STEM–HAADF images, parallel to the mesopores of the lipase-PMO and laccase-PMA samples. They show the organosiliceous matrix where the cavities appear in black, presenting a hexagonal array (*p6mm* symmetry) with good crystallinity. The crystal has to be perfectly orientated in respect to the electron beam for a reliable analysis of the pores and the mesoporous walls.

Figure 19.10c represents the relative carbon/oxygen composition in lipase-PMO. As expected, the relationship between C and O is different whether the beam illuminates the pore with the enzyme or the walls that contain $SiO_2+(—CH_2—CH_2—)$ groups; darker areas correspond to the higher C content. Such a big difference makes possible to distinguish the presence of lipase inside the pores. The fact that the nitrogen per molecule is so low made not possible to construct a map based on the N signal. However, a punctual analysis (black dashed square in Fig. 19.10c) allowed observing the nitrogen K edge at 401 eV. In Figure 19.10d the EELS analysis of the pore is shown after background subtraction where the oxygen and nitrogen edges are clearly identified.

A closer image of the pore system in the laccase-PMA sample is shown in Figure 19.11b, in which a spectrum image collection analysis was performed inside one of the pores (square). The extracted spectrum profile (raw data; Fig. 19.11c) exhibits the O and N edges corresponding to the enzyme.

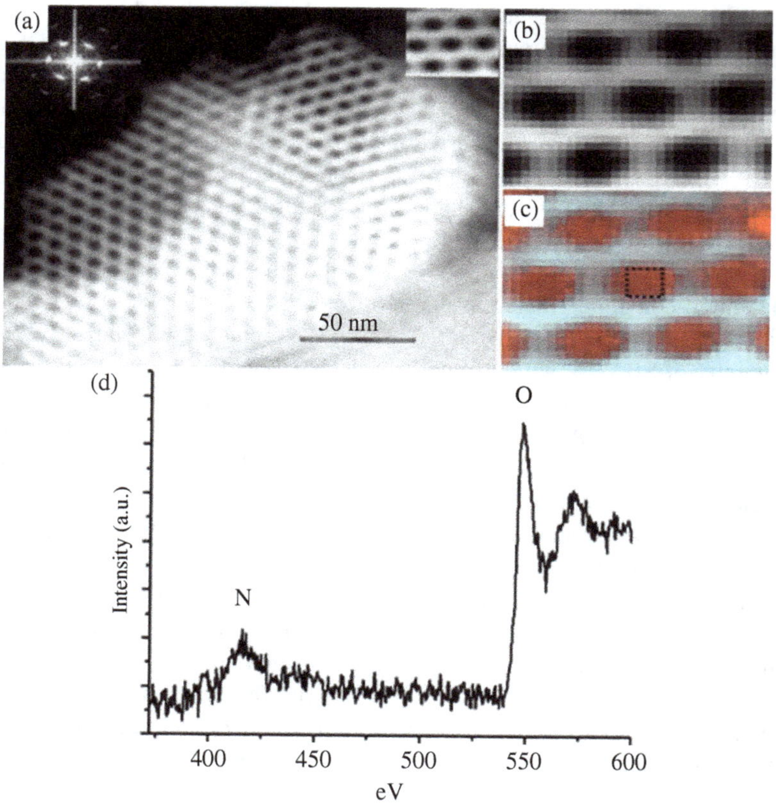

FIGURE 19.10 (a) STEM–HAADF image of a PMO crystal loaded with lipase along the [001] orientation showing the hexagonal array. (b) STEM–HAADF signal of the area analyzed. (c) Map obtained from the EELS signal of the relative C/O composition. (d) Sum of EEL spectra, after background subtraction, recorded in the dashed square marked in (c). Source: Mayoral *et al.* [89]. Reproduced with permission from Wiley.

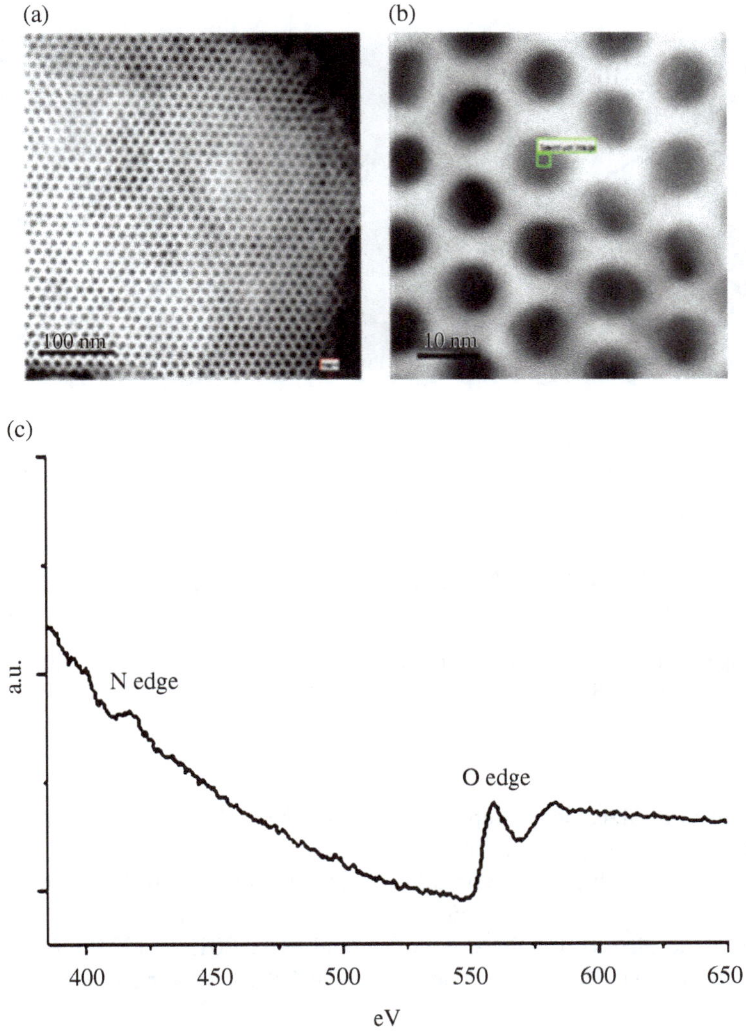

FIGURE 19.11 (a) STEM–HAADF image of the PMA loaded with laccase. (b) High-resolution image of the pores. The rectangle represents the area where the EELS analysis was carried out. (c) EELS profile collected from (b) showing the three edges that confirm the presence of laccase. Source: Reproduced with permission from Ref. [89]. © 2013 Wiley-VCH Verlag GmbH & Co. KGaA, Weinheim.

19.4.4 Stability: PMO-Enzyme Catalysts in Organic Solvents

Stabilization of the enzymes immobilized in hybrid PMO (aminosilica) deserves a separate study. As described above, in these materials the functional groups are not pendant groups anchored to the pore surface, but they are inserted within the framework bridging two silicon atoms. PMO are ethylene-bridged PMO, and PMA are aminodipropyl-bridged (also containing ethylene bridging groups) periodic mesoporous aminosilica. In both cases there is a very close match between sizes of enzyme and pore diameter.

For stability tests, lipase-PMO was suspended in methanol, and laccase-PMA was suspended in ethanol. Methanol and ethanol are solvents capable to interfere hydrogen bonds of the protein and to promote unfolding by changing the orientation of hydrophobic side groups of the peptide chain toward the surface of the protein. The close presence of ethylene groups in the pore wall promotes the formation of new enzyme–support interactions in these partially unfolded conformations of the respective enzymes. The degree of unfolding is limited by the confined space available, so these partially unfolded and still active conformations become reinforced by the new enzyme–support links formed, and consequently these structures are more rigid. This can explain the stability of laccase in ethanol (Fig. 19.12a) and especially the high resistance of lipase in such a hostile medium as methanol: even in 100% methanol, 70% lipase activity is maintained after 1 h (Fig. 19.12b) [88].

19.5 FINAL COMMENTS

The properties of non-covalently immobilized enzymes may be preserved and improved by a careful design of the support material. Hybrid organosiliceous ordered mesoporous materials have been tailor-made to immobilize lipase and laccase. The characteristics of these materials

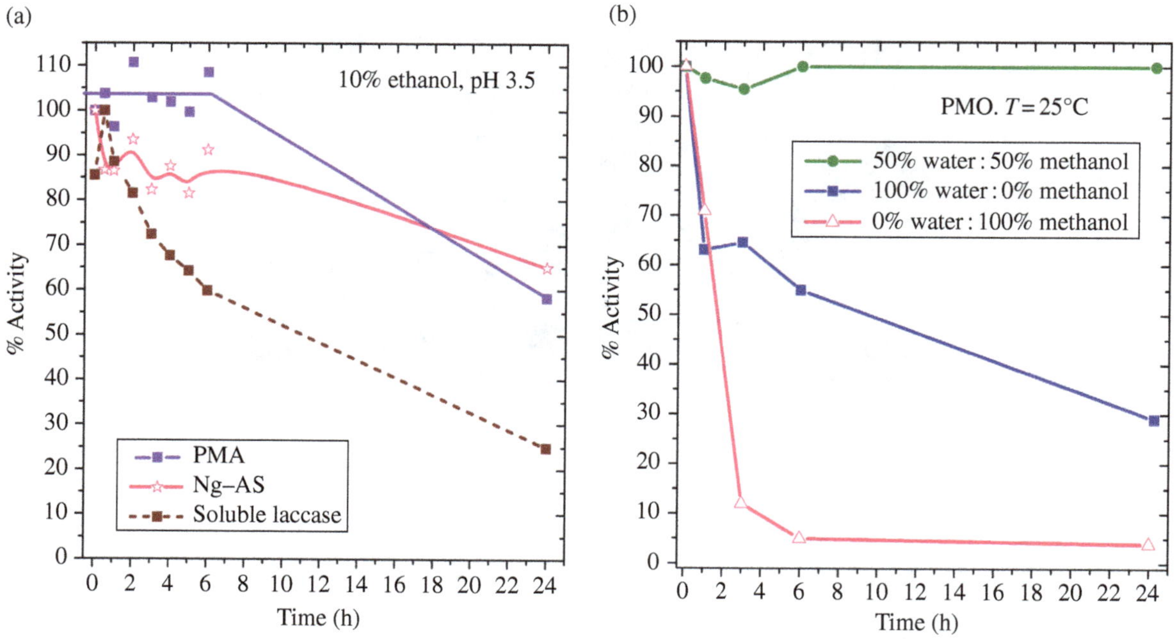

FIGURE 19.12 Stability in organic solvents. (a) Laccase catalysts in ethanol solution. (b) Lipase catalysts in methanol solutions. Source: Reproduced with permission from the Royal Society of Chemistry from Ref. [88].

such as uniform and tunable pore size, pore shape, and high surface area allow to take advantage of the interactions with enzyme surface established through the organic functionalization of the pores. Thus, the good connectivity of the porous network, the affinity for the organic groups, high surface area, and adjusted pore size lead to high enzyme loading values.

The mild nature of non-covalent interactions permits to keep high activity values of the immobilized enzymes (i.e., their catalytic efficiencies). The connectivity of pores in the uniform porous networks favors the diffusion of substrates and products. Supports having pores of high tortuosity afford lower enzyme leaching compared to materials with pores of the parallel cylinders type. In this case, the enzyme leaching can also be prevented by a close match of enzyme and pore sizes and creation of an intense non-covalent interaction. In the extreme situation, laccase could only be removed from the hybrid organosilica, PMA, at pH 14. This chemical affinity–size matching binomial has proved to be crucial also for enzyme stabilization.

Thus, the design and use of these materials permit to take advantage of the mild and nondistorting enzyme immobilization through non-covalent interactions and to improve the properties of the biocatalysts. At the same time these systems permit to achieve the main advantage of enzyme immobilization through covalent attachment, which is to avoid enzyme leaching. Unlike covalent bonded enzymes, these systems permit to reuse the support materials once the enzyme become deactivated. So, the advantages of both methods of enzyme anchoring in supports can be obtained, while the disadvantages are avoided.

ACKNOWLEDGMENTS

The authors would like to thank the Spanish MINECO for the financial support through the projects MAT2009-13569 and MAT 2012-31127. V. G. acknowledges MINECO and ICP-CSIC for the postdoc working contract.

REFERENCES

1. P. Estrada, R. Sanchezmuniz, C. Acebal, R. Arche, M. P. Castillon, *Biotechnol. Appl. Biochem.* 1991, *14*, 12.
2. A. Naidja, A. Violante, P. M. Huang, *Clays Clay Miner.* 1995, *43*, 647.
3. A. Krastanov, *J. Ind. Microbiol. Biotechnol.* 2000, *24*, 383.
4. N. Duran, M. A. Rosa, A. D'Annibale, L. Gianfreda, *Enzyme Microb. Technol.* 2002, *31*, 907.
5. B. Krajewska, *Enzyme Microb. Technol.* 2004, *35*, 126.
6. T. Jesionowski, J. Zdarta, B. Krajewska, *Adsorption* 2014, *20*, 801.
7. H. Zhao, *J. Chem. Technol. Biotechnol.* 2010, *85*, 891.
8. A. Popat, S. B. Hartono, F. Stahr, J. Liu, S. Z. Qiao, G. Qing, M. Lu, *Nanoscale* 2011, *3*, 2801.
9. N. Saifuddin, A. Z. Raziah, A. R. Junizah, *J. Chem.* 2013, 1.
10. S. A. Ansari, Q. Husain, *Biotechnol. Adv.* 2012, *30*, 512.
11. J. Liu, C. M. Li, Q. H. Yang, J. Yang, C. Li, *Langmuir* 2007, *23*, 7255.
12. V. Gascón, I. Díaz, C. Márquez-Alvarez, R. M. Blanco, *Molecules* 2014, *19*, 7057.
13. E. Serra, A. Mayoral, Y. Sakamoto, R. M. Blanco, I. Diaz, *Microporous Mesoporous Mater.* 2008, *114*, 201.

14. R. M. Blanco, P. Terreros, M. Fernandez-Perez, C. Otero, G. Diaz-Gonzalez, *J. Mol. Catal. B: Enzym.* 2004, *30*, 83.
15. H. Kawakita, K. Sugita, K. Saito, M. Tamada, T. Sugo, H. Kawamoto, *Biotechnol. Prog.* 2002, *18*, 465.
16. S. Karboune, R. Neufeld, S. Kermasha, *J. Biotechnol.* 2005, *120*, 273.
17. M. Pesic, C. Lopez, G. Alvaro, J. Lopez-Santin, *J. Mol. Catal. B: Enzym.* 2012, *84*, 144.
18. R. M. Blanco, J. M. Guisan, *Enzyme Microb. Technol.* 1989, *11*, 360.
19. V. Gascón, C. Márquez-Álvarez, R. M. Blanco, *Appl. Catal. A Gen.* 2014, *482*, 116.
20. J. J. Karimpil, J. S. Melo, S. F. D'Souza, *Int. J. Biol. Macromol.* 2012, *50*, 300.
21. F. Xu, W. H. Wang, Y. J. Tan, M. L. Bruening, *Anal. Chem.* 2010, *82*, 10045.
22. L. Cao, *Curr. Opin. Chem. Biol.* 2005, *9*, 217.
23. R. M. Blanco, J. J. Calvete, J. M. Guisan, *Enzym. Microb. Technol.* 1989, *11*, 353.
24. D. I. Fried, F. J. Brieler, M. Fröba, *ChemCatChem* 2013, *5*, 862.
25. F. Jia, B. Narasimhan, S. Mallapragada, *Biotechnol. Bioeng.* 2014, *111*, 209.
26. W. Liu, L. Wang, R. Jiang, *Top. Catal.* 2012, *55*, 1146.
27. P. C. Weber, D. H. Ohlendorf, J. J. Wendoloski, F. R. Salemme, *Science* 1989, *243*, 85.
28. C. C. Yu, Y. Y. Kuo, C. F. Liang, W. T. Chien, H. T. Wu, T. C. Chang, F. D. Jan, C. C. Lin, *Bioconjug. Chem.* 2012, *23*, 714.
29. Z. Zhou, M. Hartmann, *Chem. Soc. Rev.* 2013, *42*, 3894.
30. J. S. Beck, J. C. Vartuli, W. J. Roth, M. E. Leonowicz, C. T. Kresge, K. D. Schmitt, C. T. W. Chu, D. H. Olson, E. W. Sheppard, *J. Am. Chem. Soc.* 1992, *114*, 10834.
31. C. T. Kresge, M. E. Leonowicz, W. J. Roth, J. C. Vartuli, J. S. Beck, *Nature* 1992, *359*, 710.
32. D. Y. Zhao, Q. S. Huo, J. L. Feng, B. F. Chmelka, G. D. Stucky, *J. Am. Chem. Soc.* 1998, *120*, 6024.
33. D. Y. Zhao, J. L. Feng, Q. S. Huo, N. Melosh, G. H. Fredrickson, B. F. Chmelka, G. D. Stucky, *Science* 1998, *279*, 548.
34. G. Oye, J. Sjoblom, M. Stocker, *Adv. Colloid Interface Sci.* 2001, *89*, 439.
35. Y. Han, D. Zhang, *Curr. Opin. Chem. Eng.* 2012, *1*, 129.
36. U. Ciesla, F. Schuth, *Microporous Mesoporous Mater.* 1999, *27*, 131.
37. M. Kruk, M. Jaroniec, C. H. Ko, R. Ryoo, *Chem. Mater.* 2000, *12*, 1961.
38. A. Taguchi, F. Schüth, *Microporous Mesoporous Mater.* 2005, *77*, 1.
39. M. Hartmann, *Chem. Mater.* 2005, *17*, 4577.
40. Z. Wu, D. Zhao, *Chem. Commun.* 2011, *47*, 3332.
41. C. T. Kresge, W. J. Roth, *Chem. Soc. Rev.* 2013, *42*, 3663.
42. S. J. Holder, N. A. J. M. Sommerdijk, *Polym. Chem.* 2011, *2*, 1018.
43. Y. Wan, D. Zhao, *Chem. Rev.* 2007, *107*, 2821.
44. Q. S. Huo, D. I. Margolese, U. Ciesla, D. G. Demuth, P. Y. Feng, T. E. Gier, P. Sieger, A. Firouzi, B. F. Chmelka, F. Schuth, G. D. Stucky, *Chem. Mater.* 1994, *6*, 1176.
45. V. L. Zholobenko, A. Y. Khodakov, M. Imperor-Clerc, D. Durand, I. Grillo, *Adv. Colloid Interface Sci.* 2008, *142*, 67.
46. S. Hudson, E. Magner, J. Cooney, B. K. Hodnett, *J. Phys. Chem. B* 2005, *109*, 19496.
47. S. Z. Qiao, C. Z. Yu, W. Xing, Q. H. Hu, H. Djojoputro, G. Q. Lu, *Chem. Mater.* 2005, *17*, 6172.
48. L. Bayne, R. V. Ulijn, P. J. Halling, *Chem. Soc. Rev.* 2013, *42*, 9000.
49. C.-H. Lee, T.-S. Lin, C.-Y. Mou, *Nano Today* 2009, *4*, 165.
50. S. S. Park, C. S. Ha, *Chem. Rec.* 2006, *6*, 32.
51. Q. S. Huo, R. Leon, P. M. Petroff, G. D. Stucky, *Science* 1995, *268*, 1324.
52. T. Yamada, H. S. Zhou, K. Asai, I. Honma, *Mater. Lett.* 2002, *56*, 93.
53. K. Kosuge, T. Sato, N. Kikukawa, M. Takemori, *Chem. Mater.* 2004, *16*, 899.
54. T. Klimova, A. Esquivel, J. Reyes, M. Rubio, X. Bokhimi, J. Aracil, *Microporous Mesoporous Mater.* 2006, *93*, 331.
55. Z. W. Jin, X. D. Wang, X. G. Cui, *Colloids Surf. A* 2008, *316*, 27.
56. W. H. Zhang, L. Zhang, J. H. Xiu, Z. Q. Shen, Y. Li, P. L. Ying, C. Li, *Microporous Mesoporous Mater.* 2006, *89*, 179.
57. M. Kruk, L. Cao, *Langmuir* 2007, *23*, 7247.
58. L. Cao, T. Man, M. Kruk, *Chem. Mater.* 2009, *21*, 1144.
59. L. Cao, M. Kruk, *Colloids Surf. A* 2010, *357*, 91.
60. M. Kruk, *Acc. Chem. Res.* 2012, *45*, 1678.
61. C. V. Teixeira, H. Amenitsch, P. Linton, M. Linden, V. Alfredsson, *Langmuir* 2011, *27*, 7121.
62. P. F. Fulvio, S. Pikus, M. Jaroniec, *J. Mater. Chem.* 2005, *15*, 5049.
63. A. Stein, B. J. Melde, R. C. Schroden, *Adv. Mater.* 2000, *12*, 1403.
64. M. H. Lim, A. Stein, *Chem. Mater.* 1999, *11*, 3285.
65. G. L. Athens, R. M. Shayib, B. F. Chmelka, *Curr. Opin. Colloid Interface Sci.* 2009, *14*, 281.
66. A. S. M. Chong, X. S. Zhao, A. T. Kustedjo, S. Z. Qiao, *Microporous Mesoporous Mater.* 2004, *72*, 33.
67. A. Vinu, K. Z. Hossain, K. Ariga, *J. Nanosci. Nanotechnol.* 2005, *5*, 347.
68. A. Walcarius, M. Etienne, J. Bessière, *Chem. Mater.* 2002, *14*, 2757.
69. A. Walcarius, M. Etienne, B. Lebeau, *Chem. Mater.* 2003, *15*, 2161.
70. A. S. M. Chong, X. S. Zhao, *J. Phys. Chem. B* 2003, *107*, 12650.
71. T. Asefa, M. J. MacLachlan, N. Coombs, G. A. Ozin, *Nature* 1999, *402*, 867.
72. B. J. Melde, B. T. Holland, C. F. Blanford, A. Stein, *Chem. Mater.* 1999, *11*, 3302.
73. N. Mizoshita, T. Tani, S. Inagaki, *Chem. Soc. Rev.* 2011, *40*, 789.
74. F. Hoffmann, M. Cornelius, J. Morell, M. Froba, *J. Nanosci. Nanotechnol.* 2006, *6*, 265.
75. F. Hoffmann, M. Cornelius, J. Morell, M. Froeba, *Angew. Chem. Int. Ed.* 2006, *45*, 3216.
76. H.-S. Xia, C.-H. Zhou, D. S. Tong, C. X. Lin, *J. Porous Mater.* 2010, *17*, 225.

77. P. Van der Voort, D. Esquivel, E. De Canck, F. Goethals, I. Van Driessche, F. J. Romero-Salguero, *Chem. Soc. Rev.* 2013, *42*, 3913.

78. Z. Zhou, R. N. Klupp Taylor, S. Kullmann, H. Bao, M. Hartmann, *Adv. Mater.* 2011, *23*, 2627.

79. X. Zhou, S. Qiao, N. Hao, X. Wang, C. Yu, L. Wang, D. Zhao, G. Q. Lu, *Chem. Mater.* 2007, *19*, 1870.

80. M. Mandal, M. Kruk, *J. Mater. Chem.* 2010, *20*, 7506.

81. M. Mandal, A. S. Manchanda, J. Zhuang, M. Kruk, *Langmuir* 2012, *28*, 8737.

82. L. Han, T. Ohsuna, Z. Liu, V. Alfredsson, T. Kjellman, S. Asahina, M. Suga, Y. H. Ma, P. Oleynikov, K. Miyasaka, A. Mayoral, I. Diaz, Y. Sakamoto, S. M. Stevens, M. W. Anderson, C. H. Xiao, N. Fujita, A. Garcia-Bennett, K. B. Yoon, S. N. Che, O. Terasaki, *Z. Anorg. Allg. Chem.* 2014, *640*, 521.

83. S. J. Gregg, K. S. W. Sing, Adsorption, Surface Area, and Porosity, Academic Press, London, 1991.

84. P. B. Barret, L. G. Joyner, P. P. Halenda, *J. Am. Chem. Soc.* 1951, *1*, 373

85. M. Kruk, M. Jaroniec, A. Sayari, *Langmuir* 1997, *13*, 6267.

86. M. Thommes, K. A. Cychosz, *Adsorption* 2014, *20*, 233.

87. J. Uppenberg, M. T. Hansen, S. Patkar, T. A. Jones, *Structure* 1994, *2*, 293.

88. V. Gascón, I. Díaz, R. M. Blanco, C. Márquez-Alvarez, *RCS Adv.* 2014, *4*, 34356.

89. A. Mayoral, R. Arenal, V. Gascon, C. Marquez-Alvarez, R. M. Blanco, I. Diaz, *ChemCatChem* 2013, *5*, 903.

PART V

BIORELEVANT SYNTHESES

20

MODULATION OF BIORELEVANT RADICAL REACTIONS BY NON-COVALENT INTERACTIONS

Riccardo Amorati and Luca Valgimigli
Department of Chemistry "G. Ciamician", University of Bologna, Bologna, Italy

20.1 INTRODUCTION

Radical reactions are involved in a multitude of relevant processes and their importance is increasingly recognized in different fields ranging from biochemistry to materials science. For instance, radical polymerization accounts for approximately 45% of world polymer production [1]; on the other hand, the involvement of transient radical species in enzyme regulated processes in living organisms has stimulated, in recent years, enormous research efforts to rationalize their role in key physiological processes like mitochondrial respiration and aging [2, 3] or in the pathogenesis of several diseases [4].

Perhaps the best known example of radical reaction is hydrocarbon autoxidation, a chain process that affects any organic material, from food to petrol-derived chemicals to human beings, existing under an oxygen-rich atmosphere [5]. As a consequence, antioxidants and radical-chain inhibitors are among the most important compounds used to control key radical reactions [5]. They act by trapping chain-carrying radical species, thereby competing with chain propagation, as illustrated in Scheme 20.1. Their rational design and use need to be strictly based on their reactivity with radical species. Indeed, the understanding of radical reactions and their role in biology, along with their use and control in synthetic chemistry or in medicine, requires detailed knowledge of their kinetics and how it is influenced by the reaction medium.

In this chapter it will be explained how non-covalent interactions control the rates and the products of reactions that are typical of the radicals involved in autoxidation, namely, peroxyl (ROO•), phenoxyl (PhO•), and alkoxyl (RO•) radicals. These radicals are not only of biological interest, but they are also key intermediates in green synthesis protocols, which aim at obtaining fine chemicals from renewable materials under mild conditions. For this reason, we will illustrate some selected examples of biomimetic radical reactions that are made possible by the control of these reactive intermediates by non-covalent interactions.

20.2 SOLVENT EFFECTS IN RADICAL REACTIONS

The Hughes–Ingold rules for solvent effects on reactions in solution explain the role of the solvent as providing better or poorer stabilization of the dipoles or charges formed in the activated reaction complex as compared to the reactants. Therefore, a reaction is expected to accelerate in polar solvents if the dipolarity increases along the reaction coordinate [6]. An example of the rules applied to radical chemistry is offered by the reactivity of ligated boranes (L-BH_3). These compounds, such as aminoboranes, phosphinoboranes, and carbene boranes, are interesting reducing agents acting by H-atom transfer (HAT) that have been investigated, for instance, as replacement of tributyltin hydride and similar chemicals in tin-free reductions [7, 8]. It has been shown that the rate of their reaction with alkoxyl radicals (RO•) is enhanced by a factor of 3–5 on changing the solvent from benzene to more polar pyridine, in consequence of higher solvent stabilization of the dipolar transitions state, as compared to the (uncharged) reactants (Scheme 20.2) [8].

Similarly, the rate of electron transfer (ET) from a radical species (e.g., an alkyl radical R•) to form a cation (e.g., a

Non-Covalent Interactions in the Synthesis and Design of New Compounds, First Edition.
Edited by Abel M. Maharramov, Kamran T. Mahmudov, Maximilian N. Kopylovich, and Armando J. L. Pombeiro.

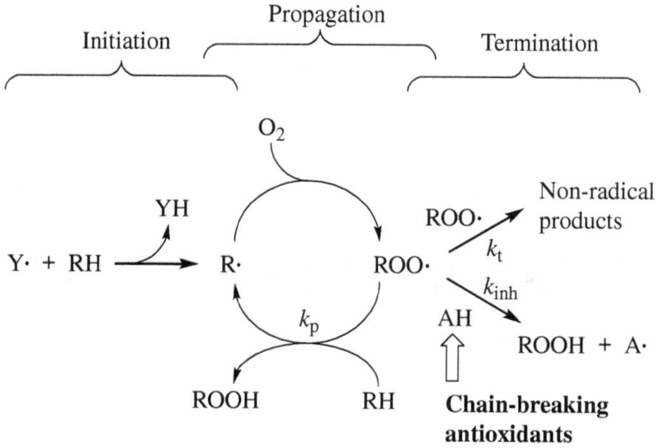

SCHEME 20.1 Scheme of the autoxidation of a hydrocarbon RH initiated by a radical Y· and inhibited by an antioxidant AH.

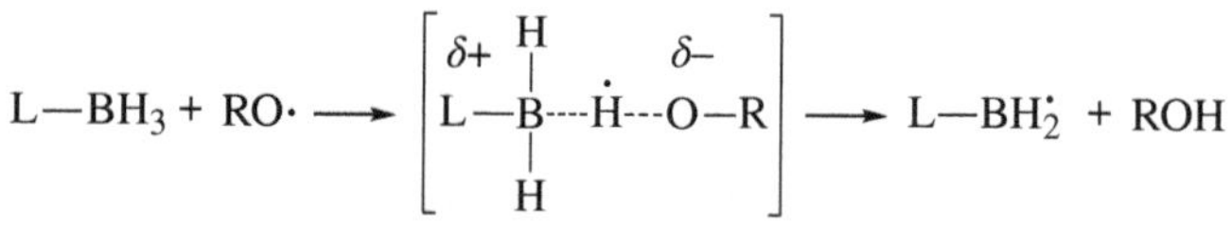

SCHEME 20.2 Reaction of ligated boranes with alkoxyl radicals.

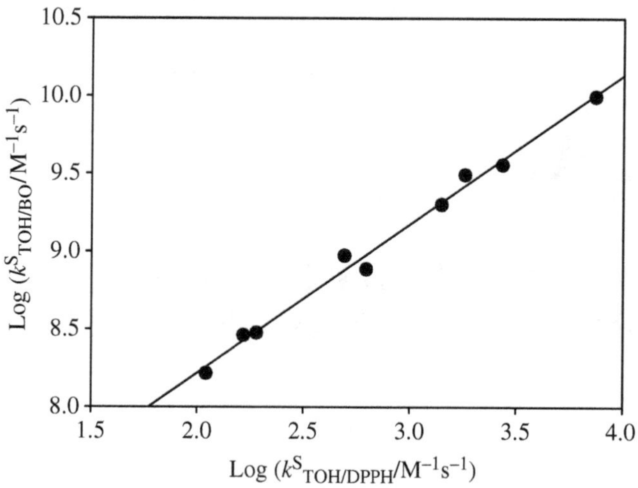

FIGURE 20.1 Kinetic solvent effect: comparison of the rate constants for reaction of α-tocopherol with ᵗBuO· and DPPH· radicals measured in different solvents at 298 K.

carbocation R^+) or to a radical species (e.g., Cl·) to form an anion (e.g., Cl^-) is expected to experience large solvent effects on changing solvent polarity [6]. Despite the above and other few examples, radical reactions involving homolytic cleavage/formation of covalent bonds and nearly isopolar transition states for long time have been generally regarded as not very sensitive to medium effects, and the lack of solvent effects has sometimes been used to infer a free radical-mediated mechanism in a novel reaction [6]. While solvent polarity may have little influence in many radical reactions, another solvent property was shown to have a major role in controlling the kinetics of radical reactions: the ability of the solvent to establish hydrogen bonding interactions with the reacting partners [9, 10]. Due to its importance, the HAT reaction has been particularly investigated in this respect [9–13].

As early as in 1964 Howard and Ingold realized that the inhibition of styrene autoxidation performed by various phenols decreased in the presence of polar solvents, the effect being more pronounced with unhindered than with 2,6-di-*tert*-butyl-substituted phenols and with acidic more than with nonacidic phenols [13]. These effects were explained as being due to H-bond formation between the phenolic —OH group (acting as H-bond donor (HBD)) and the solvent (acting as H-bond acceptor (HBA)). They suggested that only the fraction of phenol not H-bonded to a solvent molecule was able to react with ROO· radicals, thereby inhibiting the autoxidation (Scheme 20.1), while the H-bonded fraction of the phenolic antioxidant was essentially unreactive.

These important observations were somewhat overlooked at that time; however, independent confirmation came about 30 years later. When studying the reactivity of cumyloxyl (CumO·) radicals with different substrates by laser flash photolysis (LFP), Avila *et al.* realized that while the reaction with cyclohexane (H-abstraction forms a C—H bond) proceeded at nearly constant rate in a variety of solvents (i.e., in any solvent S, $k^S_{(C\text{–}H/CumO\cdot)} = (1.2 \pm 0.1) \times 10^6\,M^{-1}s^{-1}$ at 30°C), the similar reaction of CumO· with *tert*-butyl hydroperoxide or with phenol (in both cases H-abstraction forms an O—H bond) had rate constants that decreased by two orders of magnitude on changing the solvent from CCl_4 to ᵗBuOH [10]. The observed kinetic solvent effect (KSE) was attributed to the occurrence of H-bond donation from the reactive moiety in the substrate to the HBA solvent: while strong H-bonding interaction is expected to occur with the OH group of phenol or hydroperoxide, negligible interaction with the solvent is expected from hydrocarbon C—H groups [10].

Shortly later, Valgimigli *et al.* showed that the reported KSE was a general behavior of HAT reaction from a dipolar X—H moiety of any substrate to any abstracting radical [9]. Indeed, substrates like α-tocopherol and phenol, differing in reactivity by about 3 orders of magnitude, had similar KSE when reacting with abstracting radicals of largely different structures or reactivity, from the *tert*-butoxyl (ᵗBuO·) to the diphenylpicrylhydrazyl (DPPH·) radicals. For both substrates (phenol and tocopherol), it was shown that a plot of Log $k^S_{(ArOH/BuO\cdot)}$ versus Log $k^S_{(ArOH/DPPH\cdot)}$ for the reaction with ᵗBuO· and DPPH·, respectively, in a large range of solvents from hydrocarbons to γ-valerolactone, was linear with a slope of 1 (Fig. 20.1) [9]. This indicated that, to a first approximation, the solvent effect is independent of the radical species, despite the two reactions having absolute rate constants differing by 10^6-fold in the case of α-tocopherol and 10^{10}-fold in the case of phenol [9]. The KSE was again attributed to H-bonding of the substrate to the HBA solvent, as illustrated in Scheme 20.3.

According to the proposed model, to a first approximation, H-bonding interaction only affects the reactivity of the

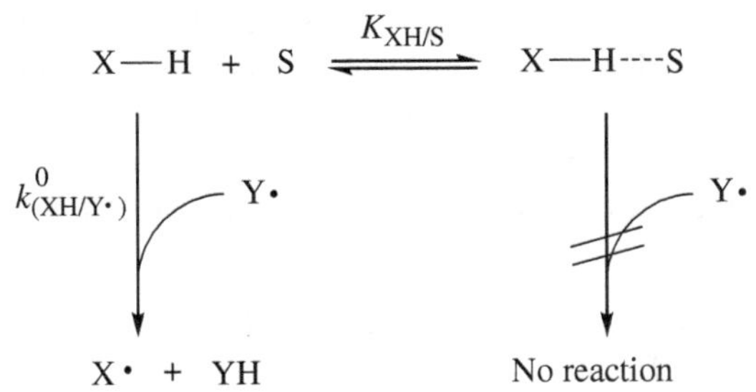

SCHEME 20.3 Model for kinetic solvent effect on formal H-atom abstraction from a polar substrate X—H by a radical Y· in the presence of the solvent S.

substrate (it is independent of the radical), and it is assumed that only the "free fraction" of the substrate not H-bonded to the solvent is available for reaction with the radical Y•, while, in the H-bonded fraction, stabilization of the XH group (i.e., increase of the X—H bond dissociation enthalpy (BDE)) impairs its reactivity. This simplified model allows predicting the rate of reaction in any solvent S according to Equation 20.1, provided the rate constant in a non-H-bonding solvent $(k^0_{XH/Y\cdot})$ is known:

$$k^S_{XH/Y\cdot} = \frac{k^0_{XH/Y\cdot}}{\left(1 + K_{XH/S} \times [S]\right)} \qquad (20.1)$$

This KSE was subsequently shown to hold for any formal HAT reaction from polar X—H groups to almost any radical, as long as the reaction has an activation barrier, that is, it is not diffusion controlled. Since the KSE is independent of the abstracting radical, the ratio $k^{S1}_{XH/Y\cdot}/k^{S2}_{XH/Y\cdot}$ for the reaction between any substrate XH and the radical Y• in two different solvents S1 and S2 is expected to be identical to the relative rate of reaction of the same antioxidant in the same solvents, with a different radical species Z• (Eq. 20.2):

$$\mathrm{KSE} = \frac{k^{S1}_{XH/Y\cdot}}{k^{S2}_{XH/Y\cdot}} = \frac{k^{S1}_{XH/Z\cdot}}{k^{S2}_{XH/Z\cdot}} \qquad (20.2)$$

20.3 QUANTITATIVE TREATMENT OF H-BOND STRENGTH IN SOLUTION: ABRAHAM'S SOLVATOCHROMIC PARAMETERS

Quantitative treatment of the KSE according to Equation 20.1 requires knowledge of the equilibrium constant for H-bond formation between the solvent (HBA) and the substrate (HBD). This can conveniently be obtained from Abraham's equation (20.3) [14]:

$$\mathrm{Log}\left(K_{HB}/\mathrm{M}^{-1}\right) = 7.354\alpha_2^{H}\beta_2^{H} - 1.094 \qquad (20.3)$$

In Abraham's equation K_{HB} (corresponding to $K_{XH/S}$) is given as a function of α_2^H and β_2^H, which are quantitative, thermodynamic-based empirical parameters obtained from solvatochromic measurements and expressing, respectively, the H-bond donating ability and the H-bond accepting ability of two generic interacting partners (e.g., substrate and solvent). Both α_2^H and β_2^H have values in the range 0 (no H-bonding) to 1. Interestingly, Ingold and coworkers showed that the KSE in HAT reactions can quantitatively be described by the empirical equation (20.4) [11], which is related to Abraham's equation:

$$\mathrm{Log}\left(k^S/\mathrm{M}^{-1}\mathrm{s}^{-1}\right) = \mathrm{Log}\left(k^0/\mathrm{M}^{-1}\mathrm{s}^{-1}\right) - 8.3\alpha_2^{H}\beta_2^{H} \qquad (20.4)$$

Since Abraham's α_2^H and β_2^H parameters refer to 1 : 1 H-bonding events in diluted solutions, Equation 20.4 implies that KSE does not depend on the dielectric constant of the solvent S, but only on its ability to form 1 : 1 complexes with the polar XH group. As hundreds of α_2^H and β_2^H values are reported in the literature [14–16], in practice Equation 20.4 allows to calculate k^S in a wide range of conditions from a single kinetic measurement.

20.4 EFFECT OF NON-COVALENT INTERACTIONS ON REACTIONS OF PHENOLIC ANTIOXIDANTS WITH PEROXYL RADICALS

Phenols represent a widespread class of organic compounds whose importance is linked to their ability to be easily oxidized to phenoxyl radicals. The amino acid tyrosine, which contains a phenolic moiety, is often present in proteins that catalyze redox reactions. The cofactor ubiquinol is a 1,4-dihydroxybenzene (1,4-hydroquinone), which is cyclically oxidized and reduced within the mitochondrial oxidative chain. Free radicals, produced in excess by respiration or by metal-catalyzed reactions inside cells, are scavenged by phenolic antioxidants, such as α-tocopherol, flavonoids, and polyphenols [17]. In most cases, their reactivity with free radicals is modulated by non-covalent interactions, which act as fine-tuning of the redox properties of phenols.

In this context, it is useful to classify the oxidation of phenols (in their neutral form) on the basis of the oxidizing agent. HAT oxidizers abstract a H^+ and a e^- in a concerted way and form a neutral phenoxyl radical, while one-electron oxidizers form a phenoxyl radical cation (Scheme 20.4). Generally speaking, HAT oxidizers are organic radicals such as alkyl (R•), alkoxyl (RO•), and peroxyl (ROO•) radicals, whereas one-electron oxidizers are based on metallic ions. However, many exceptions to this classification exist, because the actual behavior of a X• radical depends on the reduction potential of the $X\cdot/X^-$ couple and on the acidity of the corresponding X–H species.

One of the most important classes of biologically relevant oxidizers that react with phenols by HAT is alkyl peroxyl radicals, which are formed by atmospheric O_2 during the autoxidation of organic substrates, such as plastics, lubricants, foods, cosmetics, as well as biological molecules like membrane lipids. Phenols act as chain-breaking antioxidants because they formally transfer a hydrogen atom to peroxyl radicals (Eq. 20.5) much faster than the chain-propagating HAT step (Eq. 20.6). Also the phenoxyl radicals formed after the reaction with peroxyl radicals are stabilized by resonance and are unable to propagate the autoxidation [18].

$$\mathrm{ROO\bullet + ArOH \xrightarrow{k_{inh}} ROOH + ArO\bullet} \quad (20.5)$$

$$\mathrm{ROO\bullet + RH \rightarrow ROOH + R\bullet} \quad (20.6)$$

Since antioxidant activity of phenols depends on the rate of Equation 20.5, its rate constant is named "inhibition constant" (k_{inh}), which is a function of the *BDE* of the phenolic O—H bond and of the steric crowding around this group. Electron-donating groups in *ortho* and *para* position with respect to the phenolic OH lower the *BDE*(O—H) by stabilizing the phenoxyl radical, while electron-withdrawing groups have the opposite effect [18]. Weak (non-covalent) interactions modulate the antioxidant activity by changing the *BDE*(O—H) and/or the steric crowding around the reactive OH group.

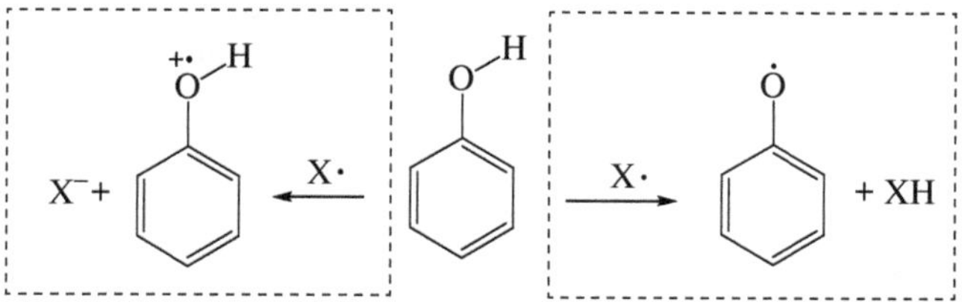

SCHEME 20.4 Oxidation pathways of phenols.

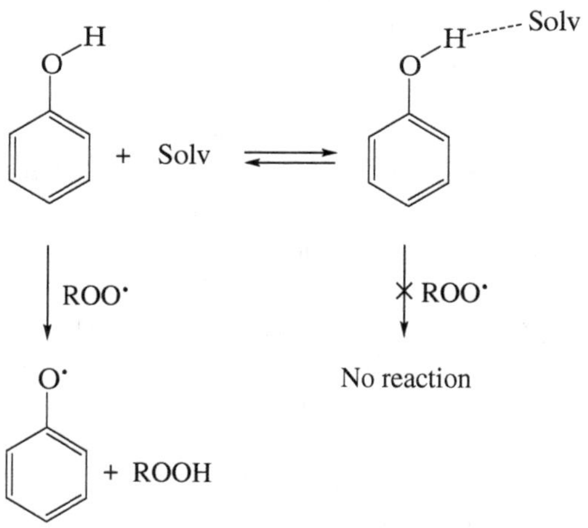

SCHEME 20.5 Kinetic solvent effect.

20.4.1 The Reactive Phenolic OH Group as H-Bond Donor

The reaction of peroxyl radicals with antioxidants progressively decreases its rate as the ability of the solvent to accept H-bonds (i.e., the solvent β_2^H value) increases. This observation was rationalized by considering that only the fraction of phenols that are not involved in H-bonds with the solvent are able to react with ROO•, as shown in Scheme 20.5 [19]. As previously noted (see Section 20.3), this mechanism of KSE is actually not specific of peroxyl radicals reacting with phenols, but holds also for any HAT oxidizer reacting with other molecules with abstractable hydrogens, able to act as H-bond donors (e.g., aromatic amines) [19]. The KSE model also predicts that the phenols that are good H-bond donors (HBD), as quantified by their α_2^H parameter, are also subject to the largest decrease of reactivity in polar H-bond accepting (HBA) solvent. Some examples are collected in Table 20.1 [20–23], where k_{inh} values in chlorobenzene (PhCl), a weak HBA solvent, are compared to those in acetonitrile (MeCN), a good HBA solvent.

The cause of KSE is both energetic and steric in nature, because the formation of H-bond with an HBA solvent causes an increase of the BDE(O-H) and because the incoming radical cannot reach the reactive OH with the same ease, when this is involved in H-bonding [19]. The existence of KSE is an experimental demonstration that the reaction between ROO• radicals and phenols occurs through the direct "contact" between the phenolic OH group and the radical. Calculations show that the reaction proceeds through the formation of a precomplex and that the electron and the proton are transferred in a concerted step [24]. The geometry of the transition state and the orbital analysis suggest that H^+ moves through the H-bond, while the electron is transferred from the π-system of the phenol to the inner oxygen of the peroxyl radical; therefore the reaction of formal HAT should more properly be considered a proton-coupled electron transfer (PCET) (Scheme 20.6) [24].

TABLE 20.1 Rate Constants for the Reaction between Phenols and Peroxyl Radicals in Two Solvents Having Different H-Bond Accepting Ability, as a Function of the H-Bond Donating Ability (α_2^H) of the Phenol

		k_{inh}/M^{-1} s^{-1}			
Phenol	α_2^H	Chlorobenzene (β_2^H=0.09) [15]	Acetonitrile (β_2^H=0.39) [20]	KSE[a]	Reference
Hydroxytyrosol	0.72	8.0×10^5	3.0×10^4	27	[21]
4-Methoxyphenol	0.57	2.6×10^5	1.6×10^4	16	[22]
α-Tocopherol	0.37	3.2×10^6	6.5×10^5	4.9	[23]
Quercetin	—	5.0×10^5	1.2×10^4	42	[21]

[a] Kinetic solvent effect, defined as k_{inh}(PhCl)/k_{inh}(MeCN).

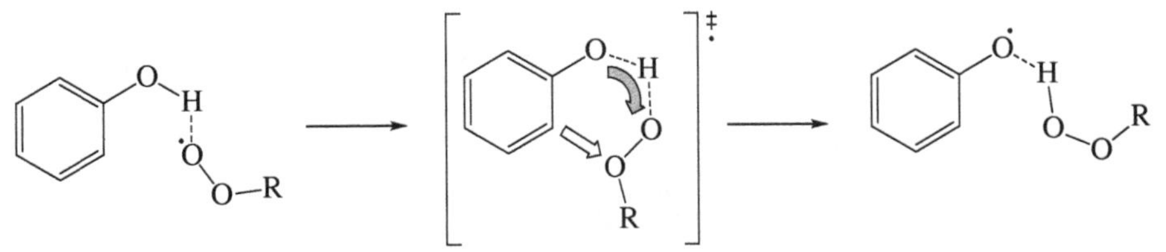

SCHEME 20.6 Transition state for the H-atom transfer mechanism. The gray and the white arrows indicate the movement of H^+ and of the electron, respectively.

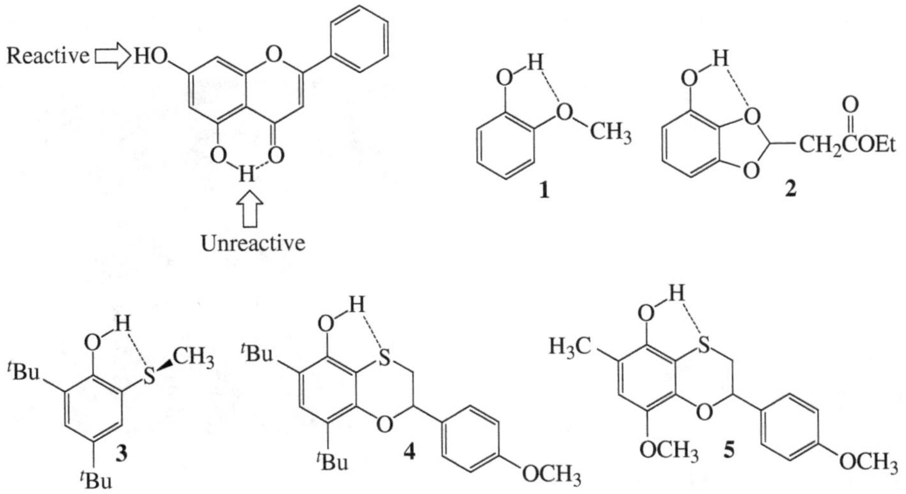

SCHEME 20.7 Examples of phenols in which the reactive OH group is an H-bond donor.

Similarly to what observed in the presence of HBA solvents, the reactivity of a phenolic OH is decreased also by the presence of HBA groups in *ortho* position, such as methoxyl [25] or carbonyl [26] groups, which are a common feature in natural polyphenols. In these systems, the decrease in reactivity is proportional to the strength of the intramolecular H-bond, so that phenols having an *ortho* OCH_3 group still react with peroxyl radicals (see compound **1**), whereas those having an *ortho* C=O group are completely unreactive (Scheme 20.7). Geometrical constrains can weaken the intramolecular H-bond, as in the case of the compound **2**, where the increase of the OH···O distance due to the penta-atomic fused ring causes an increase of k_{inh} (Table 20.2) [25].

A peculiar case is that of *ortho* XR phenols, where R is a S, Se, or Te atom, because in these compounds the formation of a H-bond requires that the dihedral angle between the aromatic ring and the X—R bond of the substituent is nearly 90° (see, for instance, compound **3** in Scheme 20.7) [27, 30]. If the *ortho* substituent is included in a cycle, as in the case of **4**, the strength of the intramolecular H-bond drops to nearly zero, as assessed by FTIR spectroscopy, and the reactivity toward ROO· drastically increases [28]. The thiaflavane antioxidant **5** was designed to exploit this feature and was found to have the same k_{inh} as α-tocopherol ($3.2\times10^6\,M^{-1}\,s^{-1}$), the most active natural lipophilic antioxidant (Table 20.2) [29].

TABLE 20.2 Rate Constants for the Reaction with Alky Peroxyl Radicals at 30°C in Chlorobenzene and Strength of the Intramolecular H-Bond

Phenol	k_{inh}/$M^{-1}\,s^{-1}$	ΔH_{HB} (kcal mol^{-1})[a]	Reference
1	4.7×10^3	~4	[25]
2	2.7×10^4	1.6	[25]
3	3.9×10^3	2.6[b]	[27]
4	3.1×10^5	~0[b]	[28]
5	3.4×10^6	~0	[29]

[a] Energy required to rotate the phenolic OH away from the *ortho* XR substituent.

[b] Corrected for the energy required to point toward a *tert*-butyl group, estimated as 1.5 kcal mol^{-1} [28].

20.4.2 Remote H-Bond Effects

The H-bond effects delineated in the previous section are all arising from interactions where the reactive phenolic —OH is the H-bond donor. However, other H-bond interactions, different from those previously described, can affect the reactivity of phenols and can be classified as "remote" effects. In such cases, the effect of non-covalent interactions on the k_{inh} is not due to steric factors, but only to modulation of the *BDE* of the phenolic O—H, which in turn is a function of the preferential stabilization either of the phenol or of the phenoxyl radical, as shown in Scheme 20.8.

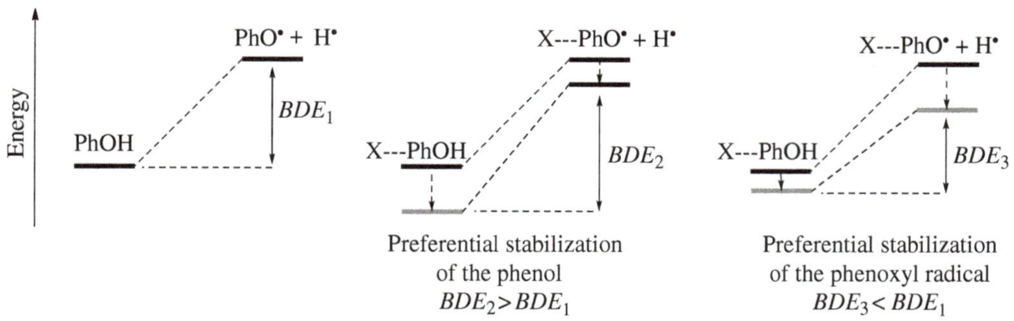

SCHEME 20.8 Effect of intra- or intermolecular remote H-bonds with **X** moiety on the *BDE* of the phenolic OH bond.

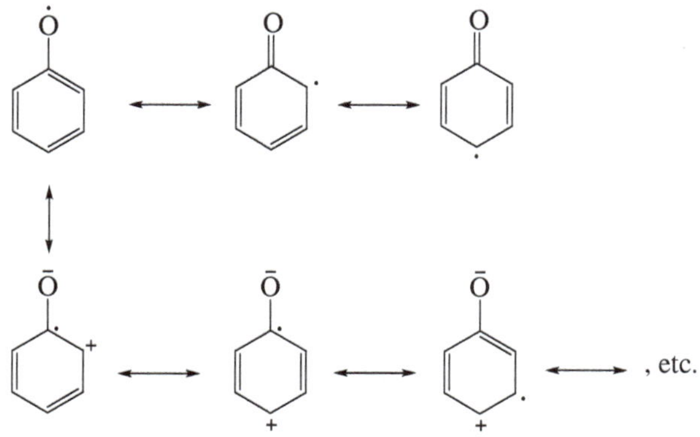

SCHEME 20.9 Resonance structures of phenoxyl radicals.

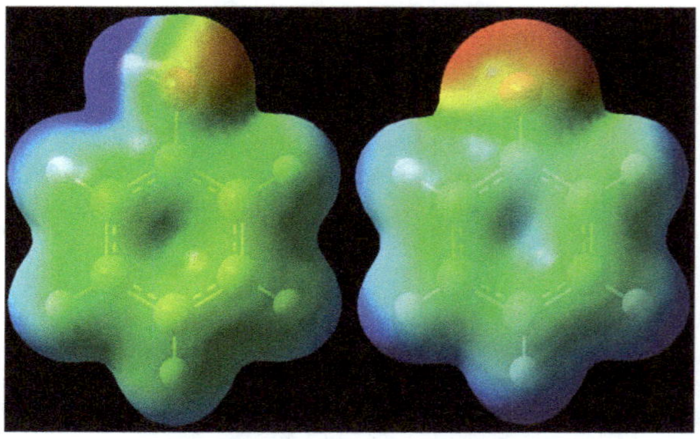

FIGURE 20.2 Electrostatic potential surface of phenol (left) and of phenoxyl radical (right), showing that the O-atom is more negative and the ring H-atoms are more positive in the radical than in the phenol. Calculated at the B3LYP/6-31 + *g*(d,p) level, red and blue colors indicate negative and positive potentials, respectively.

For a better comprehension of these effects, it is important to understand the crucial differences in the electronic distribution between phenols and their respective phenoxyl radicals, which also influence their propensity to participate in H-bonds. The first point to consider is that the phenolic —OH is an electron donor, while the —O• moiety is a strong electron acceptor, with an electron-withdrawing strength similar to that of a nitro group, as it can be visually recognized from the resonance structures with charge separation (Scheme 20.9). Electrostatic potential surfaces, shown in Figure 20.2, clearly show that, on going from the phenol to the phenoxyl radical, the oxygen atom increases its electron density, whereas the opposite occurs to the aromatic ring. The second point is that the OH group is a good H-bond donor (HBD) and a weak H-bond acceptor (HBA), whereas the phenoxyl (—O•) group is a good H-bond acceptor, having a β_2^H value similar to that of ketones [31]. Some examples of the consequences of these effects can be observed in polyphenols, such as catechols, resorcinols, and hydroquinones.

Catechols (1,2-dihydroxybenzenes) and gallols (1,2,3-trihydroxybenzenes), which are among the most common structural motives in natural polyphenols, are chain-breaking antioxidants characterized by the presence of phenolic —OH groups acting both as HBD and HBA. In such compounds, the intramolecular H-bond is not broken during the reaction with peroxyl radicals, so that the *ortho* —OH group stabilizes both the phenol and, to larger extent, the phenoxyl radical as a consequence of electron polarization in the radical. Therefore, catechol (**6**) has a *BDE*(O—H) lower than 2-methoxyphenol (**1**) and 1,4-hydroquinone (1,4-dihydroxyphenol), and its reactivity with peroxyl radicals is accordingly larger. The k_{inh} in PhCl at 30°C for **6** is $5.5\times10^5\,M^{-1}\,s^{-1}$, while that of **1** is $4.7\times10^3\,M^{-1}\,s^{-1}$ and that of hydroquinone is $1.8\times10^5\,M^{-1}\,s^{-1}$ (Scheme 20.10) [12].

Hydroquinones are important both as antioxidants and as enzyme cofactors. After the H-atom abstraction from either one of the phenolic OH groups, the remaining one becomes a much better HBD; thus the phenoxyl (semiquinone) radical can be better stabilized by H-bond acceptors than the parent phenol (the hydroquinone). These effects are responsible for the reactivity increase of the model hydroquinone **7** toward free radicals after the addition of small quantities of an HBA solvent (Scheme 20.10). The kinetic acceleration (k_2/k_1) is larger with solvent having larger HBA ability: from the k_2/k_1 ratio, the *BDE* of the "free" —OH in **7S** was found to be lower than that of the —OH groups in **7** by −1.0 and −2.0 kcal mol^{-1} after H-bonding to MeCN and dimethyl sulfoxide (DMSO), respectively [32]. Importantly, this model reaction clearly demonstrates that only H-bonds that are preformed before H-atom abstraction can influence the kinetics of the reaction.

A similar effect was also detected when comparing the reactivity of the "free" OH groups of the hydroquinone **8** and that of the resorcinol derivative **9** (Scheme 20.10). Despite

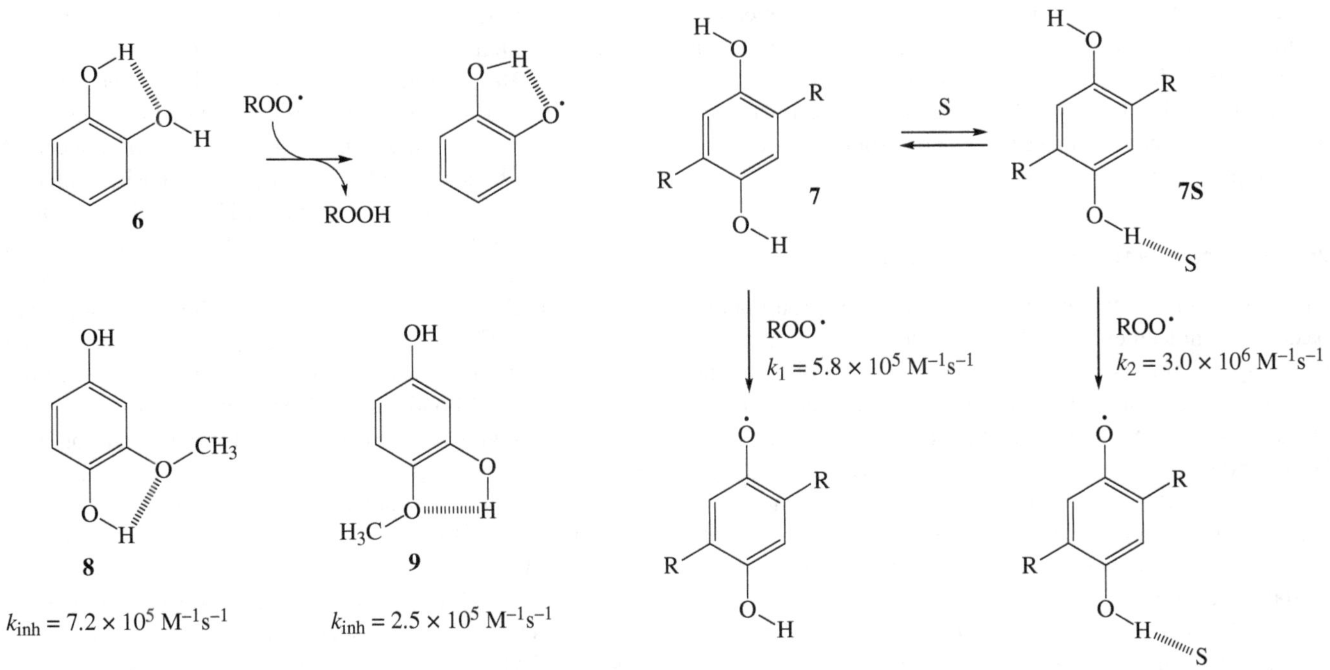

SCHEME 20.10 Remote H-bond effects in various phenols (R = isoamyl, S = HBA solvent).

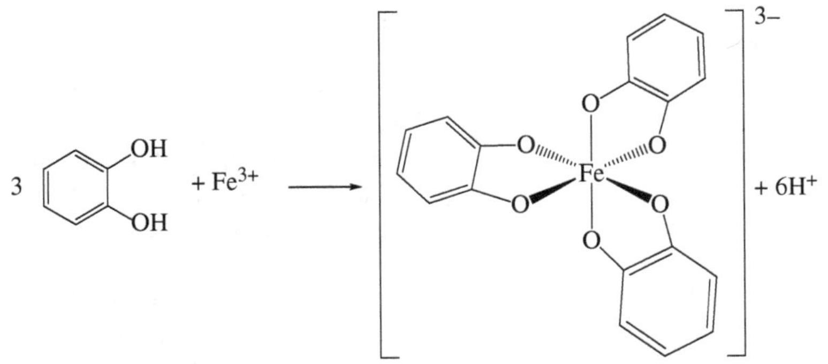

SCHEME 20.11 Interaction between iron and catechol.

the fact that the intrinsic *BDE*-lowering effect of methoxyl and hydroxyl substituents distant from the reactive —OH is about the same, **8** reacts with peroxyl radicals three times faster than **9**. The explanation is that in **8** the H-bond in the radical is stronger than that in the parent phenol, while in **9** the opposite occurs, because the electron density of the *para* –OMe group is withdrawn by the O· moiety [33].

20.4.3 Interaction with Metals

Most polyphenolic compounds are effective metal chelators, especially when they have a catechol or gallol moiety. Among the metals able to form chelate complexes, the role of iron has been investigated in detail because of its relevance in the onset of chronic neurological diseases. In the presence of hydroperoxides, Fe^{2+} ions exert prooxidant activity through the Fenton reaction [34], and this effect is increased by the presence of organic reductants (such as ascorbate) able to reduce Fe^{3+} to Fe^{2+}, thereby allowing redox cycling and extension of the prooxidant activity [34]. In the III oxidation state, iron is a strong Lewis acid and forms stable complexes with catechols or gallols after their deprotonation to the corresponding phenolates (Scheme 20.11).

On the other hand, Fe^{2+} is a borderline Lewis acid and does not bind as strongly to the oxygen atoms of polyphenol ligands. Because polyphenol ligands strongly stabilize Fe^{3+} over Fe^{2+}, catecholate and gallate complexes of Fe^{2+} rapidly oxidize in the presence of O_2 or of hydroperoxides to give Fe^{3+}–polyphenol complexes and oxygen-centered radicals. On the contrary, as Fe^{3+} complexes are more difficult to be reduced to Fe^{2+} than free Fe^{3+} ions, polyphenols impair redox cycling of iron in the presence of organic reductants, showing a preventive antioxidant behavior [34].

Concerning the reactivity of Fe^{3+}–catecholate complexes with radicals, in the case of gallocatechins, the binding with Fe^{3+} causes a complete loss of the chain-breaking antioxidant activity of the polyphenols [35], due to electron withdrawal by the metal ion. Indeed, it was shown that in water at pH 11.5 deprotonation of the Fe^{3+} complexes of gallocatechins makes them good one-electron reducers, at variance with their behavior in the less electron-rich undissociated form at pH 7 [35].

20.4.4 One-Electron Oxidation Systems

Non-covalent interactions between phenolic OH groups and a basic HBA molecules play a fundamental role in boosting the reactivity of phenols toward one-electron oxidizers. The phenoxyl radical cation that is formed after the removal of one electron is a high-energy intermediate, characterized by a large acidity ($pK_{a,H_2O} = -2$) [36]. The reaction is strongly facilitated by the presence of a base at H-bond distance from the phenolic OH group, which accepts the H^+, while the electron is transferred (Scheme 20.12). Therefore, one-electron oxidation of phenols is most often a concerted electron transfer (ET) and proton transfer (PT) process, where the electron moves from the phenol to the oxidizer, while the proton moves from the phenol to a base, such as the solvent.

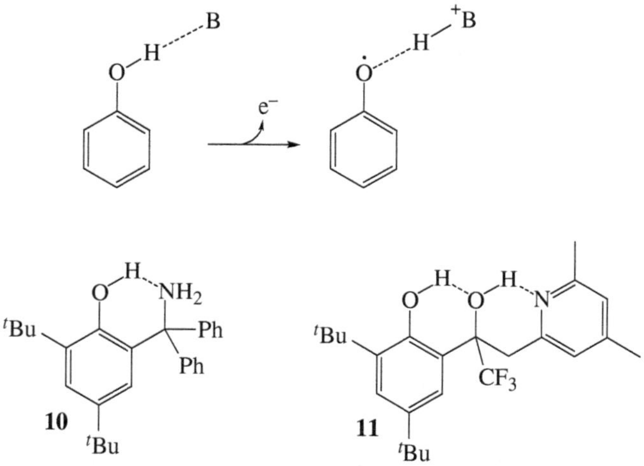

SCHEME 20.12 One-electron oxidation of phenol H-bonded to a base (**B**) by a concerted ET–PT process.

Indeed, kinetic studies of the reactions between phenols having pendant bases in *ortho* position (see, for instance, compound **10**; Scheme 20.12) and several one-electron oxidizers with varying redox potentials have demonstrated that e^- and H^+ transfers are concerted [36, 37]. The concerted ET–PT mechanism holds also when the basic center is connected to the reactive phenolic OH through a chain of H-bonds, as in the case of compound **11** [38]. The acceleration of $1e^-$ oxidation of phenols is also observed when the base is not covalently linked to it. The study of the quenching reaction of triplet C_{60} with phenols in the presence of increasing amounts of pyridines showed that only the fraction of phenols forming H-bond interactions with the base can efficiently react with $1e^-$ oxidizers [39]. H-bond formation with nonbasic solvents gives less evident but still detectable acceleration on $1e^-$ oxidation of phenols, as demonstrated by Galian *et al.* by studying the reaction of an aromatic ketone with phenol [40].

While H-bond formation with a base boosts the reactivity of phenols toward $1e^-$ oxidizers, this is no longer true in the case of HAT oxidizers, such as peroxyl radicals, which follow the KSE described in Section 20.4.1 also in the presence of bases. There is, however, a gray zone represented by radicals that behave differently depending on the polarity of the solvent, such as the diphenylpicrylhydrazyl (DPPH•) radical (see Scheme 20.13).

DPPH• is a popular probe for the assessment of the radical trapping ability of putative antioxidants. In apolar solvents, DPPH• reacts with phenols by HAT and follows the KSE also in the presence of bases. On the other hand, in polar solvents such as acetonitrile, DPPH• behaves as a sequential $1e^-/H^+$ oxidizer in the presence of bases, so that H-bonded bases increase the rate of reaction between DPPH• and phenols (Scheme 20.13) [22]. The sequential $1e^-/H^+$ oxidation by DPPH•, which is named sequential proton loss electron transfer (SPLET), occurs also in the absence of real bases in polar protic solvents like alcohols (where the solvent accepts the proton); this change in mechanism being responsible for the massive acceleration observed for the reaction of DPPH• with phenols on going from aprotic solvents like dioxane to methanol [41].

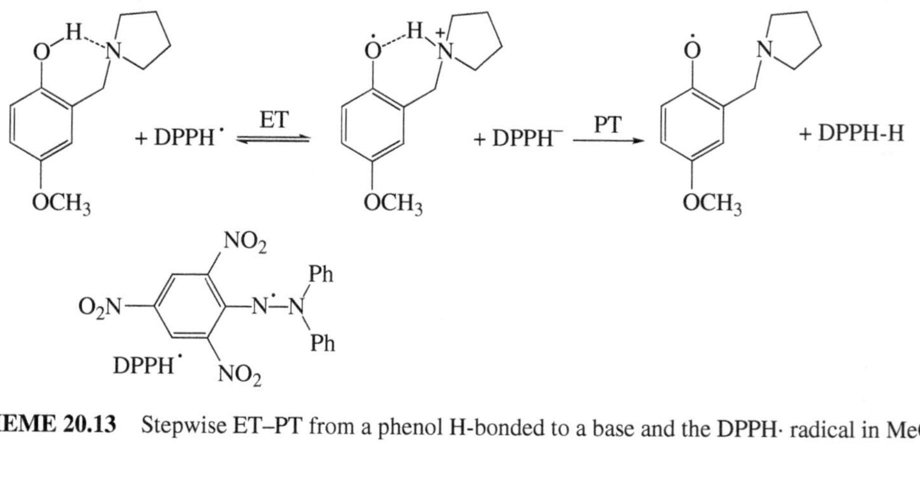

SCHEME 20.13 Stepwise ET–PT from a phenol H-bonded to a base and the DPPH· radical in MeCN.

20.5 EFFECTS OF NON-COVALENT INTERACTIONS ON THE REACTIVITY OF PHENOXYL RADICALS

Non-covalent interactions represent the way used by Nature to harness the reactivity of phenoxyl radicals, in order to perform complex redox reactions, for instance, for the photochemical splitting of water inside Photosystem II [17] or for energy transduction via the ubiquinol/ubiquinone shuttle [32]. The study of these interactions is gaining increasing interest among researchers not only to understand in deeper detail the functioning of these natural "power stations" but also to find new ways to exploit radical reactions of phenols as a novel sustainable way to obtain new pharmacologically active compounds.

Phenoxyl radicals are easily accessible by oxidation of phenols, which in turn are abundantly available from natural sources, often as by-products of the food industry. In this section, chemical–physical studies aimed at quantifying the non-covalent interactions between phenoxyl radicals and H-bond donating or accepting molecules or metals acting as Lewis acids are reported. Some examples of the applications of these concepts to the field of bioorganic and synthetic chemistry are also included.

20.5.1 Stabilization of Phenoxyl Radicals by H-Bonds

In 2003 Lucarini *et al.* quantified by EPR spectroscopy the H-bond accepting ability of the phenoxyl —O• moiety by studying the changes of the hyperfine splitting constants after the addition of increasing amounts of 1,1,1,3,3,3-hexafluoropropan-2-ol (HFP), which is a strong H-bond donating solvent, having a α_2^H value of 0.77 and a negligible tendency to accept H-bonds (β_2^H=0.03) (Scheme 20.14). From the equilibrium constants reported in that work, the β_2^H value for some phenoxyl radicals can be estimated as reported in Table 20.3 [42]. These results show that the —O• group is a moderately strong HB acceptor, having a β_2^H value similar to that of ketones (β_2^H of acetone is 0.50) [15].

Complexation of phenoxyl radicals by HFP greatly increases their persistency, because it reduces the participation of the —O• moiety in disproportionation reactions (Scheme 20.14). This effect is particularly evident in the case of the radical derived from α-tocopherol (α-TO•), as its EPR signal persists at room temperature for several hours [42]. Also the H-bond with a protonated alkylamine group in *ortho* position to the phenoxyl oxygen greatly increases the persistency of phenoxyl radicals [43]. Thanks to its negligible β_2^H value, HFP is able to selectively stabilize the phenoxyl radical with respect to the parent phenol, thus causing a marked decrease of the *BDE* of the phenolic O—H bond [42]. Further studies clarified that H-bonds are the main determinant of *BDE* changes observed on moving from a polar solvent to an apolar one. It was demonstrated by DFT calculations that the difference between the *BDE* values measured in benzene or water can be explained by keeping into account exclusively the H-bond formation with H_2O molecules [44].

The examples reported so far deal with H-bond interactions that involve the —O• moiety. In the case of phenols having polar substituents, other functional groups may play a role, for instance, the *para* OH group in the case of semiquinones. The ability of 2,6-di-*tert*-butyl-4-hydroxyphenoxyl radical to donate H-bonds to various HBA solvents has been investigated by EPR spectroscopy and by DFT calculations (Scheme 20.15) [45]. The equilibrium constants in benzene ranged from 5.1 M^{-1} for EtCN (β_2^H=0.44) to 810 M^{-1} for

TABLE 20.3 Equilibrium Constant for the H-Bond with 1,1,1,3,3,3-Hexafluoropropan-2-ol (HFP) and H-Bond Accepting Ability of Phenoxyl Radicals [42]

Phenoxyl Radical	K_{eq} [a]	β_2^H [b]
2,4,6-Tri-Me PhO·	21.0	0.42
4-OMe PhO·	43.6	0.48
α-TO·	64.2	0.51

[a] Measured in benzene at 293 K.
[b] Calculated herein from K_{eq} values.

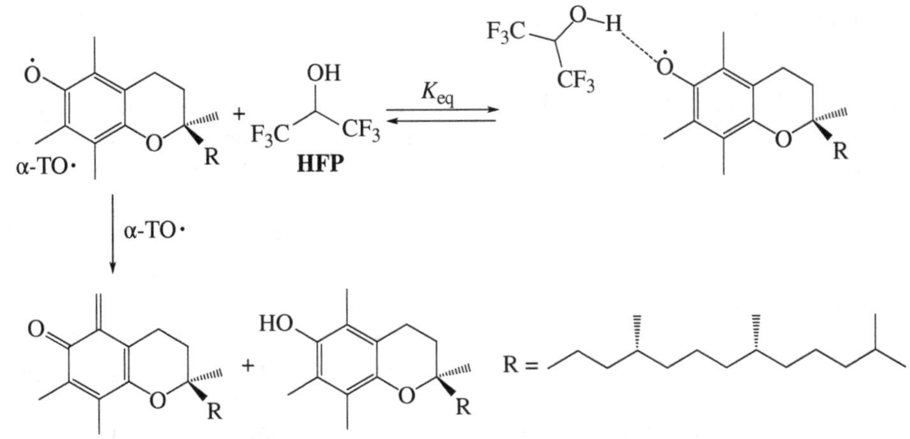

SCHEME 20.14 Hydrogen-bond formation between the phenoxyl radical of α-tocopherol (α-TO·) and HFP and mechanism for the bimolecular decay of the α-TO· radical.

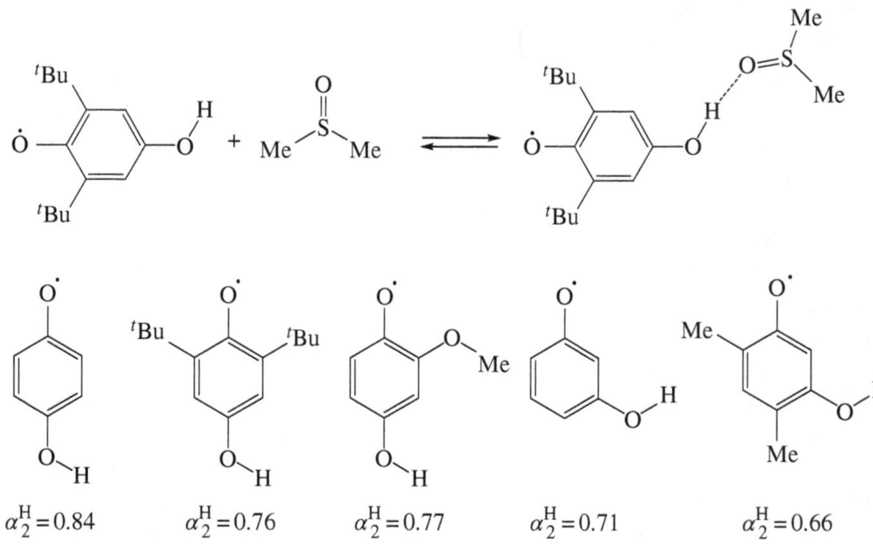

SCHEME 20.15 Hydrogen-bond equilibrium between a 4-OH-substituted phenoxyl radical and dimethyl sulfoxide and α_2^H values for the OH group of some hydroxy-substituted phenoxyl radicals.

DMSO (β_2^H = 0.78) and reached the exceptionally large value of $4 \times 10^4 M^{-1}$ for the very strong HBA solvent hexamethylphosphoramide (β_2^H = 1.0) [15]. From these measurements, it was possible to obtain the α_2^H value for the investigated radical, as 0.79 (by means of Abraham's equation; Section 20.3). H-bond formation also increased the persistency of the phenoxyl radicals, because it prevents the disproportionation 2 QH• = Q + QH_2, which is the main decay pathway of semiquinones.

Theoretical calculations allowed to investigate the H-bond donating (HBD) ability of the —OH group in different hydroxy-substituted phenoxyl radicals and to explore the substituent effect on such interaction. The results, collected in Scheme 20.15, show that the HBD ability of semiquinones is reduced by the presence of electron-donating substituents on the aromatic ring and that the 3-OH-substituted phenoxyl radicals (from resorcinols) have a lower ability to form H-bonds than 4-OH-substituted ones (from hydroquinones) [45].

20.5.2 Biosynthetic Coupling of Phenoxyl Radicals: The Dirigent Proteins

A spectacular example of how hydrogen bonds are able to direct the reactivity of phenoxyl radicals in a biological system is that represented by the dirigent proteins (DPs) in the biosynthesis of lignans in plant cells [46]. Lignans are a class of secondary metabolites found mainly in vascular plants, produced by oxidative dimerization of substituted hydroxycinnamic units: *p*-coumaryl, coniferyl, and sinapyl alcohols, also known as monolignols [46]. The interest toward this class of compounds is boosted by their remarkable biological effects, including anticancer, antioxidant, antimicrobial, anti-inflammatory, and immunosuppressive activities [47]. The biosynthesis of lignans is a two-step oxidative dimerization. First, an H-atom is abstracted from monolignols by oxidase enzymes (such as laccase) to form the corresponding phenoxyl radical. Second, two phenoxyl radicals react together to afford the final product. While in the *in vitro* experiments a complex mixture of products with no stereochemical control is always obtained, in plant cells lignans are produced with complete regiochemical and stereochemical control, as effect of the presence of DPs [46].

The first DP, discovered in *Forsythia intermedia* by Lewis and coworkers, directs the stereoselective biosynthesis of (+)-pinoresinol from coniferyl alcohol monomers [48]. It was shown that when reacting coniferyl alcohol and a laccase in the absence of DP, (+)-pinoresinol is a relatively minor product, while after the addition of DP, it becomes predominant (Scheme 20.16). It was assessed that this DP has no oxidizing activity and has little affinity for coniferyl alcohol, so it was concluded that the action of DP is to bind two coniferyl phenoxyl radicals and to catalyze the coupling in a geometrically constrained fashion. Interestingly, an enantiocomplementary DP able to mediate the formation of (−)-pinoresinol was later isolated from *Arabidopsis thaliana* [49].

The crystal structure of the (+)-pinoresinol-forming DP revealed the presence in the active site of polar amino acids able to form H-bonds with the coniferyl phenoxyl radical (Arg 141, Ser 111, Tyr 101, Asp 40), as well as lipophilic residues aimed at interacting with the aromatic cinnamic skeleton (Ile 161, Leu 113, Phe 115, Phe 79, Phe 38, Leu 174) [50]. Docking studies of DP with a substrate molecule suggest that the —O• group of the coniferyl phenoxyl radical points toward the side chain of arginine 141, which is able to donate a strong H-bond through the guanidine moiety [50]. The study of DPs is boosted by the interest in new synthetic protocols for the regio- and stereospecific coupling of phenoxyl radicals, which could provide synthetic access to many compounds with pharmaceutical activity [46].

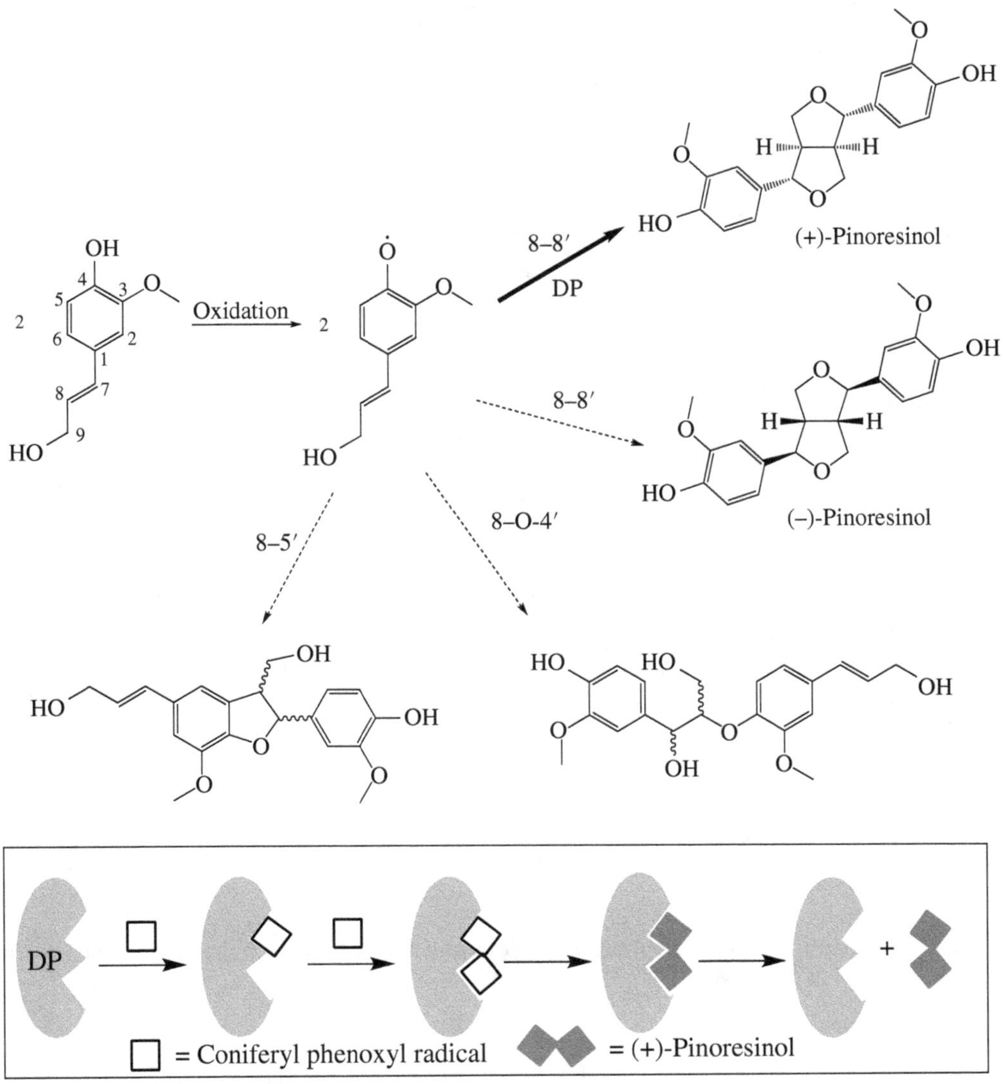

SCHEME 20.16 Main products obtained by bimolecular radical coupling of the phenoxyl radical from coniferyl alcohol. In the presence of the dirigent protein, instead of a complex mixture of dimers, only (+)-pinoresinol is obtained. In the insert, proposed reaction mechanism of DP-mediated coupling.

20.5.3 Synthesis through Phenoxyl Radicals Stabilized by H-Bonds

The sustainable synthesis of bisphenols, which are common moieties in natural products, is difficult despite their rather simple structure. Although bisphenols could be obtained directly by oxidative coupling of two phenoxyl radicals, this reaction is generally subject to side reactions and often results in polycyclic structures and in intractable mixtures of products. The group of Prof. Waldvogel discovered that by performing the oxidation of 2,4-disubstituted phenols by a boron-doped diamond electrode in a strong H-bond donating solvent—HFP—symmetric bisphenols can be obtained in high yields [51]. They extended the scope of this procedure by showing that, in the presence of HFP, also phenol–arene cross-coupling [52] or cross-coupling between two different phenols [53] could be achieved. This electrochemical method is particularly attractive because it avoids the use of oxidizing reagents or metal catalysts.

The proposed reaction mechanism (Scheme 20.17) is based on the formation at the electrode surface of highly reactive hydroxyl or alkoxyl radicals, which are stabilized by HFP and selectively react with phenols by H-atom abstraction. Resulting phenoxyl radicals are then stabilized by HFP, as demonstrated by EPR experiments [42]; thus the random coupling reactions that are typical of these radicals are strongly disfavored. The phenoxyl radicals have electrophilic nature and undergo nucleophilic attack by electron-rich arenes affording, after the final oxidation and aromatization steps, the desired products [52]. This procedure is noteworthy not only because it uses phenoxyl radicals and makes them useful reaction intermediates but also because it represents a green and sustainable way to employ phenols—which are abundantly present in plants—as building blocks for the production of fine chemicals.

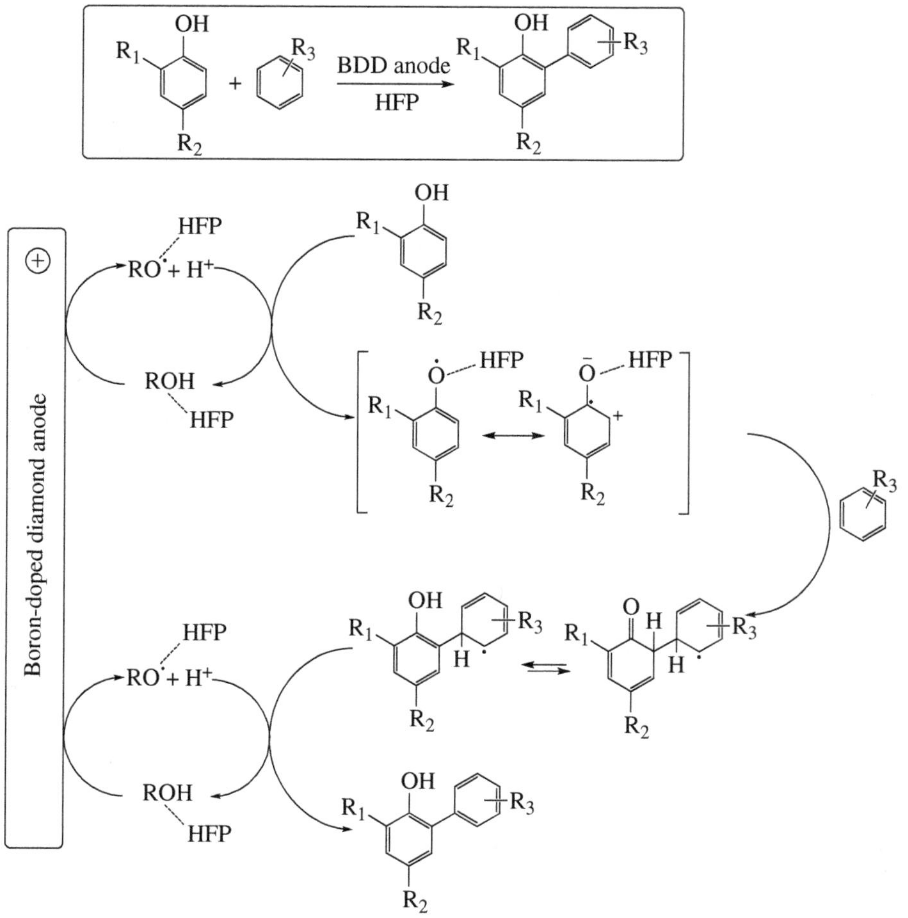

SCHEME 20.17 Proposed mechanism for the phenol–arene anodic cross-coupling by a boron-doped diamond electrode. R_1, R_2, and R_3 are electron-donating substituents (alkyl or OMe).

20.5.4 Metal Complexation Effects

An important way that leads to the stabilization of phenoxyl radicals is the interaction with metal cations acting as Lewis acids. Mukai and coworkers investigated in detail the variations of the UV–Vis spectrum and the decay kinetics of α-TO• in acetonitrile solutions after the addition of Li^+, Na^+, and Mg^{2+} ions. The equilibrium constant for the interaction between the metal ions and α-TO• was measured as 9.2, 2.8, and 45 M^{-1} for Li^+, Na^+, and Mg^{2+}, respectively [54]. At the same time, addition of Mg^{2+} ions reduced the rate constant for the bimolecular decay of α-TO• by 14-fold, indicating stabilization of the radical, which leads to results analogous to those observed in the case of HFP [55]. These results are particularly interesting because they show that non-covalent interactions can be significant also in a polar solvent such as MeCN, which competes with the phenoxyl radical for the metal cation.

A recent example in which the lack of selectivity in radical dimerization of a natural phenol has been circumvented by using the non-covalent interaction with a metallic ion has been reported by Daquino *et al.* [56]. Caffeic acid derivatives dimerize to unusual benzo[kl]xanthenes with interesting antioxidant activity [57] when treated with MnO_2, whereas a different product is formed if Ag_2O is used as oxidizing agent. The products formed by reaction with MnO_2 and Ag_2O derive, respectively, from the 8–8′ and 5–8′ coupling of the phenoxyl radical of caffeic acid (Scheme 20.18). The authors showed that when performing the Ag_2O oxidation in the presence of Mn^{2+} ions, the 8–8′ coupling product is formed in good yields, therefore demonstrating that Mn^{2+} ions have the peculiar ability to direct the reaction of the phenoxyl radicals of caffeic acid.

The mechanism of the oxidative dimerization of caffeic acid esters is based on the formation of a phenoxyl radical at the 4 position, in which the spin density is delocalized also on the carbon atoms in 5 and 8. Theoretical calculations showed that the interaction between the phenoxyl radical of caffeic acid esters and the Mn^{2+} ion reduces the spin density at the 5 position while increases that at the 8 position, therefore making the 8–8′ dimerization more favorable [56].

The biosynthetic route discovered by Daquino *et al.* [56, 57] has surprising similarities with one of the proposed pathways for the biosynthesis of lignin. Lignin is a structural component of plant cell wall, being responsible for the strength of vascular plants, and it is one of the most abundant

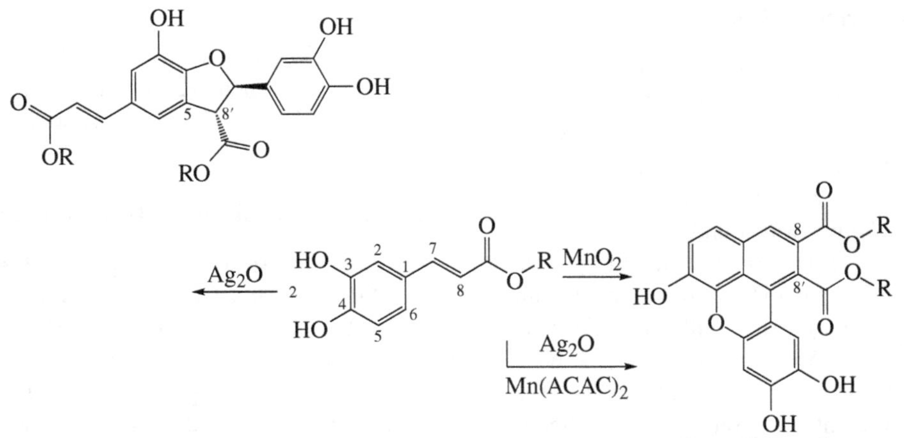

SCHEME 20.18 Oxidative dimerization of caffeic acid esters to the 5–8′ (left) and 8–8′ (right) coupling products. R=CH_3 or CH_2CH_2Ph; ACAC = acetylacetonate.

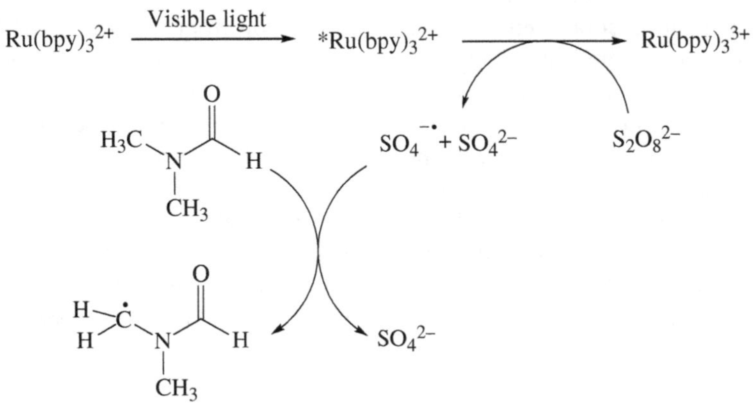

SCHEME 20.19 H-atom transfer in the alkylation of amides, catalyzed by ruthenium complexes.

biopolymers on Earth. It is formed by a radical polymerization of monolignols catalyzed by peroxidases, laccases, or other phenol oxidases. However, the exact mechanism is still debated since synthetic lignin usually displays a different ratio of covalent linkages and contains some bonds that do not occur in native lignin. Interestingly, when using Mn^{3+}/Mn^{2+} ions as redox shuttles between oxidizing enzymes and monolignols, a synthetic lignin very similar to the natural one could be obtained [58].

20.6 EFFECTS OF NON-COVALENT INTERACTIONS ON THE REACTIONS OF ALKOXYL RADICALS WITH AMINES OR AMIDES

The reactions of free radicals with the C—H bonds of alkyl amines and amides play an important role in the oxidative damage of proteins [59]. Hydroxyl (HO•), alkoxyl (RO•), and other radicals are able to strip H-atoms from both the backbone of proteins or from the side chain of amino acids [60]. After the initial H-atom abstraction, proteins undergo a series of structural modifications, including cross-linking, fragmentation, and loss of activity [59]. HAT from the C—H bonds of amines and amides is also a key step in the synthesis of substituted N-containing compounds under photoredox [61, 62] or oxidative conditions [63, 64]. For instance, the alkylation of dimethylformamide was achieved by radical oxidation using persulfate ($S_2O_8^{2-}$) in the presence of a visible light ruthenium-based catalyst (Scheme 20.19). The reaction is triggered by formation of an alkyl radical in α-position with respect to the amide N-atom [62].

Detailed investigations on the reactions of radicals with amines and amides have been performed by the group of Prof. Bietti in the case of alkoxyl (RO•) radicals, because their kinetics of formation and decay are amenable to be precisely followed by LFP technique. These reactions have been recently found to be very sensitive to non-covalent interactions, which can exert their effects mainly in four different ways: (i) radical–substrate H-bonds, (ii) intramolecular H-bonds in the substrate, (iii) intermolecular non-covalent interactions of the substrate with the solvent or with dissolved salts, and (iv) intermolecular non-covalent interactions of the abstracting radical with the solvent. These different cases will be discussed in the next paragraphs.

20.6.1 Radical–Substrate H-Bonds

The reactions of cumyloxyl (CumO•) and benzyloxy (BnO•) radicals with branched alkylamines were studied using the LFP technique [65]. These two radicals possess a very similar reactivity toward hydrocarbon substrates, indicating that the steric hindrance plays a minor role in such reactions [65]. The reaction of CumO• was found to occur essentially at the C—H bonds in α-position to the nitrogen atom ($k_{C-H} = 2.2-2.7 \times 10^7 M^{-1} s^{-1}$ in the case of secondary amines in MeCN) and only marginally with the N—H bond ($k_{N-H} = 3.1 \times 10^6 M^{-1} s^{-1}$ in MeCN), as the N—H bond is stronger than the α-C—H bond. The reason why H-atom abstraction occurs preferentially from hydrogens bound to the α-carbon is that during the HAT to RO• radicals, the incipient alkyl radical develops a partial positive charge that is stabilized by electron donation by the nitrogen lone pair, as shown in Scheme 20.20 [66].

In contrast, the BnO• radical reacts much faster than CumO• with amines, and H-atom abstraction is not selective for the C—H bonds in α-position. Theoretical calculations have shown that the reaction of BnO• with amines occurs through a rate-determining prereaction complex, which involves an uncommon C—H···N interaction between the CH_2 group of the BnO• radical and the amine N-atom (Scheme 20.20) [65]. The fact that the C—H group in BnO• can act as H-bond donor is explained in terms of the electron-withdrawing effect of the O• moiety, which increases the positive charge on the H-atom. This view is also supported by the large acidity in water ($pK_{a,H_2O} = -3$) of the benzylic CH_2 group in this radical [67]. Once the prereaction complex is formed, H-atom abstraction occurs from any nearby C—H or N—H groups, without selectivity.

20.6.2 Intramolecular H-Bonds in the Substrate

The assistance of the N lone pair to H-atom abstractions by CumO• from the α-CH bond in amines is the basis to rationalize the effects of intramolecular H-bonds observed in amino-substituted phenols, which represent a model for tyrosine residues in proteins [68]. The results from LFP studies of some piperidinylmethyl-substituted *para* methoxy phenols are presented (Scheme 20.21). By comparing the rates of the reactions (a) and (b) in Scheme 20.21, it can be inferred that the reaction of CumO• with the phenolic OH group is about twofold slower than that from the amine α-CH bonds. Although it may appear counterintuitive, H-abstraction from benzylic C—H bonds in α-position to the amine is slower than from α-C—H bonds in the piperidinyl moiety, as it can be inferred from the rate constant for the reaction of CumO• with benzylamine ($1.8 \times 10^7 M^{-1} s^{-1}$) [65].

When both groups are present in *meta* position on the same aromatic ring (entry c in Scheme 20.21), the rate constant is about the sum of those of each separate reactive moieties reported in entries a and b, indicating that the phenolic OH and the amine α-C—H bonds react independently with CumO•. However, when the two groups are in *ortho*

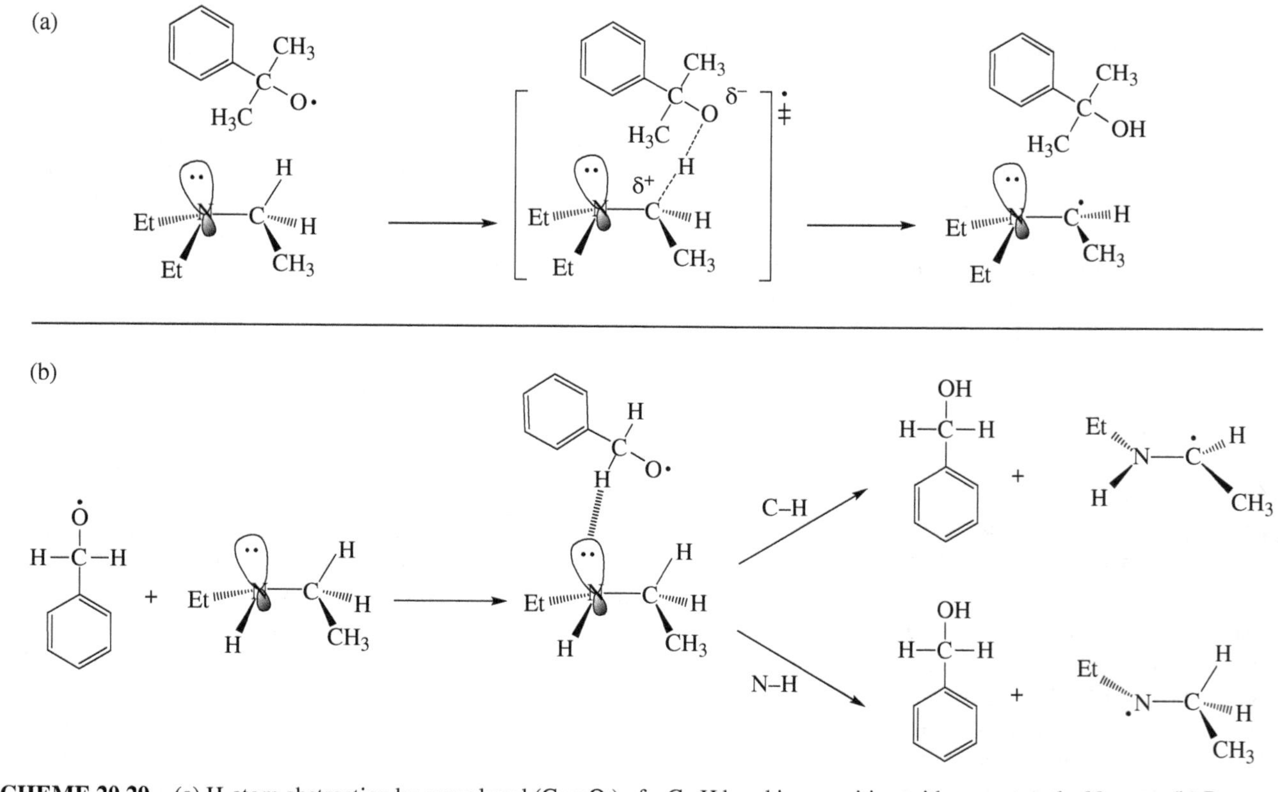

SCHEME 20.20 (a) H-atom abstraction by cumyloxyl (CumO·) of a C—H bond in α-position with respect to the N-atom. (b) Prereaction complex in the reaction between benzyloxyl radicals (BnO·) and amines and the products of H-atom abstraction from C—H or N—H bonds.

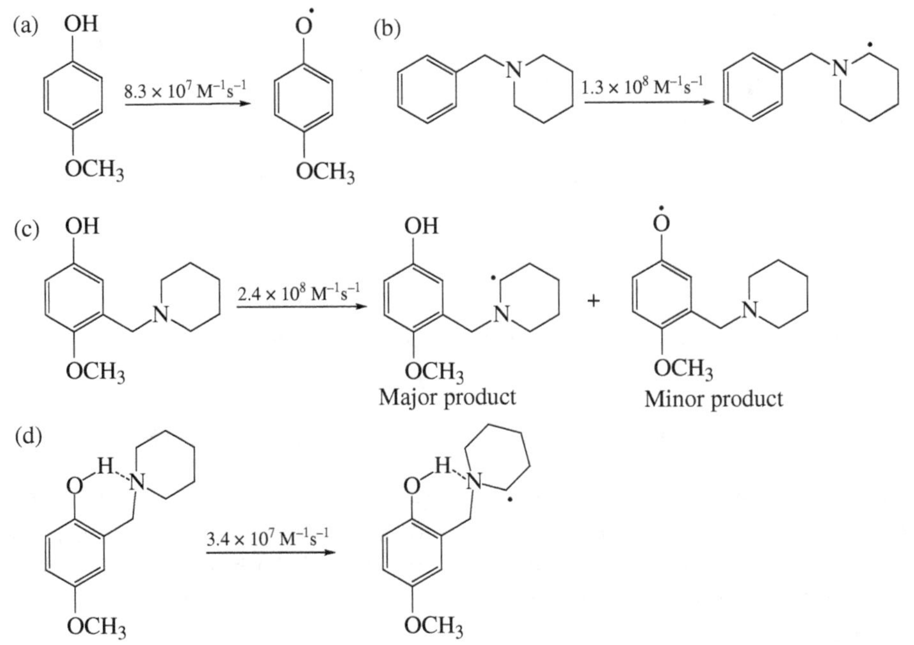

SCHEME 20.21 Rate constants and main products for the reaction of CumO· with amino-substituted phenols in MeCN at room temperature.

position (entry d in Scheme 20.21), a reactivity drop and the exclusive formation of alkyl radicals are observed. This result has been explained as being due to the intramolecular H-bond between the phenolic OH and the amine N-atom, which has a twofold effect: (i) it makes the phenolic OH group unavailable for the HAT, similarly to the case of the reaction of peroxyl radicals with intramolecularly H-bonded phenols reported in Section 20.4.1, and (ii) it reduces the charge on the nitrogen lone pair, thus diminishing the stabilization of the incipient alkyl radical [68].

20.6.3 Intermolecular Non-covalent Interactions of the Substrate with the Solvent or with Dissolved Salts

The reaction of CumO· with hydrocarbons occurs with very similar rate constants in a wide range of solvents, because neither the reactants nor the transition state is strongly solvated [69]. On the other hand, the reaction of CumO· with triethylamine in methanol is sixfold slower than in MeCN [69]. Also the addition of metal cations acting as Lewis acids to MeCN causes a marked decrease of the reactivity, in particular in the case of $Mg(ClO_4)_2$, which causes the rate constant to drop below the detection limit of the LFP technique. The relevant rate constants for the reactions of Et_3N with CumO· are collected in Table 20.4 [66, 69].

The explanation for these results is that H-bond donors such as MeOH and Lewis acids interact with the lone pair on nitrogen in Et_3N, reducing its ability to stabilize the incipient alkyl radicals during the HAT. The examples reported clearly show that non-covalent interactions play a major role in directing the reactivity of alkoxyl radicals with specific functional groups. A detailed knowledge of these effects is needed to understand radical-mediated oxidation in biological systems and to optimize the yield of radical synthesis [70].

TABLE 20.4 Rate Constants for the Reaction of CumO· and Et_3N Measured by the LFP Technique

Solvent	k_H ($M^{-1}s^{-1}$)	Reference
Benzene	2.8×10^8	[69]
MeCN	2.0×10^8	[69]
MeOH	3.8×10^7	[69]
MeCN + $LiClO_4$ (1.0 M)	9.4×10^7	[66]
MeCN + $Mg(ClO_4)_2$ (5 mM)	$<1 \times 10^6$	[66]

20.6.4 Intermolecular Non-covalent Interactions of the Abstracting Radical with the Solvent

We have previously indicated (Section 20.3) that alkoxyl radicals, such as cumyloxyl, do not change their intrinsic reactivity upon interaction with different solvents and that they experience limited influence by the solvent when reacting with a hydrocarbon [10]. However, while the rates of the reaction between CumO· (or BnO·) radicals and 1,4-cyclohexadiene or cyclohexane were identical within experimental error in solvents of largely different polarity (from isooctane to acetonitrile), the rate constant *increased* slightly (by a factor or 2 or less) but significantly in alcohols (*tert*-butanol or methanol), and the acceleration became more meaningful (a factor of 3.5–4) in the strong HBD solvent 2,2,2-trifluoroethanol [71]. The higher reaction rate in HBD (alcoholic) solvents was attributed to H-bonding from the solvent's OH to the oxygen of the alkoxyl radical, since such an H-bonding

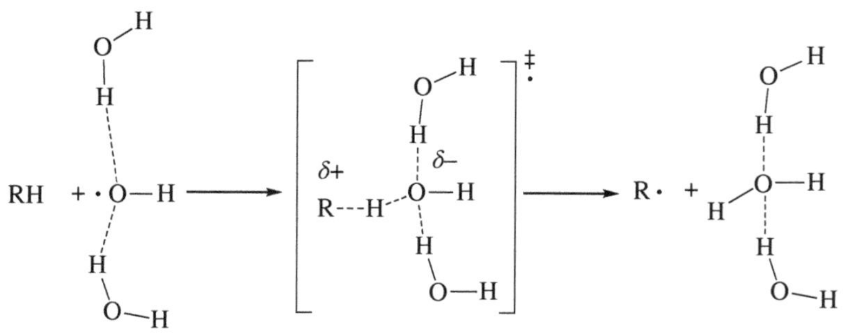

SCHEME 20.22 Acceleration of the reaction of hydroxyl or alkoxyl radicals by H-bond formation with an H-bond donating solvent.

interaction would become stronger as the reaction progresses, thereby stabilizing the TS more than the reactants.

A qualitatively similar KSE was also documented on C—H atom abstraction by hydroxyl radicals. Tanko and coworkers showed that the rate of reaction of hydroxyl radicals with a variety of hydrocarbons was larger by as much as two orders of magnitude in water than in polar aprotic solvents like acetonitrile, and attributed this enhanced reactivity to H-bonding of water (the HBD) to hydroxyl radical (the HBA) (Scheme 20.22) [72]. The strengths of such H-bond interactions are expected to increase on moving from the reactants to the transition state due to its dipolar nature with partial negative charge developing on the oxygen atom of the abstracting radical, which would explain the increased rate of reaction [72].

In summary, modulation of the reactivity of the abstracting radical by non-covalent interactions can be expected in protic solvents; however, due to the limited magnitude in the case of alkoxyl radicals, they might be relevant only in the presence of strong HBD solvents/chemicals like 2,2,2-trifluoroethanol. Conversely, they could have major relevance in the case of hydroxyl radicals.

20.7 FINAL COMMENTS

The examples reported in this chapter demonstrate that non-covalent interactions are main determinants of the rates of the reactions of peroxyl, phenoxyl, and alkoxyl radicals with many substrates. Interestingly, stabilization of phenoxyl radicals by non-covalent interactions leads to the formation of specific products in good yield, instead of complex mixtures, as it is often observed with these radicals. Clearly, this offers unique opportunities to control and manipulate these reactions to fully exploit their potential in synthetic processes. On the other hand, knowledge and control of the role of non-covalent interactions in radical chemistry allow designing radical inhibitors, such as antioxidants, suited to specific needs like biomedical applications, or it allows fine-tuning of their use in industrially relevant processes. Further studies are needed to extend the number of Abraham's α_2^H and β_2^H values (Section 20.3) available for radicals, as these solvatochromic parameters have a powerful predictive ability on the strength of non-covalent interactions.

ACKNOWLEDGMENTS

This work has been supported by MIUR (PRIN 2010–2011 2010PFLRJR, PROxi), by the University of Bologna (FARB project NANOX), by CINECA (IsC29 project WINTER), and by COST Action CM1201 (Biomimetic Radical Chemistry).

REFERENCES

1. P. Nesvadba, Radical Polymerization in Industry. In: *Encyclopedia of Radicals in Chemistry, Biology and Materials*; C. Chatgilialoglu, A. Studer, Eds.; John Wiley & Sons, Ltd, Chichester, 2012, Vol. *4*, p 1701.
2. D. Canistro, C. Boccia, R. Falconi, B. Bonamassa, L. Valgimigli, F. Vivarelli, A. Soleti, M. L. Genova, G. Lenaz, A. Sapone, F. Zaccanti, S. Z. Abdel-Rahman, M. Paolini, *J. Gerontol. A Biol. Sci. Med. Sci.* 2015, *70*, 936.
3. L. Valgimigli, A. Sapone, D. Canistro, M. Broccoli, L. Gatta, A. Soleti, M. Paolini, *Aging Clin. Exp. Res.* 2015, *27*, 235.
4. M. Valgimigli, L. Valgimigli, D. Trerè, S. Gaiani, G. F. Pedulli, L. Gramantieri, L. Bolondi, *Free Radic. Res.* 2002, *36*, 939.
5. R. Amorati, L. Valgimigli, *Free Radic. Res.* 2015, *49*, 633.
6. C. Reichardt, *Solvents and Solvent Effects in Organic Chemistry*, Wiley-VCH, Weinheim, 2003.
7. M. Lucarini, G. F. Pedulli, L. Valgimigli, *J. Org. Chem.* 1996, *61*, 1161.
8. A. Solovyev, S.-H. Ueng, J. Monot, L. Fensterbank, M. Malacria, E. Lacôte, D. P. Curran, *Org. Lett.* 2010, *12*, 2998.
9. L. Valgimigli, J. T. Banks, K. U. Ingold, J. Lusztyk, *J. Am. Chem. Soc.* 1995, *117*, 9966.
10. D. V. Avila, K. U. Ingold, J. Lusztyk, W. H. Green, D. R. Procopio, *J. Am. Chem. Soc.* 1995, *117*, 2929.
11. D. W. Snelgrove, J. Lusztyk, J. T. Banks, P. Mulder, K. U. Ingold, *J. Am. Chem. Soc.* 2001, *123*, 469.
12. R. Amorati, L. Valgimigli, *Org. Biomol. Chem.* 2012, *10*, 4147.
13. J. A. Howard, K. U. Ingold, *Can. J. Chem.* 1964, *42*, 1044.
14. M. H. Abraham, P. L. Grellier, D. V. Prior, P. P. Duce, J. J. Morris, P. J. J. Taylor, *J. Chem. Soc., Perkin Trans. 2* 1989, 699.
15. M. H. Abraham, P. L. Grellier, D. V. Prior, J. J. Morris, P. J. Taylor, *J. Chem. Soc., Perkin Trans. 2* 1990, 521.
16. M. H. Abraham, P. L. Grellier, D. V. Prior, R. W. Taft, J. J. Morris, P. J. Taylor, C. Laurence, M. Berthelot, R. M. Doherty,

M. J. Kamlet, J. L. M. Abboud, K. Sraidi, G. Guiheneuf, *J. Am. Chem. Soc.* 1988, *110*, 8534.

17. D. R. Weinberg, C. J. Gagliardi, J. F. Hull, C. F. Murphy, C. A. Kent, B. C. Westlake, A. Paul, D. H. Ess, D. G. McCafferty, T. J. Meyer, *Chem. Rev.* 2012, *112*, 4016.
18. L. Valgimigli, D. A. Pratt, Antioxidants in Chemistry and Biology. In: *Encyclopedia of Radicals in Chemistry, Biology and Materials*; C. Chatgilialoglu, A. Studer, Eds.; John Wiley & Sons, Ltd, Chichester, 2012, Vol. *3*, p. 1623.
19. L. Valgimigli, K. U. Ingold, J. Lusztyk, *J. Am. Chem. Soc.* 1996, *118*, 3545.
20. L. Valgimigli, D. Bartolomei, R. Amorati, E. Haidasz, J. J. Hanthorn, S. J. Nara, J. Brinkhorst, D. A. Pratt, *Beilstein J. Org. Chem.* 2013, *9*, 2781.
21. R. Amorati, L. Valgimigli, L. Panzella, A. Napolitano, M. d'Ischia, *J. Org. Chem.* 2013, *78*, 9857.
22. R. Amorati, S. Menichetti, C. Viglianisi, M. C. Foti, *Chem. Commun.* 2012, *48*, 11904.
23. L. Valgimigli, R. Amorati, S. Petrucci, G. F. Pedulli, D. Hu, J. J. Hanthorn, D. A. Pratt, *Angew. Chem. Int. Ed.* 2009, *48*, 8348.
24. G. A. DiLabio, E. R. Johnson, *J. Am. Chem. Soc.* 2007, *129*, 6199.
25. R. Amorati, S. Menichetti, E. Mileo, G. F. Pedulli, C. Viglianisi, *Chem. Eur. J.* 2009, *15*, 4402.
26. M. Musialik, R. Kuzmicz, T. S. Pawlowski, G. Litwinienko, *J. Org. Chem.* 2009, *74*, 2699.
27. R. Amorati, M. G. Fumo, S. Menichetti, V. Mugnaini, G. F. Pedulli, *J. Org. Chem.* 2006, *71*, 6325.
28. R. Amorati, F. Catarzi, S. Menichetti, G. F. Pedulli, C. Viglianisi, *J. Am. Chem. Soc.* 2008, *130*, 237.
29. C. Viglianisi, M. G. Bartolozzi, G. F. Pedulli, R. Amorati, S. Menichetti, *Chem. Eur. J.* 2011, *17*, 12396.
30. R. Amorati, G. F. Pedulli, L. Valgimigli, H. Johansson, L. Engman, *Org. Lett.* 2010, *12*, 2326.
31. R. Amorati, G. F. Pedulli, M. Guerra, *Org. Biomol. Chem.* 2010, *8*, 3136.
32. R. Amorati, P. Franchi, G. F. Pedulli, *Angew. Chem. Int. Ed.* 2007, *46*, 6336
33. M. C. Foti, R. Amorati, G. F. Pedulli, C. Daquino, D. A. Pratt, K. U. Ingold, *J. Org. Chem.* 2010, *75*, 4434.
34. N. R. Perron, J. L. Brumaghim, *Cell Biochem. Biophys.* 2009, *53*, 75.
35. S. V. Jovanovic, S. Steenken, Y. Hara, M. G. Simic, *J. Chem. Soc., Perkin Trans. 2* 1996, 2497.
36. J. J. Warren, T. A. Tronic, J. M. Mayer, *Chem. Rev.* 2010, *110*, 6961.
37. I. J. Rhile, T. F. Markle, H. Nagao, A. G. DiPasquale, O. P. Lam, M. A. Lockwood, K. Rotter, J. M. Mayer, *J. Am. Chem. Soc.* 2006, *128*, 6075.
38. C. Costentin, M. Robert, J.-M. Savéant, C. Tard, *Angew. Chem. Int. Ed.* 2010, *49*, 3803.
39. L. Biczók, N. Gupta, H. Linschitz, *J. Am. Chem. Soc.* 1997, *119*, 12601.
40. R. E. Galian, G. Litwinienko, J. Pérez-Prieto, K. U. Ingold, *J. Am. Chem. Soc.* 2007, *129*, 9280.
41. M. Musialik, G. Litwinienko, *Org. Lett.* 2005, *7*, 4951.
42. M. Lucarini, V. Mugnaini, G. F. Pedulli, M. Guerra, *J. Am. Chem. Soc.* 2003, *125*, 8318.
43. T. Maki, Y. Araki, Y. Ishida, O. Onomura, Y. Matsumura, *J. Am. Chem. Soc.* 2001, *123*, 3371.
44. M. Guerra, R. Amorati, G. F. Pedulli, *J. Org. Chem.* 2004, *69*, 5460.
45. R. Amorati, G. F. Pedulli, *Org. Biomol. Chem.* 2012, *10*, 814.
46. B. Pickel, A. Schaller, *Appl. Microbiol. Biotechnol.* 2013, *97*, 8427.
47. M. Saleem, H. J. Kim, M. S. Ali, Y. S. Lee, *Nat. Prod. Rep.* 2005, *22*, 696.
48. L. B. Davin, H.-B. Wang, A. L. Crowell, D. L. Bedgar, D. M. Martin, S. Sarkanen, N. G. Lewis, *Science* 1997, *275*, 362.
49. B. Pickel, M.-A. Constantin, J. Pfannstiel, J. Conrad, U. Beifuss, A. Schaller, *Angew. Chem. Int. Ed.* 2010, *49*, 202.
50. K.-W. Kim, C. A. Smith, M. D. Daily, J. R. Cort, L. B. Davin, N. G. Lewis, *J. Biol. Chem.* 2015, *290*, 1308.
51. A. Kirste, M. Nieger, I. M. Malkowsky, F. Stecker, A. Fischer, S. R. Waldvogel, *Chem. Eur. J.* 2009, *15*, 2273.
52. A. Kirste, B. Elsler, G. Schnakenburg, S. R. Waldvogel, *J. Am. Chem. Soc.* 2012, *134*, 3571.
53. B. Elsler, D. Schollmeyer, K. M. Dyballa, R. Franke, S. R. Waldvogel, *Angew. Chem. Int. Ed.* 2014, *53*, 5210.
54. K. Mukai, Y. Kohno, A. Ouchi, S. Nagaoka, *J. Phys. Chem. B* 2012, *116*, 8930.
55. Y. Kohno, M. Fujii, C. Matsuoka, H. Hashimoto, A. Ouchi, S. Nagaoka, K. Mukai, *J. Phys. Chem. B* 2011, *115*, 9880.
56. C. Daquino, A. Rescifina, C. Spatafora, C. Tringali, *Eur. J. Org. Chem.* 2009, 6289.
57. C. Spatafora, C. Daquino, C. Tringali, R. Amorati, *Org. Biomol. Chem.* 2013, *11*, 4291.
58. H. Onnerud, L. Zhang, G. Gellerstedt, G. Henriksson, *Plant Cell* 2002, *14*, 1953.
59. M. J. Davies, Oxidative Damage to Proteins. In: *Encyclopedia of Radicals in Chemistry, Biology and Materials*; C. Chatgilialoglu, A. Studer, Eds.; Wiley-VCH, Weinheim, 2012, Vol. *3*, p. 1425.
60. M. Salamone, L. Mangiacapra, M. Bietti, *J. Org. Chem.* 2015, *80*, 1149.
61. C. K. Prier, D. A. Rankic, D. W. C. MacMillan, *Chem. Rev.* 2013, *113*, 5322.
62. C. Dai, F. Meschini, J. M. R. Narayanam, C. R. J. Stephenson, *J. Org. Chem.* 2012, *77*, 4425.
63. W. Wang, X. Zhao, L. Tong, J. Chen, X. Zhang, M. Yan, *J. Org. Chem.* 2014, *79*, 8557.
64. H. Ueda, K. Yoshida, H. Tokuyama, *Org. Lett.* 2014, *16*, 4194.
65. M. Salamone, G. A. DiLabio, M. Bietti, *J. Am. Chem. Soc.* 2011, *133*, 16625.
66. M. Salamone, L. Mangiacapra, G. A. DiLabio, M. Bietti *J. Am. Chem. Soc.* 2013, *135*, 415.
67. M. Salamone, G. Anastasi, M. Bietti, G. A. DiLabio, *Org. Lett.* 2011, *13*, 260.
68. M. Salamone, R. Amorati, S. Menichetti, C. Viglianisi, M. Bietti, *J. Org. Chem.* 2014, *79*, 6196.
69. M. Bietti, M. Salamone, *Org. Lett.* 2010, *12*, 3654.
70. G. Litwinienko, A. L. J. Beckwith, K. U. Ingold, *Chem. Soc. Rev.* 2011, *40*, 2157.
71. M. Salamone, I. Giammarioli, M. Bietti, *J. Org. Chem.* 2011, *76*, 4645.
72. S. Mitroka, S. Zimmeck, D. Troya, J. M. Tanko, *J. Am. Chem. Soc.* 2010, *132*, 2907.

21

CURRENT UNDERSTANDING OF π–π INTERACTIONS AND THE APPLICATIONS IN PROTEIN DESIGN

JIANMIN GAO AND AZADE S. HOSSEINI

Department of Chemistry, Merkert Chemistry Center, Boston College, Chestnut Hill, MA, USA

21.1 INTRODUCTION

It is well known that aromatic molecules can interact with one another via electrostatic mechanisms [1]. Take benzene dimers as an example—it has been shown that benzene dimers prefer the edge–face and offset–stacked geometries over the face–face stacked geometry (Fig. 21.1). This phenomenon arises from the electrostatic potential of benzene, which possesses a positive edge and a negative center. The edge–face and offset–stacked geometries present favorable electrostatic interactions between two benzene molecules, while the face–face stacked geometry results in electrostatic repulsion [1]. The electrostatic interaction between aromatic molecules is further evidenced by the mixture of benzene and perfluorobenzene, which were found to pack against each other in the face–face stacked geometry [1]. As perfluorination reverses the electron distribution of the benzene ring, face–face stacking best matches the electrostatic potential of benzene to that of perfluorobenzene. This chapter will discern the significance of such electrostatic interactions in the context of stabilizing protein structures and facilitating ordered assembly of designed peptides.

A statistical survey of protein structures in the Protein Data Bank revealed that ~60% of aromatic side chains form a close contact with another aromatic residue [2], suggesting the significance of such interactions in Nature. In protein folding and assembly, packing of aromatic residues inevitably involves steric, hydrophobic, and electronic factors that contribute to the stability of relevant structures. To avoid confusion, we use π–π interaction to specifically refer to the electrostatic interaction between aromatic moieties. In contrast, the term "aromatic packing" will be used to describe the appearance of aromatic residues physically contacting one another.

Although aromatic molecules are clearly able to interact electrostatically, the strength of such π–π interactions had been controversial. For example, literature data range from one-tenth of a kcal/mol to well over a kcal/mol for the edge–face interaction between two benzene molecules [1]. This uncertainty casts doubt on the significance of π–π interactions in the context of complex protein structures, where numerous covalent and non-covalent interactions exist and compete with one another. It had been difficult to dissect the electrostatic π–π interactions from other factors as mutations within the natural amino acid set inevitably perturb multiple factors at the same time. With recent advances in synthetic chemistry, peptides and proteins can now be made chemically to incorporate properly designed unnatural building blocks, allowing π–π interactions to be directly probed in the context of proteins [3].

21.2 SIGNIFICANCE OF π–π INTERACTIONS IN PROTEIN FOLDING

Energetic analysis of π–π interactions in proteins is greatly facilitated by the incorporation of steric mimics of native aromatic residues into protein structures of interest. For example, the unnatural amino acids cyclohexylalanine (Cha) and perfluorophenylalanine (f_5F) have been used as analogues of phenylalanine (F or Phe). While being slightly larger than F, Cha does not have an aromatic side chain and is therefore unable to engage in π–π interactions. As an even

Non-Covalent Interactions in the Synthesis and Design of New Compounds, First Edition.
Edited by Abel M. Maharramov, Kamran T. Mahmudov, Maximilian N. Kopylovich, and Armando J. L. Pombeiro.

closer mimic of F, f_5F displays an electron-deficient aromatic ring as a consequence of perfluorination. Mutation of a Phe to Cha or f_5F can dramatically perturb π–π interactions that the Phe residue is involved in.

By comparing a phenylalanine pair (F–F) to a phenylalanine–perfluorophenylalanine pair (F–f_5F) in the helical model peptide Ala–Lys, Waters and coworkers investigated the effect of π–π interactions to the stabilization of α-helices. This report shows that a F–F pair gives better stabilized helices when incorporated at i and i+4 positions in contrast to the i−i+5 combination, indicating that F–F packing is energetically favorable [4]. Placing the F–F pair at the C-terminus of the peptide increases the energetic gain by a factor of three compared to an internal position (−0.80 kcal/mol vs. −0.26 kcal/mol), presumably due to the structural flexibility of the peptide terminus that allows optimal geometry of the F–F pair. To show that the energetic gain is not purely from hydrophobic interactions, the authors compared the F–F pair to the corresponding F–f_5F pair. The results show that, at the C-terminus of the peptide helix, F–f_5F gives energetic gain of smaller magnitude (−0.55 kcal/mol vs. −0.80 kcal/mol) despite the fact that f_5F is significantly more hydrophobic than F. This indicates that the F–F interaction is at least partly electrostatic in nature, presumably in the edge–face geometry.

Waters and coworkers have also investigated π–π interactions in the context of a β-sheet [5]. Specifically, modified β-hairpin peptides were used to study the contribution of electronic and hydrophobic effects to aromatic packing. The hairpins display a generic sequence of Ac–Arg–X_1–Val–Orn–Val–Asn–Gly–Lys–Glu–Ile–X_2–Gln–NH_2, in which the cross-strand residues X_1 and X_2 pack against each other and are varied to be either a Phe or a Cha. Again, Cha is used as a comparison to Phe because of their similarity in size and dramatic difference in electronic properties. The X_1–X_2 pair is expected to stabilize the β-hairpin structure via hydrophobic and possibly π–π interactions. Interestingly, the peptide variants with Phe–Phe and Cha–Cha pairing show comparable stabilities. However, the peptides with Phe–Cha or Cha–Phe pairs exhibit lower stabilities, indicating that Phe has a preference for self-association. Thermodynamic denaturation revealed that peptide with a Phe–Phe pair exhibits both greater enthalpic driving force and a large entropic cost, as expected for the Phe–Phe pair to engage in favorable electrostatic interactions, which restrict the Phe residues in an offset–stacked or edge–face geometry. Indeed, analysis of the aromatic region by NMR revealed a tight contact between the two rings, which adopt an edge–face geometry. The preference of self-association of Phe also suggests that, as expected for its electrostatic nature, π–π interactions may afford selectivity similar to that of a hydrogen bond.

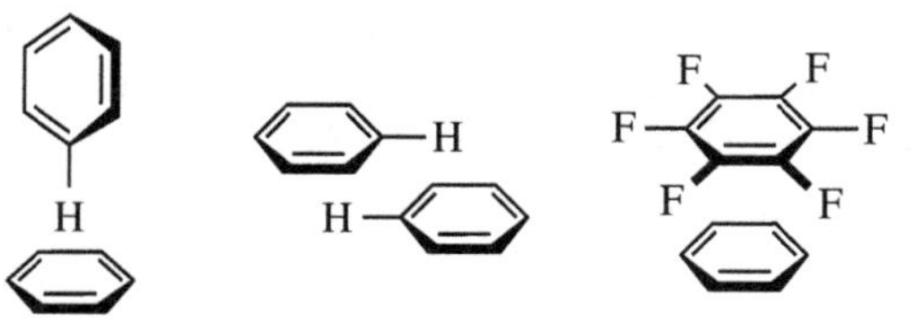

FIGURE 21.1 Possible geometries of π–π interactions. Left: edge–face. Middle: offset-stacked. Right: face–face stacked.

A report by Nowick and coworkers describes a similar study, which uses Phe and Cha in a different model β-sheet to explore the significance of π–π interactions [6]. Specifically, a series of peptides were synthesized based on a general β-sheet template. Upon dimerization, this peptide template folds into an antiparallel β-sheet, in which R1 from one monomer and R5 from the other monomer form a non-hydrogen-bonded cross-strand pair (Fig. 21.2). When different peptide variants are mixed, formation of homodimers and/or heterodimers is expected depending on the relative strength of the pairwise interaction between R1 and R5. The authors assessed the significance of π–π interactions by determining the dimerization selectivity by mixing two peptides with both R1 and R5 being Phe and Cha, respectively.

Contrary to the findings of Waters and coworkers, the Nowick system revealed no preferences for the homodimers that display Phe–Phe and Cha–Cha pairs. Instead, all four possible dimers were found to exist at comparable percentages. This finding, together with the Waters report, suggests that π–π interactions may contribute to protein stability in a context-dependent manner. Indeed, a survey of the statistical frequency of Phe–Phe pairing within β-sheet proteins showed that this interaction occurs with high frequency in hydrogen-bonded cross-strand pairs of antiparallel β-sheets but is almost absent in nonhydrogen-bonded cross-strand

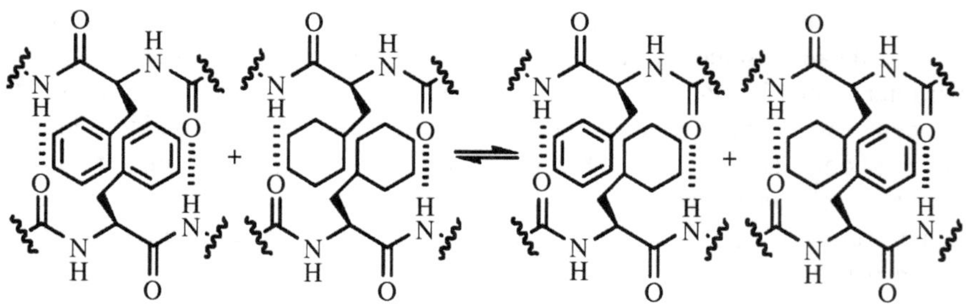

FIGURE 21.2 Equilibration of the homodimers and heterodimers of the templated β-strand reported by Nowick and coworkers. Source: Chung *et al.* [6]. Reproduced with permission of American Chemical Society.

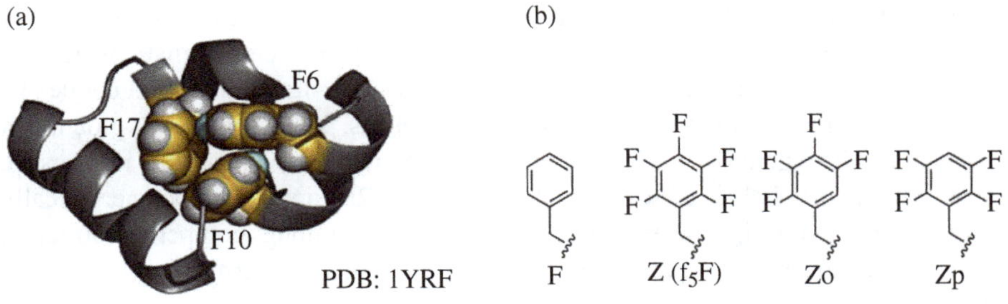

HP35:
LSDEDFKAVFGMTRSAFANLPLWKQQHLKKEKGLF

FIGURE 21.3 Structures of HP35 (a) and amino acid side chains (b) used to probe edge–face interactions.

pairs [7, 8], suggesting that the Phe–Phe contacts may not readily occur in the latter system. This statistical observation is consistent with the experimental results of Nowick and coworkers.

In contrast to the work of Waters and Nowick, where the interacting aromatic pairs reside at solvent-exposed positions of the model proteins, Gao and coworkers studied the energetic significance of π–π interactions in the solvent-excluded cores of proteins. The hydrophobic core of a protein presents a low dielectric environment where the electrostatic interactions between aromatic residues should be most significant. One of the model proteins used is the villin headpiece subdomain HP35, the core of which consists of three Phe residues (Fig. 21.3), with the edge of F10 packing against the face of F6 and the edge of F6 packing against the face of F17.

An earlier report by Gellman and coworkers shows that perfluorination of a Phe side chain (F-to-f_5F mutation) at position 10 is stabilizing to the folded structure of HP35 but destabilizing at position 6 or 17 [9]. Intrigued by this observation, Gao and coworkers postulated that the F-to-f_5F mutations may destabilize protein structures by perturbing π–π interactions.

To test the hypothesis, Gao and coworkers examined HP35 mutants that incorporate tetrafluorinated phenylalanines (Fig. 21.3) [10]. Specifically, two tetrafluorinated phenylalanine analogues, 2,3,4,5-tetrafluoro-L-phenylalanine (Zo) and 2,3,5,6-tetrafluoro-L-phenylalanine (Zp), were synthesized, and their effect on HP35 stability was compared to that of pentafluorophenylalanine, dubbed Z in this study. The results show that the F10Zo mutant exhibits a higher stability than its F10Z counterpart by −1.1 kcal/mol. The only difference between F10Zo and F10Z is a single atom (hydrogen vs. fluorine), the atom that forms the edge–face contact between residue 10 and F6. The fact that F10Zo exhibits higher stability demonstrates the importance of the edge–face interaction as it relates to protein stability. The decrease of protein stability observed for the F6Z mutant can also be explained by the fact that F6 is involved in two favorable edge–face interactions; both are eliminated by perfluorination of this Phe residue. This study clearly demonstrates that π–π interactions, specifically the edge–face interaction

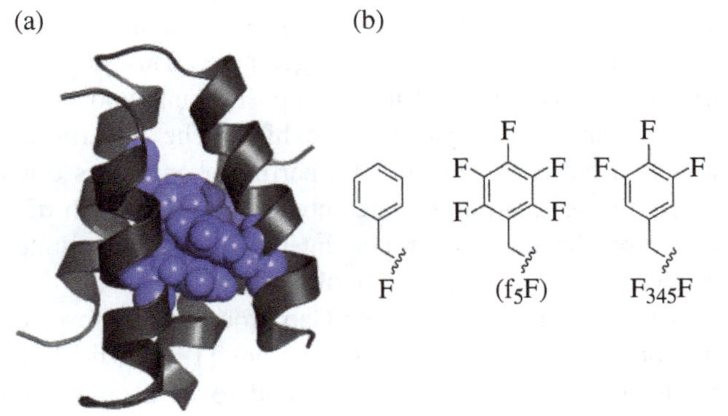

FIGURE 21.4 Cartoon representation of α2D dimer (PDB 1PQ6) highlighting the face–face stacked aromatic pairs (a) and structures of selected amino acids used to probe π–π interactions in α2D (b).

between Phe residues, can significantly contribute to protein stability (over 1 kcal/mol). The results also showcase that side-chain fluorination can be a powerful strategy for optimizing protein stabilities with consideration of not compromising edge–face interactions.

The example demonstrated above shows the benefit of exploiting edge–face interaction within a protein core. The face–face stacking interaction, however, is scarce in proteins, but there are instances where this phenomenon can be favorable. One example is α2D, a de novo designed 35-residue polypeptide that consists of aromatic residues within its core [11]. When folded, this peptide forms a dimeric helix bundle, where the F10 side chain of each monomer forms an intermolecular interaction with F29 of the other monomer in the face–face stacked geometry (Fig. 21.4).

By using α2D as a tractable system, Gao and coworkers investigated the energetics of π–π stacking and their role in stabilizing protein structures [12, 13]. Researchers incorporated various fluorinated phenylalanine analogues into positions 10 and/or 29 of the α2D sequence. The WT contains a phenylalanine at both positions and is called (F, F), while the fluorinated mutants are represented by (f_5F, f_5F). Experimental analysis indicated that upon fully fluorinating the core, melting temperatures increased from 29°C for (F, F) to 80°C for (f_5F, f_5F). Additionally, the structural stability was improved by nearly 6 kcal/mol, a phenomenon that can be explained by the increased hydrophobicity of (f_5F, f_5F). Interestingly, while both (F, F) and (f_5F, f_5F) individually form stable homodimers, the homodimers completely exchange to give a heterodimer, the hydrophobic core of which consists of two aromatic pairs where F and f_5F stack in the face–face geometry. The hydrophobic effect cannot account for the preference for the F–f_5F hetero-pairing because all these aromatic side chains are completely buried regardless of whether the proteins exist as homodimers or the heterodimer. Instead, the selective heterodimer formation results from the favorable electrostatic interaction between F and f_5F when they are stacked in the face-to-face geometry.

The core structure of the α2D heterodimer closely resembles the crystal structure of benzene and perfluorobenzene, where the two molecules are arranged at alternating positions and in a face–face stacked geometry. This structural arrangement is attributed to the opposite quadrupoles that benzene and perfluorobenzene exhibit. The electrostatic attraction between benzene and perfluorobenzene is often referred to as quadrupole interaction. The study with α2D showcases that such quadrupole interaction can be exploited to afford selective dimerization of proteins.

A follow-up report from the Gao group further revealed the complex origins of π–π interactions [13]. Particularly, dipole–dipole interactions were found to be energetically comparable to quadrupole interactions. In fact, the α2D variant ($F_{345}F$, $F_{345}F$) displays even higher stability than (f_5F, f_5F), even though f_5F is significantly more hydrophobic than $F_{345}F$ (Fig. 21.4). The superb stability of ($F_{345}F$, $F_{345}F$) is ascribed to the large dipole of $F_{345}F$ (3.5 Debye), which aligns in an antiparallel fashion in the ($F_{345}F$, $F_{345}F$) homodimer. Based on these mechanistic insights, Gao and coworkers showed that peptide self-sorting can be accomplished by taking advantage of π–π interactions alone.

Collectively, the literature data presented here provide an increasingly clear understanding of π–π interactions in the context of protein structure and stability. The electrostatic interactions between aromatic residues appear to be capable of providing significant free energies to the folding of peptide helices, β-sheet, as well as globular proteins. Importantly, π–π interactions, when used in proper contexts such as the core of α2D, appears to be energetically sufficient to afford selectivity among competing structures as demonstrated with the selective formation of α2D heterodimers.

21.3 π–π INTERACTIONS IN PEPTIDE ASSEMBLY

One area, where π–π interactions have already seen applications, is peptide self-assembly to give functional materials, such as hydrogels. We note that there is a substantial collection of publications on peptide assembly and it is not possible to include them all in this chapter. Instead, we will discuss the selected examples that present detailed structure–activity characterizations, which best illustrate the importance of π–π interactions in peptide assembly.

Peptides and amino acids are capable of forming hydrogels through spontaneous and dynamic self-assembly processes, which arise from non-covalent interactions between monomers to form amyloid-like fibrillar structures [14]. For example, both Fmoc-Phe and Fmoc-Tyr can promote the formation of hydrogels in various contexts; however, the former does not readily self-assemble in water to form hydrogels [15, 16]. The Nilsson group undertook a study to probe whether aromatic packing of Fmoc-Phe could be optimized to enhance hydrogelation behavior [17]. The initial experiments focused on probing the electronic and hydrophobic character of the phenyl side chain on self-assembly and hydrogelation (Fig. 21.5).

Fmoc-Tyr is an efficient hydrogelator, while Fmoc-Phe does not readily self-assemble in water. Fmoc-f_5F was used in order to probe the possible effects of hydrophobicity and

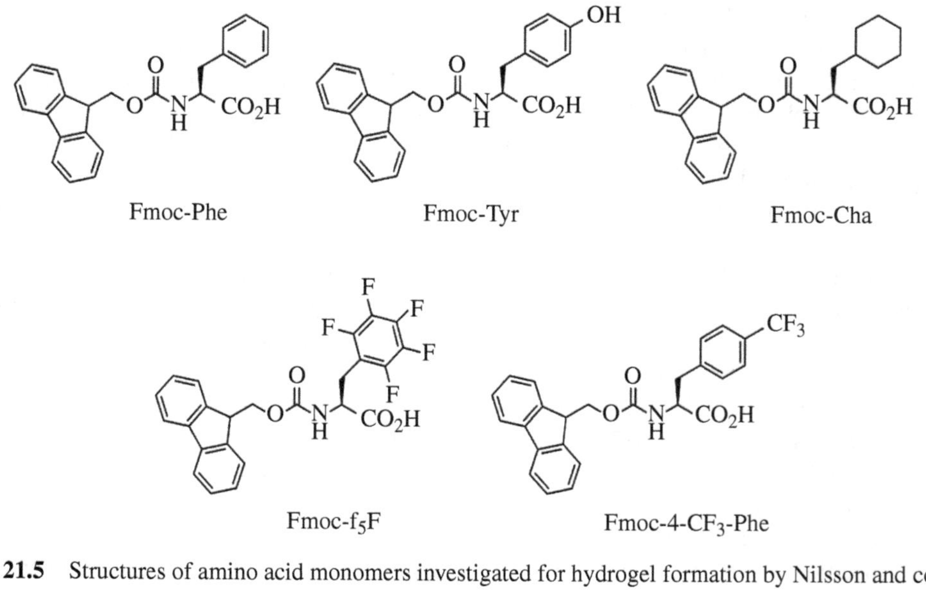

FIGURE 21.5 Structures of amino acid monomers investigated for hydrogel formation by Nilsson and coworkers.

electronics on self-assembly. If the increased hydrophobicity of Fmoc-Phe relative to Fmoc-Tyr caused the poor self-assembly behavior, then Fmoc-f_5F would be even worse. Conversely, if the electron-deficient side chain elicits favorable π–π interaction with the Fmoc moiety, then Fmoc-f_5F could be an efficient hydrogelator. Fmoc-4-CF_3-Phe was included because its hydrophobicity falls in between Fmoc-Phe and Fmoc-f_5F and it incorporates an electron-withdrawing functionality at the *para* position. This would allow perturbation of the electronics and hydrophobicity without having to directly fluorinate the ring.

Characterization results of these mutants show that Fmoc-f_5F forms a hydrogel after 5 min, while Fmoc-Tyr forms a gel after 30 min. Additionally, the authors report a critical gelation concentration (remaining monomer concentration) of 2.1 mM for Fmoc-f_5F in pure water, which is half of that for Fmoc-Tyr. Overall, Fmoc-f_5F promotes spontaneous gelation at lower concentrations and enhances rates when compared with Fmoc-Tyr. Microscopic analysis of the hydrogels formed from Fmoc-Tyr and Fmoc-f_5F was utilized to investigate the structural makeup of the assembled materials, and it was found that the modes of self-assembly for these two molecules are similar, despite their differences in hydrogelation rates and critical gelation concentrations. Fmoc-Phe and Fmoc-4-CF_3-Phe, however, showed no evidence of fibrillar assemblies.

The superior hydrogelation propensity of Fmoc-f_5F suggests that the π–π interaction between f_5F and Fmoc may be critical to the assembly of the amino acid. This statement is further supported by the comparison of Fmoc-f_5F and Fmoc-Cha. The side chain of f_5F and Cha exhibits comparable volumes and solvent accessible surface areas. However, Fmoc-Cha fails to self-assemble, presumably due to the lack of the favorable π–π interactions that Fmoc-f_5F is able to provide.

Several research groups have demonstrated the utility of short peptides containing aromatic groups toward hydrogel formation [14]. For example, both Ulijn and Gazit labs reported that under physiological conditions Fmoc-FF-OH peptides undergo spontaneous self-assembly to form fibrous hydrogels with the resulting structures exhibiting superior mechanical rigidity to those formed by building blocks with longer peptide sequences [18, 19].

Toward the utilization of such peptide hydrogels in tissue engineering, Gazit and coworkers investigated a series of hydrogelating peptides that incorporate the RGD motif [20], which can mimic the interaction between cell surface adhesion molecules and extracellular matrix proteins. The hydrogelation behavior of these peptides indicates that different spatial arrangement of aromatic moieties dramatically alters the peptide assembly (Fig. 21.6). Both Fmoc-FRGD-OH and Fmoc-RGDF-OH, which contain two aromatic moieties,

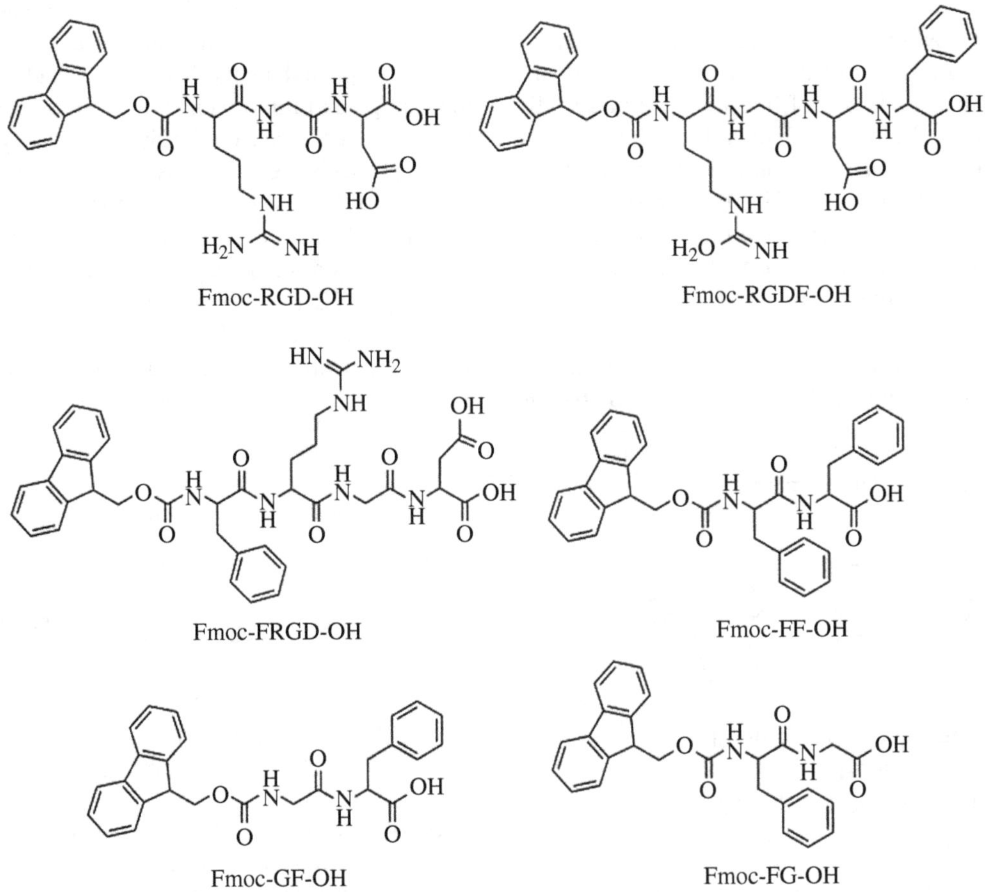

FIGURE 21.6 Structures of self-assembling peptides studied by Gazit and coworkers.

formed fibrillar structures. Removal of the Phe residue resulted in peptide Fmoc-RGD-OH, which could only form nanotubes with lower yields.

Interestingly, the shorter control peptides Fmoc-FG-OH and Fmoc-GF-OH assembled into different products: while Fmoc-FG-OH afforded fibrous networks, Fmoc-GF-OH formed much larger tubes with a diameter of several micrometers. Although the two peptides contain the same amino acid composition, the positioning of aromatic residues has a direct influence on the resulting structures. Sequences that formed hydrogels the fastest were Fmoc-FF-OH and Fmoc-FG-OH.

Rheological analysis revealed that both the numbers of aromatic residues on a given peptide backbone, along with their location, can dramatically affect the mechanical properties of the resulting hydrogel. Fmoc-FF-OH, with its three adjacent aromatic residues, exhibited the stiffest properties, while Fmoc-FG-OH and Fmoc-FRGD-OH, both containing two adjacent aromatic groups, had similar properties to each other. Not surprisingly, Fmoc-RGDF-OH formed the least stiff hydrogel, presumably due to the increased space between its aromatic residues.

Another example of engineering π–π interactions for peptide assembly comes from the collagen model peptides (CMPs). CMPs can self-assemble to form supramolecular fibril structures with collagen-like activity. One benefit is that peptide chains do not require preorganization by covalent linkages.

Maryanoff and coworkers describe a method to enhance the self-assembly properties of this system by introducing Phe and its fluorinated analogue f_5F at the ends of a CMP (Fig. 21.7) [21]. To probe the benefit of aromatic interactions in collagen assembly, the authors synthesized four different CMPs with the sequence $(Gly–Pro–HyP)_n$. Four 32-mer peptides CMPs were compared, each containing a different end group. The assembly of peptide 1a (F/f_5F) gives a bioactive collagen-like material, while peptide 1b (F/F) exhibits similar behavior.

The morphology of the resulting materials was examined by atomic force microscopy (AFM). Peptides 1a and 1b both form fibrous aggregates, while peptide 1c (Phe–Leu) assembles into smaller, more spherical aggregates. Peptide 1d (H/H), however, results in sheets with irregular shaped holes. These results can be rationalized by the fact that peptides 1a and 1b allow favorable end-to-end interactions because of the aromatic residues, although different π–π interactions are involved in 1a and 1b. Peptide 1c, which was meant to test the adequacy of hydrophobic interactions alone (Phe–Leu), failed to give collagen-like products, highlighting the importance of the aromatic residues. Not surprisingly, peptide 1d, which contained no side-chain functionalities at the R_1 and R_2 positions, also fails to assemble into collagen-like material. It remains to be seen how peptide variants with (f_5F, f_5F) or (Cha, Cha) combinations behave in self-assembly. These peptides would help better delineate the role of π–π interactions in collagen formation.

Recent work from the Yang group utilized aromatic residue-promoted peptide assembly in the design of fluorescence probes of enzymes [22]. In this study, authors described their efforts to optimize quenched probes to achieve low background fluorescence while maintaining good analyte recognition properties. In order to achieve this goal, conventional approaches utilize a FRET pair by integrating a fluorescent dye with its quencher. In contrast, Yang and coworkers take advantage of peptide assembly that results in aggregation-caused quenching (Fig. 21.8).

Authors designed compounds of dabcyl-$GF_nG_{(3-n)}$DEVDGK(FITC) as possible self-assembled probes for the enzyme caspase-3 and subsequently tested the effect that the number of phenylalanine (F) moieties had on self-assembly, the background fluorescence (no caspase cleavage), and the recovered fluorescence intensities after enzymatic cleavage. These probes were named as FITC-F_n with n denoting the number of Phe residues as shown above in the peptide sequence. Dynamic light scattering (DLS) was used to measure the critical micelle concentration (CMC) values of FITC-F_0, FITC-F_1, FITC-F_2, and FITC-F_3, which were 508, 181, 140, and 267 μg/ml, respectively. The lower CMC values for peptides containing F compared to that without F indicated that the introduction of an aromatic F could

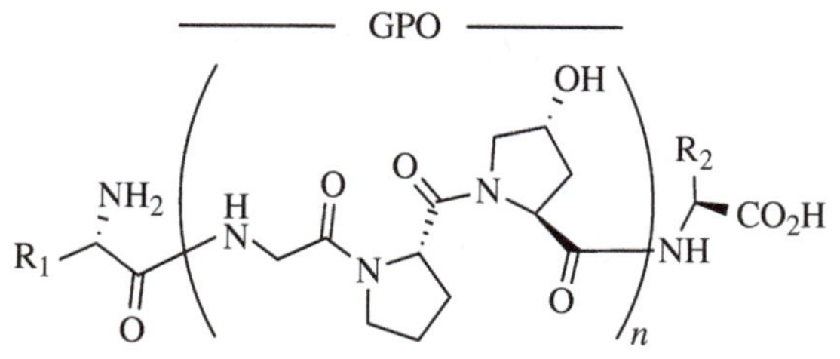

1a: $n = 10$; $R_1 = (C_6F_5)CH_2–$, $R_2 = PhCH_2–$
1b: $n = 10$; R_1, $R_2 = PhCH_2–$
1c: $n = 10$; $R_1 = Me_2CHCH_2–$, $R_2 = PhCH_2–$
1d: $n = 10$; R_1, $R_2 = H$

FIGURE 21.7 Structures of the collagen model peptides reported by Maryanoff and coworkers.

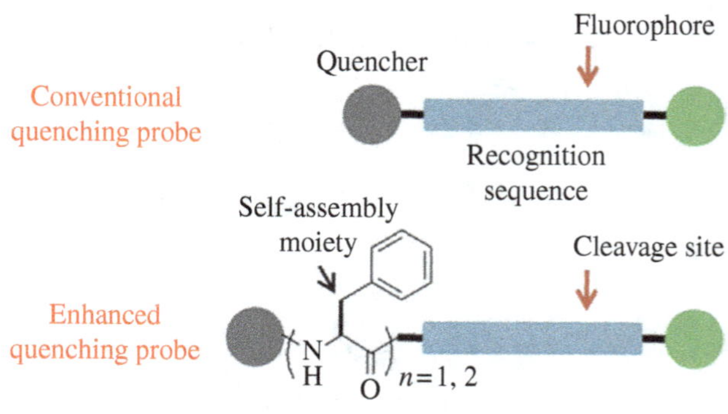

FIGURE 21.8 Design principles for probes with low background fluorescence. Top: conventional quenching probes. Bottom: enhanced quenching probes by peptide assembly. Source: Ren et al. [22]. Reproduced with permission of John Wiley & Sons.

enhance self-assembly propensity of peptides, possibly due to the enhanced hydrophobicity and π–π interactions. As expected, peptide FITC-F_0 exhibited the highest background fluorescence intensity, indicating that its poor self-assembly properties led to insufficient quenching. Interestingly, FITC-F_2 displayed lower background fluorescence in comparison to FITC-F_1 and FITC-F_3. In addition, upon enzymatic cleavage by caspase-3, FITC-F_1 and FITC-F_3 had better recovery of fluorescence than FITC-F_2, which indicates that FITC-F_2 forms tighter and less accessible aggregates than FITC-F_1 and FITC-F_3.

The fact that FITC-F_2 is more aggregation prone than FITC-F_3 suggests that hydrophobicity is not a major factor in the assembly of these peptide probes. Instead, π–π interactions are likely to play important roles in this case. Out of the four probes, FITC-F_1 exhibited the highest fluorescence increase upon enzyme cleavage, and its use as a molecular probe for imaging caspase activity in cells has been demonstrated.

Yang and coworkers further investigated an analogue of FITC-F_1 that incorporates *para*-fluorophenylalanines instead of phenylalanines. Upon self-assembly, the fluorinated analogue displayed lower background fluorescence than FITC-F_1, which makes it an even better probe for cell imaging. This result nicely showcases the use of fluorination as an engineering strategy to optimize self-assembling peptides for biological applications.

21.4 π–π INTERACTIONS IN PEPTIDE THERAPEUTICS

It is important to understand π–π interactions not only because of their relevance to protein folding and assembly but also because of their prevalence in peptide hormones, which serve as a major class of therapeutic agents in medicine. Somatostatin, for example, is a 14-residue peptide hormone involved in interactions with at least five characterized G-protein-coupled receptors, SSTR1-5. The sequence of somatostatin contains several aromatic residues (Fig. 21.9a).

Hirschmann and coworkers proposed the existence of a π–π interaction between Phe6 and Phe11 [24]. Soon after, the same group reported that the two residues form contact in an edge–face rather than face–face geometry [25, 26]. Conversely, van Binst and coworkers reported that they were unable to observe the Phe6–Phe11 aromatic interaction, but instead observed it between Phe6 and Phe7 [27, 28]. The conflicting results originate from the structural instability of somatostatin that gives rise to multiple conformations.

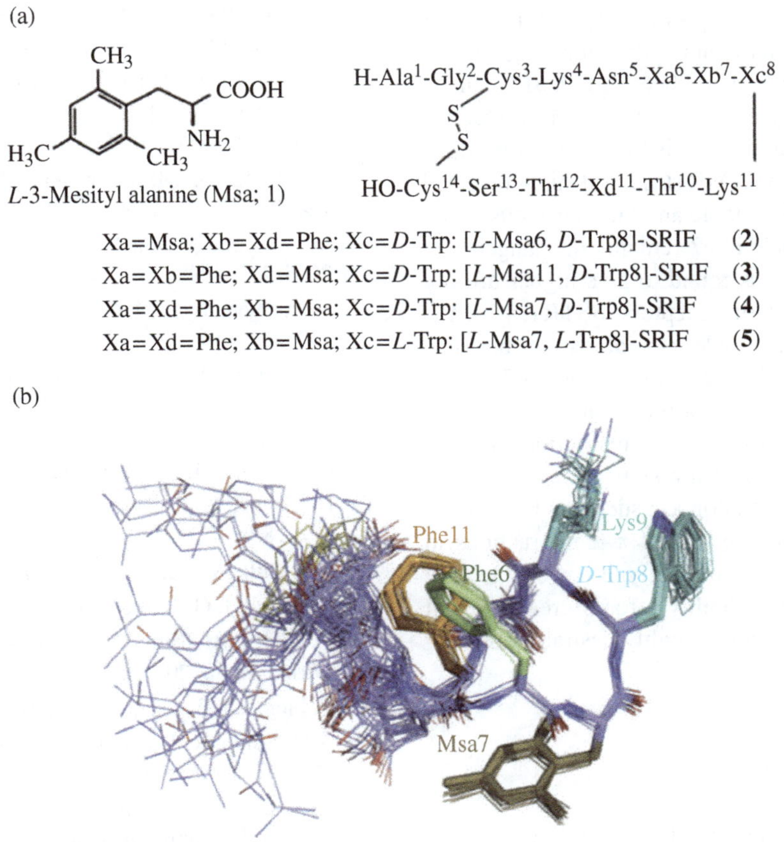

FIGURE 21.9 New somatostatin analogues incorporating Msa residues (a) and the stable structures of the peptide **4** as revealed by NMR spectroscopy (b). Source: Martin-Gago *et al.* [23]. Reproduced with permission of John Wiley & Sons.

The Riera group embarked on a study to substitute key amino acids in somatostatin with unnatural residues in an effort to seek new analogues that are conformationally stabilized by π–π interactions yet still retain a similar structure to the parent compound [23]. Various analogues were prepared by replacing the Phe residues with a more electron-rich amino acid (Msa; Fig. 21.9a) that displays a mesityl group as its side chain. It was postulated that the π–π interactions between Msa and the remaining Phe residues would be stronger than those among the Phe groups of the parent compound.

Two peptides (peptides 2 and 3; Fig. 21.9a) were initially prepared to display Msa instead of Phe at positions 6 and 11, respectively. Additionally, D-Trp was used instead of L-Trp at position 8 in order to increase physiological stability while maintaining biological activity. Based on the presumption that an aromatic interaction could occur either between residues 6 and 11 [24] or between residues 6 and 7 [27], Phe7 was also substituted with Msa (peptide 4). Interestingly, this peptide was found to exhibit outstanding receptor affinity. Consequently, the authors synthesized the same sequence with L-Trp (peptide 5) to study the effects of D-Trp substitution in this compound.

NMR spectroscopy was used to evaluate the structural preferences of the four peptides. Interestingly, peptide 4 was found to exhibit much stabilized and homogeneous structures (Fig. 21.9b). The NMR structures show that, on the top face of the molecule, Phe6 and Phe 11 pack against each other in an edge–face geometry. In contrast, Msa7 resides on the other face of the molecule, presumably stabilizing the somatostatin structure by steric and hydrophobic factors.

The π–π interaction between Phe6 and Phe11 can be potentially strengthened by their mutation to Msa, which is more electron rich and serves as a better "face" for edge–face interactions. Indeed, the NMR structural study revealed that peptides 2 and 3 exhibit stable and homogeneous structures as well, although their Phe7 remains unchanged. On the other hand, peptide 4 was found to give higher affinity toward one of the somatostatin receptors, SSTR1, which is potentially owing to the favorable π–π interaction between Msa7 of the peptide and a Phe present in SSTR1. While this is consistent with the enhanced π–π interactions, the stabilization could also benefit from steric restriction and the greater hydrophobicity of the Msa residue alone. Mutation of Phe6 and Phe 11 to other analogues, such as f_5F and Cha, would help to better define the role of π–π interactions in somatostatin structure formation. Finally, it is worth to note that the enhanced structural stability affords greater serum stability of peptides 2–4, which is highly desirable for peptide therapeutics.

21.5 FINAL COMMENTS

This chapter presents our current understandings of π–π interactions in the context of peptide and proteins. As we stated at the beginning of the chapter, we use the term π–π interactions to specifically refer to the electrostatic interaction between aromatic molecules. The recent advances of synthetic peptide/protein chemistry make it possible to directly probe π–π interactions in protein structures via the use of proper unnatural amino acids. Accumulating data indicate π–π interactions can play an important role in stabilizing protein structures and programming protein dimerization selectivities. Furthermore, guided by the consideration of π–π interactions, much success has been achieved in designing self-assembling peptides for biomedical applications, as well as in optimizing peptide therapeutics. We anticipate that further understanding and engineering π–π interactions will enable an increasing number of applications in chemistry and life sciences.

ACKNOWLEDGMENTS

We gratefully acknowledge the Smith Family Foundation, the National Science Foundation, and the National Institutes of Health for financially supporting the research of the Gao group to gain fundamental understanding of molecular interactions as described in this chapter.

REFERENCES

1. L. M. Salonen, M. Ellermann, F. Diederich, *Angew. Chem. Int. Ed.* 2011, *50*, 4808.
2. S. K. Burley, G. A. Petsko, *Science* 1985, *229*, 23.
3. C. J. Pace, J. Gao, *Acc. Chem. Res.* 2013, *46*, 907.
4. S. M. Butterfield, P. R. Patel, M. L. Waters, *J. Am. Chem. Soc.* 2002, *124*, 9751.
5. C. D. Tatko, M. L. Waters, *J. Am. Chem. Soc.* 2002, *124*, 9372.
6. D. M. Chung, Y. Dou, P. Baldi, J. S. Nowick, *J. Am. Chem. Soc.* 2005, *127*, 9998.
7. M. A. Wouters, P. M. G. Curni, *Proteins Struct. Funct. Bioinf.* 1995, *22*, 119.
8. E. G. Hutchinson, R. B. Sessions, J. M. Thornton, D. N. Woolfson, *Protein Sci.* 1998, *7*, 2287.
9. M. G. Woll, E. B. Hadley, S. Mecozzi, S. H. Gellman, *J. Am. Chem. Soc.* 2006, *128*, 15932.
10. H. Zheng, K. Comeforo, J. Gao, *J. Am. Chem. Soc.* 2009, *131*, 18.
11. R. B. Hill, W. F. DeGrado, *J. Am. Chem. Soc.* 1998, *120*, 1138.
12. H. Zheng, J. Gao, *Angew. Chem. Int. Ed.* 2010, *49*, 8635.
13. C. J. Pace, H. Zheng, R. Mylvaganam, D. Kim, J. Gao, *Angew. Chem. Int. Ed.* 2012, *51*, 103.
14. G. Fichman, E. Gazit, *Acta Biomater.* 2014, *10*, 1671.
15. A.M. Smith, R. J. Williams, C. Tang, P. Coppo, R. F. Collins, M. L. Turner, A. Saiani, R. V. Ulijn, *Adv. Mater.* 2008, *20*, 37.
16. S. Sutton, N. L. Campbell, A. I. Cooper, M. Kirkland, W. J. Frith, D. J. Adams, *Langmuir* 2009, *25*, 10285.
17. D. M. Ryan, S. B. Anderson, T. Senguen, R. E. Youngman, B. L. Nilsson, *Soft Matter* 2010, *6*, 475.

18. V. Jayawarna, M. Ali, T. Jowitt, A. F. Miller, A. Saiani, J. E. Gough, R. V. Ulijn, *Adv. Mater.* 2006, *18*, 611.
19. A. Mahler, M. Reches, M. Rechter, S. Cohen, E. Gazit, *Adv. Mater.* 2006, *18*, 1365.
20. R. Orbach, L. Adler-Abramovich, S. Zigerson, I. Mironi-Harpaz, D. Seliktar, E. Gazit, *Biomacromolecules* 2009, *10*, 2626.
21. M. A. Cejas, M. A. Kinney, C. Chen, J. G. Vinter, H. R. Almond, Jr., K. M. Balss, C. A. Maryanoff, U. Schmidt, M. Breslav, A. Mahan, E. Lacy, B. E. Maryanoff, *Proc. Natl. Acad. Sci. U. S. A.* 2008, *105*, 8513.
22. C. Ren, H. Wang, D. Mao, X. Zhang, Q. Fengzhao, Y. Shi, D. Ding, D. Kong, L. Wang, Z. Yang, *Angew. Chem. Int. Ed.* 2015, *54*, 1.
23. P. Martin-Gago, M. Gomez-Caminals, R. Ramon, X. Verdaguer, P. Martin-Malpartida, E. Aragon, J. Fernandez-Carneado, B. Ponsati, P. Lopez-Ruiz, M. A. Cortes, B. Colas, M. J. Macias, A. Riera, *Angew. Chem. Int. Ed.* 2012, *51*, 1820.
24. D. F. Veber, F. W. Holly, W. J. Paleveda, R. F. Nutt, S. J. Bergstrand, M. Torchiana, M. S. Glitzer, R. Saperstein, R. Hirschmann, *Proc. Natl. Acad. Sci. U.S.A.* 1978, *75*, 2636.
25. B. H. Arison, R. Hirschmann, W. J. Paleveda, S. F. Brady, D. F. Veber, *Biochem. Biophys. Res. Commun.* 1981, *100*, 1148.
26. S. A. Neelamkavil, B. Arison, E. Birzin, J. J. Feng, K. H. Chen, A. Lin, F. C. Cheng, L. Taylor, E. R. Thornton, A. B. Smith 3rd, R. Hirschmann, *J. Med. Chem.* 2005, *48*, 4025.
27. A. Jans, K. Hallenga, G. Van Binst, A. Michel, A. Scarso, J. Zanen, *Biochim. Biophys. Acta Protein Struct. Mol. Enzymol.* 1985, *827*, 447.
28. E. M. Van den Berg, A. W. Jans, G. Van Binst, *Biopolymers* 1986, *25*, 1895.

22

NON-COVALENT INTERACTIONS FOR THE PREPARATION OF PSEUDOPEPTIDIC SYNTHETIC COMPOUNDS AND MATERIALS

JORDI SOLÀ AND IGNACIO ALFONSO

Department of Biological Chemistry and Molecular Modeling, Institute of Advanced Chemistry of Catalonia, (IQAC-CSIC), Barcelona, Spain

22.1 INTRODUCTION

Pseudopeptidic molecules [1] are synthetic structures generally built from natural amino acid-derived and nonnatural moieties (either nonnatural amino acids or fully artificial fragments; Scheme 22.1). The combination of both substructures allows a good control of the molecular properties regarding the possibility for establishing non-covalent interactions. Thus, the presence of amino acid-based side chains allows the implementation of functional groups with acid–base character, aromatic rings, polar and nonpolar groups, H-bonding donors and acceptors, etc. Moreover, the peptidic bond itself can work as H-bonding donor and acceptor. Overall, the peptide moiety of pseudopeptides is a fantastic source of potential non-covalent interactions. On the other hand, the nonnatural (or nonpeptidic) moieties can be selected to introduce additional non-covalent interaction sites, as well as to increase rigidity, hydrophobicity, or stability. This last property is especially important if the final pseudopeptidic molecules are intended to be used in biological applications. For instance, the nonpeptidic/nonnatural fragments of the pseudopeptides increase their resistance to hydrolysis by proteases.

Regarding the potential applications of this family of compounds, many uses can arise from their structural and functional diversity. One obvious application should be as mimetic of peptide sequences for biological uses. The molecular recognition properties of the pseudopeptides have supported their applications as molecular receptors in supramolecular chemistry or as catalysts in synthetic processes. Besides, pseudopeptides have very interesting self-assembling properties that allow their use in the design of soft materials for tissue engineering or other industrial (cosmetic, food, etc.) applications.

The goal of this chapter is to serve as an illustrative overview of the importance of non-covalent interactions for the preparation of structurally defined pseudopeptidic compounds and materials. This chapter does not pretend to deeply review the state of the art within the field, because this thorough revision will be much longer than the intended one for this chapter. Representative examples were selected, where the non-covalent interactions play a significant role for the synthesis of the pseudopeptide molecule or material, as well as for the integrity of the defined structure (with a given conformation).

22.2 NON-COVALENT INTERACTIONS IN THE PREPARATION OF CYCLIC PSEUDOPEPTIDES

As stated in the introduction, pseudopeptidic molecules are privileged compounds for establishing different non-covalent interactions. Chemists have taken advantage of these interactions for preparing challenging structures that would be very difficult to obtain in the absence of the non-covalent contacts. This is especially important for the preparation of macrocyclic and macrobicyclic pseudopeptides. In the following sections, representative examples of such cyclization processes guided by non-covalent interactions are discussed.

Non-Covalent Interactions in the Synthesis and Design of New Compounds, First Edition.
Edited by Abel M. Maharramov, Kamran T. Mahmudov, Maximilian N. Kopylovich, and Armando J. L. Pombeiro.

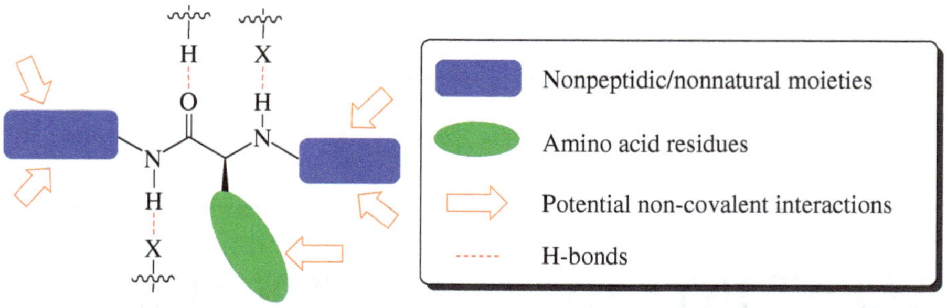

SCHEME 22.1 General structure of a pseudopeptidic molecule.

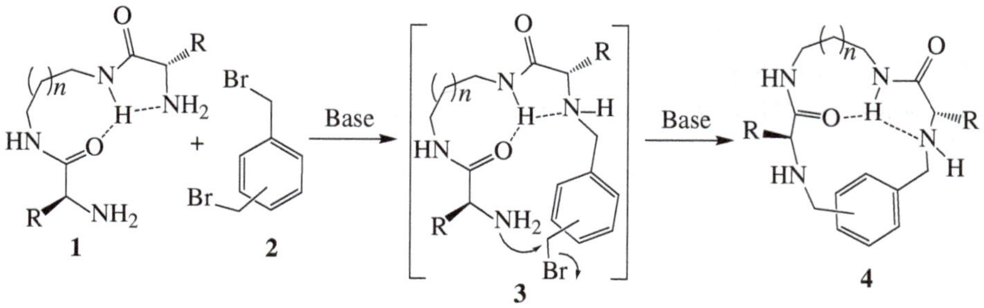

SCHEME 22.2 Efficient macrocyclization of a pseudopeptide through a U-turn preorganized intermediate.

22.2.1 Pseudopeptidic Macrocycles

Pseudopeptidic macrocycles are interesting scaffolds for several reasons [2]. The cyclic molecule constricts the conformational space available for the residues within the pseudopeptide, which in turn produces a much better defined three-dimensional structure of the systems. This specific characteristic makes the macrocyclic species very interesting for molecular recognition processes and for biomimetic uses [3]. Unfortunately, very often, the macrocyclization step is hampered by low yields and the formation of linear and cyclic oligomeric by-products [4]. In order to avoid that, the suitable preorganization of the open-chain linear precursor is necessary. This preorganization can be obtained by the correct design and implementation of non-covalent contacts that approach both ends of the linear precursor or intermediate during the cyclization process. An illustrating example is shown in Scheme 22.2 [5]. The minimalistic linear pseudopeptide precursor **1** is formed by the coupling of two amino acids with a simple aliphatic diamine. NMR and molecular modeling showed that these peptides are able to form different intramolecular H-bonds between amide and amine functional groups.

The reaction of these pseudopeptidic diamines with a dielectrophile such as **2** in acetonitrile and using potassium carbonate as a base rendered the corresponding macrocycle (**4**) in very good isolated yields (52–69%) and high selectivity considering a macrocyclization process. Besides, the reaction can be performed with different amino acids (R) and different lengths of the spacers (n). The key open-chain intermediate (**3**) formed during the reaction is preorganized in a folded U-turn conformation by the combined action of solvophobic interactions and intramolecular H-bonds.

Other clever approaches for the preorganization of linear peptides for macrocyclization reactions have been reported in the literature [6]. One interesting example has been described by Yudin and coworkers [7]. Their approach is based on simple electrostatic rational: they propose that a linear peptide should be naturally folded in organic solvents, since both ends would have complementary charges due to the zwitterionic form of the C- and N-terminus. Taking advantage of that, they used a multicomponent reaction to cyclize linear peptides (Scheme 22.3), the well-known Ugi reaction that comprises a carboxylic acid, an amine, an aldehyde, and an isocyanide.

The key for the success of the process is the participation of the amphoteric amino aldehyde (**5**) that, upon iminium formation with the N-terminus of the linear peptide, renders intermediate **7** still zwitterionic and thus folded. The approach of the isocyanide must be also dictated by coulombic attractive forces in a well-defined structure. The transient intermediate **8** evolves rapidly by an intramolecular attack of the aziridine nitrogen and the concomitant rearrangement to the final macrocycle **9**. An important consequence of the structurally well-defined intermediates is the complete stereoselectivity of the process, which is very uncommon for the Ugi reaction. A further support for the importance of the non-covalent interactions for the process is the nonobvious election of the reaction solvent. Fluorinated alcohols (mainly trifluoroethanol) are the best

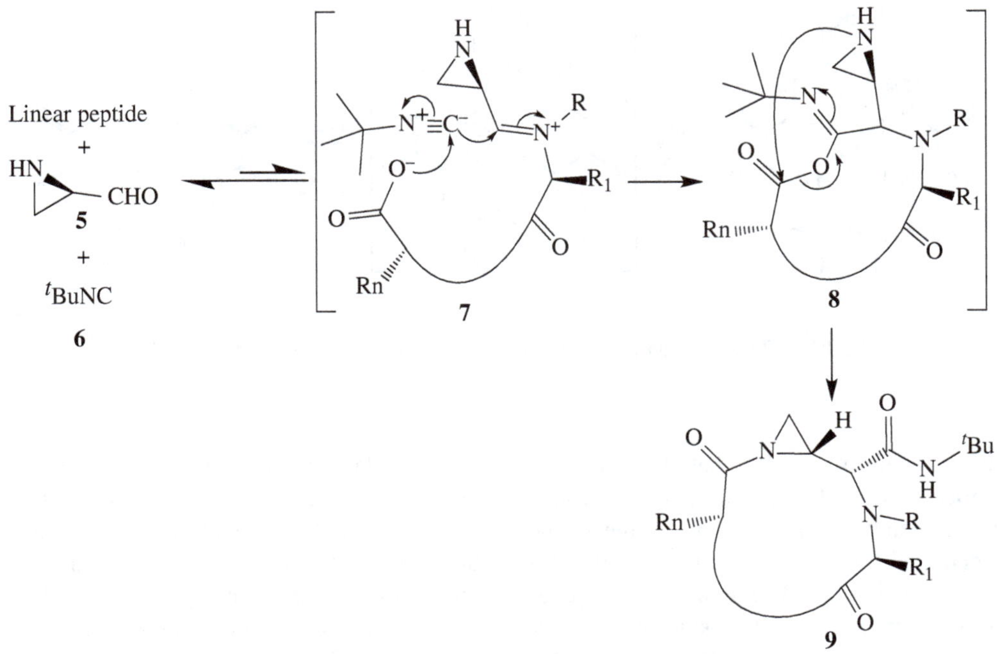

SCHEME 22.3 Macrocyclization of a linear peptide favored by polar electrostatic interactions.

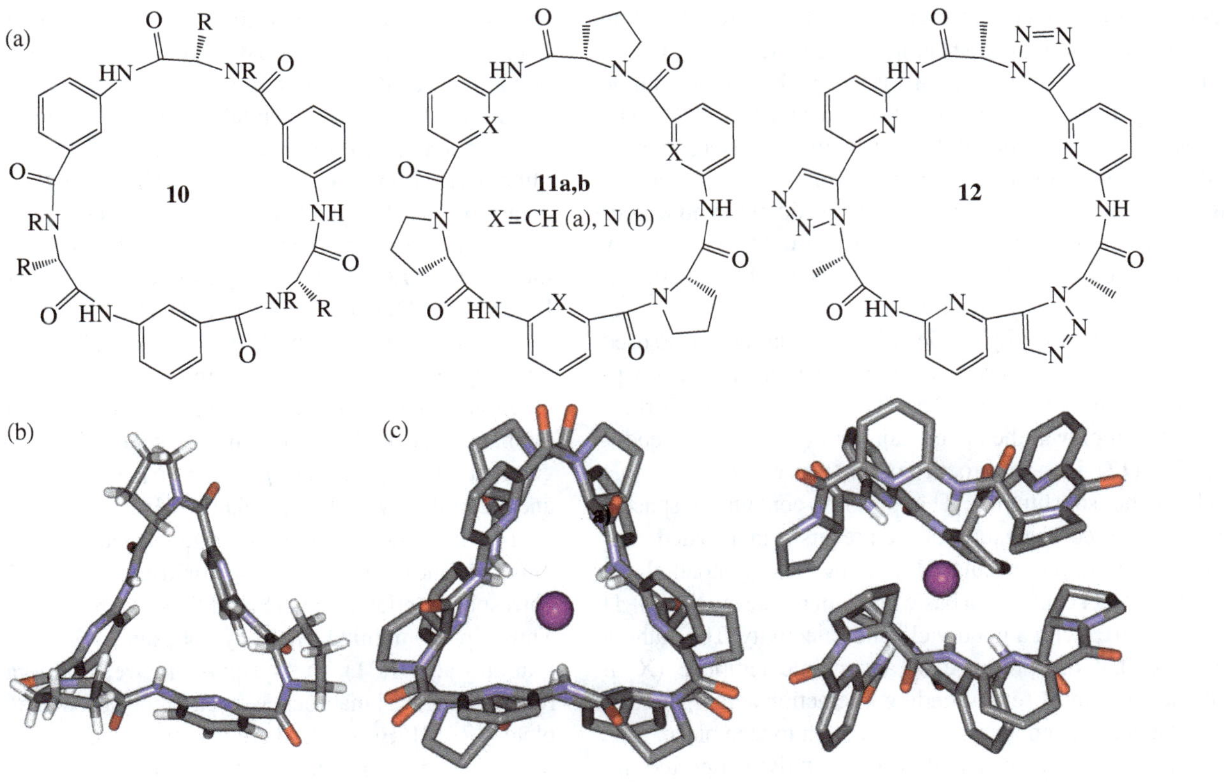

SCHEME 22.4 (a) Macrocyclic pseudopeptides described by Kubik *et al.* [10–12]. (b) Crystal structure of macrocycle **11b** (X = N). (c) Upper and side views of the crystal structure of the iodide (purple sphere) complex of **11b** (nonpolar H-atoms have been omitted for clarity).

media because they show a perfect combination of acidity (for the iminium protonation) and low H-bonding competition. This methodology has proved to be highly efficient and modular [8, 9].

Other macrocyclic pseudopeptides have been synthesized by combining natural amino acids with nonnatural moieties. For instance, the cyclic pseudopeptides developed by Kubik and coworkers [10–12] represent a very good example of how the molecular design has evolved for the optimization of the properties in the molecular recognition of different guests (Scheme 22.4) even in a very competitive medium, like aqueous solution.

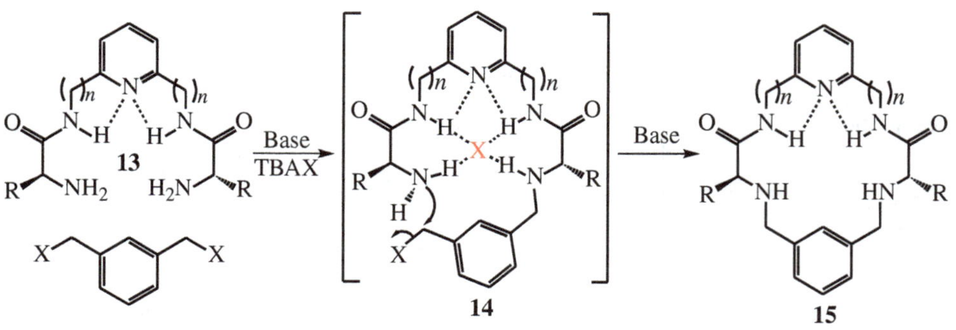

SCHEME 22.5 Anion-promoted [1 + 1] macrocyclization of pseudopeptides.

Starting from a conformationally based design of alternating 3-aminobenzoic acid and natural *N*-methylated amino acids (**10**), the better performance of proline as the amino acid was identified [10]. Then, the introduction of a nitrogen atom in the aromatic rings (**11b**) improved the recognition abilities due to the preorganized disposition of the polar amide NH atoms, as observed in its solid-state (Scheme 22.4b) and solution conformation. Binding studies suggested the formation of supramolecular complexes with a 2 : 1 macrocycle/anion stoichiometry, also confirmed by X-ray diffraction of the corresponding iodide complex (Scheme 22.4c) [11].

Later on, the isosteric substitution of the tertiary amide bonds by triazole rings has been developed [12]. The corresponding pseudopeptide also interacts very efficiently with anions in competitive solvent mixtures, with a somehow higher efficiency. In all the cases, the efficient synthesis of the apparently strained macrocyclic ring might be favored by a combination of conformational effects and intramolecular H-bonds in the key cyclization step.

Anion recognition by amide H-bonding has been also used for promoting the macrocyclization reaction in pseudopeptidic molecules (Scheme 22.5). Thus, Luis and coworkers [13] have employed the open-chain family of model pseudopeptides (**13**) in a macrocyclization process by a double nucleophilic substitution. The pyridine-containing spacers help on the preorganization of the precursor in a folded conformation by intramolecular H-bonding. The monoalkylated intermediate **14** cyclized in basic medium rendering the [1 + 1] macrocycle **15** with a good yield and selectivity. The authors proposed that the presence of an anionic template (X in Scheme 22.5) must set H-bonding interactions preorganizing **14** in the correct geometry for the reaction to take place.

A thorough experimental kinetic study supported by molecular modeling calculations clearly showed that the anion was able to stabilize the corresponding transition states (from **14** to **15**), thus producing an important kinetic anion template effect. In a later study, the same authors extended the work to other pseudopeptides with a more flexible aliphatic spacer and different side chains (R) derived from both natural and nonnatural amino acids [14]. Overall, the kinetic anion templation seemed to be a general effect in the macrocyclization reactions.

A similar rationale has been followed to prepare more elaborated larger [2+2] pseudopeptidic macrocycles (Scheme 22.6) [15]. To this aim, a reductive amination reaction was used, for which the imine macrocycle is the key intermediate. The advantage of imine bond is its reversibility, allowing self-correction and the final formation of the most stable oligomeric macrocycle. This dynamic covalent approach is the base of dynamic combinatorial libraries that will be discussed more deeply in Section 22.3. Scheme 22.6 shows how pseudopeptidic diamines **16** react with the aromatic dialdehyde (terephthaldehyde, **17**), and the reaction being monitored by NMR and ESI-MS.

In the absence of the suitable organic anionic template, the reaction led to a complex mixtures of open-chain and oligomeric imines. However, the addition of the aromatic dicarboxylate **18** stabilized the macrocyclic [2+2] imine **19** by amide–carboxylate H-bonds. This supramolecular intermediate **19** was fully characterized by NMR (including ROESY and DOSY), ESI-MS, and CD.

The *in situ* reduction of **19** rendered the corresponding macrocyclic pseudopeptide as the isolable amine **20**. The reaction can be carried out with different lengths of the aliphatic spacers (n), different side chains (R), or different geometrical dispositions (*para* or *meta*) of the dialdehyde and, accordingly, of the template [16].

The introduction of more rigid spacers in the central moiety of the open-chain precursor affects the stability of the corresponding intermediate and, thus, the reaction outcome. Thus, by combining (*R*,*R*)-cyclohexane-1,2-diamine with L-amino acids (**21**), the systems are well preorganized to render the [2+2] macrocycle even in the absence of the template [17, 18]. However, there is a *match/mismatch* effect of the stereogenic centers of the pseudopeptides, since the (*R*,*R*)-diamine/D-amino acid combination (**22**) proved to be strongly disfavored for the cyclization. Very interestingly, the anion-templation mechanism was also efficient in the *mismatch* combination of stereocenters, promoting the cyclization of the molecules [19]. This specific example represents a very remarkable example of how non-covalent interactions can guide the reactivity of the process, improving the efficiency and selectivity on the formation of a given macrocycle.

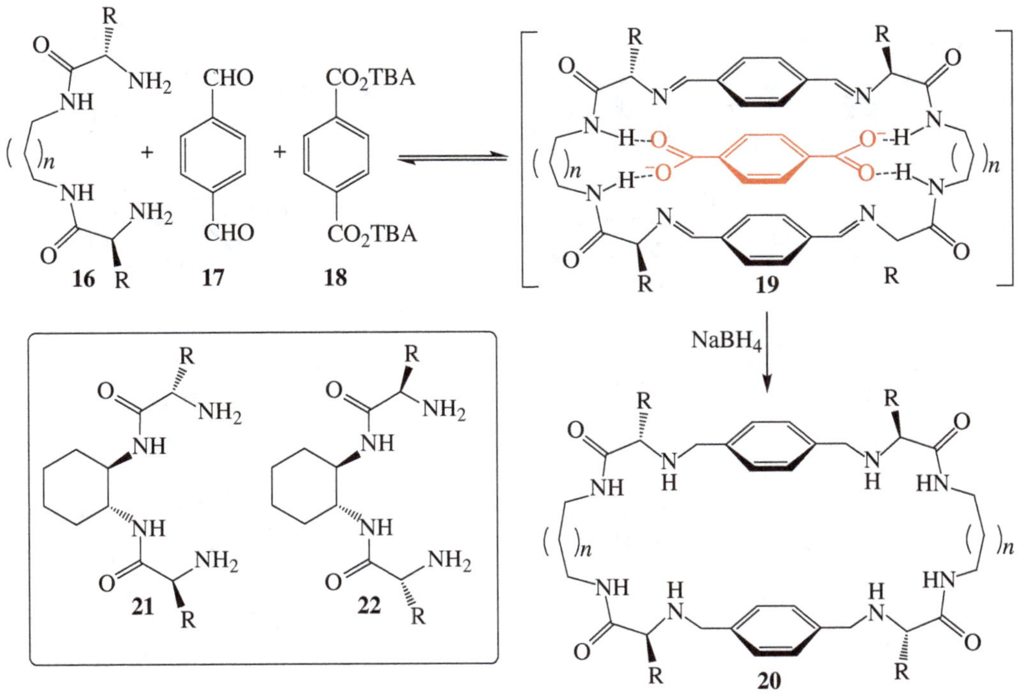

SCHEME 22.6 Anion-templated [2+2] macrocyclization of pseudopeptides.

22.2.2 Pseudopeptidic Cages

Macrobicyclic pseudopeptides, also commonly named as cages, are important molecules as constrained scaffolds and receptors. However, the formation of the three-dimensional architecture is even more challenging, since the problems of the macrocyclization reaction are multiplied upon the formation of a macrobicycle. However, similar approaches have been applied for the improvement of the yields and selectivity of their synthesis. Some illustrative examples from the recent literature have been selected here. One obvious approach to obtain a cage topology should be starting from a tripodal pseudopeptide formed by a symmetric molecule with three arms (**23**) (Scheme 22.7). The reaction of these triamines with a trifunctional electrophile **24** would render the [1+1] cage if a suitable preorganization of the precursor **23** and the reaction intermediates **25** is achieved [20].

When the tripodal spacer is the tris(ethylenetriamine) (Tren, **23a**), the process led to the expected cage (**26**) in good isolated yields and excellent selectivity. However, when the spacer is a more rigid trisubstituted aromatic ring (**23b,c**), the reaction rendered a complex mixture of different oligomers, as observed by UPLC coupled with ESI-MS. These mixtures were impossible to separate. The success of the reaction is due to the anion assistance in the preorganization of the reaction intermediates for closing the cage by anion–amide H-bonding (Scheme 22.7).

Actually, some of the obtained cages showed to be excellent receptors for inorganic anions with an extraordinary selectivity for chloride over other halides (more than two orders of magnitude in the binding constants) even in competitive media [21]. Also in this case the synthetic procedure is highly modular, allowing different substituents in both partners, the tripodal nucleophile (R_1) and the tripodal electrophile (R_2).

An anion template approach to larger cages based on a reductive amination reaction (Scheme 22.8) was also reported [22], in a clear parallelism with the previously commented procedure for the corresponding macrocycles. To that, three equivalents of the open-chain pseudopeptides were reacted with two equivalents of the symmetric aromatic trialdehyde **27** to render the [3+2] macrobicyclic hexaimine (**28**) that was reduced to the final amine **29** in good yields. Also in this case, the preorganization of the precursors is crucial for the success of the process: the same structural factors and anion-templated rules can be applied.

The key intermediate was fully characterized with different experimental techniques that allow proposing the non-covalent complex shown in Scheme 22.8, where the template–pseudopeptide interactions stabilize the cage structure and promotes the selective macrobicyclization. The procedure is rather general and the final cages showed to be excellent receptors for *N*-protected dipeptides in competitive media [23].

Other pseudopeptidic cages have been reported by different authors [24, 25], but among all the one from Kubik's group is specially remarkable [26]. Inspired by the dimeric structure of the supramolecular iodide complex of macrocycle **11b** (Scheme 22.4), the authors connected two units of the cycle with three equivalent arms using a triazole linker. The obtained pseudopeptidic cage **30** (Scheme 22.9) showed

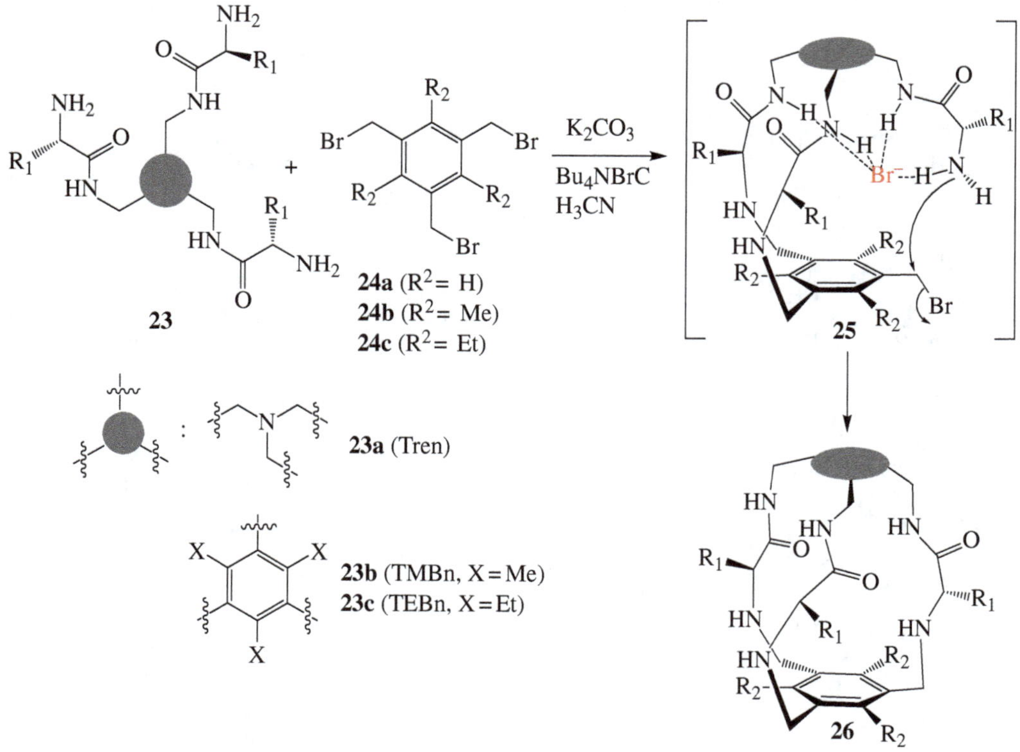

SCHEME 22.7 Synthesis of macrobicyclic cages, facilitated by H-bonding with the anion.

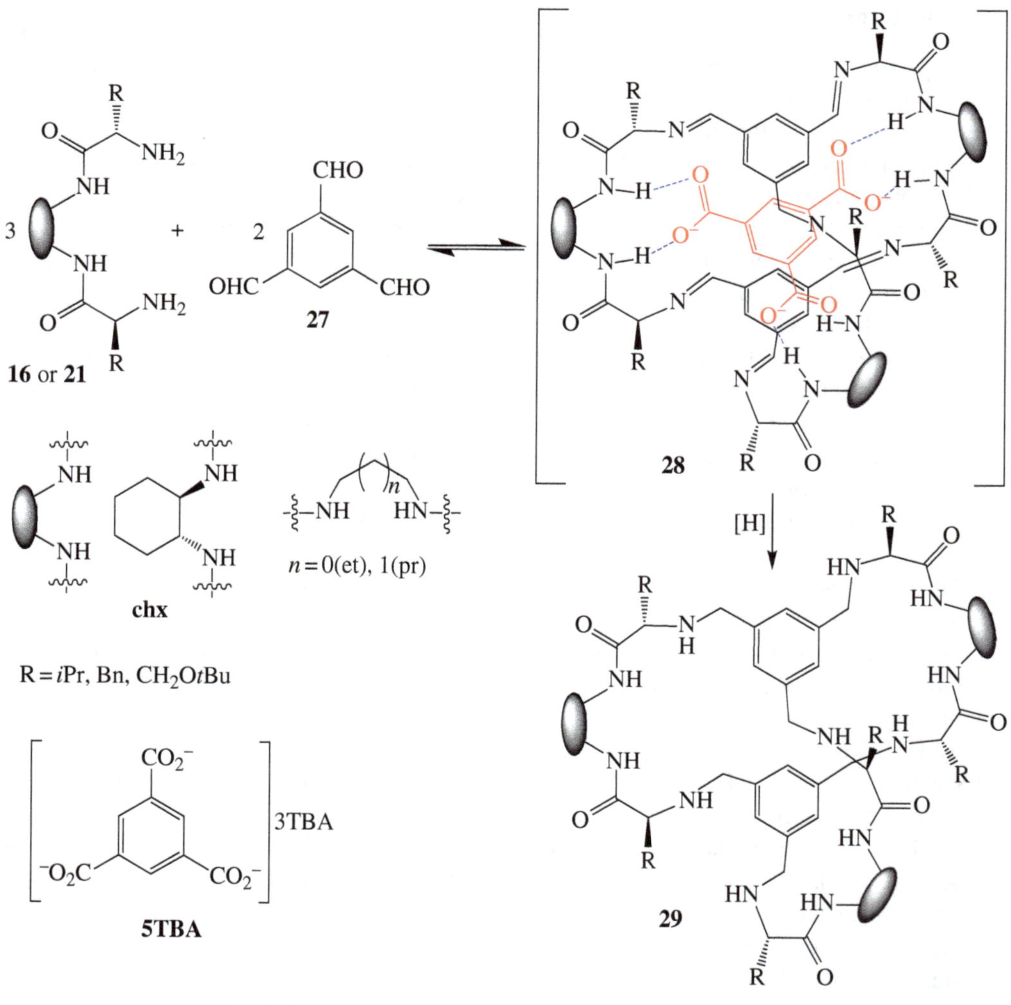

SCHEME 22.8 Anion-templated [3 + 2] macrobicyclization of pseudopeptidic cages.

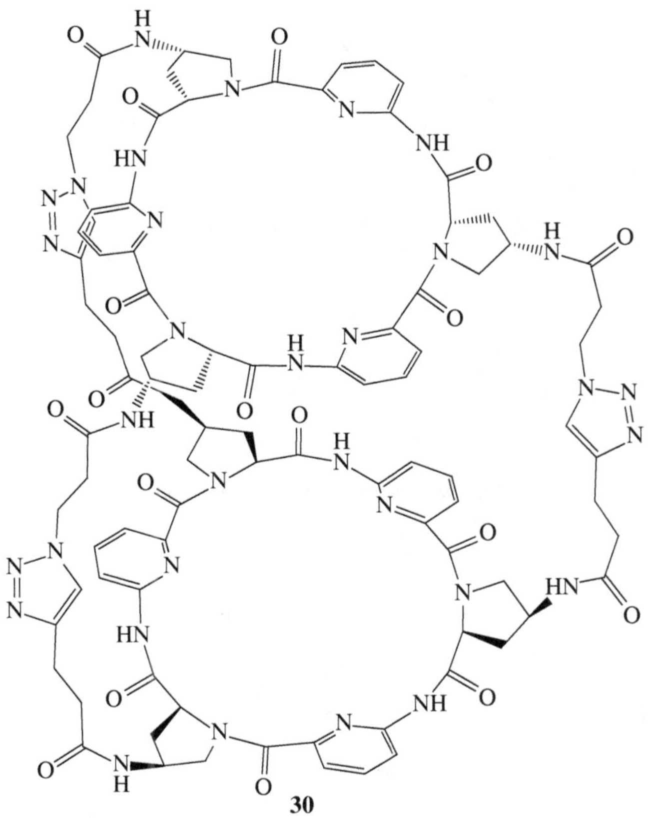

30

SCHEME 22.9 Macrobicyclic cage described by Kubik and coworkers.

an extraordinary affinity for the sulfate anion in aqueous methanol (by ITC and NMR). The molecular models of the supramolecular complex revealed a very efficient H-bonding pattern within the inner cavity of the cage for the sulfate binding. Moreover, the hydrophobic contacts between the two macrocyclic pseudopeptides additionally helped for the closing of a compact complex, tightly wrapping the sulfate anion.

22.3 PSEUDOPEPTIDES IN DYNAMIC COVALENT CHEMISTRY

The improvement of the preparation of pseudopeptidic molecules by taking advantage of the non-covalent interactions is specially expressed in dynamic covalent chemistry studies [27]. This concept is based on the connection of building blocks (BBs) through dynamic bonds that produce reversible species in solution. In this way, the system can self-correct and be adapted to the reaction conditions [28]. The mixture of species generated by the combination of the BBs is commonly named dynamic combinatorial library (DCL) to highlight its combinatorial and adaptive nature. The prevalence of one member of the library over the others in equilibrium will be dictated by the relative energy of the competing species. More specifically, the whole system will evolve to the composition that produces the lowest energy of the whole system. Therefore, the non-covalent interactions established within the library will have an impact on the stability of the whole system and accordingly on its final composition (since the system is under thermodynamic control). These non-covalent interactions can be established between the BBs, leading to the self-synthesis of the most favored member, between the different library species allowing the preparation of self-assembled materials, or between the members of the library and an external species (template) leading to the amplification and identification of the best receptor for that template.

This methodology has found applications in the field of pseudopeptides for several reasons. The relatively simple synthesis of amino acid-based BBs and the many potential non-covalent interactions are obvious factors. Besides, and more importantly, pseudopeptides are excellent molecules to establish bioinspired systems that could render important biologically related conclusions in stability studies, molecular recognition processes, and evolutionary issues.

One of the pioneering works in the field was reported by Sanders and coworkers [29]. The authors prepared a simple BB from proline by attaching hydrazine at the carboxylic end and an aromatic acyl at the amino-terminus **31** (Scheme 22.10). The aromatic ring had a dimethyl acetal group in *meta* position, which can be considered as a masked aldehyde. In acidic medium, the acetal is readily hydrolyzed, and the generated aldehyde quickly reacts with the acylhydrazine end to render a hydrazone bond. This process was reversible at acidic conditions, and the system produced a mixture (library) of dynamic pseudopeptidic macrocycles (Scheme 22.10). The addition of organic cations like a quaternary ammonium salt induced the rearrangement of the species in solution by altering the composition of the mixture. The macrocyclic ring that interacted more efficiently with the added template was highly amplified, being in this case the corresponding trimer (**32**). The increase of the pH switched off the dynamic process, allowing the isolation and characterization of the amplified species.

The same group reported a more elaborated system by the introduction of a Pro–Phe dipeptide and moving the acetal group to the *para* position of the aromatic ring in **33** (Scheme 22.11). In this case the acidic medium rendered a more complex dynamic library of pseudopeptidic hydrazones. This DCL responded to several cations through the amplification of different macrocyclic sizes. Thus, Na^+ and specially Li^+ led to the almost exclusive formation of the corresponding macrocyclic trimer **34** [30]. The observed amplification was due to the cation coordination within the macrocyclic pseudopeptide as proposed by the observation of key changes in the FTIR, ^1H, and ^{13}C NMR spectra of **34** in the presence of LiI salt. The binding was additionally supported by ^7Li NMR and ITC experiments [31]. The authors proposed the combined action of intramolecular H-bonding and metal coordination through the carbonyls and the aromatic rings of the pseudopeptide. Interestingly, the corresponding BB with a *meta* disposition in the aromatic

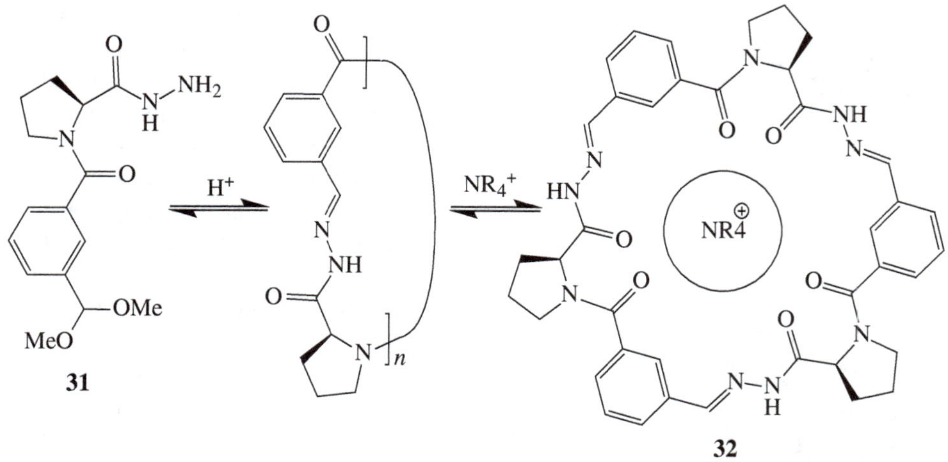

SCHEME 22.10 Amplification of macrocyclic trimer from a dynamic library of pseudopeptidic hydrazones.

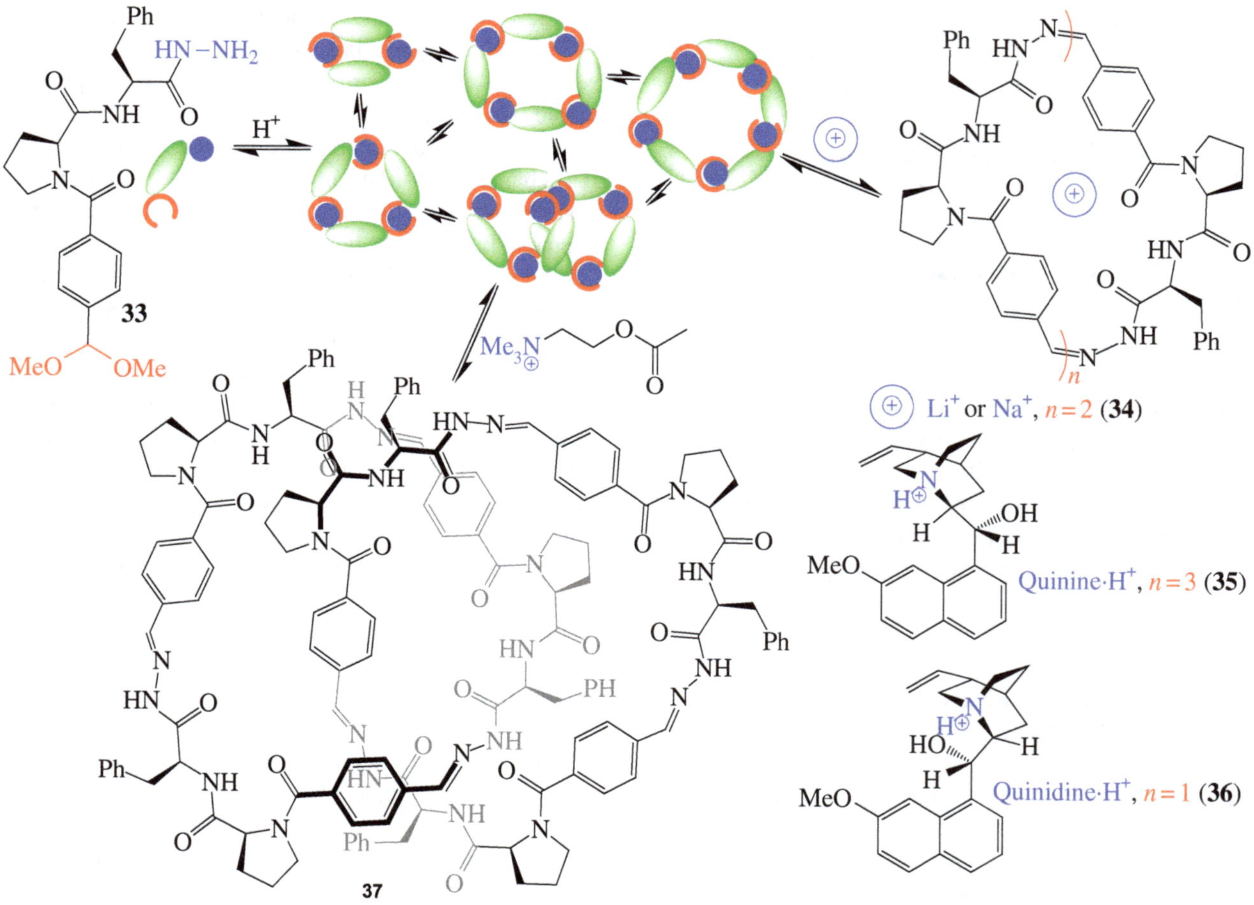

SCHEME 22.11 Different cationic templates amplified different species from a dynamic library of pseudopeptidic hydrazones.

backbone mainly formed the corresponding dimer in the presence of the same metal salts, underscoring the importance of the structure of the BBs on the final library composition.

The versatility of the methodology was exemplified when different organic cationic templates were added to the same library (Scheme 22.11). Thus, the addition of diastereomeric alkaloids amplified completely different species from the mixture: quinine promoted the tetramer **35**, while quinidine led to the formation of the dimer (**36**) [32].

Hence, the dynamic covalent chemistry approach is an excellent method for the discovery of unexpected species able to interact with a given target substrate. Thus, even the members initially absent or present in very low concentration

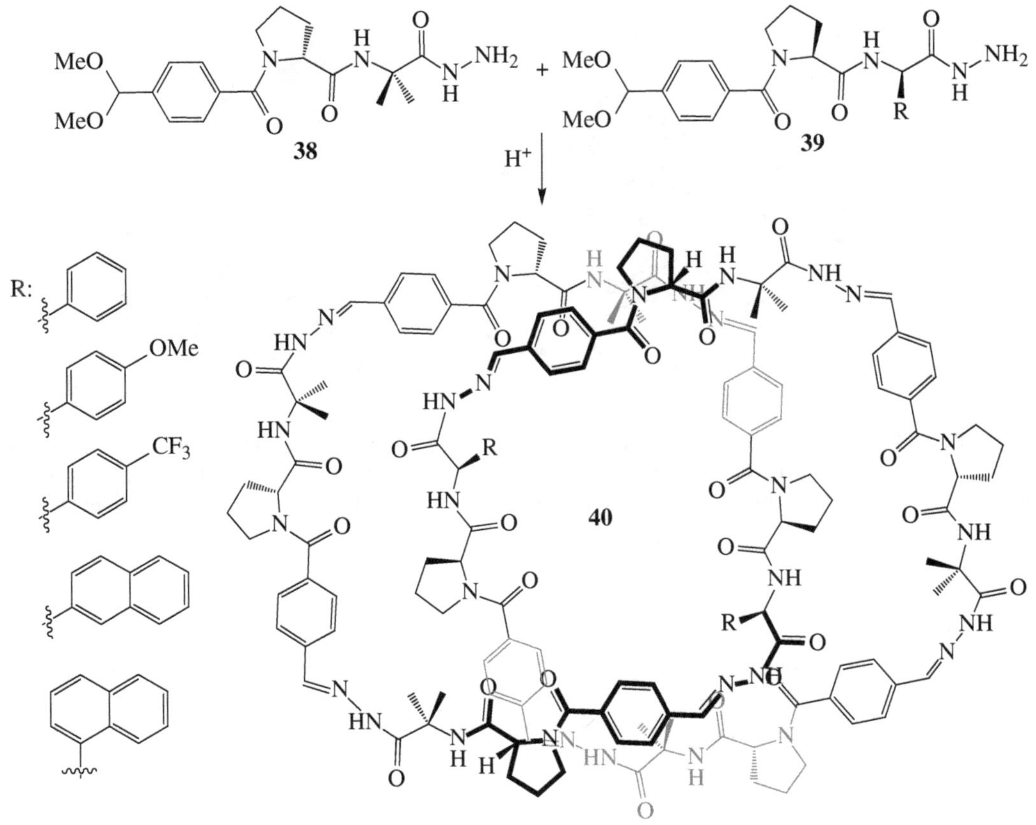

SCHEME 22.12 Formation of catenated tetramers from a DCL of pseudopeptidic hydrazones.

in the DCL can be amplified and identified upon the addition of a template. For these reasons, the DCLs have been also called virtual combinatorial libraries [33].

A very remarkable example was described by Sanders and Otto [34] also starting from the same BB **33** (Scheme 22.11). They added acetylcholine, a biologically relevant molecule, and observed the amplification of a new species with a much higher molecular weight, corresponding to a cyclic hexamer. The new species was obtained with a high yield (67%) in a template-induced preparative-scale synthesis. After the suitable isolation and structural characterization, they concluded that the hexamer was actually a [2]catenane formed by the threading of two macrocyclic trimers (**37**). Thermodynamic studies showed that this catenane was an extraordinarily efficient receptor for acetylcholine, with an affinity in the nanomolar range. Moreover, NMR experiments confirmed that the acetylcholine was deeply included within the interlocked structure. The results from this pioneering work demonstrated the power of the DCC procedure and that catenanes can be spontaneously and efficiently formed under certain experimental conditions.

Another remarkable example of interlocked pseudopeptides from a dynamic covalent mixture of hydrazones has been reported by Gagné and coworkers [35]. They used a combination of pseudopeptidic BBs (**38** and **39** in Scheme 22.12) with a defined stereochemistry of the corresponding chiral centers, and, after equilibration at acidic conditions, they observed the spontaneous main formation of octamers containing a combination of given BBs ($\mathbf{38}_6$–$\mathbf{39}_2$). These octamers were actually [2]catenanes (**40**) formed by the threading of two identical macrocyclic tetramers, as confirmed by NMR in solution and X-ray diffraction of crystal structures. The data suggested that the R residue in one of the BB was very important for the formation of the interlocked pseudopeptides. A further deep structural study of a large variety of examples [36] allowed the identification of the non-covalent interactions responsible for the formation of the catenanes (Fig. 22.1). The authors were able to obtain crystal structures of several examples. They observed many H-bonding contacts within and between catenated rings and, more importantly, several π–π and CH–π interactions that are crucial for the formation of the observed species. These important contacts were retained in solution as confirmed by NMR studies.

The same authors studied the stereoselective amplification of pseudopeptidic receptors for different nucleosides. A DCL formed by the racemic BB **38** rendered a mixture of cyclic oligomers of different sizes [37] and bearing all the possible stereochemical combinations of chiral centers. The addition of (−)-adenosine induced the amplification of the corresponding dimer with a preferred configuration. The HPLC analysis coupled with a laser polarimeter detector and complementary studies with isotopically labeled

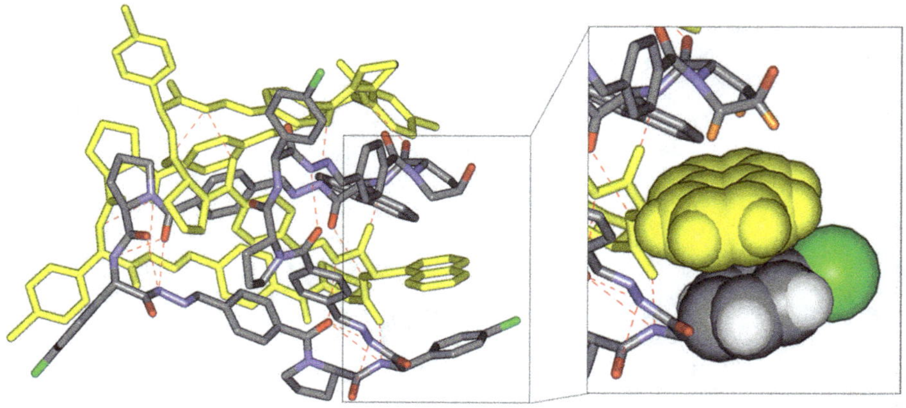

FIGURE 22.1 Crystal structure of one catenane from Gagné's group [36]. For a clearer representation, one of the symmetrically equivalent rings has been colored in yellow, and the H-atoms have been omitted. The hydrogen bonding interactions are highlighted in red dotted lines. The zoomed region on the right shows the important interactions between the catenated rings, with a π–π and a CH–π contacts that stabilize the interlocked structure.

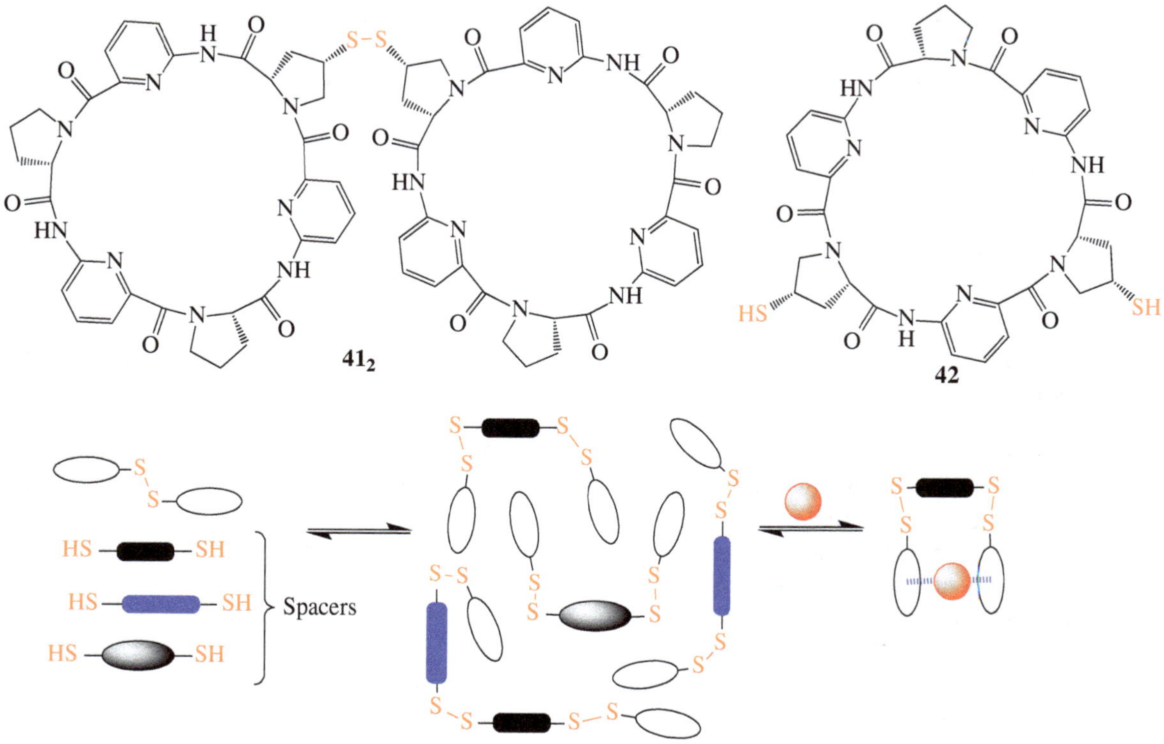

SCHEME 22.13 Amplification of pseudopeptidic receptors for anions (spheres) from a DCL of disulfides of pseudopeptidic macrocycles.

pseudoenantiomers confirmed that the amplified dimer had (*S*,*S*) configuration [38]. The versatility of the DCC procedure was further illustrated by adding other templates to the same racemic DCL. Thus, the corresponding deracemization of the mixture amplified two different macrocyclic pseudopeptides: the (*S*,*S*)-dimer from (−)-cytidine while the (*R*,*R*,*R*,*R*)-tetramer from (−)-2-thiocytidine [39].

Another reversible covalent bond widely used in the DCC field has been the disulfide group. Disulfides can be formed in aqueous solution and neutral or slightly basic pH; the reversible disulfide exchange is also compatible with these reaction conditions and many other functional groups. Moreover, thiols can be easily implemented in amino acid-derived structures using relatively simple chemistry. This fact has made the disulfide exchange one of the most popular reactions for the generation of DCLs of pseudopeptides in aqueous medium.

Otto and Kubik have collaboratively used this approach for the discovery of a pseudopeptidic receptor for anions in aqueous medium [40]. Considering the binding abilities of macrocyclic pseudopeptides from Kubik's lab and their trend to form dimeric supramolecular complexes (both in solution and in the solid state), they decided to prepare a DCL of dimers joined by disulfide bonds (Scheme 22.13).

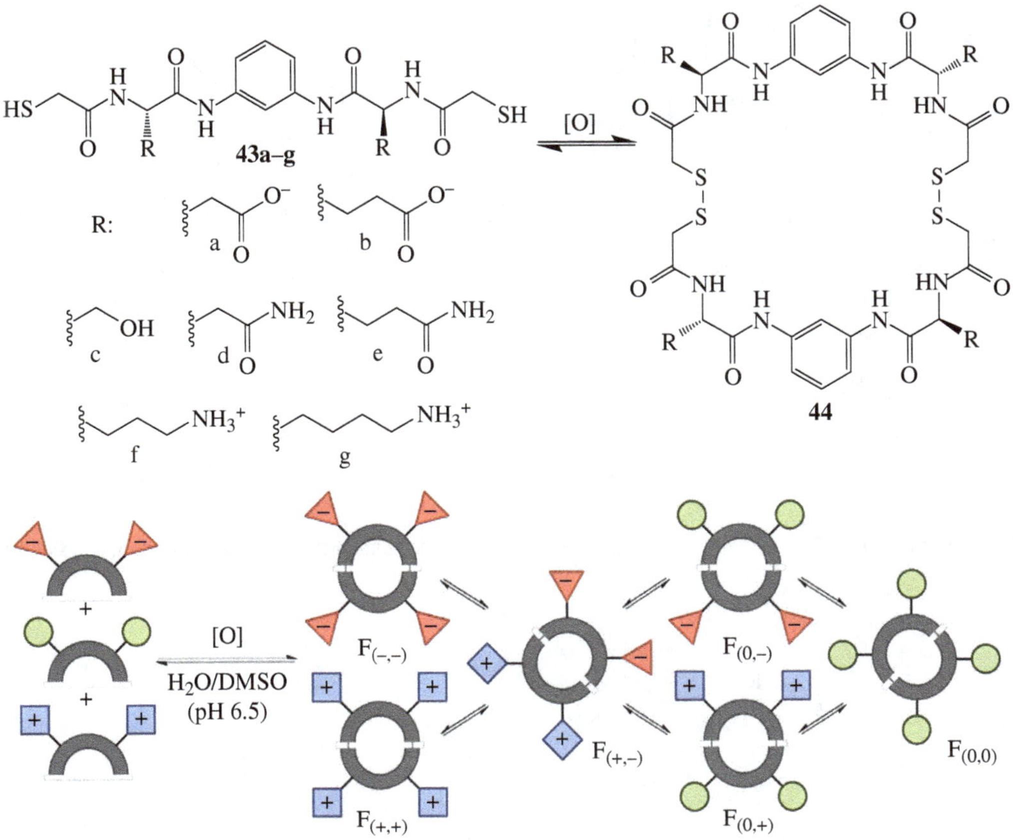

SCHEME 22.14 DCLs of pseudopeptidic macrocycles displaying electrostatic interactions between side chains.

The library was designed to generate different potential receptors by adding different dithiols as potential spacers between the macrocycles with an appended thiol (**41**). The addition of either iodide or sulfate amplified two different species, where the nature of the spacer maximized the binding, as further confirmed by ITC titration with the isolated receptors. The same researchers reported an improvement of the DCL system by attaching two thiols in each macrocycle (**42**). In this case, the mixture mainly produced macrobicycles by the connection of two BBs with two dithiol arms. From this library, a receptor with nanomolar affinity for sulfate anion in aqueous acetonitrile was identified [41]. This result is quite remarkable for a receptor based on neutral H-bond donors to bind the very elusive sulfate anion in aqueous medium.

The non-covalent interactions in disulfide-based DCLs can be also established between the BBs themselves. A recent example from our group is depicted in Scheme 22.14 [42, 43]. We prepared a bioinspired DCL of simple pseudopeptidic dithiols (**43**) with C_2 symmetry. They shared a common central *meta* phenylene spacer for promoting macrocyclization and two terminal mercaptoacetyl residues for the disulfide formation. The implementation of an amino acid moiety in each arm provided the BBs with peptide-like chemical and structural information. This design and the addition of small amounts of DMSO as organic cosolvent [42] allowed the generation of DCLs of macrocyclic pseudopeptides (mainly dimers like **44**; Scheme 22.14) with residues of different polarity and molecular charges at pH values close to neutrality. From these BBs, a library for mimicking the evolutionary patterns of halophilic proteins was designed and generated. The members of the library concentrating acidic residues (from Asp and Glu, **44a,b**) were amplified by the increase of salt concentration [43].

It is known that halophilic organisms have modified their proteins (after biological evolution) by increasing the acidic residues on the surface, thus retaining the structure and function of their biological machinery at high salt concentration. Following this study, a more diverse and complex library was prepared containing positive, negative, and neutral residues within the same backbone pseudopeptide [44]. The adaptation ability of this complex molecular network to the increase of salt gave important information about the electrostatic interactions in aqueous medium. Thus, for instance, we demonstrated that the anionic residues produced library members more sensitive to the changes of the ionic strength. Complementary NMR and molecular dynamics simulations suggested that these differences are due to a more efficient folding of the anionic species by intramolecular (carboxylate–amide and amide–amide) H-bonding.

Another remarkable example of molecular selection through non-covalent interactions was also recently reported by our group (Scheme 22.15) [45]. In this work a DCL formed by mixing BBs with different valences was generated. To that, the previously reported bipodal BB **43d** was

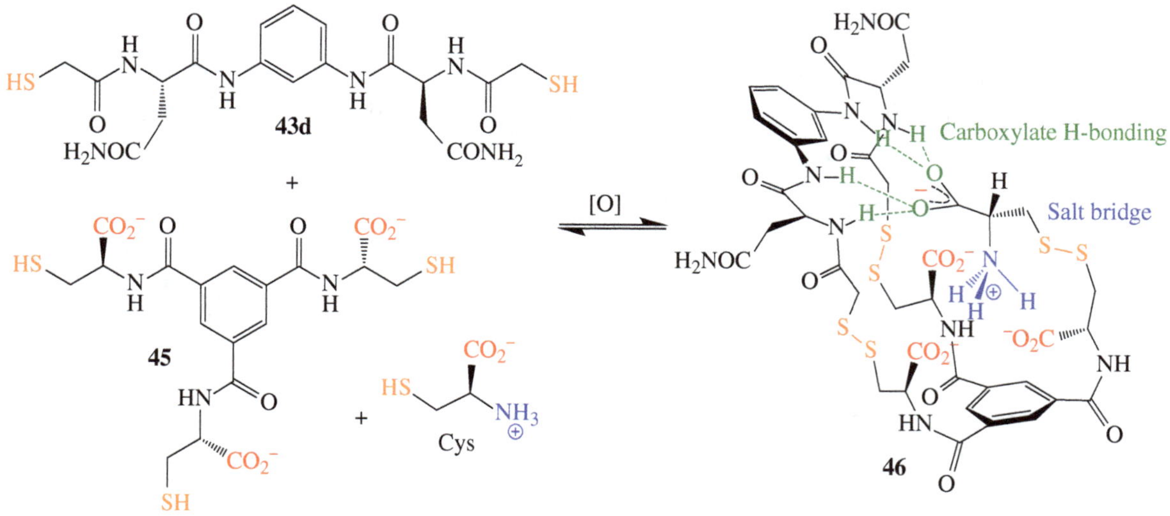

SCHEME 22.15 Constitutional self-selection of a heterotrimeric pseudopeptide from intramolecular H-bonding and salt bridge interactions.

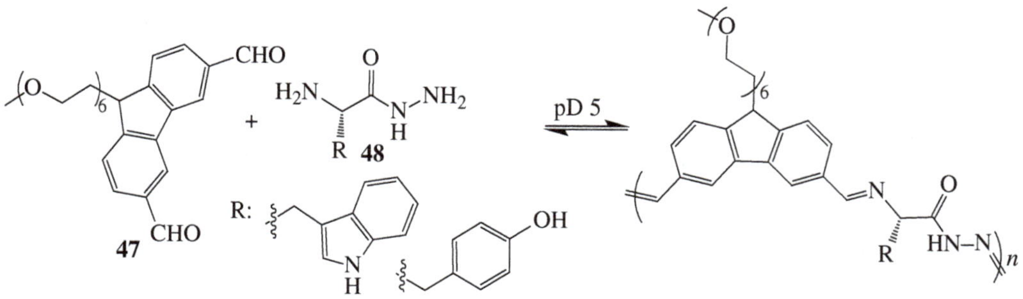

SCHEME 22.16 Dynamic polymers from hybrid imine–hydrazone pseudopeptides.

reacted with the tripodal BB **45**, formed by attaching three Cys amino acids to a central trimesic acid. The corresponding disulfides library rendered a complex mixture of homo- and hetero-oligomers. However, when performing the reaction in the presence of Cys as a monopodal thiol, a practically unique species was formed from the virtually very diverse possibilities of different topologies. The species was isolated at preparative scale and fully characterized to be the corresponding heterotrimer **46** obtained by the macrocyclic disulfide formation between **43d** and **45** and the attachment of the cysteine to the free additional thiol of **45**. The NMR and modeling data suggested that this species was obtained due to the establishment of several non-covalent interactions in a cooperative fashion. Thus, the pending zwitterionic Cys arm would fold onto the macrocyclic cavity, setting amide–carboxylate H-bonds and ammonium-carboxylate salt bridges, as depicted in Scheme 22.15. The formation of **46** occurred through error-correction processes and required the zwitterionic form of the Cys to set the polar and electrostatic interactions. Control experiments at different environmental conditions and with other BBs strongly supported this interactional model [46].

The DCC methodology has been also used for the preparation of pseudopeptidic polymeric materials that are dynamic in essence (peptide-like dynamers; Scheme 22.16). Thus, Lehn and coworkers described the generation of these species from a central aromatic dialdehyde **47** and aminoacylhydrazines **48** in acetate aqueous buffer (pD 5) [47]. Under these conditions both imine and hydrazone bonds are dynamic, leading to the formation of polymeric species with relatively good solubility in aqueous medium thanks to the polyether chains of **47**. The polymers thus formed tend to self-organize leading to globular particles sustained by hydrophobic interactions (as studied by cryo-TEM, LS, DOSY NMR, and SANS studies). These polymers displayed surprisingly low polydispersity retaining the behavior of the two dynamic covalent bonds that allow amino acidic exchange. The particles thus formed are reminiscent of globular proteins with nucleation–elongation behavior driven by hydrophobic effects, which was proposed by the authors as a simple model for the evolutionary processes of complex matter.

A final example of the application of DCC protocol within the pseudopeptidic field describes the discovery of self-replicating species. Self-replication is a very important property of living organisms that is also present at the molecular level. When a given molecular entity promotes its own formation, this species can be classified as a

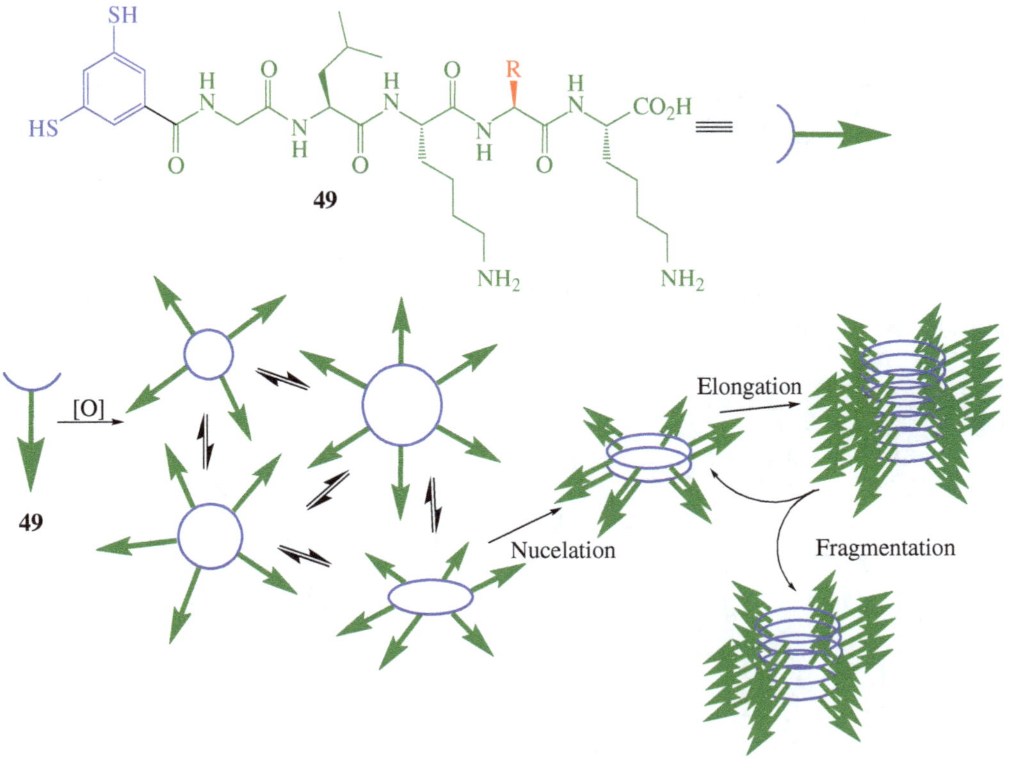

SCHEME 22.17 Design of self-assembling self-replicators from a DCL of macrocyclic pseudopeptides.

self-replicator. Otto and coworkers [48] described simple BBs **49** (Scheme 22.17) for displaying self-assembling properties in a DCL. They were composed by a short amphiphilic peptide sequence (Scheme 22.17) expected to form a β-sheet secondary structure by intermolecular non-covalent H-bonding and hydrophobic interactions. At the N-terminus, they attached a 3,5-dimercaptobenzoic acid, which was the moiety for establishing dynamic disulfide bonds leading to oligomeric macrocycles. Upon the DCL setup, the system formed macrocycles of different sizes in dynamic equilibrium in aqueous buffer. These species formed self-assembled aggregates by a nucleation mechanism assisted by the stacked rings and the β-sheet secondary structure between the peptide arms (see Scheme 22.17). These aggregates elongated toward nanometer-sized fibrils, where the fibrils ends act as templates for the formation of additional macrocycles (self-replication). The authors discovered that DCLs seeded with a given macrocyclic size (hexamer or heptamer) experienced an exponential growth of that species from the DCL of different species. Moreover, they observed that the type of mechanical agitation also affected the outcome of the system: stirring mainly produced heptamers, while shaking promoted hexamers, the authors giving convincing explanation for this difference. In further studies, the same group described the behavior of structurally different replicators depending on the peptide sequence (R in Scheme 22.17) [49] or the effect of the presence of fluorinated alcohols (trifluoroethanol) as organic cosolvents [50].

22.4 NON-COVALENT INTERACTIONS IN PEPTIDE-LIKE FOLDAMERS

Foldamers [51] are a class of molecules with a defined secondary structure akin of those encountered in natural biomolecules such as DNA or proteins. They are formed by relatively small sequences (oligomers) of repeating units (monomers). In the last years, a growing interest in the field of foldamers has been translated on the synthesis of a vast variety of peptidomimetic structures. Some of the examples include different kinds of α-, β-, and γ-peptides, aromatic amides, ureas, or peptoids.

Despite the great number of structurally diverse monomers that can be found in the literature, foldamers form only a few types of structures such as helices, linear strands, turns, and sheets. However, helical architectures are by far the most common and almost exclusive in peptide-related structures. The ability to fold into well-defined structures depends on several factors, mainly on the establishment of non-covalent interactions between usually nonconsecutive residues that can stabilize a given conformation. Thus, hydrogen bonding is fundamental in the stabilization of many helical scaffolds. In natural proteins two main helices, the α-helix and the 3_{10}-helix, are found, stabilized by $(i, i+4)$ and $(i, i+3)$ C═O···H—N hydrogen bonding, respectively. Not surprisingly, nonnatural structures that form folded structures based on amino acids have also been described [52]. For example, oligomers of the achiral α-aminoisobutyric acid (Aib) and related amino acids adopt a 3_{10} helical

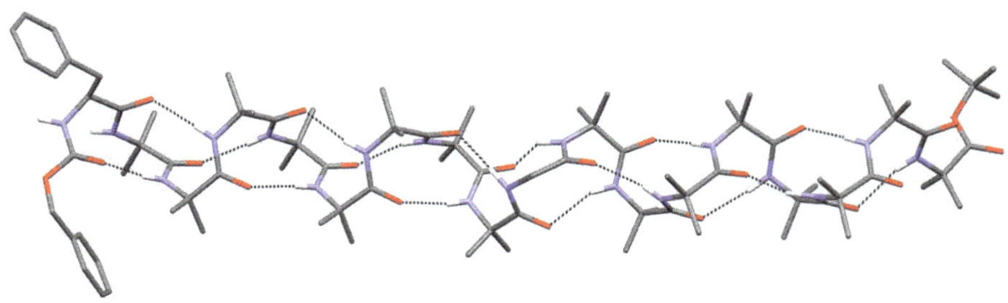

FIGURE 22.2 X-ray structure of CbzPheAib$_8$GlyAib$_8$Ot-Bu described by Clayden's group showing the stabilizing hydrogen bond contacts. Only amide hydrogens are shown for clarity.

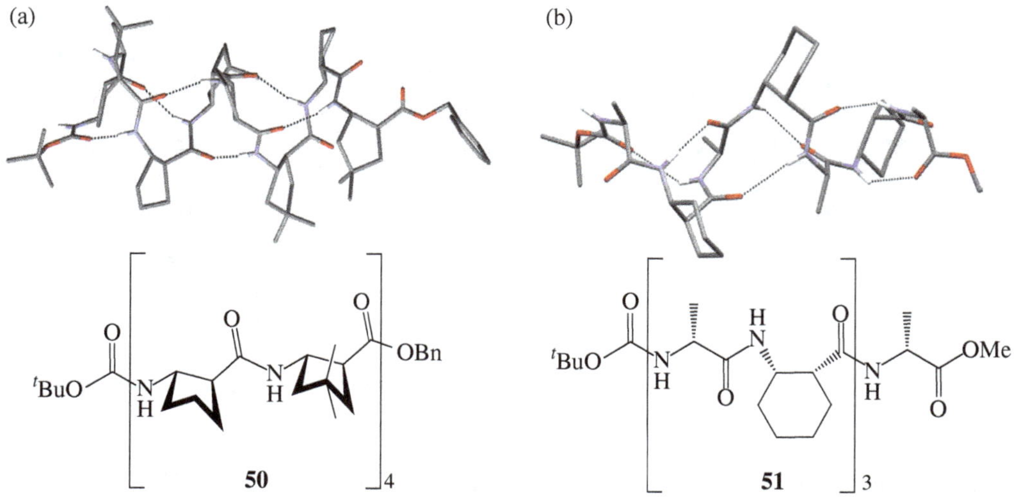

FIGURE 22.3 (a) X-ray (top) and chemical structure (bottom) of a 12-helical β-peptide described by Gellman [57]. (b) X ray structure (top) of an alternating α/β-peptide (bottom) with a 11/9 mixed helix described by Choi [58]. Only amide hydrogens are shown for clarity.

structure, which is favored by an arrangement of hydrogen bonds [53]. The longest crystalline 3_{10}-helix described is formed by the sequence CbzPheAib$_8$GlyAib$_8$Ot-Bu and involves 16 hydrogen bonds to complete 5 full turns of the helix (Fig. 22.2) [54].

β-Peptides [55] are also known to fold into stable helical conformations; the incorporation of a flexible methylene unit, however, leads to larger diversity in terms of secondary structure that arise from the different substitution possibilities. Multiple hydrogen bond patterns can be formed with different orientations: 14-, 12-, and 8-helix and 10/12-mixed helices have been described [51a, 56]. While the 8- and 12-helices have hydrogen bond orientations analogous of α-peptides, the other helices present orientations not found in natural peptides, giving rise to a different macrodipole. In this context, cyclic amino acids are especially interesting due to their conformational restrictions and the orientation of both carbonyl and amino groups.

Figure 22.3a shows an example of a 12-helix based on *trans*-2-aminocyclopentanecarboxylic acid obtained by X-ray diffraction [57]. This helix presents an (i, $i+3$) C═O···H—N hydrogen bonding arrangement comparable to α-peptides. When longer homologous γ- and δ-amino acids are used, more helical conformations and hydrogen bonding patterns are accessible, giving rise to a bigger variety of structures [59]. Moreover, peptides with heterogeneous backbones offer even further possibilities [60]. In Figure 22.3b we show an example of an α/β-peptide that forms a mixed 11/9 helicity with two types of hydrogen bonding: a hydrogen bond between the C═O of β-residue (i) and the N—H of α-residue ($i+3$) and a hydrogen bond between the C═O of α-residue (i) and the N—H of β-residue ($i-1$) [58].

Another kind of relevant foldameric structures is that composed by aromatic amino acids [61]. Hamilton showed that a series of oligoamides based on anthranilic acid subunits can adopt helical structures stabilized in part by hydrogen bonding (Fig. 22.4a) [62]. Analogously, curved folded oligoamides that form a crescent structure were described by Gong, rigidified by hydrogen bonds [65]. Ahn and coworkers designed another kind of rigid backbone based on benzamide residues held as well by hydrogen bonds [66]. Taking advantage of the intramolecular hydrogen bonding within 2′-pyridyl-2-pyridinecarboxamide units described by Boese [67], Lehn *et al.* synthesized longer oligomers that adopt helical shapes, stabilized by hydrogen bonding between the amide protons and the pyridyl moieties [63]. Further contacts are established by water molecules within the cavity that may contribute to

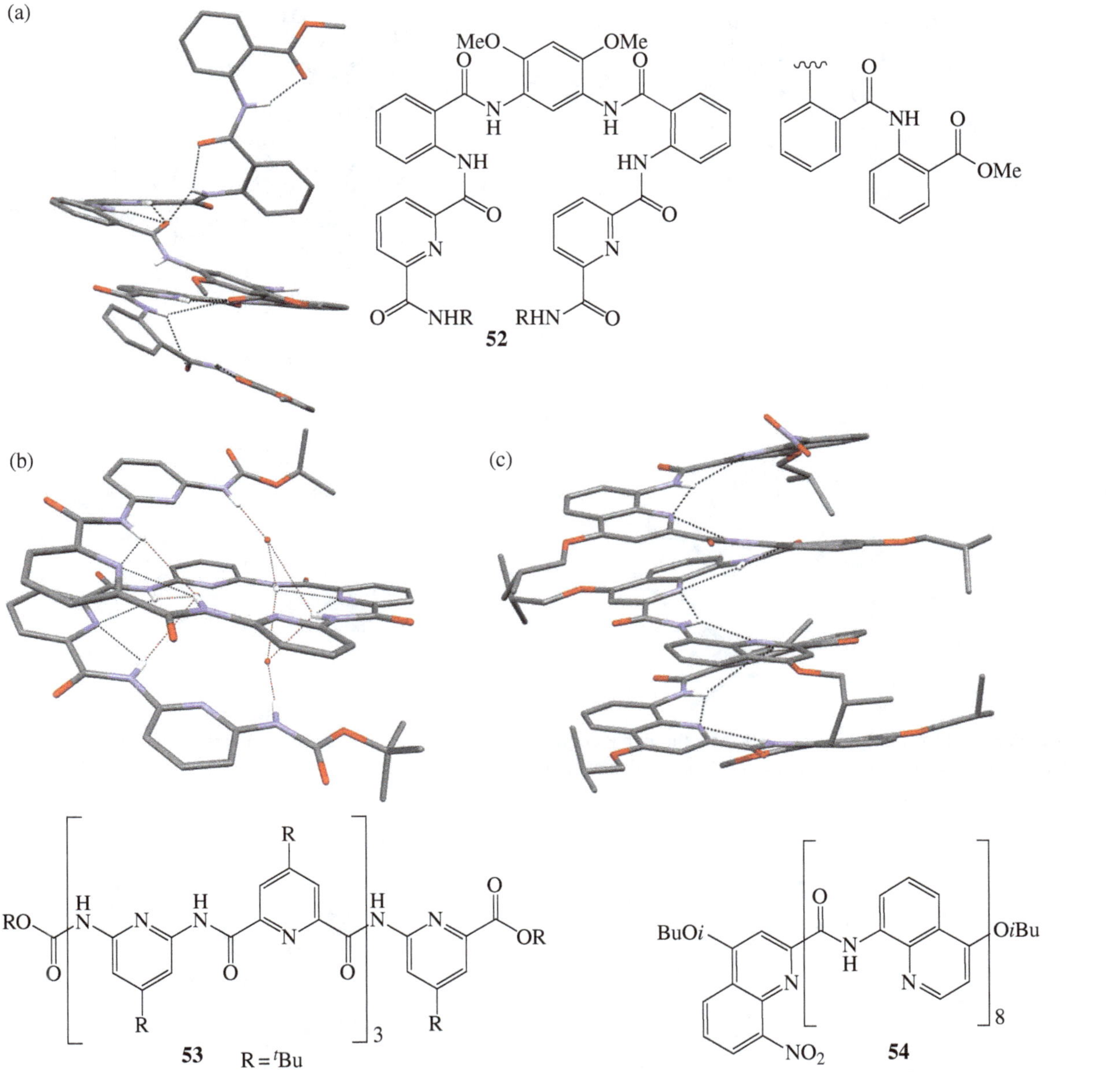

FIGURE 22.4 (a) X-ray structure and chemical structure of aromatic oligoamides. (a) Oligoanthranilamides described by Hamilton [62a]. (b) Lehn's 2′-pyridyl-2-pyridinecarboxamide oligomers [63]. (c) Quinoline-based strands synthesized by Huc [64b]. Only amide protons are shown for clarity.

the stabilization of the structure. In addition, π–π stacking contacts are possible over half a helix turn (Fig. 22.4b). The conjugation of π-orbitals along the oligomer reduces the conformational freedom, establishing contacts between the aromatic rings and the amide bonds, enabling the formation of helices even in water [68].

A more rigid helical conformation was synthesized employing quinoline BB. In this case the amide protons fill the helix hollow and are involved in two hydrogen bonds with the adjacent quinoline (Fig. 22.4c) [64]. Again, combination of different kinds of BB results in unconventional helices by combination of hydrogen bonding and π–π interactions [69].

Despite the prevalence of helical structure in peptide-like foldamers, other motifs have been formed in related compounds. For example, Vögtle *et al.* described the synthesis of a trefoil knot in oligoamides through an "internal" recognition by hydrogen bonding [70]. Hunter and Tomas reported a folded oligoamide also stabilized through hydrogen bonding interactions forming a "tail-biter" structure [71]. Nowick described the synthesis and incorporation of the Hao derivative (hydrazine, 5-amino-2-methoxybenzoic acid) into peptide sequences. The presence of this unnatural amino acid results in the formation of four hydrogen bonds that promote the mimics of a β-sheet structure (Fig. 22.5a) [72]. Moreover, cyclization through two ornithine side-chain

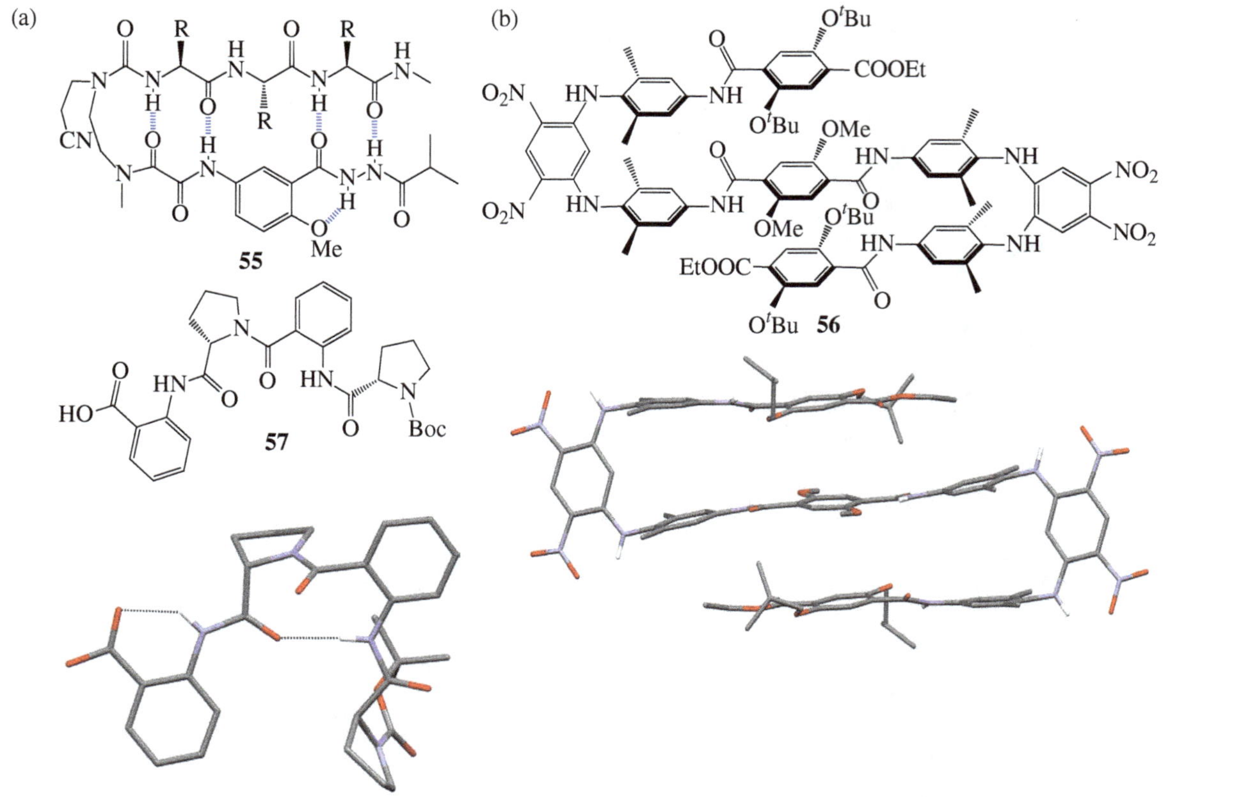

FIGURE 22.5 (a) Nowick's mimic of a β-sheet through incorporation of Hao residue [72a]. (b) Aromatic β-sheet foldamers described by Huc and crystal structure [73]. (c) Sanjayan's β-turn foldamers [74].

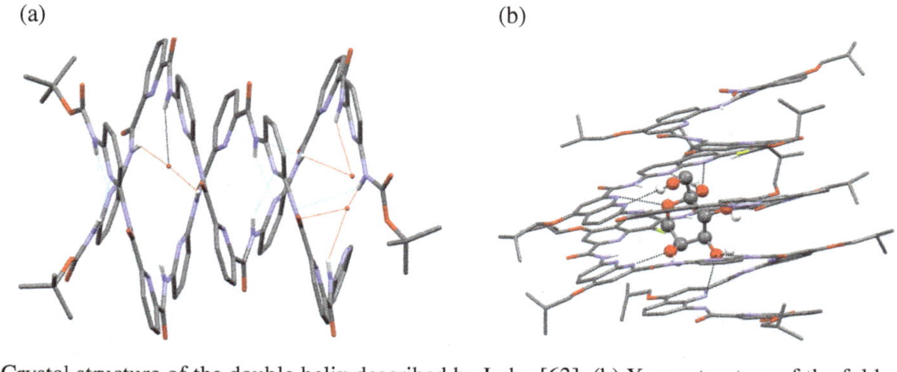

FIGURE 22.6 (a) Crystal structure of the double helix described by Lehn [63]. (b) X-ray structure of the foldameric receptor described by Huc with an encapsulated molecule of β-$^2C_5$-D-fructopyranose (ball and stick representation) [79]. Only amide and hydroxyl protons are displayed for clarity.

residues and incorporation of the Hao derivative promote the formation of β-sheet-like macrocycles [75].

The formation of pseudo-β-turn-folded peptides was described by Sanjayan in peptides consisting of repeating L-proline-anthranilic acid BB, with a $1 \rightarrow 2$ hydrogen bond pattern [74]. β-Sheet formation though hydrogen bonding has been described with β- [76] and γ-peptides [77]. Importantly, the formation of flat cyclic peptides allows the formation of peptide nanotubes through hydrogen bonding self-assembly of peptide monolayers [78]. More recently, Huc showed that the formation of artificial β-sheets based on π–π aromatic stacking interactions is possible if aggregation is prevented (Fig. 22.5b) [73].

External factors such as solvophobic effects, aggregation, or the presence of a molecule that can act as a guest may affect the folding process. Folding is sensitive to the nature of the media, and solvent composition is crucial as can favor or preclude folding processes, especially those depending on hydrogen bonding. A clear example takes place with compound **53**: as shown in Figure 22.4b it crystallizes as a single helix in polar solvents. However, while diluted samples show a single species in chloroform-*d*, upon concentration a second set of signals in ^{1}H NMR can be detected, which correspond to a dimeric structure of the compound forming a double helix (Fig. 22.6). The X-ray structure obtained by crystallization in nonpolar solvents

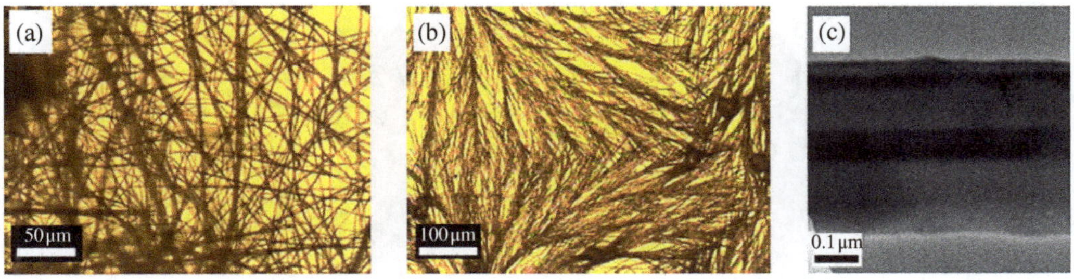

FIGURE 22.7 (a) An unordered mesh of diphenylalanine nanotubes on a gold surface. (b) A network of cast silver peptide nanotubes nanowires. (c) A single silver cast peptide nanotube. Source: Hendler *et al.* [89b]. Reproduced with permission of John Wiley & Sons.

confirms the formation of a double strand, stabilized by extensive π–π interactions and hydrogen bonds [63, 80].

Triple [81] and even quadruple [82] intertwined helices have been described for aromatic oligoamides. Moreover, as stated before, small polar molecules like water can be accommodated in the inside of the helix forming a host–guest complex stabilized by hydrogen bonding [83], and sometimes the presence of a guest can be crucial to induce the folding [84]. Furthermore, encapsulation can be diastereoselective, thus favoring one screw sense upon the incorporation of a chiral guest such as tartaric acid [85].

The presence of molecules that can be encapsulated by the helical foldamers can shift them between single and double helices. In this context, Huc *et al.* described the winding of helical oligoamides around rodlike guests with the appropriate length and hydrogen bond acceptors, forming host–guest complex that could be described as foldaxanes [86]. Selective encapsulation after conscious tailoring of foldamers can lead to remarkable results. An iterative reduction of the inner space in aromatic foldamers conduced to the synthesis of a strong binder of fructose in organic solvents, with a selectivity 300 times better than for the related mannose (Fig. 22.6) [79].

22.5 SELF-ASSEMBLED PSEUDOPEPTIDES FOR THE PREPARATION OF NANOSTRUCTURES

Molecular BBs able to self-assemble into supramolecular structures are of the utmost importance in the preparation of biological materials, from DNA structures to membrane bilayers. Self-assembly depends on the molecular information of the components (shape, distribution of polar groups, dipolar moment, etc.) that will determine the interactions formed [87]. In this context, peptides able to form nanostructures are receiving growing interest in different fields as they are interesting for biological, medical, photonic, and nanotechnological applications. The endless diversity that arises from combinatorial combination of the 20 amino acids permits the formation of a broad range of compounds too large to be discussed here; extensive reviews can be found in the literature [88].

However, some notorious examples related to the formation of nanostructures will be mentioned here. The pioneering work from Gazit *et al.* showed that peptides as short as diphenylalanine form homogenous fibril nanotubes in water, assembled by aromatic stacking [89]. When silver ion is added to the solution, the nanotubes are able to mold the reduction of silver resulting in the formation of nanowires (Fig. 22.7) [89].

Small structural changes in the molecular components can have a big influence in the supramolecular assembly. Diphenylglycine, a dipeptide closely related to diphenylalanine, forms completely different structures as it assembles into discrete, ordered nanospheres [90].

Supramolecular gels represent one of the most studied self-assembled materials. The use of molecular gels is of great interest due to its potential applicability in a broad range of fields [91]. Organic gelators can be assembled via non-covalent interactions (H-bonding, van der Waals, solvophobic, etc.) forming organized one-dimensional aggregates that can further organize into fibrils [92]. Thus, the solvent that represents by far the major component of the gel gets trapped in the network of fibrils and it is not able to flow freely.

Usually gels are formed by dissolution of the gelator in hot solvent followed by cooling to a given temperature. The morphology and stability of the aggregates may be affected by the cooling rate; depending on the energy barrier, metastable aggregates may reorder into more thermodynamically stable structures [93]. Of course, solubility plays a key role in gelation and affects both the thermal stability and the minimal gelator concentration, which is the minimum amount of compound to aggregate [94].

In water, the assembly is dominated by hydrophobic interactions. Most examples of self-assembled peptides in water contain hydrophobic residues. Stupp and coworkers described the use of peptide amphiphiles that assemble into fibrils by electrostatic interaction [95]. Interestingly, cells can be encapsulated within the fibrous network inducing the differentiation of cells into neurons [96]. The incorporation of aromatic components in the peptides allows the use of much smaller molecules thanks to the π–π interactions established; in particular, extensive work has been done with the use of Fmoc *N*-protected peptides (Fig. 22.8) [97].

Controlling self-assembly on demand is especially important. Several stimuli have been used to control assembly such as pH, changes of temperature, or addition of chemical

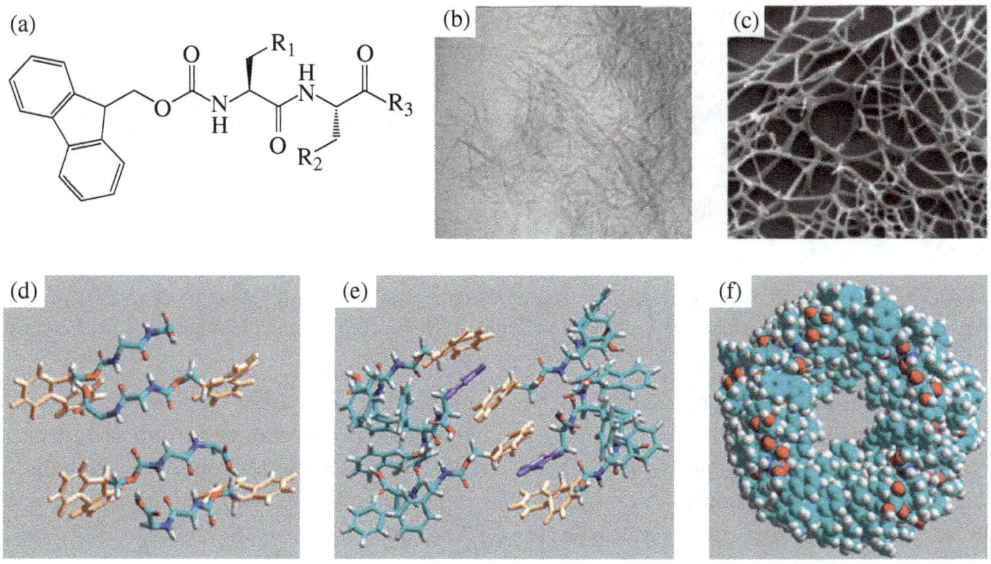

FIGURE 22.8 (a) General structure for Fmoc peptide hydrogelators. (b) Light microscopy of the nanofibrous structures made by Fmoc–Phe–Phe–OH (R_1, R_2 = Phe, R_3 = OH). (c) Microscopic structure obtained by cryo-SEM. (d) Model structure showing an antiparallel β-sheet alignment. (e) Due to the twist of β-sheets, the next sheet is rotated to maintain the interaction between the aromatic Fmoc groups, creating a cylindrical structure as shown in (f). Source: Smith *et al.* [97b]. Reproduced with permission of John Wiley & Sons.

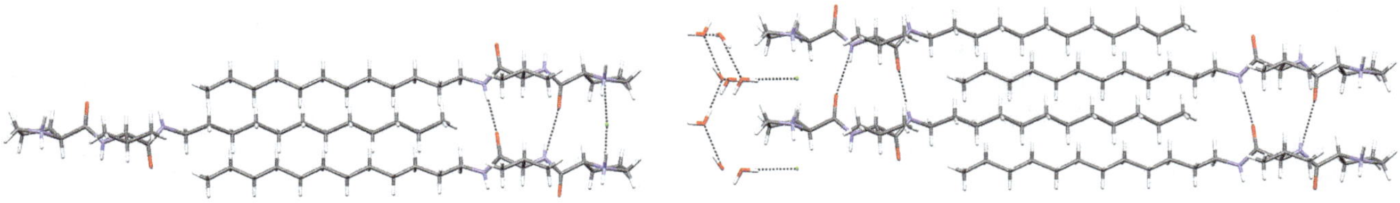

FIGURE 22.9 X-ray crystal structure of Pro–Val–NH($C_{12}H_{25}$)·HCl·2H_2O described by Escuder and Miravet [106].

reagents [98]. Ulijn demonstrated that proteases can be used to selectively trigger the self-assembly of peptide hydrogels via reversed hydrolysis [99]. Amphiphilic Fmoc-protected tripeptides are formed from the nongelating amino acids under equilibrium conditions. Normally, in water media the equilibrium should be shifted toward hydrolysis, but, as self-assembly is driven by π-stacking, a driving force is provided to the formation of the tripeptides resulting in an enzyme-promoted gelation. The nanotubular structures obtained possess a tunable diameter depending on the Fmoc peptide BB employed [100]. Enzyme action can be used also in the opposite direction. Thus, it is possible to use peptide hydrogels to selectively release particles contained within the structure only when the correct enzymes are present. Upon hydrolysis, repulsion between charges results in disassembly of the structure and release of the cargo [101], allowing, for example, the detection of proteases [102].

Changes in pH are also important in the selective formation of vesicles. The natural tripeptide Lys–Phe–Gly forms vesicles at concentrations about 0.5 mg/ml and nanotubes in more concentrated samples. These suprastructures are stable at pH 7.4 but break at slightly more acidic pH (~6.0). Modeling studies suggested that peptide assembly is driven by interplay of electrostatic, π-stacking, and H-bonding interactions [103]. The pH dependence of these structures allows the efficient delivery of anticancer drugs resulting in increased cytotoxicity.

Stimuli-responsive aggregation can lead to a tunable catalytic behavior. L-Proline-derived peptides show an important change in activity associated with gelation. Peptides containing a Pro–Val moiety form gels in organic solvents driven by formation of hydrogen bonding networks. A controlled cooling leads to more ordered and thermodynamically stable aggregates [93b]. Moreover, the aggregation process results in a basicity boost of L-proline secondary amine [104], allowing to catalyze a Henry reaction when aggregated but not when on solution [105]. Analogous peptide hydrogels based on Pro–Val peptide amides with a hydrocarbonated chain are assembled driven by hydrophobic effect and show the same catalytic behavior upon gelation [106].

In this latter case, a protonated sample of the peptides could by crystallized, giving some hints onto the gelation mechanism (Fig. 22.9). An intercalated bilayer structure is formed in which the alkyl chains are packed by van der Waals interactions. Intermolecular H-bonds between the alkylamide NH and valine carbonyl group and between the valine amide NH and proline carbonyl group provide further stability. Chloride counterions and water molecules complete the packed structure. Although nonprotonated gel probably presents a different packing, especially on the polar heads, this X-ray shows a nice picture of the interactions responsible for the formation of aggregates. The use of this compound as a preformed hydrogel allows catalyzing an aldol reaction in a biphasic system. Thus, the hydrogel containing the catalyst can be easily recovered and reused [107].

To predict the supramolecular behavior of these compounds is not an easy task. The discovery of the assembly properties has been usually dictated by serendipity or following chemists' intuition. Therefore, systematic exploration that leads to the synthesis of materials with tailored properties is of the utmost importance. In this context, Ulijin's group explored the behavior of 8000 peptides made of sequence of three amino acids using computational tools [108]. Thus, they were able to predict the supramolecular properties of the peptides and elaborate products with the required properties.

22.6 FINAL COMMENTS

The synthetic molecular entities bearing natural amino acids combined with nonnatural moieties are highly appealing for the chemical research. Most of the structural properties and potential applications of this family of molecules are related to their ability to set up non-covalent interactions of varied nature. These weak contacts can play important roles within this field, from the synthesis of challenging molecules to the stabilization of three-dimensional structures or the preparation of supramolecular entities. Representative and illustrative examples of each case have been described here. Despite much work has been already done, the large structural diversity of the amino acid structures combined with the endless number of possibilities of the nonnatural/nonpeptidic moieties allows envisioning a promising future for these fascinating and beautiful molecules. Undoubtedly, the pseudopeptides will surely walk into their bright future holding their hands by non-covalent interactions.

ACKNOWLEDGMENTS

This work has been partially supported by the Spanish Ministry of Economy and Competitiveness (MINECO, CTQ2012-38543-C03-03) and Generalitat de Catalunya (2014 SGR 231). J.S. acknowledges MINECO (Ramon y Cajal contract) and EU (FP7-PEOPLE-2012-CIG-321659) for funding fellowships.

REFERENCES

1. S. V. Luis, I. Alfonso, *Acc. Chem. Res.* 2014, *47*, 112.
2. S. E. Gibson, C. Lecci, *Angew. Chem. Int. Ed.* 2006, *45*, 1364.
3. W. A. Loughlin, J. D. A. Tyndall, M. P. Glenn, D. P. Fairlie, *Chem. Rev.* 2004, *104*, 6085.
4. N. Sokolenko, G. Abbenante, M. J. Scanlon, A. Jones, L. R. Gahan, G. R. Hanson, D. P. Fairlie, *J. Am. Chem. Soc.* 1999, *121*, 2603.
5. J. Becerril, M. Bolte, M. I. Burguete, F. Galindo, E. García-España, S. V. Luis, J. F. Miravet, *J. Am. Chem. Soc.* 2003, *125*, 6677.
6. C. J. White, A. K. Yudin, *Nat. Chem.* 2011, *3*, 509.
7. R. Hili, V. Rai, A. K. Yudin, *J. Am. Chem. Soc.* 2010, *132*, 2889.
8. S. Zaretsky, C. C. G. Scully, A. J. Lough, A. K. Yudin, *Chem. Eur. J.* 2013, *19*, 17668.
9. C. J. White, J. L. Hickey, C. C. G. Scully, A. K. Yudin, *J. Am. Chem. Soc.* 2014, *136*, 3728.
10. S. Kubik, R. Goddard, *J. Org. Chem.* 1999, *64*, 9475.
11. S. Kubik, R. Goddard, R. Kirchner, D. Nolting, J. Seidel, *Angew. Chem. Int. Ed.* 2001, *40*, 2648.
12. M. R. Krause, R. Goddard, S. Kubik, *J. Org. Chem.* 2011, *76*, 7084.
13. V. Martí-Centelles, M. I. Burguete, S. V. Luis, *Chem. Eur. J.* 2012, *18*, 2409.
14. V. Martí-Centelles, M. I. Burguete, C. Cativiela, S. V. Luis, *J. Org. Chem.* 2014, *79*, 559.
15. M. Bru, I. Alfonso, M. I. Burguete, S. V. Luis, *Angew. Chem. Int. Ed.* 2006, *45*, 6155.
16. I. Alfonso, M. Bolte, M. Bru, M. I. Burguete, S. V. Luis, J. Rubio *J. Am. Chem. Soc.* 2008, *130*, 6137.
17. M. Bru, I. Alfonso, M. I. Burguete, S. V. Luis, *Tetrahedron Lett.* 2005, *46*, 7781.
18. I. Alfonso, M. Bolte, M. Bru, M. I. Burguete, S. V. Luis, *Chem. Eur. J.* 2008, *14*, 8879.
19. M. Bru, I. Alfonso, M. Bolte, M. I. Burguete, S. V. Luis, *Chem. Commun.* 2011, *47*, 283.
20. I. Martí, J. Rubio, M. Bolte, M. I. Burguete, C. Vicent, R. Quesada, I. Alfonso, S. V. Luis, *Chem. Eur. J.* 2012, *18*, 16728.
21. I. Martí, M. Bolte, M. I. Burguete, C. Vicent, I. Alfonso, S. V. Luis, *Chem. Eur. J.* 2014, *20*, 7458.
22. A. Moure, S. V. Luis, I. Alfonso, *Chem. Eur. J.* 2012, *18*, 5496.
23. E. Faggi, A. Moure, M. Bolte, C. Vicent, S. V. Luis, I. Alfonso, *J. Org. Chem.* 2014, *79*, 4590.
24. R. B. P. Elmes, K. A. Jolliffe, *Chem. Commun.* 2015, *51*, 4951.
25. G. Haberhauer, Á. Pintér, S. Woitschetzki, *Nat. Commun.* 2013, *4*, 2945.
26. T. Fiehn, R. Goddard, R. W. Seidel, S. Kubik, *Chem. Eur. J.* 2010, *16*, 7241.
27. For some recent reviews in the field, see: (a) S. Otto, *Acc. Chem. Res.* 2012, *45*, 2200; (b) E. Moulin, G. Cormos, N. Giuseppone, *Chem. Soc. Rev.* 2012, *41*, 1031; (c) F. B. L. Cougnon, J. K. M. Sanders, *Acc. Chem. Res.* 2011, *45*, 2211; (d) J. W. Sadownik, R. V. Ulijn, *Curr. Opin. Biotechnol.* 2010,

21, 401; (e) J. N. Reek, S. Otto, Dynamic Combinatorial Chemistry, John Wiley & Sons, Inc., Hoboken, 2010; (f) S. Ladame, *Org. Biomol. Chem.* 2008, *6*, 219; (g) J.-M. Lehn, *Chem. Soc. Rev.* 2007, *36*, 151; (h) P. T. Corbett, J. Leclaire, L. Vial, K. R. West, J.-L. Wietor, J. K. M. Sanders, S. Otto, *Chem. Rev.* 2006, *106*, 3652.

28. J. Li, P. Nowak, S. Otto, *J. Am. Chem. Soc.* 2013, *135*, 9222.
29. G. R. L. Cousins, R. L. E. Furlan, Y.-F. Ng, J. E. Redman, J. K. M. Sanders, *Angew. Chem. Int. Ed.* 2001, *40*, 423.
30. R. L. E. Furlan, Y.-F. Ng, S. Otto, J. K. M. Sanders, *J. Am. Chem. Soc.* 2001, *123*, 8876.
31. S. L. Roberts, R. L. E. Furlan, S. Otto, J. K. M. Sanders, *Org. Biomol. Chem.* 2003, *1*, 1625.
32. F. Bulos, S. L. Roberts, R. L. E. Furlan, J. K. M. Sanders, *Chem. Commun.* 2007, 3092.
33. J.-M. Lehn, *Chem. Eur. J.* 1999, *5*, 2455.
34. R. T. S. Lam, A. Belenguer, S. L. Roberts, C. Naumann, T. Jarrosson, S. Otto, J. K. M. Sanders, *Science* 2005, *308*, 667.
35. M.-K. Chung, P. S. White, S. J. Lee, M. R. Gagné, *Angew. Chem. Int. Ed.* 2009, *48*, 8683.
36. M.-K. Chung, P. S. White, S. J. Lee, M. L. Waters, M. R. Gagné, *J. Am. Chem. Soc.* 2012, *134*, 11415.
37. H. Schiltz, M.-K. Chung, S. J. Leeb, M. R. Gagné, *Org. Biomol. Chem.* 2008, *6*, 3597.
38. S. M. Voshell, S. J. Lee, M. R. Gagné, *J. Am. Chem. Soc.* 2006, *128*, 12422.
39. M.-K. Chung, C. M. Hebling, J. W. Jorgenson, K. Severin, S. J. Lee, M. R. Gagne, *J. Am. Chem. Soc.* 2008, *130*, 11819.
40. S. Otto, S. Kubik, *J. Am. Chem. Soc.* 2003, *125*, 7804.
41. Z. Rodríguez-Docampo, E. Eugenieva-Ilieva, C. Reyheller, A. M. Belenguer, S. Kubik, S. Otto, *Chem. Commun.* 2011, *47*, 9798.
42. J. Atcher, I. Alfonso, *RSC Adv.* 2013, *3*, 25605.
43. J. Atcher, A. Moure, I. Alfonso, *Chem. Commun.* 2013, *49*, 487.
44. J. Atcher, A. Moure, J. Bujons, I. Alfonso, *Chem. Eur. J.* 2015, *21*, 6869.
45. J. Solà, M. Lafuente, J. Atcher, I. Alfonso, *Chem. Commun.* 2014, *50*, 4564.
46. M. Lafuente, J. Atcher, J. Solà, I. Alfonso, *Chem. Eur. J.* 2015, *21*, 17002.
47. A. K. H. Hirsch, E. Buhler, J.-M. Lehn, *J. Am. Chem. Soc.* 2012, *134*, 4177.
48. J. M. A. Carnall, C. A. Waudby, A. M. Belenguer, M. C. A. Stuart, J. J.-P. Peyralans, S. Otto, *Science* 2010, *327*, 1502.
49. M. Malakoutikhah, J. J.-P. Peyralans, M. Colomb-Delsuc, H. Fanlo-Virgós, M. C. A. Stuart, S. Otto, *J. Am. Chem. Soc.* 2013, *135*, 18406.
50. G. Leonetti, S. Otto, *J. Am. Chem. Soc.* 2015, *137*, 2067.
51. (a) S. H. Gellman, *Acc. Chem. Res.* 1998, *31*, 173180; (b) D. J. Hill, M. J. Mio, R. B. Prince, T. S. Hughes, J. S. Moore, *Chem. Rev.* 2001, *101*, 3893; (c) I. Huc, S. Hecht, Foldamers: Structure, Properties and Applications, Wiley–VCH, Weinheim, 2007; (d) G. Guichard, I. Huc, *Chem. Commun.* 2011, *47*, 5933.
52. (a) I. M. Mandity, E. Weber, T. A. Martinek, G. Olajos, G. K. Toth, E. Vass, F. Fueloep, *Angew. Chem. Int. Ed.* 2009, *48*, 2171; (b) T. A. Martinek, F. Fueloep, *Chem. Soc. Rev.* 2012, *41*, 687.
53. (a) R. P. Hummel, C. Toniolo, G. Jung, *Angew. Chem. Int. Ed.* 1987, *26*, 1150; (b) C. Toniolo, E. Benedetti, *Trends Biochem. Sci.* 1991, *16*, 350; (c) B. Pengo, F. Formaggio, M. Crisma, C. Toniolo, G. M. Bonora, Q. B. Broxterman, J. Kamphuis, M. Saviano, R. Iacovino, F. Rossi, E. Benedetti, *J. Chem. Soc. Perk. Trans. 2* 1998, 1651; (d) J. Venkatraman, S. C. Shankaramma, P. Balaram, *Chem. Rev.* 2001, *101*, 3131; (e) J. Clayden, A. Castellanos, J. Sola, G. A. Morris, *Angew. Chem. Int. Ed.* 2009, *48*, 5962.
54. (a)J. Sola, M. Helliwell, J. Clayden, *J. Am. Chem. Soc.* 2010, *132*, 4548; (b) J. Sola, M. Helliwell, J. Clayden, *Biopolymers* 2011, *95*, 62.
55. (a) D. H. Appella, L. A. Christianson, I. L. Karle, D. R. Powell, S. H. Gellman, *J. Am. Chem. Soc.* 1996, *118*, 13071; (b) D. H. Appella, L. A. Christianson, D. A. Klein, D. R. Powell, X. Huang, J. J. Barchi, Jr., S. H. Gellman, *Nature* 1997, *387*, 381; (c) D. Seebach, J. L. Matthews, *Chem. Commun.* 1997, 2015.
56. (a) R. P. Cheng, S. H. Gellman, W. F. DeGrado, *Chem. Rev.* 2001, *101*, 3219; (b) D. Seebach, D. F. Hook, A. Glättli, *Pept. Sci.* 2006, *84*, 23.
57. S. H. Choi, I. A. Guzei, L. C. Spencer, S. H. Gellman, *J. Am. Chem. Soc.* 2010, *132*, 13879.
58. M. Lee, J. Shim, P. Kang, I. A. Guzei, S. H. Choi, *Angew. Chem. Int. Ed.* 2013, *52*, 12564.
59. F. Bouillère, S. Thétiot-Laurent, C. Kouklovsky, V. Alezra, *Amino Acids* 2011, *41*, 687.
60. (a) A. Hayen, M. A. Schmitt, F. N. Ngassa, K. A. Thomasson, S. H. Gellman, *Angew. Chem. Int. Ed.* 2004, *43*, 505; (b) G. V. M. Sharma, P. Nagendar, P. Jayaprakash, P. R. Krishna, K. V. S. Ramakrishna, A. C. Kunwar, *Angew. Chem. Int. Ed.* 2005, *44*, 5878; (c) S. H. Choi, I. A. Guzei, L. C. Spencer, S. H. Gellman, *J. Am. Chem. Soc.* 2008, *130*, 6544; (d) W. S. Horne, S. H. Gellman, *Acc. Chem. Res.* 2008, *41*, 1399; (e) G. V. M. Sharma, N. Chandramouli, M. Choudhary, P. Nagendar, K. V. S. Ramakrishna, A. C. Kunwar, P. Schramm, H.-J. Hofmann, *J. Am. Chem. Soc.* 2009, *131*, 17335; (f) L. K. A. Pilsl, O. Reiser, *Amino Acids* 2011, *41*, 709; (g) B. Legrand, C. André, L. Moulat, E. Wenger, C. Didierjean, E. Aubert, M. C. Averlant-Petit, J. Martinez, M. Calmes, M. Amblard, *Angew. Chem. Int. Ed.* 2014, *53*, 13131.
61. (a) I. Huc, *Eur. J. Org. Chem.* 2004, 17; (b) Z.-T. Li, J.-L. Hou, C. Li, *Acc. Chem. Res.* 2008, *41*, 1343; (c) D.-W. Zhang, X. Zhao, J.-L. Hou, Z.-T. Li, *Chem. Rev.* 2012, *112*, 5271.
62. (a) Y. Hamuro, S. J. Geib, A. D. Hamilton, *J. Am. Chem. Soc.* 1996, *118*, 7529; (b) Y. Hamuro, S. J. Geib, A. D. Hamilton, *J. Am. Chem. Soc.* 1997, *119*, 10587.
63. V. Berl, I. Huc, R. G. Khoury, M. J. Krische, J. M. Lehn, *Nature* 2000, *407*, 720.
64. (a) H. Jiang, J.-M. Leger, C. Dolain, P. Guionneau, I. Huc, *Tetrahedron* 2003, *59*, 8365; (b) H. Jiang, J.-M. Léger, I. Huc, *J. Am. Chem. Soc.* 2003, *125*, 3448.

65. J. Zhu, R. D. Parra, H. Zeng, E. Skrzypczak-Jankun, X. C. Zeng, B. Gong, *J. Am. Chem. Soc.* 2000, *122*, 4219.
66. S. Marimganti, M. N. Cheemala, J.-M. Ahn, *Org. Lett.* 2009, *11*, 4418.
67. M. Mazik, D. Bläser, R. Boese, *Tetrahedron* 1999, *55*, 12771.
68. (a) V. Berl, I. Huc, R. G. Khoury, J. M. Lehn, *Chem. Eur. J.* 2001, *7*, 2798; (b) I. Huc, V. Maurizot, H. Gornitzka, J. M. Leger, *Chem. Commun.* 2002, 578.
69. (a) N. Delsuc, F. Godde, B. Kauffmann, J.-M. Léger, I. Huc, *J. Am. Chem. Soc.* 2007, *129*, 11348; (b) M. Kudo, V. Maurizot, B. Kauffmann, A. Tanatani, I. Huc, *J. Am. Chem. Soc.* 2013, *135*, 9628.
70. J. Brüggemann, S. Bitter, S. Müller, W. M. Müller, U. Müller, N. M. Maier, W. Lindner, F. Vögtle, *Angew. Chem. Int. Ed.* 2007, *46*, 254.
71. C. A. Hunter, A. Spitaleri, S. Tomas, *Chem. Commun.* 2005, 3691.
72. (a) J. S. Nowick, D. M. Chung, K. Maitra, S. Maitra, K. D. Stigers, Y. Sun, *J. Am. Chem. Soc.* 2000, *122*, 7654; (b) J. S. Nowick, K. S. Lam, T. V. Khasanova, W. E. Kemnitzer, S. Maitra, H. T. Mee, R. Liu, *J. Am. Chem. Soc.* 2002, *124*, 4972; (c) P.-N. Cheng, C. Liu, M. Zhao, D. Eisenberg, J. S. Nowick, *Nat. Chem.* 2012, *4*, 927.
73. L. Sebaoun, V. Maurizot, T. Granier, B. Kauffmann, I. Huc, *J. Am. Chem. Soc.* 2014, *136*, 2168.
74. P. Prabhakaran, S. S. Kale, V. G. Puranik, P. R. Rajamohanan, O. Chetina, J. A. K. Howard, H.-J. Hofmann, G. J. Sanjayan, *J. Am. Chem. Soc.* 2008, *130*, 17743.
75. R. J. Woods, J. O. Brower, E. Castellanos, M. Hashemzadeh, O. Khakshoor, W. A. Russu, J. S. Nowick, *J. Am. Chem. Soc.* 2007, *129*, 2548.
76. (a) S. Krauthauser, L. A. Christianson, D. R. Powell, S. H. Gellman, *J. Am. Chem. Soc.* 1997, *119*, 11719; (b) D. Seebach, S. Abele, K. Gademann, B. Jaun, *Angew. Chem. Int. Ed.* 1999, *38*, 1595; (c) J. M. Langenhan, I. A. Guzei, S. H. Gellman, *Angew. Chem. Int. Ed.* 2003, *42*, 2402.
77. (a) M. G. Woll, J. R. Lai, I. A. Guzei, S. J. C. Taylor, M. E. B. Smith, S. H. Gellman, *J. Am. Chem. Soc.* 2001, *123*, 11077; (b) C. R. Jones, M. K. N. Qureshi, F. R. Truscott, S.-T. D. Hsu, A. J. Morrison, M. D. Smith, *Angew. Chem. Int. Ed.* 2008, *47*, 7099.
78. (a) M. R. Ghadiri, J. R. Granja, R. A. Milligan, D. E. McRee, N. Khazanovich, *Nature* 1993, *366*, 324; (b) N. Khazanovich, J. R. Granja, D. E. McRee, R. A. Milligan, M. R. Ghadiri, *J. Am. Chem. Soc.* 1994, *116*, 6011; (c) R. J. Brea, C. Reiriz, J. R. Granja, *Chem. Soc. Rev.* 2010, *39*, 1448.
79. N. Chandramouli, Y. Ferrand, G. Lautrette, B. Kauffmann, C. D. Mackereth, M. Laguerre, D. Dubreuil, I. Huc, *Nat. Chem.* 2015, *7*, 334.
80. A. Acocella, A. Venturini, F. Zerbetto, *J. Am. Chem. Soc.* 2004, *126*, 2362.
81. Y. Ferrand, A. M. Kendhale, J. Garric, B. Kauffmann, I. Huc, *Angew. Chem. Int. Ed.* 2010, *49*, 1778.
82. Q. Gan, C. Bao, B. Kauffmann, A. Grélard, J. Xiang, S. Liu, I. Huc, H. Jiang, *Angew. Chem. Int. Ed.* 2008, *47*, 1715.
83. J. Garric, J.-M. Leger, I. Huc, *Chem. Eur. J.* 2007, *13*, 8454.
84. Y.-X. Xu, G.-T. Wang, X. Zhao, X.-K. Jiang, Z.-T. Li, *J. Org. Chem.* 2009, *74*, 7267.
85. Y. Ferrand, A. M. Kendhale, B. Kauffmann, A. Grelard, C. Marie, V. Blot, M. Pipelier, D. Dubreuil, I. Huc, *J. Am. Chem. Soc.* 2010, *132*, 7858.
86. (a) Q. Gan, Y. Ferrand, C. Bao, B. Kauffmann, A. Grélard, H. Jiang, I. Huc, *Science* 2011, *331*, 1172; (b) Q. Gan, Y. Ferrand, N. Chandramouli, B. Kauffmann, C. Aube, D. Dubreuil, I. Huc, *J. Am. Chem. Soc.* 2012, *134*, 15656.
87. G. M. Whitesides, B. Grzybowski, *Science* 2002, *295*, 2418.
88. (a) E. Gazit, *Chem. Soc. Rev.* 2007, *36*, 1263; (b) I. W. Hamley, *Angew. Chem. Int. Ed.* 2007, *46*, 8128; (c) R. V. Ulijn, A. M. Smith, *Chem. Soc. Rev.* 2008, *37*, 664; (d) M. Zelzer, R. V. Ulijn, *Chem. Soc. Rev.* 2010, *39*, 3351; (e) X. Zhao, F. Pan, H. Xu, M. Yaseen, H. Shan, C. A. E. Hauser, S. Zhang, J. R. Lu, *Chem. Soc. Rev.* 2010, *39*, 3480; (f) S. Fleming, R. V. Ulijn, *Chem. Soc. Rev.* 2014, *43*, 8150.
89. (a) M. Reches, E. Gazit, *Science* 2003, *300*, 625; (b) N. Hendler, N. Sidelman, M. Reches, E. Gazit, Y. Rosenberg, S. Richter, *Adv. Mater.* 2007, *19*, 1485.
90. M. Reches, E. Gazit, *Nano Lett.* 2004, *4*, 581.
91. (a) J. B. Matson, S. I. Stupp, *Chem. Commun.* 2012, *48*, 26; (b) B. Escuder, J. F. Miravet, Functional Molecular Gels, Royal Society of Chemistry, Oxford, 2013.
92. R. G. Weiss, P. Térech, Molecular Gels: Materials with Self-Assembled Fibrillar Networks, Springer, Dordrecht, 2006.
93. (a) J. Becerril, M. I. Burguete, B. Escuder, F. Galindo, R. Gavara, J. F. Miravet, S. V. Luis, G. Peris, *Chem. Eur. J.* 2004, *10*, 3879; (b) F. Rodriguez-Llansola, J. F. Miravet, B. Escuder, *Chem. Commun.* 2009, 209.
94. A. R. Hirst, I. A. Coates, T. R. Boucheteau, J. F. Miravet, B. Escuder, V. Castelletto, I. W. Hamley, D. K. Smith, *J. Am. Chem. Soc.* 2008, *130*, 9113.
95. (a) J. D. Hartgerink, E. Beniash, S. I. Stupp, *Proc. Natl. Acad. Sci. U. S. A.* 2002, *99*, 5133; (b) K. L. Niece, J. D. Hartgerink, J. J. J. M. Donners, S. I. Stupp, *J. Am. Chem. Soc.* 2003, *125*, 7146.
96. G. A. Silva, C. Czeisler, K. L. Niece, E. Beniash, D. A. Harrington, J. A. Kessler, S. I. Stupp, *Science* 2004, *303*, 1352.
97. (a) Z. Yang, H. Gu, Y. Zhang, L. Wang, B. Xu, *Chem. Commun.* 2004, 208; (b) A. M. Smith, R. J. Williams, C. Tang, P. Coppo, R. F. Collins, M. L. Turner, A. Saiani, R. V. Ulijn, *Adv. Mater.* 2008, *20*, 37; (c) R. Orbach, L. Adler-Abramovich, S. Zigerson, I. Mironi-Harpaz, D. Seliktar, E. Gazit, *Biomacromolecules* 2009, *10*, 2646; (d) D. J. Adams, P. D. Topham, *Soft Matter* 2010, *6*, 3707; (e) R. Orbach, I. Mironi-Harpaz, L. Adler-Abramovich, E. Mossou, E. P. Mitchell, V. T. Forsyth, E. Gazit, D. Seliktar, *Langmuir* 2012, *28*, 2015.
98. R. J. Mart, R. D. Osborne, M. M. Stevens, R. V. Ulijn, *Soft Matter* 2006, *2*, 822.
99. S. Toledano, R. J. Williams, V. Jayawarna, R. V. Ulijn, *J. Am. Chem. Soc.* 2006, *128*, 1070.

100. A. K. Das, R. Collins, R. V. Ulijn, *Small* 2008, *4*, 279.
101. P. D. Thornton, R. J. Mart, R. V. Ulijn, *Adv. Mater.* 2007, *19*, 1252.
102. A. Laromaine, L. Koh, M. Murugesan, R. V. Ulijn, M. M. Stevens, *J. Am. Chem. Soc.* 2007, *129*, 4156.
103. P. Moitra, K. Kumar, P. Kondaiah, S. Bhattacharya, *Angew. Chem. Int. Ed.* 2014, *53*, 1113.
104. F. Rodriguez-Llansola, B. Escuder, J. F. Miravet, *Org. Biomol. Chem.* 2009, *7*, 3091.
105. F. Rodríguez-Llansola, B. Escuder, J. F. Miravet, *J. Am. Chem. Soc.* 2009, *131*, 11478.
106. C. Berdugo, J. F. Miravet, B. Escuder, *Chem. Commun.* 2013, *49*, 10608.
107. F. Rodriguez-Llansola, J. F. Miravet, B. Escuder, *Chem. Commun.* 2009, 7303.
108. W. J. M. FrederixPim, G. G. Scott, Y. M. Abul-Haija, D. Kalafatovic, C. G. Pappas, N. Javid, N. T. Hunt, R. V. Ulijn, T. Tuttle, *Nat. Chem.* 2015, *7*, 30.

PART VI

MATERIAL CHEMISTRY

23

NON-COVALENT EXFOLIATION OF GRAPHITE TO PRODUCE GRAPHENE

YINGKUI YANG, DEAN SHI, AND TAO JIANG

MOE Key Laboratory for Green Preparation and Application of Functional Materials, Hubei Collaborative Innovation Center for Advanced Organic Chemical Materials, and Faculty of Materials Science and Engineering, Hubei University, Wuhan, China

23.1 INTRODUCTION

Graphene, initially labeled as an "academic material," has quickly become a real "miracle material" since its first discovery by Novoselov and Geim in 2004 [1]. Structurally, graphene is a one-atom-thick planar sheet of sp^2-bonded carbon atoms densely packed into a 2D honeycomb lattice [1]. In its hexagonal crystal lattice, each carbon atom covalently connects each other by three σ bonds (in plane), while remaining single π-orbital (out of plane) contributes to a delocalized network of π-electrons. Both armchair and zigzag edges usually coexist in graphene although the zigzag configuration is more stable [2]. A graphene nanosheet can range in the lateral size from several nanometers up to centimeters [3–5].

Graphene is conceptually a basic building block for all graphitic materials, where it can be wrapped up into 0D fullerene, rolled into 1D carbon nanotubes (CNTs) and stacked into 3D graphite [6]. Due to its unique structure, single-layer graphene exhibits a wide range of exceptional intrinsic properties including ultrahigh carrier mobility (up to $2 \times 10^5\,cm^2/Vs$) [7, 8], excellent thermal conductivity ($5 \times 10^3\,W m/K$) [9–11], negligible reflectance (<0.1%) and high optical transparency (97.7%) [12], superior Young's modulus (1.1 TPa) and fracture strength (125 GPa) [13], extremely high lateral-to-thickness ratio, and hence ultralarge specific surface area (SSA, $2630\,m^2/g$) [14]. Graphene also shows extraordinary mechanical flexibility, high thermal stability, and outstanding resistance to gas permeation [15, 16].

All the properties combined in one material of graphene have potentially enabled a wide spectrum of multifunctional applications. Over the past years, graphene conducting films have been fabricated for the use of photoanode in polymer solar cells [17–20] and counter electrode in dye-sensitized solar cells [21] as well as flexible optoelectronics by combination with its mechanical flexibility and elasticity [22–24]. Graphene with tunable work function and bandgap can also behave as highly efficient hole transport and electron transport layers, which have the ability to facilitate charge transport and collection and, hence, enhancing power conversion efficiency of solar cells [25–27].

Meanwhile, with excellent electrocatalytic activity, high electrochemical stability, and low cost, graphene-based materials can be considered as potential competitors to replace or reduce the use of expensive metallic catalysts in fuel cells [28–30]. Moreover, high theoretical capacity (744 mAh/g) and specific capacitance (550 F/g), when combined with fascinating properties mentioned above, make graphene electrodes capable of enhancing energy storage, power delivery, and cycle life toward lithium batteries [31–33] and supercapacitors [34–36]. Therefore, graphene has demonstrated its great advantages over conventional materials in energy conversion and storage applications, and it is highly expected to address ever-growing energy demands today [37–39]. In addition, graphene has potential applications in corrosion protection [40], biomedical fields [41–45], and polymer-based composites with improved mechanical, electrical, and thermal properties [46–48].

Despite promising versatility and benefits, graphene synthesis is the first step toward practical applications. Until now, two methodologies including top-down and bottom-up synthesis have been successfully applied to produce graphene

Non-Covalent Interactions in the Synthesis and Design of New Compounds, First Edition.
Edited by Abel M. Maharramov, Kamran T. Mahmudov, Maximilian N. Kopylovich, and Armando J. L. Pombeiro.

materials [49]. Top-down methods involve (i) physically exfoliating graphene from pristine graphite and graphite intercalation compounds (GICs) in the solid phase or in a liquid [50–52], (ii) electrochemical exfoliation using graphite as the sacrificial electrode in an electrolysis cell followed by collecting the isolated graphene sheets from the electrolyte solution [53–56], and (iii) chemically oxidizing graphite to graphite oxide followed by exfoliation to graphene oxide (GO) and subsequent chemical, thermal, and electrochemical reduction to graphene [5, 57, 58]. Bottom-up approaches involve building graphene from hydrocarbon precursors including chemical vapor deposition (CVD) [59–61], epitaxial growth [62–65], chemical coupling/polymerization [66–72], and polymer pyrolysis [73]. Each above method adopted for graphene production has its own advantages and disadvantages in the crystal quality (thickness, lateral size, purity, topology defects, edges, etc.), intrinsic property (electrical, optical, thermal, mechanical, surface chemistry, dispersibility, etc.), accessibility and scalability (cost, yield, etc.), and targeted applications depending on synthetic parameters [74, 75].

Among these approaches, CVD is an industrial technique that produces large-area high-quality graphene films and oriented graphene sheets by catalytic decomposition of hydrocarbons (typically CH_4) on the catalytic support (usually Cu, Ni, or alloy) at high temperatures [60]. CVD-grown graphene films can be transferred to transparent substrates by chemical etching away the metal and serve as transparent conductive electrodes (TCEs) for solar cells and flexible electronics [4]. CVD graphene is a promising method if its production cost and subsequent transfer can be further optimized in the future [76]. Bottom-up synthesis from epitaxial growth on SiC and molecular precursors is limited due to low yield, relative low quality, and high cost [63]. The top-down synthesis by reduction of GO is the most widely used method for low cost, large-scale production of graphene (usually referred as reduced GO, RGO) in the laboratory [57] and industry [77]. Intrinsic defects (edges, vacancies, distortions, etc.) and extrinsic defects (residual functional groups) in RGO can serve as active sites favorable for energy conversion and storage [21, 29, 39]. However, the oxidation process induces severe structural damage to graphene and also leads to a pollution problem because of the release of waste liquid, toxic and/or explosive gases [78]. Note that the above methods produce graphene materials based on various chemical reactions.

Since the first isolation of graphene by mechanical exfoliation from graphite [1], this solid-phase exfoliation method in a nonchemical manner has been extended to the liquid-phase exfoliation of graphite into graphene [3, 79, 80]. Such non-covalently exfoliated graphene can retain a perfect, defect-free 2D crystal structure, which is very suitable for fundamental research of its intrinsic properties. Commercial graphene materials were provided by direct liquid-phase exfoliation of graphite [77]. Liquid-exfoliated graphene suspensions can also enable the realization of solution-processed graphene films onto the substrate of interest and the production of conducting inks and top-down approaches to print electronics [81]. In addition, liquid exfoliation of graphite can allow graphene for wet chemical functionalization and hybridization with other components, forming hybrid electrodes for high-performance batteries and supercapacitors [82]. Therefore, non-covalently exfoliated graphene materials are highly desirable for both academic and industrial purposes. In this chapter, current status of non-covalent exfoliation of graphite into graphene is systematically summarized, with special attention on liquid exfoliation of graphite. We highlight benefits and address potential issues from these non-covalent methods and also forecast future research directions.

23.2 SOLID-PHASE EXFOLIATION

23.2.1 Micromechanical Cleavage

With an interlayer van der Waals (vdW) binding energy of approximately 2 eV/nm^2 within graphite [83], the force required to exfoliate graphite is in the order of magnitude of approximately 300 nN/μm^2 [84]. The weak force seems to be easily achieved; however, until 2004 Geim and Novoselov used an everyday Scotch tape to repeatedly cleave layers apart of highly oriented pyrolytic graphite (HOPG) and experimentally isolated monolayer graphene sheet for the first time [1]. This process is also accepted terminology *Scotch tape* method, as shown in Figure 23.1 [85].

The cleaved individual sheets can reach tens of microns in lateral size with a thickness of approximately 0.4 nm, and also can be pressed onto an appropriate substrate, typically but not only to the oxidized silicon [86].

The method is quite simple yet efficient and has been widely employed in laboratory to prepare highly crystalline graphene samples for fundamental study of intrinsic properties

FIGURE 23.1 Scotch tape procedure to produce graphene. Reproduced with permission from Ref. [85], copyright 2012 Macmillan Publishers Limited.

such as carrier mobility [87, 88], ballistic transport [7], thermal conductivity [10], mechanical properties [13], and Raman spectra [89]. However, such micromechanical cleavage of bulk graphite suffers from low throughput, which is not well suited for commercial applications.

23.2.2 Dry Ball Milling

The ball milling opens up an alternative and powerful process to exfoliate graphite into graphene with low-cost and high-yield production. Recent work has been performed by dry ball milling of solid mixtures of graphite and commercial melamines at 100 rpm for 30 min under N_2 or air atmosphere [90–92]. Driven by the π-conjugation of aromatic melamines and multipoint interactions of extensive hydrogen bonding 2D networks on the graphene surfaces, this milling process allows fast exfoliation of graphite into large quantities of few-layer graphene sheets with low defects (Fig. 23.2). The equivalent benzene derivatives, however, were found to not show comparable exfoliation and stabilization capability toward graphite compared to aminotriazines. Theoretic calculations further reveal that aminotriazines have a strong affinity on graphite by molecular adsorption and interadsorbate hydrogen bonding [90–92]. Solid ball milling of graphite with ammonia borane (NH_3BH_3) can also produce single (25%) and few-layered (≤5) graphene [93].

Inspired by the "Scotch tape" method, an industrial three-roll mill machine was used applicable for the continuous mechanical delamination of natural graphite by addition of a polymer adhesive during milling [94]. Single and few-layer (1.13–1.41 nm in thickness) graphene sheets with lateral sizes of 2.7–5.0 μm were obtained but showed a slightly lower electrical conductivity of 7500 S/m due to the presence of edge defects.

It is worth mentioning that edge-functionalized graphene can be produced during dry ball milling by choice of the exfoliating additives. For instance, carboxylated graphene sheets were easily achieved by ball milling of graphite in the presence of dry ice [95]. Hydroxyl-, hydrogen-, carboxylic-, sulfonic-, and carboxylic/sulfonic acid-functionalized graphene sheets were also obtained from the exfoliation of graphite by ball milling in the presence of potassium hydroxide (KOH) [96], hydrogen, carbon dioxide, sulfur trioxide, or carbon dioxide/sulfur trioxide mixture [97], respectively. This ball milling process is a promising method for direct, massive, and fast fabrication of graphene/polymer [98] and graphene inorganic composites [99] for versatile applications.

23.3 LIQUID-PHASE EXFOLIATION

Liquid-phase exfoliation of layered materials to 2D nanosheets has recently drawn increasing attention [100–103]. Compared to solid-phase exfoliated graphene, liquid-phase exfoliation of graphite has owned its superiority for both scaling-up production of high-quality graphene and allowing wet chemical functionalization [104]. In addition, mechanically split graphene sheets usually tend to reaggregate in order to restack graphitic structure and to minimize

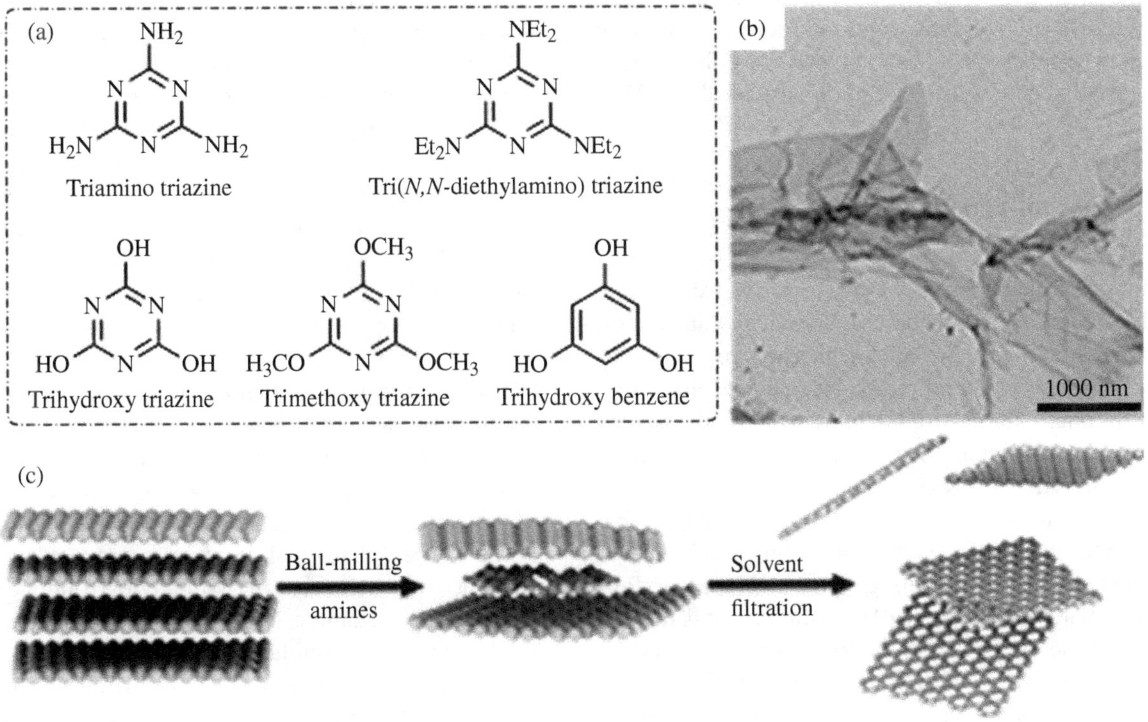

FIGURE 23.2 (a) Chemical structures of the exfoliating agents, (b) a typical TEM image of the exfoliated graphene sheets, and (c) a proposed process to produce graphene sheets. Reproduced with permission from Ref. [90], copyright 2014 American Chemical Society.

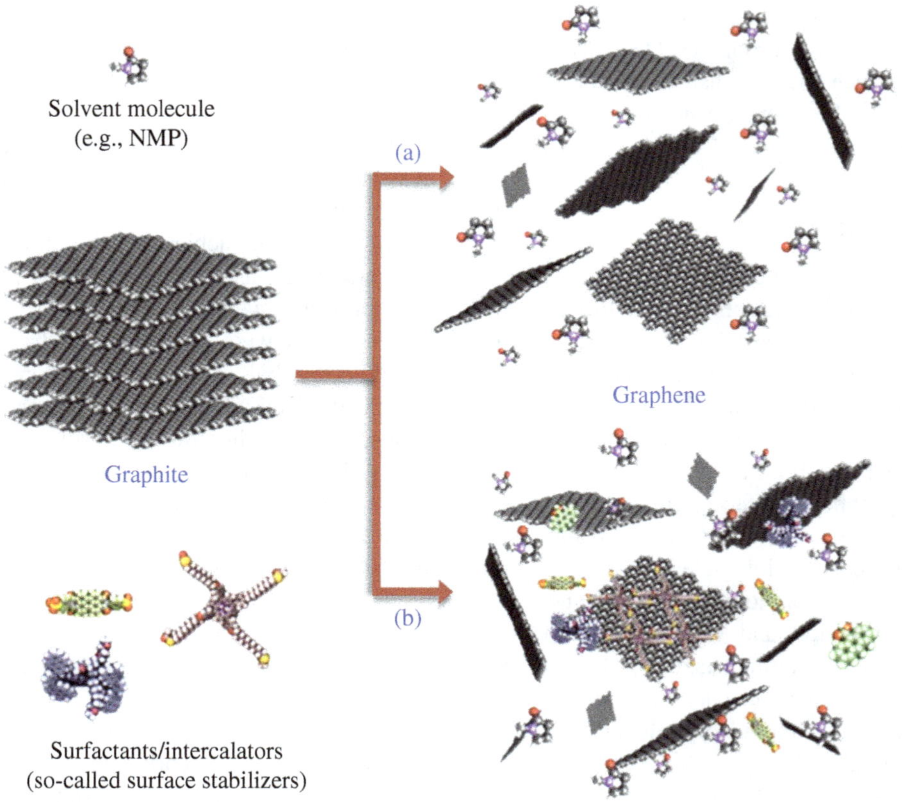

FIGURE 23.3 Schematic illustration for liquid-phase exfoliation of graphite into graphene by (a) surfactant-free and (b) surfactant-assisted exfoliation methods. Reproduced with permission from Ref. [51], copyright 2014 Royal Society of Chemistry.

surface free energy. Liquid-exfoliated graphene is relatively long-term stable and highly desirable for the production of conducting inks and top-down approaches to print electronics [105–107]. To date, liquid-phase exfoliation of graphite into graphene has been achieved in either special solvents or surfactant-assisted solution systems (Fig. 23.3) [3, 79, 80].

23.3.1 Surfactant-Free Exfoliation in Organic Solvents

Liquid-phase exfoliation involves chemical wet dispersion followed by sonication-assisted exfoliation in suitable solvents, as shown in Figure 23.3a. Note that solvent molecules themselves cannot inherently dissolve and exfoliate graphene, additional external forces, most commonly sonication treatment, must be applied to overcome the enormous vdW interactions that always operate between sheets. During sonication in liquid phase, shear forces and cavitation, that is, the growth and collapse of bubbles and voids due to pressure fluctuations, act on bulk graphite and induce exfoliation gradually [50, 51]. Another key requirement is that the solvent–graphene interactions need to balance the intersheet attractive forces after exfoliation. Liquid-exfoliated graphene sheets, otherwise, restack back to graphite-like structure. Therefore, the selection of solvents is absolutely vital for effectively exfoliating graphite.

In general, it is definitely sufficient to identify solvents for given nonpolar solutes with their Hildebrand solubility parameters (δ_T) very close to each other. For polar solutes, an additional requirement for good solvents is that both solvent and solute should have similar values of all three Hansen solubility parameters including dispersive (δ_D), polar (δ_P), and H bonding (δ_H) components. However, graphene is not a molecular solute but a 2D nanomaterial with large lateral size, and hence these criteria are clearly deviated. Pioneering work from Coleman and coworkers [108, 109] has shown that efficient solvents for graphene should have a δ_T of approximately 23 MPa$^{1/2}$ and consequently their surface tensions (γ) are approximately 40 mJ/m^2 (equivalent to surface energy of ~70 mJ/m^2), which is close to that of graphite (68 mJ/m^2) in order to minimize the entropy of mixing [3]. The proposed semiempirical criterion has provided very useful information for the choice of solvents. At present, dozens of organic solvents (see Fig. 23.4) have been tried so far to exfoliate graphite to graphene.

The first exfoliation of graphite to produce single-layer graphene was achieved in *N*-methylpyrrolidone (NMP), *N,N*-dimethylacetamide (DMA), γ-butyrolactone (GBL), and 1,3-dimethyl-2-imidazolidinone (DMEU) under ultrasonication (Fig. 23.5) [109]. After centrifugation to remove the unexfoliated graphite in NMP, the collected supernatant dispersions consist of 28% monolayer and nearly 100% few-layer (≤5) pristine graphene with lateral sizes of a few

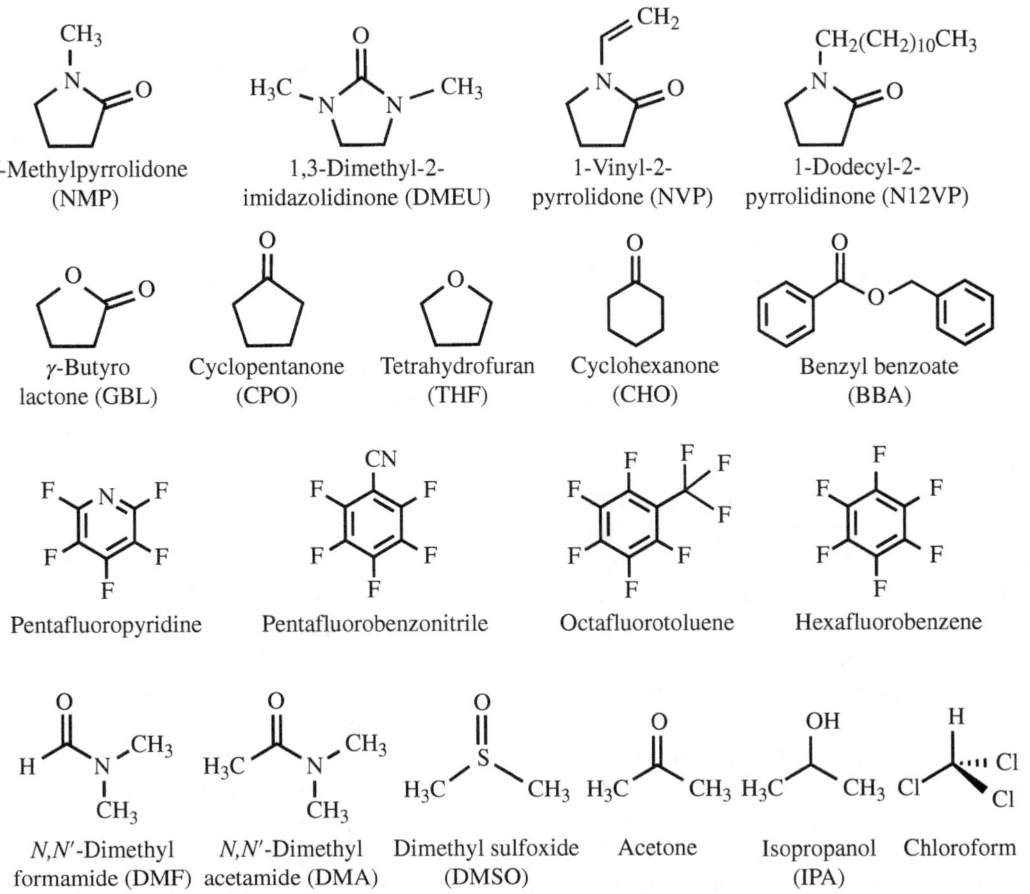

FIGURE 23.4 Chemical structures of currently reported organic solvents for the exfoliation of graphite to graphene [108–116].

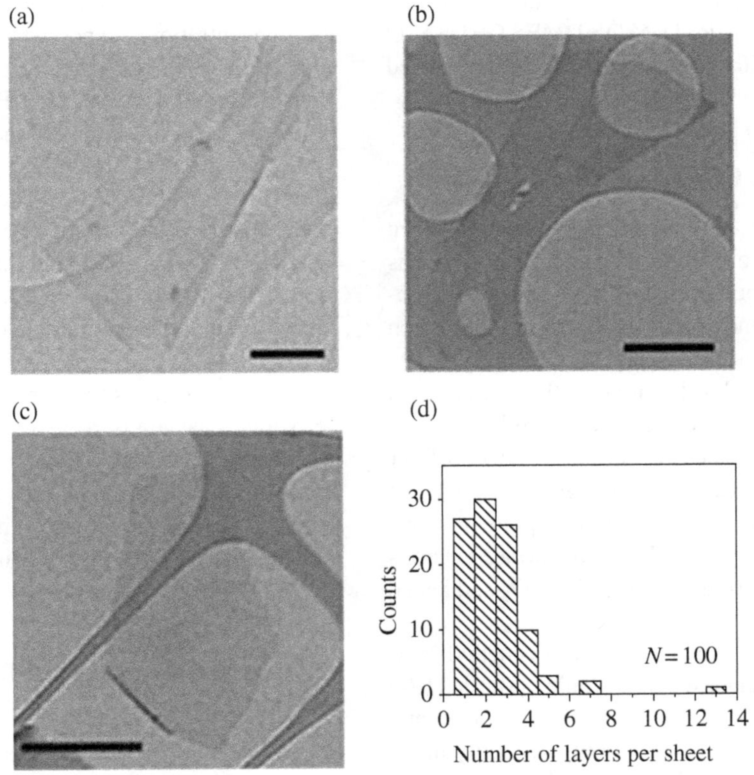

FIGURE 23.5 Bright-field TEM images of single-layer graphene sheets deposited from GBL (a), DMEU (b), and NMP (c), respectively (scale bars: 500 nm). (d) Histogram of the number of visual observations of flakes as a function of the number of monolayers per flake for NMP dispersions. Reproduced with permission from Ref. [109], copyright 2008 Macmillan Publishers Limited.

microns. The monolayer yield relative to total graphite is approximately 1 wt%, which can be gradually improved to 12 wt% with increasing sonication time of up to hundreds of hours [109, 110]. Repeated exfoliation of the recovered partially exfoliated graphite for several rounds was found to produce few-layer graphene dispersions with concentrations of drastically increasing to 63 mg/ml from 0.01 mg/ml [109, 111].

By solvothermal treatment [117] or using supercritical fluid conditions [112], the yield of monolayer and few-layer graphene can be improved to some extent due to the enhanced high diffusivity and solvating power. The lateral size of exfoliated graphene flakes can be further separated from each other by controlling the centrifugation rate [118]. The films made by these graphene sheets have high conductivity of approximately 6500 S/m [109] and low sheet resistance ($R_S \approx 1.5\,k\Omega$) [112]. However, under the same processing conditions, NMP is suggested to be the best solvent in terms of the percentage of monolayer and few-layer graphene due to its values of δ_T and γ matching graphene, while cyclopentanone (CPO) gives the highest absolute concentration of graphene dispersions and benzyl benzoate (BBA) induces a high yield of stable graphene after centrifugation.

Common polar solvents such as *N,N′*-dimethylformamide (DMF) [112–114] and dimethyl sulfoxide (DMSO) [108, 109] as well as fluorinated aromatic solvents [115] exhibit an intermediate efficiency in exfoliating graphite. In contrast, it is low efficient to exfoliate graphite in H_2O due to its excessive values of both δ_T and γ. Theoretical simulations have further demonstrated the ability of several solvents to disperse graphene as follows: NMP $\approx$ DMSO > DMF > GBL > H_2O [116], in fair agreement with the experimental results reported above.

23.3.2 Surfactant-Free Exfoliation in Ionic Liquids

Ionic liquids (ILs) as low-melting organic salts have been widely used as "green" solvents in the materials synthesis because of negligible vapor pressure, low toxicity, incombustibility, high stability, and product recovery in conjunction with tunable solubility and designable functionality [119–121]. In the first report, imidazolium-based ILs were ground with CNTs, and the heavily entangled nanotube bundles could be effectively exfoliated and disentangled into finer bundles even individuals [122]. By analogy, ILs should be alternative candidates for the exfoliation of graphite into graphene. More importantly, many of ILs have surface tensions closely matching with that of graphene [123–127] and hence are potentially available.

Several ILs (Fig. 23.6a) such as 1-butyl-3-methylimidazolium bis(trifluoromethanesulfonyl)imide ($BmimTf_2N$) [128], 1,1′-didodecyl-4,4′-bipyridinium bis(trifluoromethane sulfonyl) imide ($DdimTf_2N$) [129], 1-butyl-3-methylimidazolium hexafluorophosphate ($BmimPF_6$) [130], 1-hexyl-3-methylimidazolium hexafluorophosphate ($HmimPF_6$) [131], and 1-butyl-3-methyl imidazolium tetrafluoroborate ($BmimBF_4$) [132] have been used to produce stable dispersions (Fig. 23.6b) of exfoliated graphene with 2–7 layers and several microns in lateral size. The supernatant concentrations of graphene were reported to be approximately 0.95 mg/mL [128] and increased to be 5.33 mg/mL by pregrinding graphite with ILs before sonication treatment [130].

The exfoliation of graphite and subsequent stabilization of graphene in ILs are attributable to the π–π and/or cation–π interactions between graphene and ILs as well as strong charge polarization of graphene by ILs [122, 133, 134]. Therefore ILs are more effective in exfoliating graphite, dispersing and stabilizing graphene compared to common organic solvents.

However, organic solvents and ILs are very difficult to be fully removed when processing graphene dispersions into conductive films or composites due to their high boiling point (BP) [109, 115, 135]. It is also not easy to deposit individual graphene sheets onto the substrate for fabricating electronics due to the reaggregation during slow evaporation of high BP solvents. Meanwhile, it is virtually impossible to fully separate pure graphene from ILs due to extremely low vapor pressure of ILs and high affinity. However, good solvents for graphene generally have high BPs such as NMP (203°C), DMEU (225°C), GBL (204.5°C), BBA (323.5°C), DMSO (189°C), DMA (165.1°C), and DMF (155°C) [108, 109]. This matches the Trouton's rule that links solubility parameter to BP through the enthalpy of vaporization [136]. This logically means that poor solvents with low solubility parameters and low BPs produce graphene with low efficiency of exfoliation and consequently low concentration and poor stability. However, the extended sonication (48 h) of graphite in low BP solvents such as chloroform and isopropanol (IPA) can improve the exfoliation efficiency to achieve concentrations over 0.4 mg/mL [136], under half of that (1.2 mg/ml) reported in NMP after sonication for 460 h [110]. The dispersions consist of graphene flakes a thickness of <10 layers in chloroform and <5 layers in IPA (Fig. 23.7), and their stable concentrations can increase with the square root of ultrasonication time. Moreover, both dispersions are reasonably stable with >80% of the dispersed phase remaining after 100 h. The advantage of low BP solvents is that they allow individual graphene to be spray cast onto a given substrate while the flake aggregation is clearly observed when spraying from high BP solvents. The disadvantage is that prolonged sonication treatment usually decreases the lateral size of graphene accompanied by an increase in defect concentration [51, 110, 111, 136].

23.3.3 Surfactant-Assisted Exfoliation in Organic Solutions

When graphite is exposed to sonication in organic solvents, the as-exfoliated graphene sheets can be stabilized against aggregation by solvent molecules themselves. A key requirement is the capability to stabilize graphene for a long time at

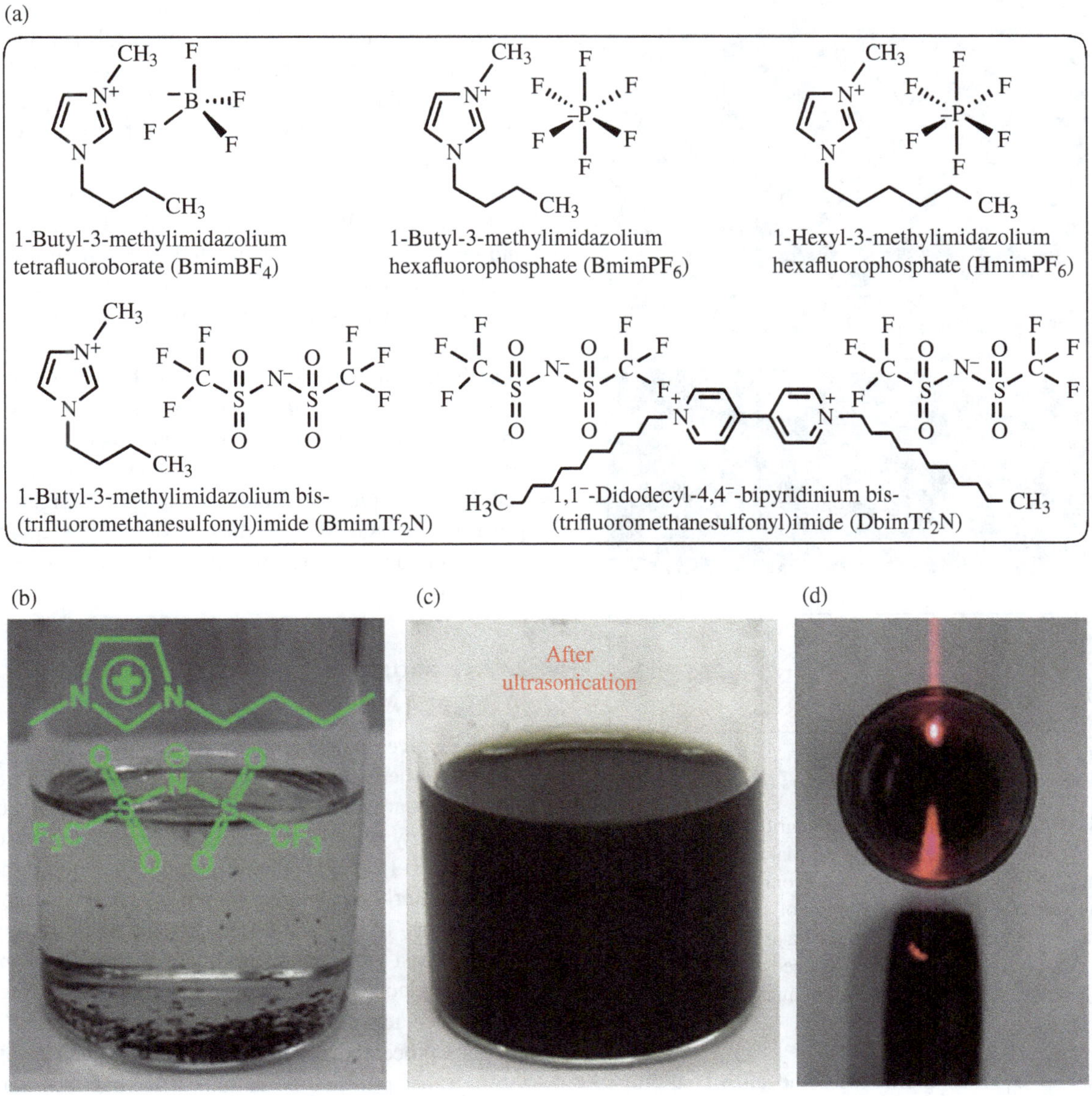

FIGURE 23.6 (a) Chemical structures of currently used ionic liquids, which are favorable for the exfoliation of graphite to graphene [128–132], and photographs of the dispersion of graphite in $BmimTf_2N$ before (b) and after (c) ultrasonication. (d) Tyndall effect of a diluted graphene suspension using a laser pointer. Reproduced with permission from Ref. [128], copyright 2010 by Royal Society of Chemistry.

high concentrations. However, graphene–solvent interactions are not sufficiently strong to compensate for vdWs attractive interactions between layers [116], as described above. Surface molecular adsorption has been suggested to prevent graphene from reaggregation [137]. Therefore, the use of functional additives in solvents is expected to promote fast exfoliation of graphite and subsequent stabilization of graphene suspensions by their strong interactions on the basal plane and edges.

Cationic surfactant-mediated exfoliation of HOPG in anhydrous acetic acid containing cetyltrimethylammonium bromide (CTAB) was reported to give an approximately 10% graphene yield [138]. In this case, CTAB molecules get adsorbed onto the electron-rich graphene surface through electrostatic interactions, while their hydrophobic alkyl chains can prevent the restacking and agglomeration of the exfoliated graphene flakes.

Based on the π–π interactions between graphene and organic porphyrins, monolayer and few-layer graphene flakes can be obtained by exfoliating graphite in NMP [139]. Recently, amphiphilic pyrene molecules were found to promote the exfoliation of edge-functionalized expanded graphite into about sub-7-nm-thick graphene flakes with low defect [140]. Such exfoliated graphene sheets can form a stable dispersion in NMP for 9 months *via* zwitterionic interaction and readily obtain a transparent conductive film on a quartz substrate without noticeable agglomeration by spray coating. The exfoliated graphene in NMP can be also

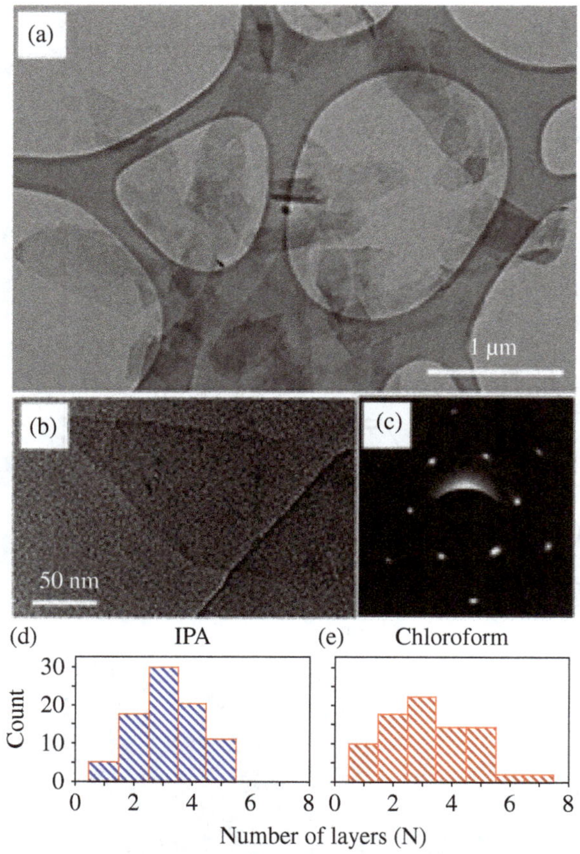

FIGURE 23.7 (a) A TEM image of a grid coated with a large number of few-layer graphene flakes exfoliated in IPA. (b) A high-resolution TEM image of a monolayer as confirmed by the electron diffraction pattern (c). Statistical analysis of TEM images showing histograms of the number of layers per flake dispersed in IPA (d) and chloroform (e), respectively. Reproduced with permission from Ref. [136], copyright 2011 American Chemical Society.

transferred to aqueous media with the aid of amphiphilic poly[styrene-*b*-(2-vinylpyridine)] (PS-*b*-P2VP) and poly (isoprene-*b*-acrylic acid) (PI-*b*-PAA) [141].

A particular benefit using additives lies in the efficient exfoliation of graphite in low BP solvents. For instance, porphyrin-assisted exfoliation of graphite in tetrahydrofuran (THF) was reported to create functional graphene nanohybrids [142]. Zinc phthalocyanines based on oligo-*para*-phenylene vinylene (ZnPc-oPPV) in THF can interact tightly with graphite affording stable and finely dispersed graphene suspensions [143]. The p-type character of oPPV backbone facilitates the π–π interactions with graphene layers, while repulsive forces between oPPV chains and steric hindrance prevent individual systems to interact with each other. Consequently, higher exfoliation degree correlates with more efficient charge transfer between ZnPc and graphene.

Compared to organic additives, functional polymers with higher molecular weight can be more easily adsorbed onto the surface of the newly exfoliated graphene flakes, therefore providing larger steric stabilization against their restacking [144–146]. A conjugated quinquethiophene-terminated PEG was used to effectively exfoliate graphite into graphene in ethanol under bath-type sonication [147]. The high-quality graphene film can be fabricated from its dispersion by means of vacuum filtration and gives a R_S of 0.3 kΩ/sq and a transparency of 74% at 550 nm after further treatment with HNO_3 and $SOCl_2$. By using a hyperbranched polyethylene as the stabilizer, graphene dispersions with concentrations of 0.18 mg/ml in $CHCl_3$ and 0.045 mg/ml in THF were obtained and could be further concentrated to 3.4 mg/ml by solvent evaporation [144]. However, the exfoliation process is completely ineffective using heptane and toluene as solvents.

Efficient graphene exfoliation was also achieved in ethanol through polymeric stabilization of ethyl cellulose [145, 146]. The resulting graphene dispersions can be further concentrated to form stable graphene inks of >1 mg/ml by iterative solvent exchange. Special benefits are that polymer-stabilized graphene solutions can be straightforwardly processed into GBMs including flexible free-standing transparent conductive thin films and graphene/polymer composites [144–146].

23.3.4 Surfactant-Assisted Exfoliation in Aqueous Media

Organic solvents with suitable values of δ_T and γ have demonstrated their own advantages in exfoliating graphite. However, many of high BP solvents such as NMP and DMF are relatively expensive and also toxic to organs to some extent [148–150]. Exfoliating graphite in water is desirable but particularly challenging due to very high surface tension (72.8 mJ/m^2) of water and hydrophobic nature of graphene. Over the past years, many commercial surfactants including organic molecules and polymers have been employed to tune the surface tension of water phase to a proper level for aqueous processed exfoliation and subsequent stabilization through the repulsive potential barrier between surfactant-adsorbed graphene sheets (Fig. 23.3b). Figure 23.8 shows their chemical structures, names, and corresponding abbreviations in the text.

Coleman and coworkers [151] first reported the large-scale exfoliation of graphite in an aqueous solution of SDBS with the aid of ultrasound. The exfoliated graphene suspensions consist of >40% multilayer (<5) and approximately 3% monolayer graphene, and they are well stabilized against reaggregation for approximately 6 weeks by large potential barrier originating in the Coulomb repulsion between surfactant-coated sheets. Due to the persistence of SDBS molecules, monolayer height is approximately 1 nm, and the vacuum deposited films (~30 nm in thickness) give a low DC conductivity of 35 S/m, which can increase to 1500 S/m by thermal annealing to remove residual SDBS. Later, this aqueous phase exfoliation of graphite is enabled by a planar surfactant of SC [152], which forms a stable encapsulation layer on each side of the suspended graphene sheets. The exfoliated graphene dispersion is stable for several weeks, with concentrations up to 0.3 mg/mL with low-power sonication for long times (up to

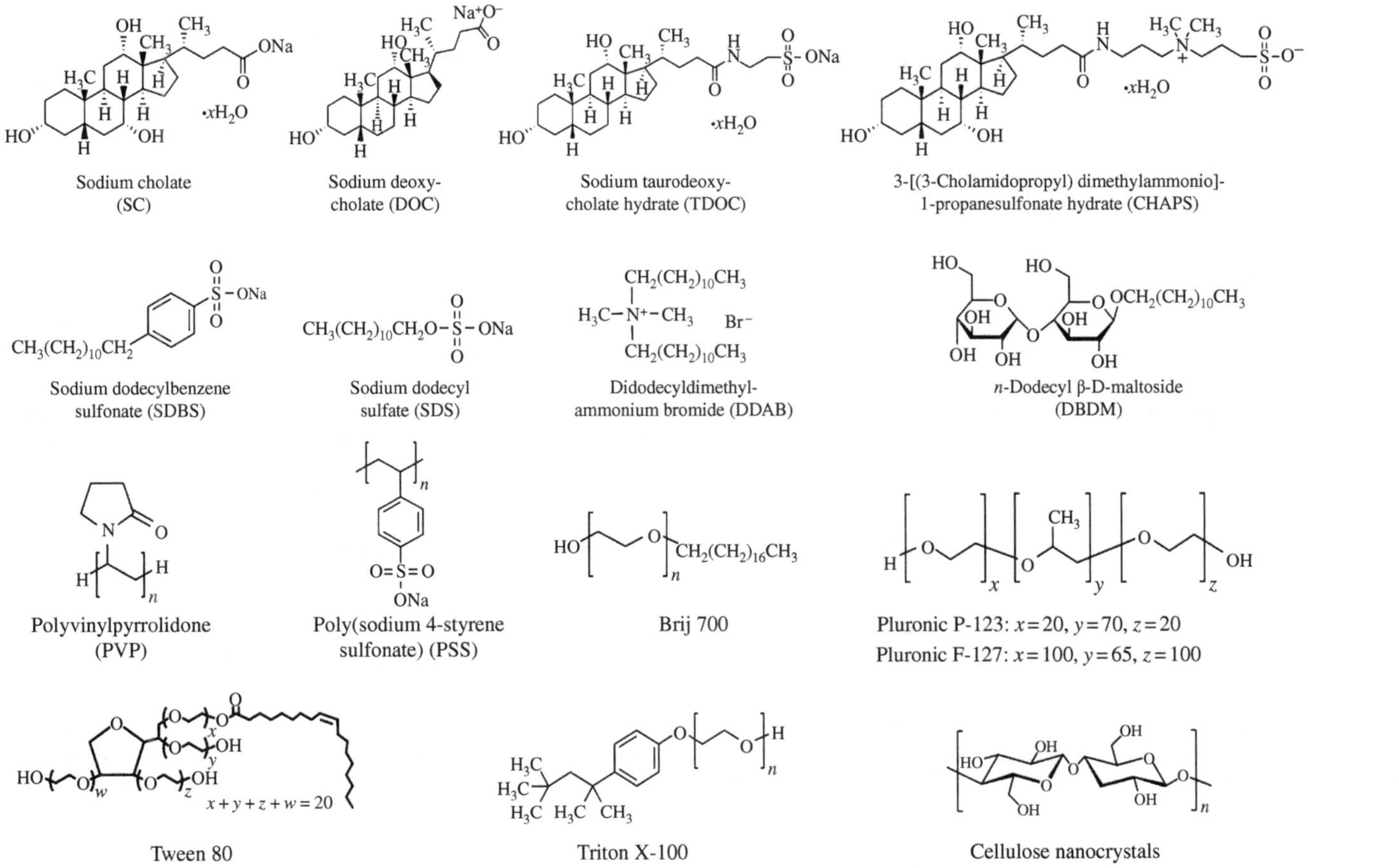

FIGURE 23.8 Chemical structures, corresponding names, and abbreviations of organic and polymer surfactants used as exfoliating agents and stabilizers in the aqueous processing of graphite toward graphene sheets [151–158].

400h) [153]. Moreover, stable dispersions can be cast into high-quality, free-standing films with excellent electrical and mechanical properties or processed into flexible, transparent, and conducting films on substrates [154].

In addition to SDBS and SC, other surfactants including polymers and biomolecules (Fig. 23.8) have been used to assist the exfoliation of graphite during aqueous phase processing [155–158]. These have usually produced molecule-coated monolayer graphene with yields up to approximately 10% for specific surfactants and stable dispersions at concentrations on the order of 0.1mg/mL. However, there are significant differences in the capability of surfactants to exfoliate and disperse graphene in water.

In general, under comparable processing conditions, nonionic surfactants appear to be more effective in exfoliating and stabilizing graphene compared to ionic ones [155, 156]. The stable concentrations of graphene dispersions are usually above 0.10mg/ml whereas ionic surfactants yield values lie in 0.01–0.10mg/ml in most cases. Particularly remarkable is a triblock copolymer of Pluronic P-123, which gives the highest concentration up to 1.5mg/ml after sonication for 5h [155]. It has been proposed that ionic surfactants are non-covalently absorbed onto graphene, resulting in the formation of charged sheets, which provide electrostatic repulsion to prevent them from aggregation.

There are no statistical correlations of graphene concentrations with the polarity of head groups of surfactants. In contrast, nonionic surfactants generally consist of a hydrophobic tail and a long hydrophilic part with high molecular weight and thus large steric repulsions stabilize graphene. The additional repulsive and stabilizing effect possibly originates from the presence of terminal hydroxyl groups and ether linkages interacting with water by H bonding.

A recent study has suggested that the nature of hydrophobic part of surfactants is critical for an efficient dispersion [156]. A nonionic surfactant Triton X-100 was found to give the highest concentration of graphene dispersions due to strong π–π interaction between graphene and benzene rings in Triton X-100. Additional incorporation of metallic salts such as $FeCl_2$ [159], NaCl, and $CuCl_2$ [160] into aqueous media containing surfactants can further facilitate the exfoliation of graphite due to the ion intercalation-induced expansion of graphite.

Pyrene, perylene, and their derivatives containing hydrophobic aromatic rings and hydrophilic functional groups (Fig. 23.9) have been frequently used as surfactant additives to exfoliate graphite and stabilize graphene dispersions [161]. Generally, while sonication serves to overcome strong vdW interactions between layers in graphite, these conjugated

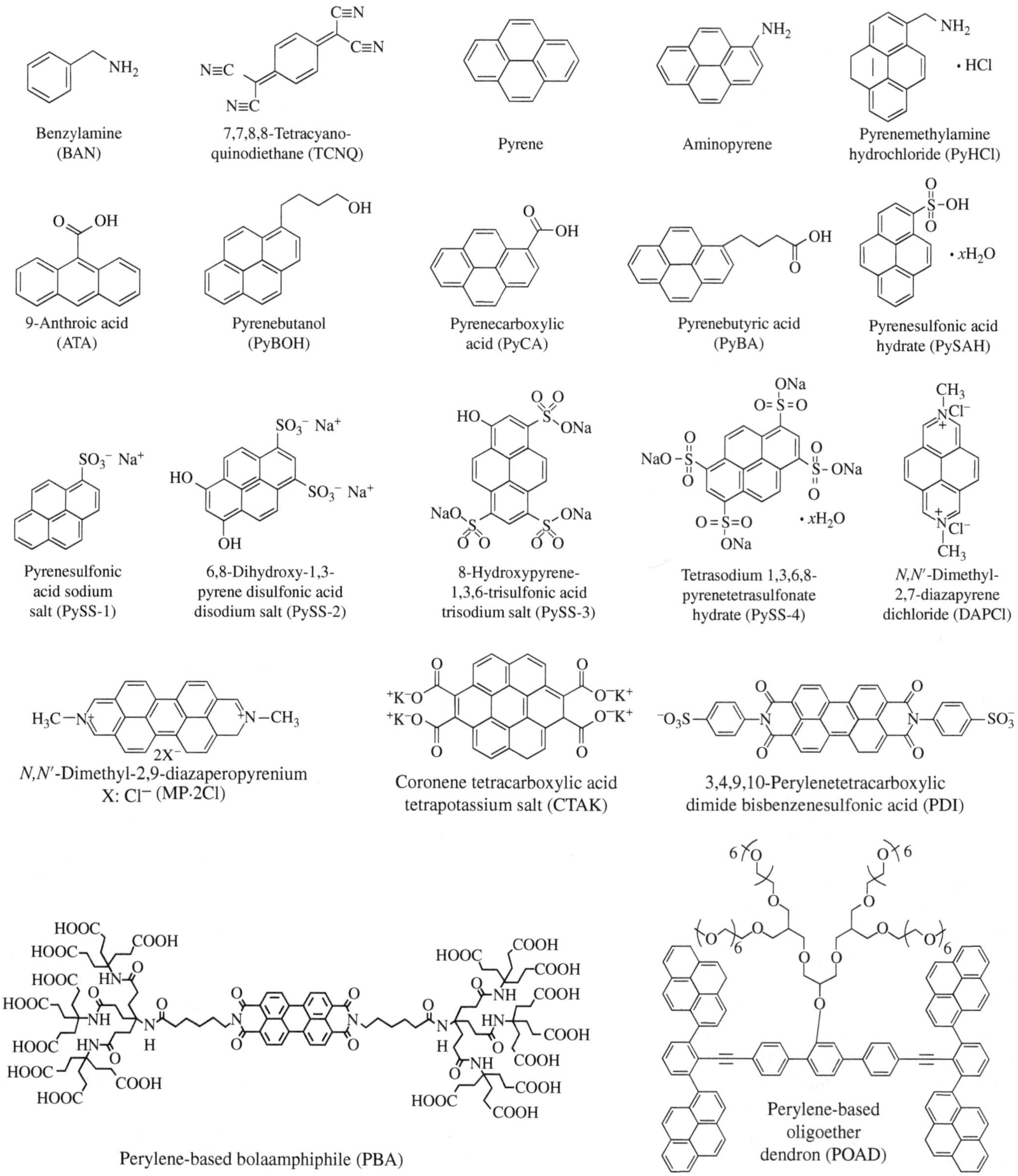

FIGURE 23.9 Chemical structures, corresponding names, and abbreviations of pyrene- and perylene-based derivatives used as surfactants in the aqueous processing of graphite toward graphene [161–172].

molecules tend to intercalate into interlayers and then adsorb *in-situ* onto the graphene surface driven by hydrophobic interactions and π–π stacking between the two [173]. This process guarantees stabilization of the exfoliated graphene in water by reducing surface free energy. However, both exfoliation efficiency and stabilization ability of these aromatic molecules strongly depend on their molecular structure, solubility, and interaction energy with graphene or water.

Several comparative studies have been recently performed on pyrene derivatives [162–166]. Pyrene, aminopyrene, and

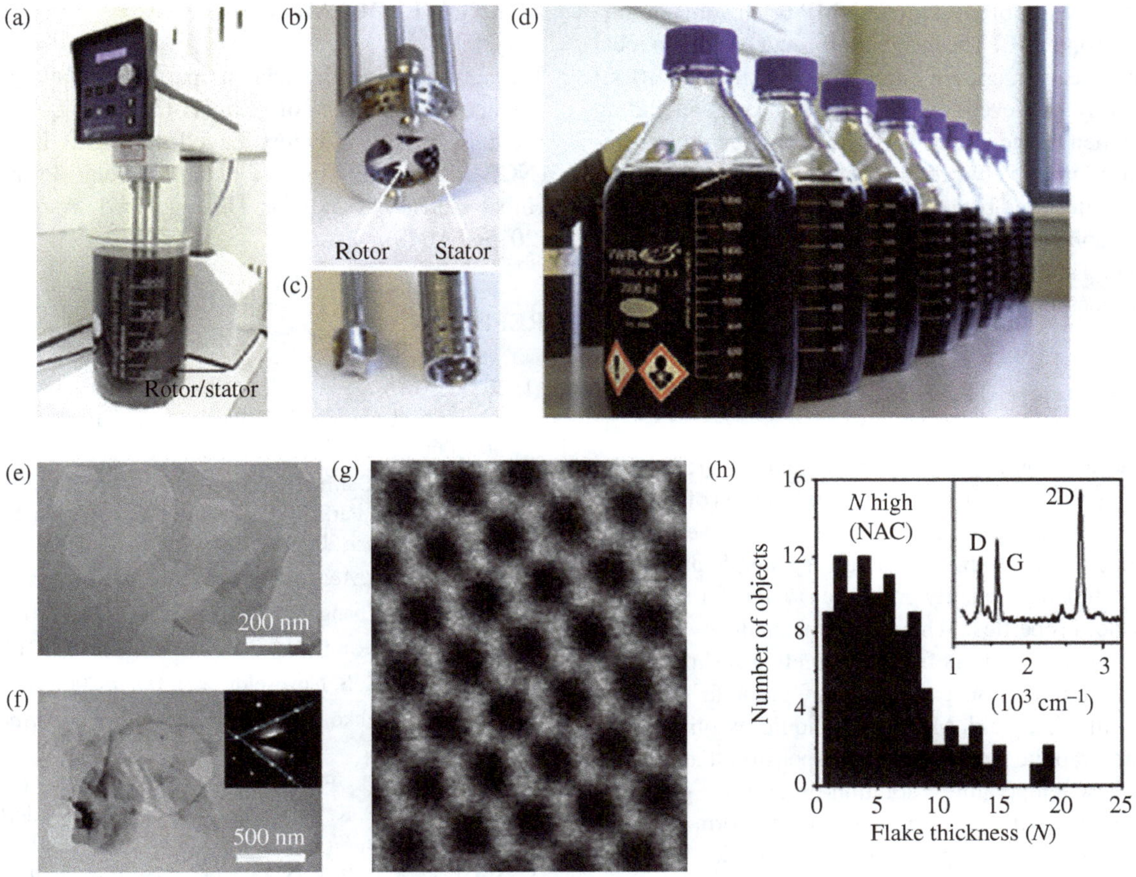

FIGURE 23.10 Production of graphene by shear mixing. (a) A Silverson model high-shear mixer with mixing head in a 5 L beaker of graphene dispersion. Close-up view of a $D = 32$ mm mixing head (b) and a $D = 16$ mm mixing head with rotor (left) separated from stator (c). (d) Photograph of graphene NMP dispersions produced by shear exfoliation. (e, f) TEM images of individual monolayer graphene as evidenced by its electron diffraction pattern (f, inset) and a monolayer imaged by high-resolution scanning TEM (g). (h) Histogram of graphene thickness as measured by AFM on a SC-exfoliated sample. Raman characterization (inset) confirms the presence of monolayer graphene. Reproduced with permission from Ref. [175], copyright 2014 Macmillan Publishers Limited.

PyBOH are insoluble in water and cannot induce the exfoliation of graphite, whereas both PySAH and PyHCl are partially soluble in water. Excess PySAH and PyHCl therefore precipitate and the exfoliated graphene concentration remains constant beyond their saturation concentrations. However, stable concentration of graphene goes through a maximum with increasing PySS-1, PySS-4, PyBA, and PyCA in solutions. Moreover, sulfate group-containing pyrene derivatives such as PySS-1 show higher ability in stabilizing graphene as compared with PySAH possibly due to the electrostatic effect in which both ionic strength and size of Na^+ in PySS-1 are greater than those of H^+ in Py-SAH.

Among pyrene derivatives, PySS-1 was reported to be the most effective one to yield approximately 70% few-layer graphene with concentrations up to 0.8–1 mg/ml. Graphite can be also exfoliated in aqueous solutions of perylene-based ionic salts (PDI [167], MP·2Cl [168]) and nonionic amphiphilic oligomers (PBA [169], POAD [170]) to obtain highly stable graphene dispersions with concentrations up to 1.5 mg/mL. Pyrene-labeled single-stranded DNAs are effective in exfoliating graphite and stabilizing graphene [174]. Benzylamine (BAN) [171] and anionic coronene derivative (CTAK [172])-assisted exfoliation of graphite have been achieved successfully. In addition, these aromatic conjugates are usually electron-rich and can synergistically promote the exfoliation, dispersion, and stabilization of graphene in water by charge-transfer interactions with graphene.

Till now, liquid-phase exfoliation of graphite has become an important technology to produce high-quality graphene for many fundamental applications. However, these methods tend to exhibit either high production rates through long-time sonication or low defect concentration along with low yield but usually not both. The processing scalability is therefore limited to the use of sonication as an energy source. At present, an overwhelming majority of published papers describe graphene production rates below 0.04 g/h (see *Supporting Information* of Ref. [175]), far too low for commercial production. However, Coleman and coworkers [175] demonstrated that high-shear mixing (Fig. 23.10) as a scalable alternative to sonication can achieve the highly efficient exfoliation of graphite in NMP or an aqueous surfactant solution of SC. The exfoliation occurs once the local shear rate exceeds a critical value of approximately $10^4\,s^{-1}$ and is readily achieved in a range of mixers including simple kitchen blenders. In this situation, few-layer graphene can be massively produced in liquid volumes up to hundreds of liters and beyond. Raman spectra also show the resulting graphene to be unoxidized and defect free in the basal plane. This technology may open up a new way for solution-processed graphene on a large scale.

23.4 CONCLUSIONS AND OUTLOOK

The past decade has witnessed considerable progress in the synthesis of graphene by non-covalent exfoliation of graphite into graphene in a liquid or in the solid state. The original "Scotch tape method" appears to be very simple and effective in providing high-quality graphene crystals for researching intrinsic properties. This micromechanical cleavage technique, however, suffers from extremely low throughput. In contrast, liquid-phase exfoliation of graphite produces graphene with high yield. Meanwhile, liquid-exfoliated graphene from graphite forms a stable suspension that enables realizing low-temperature solution-processed graphene films or mixing with other components to form hybrids, which can serve as high-performance electrodes for solar cells, batteries, and supercapacitors. This also allows the production of conducting inks and top-down approaches to print electronics. However, the liquid-exfoliation process usually produces a mixture of monolayer, few-layer graphene sheets, and unexfoliated graphite flakes, and they need to be carefully separated from each other. Moreover, specific surfactants may be required in order to improve the exfoliation efficiency and yield, thus suffering from solvent or surfactant contamination.

Based on the above considerations, future research should concentrate on (i) optimizing the exfoliation process to obtain large-area mono-/few-layer graphene with few defect (or defect free) and high yield (e.g., by improving the ball milling or shearing-exfoliation technique); (ii) the appropriate choice of green solvents and dispersants to achieve stable suspensions with high concentration; (iii) removing/recycling residual solvents and stabilizers from final product; (iv) pretreating graphite to improve the exfoliation efficiency by intercalating compounds in combination with thermal expansion; and (v) clarifying the mechanisms of dispersion, exfoliation, and subsequent stabilization of graphene in various solvents or solutions. More importantly, to have commercially potential, non-covalently exfoliated graphene materials should demonstrate their properties beyond those of existing materials in addition to the mass production capacity and comparable costs. There are great challenges for achieving such goals and therefore a wide range of technical innovations is still necessary in the future.

ACKNOWLEDGEMENTS

The authors thank financial support from National Natural Science Foundation of China (51273057, 51073050), Program for New Century Excellent Talents in University (NCET-12-0709), Hubei Chutian Scholar Program, and Hubei Science Fund for Distinguished Young Scholars (2015CFA048).

REFERENCES

1. K. S. Novoselov, A. K. Geim, S. Morozov, D. Jiang, Y. Zhang, S. Dubonos, I. Grigorieva, A. Firsov, *Science* 2004, *306*, 666.
2. Ç. Ö. Girit, J. C. Meyer, R. Erni, M. D. Rossell, C. Kisielowski, L. Yang, C.-H. Park, M. F. Crommie, M. L. Cohen, S. G. Louie, A. Zettl, *Science* 2009, *323*, 1705.
3. J. N. Coleman, *Acc. Chem. Res.* 2013, *46*, 14.
4. Y. Zhang, L. Zhang, C. Zhou, *Acc. Chem. Res.* 2013, *46*, 2329.
5. S. Park, R. S. Ruoff, *Nat. Nanotechnol.* 2009, *4*, 217.
6. A. K. Geim, K. S. Novoselov, *Nat. Mater.* 2007, *6*, 183.
7. X. Du, I. Skachko, A. Barker, E. Y. Andrei, *Nat. Nanotechnol.* 2008, *3*, 491.
8. M. Orlita, C. Faugeras, P. Plochocka, P. Neugebauer, G. Martinez, D. K. Maude, A.-L. Barra, M. Sprinkle, C. Berger, W. A. De Heer, *Phys. Rev. Lett.* 2008, *101*, 267601.
9. S. Ghosh, I. Calizo, D. Teweldebrhan, E. Pokatilov, D. Nika, A. Balandin, W. Bao, F. Miao, C. N. Lau, *Appl. Phys. Lett.* 2008, *92*, 151911.
10. A. A. Balandin, S. Ghosh, W. Bao, I. Calizo, D. Teweldebrhan, F. Miao, C. N. Lau, *Nano Lett.* 2008, *8*, 902.
11. I. Calizo, A. Balandin, W. Bao, F. Miao, C. Lau, *Nano Lett.* 2007, *7*, 2645.
12. R. R. Nair, P. Blake, A. N. Grigorenko, K. S. Novoselov, T. J. Booth, T. Stauber, N. M. R. Peres, A. K. Geim, *Science* 2008, *320*, 1308.
13. C. Lee, X. Wei, J. W. Kysar, J. Hone, *Science* 2008, *321*, 385.
14. M. D. Stoller, S. Park, Y. Zhu, J. An, R. S. Ruoff, *Nano Lett.* 2008, *8*, 3498.
15. A. K. Geim, *Science* 2009, *324*, 1530.
16. K. S. Novoselov, V. I. Fal'ko, L. Colombo, P. R. Gellert, M. G. Schwab, K. Kim, *Nature* 2012, *490*, 192.
17. M. He, J. Jung, F. Qiu, Z. Lin, *J. Mater. Chem.* 2012, *22*, 24254.
18. A. Iwan, A. Chuchmala, *Prog. Polym. Sci.* 2012, *37*, 1805.
19. D. S. Hecht, L. Hu, G. Irvin, *Adv. Mater.* 2011, *23*, 1482.
20. S. Pang, Y. Hernandez, X. Feng, K. Müllen, *Adv. Mater.* 2011, *23*, 2779.
21. J. D. Roy-Mayhew, I. A. Aksay, *Chem. Rev.* 2014, *114*, 6323.
22. T.-H. Han, Y. Lee, M.-R. Choi, S.-H. Woo, S.-H. Bae, B. H. Hong, J.-H. Ahn, T.-W. Lee, *Nat. Photonics* 2012, *6*, 105.
23. F. Bonaccorso, Z. Sun, T. Hasan, A. C. Ferrari, *Nat. Photonics* 2010, *4*, 611.
24. G. Eda, M. Chhowalla, *Adv. Mater.* 2010, *22*, 2392.
25. J. Liu, M. Durstock, L. Dai, *Energy Environ. Sci.* 2014, *7*, 1297.

26. L. Dai, *Acc. Chem. Res.* 2013, *46*, 31.
27. Z. Yin, J. Zhu, Q. He, X. Cao, C. Tan, H. Chen, Q. Yan, H. Zhang, *Adv. Eng. Mater.* 2014, *4*, 1300574.
28. X.-K. Kong, C.-L. Chen, Q.-W. Chen, *Chem. Soc. Rev.* 2014, *43*, 2841.
29. M. Liu, R. Zhang, W. Chen, *Chem. Rev.* 2014, *114*, 5117.
30. M. Kaukonen, A. V. Krasheninnikov, E. Kauppinen, R. M. Nieminen, *ACS Catal.* 2013, *3*, 159.
31. M. Srivastava, J. Singh, T. Kuila, R. K. Layek, N. H. Kim, J. H. Lee, *Nanoscale* 2015, *7*, 4820.
32. R. Raccichini, A. Varzi, S. Passerini, B. Scrosati, *Nat. Mater.* 2015, *14*, 271.
33. S. Han, D. Wu, S. Li, F. Zhang, X. Feng, *Small* 2013, *9*, 1173.
34. J. Zhu, D. Yang, Z. Yin, Q. Yan, H. Zhang, *Small* 2014, *10*, 3480.
35. F. Beguin, V. Presser, A. Balducci, E. Frackowiak, *Adv. Mater.* 2014, *26*, 2219.
36. J. Yan, Q. Wang, T. Wei, Z. Fan, *Adv. Eng. Mater.* 2014, *4*, 13300816.
37. F. Bonaccorso, L. Colombo, G. Yu, M. Stoller, V. Tozzini, A. C. Ferrari, R. S. Ruoff, V. Pellegrini, *Science* 2015, *347*, 1246501.
38. I. V. Lightcap, P. V. Kamat, *Acc. Chem. Res.* 2013, *46*, 2235.
39. Y. Sun, Q. Wu, G. Shi, *Energy Environ. Sci.* 2011, *4*, 1113.
40. S. Böhm, *Nat. Nanotechnol.* 2014, *9*, 741.
41. M. Drndić, *Nat. Nanotechnol.* 2014, *9*, 743.
42. D. Bitounis, H. Ali-Boucetta, B. H. Hong, D.-H. Min, K. Kostarelos, *Adv. Mater.* 2013, *25*, 2258.
43. C. Chung, Y.-K. Kim, D. Shin, S.-R. Ryoo, B. H. Hong, D.-H. Min, *Acc. Chem. Res.* 2013, *46*, 2211.
44. A. Bianco, *Angew. Chem. Int. Ed.* 2013, *52*, 4986.
45. H. Y. Mao, S. Laurent, W. Chen, O. Akhavan, M. Imani, A. A. Ashkarran, M. Mahmoudi, *Chem. Rev.* 2013, *113*, 3407.
46. Y. Yang, C. E. He, R. G. Peng, A. Baji, X. Du, Y. L. Huang, X. Xie, Y.-W. Mai, *J. Mater. Chem.* 2012, *22*, 5666.
47. Y. Yang, W. Zhan, R. Peng, C. He, X. Pang, D. Shi, T. Jiang, Z. Lin, *Adv. Mater.* 2015, *27*, 6376.
48. Y. Yang, C. E. He, W. Tang, C. P. Tsui, D. Shi, Z. Sun, T. Jiang, X. Xie, *J. Mater. Chem. A* 2014, *2*, 20038.
49. M. J. Allen, V. C. Tung, R. B. Kaner, *Chem. Rev.* 2010, *110*, 132.
50. M. Cai, D. Thorpe, D. H. Adamson, H. C. Schniepp, *J. Mater. Chem.* 2012, *22*, 24992.
51. A. Ciesielski, P. Samori, *Chem. Soc. Rev.* 2014, *43*, 381.
52. M. Inagaki, Y. A. Kim, M. Endo, *J. Mater. Chem.* 2011, *21*, 3280.
53. K. S. Rao, J. Sentilnathan, H. W. Cho, J. J. Wu, M. Yoshimura, *Adv. Funct. Mater.* 2015, *25*, 298.
54. A. Abdelkader, A. Cooper, R. Dryfe, I. Kinloch, *Nanoscale* 2015, *7*, 6944.
55. J. Wang, J. Huang, R. Yan, F. Wang, W. Cheng, Q. Guo, J. Wang, *J. Mater. Chem. A* 2015. *3*, 3144.
56. M. P. Kumar, M. Nidhi, C. Srivastava, *RSC Adv.* 2015, *5*, 24846.
57. C. K. Chua, M. Pumera, *Chem. Soc. Rev.* 2014, *43*, 291.
58. S. Pei, H.-M. Cheng, *Carbon* 2012, *50*, 3210.
59. A. W. Tsen, L. Brown, R. W. Havener, J. Park, *Acc. Chem. Res.* 2013, *46*, 2286.
60. D. Wei, B. Wu, Y. Guo, G. Yu, Y. Liu, *Acc. Chem. Res.* 2013, *46*, 106.
61. Z. Yan, Z. Peng, J. M. Tour, *Acc. Chem. Res.* 2014, *47*, 1327.
62. L. Chen, Y. Hernandez, X. Feng, K. Muellen, *Angew. Chem. Int. Ed.* 2012, *51*, 7640.
63. P. W. Sutter, J.-I. Flege, E. A. Sutter, *Nat. Mater.* 2008, *7*, 406.
64. W. Yang, G. Chen, Z. Shi, C.-C. Liu, L. Zhang, G. Xie, M. Cheng, D. Wang, R. Yang, D. Shi, *Nat. Mater.* 2013, *12*, 792.
65. P. C. Rogge, K. Thürmer, M. E. Foster, K. F. McCarty, O. D. Dubon, N. C. Bartelt, *Nat. Commun.* 2015, *6*, 6880.
66. J. Cai, P. Ruffieux, R. Jaafar, M. Bieri, T. Braun, S. Blankenburg, M. Muoth, A. P. Seitsonen, M. Saleh, X. Feng, K. Mullen, R. Fasel, *Nature* 2010, *466*, 470.
67. W. Pisula, X. Feng, K. Muellen, *Chem. Mater.* 2011, *23*, 554.
68. L. Dössel, L. Gherghel, X. Feng, K. Müllen, *Angew. Chem. Int. Ed.* 2011, *50*, 2540.
69. P. Ruffieux, J. Cai, N. C. Plumb, L. Patthey, D. Prezzi, A. Ferretti, E. Molinari, X. Feng, K. Müllen, C. A. Pignedoli, R. Fasel, *ACS Nano* 2012, *6*, 6930.
70. X.-H. Li, M. Antonietti, *Angew. Chem. Int. Ed.* 2013, *52*, 4572.
71. X. H. Li, S. Kurasch, U. Kaiser, M. Antonietti, *Angew. Chem. Int. Ed.* 2012, *51*, 9689.
72. X. Liu, M. Antonietti, *Adv. Mater.* 2013, *25*, 6284.
73. Y. Li, Z. Li, P. K. Shen, *Adv. Mater.* 2013, *25*, 2474.
74. R. S. Edwards, K. S. Coleman, *Nanoscale* 2013, *5*, 38.
75. S. Bae, H. Kim, Y. Lee, X. Xu, J.-S. Park, Y. Zheng, J. Balakrishnan, T. Lei, H. Ri Kim, Y. I. Song, Y.-J. Kim, K. S. Kim, B. Ozyilmaz, J.-H. Ahn, B. H. Hong, S. Iijima, *Nat. Nanotechnol.* 2010, *5*, 574.
76. K. S. Kim, Y. Zhao, H. Jang, S. Y. Lee, J. M. Kim, K. S. Kim, J.-H. Ahn, P. Kim, J.-Y. Choi, B. H. Hong, *Nature* 2009, *457*, 706.
77. W. Ren, H.-M. Cheng, *Nat. Nanotechnol.* 2014, *9*, 726.
78. C. Galande, W. Gao, A. Mathkar, A. M. Dattelbaum, T. N. Narayanan, A. D. Mohite, P. M. Ajayan, *Part. Part. Syst. Charact.* 2014, *31*, 619.
79. J. N. Coleman, *Adv. Funct. Mater.* 2009, *19*, 3680.
80. X. Cui, C. Zhang, R. Hao, Y. Hou, *Nanoscale* 2011, *3*, 2118.
81. X. Wan, Y. Huang, Y. Chen, *Acc. Chem. Res.* 2012, *45*, 598.
82. Z.-S. Wu, G. Zhou, L.-C. Yin, W. Ren, F. Li, H.-M. Cheng, *Nano Energy* 2012, *1*, 107.
83. L. A. Girifalco, R. A. Lad, *J. Chem. Phys.* 1956, *25*, 693.
84. Y. Zhang, J. P. Small, W. V. Pontius, P. Kim, *Appl. Phys. Lett.* 2005, *86*, 073104.
85. R. Van Noorden, *Nature* 2012, *483*, S32.
86. P. Blake, E. W. Hill, A. H. Castro Neto, K. S. Novoselov, D. Jiang, R. Yang, T. J. Booth, A. K. Geim, *Appl. Phys. Lett.* 2007, 91.
87. S. Morozov, K. Novoselov, M. Katsnelson, F. Schedin, D. Elias, J. Jaszczak, A. Geim, *Phys. Rev. Lett.* 2008, *100*, 016602.

88. E. V. Castro, H. Ochoa, M. Katsnelson, R. Gorbachev, D. Elias, K. Novoselov, A. Geim, F. Guinea, *Phys. Rev. Lett.* 2010, *105*, 266601.
89. S. Berciaud, S. Ryu, L. E. Brus, T. F. Heinz, *Nano Lett.* 2009, *9*, 346.
90. V. Leon, A. M. Rodriguez, P. Prieto, M. Prato, E. Vazquez, *ACS Nano* 2014, *8*, 563.
91. V. Leon, M. Quintana, M. A. Herrero, J. L. G. Fierro, A. de la Hoz, M. Prato, E. Vazquez, *Chem. Commun.* 2011, *47*, 10936.
92. J. D. Wuest, A. Rochefort, *Chem. Commun.* 2010, *46*, 2923.
93. L. Liu, Z. Xiong, D. Hu, G. Wu, P. Chen, *Chem. Commun.* 2013, *49*, 7890.
94. J. Chen, M. Duan, G. Chen, *J. Mater. Chem.* 2012, *22*, 19625.
95. I.-Y. Jeon, Y.-R. Shin, G.-J. Sohn, H.-J. Choi, S.-Y. Bae, J. Mahmood, S.-M. Jung, J.-M. Seo, M.-J. Kim, D. Wook Chang, L. Dai, J.-B. Baek, *Proc. Natl. Acad. Sci. U. S. A.* 2012, *109*, 5588.
96. L. Yan, M. Lin, C. Zeng, Z. Chen, S. Zhang, X. Zhao, A. Wu, Y. Wang, L. Dai, J. Qu, M. Guo, Y. Liu, *J. Mater. Chem.* 2012, *22*, 8367.
97. I.-Y. Jeon, H.-J. Choi, S.-M. Jung, J.-M. Seo, M.-J. Kim, L. Dai, J.-B. Baek, *J. Am. Chem. Soc.* 2013, *135*, 1386.
98. H. Wu, W. Zhao, H. Hu, G. Chen, *J. Mater. Chem.* 2011, *21*, 8626.
99. S. Yang, P. Gao, D. Bao, Y. Chen, L. Wang, P. Yang, G. Li, Y. Sun, *J. Mater. Chem. A* 2013, *1*, 6731.
100. R. J. Smith, P. J. King, M. Lotya, C. Wirtz, U. Khan, S. De, A. O'Neill, G. S. Duesberg, J. C. Grunlan, G. Moriarty, J. Chen, J. Wang, A. I. Minett, V. Nicolosi, J. N. Coleman, *Adv. Mater.* 2011, *23*, 3944.
101. J. N. Coleman, M. Lotya, A. O'Neill, S. D. Bergin, P. J. King, U. Khan, K. Young, A. Gaucher, S. De, R. J. Smith, I. V. Shvets, S. K. Arora, G. Stanton, H.-Y. Kim, K. Lee, G. T. Kim, G. S. Duesberg, T. Hallam, J. J. Boland, J. J. Wang, J. F. Donegan, J. C. Grunlan, G. Moriarty, A. Shmeliov, R. J. Nicholls, J. M. Perkins, E. M. Grieveson, K. Theuwissen, D. W. McComb, P. D. Nellist, V. Nicolosi, *Science* 2011, *331*, 568.
102. V. Nicolosi, M. Chhowalla, M. G. Kanatzidis, M. S. Strano, J. N. Coleman, *Science* 2013, *340*, 1420.
103. K. Lee, H.-Y. Kim, M. Lotya, J. N. Coleman, G.-T. Kim, G. S. Duesberg, *Adv. Mater.* 2011, *23*, 4178.
104. A. Hirsch, J. M. Englert, F. Hauke, *Acc. Chem. Res.* 2013, *46*, 87.
105. F. Torrisi, T. Hasan, W. Wu, Z. Sun, A. Lombardo, T. S. Kulmala, G.-W. Hsieh, S. Jung, F. Bonaccorso, P. J. Paul, D. Chu, A. C. Ferrari, *ACS Nano* 2012, *6*, 2992.
106. K.-Y. Shin, J.-Y. Hong, J. Jang, *Adv. Mater.* 2011, *23*, 2113.
107. L. Huang, Y. Huang, J. Liang, X. Wan, Y. Chen, *Nano Res.* 2011, *4*, 675.
108. Y. Hernandez, M. Lotya, D. Rickard, S. D. Bergin, J. N. Coleman, *Langmuir* 2009, *26*, 3208.
109. Y. Hernandez, V. Nicolosi, M. Lotya, F. M. Blighe, Z. Sun, S. De, I. T. McGovern, B. Holland, M. Byrne, Y. K. Gun'ko, J. J. Boland, P. Niraj, G. Duesberg, S. Krishnamurthy, R. Goodhue, J. Hutchison, V. Scardaci, A. C. Ferrari, J. N. Coleman, *Nat. Nanotechnol.* 2008, *3*, 563.
110. U. Khan, A. O'Neill, M. Lotya, S. De, J. N. Coleman, *Small* 2010, *6*, 864.
111. U. Khan, H. Porwal, A. O'Neill, K. Nawaz, P. May, J. N. Coleman, *Langmuir* 2011, *27*, 9077.
112. D. Rangappa, K. Sone, M. Wang, U. K. Gautam, D. Golberg, H. Itoh, M. Ichihara, I. Honma, *Chem. Eur. J.* 2010, *16*, 6488.
113. A. V. Alaferdov, A. Gholamipour-Shirazi, M. A. Canesqui, Y. A. Danilov, S. A. Moshkalev, *Carbon* 2014, *69*, 525.
114. P. Blake, P. D. Brimicombe, R. R. Nair, T. J. Booth, D. Jiang, F. Schedin, L. A. Ponomarenko, S. V. Morozov, H. F. Gleeson, E. W. Hill, A. K. Geim, K. S. Novoselov, *Nano Lett.* 2008, *8*, 1704.
115. A. B. Bourlinos, V. Georgakilas, R. Zboril, T. A. Steriotis, A. K. Stubos, *Small* 2009, *5*, 1841.
116. C.-J. Shih, S. Lin, M. S. Strano, D. Blankschtein, *J. Am. Chem. Soc.* 2010, *132*, 14638.
117. Z. Tang, J. Zhuang, X. Wang, *Langmuir* 2010, *26*, 9045.
118. U. Khan, A. O'Neill, H. Porwal, P. May, K. Nawaz, J. N. Coleman, *Carbon* 2012, *50*, 470.
119. H. Weingärtner, *Angew. Chem. Int. Ed.* 2008, *47*, 654.
120. T. L. Greaves, C. J. Drummond, *Chem. Rev.* 2007, *108*, 206.
121. R. Peng, Y. Wang, W. Tang, Y. Yang, X. Xie, *Polymers* 2013, *5*, 847.
122. T. Fukushima, A. Kosaka, Y. Ishimura, T. Yamamoto, T. Takigawa, N. Ishii, T. Aida, *Science* 2003, *300*, 2072.
123. J. Restolho, J. L. Mata, B. Saramago, *J. Colloid Interface Sci.* 2009, *340*, 82.
124. M. Tariq, M. G. Freire, B. Saramago, J. A. P. Coutinho, J. N. C. Lopes, L. P. N. Rebelo, *Chem. Soc. Rev.* 2012, *41*, 829.
125. R. Sedev, *Curr. Opin. Colloid Interface Sci.* 2011, *16*, 310.
126. S. Zhang, N. Sun, X. He, X. Lu, X. Zhang, *J. Phys. Chem. Ref. Data* 2006, *35*, 1475.
127. C. Kolbeck, J. Lehmann, K. R. J. Lovelock, T. Cremer, N. Paape, P. Wasserscheid, A. P. Fröba, F. Maier, H. P. Steinrück, *J. Phys. Chem. B* 2010, *114*, 17025.
128. X. Q. Wang, P. F. Fulvio, G. A. Baker, G. M. Veith, R. R. Unocic, S. M. Mahurin, M. F. Chi, S. Dai, *Chem. Commun.* 2010, *46*, 4487.
129. A. Safavi, M. Tohidi, F. A. Mahyari, H. Shahbaazi, *J. Mater. Chem.* 2012, 22.
130. D. Nuvoli, L. Valentini, V. Alzari, S. Scognamillo, S. B. Bon, M. Piccinini, J. Illescas, A. Mariani, *J. Mater. Chem.* 2011, *21*, 3428.
131. N. G. Shang, P. Papakonstantinou, S. Sharma, G. Lubarsky, M. Li, D. W. McNeill, A. J. Quinn, W. Zhou, R. Blackley, *Chem. Commun.* 2012, *48*, 1877.
132. J. Lu, J.-X. Yang, J. Wang, A. Lim, S. Wang, K. P. Loh, *ACS Nano* 2009, *3*, 2367.
133. Y. Shim, H. J. Kim, *ACS Nano* 2009, *3*, 1693.
134. J. Y. Wang, H. B. Chu, Y. Li, *ACS Nano* 2008, *2*, 2540.
135. U. Khan, P. May, A. O'Neill, J. N. Coleman, *Carbon* 2010, *48*, 4035.
136. A. O'Neill, U. Khan, P. N. Nirmalraj, J. Boland, J. N. Coleman, *J. Phys. Chem. C* 2011, *115*, 5422.
137. C. Backes, F. Hauke, A. Hirsch, *Adv. Mater.* 2011, *23*, 2588.

138. S. Vadukumpully, J. Paul, S. Valiyaveettil, *Carbon* 2009, *47*, 3288.

139. J. Geng, B.-S. Kong, S. B. Yang, H.-T. Jung, *Chem. Commun.* 2010, *46*, 5091.

140. J. S. Park, L. Yu, C. S. Lee, K. Shin, J. H. Han, *J. Colloid Interface Sci.* 2014, *417*, 379.

141. T. Skaltsas, N. Karousis, H.-J. Yan, C.-R. Wang, S. Pispas, N. Tagmatarchis, *J. Mater. Chem.* 2012, *22*, 21507.

142. J. Malig, A. W. I. Stephenson, P. Wagner, G. G. Wallace, D. L. Officer, D. M. Guldi, *Chem. Commun.* 2012, *48*, 8745.

143. L. Brinkhaus, G. Katsukis, J. Malig, R. D. Costa, M. Garcia-Iglesias, P. Vázquez, T. Torres, D. M. Guldi, *Small* 2013, *9*, 2348.

144. L. Xu, J.-W. McGraw, F. Gao, M. Grundy, Z. Ye, Z. Gu, J. L. Shepherd, *J. Phys. Chem. C* 2013, *117*, 10730.

145. E. B. Secor, P. L. Prabhumirashi, K. Puntambekar, M. L. Geier, M. C. Hersam, *J. Phys. Chem. Lett.* 2013, *4*, 1347.

146. Y. T. Liang, M. C. Hersam, *J. Am. Chem. Soc.* 2010, *132*, 17661.

147. M. S. Kang, K. T. Kim, J. U. Lee, W. H. Jo, *J. Mater. Chem. C* 2013, *1*, 1870.

148. G. L. Kennedy, *Drug Chem. Toxicol.* 1986, *9*, 147.

149. H. M. Solomon, B. A. Burgess, G. L. Kennedy, R. E. Staples, *Drug Chem. Toxicol.* 1995, *18*, 271.

150. H. J. Heipieper, F. J. Weber, J. Sikkema, H. Keweloh, J. A. M. de Bont, *Trends Biotechnol.* 1994, *12*, 409.

151. M. Lotya, Y. Hernandez, P. J. King, R. J. Smith, V. Nicolosi, L. S. Karlsson, F. M. Blighe, S. De, Z. Wang, I. T. McGovern, G. S. Duesberg, J. N. Coleman, *J. Am. Chem. Soc.* 2009, *131*, 3611.

152. A. A. Green, M. C. Hersam, *Nano Lett.* 2009, *9*, 4031.

153. M. Lotya, P. J. King, U. Khan, S. De, J. N. Coleman, *ACS Nano* 2010, *4*, 3155.

154. S. De, P. J. King, M. Lotya, A. O'Neill, E. M. Doherty, Y. Hernandez, G. S. Duesberg, J. N. Coleman, *Small* 2010, *6*, 458.

155. L. Guardia, M. J. Fernández-Merino, J. I. Paredes, P. Solís-Fernández, S. Villar-Rodil, A. Martínez-Alonso, J. M. D. Tascón, *Carbon* 2011, *49*, 1653.

156. M. Buzaglo, M. Shtein, S. Kober, R. Lovrincic, A. Vilan, O. Regev, *Phys. Chem. Chem. Phys.* 2013, *15*, 4428.

157. P. M. Carrasco, S. Montes, I. García, M. Borghei, H. Jiang, I. Odriozola, G. Cabañero, V. Ruiz, *Carbon* 2014, *70*, 157.

158. J.-W. T. Seo, A. A. Green, A. L. Antaris, M. C. Hersam, *J. Phys. Chem. Lett.* 2011, *2*, 1004.

159. X. Liu, M. Zheng, K. Xiao, Y. Xiao, C. He, H. Dong, B. Lei, Y. Liu, *Nanoscale* 2014, *6*, 4598.

160. L. Niu, M. Li, X. Tao, Z. Xie, X. Zhou, A. P. A. Raju, R. J. Young, Z. Zheng, *Nanoscale* 2013, *5*, 7202.

161. Y. Xu, H. Bai, G. Lu, C. Li, G. Shi, *J. Am. Chem. Soc.* 2008, *130*, 5856.

162. D. Parviz, S. Das, H. S. T. Ahmed, F. Irin, S. Bhattacharia, M. J. Green, *ACS Nano* 2012, *6*, 8857.

163. H. Yang, Y. Hernandez, A. Schlierf, A. Felten, A. Eckmann, S. Johal, P. Louette, J. J. Pireaux, X. Feng, K. Mullen, V. Palermo, C. Casiraghi, *Carbon* 2013, *53*, 357.

164. J.-H. Jang, D. Rangappa, Y.-U. Kwon, I. Honma, *J. Mater. Chem.* 2011, *21*, 3462.

165. A. Schlierf, H. Yang, E. Gebremedhn, E. Treossi, L. Ortolani, L. Chen, A. Minoia, V. Morandi, P. Samori, C. Casiraghi, D. Beljonne, V. Palermo, *Nanoscale* 2013, *5*, 4205.

166. M. Zhang, R. R. Parajuli, D. Mastrogiovanni, B. Dai, P. Lo, W. Cheung, R. Brukh, P. L. Chiu, T. Zhou, Z. Liu, E. Garfunkel, H. He, *Small* 2010, *6*, 1100.

167. Q. Su, S. Pang, V. Alijani, C. Li, X. Feng, K. Müllen, *Adv. Mater.* 2009, *21*, 3191.

168. S. Sampath, A. N. Basuray, K. J. Hartlieb, T. Aytun, S. I. Stupp, J. F. Stoddart, *Adv. Mater.* 2013, *25*, 2740.

169. J. M. Englert, J. Röhrl, C. D. Schmidt, R. Graupner, M. Hundhausen, F. Hauke, A. Hirsch, *Adv. Mater.* 2009, *21*, 4265.

170. D.-W. Lee, T. Kim, M. Lee, *Chem. Commun.* 2011, *47*, 8259.

171. C.-H. Hsu, H.-Y. Liao, Y.-F. Wu, P.-L. Kuo, *ACS Appl. Mater. Interfaces* 2011, *3*, 2169.

172. A. Ghosh, K. V. Rao, S. J. George, C. N. R. Rao, *Chem. Eur. J.* 2010, *16*, 2700.

173. J. Björk, F. Hanke, C.-A. Palma, P. Samori, M. Cecchini, M. Persson, *J. Phys. Chem. Lett.* 2010, *1*, 3407.

174. F. Liu, J. Y. Choi, T. S. Seo, *Chem. Commun.* 2010, *46*, 2844.

175. K. R. Paton, E. Varrla, C. Backes, R. J. Smith, U. Khan, A. O'Neill, C. Boland, M. Lotya, O. M. Istrate, P. King, T. Higgins, S. Barwich, P. May, P. Puczkarski, I. Ahmed, M. Moebius, H. Pettersson, E. Long, J. Coelho, S. E. O'Brien, E. K. McGuire, B. M. Sanchez, G. S. Duesberg, N. McEvoy, T. J. Pennycook, C. Downing, A. Crossley, V. Nicolosi, J. N. Coleman, *Nat. Mater.* 2014, *13*, 624.

24

ELECTROSTATIC INTERACTIONS IN THE DESIGN OF POLYMERIC PRODUCTS

LORENZO M. POLGAR AND FRANCESCO PICCHIONI
Department of Chemical Engineering, University of Groningen, Groningen, the Netherlands

24.1 INTRODUCTION

Polymeric products have been playing an important role in our daily life since the beginning of the twentieth century. While covalent bonds constitute most often to the interaction along the polymeric backbone, a large variety of physical and chemical interactions between the chains can be used in order to modulate the product's final properties. Among all possibilities, electrostatic interactions have been the subject of many studies at both academic and industrial level with corresponding products being currently present in the market. While initial investigations were mainly dealing with elastomeric materials [1], subsequent studies were mainly boosted by research in polyurethanes [2] (for coating applications) as well as in fuel cells. The latter has been generating a constant demand for suitable polymeric materials for membrane applications [3]. Indeed, many papers have been focusing at describing and reviewing experimental and theoretical studies on the transport properties of several different ionomeric membranes generally for fuel cells applications [4]. In particular, many studies tried to tackle the issue of structure–property relationship for membranes [5], a necessary tool for design of new materials for any given application.

In many cases, the focus lies on the cross-linking strategy [6] and, almost exclusively, on the proton conductivity. The latter is a function of the kind of electrostatic interaction present (e.g., carboxylic vs. sulfonic groups, the most commonly used for industrial applications), the amount and distribution of ions along the backbone as well as the morphology of the material [7]. Besides pure materials, electrostatic interactions have been also used in order to selectively modify properties of polymeric blends and composites [8].

In this chapter, we focus in first instance on the thermal and mechanical behavior of ionomers in the solid state. This is quite relevant for the use of these materials as bulk polymeric products but also for their specific membrane application in view of the mechanical stability required [9]. In doing so, we will briefly review some synthetic strategies that can be used to introduce electrostatic interactions in polymeric systems, with attention to recent examples for different applications. Subsequently we will focus on the thermal and mechanical behavior in the solid state. Finally, we will also provide some perspectives for solution applications as they are becoming more and more relevant especially for industrial applications.

24.2 SYNTHETIC STRATEGIES

Electrostatic interactions are inherently associated with the presence of polar groups along the polymeric backbone. Such interactions become relevant when they are present in a matrix with a relatively low dielectric constant. The effect can be simply deduced from the Coulomb law describing the energy of interaction (V) as function of the relative distance between the ions (R), the magnitude of the charges interacting (q_1 and q_2), and the dielectric constant of the medium (ε):

$$V = \frac{q_1 q_2}{4\pi\varepsilon R} \tag{24.1}$$

Consequently, a lower the dielectric constant of the medium results in a higher interaction energy. This entails the necessity to embed polar groups in a relatively apolar

Non-Covalent Interactions in the Synthesis and Design of New Compounds, First Edition.
Edited by Abel M. Maharramov, Kamran T. Mahmudov, Maximilian N. Kopylovich, and Armando J. L. Pombeiro.

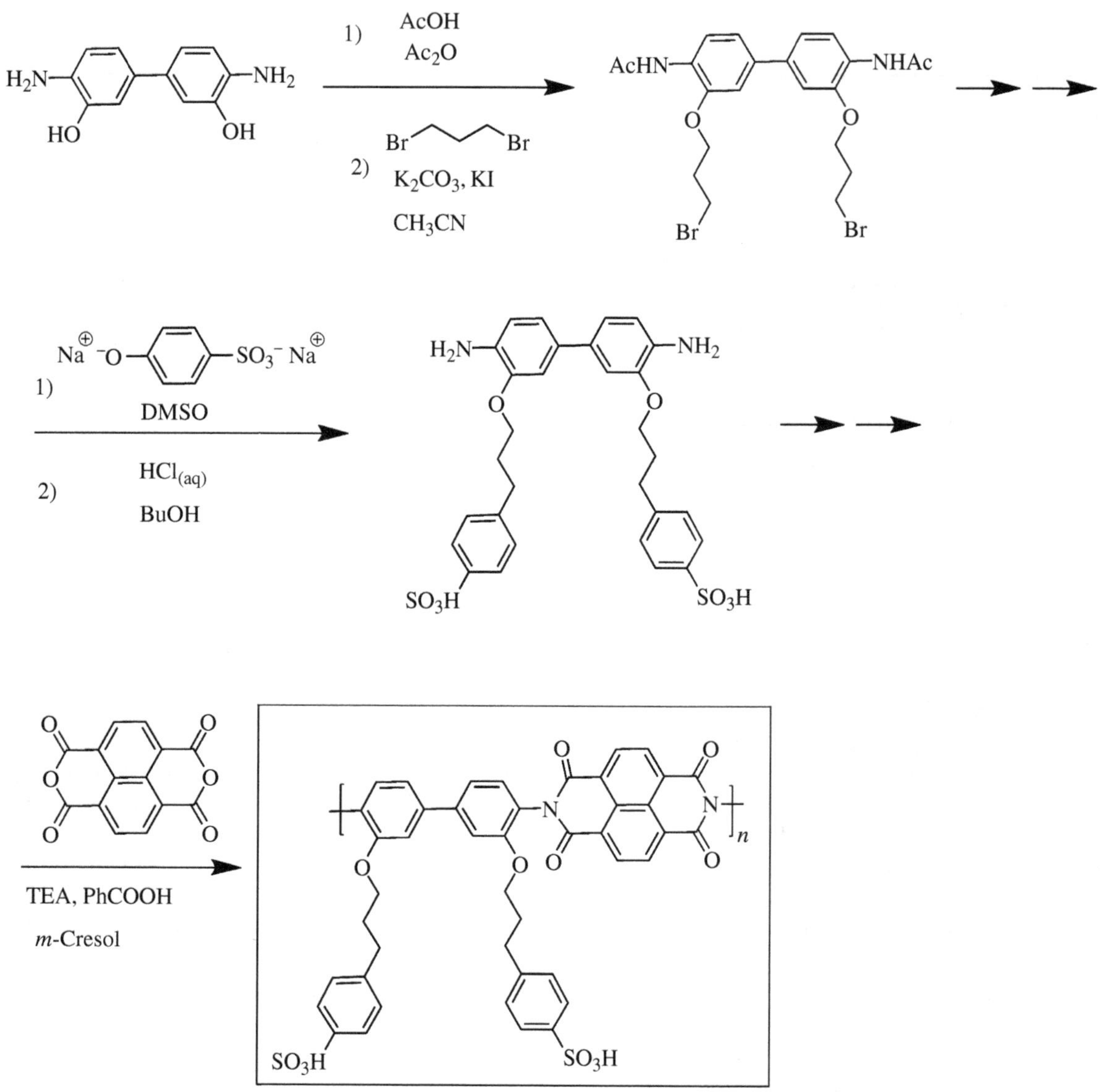

FIGURE 24.1 Paradigmatic example of a synthetic route required for the copolymerization of a polyamide. Redrawn from Ref. [10d].

matrix and thus, from a macromolecular point of view, the presence of polar and apolar monomeric units along the same polymeric backbone. This can be achieved either by *in situ* copolymerization of these different monomers or via postpolymerization modification.

24.2.1 Copolymerization

Most often, this synthetic strategy provides very well-characterized polymers and as such is crucial in studies dealing with ionomers application in membranes for fuel cells [10]. The synthetic routes are often cumbersome and consisting of several subsequent steps [10d, 11]. In many cases such lengthy synthetic steps are directly related to the peculiar chemical structure of the ionomers to be prepared. Fuel cell application requires electrical conductivity (ensured by the salt/acid group) as well as exceptional long-term stability in terms of mechanical behavior. Besides perfluorinated polymers [3a], the presence of aromatic groups represents a viable option to achieve the properties combination described above [4c, 12]. A paradigmatic example for such synthetic pathways clearly illustrates the aspects outlined above (Fig. 24.1) [10d].

It must be noticed here how such synthetic efforts (e.g., all steps above are carried out in solution and require accurate control in order to avoid side reactions) are often counterbalanced by the possibility to establish a reliable structure–property relationship (see *below*) [13]. Furthermore, from an applicative point of view, this also yields materials with superior performance (in terms of conductivity) with respect to commercially available counterparts [10d].

24.2.2 Postpolymerization Modification

The introduction of electrostatic interactions in existing polymeric materials by chemical modification is often perceived as a simple method, conceivable for upscaling at industrial level. One of the first examples in this respect is represented by the sulfonation of aromatic polymers such as, for example, polystyrene (Fig. 24.2) [1, 14] or related copolymers [15].

The methods used rely on the use of sulfonating agents derived from sulfonic acid and are generally not easy to control in terms of regioselectivity as well as the presence of side reactions [14c]. Recent developments focused on the study of relatively mild methods for the sulfonation step [16]. This makes it possible to carry out the reaction in the presence of other functional groups (Fig. 24.3).

Recent developments in controlled radical polymerization techniques (e.g., atom transfer radical polymerization) allow the use of very well-defined precursors [17]. Even with the limitation outlined above, this yields ionomers with narrow molecular weight distribution. In this context, attention has been paid to reversible addition–fragmentation techniques (RAFT) for the preparation of ionomeric initiators (Fig. 24.4) [18].

These recent developments represent a clear novelty as they pave the way toward the relatively easy preparation of different polymers displaying a well-defined structure in terms of comonomer (block vs. random) distribution, which is paramount in obtaining reliable structure–property relationships.

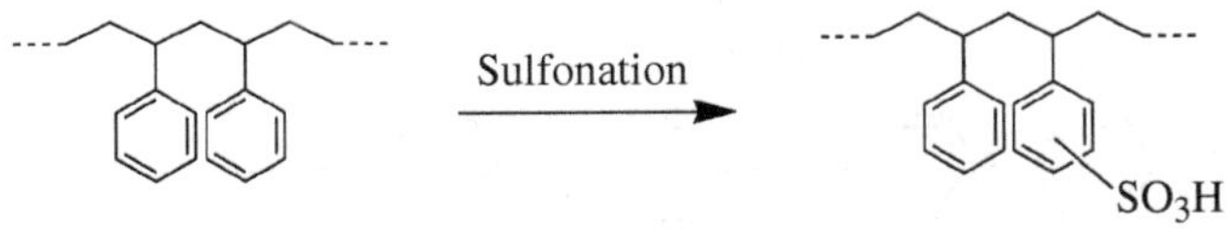

FIGURE 24.2 Sulfonation of polystyrene as a paradigmatic example of postpolymerization modification. Redrawn from Ref. [14c].

24.3 PROPERTIES OF IONOMERS

The presence of electrostatic interactions (both intra- and intermolecular) has a clear and distinctive effect on the properties of a solid polymeric materials. We conveniently divide the discussion according to the kind of properties.

24.3.1 Solubility and Water Uptake

The strength of ionic interactions is often enough to change the solubility (often in water) of the parent neutral polymer in a dramatic way [19]. The effect is usually so large that it might overcome not only the apolar character of the backbone, but also its topological characteristics (e.g., the presence of branches) [20]. More important than solubility is the water uptake, which is often an unwanted phenomenon for application solid polymers. Water uptake is a crucial factor for ionomers in general and membranes for fuel cells in particular since it influences the stability of the membrane [21]. Water uptake is generally a function of the amount of ionic groups along the backbone (Fig. 24.5) [16].

This relationship is rather intuitive and can also be found for other systems [22]. On the other hand, the distribution of the ionic groups also has a relevant role in the water uptake behavior. The diffusion coefficient of water for three different sulfonated aromatic polymers (Fig. 24.6) [23] was found to depend on the relative humidity applied during the measurements, the distribution of sulfonic groups along the chain, and the corresponding morphology.

If polymer A is characterized by random distribution of sulfonic groups and a relatively small domain size. Polymer B is an alternating copolymer with a very regular distribution of sulfonic groups resulting in relatively large but isolated domains. Finally, polymer C showed clear phase separation

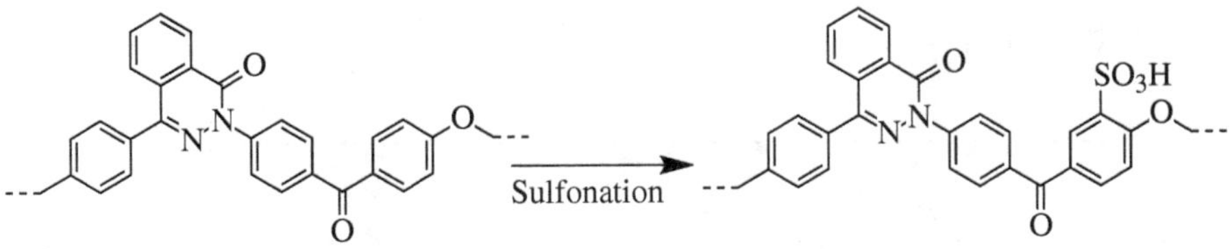

FIGURE 24.3 Sulfonation of aromatic polymers containing functional groups. Redrawn from Ref. [16].

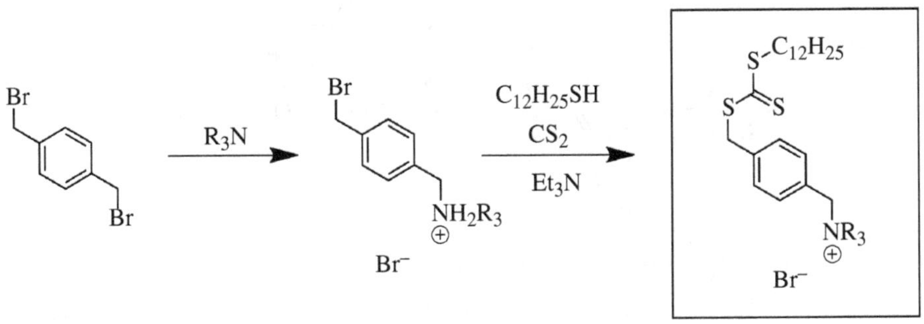

FIGURE 24.4 Synthesis of ionomeric RAFT initiator (4-(bromomethyl)-*N,N,N*-trialkylbenzyl ammonium compounds as example). Redrawn from Ref. [18].

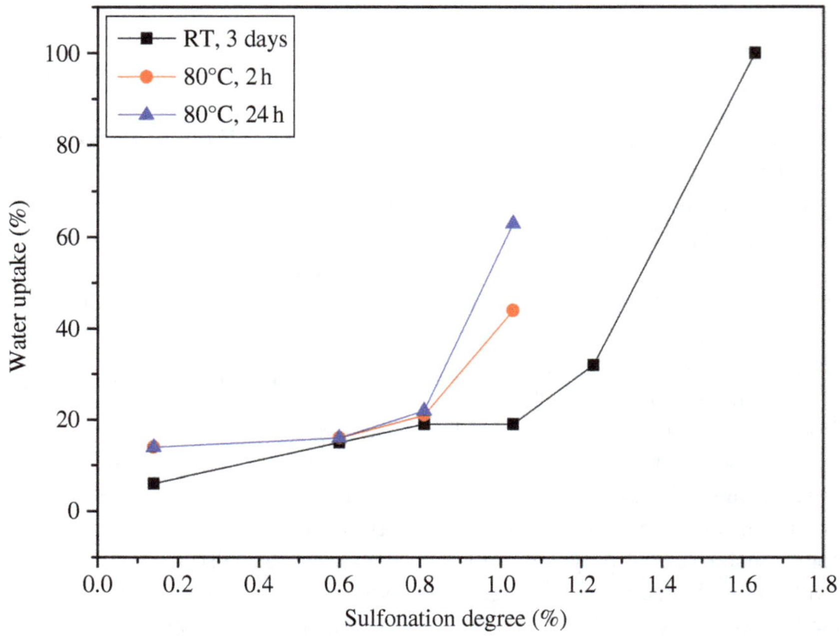

FIGURE 24.5 Water uptake as function of the degree of sulfonation (chemical structure in Fig. 24.3) at different temperatures. Data from Ref. [16].

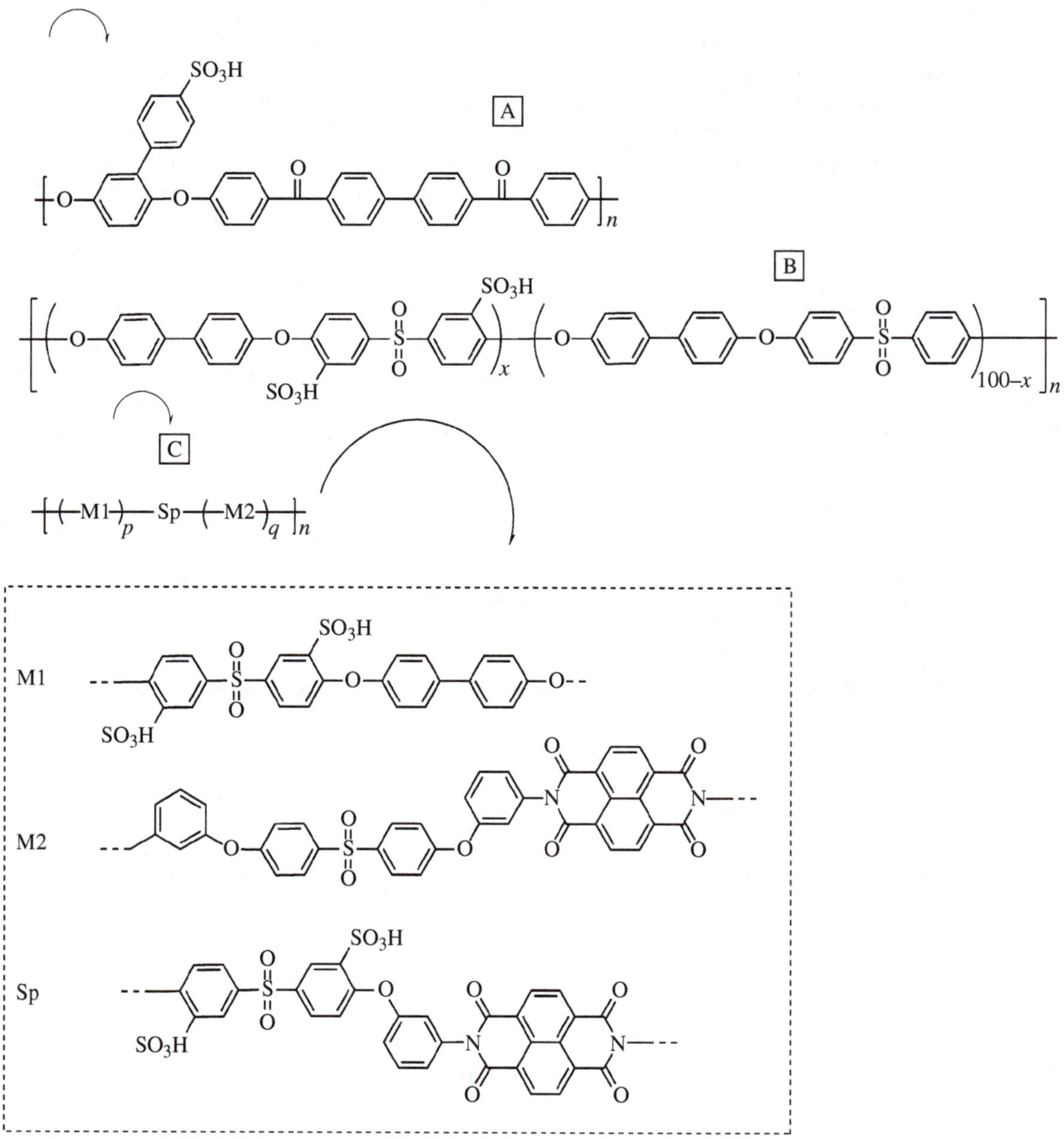

FIGURE 24.6 Structure of aromatic sulfonated polymers used for water uptake measurements. Redrawn from Ref. [23].

between the hydrophilic and hydrophobic blocks. The water diffusion coefficient was found to decrease in the following order: B>A>C. This might be tentatively related to the number and size of ionic aggregates formed in this system, in turn dependent on the group distribution along the chain.

24.3.2 Thermal Properties

For amorphous polymers in general, the introduction of ionic groups leads to higher T_g as measured by DSC. This has been observed for several systems including (meth) acrylic [24] and styrene copolymers [8e, 24b, 25]. The effect is clearly related to the kind of ionic groups present. For copolymers of styrene and vinyl pyridine (quaternized with methyl iodide) [25], this increase in T_g is a function of both the ionic content and the kind of pyridine derivative used (Fig. 24.7).

The data represent a deviation from the expectations. The slope (dT_g/dc, with c being the ionic content) of about 3.5°C/(mol %) that is observed is in agreement with the one measured for anionic PS-based copolymers (e.g., styrene/sodium methacrylate) [25]. Since the T_g variation was expected to be inversely proportional to the interionic distance (bigger in the case of pyridinium cations than in styrene anions), the measured value seemed to be overestimated. The authors attributed this deviation to the morphology of the sample as deducted, among others, from DMTA measurements. From these, the absence of a relevant phase separation (ionic clusters and the polymer matrix) could be deduced. Such phase separation is present in most ionomeric systems (*see below*) and actually entails the appearance of an additional thermal transition (T_i, order–disorder or "melting" of the ionic domains). This is a function of the amount of ionic groups as well as the kind of counterion, as is clearly demonstrated for ethylene/methacrylic acid copolymers (Fig. 24.8) [26].

It seems clear that, for amorphous polymers, the changes in thermal behavior are also dependent on the chemical structure of the nonionic moieties along the polymer backbone. This was also confirmed for ionomeric polyurethanes [27]. For ionomers that are displaying crystalline domains, the final thermal behavior is also dependent on the crystallinity. These crystalline domains also represent, in striking similar way to electrostatic interactions, a kind a physical cross-linking [28]. Furthermore, the presence of ionic domains has a clear influence on the crystallization behavior as demonstrated for ethylene/methacrylic acid copolymers (Fig. 24.9) [26].

It is worth noticing how the T_c generally decreases with the amount of ionic groups, thus suggesting that the presence of strong electrostatic interactions might severely hinder the crystallization of the ethylene segments. For relatively weaker interactions (i.e., low ionic content and rare earth metal counterions such as Cu, Mn, and Co), a slight increase in low ionic content might be detected. This is probably due to nucleating effects of the small and weak ionic domains. On the other hand, the decrease in T_c at high ion content is much less pronounced than for alkaline and alkaline earth metal slats.

Besides the specific details for this system, these considerations highlight the importance of the morphology on the thermal and mechanical behavior, as well as the necessity to combine thermal analysis with a dynamic and mechanical one in order to elucidate the peculiar characteristics of these systems. These aspects are discussed in the next sections.

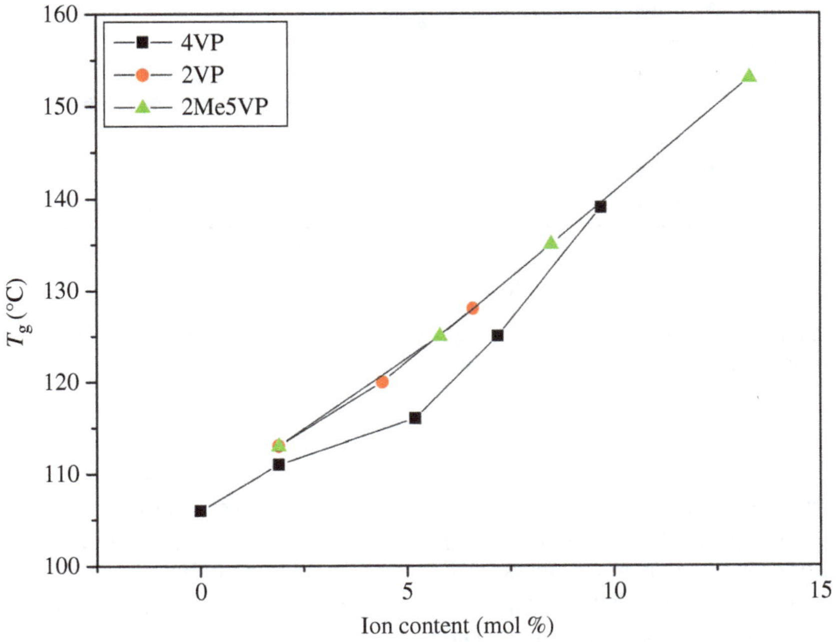

FIGURE 24.7 T_g of styrene-vinyl pyridine copolymers as function of ion content and kind of pyridine. 4VP=4-vinyl pyridine, 2VP=2-vinyl pyridine, 2Me5VP=2-methyl-5-vinyl pyridine. Data from Ref. [25].

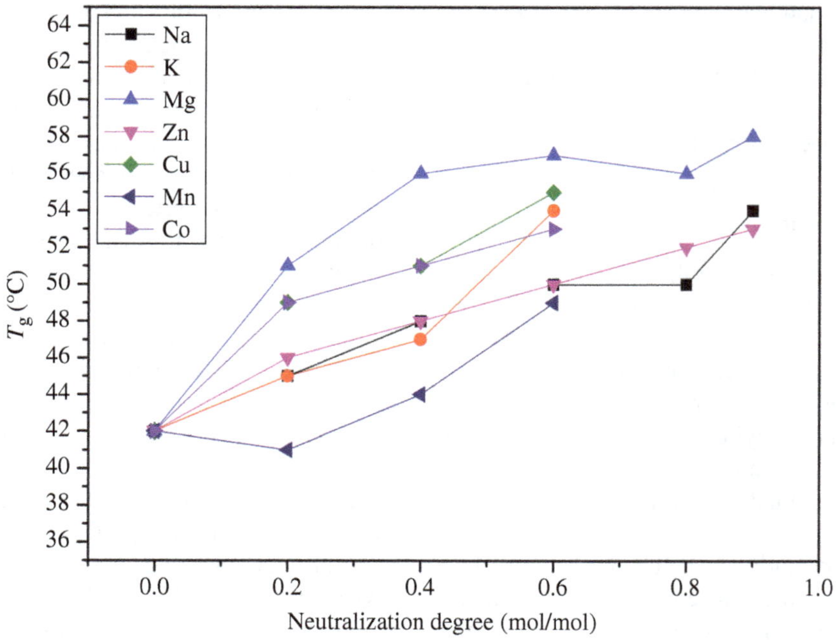

FIGURE 24.8 "Melting" temperature of ionic domains in ethylene/methacrylic acid copolymers as function of the degree of neutralization of the acid groups and the kind of counterion. Data from Ref. [26].

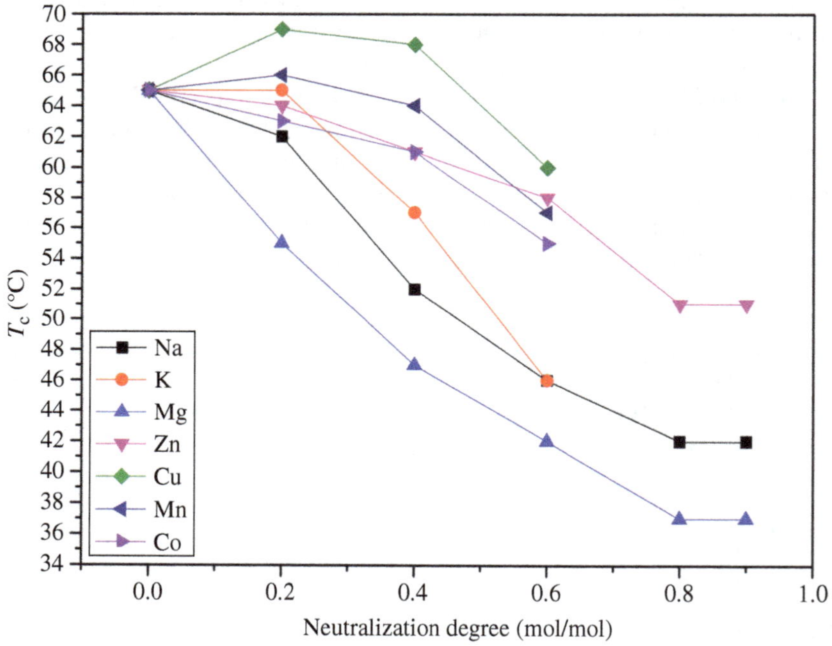

FIGURE 24.9 Crystallization temperature (T_c) of ethylene segments as function of the degree of neutralization and the kind of cation (ethylene/methacrylic acid copolymers as example). Data from Ref. [26].

24.3.3 Morphology

The morphological characteristics of ionomers have been extensively studied in the past 50 years [29]. The basic idea [1] is that in first instance cations and anions in ionomers form ion pairs, separated only according to a distance equal to their ionic radius. These ion pairs can then aggregate (depending on steric and energetic considerations) to form a multiplet while, at larger concentrations, clusters (i.e., loose aggregates of multiplets) are invoked. The fundamental difference between clusters and multiplets is the presence of a hydrocarbon component (derived from segments of the corresponding polymer). Hydrocarbon components are present in clusters but not in multiplets.

The size of clusters is a typical trade-off between attraction (electrostatic interactions between multiplets) and repulsion (elastic energy of the polymeric chain) and represents a crucial parameter in determining properties such as electrical conductivity. Basic calculations [1] result in equations describing the cluster size as follows:

$$n = \rho \frac{N}{M_c}\left[\frac{4l^2}{3kT_s}\frac{k'}{K}\frac{\overline{h}^2}{\overline{h_0}^2}\frac{e^2}{4\pi\varepsilon r} + 2\left(\frac{M_0 M_c}{\rho N}\right)^{\frac{2}{3}}\right]^{\frac{3}{2}} \quad (24.2)$$

where ρ is the density, n the number of ion pairs per cluster, N the Avogadro number, ε the electrical permittivity, T_s the temperature above which the cluster collapses, l the length of a backbone bond, M_c the average molecular weight between two ions on the chain, M_0 the molecular weight of the monomeric unit, e the electronic charge, r the distance between the positive and negative centers in an ion pair, K the dielectric constant, k' a parameter related to the energy release after cluster collapse, and $\overline{h}^2$ and $\overline{h_0}^2$ the mean square end-to-end distance for the free chain and freely jointed chain, respectively. k' depends on the cluster geometry and must be estimated to predict the n value. Equation 24.2 generally lacks predictive power. Its importance resides mainly in the definition of clusters and multiplets and of their stability (T_s) [30].

Subsequent models, with respect to the pioneering ones, basically rely on the same concepts while introducing additional factors such as the presence of unneutralized acid groups (especially relevant for carboxyl acids [31]). These improvements are quite essential since the presence of acid groups in their neutral form also contributes to the final mechanical behavior [32], probably via hydrogen bonding formation. The formation of clusters [33] and their stability is widely recognized as a crucial factor in determining the mechanical behavior, as will be discussed in the following.

24.3.4 Mechanical Properties

Mechanical properties of ionomers are a function of the level of humidity [34], and attention should be paid to comparisons in this respect. On a general level, the typical DMTA curves for ionomers (Fig. 24.10) display the presence of an additional transition plateau with respect to the corresponding neutral polymers.

The position (loss tangent peak) of this additional transition is related to the collapse of the ionic aggregates and is a function of the phase separation in the system (i.e., how well defined the clusters are in the matrix) [35]. This is in turn dependent on the amount of ionic groups [1, 36] as well as (slightly) on the kind of counterions. In many systems [24b, 26], there is no clear distinction between these two transitions up to the level of no clear detection. This happens when multiplets remain soluble in the matrix and, thus, give no aggregation [25] or for very rigid polymeric backbones [33b]. Besides the position of this additional relaxation, electrostatic interactions are generally responsible for a relevant modulus increase in the rubbery plateau region [1]. The same is, to a lesser extent, true for the glassy modulus [24b]. In both cases, the underlying idea is that ionic clusters represent physical cross-linking points for the material [37].

Similar to DMTA measurements, tensile strength properties are deeply influenced by the presence of ions. This generally results in a higher modulus and tensile strength and a lower elongation at break. This is predictable on the basis of the physical network induced by the electrostatic

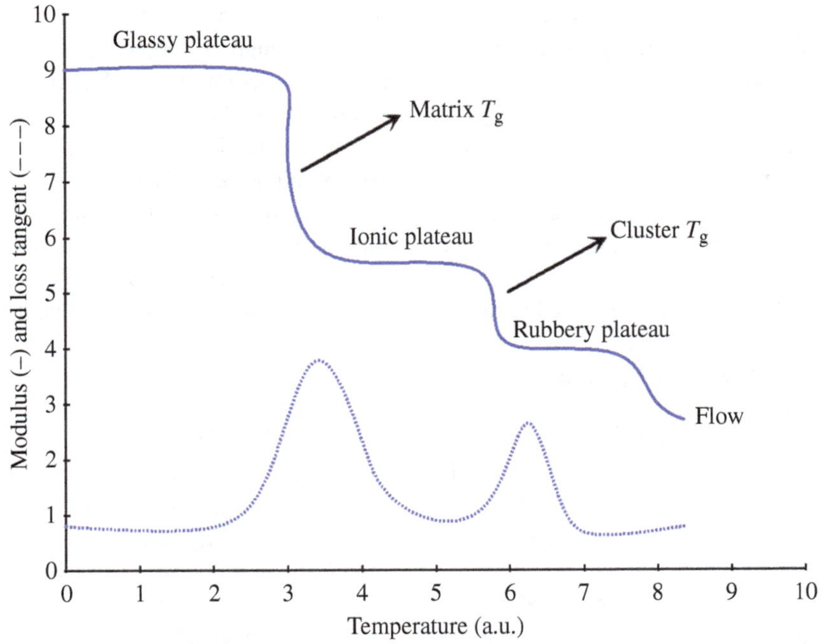

FIGURE 24.10 Schematic representation of DMTA measurements for ionomers.

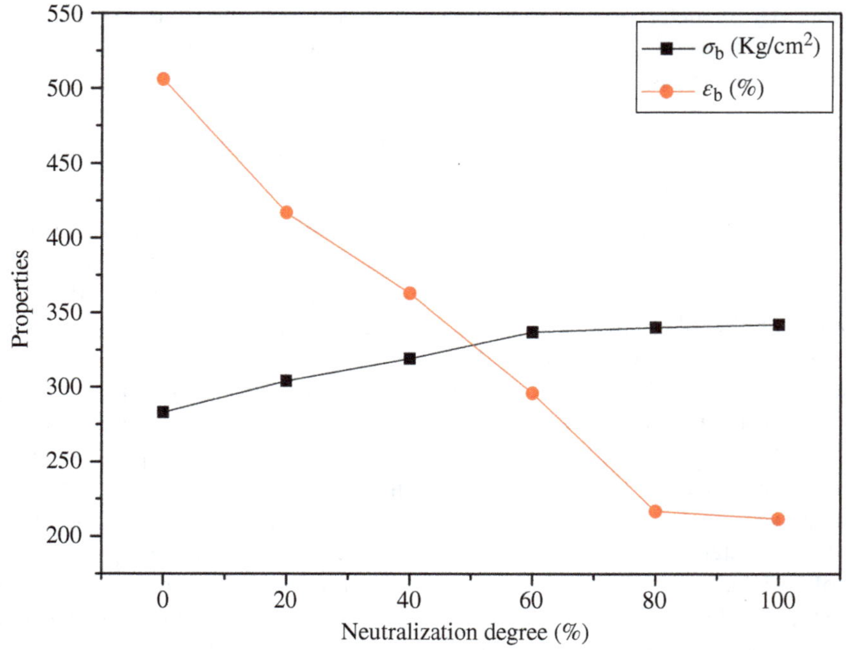

FIGURE 24.11 Tensile strength (σ_b) and elongation at break (ε_b) as a function of the degree of neutralization (i.e., amount of ionic groups) (poly(urethane) ionomers as example). Data from Ref. [39].

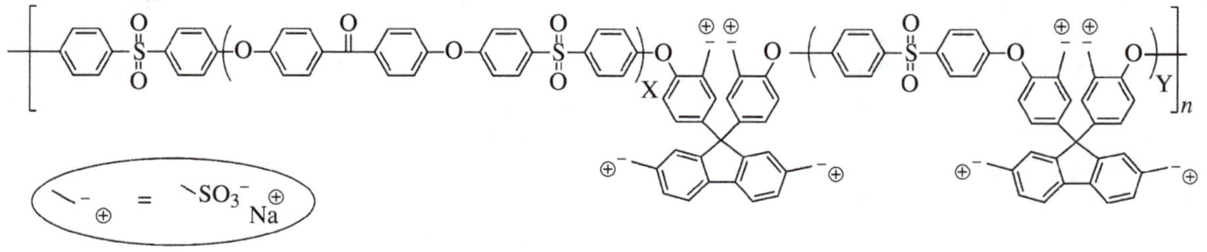

FIGURE 24.12 Chemical structure of poly(arylene ether sulfone ketone) membranes. Redrawn from Ref. [40].

interactions. The mechanical properties are therefore clearly a function of the chemical structure, namely, the amount of ionic groups [24a, 38], as illustrated for poly(urethane) ionomers (Fig. 24.11) [39].

Exceptions to this general trend have also been reported. This can be illustrated for membranes from poly(arylene ether sulfone ketone) (Fig. 24.12) [40]. For these materials, the relative ratio of X/Y represents the amount of ionic groups.

The corresponding tensile properties (Fig. 24.13) nicely illustrate some general principles [40].

It is possible to notice how, at fixed X, an increase in Y (more ionic groups) leads to an overall decrease in the elongation at break (ε_b), tensile strength (σ_b) and modulus (E). This is probably related to the specific (i.e., fully aromatic and rigid) nature of the nonionic blocks. This trend is then again reversed when introducing aliphatic spacers along the backbone [10a].

Moreover, the position of the ionomeric group along the chain and its distance from the backbone are also important for the final properties. This was clearly illustrated for poly(imide)-based membranes (Fig. 24.14) [10b]. In this case, although differences are clearly observed, no clear trend as function of the $-SO_3H$ position on the ring and its distance from the backbone could be detected.

The mechanical behavior is also clearly a function of the type of ion present. For example, ionomers based on sulfonic groups were reported to display higher tensile strength than ionomers based on carboxylic ones [41]. This can, again, be related to the acidity of the group and ultimately to the physical network density as result of the different clusters structure.

24.3.5 Rheology

It is worth mentioning how the rheological behavior of ionomers has been the subject of many scientific papers [1, 26]. Generally speaking, a clear relationship can be found between viscosity in the melt and the kind and amount of ionic groups, with the latter being much more relevant than the former [42]. Also in this case, the inferences provided to explain the different experimental data rely on

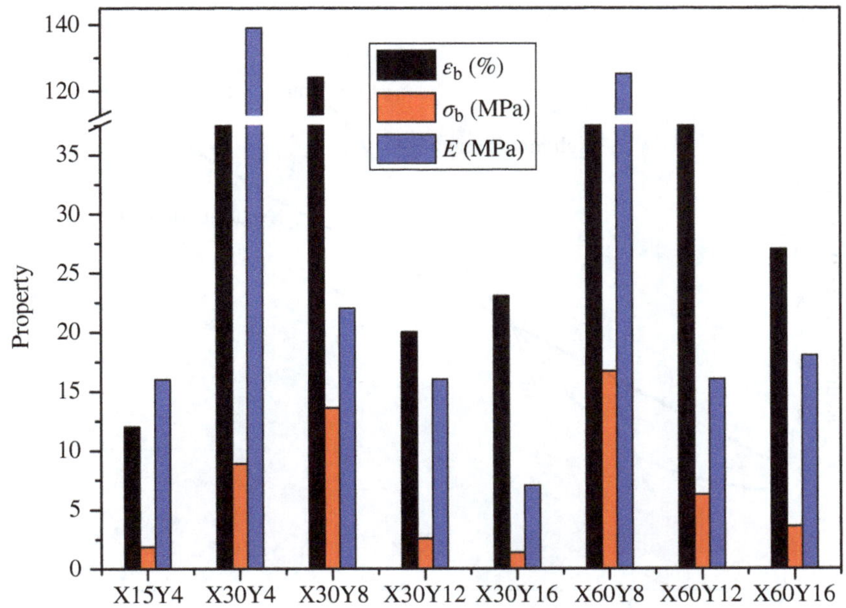

FIGURE 24.13 Mechanical properties of membranes based on poly(arylene ether sulfone ketone). Data from Ref. [40].

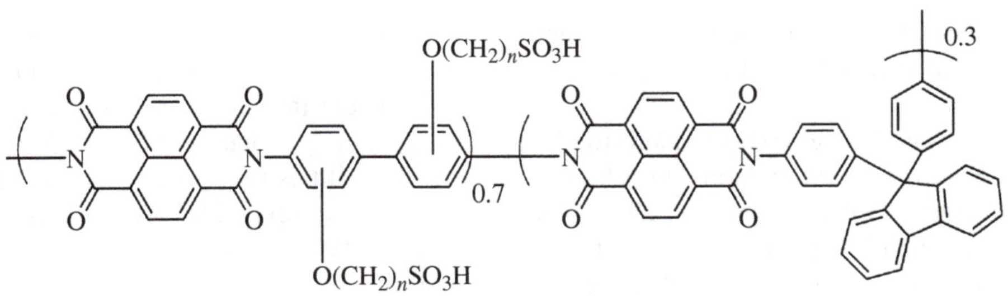

FIGURE 24.14 Polyimide-based membranes. Redrawn from Ref. [10b].

the theory of cluster formation [1, 43], possibly with distinctions between ion-rich and ion-poor domains [43, 44]. Moreover, several models (such as ion hopping [45]) have been invoked to explain peculiar relaxation behavior of ionomers in the melt.

24.3.6 Properties in Solution

The presence of different monomers (characterized by different polarity) along the polymer backbone also renders ionomers attractive for their solution properties in a way very much similar to polyelectrolytes. For example, the association behavior of such polymers in micelles is strongly associated with the interaction of the different blocks/segments with the solvent. The presence of electrostatic interactions changes this in a dramatic way [46], thus resulting in a variety of different kinds of aggregates [47]. This has, in turn, a clear influence on the rheological and interfacial properties of such materials [48].

The rheological properties are generally dependent on the amount of ionic groups and their distribution, as well as on the kind of solvent. This was clearly demonstrated for styrene/sulfonated styrene copolymers (sodium salts) [14b]. The corresponding reduced viscosity curves (Fig. 24.15) display some general features.

A decrease in viscosity at low concentration is observed while a subsequent increase is also detected, both becoming more pronounced with an increase in ion content. This can be easily interpreted in terms of intra- and intermacromolecular electrostatic interactions. At low concentrations, intramolecular interactions are dominant and result in lower reduced viscosity. With increasing concentration, intermolecular interactions become dominant and induce aggregation between chains and thus, ultimately, higher reduced viscosity values. This behavior changes as a function of the kind of solvent used, as well as the average molecular weight of the chains. The general increase of viscosity with concentration has been also

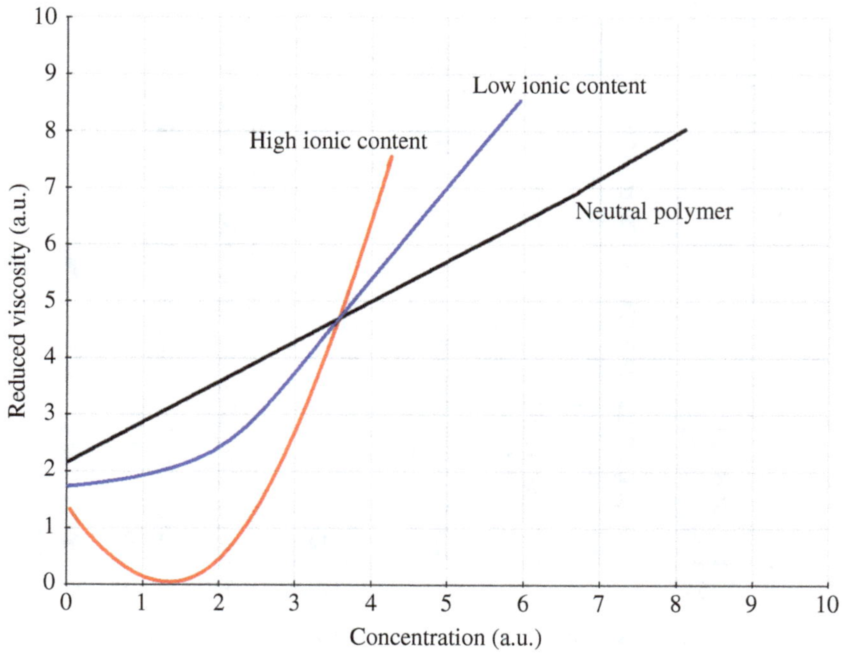

FIGURE 24.15 Schematic representation of rheological behavior for ionomers in solution (styrene/sulfonated styrene copolymers as example). Based on Ref. [14b].

observed for other systems [49], while the initial decrease (magnified and exaggerated in Fig. 24.15) is most often left unreported.

Quite recently, attention has also been paid to the consequence of such electrostatic interactions (in solution) in terms of polymer/surfactant interactions [50]. This is not only interesting from a scientific point of view, but also quite relevant for industrial applications. Enhanced oil recovery (EOR), a family of techniques aimed at fully exploiting oil reservoirs [51], is a good example. EOR makes use, among others, of surfactant and polymeric solutions in water, both components often displaying the presence of ionic groups. The interplay between electrostatic interactions (polymer/polymer, surfactant/surfactant, and polymer/surfactant) is crucial in determining the rheological behavior and, consequently, in the overall EOR operation [48].

24.4 PERSPECTIVE IN FUTURE RESEARCH

Although research on ionomers dates back to the middle of the previous century, many issues remain open (*vide supra*) in terms of structure–property relationships. These are very useful in the design of new polymers, but are often system dependent, hindering thus an *a priori* theoretical design. A possible exception, stimulated by the corresponding application for fuel cells, is represented by electrical conductivity in membranes. On the other hand, mechanical and thermal behavior, although fundamentally understood in terms of ionic aggregates and characterized by quite general qualitative trends (*vide supra*), often requires *ad hoc* analyses (i.e., dependent on the particular system under investigation) for a refined fine tuning of the corresponding properties. On the other hand, the rise of new possibilities for property tailoring (such as use of dedicated plasticizers [14a] as well as of ampholitic ionic groups [10c]) offers the designer a wider range of possibilities.

From a purely scientific point of view, many reports have been relating electrostatic interactions in ionomers to relatively new effects and properties. Relevant examples include shape memory effects in polyurethanes [2c], use of rare earth metals as counterions inducing fluorescent behavior [52], the use of organic counterions stimulated by the corresponding recent developments in ionic liquids [53], and the synthesis of redox-active ionomers [54]. These developments suggest even more fundamentally oriented research in the coming years.

REFERENCES

1. W. J. Macknight, T. R. Earnest, *J. Poly. Sci. Macromol. Rev. D* 1981, *16*, 41.

2. (a) S. L. Hsu, H. X. Xiao, H. H. Szmant, K. C. Frisch, *J. App. Polym. Sci.* 1984, *29*, 2467; (b) B. K. Kim, J. C. Lee, *J. Polym. Sci. A, Polym. Chem.* 1996, *34*, 1095; (c) B. K. Kim, S. Y. Lee, J. S. Lee, S. H. Baek, Y. J. Choi, J. O. Lee, M. Xu, *Polymer* 1998, *39*, 2803; (d) M. C. Delpech, F. M. B. Coutinho, *Polym. Test.* 2000, *19*, 939; (e) C. Z. Yang, T. G. Grasel, J. L. Bell, R. A. Register, S. L. Cooper, *J. Polym. Sci. B Polym. Phys.* 1991, *29*,

581; (f) H. S. Xu, C. Z. Yang, *J. Polym. Sci. B Polym. Phys.* 1995, *33*, 745; (g) O. Lorenz, F. Haulena, D. Kleborn, *Angew. Makromol. Chem.* 1973, *33*, 159; (h) S. Y. Lee, J. S. Lee, B. K. Kim, *Polym. Inter.* 1997, *42*, 67; (i) D. J. Hourston, G. D. Williams, R. Satguru, J. C. Padget, D. Pears, *J. Appl. Polym. Sci.* 1999, *74*, 556; (j) F. M. B. Coutinho, M. C. Delpech, *Polym. Test.* 1996, *15*, 103; (k) B. K. Kim, T. K. Kim, *J. Appl. Polym. Sci.* 1991, *43*, 393.

3. (a) C. Heitner-Wirguin, *J. Membr. Sci.* 1996, *120*, 1; (b) N. W. Deluca, Y. A. Elabd, *J. Polym. Sci. B Polym. Phys.* 2006, *44*, 2201.
4. (a) F. M. Collette, F. Thominette, H. Mendil-Jakani, G. Gebel, *J. Membr. Sci.* 2013, *435*, 242; (b) N. P. Berezina, N. A. Kononenko, O. A. Dyomina, N. P. Gnusin, *Adv. Colloid. Interf. Sci.* 2008, *139*, 3; (c) B. Bae, K. Miyatake, M. Watanabe, *ACS Appl. Mater. Interfaces* 2009, *1*, 1279; (d) P. Jannasch, *Curr. Opin. Colloid. Interface Sci.* 2003, *8*, 96; (e) R. B. Moore, C. R. Martin, *Macromolecules* 1988, *21*, 1334.
5. (a) C. Lee, S. Sundar, J. Kwon, H. Han, *J. Polym. Sci. A Polym. Chem.* 2004, *42*, 3612; (b) J. R. Varcoe, R. C. T. Slade, E. L. H. Yee, S. D. Poynton, D. J. Driscoll, D. C. Apperley, *Chem. Mater.* 2007, *19*, 2686.
6. (a) A. Jasti, V. K. Shahi, *RSC Adv.* 2014, *4*, 19238; (b) A. Jasti, V. K. Shahi, *J. Power Sources* 2014, *267*, 714; (c) T. A. Witten, M. H. Cohen, *Macromolecules* 1985, *18*, 1915.
7. (a) S. J. Paddison, *Annu. Rev. Mater. Res.* 2003, *33*, 289; (b) E. D. Gomez, A. Panday, E. H. Feng, V. Chen, G. M. Stone, A. M. Minor, C. Kisielowski, K. H. Downing, O. Borodin, G. D. Smith, N. P. Balsara, *Nano Lett.* 2009, *9*, 1212.
8. (a) P. Datta, C. Guha, G. Sarkhel, *Polym. Adv. Technol.* 2014, *25*, 1454; (b) P. Datta, C. Guha, G. Sarkhel, *Poly.-Plast. Technol. Eng.* 2014, *53*, 80; (c) P. Datta, C. Guha, G. Sarkhel, *J. Macromol. Sci. A Pure Appl. Chem.* 2014, *51*, 820; (d) R. K. Shah, D. L. Hunter, D. R. Paul, *Polymer* 2005, *46*, 2646; (e) F. Picchioni, I. Giorgi, E. Passaglia, G. Ruggeri, M. Aglietto, *Polym. Int.* 2001, *50*, 714; (f) A. Molnar, A. Eisenberg, *Macromolecules* 1992, *25*, 5774; (g) J.-M. Yeh, C.-T. Yao, C.-F. Hsieh, H.-C. Yang, C.-P. Wu, *Eur. Polym. J.* 2008, *44*, 2777.
9. X. Huang, R. Solasi, Y. Zou, M. Feshler, K. Reifsnider, D. Condit, S. Burlatsky, T. Madden, *J. Polym. Sci. B Polym. Phys.* 2006, *44*, 2346.
10. (a) N. Asano, M. Aoki, S. Suzuki, K. Miyatake, H. Uchida, M. Watanabe, *J. Am. Chem. Soc.* 2006, *128*, 1762; (b) T. Yasuda, K. Miyatake, M. Hirai, M. Nanasawa, M. Watanabe, *J. Polym. Sci. A Polym. Chem.* 2005, *43*, 4439; (c) D. G. Peiffer, R. D. Lundberg, *Polymer* 1985, *26*, 1058; (d) K. Miyatake, T. Yasuda, M. Hirai, M. Nanasawa, M. Watanabe, *J. Polym. Sci. A Polym. Chem.* 2007, *45*, 157.
11. M. Tanaka, M. Koike, K. Miyatake, M. Watanabe, *Polym. Chem.* 2011, *2*, 99.
12. (a) K. Miyatake, A. S. Hay, *J. Polym. Sci. A Polym. Chem.* 2001, *39*, 3211; (b) K. Miyatake, N. Asano, M. Watanabe, *J. Polym. Sci. A Polym. Chem.* 2003, *41*, 3901; (c) K. Miyatake, Y. Chikashige, E. Higuchi, M. Watanabe, *J. Am. Chem. Soc.* 2007, *129*, 3879; (d) Y. Chikashige, Y. Chikyu, K. Miyatake, M. Watanabe, *Macromolecules* 2005, *38*, 7121; (e) H.-S. Lee, A. S. Badami, A. Roy, J. E. McGrath, *J. Polym. Sci. A Polym. Chem.* 2007, *45*, 4879; (f) R. Mukherjee, A. K. Mohanty, S. Banerjee, H. Komber, B. Voit, *J. Membr. Sci.* 2013, *435*, 145; (g) P. X. Xing, G. P. Robertson, M. D. Guiver, S. D. Mikhailenko, S. Kaliaguine, *J. Polym. Sci. A Polym. Chem.* 2004, *42*, 2866; (h) M. Tanaka, K. Fukasawa, E. Nishino, S. Yamaguchi, K. Yamada, H. Tanaka, B. Bae, K. Miyatake, M. Watanabe, *J. Am. Chem. Soc.* 2011, *133*, 10646; (i) S. Matsumura, A. R. Hlil, C. Lepiller, J. Gaudet, D. Guay, Z. Shi, S. Holdcroft, A. S. Hay, *Macromolecules* 2008, *41*, 281; (j) B. Lafitte, P. Jannasch, *Adv. Funct. Mater.* 2007, *17*, 2823; (k) J. Saito, K. Miyatake, M. Watanabe, *Macromolecules* 2008, *41*, 2415; (l) Y. Lim, D. Seo, S. Lee, M. A. Hossain, J. Lim, S. Lee, T. Hong, W. Kim, *J. Nanosci. Nanotechnol.* 2014, *14*, 7948; (m) Y. Lim, D. Seo, M. A. Hossain, S. Lee, J. Lim, H. Jang, T. Hong, W. Kim, *Electrochim. Acta* 2014, *118*, 18; (n) Y. Lim, S. Lee, D. Seo, H. Jang, M. A. Hossain, H. Ju, T. Hong, W. Kim, *Int. J. Hydrog. Energy* 2014, *39*, 1539; (o) Y. Lim, H. Lee, S. Lee, H. Jang, M. A. Hossain, Y. Cho, T. Kim, Y. Hong, W. Kim, *Electrochim. Acta* 2014, *119*, 16; (p) B. R. Einsla, Y. T. Hong, Y. S. Kim, F. Wang, N. Gunduz, J. E. McGrath, *J. Polym. Sci. A Polym. Chem.* 2004, *42*, 862.
13. L. E. Karlsson, P. Jannasch, *J. Membr. Sci.* 2004, *230*, 61.
14. (a) C. G. Bazuin, A. Eisenberg, *J. Polym. Sci. B Polym. Phys.* 1986, *24*, 1137; (b) M. Hara, J. L. Wu, A. H. Lee, *Macromolecules* 1988, *21*, 2214; (c) N. Carretta, V. Tricoli, F. Picchioni, *J. Membr. Sci.* 2000, *166*, 189.
15. D. Suleiman, G. Carreras, Y. Soto, *J. Appl. Polym. Sci.* 2013, *128*, 2297.
16. Y. Gao, G. P. Robertson, M. D. Guiver, X. G. Jian, *J. Polym. Sci. A Polym. Chem.* 2003, *41*, 497.
17. D. Baigl, T. A. P. Seery, C. E. Williams, *Macromolecules* 2002, *35*, 2318.
18. L. H. Zhang, L. R. Cool, C. Wesdemiotis, R. A. Weiss, K. A. Cavicchi, *Polym. Chem.* 2014, *5*, 6108.
19. W. F. Lee, C. C. Tsai, *Polymer* 1994, *35*, 2210.
20. A. Asif, W. F. Shi, *Eur. Polym. J.* 2003, *39*, 933.
21. (a) N. Li, S. Zhang, J. Liu, F. Zhang, *Macromolecules* 2008, *41*, 4165; (b) B. Smitha, S. Sridhar, A. A. Khan, *J. Membr. Sci.* 2005, *259*, 10.
22. Y. A. Elabd, E. Napadensky, *Polymer* 2004, *45*, 3037.
23. M. L. Einsla, Y. S. Kim, M. Hawley, H.-S. Lee, J. E. McGrath, B. Liu, M. D. Guiver, B. S. Pivovar, *Chem. Mater.* 2008, *20*, 5636.
24. (a) E. Hirasawa, Y. Yamamoto, K. Tadano, S. Yano, *Macromolecules* 1989, *22*, 2776; (b) X. Ma, J. A. Sauer, M. Hara, *Macromolecules* 1995, *28*, 3953.
25. S. Gauthier, D. Duchesne, A. Eisenberg, *Macromolecules* 1987, *20*, 753.
26. E. Hirasawa, Y. Yamamoto, K. Tadano, S. Yano, *J. Appl. Polym. Sci.* 1991, *42*, 351.
27. Y. M. Lee, J. C. Lee, B. K. Kim, *Polymer* 1994, *35*, 1095.
28. (a) A. C. C. Yang, R. Narimani, Z. Zhang, B. J. Frisken, S. Holdcroft, *Chem. Mater.* 2013, *25*, 1935; (b) H. W. Starkweather, *Macromolecules* 1982, *15*, 320.
29. (a) C. L. Marx, D. F. Caulfiel, S. L. Cooper, *Macromolecules* 1973, *6*, 344; (b) M. Lee, J. K. Park, H.-S. Lee, O. Lane, R. B. Moore, J. E. McGrath, D. G. Baird, *Polymer* 2009, *50*, 6129; (c) R. B. Moore, C. R. Martin, *Macromolecules* 1989, *22*, 3594; (d) J. M. Serpico, S. G. Ehrenberg, J. J. Fontanella, X. Jiao, D. Perahia, K. A. McGrady, E. H. Sanders, G. E. Kellogg, G. E. Wnek, *Macromolecules* 2002, *35*, 5916.

30. B. Hird, A. Eisenberg, *Macromolecules* 1992, *25*, 6466.
31. M. M. Coleman, J. Y. Lee, P. C. Painter, *Macromolecules* 1990, *23*, 2339.
32. (a) K. A. Mauritz, R. I. Blackwell, F. L. Beyer, *Polymer* 2004, *45*, 3001; (b) P. Cousin, P. Smith, *J. Polym. Sci. B Polym. Phys.* 1994, *32*, 459.
33. (a) J. S. Kim, G. M. Wu, A. Eisenberg, *Macromolecules* 1994, *27*, 814; (b) W. Essafi, G. Gebel, R. Mercier, *Macromolecules* 2004, *37*, 1431.
34. (a) S. B. Mitra, B. L. Kedrowski, *Dent. Mater.* 1994, *10*, 78; (b) S. Uno, W. J. Finger, U. Fritz, *Dent. Mater.* 1996, *12*, 64.
35. K.-H. Ko, H. R. Park, J.-S. Kim, Y.-W. Kim, *J. Appl. Polym. Sci.* 2013, *129*, 2443.
36. M. Luqman, K.-H. Ko, J.-S. Kim, K. Shin, *J. Appl. Polym. Sci.* 2013, *128*, 1764.
37. R. A. Weiss, J. J. Fitzgerald, D. Kim, *Macromolecules* 1991, *24*, 1071.
38. P. Krol, *Prog. Mater. Sci.* 2007, *52*, 915.
39. B. K. Kim, Y. M. Lee, *Colloid Polym. Sci.* 1992, *270*, 956.
40. B. Bae, K. Miyatake, M. Watanabe, *Macromolecules* 2010, *43*, 2684.
41. S. A. Visser, S. L. Cooper, *Macromolecules* 1991, *24*, 2576.
42. T. C. Ward, A. V. Tobolsky, *J. Appl. Polym. Sci.* 1967, *11*, 2403.
43. M. Rigdahl, A. Eisenberg, *J. Polym. Sci. B Polym. Phys.* 1981, *19*, 1641.
44. M. R. Tant, G. L. Wilkes, *J. Macromol. Sci. Rev. Macromol. Chem. Phys.* 1988, *C28*, 1.
45. P. Vanhoorne, R. A. Register, *Macromolecules* 1996, *29*, 598.
46. (a) Y. S. Yu, L. F. Zhang, A. Eisenberg, *Macromolecules* 1998, *31*, 1144; (b) J. Zhu, A. Eisenberg, R. B. Lennox, *Macromolecules* 1992, *25*, 6547; (c) J. V. M. Weaver, Y. Q. Tang, S. Y. Liu, P. D. Iddon, R. Grigg, N. C. Billingham, S. P. Armes, R. Hunter, S. P. Rannard, *Angew. Chem. Int. Ed.* 2004, *43*, 1389.
47. (a) O. Terreau, L. B. Luo, A. Eisenberg, *Langmuir* 2003, *19*, 5601; (b) A. Harada, K. Kataoka, *Macromolecules* 2003, *36*, 4995; (c) A. Desjardins, T. G. M. Vandeven, A. Eisenberg, *Macromolecules* 1992, *25*, 2412.
48. (a) P. Raffa, P. Brandenburg, D. A. Z. Wever, A. A. Broekhuis, F. Picchioni, *Macromolecules* 2013, *46*, 7106; (b) P. Raffa, M. C. A. Stuart, A. A. Broekhuis, F. Picchioni, *J. Colloid. Interface Sci.* 2014, *428*, 152.
49. R. A. Weiss, A. Sen, C. L. Willis, L. A. Pottick, *Polymer* 1991, *32*, 1867.
50. V. G. Vasil'ev, L. A. Wasserman, G. G. Nikiforova, L. I. Komarova, G. I. Timofeeva, I. G. Plashchina, S. N. Salazkin, V. S. Papkov, *Polym. Sci. Ser. A* 2014, *56*, 17.
51. D. A. Z. Wever, F. Picchioni, A. A. Broekhuis, *Prog. Polym. Sci.* 2011, *36*, 1558.
52. E. Banks, Y. Okamoto, Y. Ueba, *J. Appl. Polym. Sci.* 1980, *25*, 359.
53. Z. H. Si, Z. Sun, F. L. Gu, L. H. Qiu, F. Yan, *J. Mater. Chem. A* 2014, *2*, 4413.
54. A. Wiesemann, R. Zentel, T. Pakula, *Polymer* 1992, *33*, 5315.

25

SUPRAMOLECULAR STERIC HINDRANCE AT BULKY ORGANIC/POLYMER SEMICONDUCTORS AND DEVICES

Ling-Hai Xie[1] and Wei Huang[1,2]

[1] *Center for Molecular System & Organic Devices (CMSOD), Key Laboratory for Organic Electronics and Information Displays & Institute of Advanced Materials (IAM), Jiangsu National Synergetic Innovation Center for Advanced Materials (SICAM), Nanjing University of Posts & Telecommunications, Nanjing, China*

[2] *Key Laboratory of Flexible Electronics (KLOFE) & Institute of Advanced Materials (IAM), Jiangsu National Synergetic Innovation Center for Advanced Materials (SICAM), Nanjing Tech University (NanjingTech), Nanjing, China*

25.1 INTRODUCTION

Advanced (opto)electronics concerns versatile organic/polymer materials in mobile information technology and sustainable energy because of their unique properties, such as flexible molecular tolerability, excellent mechanical features, and low-cost fabrication over inorganic counterparts [1, 2]. In the last decade, there were made many organic devices, such as organic light-emitting diodes (OLEDs) [3–5], organic photovoltaics (OPVs) [6, 7], organic field-effect transistors (OFETs) [8–11], information memory [12–16], and organic lasing [17] units, in terms of device efficiency, performance, stability, and reliability, as well as fabrication methods. OLED is a particularly impressive example that has been commercialized as the third-generation display technology. White OLED lighting sources have reached the power efficiency of 100 lm/W, which allows them to be applied in the near future. However, organic devices still face unprecedented challenge on the road to commercialization [18–26].

The era of consciousness, labeled with intelligent robots, requires the achievement of smart devices, the integrated entities of various computing devices for information processing, sensor and actuator devices for action control, generator and photovoltaic devices for energy supply, and ionic devices for matter transport as well as monitor devices for the environment [27]. The prerequisite of surface-equals-device engineering is to fabricate film devices with robust merits of high efficiency, reliable performance, and repeatable procedure, as well as long lifetime via molecular manufacturing technology.

In order to solve all the device technology issues that hindered the commercialization, except for the technique breakthrough of substrates and electrodes [28, 29], it is necessary to completely decode organic devices that resembled cells in the human body from the complex system point of view. The forbidden challenges stem from the complexity of organic devices that in turn make it possible to create molecular conscious systems dominated by multistate structures and involve hierarchically sandwiched devices, mesoscale films, nanoscale morphology, and supramolecular aggregates as well as molecular building blocks from top to bottom, if one defines the big as top/up and small as bottom/down (Fig. 25.1).

As a result, it is inevitable to systematically decode, analyze and evaluate organic devices starting from atoms in the periodic table of elements. In the last decade, we initiated a hierarchical chemistry of organic semiconductors, which are regarded as a molecular system approach. Hierarchical chemistry involves the integration of all chemical toolboxes and includes covalent synthesis, supramolecular chemistry, and molecular nanotechnology, as well as mesoscale assembly to explore one concrete specific molecule, material, or device, affording the principle of the multitarget-directed systems molecular design.

We observed the supramolecular interactions at the bulky spiro-configuration, which inspired us to define the concept on supramolecular steric hindrance (SSH) and to

Non-Covalent Interactions in the Synthesis and Design of New Compounds, First Edition.
Edited by Abel M. Maharramov, Kamran T. Mahmudov, Maximilian N. Kopylovich, and Armando J. L. Pombeiro.

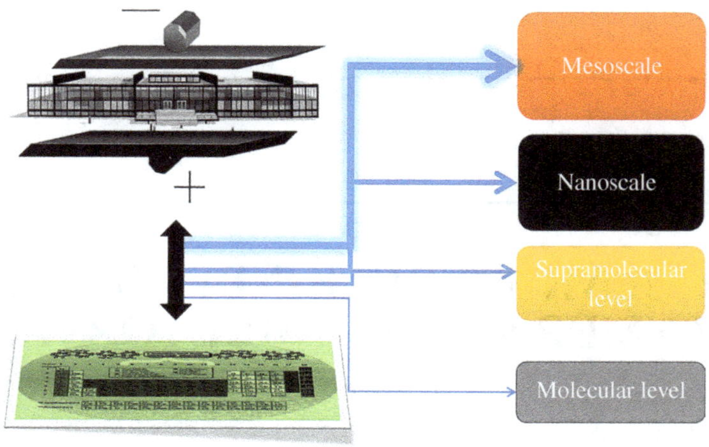

FIGURE 25.1 Schematic landscape of organic semiconductors and devices from atom at periodic table to sandwiched film devices at the microscopic scale.

make molecular design of supramolecular bulk motifs [30]. In this chapter, we will introduce the phenomena of SSH and molecular design of SSH in organic/polymer optoelectronic materials by jointing covalent synthesis, supramolecular chemistry, nanocrystal engineering, and theoretical tools as well as film and device fabrication and evaluation. Four-element principle of the design of organic materials is introduced in the second section. After this, the detailed discussions on SSH and its effect on OLEDs, π-stacked bulk supramolecules, basketweave motifs for the assembly of organic nanomosaics, a hierarchical chemistry starting from molecular design of SSH, and supramolecular hindrance polymer (SHP) semiconductors are reported. Both the steric hindrance with repulsive interaction and non-covalent bonds with attractive interaction become the key pair of opposite factors in the four-element design principle.

25.2 FOUR-ELEMENT DESIGN PRINCIPLE

Modern four-element philosophy involves the four major concepts of the universe, which are matter, energy, information, and consciousness (MEIC) [31], which is different from the counterpart theory from Aristotle. This MEIC philosophy allows the whole-picture observation of objects in the universe. The MEIC philosophy guides us not only to enter the era of consciousness in the next half of the twenty-first century but also to investigate the chemistry of any molecule, material, device, cell, or any other system.

From the chemistry point of view, organic semiconductors belong to molecular π-system where charge carriers, hole and electron, could transport or hop within the limit in π-tube materials. These materials have four characteristics, which are electronic structure, steric hindrance, conformational topology, and non-covalent interaction at the molecular level (Fig. 25.2) [32].

The basic four characteristics of organic/polymer semiconductors are extracted from the theories of organic physical chemistry. The first element is electronic structure, which can be expressed by several parameters such as HOMO and LUOMO, bandgaps, dipoles, exciton binding energy, triplet energy level, EPS, and so on according to molecular orbital theory and frontier orbital theory. Some parameters can be calculated theoretically by means of Gaussian software package and measured experimentally by UPS, electrochemical CV methods, and others. Electronic structures are determined by several parameters, including valence state of consisted atoms, electronegativity of atom components, and atom linkage pattern via covalent bonds. Electronic structures determine a series of key charge processes, such as electronic transfer, exciton behaviors, energy transfer, charge injection and transports, and so on. Molecular electronic structures determine the device electrical performances via several scales ranging from supramolecular, nanoscale, and mesoscale to device levels. Until now, electronic structures can be tuned by several tools, including donor–acceptor concept [33], metal or heteroatoms doping, intermediate species, resonance structure and aromaticity, etc.

The second characteristic is steric hindrance, which means nonactive segments, such as sp^3 carbon in organic semiconductors that serve as insulator elements. Steric effect includes a series of concepts and phenomena where steric strain affords drive forces for the organization of intramolecular conformation and configuration as well as molecular packing. With regard to function core, steric hindrance serves as one organizer with a capability of either repulsion or shield. Steric hindrance/bulk groups can be examined by the measurement of electronic structures that could not change in terms of frontier molecular orbitals such as HOMO, LUMO, and bandgap obviously. For devices, steric bulk could improve device stability and procedure reliability by means of resisting the interaction from intermolecular force and external condition or environments.

The third characteristic is conformational topology, which evaluates the softness or flexibility of molecules. Generally, conformation complexity is higher in polymers

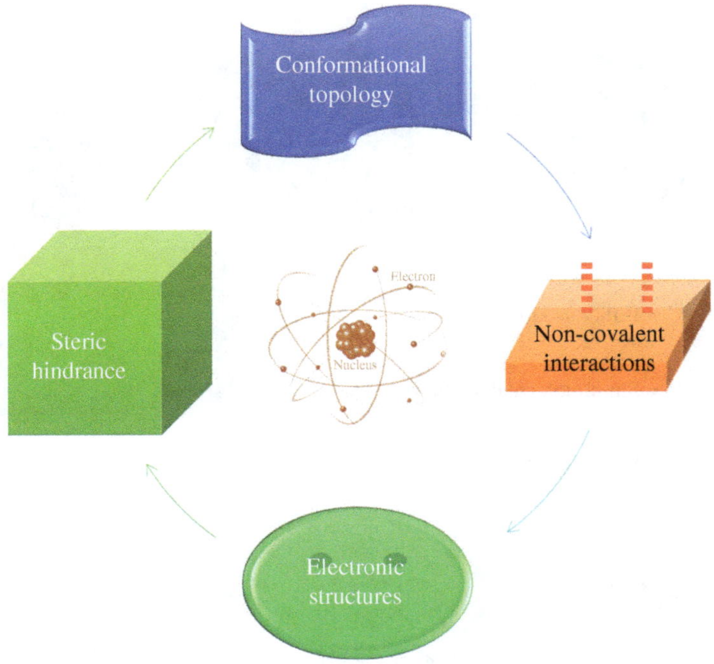

FIGURE 25.2 Four-element structure of covalent entities in organic semiconductors.

than in small molecules. Diverse polymer chains lead to the changes in terms of viscosity, phase behaviors, and mechanical properties. There are several pairs of conformers, such as linear versus multiarm, open versus cycle, rod versus coil, and rigid versus soft. Reversible conformational behaviors could serve as the special function in organic devices, such as conformation memory and memristor.

The fourth characteristic is non-covalent interactions, which are basic self-recognizing forces in molecular semiconductors. Non-covalent forces are of several kinds, such as hydrogen bonding, electrostatic interactions, and π–π interactions, as well as coordinative forces. Non-covalent forces in molecules can be identified by several characterization techniques such as single-crystal X-ray diffractions, nuclear magnetic resonance (NMR), STM characterization, CD spectra, and terahertz time-domain spectroscopy (THz-TDS) [34] and visualized by theoretical methods. In different vacuum evaporation deposition technologies, the more complex non-covalent processes occur in low-cost printable procedures on semiconductor films. As a result, it is necessary to uncover the non-covalent forces in molecular packing and film morphology as well as phase change under the external condition change. Unclarified non-covalent organizations are a crucial factor to make the failure of the commercialization in organic/polymer devices such as O/PLEDs, OPVs, transistors, and photonic devices when compared to the inorganic counterparts. Well-directed supramolecular bonds could drive molecules into well-organized nanostructures derived from organic semiconductors.

The four above aspects dominate the up-level molecular behavior, such as supramolecular aggregation, nanoscale assembly, and film-forming and device performance. When one atom or molecular segment is introduced into molecular backbones, at least one aspect would be changed, followed by intermolecular supramolecular effects, nanoscale, and mesoscale, and then influences the thin film and device. Under the comprehensive framework, the complexity of organic devices stems from both intramolecular coupling of four aspects in molecular entities and intercoupling between different levels or scales.

There exists the interplay between mesoscale, nanoscale, supramolecular level, and molecular scale in organic devices inside the microscopic level. Generally, if an atom or molecular segment is introduced into molecular entities, several properties are changed, even all aspects that explain the complexity of molecular world starting from a periodic table of elements. As a result, one of the main tasks is to make the four-element analysis useful to clarify the multirole of any atom or molecular segment. Another task is to design the modeling single-element building blocks in view of the multiscale contribution at each level from atom to device. Four-element design at the molecular level is the starting entry to fabricate the repeatable and reliable devices on need.

25.3 SSH IN ORGANIC SEMICONDUCTORS

25.3.1 SSH in Bulk Semiconductors and OLEDs

Organic bulk semiconductors have excellent thermal and morphological stability by means of high glassy transition temperature T_g. They are competitive candidates to commercially available OLED displays and lighting sources. In general, bulky semiconductors have functional π-segments tethered with the bulky groups via covalent linkages. Typical bulky groups include cruciform-shaped spirofluorenes or

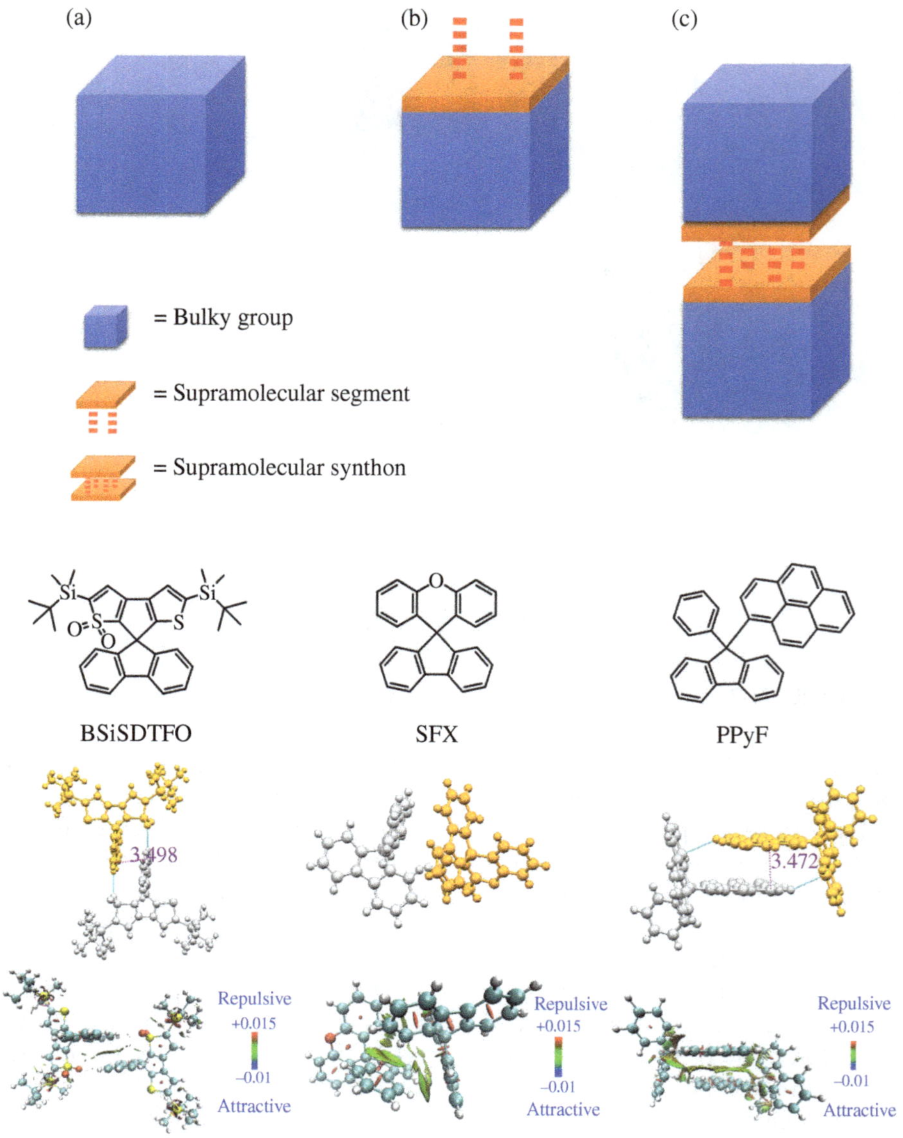

FIGURE 25.3 Schematic cartoon of bulky moieties, bulky groups with supramolecular forces, and their dimers as well as a series of model compounds with dimer motifs and non-covalent visualization graphs.

nonplanar diarylfluorenes that have sp^3 carbon serving as an electron insulator. sp^3 carbons offer cubic configuration with the steric hindrance effect that enable either the distortion or strain of intramolecular π-segments or separation of the intermolecular interactions (Fig. 25.3).

In 1996, Salbeck *et al.* introduced spiro-configuration into hole-transporting materials for the improvement of thermal stability and amorphous state [35]. For light-emitting materials, the bulky groups afford an obstacle to hinder π–π interactions of chromophores, facilitating the pure molecular light-emitting behaviors. Following this strategy, during the last 10 years, we discovered one-pot protocol to spirofluorenes such as spiro(fluorene-9,9′-xanthene) (SFX) [36–40], SFDBX [41], SFDBA, and SFDBAO, followed by complex diarylfluorenes (CDAFs) via carbinol's Friedel–Crafts [42] and C—C coupling [43] reactions. We especially explored their single crystals by X-ray diffraction to observe the intermolecular weak interactions at the bulk semiconductors.

In general, bulky groups offer repulsive forces that are opposed to the attractive non-covalent force at supramolecular motifs. However, the weak interactions can occur at the bulky spiro-scaffolds and diarylfluorenes with the steric effect that generally damage the intermolecular interaction (Fig. 25.3a). In BSiSDTFO, one supramolecular motif is the π–π stacking interaction with distances of 3.498Å between spirofluorene segments, together with two pairs of C—H···O weak interactions [30].

We defined the bulky groups with well-defined non-covalent forces as SSH. One key effect of SSH is packing bulky groups into supramolecular motifs via dimer, trimer, tetramer, and so on. Supramolecular dimer of bulky BSiSDTFO determined the type of exciton with single-molecule behaviors rather than excimer emission. BSiSDTFO

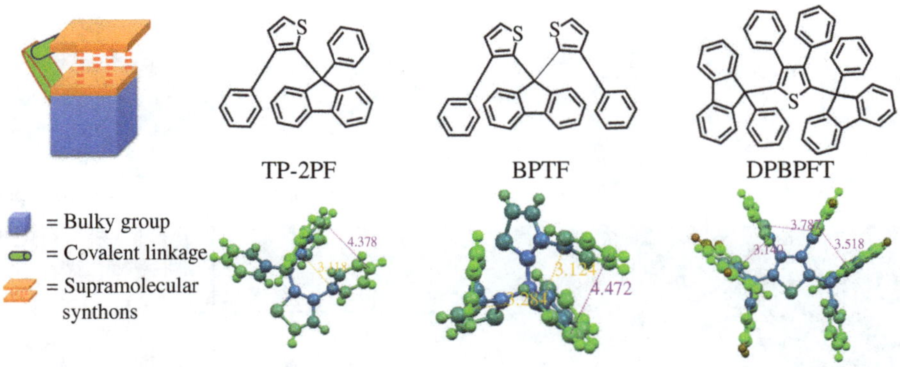

FIGURE 25.4 Complex diarylfluorenes as bulky supramolecular semiconductors.

also exhibited higher fluorescent quantum yields of more than 20% in solid state. The anomalous photophysical properties are ascribed to the synergic combination of steric hindrance with non-covalent interactions.

Meanwhile, among rigid and bulky moieties, we discover a similar phenomenon from another spiro-arene, SFX. Single-crystal X-ray diffraction indicated that non-covalent forces do not exist between SFXs. In order to get insight into the secondary forces, we utilized non-covalent visualization tools and the theoretical calculation of interaction energy (IE) [44]. As a result, non-covalent interactions between SFX molecules with a large IE value were visualized by NCIPLOT software (Fig. 25.3b). The observation inspired us to demonstrate the complex relation between bulky groups with steric hindrance and non-covalent forces in organic semiconductors.

After this observation of spiro-based compounds, we introduced the phenylfluorenyl moieties (PFMs) as another bulky group to suppress π–π stacking interaction in the design of deep-blue light-emitting materials. When PFM capped at one end of pyrenes were studied, diffraction data indicated that π–π interactions between pyrenes still exist, without being completely suppressed. We also observed the tight pairs of pyrene chromophores combined via strong π–π stacking interactions [42]. As a result, PPyF exhibits typical SSH features. If only increasing the number of bulky PFMs, the larger distance between pyrenes is realized by such two-end substituents.

SSH leads to different performances in OLED materials [45, 46]. We compared the difference between spiro-configuration and alkyl side chain at 9-position of fluorene in phosphorescent metal complex. The rather low turn-on voltage of approximately 5.2V was observed in spiro-Ir OLED, which is lower than that in alkyl-Ir OLED. The result can be explained by the molecular packing between spirofluorene segments or between host and spirofluorenes, which leads to the charge hopping chance more than with the isolating alkyl side chains. The final device exhibits luminance efficiency of 44.3 cd/A[1] and power efficiency of 34.8 lm/W with yellow colors. In addition, from the electronic structures and supramolecular interaction point of view, we designed a series of next-generation bulky groups with steric hindrance, including spirothiophene [47], 9-phenylfluorenes [48], 9-pyridylfluorenes [49], and 9-phenyldiazafluorenyl moieties with C—H···N interaction.

Non-covalent forces at bulky groups offer an effective tool to change molecular behavior, which is completely different from either the steric hindrance or weak forces alone. Supramolecular analysis of bulky organic/polymer semiconductors offers a way to understand the interplay between steric hindrance and non-covalent interactions to influence the optoelectronic features and the film morphology.

25.3.2 Bulk Supramolecules

Non-covalent interaction at bulky groups can be fixed by covalent bond to create supramolecules. Bulky π-stacked molecules are an alternative to organic/polymer semiconductors, intrinsically different from π-conjugated organics/polymers in terms of charge carrier transport and injection. One example is complex diarylfluorenes (CDAFs) (Fig. 25.4).

Intermolecular forces in diarylfluorenes with SSH are easily tuned by introducing *ortho*-substituents. As a result, a series of CDAFs as the model of bulky π-stacked molecules are designed [47, 50, 51]. Due to π-stacking in CDAFs, these supramolecular organic semiconductors (SOSs) have intramolecular electron transfer mechanism. The stacking degree can be identified by single-crystal X-ray diffraction and NMR spectroscopy, besides other non-covalent visualization tools. For example, congested TP-2PF (CCDC 817167) has an obvious intramolecular π-stack with a shortest distance of 3.118 Å. TP-2PF's fluorene plane exhibits obvious out-of-plane bending conformation owing to steric strain. For BPTF, π-stacking interaction occurs owing to two *ortho*-substituent steric hindrances, probably offering an extraordinary pathway for charge carriers beyond fluorene conjugation backbone. They are valuable tools to modulate photophysical and electronic behaviors in the design of organic semiconductors.

DPBPFT (CCDC 817168) was investigated as a π-stacked semiconductor by UV–Vis absorption, photoluminescence, and electrochemical methods. A supramolecular light-emitting device was fabricated with a configuration of ITO/MoOx(2 nm)/NPB(30 nm)/DPBPFT(30 nm)/TPBi(10 nm)/

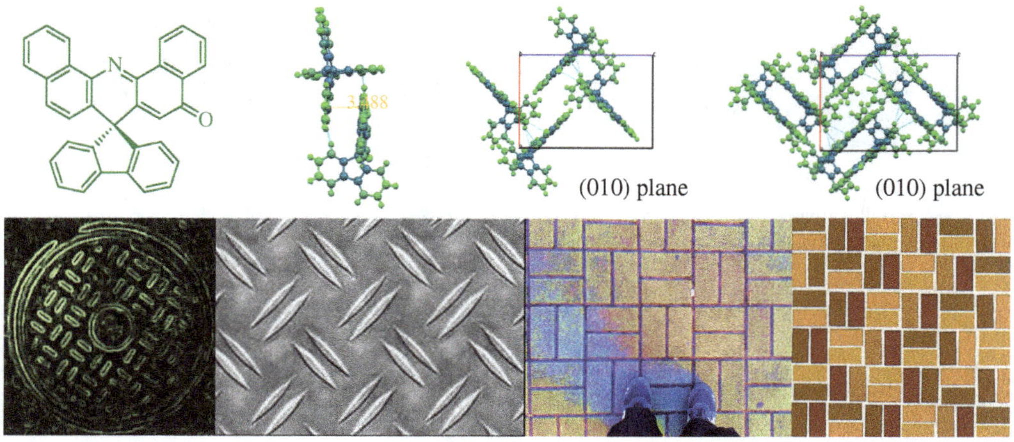

FIGURE 25.5 Molecular structure of SFDBAO, its dimer, tetrad, and octet motifs, together with their similar patterns in macroscopic world.

Bebq2(30 nm)/LiF(1 nm)/Al. The supramolecular devices exhibited an anomalous electroluminescence with a turn-on voltage of 9 V, a maximum luminance of *ca.* 1282 cd/m^2, and a luminous efficiency of 0.24 cd/A. Weak interactions occur inside molecules that afford potential function behavior to couple the charge with exciton, spin, and molecular mechanical motion. Nevertheless, stacked molecules afford one potential kind of multifunctional behavior for molecular devices and film electronics that will be investigated thoroughly in CMSOD@IAM.

25.3.3 Basketweave Motifs in the Assembly of Organic Nanocrystals

Molecular packing modes influence optoelectronic properties and device performances significantly. We carefully examined a series of spiro-based organic semiconductors. We found a four-element motif in terspirofluorene double-stranded supramolecular polymer. SSH in several spirocyclic arenes not only created bulky supramolecular dimers but often generated tetramolecular motifs, suggesting that they are excellent tectons. Impressively, a series of unique molecular packing modes in SFDBAO was extracted as shown in Figure 25.5 [52].

SFDBAO dimers are fabricated by π–π stacking with a distance of 3.488 Å and two pairs of C—H···O bonds. Spiro-tetrads are also observed in single crystals of SFBAO. The spiro-tetrad consists of four SFDBAO via C—H···O and C—H···π bonds. From further supramolecular analysis of SFDBAO, the unique supramolecular octet motifs lie at the plane of (010). Complex four-dimer motifs consist of eight SFDBAO with the non-covalent forces, including the above-mentioned π–π stacking interaction and C—H···O and C—H···π forces. To the end, a series of rare quadrats are identified with a molecular packing pattern of double basketweave mode at the molecular levels, which are different from the other molecular packing modes such as herringbone, stretcher, and stack in high-mobility materials.

These basketweave molecular packing patterns/motifs are universal in macroscopic world. Such patterns appear on the sidewalk pavement, hotel wall, tread plate of the bus or car, and manhole cover in campus of NUPT as well as in mats. The pattern is also similar to the logo or trademark of the Columbia Sportswear Company and Sun Microsystems. Such patterns can be made by jigsaw pieces clip art, bricks, tiles, mosaics, or puzzle blocks. SFDBAO's supramolecular basketweave motifs are not only aesthetic but also functional, serving as molecular tectons in molecular nanotectonics.

We utilized surfactant-assisted reprecipitation to explore supramolecular self-assembly of SFDBAO toward molecular nanocrystal engineering. SFDBAO could enter into various monodispersed nanocrystal polyhedral shapes such as hexahedra, octahedra, and decahedra by changing the types of surfactants, for example, Pluronic 123 (P123) or cetyltrimethyl ammonium bromide (CTAB). Organic polyhedral micro-/nanocrystals were fully characterized by the SEM, SAED, TEM, and XRD. The various two- or three-dimensional nanocrystalline polyhedra and nanosheets offer the state-of-the-art models to get insight into crystal growth kinetics and supramolecular polymerization mechanism. Bulky spiro-arenes are also promising bulky tectons for semiconducting nanocrystals with the potential application of printing electronics.

25.3.4 A Hierarchical Chemistry of Organic Devices

Rational design is the goal of molecular engineering chemistry that is hindered by multiscale relevancy. We attempted to design the organic devices via nanocrystals and nanofilms by manipulating non-covalent interactions. We designed SSH by using covalently linking steric repulsive groups with supramolecular attractive groups into one molecular entity (also called repulsor–attractor design). Following this strategy, any bulky groups or any supramolecular segments can be introduced into one molecule in order to create one kind of supramolecular hindrance molecules. Bulky groups are one major kind of building blocks with the sp^3 carbon, such as the above spiro-arenes, diarylfluorenes, iptycenes [53], adamantanes, and other rigid moieties. Non-covalent interaction can be hydrogen bonding and electrostatic forces, as well as host–guest bonding, which are stronger than others. Interplay between steric hindrance and non-covalent interactions can change growth direction or supramolecular polymerization kinetics. Some efforts have been made by us to synthesize a series of hindrance-capped supramolecular synthons.

We examine SSH as an effective tool for the control over the geometric shape of nanocrystals from solution [54]. One proof-of-concept modeling bulky tecton is Py-SFDBAO, where SFDBAO is regarded as bulky groups and Py as a strong π–π stacking synthon. As a result, we observed the perfect dimers of Py-SFDBAO with a longitude of 1.25 nm that assemble into a supramolecular octet with the basketweave motifs. In consideration of the anisotropic supramolecular interactions at horizontal and vertical direction, Py-SFDBAO could assemble into a two-dimensional nanosheet that resembles the bilayer lipid entity in cells.

Py-SFDBAO assembles into a nanosheet via surfactant-assisted reprecipitation; the nanostructures were characterized by SEM, TEM, XRD, and AFM. From SEM, the area of nanosheets with regular square edge is about 4 μm^2. XRD and SEAD indicated that the plane of nanosheets is (100), where Py-SFDBAOs stand up in nanomosaics. The thickness of about 12.9 nm is equivalent to 10 molecular layers with an interdigital dimer. Unambiguously, the growth of Py-SFDBAO in plane of (200) and supramolecular octet with regular shape determined the near square-shaped edge in nanocrystals. Py-SFDBAOs are perfect molecular building blocks with well-defined repulsor–attractor feature of bulky supramolecular motifs. Under the regime, the basic picture is that SFDBAOs first aggregate into dimer with the length of approximately 1.25 nm, followed by the octamers, and the quadrat grows into nanomosaics that finally resemble the lipid membranes.

The organic nanocrystalline nanomosaics offer a state-of-the-art model to get insight into the kinetic and thermodynamic processes of supramolecular polymerization that are the key issues in supramolecular chemistry. The above hypothesis and deduction make easier to understand the effect of specific geometric shapes and non-covalent interactions in the controllable shape of nanocrystals from zero, one, two, and three dimensionality, affording the guideline for molecular design of supramolecular architectures. Nanosheets could form the uniform films with crystalline state by spin coating from the water-phase dispersed nanoinks (Fig. 25.6).

GIXD characterization further confirmed the film morphology and molecular orientation in crystalline state. Two-dimensional nanosheets offer a platform to eco-friendly fabrication of nanocrystalline heterojunction that are the key components to tune charge carriers for the application of LEDs, OPVs, and photodetectors in organic electronics. Nanomosaics could be other excellent kinds of nanobuilding blocks for the nanoassembly or mesoscale assembly for complex surfaces or films.

Finally, a sandwiched film diode was fabricated with a configuration of Si/nanomosaics/Au and ITO/nanomosaics/Al, exhibiting multifunctionality with both photoswitching effect and WORM-type memory. A single-cell "AND" logic gate circuit was achieved by the nanocrystalline film devices. The mechanism of molecular nanocrystal devices involves the trapped charge-limited current and F–N tunneling. In a word, SSH design in hierarchical chemistry is exploited in organic devices via basketweave motifs and nanosheets or nanomosaics, affording the eco-friendly optical nanoinks in organic electronics. SSH effects make a starting point of evolution from solution, via thin films, to devices in organic electronics, transforming organic semiconductors into supramolecular nanofilm electronics that lie between solid film electronics and nanoelectronics.

25.4 EXTENSION OF SSH CONCEPT TO POLYMER SEMICONDUCTORS

25.4.1 SHPs

SSH can be introduced into polymers and to make a new kind of materials that can be coined as SHPs (Fig. 25.7). However, the SSH effect in conjugated polymers are difficult to observe, because conformation is more complex in polymer semiconductors than those in small organic semiconductors. In general, the introduction of bulky groups with steric hindrance into conjugated polymers [55–57], conjugation-interrupted polymers [58], and stacked polymers [59–63] could improve their thermal and morphological stability with higher T_g.

Spiro-based bulky conjugated polymers are in amorphous state, in which the backbone chain is controlled with difficulty by external conditions [64, 65]. Different from conjugated polymers, the stacked polymers can be regarded as the models of supramolecular semiconductors owing to the π–π hopping channels. Stacked polymers have three types of possible conformations, including full face-to-face, offset stacked modes and molecular tweeter mode [66].

Under the SSH concept, we designed hindrance-functionalized stacked polymers to modulate the conformation, a key parameter in polymers. From UV–Vis absorption and photoluminescence spectra, we can deduce that polymers possess more partially stacked conformations in comparison with initial PVK in both solution and films. PVK–PF exhibits the fluorescence spectra with narrow profile. Owing to steric hindrance effects, the blue-shift PL spectra occur, indicating the face-to-face π–π stacking motifs were suppressed. We also examine electrically triggered conformational changes by measuring current–voltage curve in diode with a sandwiched configuration of ITO/PVK–PF/Al that was fabricated by spin coating procedures. The diode exhibits electrical hysteresis with bistability, suggesting that phase change occurs in the PVK–PF film by means of the SSH effects in stacked polymer semiconductors, although the

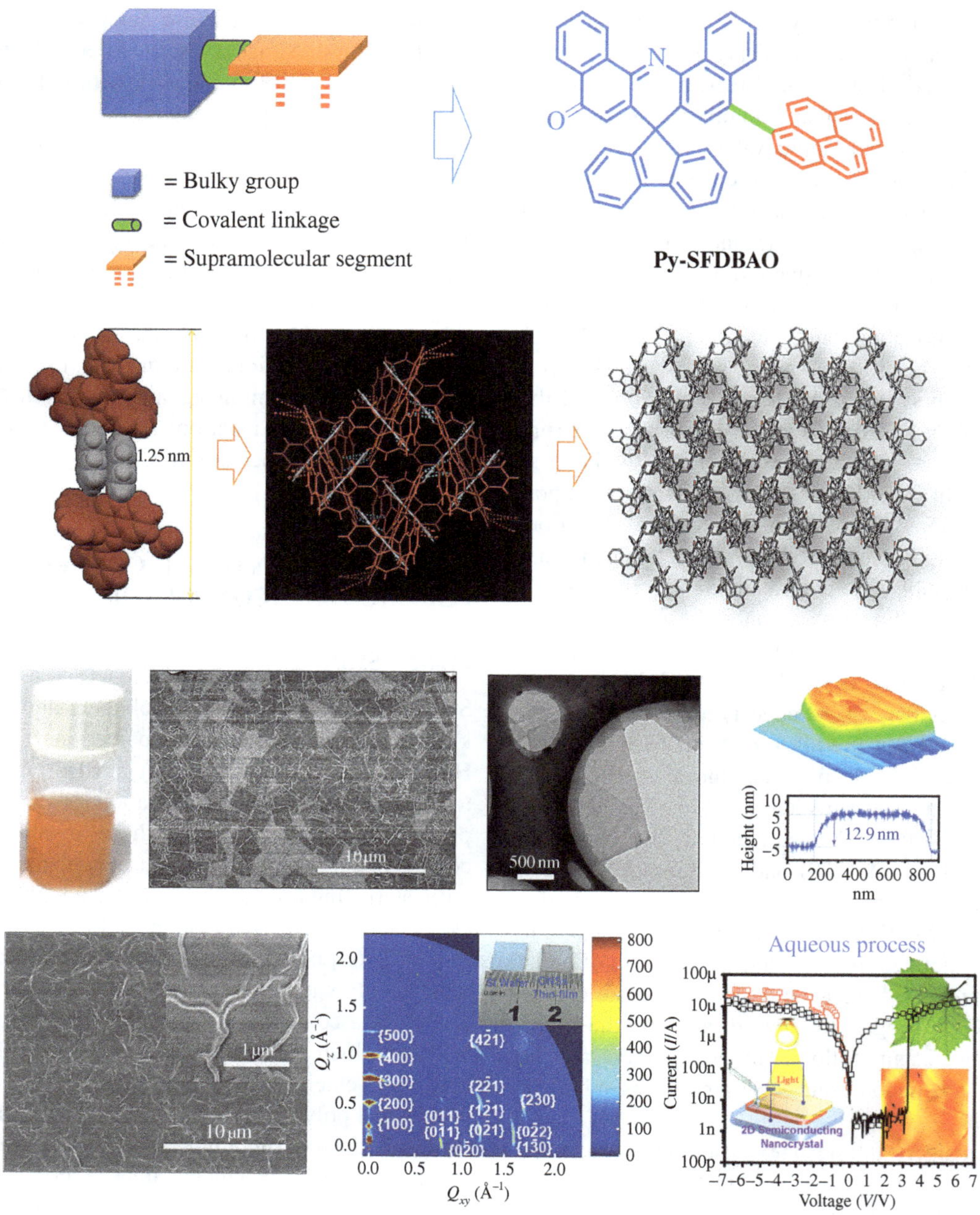

FIGURE 25.6 A hierarchical chemistry to devices from SSH design with a repulsor–attractor concept, via organic synthesis of modeling molecules, supramolecular motifs, and nanocrystals, to nanocrystalline films and nanofilm devices.

filament mechanism should be excluded carefully. Stacked bulk polymers will be promising polymers with memory effect [67].

On the other hand, conjugated polymers are the main kind of polymer semiconductors where the control over conformation is required in order to explore the multifunction or improve the device performance, reliability, and lifetime. Conformational changes are the origin of phase transition and polymorphism in printing electronics or solution-processable devices. Among them, polyfluorenes [68] have several metaphases, including beta phase, gamma phase, and gel phase as well as green band-related phase that is still waiting for the clarification of origin. For example, the beta phase is one with high performance, quantum efficiency, and better charge transporting ability. The desired materials are the bulky polyfluorenes with beta phase, which have both thermal stability and high charge delocalization ability [69, 70].

In order to improve the tight molecular packing, SSH becomes an intermediate tool to balance the contradiction between distorted chains and crystalline chains. In order to obtain the beta phase of polydiarylfluorenes, we introduced alkyl side chain at the crossing 4-position of fluorene. We proposed a monomer with SSH that has the bulky diaryl moieties at 9,9-position of fluorene and the alkyl side chain with attractive van der Waals force at crossing 4-position of

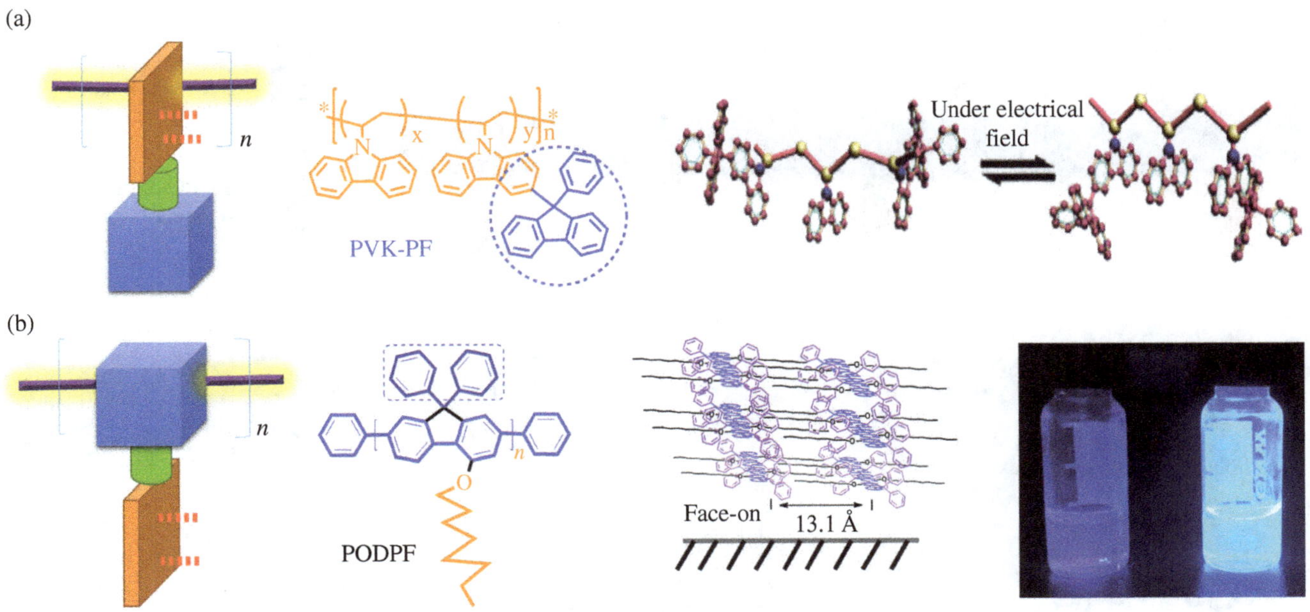

FIGURE 25.7 Two types of supramolecular hindrance polymers (SHPs), including a stacked polymer with the bulky moieties and a bulky poly(diarylfluorene) with non-covalent force.

fluorene [71]. As a result, PDOPF could adopt a beta phase of polyfluorene with a dihedral angle of 165° in both solution and film. Synergic inductions are favorable for β-chain conformations via the two-side balanced introduction of bulky groups with steric repulsion and alkyl chains with the van der Waals forces. 2D-GIXD suggest that backbone lies at a face-on mode with the interchain distance of about 13.1 Å. Beta-phased bulky polyfluorenes are more stable for the application of lasing materials. SSH will become crucial tool to realize controllable phase in polymer semiconductors.

SSHs are easily integrated into supramolecular conjugated polymers (SCPs) that are promising advanced multifunctional materials with the merits of both semiconductors and supramolecules, such as reversible and stimulus-responsive behaviors, self-assembly, and self-healing, offering a promising platform to construct smart semiconductors and devices [72]. In this aspect, we created a series of SCPs by inducing the hydrogen bonding, coordination, and electrostatic forces in polymer backbones or side chains that have been synthesized to examine the dual nature of SCPs. One SCP is poly(diazafluorene)s [73–75]; the other is PPFOH with SSH feature [76, 77] (Fig. 25.8).

We classified solvents into two basic types according to whether it is gelation or not. The basic supramolecular motifs of PPFOH and poly(diazafluorenes) in different solvents were deduced by means of NMR characterization (Fig. 25.8), although supramolecular interactions are more complex in conjugated polymers than in small molecules owing to the unobtainable single crystals. NMR spectra indicated that PPFOH or polydiazafluorene in one miscible solvent exhibits single-chain behaviors with the solvent–polymer complexes, which generally hinder polymer chain condensation into networks at gel state. In the nonmiscible solvents, non-covalent interactions among polymer chains could be destroyed, which is confirmed by the broad peaks in NMR, leading a typical polymer–polymer system. Furthermore, we found that the emission color of poly(diazafluorenes) can be modulated by the coordination to metal ions and boron reagent as well as solvents from blue to green yellow.

In order to further amplify the smart luminescence, we proposed the attractor–repulsor design of SHPs. In comparison to poly(diazafluorenes), PPFOH with SSH exhibit the big solvent and molecular weight effects on fluorescence behaviors. The enormous multicolor films of PPFOH probably are contributed by hydrogen bonding of flurorenol, which was identified by X-ray diffraction [78].

9-Phenylfluorenol is the typical monomer with an SSH effect that only can be combined in dimer rather than tetramer by the H-bonding. We also studied the difluorenols for the evidence of H-bonding between polymer–polymer chains. Color changes probably stem from the synergetic interaction between chain packing and conformational change or entanglement under the solvent-dependent condensation that represent the complex nature of functional supramolecular polymer soft matters.

It should be noted that the big change in different solvents was amplified by conjugated polymer chains, compared to small organic molecules. PPFOH also exhibits the large solvent effect on electroluminescence. Finally, PLEDs with a configuration of ITO/PEDOT:PSS/PPFOH/TPBi/LiF/Al exhibit the full color emission from blue, green, yellow, and white by means of only change of solvent and molecular weight. Supramolecular electroluminescence from PPFOH facilitates a potential application of stimulus-responsive sensor and actuators as well as their multifunctional MIMO devices.

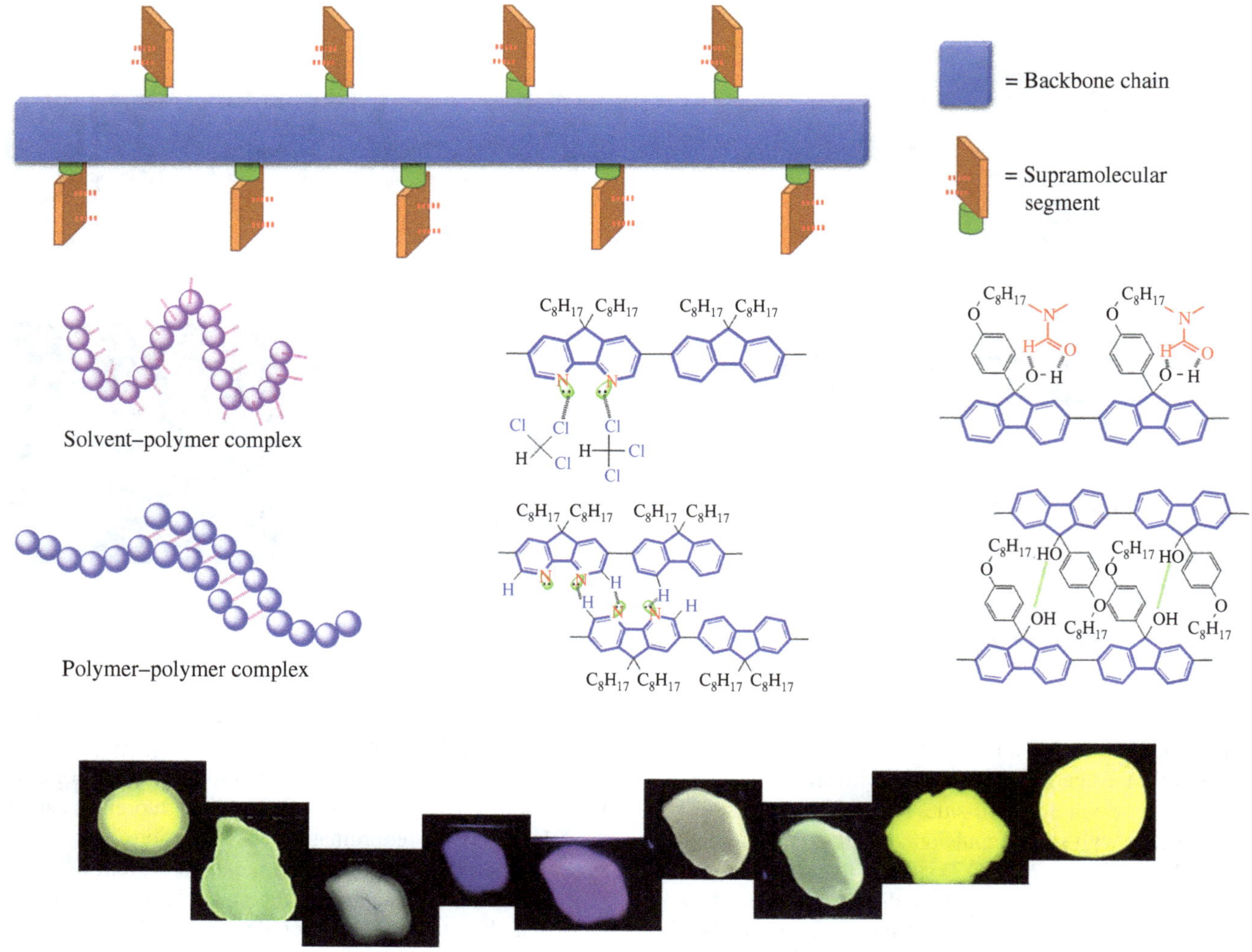

FIGURE 25.8 Conceptual models of polyfluorene-based supramolecular conjugated polymers.

25.5 FINAL COMMENTS

In summary, both hierarchical chemistry and four-element approach have been proposed to challenge the complexity of organic devices. From the observation of supramolecular interactions at bulky semiconductors, the concept on SSH was extracted for the explanation of low turn-on voltage in OLEDs. Following this concept, bulky supramolecules and molecular tectons were designed with a feature of attractor–repulsor via molecular engineering, resulting in impressive basketweave motifs that were discovered in spiro-based organic semiconductors.

Molecular design of SSH offers a series of advanced molecular tectons for the assembly of well-defined two- or three-dimensional micro-/nanocrystals, such as nanosheets, polyhedral, and nanomosaics, via surfactant-assisted reprecipitation, affording a state-of-the-art platform to molecular nanotectonics. Non-covalent interactions innovate the eco-friendly optical/semiconducting nanoinks for the application of wearable electronics or E-textiles, representing system chemistry to organic devices.

Conformational changes of PVK–PF have been tuned by SSH that become a key mechanism to design polymer memory materials for information storage. Bulky poly(diarylfluorenes), PODPF, can be transformed from alpha phase into beta phase by the cooperative repulsive and attractive interaction under the specific solvent or thermal conditions in spin-coated films. Bulky PPFOH as a paradigm of SCPs exhibits the rarely big solvent and molecular weight effects on photophysics. Supramolecular electroluminescence with abundant colors strongly depends on external conditions, molecular weight, or electrical fields, affording alternative possibility to create smart sensors and actuators.

The four-element theory allows not only to guide the molecular design of organic molecules, polymers, and nanocrystals but also to design the nanomolecules and nanopolymers [17, 18, 79–82], such as nanogrids, poly(grid)s, and covalent molecular motifs (Fig. 25.9).

Different from graphene oxides [83–89], supramolecular nanomolecules and nanopolymers serve as atom-precise supramolecular nanomaterials that combine both features of

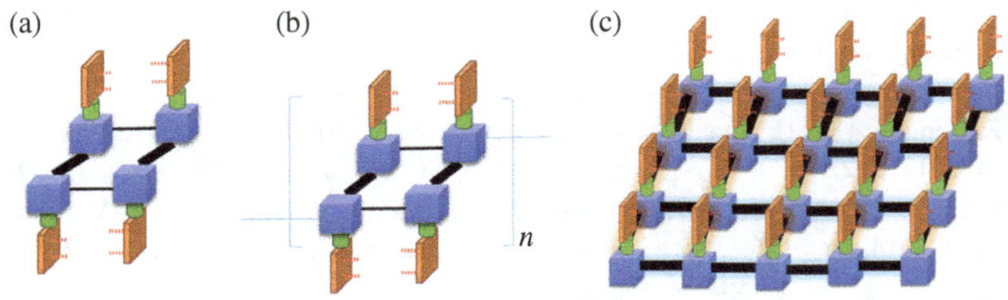

FIGURE 25.9 Supramolecular nanogrids and supramolecular nanopoly(grid)s.

hard materials and soft materials into one chemical nanoentity, affording a robust platform to challenge the complexity and functionality of organic devices. Grid functions make the unprecedented chance to activate the molecular behavior and to create the versatile nanosystems via the molecular manufacturing by integrating various functions, including electron, photon and exciton, quantum, plasmon, polariton, phonon, magnet, ion, mechanon, and so on. Hierarchical chemistry organizes grid function into systems via noncovalent synthesis, polymerization engineering, supramolecular approaches, nanotechnology, and mesoscale toolbox, starting from periodic table elements. As a result, grid function makes organic smart systems by the controllable motion at all scales. Ultimately, a dream of chemical robots could be true that will release scientist from lab at the second half of the twenty-first century.

ACKNOWLEDGMENTS

This work was partially supported by the National Natural Science Funds for Excellent Young Scholar (21322402), National Key Basic Research Program of China (973) (2015CB932200), National Natural Science Foundation of China (U1301243, 21274064), The Program for New Century Excellent Talents in University (NCET-11-0992), Doctoral Fund of Ministry of Education of China (20133223110007), and Excellent Science and Technology Innovation Team of Jiangsu Higher Education Institutions (2013) and Open Project from State Key Laboratory of Supramolecular Structure and Materials at Jilin University (sklssm2015022). Project was funded by the Priority Academic Program Development of Jiangsu Higher Education Institutions.

REFERENCES

1. S. R. Forrest, *Nature* 2004, *428*, 911.
2. L.-H. Xie, C.-R. Yin, W.-Y. Lai, Q.-L. Fan, W. Huang, *Prog. Polym. Sci.* 2012, *37*, 1192.
3. S. Chen, L. Deng, J. Xie, L. Peng, L. Xie, Q. Fan, W. Huang, *Adv. Mater.* 2010, *22*, 5227.
4. L. Deng, S. Chen, J. Xie, Y. Qian, L. Xie, N. Shi, B. Liu, W. Huang, *Org. Electron.* 2013, *14*, 423.
5. Y. Ni, H. Su, W. Huang, J. Zhao, Y. Qian, L. Xie, G. Xe, Y. Zhao, A Spiro [Fluorene-9, 9'-Xanthene]-Based Host Material for Efficient Green and Blue Phosphorescent OLED, in International Conference on Process Equipment, Mechatronics Engineering and Material Science (PEME 2013), June 15–16, 2013, Wuhan, China.
6. C. Ou, D. Zhao, C. Zhang, F. Wang, L. Wang, L. Xie, B. Ren, X. Sun, S. Li, G. Zhang, W. Huang, *Synth. Met.* 2014, *187*, 118.
7. B.-Y. Ren, C.-J. Ou, C. Zhang, Y.-Z. Chang, M.-D. Yi, J.-Q. Liu, L.-H. Xie, G.-W. Zhang, X.-Y. Deng, S.-B. Li, W. Wei, W. Huang, *J. Phys. Chem. C* 2012, *116*, 8881.
8. W.-W. Shi, W. Li, M.-D. Yi, L.-H. Xie, W. Wei, W. Huang, *Acta Phys. Sin.* 2012, *61*, 228502.
9. J. Dong, Y.-H. Chai, Y.-Z. Zhao, W.-W. Shi, Y.-X. Guo, M.-D. Yi, L.-H. Xie, W. Huang, *Acta Phys. Sin.* 2013, *62*, 047301.
10. M. Yi, X. Xia, T. Yang, Y. Liu, L. Xie, X. Zhou, W. Huang, *Appl. Phys. Lett.* 2011, *98*, 073309.
11. M. Yi, Y. Guo, J. Guo, T. Yang, Y. Chai, Q. Fan, L. Xie, W. Huang, *J. Mater. Chem. C* 2014, *2*, 2998.
12. J. Liu, Z. Yin, X. Cao, F. Zhao, A. Lin, L. Xie, Q. Fan, F. Boey, H. Zhang, W. Huang, *ACS Nano* 2010, *4*, 3987.
13. M. Yi, L. Zhao, Q. Fan, X. Xia, W. Ai, L. Xie, X. Liu, N. Shi, W. Wang, Y. Wang, W. Huang, *J. Appl. Phys.* 2011, *110*, 063709.
14. Z.-Z. Du, W. Li, W. Ai, Q. Tai, L.-H. Xie, Y. Cao, J.-Q. Liu, M.-D. Yi, H.-F. Ling, Z.-H. Li, W. Huang, *RSC Adv.* 2013, *3*, 25788.
15. J. Lin, W. Li, Z. Yu, M. Yi, H. Ling, L. Xie, S. Li, W. Huang, *J. Mater. Chem. C* 2014, *2*, 3738.
16. Y.-H. Chai, Y.-X. Guo, W. Bian, W. Li, T. Yang, M.-D. Yi, Q.-L. Fan, L.-H. Xie, W. Huang, *Acta Phys. Sin.* 2014, *63*, 27302.
17. Y. Qian, Q. Wei, G. Del Pozo, M. M. Mroz, L. Lueer, S. Casado, J. Cabanillas-Gonzalez, Q. Zhang, L. Xie, R. Xia, W. Huang, *Adv. Mater.* 2014, *26*, 2937.
18. L. H. Xie, X. Y. Hou, C. Tang, Y. R. Hua, R. J. Wang, R. F. Chen, Q. L. Fan, L. H. Wang, W. Wei, B. Peng, W. Huang, *Org. Lett.* 2006, *8*, 1363.
19. L.-H. Xie, S.-H. Yang, J.-Y. Lin, M.-D. Yi, W. Huang, *Philos. Trans. R. Soc. A* 2013, *371*, 20120337.
20. M.-L. Sun, W.-S. Zhu, Z.-S. Zhang, C.-J. Ou, L.-H. Xie, Y. Yang, Y. Qian, Y. Zhao, W. Huang, *J. Mater. Chem. C* 2015, *3*, 94.
21. J. Liu, Z. Lin, T. Liu, Z. Yin, X. Zhou, S. Chen, L. Xie, F. Boey, H. Zhang, W. Huang, *Small* 2010, *6*, 1536.
22. Y. Yang, Z. Liu, Z. Yin, Z. Du, L. Xie, M. Yi, J. Liu, W. Huang, *RSC Adv.* 2014, *4*, 55671.

23. W. Shi, Q. Tai, X.-H. Xia, M.-D. Yi, L.-H. Xie, Q.-L. Fan, L.-H. Wang, A. Wei, W. Huang, *Chin. Phys. Lett.* 2012, *29*, 087201.
24. W. Ai, L. Xie, Z. Du, Z. Zeng, J. Liu, H. Zhang, Y. Huang, W. Huang, T. Yu, *Sci. Rep.* 2013, *3*, 2341.
25. W. Ai, Z. Luo, J. Jiang, J. Zhu, Z. Du, Z. Fan, L. Xie, H. Zhang, W. Huang, T. Yu, *Adv. Mater.* 2014, *26*, 6186.
26. W. Ai, X. Cao, Z. Sun, J. Jiang, Z. Du, L. Xie, Y. Wang, X. Wang, H. Zhang, W. Huang, T. Yu, *J. Mater. Chem. A* 2014, *2*, 12924.
27. L.-H. Xie, W. Huang, *Advances in Optoelectronics Research*; Nova Science Publishers: Hauppauge, 2014; 29, Chapter 2.
28. Z.-D. Liu, Z.-Y. Yin, Z.-H. Du, Y. Yang, M.-M. Zhu, L.-H. Xie, W. Huang, *Nanoscale* 2014, *6*, 5110.
29. J. Liu, L. Xie, W. Huang, *Integr. Ferroelectr.* 2011, *128*, 105.
30. L.-H. Xie, X.-Y. Hou, Y.-R. Hua, Y.-Q. Huang, B.-M. Zhao, F. Liu, B. Peng, W. Wei, W. Huang, *Org. Lett.* 2007, *9*, 1619.
31. L.-H. Xie, W. Huang, *Advances in Science*; Science Network: Canada, 2013; Vol. *1*, 104, Chapter 4.
32. L.-H. Xie, Y.-Z. Chang, J.-F. Gu, R.-J. Sun, J.-W. Li, X.-H. Zhao, W. Huang, *Acta Phys. -Chim. Sin.* 2010, *26*, 1784.
33. W. L. Yu, H. Meng, J. Pei, W. Huang, *J. Am. Chem. Soc.* 1998, *120*, 11808.
34. L.-H. Xie, H. Dai, B.-B. Jin, Y. Han, Q. Tai, M.-D. Yi, T. Yang, P.-H. Wu, W. Huang, *Macromol. Chem. Phys.* 2012, *213*, 1441.
35. J. Salbeck, New low Molecular organic compounds with high glass transition temperature as materials for blue electroluminescence, in, *Inorganic and Organic Electroluminescence*, edited by R. H. Mauch and H.-E. Gumlich; Wissenschaft und Technik: Berlin, 1966), p. 243.
36. L.-H. Xie, F. Liu, C. Tang, X.-Y. Hou, Y.-R. Hua, Q.-L. Fan, W. Huang, *Org. Lett.* 2006, *8*, 2787.
37. J.-F. Gu, G.-H. Xie, L. Zhang, S.-F. Chen, Z.-Q. Lin, Z.-S. Zhang, J.-F. Zhao, L.-H. Xie, C. Tang, Y. Zhao, S.-Y. Liu, W. Huang, *J. Phys. Chem. Lett.* 2010, *1*, 2849.
38. Y. Qian, G. Xie, S. Chen, Z. Liu, Y. Ni, X. Zhou, L. Xie, J. Liang, Y. Zhao, M. Yi, Y. Zhao, W. Wei, W. Huang, *Org. Electron.* 2012, *13*, 2741.
39. M.-L. Sun, S.-Z. Yue, J.-R. Lin, C.-J. Ou, Y. Qian, Y. Zhang, Y. Li, Q. Wei, Y. Zhao, L.-H. Xie, W. Huang, *Synth. Met.* 2014, *195*, 321.
40. C.-J. Ou, Z.-F. Lei, M.-L. Sun, L.-H. Xie, Y. Qian, X.-W. Zhang, W. Huang, *Synth. Met.* 2015, *200*, 135.
41. F. Liu, L.-H. Xie, C. Tang, J. Liang, Q.-Q. Chen, B. Peng, W. Wei, Y. Cao, W. Huang, *Org. Lett.* 2009, *11*, 3850.
42. L.-H. Xie, X.-Y. Hou, Y.-R. Hua, C. Tang, F. Liu, Q.-L. Fan, W. Huang, *Org. Lett.* 2006, *8*, 3701.
43. X. Cao, W. Yang, C. Liu, F. Wei, K. Wu, W. Sun, J. Song, L. Xie, W. Huang, *Org. Lett.* 2013, *15*, 3102.
44. J. Li, Y. Liu, Y. Qian, L. Li, L. Xie, J. Shang, T. Yu, M. Yi, W. Huang, *Phys. Chem. Chem. Phys.* 2013, *15*, 12694.
45. L.-H. Xie, R. Zhu, Y. Qian, R.-R. Liu, S.-F. Chen, J. Lin, W. Huang, *J. Phys. Chem. Lett.* 2010, *1*, 272.
46. X.-H. Zhao, G.-H. Xie, Z.-D. Liu, W.-J. Li, M.-D. Yi, L.-H. Xie, C.-P. Hu, R. Zhu, Q. Zhao, Y. Zhao, J.-F. Zhao, Y. Qian, W. Huang, *Chem. Commun.* 2012, *48*, 3854.
47. L.-H. Xie, T. Fu, X.-Y. Hou, C. Tang, Y.-R. Hua, R.-J. Wang, Q.-L. Fan, B. Peng, W. Wei, W. Huang, *Tetrahedron Lett.* 2006, *47*, 6421.
48. X.-Y. Hou, T. C. Li, C.-R. Yin, H. Xu, J. Lin, Y.-R. Hua, D.-Y. Chen, L.-H. Xie, W. Huang, *Synth. Met.* 2009, *159*, 1055.
49. X.-H. Zhao, Z.-S. Zhang, Y. Qian, M.-D. Yi, L.-H. Xie, C.-P. Hu, G.-H. Xie, H. Xu, C.-M. Han, Y. Zhao, W. Huang, *J. Mater. Chem. C* 2013, *1*, 3482.
50. Y. Yang, J.-F. Zhao, R.-R. Liu, J.-W. Li, M.-D. Yi, G.-H. Xie, L.-H. Xie, Y.-Z. Chang, C.-R. Yin, X.-H. Zhou, Y. Zhao, Y. Qian, W. Huang, *Tetrahedron* 2013, *69*, 6317.
51. C.-P. Hu, Y.-Y. Liu, L.-H. Xie, J.-W. Li, Y.-R. Li, Q. Tai, M.-D. Yi, W. Huang, *Chem. Phys. Lett.* 2013, *578*, 150.
52. Z.-Q. Lin, P.-J. Sun, Y.-Y. Tay, J. Liang, Y. Liu, N.-E. Shi, L.-H. Xie, M.-D. Yi, Y. Qian, Q.-L. Fan, H. Zhang, H. H. Hng, J. Ma, Q. Zhang, W. Huang, *ACS Nano* 2012, *6*, 5309.
53. S.-H. Yang, Z.-Q. Lin, N.-E. Shi, L.-Z. Jin, M.-N. Yu, L.-H. Xie, M.-D. Yi, W. Huang, *CrystEngComm.* 2015, *17*, 1448.
54. Z.-Q. Lin, J. Liang, P.-J. Sun, F. Liu, Y.-Y. Tay, M.-D. Yi, K. Peng, X.-H. Xia, L.-H. Xie, X.-H. Zhou, J.-F. Zhao, W. Huang, *Adv. Mater.* 2013, *25*, 3664.
55. W. L. Yu, J. Pei, W. Huang, A. J. Heeger, *Adv. Mater.* 2000, *12*, 828.
56. X.-Y. Hou, T.-C. Li, J. Liang, D.-Y. Chen, L.-H. Xie, W. Huang, *Acta Chim. Sin.* 2008, *66*, 2575.
57. S. Si, L. Xie, W. Wei, B. Peng, W. Huang, *Acta Polym. Sin.* 2007, 148.
58. F. Liu, J.-Q. Liu, R.-R. Liu, X.-Y. Hou, L.-H. Xie, H.-B. Wu, C. Tang, W. Wei, Y. Cao, W. Huang, *J. Polym. Sci. A Polym. Chem.* 2009, *47*, 6451.
59. L.-H. Xie, X.-Y. Deng, L. Chen, S.-F. Chen, R.-R. Liu, X.-Y. Hou, K.-Y. Wong, Q.-D. Ling, W. Huang, *J. Polym. Sci. A Polym. Chem.* 2009, *47*, 5221.
60. C.-R. Yin, S.-H. Ye, J. Zhao, M.-D. Yi, L.-H. Xie, Z.-Q. Lin, Y.-Z. Chang, F. Liu, H. Xu, N.-E. Shi, Y. Qian, W. Huang, *Macromolecules* 2011, *44*, 4589.
61. C.-R. Yin, Y. Han, L. Li, S.-H. Ye, W.-W. Mao, M.-D. Yi, H.-F. Ling, L.-H. Xie, G.-W. Zhang, W. Huang, *Polym. Chem.* 2013, *4*, 2540.
62. S.-H. Ye, C.-R. Yin, Z. Zhou, T.-Q. Hu, Y.-H. Li, L. Li, L.-H. Xie, W. Huang, *J. Polym. Sci. B Polym. Chem.* 2014, *52*, 587.
63. L. Li, T.-Q. Hu, C.-R. Yin, L.-H. Xie, Y. Yang, C. Wang, J.-Y. Lin, M.-D. Yi, S.-H. Ye, W. Huang, *Polym. Chem.* 2015, *6*, 983.
64. J. Liang, Y. Qian, L.-H. Xie, N.-E. Shi, S.-F. Chen, X.-Y. Deng, W. Huang, *Acta Phys. -Chim. Sin.* 2010, *26*, 946.
65. L. Zhang, Z.-Q. Lin, J.-F. Gu, C.-R. Yin, X.-Y. Hou, F. Liu, Y.-Y. Liu, L.-H. Xie, S.-F. Chen, W. Huang, *Acta Phys. -Chim. Sin.* 2010, *26*, 1934.
66. L.-H. Xie, Q.-D. Ling, X.-Y. Hou, W. Huang, *J. Am. Chem. Soc.* 2008, *130*, 2120.
67. *Nat. Mater.* 2008, *7*, 267.
68. Z.-Q. Lin, N.-E. Shi, Y.-B. Li, D. Qiu, L. Zhang, J.-Y. Lin, J.-F. Zhao, C. Wang, L.-H. Xie, W. Huang, *J. Phys. Chem. C* 2011, *115*, 4418.
69. X. Zhang, Q. Hu, J. Lin, Z. Lei, X. Guo, L. Xie, W. Lai, W. Huang, *Appl. Phys. Lett.* 2013, *103*, 153301.

70. X. Zhang, Z. Lei, Q. Hui, J. Lin, Y. Chen, L. Xie, W. Lai, W. Huang, *Appl. Phys. Express* 2014, *7*, 101601.
71. J.-Y. Lin, W.-S. Zhu, F. Liu, L.-H. Xie, L. Zhang, R. Xia, G.-C. Xing, W. Huang, *Macromolecules* 2014, *47*, 1001.
72. G.-W. Zhang, L. Wang, L.-H. Xie, J.-Y. Lin, W. Huang, *Int. J. Mol. Sci.* 2013, *14*, 22368.
73. J.-F. Zhao, L. Chen, P.-J. Sun, X.-Y. Hou, X.-H. Zhao, W.-J. Li, L.-H. Xie, Y. Qian, N.-E. Shi, W.-Y. Lai, Q.-L. Fan, W. Huang, *Tetrahedron* 2011, *67*, 1977.
74. W.-J. Li, H.-M. Wu, Y.-B. Li, C.-P. Hu, M.-D. Yi, L.-H. Xie, L. Chen, J.-F. Zhao, X.-H. Zhao, N.-E. Shi, Y. Qian, C. Wang, W. Wei, W. Huang, *Tetrahedron* 2012, *68*, 8216.
75. W.-J. Li, B. Liu, Y. Qian, L.-H. Xie, J. Wang, S.-B. Li, W. Huang, *Polym. Chem.* 2013, *4*, 1796.
76. J. Lin, Z. Yu, W. Zhu, G. Xing, Z. Lin, S. Yang, L. Xie, C. Niu, W. Huang, *Polym. Chem.* 2013, *4*, 477.
77. J.-Y. Lin, J. Wong, L.-H. Xie, X.-C. Dong, H. Y. Yang, W. Huang, *Macromol. Rapid Commun.* 2014, *35*, 895.
78. I. Csoregh, M. Czugler, E. Weber, *J. Phys. Org. Chem.* 1993, *6*, 171.
79. B.-M. Zhao, N.-N. Fu, L.-H. Xie, L.-H. Wang, W. Huang, *Chem. Lett.* 2008, *37*, 622.
80. Y.-Z. Chang, Q. Shao, L.-Y. Bai, C.-J. Ou, J.-Y. Lin, L.-H. Xie, Z.-D. Liu, X. Chen, G.-W. Zhang, W. Huang, *Small* 2013, *9*, 3218.
81. L. Wang, G.-W. Zhang, C.-J. Ou, L.-H. Xie, J.-Y. Lin, Y.-Y. Liu, W. Huang, *Org. Lett.* 2014, *16*, 1748.
82. G.-W. Zhang, L. Wang, L.-H. Xie, X.-Y. Hou, Z.-D. Liu, W. Huang, *J Nanomater.* 2013, 368202.
83. W. Ai, J.-Q. Liu, Z.-Z. Du, X.-X. Liu, J.-Z. Shang, M.-D. Yi, L.-H. Xie, J.-J. Zhang, H.-F. Lin, T. Yu, W. Huang, *RSC Adv.* 2013, *3*, 45.
84. W. Ai, W. Zhou, Z. Du, Y. Du, H. Zhang, X. Jia, L. Xie, M. Yi, T. Yu, W. Huang, *J. Mater. Chem.* 2012, *22*, 23439.
85. W. Ai, Z.-Z. Du, J.-Q. Liu, F. Zhao, M.-D. Yi, L.-H. Xie, N.-E. Shi, Y.-W. Ma, Y. Qian, Q.-L. Fan, T. Yu, W. Huang, *RSC Adv.* 2012, *2*, 12204.
86. W Ai, Z. Du, Z. Fan, J. Jiang, Y. Wang, H. Zhang, L. Xie, W. Huang, T. Yu, *Carbon* 2014, 76, 148.
87. Z. Du, W. Ai, L. Xie, W. Huang, *J. Mater. Chem. A* 2014, *2*, 9164.
88. Z. Du, W. Ai, J. Zhao, L. Xie, W. Huang, *Sci. China, Ser. E* 2014, *57*, 244.
89. F. Zhao, J. Liu, X. Huang, X. Zou, G. Lu, P. Sun, S. Wu, W. Ai, M. Yi, X. Qi, L. Xie, J. Wang, H. Zhang, W. Huang, *ACS Nano* 2012, *6*, 3027.

INDEX

Abraham's equation, 365, 372, 378
acetylacetone dioxygenase (Dke1), 4, 5
adamantylamine, 231, 233
alcohol oxidation, 324
aldol reaction, 14, 409
allylation, 334, 336
amino acids, 10, 11, 74–76, 221, 264, 265, 278, 347, 372, 375, 381, 383, 384, 388, 391–394, 402–404, 407–410
amplification, 65, 72, 397–400
anion–π complexes, 201, 202
anion–π interaction, 199–212
anion-templated synthesis, 64–67, 69, 71–73, 80
anticancer, 92, 136, 143, 267, 372, 408
anti-ferromagnetic coupling, 106–108, 177
anti-Hofmeister, 209
anti-inflammatory, 372
antimicrobial, 372
antioxidant, 363–370, 372, 374, 378, 379
aromatic melamines, 417
aromatic oligoamides, 405, 407
aromatic ring, 23, 49–52, 54, 58–60, 101, 102, 159, 172, 173, 187, 195, 199–202, 206, 209, 217, 223, 244, 245, 247–251, 253, 255, 257, 318, 322, 350, 367, 368, 372, 376, 382, 391, 394, 395, 397, 405, 423
aromatic solvents, 420
aspartic proteases, 9
asymmetric Pummerer reaction, 18, 19
autoxidation, 363–366
azaborine, 24–28
azacyclophane, 49, 52, 54, 58–61
aza-Henry reaction, 10, 11
azo dyes, 229

ball milling, 417, 426
basketweave motifs, 444, 448, 449, 452
Baylis–Hillman reaction, 17, 18
beads, 231, 237, 345
benzylazacyclophane synthesis, 59–61
benzyloxyl radicals, 376
binding constant, 214, 284, 287, 395, K 61
BINOL, 213, 214
biocatalyst, 345, 352, 354, 355, 358
biomedical materials, 304, 306, 309
biomimetic radical, 363, 378
biomolecules, 32, 36, 318, 348, 403, 423
bipyridine, BPY, 34, 146, 171–174, 177, 179, 180, 182, 193, 280, 286, 287, 297, 318, 322
2,6-bis(*N*-methylbenzimidazolyl)pyridine, BIP, 284
BN bond, 23–26, 29, 30, 33, 34, 36, 38, 43
BN interaction, 23, 33, 34, 37, 39–42, 44
BN isosterism, 24
BODIPY, 31, 32
boomerang effect, 319, 320, 324
borazine, 24–26, 34, 35
boron neutron capture therapy, 30
bovine serum albumin (BSA), 265
Brønsted acids, 15, 328
β-sheet, 382, 384, 403, 405, 406, 408
bulk water, 85, 91–93, 97
bulky semiconductors, 445, 452

calixarene-like chiral macrocyclic amine, 222, 224
calixarenes, 81, 202, 213, 230, 261, 262, 264–271
Cambridge Structural Database, 88, 182, 208, 242, 244, 248, 250, 253
camptothecin, 267
carbon nanotubes (CNTs), 103, 111, 208, 318, 346, 415
catechols, 23, 33, 34, 368, 369
cation–π interaction, 87, 199–201, 266, 321, 342, 343, 420
cation-templated synthesis, 71, 73
C–C cleavage, 4–6
CCSD(T), 116, 119, 120, 245, 246, 248, 250, 252, 254

Non-Covalent Interactions in the Synthesis and Design of New Compounds, First Edition.
Edited by Abel M. Maharramov, Kamran T. Mahmudov, Maximilian N. Kopylovich, and Armando J. L. Pombeiro.

chalcogen bond, 18, 155, 157
charge transfer, 34, 41, 86, 93, 96, 134, 158, 188, 190, 191, 193–195, 200, 202, 245, 283, 290, 422, 425
chelation, 126, 347
chemical shift difference, 216, 218, 220, 222, 223, 226
chiral aza-crown macrocycle, 226
chiral derivatizing agent, 13, 213
chiral discrimination, 221, 223
chiral macrocycle, 213, 226
chiral macrocyclic amide, 214, 215
chiral macrocyclic amine, 213, 214, 216, 217, 221–225
chiral rhombamine macrocycle, 218, 219
chiral shift experiment, 218, 222
chiral solvating agent, 213, 214, 216, 218, 220, 222, 224, 226
CH/O interactions, 254
CH/π interaction, 241, 250, 253, 255, 257
chromium, 123, 127
*cis*platin, 92, 93, 136, 143
CMI-1, 348
co-condensation, 346, 349, 354
colloidosomes, 231, 237
colloids, 117, 231, 232, 327, 359
colorimetric sensor, 42, 44
complex diarylfluorenes (CDAFs), 446, 447
cooperative catalysis, 327, 328, 343
cooperativity, 87, 88, 207, 208, 246, 247, 329
coordination interaction, 283, 285, 286, 297–299, 302, 303, 309
coordination polymers, 153, 163, 164
coordination self-assembly, 285, 288, 296
coordination shell, 91
Cope-type hydroamination, 6
copolymers, 231, 290, 291, 294, 305, 350, 433, 435, 436, 439, 440
Coulomb law, 431
covalent interaction, 116, 121, 129, 163, 284–286
cross-coupling reaction, 28, 294, 305, 339
cross-linked polymer, 284, 305
crystal design, 16, 185
crystal structures, 86, 88, 94, 150, 157, 158, 166, 167, 173, 187, 195, 201–207, 209, 242–245, 247–252, 255, 257, 266, 399
Cu(II) complexes, 106, 107
cucurbituril, 10, 68, 229, 230, 232, 234, 236, 238, 261, 269
[3+2] cycloaddition, 330, 331, 333, 334
cyclodextrin, 111, 213, 230, 261, 269, 271, 277, 299
cyclo-oligomerization, 71, 72, 74, 75
cytotoxicity, 136, 230, 306, 408

defect-free, 416, 426
degree of polymerization, DP, 271, 284
delocalization index, 26, 190, 191
density functional theory, 96, 108, 116, 200, 350
DFT calculations, 94, 113, 117, 119, 123, 200, 244, 251, 253, 371
α,β-diamino acids, 10, 11
1,2-diaminocyclohexane, 216, 226
diastereomeric complex, 213, 216, 222
dibucaine, 267, 268
dihalogen, 187–190, 192, 193, 195, 196
diphenylglycine, 407
1,3-dipolar cycloaddition, 12, 13
discrimination ability, 216
dispersion, 24, 88, 93, 115–121, 123, 126, 129–132, 134–138, 158, 190, 245, 246, 251, 252, 255, 257, 418–423, 425, 426
dispersion (London) force, 88, 115, 117–121, 134
donor-acceptor, 5, 87, 101, 125, 195, 205, 267, 270, 331, 444
donor-acceptor interactions, 267
dopamine, 36, 40, 49–52, 58–61
dynamic library, 397, 398

edge–face interaction, 381, 383, 388
edge-to-face (T-shaped), 101, 102, 104, 105
EELS, 356, 357
electrochromic materials, 304
electrostatic combination, 290
electrostatic interaction, 58, 85, 87, 88, 103, 189, 229, 277, 284–286, 298–300, 316, 321, 327, 346, 347, 352, 355, 381–384, 388, 393, 401, 402, 407, 421, 431–433, 435, 437–440, 442, 445
electrostatic self-assembly, 290, 292, 298
empirical potential structure refinement (EPSR), 90, 93, 96–98
enantiodiscrimination, 214, 218, 226
enantiomeric purity, 213, 216
enantioselective cyclization, 8
enzyme, 4, 9, 15, 16, 27, 36, 63, 171, 199, 271, 272, 277, 321, 327, 345–350, 352–357, 359, 363, 368, 372, 375, 386, 387, 408
enzyme leaching, 346, 347, 352, 354, 355, 358
enzyme loading, 346, 347, 352–355, 358
enzyme retention, 355
EPSR and time-dependent density-funtional theory (TD-DFT), 90, 93, 96–98
era of consciousness, 443, 444
Eschweiler–Clarke methylation, 59
ethylene/methacrylic acid, 435, 436

face-to-face (sandwich), 52, 76, 101, 102, 110, 111, 134, 252, 255, 384, 449
FDU-5, 348
FDU-12, 348, 350–353
ferromagnetic coupling, 106–108
few-layer grapheme, 417, 420–422, 426
formaldehyde, 51–61, 70, 71, 77, 79
four-element theory, 452
Friedel–Crafts reaction, 77, 446
frustrated Lewis pairs, 30, 31
FSM-16, 348
functionalization, 6, 238, 283, 315, 316, 318, 325, 346, 349, 352, 354, 355, 358, 416, 417

gallol, 368, 369
gold(I), 105, 134
grafting, 285, 346, 349, 353, 354
grapheme, 24, 106, 315–325, 415–428, 452
graphene oxide, 317, 322, 416
graphite, 24, 25, 316–318, 415–428

Hajos–Parrish reaction, 15
halogen bond, 3, 16, 17, 20, 86, 145–163, 182, 187–196, 327, 337–341, 343, 344
halonium cation (ion), 190, 192
H-atom transfer (HAT), 363, 366, 367, 375
H-bond accepting, 365, 366, 371
H-bonds, 167, 170, 174, 176, 366, 368, 370–373, 375, 376, 392, 394, 402, 409

Heck reaction, 321
hemichelation, 126, 127, 130, 138
heterogeneous catalysis, 268
heteronuclear dihalogens, 189, 190, 192, 193, 195
hexafluorobenzene, 201, 419
hierarchical chemistry, 443, 444, 448, 450, 452, 453
hierarchical self-assembly, 283, 292, 297, 299
high boiling points, 420
HMS, 348
homolytic cleavage, 95, 364
horizontal displacement, 241–258, 260
host–guest 68, 72, 111, 138, 185, 199, 200, 202, 204, 205, 217, 229–237, 261, 263, 264, 266, 269–271, 274–277, 279, 283, 292, 297, 299, 300, 302, 306, 308, 407, 449
host–guest chemistry, 199, 229, 230, 232, 235–237, 263, 266
host–guest interaction, 68, 217, 229, 233, 234, 236, 269–271, 274, 275, 297, 300, 302, 306, 308
host–guest recognition, 283, 292, 299, 302, 306
Hughes–Ingold rules, 363
hybrid materials, 163, 315, 316, 319, 320, 322, 323, 325, 347
hybrid ordered mesoporous material, 345, 346, 348, 350, 352, 354, 356, 358, 360
hybrid organosiliceous material, 346
hydrodefluorination, 324, 325, 343
hydroformylation, 331–333
hydrogelation, 384, 385
hydrogelators, 408
hydrogels, 111, 231, 237, 266, 384–386, 408
hydrogenation, 10, 103, 104, 320, 324, 333, 337, 338, 343
hydrogen bond, 3–18, 20, 21, 29, 30, 38, 42, 52–55, 58–61, 63, 64, 68–70, 72, 74, 76, 78, 81, 85–88, 91–95, 98, 108, 109, 117, 120, 145, 146, 148–150, 153, 155, 157, 158, 163, 164, 168, 170, 172, 178, 180–182, 187, 192–195, 199, 200, 205, 207, 209, 216, 218, 219, 222, 226, 229, 230, 248–250, 253, 255, 263, 266, 283, 292, 297, 302, 303, 315, 318, 328–331, 333–337, 339–341, 343, 346, 347, 357, 364, 371, 372, 382, 403–408, 417, 437, 445, 449, 451
hydrogen bond-directed synthesis, 63, 74, 76, 78
hydrogen storage, 24, 32, 33, 43
hydrolysis of ethers, 14
hydrophilic–hydrophobic balance, 277
hydrophobic, 16, 24, 27, 68, 71, 85, 96, 111, 163, 168, 182, 229, 231, 261, 264, 266, 274–277, 298, 318, 327, 346–349, 352–355, 357, 381–388, 391, 397, 402, 403, 407, 408, 421–424, 435
hydrophobic effects, 85, 275, 382, 402
hydrophobic interaction, 163, 318, 346, 347, 352, 354, 382, 386, 402, 403, 407, 424
hydroquinones, 368, 372
hyperbranched MSP, 284, 292, 294, 297, 299, 300, 302, 309, 422

imidazolium ionic-liquids, 420
immobilization, 37, 103, 105, 111, 205, 315, 316, 318, 319, 321–325, 345–350, 352–355, 358
immunosuppressive, 372
interaction energy (IE), 126, 137, 192, 200, 201, 203, 241, 245, 246, 248, 250, 252, 254, 255, 257, 424, 431, 447
intramolecular hydrogen bonds, 16, 63, 68, 74, 76, 81, 218, 219
intramolecular proton transfer, 13, 336
in vitro, 230, 231, 238, 261, 267, 306, 372
ion-dipole interactions, 229
ionomers, 431–433, 435–440
iridium, 109–111, 125, 126, 142
iridium(III) complexes, 109
irreversible bonds, 63, 64, 81
isomerization, 7–10, 18, 19, 105, 125, 237, 273, 337
isoquinoline, 34, 49–52, 58–61

Job plot, 221, 223, 224

Keggin-type, 209
keto–enol tautomerism, 16
kinetic control, 63, 64, 73, 74, 269
kinetic solvent effect, 364–366
KIT-6, 348, 352–354
Krebs cycle, 15, 16

laccase, 352, 354–358, 372, 375
Langmuir–Blodgett films, 167
lanthanide cations, 284
lanthanides, 283
Lewis acid, 12, 18, 23, 27, 28, 30, 32, 71, 79, 117, 123, 125, 126, 145, 187, 190, 329, 339, 340, 369, 371, 374, 377
Lewis base, 16, 18, 23, 28, 30, 31, 118, 145, 187, 188, 195, 328, 337, 338
Lewis pairs, 30, 31
lignin, 374, 375
linear MSP, 284–286, 290, 298, 302, 303, 309
lipase, 352–358
liquid exfoliation, 416, 426
lone-pair/π interactions, 241, 242, 247
L-proline, 406, 408

macrocycle, 33, 34, 47, 51, 52, 54, 58–60, 63–74, 76–82, 213, 214, 216, 218–222, 224–226, 230, 261, 274, 392–395, 397, 400, 401, 403, 406
macrocyclic pseudopeptides, 393, 397, 400, 401, 403
macrocyclization, 52–55, 58–61, 63–69, 74, 77, 79, 81, 392–395, 401
magnetic coupling, 101, 106–108, 111
manganese, 108, 109, 130, 177
Mannich reaction, 10, 15, 52, 60, 69
McConnell I spin polarization, 107, 108
MCF, 230, 306, 348
MCM-41, 347, 348
MCM-48, 347, 348
mesostructure, 347–350
mesostructured material, 348
metal ion coordination, 298
metallacycle, 117, 130, 147, 149, 204, 299
metal–ligand coordination, 283, 284, 293, 297, 298, 328
metallo-supramolecular chemistry, 283
micelles, 231, 233, 237, 268, 274, 276, 277, 291, 292, 347, 348, 439
Michael addition, 6, 331, 332
micromechanical cleavage, 416, 417, 426
Mn(II) and Mn(III) complexes, 108
molecular nanotectonics, 448, 452
molybdenum, 121, 122
monolayer grapheme, 416, 422, 423, 425
morphology, 270, 273, 274, 276, 323, 346, 350, 386, 407, 431, 433, 435, 436, 443, 445, 447, 449

M41S, 347
MSP network, 292, 294, 296–302, 305, 308, 309
MSU, 348
multi-walled carbon nanotubes (MWCNTs), 103, 105, 106

nanomaterial, 345, 347, 418, 452
nanomosaics, 444, 449, 452
nanoparticles, 90, 117, 230, 231, 234–238, 266, 272, 278, 279, 321, 323, 324, 345–347
nanostructures, 229–231, 235, 237, 238, 291, 292, 407, 445, 449
nanotubes, 24, 103, 111, 208, 241, 318, 345, 346, 386, 406–408, 415
nanowires, 208, 407
N-arylaziridine, 17
N-benzhydryl acetamide, 17, 18
Ni(II) complex, 10, 106, 108, 113
NMR spectroscopy, 33, 89, 94, 122, 123, 134–136, 213, 216, 225, 226, 262, 263, 325, 387, 388, 447
non-covalent interaction of water, 90, 92, 94, 96–98
non-covalent visualization, 446, 447
nonpolar interactions, 166, 167
nonracemic 1,3-diamines, 10, 11

offset face-to-face (parallel-displaced), 101, 102
OH/π interactions, 242, 243, 245, 247, 249, 257
optoelectronic material, 304, 444
ordered mesoporous material, 345–358, 360
organocatalysis, 327, 343
orthogonal combination, 297, 302, 303
orthogonal self-assembly, 297, 299, 302, 309

palladium, 4, 28, 118, 128, 136, 321–325, 334, 336
parallel interactions, 241–246, 248–258, 260
peptide, 9, 10, 33, 37, 63, 76, 78, 234, 235, 238, 355, 357, 381–388, 391–393, 401–409
peptide assembly, 384–386, 408
peptide hormones, 387
peptide therapeutics, 387, 388
Perylene, 274, 275, 423–425
Phenanthroline, 39, 106, 107, 155, 172, 173, 286, 287, 318, 322
Phenylalanine, 29, 221, 250, 251, 381–383, 386, 387
phenylethylamines, 49, 50, 59
2-phenylquinoline, 16, 17
photoluminescent, 101, 109–111, 304
photoluminescent material, 304
photophysical properties, 31, 109, 135, 447
Pictet–Spengler reaction, 49–52, 58–60
π-acidic, 203, 204, 207, 209
π-effects, 3, 85, 327
π-electron acceptor, 263
π-electron-rich, 3, 12, 23, 32, 42, 101, 145, 187, 188, 193, 199, 201, 202, 230, 261, 270, 304, 337, 340, 343, 370, 373, 388, 421, 425
π–π interaction, 63, 101–114, 120, 155, 166, 170, 173, 174, 176, 182, 208, 222, 226, 262, 275, 318–320, 322, 323, 341, 343, 381–388, 399, 407, 421–423, 445–448
π–π stacking, 101, 103, 106–111, 113, 163, 170–173, 175, 179, 180, 182, 205, 207, 216, 277, 278, 283, 290, 292, 383, 405, 406, 424, 446–449
π-stacking, 54, 101, 110, 111, 261, 262, 274, 275, 315, 316, 318, 322, 325, 408, 447
π-systems, 101, 102, 199, 202, 241
platinum, 92, 93, 111
platinum(II) complexes, 93, 111, 113, 290–292, 299
PMA, 276, 354–358
PMO, 348–350, 352–358
pnicogen bond, 18, 20, 340
polarization, 25, 87, 92, 93, 102, 107–109, 147, 158, 187, 188, 190, 193, 194, 200, 201, 368, 420
polar solvents, 58, 66, 174, 303, 317, 319, 320, 363, 364, 370, 406, 420
poly(diazafluorenes), 451
polyamide, 216, 432
polyaromatic molecules, 318
polyazamacrocycle, 216
polydiarylfluorenes, 450
polymer, 24, 28, 34, 35, 42–44, 52, 54, 64, 134, 153, 163, 164, 171, 178, 180, 181, 183, 231, 233, 236, 237, 269–274, 283–286, 288, 290–294, 296–298, 300, 302–306, 308, 309, 318, 331, 345, 346, 349, 363, 402, 415–417, 422, 423, 432–437, 439, 450–454
polymer network, 294
polypeptide hydrolysis, 9
polyphenols, 365, 367–370
polyurethane, 431, 435, 440
pore shape, 352, 354, 355, 358
pore size, 316, 346–355, 358
porphyrins, 64, 154, 213, 269, 270, 421
pristine graphene, 316–318, 325, 418
protein, 23, 27, 29, 32, 33, 36, 37, 85, 199, 231, 232, 238, 241, 243, 249, 250, 265, 266, 277, 279, 306, 318, 321, 345–348, 350, 355, 357, 365, 372, 373, 375, 376, 381–384, 386–388, 401–403
protein data bank, 250, 257, 381
protein design, 381, 382, 384, 386, 388
protein folding, 85, 381, 383, 387
proton-coupled electron transfer, 366
proton transfer, 13, 23, 329, 333, 335–338, 370
pseudopeptides, 391, 393–395, 397, 399–403, 407, 409
pseudopeptidic molecules, 391, 394, 397
pyrenes, 27, 277, 447
pyridine, 34, 35, 68, 69, 78, 103, 106, 109, 111, 150, 154, 156, 170, 188, 190, 200, 201, 205, 253–258, 284, 286, 292, 363, 370, 394, 435

QTAIM analysis, 194, 195

radical dimerization, 374
radical ion, 207, 208
radical reactions, 363, 364, 366, 368, 370–372, 374, 376, 378
receptors for anions, 199, 209, 400
recycling and reuse, 103
reduced graphene oxide, 316, 317
regioselectivity, 5, 50, 51, 58, 60, 262, 345, 433
resorcinols, 368, 372
retro-Claisen reaction, 4, 5
rhenium, 130
rhodium, 103, 126, 129, 142, 319, 320, 322, 329, 333
rhodium(I) complexes, 103, 329
ruthenium, 95, 96, 103, 104, 142, 319, 320, 322–325, 336, 337
ruthenium(II), 103, 329
ruthenium(II) complexes, 103, 329

salt-bridge interactions, 402
SBA-3, 348
SBA-12, 348
SBA-15, 348, 350–355
SBA-16, 348, 352, 353
Schiff base, 33, 49–51, 59, 61, 64, 208, 286
Schmidt cyclization, 343
Schwartz's reagent, 329
Scotch-tape method, 416, 417, 426
self-assembling, 269, 274, 291, 292, 391, 403
self-assembling peptides, 385, 387, 388
self-healing, 231, 238, 269, 283, 305, 307, 309, 451
self-replicators, 403
SEM, 291, 292, 350, 352, 408, 448, 449
sequential/relay catalysis, 327, 328
Shvo's catalyst, 336–338
σ-hole, 145, 159, 187, 188, 193, 199, 338 (AU: Pls insert the symbol)
silane, 343, 346
silica, 28, 208, 231, 232, 236, 237, 345–350, 352–355
single-walled carbon nanotubes (SWCNTs), 103
solar cell, 208, 304, 305, 415, 416, 426
solubility parameters, 418, 420
solution properties, 439
solvent effects, 363, 364
somatostatin, 36, 387, 388
spider, 146, 149
spirofluorenes, 445–447
stability constants, 70, 262–265
stabilization, 3, 9, 12, 15, 24, 26, 28, 115, 116, 120, 122, 124–126, 128, 130, 132, 134, 136, 138, 140, 142, 148–150, 178, 205, 207, 241, 247, 253, 275, 324, 328, 346, 357, 358, 363, 365, 367, 368, 371, 374, 377, 378, 382, 388, 403, 405, 409, 417, 420–422, 424–426
stacked polymers, 449
star-shaped MSP, 292–294, 309
STEM–HAADF, 356, 357
stereoselectivity, 7, 15, 28, 334, 335, 392
stimuli-responsive, 229–231, 235–238, 271, 277, 290, 303, 408
structure–property relationships, 433, 440
subphthalocyanine, 32
sulfonation, 433, 434
p-sulfonatocalix[n]arenes, 261, 262, 264, 266, 268, 270, 272, 274, 276, 278
support, 6, 12, 37, 45, 90, 91, 93, 103, 113, 127, 138, 184, 210, 259, 279, 315, 316, 319, 320, 322–324, 331, 345–350, 352–355, 357, 358, 392, 416, 426
supramolecular hindrance polymers (SHPs), 451
supramolecular polymer, 236, 269, 270, 271S, 272–274, 277, 278, 283, 284, 286, 288, 290, 292, 294, 296, 298, 300, 302, 304, 306, 308, 448, 449, 451
supramolecular semiconductors, 447, 449
supramolecular steric hindrance, 443, 444, 446, 448, 450, 452
supramolecular structures, 24, 33, 150, 157, 247, 249, 252, 253, 255, 257, 407
surfactant-assisted exfoliation, 420, 422
surfactants, 274–276, 318, 347, 348, 418, 422–424, 426, 448
synergistic catalysis, 327, 328

TEM, 274, 276, 291, 292, 350, 351, 356, 417, 419, 422, 425, 448, 449
template, 52–55, 59–61, 63–74, 77, 79–81, 101, 148, 149, 204, 206, 316, 347–349, 382, 394–400, 403
terpyridine (TPY), 284, 288
tert-butyl hydroperoxide, 364
tetracaine, 267, 268
1,2,4,5-tetrazine, 201, 202, 204
tetrazoles, 12
theoretical calculations, 94, 146
thermodynamic control, 63, 64, 73, 79–81, 269, 397
topological analysis, 189, 194
topotecan, 267
transition metal, 52, 63, 71, 88, 91, 92, 101, 103, 115–122, 124, 126, 128, 130–132, 134, 136–138, 141, 148, 177, 201–203, 264, 284, 288, 290, 293, 296, 316, 321, 322, 327–329, 338, 343
transition metal complex, 92, 101, 116, 117, 119–121, 134, 137, 203, 316, 328
trianglamine, 216, 217
Tsuji–Trost cyclization, 334
Tyndall effect, 276, 421
tyramine, 50–52, 54, 55, 58–61
tyrosine, 49, 51–58, 60, 61, 365, 376
2D-GIXD, 449, 451
2D NOESY, 222, 224

α,β-unsaturated ketones, 330, 331

van der Waals, 18, 24, 77, 85–89, 115–117, 119, 122, 123, 126, 131, 134, 138, 145, 146, 154, 159, 170, 178, 179, 229, 230, 262, 264, 292, 298, 315, 318, 327, 347, 407, 409, 416, 450, 451
  forces, 3, 85, 86, 88, 170, 315, 318, 347, 451
  interactions, 86–89, 116, 134, 138, 230, 262, 264, 409
  radii, 18, 86, 88, 89, 122, 131, 145, 146, 159
viologen, 229–231, 234–236, 263–265, 270–275, 277, 280

Wulff-type boronic acids, 23, 37, 39, 42

x-ray diffraction (XRD), 89, 121–123, 129, 134, 138, 166, 169, 350, 448, 449

zwitterionic, 9, 28, 37, 42, 164, 165, 336, 347, 392, 402, 421

www.ingramcontent.com/pod-product-compliance
Lightning Source LLC
LaVergne TN
LVHW082108200926
848484LV00008B/1926